BIOLOGY

THE UNITY AND DIVERSITY OF LIFE

FIFTH EDITION

CECIE STARR
Belmont, California

RALPH TAGGART
Michigan State University

Advisors and Contributors

John Alcock, Arizona State University, *Chapters 48, 49*
Robert Colwell, University of California, Berkeley, *Chapters 43, 44*
Cleon Ross, Colorado State University, *Chapters 20, 22*
Samuel Sweet, University of California, Santa Barbara, *Chapter 37*

Wadsworth Publishing Company
Belmont, California
A Division of Wadsworth, Inc.

We are no more than
sunlight dancing on the stream—
And no less.

Cover photograph, snowy egret at sunrise, by Glenn Van Nimwegen. Like all organisms on earth today, this egret is a living representative of an evolutionary line extending out of the distant past; and like nearly all organisms, it depends ultimately on energy flow from the sun. Evolution and energy flow—these are the unifying themes of this book.

Biology Editor: Jack C. Carey

Editorial Assistant: Sue Belmessieri

Production Manager: Mary Forkner Douglas

Art Director: Stephen Rapley

Copy Editor: Elizabeth Judd

Permissions and Photo Research: Marion Hansen

Production Assistants: Cornelia Lovette, Kathy McCann, Martha Simmons

Artists: Lewis Calver, Joan Carol, Raychel Ciemma, Ron Ervin, Darwen Hennings, Vally Hennings, Joel Ito, Keith Kasnot, Julie Leech, Laszlo Meszoly, Victor Royer

Graphic Artists: Susan Breitbard, Alan Noyes, Jeanne Schreiber, Salinda Tyson

Marketing: Theresa Coyne, Joy Westberg

Typography: Jonathan Peck Typographers, Ltd.

Color Processing: Williams Litho Service

Printing: Rand McNally

Printed in the United States of America

2 3 4 5 6 7 8 9 10—93 92 91 90 89

Library of Congress Cataloging in Publication Data

Starr, Cecie.
    Biology: the unity and diversity of life.

    Includes index.
    1. Biology. I. Taggart, Ralph. II. Title.
QH308.2.K57 1989   574   88-33774
ISBN 0-534-09180-6

PREFACE

More than twelve years ago, we wrote the first edition of *Biology: The Unity and Diversity of Life*. Our approach was novel at the time; we introduced the principles of evolution and energy flow early on and then *used* them as a way of interpreting the observations and experiments described throughout the book. We believed then, as we do now, that students can make more sense of life in general and their own lives in particular when they truly understand the integrative power of those two great principles.

Why will it be difficult to produce a vaccine against HIV, the causative agent of AIDS? How can we explain the similarities and differences between a human and a chimp? What repercussions can we expect from the rapid and wholesale destruction of tropical rain forests? What will happen if the average temperature near the earth's surface continues to rise through amplification of the greenhouse effect? With the principles of evolution and energy flow to guide them, students can think their way through any number of issues or observations about the natural world; and they can better evaluate what they hear about and read every day of their lives.

Several thousand instructors apparently liked our approach; over the years, close to a million of their students have used our book in its new and previously owned forms. The acceptance has been gratifying, but it also has reinforced our deep sense of responsibility to those who have placed their trust in our efforts. That is why, as soon as one edition is published, we begin work on the next, using the following standards as our guide:

1. Use the principles of evolution and energy flow as a conceptual framework for each chapter.

2. Distill the main concepts and outline the research trends in all major fields.

3. Give enough examples of problem solving and experiments to provide familiarity with a scientific approach to interpreting the world.

4. Include enough comparative biology to convey a sense of the unity and diversity among organisms.

5. Include enough human biology to enhance understanding of our evolution, behavior, and ecology as well as our body structure and functioning.

6. Be selective in developing the vocabulary necessary to comprehend what is being talked about today in each field.

7. Present the material accurately but not at a high level.

8. Write clearly, in a style that is not boring or patronizing.

9. Create easy-to-follow line illustrations and select photographs that are informative as well as beautiful.

10. Keep the wrtiting free of bias, thereby giving students the chance to form their own opinions about the material being presented.

REVISION HIGHLIGHTS

This is a major revision effort, with eighty percent of the book being updated and rewritten for greater clarity. We researched the current literature and corresponded with specialists in different fields to identify new information or ideas that called for new approaches or shifts in emphasis. Simultaneously, thirty adopters throughout the country were preparing "diary" reviews of the fourth edition. They evaluated the effectiveness of each chapter and let us know whenever a portion of their students had trouble with some inherently difficult topic. This drove us to write draft after draft manuscript, until instructors who said their students had *liked* the material before were actually enthusiastic about how much more effective it has become.

The revision is too extensive to describe completely, but the following examples provide insight into the types of changes that were made:

1. Chapter 5, which starts students thinking about the relationships between cell structure and function, is more tightly written; see, for example, the sections on chromosomes and the cytomembrane system. The section on the cytoskeleton was revised to reflect emerging ideas about how the cytoskeleton organizes the cell division planes during embryonic development. Similarly, Chapter 6 is updated to reflect advances in our understanding of membrane structure and function; Figure 6.2 is an example of how the recent data are incorporated into illustrations as well as into the text.

2. The unit on metabolism has been simplified without distorting the science. Chapter 7, for example, has sharper descriptions of metabolic pathways, enzyme structure and function, the ADP-ATP cycle, coenzymes,

and electron transport systems. Because the basics are covered in this separate chapter, the subsequent chapters on photosynthesis and aerobic respiration are kept uncluttered. The entire unit benefits from many new, informative illustrations.

3. The chapters on mitosis and meiosis (10 and 11) are reorganized and rewritten. Inherently confusing terms (chromosome and chromatid, haploidy, diploidy, pairs of homologues) are explained at the outset in words and simple pictures (pages 140 and 141). By the time students start Chapter 11, a few simple "refresher" paragraphs on terminology (page 151) set the stage for the discussion. A word-picture overview of meiosis follows (page 152). With this preparation, students can focus on *details* of the division mechanisms, these being the critical foundation for the remaining chapters in the unit.

4. Take a moment to look at the many new illustrations throughout the genetics unit, particularly the three-dimensional paintings. More than thirty instructors and specialists—and here we must acknowledge the valuable ideas of Joseph Bonfiglio—worked with us to make the new art accurate and pedagogically effective. This may be one of the reasons why the book works so well in the classroom: while each chapter is being worked on, the illustrations receive as much attention as the text. Over the years, students have let us know how much they appreciate the results.

5. Chapter 12 (Mendelian genetics) has been called "the" model chapter for giving students a sense of how to think critically; even so, we tightened the writing and improved the illustrations. Chapters 12 and 13 are updated and rewritten for greater clarity. The human genetics chapter (14) is now clearly organized by major patterns of inheritance (autosomal recessive, autosomal dominant, and X-linked); it also has a new section on RFLPs. Gordon Edlin deserves special thanks for his insights concerning this and other genetics chapters.

6. Chapter 17 has been reorganized; prokaryotic control mechanisms are described simply and accurately, and there are better examples of gene control in eukaryotes. The chapter on recombinant DNA and genetic engineering (18) has been completely rewritten. It starts with a nonthreatening, historical perspective before getting into the new technologies, and it concludes with a balanced look at the social implications of genetic engineering.

7. At the suggestion of Lisa Wagner, we reorganized and simplified Chapter 19 (plant structure). A brief overview of the plant body precedes the discussion of tissue types, primary growth, and secondary growth. Related art is grouped together (see the comparison of monocot and dicot stems on page 254, for example).

8. Similarly, Chapter 23 (animal tissues and organ systems) now begins with an overview of the levels of tissue organization and a "road map" (Figure 23.4) of the human organ systems to be covered in subsequent chapters. The entire animal anatomy-physiology unit has more micrographs and accompanying diagrams of tissues.

9. Nearly all students approach the topics of neural and endocrine function with justifiable trepidation. We worked hard to simplify as well as update our coverage of these topics (in Chapters 24 and 26) and to develop new illustrations that make the concepts more accessible (Figures 24.6 and 26.12, for example). Chapters 28 (skeletal-muscular systems) and 29 (circulatory systems) have valuable new art, and both have updated, simpler text descriptions. The Commentary on cardiovascular disorders (page 408) is expanded and includes new illustrations.

10. Chapter 30, on immunology, has been reorganized and rewritten to keep pace with this rapidly changing field and to make the material more accessible to students; the illustrations also reflect new information. The section on antibody diversity is much improved; the Commentary on AIDS is updated and expanded. Chapter 33 (temperature control, fluid regulation) is rewritten. The descriptions of renal function are clarified and there is a new section on acid-base balance.

11. Chapter 34 (animal reproduction, embryonic development) is reorganized into two parts: first a description of the main developmental stages, then a discussion of mechanisms underlying development. This organization, recommended by William Bradshaw, affords a clear intellectual path through complex material; it also affords more flexibility in assigning only the basics. Chapter 35 (human reproduction, development) is updated and expanded, with refined illustrations (Figure 35.8 is an example). Figure 35.21, a new figure, diagrams the critical periods of human embryonic development. Also new is a Commentary on cancer in the reproductive tract (page 526).

12. All chapters in the evolution and diversity unit are greatly improved. Chapter 39 has better coverage of viruses and bacteria, including new life-cycle illustrations, photographs, and paintings. Chapter 40 (fungi and plants) is rewritten to give more balanced coverage of the different groups. It has more examples of interest to students, such as the Commentary on fungi (page 620) and commercial uses of red and brown algae. The tables are simplified. Working with several specialists in botany, we developed beautiful, accurate paintings of life cycles (pages 618, 628, 631, and 639, for example). The new life cycle on monocots (the lily) is included to

round out the picture for flowering plants (Chapter 21 has a life cycle for dicots).

13. Eugene Kozloff worked with us to improve the invertebrate section in Chapter 41 (animal diversity). The vertebrate section is expanded and has a much clearer organization. The new photographs and line illustrations here are truly spectacular. After consulting with prominent physical anthropologists, we rewrote the chapter on human origins and evolution (42). It has a new introduction to mammalian and then primate characteristics, an updated picture of hominid evolution, and many new, excellent illustrations.

14. The ecology unit is revised considerably. The text and illustrations of human population growth (Chapter 43) have been updated. Ecosystems concepts are now in sharp focus, and there is a new Commentary on the greenhouse effect (Chapter 45). The global forces that shape the distribution of ecosystems are now a major theme in Chapter 46 (biosphere). The classification of biomes is improved, and there is now a balanced coverage of aquatic ecosystems, including more on marine ecosystems. Concepts are given practical application in the new Commentary at the chapter's end, which describes how the El Niño Southern Oscillation profoundly disrupts the world's climates and human affairs. Chapter 47 (human impact on the biosphere) is rewritten. It is now global rather than regional in focus, with attention given to acid deposition, the ozone hole, tropical forest destruction, desertification, and depletion of energy reserves, all in the context of human population growth.

15. The classification scheme followed throughout the book (and summarized in an appendix) has been brought in line with current thinking. There is no consensus here, but responses to a detailed questionnaire told us which groupings are most acceptable to the most instructors and specialists. Page numbers after the entries refer to text pages on which the different organisms are described and illustrated.

THE ILLUSTRATION PROGRAM

We worked directly with outstanding artists, medical illustrators, and photographers in developing an additional 400 illustrations. Moreover, existing art has been recast in a three-dimensional style. We personally blocked out the placement of text and illustrations on each page and sized and color-coded the art (we did this to make it easier for students to see how the text and illustrations, as well as the parts within an illustration, fit together).

New to this edition are three-dimensional paintings that help make hard-to-visualize topics more tangible. Also new are the full-color life cycles of plants, as well as paintings that detail the structure and function of viruses, bacteria, fungi, plants, and animals.

We expanded the number of illustrations in which visual and written summaries are combined to make concepts easier to grasp. Wherever feasible, we broke down the information into a series of steps, which are far less threatening to students than one large, complicated diagram.

STUDY AIDS

Students taking introductory biology will not already know enough about biology to spot all key concepts in page after text page. *Major summary statements* are highlighted with blue lines and printed in boldface for emphasis. Taken together, the statements are an easy-to-identify, *in-context summary* of key concepts. Many are in list form (see, for example, pages 88, 141, and 163). Concepts are also reinforced by *end-of-chapter summaries* as well as *summary illustrations* (see pages 160 and 218 for examples of the latter). This edition also has more *summary tables*.

Review questions are keyed to italic and boldfaced sentences as a way of reinforcing the main concepts. Italicized numerals at the end of each question refer to pages on which students can find the answers.

Chapters 12–14 have *genetics problems*, with *detailed solutions* in Appendix II. The *glossary* brings together the text's main definitions. It includes pronunciation guides and origins of words, when such information will make seemingly formidable words less so. The *index* is extensive, simply because students may find a door to the text more quickly through finer divisions of topics.

APPLICATIONS

Within the chapters, we address numerous topics of social concern from a biological perspective, including the effects of smoking (covered in the respiration chapter and in the discussion of human embryonic development), anorexia nervosa, bulimia, fetal alcohol syndrome (FAS), abortion, and genetic engineering. Many more topics important to students have been added.

Many chapters also have *Case Studies* that show how general concepts apply to specific situations. For example, after the discussion of immunity, a case study shows students how their own body would mount an immune response to a bacterial attack (page 429).

The *Commentaries* are "outside readings" built into the book. They explore such thought-provoking topics as drug abuse, cancer, death, cardiovascular and lung disorders, AIDS, sexually transmitted diseases, the greenhouse effect, tropical forest destruction, and drought prediction based on studies of interactions between the oceans and the atmosphere.

The *Perspectives* are end-of-chapter sections that encourage students to take a moment to think about the connections between chapters and units. Many also bridge chapter topics and the student's world, inviting reflections on the past and possible futures. Two examples appear on pages 133 and 805.

SUPPLEMENTS

An extensive package of 14 supplements accompanies this text. There are now 140 illustrations on full-color *transparencies*. These figures also are available on 35mm slides. An additional 200 figures are available on transparency masters. All transparencies and slides are labeled with large, bold type.

There is a new *Test Items* booklet, with 2,500 questions. The questions are also available on IBM, Apple IIe, and Macintosh disks. Accompanying the disks is a database manager and word processor that helps instructors prepare tests. The disks also include a chapter objectives data bank that allows instructors to select and modify learning objectives.

New for this edition is a comprehensive *Instructor's Resource Manual* by Larry Lewis, Susan Feldkamp, Larry Sellers, and Jane Taylor. Each chapter includes revision highlights, a chapter outline, a list of boldface or italic terms, chapter objectives, a detailed lecture outline correlated with the transparencies, suggestions for presenting the material, suggestions for classroom and laboratory demonstrations, suggested discussion questions, research paper topics and library activities, and annotated lists of filmstrips and videos. There also are three valuable appendices: a list of library sources for student research, a bibliography of filmstrip and video sources, and a ten-page guide that aids student in writing research papers and reports.

Jane Taylor revised the *Study Guide* to correspond with the new edition. In this supplement, learning aids are organized by chapter section, allowing students to focus on smaller amounts of material and to skip over unassigned sections. Because students learn in different ways, there is a variety of pedagogical aids to help them. Each chapter section has a detailed summary, list of key terms, learning objectives, and different kinds of self-quiz questions. The new edition has a multiple-choice test at the end of each chapter. Also available is an *Answer Book* by David Cotter that provides detailed responses for end-of-chapter Review Questions found in the book.

Although the new edition contains many more applications, *Current Readings in Biology*, an anthology compiled by John Crane, provides additional applications of biological concepts. Another rich anthology, *Science and the Human Spirit: Context for Writing and Learning* by Fred O. White, helps students learn how to read and write about the sciences. Topics include ethical issues, significant discoveries and breakthroughs, and women in science.

Finally, Jim Perry and David Morton revised their *Laboratory Manual* to correspond with the new edition and to take advantage of feedback from ten instructors who class-tested it. This well-received manual has thirty-three experiments and exercises, each divided into distinct parts that can be assigned individually, depending on the time available in the laboratory. All experiments and exercises have the same format, with objectives, a discussion (introduction, background, and relevance), a list of materials for each part of the experiment, procedure, pre-lab questions, and post-lab questions. Each procedure is in list form, with each step numbered. The procedures are detailed enough so each exercise is self-explanatory. An *Instructor's Manual* by Joy Perry accompanies the laboratory manual. It covers quantities, procedures for preparing reagents, time requirements for each portion of the exercise, notes to help make the lab a success, and vendors of materials, with item numbers.

A COMMUNITY EFFORT

It would be rather fatuous of any author to pretend that a general biology book can be written in a vacuum; biology simply has diverged too rapidly in too many directions for any one person to keep up with all of it. Over the years, more than 1,000 instructors and research specialists have given us the benefit of their insights, and they must be given credit for their contribution. We owe special thanks to John Alcock, Robert Colwell, Cleon Ross, and Samuel Sweet—not only for contributing chapters in their specialized areas but also for their good grace in allowing us to rewrite the material in the style and at the level of the entire book. We owe special thanks also to James Bonner, Eugene Kozloff, George Lefevre, and William Parson for their warm advice, support, and contributions over the years.

We owe thanks to William Fennel, Stephen Hedman, John Jackson, Kenneth Jones, Joyce Maxwell, Heather McKean, Douglas Reynolds, John Rickett, and Pat Starr. They all transcend the "reviewer" designation; they have made detailed criticisms of the entire manuscript, edition after edition. This time around, Thomas Gray, Michael Tansey, and Larry Sellers join them. The advice of Gary

Atchison, George Cox, William Schlesinger, and David Tilman was invaluable in revising the ecology unit.

Some of the biologists mentioned earlier in the preface were diary reviewers, along with Samuel Bieber, Jean Bowles, William Bradshaw, William Brunckhorst, Lynn Clark, R. Dean Decker, Dorothy Dunning, Thomas Eickbush, Charles Flora, Robert Kull, Kim Mattson, Mary McKitrick, William Morrison, Diane Nelson, Tom Palko, and Lisa Wagner. By helping us class-test the material, all were instrumental in shaping the revision. They and the other reviewers listed on the next two pages will know, in reading through this new edition, where they have left their imprint.

A word for Wadsworth. Once again, we lured Mary Forkner Douglas into becoming our production manager; once again, Stephen Rapley was our art director. Anyone looking through the text and sophisticated graphics of this book will know at once how great these two individuals are. They have been absolutely unwavering in their personal and professional support. And once again, Dick Greenberg, Kathie Head, and Steve Rutter gave their unqualified support to the revision effort.

After the last time, we didn't think Marion Hansen would put up with our demands this time; but she did, and students can thank her for finding the beautiful new photographs in this edition. Sue Belmessieri kept the revision effort going smoothly with her organizational skills and good humor. Deborah Cogan came through in the eleventh hour. Lewis Calver, Raychel Ciemma, Darwen Hennings, Vally Hennings, and Victor Royer deserve special recognition; these outstanding artists are deservedly well known, and they have done their best for us over the years. This edition, Joan Carol and Keith Kasnot made valuable artistic contributions. Susan Breitbard, Alan Noyes, and Jeannie Schreiber played a major role in developing complex pieces of art and in making soothing noises for exhausted authors. The staff at Jonathan Peck Typographers did their usual superb job.

Jack Carey has guided the development of this book through all of its editions. Today, biology textbooks must do so much for so many diverse students; and it takes a serious, knowledgeable editor to help authors produce a good one. Over the past twenty-five years, Jack has met with instructors and researchers throughout North America and has developed what is possibly the most comprehensive list of respected reviewers in biology education. He is the catalyst for the intellectual actions and reactions that are the foundation of this book.

FIFTH EDITION REVIEWERS

Thomas Adams, Michigan State University
Peter Armstrong, University of California, Davis
Gary Atchison, Iowa State University
Andrew Bajer, University of Oregon
Aimee Bakken, University of Washington
Barry Batzing, State University of New York, Cortland
Suzanne Bayley, Freshwater Institute
Loren Bieber, Michigan State University
Samuel Bieber, Old Dominion University
Antonie Blackler, Cornell University
Robert Bland, College of St. Thomas
Joe Bonfiglio, West Valley College
Jean Bowles, Metropolitan State College
William Bradshaw, Brigham Young University
Mark Brinson, East Carolina University
Leon Browder, University of Calgary
William Brunckhorst, University of Wisconsin, Whitewater
Robert Burnes, University of Arkansas for Medical Sciences
Clyde Calvin, Portland State University
Christine Case, Skyline College
A. Kent Christensen, University of Michigan Medical School
Donald Christian, University of Minnesota
Lynn Clark, Iowa State University
Richard Coles, Washington University
Edwin Cooper, University of California, Los Angeles
George Cox, San Diego State University
John Crowe, University of California, Davis
R. Dean Decker, University of Richmond
Fred Delcomyn, University of Illinois, Urbana
Nancy Dengler, University of Toronto
Ronald Dengler, University of Toronto
Katherine Denniston, Towson State University
William Dentler, University of Kansas
Jean DeRousseau, New York University
Dean Dluzen, University of Illinois, Urbana
Dorothy Dunning, West Virginia University
Gordon Edlin, University of California, Davis
Thomas Eickbush, University of Rochester
James Estes, University of Oklahoma
William Fennel, Eastern Michigan University
Phillip Fields, University of South Alabama, College of Medicine
Harvey Fineberg, Harvard School of Public Health
D. J. Finnegan, University of Edinburgh

Charles Flora, Western Washington University
Christine Foyer, The University of Sheffield, UK
Michael Freeling, University of California, Berkeley
Theodore Friedmann, University of California, San Diego
Jeffery Froehlich, University of New Mexico
Fred Funk, Northern Arizona University
Chris George, California State Polytechnic University
Christopher Goff, Haverford College
Richard Gonzalez, United States Army Research Institute of Environmental Medicine
H. Maurice Goodman, University of Massachusetts Medical School, Worcester
Aubrey Gorbman, University of Washington
Richard Gordon, The University of Manitoba
Linda Graham, University of Wisconsin
Thomas Gray, University of Kentucky
Mac Hadley, University of Arizona
Janet Dunaif-Hattis, Northwestern University
Stephen Hedman, University of Minnesota
Craig Heller, Stanford University
Tom Hellier, University of Texas, Arlington
Thomas Herbert, University of Miami
John Hershey, University of California, Davis
Milton Hildebrand, University of California, Davis
Ann Hobbs, National Institute of Health
Tom Hollinger, University of Florida
Edwin House, Idaho State University
John Jackson, North Hennepin Community College
Eric Jacobsson, University of Illinois, Urbana
Duane Jeffery, Brigham Young University
Kenneth Jones, California State University, Northridge
Patricia Jones, Stanford University
Gordon Kaye, Albany Medical School
Douglas Kelly, University of Southern California, School of Medicine
Regis Kelly, University of California, San Francisco
Bryce Kendrick, University of Waterloo, Ontario
Richard Kessel, University of Iowa
Jack Keyes, University of Oregon
John W. Kimball, Andover, Massachusetts
Arnold Kluge, University of Michigan
Richard Korf, Cornell University
Eugene Kozloff, Friday Harbor Laboratory
Robert Kull, United States Air Force Academy
Armand Kuris, University of California, Santa Barbara

John Lammert, Gustavus Adolphus College
William Lassiter, School of Medicine, University of North Carolina, Chapel Hill
C. Roland Leeson, University of Illinois, Urbana
George Lefevre, California State University, Northridge
Matthew Levy, Mt. Sinai School of Medicine, City University of New York
Harvey Lillywhite, University of Florida
Randy Linde, Palo Alto Medical Foundation
Robert Little, Medical College of Georgia
Reuben Lotan, University of Texas, M. D. Anderson Cancer Center
Roderick MacLeod, University of Illinois, Urbana
Richard MacMillen, University of California, Irvine
Michael Madigan, Southern Illinois University
Alan Mann, University of Pennsylvania
Kim Mattson, West Virginia University
Linda Maxson, University of Illinois, Urbana
Joyce Maxwell, California State University, Northridge
Heather McKean, Eastern Washington State University
Mary McKitrick, University of Michigan, Ann Arbor
F. M. Anne McNabb, Virginia Polytechnic Institute and State University
Thomas Mertens, Ball State University
G. Tyler Miller, St. Andrew's Presbyterian College
Hylan Moises, University of Michigan
Elizabeth Moore-Landecker, Glassboro State College
William Morrison, Shippensburg University of Pennsylvania
David Morton, Frostburg State University
David Mount, University of Arizona
Richard Murphy, University of Virginia Medical School
Diane Nelson, East Tennessee State University
David Norris, University of Colorado, Boulder
John Obringer, United States Air Force Academy
Merle Olson, University of Texas Health Science Center
Anna Pai, Montclair State College
Tom Palko, Arkansas Tech University
Jane Parnes, Stanford University
Gordon Parry, University of California, Berkeley
William Parson, University of Washington
Thomas Parsons, University of Toronto
David Patterson, Eleanor Roosevelt Institute for Cancer Research
Jan Pechenik, Tufts University
Harry Peery, University of Toronto
Jane Phillips, University of Wisconsin, Madison
Carl Pierce, Washington University
Marian Reeve, Merritt College
Louis F. Reichardt, University of California, San Francisco
Doug Reynolds, Eastern Kentucky University
John Rickett, University of Arkansas

Robert Robbins, Michigan State University
William Roberts, University of California, San Francisco
Cleon Ross, Colorado State University
Gordon Ross, University of California Medical Center, Los Angeles
Thomas Rost, University of California, Davis
Rudolfo Ruibal, University of California, Riverside
James Sackett, University of California, Los Angeles
Richard Sayre, Ohio State University
Stephen Scheckler, Virginia Polytechnic Institute and State University
David Schindler, Freshwater Institute
William Schlesinger, Duke University
Rudi Schmid, University of California, Berkeley
Donald Schneider, Dartmouth Medical School
Jurgen Schnermann, University of Michigan, School of Medicine
Richard Searles, Duke University
Larry Sellers, Louisiana Tech University
Brian Shea, Northwestern University
Gordon Shepherd, Yale University Medical School
Steven Shimada, Yale University Medical School
Roger Sloboda, Dartmouth College
Ralph Smith, University of California, Berkeley
Val Smith, University of North Carolina, Chapel Hill
Wayne Smith, University of Tampa
Frank Solomon, Massachusetts Institute of Technology
Pat Starr, Mount Hood Community College
Janet Stein, formerly of University of British Columbia
Samuel Sweet, University of California, Santa Barbara
Michael Tansey, Indiana University, Bloomington
David Tilman, University of Minnesota, Minneapolis-St. Paul
Kenneth Todar, University of Wisconsin, Madison
Heinz Valtin, Dartmouth Medical School
Mary Jane Voll, University of Maryland, College Park
Edward Voss, Jr.; University of Illinois
Robert Waaland, University of Washington
David Wade, Southern Illinois University
Lisa Wagner, Georgia Southern College
Robert Warmbrodt, University of Maryland
Wesley Weathers, University of California, Davis
Philippa Webster, Stanford University
Mark Weiss, Wayne State University
Robert Wetzel, University of Michigan
Mark Wheelis, University of California, Davis
Steven White, San Jose State University
Warren Wickelgren, University of Colorado, Health Science Center
Sandra Winicur, Indiana University, South Bend
Dana Wrensch, Ohio State University
John Wright, Gull Foundation for Medical Research
Adrienne Zihlman, University of California, Santa Cruz

CONTENTS IN BRIEF

CONTENTS

UNIT FOUR / PLANT SYSTEMS AND THEIR CONTROL

UNIT FIVE / ANIMAL SYSTEMS AND THEIR CONTROL

UNIT SEVEN / EVOLUTION AND DIVERSITY

UNIT ONE

UNIFYING CONCEPTS IN BIOLOGY

1

ON THE UNITY
AND DIVERSITY
OF LIFE

Buried somewhere in that mass of nerve tissue just above and behind your eyes are memories of first encounters with the living world. Still in residence are sensations of discovering your own two hands and feet, your family, the change of seasons, the smell of rain-drenched earth and grass. In that brain are traces of early introductions to a great disorganized parade of insects, flowers, frogs, and furred things, mostly living, sometimes dead. There, too, are memories of questions—"*What is life?*" and, inevitably, "*What is death?*" There are memories of answers, some satisfying, others less so.

Observing, asking questions, accumulating answers—in this manner you have acquired a store of approximate knowledge about the world of life. During the journey to maturity, experience and education have been refining your questions, and no doubt the answers are more difficult to come by. What *is* life? What characterizes the living state? The answers you get may vary, depending, for example, on whether they come from a physician, a clergyman, or a parent of a severely injured person who is maintained by mechanical life support systems because the brain is no longer functioning.

Figure 1.1 A common egret on the wing, against a background of bald cypress and Spanish moss. These three kinds of organisms are diverse in appearance, yet they have much in common. They illustrate the unity and diversity inherent in *all* of life, which is the subject of this chapter.

Yet despite the changing character of the questions, the world of living things persists much as it did before. Leaves still unfurl during the spring rains. Animals are born, they grow, reproduce, and die even as new individuals of their kind replace them. *The most important difference is in the degree of insight you now bring to observations, questions, and answers about such events.*

It is scarcely appropriate, then, for a book to proclaim that it is your introduction to biology—"the study of life"—when you have been studying life ever since awareness of the world began penetrating your brain. The subject is the same familiar world that you have already thought about to no small extent. That is why this book proclaims only to be biology *revisited*, in ways that may help carry your thoughts about life to deeper, more organized levels of understanding.

Let us return to the question, What is life? The answer has yet to be reduced to a simple definition. The word embodies a story that has been unfolding in countless directions for several billion years! To biologists, "life" is what it is by virtue of its ancient molecular origins and its degree of organization. "Life" is a way of capturing and systematically using energy and materials. "Life" is a commitment to a specific program of growth and development; it is a capacity for reproduction. "Life" is adjustment to changing conditions—it is *adaptive* to environmental change, both in the short term and through successive generations. Clearly, a short list of definitions only hints at all that the word conveys. Deeper insight into its meaning comes with wide-ranging study of its characteristics, *for life cannot be understood in isolation from its history and its adaptive potential.*

Throughout this book you will encounter examples of living things—how they are constructed, how they function, where they live, what they do. You also will come across statements about those examples and the generalizations that can be drawn from them. Such statements are highlighted with dark type and separated by lines from the text. All the statements, taken together, will give you a sense of what "life" is.

With this in mind, let us turn to a few examples that can illustrate the most general concepts of all. Although the concepts are explored more fully in later chapters, they are summarized here and in Table 1.1 to provide perspective on things to come. You may also find it useful to refer to these concept summaries later on, as a way of reinforcing your grasp of details.

ORIGINS AND ORGANIZATION

Suppose someone asks you to point out the difference between a frog and a rock. The frog, you might say, has a body of truly complex organization. Its hundreds of thousands of individual cells are organized into tissues.

Its tissues are arranged into organs such as a heart and a stomach. The frog can move about on its own. And sooner or later (given a receptive member of the opposite sex), it can reproduce. A rock shows no such ordered complexity, it cannot move by itself, and it certainly cannot reproduce either on its own or in the company of another rock. If you deduce from this that a living organism has complex regional organization, the capacity to move, and the capacity to reproduce, then the frog is alive and the rock is not.

Now suppose someone asks you to point out the difference between a bacterium and a rock. A bacterium is one of the simplest organisms, no more than a single cell. Yet microscopic examination shows that a bacterial body is regionally organized in complex ways. All but a few bacteria have an outer wall. All have a plasma membrane (a saclike outer membrane that helps control the kinds of substances moving into and out of the cell). The membrane encloses a semifluid substance in which spe-

Table 1.1 Some Characteristics of Life

1. All organisms are assembled from the same raw materials, according to the same laws of energy, as nonliving things—but they show more complex organization.

2. All organisms are part of webs of organization in nature, in that they depend directly or indirectly on one another for materials and energy.

3. All organisms show metabolic activity: they have the capacity to acquire and use energy to stockpile, tear down, build, and eliminate materials in ways that promote their survival and reproduction.

4. All organisms use homeostatic controls to respond to environmental changes, in ways that maintain favorable operating conditions for the body.

5. All organisms have the capacity for growth, development, and reproduction.

6. In all organisms, DNA is the molecule of inheritance; its instructions for reproducing heritable traits are passed on from parents to offspring.

7. All organisms show adaptive potential: heritable variations in their form, functioning, and behavior may allow them to adjust to changes in their environment, both over the short term and through successive generations.

8. Organisms show variations in form, functioning, and behavior that have accumulated as a result of natural selection and other evolutionary forces. The diversity of life is the sum total of those variations.

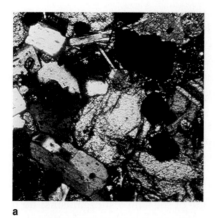

a

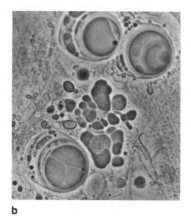

b

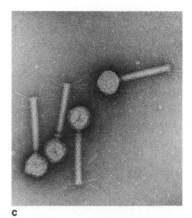

c

Figure 1.2 A hierarchy of structural organization between a frog and a rock. The sizes of the subjects in these photographs are not to the same scale relative to one another. (**a**) Basalt rock, thin-section, from Marianas Trench. (**b**) Liposomes, artificial lipid spheres with membrane properties that mimic living cell membranes. (**c**) Virus particles, with outer layers enclosing hereditary instructions. (**d**) Bacterial cell, sliced to show the inside. (**e**) Some of the complex structures within a single cell in a multicellular plant. (**f**) The many-celled frog body.

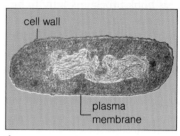

cell wall

plasma membrane

d

cific structures are embedded (Figure 1.2). All bacteria can divide and reproduce. Some can move on their own through the surroundings; others cannot. The "movement" criterion is becoming a little fuzzy now. However, by the other two standards (complex organization and the capacity to reproduce), a bacterium is alive.

Suppose you are now asked to compare a virus with a rock. A virus is a peculiar particle in the shadowy world between the living and the nonliving. Viruses do have a distinctive organization. Some, for instance, have a "head" end, a sheathlike midsection, and a "tail" end. A virus has no means whatsoever of moving on its own. It cannot reproduce on its own. Yet all viruses contain instructions for producing copies of themselves. To be carried out, the instructions must become incorporated into a living host cell. In effect, a virus takes over the cellular machinery, to the extent that its host starts following *viral* instructions. The machinery starts churning out parts that cannot be used by the cell, but that can be used for building new viruses! By your initial criteria—complex organization, movement, reproduction—is a virus alive? In some respects yes, in others no.

Somewhere below the boundary to the living world are microscopically small, water-filled sacs called liposomes (Figure 1.2b). Under the right conditions, they assemble spontaneously from simple lipid molecules. (You can observe this kind of assembly by stirring some oil into a glass of water. Oil molecules will not mix with water molecules; they cluster into droplets.) Intriguingly, lipids also are the main structural component of cell membranes—and many properties of liposomes correspond to properties of cell membranes. Both self-

assemble, and both have the capacity for self-repair (for example, they seal themselves when punctured). Both let water molecules enter freely but keep certain ions out. Thus liposomes show at least some *potential* for organization.

What about reproduction? When lipid molecules are added to them, liposomes grow in size. They can even grow to the extent that parts break off and form new spheres, which grow in turn. But this is not reproduction; this is only random chemical growth and fragmentation into (generally) nonidentical parts. So you are left to conclude that liposomes are not alive. Still, the way they are organized seems more intriguing than the organization of a rock.

And what about that rock? Although it seems so different from a bacterium, at the levels of viruses and liposomes the difference begins to blur. At a still deeper level, the difference becomes nonexistent. Frog, bacterium, virus, liposome, rock—all turn out to be composed of the same raw materials (particles called protons, electrons, and neutrons). And those materials become organized relative to one another according to the same physical laws. At the heart of those laws is something called **energy**—a capacity for interaction between particles, a capacity to make things happen, to do work. Energetic interactions join particles together, in predictable ways, and form atoms. They bind atom to atom in predictable patterns, thereby giving rise to molecules that form (for example) all frogs and rocks. Energetic interactions hold a frog and a rock together; the flow of energy organizes and holds entire communities of organisms together. Thus we have a profound concept:

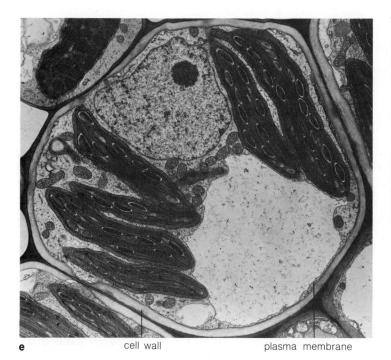

e   cell wall   plasma membrane

f

The structure and organization of the nonliving and the living world arise from the same fundamental properties of matter and energy.

This concept is even used to explain the origin of life. It now appears that the first cells emerged through the evolution of complex systems of molecules. Look carefully at Figure 1.3. This figure outlines the levels of organization in nature that we will be considering throughout the book. For now, consider the idea that these levels echo stages in the history of life. In the beginning, atoms and molecules interacted under certain conditions and formed the large molecules that are now found in all living things. The molecules became organized into systems—spheres, for example, much like liposomes. Inside the spheres, some molecules became organized in ways that allowed them to *duplicate* themselves—and to lay the foundation for reproduction. This capacity led to the *cell*—the smallest unit of life that still retains the properties of life.

What we are unfolding here is a picture of increasingly ordered patterns in the use of materials and energy. It accounts, as any speculation about the origin of life must do, for this apparent fact:

The "difference" between the living state and the nonliving state lies in the degree to which energy is used and materials are organized.

UNITY IN BASIC LIFE PROCESSES

Metabolism

Raw materials do not assemble on their own to form each new organism. The assembly processes—indeed, all processes associated with life—are the result of *energy transfers* between substances. For example, in a process called **photosynthesis**, plants absorb sunlight energy and use it to form compounds such as adenosine triphosphate, or ATP; then the energy of ATP is used to build sugar, starch, and other molecules. Here, energy is transferred from the sun, to ATP, and then on to molecules that the cell uses as building blocks or tucks away as energy reserves. In another process, called **aerobic respiration**, cells tap their energy reserves by breaking apart molecules so that some energy becomes available to form ATP—which in turn delivers energy to sites where specific cellular activities take place.

This example tells us something about the nature of energy transfers. Energy stores can be used up. Organisms cannot create "new" energy from nothing; to stay alive, they must tap an existing energy source from their surroundings (the sun, nectar from a flower, a chicken dinner), then transform the acquired energy into forms that can be stored and used to do cellular work.

All forms of life extract and transform energy from their surroundings, and they use it for manipulating materials in ways that assure maintenance, growth, and reproduction. More briefly, they show what is called "metabolic activity."

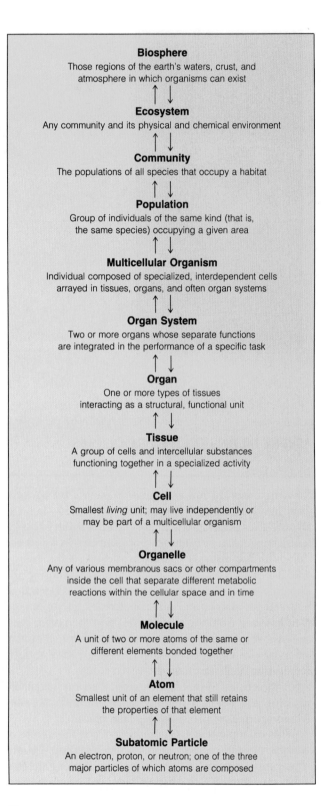

Biosphere
Those regions of the earth's waters, crust, and
atmosphere in which organisms can exist

↑ ↓

Ecosystem
Any community and its physical and chemical environment

↑ ↓

Community
The populations of all species that occupy a habitat

↑ ↓

Population
Group of individuals of the same kind (that is,
the same species) occupying a given area

↑ ↓

Multicellular Organism
Individual composed of specialized, interdependent cells
arrayed in tissues, organs, and often organ systems

↑ ↓

Organ System
Two or more organs whose separate functions
are integrated in the performance of a specific task

↑ ↓

Organ
One or more types of tissues
interacting as a structural, functional unit

↑ ↓

Tissue
A group of cells and intercellular substances
functioning together in a specialized activity

↑ ↓

Cell
Smallest *living* unit; may live independently or
may be part of a multicellular organism

↑ ↓

Organelle
Any of various membranous sacs or other compartments
inside the cell that separate different metabolic
reactions within the cellular space and in time

↑ ↓

Molecule
A unit of two or more atoms of the same or
different elements bonded together

↑ ↓

Atom
Smallest unit of an element that still retains
the properties of that element

↑ ↓

Subatomic Particle
An electron, proton, or neutron; one of the three
major particles of which atoms are composed

Figure 1.3 Levels of organization in nature.

All forms of life can acquire and use energy to stockpile, tear down, build, and eliminate materials in ways that assure survival and reproduction. This capacity is called <u>metabolism</u>.

Growth, Development, and Reproduction

Through metabolic events, living things come into the world, grow and develop, and reproduce. Most then move on through decline and death according to a timetable for their kind. Even as individual organisms die, reproduction assures that new organisms having much the same traits will replace them.

Yet "an organism" is much more than a single organized form having a single set of functions during its lifetime. One example will make the point, even though actual details vary considerably among organisms.

A tiny egg deposited on a branch by a female moth is a compact transitional form (Figure 1.4). It contains all the instructions necessary to become an adult moth. Before becoming a moth, developmental events inside the egg lead to an entirely different form: a wingless, many-legged larva called a caterpillar.

The caterpillar hatches during a warm season when tender new leaves unfold. Not coincidentally, the caterpillar is a streamlined "eating machine" able to tear and chew tender plant tissues. It has a capacity for extremely rapid growth. It eats and grows until some internal alarm clock goes off, setting in motion events that lead to profound changes in form. Some cells are disassembled, other cells multiply and are assembled into entirely different patterns. Tissues, too, are moved about during this wholesale remodeling, the so-called pupal stage. From the pupa, the adult moth emerges.

The moth is the "reproductive machine" stage. Its head has a tubelike extension (a proboscis) that draws nectar from flowers. From the nectar comes energy that powers free-wheeling flights. For this insect, wings are emblazoned with colors and move at a frequency that can attract a potential mate. The moth has organs in which egg or sperm develop, and which enhance fertilization of an egg and the production of offspring.

None of these stages is "the insect." "The insect" is a series of stages in organization, with different adaptive properties emerging at each stage.

Each organism arises through <u>reproduction</u>—that is, the production of offspring by one or more preexisting parent organisms.

Each organism undergoes <u>development</u>: it proceeds through a series of changes in form and behavior over time. The changes occur at about the same rate and in the same way for all organisms of a given type.

Homeostasis

Any attempt to define the nature of life cannot focus only on the organism, for an organism cannot exist apart from its surroundings. The living state is maintained within rather narrow limits. Concentrations of substances such as carbon dioxide and oxygen must not rise above or fall below certain levels. Toxic substances must be avoided or eliminated. Certain kinds of foods must be available, in certain amounts. Water, oxygen, carbon dioxide, light, temperature—such environmental factors dictate the terms of survival. *And such terms are subject to change.*

How do living things respond to changes in the environment? They respond in two ways. First, all organisms have built-in means of making internal adjustments to outside changes. The adjustments help maintain operating conditions within some tolerable range. This capacity for maintaining the "internal environment" is known as **homeostasis**. Individual cells have homeostatic controls. (For instance, they have mechanisms for bringing in substances that are in short supply and for eliminating other substances.) Multicelled organisms also have homeostatic controls. (Birds, for instance, have sensors that signal the brain when the outside temperature drops. The brain may send signals to cells that control feather movements. Special movements lead to feather fluffing, a behavior that retains heat and helps maintain body temperature.)

Homeostasis implies constancy, a sort of perpetual bouncing back to some limited set of operating conditions. In some respects, constancy is indeed vital. Your red blood cells will not function unless they are bathed in water that contains fairly exact amounts of dissolved components. Your body works so that the bathwater, so to speak, is always much the same.

Yet living things also respond in a second way to changing conditions. All organisms adjust to certain *directional* changes in the internal and external environments. We might call this **dynamic homeostasis**, for the living state is maintained through adjustments that shift the form and function of the organism over time.

A simple example will do here. In humans, irreversible chemical changes trigger puberty, the age at which sexual reproductive structures mature and become functional. At puberty, the body steps up its secretions of such hormones as androgens (in males) and estrogens (in females). The increased secretions are necessary for sexual maturation. They call for entirely new events such as the menstrual cycle. This cycle includes a rhythmic accumulation of substances that prepare the female body for pregnancy, followed by disposal of substances when pregnancy does not occur. It is not that homeostasis no longer operates. It is that developmental events now demand new kinds of adjustments in the internal state.

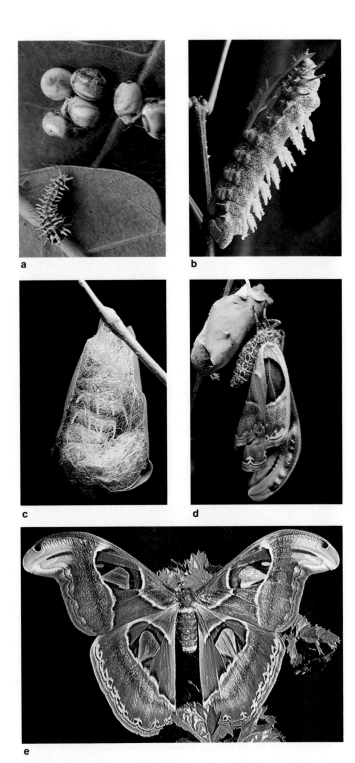

Figure 1.4 "The insect"—a continuum of developmental stages, with new adaptive properties emerging at each stage. Shown here: the development of a giant moth, from egg (**a**) to larval stage (**b**), to pupal form (**c**), to emergence of the resplendent moth form (**d, e**).

Figure 1.5 Underwater tropical reef off the Florida Keys. Elkhorn coral rises above the forestlike growth of staghorn coral.

All forms of life depend on homeostatic controls, which maintain the living state as internal and external conditions change.

Some homeostatic controls keep internal conditions within some tolerable range throughout the life cycle.

Some homeostatic controls govern new adjustments in the internal state as the life cycle unfolds.

DNA: Storehouse of Constancy and Change

Upon thinking about the preceding examples of development, you might wonder what could be responsible for **inheritance**—the transmission, from parent to offspring, of structural and functional patterns characteristic of each kind of organism. How is it that a bacterium can divide and develop into two fairly exact copies of itself? How is it that corn seeds can germinate and develop into fairly exact replicas of parent corn plants? Within each individual, there must be a storehouse of hereditary information.

This storehouse of information has a remarkable characteristic. Although offspring resemble their parents in form and behavior, *variations* can exist on the basic plan. A newly produced bacterium might not be able to assemble (as it is supposed to) some molecule that is vital to its functioning. Some humans are born with six digits on each hand instead of five. *Overall, hereditary instructions must remain intact to assure faithful transmission of traits—yet they also must be subject to change in some details!*

As you have probably learned by now, we know where the instructions reside. In all living cells, they are encoded in molecules of deoxyribonucleic acid, or DNA. We also know that changes can occur in the kind, structure, sequence, or number of component parts of DNA. These changes are **mutations**. Most mutations are harmful, for the DNA of each kind of organism is a package of information that is finely tuned to a given environment. In addition, its separate bits of information are part of a coordinated whole. When one crucial part changes, the whole living system may be thrown off balance. Such is the fate of the bacterium mentioned in the preceding paragraph, for the change probably means that it is doomed.

However, sometimes a mutation may prove to be harmless, even beneficial, under prevailing conditions. For example, mutation produces a dark-colored form of a moth (*Biston betularia*) that otherwise is light-colored. When the mutant moth rests on soot-covered trees, bird predators simply do not see it. In places where there happen to be lots of soot-covered trees (as in industrial regions), the mutant stands a better chance of not being eaten—hence surviving and reproducing—than its light-colored kin.

DNA is a storehouse of patterns for all heritable traits.

Mutations introduce variations in the patterns.

The environment—internal and external—is the testing ground for the combination of patterns that come to be expressed in each individual.

DIVERSITY IN FORM AND FUNCTION

Until now, we have focused on the unity of life—on characteristics shared by all organisms. We touched on the obvious—that all organisms have the potential for reproduction and development. Moreover, we suggested that the structure and organization of all living things arise from basic properties of matter and energy, that all organisms rely on metabolic and homeostatic processes, and that all organisms have the same molecular basis of inheritance. At one time people had no idea that living things hold these other characteristics in common. What *was* apparent, and difficult to explain, was the tremendous *diversity* of life. Why is it that almost every environment is host to an astonishing array of different organisms? Before trying to answer this question, let's briefly consider some aspects of diversity in two different settings.

The Tropical Reef

Imagine yourself exploring a tropical reef of the sort shown in Figures 1.5 and 1.6. Long ago, small animals called corals began to grow and reproduce in the warm, nearshore waters of tropical seas (Figure 1.7). The skeletons they left behind served as a foundation for more corals to build upon. As skeletons and residues accumulated, the reef grew. All the while, tides and currents carved ledges and caverns into it. Today, a reef's spine can be decked out with any number of *750 different kinds* of animals called corals.

Plants called red algae typically encrust the coral foundation. In shallow waters behind the reef, red algae give way to blue-green forms. Many small, transparent animals feed on algae and other plants. These animals in turn are food for still larger animals, including some of the world's *20,000 different kinds* of fishes. Squatting

Figure 1.6 Who eats whom on coral reefs. (**a**) Crown-of-thorns, a sea star that feasts on tiny corals. (**b**) Sea anemone, an animal with weapon-studded tentacles, which ensnare tiny animals floating past. (**c**) Sponges, with pores opened toward the oncoming food-laden currents. (**d**) Clownfish, curiously at home above the mouth of a sea anemone—a mouth through which other kinds of edible fish quickly disappear. (**e**) Green algae, plants that are food for various reef organisms. (**f**) Red algae, food for various animals (but not for this chambered nautilus, a shelled animal that swims expertly after shrimp and other prey).

(**g**) A school of goatfish, which feed on small, spineless animals on the sea floor. Goatfish are tasty to humans, also to large fish. (**h**) Some fish are not on the general menu. Here, a blue wrasse safely picks off and dines on parasites that live on this large predatory fish. (**i**) Stone crab. Depending on the species, crabs eat plants, animals, and organic remains. The moray (**l**) prefers meat. (**j**) Lion fish, with its fanned, poison-tipped spines warning away intruders. (**k**) Find the scorpion fish—a dangerous animal that lies camouflaged and motionless on the sea bottom, the better to surprise unsuspecting prey.

a

b

c

Figure 1.7 A few master builders of coral reefs: (**a**) pillar coral, (**b**) daisy coral, and (**c**) green tube coral.

Figure 1.8 (right) The East African Rift Valley, some 6,400 kilometers (4,000 miles) long. The sparsely wooded grasslands in this valley are home for a diverse array of animal life and were the probable birthplace of the human species.

on the reef are predatory sea anemones, each with a mouth fringed with tentacles that capture smaller fish. Yet, hovering above the tentacles *is* a certain kind of fish! It is as edible as most others, but somehow it is not recognized as prey. The fish moves away, captures food, and returns to the anemone's tentacles—which give it protection from other predators. The anemone eats food scraps that fall from the mouth of the fish. In effect, the animals are allies: one receives protection, the other receives food (Figure 1.6d).

The reef is also home for different kinds of sea stars. When feeding, a sea star extends its stomach from its body and puts it into coral chambers. Each chamber resident is digested in place before the stomach is pulled out. When sea stars reproduce, millions of larvae emerge and feed on microscopic algae. Then, as the larvae grow, they become food for the meat-eating corals! It is the corals that now grow and reproduce. In time they repopulate the reef regions that the earlier generation of sea stars had stripped clean. Sea star larvae that do escape grow to become diner instead of dinner, and thereby initiate a new cycle of death and life.

Before speculating on what could account for the diversity of reef organisms, imagine yourself in another setting to see whether a comparison yields any similarities or differences that might provide added insight.

The Savanna

In the shadow of Kilimanjaro, a volcanic peak rising above the edge of the East African Rift Valley, grasslands sweep out to the northeast. This is the African *savanna*, a region of warm grasslands punctuated by scattered stands of shrubs and trees (Figure 1.8). More large ungulates (hoofed, plant-eating animals) live here than anywhere else. One form, the giraffe, browses on leaves beyond the reach of other ungulates. Another form is the Cape buffalo (Figure 1.9). An adult male can weigh a ton, it has formidable horns, and its behavior is unpredictable. It is rarely troubled by predators. Other forms include zebra and impala, which are smaller, more vulnerable, and more abundant than Cape buffalo. They are constantly troubled by such predators as lions and cheetahs. Their remains (as well as the remains of lions and cheetahs) are picked over by scavengers—hyenas, jackals, vultures, and marabou storks (Figure 1.10).

Buffalo, zebra, impala, rhinoceros, giraffe—these and *eighty-five other kinds* of large, plant-eating animals live in the immense valley, as do different kinds of animals that feed on them. They exist side by side in time, moving westward, southward, and back again as dry seasons follow rains, as scorched earth gives way to a resurgence of plant growth.

A Definition of Diversity

When you compare the diversity of the reef with that of the savanna, what conclusions can be drawn? One thing their diverse occupants have in common is some specialization in "who eats whom," beginning with plants and proceeding through different forms of animals that eat the plants and one another. You could spend years observing organisms in any setting, at microscopic as well as macroscopic levels, and you would find that they speak eloquently of the same challenge. *All organisms must be equipped to obtain a share of available resources.* In large part, diversity in form, function, and behavior represents specialized ways to get and use resources—and to avoid becoming a resource for some other organism. In light of this observation, let us now address the question of how such diversity could have come about.

Since the time of their origin, living things have required a constant supply of energy and materials. Yet think about the times when you yourself have encountered shortages (for example, of water, gasoline, electricity, or lettuce). In the past, as today, resources were not always abundant. More often than not, *members of every group of organisms must have been demanding a share of limited resources.*

Imagine, next, that *variant* members occasionally appeared (perhaps through DNA mutations). Some variant forms might have been better equipped for securing resources. Some might have been better equipped for responding to predators, prey, or inadvertent allies around them. Accordingly, they would have tended to be the ones that survived and reproduced. Through reproduction, the heritable basis for the variation would have been transmitted to offspring. Given that such offspring were more likely to survive, they would have occurred with increasing frequency in each new generation. There would have been *natural selection,* within the group, of those individuals better adapted to prevailing conditions.

This line of thought amounts to one view of a main road to diversity. Other roads opened up also, but whatever the pathways taken, a strong argument can be made that the following is true:

Diversity is the sum total of variations in form, functioning, and behavior that have accumulated in different lines of organisms.

Such variations tend to be adaptive (or at least not harmful) under prevailing conditions.

a

b

c

Figure 1.9 A sampling of the ninety kinds of large plant-eating animals that live in the savanna—a clear example of diversity in a single environment. (**a**) A herd of Cape buffalo. Imagine yourself a predator this close to the herd and you get an idea of one of the benefits of group living. (**b**) Zebra mother and offspring. (**c**) Male and female impala on the alert, ready to take cover in the nearby woods. (**d**) The rhinoceros, another formidably decked-out plant eater. (**e**) The giraffe, browsing on vegetation high up.

ENERGY FLOW AND THE CYCLING OF RESOURCES

Let's now put this view of diversity in the context of ecological interactions. The geologic record suggests that the first forms of life on earth lived independently of one another, perhaps in tidal flats along the margins of ancient seas. They must have fed on already existing substances, such as simple carbon-containing compounds that had accumulated through volcanic eruptions and other geologic processes. Eventually, perhaps when food supplies began to dwindle, those organisms

d

e

began relying more *on each other* as sources of energy and materials. Thus, by necessity, community interactions such as predation began and have continued in ever richer diversity. Through these interactions, few existing energy sources are unexploited. One example will make the point, even though the cast of characters seems of a most improbable sort.

First we have the adult male elephant of the African savanna (Figure 1.11). It stands almost two stories high at the shoulder and weighs more than eight tons. This grazing animal eats quantities of plants, the remains of which leave its body as droppings of considerable size. Appearances to the contrary, locked in the droppings are substantial stores of unused nutrients. With resource availability being what it is, even waste products from one kind of organism are food for another.

And so we next have little dung beetles rushing to the scene almost simultaneously with the uplifting of the

a

b

Figure 1.10 Predators and scavengers of the savanna. (**a**) An adult lioness standing over a fresh kill. These large cats stalk the herds at dusk or afterward, typically concealing themselves in dense or low-lying vegetation. (**b**) Vultures, together with marabou storks, feed on locusts, small birds, and small mammals—but they also clean up whatever carrion becomes available to them. In this dual predator-scavenger role, they are like other diverse animals of the savanna, including hyenas and jackals.

a

b

c

d

Figure 1.11 An ecological interdependency of a most improbable sort, beginning with the plants that feed the elephants (**a**), the dung that leaves the elephant (**b**), the beetles that roll dung balls away and bury them (**c**), ending with the beetle larva (**d**) that hatches in the dung—and the remains of the dung itself, enriching the soil in which plants grow, eventually to feed the elephants.

elephant tail. With great precision they carve out fragments of the dung into round balls. The dung balls are rolled off and buried underground in burrows, where they serve as compact food supplies. In these balls the beetles lay eggs, a reproductive behavior that assures the forthcoming offspring of a food supply. Also assured is an uncluttered environment. If the dung were to remain aboveground, it would dry out and pile up beneath the hot African sun. Instead, the surface of the land is tidied up, the beetle has its resource, and the remains of the dung are left to decay in burrows—there to enrich the soil that nourishes the plants that sustain (among others) the elephants.

Such interactions of organisms with their environment and with one another are the focus of **ecology**. Everywhere you look you will find different organisms locked in ecological interdependency. Some interdependencies are simple and others complex, and some seem outlandish—yet the underlying principle is the same:

Most forms of life depend directly or indirectly on one another for materials and energy.

At its most inclusive level, ecological interdependency encompasses the biosphere (the entire zone of earth, water, and atmosphere in which life can exist). With few exceptions, living organisms are linked together by a one-way flow of energy from the sun and by a cycling of materials (such as carbon dioxide and oxygen) on a global scale. Plants (and some photosynthetic microorganisms) harness sunlight energy. They use this energy in constructing and maintaining the plant body. Directly or indirectly, other organisms feed on energy stored in plant parts. Microscopic decomposers obtain energy by breaking apart molecules of plant and animal remains. Through their activity, they help recycle many materials for new generations of life (Figure 1.12).

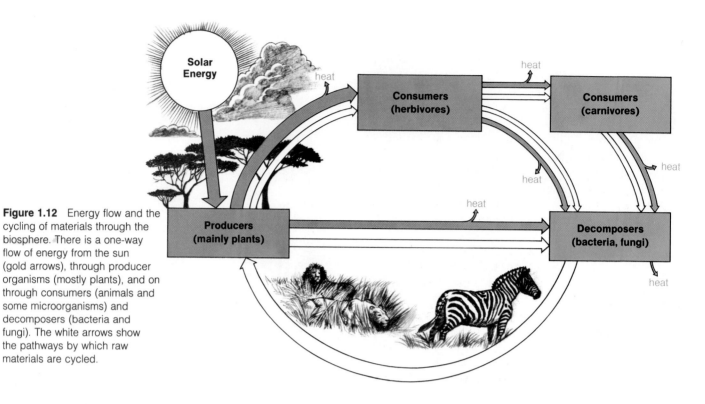

Figure 1.12 Energy flow and the cycling of materials through the biosphere. There is a one-way flow of energy from the sun (gold arrows), through producer organisms (mostly plants), and on through consumers (animals and some microorganisms) and decomposers (bacteria and fungi). The white arrows show the pathways by which raw materials are cycled.

PERSPECTIVE

This chapter has touched on two fundamental aspects of life: its unity and diversity. All organisms are *alike* in sharing common origins, in adhering to the same rules that govern the organization of matter and energy, in relying on metabolic and homeostatic processes, and in having the same molecular basis of inheritance. They are also dramatically *unalike* in appearance and behavior.

There are *many millions* of different organisms, and only a fraction have been catalogued and studied. Many millions more once existed and became extinct. Explaining how such immense diversity arose, and at the same time accounting for its underlying unity, would be quite an accomplishment. In biology, such an explanation was put together. It is called the principle of evolution by means of natural selection. It is, in fact, the formal statement of the informal picture of natural selection presented earlier in the chapter.

A principle is an idea whose validity holds up even when many different observations and experiments are used to test it. The one we are talking about here will be described in the next chapter, together with examples of observations that led to its formulation. This powerful concept will help you interpret many of the observations and experimental results you will encounter throughout this book. In turn, you will see how observations and experiments from many lines of inquiry lend support to the principle. It is because of this depth and range of supporting evidence that biologists in general view the

principle as the most logical explanation of the seeming contradiction inherent in life: its unity *and* diversity.

Review Questions

1. For this and subsequent chapters, make a list of the **boldface** terms that occurred in the text. Write a definition next to each, then check it against the one in the text. (You will be using these terms later on.)

2. Why is it difficult to give a simple definition of life? *2* (For this and subsequent chapters, *italic numbers* following review questions indicate the pages on which the answers may be found.)

3. What is meant by "adaptive"? Give some examples of environmental conditions to which plants and animals must be adapted in order to stay alive. *2, 13*

4. If the structure and organization of all things arise from the basic nature of matter and energy, then what is the essential difference between living and nonliving things? *5*

5. Study Figure 1.3. Then, on your own, arrange and define the levels of biological organization. What concept ties this organization to the history of life, from the time of origin to the present? *5*

6. In what fundamental ways are all organisms alike? *2, 9*

7. What is metabolic activity? *5–6*

8. What aspect of life is being overlooked when you talk about "the animal" called a frog? (Hint: What's a tadpole?) *6*

9. What is DNA? What is a mutation? Why are most mutations likely to be harmful? *8*

10. Outline the one-way flow of energy and the cycling of materials through the biosphere. *17–18*

ON SCIENTIFIC PRINCIPLES

The preceding chapter claimed that this book can help carry your thinking about life to deeper, more organized levels of understanding. By itself that is a presumptuous claim, for success depends partly on how open you are to thinking about things in light of the principle of evolution. This principle is so powerful it can be used to make sense of observations at all levels of biological organization, from molecules to the biosphere. That is why it is being introduced here. Details of observations and experiments are a necessary part of the chapters that follow, for they represent evidence in support of the statements being made.

What, exactly, is a "principle"? The question is important, for the answer provides insight into why the principle of evolution is used with such confidence. A **principle** is a way of explaining a major phenomenon of nature, one that has been synthesized from a large body of information. Thus the idea of evolution developed over centuries, as naturalists and travelers observed and collected specimens of living and extinct forms, then asked questions about the remarkable diversity those specimens represented. It became clear that almost all organisms alive today are very different from organisms of the remote past. Eventually there was overwhelming evidence that the difference was a consequence of evolution—of changes in lines of descent that have accrued since life began.

If we were to idealize the route from a question about such a major aspect of the world to a fundamental explanation for it, we might end up with a list like this:

1. Ask a question (or identify a problem).

2. Make one or more **hypotheses**, or educated guesses, about what the answer (or solution) might be. This means using the process of **induction**: sorting through clues, hunches, and observations, then combining bits of information and logic to produce a general statement (the hypothesis).

3. Predict what the consequences might be if a hypothesis is valid. This process of reasoning from a general statement to predicting consequences is called **deduction** (and sometimes the "if-then" process).

4. Devise ways to *test* those deductions by making observations, developing models, or performing experiments.

5. Repeat the tests as often as necessary to determine whether results will be consistent and as predicted.

6. Report objectively on the tests and on conclusions drawn from them.

7. Examine alternative hypotheses in the same manner.

2
METHODS AND ORGANIZING CONCEPTS IN BIOLOGY

Figure 2.1 Galápagos tortoises, one of many diverse organisms that Charles Darwin encountered on the Galápagos Islands.

This route represents what might be called a scientific approach to interpreting the natural world. Hypotheses are proposed, then deductions are made and tested (see *Commentary*). There is no such thing as any one "scientific method" of doing this. In practice, insights arise from accident and intuition as well as from methodical search. Some individuals adhere to existing procedures, others improvise as they go. No matter what the individual "method," however, the bottom line in science is this: *Hypotheses must be testable—and the tests must not be so loosely conceived that they cannot be duplicated or verified.*

No scientist can put forward an idea and demand that it be believed as true, no questions asked. In science, there are no absolute truths. There are only high probabilities that an idea is correct in the context of observations and tests made so far. Instead of absolutes, there is **suspended judgment**. This means a hypothesis is tentatively said to be valid if it is consistent with observations at hand. You won't (or shouldn't) hear a scientist say, "There is no other explanation!" More likely you will hear, "Based on present knowledge, this explanation is our best judgment at the moment."

Often the weight of evidence is so convincing that the hypothesis becomes accepted as a **theory**: a coherent set of ideas that form a general frame of reference for further studies. In science, the word "theory" is not used lightly. It is bestowed only on hypotheses that can be relied upon with a very high degree of confidence.

Sometimes substantiating evidence seems so overwhelming that a theory is elevated to the status of principle and serves as a doctrine from which other concepts are drawn. Even a principle is not beyond scrutiny, for new observations and test results may call for its modification or replacement. Far from being a disaster, this activity stimulates the development of even more adequate explanations.

Obviously, individual scientists would rather have their name associated with useful explanations than with useless ones. But they must be objective and keep asking themselves: "Is my thinking consistent with available observations and tests of what I hope to explain?" This does not mean all scientists are objective all of the time or even most of the time; no one can lay claim to that. It means only that they are expected as individuals to forsake pride and prejudice by testing their beliefs, even in ways that might prove them wrong. Even if an individual scientist doesn't, or won't, *others will*—for science proceeds as a community that is both cooperative and competitive. Ideas are shared, with the understanding that it is just as important to expose errors as it is to applaud insights.

This call for objectivity strengthens the theories and principles that do emerge from scientific studies. Yet it also puts limits on the kinds of studies that can be carried out. Beyond the realm of what can be analyzed with the methods and technology available, certain events remain unexplained. Why do we exist, for what purpose? Why does any one of us have to die at a particular moment and not another? Why do we sense beauty in some things and recoil in horror from others?

Answers to such questions are *subjective*; they come from within, as a consequence of all those factors shaping the consciousness of each individual. Because these factors can be infinitely variable, they do not readily lend themselves to scientific analysis.

This is not to say that subjective answers are without value. No human society can function without a shared commitment to standards for making judgments, however subjective those judgments might be. Moral, aesthetic, economic, and philosophical standards vary from one society to the next. But all guide their members in deciding what is important and good, and what is not. All attempt to give meaning to what we do.

Every so often, scientists stir up controversy when they explain part of the world that was previously considered beyond natural explanation, or belonging to the *supernatural*. This is sometimes true when moral codes are interwoven with religious narratives, which grew out of observations by ancestors. Questioning some long-standing view of the world may be misinterpreted as questioning morality, even though the two are not remotely synonymous.

For example, centuries ago Nicolaus Copernicus studied the movements of planets and stated that the earth circles the sun. Today the statement seems obvious. Back then, it was heresy. The prevailing belief was that the Creator had made the earth (and, by extension, humankind) the immovable center of the universe! Not long afterward a respected professor, Galileo Galilei, studied the Copernican model of the solar system. He thought it was a good one and said so. He was forced to retract his statement publicly, on his knees, and to put the earth back as the fixed center of things. (Word has it that when he stood up he muttered, "But it moves nevertheless.")

Today, as then, society has its sets of standards. Today, as then, those standards may be called into question when a new, natural explanation runs counter to a supernatural belief. When this happens it doesn't mean that scientists as a group are less moral, less lawful, less sensitive, or less caring than any other group. It means only that their individual and collective work has been guided by one additional standard: *The external world, not internal conviction, must be the testing ground for scientific beliefs.*

Systematic observations, hypotheses, predictions, tests—in all these ways, science differs from systems of belief that are based on faith, force, authority, or simple

Testing the Hypothesis Through Experiments

William H. Leonard, Clemson University

How is it that scientists probe so skillfully into the monument of life and discover so much about its foundations? What is it about their manner of thinking that yields such precise results? Simply put, scientific inquiry routinely depends on systematic observation and test.

Observations can be made directly, through systems of vision, hearing, taste, olfaction, and touch. They can be made indirectly, through use of special equipment (such as a microscope) that extends the range of perception. With practice, we can become skilled at *making systematic observations*. This means focusing one or more senses on a particular object or event in the environment, and screening out the "background noise" of information that probably has no bearing on our focus.

Hypothesizing means putting together a tentative explanation to account for an observation. When a hypothesis is scientific, it is *testable* through experiments. Experiments are devised to test whether predictions that can be derived from the hypothesis are correct. Thus the hypothesis must be constructed so that it provides a framework for stating the results of an experiment. Its content must be more specific than a problem statement, and often it is worded in the negative. Why is this so? Scientists tend to accept tentatively a plausible idea until it is shown to be false. It is difficult to prove experimentally that a hypothesis is true, because its validity would have to be demonstrated for all possible cases and under all possible conditions. Scientists therefore continue to test hypotheses by devising experiments that might show them to be false. If they succeed, then the hypothesis must be modified or discarded. That is why hypotheses are expressed in the negative.

For example, "DDT concentrations of 0.0001 percent by weight in the food of laboratory rats will not have harmful effects on the maintenance of the rat population over five years." If experiments reveal harmful effects at that dosage, then the hypothesis is not correct, and support is given to the idea that DDT is harmful.

Testing the hypothesis through experiments is at the heart of scientific inquiry. The goal is to control all variables except the one under study. Variables are events or conditions subject to change. For example, variables that are common to many biological experiments are the amount of light, temperature, and moisture. Others are concentrations of substances and numbers of organisms (population density) in a defined space. There are three general categories of variables:

independent variables	*the condition or event under study*
dependent variables	*ones that can possibly change because of the presence of, or change in, an independent variable*
controlled variables	*conditions that could affect the outcome of an experiment but that do not, because they are held constant*

An experimenter observes or manipulates one independent variable at a time, to identify any effects it has on dependent variables. If more than one independent variable were studied simultaneously, it would not be clear which one was responsible for the observed experimental results.

In one classic experimental design, a population of organisms is divided into two groups. The experimental group is the one subjected to the independent variable; the control group is not. All other variables are held the same in both groups. Thus, any differences that show up in test results for the two groups can be attributed to the independent variable. The illustration on the next page is an example of the use of experimental and control groups. This experiment has been used to test the hypothesis that laboratory rats ingesting DDT with normal food will lose weight, show less resistance to disease, and have a lower reproductive rate than rats not ingesting DDT. Notice

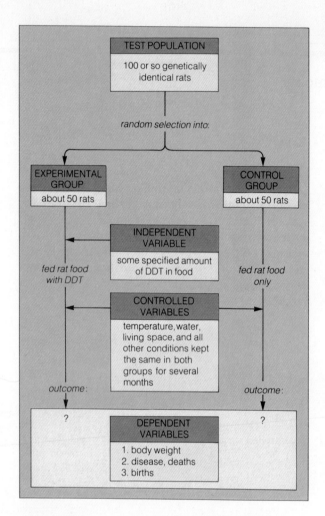

An example of a classic experimental design in biology. The experiment is designed to test the hypothesis that DDT ingested with food will not have harmful effects on laboratory rats over a period of time. With all other variables held constant, test results should refute or support the hypothesis.

that the rats were randomly assorted into either the experimental group or the control group. *Randomization* ensures that both groups are representative (or equivalent) samples of the original population. When any test group is not equivalent to a natural population, *sampling error* is introduced into the experiment. Then one could argue that any experimental results were due to differences in the composition of the different test groups, instead of a result of the independent variable.

Collecting and organizing test results is a necessary process in biological experiments. Data tables or graphs are used to organize and display information for analysis. Graphs are especially useful in illustrating trends or patterns. Data analysis is less mechanical and more conceptual than collecting and organizing the information. Often, statistical tests are used to determine if differences between experimental data and control data are *significant* or are likely due only to chance. If it can be argued that the differences are due to chance only, then it can also be argued that the independent variable had no effect. For example, say that at the end of an experiment on DDT effects, the average adult rat weight was 187.4 grams in the DDT-fed group, and 206.7 grams in the control group. Is this difference significant enough to suggest that there was an actual effect? The use of mathematical tools characteristic of statistical analysis could help in finding an answer.

Generalizing from test results requires careful and objective analysis of the data gathered. Usually, the hypothesis under test is accepted or rejected on the basis of conclusions drawn. A statement is written about what new insights (if any) have been gained into the original problem. Apparent trends are noted when the same data appear in test results gathered over a period of time. Often, further questions and hypotheses are posed in an attempt to guide additional studies of the problem.

consensus. It is not any "law" that is the focus of science. Rather the focus is on the observations that the "law" attempts to explain. A "law" can be invalidated by new evidence, gathered through ongoing tests and clarification of what those observations really mean.

There are, in the history of science, a few individuals who challenged beliefs held not only by society but by the scientific community within it. In biology, Charles

Darwin is among them. More than a century ago, he put together a theory and started lines of investigation that are still flourishing. Tracing Darwin's story and its antecedents will show that he relied on the scientific approach. In his words, "The line of argument pursued through my theory is to establish a point as a probability by deduction and to apply it as hypothesis to other points to see whether it will solve them."

EMERGENCE OF EVOLUTIONARY THOUGHT

More than two thousand years ago, the seeds of biological inquiry were taking hold among the ancient Greeks. This was a time when popular belief held that supernatural beings intervened directly in human affairs. For example, the gods were said to cause a common ailment known as the sacred disease. Yet from a physician of the school of Hippocrates, these thoughts come down to us:

It seems to me that the disease called sacred . . . has a natural cause, just as other diseases have. Men think it divine merely because they do not understand it. But if they called everything divine that they did not understand, there would be no end of divine things! . . . If you watch these fellows treating the disease, you see them use all kinds of incantations and magic—but they are also very careful in regulating diet. Now if food makes the disease better or worse, how can they say it is the gods who do this? . . . It does not really matter whether you call such things divine or not. In Nature, all things are alike in this, in that they can be traced to preceding causes.

—On the Sacred Disease (400 B.C.)

Such passages reflect the start of a commitment to finding natural explanations for observable events.

Aristotle was foremost among the early naturalists and described the world around him in excellent detail. He had no reference books or instruments to guide him, for biological science in the Western world *began* with the great thinkers of this age. Yet here was a man who was no mere collector of random bits of information. In his descriptions is evidence of a mind perceiving connections between observations and making hypotheses to explain the order of things.

When Aristotle began his studies, he believed (as did others) that each kind of organism was distinct from all the rest. Later he wondered about bizarre forms that could not be readily classified. In structure or function, they so resembled other forms that their place in nature seemed blurred. (For example, to Aristotle some sponges looked like plants but in their feeding habits they were animals.) He came to view nature as organized gradually from lifeless matter through complex forms of plant and animal life. This view is reflected in his model of biological organization (Figure 2.2), the first theoretical framework in the history of biology.

By the fourteenth century, this line of thought had become transformed into a rigid view of life. A great Chain of Being was seen to extend from the lowest forms to humans and on to spiritual beings. Each kind of being,

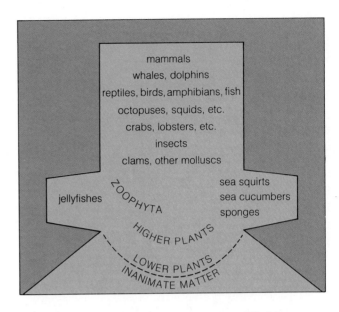

Figure 2.2 Scala Naturae—Aristotle's "ladder of life," the prototype of modern classification schemes.

or **species**, as it was called, was seen to have a fixed, separate place in the divine order of things. Each had remained unchanged since the time of creation, a permanent link in the chain. Scholars thought that once they had discovered, named, and described all the links, the meaning of life would be revealed to them. Contradictory views were not encouraged; scientific inquiry was channeled into an encyclopedic recording of the links.

As long as the world of living things meant mostly those forms in Europe, the task seemed manageable. With the global explorations of the sixteenth century, however, "the world" expanded enormously. Naturalists were soon overwhelmed by descriptions of thousands of plants and animals discovered in Asia, Africa, the Pacific islands, and the New World. Some specimens appeared to be quite similar to common European forms, but some were unique to different lands. *How could those extraordinarily diverse organisms be classified?*

The naturalist Thomas Moufet, in attempting to sort through the bewildering array, simply gave up and recorded such gems as this description of grasshoppers and locusts: "Some are green, some black, some blue. Some fly with one pair of wings; others with more; those that have no wings they leap; those that cannot fly or leap they walk; some have long shanks, some shorter. Some there are that sing, others are silent. . . ." It was not exactly a time of subtle distinctions.

Table 2.1	Classification of Three Organisms According to a Linnean Scheme		
Category (taxon)	Corn	Housefly	Human
Kingdom	Plantae	Animalia	Animalia
Phylum (or division, in botanical schemes)	Anthophyta	Arthropoda	Chordata
Class	Monocotyledonae	Insecta	Mammalia
Order	Commelinales	Diptera	Primates
Family	Poaceae	Muscidae	Hominidae
Genus	*Zea*	*Musca*	*Homo*
Species	*mays*	*domestica*	*sapiens*

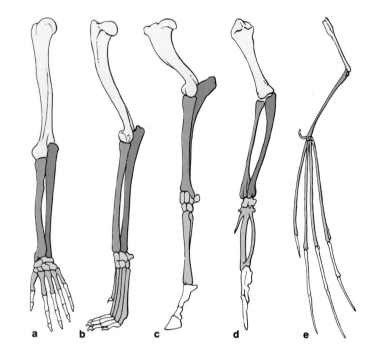

Figure 2.3 Similarities in forelimb structure among a few vertebrates: (**a**) human, (**b**) dog, (**c**) horse, (**d**) bird, and (**e**) bat. The drawings are not to scale relative to one another. Similar structures are shaded the same way from one animal to the next.

Linnean System of Classification of Organisms

The first widely accepted method of classification is attributed to Carl von Linné, now known by his latinized name, Linnaeus. This man was an eighteenth-century naturalist whose enthusiasm knew no bounds. He sent ill-prepared students around the world to gather specimens of plants and animals, and is said to have lost a third of his collectors to the rigors of their expeditions. Although perhaps not very commendable as a student adviser, Linnaeus did go on to develop the **binomial system of nomenclature**. With this system, each organism could be classified by assigning it a Latin name consisting of two parts.

For instance, *Ursus maritimus* is the scientific name for the polar bear. The first name refers to the **genus** (plural, genera), and the first letter of that name is capitalized. Distinct but obviously similar species are grouped in the same genus. Thus other bears are *Ursus arctos*, the Alaskan brown bear; and *Ursus americanus*, the black bear. The second, uncapitalized name is the **specific epithet**. The specific epithet is never used without the full or abbreviated generic name preceding it, for it also can be the second name of a species found in an entirely different genus. For instance, the Atlantic lobster is called *Homarus americanus;* the American toad, *Bufo americanus*. (Hence one would not order *americanus* for dinner unless one is willing to take what one gets.)

The binomial system was at the heart of a scheme that was thought to mirror the patterns of links in the great Chain of Being. This classification scheme was based on perceived similarities or differences in physical features (coloration, number of legs, body size, and so forth). Eventually it became structured into a system with more inclusive levels of organization. For example, a **family** was said to include all genera that resemble one another more than they do the genera of other families; an **order** includes all families that resemble one another; and so on (Table 2.1).

In retrospect, we can say that the Linnean system was the basis of the first widely accepted, shared language for naming and classifying organisms. It came at a time when ordering was desperately needed. Yet we must also say that the Linnean system reinforced the prevailing view—that species were unique and *unchanging* kinds of organisms, each locked in place in the Chain of Being. To this day, the use of rigid categories for classifying organisms works in subtle ways on our perceptions of the diversity of living things.

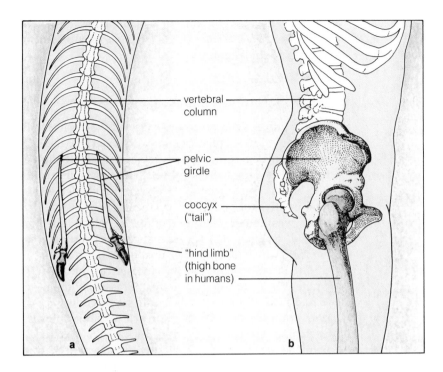

vertebral column

pelvic girdle

coccyx ("tail")

"hind limb" (thigh bone in humans)

a

b

Figure 2.4 (**a**) Bony structures in a python that correspond to the pelvic girdle of vertebrates. Small "hind limbs" protrude through the skin on the underside (ventral surface). (**b**) Pelvic girdle in a human.

Challenges to the Theory of Unchanging Life

By the late eighteenth and early nineteenth centuries, the somewhat passive cataloging of life was disrupted. Puzzling information was emerging from comparative anatomy (the dissection and comparison of body structure and patterning in major groups of animals). For example, most mammals have two forelimbs. Whales have two flippers, humans have two arms, bats have two wings, and so forth (Figure 2.3). These structures vary in size, shape, and how they are put to use (as in flying versus swimming). Yet when early anatomists analyzed forelimbs of different mammals, they found them to be similar in general body position, in component materials, and in how they developed in the embryo. What meaning could be assigned to the unmistakable resemblances and variations in body plan?

According to one explanation, at the time of creation there was no need to come up with completely new body patterns for each organism, because some patterns were perfect and worked so well. Yet if that were true, then what could explain another discovery—that some body parts have no apparent function at all?

For example, snakes and humans belong to the major group called vertebrates (animals with backbones).

Unlike humans, snakes have no limbs. However, if snakes had been created in a state of limbless perfection, then why do some have bony parts that correspond to a pelvic girdle? A pelvic girdle (Figure 2.4) is a set of bones to which *hind limbs* attach! As another example, at the tip of the human backbone, anatomists discovered bony parts that looked exactly like bones in an animal tail. What would the remnants of a tail be doing in a perfectly constructed human body? It was not a time of easy answers.

Also puzzling was the world distribution of plants and animals. For instance, naturalists found that marsupials (pouched mammals such as kangaroos) were not at all common to most places but abounded in Australia. Cactus plants were observed in North and South American deserts yet were nowhere to be seen in Australian and Asian deserts. If all species had been created at the same time in the same place, as most scholars then believed, *how could so many be restricted to one part of the world or another?*

In the late eighteenth century, two hypotheses were advanced to explain these observations. Both are credited to the zoologist Georges-Louis Leclerc de Buffon. If there had been only a single center of creation, thought Buffon, then species spreading out from it would have been stopped sooner or later by mountain barriers and oceans. But what if there had been *several "centers of creation"*? *Perhaps the creation of species had been "spread out" in space.*

Buffon's work in zoology also led him to suggest that species might not have been created in a perfect state. For instance, if the pig had been so perfectly constructed, then why did it have lateral toes—which are too high on the leg to reach the ground? *Perhaps species became modified over time.*

Evidence favoring these two hypotheses came from fossils: the remains or impressions of once-living organisms entombed in the earth. Fossils of different types were buried in a progression of distinct layers under surface rocks and soil. In underlying (earliest deposited) rock layers, fossils of marine organisms were somewhat simple in structure. In layers above them, fossils of similar structure showed more complexity. Finally, in the uppermost (most recently deposited) layers, they closely resembled living marine organisms. Some naturalists thought these patterns were a record of successive changes in various forms of life. The reasons for change and the time required for it were not known—*but the very concept of change ran counter to the concept of the fixity of species.*

Many naturalists now tried to reconcile the new observations of changing fossil patterns with a traditional conceptual framework that did not allow for change. The nineteenth-century anatomist Georges Cuvier had spent twenty-five years comparing fossils with living organisms. He perceived that the fossil record changed abruptly in certain rock layers. The layers were so distinct from ones that had been deposited before them that they seemed to be boundaries for dramatic change in ancient environments. Four broad eras were recognized, each with a distinctive array of fossils. (These eras are now called the Proterozoic, Paleozoic, Mesozoic, and Cenozoic.)

Cuvier actually had made some astute inferences about past episodes of catastrophic change. He attempted to explain these changes with his concept of **catastrophism**. There was only one time of creation, said Cuvier, which had populated the world with all species. Many species had been destroyed in a global catastrophe. The few survivors repopulated the world. It was not that the survivors were *new* species. Naturalists simply hadn't got around to discovering earlier fossils of them, fossils that *would* date to the time of creation. Another catastrophe wiped out more species and led to repopulation by the survivors, and so on through various catastrophes.

Investigations never have turned up the fossils needed to support Cuvier's concept. Rather, they have turned up considerable fossil evidence against it (Chap-

ter Thirty-Seven). The concept is illuminating, however, *for it shows how prevailing beliefs may influence explanations of what is being observed.*

Lamarck's Theory of Evolution

One of Cuvier's contemporaries viewed the fossil record differently. Jean-Baptiste Lamarck believed that life had been created long ago in a simple state. He believed further that it gradually improved and changed into complex levels of organization. The force for change was a built-in drive for perfection, up the Chain of Being. The drive was centered in nerve fibers, which directed "fluida" (vaguely defined substances) to body parts in need of change (in a manner unspecified).

For instance, suppose the ancestor of the modern giraffe was a short-necked animal. Pressed by the need to find food, this animal constantly stretched its neck to browse on leaves beyond the reach of other animals. Stretching directed fluida to its neck, making the neck permanently longer. The slightly stretched neck was bestowed on offspring, which stretched their necks also. Thus generations of animals desiring to reach higher leaves led to the modern giraffe. Conversely, a vestigial structure was an organ no longer being exercised enough. It was withering away from disuse, and each newly withered form was passed on to offspring.

Such was the Lamarckian hypothesis of **inheritance of acquired characteristics**—the notion that changes acquired during an individual's life are brought about by environmental pressure and internal "desires," and that offspring inherit the desired changes.

Lamarck's contemporaries considered the hypothesis a wretched piece of science, largely because Lamarck habitually made sweeping assertions but saw no need to support them with observations and tests. In retrospect, perhaps we can find kinder words for the man. His work in zoology was respected. And he did indeed piece together a foundation for an evolutionary theory: *Species change over time, and the environment is a factor in that change.* It was his misfortune that he made some crucial observations but came up with a hypothesis that has never been supportable by tests.

EMERGENCE OF THE PRINCIPLE OF EVOLUTION

Naturalist Inclinations of the Young Darwin

Charles Darwin (Figure 2.5) was to develop an evolutionary theory that has had repercussions throughout Western civilization. His early environment may have influenced that destiny. His grandfather, a physician and naturalist, was one of the first to propose that all organisms are related by descent. Being from a wealthy family, Darwin had the means to indulge his interests. When he was eight years old, he was an enthusiastic but haphazard shell collector. At ten, he focused on the habits of insects and birds. At fifteen, he found schoolwork boring compared to the pursuit of hunting, fishing, and observing the natural world.

At college, Darwin attempted to study medicine. He abandoned the study after realizing he never could practice surgery on his fellow humans, given the crude and painful procedures available. For a while he followed his own inclinations toward natural history. Then his father suggested that a career as a clergyman might be more to his liking, so Darwin packed for Cambridge. His grades were good enough to earn him a degree. But most of his time was spent among faculty members with leanings toward natural history. It was the botanist John Henslow who perceived and respected Darwin's real interests. Henslow arranged for him to take part in a training expedition led by an eminent geologist. At the pivotal moment when Darwin had to decide on a career, Henslow arranged that he be offered the position of ship's naturalist aboard H.M.S. *Beagle*.

Figure 2.5 Charles Darwin, at about the time he accepted the position of ship's naturalist aboard H.M.S. *Beagle*.

Voyage of the *Beagle*

The *Beagle* was about to sail for South America to complete earlier work on mapping the coastline. Prolonged stops at islands, near mountain ranges, and along rivers would give Darwin a chance to study many diverse forms of life. Almost from the start of the voyage, the young man who had hated work suddenly began to work with enthusiasm, despite lack of formal training. Throughout the journey to South America, he collected and examined marine life. And he read Henslow's parting gift, the first volume of Charles Lyell's *Principles of Geology*.

Amplifying earlier ideas of the geologist James Hutton, Lyell argued that processes now molding the earth's surface—volcanic activity, the gradual uplifting of mountain ranges, erosion by wind and water—had also been at work in the past. This concept is called **uniformitarianism**. It called into question prevailing ideas about the age of the earth. (For example, Jewish calendar-years were based on the concept that the earth was less than 6,000 years old.) Given the rates at which known geologic processes proceed, Lyell had reckoned that it would have taken not a few thousand years but millions of years to mold the land into its current configurations.

The implications of this concept were staggering. One of the reasons that evolution initially seemed so implausible was that it was hard to imagine all of life's exuberant diversity evolving in the space of a few millenia. If Lyell were interpreting the geologic record correctly, *then there had indeed been time enough for evolution.*

Darwin and Wallace: The Theory Takes Form

Clues From the Voyage. Darwin returned to England in 1836, after nearly five years at sea. In the years to follow, his writings established him as a respected figure in natural history. All the while, his consuming interest was the "species problem." *What could explain the remarkable diversity among organisms?* As it turned out, field observations he had made during his voyage enabled him later to recognize two clues that pointed to the answer.

First, while the Argentine coast was being mapped, Darwin repeatedly got off the boat (he was prone to seasickness). He made many excursions inland, where he made detailed field observations and collected various

Figure 2.6 The armadillo. This living animal, along with fossils of animals that were very similar to it (fossils of glyptodont, shown in the reconstruction at left), provided Darwin with a clue that helped him develop his theory of evolution.

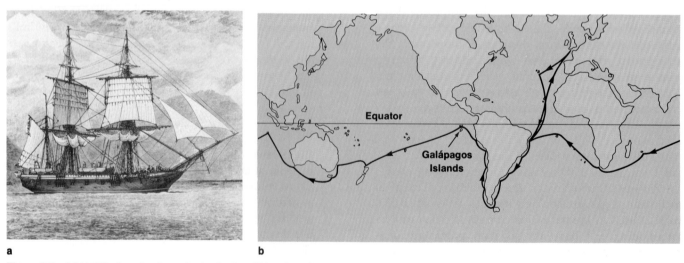

a

b

Figure 2.7 (**a**) H.M.S. *Beagle,* shown in the Straits of Magellan. (**b**) The route of the *Beagle*'s five-year voyage around the world. The Galápagos Islands, about 1,000 kilometers (600 miles) off the coast of Ecuador, support a number of unique plant and animal species. The diversity of life on these isolated islands profoundly influenced Darwin's thinking about the evolution of species.

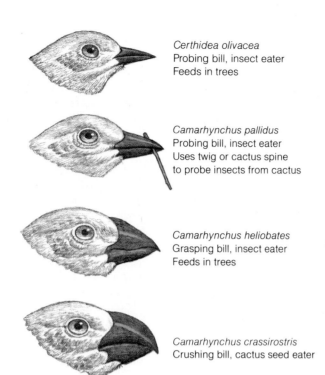

Certhidea olivacea
Probing bill, insect eater
Feeds in trees

Camarhynchus pallidus
Probing bill, insect eater
Uses twig or cactus spine
to probe insects from cactus

Camarhynchus heliobates
Grasping bill, insect eater
Feeds in trees

Camarhynchus crassirostris
Crushing bill, cactus seed eater

a

b

Figure 2.8 (**a**) Examples of variation in beak shape among different species of Darwin's finches, as correlated with feeding habits. (**b**) Woodpecker finch (*C. pallidus*), using a cactus spine to pry out wood-boring insects.

fossils. He saw for the first time many unusual species, including the armadillo (Figure 2.6). Among the fossils were remains of glyptodonts: extinct animals that looked suspiciously like living armadillos. Here were two bizarre but very similar animals, separated in time but confined to the same part of the world. If both had been created separately at the same time, then why were living armadillos lumbering about the very spot where fossil remains of others like them lay buried? *Wouldn't it make more sense to assume that one form evolved from the other?*

Second, Darwin had observed local populations of the same species at many different places along the east coast of South America. It later struck him that individuals within those populations varied in such traits as size, form, coloration, and behavior. When two populations were geographically close, the differences often didn't amount to much. But when populations were separated by some distance, the differences were so pronounced that they might almost be judged as separate species. Darwin had seen nothing like this geographic variation in England—a small island where environmental conditions do not vary much.

The phenomenon of geographic variation was pronounced in the Galápagos Islands, about 1,000 kilometers off the coast of Ecuador (Figure 2.7). These tiny islands arose long ago as volcanoes from the sea floor, hence originally had been devoid of life. Every island or island cluster housed diverse organisms, including its own species of finch. Each finch species had a distinct beak shape, which seemed to be related to particular types of food in the local environment (Figure 2.8). Now, if all those species had not changed since the moment of their creation, then a slightly different finch must have been created for each speck of land in the Galápagos group! *Would it not be simpler to assume that geographic variants had evolved from a single ancestral species of finch?*

Darwin's Deductions. If Darwin were correct in his hypothesis that species evolve, then what mechanism could account for their evolution? No one had ever seen one species change into another. Darwin thought no one would because he believed evolution was exceedingly slow. Even so, he did come up with a plausible mech-

Figure 2.9 A few examples of the more than 300 varieties of domesticated pigeons. Such forms have been derived, by selective breeding, from the wild rock dove (**a**).

anism after reading *Essay on the Principle of Population* by Thomas Malthus—a clergyman and economist.

Malthus stated that "nature has scattered the seeds of life abroad with the most profuse and liberal hand [but] has been comparatively sparing in the room and the nourishment necessary to rear them." Thus, any population tends to outgrow its resources, and its individual members must compete for what is available.

When Darwin reflected on his observations of the natural world, it dawned on him that the normal variation in local populations of a species could include differences in the ability to acquire resources. If there were indeed a struggle for existence, then some *variant* individuals might have a competitive edge in surviving and reproducing. Nature would *select* certain traits and eliminate others—and slowly the population could change! Thus Darwin deduced that "natural selection" among variant individuals could be a mechanism of evolution.

The Theory of Natural Selection. Now the problem became one of compiling evidence of natural selection. Darwin knew he had to find a way to convince the scientific community that the normal variation within species could have occurred without divine creation. He did this by turning to a familiar practice among plant and animal breeders of the day: the development of new varieties within a species.

A compelling example was the flamboyant variation among domesticated pigeons (Figure 2.9). Then, as now, breeders decided which traits were desired, such as a black tail, and they selected individual variants having the most black coloration in their tail feathers. By permitting only those birds to mate, they fostered the trait and eliminated others from the population.

Thus Darwin used *artificial* selection as a model for *natural* selection. He also put together ample examples of variation in natural populations and thereby provided compelling evidence for one of his key deductions: *If evolution occurs by natural selection, then there must be variation in populations.*

For natural selection to operate, more offspring must be produced than can survive; otherwise there would be no competition among variant individuals, hence no requirement for selection among them. Darwin argued that this deduction is probably valid, given that populations typically remain much the same in size. (To give a simple example, a single sea star can release 2,500,000 eggs every year, but the oceans obviously do not fill with sea stars.)

On the basis of such reasoning, Darwin put together his **theory of natural selection**, which is expressed here in modern form:

Figure 2.10 Alfred Wallace. Although Darwin and Wallace had worked independently, they both arrived at the same concept of natural selection. Darwin tried to insist that Wallace be credited as originator of the theory, being the first to circulate a report of his work. Wallace refused; he would not ignore the decades of work Darwin had invested in accumulating supporting evidence.

1. In any population, more offspring tend to be produced than can survive to reproductive age.

2. Members of a population vary in form and behavior. Much of the variation is heritable.

3. Some varieties of heritable traits improve chances of surviving and reproducing under prevailing conditions.

4. Because bearers of such traits are more likely to reproduce, their offspring tend to make up more of the reproductive base for each new generation. This tendency is called *differential reproduction*.

5. *Natural selection is the result of differential reproduction.* Some traits show up (are selected for) with increased frequency because their bearers contribute proportionally more offspring to succeeding generations.

Having formulated his concept of natural selection, Darwin continued his research and sifted his data for flaws in his reasoning. Then, in 1858, he received a paper from Alfred Wallace outlining the same concept! Like Darwin, Wallace was a respected naturalist, with thirteen years of research in South America and the Malay Archipelago to his credit. However, Darwin had only circulated a letter outlining the main points of his ideas even though he had been working for more than twenty years on the problem; the same ideas flashed into Wallace's mind and he had written out his paper in two days (Figure 2.10).

Despite the shock, Darwin sent Wallace's paper to colleagues and suggested that it be published. His colleagues prevailed on him to gather his own notes into a paper that could be presented along with Wallace's. In 1858, both papers were presented to the Linnean Society. The next year Darwin published his monumental book, *On the Origin of Species by Means of Natural Selection.*

On the Origin of Species. Darwin's theory still faced a crucial test. *If evolution occurs, then there should be evidence of one kind of organism changing into a different kind.* The fossil record seemed to contain no transitional forms, the so-called "missing links" between one major type of organism and another. Oddly enough, two years after Darwin's book was published such a fossil did turn up—and no one paid much attention to it. The fossil speci-

Figure 2.11 *Archaeopteryx*, a link between reptiles and birds. To the left, a restoration based on the fossil shown above.

men, named *Archaeopteryx*, had traits reminiscent of small, two-legged reptiles and modern birds (Figure 2.11). Like reptiles, it had teeth and a long, bony tail. Yet its entire body was covered with feathers!

No other evidence was forthcoming in Darwin's time. Almost seventy years would pass before solid evidence would show that his idea of evolution holds under many different tests, on many different levels of biological organization. Ironically, even though the idea itself had at last gained respectability, it would be that long before most of the scientific community would agree with Darwin and Wallace's remarkable insight—that *natural selection* is a major mechanism by which evolution occurs. In the meantime, their names would be associated mostly with the concept that life evolves—something that others had proposed before them.

AN EVOLUTIONARY VIEW OF DIVERSITY

With widespread acceptance of evolutionary thought, fresh winds began blowing through the rigid framework of biological classification. It was not that classification schemes were blown away; they were and continue to be useful ways of storing and retrieving information.

Rather, the emphasis shifted to identifying evolutionary links between species, including those already categorized.

Today, particular characteristics are viewed as being indicative of different lines of descent. A *genus* includes only those species related by descent from a fairly recent, common ancestral form. A *family* includes all genera related by descent from a more remote common ancestor, and so on up to the highest (and most inclusive) levels of classification: *phylum* and *kingdom*. A scheme that takes into account the evolution of major lines of descent is known as a natural system, or **phylogenetic system of classification**.

In this book we use a modified version of Robert Whittaker's phylogenetic system of classification. In this scheme, all organisms are assigned to one of five kingdoms: Monera, Protista, Fungi, Plantae, and Animalia (Table 2.2).

Regardless of its strengths, no classification system should be viewed as *the* system. As long as observations continue to be made, different people will interpret relationships among organisms in different ways. (For example, some researchers group all forms of life into two kingdoms; others group them into as many as twenty.) The system used here helps summarize knowledge about the diversity of life but it, too, is subject to modification as new evidence turns up.

Table 2.2 Classification Scheme Used in This Book

Kingdom	Body Plan	Cell Type	Mode of Nutrition	Representatives
Monera	Single cells	*Prokaryotic*; the cell body does not contain a nucleus or other membrane-bound organelle.	Some are *autotrophs*; they synthesize their own food from simple raw materials (such as carbon dioxide and water) using an environmental energy source (such as sunlight). The *photosynthetic* autotrophs use sunlight. Other species are *heterotrophs*; they feed on tissues or wastes of other organisms.	All are bacteria. (Cyanobacteria, formerly called blue-green algae, are now in this category.)
Protista	Single cells, mostly	*Eukaryotic*; the cell body has a nucleus and other organelles.	Some species are photosynthetic autotrophs; others are heterotrophs.	Euglenids, golden algae and diatoms, dinoflagellates, protozoans, and slime molds.
Fungi	Multicelled, mostly	*Eukaryotic*	All are heterotrophs, either saprobic (feeding on nonliving matter) or parasitic. All rely on extracellular digestion (they secrete substances that break down food, then fungal cells absorb breakdown products).	Chytrids, water molds, zygospore-forming fungi, sac fungi, club fungi, imperfect fungi.
Plantae	Multicelled, mostly	*Eukaryotic*	All are photosynthetic autotrophs.	Red algae, brown algae, green algae, mosses, horsetails, ferns, seed plants (such as cycads, ginkgo, conifers, and flowering plants).
Animalia	Multicelled	*Eukaryotic*	Diverse heterotrophs, including predators and parasites.	Sponges, jellyfishes, comb jellies, flatworms, roundworms, segmented worms, arthropods (such as insects and lobsters), mollusks (such as snails), echinoderms (such as sea stars), and chordates (fishes, amphibians, reptiles, birds, mammals).

PERSPECTIVE

When Darwin disembarked from the *Beagle* with his observations and thoughts on diversity, he set in motion a chain of events that made the study of life simpler— and, at the same time, more complex. As you will discover in chapters to follow, the concept of evolving life provides a clear path through the seeming maze of species diversity. However, even though species are no longer generally regarded as unchanging, major questions have been raised concerning exactly *how* they change. *As in all of science, Darwin and Wallace's principle remains open to test, open to revision.* Is evolution always as extremely gradual as Darwin envisioned it? Or do some evolutionary changes occur rapidly (as some current investigators believe)? Is natural selection the only evolutionary process? What, exactly, causes the variation that selective agents act upon? Do species change only in directions corresponding to environmental pressures, or are there also built-in limits on which ways they *can* go?

We will be returning to the more recent questions, hypotheses, and tests later in the book.

SUMMARY

1. A scientific *principle* is a way of explaining a major aspect of nature, put together from a large body of information. An example is the principle of evolution, which was developed from information gathered over centuries.

2. Although there is no such thing as "the scientific method" of gathering information, scientists do use these processes:

   a. *Hypothesizing* about the meaning of some aspect of the world (for example, proposing that the tremendous diversity among organisms, both living and extinct, came about by evolution.)

   b. *Predicting* what the consequences will be if a hypothesis is valid (the "if-then" process).

   c. *Testing* the hypothesis in ways that can be duplicated or verified by others, as through systematic observation and through experimentation.

3. With enough substantiating tests, a hypothesis may become a *theory* (a coherent set of ideas that form a general frame of reference for further studies and that can

be used with a high degree of confidence). When overwhelming evidence favors it, a theory may become a principle. In science, theories and principles are always open to further test and modification.

4. Attempts to find meaning in the diversity of organisms led to classification schemes based on perceived similarities and differences in physical traits. Thus each distinct kind of organism was called a *species*, distinct species resembling one another more than they resembled other species were grouped into the same *genus*, and so on with increasingly inclusive groupings into *family*, *order*, *class*, *phylum*, and *kingdom*.

5. The idea that species *evolve* (change over time) emerged through:

   a. Comparisons of the body structure and patterning among major groups of animals.

   b. Questions about the world distribution of plants and animals.

   c. Observations of fossils of different types (structurally simple to complex) buried in a series of distinct layers of the earth (most ancient layers to more recent layers).

6. Darwin and then Wallace observed that individuals in local populations of a species vary in size, form, and other traits that might influence their ability to acquire resources—hence to survive and reproduce. If those traits had a heritable basis, then nature would "select" certain varieties of traits and eliminate others through the generations, and the population would change (evolve). Darwin further observed that for evolution to occur, more offspring must be produced than can survive to reproductive age; otherwise there would be no need for selection among them. These concepts are central to the *theory of evolution by natural selection*.

7. One "test" of Darwin's theory would be evidence of one major kind of organism changing into another kind. *Archaeopteryx* provided early evidence; later work at many levels of biological organization has provided a large body of substantiating evidence.

8. Classification schemes are now constructed to indicate possible evolutionary links between organisms. In one current scheme, the most inclusive evolutionary categories are the kingdoms Monera, Protista, Fungi, Plantae, and Animalia.

Review Questions

1. Define inductive reasoning and deductive reasoning. *19*

2. How do beliefs derived from a scientific approach differ from beliefs based on faith, force, authority, or simple consensus? *20–21*

3. The following terms are important in scientific testing: control group, experimental group, independent variables, dependent variables, controlled variables, randomization, sampling error. Can you define them? *21–22*

4. Design a test to support or refute the following hypothesis: The body fat in rabbits appears yellow in certain mutant individuals—but only when those mutants also eat leafy plants containing a yellow pigment molecule called xanthophyll. *Refer to pages 21–22*

5. Witnesses in a court of law are asked to "swear to tell the truth, the whole truth, and nothing but the truth." What are some of the problems inherent in the question? *(20)* Can you think of a better alternative?

6. Spend some time watching television (commercials, news broadcasts, documentaries), and write down examples of statements presented as facts. Also write down why you think each statement is either plausible or nonsense. (For example, when "leading doctors" are said to recommend something, what kind of doctors are they, how many are they, are they doctors everywhere or in a village in Samoa, and what, exactly, are they leading in?)

7. List and define the main categories in the Linnean system of classification. *(24)* How do you suppose this system influences our perceptions of the diversity of living things?

8. What is a phylogenetic system of classification? *32–33*

9. State the key points of the theory of natural selection. What is meant by differential reproduction? *31*

Readings

Darwin, C. 1957. *Voyage of the Beagle.* New York: Dutton. In his own words, what Darwin saw and thought about during his global voyage.

Dobzhansky, T. 1973. "Nothing in Biology Makes Sense Except in the Light of Evolution." *The American Biology Teacher* 35(3):125–129. Personal views of one of the world's leading geneticists, who argues that the principle of evolution does not clash with religious faith.

Gould, S. 1982. "The Importance of Trifles." *Natural History* 91(4):16–23. Gould discusses the principles of reasoning that are evident in one of Darwin's last books (on worms). He argues persuasively that Darwin was indeed one of the great thinkers and not "a great assembler of facts and a poor joiner of ideas," as a detractor once wrote.

Mayr, E. 1976. *Evolution and the Diversity of Life.* Cambridge, Massachusetts: Belknap Press. Insights into a mind searching for ways to untangle the knot of biological diversity.

Moore, J. 1984. "Science as a Way of Knowing—Evolutionary Biology." *American Zoologist* 24:467. Recommended for those who would like more evidence for the line of thought presented in this chapter. An excellent essay.

Moorhead, A. 1969. *Darwin and the Beagle.* New York: Harper and Row. Well-illustrated account of the places Darwin visited, what he observed, and the home to which he returned.

Singer, C. 1962. *A History of Biology to About the Year 1900.* New York: Abelard-Schuman. Out of date, but contains absorbing portrayals of the men and women who led the way in developing basic biological concepts.

UNIT TWO

THE CELLULAR BASIS OF LIFE

3

CHEMICAL FOUNDATIONS FOR CELLS

Sunlight energy enters a photosynthetic cell in a blade of grass and sets up a disturbance among some of its molecules. Because of the way those molecules interact with one another, the disturbance is highly channeled: energy is absorbed, converted, transferred to other molecules, and used in building sugar. A bacterium present in your gut depends on energy stored in the sugar present in milk. It can extract that energy because some of the enzymes it produces can dismantle milk-sugar molecules quickly and specifically. What happens if you stop drinking milk? If the bacterium does not have other enzymes able to handle other foods, that is the end of the bacterium.

No matter what examples come to mind, all events in the living world begin with the organization and behavior of atoms and molecules. It is here that energy transfer within and between living things originates. It is here that the shape and function of cells, and the multicelled body, are determined. What is it about carbohydrates, lipids, and proteins that makes them such suitable building blocks in cell architecture? How can some molecules change shape or be zipped open so that vital reactions can be played out on their surfaces? Why is it that some molecules can be broken apart more easily than others? Answers to such questions are found in rules governing (1) the internal organization of atoms and (2) how atoms and molecules behave relative to one another.

ORGANIZATION OF MATTER

All the diverse substances that occur naturally on earth are alike in two respects: they occupy space and have mass. All contain one or more types of about ninety naturally occurring **elements**, which are materials that cannot be decomposed into substances with different properties.

By international agreement, a one- or two-letter chemical symbol stands for each element, regardless of the element's name in different countries. For example, what we call *nitrogen* is called *azoto* in Italian and *stickstoff* in German. But the symbol for this element is always N. Similarly, the symbol for the element sodium is always Na (from the Latin *natrium*). Table 3.1 lists elements that are most common in living things, along with their chemical symbols.

Different elements combine in proportions that are fixed and unvarying to form **compounds**. For example, the compound water has a fixed proportion of two elements: 11.9 percent hydrogen to 88.1 percent oxygen by mass. Compounds are unlike **mixtures**, in which two or more elements can be present in varying proportions. Seawater is a mixture. It consists of sodium, chlorine,

Table 3.1 Atomic Number and Mass Number of Elements Commonly Found in Living Things				
Element	Symbol	Atomic Number	Most Common Mass Number	Abundance in Human Body* (% Weight)
hydrogen	H	1	1	10.0
carbon	C	6	12	18.0
nitrogen	N	7	14	3.0
oxygen	O	8	16	65.0
sodium	Na	11	23	0.15
magnesium	Mg	12	24	0.05
phosphorus	P	15	31	1.1
sulfur	S	16	32	0.25
chlorine	Cl	17	35	0.15
potassium	K	19	39	0.35
calcium	Ca	20	40	2.0
iron	Fe	26	56	0.004
iodine	I	53	127	0.0004

*Approximate values.

(hydrogen, carbon, nitrogen, oxygen together 96%)

potassium, calcium, sulfur, magnesium, and other substances dissolved in water, but the percentage of each substance varies from place to place. The compound water is 11.9 percent hydrogen and 88.1 percent oxygen no matter where you find it.

Atoms and Ions

In even a small sample of an element such as carbon (C), a gigantic number of atoms are massed together. By definition, an **atom** is the smallest portion of an element that still retains the properties of the element. Atoms contain no more than three major kinds of particles, called *protons, neutrons, and electrons*. Yet if this is true, how is it that each element displays unique properties? We can approach the question through a simple generalization:

Atoms of each element differ from those of all other elements in their number of protons and electrons, and in their electron arrangements.

The number and arrangement of subatomic parts dictate how atoms can combine to form **molecules**, which are units of two or more atoms of the same or different elements bonded together. They dictate what the properties of molecules will be and how (if at all) different molecules will interact. Thus a look at what goes on inside atoms will help explain why substances in living cells behave as they do.

Let's begin with the fact that an atom has one or more protons, each of which has a positive (+) charge. (Electric charges are positive or negative. Two opposite charges attract each other; identical charges repel each other.) Every atom except a form of hydrogen also has one or more neutrons, which are electrically neutral.

Protons and neutrons form the **atomic nucleus**, the core of the atom that accounts for almost all of its mass. Electrons, which have a negative (−) charge, are attracted to the positively charged nucleus and move rapidly around it. They occupy most of the volume of an atom.

The number of protons in the nucleus is called the **atomic number**, and it is different for each element. For instance, a hydrogen atom has one proton; its atomic number is 1. A carbon atom has six protons; its atomic number is 6.

The total number of protons *and* neutrons in the nucleus is called the **mass number**. For example, the most common form of carbon atom has six protons and six neutrons; its mass number is 12. The masses of different types of atoms relative to one another are also called *atomic weights*. This term is not really precise (mass is not quite the same thing as weight), but it has been

entrenched for more than a century and its use continues.

Suppose you managed to isolate one atom from all others. You would find that its number of protons (positive charges) is exactly balanced by its number of electrons (negative charges). Thus an atom as a whole is electrically neutral; it has a *net* charge of zero.

Of course, most atoms in the world do not float about all by themselves. They bump into other atoms all the time. Often they are bombarded by sunlight. Disturbances of this sort sometimes upset the balance between protons and electrons. The number of protons in an atom remains fixed, but one or more *electrons* can be knocked out of it, pulled away from it, or added to it. An atom that loses or gains one or more electrons ends up being positively or negatively charged; and in this state it is called an **ion**.

For example, a sodium (Na) atom has eleven protons and eleven electrons. It can lose an electron, thereby becoming a positively charged sodium ion (Na^+). A chlorine (Cl) atom has seventeen protons and seventeen electrons. It can gain another electron, thereby becoming a negatively charged chloride ion (Cl^-).

An isolated atom has a net electric charge of zero.

An ion is an atom (or a compound) that has gained or lost one or more electrons, hence has acquired an overall positive or negative charge.

Isotopes

All atoms of an element have the same number of protons, but they can vary slightly in the number of neutrons. Most elements have these variant forms, which are called **isotopes**. For example, "a carbon atom" might be carbon 12 (containing six protons, six neutrons), carbon 13 (six protons, seven neutrons), or carbon 14 (six protons, eight neutrons). These can be written as ^{12}C, ^{13}C, and ^{14}C.

Some isotopes are stable, in that they do not change into other atomic forms. **Radioactive isotopes** have unstable nuclei. Over a given period, they spontaneously give off subatomic particles and energy, and they break down (decay) into atoms of different types. For example, carbon 14 has too many neutrons for stability and it decays into nitrogen 14. It takes 5,730 years for half the carbon 14 atoms in a sample of material to do this. Such unvarying rates of decay are used to determine the age of rock samples and fossils (Chapter Thirty-Seven).

Radioactivity can be detected by various methods, hence radioactive isotopes can be used as **tracers**. They can be introduced into some system in order to identify the pathways or destination of a particular substance in that system. Such "systems" might be cells, the human body, even whole ecosystems.

For example, plants use carbon as a building block in photosynthesis. By putting plant cells in a medium enriched in carbon 14, researchers were able to identify the exact steps by which plants take up carbon and use it in assembling complex food molecules. As another example, by injecting a tiny amount of the isotope iodine 131 into a patient's bloodstream, physicians can determine whether the patient has a normal thyroid gland. A normal gland will quickly take up all iodine—including that used as a tracer—from the bloodstream. Isotopes were also used to help determine the location of hereditary instructions inside the cell.

All atoms of an element have the same number of protons but can vary slightly in the number of neutrons. The variant forms are isotopes.

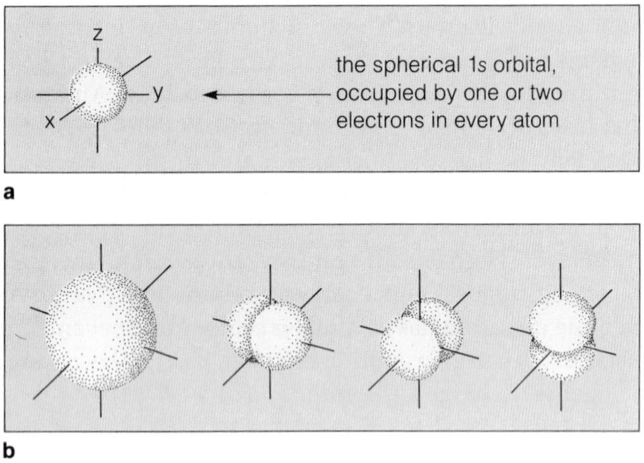

---

a

b

Figure 3.1 Models of electron orbitals. In such models, the nucleus is located where the lines of three axes (x, y, z) intersect. At the first energy level, one or two electrons in every atom occupy the spherelike 1s orbital shown in (**a**). At the second energy level, electrons can occupy the four different orbitals shown in (**b**). The dumbbell-shaped orbitals in this group have somewhat higher energies than the spherical one.

How Electrons Are Arranged in Atoms

Energy Levels for Electrons. No matter how many electrons there are in an atom, each is attracted to the positively charged protons but repelled by other electrons that may be present. How do they spend as much time as possible near the nucleus and, simultaneously, keep far away from each other? They do so by moving in different **orbitals**, which are regions of space around the nucleus in which electrons are likely to be at any instant (Figure 3.1). Each orbital can be occupied by one or, at most, two electrons.

The single electron of a hydrogen atom occupies the 1s orbital (the s stands for spherical), which is the one closest to the nucleus. So do the two electrons of the helium atom. It takes a lot of energy to remove an electron from that orbital, so the electron is said to be at the *lowest energy level.*

What happens in atoms with more than two electrons? Because there are no vacancies in the 1s orbital, other electrons occupy different orbitals that take them farther away, on the average, from the nucleus. It takes less energy to remove an electron that is at a greater distance from the nucleus; hence these electrons are at *higher energy levels.* At the second level, there can be as many as eight electrons in four different orbitals (Table 3.2).

(Another, although not quite accurate way to think about energy levels is to visualize the nucleus surrounded by circles, or shells. The innermost shell would correspond to the lowest energy level and successive shells to higher energy levels, as Figure 3.2 indicates.)

Electrons occupy different orbitals (regions of space around the nucleus in which electrons are likely to be at any instant).

Each orbital can be occupied by one or at most two electrons.

Electrons closest to the nucleus are at the lowest energy level; those that tend to be at a greater distance from it are at higher energy levels.

Table 3.2 Electron Distribution Among Energy Levels for a Few Elements			Energy Level*		
Element	Chemical Symbol	Atomic Number	First	Second	Third
hydrogen	H	1	1	—	—
helium	He	2	2	—	—
carbon	C	6	2	4	—
nitrogen	N	7	2	5	—
oxygen	O	8	2	6	—
neon	Ne	10	2	8	—
sodium	Na	11	2	8	1
magnesium	Mg	12	2	8	2
phosphorus	P	15	2	8	5
sulfur	S	16	2	8	6
chlorine	Cl	17	2	8	7

*There can be a maximum of two electrons at the first energy level, eight at the second, and eighteen at the third.

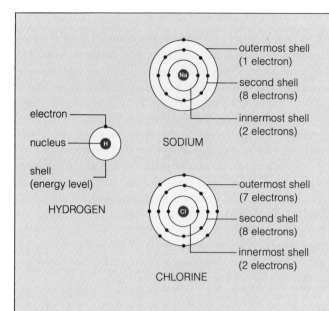

Figure 3.2 Keeping track of electrons: some options.

Atomic structure is most accurately represented by the *orbital model*, in which one or more rapidly moving electrons occupy a volume of space around the nucleus. Electron orbitals of various shapes are associated with each energy level (Figure 3.1 and Table 3.2).

The orbital model is not the most convenient one to use when tracking interactions between atoms. A simplified (if not quite accurate) model is often used instead. In the *shell model*, energy levels that electrons occupy are represented by circles (shells) around the nucleus. Electrons are drawn as dots somewhere on the circles.

An even more stripped-down convention uses only the electron "dots" of the outermost energy level (or shell) with the chemical symbol for the atom. For example, hydrogen and sodium both have only one electron in the outermost shell; chlorine has seven. They would be represented as

$$ \text{H·} \qquad \text{Na·} \qquad \text{:} \ddot{\text{C}}\text{l·} $$

The dots can be grouped in pairs to convey information on how the electrons are distributed among the different orbitals of the shell.

Electron Excitation. When an atom (or molecule) is hit with sunlight, it can *absorb* energy. When that happens, an electron may move briefly into an orbital at a higher energy level, farther from the nucleus (Figure 3.3). Within a fraction of a second the electron returns to the lowest available energy level, releasing energy when it does. Many metabolic events, including photosynthesis and visual perception, depend on such electron boosts.

To reach a higher energy level, an electron must get energy boosts of specific sizes. It can move to one level, or the next level, or the next, but never in between. It's something like standing on a ladder. You can stand on any of the rungs, and you can move your feet from rung to rung. But you certainly can't stand between rungs and neither, by analogy, can electrons.

For example, electrons in certain pigment (light-absorbing) molecules in photosynthetic cells can be excited to higher energy levels by absorbing wavelengths of sunlight energy. But not just any wavelength will do. Different pigments require *specific* wavelengths to boost their electrons to their elevated energy levels (light of different wavelengths varies in how energetic it is).

In looking at Table 3.2, you can see that some atoms have vacancies at the highest occupied energy level. Such atoms tend to form bonds with other elements. This is notably true of hydrogen, carbon, nitrogen, and oxygen—the main building blocks of organisms.

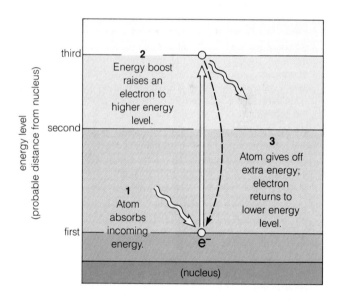

Figure 3.3 Movement of an electron to a higher energy level when an atom (or molecule) absorbs a precise amount of incoming energy, and return of the electron to a lower energy level when the atom releases the extra energy. Photosynthesis and other important biological events are powered by the energy released, in controlled ways, from excited molecules.

Figure 3.4 Chemical bookkeeping.

Symbols for elements are used in writing *formulas*, which identify the composition of compounds. (For example, water has the formula H_2O. The subscript indicates two hydrogen atoms are present for every oxygen atom.) Symbols and formulas are used in *chemical equations*: representations of reactions among atoms and molecules.

In written chemical reactions, an arrow means "yields." Substances entering a reaction (*reactants*) are to the left of the arrow. Products of the reaction are to the right. For example, the overall process of photosynthesis is often written this way:

$$6CO_2 \quad + \quad 6H_2O \longrightarrow C_6H_{12}O_6 \quad + \quad 6O_2$$

6 carbons	12 hydrogens	6 carbons	12 oxygens
12 oxygens	6 oxygens	12 hydrogens	
		6 oxygens	

Notice there are as many atoms of each element to the right of the arrow as there are to the left (even though they are combined in different forms). Atoms taking part in chemical reactions may be rearranged but they are never destroyed. The *law of conservation of mass* states that the total mass of all materials entering a reaction equals the total mass of all the products.

When thinking about cellular reactions, keep in mind that no atoms are lost, so the equations you use to represent them must be balanced in this manner.

Both the reactants and products can be expressed in moles. A "mole" is a certain number of atoms or molecules of any substance, just as "a dozen" can refer to any twelve cats, roses, and so forth. Its weight (in grams) equals the total atomic weight of the atoms that compose the substance.

For example, the atomic weight of carbon is 12; hence one mole of carbon weighs 12 grams. A mole of oxygen (atomic weight 16) weighs 16 grams. Can you show why a mole of water (H_2O) weighs 18 grams, and why a mole of glucose ($C_6H_{12}O_6$) weighs 180 grams?

BONDS BETWEEN ATOMS

Let us turn now to the basis of reactions among atoms. In case you are not familiar with such reactions, take a moment to review Figure 3.4, which summarizes a few conventions used in describing them.

The Nature of Chemical Bonds

A **chemical bond** is a union between the electron structures of two or more atoms or ions. Many such bonds depend on an atom giving up, gaining, or sharing one or more electrons. *Hence a chemical bond is not an object; it is an energy relationship.* Atoms tend to enter into these relationships when there are vacancies at their highest occupied energy levels. In this state, they are more readily disposed to donate, accept, or share electrons.

Certain types of bonds predominate at the temperatures, pressures, and moisture levels characteristic of the cellular world. Strong bonds occur *within* molecules. Weak interactions also occur *between* molecules or between parts of a single molecule. As you will see, thousands of such interactions stabilize the shape of many biological molecules, and they influence the organization of molecules within cells.

Ionic Bonding

An ion forms when an atom loses or gains one or more electrons. For this to happen, another atom of the right kind must be nearby to accept or donate the electrons. Since one loses and one gains electrons, *both* become ionized. Depending on the environment in which the transfer is made, the two ions can go their separate ways or remain together as a result of the mutual attraction of their opposite charges. An association of two oppositely charged ions is an **ionic bond**.

For example, only one electron is present at the highest occupied energy level of a sodium atom (Table 3.2). There is only one vacancy left at the highest occupied energy level of a chlorine atom. When these two atoms encounter each other, sodium tends to transfer an electron to chlorine (Figure 3.5). The two atoms thus become oppositely charged ions (Na^+ and Cl^-) and remain in association as $NaCl$, or sodium chloride, which is our familiar table salt.

In an ionic bond, a positive and a negative ion are linked by the mutual attraction of opposite charges.

Covalent Bonding

Nonpolar and Polar Covalent Bonds. Sometimes an attraction between two atoms is almost but not quite enough for one to pull electrons away from the other. The atoms end up sharing electrons, in what is called a **covalent bond**.

A single covalent bond between two hydrogen atoms may be written as H—H. (Such representations, in which a line signifies a bond, are called structural for-

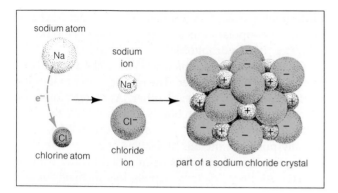

Figure 3.5 Ionic bonding in sodium chloride, or table salt. Strictly speaking, there is no such thing as a "molecule" of NaCl in table salt, for the attraction is not restricted to a single pair of ions. Attractions organize many NaCl units into a crystalline structure, in which units are repeated over and over in three dimensions in a regular, latticelike pattern.

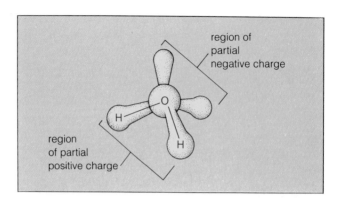

Figure 3.6 Polarity of a water molecule (H₂O). Because of the arrangement of electron orbitals and the bonding angles between oxygen and hydrogen, the molecule as a whole is polar (it carries a slight negative charge at one end and a slight positive charge at the other). Many properties of liquid water—indeed, of organisms—can be traced to this polarity.

mulas.) Each hydrogen atom shares its electron with the other:

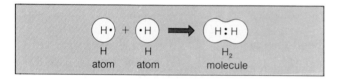

where the dots signify electrons. Similarly, in a double covalent bond, two atoms share two pairs of electrons. An example is the O_2 molecule, or O=O. In a triple covalent bond (such as N≡N), two atoms share three pairs of electrons.

In *nonpolar covalent bonds*, both atoms exert the same pull on shared electrons. The word nonpolar implies that there is no difference between the two "ends" (or poles) of the bond. An example is the H—H molecule. Each hydrogen atom has only one proton, hence the same positive charge and the same pull on shared electrons.

In *polar covalent bonds*, one atom exerts more of a pull on shared electrons. This happens when atoms of two different elements share electrons. Because unlike elements do not have the same number of protons, they are not equally attractive to electrons—which end up associating with one atom more than with the other. For example, electrons shared in a water molecule (H—O—H) are less attracted to the hydrogens than to the oxygen, which has more protons in its nucleus.

In a polar covalent bond, the atom exerting the greater pull on the electrons ends up carrying a slight negative charge (it is more "electronegative"). But the other atom carries a slight positive charge, so the bonding arrangement as a whole has no *net* charge.

In a nonpolar covalent bond, atoms share electrons equally, hence there is no difference in charge between the two poles of the bond.

In a polar covalent bond, atoms share electrons unequally, hence there is a slight difference in charge between the two poles of the bond.

Polarity of a Water Molecule. Oxygen has two electrons at the first energy level and six at the second. When oxygen interacts with hydrogen to form a water molecule, electrons at the second energy level jockey for position, and four of them end up in orbitals that vaguely resemble teardrops. Two of those electrons are shared with the two hydrogen atoms, which are associated with two of the teardrops (Figure 3.6). Other unpaired electrons take up positions on the other side of the oxygen nucleus.

With this arrangement, a water molecule carries a slight negative charge near the oxygen relative to the hydrogens. Also, because of repulsive forces between the protons, the two teardrops associated with the hydrogen atoms tend to be pushed slightly apart. Overall, it is possible to identify two regions of opposite charge; *the water molecule itself is polar.* Many of the properties of water arise from this polarity.

To keep things simple in sketches that follow, the polar water molecule will be illustrated in this fashion:

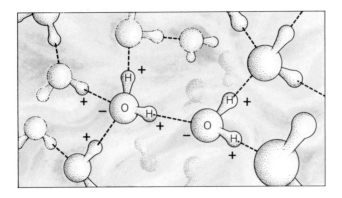

Figure 3.7 Hydrogen bonds between water molecules in liquid water.

Hydrogen Bonding

An atom taking part in a polar covalent bond may also come under the influence of other atoms and molecules. For example, because each hydrogen atom in a water molecule carries a slight positive charge, it can be somewhat attracted to another oxygen atom or to a nitrogen atom in the vicinity. These atoms carry a slight negative charge. Thus we have a **hydrogen bond**, in which an electronegative atom interacts weakly with a hydrogen atom that is already participating in a polar covalent bond:

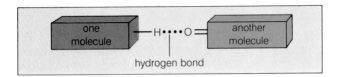

Hydrogen bonds help stabilize the structure of many large biological molecules. They also impart some structure to liquid water. Because a water molecule is polar, each of its hydrogen atoms can form a weak hydrogen bond with the oxygen of a neighboring water molecule. Among neighbors, oppositely charged parts tend to orient toward each other, and parts of like charge away from each other. The bond is strongest when the hydrogen lies in a straight line with two oxygens (Figure 3.7).

In a hydrogen bond, an electronegative atom weakly attracts a hydrogen atom that is covalently bonded to a different atom.

Hydrophobic Interactions

Given that water molecules show polarity, how do other substances act in their presence? If a substance is polar, it is attracted to water molecules and is called **hydro-** **philic** (water-loving). In contrast, if a substance is wholly or largely nonpolar at its surface, it tends to be repelled by water. It is **hydrophobic** (water-dreading).

Consider what happens when you shake a bottle containing salad oil and water. The oil and water molecules hold little attraction for each other. As water molecules are reunited by hydrogen bonds (which replace the ones that were broken when you shook the bottle), they actually push out the oil. Molecules of oil are forced to cluster together in droplets or in a film on the water's surface. As they do, they expose less of their surface area to water—so the oil interferes less with the attraction that water molecules hold for one another.

In a hydrophobic interaction, nonpolar groups cluster together in water. The clustering is not a true bond; the surrounding water molecules simply push out the nonpolar groups.

Although hydrophobic interactions are not bonds, the clustering effect does influence the shapes of many large molecules, such as proteins and lipid compounds. As such, it also influences the architecture of membranes and other cell structures.

Bond Energies

How much energy is involved in the different kinds of bonds described here? The values have been determined by measuring the amounts of energy needed to break the bonds apart.

Bond energy is measured in terms of kilocalories per mole. A **kilocalorie** is the same thing as a thousand calories—the amount of energy needed to heat 1,000 grams of water from 14.5°C to 15.5°C at standard pressure. (Because energy can be converted from one form to another, bond energies can be expressed in kilocalories even though the energy is used for something other than heating water.)

For covalent bonds between atoms of carbon, nitrogen, oxygen, and hydrogen, bond energies typically range between 80 and 110 kilocalories per mole. Under cellular conditions, ionic bonds are much weaker. Typically the bond energy is about 5 kilocalories per mole. Bond energies average 4 to 6 kilocalories per mole for hydrogen bonds.

Your cells break apart covalent bonds and form new bonds when they degrade the sugar glucose. The overall reaction can yield as much as 686 kilocalories per mole. Even with the energy inherent in sugars (and other substances) stored in your body, it still takes additional energy to keep your body active and functioning. On the average, you take in about 2,000 to 2,800 kilocalories each day to make up the difference.

ACIDS, BASES, AND SALTS

Acids and Bases

Hydrogen takes part in many cellular events, being donated by one molecule and accepted by another when substances are put together and pulled apart. When a hydrogen atom is pulled away from a molecule, it becomes for one fleeting moment a naked proton, stripped of its electron. It is a **hydrogen ion**, or H^+.

Most cellular reactions occur in solution (in some kind of liquid mixture). A substance that releases a hydrogen ion in solution is called an **acid**. A substance that combines with a hydrogen ion in solution is a **base**.

Upon thinking about it, you can deduce that "acid" and "base" can refer to two different states of what is essentially the same molecule. Once stripped of a hydrogen ion, an acid may be free to accept another (serve as a base). Similarly, once a base accepts a hydrogen ion, it may be free to donate it elsewhere (serve as an acid).

Suppose you add hydrochloric acid (HCl) to water. The molecules of this acid dissociate (separate) into ionized parts (H^+ and Cl^-). The hydrogen ion is attracted to any neighboring water molecule, which thereby becomes a hydronium ion:

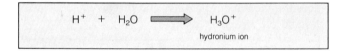

In this reaction, water acts as a base. However, other basic substances besides water are present in cells, and they tend to be more powerful acceptors of hydrogen ions. Hence the hydronium ion quickly does an about-face and acts as an acid, donating its extra hydrogen to its stronger neighbors.

The most powerful hydrogen ion acceptors are **hydroxide ions**, which can be written as OH^-. Many compounds are sources of hydroxide ions. For example:

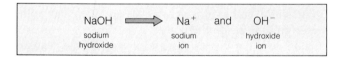

The pH Scale

Depending on how many molecules give up and latch onto hydrogen ions, cells are subject to slight shifts in hydrogen ion concentrations. Changes in acidity can be measured in terms of the **pH scale**, which is based on the concentration of hydrogen ions in solution. The most useful part of the scale ranges from 0 (most acidic) to 14 (most basic). The midpoint of this range, 7, represents a neutral solution in which the H^+ concentration equals

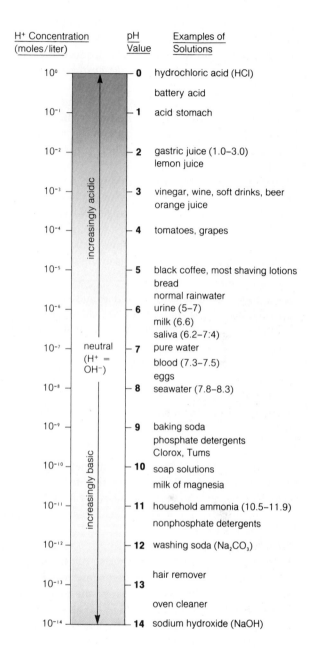

Figure 3.8 The pH scale, in which a fluid is assigned a number according to the number of hydrogen ions present in a liter of that fluid. The most useful part of the scale ranges from 0 (most acidic) to 14 (most basic), with 7 representing the point of neutrality.

A change of only 1 on the pH scale means a tenfold change in hydrogen ion concentration. Thus, for example, the gastric juice in your stomach is ten times more acidic than vinegar, and vinegar is ten times more acidic than tomatoes.

the OH⁻ concentration. A change in one unit in pH means a tenfold change in hydrogen ion concentration.

Pure water has a pH of 7. Any solution with a pH of less than 7 has more H⁺ than OH⁻ ions. The converse is true of any solution with a pH of more than 7. As Figure 3.8 indicates, *the greater the hydrogen ion concentration, the lower the pH value.*

Each living cell is sensitive to pH, and its interior usually will not range far from neutrality. But the pH of the surroundings may be quite different from that inside the cell. For instance, cells of sphagnum mosses grow in peat bogs, where the pH is 3.2 to 4.6 (highly acidic). Some nematodes (a type of worm) thrive in places where the pH is 3.4. For plants and animals living in rivers, the pH of the surrounding water ranges between 6.8 and 8.6. The pH of fluids bathing most cells of your body is between 7.35 and 7.45. (Exceptions include the highly acidic gastric juices that bathe cells lining your stomach.)

Although the cell interior usually does not range far from neutrality, each cell is adapted to a particular environmental pH range. That range can be quite different for different kinds of cells.

Buffers

Given that hydrogen ions are continually being produced and used in cells, why is it that the cell interior normally does not show drastic shifts in pH? The answer is that most cells have several mechanisms for keeping the internal pH fairly constant.

For example, some substances in the cell act as **buffers**: they combine with and/or release hydrogen ions in response to changes in pH. When some reactions produce an excess of H⁺, buffers accept the excess. When other reactions deplete H⁺, buffers are able to dole out reserves. Carbonic acid is one of the major buffers. This acid can dissociate into a bicarbonate ion and H⁺ in water:

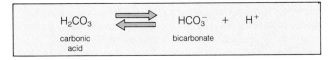

In turn, bicarbonate can combine with hydrogen ions to form carbonic acid, as the reverse arrow indicates in the above equation.

Buffer molecules combine with or release hydrogen ions in response to changes in cellular pH.

Dissolved Salts

A **salt** is an ionic compound formed by the reaction between an acid and a base. Sodium chloride (NaCl) is a salt:

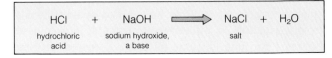

Many salts dissociate into their positively and negatively charged ions when placed in water. In cells, the ions formed from salts include positively charged ions of potassium (K^+), sodium (Na^+), calcium (Ca^{++}), and magnesium (Mg^{++}). They also include the negatively charged chloride ions (Cl^-).

Ions play vital roles in cells. For example, potassium ions help activate many enzymes. In plants, they also seem to be involved in transporting nitrates and phosphates. In animals, signals cannot travel through the nervous system without sodium and potassium ions. Calcium ions take part in cell movements, cell division, nerve functioning, muscle contraction, and blood clotting. In the human body, sodium and chloride ions make up about ninety percent of all ions dissolved in the fluid around cells. The ionic composition of this fluid influences the movement of water and dissolved substances into and out of cells.

Salts are ionic compounds that are formed by the reaction between an acid and a base, and that dissociate into positively and negatively charged ions in water.

Water Molecules and Cell Organization

Water makes up about seventy-five to eighty-five percent of the weight of an active cell, on the average. Water also bathes the cell surface. Surface parts of multicelled plants, animals, and fungi may be exposed to dry soil or air, but even here the active cells are in contact with water present inside the body.

Life does not necessarily end if most of the cellular water is removed. Some plant spores can resume active growth after being dried out for centuries. The water bear also can survive dry spells (Figure 3.9). Water bears, nematodes, rotifers, lichens, mosses, spores, and seeds —all are well matched to environments in which water periodically becomes scarce. Yet with their seeming independence of free-flowing water, the key word in all of this is *active*:

None of the activities associated with the term "living" proceeds without water.

What is it about water that makes it so central to cell functioning? As you will now see, cells depend especially on three properties of water: its internal cohesion, its ability to stabilize temperature, and its capacity to dissolve many substances.

Hydrogen Bonding in Liquid Water

Cohesion. When a substance is overly stretched (placed under too much tension), it can rupture. *Cohesion* is the capacity to resist rupturing under tension. It results from attractions between molecules that make up the substance, such as the hydrogen bonds between water molecules in liquid water.

At interfaces between air and water, hydrogen bonds exert a constant inward pull on molecules at the water's surface and impart a high surface tension to it. The cohesive forces of water resist surface breakthrough. That is why beads of water form and a pond surface resists penetration by small leaves and insects (Figure 3.10). Cohesion, combined with other factors, allows whole columns of water to be pulled through narrow, cellular pipelines to the tops of even the tallest trees.

Numerous hydrogen bonds between water molecules give liquid water high cohesion. The bonds (which individually are weak) resist breaking even when an external force puts the water surface under tension.

Temperature Stabilization. Water helps stabilize temperatures inside and outside the cell because of its high specific heat, high heat of vaporization, and high heat of fusion. All three properties arise largely through hydrogen bonds among water molecules.

Specific heat is the amount of heat energy needed to increase the temperature of one gram of a substance by 1°C. (As you probably know, temperature is a measure of the rate of molecular motion.) Some substances require more energy than others to reach a given level of molecular motion. When liquid water is heated, its individual molecules cannot move faster until the hydrogen bonds among them are broken. Thus water absorbs considerable heat before its temperature increases markedly.

Heat of vaporization is the amount of heat energy that one gram of liquid at its boiling point must absorb before it is converted to gaseous form. It takes 539 calories of heat energy to convert a gram of liquid water into steam at the boiling point of 100°C. (The heat of vaporization for water is twice that for ethanol.)

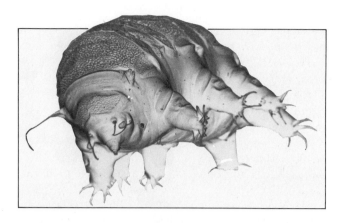

Figure 3.9 A tardigrade (water bear). This tiny animal survives in ponds and damp soil even when its home freezes over or bakes in the sun. When liquid water becomes scarce, the tardigrade dries out systematically and enters a state of suspended animation. Body parts through which water normally escapes are withdrawn from exposure to air. As the body contracts, its cells produce a compound that replaces much of the water normally surrounding large biological molecules. The compound holds the molecules in place and protects them from mechanical damage. When moist conditions return, the water bear revives and actively goes about its business.

Figure 3.10 A water strider, able to walk on water because of water's high surface tension. This surface tension results from the tenacity with which water molecules cling to one another through hydrogen bonds.

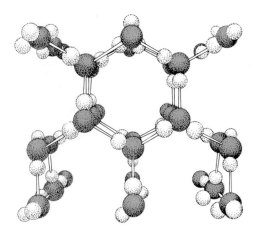

Figure 3.11 Crystal lattice structure of ice. The brown spheres represent oxygen; the white spheres, hydrogen. The "sticks" connecting them represent hydrogen bonds. Below 0°C, each water molecule becomes locked by four hydrogen bonds into a crystal lattice. In this bonding pattern, molecules are spaced farther apart than they would be in liquid water at room temperature (where constant molecular motion usually prevents the maximum number of hydrogen bonds from forming).

Because of the extended bonding pattern, ice is less dense than liquid water and is able to float on it. During winter freezes, sheets of ice that form on surfaces of ponds, lakes, and streams act like a blanket that holds in the water's heat.

Because of hydrogen bonding, water molecules resist separating from each other in liquid water. When a molecule absorbs enough heat energy, however, the bonds can be broken and it can escape from the water's surface. This process is called **evaporation**. Most of the energy needed for the escape is obtained from the surrounding liquid, hence evaporation lowers the surface temperature of water. Oasis plants and many animals cool off through evaporative water loss.

Water also has a high *heat of fusion*: it resists changing from liquid to solid when heat is lost at the freezing point. At room temperature, water is fluid because its relatively weak hydrogen bonds are constantly breaking and rapidly forming again, which allows the molecules some freedom of movement. Only when the temperature drops below 0°C do the molecules become locked in the rigid bonding pattern characteristic of ice (Figure 3.11).

Hydrogen bonds between water molecules give water the capacity to help stabilize temperatures inside and outside the cell.

Solvent Properties of Water

Because of the polar nature of its molecules, water is an excellent solvent for ions and polar molecules. A **solvent** is any fluid in which one or more substances can be dissolved. To understand what dissolving means, consider the following example.

Crystals of table salt (NaCl) separate into Na^+ and Cl^- when they are placed in water. Water molecules tend to cluster around each positively charged ion with their "negative" ends pointing toward it:

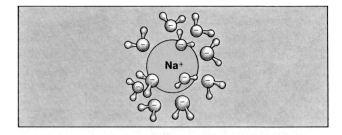

Similarly, water molecules tend to cluster around each negatively charged ion with their "positive" ends pointing toward it:

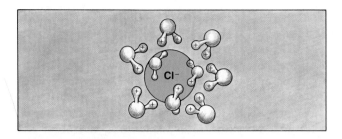

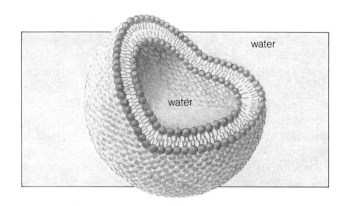

These so-called **spheres of hydration** shield charged ions and keep them from interacting. Thus they force ions to remain dispersed in water.

A charged substance is said to be *dissolved* in water when spheres of hydration form around its individual molecules. The dissolved substances are known as **solutes**. Most ionic substances are at least partially soluble in water. So are most polar molecules, especially those which tend to form hydrogen bonds with water.

Water and the Organization Underlying the Living State

The organization of almost all large biological molecules is influenced by their interaction with water. For example, proteins have many acidic and basic groups on their surfaces. Depending on cellular pH, the protein surface can be positively or negatively charged overall. The charged regions and polar groups attract water molecules. They also attract ions—which in turn attract water. In this manner, an electrically charged "cushion" of ions and water forms around the protein surface:

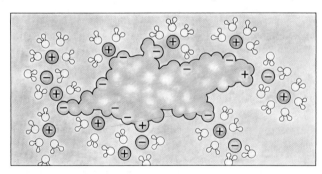

Thus, through interactions with water and ions, proteins can remain dispersed in cellular fluid. (In contrast, particles in a suspension are large enough to settle out.)

As another example, many lipids have hydrophobic tails (which repel water) attached to a hydrophilic head (which, being polar, attracts water). In the presence of water, these lipid molecules tend to cluster, with all of the hydrophilic heads facing the water and all of the hydrophobic tails packed in a sheetlike array that excludes water. Imagine two such sheets arranged into a sphere that has water on the inside and out:

where orange indicates the hydrophilic heads and yellow, the hydrophobic tails. *This "lipid bilayer" arrangement is the framework for all cell membranes.*

Many more examples could be given of interactions between the polar water molecule and other substances characteristic of the cellular world. For now, the point to keep in mind is this:

The properties of water profoundly influence the organization and behavior of substances that make up cells and the cellular environment.

SUMMARY

1. Matter is composed of elements. The atoms of each element have a unique number of protons in the nucleus. Except in the case of a form of hydrogen, the nucleus also contains neutrons. The number of electrons arranged around the nucleus equals the number of protons. The number and arrangement of electrons dictate how atoms can form molecules (units of two or more atoms of the same or different elements bonded together). They dictate the properties of molecules and how (or whether) molecules will interact.

2. In compounds (such as water), certain elements are always present in fixed, unvarying proportions. In mixtures (such as seawater), two or more elements can be present in varying proportions.

3. An isolated atom has a *net* electric charge of zero. An ion is an atom or compound that has gained or lost one or more electrons (it has acquired an overall positive or negative charge).

4. *All* atoms of an element have the same number of protons but they can vary slightly in the number of neutrons. Variant forms of atoms of an element are called isotopes. Radioactive isotopes have unstable nuclei; over a specific period they spontaneously decay into atoms of different types. They are used as tracers in biological systems and are used in determining the age of fossils (remains of once-living organisms).

5. Electrons can be at low energy levels (close to the nucleus) or at higher ones (farther away from it). They occupy orbitals at these different levels. An orbital may be vacant or may have one or two electrons, but never more than two. Energy inputs can boost electrons to higher energy levels (or out of the atom entirely). When electrons return to a lower energy level, atoms release energy. Photosynthesis and other important biological events are powered by energy released from molecules that have absorbed extra energy.

6. Atoms of hydrogen, carbon, nitrogen, and oxygen (the main structural elements of cells) have vacancies at their highest occupied energy levels. They tend to form bonds with other elements, thereby gaining enough electrons to fill their highest occupied energy level.

7. A chemical bond is a union between the electron structures of two or more atoms or ions:

   a. Ionic bond: a positive and a negative ion remain together by the mutual attraction of opposite charges.

   b. Nonpolar covalent bond: atoms share one or more electrons equally; there is no difference in charge between the two poles of the bond.

   c. Polar covalent bond: atoms share one or more electrons unequally; there is a slight difference in charge between the two poles of the bond (one end is electronegative).

   d. Hydrogen bond: an electronegative atom weakly attracts a hydrogen atom that is already covalently bonded to a different atom.

8. Acids are substances that release hydrogen ions (H^+) in solution; bases are substances that combine with hydrogen ions. Typically, the H^+ concentration in cells equals the hydroxide ion (OH^-) concentration; this represents neutrality on the pH scale. Cells are sensitive to changes in pH.

9. A salt is an ionic compound (such as NaCl) formed by a reaction between an acid and a base. Salts dissociate into positively and negatively charged ions in solution, and these ions play vital roles in cell functions.

10. A water molecule shows polarity; due to its electron arrangement, one end is electronegative. Other polar molecules are attracted to water (they are hydrophilic); nonpolar molecules are repelled by it (they are hydrophobic).

11. The properties of water profoundly influence the organization and behavior of substances that make up cells and cellular environments.

   a. Because of hydrogen bonding between its molecules, water has the capacity to help stabilize the temperatures inside and outside the cell.

   b. Because of water's capacity for hydrogen bonding, its molecules show cohesion (a capacity to resist rupturing when placed under tension, as when water

molecules are pulled up from roots to leaves at the tops of trees).

   c. Because of the polar nature of its molecules, water is an excellent solvent for ions and polar molecules.

   d. Because of the polar nature of water molecules, hydrophobic and hydrophilic substances interact with water in ways that influence the shapes of large molecules and the structure of membranes and other cell components.

Review Questions

1. Define the following: element, atom, molecule, compound. What are the six main elements (and their symbols) in most organisms? 36–37

2. Define proton, neutron, and electron. How are electrons arranged around the nucleus of an atom? 37

3. Explain the difference between an atom, an ion, and an isotope. 37–38

4. Each orbital around an atomic nucleus can be occupied by one or (at most) two electrons. This means that all the orbitals at the highest energy level for an atom (in other words, the ones farthest from the nucleus) might or might not be filled at a given time. If there is one or more "vacancies" at the highest occupied energy level, is the atom *more* or *less* likely to form bonds with other elements? 39

5. Explain the difference between covalent, ionic, and hydrogen bonds. 40–42

6. What is the difference between a hydrophilic and a hydrophobic interaction? Is a film of oil on water an outcome of bonding between the molecules making up the oil? 42

7. Define an acid, a base, and a salt. On a pH scale from 0 to 14, what is the acid range? Why are buffers important in living cells? 43–44

8. Cell functioning depends on the properties of water, such as its internal cohesion. What type of chemical bond represents the basis of these properties? 45–46

9. Describe how the polarity of the water molecule is the basis of the solvent properties of water. (As part of your answer, describe how spheres of hydration form around a positive or negative ion.) 46–47

10. How do water molecules influence the organization of proteins and lipids in a cell? 47

Readings

Breed, A., et al. 1982. *Through the Molecular Maze*. Los Altos, California: Kaufmann. A short guide to basic chemical concepts.

Lehninger, A. 1982. *Principles of Biochemistry*. New York: Worth. Classic reference book in the field.

Mertz, W. 1981. "The Essential Elements." *Science* 213: 1332–1338.

Miller, G. 1987. *Chemistry: A Basic Introduction*. Fourth edition. Belmont, California: Wadsworth. Well-written, accessible introduction to chemical principles.

The preceding chapter included a rather odd statement, that living cells are between seventy-five and eighty-five percent water. Whether or not you accepted the statement without giving it a second thought may be a test of how critically you are evaluating what you read. For example, given the many trillions of cells comprising your body, might you not wonder why you do not ooze forward when you walk or quiver like Jell-O, which also is mostly water? Obviously there must be more than free water molecules in your cells.

By far, the three most abundant elements in your body are oxygen, hydrogen, and carbon. (As Table 3.1 indicated, they represent ninety-three percent of its weight. Nitrogen, calcium, and phosphorus account for a little over six percent, and very small amounts of other elements make up the rest.) Much of the oxygen and hydrogen is indeed linked together in the form of water. But these two elements also are linked in significant amounts to carbon—the most important structural element in the body.

THE ROLE OF CARBON IN CELL STRUCTURE AND FUNCTION

A carbon atom can form as many as four covalent bonds with other carbon atoms as well as with other elements. In cells, carbon atoms linked one after another in chains or rings form the backbones (or skeletons) for diverse compounds. These backbones occur in strandlike, globular, and sheetlike molecules, some of which contain thousands, even millions, of atoms. The carbon compounds assembled in cells are *organic* molecules. (The term distinguishes them from the simple *inorganic* compounds, such as water and carbon dioxide, which have no carbon chains or rings.)

Families of Small Organic Molecules

Compounds having no more than twenty or so carbon atoms are considered to be small organic molecules. The four main families of these small molecules are called simple sugars, fatty acids, amino acids, and nucleotides. Usually they form pools of dissolved materials that the cell can tap for energy or for building blocks in the synthesis of large molecules (macromolecules). The main macromolecules in cells are polysaccharides, lipids, proteins, and nucleic acids.

Four families of small carbon compounds—simple sugars, fatty acids, amino acids, and nucleotides—serve as energy sources and as building blocks for the macromolecules present in cells.

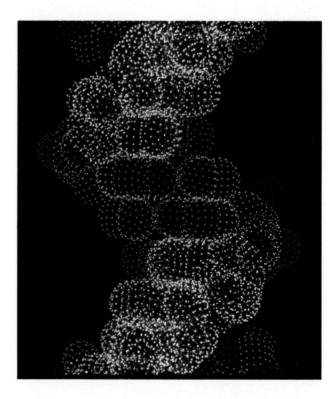

Figure 4.1 Model of a segment of DNA, a molecule that is central to maintaining and reproducing the cell. Carbon atoms are in green. Other atoms represented are nitrogen (blue), oxygen (red), and phosphorus (yellow). Most of the carbon atoms are bonded in ring structures that interconnect to form the carbon skeleton of DNA.

Table 4.1 Some Classes of Organic Compounds

Class	Some Characteristic Functional Groups		Occur in:
Hydrocarbons	methyl ($-CH_3$):	H | $-C-H$ | H	fats, oils, waxes
Alcohols	hydroxyl:	$-OH$	sugars
Carbonyls	aldehyde ($-CHO$):	O || $-C-H$	sugars
	ketone ($>C=O$):	O || $-C-C-C$	sugars
	carboxyl ($-COOH$):	O || $-C-OH$	sugars, fats, amino acids
Amines	amino ($-N$, $-NH$, $-NH_2$):	H | $-N-H$	amino acids, proteins
Phosphate compounds	phosphate ($-$ⓟ):	O || $-P-O$ | O	energy carriers (such as ATP)

Properties Conferred by Functional Groups

All organic molecules have a carbon backbone, but they differ in what is attached to it. Atoms or groups of atoms covalently bonded to a carbon backbone are called **functional groups** (Table 4.1). Organic molecules are distinctive in their structure and properties because of their characteristic functional groups, as a few examples will illustrate.

Carbon-Hydrogen Compounds. There is a TinkerToy quality to carbon compounds, in that a single carbon atom can be the start of truly diverse molecules assembled from "straight-stick" covalent bonds. Consider the *hydrocarbons*, which consist only of hydrogen and carbon. Methane (CH_4) is the simplest hydrocarbon. If you were to strip one hydrogen from methane, the result would be a *methyl group*:

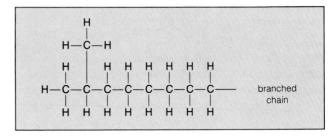

Now imagine that two methane molecules are each stripped of a hydrogen atom and bonded together. If the resulting structure were to lose a hydrogen atom, you would end up with an *ethyl group*:

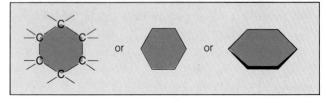

In this mental exercise, you could go on building a continuous chain, with all the carbon atoms arranged in a line:

You might even have chains coiled back on themselves into rings:

All of the structures illustrated so far have nonpolar bonds and will not dissolve in water; they are hydrophobic. These are properties of fats, oils, and waxes—which are largely hydrocarbon.

Carbon-Hydrogen-Oxygen Compounds. Compounds that we call *alcohols* are mostly carbon and hydrogen, but they also incorporate one or more *hydroxyl groups* ($-OH$). Methyl alcohol can cause blindness and death, even in small amounts. Yet ethyl alcohol occurs in alcoholic beverages. Glycerol, one of the building blocks of fats and oils, is an alcohol. So are all sugars. All of these compounds will dissolve in water because of hydrogen bonding at their $-OH$ groups.

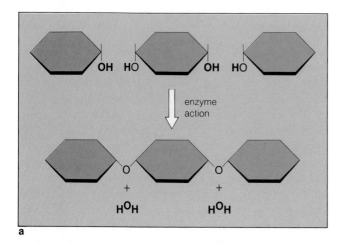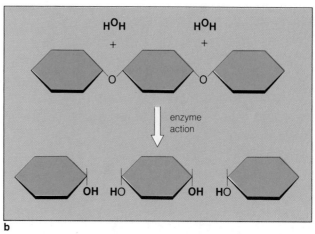

Figure 4.2 (**a**) Condensation of three subunits into a larger molecule. Water can be formed during the reaction. (**b**) Hydrolysis of a molecule into three subunits.

Condensation and Hydrolysis

How do small organic compounds combine to form macromolecules? They don't just get together on their own. Their union depends on the action of enzymes, which are a special class of proteins that speed up reactions between specific substances.

For example, in **condensation**, enzymes speed up the covalent linkage of small molecules (such as short carbon-based chains and rings) in a reaction in which water also may form. Essentially, one molecule is stripped of a hydrogen ion and another is stripped of an —OH group, then the two molecules become joined. At the same time, the H^+ and OH^- that are released can combine to form a water molecule (Figure 4.2).

Condensation reactions can produce a **polymer**: a molecule composed of anywhere from three to millions of relatively small subunits, which may or may not be identical. The individual subunits incorporated in polymers are called **monomers**.

Often the substances used in assembling macromolecules are obtained by breaking down other molecules. A common breakdown process, called **hydrolysis**, is like condensation in reverse. Covalent bonds between parts of molecules are broken and an H^+ ion and an OH^- group derived from water become attached to the fragments (Figure 4.2).

Condensation is the covalent linkage of small molecules in a reaction that can also involve the formation of water.

Hydrolysis is the cleavage of a molecule into two or more parts by reaction with water.

CARBOHYDRATES

Almost all eukaryotic cells use carbohydrates directly or indirectly for energy. Bacteria, plants, and fungi also use carbohydrates as major structural materials. **Carbohydrates** are monomers or polymers of a sugar. A *sugar* is a compound in which carbon, hydrogen, and oxygen atoms are combined in about a 1:2:1 ratio. This means that for every carbon atom present, there typically are two hydrogen atoms and one oxygen atom. Thus the composition of a sugar is $(CH_2O)_n$.

Monosaccharides

The simplest carbohydrates are sugar monomers, or **monosaccharides**. ("Saccharide" comes from a Greek word meaning sugar.) Typically, monosaccharides have a backbone of three to seven carbon atoms. They also have an aldehyde group or a ketone group and two or more hydroxyl groups (Table 4.1).

Of the more than 200 monosaccharides known, the most common have a backbone of three, five, or six carbon atoms. Those with five or more tend to form ring structures when dissolved in cellular fluids. *Ribose* and *deoxyribose* (each with five carbon atoms) are key components of nucleic acids. *Glucose* and *fructose* (each with six carbon atoms) are two of the most abundant sugar monomers.

Both glucose and fructose have the same molecular formula $(C_6H_{12}O_6)$, but they differ slightly in their structure and properties. These sugars are structural isomers: they have the same numbers of the same kinds of atoms, but the atoms are not attached to one another in the same ways (Figure 4.3).

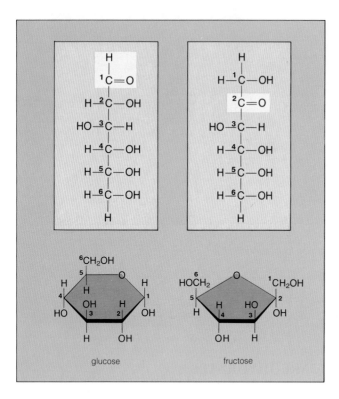

Figure 4.3 Straight-chain and ring forms of two monosaccharides: glucose and fructose. (For reference purposes, sometimes the carbon atoms of sugars are numbered in sequence, starting at the end of the molecule closest to the aldehyde or ketone group.)

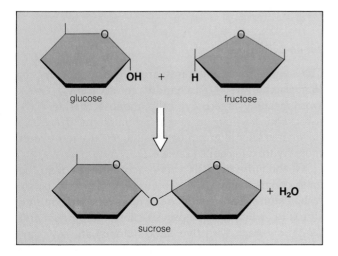

Figure 4.4 Condensation of two monosaccharides (glucose and fructose) into a disaccharide (sucrose).

Disaccharides

Monosaccharides have several free —OH groups, where bonds can be made with another sugar monomer or some other compound. A **disaccharide** results when two monosaccharides are covalently bonded (Figure 4.4).

The most abundant of all sugars is the disaccharide *sucrose*, which consists of one glucose and one fructose subunit. Sucrose is the form in which carbohydrates are transported through leafy plants. The sugar you buy at the market is sucrose that has been extracted and crystallized from plants such as sugarcane. The disaccharide *lactose* (one glucose and one galactose subunit) occurs in the milk of mammals. *Maltose* (two glucose subunits) occurs in germinating seeds. Among other things, maltose is an ingredient in beer production.

Polysaccharides

When more than two monosaccharides are bonded covalently, the result is a **polysaccharide**. In some polysaccharides the sugar subunits are all the same kind; others incorporate two or more different kinds.

Land plants have an abundance of the polysaccharide *starch* (a storage form for sugar). They also may have an abundance of the polysaccharide *cellulose* (a structural material in their cell walls). Although starch and cellulose are both assembled from glucose units, they have different properties. In plant cells, starch molecules can cluster in large granules called starch grains, which can be quickly hydrolyzed into sugar subunits used in metabolism. In contrast, cellulose is tough, fibrous, and insoluble in water.

The differences between starch and cellulose arise from differences in the bonding alignments between their glucose subunits. The oxygen atoms that link the adjacent glucose subunits are not oriented the same way in the two polysaccharides. In starch, the alignment permits the glucose chain to twist into a coiled structure that favors granule formation (Figure 4.5). In cellulose, the glucose chain is extended and many chains lie side by side, hydrogen-bonded to each other at —OH groups. The hydrogen bonds stabilize the glucose chains in tight bundles (Figure 4.6). This bonding arrangement resists most digestive enzymes except those present in termites, wood-rot fungi, and certain bacteria.

Glycogen is a highly branched polysaccharide (Figure 4.7). It is a form in which sugar is stored in fungi and certain animal tissues (such as liver and muscle tissues). *Chitin* incorporates nitrogen and is called a modified polysaccharide. Many animals and fungi have chitin-secreting cells. The chitin secretions are the main structural material in external skeletons and other hard parts of many insects and crustaceans (including crabs). It imparts firmness to the cell walls of most fungi.

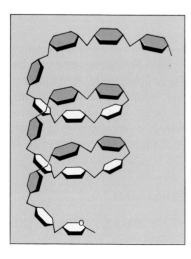

Figure 4.5 Oxygen bridges between the glucose subunits of amylose, a form of starch. The boxed inset depicts the coiling of an amylose molecule, which is stabilized by hydrogen bonds.

Figure 4.6 Structure of cellulose, which is composed of glucose subunits. Neighboring cellulose molecules link together at —OH groups to form a fine strand (microfibril). These strands can be twisted into a threadlike fibril. In some cells, fibrils are coiled into a macrofibril, which is as strong as a steel thread of the same thickness.

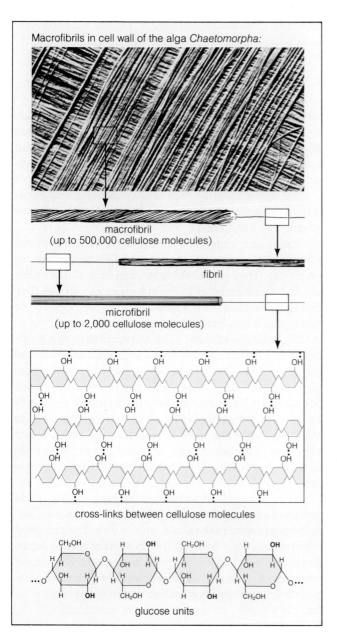

Macrofibrils in cell wall of the alga *Chaetomorpha:*

macrofibril
(up to 500,000 cellulose molecules)

fibril

microfibril
(up to 2,000 cellulose molecules)

cross-links between cellulose molecules

glucose units

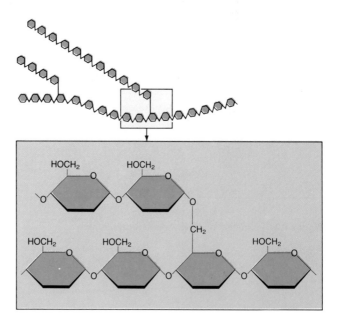

Figure 4.7 Branched structure of glycogen, a form in which sugars are stored in some animal tissues.

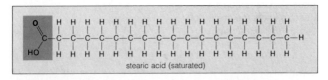

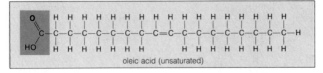

Figure 4.8 Structural formulas for a saturated and an unsaturated fatty acid.

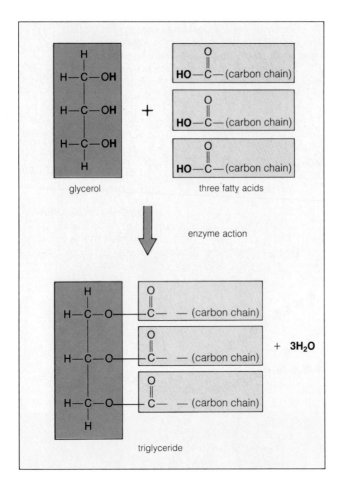

Figure 4.9 Formation of a triglyceride (a fat).

LIPIDS

Lipids are largely hydrocarbon and show little tendency to dissolve in water, but they do dissolve in nonpolar solvents (such as ether). Some lipids function in the storage and transport of energy; others are key components of membranes, protective coats, and other cell structures. Here we shall consider two subcategories of lipids: those with and those without fatty acid components.

Lipids With Fatty Acid Components

Many lipids such as butter and vegetable oils have fatty acid components. A **fatty acid** is a long, unbranched hydrocarbon with a —COOH group at the end (Figure 4.8). When a fatty acid is part of a more complex lipid molecule, it is usually stretched out like a flexible tail. Let's look briefly at three common lipids having fatty acid tails: the glycerides, phospholipids, and waxes.

Glycerides. Glycerides are the most abundant lipids and the richest source of energy in the body. A **glyceride** molecule has one, two, or three fatty acid tails attached to an alcohol backbone. In this case, the alcohol is glycerol (see Figure 4.9).

The terms *monoglyceride, diglyceride,* and *triglyceride* refer to whether one, two, or three fatty acid tails are attached to the glycerol.

Plants and animals store lipids in the form of triglycerides (Figure 4.9). Most of the fats we eat are triglycerides. They are insoluble in water and tend to clump together in fat globules in the intestine. Special digestive processes break apart the globules, then break down the triglycerides into parts that the body can absorb.

Triglycerides that are solid at room temperature are called **fats**; those that are liquid are called **oils**. Most of the fatty acid tails in butter, bacon, and other animal fats are *saturated*: they have only single covalent bonds between carbon atoms. Saturated fatty acids can snuggle up to one another in parallel arrays. Soybean, corn, and other vegetable oils have *unsaturated* fatty acids, which have one or more double covalent bonds that create kinks in the tails. Tight packing is disrupted where these bonds occur.

Some amount of certain unsaturated fatty acids appears to be important in nutrition. In one study, immature rats were placed on a fat-free diet. The rats grew abnormally, their hair fell off, their skin turned scaly, and they died young. These conditions never developed when small amounts of linoleic acid (a fatty acid with two double bonds) were added to the diet.

Phospholipids. Membranes of plant and animal cells have abundant phospholipids. A **phospholipid** molecule has a glycerol backbone, two fatty acid tails, and a phosphate group to which a number of different small hydrophilic groups may be attached. One example is shown in Figure 6.1.

Waxes. In **waxes**, long-chain fatty acids are linked to long-chain alcohols or to carbon rings. Wax secretions help form coats on leaves, fruits, animal skin, feathers, and fur. Beeswax is a structural material used to construct the honeycomb of beehives. In some plants, waxes are important components of *cutin*, a secretion that covers and waterproofs stem, leaf, and fruit surfaces and that may impart some resistance to disease-causing organisms. Figure 4.10 shows the waxy surface coat of a blade of grass.

Lipids Without Fatty Acid Components

Lipids that have no fatty acid tails are less abundant than the ones described so far, but many play important roles in cell membranes and in the regulation of metabolism. Some (such as the terpenes) are long, water-insoluble chains; others (such as the steroids) have ring structures.

For example, all **steroids** start out with the same backbone of four carbon rings:

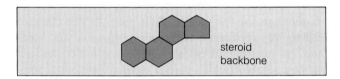

steroid backbone

However, they differ in the number and location of double bonds in the backbone, and in the number, position, and type of functional groups attached to it.

Cholesterol, the most common steroid in animal tissues, is a component of cell membranes. A cholesterol molecule can undergo rearrangements that lead to the formation of such substances as sex hormones and bile acids (which function in digestion). Plant tissues do not contain cholesterol; their steroids are called phytosterols.

PROTEINS

In addition to carbon, hydrogen, and oxygen, the molecules called **proteins** also contain nitrogen and (usually) sulfur. Of all biological molecules, proteins are the most diverse. In this class of molecules are thousands of dif-

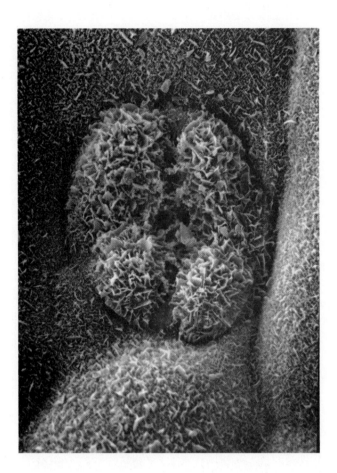

Figure 4.10 Scanning electron micrograph of wax deposits on cells making up the upper epidermal surface of a blade of grass. Here, the shredlike wax depositions surround a stoma, one of the openings that help control movements of gases and water across the leaf epidermis.

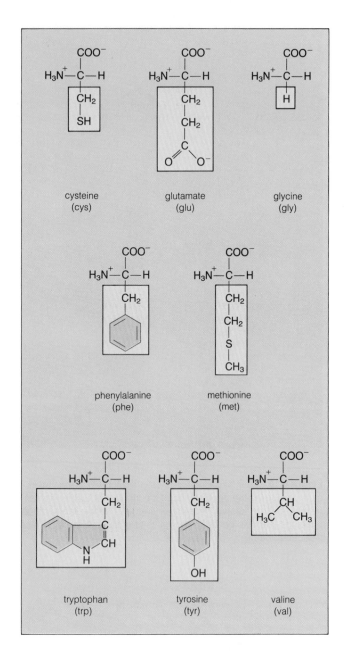

Figure 4.11 Structural formulas for eight of the twenty common amino acids. The R groups are within the shaded boxes.

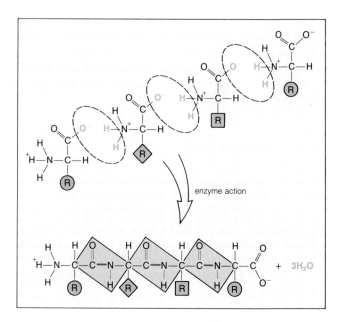

Figure 4.12 Condensation of a polypeptide chain from four amino acid units.

ferent enzymes, which make specific reactions proceed faster than they would on their own. In this class are substances concerned with cellular movements. Here, too, are storage molecules and transport molecules such as hemoglobin (which carries oxygen in blood). Some proteins are hormones; some help defend the vertebrate body against damage or attack. Other proteins are structural materials of the first rank, the stuff of cell walls and membranes, bone and cartilage, hoof and claw.

Primary Structure: A String of Amino Acids

Despite their diversity, proteins typically contain only twenty different kinds of amino acids. An **amino acid** has four parts, all bonded covalently to one carbon atom. The parts are an amino group (—NH₂), a carboxyl group (—COOH), a hydrogen atom, and some distinct atom or cluster of atoms designated the *R group* (Figure 4.11). Under cellular conditions, the amino and the carboxyl parts are ionized as shown here:

How do amino acids become linked to form proteins? A covalent bond forms between the amino group of one amino acid and the carboxyl group of another (Figure 4.12). This covalent linkage, called a **peptide bond**, results in a molecule called a dipeptide. Three or more amino acids linked together form a **polypeptide chain**.

Which kind of amino acid follows another in the chain is always the same for all proteins of a given type. For example, the two chains making up the protein insulin always have the sequences shown in Figure 4.13. The specific sequence of amino acids in a polypeptide chain constitutes the **primary structure** of a protein.

Spatial Patterns of Protein Structure

The sequence of amino acids influences the shape that a protein can assume, what its function will be, and how it will interact with other substances. It does so in two major ways. First, it influences the patterns of hydrogen bonding along the chain. Second, the R groups in the sequence interact and determine the way the chain can bend and twist into its three-dimensional shape.

In Figure 4.12, notice the so-called peptide groups, which are indicated by the tan-shaded squares. Because of the way electrons are shared in each square, the atoms linked by the red bonds tend to be positioned rigidly in the same plane (the square). Only the atoms outside the squares have some freedom in how they become oriented.

These bonding patterns impose some limits on the protein structures possible. Most often, hydrogen bonds form between every fourth amino acid and hold the chain in a helical coil about its own axis (Figure 4.14). Coils of this sort occur in hemoglobin, for example. In other cases, the chain is almost fully extended, and hydrogen bonds form between *different* chains (Figure 4.14). These bonds hold many chains side by side in a sheetlike structure, as they do in the protein that makes

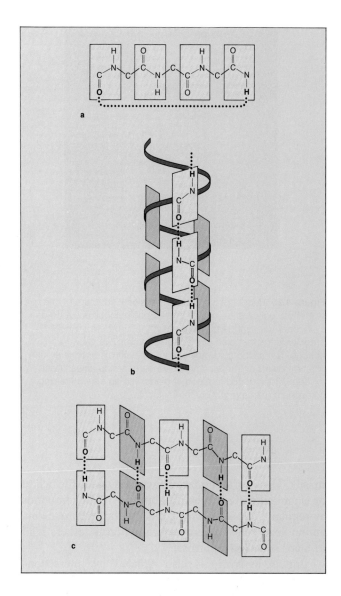

Figure 4.14 (**a**) Hydrogen bonds (dotted lines) in a polypeptide chain, which can produce a coiled chain (**b**) or a sheetlike array of chains (**c**).

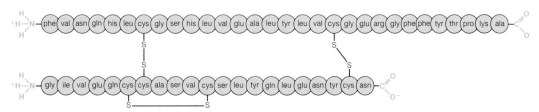

Figure 4.13 Linear sequence of amino acids in bovine (cattle) insulin, as determined by Frederick Sanger in 1953. This protein is composed of two polypeptide chains, linked together by disulfide bridges (—S—S—).

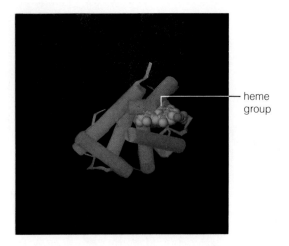

heme
group

Figure 4.15 Model of the three-dimensional structure of one of the four polypeptide chains in a hemoglobin molecule. Red cylinders represent parts of the chain that are helically coiled. Irregular loops that disrupt the regular coiling are shown as ribbons. The amino acid proline causes such disruptions at four different points in the primary sequence of amino acids. When oxygen is transported in blood, it binds with an iron-containing component (heme group) of the molecule.

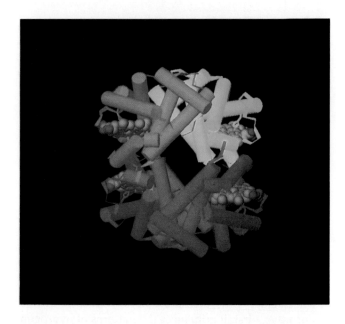

Figure 4.16 Quaternary structure of the protein hemoglobin in human blood. Hemoglobin is a red pigment circulating in animal blood and carrying vital oxygen to tissues. The hemoglobin molecule consists of four polypeptide chains, held tightly together by numerous weak bonds. (Compare Figure 4.15.)

up silk. The term **secondary structure** refers to the helical or extended pattern brought about by hydrogen bonds at regular intervals along a polypeptide chain. Some amino acids tend to favor helical patterns, others tend to favor sheetlike patterns.

Protein structure is also affected by interactions among R groups. Most helically coiled chains become further folded into some characteristic shape when one R group interacts with another R group some distance away, with the backbone itself, or with other substances present in the cell. The term **tertiary structure** refers to the folding that arises through interactions among R groups of a polypeptide chain. Figure 4.15 shows one of the diverse shapes achieved through such interactions.

Quaternary structure, the fourth level of protein architecture, results from interactions between two or more polypeptide chains in some proteins. The resulting protein can be globular, fiberlike, or some combination of the two shapes. For example, hemoglobin is globular, overall (Figure 4.16). Collagen, the most common animal protein, is fibrous (Figure 4.17). Skin, bone, tendons, cartilage, blood vessels, heart valves, corneas—these and other structures depend on the strength inherent in collagen.

Protein Denaturation

Many experiments support the idea that primary structure dictates the shape that a protein maintains under normal conditions. Among these are studies of proteins that have undergone **denaturation**: a disruption of the interactions holding a molecule in its three-dimensional form. When denatured, the polypeptide chain unwinds or changes shape.

Denaturation can be brought about by exposure to high temperatures (typically above 60°C) or to chemical agents that disrupt the bonds on which secondary, tertiary, and quaternary structure are based. Following the drastic structural changes brought about by denaturation, the protein no longer can perform its biological functions.

For example, the white portion of an uncooked chicken egg is a concentrated solution of the protein albumin. When you cook an egg, the heat does not affect the strong covalent bonds of albumin's primary structure, but it destroys the weaker bonds that maintain secondary and tertiary structure. Although denaturation can be reversed for some kinds of proteins when normal conditions are restored, albumin isn't one of them. There is no way to uncook a cooked egg.

In itself, a stretched-out polypeptide chain plays no functional role in a cell. However, its amino acid sequence dictates the final three-dimensional structure of a protein—and this structure dictates how the protein will interact with other cell substances.

NUCLEOTIDES AND NUCLEIC ACIDS

Nucleotides are vital to the operation and the reproduction of cells. Each **nucleotide** contains three different kinds of components: a five-carbon sugar (ribose or deoxyribose), a nitrogen-containing base (either a single-ringed pyrimidine or a double-ringed purine), and a phosphate group. For example, one nucleotide has the three components hooked together in this way:

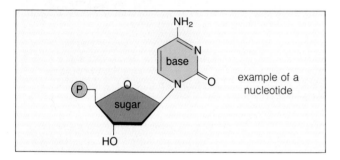

Three kinds of nucleotides or nucleotide-based molecules are the adenosine phosphates, the nucleotide coenzymes, and the nucleic acids. Later chapters will explore the structure and function of these molecules. Here, we will simply summarize their functions.

Adenosine phosphates are relatively small molecules that function as chemical messengers within and between cells, and as energy carriers. Cyclic adenosine monophosphate (cAMP) is a chemical messenger. Adenosine triphosphate (ATP) is a nucleotide that serves as an energy carrier.

Nucleotide coenzymes transport the hydrogen atoms and electrons necessary in metabolism. Nicotinamide adenine dinucleotide (NAD$^+$) and flavin adenine dinucleotide (FAD) are two of these coenzymes.

Nucleic acids, which are nucleotide-based molecules, are single- or double-stranded. A single strand consists of nucleotide units strung into long chains, with a phosphate bridge connecting sugars and with bases sticking out to the side (Figure 4.18). The sequence in which the four kinds of bases follow one another varies among nucleic acids.

Deoxyribonucleic acid (DNA) and the ribonucleic acids (RNAs) are built according to the plan just outlined. DNA is usually a double-stranded molecule that twists

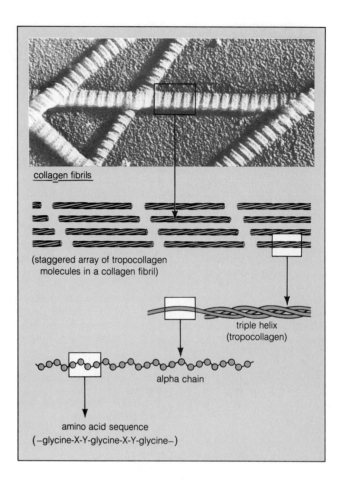

Figure 4.17 Structure of collagen, from fibrils down to its amino acid sequence. The X designates any amino acid; the Y designates either a proline or a modified proline (with an –OH group attached).

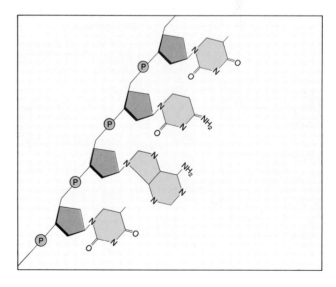

Figure 4.18 Example of bonds between nucleotides in a nucleic acid molecule.

Table 4.2 Summary of the Main Biological Molecules

Category	Main Subcategories	Some Examples and Their Functions	
CARBOHYDRATES *contain an aldehyde or a ketone group, one or more hydroxyl groups*	**Monosaccharides**	Glucose	Energy reservoir for metabolism, structural unit for biosynthesis
	Disaccharides	Sucrose	Form of sugar transported in complex plants
	Polysaccharides	Starch	Food storage
		Cellulose	Structural roles
LIPIDS *are largely hydrocarbon, generally do not dissolve in water but dissolve in nonpolar substances*	**Lipids With Fatty Acid Components:**		
	Glycerides: one, two, or three fatty acid tails attached to glycerol backbone	Fats (e.g., butter) Oils (e.g., corn oil)	Forms in which energy is stored
	Phospholipids: phosphate group and (often) two fatty acids attached to glycerol backbone	Phosphatidylcholine	Key component of cell membranes
	Waxes: long-chain fatty acid tails attached to alcohol	Earwax Waxes in cutin	Protective barrier to inner ear Help reduce water loss from some plant parts
	Lipids With No Fatty Acid Components:		
	Steroids: four carbon rings; the number, position, and type of functional groups vary	Cholesterol	Component of animal cell membranes; can be rearranged into other steroids (e.g., bile acids, male and female sex hormones)
PROTEINS *are polypeptides (up to several thousand amino acids, covalently linked)*	**Fibrous Proteins:** Individual polypeptide chains, often linked into tough, water-insoluble molecules	Keratin Collagen	Structural element of hair, nails Structural element of bones
	Globular Proteins: One or more polypeptide chains folded and linked into globular shapes; many roles in cell activities	Polymerases Hemoglobin Insulin Antibodies	Serve as enzymes Transport of oxygen Hormone that influences metabolism Tissue defense
NUCLEOTIDES *are derived from monomers having a five-carbon sugar, a phosphate group, and a nitrogen-containing base*	**Adenosine Phosphates**	ATP	Energy carrier
	Nucleotide Coenzymes	NAD^+, $NADP^+$	Transport of protons (H^+) and electrons from one reaction site to another
	Nucleic Acids (chains of nucleotide monomers)	DNA, RNAs	Storage, transmission, translation of genetic information

helically about its own axis (Figure 4.1). The bases of one strand are connected by hydrogen bonds to bases of the other strand. You will be reading more about these molecules in chapters to come. For now, it is enough to know the following: (1) Genetic instructions are encoded in the sequence of bases in DNA, and (2) RNA molecules function in the processes by which genetic instructions are used in building proteins.

SUMMARY OF THE MAIN BIOLOGICAL MOLECULES

Table 4.2 summarizes the main categories of biological molecules that have been described in this chapter. Included in this table are the most common classes of molecules within each category. We will have occasion to return to their nature and roles in diverse life processes.

Review Questions

1. What are the four main families of small organic molecules used in cells for the assembly of carbohydrates, lipids, proteins, and nucleic acids (the large biological molecules)? *49*

2. Identify which of the following is the carbohydrate, fatty acid, amino acid, and polypeptide (*refer to pages 51, 54, 56*):

    a. $^+NH_3$—CHR—COO$^-$
    b. $C_6H_{12}O_6$
    c. (glycine)$_{20}$
    d. $CH_3(CH_2)_{16}COOH$

3. Amylose (a form of starch) and cellulose are both assembled from glucose subunits, but they have very different properties. Can you explain how these properties arise? *52*

4. Explain the difference between saturated and unsaturated fats in terms of the carbon skeletons of their fatty acid tails. Is butter a saturated or unsaturated fat? *54*

5. Is this statement true or false? Enzymes are proteins, but not all proteins are enzymes. *55–56*

6. Describe the four levels of protein structure. How do the side groups of a protein molecule influence its interactions with other substances? Give an example of what happens when the bonds holding a protein together are disrupted. *57–58*

7. Distinguish between the following:

    a. monosaccharide, polysaccharide *51–52*
    b. peptide, polypeptide *56–57*
    c. glycerol, fatty acid *54*
    d. nucleotide, nucleic acid *59*

8. Define the general structure and function of three kinds of nucleotides or nucleotide-based molecules that are important to cell functioning and reproduction. *59–61*

Readings

Alberts, B., et al. 1983. *Molecular Biology of the Cell.* New York: Garland Publishing. Extraordinarily clear descriptions of the biological molecules; good illustrations.

Dickerson, R., and I. Geis. 1976. *Chemistry, Matter, and the Universe.* Menlo Park, California: Benjamin/Cummings. Introductory text with a biological point of view.

Eyre, D. 1980. "Collagen: Molecular Diversity in the Body's Protein Scaffold." *Science* 207:1315–1322.

Karplus, Martin, and J. Andrew McCammon. April 1986. "The Dynamics of Proteins." *Scientific American* 254(4):42–51.

"The Molecules of Life." October 1985. *Scientific American.* This entire issue is devoted to articles on current insights into DNA, proteins, and other biological molecules. Excellent illustrations.

Sharon, N. November 1980. "Carbohydrates." *Scientific American* 243(15):90–116. Describes carbohydrates and the roles they play in organisms.

Weinberg, R. October 1985. "The Molecules of Life." *Scientific American* 253:48–57. Survey of the new techniques and discoveries of molecular biology.

5

CELL STRUCTURE AND FUNCTION: AN OVERVIEW

GENERALIZED PICTURE OF THE CELL

Emergence of the Cell Theory

Early in the seventeenth century, Galileo Galilei arranged two glass lenses in a cylinder. With this instrument he happened to look at an insect and thereby came to describe the stunning geometric patterns of its tiny eyes. Thus Galileo, who was not a biologist, was the first to record a biological observation made through a microscope. The study of the cellular basis of life was about to begin. First in Italy, then in France and England, biologists set out to explore a world whose existence had not even been suspected.

At mid-century Robert Hooke, "Curator of Instruments" for the Royal Society of England, was at the forefront of the studies. When Hooke first turned one of his microscopes to a thinly sliced piece of cork from a mature tree, he observed tiny, empty compartments. He gave them the Latin name *cellulae* (meaning small rooms); hence the origin of the biological term "cell." They were actually walls of dead cells, which is what cork is made of, although Hooke did not think of them as being dead because he did not know that cells could be alive. He also noted that cells in other plant materials contained "juices."

Given the simplicity of their instruments, it is amazing that the pioneers in microscopy saw as much as they did. Antony van Leeuwenhoek, a shopkeeper, had great skill in constructing lenses and, possibly, the keenest vision. He even observed a single bacterium—a type of organism so small it would not be seen again for another two centuries! Yet this was mostly an age of exploration, not of interpretation. Once the limits of their simple instruments had been reached, the early microscopists gave up interest in cell structure without having been able to explain what they had seen.

Then, in the 1820s, improvements in lens design brought cells into sharper focus. Robert Brown, a botanist, observed a spherelike structure in every plant cell he examined; he called the structure a "nucleus." By 1839, the zoologist Theodor Schwann reported the presence of cells in animal tissues. He began working with Matthias Schleiden, a botanist who had concluded that all plant tissues are composed of cells and that the nucleus is somehow paramount in a cell's development. Both investigators proposed that each cell develops as an independent unit, even though its life is influenced by the organism as a whole. Schwann distilled the meaning of the new observations into what became known as the first two generalizations of the **cell theory**:

All organisms are composed of one or more cells.

The cell is the basic <u>living</u> unit of organization for all organisms.

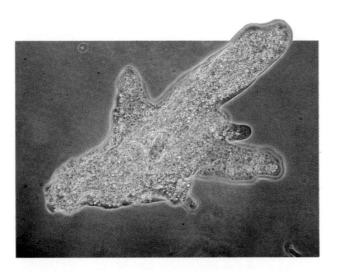

Figure 5.1 Light micrograph of a "cell"—a rather vague name that does not even hint at the astonishing events going on inside this living form, *Amoeba proteus*, as it moves through a water droplet.

Yet a question remained: Where do cells come from? A decade later the physiologist Rudolf Virchow completed studies of growth and reproduction (their division into two cells). He reached this conclusion, which became the third generalization of the cell theory:

All cells arise from preexisting cells.

Not only was a cell viewed as the smallest living unit, the continuity of life was now seen to arise directly from the division and growth of single cells. Within each tiny cell, events were going on that had profound implications for all levels of biological organization!

Basic Aspects of Cell Structure and Function

Cells vary in size, shape, and complexity. Even so, they all have three general features in common: a *nucleus* (or nucleoid), surrounded by *cytoplasm*, which in turn is surrounded by a *plasma membrane*:

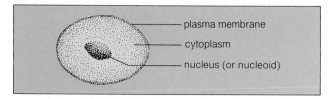

These features, which are the structural basis of all cellular events, can be defined in this way:

1. Plasma membrane. The cell's outermost membrane separates internal events from the environment and allows them to proceed in organized, controlled ways. It does not totally isolate the interior; several substances diffuse across the membrane and others move through channels built into it. Some membrane components serve as docks (receptors) for substances that can affect the cell's activities.

2. Nucleus. This membrane-bound compartment contains hereditary instructions (DNA) and other molecules that function in how the instructions are read, modified, and dispersed. In bacterial cells only, DNA occurs in a region called the *nucleoid*; there is no surrounding membrane.

3. Cytoplasm. The cytoplasm includes everything between the plasma membrane and the nucleus. It has internal membrane systems, particles, and filaments bathed in a semifluid matrix (the cytosol). Some components synthesize, store, or degrade substances. The filaments form a tiny skeleton that imparts shape and permits movement.

Cell Size and Cell Membranes

Can any cell be seen with the naked eye? There are a few, including the "yolks" of bird eggs, cells in the red part of a watermelon, and the fish eggs we call caviar. Generally, however, cells and their structures are too small to be observed without microscopes; they are measured in micrometers and nanometers. Just how small is a micrometer? Your red blood cells are about 6 to 8 micrometers across—and a string of about 2,000 would only be as long as your thumbnail is wide! We use light microscopes to explore cellular details down to about 0.2 micrometer and electron microscopes for details smaller than this (Figures 5.1 through 5.4).

Figure 5.2 Units of measure used in microscopy. The micrometer is used in describing whole cells or large cell structures. The nanometer is used in describing cell ultrastructures and large organic molecules.

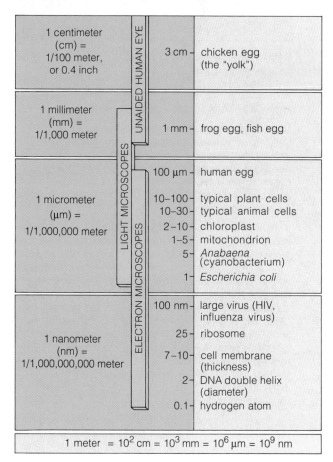

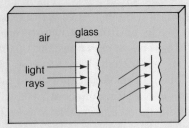

air　glass

light
rays

a *Refraction of light rays
(The angle of entry and the
molecular structure of the
glass determine how much
the rays will bend)*

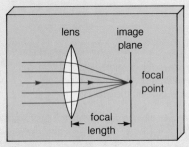

lens　image
plane

focal
point

focal
length

b *Focusing of light rays*

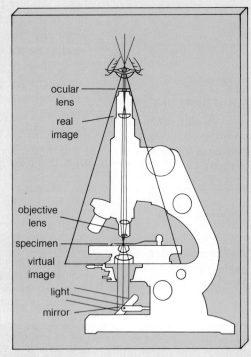

ocular
lens

real
image

objective
lens

specimen

virtual
image

light

mirror

c *Compound light microscope*

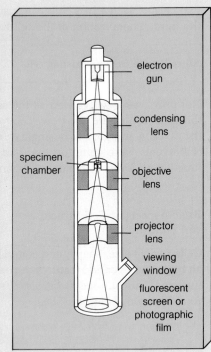

electron
gun

condensing
lens

specimen
chamber

objective
lens

projector
lens

viewing
window

fluorescent
screen or
photographic
film

d *Transmission electron microscope*

Figure 5.3　Microscopes—gateways to the cell.

Light Microscopes　(a) Light microscopy relies on the bending, or refraction, of light rays. Light rays pass straight through the center of a curved lens. The farther they are from the center, the more they bend. (**b, c**) The *compound light microscope* is a two-lens system. All rays coming from the object being viewed are channeled to a single place behind the system of curved lenses.

If you wish to observe a *living* cell, it must be small or thin enough for light to pass through. Also, structures inside cells can be seen only if they differ in color and density from their surroundings—but most are nearly colorless and optically uniform in density. Specimens can be stained (exposed to dyes that react with some cell structures but not others), but staining usually alters the structures and kills the cells. Finally, dead cells begin to break down at once, so they must be pickled or preserved (fixed) before staining. Most observations have been made of dead, pickled, or stained cells. Largely transparent living cells can be observed through the *phase-contrast microscope*. Here, small differences in the way different structures refract light are converted to larger variations in brightness.

No matter how good a glass lens system may be, when magnification exceeds 2,000× (when the image diameter is 2,000 times as large as the object's diameter), cell structures appear large but are not clearer. By analogy, when you hold a

magnifying glass close to a newspaper photograph, you see only black dots. You cannot see a detail as small as or smaller than a dot; the dot would cover it up. In microscopy, something like dot size intervenes to limit *resolution* (the property that determines whether small objects close together can be seen as separate things). That limiting factor is the physical size of wavelengths of visible light.

Light comes in different wavelengths, or colors. Red wavelengths are about 750 nanometers and violet wavelengths are about 400 nanometers; all other colors fall in between. If an object is smaller than about one-half the wavelength, light rays passing by it will overlap so much that the object won't be visible. The best light microscopes resolve detail only to about 200 nanometers.

Transmission Electron Microscopes　Electrons are usually thought of as particles, but they also behave like waves. Electron wavelengths are about 0.005 nanometer—about 100,000 times shorter than those of visible light! Ordinary lenses cannot be used to focus such accelerated streams of electrons, because glass scatters them. But each electron carries an electric charge, which responds to magnetic force. A magnetic field can divert electrons along defined paths and channel them to a focal point. Magnetic lenses are used in *transmission electron microscopes* (**d**).

Electrons must travel in a vacuum, otherwise they would be randomly scattered by molecules in the air. Cells can't live in

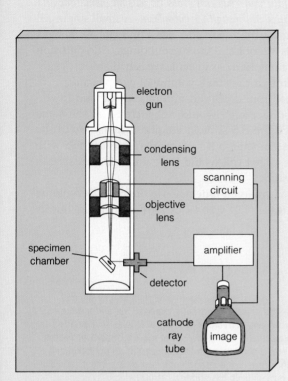

e *Scanning electron microscope*

a vacuum, so living cells cannot be observed at this higher magnification. In addition, specimens must be sliced extremely thin so that electron scattering corresponds to the density of different structures. (The more dense the structure, the greater the scattering and the darker the area in the final image formed.) Specimen fixation is crucial. Fine cell structures are the first to fall apart when cells die, and artifacts (structures that do not really exist in cells) may result. Because most cell materials are somewhat transparent to electrons, they must be stained with heavy metal "dyes," which can create more artifacts.

With *high-voltage electron microscopes*, electrons can be made ten times more energetic than with the standard electron microscope. With the energy boost, intact cells several micrometers thick can be penetrated. The image produced is something like an x-ray plate and reveals the three-dimensional internal organization of cells (see, for example, Figure 5.20c).

Scanning Electron Microscopes (e) With a *scanning electron microscope*, a narrow electron beam is played back and forth across a specimen's surface, which has been coated with a thin metal layer. Electron energy triggers the emission of secondary electrons in the metal. Equipment similar to a television camera detects the emission patterns, and an image is formed. Scanning electron microscopy does not approach the high resolution of transmission instruments. However, its images have fantastic depth.

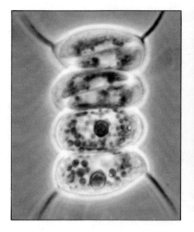

a Light micrograph (phase-contrast).

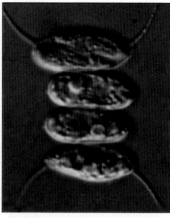

b Light micrograph (Nomarski process).

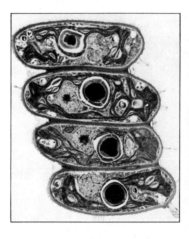

c Transmission electron micrograph, thin section.

d Scanning electron micrograph

Figure 5.4 Comparison of how different types of microscopes reveal cellular details. The specimen is the green alga *Scenedesmus*; the magnification (1000×) is the same in all cases. (**a**) Phase-contrast and (**b**) Nomarski techniques create optical contrasts without staining the cells; both have enhanced the value of light microscopes. (**c**) Details of a cell's internal structure show up best with transmission electron microscopy. (**d**) Scanning electron microscopy provides a three-dimensional view of surface features.

diameter (cm):	0.5	1.0	1.5
surface area (cm²):	0.79	3.14	7.07
volume (cm³):	0.06	0.52	1.77
surface-to-volume ratio:	13.17:1	6.04:1	3.99:1

Figure 5.5 Relationship between the surface area and volume when a sphere is enlarged. Notice that as the diameter increases, the volume is increasing more than the surface area.

Why are most cells so small? *The smaller the cell, the more efficiently materials can cross its plasma membrane and become distributed through the cytoplasm.* A small single cell has enough surface area to accommodate exchanges between its interior and the surroundings. And diffusion (the random motion of molecules) easily distributes substances through the small volume of cytoplasm.

If a small round cell did nothing but expand in volume, its surface area would not increase at the same rate. (As Figure 5.5 shows, volume increases with the cube of the diameter, but surface area increases only with the square.) That cell would have an inefficient *surface-to-volume ratio.*

If the cell enlarged four times in diameter, its volume would increase (4 × 4 × 4), or sixty-four times. But its

cytoplasm   microvillus   intestinal lumen

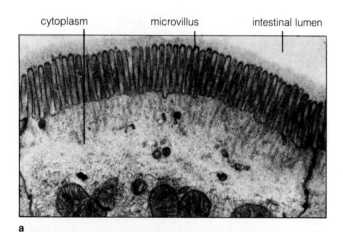

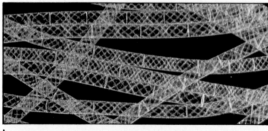

a

b

c

d

Figure 5.6 (**a**) Improving the surface-to-volume ratio in a single cell. Extensions of the plasma membrane greatly increase the surface area exposed to the environment. A *microvillus* is an example. This slender, cylindrical extension occurs in animal cells that rapidly absorb or secrete molecules. Shown here, a slice through the top part of an epithelial cell that faces the intestinal lumen (the space enclosed by the intestinal wall). Numerous microvilli project from the surface. 21,000×.

Multicellular responses to the constraints of the surface-to-volume ratio. (**b**) Threadlike body plan of a green alga, *Spirogyra*. (**c**) Sheetlike body plan of a red alga. (**d**) Massive body plan of a whale, which requires internal transport systems that cut through the volume of tissue masses and so shrink the diffusion distance between individual cells and the environment.

surface area would increase by only sixteen times—so each unit of plasma membrane would have to serve four times as much cytoplasm as before! Past a certain point, the inward flow of nutrients and outward flow of wastes (some toxic) would not be fast enough, and the cell would die. So we can see why most cells are small—or long and thin, or with outfoldings and infoldings that increase the membrane surface relative to the volume of cytoplasm (Figure 5.6).

We see the effects of surface-to-volume constraints on multicelled body plans, also (Figure 5.6). Some algae are strandlike, with cells attached end to end and each interacting directly with the environment. Other algae and a few protistans are sheetlike, with cells at or near the body surface. Massive plants and animals have transport systems that move materials to and from millions, billions, even trillions of cells packed together in tissues. For example, the animal circulatory system quickly delivers materials from the environment to all living cells and sweeps wastes away from them. Its "highways" cut through the volume of tissue and so shrink the distance that would otherwise have to be traversed by diffusion.

PROKARYOTIC CELLS— THE BACTERIA

Let's begin our survey of cells with bacteria. They are the smallest cells of all, and in structural terms, they are the simplest to think about. Most bacteria have a rigid or semirigid **cell wall**, which supports the cell and imparts shape to it (Figure 5.7). The wall is composed of secretions from the bacterium; sometimes it has a protective coating. The wall surrounds the plasma membrane, which incorporates many enzymes and other molecules used in cell activities. Inside the cell, the DNA is concentrated in the irregularly shaped region called the nucleoid. There is only a small volume of cytoplasm, and numerous ribosomes are dispersed through it.

As Figure 5.2 indicates, a **ribosome** is a tiny particle. It is composed of two subunits, each being an aggregate of RNA and protein molecules. In all cells, not just bacteria, ribosomes are the workbenches for protein synthesis; new polypeptide chains are constructed at their surface (page 73).

Bacterial cells are said to be **prokaryotic** because they do not have a nucleus. The word means "before the nucleus," and it implies that some forms of bacteria existed on earth before the evolution of nucleated cells. All bacteria are prokaryotic, including the cyanobacteria (which formerly were called blue-green algae). Chapters Thirty-Eight and Thirty-Nine provide closer looks at the evolution, structure, and functioning of bacteria. Here our focus will be on the nucleated cells, the eukaryotes.

EUKARYOTIC CELLS

Function of Organelles

All cells except bacteria are **eukaryotic**. They contain many **organelles**, which are membranous sacs, envelopes, and other compartmented portions of the cytoplasm. The most conspicuous organelle is the nucleus (hence the name eukaryotic, which means "true nucleus").

No chemical apparatus in the world can match the eukaryotic cell for the sheer number of chemical reactions that proceed in so small a space. Many of the reac-

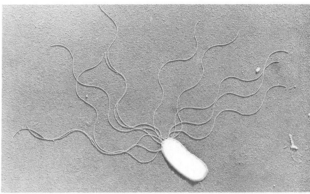

a

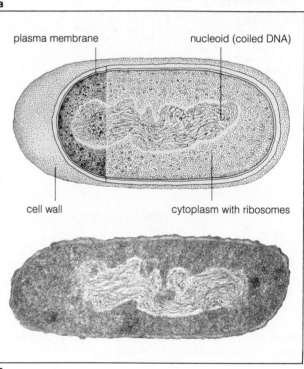

b

Figure 5.7 Bacterial body plans. (**a**) Surface view of *Pseudomonas marginalis*, which is equipped with motile structures called bacterial flagella, 12,200×. (**b**) Sketch and transmission electron micrograph of *Escherichia coli*, 36,500×.

plasma membrane   nucleoid (coiled DNA)

cell wall   cytoplasm with ribosomes

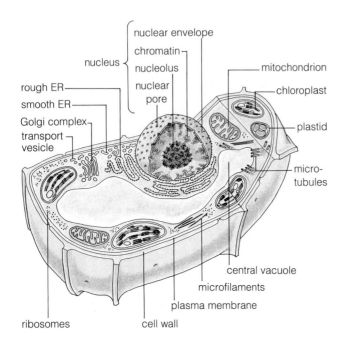

Figure 5.8 Generalized sketch of a plant cell, showing the types of organelles that may be present.

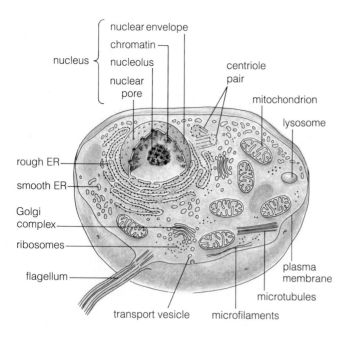

Figure 5.9 Generalized sketch of an animal cell, showing the types of organelles that may be present.

tions are incompatible. (For example, a starch molecule can be synthesized by some reactions and degraded by others—but a cell would gain nothing if both types of reactions proceeded simultaneously on the same molecule.) Reactions proceed smoothly in cells, largely for these reasons:

Organelles physically separate chemical reactions (many of which are incompatible) in the space of the cytoplasm.

Organelles also separate different reactions in time, as when molecules are produced or processed in one organelle, then are used later in other reaction sequences.

Keep these two points in mind as you read through the rest of this unit, for they give insight into why many cell activities proceed as they do.

Typical Components of Eukaryotic Cells

In general, eukaryotic cells contain the following organelles, each with specific functions:

nucleus	*physical isolation and organization of DNA*
endoplasmic reticulum	*modification of polypeptide chains into mature proteins; lipid synthesis*
Golgi bodies	*further modification, sorting, and shipping of proteins and lipids for secretion or for use in cell*
lysosomes	*intracellular digestion*
transport vesicles	*transport of a variety of materials to and from organelles and plasma membrane*
mitochondria	*ATP formation*

In addition to the organelles just listed, eukaryotic cells have many thousands of *ribosomes*, either "free" in the cytoplasm or attached to particular membranes. And they have a *cytoskeleton*, an internal network of protein filaments responsible for the cell's shape, internal organization, movement of structures and organelles, and often movement through the environment.

Photosynthetic cells alone have one or more *plastids*, organelles that function in food production and storage. Cells of many fungi and plants have one or more fluid-filled sacs called *central vacuoles*. Cells of many protistans, fungi, and plants have a *cell wall* surrounding the plasma membrane.

Figures 5.8 through 5.11 show how some of these components might be arranged in a typical plant and animal cell. Keep in mind that calling them "typical" is like calling a squid or cactus a "typical" animal or plant; mind-boggling variation exists on the basic plan.

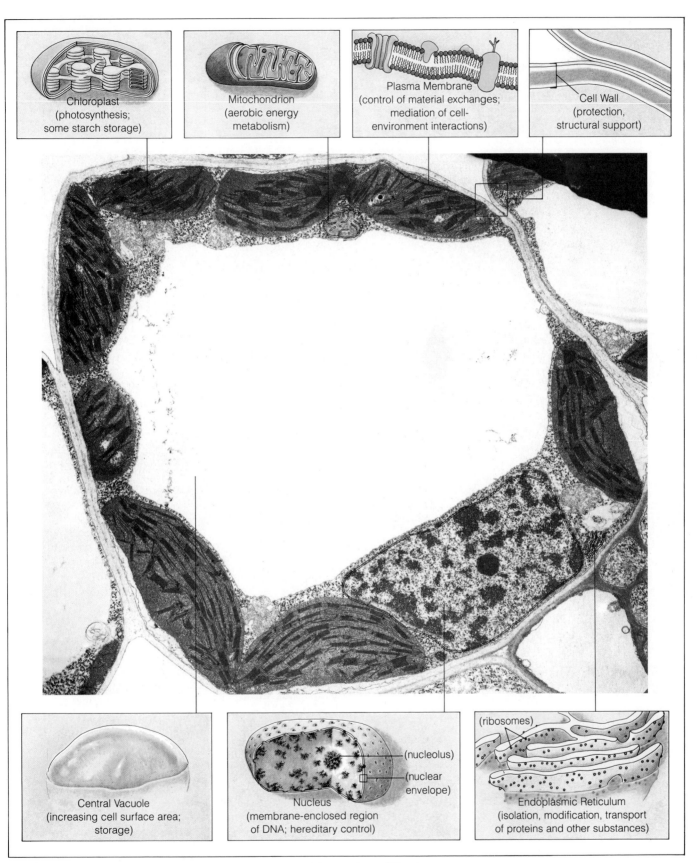

Chloroplast
(photosynthesis; some starch storage)

Mitochondrion
(aerobic energy metabolism)

Plasma Membrane
(control of material exchanges; mediation of cell-environment interactions)

Cell Wall
(protection, structural support)

Central Vacuole
(increasing cell surface area; storage)

Nucleus
(membrane-enclosed region of DNA; hereditary control)

(nucleolus)

(nuclear envelope)

(ribosomes)

Endoplasmic Reticulum
(isolation, modification, transport of proteins and other substances)

Figure 5.10 Transmission electron micrograph of a plant cell from a blade of Timothy grass, cross-section, 11,300×.

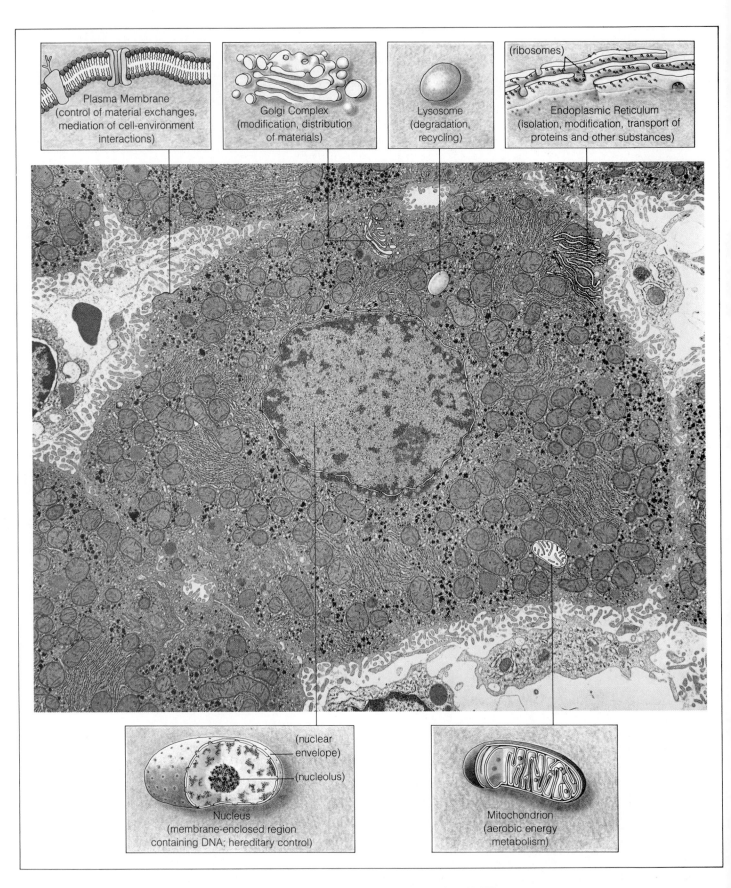

Plasma Membrane
(control of material exchanges,
mediation of cell-environment
interactions)

Golgi Complex
(modification, distribution
of materials)

Lysosome
(degradation,
recycling)

(ribosomes)

Endoplasmic Reticulum
(isolation, modification, transport of
proteins and other substances)

(nuclear
envelope)

(nucleolus)

Nucleus
(membrane-enclosed region
containing DNA; hereditary control)

Mitochondrion
(aerobic energy
metabolism)

Figure 5.11 Transmission electron micrograph of an animal cell from a rat liver, cross-section, 15,000×.

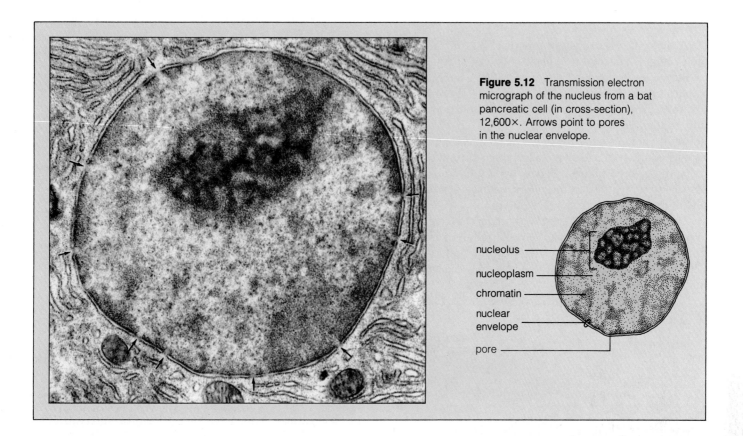

Figure 5.12 Transmission electron micrograph of the nucleus from a bat pancreatic cell (in cross-section), 12,600×. Arrows point to pores in the nuclear envelope.

nucleolus

nucleoplasm

chromatin

nuclear envelope

pore

THE NUCLEUS

Recall that carbohydrates, lipids, proteins, and nucleic acids are the main building blocks of cell architecture. It takes the special class of proteins called enzymes to build and use those molecules as a cell grows, maintains itself, and reproduces. Thus, *cell structure and function begin with proteins—and instructions for building the proteins themselves are contained in DNA.*

Especially among multicelled eukaryotes, which have complex programs of growth and development, DNA instructions are used in intricate ways. Also, eukaryotes have a great deal of DNA, which must be duplicated, sorted out, and condensed into compact structures before a cell reproduces itself by dividing in two. The isolation of eukaryotic DNA in its own membrane-bound compartment, the **nucleus**, serves two functions: it helps control access to the DNA and it simplifies DNA packaging for cell division. Every nucleus has the components that are listed in Table 5.1.

Although the nucleus contains the instructions for protein synthesis, the proteins actually are assembled elsewhere. First, an RNA molecule is assembled on a DNA region that codes for a specific protein. Then the RNA moves into the cytoplasm, where its coded message is translated into a polypeptide chain (the protein's primary structure).

Table 5.1 Components of the Nucleus	
Component	Definition
Chromosomes	*DNA molecules and numerous proteins attached to them*
Nucleoplasm	*fluid portion of the nucleus*
Nucleolus	*dense cluster of the types of RNA and proteins used to assemble ribosomal subunits*
Nuclear envelope	*membranous, pore-riddled boundary between the nuclear interior and the cytoplasm*

Chromosomes

Eukaryotic DNA is not a "naked" molecule; a great number of different proteins are closely associated with it at all times. Some proteins are enzymes that take part in RNA assembly and other important tasks. Others form a scaffold that organizes the DNA during cell division.

Sometimes the DNA and proteins in the nucleus of a nondividing cell look rather grainy (Figure 5.12). The word **chromatin** is used to describe the grainy appearance of these two substances in micrographs, but don't let appearances mislead you. At high magnification, a DNA molecule clearly is not granular. It is threadlike, with proteins attached to it like beads on a chain.

Before a cell divides in two, each DNA molecule is duplicated (both new cells get a complete set of DNA this way). Then the duplicated molecules fold and twist into condensed structures, proteins and all. The condensed structures were the only ones that early microscopists could see, and they called them **chromosomes** (meaning colored bodies). Today, we call DNA and its associated proteins a chromosome regardless of whether it is in its threadlike or condensed form.

The point is, *chromosomes do not always look the same during the life of a cell*. We will consider different aspects of "the chromosome" in chapters to come, and it will help to keep this point—and the following sketch—in mind:

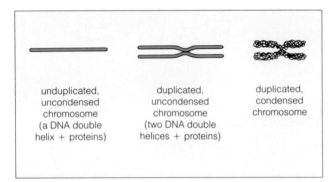

unduplicated, uncondensed chromosome (a DNA double helix + proteins)

duplicated, uncondensed chromosome (two DNA double helices + proteins)

duplicated, condensed chromosome

Nucleolus

As cells grow and develop, dense masses of irregular size and shape form in the nucleoplasm. They consist largely of RNA and proteins. Two or three masses form in most cells, but there can be a thousand or more in cells that develop into amphibian eggs. Each mass is a **nucleolus** (plural, nucleoli), and it retains its distinct appearance only in nondividing cells (Figure 5.12).

A nucleolus is the site where subunits of ribosomes are assembled. It is located on chromosomal regions that contain the instructions for synthesizing certain types of RNA molecules. Here, RNA is assembled, processed, and packaged with proteins into ribosomal subunits, which are then shipped to destinations in the cytoplasm. A complete, two-part ribosome is never found in the nucleus.

Nuclear Envelope

Figures 5.12 and 5.13 show details of the **nuclear envelope**. This two-membrane envelope is the outermost component of the nucleus. Ribosomes are attached to its outer surface, which faces the cytoplasm, and pores extend across the envelope at regular intervals. The structure of the nuclear envelope suggests that it is the boundary for controlled exchanges between the nucleus and the cytoplasm, with its pores being passageways.

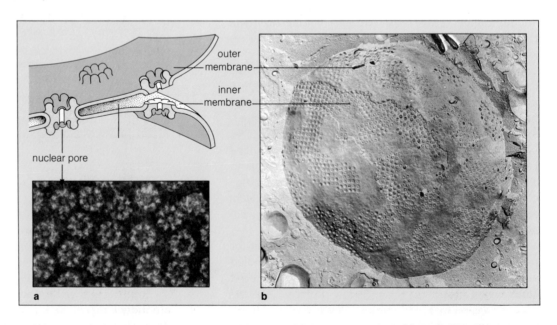

outer membrane

inner membrane

nuclear pore

a

b

Figure 5.13 Nuclear envelope structure. (**a**) Electron micrograph of the pores that span the envelope. (**b**) Electron micrograph of a freeze-fractured nucleus (page 88), revealing both membranes of the envelope, 10,300×. Each membrane has a lipid bilayer structure, as described on page 86.

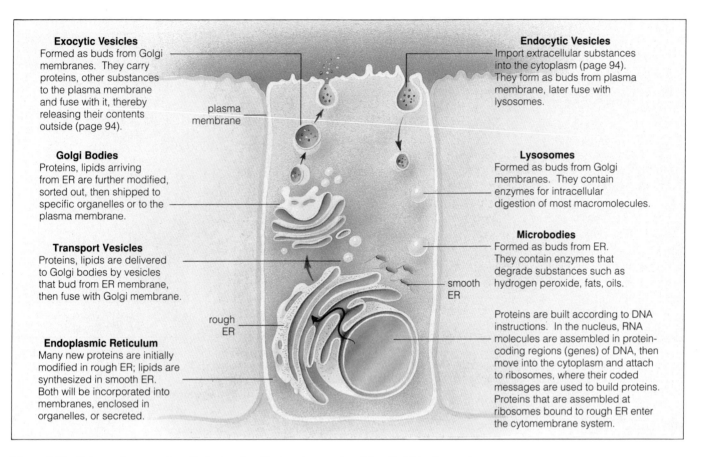

Exocytic Vesicles
Formed as buds from Golgi membranes. They carry proteins, other substances to the plasma membrane and fuse with it, thereby releasing their contents outside (page 94).

Golgi Bodies
Proteins, lipids arriving from ER are further modified, sorted out, then shipped to specific organelles or to the plasma membrane.

Transport Vesicles
Proteins, lipids are delivered to Golgi bodies by vesicles that bud from ER membrane, then fuse with Golgi membrane.

Endoplasmic Reticulum
Many new proteins are initially modified in rough ER; lipids are synthesized in smooth ER. Both will be incorporated into membranes, enclosed in organelles, or secreted.

plasma membrane

rough ER

smooth ER

Endocytic Vesicles
Import extracellular substances into the cytoplasm (page 94). They form as buds from plasma membrane, later fuse with lysosomes.

Lysosomes
Formed as buds from Golgi membranes. They contain enzymes for intracellular digestion of most macromolecules.

Microbodies
Formed as buds from ER. They contain enzymes that degrade substances such as hydrogen peroxide, fats, oils.

Proteins are built according to DNA instructions. In the nucleus, RNA molecules are assembled in protein-coding regions (genes) of DNA, then move into the cytoplasm and attach to ribosomes, where their coded messages are used to build proteins. Proteins that are assembled at ribosomes bound to rough ER enter the cytomembrane system.

Figure 5.14 Cytomembrane system. Endoplasmic reticulum, transport vesicles, Golgi bodies, and endocytic vesicles are components of the secretory pathway of this system (purple arrows).

CYTOMEMBRANE SYSTEM

In our introduction to the nucleus, we outlined the sequence by which DNA instructions for building proteins are entrusted to RNA, which carries them to ribosomes in the cytoplasm for translation into polypeptide chains. What happens to the newly formed chains? Many end up in the pool of proteins dissolved in the cytoplasm; others enter the cytomembrane system.

The **cytomembrane system** includes the endoplasmic reticulum (ER), Golgi bodies, lysosomes, and a variety of vesicles. We call it a system because of the membrane "traffic" to and from its components (Figure 5.14). Briefly, cell membranes are composed mostly of lipid and protein molecules that are modified or assembled in the ER and Golgi bodies. Those two organelles grow and maintain themselves by adding some of the molecules to their own membrane. Parts of their membrane "bud" and break away as vesicles that move through the cytoplasm, then fuse with other organelles or with the plasma membrane. The plasma membrane itself buds inward to form vesicles that later fuse with other organelles.

Membrane production and distribution is only one function of the cytomembrane system. If we were to construct a flowchart for the proteins and lipids produced and used by the system, it would look like this:

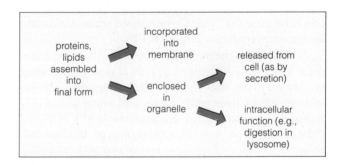

Endoplasmic Reticulum and Ribosomes

The membrane of **endoplasmic reticulum** (ER) is folded into tubes and flattened sacs, the interiors of which form a single, continuous space that is separate from the cytoplasm. The membrane has rough and smooth regions, due mainly to the presence or absence of ribosomes on

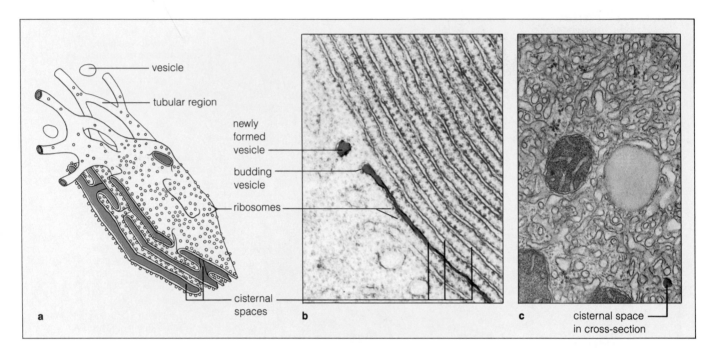

vesicle

tubular region

newly formed vesicle

budding vesicle

ribosomes

cisternal spaces

a

b

c

cisternal space in cross-section

Figure 5.15 (**a, b**) Rough endoplasmic reticulum, showing how the membrane surface facing the cytoplasm is studded with ribosomes, 25,000×. (**c**) Smooth endoplasmic reticulum, in cross-section, 35,000×.

the side of the membrane facing the cytoplasm. *Rough ER* has many ribosomes attached, and they give it a grainy appearance in electron micrographs; *smooth ER* is free of ribosomes (Figure 5.15). Both occur in most cells, although the relative proportion varies among different cell types.

Smooth ER. This part of the ER membrane curves through the cytoplasm like interconnecting pipelines. Often it looks like clustered sacs in micrographs, but this is because the "pipes" were sliced crosswise during specimen preparation (Figure 5.15c).

Smooth ER is the major site of lipid synthesis in many cells. It is highly developed in cells of developing seeds and animal cells that secrete steroid hormones. In the smooth ER of liver cells, drugs and some potentially harmful by-products of metabolism are inactivated ("detoxified"). A specialized version of this smooth membrane (the sarcoplasmic reticulum) occurs in skeletal muscle; it stores and releases calcium ions, which play a role in muscle contraction.

Rough ER. Newly minted polypeptide chains destined for secretion or for delivery to several organelles begin their journey in rough ER, which typically is organized into stacked, flattened sacs. Apparently, only chains hav-

ing a "signal" (a sequence of about fifteen to twenty specific amino acids) can enter the sacs; chains without the signal join the cytoplasmic pool of proteins.

A major function of rough ER is to synthesize and then attach a special polysaccharide to most of the chains passing through. Later, in the Golgi bodies, the polysaccharide is modified into one of several "shipping labels" that send each chain to its ultimate destination.

Rough ER is especially abundant in secretory cells (such as pancreatic cells that produce and secrete digestive enzymes) and in immature egg cells that grow rapidly in size (frog and other amphibian eggs are like this).

Golgi Bodies

In **Golgi bodies**, many proteins and lipids undergo final modification, then they are sorted out and packaged for specific destinations. Here, polysaccharides that were attached to each protein in the ER are trimmed and elongated in specific ways. The final results allow components of the Golgi membrane to "recognize" differences among many products and to form vesicles with special mailing tags around them.

All eukaryotic cells have one or more Golgi bodies. In outward appearance, a Golgi body resembles a stack

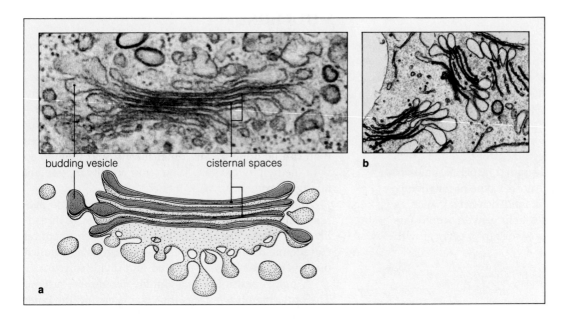

budding vesicle   cisternal spaces

b

a

Figure 5.16 (**a**) Electron micrograph and sketch of a Golgi body from an animal cell, 53,000×. (**b**) Golgi bodies in a root cap cell of corn, 14,000×.

of pancakes—usually eight or less, and usually curled at the edges (Figure 5.16). The topmost pancakes (the ones farthest from the nucleus in animal cells) bulge at the edges, then the bulges break away as vesicles. Some secretory cells concentrate and store one or more products in such vesicles until the cell is signaled to release them. Figure 5.14 shows the secretory pathway, which proceeds through the ER and a Golgi body, and then across the plasma membrane.

Lysosomes

In all eukaryotic cells, some vesicles that bud from Golgi bodies become **lysosomes**, the main organelles of intracellular digestion. Forty or so enzymes in these membrane bags can break down every polysaccharide, nucleic acid, and protein, along with some lipids.

Lysosomes act on substances that have already become wrapped in cell membrane. For example, when cholesterol molecules are transported in the blood, they become complexed with other substances. The complexes bind at the surface of cells and an endocytic vesicle forms around them (Figures 5.14 and 6.10). When the vesicle fuses with a lysosome, enzymes break down the complex and free the cholesterol, which becomes available for cellular use. Lysosomal enzymes also degrade organelles that are worn out or that have served their function (Figure 5.17), and they can destroy bacteria and foreign particles.

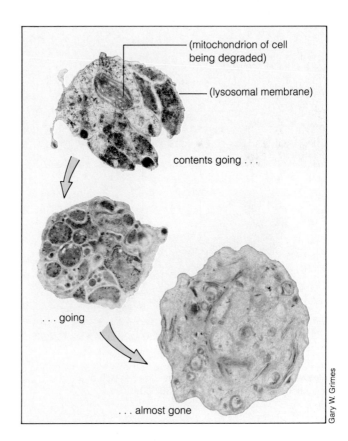

(mitochondrion of cell being degraded)

(lysosomal membrane)

contents going . . .

. . . going

. . . almost gone

Gary W. Grimes

Figure 5.17 Digestion of organelles from a destroyed cell, as seen in a lysosome, 11,000×.

Microbodies

Some bags of enzymes that bud from the ER membrane become **microbodies**, a type of vesicle found in at least some cells of every eukaryote. The ones called "peroxisomes" have enzymes that use oxygen to degrade fatty acids and amino acids. Hydrogen peroxide, a potentially harmful substance, is a by-product of the reactions. However, another enzyme converts the hydrogen peroxide to water or uses it to degrade alcohol. If you drink alcohol, nearly half of it is degraded in peroxisomes of your liver and kidney cells. Another type of microbody (glyoxysomes) is abundant in lipid-rich seeds, such as those of peanuts. Its enzymes help convert stored fats and oils to the sugars necessary for rapid, early growth of the plant.

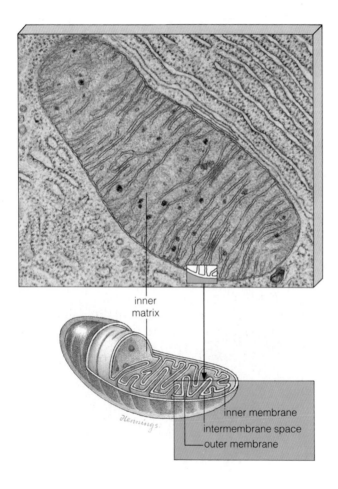

Figure 5.18 Micrograph and generalized sketch of a mitochondrial membrane system, 36,500×.

MITOCHONDRIA

As we have seen, cells have the means to synthesize and distribute proteins and lipids. They also have the means to use energy stored in carbohydrates to form molecules such as ATP—which provides energy to drive a variety of cellular reactions. The **mitochondrion** (plural, mitochondria) is an organelle specialized for this task. Mitochondria can extract far more energy from carbohydrates than can be done by any other means, and they do so with the help of oxygen. When you breathe in, you are taking in oxygen for your mitochondria.

Mitochondria are often about the size of some bacteria. They are round, potato-shaped, tubelike, or threadlike, but these shapes are not rigid. When chemical conditions shift in the cell, mitochondria grow and branch out, even fuse with one another and divide in two.

A mitochondrion has a double-membrane system. The outer membrane faces the cytoplasm and the inner membrane usually has many deep, inward folds. As Figure 5.18 shows, the system creates two compartments: an intermembrane space (between the two membranes) and a mitochondrial matrix (the innermost space). Hydrogen ions and electrons are moved from one compartment to the other in ways that cause ATP to form; then oxygen combines with the "spent" electrons and hydrogen to form water (pages 129-131).

A eukaryotic cell typically contains anywhere from a dozen to a thousand mitochondria. Animal cells generally have more mitochondria than plant cells do. Mitochondria are abundant in muscle cells, parts of nerve cells, and near the surface of cells that specialize in absorbing or secreting substances. All of these cells have high energy demands.

SPECIALIZED PLANT ORGANELLES

Chloroplasts and Other Plastids

Most plant cells contain organelles specialized for photosynthesis and storage, these being classified as "plastids." Three kinds are common:

chloroplasts	*with photosynthetic pigments and starch-storing capacity*
chromoplasts	*with pigments that may function in pollination and seed dispersal*
amyloplasts	*with starch-storing capacity; no pigments*

Chloroplasts specialize in photosynthesis and, like mitochondria, they have a double-membrane system. Parts of the inner membrane commonly are organized as stacked disks (called grana). Pigments, enzymes, and other molecules embedded in the inner membrane trap

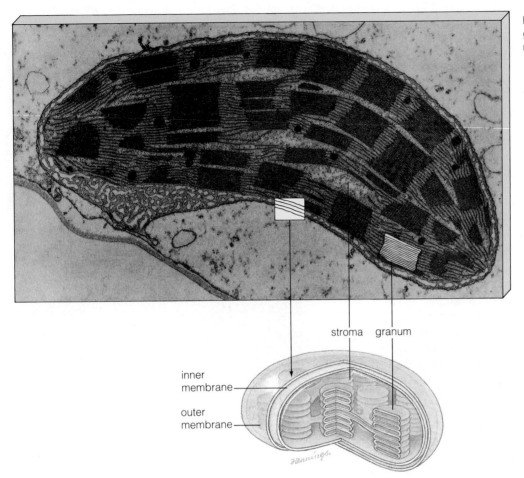

Figure 5.19 Micrograph and generalized sketch of a chloroplast membrane system, 19,800×.

inner
membrane

outer
membrane

stroma   granum

sunlight energy and take part in ATP formation. A semi-fluid matrix (the stroma) surrounds the grana; here, starch and proteins are synthesized (Figures 5.19 and 8.2c). Often, aggregates of new starch molecules are temporarily stored as "starch grains" in chloroplasts.

Chloroplasts typically are oval or disk-shaped. They appear green in land plants and green, yellow-green, or golden brown among algae. Their color depends on the kinds and numbers of light-absorbing pigment molecules in their membranes. *Chlorophyll*, a major pigment, appears green because it absorbs most wavelengths (colors) of visible light except those corresponding to green light, which it transmits. Chlorophyll occurs in all chloroplasts, but it may be masked by other pigments, as it is in brown algae.

Chromoplasts store red or brown pigments that give flower petals, fruits, and some roots (such as carrots) their characteristic color. The colorless **amyloplasts**, which occur in plant parts exposed to little (if any) sunlight, are often storage sites for starch grains. They are abundant in cells of stems, potato tubers, and many seeds.

Central Vacuoles

Mature and still-living plant cells often have a large, fluid-filled **central vacuole** (Figure 5.10). This organelle forms when small vesicles fuse during cell growth and development. It usually occupies as much as fifty to ninety percent of the cell interior, and this leaves only a narrow zone of cytoplasm between the vacuole's membrane and the cell wall.

A central vacuole can serve as a storage area for metabolic products (such as amino acids and sugars), ions, and toxic wastes. However, its main function may be to increase cell size and surface area. Water pressure builds up inside the vacuole during plant growth and causes the vacuole to press against the cell wall. The cell enlarges permanently under this force. The extent of its enlargement depends partly on the volume possible within the confines of the wall, which is pliable at first but then becomes rigid at maturity. The increased surface area enhances the rate of absorption of the mineral salts necessary for plant nutrition, a topic that we will consider in Chapter Twenty.

THE CYTOSKELETON

Structure and Function of the Cytoskeleton

Think about all the micrographs you have ever seen of living cells—red blood cells, cells of an onion root tip, eggs. Each cell type has a characteristic shape and internal organization. Most can reposition some of their components (such as chromosomes) in predictable ways. Many cells, including the amoeba shown in Figure 5.1, also have characteristic ways of moving through the environment. What can you deduce from all of this? *The cell's very shape, internal organization, and capacity for motion probably are heritable features.*

Here again, particular aspects of cell structure and function can be traced to proteins specified by DNA. Tubulin, actin, and myosin are examples of proteins that become assembled into filaments in the cytoplasm. The filaments (and other proteins) are organized into a **cytoskeleton**: an interconnecting system of bundled-up fibers, slender threads, and lattices that crisscross through the cytoplasm, from the nucleus all the way to the plasma membrane. Figure 5.20 shows parts of the cytoskeleton from an animal cell and a plant cell.

Three main classes of cytoskeletal filaments have been identified: **microtubules** (mostly tubulin), **microfilaments** (mostly actin), and **intermediate filaments** (at least five types, each with a distinct protein such as keratin). Microtubules are the largest:

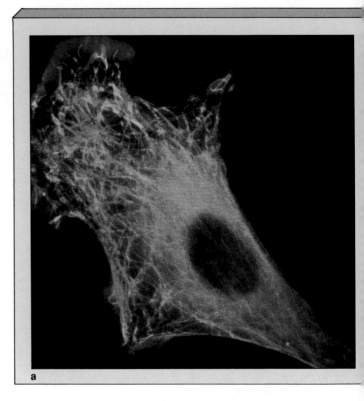

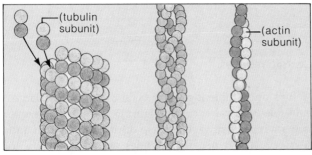

microtubule   intermediate   microfilament
                  filament

Some parts of the cytoskeleton are *permanent*. For example, a highly organized, stable array of microtubules occurs inside the flagellum (a type of motile structure) of many free-living cells. Skeletal muscle cells have actin and myosin filaments arranged in parallel to form permanent contractile units. Intermediate filaments are stable components of plant cells, muscle cells, skin, and hair.

However, some parts of the cytoskeleton are *transient*; they appear and disappear at different times in the life of a cell. Thus some microtubules are assembled when it is time to form a "spindle" that moves chro-

mosomes during cell division, then they are disassembled when the task is done.

Microtubular spindles and other aspects of the cytoskeleton are described in chapters to come. Here we will simply use the array of microtubules in flagella and cilia as an example of the cell's internal organization and capacity for motion.

Flagella and Cilia

Flagella (singular, flagellum) are microtubular structures that propel many free-living eukaryotic cells through their environment. Certain protistans, including the trypanosomes that cause African sleeping sickness and the dinoflagellates that cause "red tides," have one or more of these structures (Chapter Thirty-Nine). So do sperm and similar cells that function in the reproduction of protistans, fungi, some plants, and most animals.

Cilia (singular, cilium) are microtubular structures that many free-living cells use for propulsion and that some stationary cells use for stirring up the surrounding medium. Typically many cilia are arrayed at the cell surface (Figure 5.21a). Many thousands of ciliated cells line certain cavities and tubes in the animal body (such as the nasal cavity and bronchioles); their coordinated beating helps move fluids and dissolved particles.

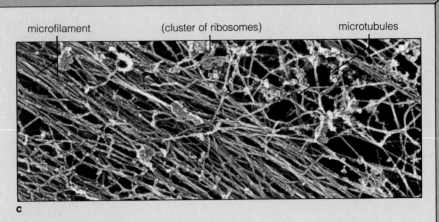

microfilament   (cluster of ribosomes)   microtubules

c

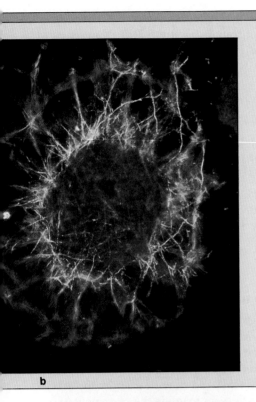

b

Figure 5.20 The cytoskeleton—the basis of a cell's shape, internal organization, and capacity for motion. (**a**) Cytoskeleton of a fibroblast (a cell that gives rise to certain connective tissues in animals), made visible by *fluorescence microscopy*. Molecules that bind only to specific proteins were labeled with fluorescent dyes, then injected into a fibroblast. The glow from the bound molecules marked the location of three different proteins. In this composite of three images, actin (blue) and vinculin (red) are associated with microfilaments, and tubulin (green) with microtubules. (**b**) Cytoskeleton of a plant cell (African blood lily). The green filaments are microtubules. They are outside the nucleus, in which chromosomes (stained purple) are clustered. (**c**) Close-up of a cytoskeleton, conveying the remarkable complexity of the cell's "skeleton."

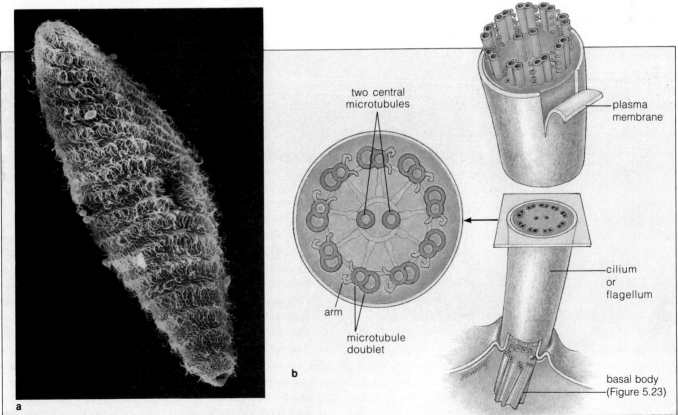

two central
microtubules

plasma
membrane

arm

microtubule
doublet

cilium
or
flagellum

basal body
(Figure 5.23)

a

b

Figure 5.21 (**a**) Scanning electron micrograph showing the hairlike cilia on the surface of *Paramecium*, 500×. (**b**) The 9+2 array of microtubules in a cilium or flagellum.

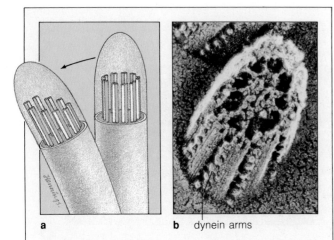

a

b dynein arms

Figure 5.22 Movement of cilia and flagella. (**a**) In an unbent cilium or flagellum, all microtubule doublets extend the same distance into the tip. With bending, the doublets on the outside of the arc are displaced farthest from the tip. (**b**) Many clawlike "arms" extend at regular intervals along the length of each microtubule doublet in the outer ring. The arms are composed of dynein, an ATP-hydrolyzing enzyme. All arms protrude in the same direction—toward the next doublet in the ring. When dynein binds and splits an ATP molecule, the angle of the arm changes with respect to the doublet in front of it. The arm bends and is strongly attracted to the doublet in front of it. On contact, the arm "unbends" with great force and causes its neighbor to move. The dynein releases its hold, after which it can grab another ATP molecule and attach to a new binding site on the doublet.

In other words, dynein arms on one doublet swing back and forth like tiny oars, displacing the neighboring doublet with each oarlike arc. The flagellum or cilium bends toward the side where the displacement is greatest.

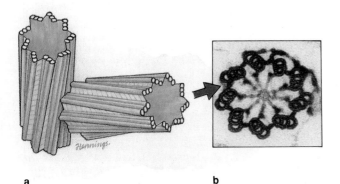

a   b

Figure 5.23 (**a**) A pair of centrioles, which occur near the nucleus of many cells. Centrioles apparently help organize the cytoskeleton; in many species they are templates for basal bodies, which give rise to the microtubular core of cilia and flagella. (**b**) Electron micrograph of a basal body, thin section, from a protistan (*Saccinobacculus*).

Flagella are longer than cilia, but both have the same cytoskeletal organization (Figure 5.21b). Nine pairs of microtubules are arranged radially around two central microtubules, in what is called a *9 + 2 array*. The array is surrounded by part of the plasma membrane. Figure 5.22 describes how interactions between microtubules cause the cilium or flagellum to bend, this being the basis of propulsion.

We should point out that some prokaryotes also have "flagella" that differ from eukaryotic flagella in structure and composition. Each is composed of protein chains of flagellin subunits, coiled into a cylinder. Although the bacterial flagellum looks something like a whip (Figure 5.7a), it is more like a propeller rotating around the attachment point.

What Organizes the Cytoskeleton?

Work to date suggests that microtubules play key roles in the cell's internal organization, movement, and division. We know, for example, that cellulose strands maintain the shape of plant cell walls—but a temporary microtubular "scaffold" laid down earlier by those cells during their growth seem to guide the placement of cellulose deposits in the newly forming walls.

If microtubules are so central to the cell's organization and functioning, then the question becomes this: *What organizes the microtubules?*

The likely candidates are **microtubule organizing centers** (MTOCs), which are small masses of proteins and other substances involved in the assembly and functioning of microtubules. In most animal cells, a prominent MTOC near the nucleus also includes a pair of centrioles. **Centrioles** are small cylinders composed of triplet microtubules (Figure 5.23). While the DNA is being duplicated before cell division, centrioles also are duplicated; a new one grows at right angles to the parent structure.

Centrioles play a vital role in ciliated or flagellated cells. When such cells are first forming, a centriole moves through the cytoplasm and becomes positioned near the plasma membrane. There, it acts as a template (that is, a structural pattern) for assembling one or more basal bodies. Each **basal body** gives rise to the microtubules that form the core structure of a cilium or flagellum (Figure 5.21).

At one time, centrioles also were thought to direct cell division. Yet division itself proceeds as usual in flowering plants, conifers, a type of fruit fly, and other species that have no centrioles. It also can proceed during the early development of a mouse embryo even if the centrioles have been removed by microsurgery from the fertilized egg.

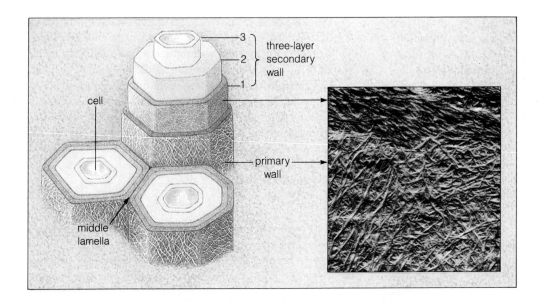

Figure 5.24 Primary and secondary walls of plant cells as they would appear in partial cross-section. The micrograph shows primary and secondary walls of a plant cell. The strands are largely cellulose.

It now appears that the organization and orientation of microtubules depend on the number, type, and location of MTOCs in the cytoplasm. Intriguingly, although the centriole-deprived cells in the mouse embryo did divide, their repeated divisions followed a disorganized pattern that produced a deformed embryo! (During development, each cell must divide at a prescribed angle relative to the other cells. The successive division planes influence the shape of the developing embryo and the adult form.) These experimental results suggest that the centriole pair of the MTOC near the nucleus governs the plane of cell division.

CELL SURFACE SPECIALIZATIONS

Cell Walls

For many cells, surface deposits outside the plasma membrane form coats, capsules, sheaths, and walls. **Cell walls** occur among bacteria, protistans, fungi, and plants. (Animal cells do not produce walls, although some secrete products to the surface layer of tissues in which they occur.)

Most cell walls have carbohydrate frameworks. They generally provide support and resist mechanical pressure, as when they keep plant cells from stretching too much while they expand with incoming water during growth. Even the most solid-looking walls have micro-

scopic spaces that make them porous, so water and solutes can move to and from the plasma membrane.

In new plant cells, cellulose strands are bundled together and added to a developing "primary" cell wall. After the main growth phase, many types of plant cells also deposit materials to form an inner, rigid "secondary" cell wall, which often consists of several layers (Figure 5.24). Cutin, suberin, and waxes commonly are embedded in cell walls at the plant's surface; they play a protective role and help reduce water loss.

The Extracellular Matrix and Cell Junctions

In multicelled plants, fungi, and animals, each living cell must interact with its physical surroundings, but it also must interact with its cellular neighbors. At the surface of tissues, cells must link tightly together so that the interior of the organism (or organ) is not indiscriminately exposed to the outside world. In all tissues, cells of the same type must recognize one another and physically stick together. Finally, in tissues where cells must act in coordinated fashion (as they do in heart muscle), the cells must share channels to exchange signals, nutrients, or both.

Extracellular Matrix. In animals, a meshwork of macromolecules, called the **extracellular matrix**, holds the cells of many tissues together. The shape of the matrix influences how cells of a given tissue will divide and what their shape will be; its composition influences cell metabolism. Its components often include collagen and other fibrous proteins, glycoproteins, and specialized polysaccharides that form a jellylike or watery "ground substance." Nutrients, hormones, and other molecules readily diffuse from cell to cell through the ground substance.

In multicelled plants, adjacent cells are cemented together at their primary walls (Figure 5.24). The cementing material (the middle lamella) has an abundance of pectin compounds.

Cell Junctions in Animals. Cell-to-cell junctions are common in animal tissues. They are illustrated in Chapter Twenty-Three, but here we can simply mention the three most common types.

Tight junctions occur between cells of epithelial tissues (which line the body's outer surface, inner cavities, and organs). Cytoskeletal strands at the plasma membrane of one cell match up and fuse with strands of its neighbor cells, forming tight seals that keep molecules from freely crossing the epithelium and entering deeper tissues.

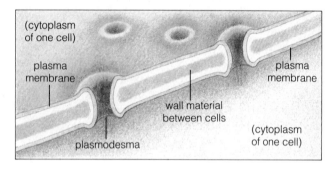

Figure 5.25 Plasmodesmata—channels that cross the wall material between plant cells. Such channels provide direct links between the cytoplasm of adjacent cells.

(Such seals keep stomach acids from leaking into other tissues, for example.) *Adhering junctions* are like spot welds at the plasma membranes of two adjacent cells. They help hold cells together in tissues that are subject to stretching, such as epithelium of the skin, heart, and stomach. At *gap junctions*, small, open channels directly link the cytoplasm of adjacent cells. In heart muscle and smooth muscle, gap junctions provide rapid communication between cells. In liver and other tissues, gap junctions allow small molecules and ions to pass directly from one cell to the next.

Cell Junctions in Plants. In land plants, living cells are linked wall-to-wall, not membrane-to-membrane. However, channels called *plasmodesmata* (singular, plasmodesma) extend across adjacent walls and connect the cytoplasm of neighboring cells (Figure 5.25). There can be 1,000 to 100,000 plasmodesmata penetrating the walls of a cell. The number affects the rate at which nutrients and other substances are transported between adjacent cells.

We will be returning to such cell-to-cell interactions. For now, the point to remember is this: *In multicelled organisms, coordinated activity among cells depends on the linkage and communication provided by the extracellular matrix and by membrane specializations of adjacent cells.*

SUMMARY

All living things are made of one or more cells; each cell is the basic living unit; and a new cell arises only from cells that already exist. Beginning with this cell theory, we looked at the internal structures of this basic unit of life, which either exists independently or has the potential to do so (as evidenced by fibroblasts and other cells that crawl around in a culture dish after being isolated from animal tissues).

We saw that, at the minimum, all cells have a nucleus (or nucleoid), cytoplasm, and a plasma membrane, which acts as a boundary for exchanges between the cell's interior and its surroundings.

Eukaryotic cells have added considerably to this basic plan. They have a variety of organelles—membranous compartments concerned with acquiring and using energy, building molecules, and tearing down molecules in controlled, specialized ways. Table 5.2 summarizes their organelles and also lists the structures common to both prokaryotic and eukaryotic cells. Details of how those organelles function will occupy our attention in chapters to follow.

Table 5.2   Summary of Typical Components of Prokaryotic and Eukaryotic Cells

Cell Component	Function	Prokaryotic Moneran	Eukaryotic Protistan	Fungus	Plant	Animal
Cell wall	Protection, structural support	✔*	✔*	✔	✔	
Plasma membrane	Regulation of substances moving into and out of cell	✔	✔	✔	✔	✔
Nucleoid	Region of DNA (hereditary control)	✔				
Nucleus	Physical isolation and organization of DNA		✔	✔	✔	✔
DNA	Encoding of hereditary information	✔	✔	✔	✔	✔
RNA	Transcription, translation of DNA messages into specific proteins	✔	✔	✔	✔	✔
Chromosome (DNA + protein)	Packaging of DNA; control of gene expression		✔	✔	✔	✔
Nucleolus	Assembly of ribosomal subunits		✔	✔	✔	✔
Ribosome	Protein synthesis	✔	✔	✔	✔	✔
Endoplasmic reticulum	Modification of many proteins into mature form; lipid synthesis		✔	✔	✔	✔
Golgi body	Final modification of proteins, lipids; sorting and packaging them for shipment inside cell or for export		✔	✔	✔	✔
Lysosome	Intracellular digestion		✔	✔*	✔	✔
Microbodies	Varied, but including hydrogen peroxide degradation		✔	✔	✔	✔
Mitochondrion	ATP formation	**	✔	✔	✔	✔
Photosynthetic pigment	Light-energy conversion	✔*	✔*		✔	
Plastid: Chloroplast	Photosynthesis, some starch storage		✔*		✔	
Chromoplast	Pigment storage				✔	
Amyloplast	Starch storage				✔	
Central vacuole	Increasing cell surface area, storage		✔*		✔	
Cytoskeleton, including microtubules, intermediate filaments, microfilaments	Cell shape, internal organization, basis of cellular motion		✔*	✔*	✔*	✔
Complex flagellum; cilium	Movement		✔*	✔*	✔*	✔
MTOCs (including centrioles)	Master organizers of cytoskeleton, possibly of orientation planes for cell division in developing embryos and regions of new growth		✔*	✔*	✔*	✔

*Known to occur in at least some groups.

**Aerobic reactions do occur in many groups, but mitochondria are not involved.

Review Questions

1. State the three principles of the cell theory. *62-63*

2. All cells share three structural features: a nucleus (or nucleoid), cytoplasm, and a plasma membrane. Can you describe the functions of each? *63, 83*

3. Why is it highly improbable that you will ever encounter a predatory two-ton living cell on the sidewalk? *66*

4. Are all cells microscopic? Is the micrometer used in describing whole cells or extremely small cell structures? Is the nanometer used in describing whole cells or cell ultrastructure? *63*

5. Suppose you want to observe details of the surface of an insect's compound eye. Would you benefit most from a compound light microscope, transmission electron microscope, or scanning electron microscope? *64-65*

6. Eukaryotic cells generally contain these organelles: nucleus, endoplasmic reticulum, Golgi bodies, lysosomes, microbodies, and mitochondria. Can you describe the function of each? *69, 83*

7. What are the components of the cytomembrane system? Sketch their general arrangement, from the nuclear envelope to the plasma membrane, and describe the role of each in the flow of materials between these two boundary layers. *73, 74-76*

8. Lysosomes dismantle and dispose of malfunctioning organelles and foreign particles. Can you describe how? *75*

9. Describe the structure and function of chloroplasts and mitochondria. Mention the ways in which they are similar. *76-77*

10. Is this statement true or false? All chloroplasts are plastids, but not all plastids are chloroplasts. *76-77*

11. What are the functions of the central vacuole in mature, living plant cells? *77*

12. What is a cytoskeleton? How do you suppose it might aid in cell functioning? *78-80*

13. Are all components of the cytoskeleton permanent? *78*

14. What gives rise to the microtubular array of cilia and flagella? Distinguish between a centriole and a basal body. *80*

15. Cell walls occur among which organisms: bacteria, protistans, plants, fungi, or animals? Are cell walls solid or porous? *81*

16. In plants, is a secondary cell wall deposited inside or outside the surface of the primary cell wall? Do all plant cells have secondary walls? *81*

17. What are some functions of the extracellular matrix in animal tissues? *81-82*

18. In multicelled organisms, coordinated interactions depend on linkages and communication between adjacent cells. What types of junctions occur between adjacent animal cells? Plant cells? *82*

19. With a sheet of paper, cover the Table 5.2 column entitled Function. Can you now name the primary functions of the cell structures listed in this table? *83*

20. Having done the preceding exercise, can you now write a paragraph describing the differences between prokaryotic and eukaryotic cells?

Readings

Birchmeier, W. April 1984. "Cytoskeleton Structure and Function." *Trends in Biochemical Sciences.* 9:192–195.

Bloom, W., and D. Fawcett. 1986. *A Textbook of Histology.* Eleventh edition. Philadelphia: Saunders. Outstanding reference book on cell structure.

deDuve, C. 1985. *A Guided Tour of the Living Cell.* New York: Freeman. Beautifully illustrated introduction to the cell; two short volumes.

Fawcett, D. 1981. *The Cell: An Atlas of Fine Structure.* Second edition. Philadelphia: Saunders. Outstanding collection of micrographs.

Karsenti, E., and B. Maro. November, 1986. "Centrosomes and the Spatial Distribution of Microtubules in Animal Cells." *Trends in Biochemical Sciences* 11(11):460–463.

Kessel, R., and C. Shih. 1974. *Scanning Electron Microscopy in Biology.* New York: Springer-Verlag. Stunning micrographs of cell structures.

Newcomb, E. 1980. "The General Cell." In *The Biochemistry of Plants,* vol. 1 (N. Tolbert, editor). New York: Academic Press. Survey of plant organelles.

Porter, K., and J. Tucker. March 1981. "The Ground Substance of the Living Cell." *Scientific American* 244(3):57–67. Summary of studies on the scaffolding inside the cell.

Rothman, J. September 1985. "The Compartmental Organization of the Golgi Apparatus." *Scientific American* 253(3):74–89. Describes the three specialized compartments of Golgi bodies.

Rothman, J., and J. Lenard. April 1984. "Membrane Traffic in Animal Cells." *Trends in Biochemical Sciences* 9:176–178.

Weber, K., and M. Osborn. October 1985. "The Molecules of the Cell Matrix." *Scientific American* 253(4):110–120. Summarizes techniques used to study the cytoskeleton.

Weibe, H. 1978. "The Significance of Plant Vacuoles." *Bioscience* 28:327–331.

As small as it may be, a cell is a *living* thing engaged in the risky business of survival. At any moment its life can be snatched from it if, for example, the concentrations of dissolved ions shoot above or plummet below prescribed levels. For most cells in most places, such concentrations are not constants. Think of the single-celled organisms living in estuaries, where seawater mixes with freshwater currents from rivers and streams. Ion concentrations change with the tides, but ion concentrations *inside* those cells must not change to the same extent. If they did, the cells would die, just as you would die if you were shipwrecked and drank nothing but salt-laden seawater.

So think of the cell for what it is: a tiny compartment in an immense world that is, by comparison, disordered, sometimes benign, and sometimes harsh. No matter what goes on outside, some things have to be kept out and others have to be brought into the cell, where organized structures and activities prevail. For this bit of the world, the bastion against disorganization is the **plasma membrane**—a thin, seemingly flimsy surface layering of little more than lipids and proteins, dotted here and there with guarded passageways and twiglike carbohydrate groups. Across this membrane, materials are exchanged between the cytoplasm and the surroundings. Then, within the cytoplasm of eukaryotic cells, exchanges are made across **internal cell membranes**, which form the compartments called organelles. This most fundamental of all cell structures, the membrane, is the focus of this chapter.

MEMBRANE STRUCTURE AND FUNCTION

FLUID MEMBRANES IN A LARGELY FLUID WORLD

The Lipid Bilayer

Water bathes the outer surface of living cells and fills most of their interior. You might be thinking that the plasma membrane would have to be a rather solid structure, given that both sides of it are immersed in fluid, yet it happens that the plasma membrane is fluid, too! When you puncture a cell with a fine needle, the cell will not lose cytoplasm when the needle is withdrawn. Rather, the cell surface seems to flow over and seal the puncture. How can a fluid cell membrane remain distinct from fluid surroundings? The answer is found in certain properties of lipid molecules.

Lipid molecules cluster spontaneously when surrounded by water. For example, consider the **phospholipids**, the most abundant type of lipid in cell membranes. Figure 6.1 shows a typical phospholipid. It

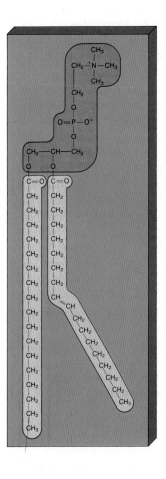

Figure 6.1 Structural formula of a typical phospholipid found in animal cell membranes. The hydrophilic head is shown in orange, and the hydrophobic tails in gold. The bend caused by the double bond in one of the tails is no more than a small kink.

has a hydrophilic (water-loving) head and two hydrophobic (water-dreading) tails:

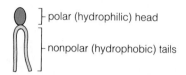

- polar (hydrophilic) head

- nonpolar (hydrophobic) tails

When many phospholipid molecules are surrounded by water, hydrophobic interactions cause their fatty acid tails to cluster together. The clustering may produce a **lipid bilayer**, with the hydrophilic heads on the outside of a two-layer sheet, facing the surrounding water, and hydrophobic tails sandwiched between them:

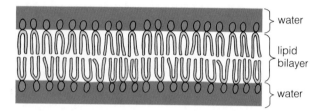

} water

} lipid bilayer

} water

Such bilayers are the structural foundation for the plasma membrane and organelle membranes.

Because lipid bilayers minimize the number of hydrophobic groups exposed to water, the fatty acid tails do not have to spend a lot of energy fighting the water molecules, so to speak. Thus the reason a punctured plasma membrane tends to seal itself is that the puncture is energetically unfavorable (it leaves too many hydrophobic groups exposed).

Ordinarily, of course, few cells are ever punctured with fine needles. But the self-sealing behavior of their membrane lipids is useful for far more than damage control. As you will see, parts of the plasma membrane sink inward and drag extracellular substances in with them. The sinks deepen into spheres that pinch off from the plasma membrane, thereby forming closed vesicles that can move into the cytoplasm. Similarly, buds form from organelle membranes, and when the buds pinch off, the self-sealing behavior of the lipid bilayer maintains the integrity of both the newly formed vesicles and the parent organelle.

Lipid molecules surrounded by water show self-assembling and self-sealing behavior. This behavior maintains the integrity of the plasma membrane and organelles.

Within the bilayer, individual lipid molecules show quite a bit of movement. They diffuse sideways, they spin about their long axis, and their tails flex back and forth. These movements keep adjacent lipids from packing into a solid layer, hence they impart fluidity to the membrane. Packing is also disrupted by lipids with short tails (which cannot interact as strongly as long-tailed neighbors do) and by unsaturated tails (which have kinks at their double bonds).

Within a bilayer, lipid molecules move rapidly and show variations in their packing tendencies, thereby contributing to membrane fluidity.

Membrane Proteins

Cell membranes are composed of more than a uniformly smooth lipid bilayer. A variety of proteins are embedded in the bilayer or positioned at its two surfaces. Most are *glycoproteins*, with short polysaccharide chains covalently bonded to them. (The membrane also contains glycolipids.) The polysaccharide chains always extend into the extracellular fluid, never into the cytoplasm.

The **fluid mosaic model** of membrane structure was put together in 1972 by S. J. Singer and G. Nicolson. Figure 6.2 shows a recent version. (Keep in mind that not all membranes are *exactly* like this; for example, there are no cholesterol molecules in plant cell membranes.) In this model, the membrane is assumed to be "fluid" because of the movements and packing variations of its lipid components. The lipids themselves give the membrane its basic structure and its relative impermeability to water-soluble molecules. The "mosaic" is the intricate composite of proteins and lipids of the membrane. The proteins are responsible for most membrane functions.

Different types of membrane proteins have now been identified. They include *channel, transport, electron-transfer, recognition,* and *receptor* proteins (Figure 6.2). Their names reflect their functions, as the following paragraphs indicate.

Channel proteins allow passage of water-soluble substances across the lipid bilayer. They are embedded in the bilayer, but there is a channel through the body of the protein, one that opens to both the cytoplasm and extracellular fluid. Some channels are open all the time; others have "gates" that open at certain times only. Examples are the channels for sodium ions that function in the excitation of nerve cells (page 324).

Transport proteins also allow passage of substances across the bilayer. Unlike the channel proteins, however, they require energy to "pump" specific substances in a specific direction. An example is the sodium-potassium pump (pages 94 and 324).

Electron-transfer proteins accept electrons from one molecule and "carry" (transfer) them to another. In some cases, they also transfer H^+ ions. The cytochromes are

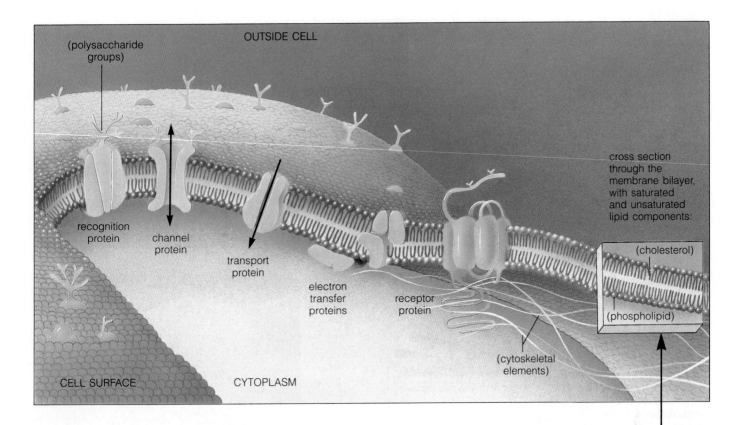

OUTSIDE CELL

(polysaccharide groups)

recognition protein

channel protein

transport protein

electron transfer proteins

receptor protein

CELL SURFACE

CYTOPLASM

(cytoskeletal elements)

cross section through the membrane bilayer, with saturated and unsaturated lipid components:

(cholesterol)

(phospholipid)

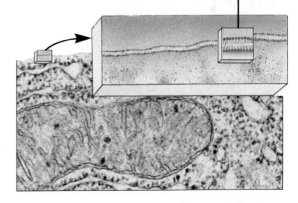

Figure 6.2 Increasingly enlarged views of the plasma membrane, based on the fluid mosaic model of membrane structure. The micrographs show the plasma membrane of a pancreatic cell, thin section.

examples of electron-transfer proteins that function in photosynthesis and aerobic respiration.

Recognition proteins function in tissue formation and, later, in cell-to-cell interactions. In multicelled animals, the polysaccharide chains of recognition proteins (and of glycolipids) form a "sugar coat" at the outer surface of the plasma membrane. Some proteins of the coat promote adhesion between cells. Others are like molecular fingerprints for each type of cell; they function in cell recognition and coordinated cell behavior within tissues.

Receptor proteins are like switches that are turned on or turned off when particular substances bind to them. They are specialized receivers of outside information that can trigger alterations in metabolism or cell behavior. Different cell types have different combinations of receptors. In vertebrates, some receptors are widespread among the body's cells, but others are restricted to only a few cell types. Most commonly, the receptors are arrayed at the plasma membrane; some are located inside the cell.

We will look closely at receptor proteins in several chapters, but we can mention a few examples here. The receptors of certain white blood cells can lock onto molecular groups found at the surface of bacterial cells and other tissue invaders; they play a key role in the body's

Figure 6.3 Freeze-fracturing and freeze-etching. (**a**) In the freeze-fracture step, specimens being prepared for electron microscopy are rapidly frozen, then fractured by a sharp blow from the edge of a fine blade. (**b**) Sometimes specimens are freeze-etched: more ice is evaporated from the fracture face to expose the outer membrane surface. (**c**) In a process called metal shadowing, the fractured surface is coated with a layer of carbon and heavy metal such as platinum. The coating is thin enough to replicate details of the exposed specimen surface. The metal replica, not the specimen itself, is used for micrographs. (**d**) Fractured membranes commonly split down the middle of the lipid bilayer. Typically, one inner surface is studded with particles and depressions, and the other is a complementary pattern of depressions and particles. The particles are membrane proteins. The micrograph shows part of a replica of a red blood cell, prepared by freeze-fracturing and freeze-etching.

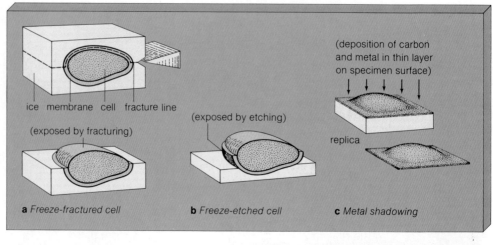

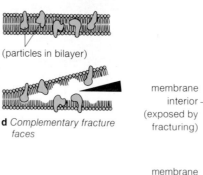

d *Complementary fracture faces*

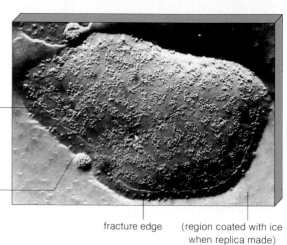

defense responses (Chapter Thirty). Receptors adjacent to channel proteins can trigger the opening or closing of the "channel." Receptors for a hormone called somatotropin switch on the enzymes that crank up metabolic machinery for cell growth and division. Mutations in those signal receptors can hinder cell growth. Other receptor malfunctions contribute to some forms of cancer and diabetes.

For any cell, the number of membrane proteins just described is much more extensive than the simplified model in Figure 6.2 might lead you to believe. Figure 6.3 provides a better picture, made possible by specimen preparation methods called freeze-fracturing and freeze-etching.

Summary of Membrane Structure and Function

Throughout this book, we will be looking at different aspects of membrane structure and function. For now, the features that all cell membranes hold in common may be summarized this way:

1. Cell membranes are composed of lipids (phospholipids especially) and proteins.

2. Membrane lipids have hydrophilic heads and hydrophobic tails, and when surrounded by water they assemble spontaneously into a bilayer. All heads are at the two outer faces of a lipid bilayer and all tails are sandwiched between them.

3. The lipid bilayer is the basic *structure* of all cell membranes and serves as a *hydrophobic barrier* between two solutions. A plasma membrane separates the cytoplasm and extracellular fluid, and organelle membranes separate different solutions in the space of the cytoplasm.

4. Membrane fluidity arises through rapid movements and packing variations among individual lipid molecules.

5. Membrane *functions* are carried out largely by proteins embedded in the bilayer or positioned at one or the other membrane surface.

DIFFUSION

Before exploring the details of membrane function, let's consider the natural, unassisted movements of water and solutes. Many membrane properties involve mechanisms that work with or against these movements.

Gradients Defined

"Concentration" refers to the number of molecules (or ions) of a substance in a given volume of space. In the absence of other forces, molecules of a given type move down their **concentration gradient**, which means they tend to move to a region where they are less concentrated. They are driven to do so because they are constantly colliding with one another millions of times a second. Random collisions do send the molecules back and forth, but the *net* movement is outward from the region of greater concentration.

Similarly, for any defined volume of space, gradients can also exist between two regions that differ in pressure, temperature, or net electric charge.

When your heart contracts, it generates fluid pressure that is greatest in the first artery leaving the heart and lowest in the last vein leading back into it. When your skis have a mind of their own and you find yourself sitting in a snowbank, heat energy flows down the thermal gradient and is transferred away from your body to the snow. The very functioning of your nervous system depends on the fact that your cells have an overall negative charge compared to extracellular fluid, and positively charged sodium ions will (when they can) follow the electric gradient and move inward. As you will see throughout this book, gradients are central to a variety of dynamic processes.

Simple Diffusion

The random movement of like molecules or ions down a concentration gradient is called **simple diffusion**. Simple diffusion of small molecules accounts for the greatest volume of substances moving across cell membranes.

The direction in which a substance diffuses depends on its own concentration gradient, not on any others. In other words, each substance diffuses *independently* of other substances that may be present. Suppose you put a few drops of red food coloring into one end of a pan filled with water. At first all of the dye molecules remain at that end, but many molecules start careening toward the other end. Even though collisions are also sending some dye molecules back, *more* molecules are leaving the concentrated region, and the net movement is down the gradient. For the same reason, water moves in the opposite direction, to the region where water molecules are less concentrated in the pan (Figure 6.4).

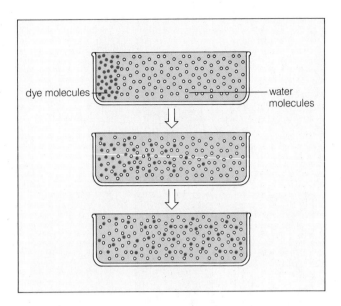

Figure 6.4 Diagram of the simple diffusion of dye molecules in one direction and water molecules in the opposite direction in a pan of water.

Even when dye molecules and water molecules are dispersed evenly in the pan, collisions still occur at random. But now there are no more gradients, hence there is no *net* movement in either direction. The molecules are said to be at dynamic equilibrium.

The *rate* of diffusion depends on several things. For example, the rate is faster when the concentration gradient is steep. It also is faster at higher temperatures, because heat energy causes molecules to move more rapidly (hence to collide more frequently). The rate is also affected by molecular size (smaller molecules move faster than large ones do at the same temperature).

Simple diffusion accounts for most of the short-distance transport of substances moving into and out of cells.

Bulk Flow

Diffusion rates are often enhanced by **bulk flow**, which is the tendency of different substances in a fluid to move together in the same direction in response to a pressure gradient.

For example, in complex plants and animals, nutrients and other substances do not diffuse slowly across large tissue masses to reach individual cells. They are moved rapidly by bulk flow through transport tubes, which occur in the respiratory and circulatory systems of animals and in the vascular systems of plants. In effect, bulk flow "shrinks" the distances that molecules must

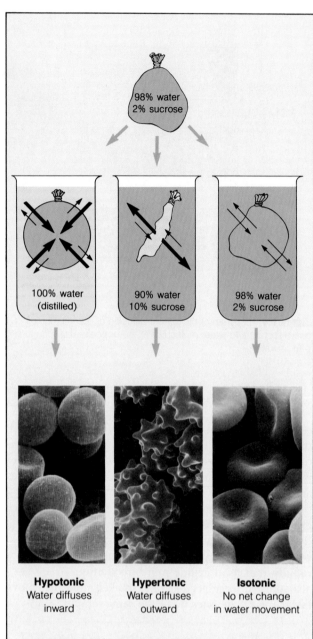

Figure 6.5 Effects of osmosis in different environments. The sketches show why it is important for cells to be matched to solute levels in their environment. (In each sketched container, arrow width represents the relative amount of water movement.)

The micrographs correspond to the sketches. They show the kinds of shapes that might be seen in red blood cells placed in *hypotonic* solutions (influx of water into the cell), *hypertonic* solutions (outward flow of water from the cell), and *isotonic* solutions (internal and external solute concentrations are matched).

Red blood cells have no special mechanisms for actively taking in or expelling water molecules. Hence they swell or shrivel up, if solute levels in their environment change.

move by diffusion as they are carried between the environment and interior cells. It also shrinks the diffusion distance between two body regions (as between leaves and roots).

OSMOSIS

Osmosis Defined

A membrane is "differentially permeable" when some substances but not others can pass through it. **Osmosis** is the movement of water across any differentially permeable membrane in response to solute concentration gradients, a pressure gradient, or both. (A solute, recall, is any dissolved substance.)

Imagine you have a plastic bag that acts like a membrane, in that it is permeable to water molecules but impermeable to larger molecules. You fill the bag with water containing a small amount of table sugar, then you put it in a container of distilled water (which is nearly free of solutes). Because the concentration of water molecules inside the bag is less than outside, water follows its gradient; there is a net movement of water into the bag (Figure 6.5). However, sugar cannot follow *its* concentration gradient and move out of the bag, because sugar molecules are too large to pass through the membrane. Soon the bag swells with water, and eventually it may spring a leak or burst.

Suppose you had immersed the bag in water having more dissolved sugar than the solution inside the bag did. The net movement of water would have been outward, and the bag would have shriveled (Figure 6.5).

Finally, suppose you had used distilled water inside and outside the bag. Because the solute concentrations would be the same, there would be no water concentration gradient—and there would be no net movement of water in either direction.

Osmosis is the passive movement of water across a differentially permeable membrane in response to solute concentration gradients, a pressure gradient, or both.

Tonicity

Osmotic movements across cell membranes are affected by **tonicity**—that is, the relative concentrations of solutes in two fluids. In this case, we are talking about the extracellular fluid and the fluid portion of cytoplasm. When solute concentrations are equal in both fluids, or *isotonic*, there is no net osmotic movement of water in either direction. When the solute concentrations are not equal, one fluid is *hypotonic* (has less solutes) and the

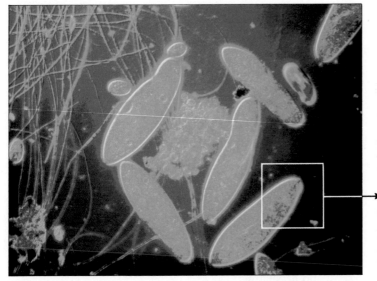

a

Figure 6.6 Osmosis and *Paramecium*, a single-celled protozoan that lives in fresh water (a hypotonic environment). Because its cell interior is hypertonic relative to the surroundings, water tends to move into the cell by osmosis. If the influx were left unchecked, the cell would become bloated and the plasma membrane would rupture. However, the excess is expelled by an energy-requiring transport process that is based on a specialized organelle called the contractile vacuole. Tubelike extensions of this organelle extend through the cytoplasm. Water collects in these extensions and drains into a central vacuolar space. When filled, the vacuole contracts and the water is forced into a small pore that empties to the outside.

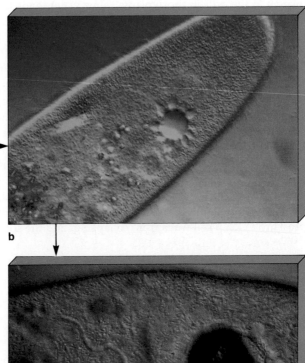

b

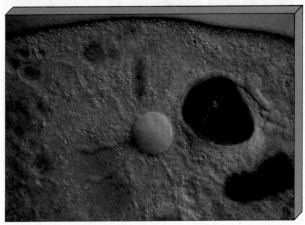

c contractile vacuole emptied

d contractile vacuole filled

other is *hypertonic* (has more solutes). Water molecules tend to move from a hypotonic fluid to a hypertonic one.

If cells did not have mechanisms for adjusting to differences in tonicity, they would shrivel or burst like the plastic bag described above. For example, many solutes are dissolved in your bloodstream and in the cytoplasm of red blood cells. But suppose you immerse a red blood cell in a hypotonic solution. That cell contains many large organic molecules, which cannot cross the plasma membrane. Although those molecules cannot move out, water can move in. Internal pressure builds up, but red blood cells have no mechanism for disposing of the excess water. The cell continues to swell until the membrane ruptures. Then the cell undergoes *lysis* (it becomes grossly "leaky") and is destroyed.

If the red blood cell had been placed in a hypertonic solution, water would have moved out of the cytoplasm and the cell would have shriveled. This particular cell maintains its volume only when solute concentrations stay much the same on both sides of the plasma membrane. Other types of cells can live in hypotonic or hypertonic environments, as Figure 6.6 illustrates.

Water Potential

The soil in which most land plants grow is hypotonic relative to the cells in those plants. Thus water tends to move by osmosis into the plant. When individual cells absorb the water, pressure increases against the cell wall, which is fairly rigid. This internal pressure on a cell wall is called **turgor pressure**.

Pressure will build up inside any walled cell when water moves inward by osmosis, but water will also be squeezed back out when turgor pressure is great enough to counter the effects of internal solutes. Both forces have the potential to cause the directional movement of water. The sum of these two opposing forces is called the **water potential**.

When the outward flow equals the rate of inward diffusion, turgor pressure is constant. There is enough pressure to keep cell walls from collapsing, and the soft parts of the plant body are maintained in erect positions. When the environment is so dry that the movement of water into the plant dwindles or stops, water moves out of the cells and soft parts of the plant body wilt. Similarly, wilting occurs when solutes reach high concentrations in the environment (Figure 6.7).

MOVEMENT OF WATER AND SOLUTES ACROSS CELL MEMBRANES

The Available Routes

Some substances can move into or out of cells only with the aid of membrane proteins; others can cross the lipid bilayer itself. Only small, electrically neutral molecules readily diffuse across the bilayer. Examples are water, oxygen, carbon dioxide, and ethanol, all of which have no *net* charge:

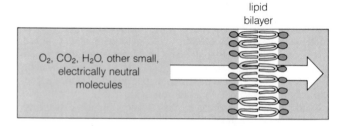

Simple diffusion of such small molecules accounts for the greatest volume of substances moving across cell membranes.

In contrast, glucose and other large molecules that are electrically neutral almost never diffuse across the lipid bilayer. Neither do positively or negatively charged ions, no matter how small they are:

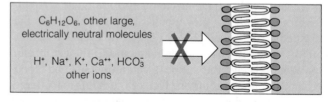

Different proteins transport these substances across the membrane. In **active transport**, a transport protein receives an energy boost that allows it to move a solute either with or against a concentration gradient. In **passive transport**, a channel protein does not require an energy boost; the solute simply moves through the protein's interior, following its concentration gradient.

Through a combination of simple diffusion, passive transport, and active transport, cells or organelles are supplied with raw materials and they are rid of wastes, at controlled rates. These mechanisms control secretions of cell products. They also help maintain pH and volume inside the cell or organelle within some functional range.

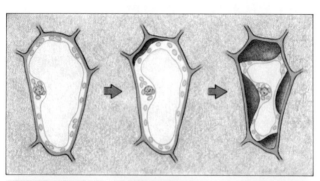

Figure 6.7 Plant wilt resulting from loss of turgor in cells. **(a)** At the start of this experiment, ten grams of salt (NaCl) in about sixty milliliters of water is added to a pot containing tomato plants. **(b)** After about five minutes, wilting is pronounced and the plant is collapsing. **(c)** After twenty-seven minutes, wilting is severe. The corresponding sketches show progressive plasmolysis (a shrinking of cytoplasm away from the cell walls).

Facilitated Diffusion

Channel proteins are particular about which solutes they will passively transport across the membrane. Because these proteins assist solutes only in the direction that simple diffusion would take them, the movement is called **facilitated diffusion**.

Each type of channel protein probably has a particular array of hydrophilic groups projecting into the space of the interior channel (Figure 6.8). When water-soluble molecules bind with the groups, the binding apparently triggers a change in the protein shape, and this change permits the solute to move through the hydrophilic interior. While the solute makes its passage, the protein closes in behind it and returns to its original shape.

Active Transport

When transport proteins move specific solutes into and out of cells, they undergo changes in shape, somewhat like the changes induced in the carriers responsible for facilitated diffusion. In this case, however, an energy boost leads to a *series* of changes, which cause the protein to pump solutes rapidly across the membrane (Figure 6.9). Most often, ATP donates energy to the protein.

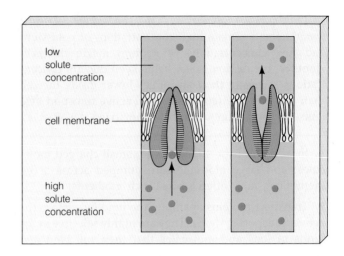

Figure 6.8 Possible mechanism of facilitated diffusion. The sketch shows a channel protein as if it were sliced down through its midsection. Water-soluble molecules bind with the hydrophilic groups present on the interior surface of the protein (the shaded areas). The binding changes the protein shape in a way that allows a bound molecule to cross the membrane.

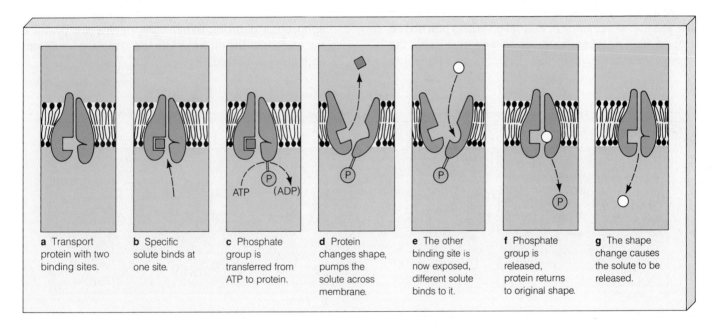

a Transport protein with two binding sites.

b Specific solute binds at one site.

c Phosphate group is transferred from ATP to protein.

d Protein changes shape, pumps the solute across membrane.

e The other binding site is now exposed, different solute binds to it.

f Phosphate group is released, protein returns to original shape.

g The shape change causes the solute to be released.

Figure 6.9 Simplified picture of an active transport system that occurs in animal cell membranes. Transport of one kind of solute across the membrane is coupled with transport of another kind in the opposite direction. The transport protein receives an energy boost from ATP and so undergoes changes in its shape that are necessary for the transport process.

One active transport system, the *sodium-potassium pump*, helps maintain high concentrations of potassium and low concentrations of sodium inside the cell. Another, the *calcium pump*, helps keep calcium concentrations at least a thousand times lower inside the cell than outside. The features that all active transport systems hold in common can be summarized this way:

1. In active transport, small ions, small charged molecules, and large molecules are pumped across a cell membrane, most often against their gradients.

2. Transport proteins spanning the lipid bilayer are the active transport systems. They are highly selective in the kinds of ions and molecules that they will bind and transport.

3. The proteins act when a specific solute is bound in place and when they receive an energy boost.

Exocytosis and Endocytosis

Exocytosis and endocytosis are processes by which the plasma membrane or organelle membranes pinch off and form transport vesicles around substances (Figure 6.10). **Exocytosis**, the process by which substances are moved *out* of a cell, calls for fusion of cytoplasmic vesicles with the plasma membrane. For example, vesicles pinched off from Golgi membranes travel to and fuse with the plasma membrane, then their contents are secreted to the outside. This process was described in the preceding chapter. In **endocytosis**, a region of the plasma membrane encloses particles at or near the cell surface, then it pinches off to form a vesicle that moves into the cytoplasm.

The amoeba relies on endocytosis. This single-celled organism is phagocytic (a "cell eater"). When it encounters a chemically "tasty" particle or cell, one or two lobes form at its surface. The lobelike extensions curve back and form a compartment around the particle, and this compartment becomes an endocytic vesicle (Figure 6.11). Most endocytic vesicles fuse with lysosomes and their contents are digested (page 75). Phagocytic white blood cells rely on endocytosis when they destroy harmful agents such as bacteria.

Endocytosis also transports liquid droplets into animal cells. (Sometimes this process is called *pinocytosis*, which means "cell drinking.") A depression forms at the surface of the plasma membrane and dimples inward around extracellular fluid. An endocytic vesicle forms and moves inside the cytoplasm, where it fuses with lysosomes.

In *receptor-mediated endocytosis*, specific molecules are brought into the cell through involvement of specialized regions of the plasma membrane that form coated pits. Each pit, or shallow depression, is coated with a dense lattice (Figure 6.12). The pit appears to be lined with surface receptors that are specific (in this example) for lipoprotein particles. When lipoproteins are bound to the receptors, the pit sinks into the cytoplasm and forms an endocytic vesicle.

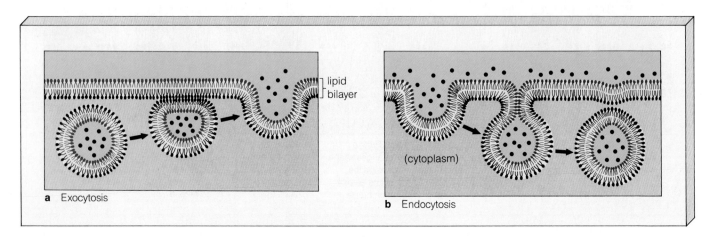

Figure 6.10 (**a**) Fusion of a transport vesicle with the plasma membrane during exocytosis and (**b**) formation of a transport vesicle during endocytosis.

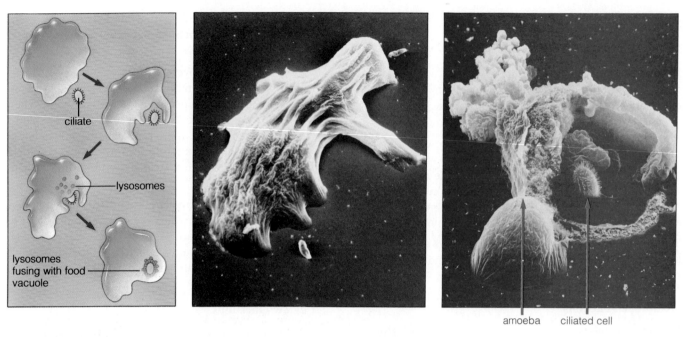

amoeba   ciliated cell

Figure 6.11 Endocytosis as demonstrated by a phagocytic cell. The scanning electron micrographs show an amoeba entrapping a ciliated cell.

ciliate

lysosomes

lysosomes fusing with food vacuole

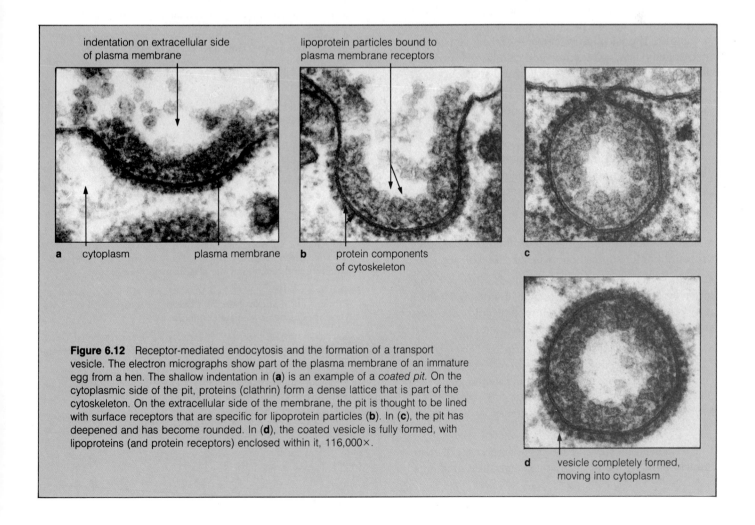

indentation on extracellular side of plasma membrane

lipoprotein particles bound to plasma membrane receptors

a cytoplasm   plasma membrane

b protein components of cytoskeleton

c

d vesicle completely formed, moving into cytoplasm

Figure 6.12 Receptor-mediated endocytosis and the formation of a transport vesicle. The electron micrographs show part of the plasma membrane of an immature egg from a hen. The shallow indentation in (**a**) is an example of a *coated pit*. On the cytoplasmic side of the pit, proteins (clathrin) form a dense lattice that is part of the cytoskeleton. On the extracellular side of the membrane, the pit is thought to be lined with surface receptors that are specific for lipoprotein particles (**b**). In (**c**), the pit has deepened and has become rounded. In (**d**), the coated vesicle is fully formed, with lipoproteins (and protein receptors) enclosed within it, 116,000×.

SUMMARY

Membrane Structure

1. Living cells are bathed in fluid of one sort or another, and their interior is also fluid. Membranes serve as boundaries between the external fluid world and the cytoplasm (or between different cytoplasmic regions).

2. The lipid bilayer is the basic structure of cell membranes. There are two layers of lipids (phospholipids especially), with the hydrophobic tails of the molecules sandwiched in between the hydrophilic heads. The membrane is rather fluid because of packing variations and rapid movements among the individual molecules.

3. Membrane functions are carried out by diverse proteins embedded in the bilayer or weakly bonded to one of its surfaces.

Membrane Functions

1. *Control of substances moving into and out of cells.* The plasma membrane is a thin, differentially permeable boundary between organized metabolic events in the cytoplasm and events in the outside world. Membrane proteins play passive or active roles in carrying or pumping substances across the lipid bilayer portion of the membrane. The selective movements of these substances affect metabolism, cell volume and cellular pH, nutrient stockpiling, and the removal of harmful substances.

2. *Compartmentalization of internal cellular space.* Within the cytoplasm, membranes form separate compartments (organelles) in which specialized activities occur. The nucleus, the grana of chloroplasts, and the inner and outer mitochondrial membranes are examples. Like the plasma membrane, these internal membranes are differentially permeable, a feature that is vital in the functions of organelles.

3. *Signal reception.* Some receptor proteins bind signaling molecules such as hormones, and when they do, they trigger alterations in cell behavior or metabolism.

4. *Cell-to-cell recognition.* In multicelled animals, recognition proteins help cells of like type identify and adhere to one another during tissue formation and, at a later stage of development, during tissue interactions.

5. *Transport of macromolecules and particles.* Through exocytosis and endocytosis, cells take in or eject large molecules or particles across the plasma membrane.

Movement of Water and Solutes Across Membranes

1. Exchanges of water and solutes occur across cell membranes, both with and against concentration gradients.

2. Simple diffusion is a natural, unassisted movement of solutes from a region of high concentration to a region of lower concentration.

3. Osmosis is the movement of water across a differentially permeable membrane in response to solute concentration gradients, a pressure gradient, or both. When conditions are isotonic (equal concentrations of solutes across the membrane), there is no osmotic movement in either direction. Water tends to move from hypotonic fluids (with fewer solutes) to hypertonic fluids (with more solutes).

4. Many solutes are actively pumped across membranes by transport proteins, which require an energy boost to do this. Other solutes are passively assisted across membranes by channel proteins, which help the solutes move in the direction that diffusion would take them.

Review Questions

1. Describe the fluid mosaic model of plasma membranes. What makes the membrane fluid? What parts constitute the mosaic? *86*

2. List the six structural features that all cell membranes have in common. *88*

3. Describe some functions of membrane proteins. *86-87*

4. Diffusion accounts for the greatest volume of substances moving into and out of cells. How does diffusion work? *89*

5. What is osmosis, and what causes its occurrence? *90-91*

6. Explain the difference between active and passive transport mechanisms. *92-93*

7. What types of substances can readily diffuse across the lipid bilayer of a cell membrane? What types of substances must be actively or passively transported across? *92-93*

8. Can you explain the difference between exocytosis and endocytosis? *94-95*

Readings

Bretscher, M. October 1985. "The Molecules of the Cell Membrane." *Scientific American* 253(4):100–108. Fairly recent description of the structure and function of the plasma membrane.

Dautry-Varsat, A., and H. Lodish. May 1984. "How Receptors Bring Proteins and Particles Into Cells." *Scientific American* 250(5):52–58. Describes receptor-mediated endocytosis.

Karp, G. 1984. *Cell Biology.* Second edition. New York: McGraw-Hill. Chapter 5 gives a detailed, dynamic picture of membrane function. This is an excellent reference book.

Singer, S., and G. Nicolson. 1972. "The Fluid Mosaic Model of the Structure of Cell Membranes." *Science* 175:720–731.

Unwin, N., and R. Henderson. February 1984. "The Structure of Proteins in Biological Membranes." *Scientific American* 250(2):78–94.

When you look at a single living cell with the aid of a microscope, you are watching a form pulsing with activity. Through its movements, it is identifying and taking in raw materials suspended in the water droplet on the slide. To power those tiny movements, the cell is extracting energy from food molecules it had stored away earlier. Even as you observe it, the cell is using materials and energy as it builds and maintains its membranes and organelles, its stores of chemical compounds, its information-storage system, its pools of enzymes. It is alive; it is growing; it may reproduce itself. Multiply this activity by *65 trillion cells* and you have an inkling of what goes on in your own body as you sit quietly, observing that single cell!

With this chapter we turn to one of the most fundamental aspects of the living cell, its reliance on metabolism. By definition, **metabolism** is the controlled capacity to acquire and use energy for stockpiling, breaking apart, building, and eliminating substances in ways that contribute to cell survival and reproduction.

7

GROUND RULES OF METABOLISM

THE NATURE OF ENERGY

Two Laws Governing Energy Transformations

Energy is a capacity to make things happen, to cause change, to do work. You use energy to put a hard wax finish on a car. When paste wax is buffed, attractions between molecules in the paste are broken, new ones form, and the molecules become repositioned in ways

Figure 7.1 All events large and small, from the birth of stars to the death of a microorganism, are governed by laws of energy. Shown here, eruptions on the sun's surface and, to the right, *Volvox*—each sphere a colony of microscopically small single cells able to capture sunlight energy that indirectly drives their life processes.

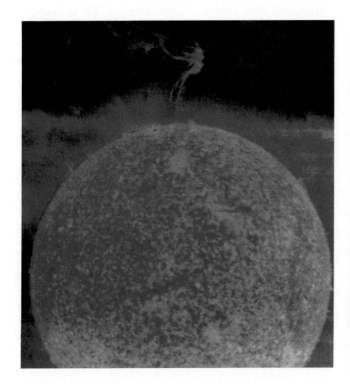

that produce a hard, glossy finish. A cell, too, uses energy to make things happen, as when it makes large molecules from smaller, simpler ones.

Can you (or the cell) create your own energy from scratch, or must it be borrowed from someplace else? You can gain insight into the question by considering two laws of thermodynamics that govern the transformation of energy.

The **first law of thermodynamics** deals with the *quantity* of energy available. It may be expressed this way:

The total amount of energy in the universe remains constant. More energy cannot be created. And existing energy cannot be destroyed; it can only undergo conversion from one form to another.

According to the first law, there is a finite amount of energy in the universe. It is distributed among various forms, such as chemical energy stored in molecules and light energy from the sun. These and other forms of energy are interconvertible. Thus plants absorb sunlight energy and convert it to chemical energy, part of which becomes stored in starch. You eat plants, and some of the stored chemical energy becomes converted in your body—for example, to mechanical energy that powers your movements. With each conversion, some energy escapes into the surroundings as heat. (Because of ongoing conversions in your cells, you body steadily gives off about the same amount of heat as a hundred-watt light bulb.) However, none of the energy *vanishes*.

More precisely, the total energy content of a system *and its surroundings* remains constant (Figure 7.2). As this figure indicates, the term "system" means all matter within a specified region. (For example, a plant is a system; so is a strand of DNA or a galaxy.) "Surroundings" can mean the entire universe or some specified region in contact with the system.

The first law means that you cannot increase the universal energy pool by making "new" energy out of nothing, and you cannot destroy any of it, either. *All you can do is channel the ways in which energy changes from one form to another, and thereby temporarily hoard or let go of what is already there.*

This brings us to the *quality* of energy available. Energy that is concentrated and organized (for example, in a starch molecule) is high quality, since it lends itself to conversions. Heat spread out in the atmosphere is an example of low-quality energy; for all practical purposes, it cannot be gathered up and converted to other forms.

According to the **second law of thermodynamics**, the spontaneous direction of energy flow is from high-quality to low-quality forms. "Spontaneous" means the natural or *most probable* direction, regardless of whether it takes a fraction of a second or billions of years.

Left to itself, any system spontaneously undergoes conversions to less organized forms. Each time that happens, some energy is randomly dispersed in a form (usually heat) that is not as readily available to do work.

If the first law suggests that we can't get something from nothing, the second law suggests that we don't have a chance of breaking even—because the amount of energy *not* available to do work is *increasing*! The reason is that no energy conversion can ever be 100 percent efficient. Each time one system transfers energy to another system, some of the energy goes off as heat.

Entropy is a measure of the degree of disorder of a system. According to the second law of thermodynamics, the entropy of an isolated system tends to increase. The ultimate destination of the universe is therefore a state of maximum entropy. It has been estimated that about 5 billion years from now, everything will be at the same temperature, and energy conversions as we know them will never happen again.

Living Systems and the Second Law

Can it be that life is one glorious pocket of resistance to the rather depressing flow of the universe toward maximum entropy? After all, with the birth and growth of each new organism, energy becomes more concentrated and organized, not less so! Yet a simple example will show that the second law of thermodynamics does indeed apply to life on earth.

The primary source of energy for almost all organisms is the sun, which is steadily losing energy. Plants intercept some of this energy, then they lose energy to other organisms that feed, directly or indirectly, on plants. At each energy transfer along the way, more heat energy is

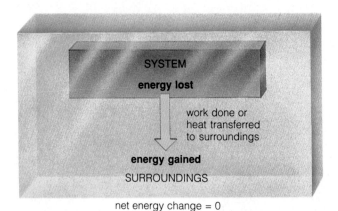

Figure 7.2 Energy content of a system and its surroundings.

added to the universal pool. Overall, then, energy is still flowing in one direction. *The world of life maintains a high degree of organization only because it is being resupplied with energy lost from someplace else.*

There is a steady flow of sunlight energy into the interconnected web of life, and this compensates for the steady flow of energy leaving it.

METABOLIC REACTIONS: THEIR NATURE AND DIRECTION

Energy Changes in Metabolic Reactions

Let's now turn to the energy transformations taking place when cells shuffle materials during metabolic reactions. At the temperature and pressure levels typical of the cellular world, the materials present at the end of a reaction (the products) may have less or more energy than the starting materials (the reactants) had.

Reactions showing a net loss in energy are called *exergonic* (the term means "energy out"):

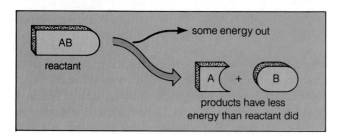

Reactions showing a net gain in energy are called *endergonic* ("energy in"). An endergonic reaction does not violate the second law of thermodynamics. It occurs only when extra energy (lost from some other system) is fed into the reaction:

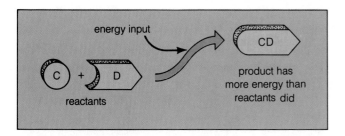

Reversible Reactions

Living cells are bathed in fluid and are largely fluid themselves. Molecules in this fluid move about constantly and bump into each other, and most often they react spontaneously in regions where they are most concentrated. (The more concentrated they are, the more their random movements put them on collision courses.)

For example, when you exercise, muscle cells give off carbon dioxide wastes that diffuse through the muscle tissue and into nearby blood vessels (where carbon dioxide is less concentrated). There, most of the carbon dioxide reacts spontaneously with water molecules to form carbonic acid, which can then dissociate into bicarbonate and hydrogen ions:

$$\text{reactants:} \qquad\qquad\qquad\qquad \text{products}$$
$$CO_2 + H_2O \implies \underset{\substack{\text{carbonic}\\\text{acid}}}{H_2CO_3} \implies \underset{\text{bicarbonate}}{HCO_3^- + H^+}$$

The arrows signify that the reaction runs in the "forward" direction. As more and more product molecules form over time, not as many reactant molecules remain. All the while, the concentration of product molecules has been increasing—which means they will collide more frequently than before. Some fraction will now have enough collision energy to be driven in the opposite direction:

$$CO_2 + H_2O \impliedby H_2CO_3 \impliedby HCO_3^- + H^+$$

where the arrows signify that this is the "reverse" reaction. This reaction occurs in your lungs, where carbon dioxide is exhaled from your body.

Most reactions are reversible. The greater the concentration of reactants, the faster the forward reaction. The greater the concentration of products, the faster the reverse reaction.

Unless other events in the cell (or body) keep it from doing so, a reversible reaction will approach **dynamic equilibrium**. This is the state at which the rates of a forward and reverse reaction are equal (Figure 7.3). Then, there is no *net* change in the concentrations of reactants and products even though molecules are still moving about. (It is like a party with as many people wandering in as wandering out of two adjoining rooms. The total number in each room stays the same—say, thirty in one and ten in the other—even though the mix of people in each room continually changes.)

When the forward and reverse reactions are proceeding at the same rate, how many reactant molecules and how many product molecules will there be? That depends on the substances involved. For example, in one reversible reaction, glucose-1-phosphate is rearranged into glucose-6-phosphate. When the concentra-

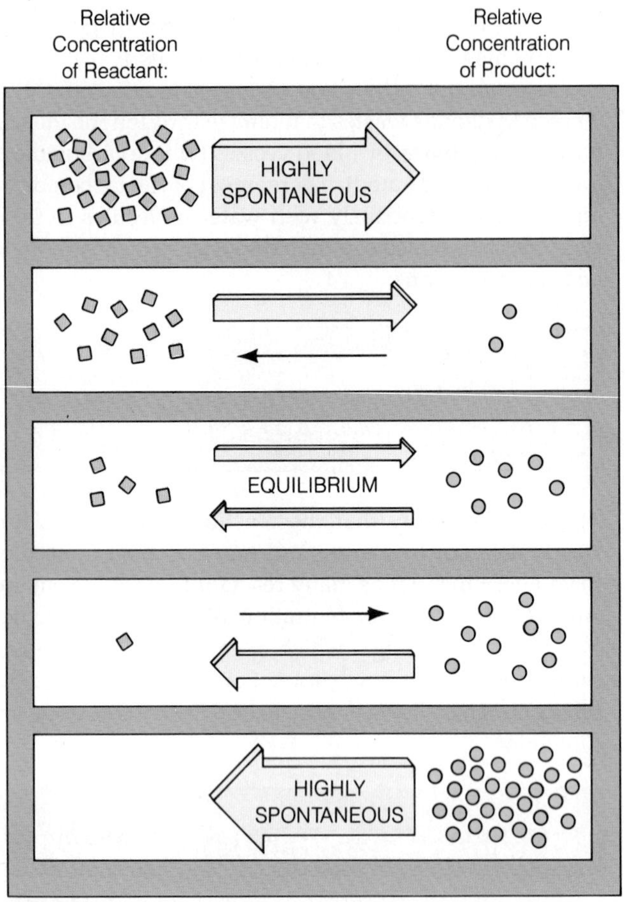

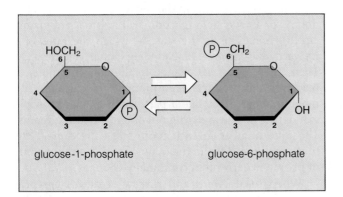

Relative Concentration of Reactant: Relative Concentration of Product:

HIGHLY SPONTANEOUS

EQUILIBRIUM

HIGHLY SPONTANEOUS

Figure 7.3 Chemical equilibrium. With high concentrations of reactant molecules, reactions generally proceed most strongly in the forward direction. With high concentrations of product molecules, they proceed most strongly in reverse. At chemical equilibrium, the *rates* of the forward and reverse reactions are equal. (Note that the *concentrations* are not necessarily the same, as described in the text.)

$HOCH_2$

glucose-1-phosphate   glucose-6-phosphate

Figure 7.4 Example of a reversible reaction.

tions of both substances are the same, this forward reaction occurs 19 times faster than the reverse (Figure 7.4). For the reverse reaction to proceed just as fast as the forward one does, there must be 19 molecules of glucose-6-phosphate for every 1 molecule of glucose-1-phosphate, or a concentration ratio of 19:1. Every reaction has its own "equilibrium constant," which is simply the ratio of reactant and product concentrations when the forward and reverse reaction rates are equal.

Metabolic Pathways

Cells can use only so many molecules of a substance at a given time, and they have only so much space to hold any excess. If cells were to produce more of the substance than they could use, put into storage, or secrete, the excess might cause problems.

In the genetic disorder phenylketonuria (PKU), a normal reaction sequence is blocked and phenylalanine accumulates in the body. The excess can enter into reactions that produce phenylketones, the accumulation of which leads to severe mental retardation. This example underscores a key aspect of metabolism: *Compounds must be produced in concentrations high enough to allow required reactions to run to completion, but low enough to prevent their being diverted into unnecessary and potentially disruptive side reactions.*

Normally, cells maintain, increase, and decrease the concentrations of different substances by coordinating a variety of metabolic pathways. A **metabolic pathway** is an orderly sequence of reactions, each step of which is quickened with the help of a specific enzyme. Each step of the pathway produces some small change in a substance, with an atom or molecule typically becoming attached to it or broken off and transferred elsewhere. As Figure 7.5 shows, most sequences are linear; some are circular. Notice the branchings leading into and out of the cyclic route. Branches often serve to link different pathways, with products of one pathway serving as reactants for others (Figure 7.6).

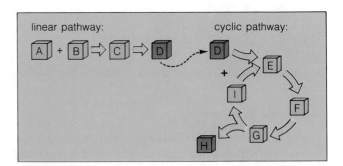

linear pathway: cyclic pathway:

Figure 7.5 Linear and cyclic metabolic pathways.

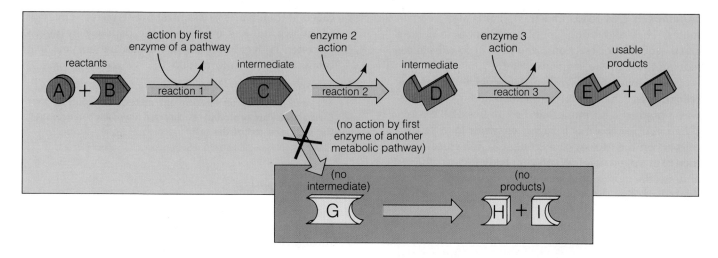

Figure 7.6 Simplified picture of two metabolic pathways. The "product" of the first reaction is not the final compound but an intermediate (C), which can enter into side reactions (a different pathway) or continue along the main pathway. Controls over enzyme activity at branch points help channel intermediate molecules into the pathway that will lead to the product required at the time.

Overall, the major metabolic pathways are either degradative or biosynthetic. In **degradative pathways**, carbohydrates, lipids, and proteins are broken down in stepwise reactions that lead to products of lower energy; often, some of the energy that is released is used to do cellular work. In **biosynthetic pathways**, small molecules (such as sugars) are assembled into large molecules, such as complex carbohydrates, lipids, and proteins.

We can define the participants in metabolic pathways in the following way:

reactants	organic and inorganic substances able to enter into a reaction; also called substrates or precursors
metabolites	compounds being funneled through a metabolic pathway; often, intermediate forms in assembly or breakdown reactions
enzymes	proteins that serve as catalysts (substances that speed up chemical reactions)
cofactors	coenzymes (such as FAD and NAD$^+$) and metal ions that help enzymes catalyze a reaction or that carry atoms or molecules stripped from the substrate to another site
energy carriers	mainly ATP, which readily donates energy to diverse reactions
end products	the substances present at the conclusion of a metabolic pathway

Let's take a look at some of these substances and at the roles they play in metabolism.

ENZYMES

Enzymes Defined

A cup of sugar left undisturbed for twenty years will change very little. But when you put some of that sugar in your mouth, it undergoes chemical change very rapidly. Enzymes secreted by some of your cells account for the difference in the rate of change. **Enzymes** are proteins with enormous catalytic power, which means they greatly enhance the rate at which specific reactions approach equilibrium.

Enzymes do not make anything happen that would not eventually happen on its own. They merely make it happen more rapidly (at least a million times faster, usually). And they make it happen again and again; enzyme molecules are not permanently altered or consumed in a reaction.

Also, an enzyme is quite selective about which reaction it will enhance and which reactants it will deal with, these being called its **substrates**. For example, thrombin (an enzyme involved in blood clotting) catalyzes the breaking of a peptide bond *only* when the bond is between two particular amino acids in this order:

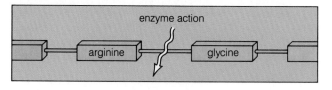

Why is the preference of enzymes for specific substrates so important? If we think of metabolic pathways as chemical roads through a cell, then enzymes are like on-ramps at crossroads and traffic lights along certain routes. They allow only specific substances to enter a given sequence of reactions, and they keep substances moving through the sequence (Figure 7.6).

Controls over the enzymes of different pathways allow cells to direct the flow of nutrients, building materials, waste products, hormones, and so on in suitable ways. When you eat too much sugar, enzymes in your liver cells act on the excess, converting it first to glucose and then to glycogen or fat. When your body uses up glucose and needs more, enzymes break down glycogen to release its glucose subunits. In this case, a hormone called glucagon acts as a control over enzyme activity. It stimulates the key enzyme in the pathway by which glycogen is degraded, and it inhibits the enzyme that catalyzes glycogen formation.

Enzymes act only on specific substrates, and controls over their activity are central to the directed flow of substances into, through, and out of the cell.

Enzyme Structure

Most enzymes are globular (globe-shaped), with at least one surface region folded in the shape of a crevice. At this crevice, which is called an **active site**, a particular reaction is catalyzed (Figure 7.7). An **enzyme-substrate complex** forms when substrate becomes bound in the crevice. The complex is short-lived, partly because only weak bonds hold it together.

As long ago as 1890, Emil Fischer suggested that the shape of some region of the enzyme's surface matches that of its substrate, much as a lock precisely matches its key. The metaphor is still useful, even though it now appears that the match is not a rigid arrangement. As first proposed by Daniel Koshland in 1973, active sites undergo changes when they interact with substrates.

According to Koshland's **induced-fit model**, an active site making contact with its substrate almost but not quite matches it (Figure 7.8). This means that the binding between them is not as strong as it could be. Even so, the interaction is enough to induce structural changes in the active site and to distort the bound substrate, such that the site and the substrate become fully complementary to each other.

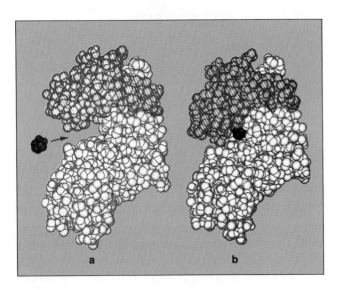

Figure 7.7 Model of the induced fit between an enzyme (hexokinase) and its bound substrate (a glucose molecule, shown here in red).

(a) The cleft into which the glucose is heading is the enzyme's active site. **(b)** In this enzyme-substrate complex, notice how the enzyme shape is altered temporarily: the upper and lower parts now close in around the substrate.

Enzyme Function

Activation Energy. How do enzymes enhance reaction rates? We can begin with a simple fact: For any reaction to occur, reactant molecules must collide with each other with some minimum amount of energy. That amount is like a hill over which the molecules must be pushed. Think about H_2 and O_2 molecules, which are not inclined to react on their own. When they absorb enough energy (as from an electric spark), they collide with enough force to get to the top of the hill. At the crest of an energy hill, the reactants are in an activated, intermediate condition called the "transition" state. Now the reaction proceeds spontaneously, just as a boulder pushed up and over the crest of a hill rolls down on its own.

For any reaction, the minimum amount of energy needed to bring all the molecules in one mole of a substance to the transition state is called the **activation energy**.

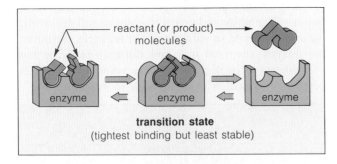

Figure 7.8 Induced-fit model of enzyme-substrate interactions. Only when the substrate is bound in place is the active site complementary to it. The most precise fit occurs during the transition state.

reactant (or product) molecules

enzyme

enzyme

enzyme

transition state
(tightest binding but least stable)

An enzyme enhances the rate of a given reaction by *lowering* the required activation energy (Figure 7.9). How can it do this? Among other things, weak but extensive bonding in its active site puts substrates in positions and at distances that promote reaction. (In contrast, reactants colliding on their own do so from random directions, hence mutually attractive chemical groups may not make contact and reaction may not occur.)

Effect of Substrate Concentrations. Keep in mind that enzymes lower the crest of an energy hill for both reactant and product molecules. That is, they make it easier for reactants to reach the crest and be converted into products, but they also make it easier for the *products* to reach the crest and return to the reactant form. Which way the reaction goes depends partly on the relative concentrations of reactant and product molecules (which can change, depending on other cellular events) and partly on the equilibrium constant.

Enzymes influence the <u>rate</u> at which forward and reverse reactions reach equilibrium. They do not change what the concentration ratio of reactant to product molecules will be at equilibrium.

Effects of Temperature and pH. Most enzymes cannot tolerate high temperatures. When the surroundings become warmer, enzyme activity increases until it reaches a maximum rate at some temperature (which varies for different enzymes). Beyond that temperature, the reaction rate decreases sharply (Figure 7.10). The increased thermal energy has disrupted weak bonds holding the enzyme molecule in its three-dimensional shape, denaturation (page 58) has occurred, and the active site is so altered that the substrate cannot bind to it. Even brief exposure to high temperatures will destroy enzymes and thereby affect metabolism. This is one of the consequences of extremely high fever; when body temperature reaches 44°C (112°F), death generally follows.

Also, most enzymes are effective when their surroundings are neutral (pH 7) or nearly so (Figure 7.10). At higher or lower pH values, hydrogen bonds and other weak attractions holding the enzyme in its three-dimensional shape are disrupted and enzyme function is affected. Pepsin is one of the exceptions; it acts in extremely acidic fluid in the stomach. Trypsin is another; it acts in a more basic fluid (about pH 8) in the small intestine.

An enzyme functions only within a limited range of temperature and pH.

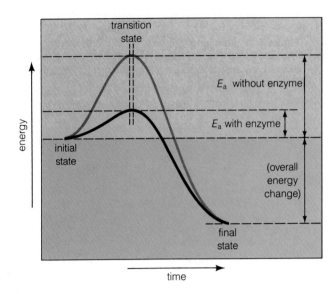

Figure 7.9 Energy hill diagram showing the effect of enzyme action. An enzyme greatly enhances the rate at which a reaction proceeds because it lowers the required activation energy (E_a). In other words, not as much collision energy is needed to boost all of the reactant molecules to the crest of the energy hill (the transition state).

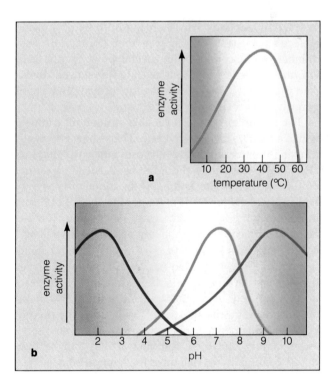

Figure 7.10 (**a**) Effect of increasing temperature on enzyme activity. (**b**) Effect of pH on enzyme activity. The brown line charts the activity of an enzyme that is fully functional in neutral solutions. The red line charts the activity of one that is functional in basic solutions; the purple line, in acidic solutions.

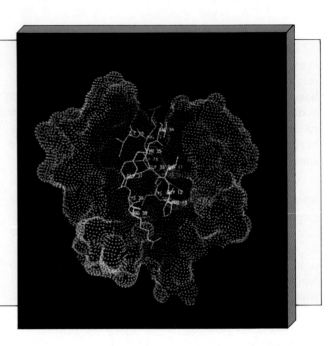

Figure 7.11 Computer-graphics model of the enzyme trypsin (blue) with an inhibitor (green) bound to the active site. The enzyme parts colored red are not exposed to the surroundings when the inhibitor is bound.

This inhibitor is almost exactly complementary in structure to the active site and dissociates slowly from it. The tight, prolonged binding is vital. Trypsin figures in the digestion of proteins in the small intestine. It is so powerful that it is assembled in inactive form in the pancreas, then isolated in membranous vesicles until it is secreted into the small intestine.

Pancreatic trypsin inhibitor serves as a safeguard: it inactivates trypsin molecules that escape packaging in the pancreas. Without this safeguard, trypsin could be unleashed against the protein components of pancreatic tissues and blood vessels. In acute pancreatitis, this and other pancreatic enzymes are activated prematurely; the results are sometimes fatal.)

Regulation of Enzyme Activity

How many product molecules will form in a given time interval? It depends partly on the number of enzyme molecules available to catalyze their formation. That number can be controlled in several ways. Synthesis of enzyme molecules can be accelerated or slowed down. And the activity of enzymes already formed can be temporarily or permanently shut down.

For example, some "reversible" inhibitors compete with a substrate for an active site. This is how pancreatic trypsin inhibitor shuts down trypsin molecules that leak into places where they shouldn't be. A substrate cannot bind to the active site as long as the inhibitor is already bound there (Figure 7.11). As another example, other inhibitors bind irreversibly with a key group on the enzyme surface, thereby making catalysis impossible.

The enzymes in metabolic pathways are not equally fast at what they do. One enzyme (or more) catalyzes the slowest step and so sets the pace of the overall reaction sequence. (Every other enzyme in the pathway can only act as fast as its substrates form during the preceding step.) Such "pacemakers" of metabolism also serve as **regulatory enzymes**, for their activity can be fine-tuned continually in response to chemical signals. *And stimulation or inhibition of the pacemakers can stoke or shut down the appearance of end products of the pathway, depending on the cell's requirements.*

One form of enzyme regulation is called allosteric control. Besides having an active site, an *allosteric* enzyme also has one or more regulatory sites at which a specific molecule binds and serves as the signal to change enzyme activity (Figure 7.12).

Often the "signal" is a molecule of the end product of the pathway. When more product molecules are being made than the cell can use, the substance binds to the allosteric enzyme and shuts it down. This is a form of **feedback inhibition**, whereby an increase in a substance inhibits the very process leading to the increase. When more product molecules are used up, the substance is released from enzyme molecules, which then can function again.

COFACTORS

Most enzymes catalyze the transfer of electrons, atoms, or functional groups. Many require the assistance of non-protein components called **cofactors**, either to help catalyze the reaction or to serve fleetingly as the transfer agents.

Some cofactors are bound so tightly to the enzyme that they are, in effect, part of the enzyme itself (in which case they are said to be prosthetic groups). Others are free-moving; they deliver atoms or parts of molecules stripped from substrates to other reaction sites.

Several large organic molecules, called *coenzymes*, serve as cofactors. **FAD** (flavin adenine dinucleotide) is an example. It is bound permanently to an enzyme that strips hydrogen atoms from its substrate. **NAD⁺** (nicotinamide adenine dinucleotide) and the related form **NADP⁺** are free-moving; each delivers a proton (H^+) and two electrons stripped from a substrate to other reaction sites. When carrying protons and elec-

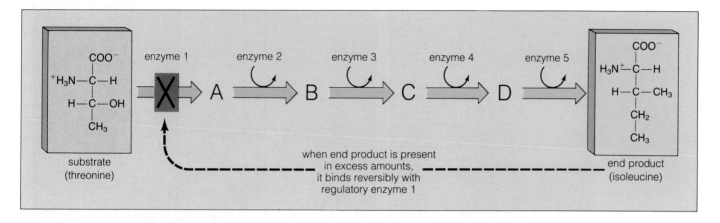

Figure 7.12 Feedback inhibition through an allosteric enzyme. Such enzymes have, in addition to the active site, one or more regulatory sites where a specific metabolite binds and signals the enzyme to shut down its catalytic activity.

Here, the end product (isoleucine) binds reversibly to the first enzyme in the pathway leading to its production. When the isoleucine concentration drops, fewer molecules are around to inhibit the regulatory enzyme present, so isoleucine production can rise again. The rapid reversibility of such forms of feedback inhibition allows cellular concentrations of substances to be adjusted quickly to metabolic needs.

trons, these compounds are called $FADH_2$, NADH, and NADPH.

FAD and NAD^+ have key roles in pathways by which carbohydrates are degraded. $NADP^+$ is an important coenzyme in photosynthesis. It also serves as a link between some of the main degradative and biosynthetic pathways (for example, the ones by which fatty acids are assembled).

Metal ions also serve as cofactors. These inorganic substances include ferrous iron (Fe^{++}), which assists cytochrome molecules. The **cytochromes** are electron-transfer proteins bound in cell membranes, such as the membranes of chloroplasts and mitochondria.

ATP: THE MAIN ENERGY CARRIER

Structure and Function of ATP

Earlier we noted that sunlight is the primary energy source for the web of life. Before that energy can be used in cell activities, it must first be transformed into the chemical energy of adenosine triphosphate, or **ATP**. Similarly, cells cannot directly use chemical energy stored in carbohydrates and other large organic molecules; they must first transform it to the energy of ATP.

As Figure 7.13 shows, ATP is composed of adenine (a nitrogen-containing compound), ribose (a five-carbon sugar), and a triphosphate (three linked phosphate groups). ATP provides energy for biosynthesis, active transport across cell membranes, and molecular dis-

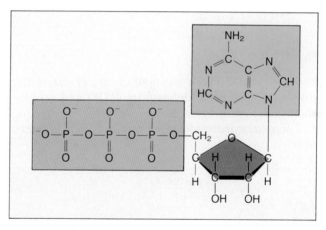

Figure 7.13 Structural formula for adenosine triphosphate, or ATP. The triphosphate group is shaded in gold, the sugar ribose in pink, and the adenine portion in blue.

placements (such as those underlying muscle contraction). It does more than link the energy-trapping reactions and building programs of photosynthesis; it also links the major degradative and biosynthetic pathways. In fact, ATP transfers energy to so many different reactions that it can be likened to a gold coin (rather than, say, a dollar bill or peso), which is accepted at once as currency all over the world.

Almost all metabolic pathways directly or indirectly run on energy supplied by ATP.

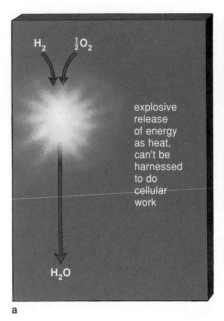
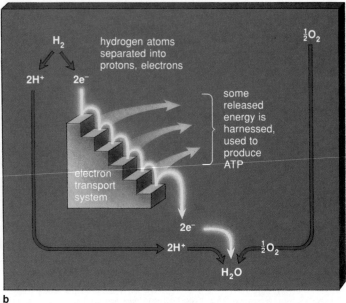

Figure 7.14 (a) When hydrogen and oxygen are made to react (say, by an electric spark), most of their energy is released as heat. (b) In cells, the same type of reaction is made to occur in many small steps that allow much of the released energy to be harnessed in usable form. These "steps" are electron transfers, often between molecules that operate together as an electron transport system.

The ATP/ADP Cycle

In the **ATP/ADP cycle**, an input of energy drives the linkage of adenosine diphosphate (ADP) and a phosphate group (or inorganic phosphate) into ATP; then the ATP donates a phosphate group elsewhere and becomes ADP:

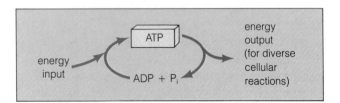

The addition of phosphate to a molecule is called **phosphorylation**. What is so important about it? When a molecule becomes phosphorylated by ATP, its store of energy generally increases *and it becomes primed to enter a specific reaction*.

With the ATP/ADP cycle, cells have a renewable means of conserving energy and transferring it to specific reactions. The ATP turnover is rather breathtaking. Typically, ATP molecules give up energy within sixty seconds of their formation. Even if you were bedridden for twenty-four hours, your cells would turn over forty kilograms (eighty-eight pounds) or so of ATP molecules simply for routine maintenance!

ELECTRON TRANSPORT SYSTEMS

In chapters to follow, we will be looking at the major metabolic pathways called photosynthesis (which is biosynthetic) and aerobic respiration (which is degradative). ATP is produced during both pathways, and its production depends on **electron transport systems**. Such systems consist of membrane-bound enzymes and cofactors, operating one after another in a highly organized sequence.

In an electron transport system, electrons are transferred from one component (the donor molecule) to another (an acceptor) in stepwise reactions. When an atom or molecule gives up one or more electrons, it is said to be *oxidized*. When it accepts one or more electrons, it is *reduced*. The term **oxidation-reduction reaction** refers to an electron transfer.

In one type of electron transport system, the first molecule accepts H^+ and electrons from NADH. Then the electrons are transferred in stepwise fashion to other membrane-bound carriers that include cytochromes. The last electron carrier in the series gives up the electron to an external acceptor molecule (one that is not part of the transport system). Oxygen is one type of final acceptor molecule.

The point of such oxidation-reduction sequences is the generation of usable forms of energy. By analogy, think of an electron transport system as a staircase (Figure 7.14).

Electrons that have been "raised" (excited) to the top of the staircase have the most energy. They drop down the staircase, one step at a time (they are transferred from one electron carrier to another). With each drop, some of the energy being released is harnessed to do work—for example, to make hydrogen ions (H^+) move in ways that establish pH and electric gradients across membranes. Such gradients, as you will discover, are central in ATP formation.

SUMMARY

1. Cells acquire and use energy to accumulate, break down, synthesize, and rid themselves of substances in controlled ways. These activities, which sustain cell growth, maintenance, and reproduction, are called metabolism.

2. Cells use different forms of energy (such as chemical and mechanical energy), each of which can be measured in terms of its capacity to do work.

3. Cellular use of energy conforms to two laws of thermodynamics. According to the first law, energy can be converted from one form to another but the total amount in the universe never changes. (No more energy can be created, and none of the existing energy ever vanishes.)

4. According to the second law, with each energy conversion, some energy is dispersed in a form that is not as readily available to do work.

5. Cells lose energy during metabolic reactions, but they also replace it with energy being lost from someplace else. In this way they maintain their high degree of organization.

6. Directly or indirectly, some of the energy being lost from the sun is the source of energy replacements for almost all organisms on earth.

7. If left undisturbed, most metabolic reactions approach a state of dynamic equilibrium (in which there is no further net change in the concentrations of reactants and products).

8. A metabolic pathway is a stepwise sequence of reactions in a cell. In *biosynthetic* pathways, organic molecules are assembled and energy becomes stored in them. In *degradative* pathways, organic molecules are broken apart and energy is released.

9. The following substances take part in metabolic reactions:
   a. Reactants: the substances that enter the reactions.
   b. Metabolites: intermediate compounds in a series of reactions.
   c. Enzymes: proteins that serve as catalysts (they speed up reactions).
   d. Cofactors: coenzymes (including NAD^+) and metal ions that help catalyze reactions or carry functional groups stripped from substrates.
   e. Energy carriers: mainly ATP, which readily donates energy to other molecules; most metabolic pathways run directly or indirectly on ATP energy.
   f. End products: the substances formed at the end of a metabolic pathway.

10. Enzymes do not change what the concentration ratio of reactant to product molecules will be at equilibrium. They only increase the rate at which a reaction approaches equilibrium (by lowering the required activation energy).

11. In oxidation-reduction reactions (electron transfers), energy is released that can be used to do work—for example, to make ATP.

Review Questions

1. State the first and second laws of thermodynamics. Which law deals with the *quality* of available energy, and which deals with the *quantity*? Can you give some examples of high-quality energy? *98*

2. Does the living state violate the second law of thermodynamics? In other words, how does the world of living things maintain a high degree of organization, even though there is a universal trend toward disorganization? *98-99*

3. In metabolic reactions, does equilibrium imply equal concentrations of reactants and products? Can you think of some cellular events that might keep a reaction from approaching equilibrium? *99-100*

4. Describe an enzyme and its role in metabolic reactions. How do enzymes affect the proportions of reactants and products that will be present at equilibrium? *101-104*

5. Define "substrate" and "active site." Why is binding at an active site a readily reversible event? *102*

6. The high temperatures associated with severe fevers can impair cell functioning. Can you explain why? *103*

7. What are the three molecular components of ATP? What is the function of ATP, and why is phosphorylation of a molecule by ATP so important? *105-106*

8. What is an oxidation-reduction reaction? What is its function in cells? *106-107*

Readings

Atkins, P. 1984. *The Second Law*. New York: Freeman.

Doolittle, R. October 1985. "Proteins." *Scientific American* 253(4):88–99.

Fenn, J. 1982. *Engines, Energy, and Entropy*. New York: Freeman. Deceptively simple introduction to thermodynamics; good analogies. Paperback.

Fersht, A. 1985. *Enzyme Structure and Mechanism*. Second edition. New York: Freeman.

8

ENERGY-ACQUIRING
PATHWAYS

Just before dawn in the Midwest the air is dry and motionless; the heat that has scorched the land for weeks still rises from the earth and hangs in the air of a new day. There are no clouds in sight. There is no promise of rain. For hundreds of miles in all directions, crops stretch out, withered or dead. All the sophisticated agricultural methods in the world can't save them now. In the absence of one vital resource—water—life in each cell of those many thousands of plants has ceased.

In Los Angeles, a student reading the morning newspaper complains that the Midwest drought will probably cause a hike in food prices. In Washington, D.C., economists calculate the crop failures in terms of decreased tonnage available for domestic consumption and export and of what it means to the nation's balance of payments. In Ethiopia, a child with bloated belly and spindly legs waits passively for death. Even if food from the vast agricultural plains of North America were to reach her now, it would be too late. Deprived of food resources too long, cells of her body will never grow normally again.

You are about to explore the ways in which cells acquire and use energy. You will be considering cellular pathways that might at first seem to be far removed from the world of your interests. Yet the food molecules on which you and almost all other organisms depend cannot be built or used without those pathways and the raw materials required for their operation.

Figure 8.1 Links between the major energy-trapping and energy-releasing pathways.

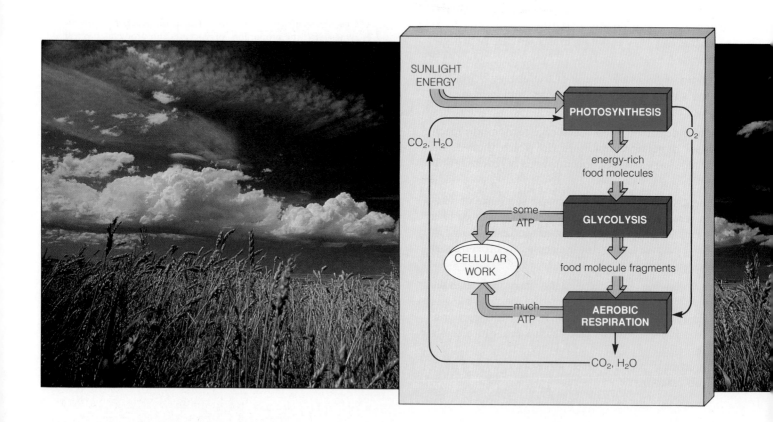

FROM SUNLIGHT TO CELLULAR WORK: PREVIEW OF THE MAIN PATHWAYS

No matter what the organism, the structural organization of its cells is based on organic compounds (which have a backbone of carbon atoms), and its cellular activities are driven by chemical energy stored in many of those compounds. The questions become these:

1. Where does the carbon come from in the first place?

2. Where does the energy come from to drive the linkage of carbon and other atoms into organic compounds?

3. How does the energy inherent in those compounds become available to do cellular work?

The answers vary, depending on whether you are talking about autotrophs or heterotrophs. **Autotrophic organisms** get their carbon and energy from the physical environment; they are "self-nourishing" (which is what autotroph means). Their carbon source is carbon dioxide (CO_2), a gaseous substance all around us in the air and dissolved in water. *Photosynthetic* autotrophs, which include all plants, some protistans, and some bacteria, harness light energy from the sun for building organic compounds. *Chemosynthetic* autotrophs, which are limited to a few bacteria, extract energy from chemical reactions involving inorganic substances (such as sulfur).

In contrast, **heterotrophic organisms** are not self-nourishing; they feed on autotrophs, each other, and organic wastes. They get carbon and energy for their building programs from organic compounds *already built* by autotrophs. Animals, fungi, many protistans, and most bacteria are heterotrophs.

It follows, from the above, that carbon and energy enter the web of life primarily through *photosynthesis*, the main biosynthetic pathway. Energy that becomes stored in organic compounds as a result of photosynthesis can be released in *glycolysis*, the first stage of the main degradative pathways used by most autotrophs and heterotrophs. The degradative pathway that yields the most energy is called *aerobic respiration*. Figure 8.1 shows the links between these energy-acquiring and energy releasing pathways, which are the focus of this chapter and the next.

PHOTOSYNTHESIS

Simplified Picture of Photosynthesis

Photosynthesis consists of two sets of reactions. In the **light-dependent reactions**, sunlight energy is absorbed and converted to chemical energy, which is stored briefly in ATP and NADPH. In the **light-independent reactions**, sugars and other organic compounds are assembled with

the help of ATP and NADPH. Often photosynthesis is summarized this way:

$$2H_2O + CO_2 \xrightarrow{\text{sunlight}} O_2 + (CH_2O) + H_2O$$

Here, hydrogen atoms obtained from water molecules are used in forming compounds based on some number of (CH_2O) units. For instance, for reactions leading to glucose formation, you would have to multiply everything by six (to get the six carbons, twelve hydrogens, and six oxygens of the glucose molecule):

$$12H_2O + 6CO_2 \xrightarrow{\text{sunlight}} 6O_2 + C_6H_{12}O_6 + 6H_2O$$

The above equation does not provide much insight into what actually goes on. For example, oxygen is shown as a by-product of photosynthesis. Did it come from the H_2O or the CO_2? Also, why is water shown as a by-product as well as a reactant? To answer these questions, we must expand the summary equation a bit:

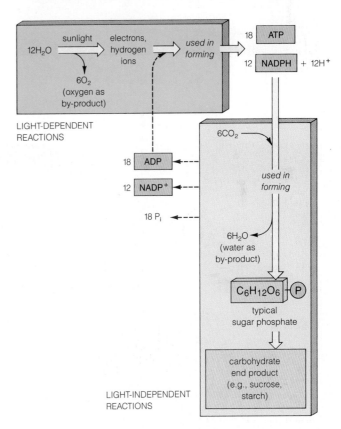

As you can see, the oxygen is obtained early on (when water molecules are split). And "new" water molecules are formed and used at different reaction steps, beginning with ATP formation, with six left over at the end of the entire sequence. Notice also how ATP and NADPH

bridge the two sets of reactions. ATP carries energy and NADPH carries electrons and hydrogen atoms for the synthesis reactions (page 105).

Finally, glucose is often used as the textbook example of a photosynthetic end product in order to keep the chemical bookkeeping simple. But notice that the light-independent reactions don't really end with glucose ("$C_6H_{12}O_6$"). Glucose and other sugar monomers do indeed form, but they have phosphate groups attached that prime them for further reaction. In other words, those sugar phosphates are *intermediates*. Almost always they are linked at once into sucrose, starch, and other carbohydrates that are the true end products of photosynthesis.

Chloroplast Structure and Function

The light-dependent reactions of photosynthesis take place at specialized cell membranes, such as the **thylakoid membrane** inside chloroplasts. That membrane is folded into a system of stacked disks, called **grana**, and flattened channels (page 77 and Figure 8.2). The interior spaces of the disks and channels are open to one another, so the membrane system actually forms a single compartment. The compartment is a reservoir for hydrogen ions, and it is tapped for ATP formation. The synthesis

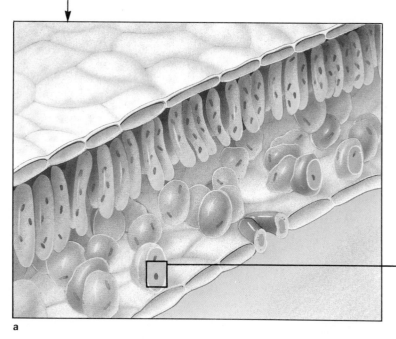

a

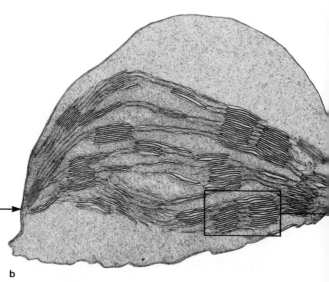

b

Figure 8.2 Functional zones of a chloroplast from the leaf of a sunflower plant (*Helianthus*). The light-dependent reactions of photosynthesis occur at thylakoid membranes, and they lead to ATP and NADPH formation. The light-independent reactions occur in the stroma. They lead to production of sugars and other carbon-containing molecules. (**a**) Section through a sunflower leaf, showing chloroplast-containing cells. (**b**) Chloroplast in cross-section, 25,000×. (**c**) Two of the grana, 93,000×. (**d**) Where photosynthetic reactions occur.

of phosphorylated sugars and starch occurs in the **stroma**, a semifluid matrix that surrounds the thylakoid compartment.

The internal organization of a chloroplast might seem to be less than memorable, until you remember that we are talking about a very small space. If you could line up 2,000 chloroplasts, one after another, the lineup would be no wider than your thumbnail. Imagine all the chloroplasts in one lettuce leaf, each a tiny factory for producing sugars and starch—and you begin to get a sense of the magnitude of metabolic events required to feed you and all other organisms on earth.

LIGHT-DEPENDENT REACTIONS

Three distinct events unfold during the light-dependent reactions. Light energy is absorbed, electron and hydrogen transfers lead to ATP and NADPH formation, and electrons are replaced in the system that originally gives up electrons for the reactions.

Light Absorption in Photosystems

Light-Trapping Pigments. Photosynthetic membranes contain light-absorbing molecules called **pigments**. A pigment molecule absorbs photons, which are individual packets of energy from the sun. Some photons have more energy than others, and the differences correspond to different wavelengths of light. The shorter the wavelength, the more energy the photon carries. Most organisms use wavelengths ranging from about 400 to 750 nanometers for light-requiring processes such as photosynthesis and vision. That is a very small part of the electromagnetic spectrum (Figure 8.3), but those are the wavelengths we perceive as different colors of light.

Each type of pigment effectively absorbs only certain wavelengths and transmits the rest. Chlorophylls readily absorb blue and red wavelengths, but they mostly transmit green. The carotenoids absorb violet and blue wavelengths but transmit yellow. Wavelengths that *can* be absorbed impart just the right amount of energy to excite one of the electrons in the pigment molecule. Figure 8.3

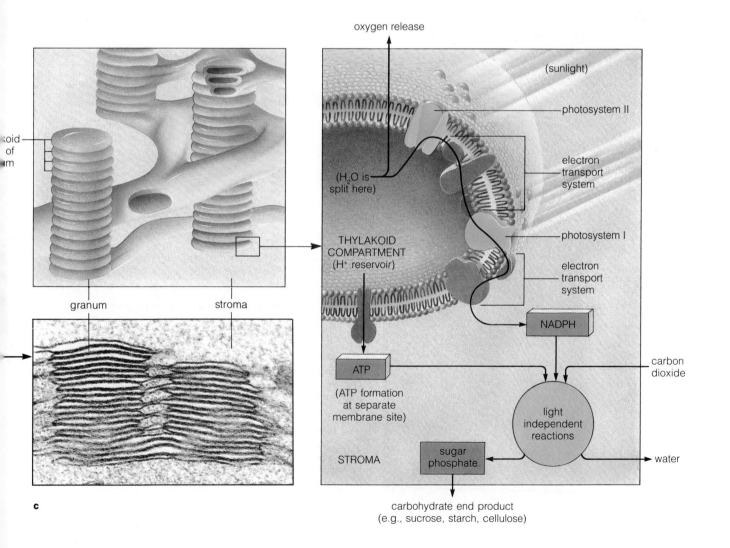

granum   stroma

c

oxygen release

(sunlight)

photosystem II

(H₂O is split here)

electron transport system

THYLAKOID COMPARTMENT (H⁺ reservoir)

photosystem I

electron transport system

NADPH

carbon dioxide

ATP

(ATP formation at separate membrane site)

STROMA

light independent reactions

water

sugar phosphate

carbohydrate end product (e.g., sucrose, starch, cellulose)

Figure 8.3 (Above) Where wavelengths of visible light occur in the electromagnetic spectrum. For most organisms, the range of photosensitivity is limited largely to wavelengths ranging from about 400 to 750 nanometers. Shorter wavelengths (such as ultraviolet and x-rays) are so energetic they can break bonds in organic molecules; hence they can destroy cells. Longer wavelengths (such as infrared) are not energetic enough to power chemical changes in the molecules required to form NADPH, which is used in biosynthesis.

(Below) Ranges of wavelength absorption for several photosynthetic pigments. Absorption peaks correspond to the measured amount of energy absorbed and used in photosynthesis. Colors used here correspond to the colors transmitted by each pigment type. (Thus chlorophylls show peak absorption of blue and red wavelengths and transmit wavelengths in between.) Together, different photosynthetic pigments can absorb most of the wavelength energy available in the spectrum of visible light.

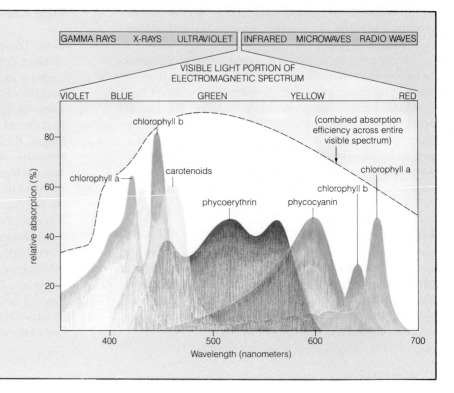

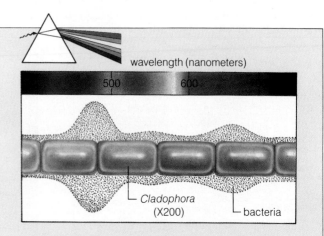

Cladophora
(X200)

bacteria

Figure 8.4 T. Englemann's 1882 experiment, which revealed the most effective wavelengths for photosynthesis by *Cladophora*, a filamentous green alga.

Oxygen is a by-product of photosynthesis, as the photograph (left) of the aquatic plant *Elodea* indicates. (The bubbles of oxygen sometimes are visible on sunny days.) Oxygen also is used by many organisms in the energy-releasing pathway called aerobic respiration. Englemann suspected that aerobic (oxygen-using) bacteria living in the same places as the alga would congregate in areas where oxygen was being produced.

Englemann used a crystal prism to cast a tiny spectrum of colors on a microscope slide. Then he positioned an algal filament to run parallel with the spectrum (as shown above). The bacteria did indeed cluster next to the filament where the most oxygen was being released. And those regions corresponded to colors (wavelengths) being absorbed most effectively—in this case, violet and red.

includes plots (absorption spectra) that show which wavelengths can make this happen in several kinds of pigments.

Figure 8.4 shows an early experiment designed to test which wavelengths are used most effectively by a green alga. In this green plant and others, the main pigments are chlorophyll *a*, chlorophyll *b*, and the carotenoids.

Photosystems. The light-absorbing pigments just described are not scattered at random in the thylakoid membrane of chloroplasts. They are organized into many clusters, each with 200 to 300 pigment molecules. Each cluster is a **photosystem**.

Over ninety percent of the pigments in a photosystem do nothing more than "harvest" light energy. When they absorb photon energy, one of their electrons becomes excited and reaches a higher energy level (page 39). When that electron returns to a lower level, the extra energy is released and hops from one pigment to another. With each hop, a little energy is usually lost (as heat). The energy remaining corresponds to longer and longer wavelengths, compared to the original photon energy.

Only a few chlorophyll molecules can respond to the longest of those wavelengths. They act like a sink, or *trap*, for all the energy being harvested by all the other pigments. And they are the only pigments that can *give up* electrons for use in photosynthesis.

When energy flows into an energy trap, an electron is excited and is rapidly transferred to an acceptor molecule embedded in the thylakoid membrane. Thus, *the first event of photosynthesis is the light-activated transfer of an electron from a special chlorophyll molecule to an acceptor molecule*. It is over in less than a billionth of a second.

Two Pathways of Electron Transfer

During their journey, electrons expelled from a chlorophyll molecule pass through one or two **electron transport systems**. Each system is a series of molecules that accept and then donate electrons to the next molecule in line. (Electron transfers are also called oxidation-reduction reactions.) With the energy released at some transfers, inorganic phosphate is attached to an ADP molecule to form ATP. We call this a *photo*phosphorylation reaction (it depends on an earlier input of light energy).

The Cyclic Pathway. In **cyclic photophosphorylation**, electrons proceed through only one transport system, then return to the photosystem that gave them up (Figure 8.5). This so-called **photosystem I** is distinguished by one of its chlorophyll molecules, P700, which absorbs wavelengths of about 700 nanometers.

The cyclic pathway is probably one of the oldest and simplest means of ATP production. The fossil record shows that early photosynthetic organisms were no larger than existing bacteria, so their body-building programs could scarcely have been enormous. Thus they could have used ATP for building organic compounds, even though synthesis reactions using ATP alone are rather inefficient. (ATP carries only energy to sites where organic compounds are built; the electrons and hydrogen atoms required must be obtained by other reaction steps.) However, the energy made available by the cyclic pathway alone would not have been enough to sustain the evolution of larger photosynthetic organisms, including land plants.

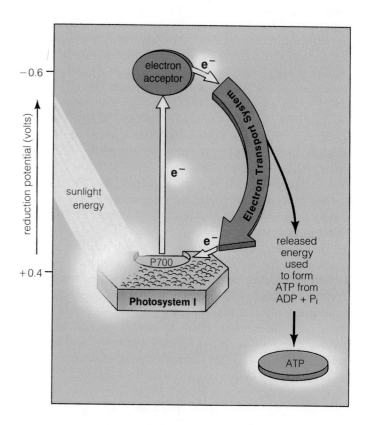

Figure 8.5 Cyclic photophosphorylation, which yields ATP. The vertical scale of this figure indicates the relative tendencies of electron-carrying molecules to take up electrons. Electrons are driven uphill by light, and they move back down the energy hill in the transport system.

The Noncyclic Pathway. Today, land plants rely mostly on **noncyclic photophosphorylation.** In this pathway, electrons are transferred through two photosystems and two electron transport systems, and ADP is still phosphorylated to form ATP. But the electrons do not move in a circle. They end up in NADPH—and hydrogen and electrons carried by this coenzyme are used directly in the synthesis of organic compounds!

As Figure 8.6 shows, the noncyclic pathway begins at **photosystem II,** which is distinguished by one of its chlorophyll molecules (P680). Absorbed light energy causes P680 to give up an electron to an acceptor molecule. The electron is transferred to a transport system, which delivers it to chlorophyll P700 of photosystem I.

When P700 absorbs light energy, electrons are boosted to a higher energy level and passed to a second transport system. Remember that transport systems are like steps on an energy staircase—and the boost places the electrons at the top of a higher staircase. There is enough energy left at the bottom of this staircase to drive the linkage of two electrons and a hydrogen ion (H^+) to $NADP^+$, the result being NADPH.

Thus there is a one-way flow of electrons to NADPH. In the meantime, the P680 molecule that gives up the electrons in the first place is getting replacements—from water. Water molecules are being split into oxygen, "naked" protons (that is, hydrogen ions, or H^+), and electrons inside the thylakoid compartment. Photon energy indirectly drives this reaction sequence, which is called **photolysis.**

The oxygen atoms split from water molecules are by-products of the noncyclic pathway. Oxygen has been accumulating ever since this pathway emerged nearly 3.5 billion years ago. It profoundly changed the earth's atmosphere. And it made possible aerobic respiration, the most efficient way of all to release energy from organic compounds. If the noncyclic pathway had not emerged, you and all other animals would not be around today, breathing in the oxygen that helps keep your cells alive.

With cyclic photophosphorylation, ATP alone forms.

With noncyclic photophosphorylation, ATP <u>and</u> NADPH form.

Oxygen, an end product of the noncyclic pathway, contributed to the evolution of the earth's atmosphere and made aerobic respiration possible.

A Closer Look at ATP Formation

So far, you have seen what happens to the electrons and oxygen released when water molecules are split at the start of the noncyclic pathway. What happens to the hydrogen ions? They accumulate inside the thylakoid compartment:

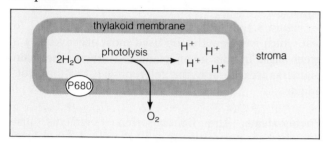

Hydrogen ions also accumulate in the compartment when electron transport systems are operating. (This is true of both the cyclic and noncyclic pathway.) During some of the electron transfers, hydrogen ions are picked up from the stroma and are released inside the thylakoid compartment:

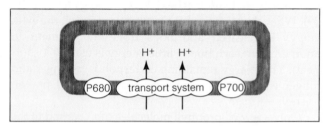

Thus, through photolysis and electron transport, hydrogen ions become much more concentrated in the compartment than in the stroma. The lopsided distribution of those positively charged ions also creates a difference in electric charge across the membrane. An electric gradient as well as a concentration gradient has become established—*and the energy inherent in those gradients can be tapped to form ATP.* The combined force of the concentration and electric gradients propels hydrogen ions out of the compartment and into the stroma. The ions flow through channel proteins, called ATP synthases, that span the membrane and that are complexed with enzyme machinery. The ion flow drives the enzyme machinery by which ADP and inorganic phosphate are linked to form ATP in the stroma:

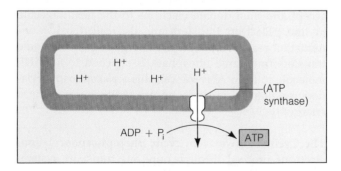

The idea that concentration and electric gradients across a membrane drive ATP formation is known as the **chemiosmotic theory.**

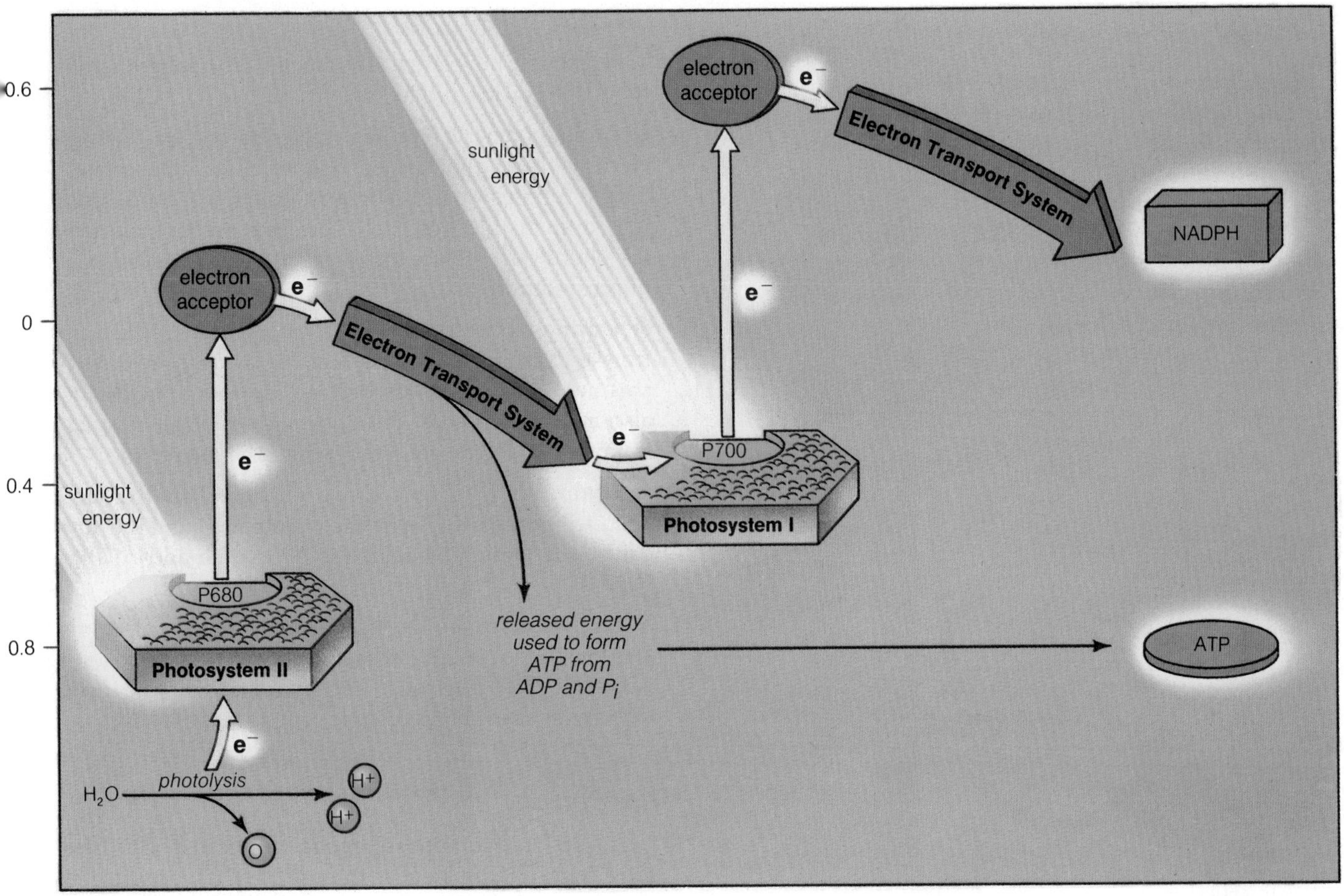

-0.6
0
0.4
0.8
electron acceptor
e−
Electron Transport System
NADPH
sunlight energy
electron acceptor
e−
Electron Transport System
e−
e−
P700
Photosystem I
sunlight energy
P680
Photosystem II
released energy used to form ATP from ADP and Pi
ATP
e−
H2O
photolysis
H+
H+
O

Figure 8.6 Noncyclic photophosphorylation, which yields NADPH as well as ATP. Electrons derived from the splitting of water molecules (photolysis) travel through two photosystems, which work together in boosting the electrons to an energy level high enough to lead to NADPH formation. For every 12 water molecules split, 24 hydrogen ions are released and play a role in the formation of 24 ATP molecules. Of those ATP molecules, 18 are used in the light-independent reactions; the other 6 can be used for other cell activities.

A total of eighteen ATP molecules form during the light-dependent reactions, and a total of eighteen water molecules are released during their formation. Of those water molecules, twelve will be used during the light-independent reactions by which sugar phosphates are assembled; the remaining six will be the "$6H_2O$" by-products remaining at the end of photosynthesis.

Summary of the Light-Dependent Reactions

So far, we have described the light-dependent reactions as separate events. We can now bring those events together into an integrated picture of the first stage of photosynthesis:

Light Absorption

1. In chloroplasts, light is absorbed by two types of photosystems (clusters of photosynthetic pigments embedded in the thylakoid membrane).

2. Light absorption causes the transfer of electrons from photosystem I or II to an acceptor molecule, which will donate them to a transport system in the membrane.

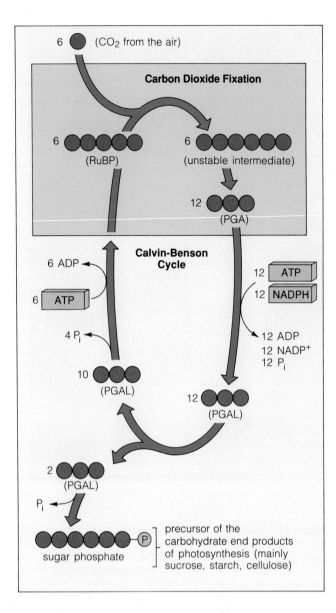

6 ● (CO₂ from the air)

Carbon Dioxide Fixation

6 ●●●●●   6 ●●●●●
(RuBP)   (unstable intermediate)

12 ●●●
(PGA)

Calvin-Benson Cycle

6 ADP ←

6 | ATP

4 Pᵢ ←

12 | ATP
12 | NADPH

12 ADP
12 NADP⁺
12 Pᵢ

10 ●●●
(PGAL)

12 ●●●
(PGAL)

2 ●●●
(PGAL)

Pᵢ ←

●●●●●●–(P)   precursor of the
sugar phosphate   carbohydrate end products
of photosynthesis (mainly
sucrose, starch, cellulose)

Figure 8.7 Summary of the light-independent reactions of photosynthesis. The carbon atoms of the different molecules are depicted in red. All of the intermediates have one or two phosphate groups; for simplicity we show only the one on the "end product" (sugar phosphate).

Noncyclic Pathway

1. Existing land plants rely mainly on noncyclic photophosphorylation, which yields ATP *and* NADPH.

2. In the noncyclic pathway, there is a one-way flow of electrons from photosystem II, through a transport system, to photosystem I, and on through a second transport system. Electrons released from water molecules (by photolysis) replace the electrons being expelled from photosystem II.

3. During electron transfers, hydrogen ions picked up from the stroma are released inside the thylakoid compartment. Hydrogen ions derived from water molecules also accumulate here. Both activities establish concentration and electric gradients across the membrane.

4. Hydrogen ions flow down the gradient (through channel proteins that span the membrane), and the flow drives the joining of inorganic phosphate and ADP to form ATP.

5. At the end of the second transport system, electrons are donated to NADP⁺, which combines with H⁺ to form NADPH. The electrons and hydrogen can be used directly in assembling organic compounds.

Cyclic Pathway

1. In cyclic photophosphorylation, excited electrons flow from photosystem I, through a transport system, then back to the photosystem. This pathway yields ATP only.

2. Operation of electron transport systems causes hydrogen ions to accumulate inside the thylakoid compartment.

3. Energy inherent in the resulting concentration and electric gradients between the compartment and the stroma is tapped to form ATP, just like it is in the noncyclic pathway.

LIGHT-INDEPENDENT REACTIONS

ATP provides energy and NADPH provides hydrogen atoms and electrons for the "synthesis" part of photosynthesis.

This second set of reactions is called "light-independent" because it does not depend directly on sunlight. (The reactions can proceed as long as ATP and NADPH are available, but those molecules are normally produced during daylight, so the light-independent reactions occur mostly during the day.)

Carbon Dioxide Fixation and the Calvin-Benson Cycle

A cyclic pathway called the **Calvin-Benson cycle** (after its discoverers, Melvin Calvin and Andrew Benson) is at the heart of the light-independent reactions. The cycle occurs in the stroma of chloroplasts, and it requires these substances:

1. ATP and NADPH, typically provided by the light-dependent reactions

2. Carbon dioxide from the air around photosynthetic cells

3. Ribulose bisphosphate (RuBP), a five-carbon sugar

4. Enzymes that catalyze each reaction step

At certain steps in the pathway, carbon is "captured" for the reactions, a sugar phosphate forms, and RuBP (needed to capture the carbon) is regenerated. Let's follow the steps by which six carbon dioxide molecules enter the cycle and are used in forming one sugar phosphate molecule. This sequence is shown in Figure 8.7.

First, carbon dioxide is attached to RuBP to form a six-carbon intermediate. The intermediate is highly unstable and is broken apart at once into two molecules of phosphoglycerate, or *PGA*, which is a three-carbon compound. This initial sequence is called **carbon dioxide fixation**—that is, the carbon present in a gaseous molecule (carbon dioxide) has become affixed to an organic compound.

Next, PGA receives a phosphate group (from ATP). The resulting intermediate receives H^+ and electrons from NADPH to form a three-carbon compound called phosphoglyceraldehyde, or *PGAL*.

For every six carbon dioxide molecules fixed at the start of the cycle, twelve PGAL are produced. Ten of the twelve now undergo complex rearrangements into new RuBP molecules—which can be used to fix more carbon dioxide. What happens to the remaining two PGAL molecules? They are combined and rearranged into an intermediate such as glucose-6-phosphate. (The name simply means a phosphate group is attached to carbon 6 of the glucose molecule; Figure 7.4). *The energy of sunlight, which excited electrons in the first stage of photosynthesis, is now stored as chemical energy in an organic compound.*

The Calvin-Benson cycle yields enough RuBP to replace the six used in carbon dioxide fixation. The ADP, $NADP^+$, and phosphate leftovers are sent back to the light-dependent reaction sites, where they are converted once more to NADPH and ATP. The sugar phosphate formed in the cycle can now serve as a building block for the plant's main carbohydrates, including sucrose, starch, and cellulose. Synthesis of those compounds by different metabolic pathways marks the conclusion of the light-independent reactions.

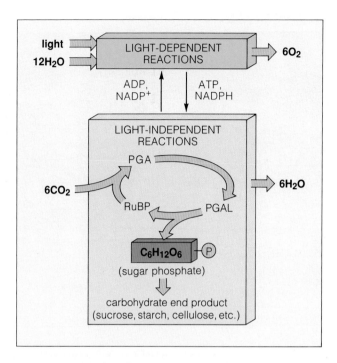

Figure 8.8 Summary of the main reactants, intermediates, and products of photosynthesis corresponding to the equation:

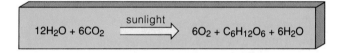

$$12H_2O + 6CO_2 \xrightarrow{\text{sunlight}} 6O_2 + C_6H_{12}O_6 + 6H_2O$$

Summary of the Light-Independent Reactions

Figure 8.8 relates the light-independent reactions to the overall events of photosynthesis. The key steps of these reactions can be summarized as follows:

1. Carbon dioxide is fixed to RuBP, making an unstable intermediate that is broken apart into two three-carbon PGA molecules.

2. PGA is phosphorylated (made more reactive) by ATP, and it receives H^+ and electrons from NADPH. The result is PGAL.

3. Through complex reactions, PGAL is rearranged into new RuBP molecules and into sugar phosphate.

4. It takes six turns of the Calvin-Benson cycle to produce one sugar phosphate (only one of every six PGAL molecules produced in the cycle is funneled into carbohydrate synthesis).

5. The sugar phosphates are intermediates in the light-independent reactions; they can enter pathways by which many different carbohydrate end products form.

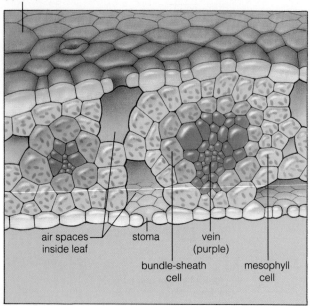

upper epidermis of leaf

air spaces inside leaf

stoma

vein (purple)

bundle-sheath cell

mesophyll cell

Figure 8.9 Internal structure of a leaf from corn (*Zea mays*), a typical C4 plant. Notice how the large, chloroplast-stuffed bundle-sheath cells surround the veins and are in turn surrounded by mesophyll cells.

How Autotrophs Use Intermediates and Products of Photosynthesis

So here we are, with sugar phosphates formed in a microscopic speck of an organelle. Visualize millions of such specks in, say, a corn plant. Where do the sugar phosphates go from here? Although a fraction of them are used at once as fuel to provide energy for cellular work, almost all are funneled into the synthesis of sucrose and starch.

Of all the carbohydrates produced by photosynthesis, sucrose is the most easily transportable. Conducting tissues carry it from sites in the leaf to living cells in all parts of the corn plant body. Starch is the main storage form of carbohydrate in the leaves, stems, and roots.

What happens in other plants? Some of the sucrose produced in leaves of potato plants is converted and stored as starch in underground stem regions called tubers (the "potatoes"). In sugar beets, onions, and sugarcane, sucrose itself is the main storage form.

Photosynthetic autotrophs also use intermediates and products of photosynthesis when they assemble lipids and amino acids. Indeed, some green algae use more than ninety percent of the carbon fixed when they construct proteins and lipids. These plants have a brief life cycle, and they live in places where sunlight and water are plentiful. They put most photosynthetic products into growth and reproduction instead of diverting them to storage forms.

C4 Plants

By now, you may be thinking that photosynthesis always proceeds at the same pace in all plants. This is not the case. The rate of photosynthesis depends critically on environmental conditions, including temperature, sunlight intensity, and water availability.

Hot, dry conditions put many leafy plants under stress. Leaves have a waxy surface covering that retards moisture loss, and any water that does escape from the leaf does so primarily through tiny passages (called stomata) through the surface layers. On hot, dry days, most passages close and water is conserved. But when they are closed, not as much carbon dioxide can diffuse into the leaf, and oxygen (a by-product of photosynthesis) builds up inside. In other words, *hot, dry conditions promote a high O2-to-CO2 ratio inside the leaf*. The stage is set for a wasteful process called photorespiration.

In **photorespiration**, oxygen instead of carbon dioxide becomes affixed to the RuBP on which the Calvin-Benson cycle turns. (This is not the same thing as aerobic respiration, an entirely different metabolic pathway.) It so happens that either dissolved oxygen *or* carbon dioxide can serve as a substrate of the enzyme that attaches carbon dioxide to RuBP. These substances compete for the enzyme's active site. When there is far more oxygen than carbon dioxide in the leaf, photorespiration wins out:

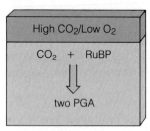

High CO_2/Low O_2

CO_2 + RuBP

two PGA

Calvin-Benson cycle predominates

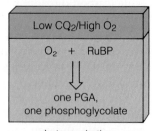

Low CO_2/High O_2

O_2 + RuBP

one PGA, one phosphoglycolate

photorespiration predominates

The formation of sugar phosphates depends on PGA. Photorespiration produces only one PGA (and a two-carbon compound called phosphoglycolate, which is rapidly broken down into carbon dioxide and water). Many plants lose as much as fifty percent of the fixed carbon through photorespiration—hence a great deal of their capacity for growth.

Not all plants are hindered by high oxygen levels. Corn, sugarcane, and crabgrass are among the so-called **C4 plants**, which have a carbon-fixing system that *precedes* the Calvin-Benson cycle. As Figure 8.9 shows, the leaves of C4 plants have two kinds of photosynthetic cells:

mesophyll cells	*preliminary fixation and rapid transport of carbon dioxide to bundle-sheath cells*
bundle-sheath cells	*carbon dioxide fixation through the Calvin-Benson cycle*

In the mesophyll cells, carbon dioxide is attached to phosphoenol pyruvate (PEP), in this way forming oxaloacetate. This first intermediate of carbon dioxide fixation has four carbon atoms, hence the name "C4 plant." (In "C3 plants," the first intermediate is three-carbon PGA.)

The oxaloacetate is transferred to an adjacent bundle-sheath cell, where the carbon dioxide is released and picked up by the Calvin-Benson cycle (Figure 8.10). Thus carbon dioxide is fixed and then released deeper in the leaf, where it is fixed again. In this way, C4 plants increase the carbon dioxide concentration and produce a higher CO_2-to-O_2 ratio in cells where the Calvin-Benson cycle operates—even under hot, dry conditions. *The higher ratio prevents photorespiration but permits photosynthesis.*

The elaborate carbon fixation in C4 plants comes at a higher cost—that is, it requires more ATP molecules than C3 plants use to fix carbon. But C4 plants end up building just as many carbohydrates with *less* water loss under hot, dry conditions. That is why a lawn of Kentucky bluegrass (a C3 plant) that thrives during cool spring weather can be overwhelmed in the hot summer by crabgrass.

Kentucky bluegrass, wheat, oats, rice, and other species have the advantage in regions where temperatures drop below 25°C because they are less sensitive to cold. But the regions of North America with the highest temperatures during the growing season have the highest percent of C4 plants—for example, eighty percent of all native species in Florida compared to zero percent in Manitoba, Canada, near Hudson Bay.

CHEMOSYNTHESIS

Photosynthesis so dominates the energy-trapping pathways that sometimes it is easy to overlook other, less common routes. The chemosynthetic autotrophs obtain energy not from sunlight but rather from the oxidation of inorganic substances (ammonium ions, iron or sulfur compounds, and so on). That is their source of energy for building organic compounds.

For example, some bacteria present in soil use ammonia (NH_3) molecules as an energy source, stripping them of protons and electrons. Nitrite ions (NO_2^-) and nitrate ions (NO_3^-) are the remnants of their activities. Compared with ammonium ions, nitrite and nitrate ions are readily washed out of the soil, so the action of these so-called nitrifying bacteria can lower the soil fertility. Farmers sometimes add chemicals to the soil that inhibit bacteria.

We will return later to the environmental effects of chemosynthetic autotrophs. Here, in this unit, we will turn next to pathways by which carbon-containing products of autotrophs can be used as energy sources for cellular work.

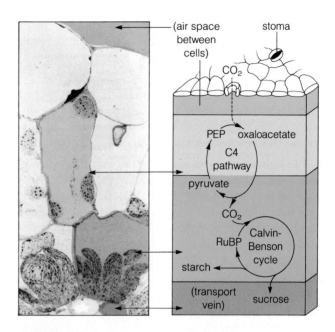

Figure 8.10 Organization of a carbon-fixing system that *precedes* the Calvin-Benson cycle in C4 plants. This system is called the C4 pathway. In the micrograph, a mesophyll cell is shaded light green; a bundle-sheath cell is shaded darker green.

SUMMARY

1. Carbohydrates and other organic compounds are used for cell architecture and as energy stores that can be tapped to drive cell activities.

2. Autotrophs ("self-feeders") use carbon dioxide from the environment as the carbon source for building organic compounds. *Photosynthetic* autotrophs (plants, for example) use sunlight as an energy source. *Chemosynthetic* autotrophs (some bacteria) use energy from chemical reactions involving sulfur or other inorganic substances.

3. Heterotrophs (animals, fungi, many protistans, and most bacteria) obtain carbon and energy from organic compounds already built by autotrophs. They do this by feeding on autotrophs, each other, and organic wastes.

4. Photosynthesis is the main biosynthetic pathway by which carbon and energy enter the web of life. It consists of two sets of reactions:

   a. *Light-dependent reactions* that produce ATP and NADPH (or ATP alone).

   b. *Light-independent reactions* that produce sugar phosphates (such as glucose-6-phosphate), which can be used in building complex carbohydrates such as sucrose, starch, and cellulose.

5. In the chloroplasts of land plants, the light-dependent reactions occur at a thylakoid membrane, which is sur-

rounded by a semifluid matrix called the stroma. Embedded in the membrane are photosystems (which contain light-absorbing pigment molecules such as chlorophyll) and electron transport systems.

6. The thylakoid membrane is organized into stacked disks (grana) and flattened channels that form a continuous compartment, which serves as a hydrogen ion reservoir. This reservoir is tapped for ATP formation.

7. The light-dependent reactions can be cyclic or noncyclic. In cyclic photophosphorylation, excited electrons from photosystem I flow through a transport system, then back to the photosystem. ATP forms; it provides energy to build carbohydrates during the light-independent reactions.

8. In noncyclic photophosphorylation, there is a one-way flow of electrons from photosystem II, through a transport system, to photosystem I, and through a second transport system. The electrons end up in NADPH. Electrons lost from photosystem II are replaced through photolysis: the splitting of water molecules into oxygen, protons (that is, hydrogen ions), and electrons.

9. The noncyclic pathway yields ATP (which provides energy) and NADPH (which provides hydrogen ions and electrons) for building carbohydrates.

10. According to the chemiosmotic theory, ATP forms as follows during the light-dependent reactions:

    a. Hydrogen ions accumulate in the thylakoid compartment (as a result of photolysis and electron transfers through transport systems). Thus H^+ concentration and electric gradients are produced between the compartment and the stroma.

    b. Hydrogen ions flow down the gradient, through channel proteins that span the thylakoid membrane. Energy inherent in the flow drives the coupling of ADP and inorganic phosphate into ATP.

11. The light-independent reactions take place in the stroma. The key pathway is the Calvin-Benson cycle, by which carbon dioxide is fixed, sugar phosphate forms, and RuBP is regenerated.

12. In carbon dioxide fixation, RuBP combines with carbon dioxide from the air, then splits into two PGA molecules. The PGA is phosphorylated by ATP. The resulting intermediate receives H^+ and electrons from NADPH to form PGAL. Five of every six PGAL molecules are used to regenerate RuBP for the cycle; the remaining PGAL can be used to form a sugar phosphate.

13. Some plants living in hot climates have an additional carbon-fixing system, the C4 pathway. This pathway helps circumvent photorespiration, in which oxygen instead of carbon dioxide becomes affixed to RuBP. (Otherwise, photorespiration undoes much of what photosynthesis accomplishes.)

Review Questions

1. Define the difference between autotrophs and heterotrophs, and give examples of each. In what category do photosynthesizers fall? *109*

2. Summarize the photosynthesis reactions in words, then as an expanded equation. Distinguish between the light-dependent and the light-independent stage of these reactions. Be sure to show the electron and hydrogen "bridges" between the two stages. *109*

3. Oxygen is a product of photolysis. What *is* photolysis, and does it occur during the first or second stage of photosynthesis? *114*

4. Describe where the light-dependent reactions occur in the chloroplast, and name the molecules formed there. Do the same for the light-independent reactions. *110–111*

5. A thylakoid compartment is a reservoir for which of the following substances: glucose, photosynthetic pigments, hydrogen ions, fatty acids? *110, 114*

6. Sketch the reaction steps of noncyclic photophosphorylation, showing where the excited electrons eventually end up. Do the same for the cyclic pathway. Which pathway has the greater energy yield? *113–115*

7. Is oxygen an end product of cyclic or noncyclic photophosphorylation? *114*

8. Describe the chemiosmotic theory of ATP formation in chloroplasts. Use sketches to show the proposed movements of hydrogen ions across thylakoid membranes. *114*

9. Which of the following substances are *not* required for the light-independent reactions: ATP, NADH, RuBP, carotenoids, free oxygen, carbon dioxide, enzymes? *117*

10. Suppose a plant carrying out photosynthesis were exposed to carbon dioxide molecules that contain radioactively labeled carbon atoms ($^{14}CO_2$). In which of the following compounds will the labeled carbon first appear? *117*
    a. NADPH   c. PGAL
    b. pyruvate   d. PGA

11. How many CO_2 molecules must enter the Calvin-Benson cycle to produce one sugar phosphate molecule? Why? *116–117*

12. Give examples of how different autotrophs use the intermediates and products of photosynthesis. *119*

Readings

Hinkle, P., and R. McCarty. March 1978. "How Cells Make ATP." *Scientific American* 238(23):104–123. How the chemiosmotic theory explains ATP formation in both chloroplasts and mitochondria.

Miller, K. 1982. "Three-Dimensional Structure of a Photosynthetic Membrane." *Nature* 300:5887.

Moore, P. 1981. "The Varied Ways Plants Tap the Sun." *New Scientist.* 12 February, pp. 394–397. Clear, simple introduction to the C4 plants.

Salisbury, F., and C. Ross. 1985. *Plant Physiology.* Third edition. Belmont, California: Wadsworth.

Zubay, G. 1988. *Biochemistry.* Second edition. Menlo Park, California: Addison-Wesley. Advanced but authoritative introduction to photosynthesis.

It is one of the quirks of the human mind that plants just aren't thought about very often as *living* organisms. Even vegetarians who become distressed at the thought of eating the flesh of an animal can relish the flesh of a peach. Perhaps it is understandable. Lacking autotrophic equipment of our own, we have to depend on something to produce energy for us, and if we carried a concern for the sanctity of life too far, we'd all starve to death.

Yet there is an undeniable unity among organisms that becomes evident at the biochemical level. Animals, plants, fungi, protistans, and most bacteria use organic compounds for energy and raw materials in much the same way you do. Each one uses ATP—the prime energy carrier for life in all of its forms. There is, in short, a remarkable similarity in the ebb and flow of energy through the individual—hence through the biosphere. We will return to this idea at the chapter's end.

9

ENERGY-RELEASING PATHWAYS

OVERVIEW OF THE MAIN ENERGY-RELEASING PATHWAYS

We can use wood as building material or as a source of fuel that can be burned quickly or stored. Cells have a similar option when it comes to using carbohydrates, lipids, and proteins for raw materials or energy.

For example, our cells need a steady supply of organic compounds simply to remain structurally intact and to carry out energy-requiring tasks. When the body takes in more carbohydrates than its cells are calling for, the excess can be stored as glycogen (the animal body's equivalent of starch), most notably in liver and muscle cells. Excess lipids are stored as glistening droplets in the cells of adipose tissue (fat). Between meals, when demands for raw materials and energy exceed dietary intake, the body draws on its stores of organic compounds.

Of the foods we eat, carbohydrates—glucose molecules especially—are the main source of energy. (Our cells use free glucose for as long as it is available, then they break down stored carbohydrates to release glucose subunits. They degrade lipids next, then proteins as a last resort.) In the main energy-releasing pathways, *chemical energy is released from glucose and other carbohydrates through phosphorylation and oxidation-reduction reactions.* Reactions of this sort were described in earlier chapters, but let's take a moment here to consider their roles in glucose breakdown.

Phosphorylation: You have probably heard the expression, "It takes money to make money." It means investing some money in an activity that will produce more money in return. Cells, too, have to make an up-front investment (of two ATP) when they break apart glucose.

a

b

c

Figure 9.1 Three energy-releasing pathways, in which electrons stripped from sugar molecules are transferred in ways that produce ATP. Each pathway is characterized by the final acceptor of the "spent" electrons.

(**a**) Bacterial residents of sulfur hot springs use *anaerobic electron transport*, in which an inorganic compound in the environment (in this case, sulfate) is the final electron acceptor. (**b**) Yeasts that live in the dustlike coating on these grapes use *fermentation*, in which the partially dismantled sugar molecule itself takes back the electrons. (**c**) Gorillas and most other organisms (including the plant upon which the gorilla is dining) use *aerobic respiration*, in which oxygen is the final electron acceptor.

Glucose simply will not give up stored energy unless it is phosphorylated (each ATP must transfer a phosphate group to it). On that original investment, cells will show a *net* return of two or thirty-six (and sometimes more) ATP for each glucose molecule degraded.

Oxidation-Reduction: Now, there is quite a difference between two and thirty-six ATP as the net energy harvest from a single molecule of glucose. The outcome depends on how completely glucose is degraded and on the use made of its hydrogen atoms—each with one proton (H^+) and one electron. The H^+ and electrons are used in oxidation-reduction reactions—that is, in electron transfers from one molecule to another. Refer to Figure 7.14, which likens electron energy levels to steps of a staircase. Obviously, there can be more transfers when the electrons are delivered to the top step for glucose breakdown and when the final electron acceptor is at a very low step! This is important to think about, for energy released at several steps of the staircase contributes to ATP formation. *When the first electron donor in a pathway is not at the top step or when the final acceptor is not at the lowest, less ATP can form.*

The main energy-releasing pathways differ in the final acceptor of electrons stripped from glucose. Figure 9.2 is an overview of the pathways, which are aerobic respiration, anaerobic electron transport, and two kinds of fermentation routes.

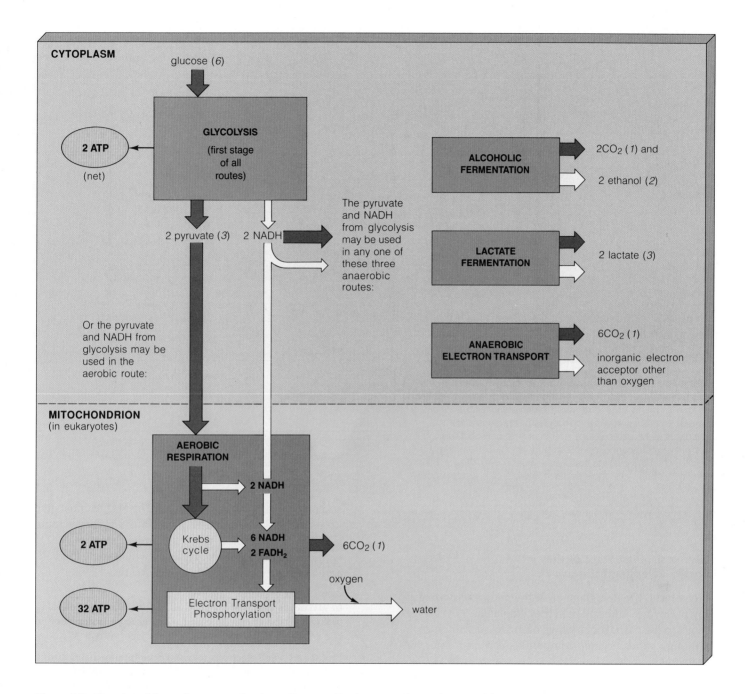

Figure 9.2 Overview of the main energy-releasing pathways, with glucose as the starting material. Depending on the organism and on environmental conditions, any one of the four routes shown might be taken.

Dark, wide arrows indicate the destination of carbon atoms derived from the glucose molecule. (Numbers in parentheses refer to the number of carbon atoms in each molecule of the compound listed.)

Yellow arrows indicate the destinations of the protons (H⁺) and electrons derived from glucose. A simple way to remember how many coenzymes (NAD⁺ or FAD) are required for the reactions is to add them up from the start—*in all cases, glycolysis*—to finish.

A simple way to remember the *net* energy yield of each route shown is to add up the ATP formed from the start (glycolysis) to finish. Neither anaerobic electron transport nor the fermentation routes yield more than two net ATP. Aerobic respiration commonly has a net energy yield of thirty-six ATP for each glucose molecule.

Figure 9.3 Typical net energy yield of aerobic respiration, the only pathway that can deliver enough ATP for strenuous activity—as long as plenty of glucose and oxygen are available. When oxygen levels become low in muscles being given a workout, glycolysis alone can produce ATP in muscle cells, but this alternative route cannot be followed for long. Muscles fatigue quickly when the glucose supplies in muscle cells (that is, their glycogen stores) are depleted; then, muscle cells lose the ability to contract.

Glycolysis: First Stage of the Energy-Releasing Pathways

Glycolysis, the first stage of all the main energy-releasing pathways, takes place in the cytoplasm. "Glycolysis" refers to the partial breakdown of sugars such as glucose, although other organic compounds can also enter the reactions.

Glucose has a backbone of six carbon atoms (Figure 4.3). During glycolysis, phosphate groups from two ATP become attached to the backbone. Then the molecule is degraded, and along the way, the H^+ and electrons stripped from intermediates are picked up by the coenzyme NAD^+ to form two NADH (page 105). The end products of glycolysis are two **pyruvate** molecules, each with a three-carbon backbone (Figure 9.2).

Energy released during glycolysis is used to form four ATP. But remember, two ATP were invested at the start of glycolysis, so the *net* yield is two ATP.

The ATP formed during glycolysis can be used in a variety of cell activities. But what does the cell do with the pyruvate and NADH? The pyruvate can travel different routes from here, depending on the type of cell and on environmental conditions. The NADH must give up its cargo of H^+ and electrons along whichever route is taken. If it did not, the cell's limited store of NAD^+ would soon be depleted, glycolysis would come to a halt, and a cellular energy crisis would follow.

In glycolysis (the first stage of all the main energy-releasing pathways), the energy released when glucose is degraded to two pyruvate molecules leads to a net yield of two ATP.

Protons (H^+) and electrons released from intermediates during glycolysis are transferred to NAD^+ to form NADH. When NADH later donates them to an acceptor molecule, NAD^+ is regenerated.

The pyruvate enters the next stage of carbohydrate breakdown. How it will be put to use depends on which pathway is taken.

Aerobic Respiration

All organisms except some microbes can use the pyruvate and NADH from glycolysis in **aerobic respiration**, an energy-releasing pathway in which oxygen is the final electron acceptor. In eukaryotes, the metabolic machinery for this pathway is housed in mitochondria (page 76).

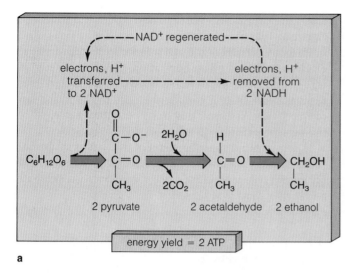
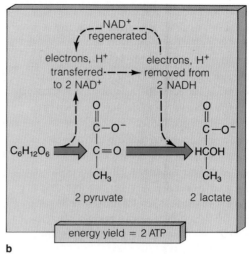

Figure 9.4 (**a**) Alcoholic fermentation and (**b**) lactate fermentation. The small energy yield is from glycolysis, the first stage of the reactions. NAD⁺ is regenerated during subsequent steps.

The two pyruvate molecules are completely dismantled in the aerobic pathway, with the carbon and oxygen atoms ending up in carbon dioxide and water molecules. (This is where the carbon dioxide exhaled from your lungs comes from.) The H⁺ and electrons released from intermediates are picked up by the NAD⁺ and another coenzyme, FAD, which donate them to an electron transport system. That system gives up the electrons to oxygen, which combines with H⁺ to form water.

Operation of the transport system leads to the formation of many ATP molecules. From glycolysis to its final stage, the aerobic pathway commonly has a net yield of thirty-six ATP (Figure 9.3).

In aerobic respiration, the pyruvate from glycolysis is completely degraded to carbon dioxide and water.

NAD⁺ and FAD accept hydrogen and electrons stripped from intermediates and donate them to an electron transport system. Oxygen is the final acceptor of those electrons.

From glycolysis (in the cytoplasm) to the final reactions (in the mitochondrion), this pathway commonly yields thirty-six ATP for every glucose molecule.

Incidentally, even some "aerobic" cells do not use oxygen all the time. Your muscle cells use the aerobic pathway when plenty of oxygen is available, but they can use other final electron acceptors when it is not (as during strenuous exercise). Denitrifying bacteria that live in soil normally use oxygen, but when the soil is saturated with standing water, they switch to nitrite (NO_2^-) as the final acceptor. These and other switch-hitting cells are "facultative" aerobes.

In contrast, "anaerobes" never use oxygen as the final electron acceptor. Some are indifferent to the presence or absence of free oxygen, but others are "strict anaerobes" that die on exposure to it. Only some microbes are anaerobic, and they degrade glucose by means of anaerobic electron transport or fermentation pathways.

Anaerobic Electron Transport

In **anaerobic electron transport**, NADH donates electrons stripped from glucose to a transport system bound in the microbe's plasma membrane, and an inorganic compound then serves as the final electron acceptor. For example, sulfate-reducing bacteria are strict anaerobes that live in soils, aquatic habitats, and the intestines. They transfer electrons to sulfate (SO_4^-) in their environment. Although the reactions regenerate NAD⁺, they do not add any more ATP to the small net yield from glycolysis.

Fermentation Pathways

In other anaerobic routes called **fermentation** pathways, the final electron acceptor is actually an intermediate or product of the glucose molecule being degraded. An "outside" acceptor (such as oxygen or sulfate) is not needed at all.

In *alcoholic fermentation*, pyruvate from glycolysis is broken down to acetaldehyde, which accepts electrons from NADH and thereby becomes ethanol (Figure 9.4a). Yeasts (single-celled fungi) use this pathway. Yeast dough "rises" when carbon dioxide, the gaseous product of the reactions, accumulates in the dough. The alcohol in beer, distilled spirits, and wine results from the fermentation activities of certain yeasts.

In *lactate fermentation*, the pyruvate is converted into lactate, a three-carbon compound (Figure 9.4b). Sometimes this compound is called lactic acid, but it is more accurate to refer to the ionized form (lactate), which predominates under cellular conditions.

One group of bacteria produces lactate exclusively as the fermentation product; milk or cream turned sour is a sign of their activity. The reactions also can occur in some animal cells that normally rely on aerobic respiration. For example, sometimes skeletal muscle cells contract so vigorously that they exceed the body's capacity to deliver oxygen to them. Under those conditions, they switch to lactate fermentation.

Keep in mind that glucose is not completely degraded by either anaerobic electron transport or fermentation, so considerable energy still remains in the products. No more ATP is produced, beyond the two molecules from glycolysis; the final steps serve only to regenerate NAD+. Also keep in mind that the low energy yield of anaerobic metabolism is quite enough for some microbes, and it even can help carry some otherwise aerobic cells through times of stress. But it is not enough to sustain the activities of large, multicelled organisms—this being one of the reasons why you never will come across an anaerobic elephant.

Anaerobic electron transport or fermentation has a net yield of two ATP molecules (from glycolysis). NAD+ is regenerated during the reactions.

A CLOSER LOOK AT AEROBIC RESPIRATION

In later chapters, we will be looking at how animals and plants function. Digestion and muscle action in humans and other animals, sugar transport through the plant body—all such activities depend absolutely on the high ATP yield of aerobic respiration. The preceding overview is enough to get you through those chapters, but the following details about the aerobic pathway will lead to better understanding of how the body can mobilize its energy resources.

As we have seen, glycolysis is the first stage of the aerobic pathway. Two more stages follow. Pyruvate is completely degraded and ATP is produced in the second stage—the **Krebs cycle** and a few reactions preceding it.

Also, the coenzymes NAD+ and FAD pick up H+ and electrons that are released during the Krebs cycle, and in this way NADH and FADH$_2$ form. In the third stage, **electron transport phosphorylation**, the coenzymes give up H+ and electrons to transport systems. Operation of the transport system commonly leads to the formation of thirty-two ATP, and this brings the total net yield of the aerobic pathway to thirty-six ATP.

Keep in mind that the steps of this pathway do not proceed all by themselves. *Enzymes catalyze each step, and the intermediate produced at a given step serves as a substrate for the next enzyme in the series.*

First Stage: Glycolysis

The initial steps of glycolysis are *energy-requiring*; they do not proceed without an energy input from ATP. After glucose receives a phosphate group from ATP, the molecule undergoes internal rearrangements, then receives another phosphate group from another ATP (Figure 9.5). The resulting intermediate is split into two molecules, one of which is phosphoglyceraldehyde, or **PGAL**. Because each intermediate can be converted easily to the other, we can say that two PGAL have formed. PGAL formation marks the beginning of the *energy-releasing* steps of glycolysis.

Each PGAL now becomes the substrate of an enzyme that makes it give up two hydrogen atoms. Both electrons and one H+ of those atoms become attached to NAD+, forming NADH. (The other H+ remains in solution.) NADH leaves the reaction site. The intermediate remaining at the site now combines with inorganic phosphate to form a rather unstable molecule—which readily gives up a phosphate group to ADP.

Thus ATP is formed by the direct, enzyme-mediated transfer of a phosphate group from a substrate to ADP. This mechanism is called **substrate-level phosphorylation**. No electron transport is required, and neither is oxygen.

Now the intermediate gives up H+ and OH− (which can combine to form a water molecule). The result is an unstable intermediate called PEP. When PEP breaks down, the energy released is enough to drive the transfer of one of its phosphate groups to ADP, forming ATP. After this second substrate-level phosphorylation, the molecule remaining is pyruvate (Figure 9.5).

In all, the initial breakdown of glucose in glycolysis leads to the formation of four ATP, two NADH, and two pyruvate molecules. (Two ATP were needed to start the reactions, however, so there is a *net* yield of only two ATP.)

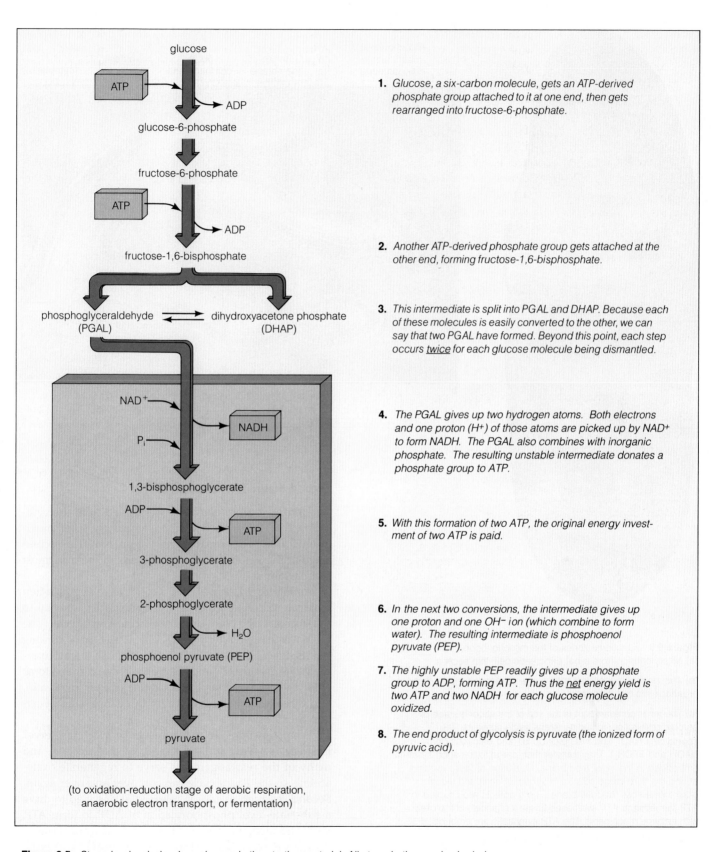

1. Glucose, a six-carbon molecule, gets an ATP-derived phosphate group attached to it at one end, then gets rearranged into fructose-6-phosphate.

2. Another ATP-derived phosphate group gets attached at the other end, forming fructose-1,6-bisphosphate.

3. This intermediate is split into PGAL and DHAP. Because each of these molecules is easily converted to the other, we can say that two PGAL have formed. Beyond this point, each step occurs _twice_ for each glucose molecule being dismantled.

4. The PGAL gives up two hydrogen atoms. Both electrons and one proton (H+) of those atoms are picked up by NAD+ to form NADH. The PGAL also combines with inorganic phosphate. The resulting unstable intermediate donates a phosphate group to ATP.

5. With this formation of two ATP, the original energy investment of two ATP is paid.

6. In the next two conversions, the intermediate gives up one proton and one OH− ion (which combine to form water). The resulting intermediate is phosphoenol pyruvate (PEP).

7. The highly unstable PEP readily gives up a phosphate group to ADP, forming ATP. Thus the _net_ energy yield is two ATP and two NADH for each glucose molecule oxidized.

8. The end product of glycolysis is pyruvate (the ionized form of pyruvic acid).

Figure 9.5 Steps in glycolysis when glucose is the starting material. All steps in the purple-shaded area occur _twice_ for each glucose molecule being degraded.

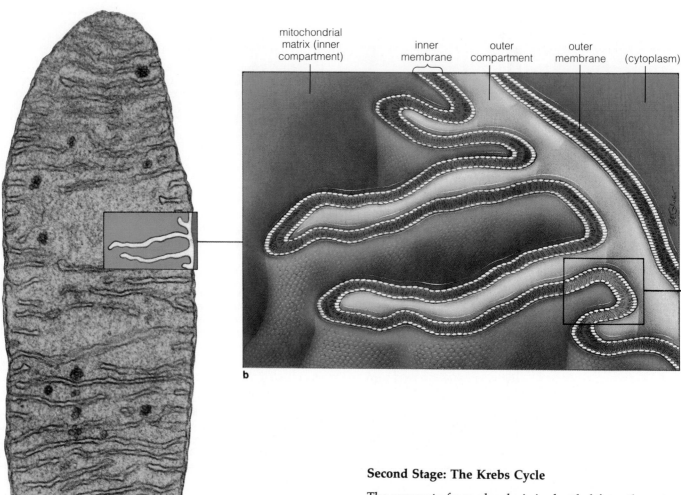

mitochondrial matrix (inner compartment)   inner membrane   outer compartment   outer membrane   (cytoplasm)

b

Figure 9.6 Functional zones of the mitochondrion.
(a) Mitochondrion from a bat pancreatic cell, thin section.

(b) The inner mitochondrial membrane divides the inside of this organelle into two compartments.

(c) The inner compartment is the site of the second stage of aerobic respiration, including Krebs cycle activities. The coenzymes NAD^+ and FAD pick up H^+ and electrons (forming NADH and $FADH_2$). They transfer their cargo to transport systems embedded in the inner membrane. Operation of these systems during the third stage sets up H^+ concentration and electric gradients across the membrane. The gradients are coupled to ATP formation at ATP synthases (protein channels complexed with enzyme machinery). The ATP synthases are embedded elsewhere in the same inner membrane.

Second Stage: The Krebs Cycle

The pyruvate from glycolysis is shuttled from the cytoplasm into the mitochondrion, where an inner membrane forms two compartments (Figure 9.6). The second stage of the aerobic pathway proceeds in the innermost compartment.

As Figure 9.7 shows, the second stage begins when pyruvate is converted to a form that can be attached to oxaloacetate, the point of entry into the Krebs cycle. (The cycle was named after Hans Krebs, who began working out its details in the 1930s.) In this cyclic pathway, intermediates give up H^+ and electrons. Three carbon atoms enter the reactions (as the backbone of each pyruvate), and three carbon atoms leave (in three molecules of carbon dioxide).

What functions do these metabolic jugglings serve? First, cells have a limited supply of oxaloacetate, and many of the rearrangements serve to regenerate oxaloacetate so the Krebs cycle can run over and over again. Second, enough energy is released at one step to drive a substrate-level phosphorylation of ADP to form ATP. Last, and most important, the H^+ and electrons released during the reactions are transferred to NAD^+ and FAD.

It takes two preparatory reaction sequences and two turns of the Krebs cycle to dismantle the two pyruvate molecules. Pyruvate breakdown adds only two more

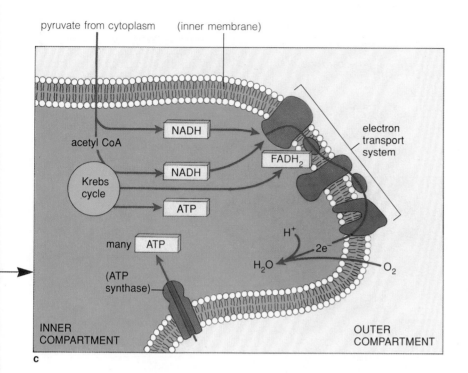

c

ATP molecules to the small net yield from glycolysis. *But it also adds many coenzymes to the number that can be used to transfer electrons to the third stage of the aerobic pathway:*

Glycolysis:		2 NADH
Pyruvate conversion preceding Krebs cycle:		2 NADH
Krebs cycle:	2 FADH$_2$	6 NADH
Total electron carriers sent to third stage of aerobic pathway:	2 FADH$_2$ +	10 NADH

Third Stage: Electron Transport Phosphorylation

We turn now to one of the most impressive of all ATP-yielding mechanisms. Consider that an adult male uses up about 2,800 kilocalories of energy during a normal day's activities. It takes about 150 kilograms of ATP daily to provide that much energy. However, there is only about 0.005 kilogram of ATP in the entire body at any one time! How does the body make up the difference? Its cells phosphorylate ADP over and over again, many thousands of times a day (page 106). This is an ATP/ADP cycle on a grand scale, and electron transport phosphorylation is the only mechanism for handling such an enormous turnover.

The third stage of the aerobic pathway is carried out by enzymes and other proteins embedded in the inner membrane of the mitochondrion (Figure 9.7). The proteins are organized into two types of reaction sites that have the following functions:

Reaction Site	Function
electron transport system	*establishes concentration and electric gradients across the membrane*
ATP synthase system	*phosphorylates ADP to form ATP*

The reactions begin in the inner compartment, when NADH and FADH$_2$ give up H$^+$ and electrons to an electron transport system (Figure 9.8). At some transfers through this system, the electrons are accepted and passed on, but the H$^+$ is left behind—in the outer compartment:

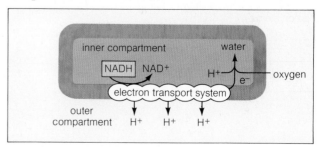

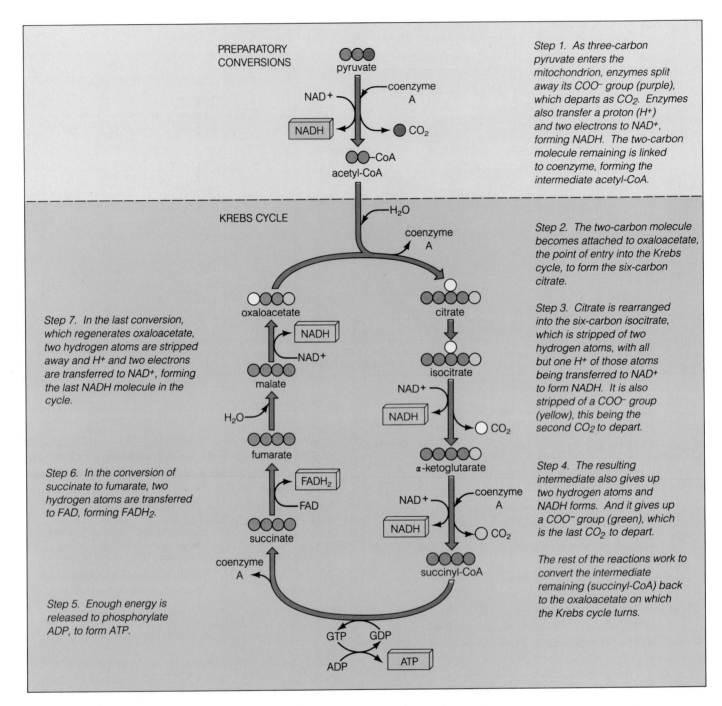

PREPARATORY CONVERSIONS

pyruvate

NAD+ ← coenzyme A

NADH

CO₂

CoA
acetyl-CoA

Step 1. As three-carbon pyruvate enters the mitochondrion, enzymes split away its COO⁻ group (purple), which departs as CO₂. Enzymes also transfer a proton (H⁺) and two electrons to NAD⁺, forming NADH. The two-carbon molecule remaining is linked to coenzyme, forming the intermediate acetyl-CoA.

KREBS CYCLE

H₂O

coenzyme A

Step 2. The two-carbon molecule becomes attached to oxaloacetate, the point of entry into the Krebs cycle, to form the six-carbon citrate.

oxaloacetate

citrate

Step 7. In the last conversion, which regenerates oxaloacetate, two hydrogen atoms are stripped away and H⁺ and two electrons are transferred to NAD⁺, forming the last NADH molecule in the cycle.

NADH
NAD⁺
malate

H₂O

fumarate

isocitrate

NAD⁺
NADH
CO₂

α-ketoglutarate

Step 3. Citrate is rearranged into the six-carbon isocitrate, which is stripped of two hydrogen atoms, with all but one H⁺ of those atoms being transferred to NAD⁺ to form NADH. It is also stripped of a COO⁻ group (yellow), this being the second CO₂ to depart.

Step 6. In the conversion of succinate to fumarate, two hydrogen atoms are transferred to FAD, forming FADH₂.

FADH₂
FAD

succinate

coenzyme A

NAD⁺ ← coenzyme A
NADH
CO₂

succinyl-CoA

Step 4. The resulting intermediate also gives up two hydrogen atoms and NADH forms. And it gives up a COO⁻ group (green), which is the last CO₂ to depart.

The rest of the reactions work to convert the intermediate remaining (succinyl-CoA) back to the oxaloacetate on which the Krebs cycle turns.

Step 5. Enough energy is released to phosphorylate ADP, to form ATP.

GTP GDP

ADP ATP

Figure 9.7 Second stage of the aerobic pathway: the Krebs cycle and the preparatory conversions immediately preceding it. This diagram shows what happens to one pyruvate molecule entering the reactions. Because two pyruvate molecules are formed during the breakdown of every glucose molecule in glycolysis, the steps shown occur twice (once for each pyruvate molecule). The circles represent carbon atoms.

The positively charged hydrogen ions accumulate in the outer compartment, and this sets up concentration and electric gradients across the membrane. The ions follow the gradients and flow inward, through the ATP synthase system that serves as a channel across the membrane. Energy associated with the flow drives ATP formation:

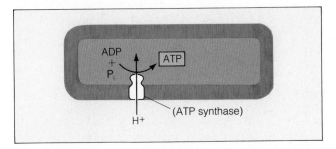

Do these events sound familiar? They should: ATP is formed in much the same way in chloroplasts of photosynthetic cells (page 114). Again, the idea that concentration and electric gradients across a membrane drive the phosphorylation of ADP is called the chemiosmotic theory of ATP formation.

Electron transport phosphorylation in mitochondria involves two events: (1) the transfer of electrons and H⁺ through a membrane-bound transport system, and (2) actual ATP formation.

Glucose Energy Yield

The net energy yield from the three stages of aerobic respiration is commonly thirty-six or thirty-eight ATP for every glucose molecule degraded (Figure 9.9). The number varies because each NADH produced in the cytoplasm (by glycolysis) can yield two *or* three ATP, for the following reasons.

Each NADH formed in mitochondria (during the second-stage reactions) delivers enough energy to form three ATP molecules. FADH₂ does not: it donates electrons at a lower step in the electron transport sequence, so it delivers only enough energy to form two ATP (Figure 9.8).

It so happens that cytoplasmic NADH only delivers electrons *to* the mitochondrion, not into them. In liver, heart, and kidney cells, a shuttle built into the outer membrane accepts the electrons, then gives them up to an NAD⁺ molecule inside. Because the resulting NADH molecule delivers electrons to the top step of the transport system, its energy can produce three ATP in these cells for an overall energy harvest of thirty-eight ATP molecules.

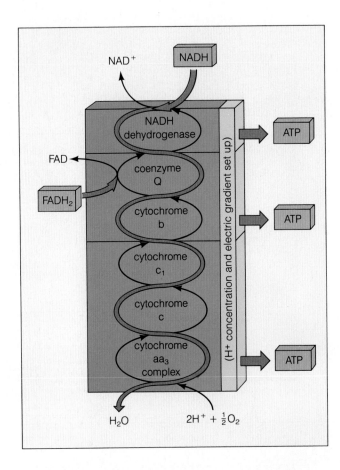

Figure 9.8 Some components of the system used in electron transport phosphorylation. The system runs on electron energy delivered to it by NADH (from glycolysis and the Krebs cycle) and FADH₂ (from the Krebs cycle).

Energy associated with three of the transfers through this electron transport system drives the formation of ATP. The NADH can deliver enough energy to form all three ATP molecules. The FADH₂ delivers enough energy to form two ATP only.

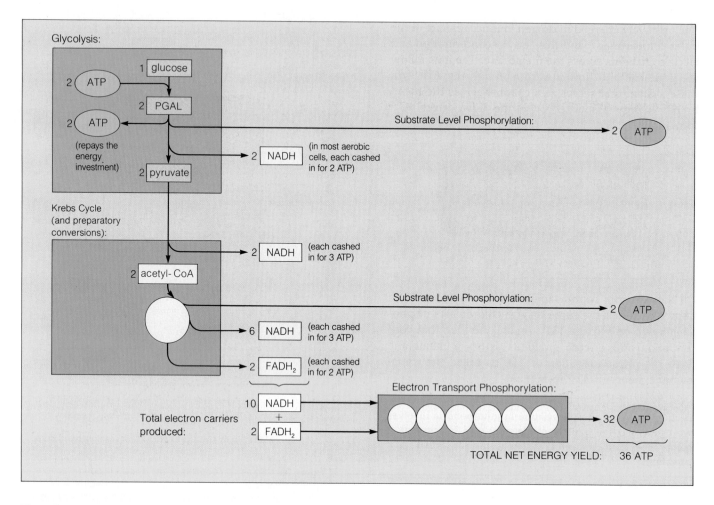

Figure 9.9 Summary of the energy harvest from one glucose molecule sent through the aerobic respiration pathway. Actual ATP yields vary, depending on cellular conditions and on the mechanism used to transfer energy from cytoplasmic NADH into the mitochondrion. Commonly, the overall energy harvest is thirty-six ATP (see page 131).

One NADH produced within the mitochondrion can yield three ATP. One cytoplasmic NADH (from glycolysis) may yield two or three ATP, depending on what kind of shuttle mechanism transfers its energy cargo to a membrane-bound transport system in the mitochondrion (see text).

More commonly, as in skeletal muscle and brain cells, a different shuttle donates electrons to FAD inside the mitochondrion. Because the $FADH_2$ so formed delivers electrons one step down in the transport system, only two ATP can form for an overall energy harvest of thirty-six ATP.

When glucose is broken down completely to carbon dioxide and water, about 686 kilocalories of energy are released. Of that, about 7.5 kilocalories become conserved for further use in each of the thirty-six ATP molecules typically formed. Thus the energy-releasing efficiency of this pathway is (36)(7.5)/(686), or thirty-nine percent. When the net yield is only two ATP (as in fermentation), then the energy-releasing efficiency of the pathway is only (2)(7.5)/(686), or two percent.

FUEL OR BUILDING BLOCKS? CONTROL OF CARBOHYDRATE METABOLISM

Glycolysis and the Krebs cycle serve two functions in metabolism. As we have seen, they can dismantle organic compounds for energy that can drive ATP formation. But the intermediates formed during the reactions also can be diverted to the synthesis of lipids, carbohydrates, proteins, and nucleic acids (refer to Figure 9.10).

What determines whether a molecule will be degraded or pulled out of energy-releasing pathways in an intermediate form that can be used in biosynthesis reactions? The answer is that the cell adjusts the activities of key enzymes according to metabolic requirements.

The main controls are over the enzymes governing steps in which so much energy is released that the reaction, once carried out, cannot be reversed.

Phosphofructokinase is an example of a metabolic control point. This enzyme adds a second phosphate group to fructose-6-phosphate during glycolysis (Figure 9.5). Once that happens, the molecule is "committed" to splitting apart. If the enzyme is inhibited, then the molecule is not degraded for its energy.

When cells have enough energy, ATP concentrations are high. At high levels, ATP actually inhibits phosphofructokinase (page 104) and thereby slows down glycolysis. When cells require energy, they draw upon their ATP stores and the ATP concentrations decrease accordingly. Then the enzyme activity increases—and glucose and other molecules can be broken apart rapidly (to yield more ATP). By coordinating the activity of many different enzymes, cells constantly adjust to the supplies of currently available resources and to the demands being made on those supplies.

Controls over enzymes that catalyze key steps in glycolysis and the Krebs cycle govern whether molecules are degraded as an ATP-generating energy source or converted to intermediate forms for use in biosynthesis.

PERSPECTIVE

In this unit, you have read about pathways by which cells trap, store, and then release energy to drive their activities. Let's conclude the unit with a glimpse of how the main pathways—photosynthesis and aerobic respiration—came to be linked on a grand scale. Although later chapters describe the evolution of these pathways in detail, this overview may give you a sense of how central the links between them have become for life as we now know it.

When life began, free oxygen was absent from the earth's atmosphere. Most likely the early cells produced ATP by pathways similar to glycolysis, and given the anaerobic conditions, fermentation routes must have predominated. Which materials were they consuming as an energy source? They could have used organic compounds already formed by physical processes. (Lightning, volcanic action, and so on can cause carbon, hydrogen, and oxygen to combine into organic compounds.)

Between about 3.5 billion and 600 million years ago, metabolic pathways apparently underwent modification. Perhaps through mutations, some of the enzyme systems used to *degrade* organic compounds were modified and extended, so that the reactions could also run in reverse. In other words, the modified systems could

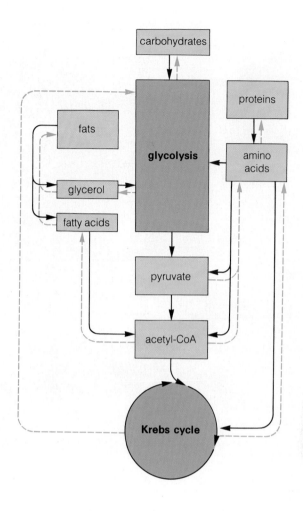

Figure 9.10 Where various substances flow into the glycolytic pathway and the Krebs cycle (black arrows). Different intermediates of glycolysis and the Krebs cycle can be pulled out of the energy-releasing pathways and used in biosynthesis (blue arrows).

build carbon compounds from carbon dioxide and water. Driving the reactions in an uphill direction would have required a steady supply of energy. That energy came from the sun; early in the history of life, photosynthesis emerged.

Photosynthesis turned out to be a profound force in evolution. Oxygen, a by-product of the reactions, began to accumulate in the atmosphere. At least some of the ancient photosynthetic cells were opportunistic about the increasing oxygen levels. In some way, mutations had produced cells that could put oxygen to use as an electron acceptor for energy-releasing reactions. Once oxygen became abundant, some of the aerobic cells apparently evolved in ways that allowed them to abandon photosynthesis entirely. Among those cells were the forerunners of animals and other organisms that could survive with aerobic machinery alone.

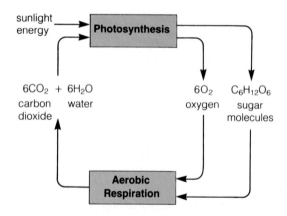

sunlight energy → **Photosynthesis**

$6CO_2$ + $6H_2O$ $6O_2$ $C_6H_{12}O_6$
carbon   water   oxygen   sugar
dioxide   molecules

Aerobic Respiration

Figure 9.11 The cycle in which molecules composed of carbon, hydrogen, and oxygen flow through all organisms on earth. In this recycling of matter through time, each birth is affirmation of our ongoing capacity for organization, each death a renewal.

With aerobic respiration, life became self-sustaining. For its final products—carbon dioxide and water—are precisely the materials used to build organic compounds in photosynthesis! Thus the flow of carbon, hydrogen, and oxygen through the energy pathways of living organisms came full circle (Figure 9.11).

Perhaps one of the most difficult connections you are asked to perceive is the link between yourself—a living, intelligent being—and such remote-sounding subjects as energy, metabolic pathways, and the cycling of oxygen, hydrogen, nitrogen, and carbon. Is this really the stuff of humanity? Think back, for a moment, on the description of a water molecule. A pair of hydrogen atoms competing with an oxygen atom for a share of the electrons joining them doesn't exactly seem close to our daily lives. But from that simple competition, the polarity of the water molecule arises. As a result of the polarity, hydrogen bonds form between water molecules. And that is a beginning for the organization of lifeless matter which leads, ultimately, to the organization of matter in all living things.

For now you can imagine other kinds of molecules interspersed in water. Many are nonpolar and resist interaction with the water molecules. Others are polar and respond by dissolving in it. And the lipids among

them (with water-soluble *and* water-insoluble regions) spontaneously form a two-layered film. Such lipid bilayers are the basis for all cell membranes, hence all cells. The cell has been, from the beginning, the fundamental *living* unit.

With the boundary afforded by a cell membrane, chemical reactions can be contained and controlled. The essence of life *is* chemical control. This "control" is not some mysterious force. It is a chemical responsiveness to energy changes and to the kinds of molecules present in the environment. It operates by "telling" a class of protein molecules—enzymes—when and what to build, and when and what to tear down. And it is not some mysterious force that creates the enzymes themselves. DNA, the slender double strand of heredity, has the chemical structure—*the chemical message*—that allows molecule faithfully to reproduce molecule, one generation after the next. Those DNA strands tell many billions of cells in your body how countless molecules must be built and torn apart for their stored energy.

So yes, oxygen, hydrogen, carbon, and nitrogen represent the stuff of you, and us, and all of life. But it takes more than molecules to complete the picture. It is because of the way molecules are organized and maintained by a constant flow of energy that you are alive. It takes outside energy from sources such as the sun to drive their formation. Once molecules are assembled into cells, it takes outside energy derived from food, water, and air to sustain their organization. Plants, animals, fungi, protistans, and bacteria are part of a web of energy use and materials cycling that interconnects all levels of biological organization. Should energy fail to reach any part of any of those levels, life there will dwindle and cease.

For energy flows through time in only one direction—from forms rich in energy to forms having less usable stores of it. Only as long as sunlight flows into the web of life—and only as long as there are molecules to recombine, rearrange, and recycle—does life have the potential to continue in all its rich diversity.

Life is, in short, no more *and no less* than a marvelously complex system of prolonging order. Sustained by energy transfusions, it continues because of a capacity for self-reproduction—the handing down of hereditary instructions. With those instructions, it has the means for organizing energy and materials generation after generation. Even with the death of the individual, life is prolonged. With death, molecules are released and can be recycled once more, providing raw materials for new generations. In this flow of matter and energy through time, each birth is affirmation of our ongoing capacity for organization, each death a renewal.

SUMMARY

1. ATP is the main energy carrier in cells. By donating a phosphate group to many kinds of enzymes, reactants, and other molecules, it primes them to enter a reaction. Photosynthesis or chemosynthesis can produce ATP. Degradative, energy-releasing pathways also produce ATP.

2. The main energy-releasing pathways are called aerobic respiration, alcoholic fermentation, lactate fermentation, and anaerobic electron transport.

3. In all of those pathways, chemical energy is released from glucose and other carbohydrates through phosphorylation and electron transfers (oxidation-reduction reactions).

4. Glycolysis, the first stage of all the main energy-releasing pathways, takes place in the cytoplasm and requires an energy input. Glucose is phosphorylated twice (two ATP each donate a phosphate group to it), then broken down into two pyruvate molecules.

5. During glycolysis, substrate-level phosphorylations provide a *net* yield of two ATP (two substrate molecules each transfer a phosphate group directly to ADP). Protons (H^+) and electrons stripped from intermediates are picked up by NAD^+ to form NADH. When this coenzyme donates the electrons to an acceptor molecule, NAD^+ is regenerated.

6. In *aerobic respiration*, oxygen is the final acceptor of electrons transferred during the reactions. The net energy yield, from the first stage (glycolysis) to the last, is commonly thirty-six ATP.

7. Aerobic respiration consists of three sets of reactions: glycolysis, the Krebs cycle (and preparatory conversions), and electron transport phosphorylation.

   a. *Glycolysis:* Glucose is broken down into two pyruvate, and two NADH form; the net yield is two ATP.

   b. *Conversions preceding Krebs cycle:* The pyruvate is converted to a form that can enter a cyclic metabolic pathway. For each pyruvate, an NADH forms and a carbon atom departs (as CO_2) before the cycle begins.

   c. *Krebs cycle:* For each pyruvate, hydrogen atoms donated by reaction intermediates yield enough H^+ and electrons to form three NADH and one $FADH_2$; also two more CO_2 form, and so does one ATP.

   d. So far (glycolysis through the Krebs cycle), one glucose (or two pyruvate) has yielded ten NADH, two $FADH_2$, and four ATP.

   e. *Electron transport phosphorylation:* NADH and $FADH_2$ donate electrons to a transport system embedded in the inner mitochondrial membrane. Operation of the system sets up H^+ and electric gradients across the mem-

brane. When H^+ moves down the gradients (through enzymes spanning the membrane), energy associated with the flow drives the phosphorylation of ADP to produce ATP. Oxygen finally accepts the electrons and combines with H^+ to form water.

8. Compared with aerobic respiration, the other main energy-releasing pathways have only a small net yield (two ATP, from glycolysis) because glucose is not completely degraded. Following glycolysis, the remaining reactions serve to regenerate NAD^+:

   a. In *alcoholic fermentation*, the pyruvate from glycolysis is converted to acetaldehyde, which accepts electrons from NADH to form ethanol.

   b. In *lactate fermentation*, the pyruvate accepts electrons from NADH and is converted to lactate.

   c. In *anaerobic electron transport*, electrons from NADH are transferred to inorganic compounds (such as sulfate) in the environment.

Review Questions

1. ATP can be produced when carbohydrates are degraded. Phosphorylation and oxidation-reduction reactions are needed to do this. Can you describe the roles of these reactions in the main energy-releasing pathways? *121–122*

2. Which energy-releasing pathways occur in the cytoplasm? In the mitochondrion of eukaryotes? *124*

3. Is the following statement true? Your muscle cells cannot function at all unless they are supplied with oxygen. *124–125*

4. Glycolysis is the first stage of all the main pathways by which glucose is degraded. Can you define those pathways in terms of the final electron acceptor for their reactions? If you include the two ATP molecules formed during glycolysis, what is the *net* energy yield from one glucose molecule for each pathway? *123, 124–125*

5. In anaerobic routes of glucose breakdown, further conversions of pyruvate do not yield any more usable energy. What, then, is the advantage of the conversions? *125*

6. Can you describe the functions of the Krebs cycle? Of electron transport phosphorylation? *128–129*

Readings

Becker, W. 1986. *The World of the Cell*. Menlo Park, California: Benjamin/Cummings. Chapters 7 and 8 are a good place to start for further readings on anaerobic and aerobic metabolism.

Brock, T., B. Smith, and M. Madigan. 1988. *Biology of Microorganisms*. Fifth edition. Englewood Cliffs, New Jersey: Prentice-Hall. Clear descriptions of the energy-releasing pathways of microbes.

Lehninger, A. 1982. *Principles of Biochemistry*. New York: Worth. Clear, accessible introduction to metabolic pathways.

UNIT THREE

THE ONGOING FLOW OF LIFE

10

CELL REPRODUCTION

Strictly speaking, reproduction means making copies. Cells reproduce by dividing in two, and the so-called "daughter cells" may indeed be exact copies of the parental cell. For example, every five days, the lining of your small intestine is entirely replaced as a result of ongoing cell divisions. All the new cells are true copies of the old (they have to be, for the body depends on this cell lineage to perform a vital task—selective absorption—in an unvarying way). Bacterial cells produce true copies of themselves most of the time—yet many also dabble in the bacterial version of sex, which introduces variation in offspring. Strawberry plants can give rise to copies or variants, depending on how they reproduce. You are not an exact copy of either of your parents.

So let's embroider the narrow definition by saying that, in biology, **reproduction** means producing a new generation of cells or multicelled individuals that may or may not be exact copies of the parents.

Reproduction is part of a **life cycle**, a recurring frame of events in which individuals grow, develop, and reproduce according to a program of instructions encoded in DNA, which they inherit from their parents. Free-living cells such as bacteria can grow and reproduce by dividing in two. Grow, divide, grow, divide—this "cell cycle" *is* their life cycle. A multicelled plant or animal has a more

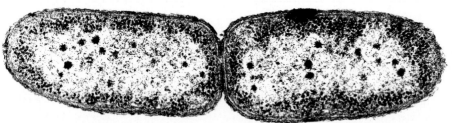

Figure 10.1 Simplified picture of prokaryotic fission, which may be like the ancient division mechanisms that preceded mitosis and meiosis. The micrograph of the bacterium *Pseudomonas* corresponds to (**e**).

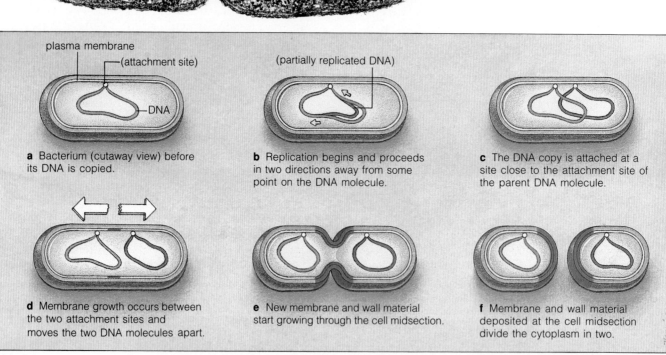

plasma membrane
(attachment site)
(partially replicated DNA)
DNA

a Bacterium (cutaway view) before its DNA is copied.

b Replication begins and proceeds in two directions away from some point on the DNA molecule.

c The DNA copy is attached at a site close to the attachment site of the parent DNA molecule.

d Membrane growth occurs between the two attachment sites and moves the two DNA molecules apart.

e New membrane and wall material start growing through the cell midsection.

f Membrane and wall material deposited at the cell midsection divide the cytoplasm in two.

intricate life cycle, with perhaps hundreds or many millions of cells growing, developing, and dividing at different times. Yet even here, when it comes time to produce offspring, *a single, dividing cell is still the bridge to the next generation:*

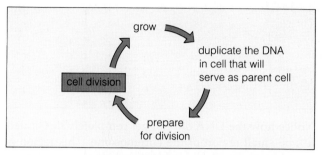

Answers to three questions about cell division are central to an understanding of reproduction. *First,* what structures and substances are necessary for inheritance? *Second,* how are they divided and distributed into daughter cells? *Third,* what are the division mechanisms themselves? We will require more than one chapter to consider the answers (and best guesses) about the reproductive process. However, the following paragraphs can help you keep the overall picture in focus.

OVERVIEW OF DIVISION MECHANISMS

Regardless of the division mechanism employed, the ground rule for cell reproduction is this: *Each cell of the new generation must receive a duplicate of all the parental DNA.* DNA, recall, contains the hereditary instructions for making specific proteins. Some of the proteins serve as structural materials; many serve as the enzymes necessary for the synthesis of carbohydrates, lipids, and other substances. These substances are used in building the membranes and other components of each new cell:

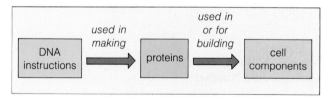

New cells simply cannot grow or function properly unless they receive all the DNA instructions that will allow them to do so.

Also, the cytoplasm of the parental cell already has operating machinery—enzymes, organelles, cytoskeletal elements, and so on. When a daughter cell inherits what looks merely like a blob of cytoplasm, it really is getting "start-up" machinery for its operation, until it has time to use its inherited DNA for growing and developing on its own.

Each cell of a new generation must receive a duplicate of all the parental DNA and enough cytoplasmic machinery to start up its own operation.

Cell reproduction occurs through three main categories of division mechanisms, which we will consider in this chapter and the next:

Mechanism:	Used By:
prokaryotic fission	*bacterial cells*
mitosis, cytokinesis	*single-celled eukaryotes (for asexual reproduction)*
	multicelled eukaryotes (for bodily growth; also for asexual reproduction in some species)
meiosis, cytokinesis	*all eukaryotes (only to form gametes for sexual reproduction)*

Prokaryotic fission is the simplest of the three; dividing the duplicated DNA *and* the cytoplasm into two daughter cells proceeds rapidly, at the same time. Mitosis and meiosis are nuclear division mechanisms only. Cytoplasmic division, or **cytokinesis,** usually takes place when the division of DNA for each new daughter cell is completed (or just about completed).

Prokaryotic Fission

Only a single DNA molecule has to be duplicated and parceled out by a dividing bacterium. Since there is no nuclear envelope around bacterial DNA (as there is in eukaryotes), cell division is straightforward, as the following example illustrates.

Suppose you place a bacterium on a nutrient-rich medium in a culture dish and allow it to grow for, say, thirty minutes. During its growth period, it also duplicates its DNA (a process called DNA replication). Both DNA molecules are attached to the plasma membrane. At the time of division, new membrane and wall material start growing through the cell midsection. The DNA molecules simply are moved apart as new material is added laterally between them (Figure 10.1). Then the membrane and wall grow across the cell midsection to divide the cell body in two. This partitioning into two equivalent parts is a type of "binary" fission.

In prokaryotic fission, both molecules of the duplicated DNA are attached to the plasma membrane, and <u>membrane growth</u> between their attachment sites pushes them apart before the cell divides in two.

The Eukaryotic Chromosome

Compared to bacteria, eukaryotic cells have more complex DNA and much more of it, so perhaps it is no surprise that their division mechanisms are more involved. We saw earlier that a eukaryotic DNA molecule has numerous proteins attached to it (page 71). The proteins generally are equal in mass to the DNA itself, and together they form a structure called the **chromosome**.

Between divisions, a chromosome is stretched out in threadlike form. It is still threadlike when it is duplicated prior to cell division, and the two threads remain attached for a while as **sister chromatids** of the chromosome. Each "thread," of course, is a DNA double helix

with its associated proteins, which we show here as a simplified version of a current model:

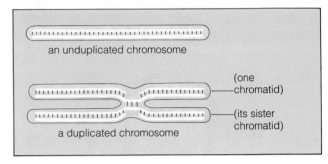

Notice how the DNA did not get completely duplicated in one small region of the chromosome. This region is the **centromere**. Here we find two **kinetochores**, which are groups of chromosomal proteins that form on the outer faces of the centromere, one for each chromatid:

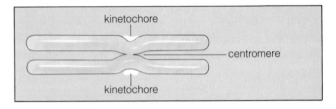

As you will see, the kinetochores become attached to microtubules that help move the chromosomes during nuclear division.

Keep in mind that the preceding model is highly simplified. The DNA double helix in each chromatid is actually two molecular strands twisted together tightly and repeatedly like a spiral staircase (that's what the "double helix" means), and it is much, much longer than can be shown here. Also, a chromatid never looks this "fat" until just before nuclear division, when its DNA and associated proteins become wound up, folded, and looped many, many thousands of times into a highly condensed form. You can see some of the major loops in Figure 10.2.

Table 10.1 Number of Chromosomes in the Somatic Cells of Some Eukaryotes*	
Mosquito, *Culex pipiens*	6
Fruit fly, *Drosophila melanogaster*	8
Garden pea, *Pisum sativum*	14
Corn, *Zea mays*	20
Lily, *Lilium*	24
Yellow pine, *Pinus ponderosa*	24
Frog, *Rana pipiens*	26
Earthworm, *Lumbricus terrestris*	36
Rhesus monkey, *Macaca mulatta*	42
Human, *Homo sapiens*	46
Orangutan, *Pongo pygmaeus*	48
Chimpanzee, *Pan troglodytes*	48
Gorilla, *Gorilla gorilla*	48
Potato, *Solanum tuberosum*	48
Amoeba, *Amoeba*	50
Horse, *Equus caballus*	64
Horsetail, *Equisetum*	216
Adder's tongue fern, *Ophioglossum reticulatum*	1,000+

*These examples are a sampling only. Chromosome number for most species falls between ten and fifty.

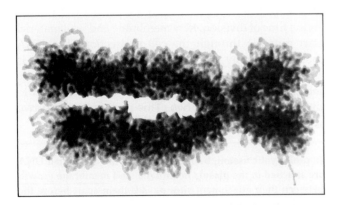

Figure 10.2 A duplicated human chromosome just before mitotic division.

Mitosis, Meiosis, and the Chromosome Number

It took billions of cell divisions to produce all the tissues of your body, and every day, millions of cells still divide every second just to replace worn-out or damaged predecessors and to keep that body running smoothly. Most are **somatic cells**; they are the ones making up all body tissues *except* for the **germ cells**, a cell lineage set aside for sexual reproduction.

Sexual reproduction begins with meiosis and proceeds through the formation of **gametes** (sex cells, such as sperm and eggs). It ends at fertilization, when the sperm nucleus and egg nucleus fuse in the first cell of the new individual (the zygote).

A zygote grows into a multicelled adult by way of mitosis, which faithfully maintains the chromosome number characteristic of the species, division after division. For example, there are 46 chromosomes in your somatic cells, 48 in a gorilla's, and 14 in a garden pea's (Table 10.1). With each mitotic cell division, that number doesn't change:

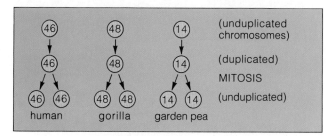

As Figure 10.3 shows, every two chromosomes in a human somatic cell have the same length and shape; their hereditary instructions also deal with the same traits. The two corresponding chromosomes are called **homologous chromosomes**. Homologues don't interact at all during mitosis, but they pair with each other during meiosis. (You may have noticed that the X and Y chromosomes don't look alike, but they pair during meiosis so we still call them homologues.)

Meiosis only occurs in germ cells that are destined to give rise to gametes. With this division mechanism, the chromosome number is *not* maintained; it is reduced by half. And not just any half—each gamete ends up with one of each pair of homologous chromosomes. (It doesn't matter which of the two it gets.) It takes two

nuclear divisions to put one chromosome of each type into each gamete:

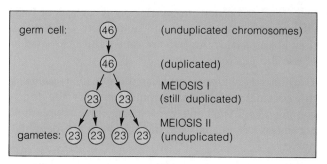

Thus gametes end up with a **haploid** number of chromosomes, meaning "half" the parental number. Then, when a sperm nucleus and an egg nucleus fuse at fertilization, the **diploid** number is restored. "Diploid" means having two chromosomes of each type (that is, pairs of homologous chromosomes) in the somatic cells of sexually reproducing species.

Now we begin our closer look at eukaryotic cell division. Before getting into the details, be sure you have a good understanding of these points:

---

1. In eukaryotes, instructions for heritable traits are distributed among a number of chromosomes. All members of a given species normally have the same number of chromosomes in their somatic cells.

2. Somatic cells commonly are diploid; they have two of each type of chromosome, called homologues. In general, two homologues have the same length and shape, and their genes deal with the same heritable traits.

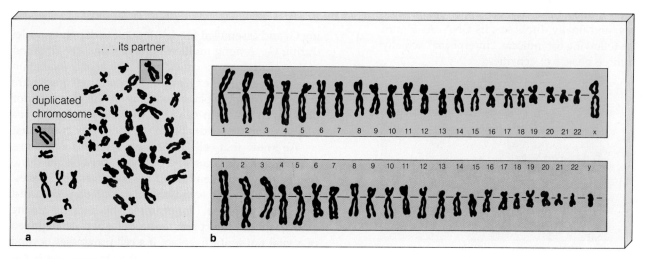

Figure 10.3 (**a**) Photograph of the 46 chromosomes from a diploid cell of a human male. All are in the duplicated state. (**b**) By cutting apart and arranging the chromosomes according to length and shape, we see that there are two sets of 23 chromosomes, with all the chromosomes in one set having a partner, or homologue, in the other. The partners don't interact at all during mitosis, but they pair up with each other during meiosis.

3. Prior to nuclear division, each chromosome is duplicated. Whereas each previously was one DNA double helix (and associated proteins), now it consists of two, which remain attached at the centromere. As long as they are attached, the two are called sister chromatids of the chromosome.

4. Mitosis is a type of nuclear division that *maintains* the parental number of chromosomes for the forthcoming cells. In just one division step, all the duplicated chromosomes are split and distributed to two daughter nuclei.

5. Mitosis and cytokinesis (cytoplasmic division) are the basis of bodily growth for multicelled eukaryotes. They also are the basis of asexual reproduction for some eukaryotes.

6. Meiosis is a type of nuclear division that *reduces* the parental chromosome number by half—to the haploid number. It puts only one chromosome of each type (no homologous partners) in each of four daughter nuclei.

7. Meiosis is the basis of gamete formation; it occurs only in germ cells set aside for sexual reproduction.

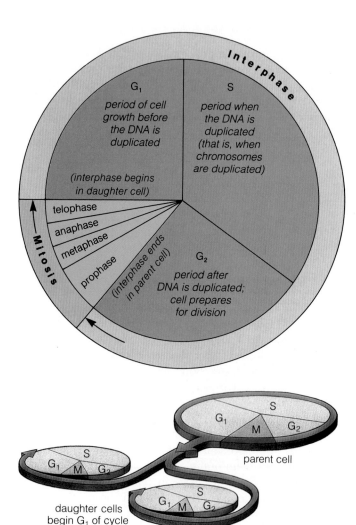

WHERE MITOSIS OCCURS IN THE CELL CYCLE

For eukaryotic cells in general, the life cycle extends from the time of the cell's formation until its own subsequent division is completed. Mitosis is a very small part of this cycle; it only lasts for a few minutes or an hour or more. Normally, the nondividing cell spends about ninety percent of its life in **interphase**, when it increases its mass, approximately doubles the number of its cytoplasmic components, and finally duplicates its DNA. As Figure 10.4 and the following list indicate, "interphase" actually consists of three phases of activities:

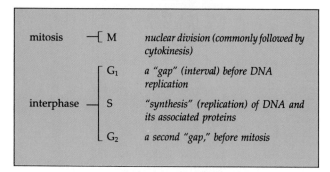

mitosis	— [M	*nuclear division (commonly followed by cytokinesis)*
	[G_1	*a "gap" (interval) before DNA replication*
interphase —	S	*"synthesis" (replication) of DNA and its associated proteins*
	[G_2	*a second "gap," before mitosis*

The "gaps" were so named before we knew much about what was going on during the cell cycle. Today we know that new cell components are synthesized dur-

Figure 10.4 Eukaryotic cell cycle. This drawing has been generalized. There is great variation in the length of different stages from one cell to the next.

ing G_1 and assembled for distribution to daughter nuclei during G_2. Among multicelled organisms, the duration of the cell cycle varies considerably. Most of the variation is associated with the G_1 phase. For example, the mature cells of a complex plant may stay in the G_1 phase for days, even years. In contrast, the cells of an early sea urchin embryo may double in number every two hours.

We know that cells must have built-in instructions regarding the duration of a cell cycle, because the duration is fairly consistent for all the members of a species. Sometimes adverse environmental conditions do arrest cells in the G_1 phase. (For example, this happens among amoebas and other protistans when they are deprived of a vital nutrient.) Even so, if a cell progresses beyond a certain point in G_1 (the so-called restriction point), the cycle normally will be completed regardless of outside conditions.

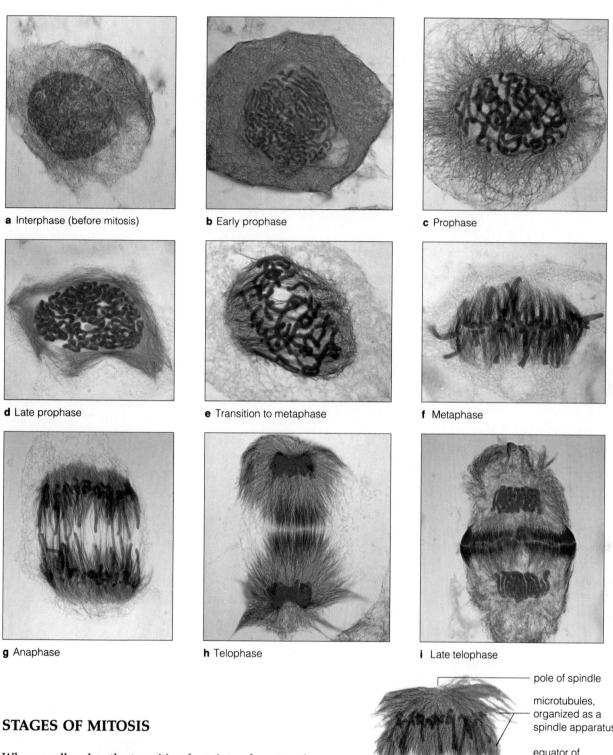

a Interphase (before mitosis)   **b** Early prophase   **c** Prophase

d Late prophase   **e** Transition to metaphase   **f** Metaphase

g Anaphase   **h** Telophase   **i** Late telophase

STAGES OF MITOSIS

When a cell makes the transition from interphase to mitosis, it stops constructing new cell parts. Profound changes now begin to occur in the nucleus. The changes proceed continuously, one after another, through four sequential stages: prophase, metaphase, anaphase, and telophase. The stages do not have distinct boundaries. If you were to watch a film on mitosis, you would see that the photomicrographs in Figure 10.5 are merely single frames of a continuous, changing drama.

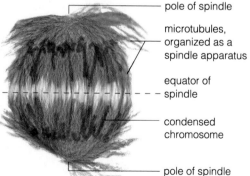

pole of spindle

microtubules, organized as a spindle apparatus

equator of spindle

condensed chromosome

pole of spindle

Figure 10.5 Mitosis in a plant cell (*Haemanthus*). The chromosomes are stained blue, and the microtubules that move them about are stained red.

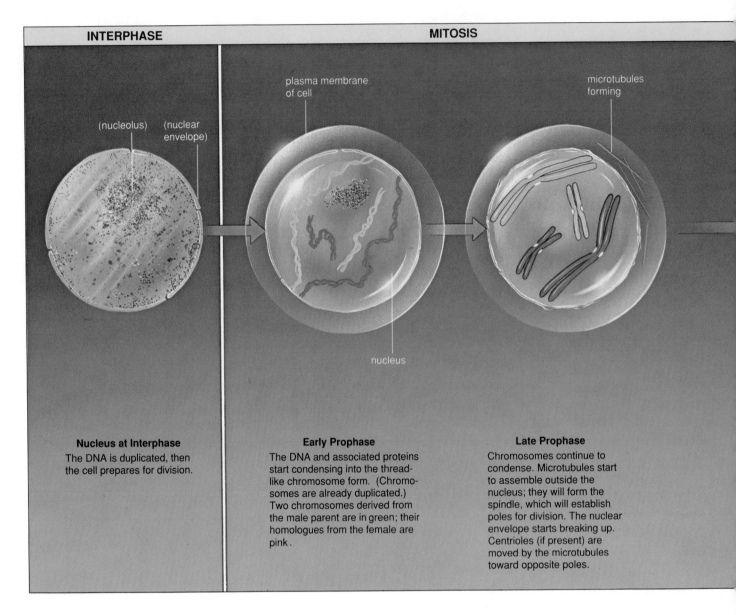

Nucleus at Interphase

The DNA is duplicated, then the cell prepares for division.

Early Prophase

The DNA and associated proteins start condensing into the thread-like chromosome form. (Chromosomes are already duplicated.) Two chromosomes derived from the male parent are in green; their homologues from the female are pink.

Late Prophase

Chromosomes continue to condense. Microtubules start to assemble outside the nucleus; they will form the spindle, which will establish poles for division. The nuclear envelope starts breaking up. Centrioles (if present) are moved by the microtubules toward opposite poles.

Figure 10.6 Mitosis: the nuclear division mechanism that maintains the parental chromosome number in each daughter nucleus. Shown here, a diploid animal cell (with pairs of homologous chromosomes, derived from two parents).

For the sake of clarity, only two pairs of homologues are shown and the spindle apparatus is simplified. With rare exceptions, the picture is more involved than this, as indicated by the micrograph to the right. It shows the duplicated chromosomes of a diploid cell from the Oregon newt, lined up at the spindle equator at metaphase.

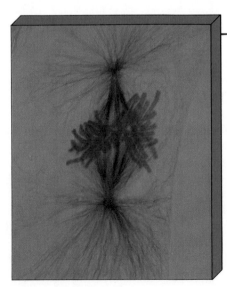

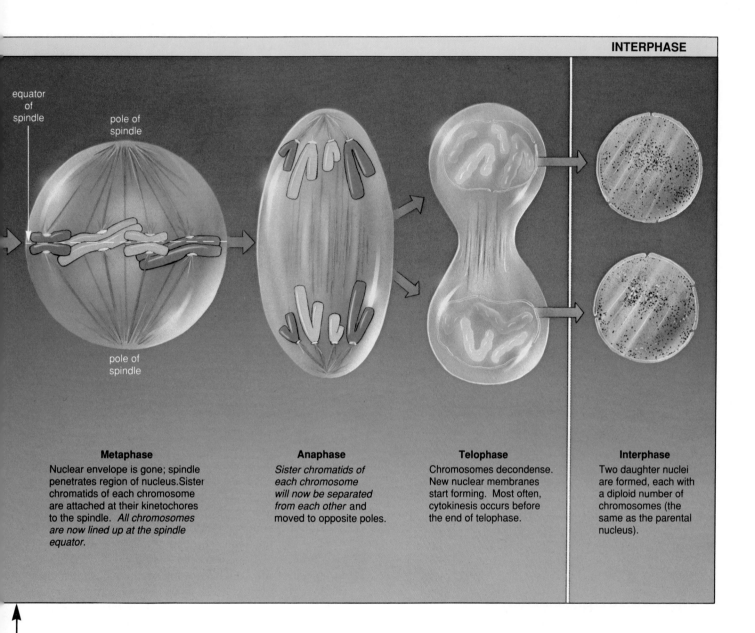

Metaphase	**Anaphase**	**Telophase**	**Interphase**
Nuclear envelope is gone; spindle penetrates region of nucleus. Sister chromatids of each chromosome are attached at their kinetochores to the spindle. *All chromosomes are now lined up at the spindle equator.*	*Sister chromatids of each chromosome will now be separated from each other* and moved to opposite poles.	Chromosomes decondense. New nuclear membranes start forming. Most often, cytokinesis occurs before the end of telophase.	Two daughter nuclei are formed, each with a diploid number of chromosomes (the same as the parental nucleus).

Prophase: Mitosis Begins

The Prophase Chromosome.   The first stage of mitosis, called *prophase*, is evident when chromosomes become visible in the light microscope as threadlike forms. (The term mitosis refers to this emergence; it comes from the Greek *mitos*, which means thread.) Each chromosome was duplicated earlier, during interphase, so it already consists of two sister chromatids joined at the centromere. In late prophase, each one becomes condensed into thicker, rodlike forms (Figures 10.5 and 10.6).

The Microtubular Spindle.   The chromosomes in a dividing nucleus cannot move on their own. Rather, they are moved in precise ways by microtubules. Remember that microtubules are pervasive components of the cyto-skeleton. When a nucleus is about to divide, those microtubules disassemble and then initially reassemble just outside the nucleus, often in parallel. Together they form a **spindle apparatus**, which is the actual machinery that moves the chromosomes for distribution to daughter nuclei. The spindle establishes two poles that are the prescribed destinations for the chromosomes.

What determines where the two poles will become established? In many cells, a microtubule organizing center near the nucleus seems responsible. Sometimes this center includes a pair of centrioles that were duplicated earlier, during interphase. As Figure 10.7 shows, a centriole pair is moved to each pole of the spindle during its formation. The positioning and orientation of centrioles seem to influence the organization of the cyto-skeleton that will form in each daughter cell (page 81).

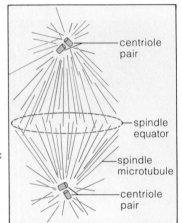

Figure 10.7 One type of mitotic spindle in eukaryotic cells having centrioles (refer to page 80). The microtubules form two "half-spindles" that meet at the spindle equator.

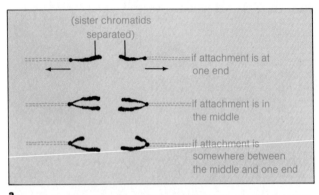

a

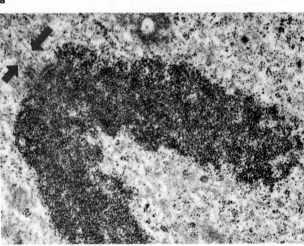

b

Figure 10.8 (a) When chromosomes are being moved to spindle poles, they can appear straight, U-shaped, or J-shaped, depending on the location of the centromere region.

(b) Micrograph of a chromosome from the alga *Oedogonium* at anaphase, as it is being pulled to a spindle pole by microtubules (red arrows point to two of them).

Metaphase

Late in prophase, the nuclear envelope breaks up. By **metaphase**, it has become fragmented into numerous small vesicles that can be detected only at high magnification. Now the spindle is free to penetrate the nuclear area.

When you observe living cells at early metaphase, you may wonder why their chromosomes seemingly go into a frenzy. This happens when microtubules of the spindle make their first pass at harnessing the chromosomes. With the first random contacts, each chromosome spins around its long axis, and it is yanked back and forth until its two kinetochores are firmly oriented toward opposite poles. Once this happens, forces associated with the spindle move the chromosomes until they all lie halfway between the poles, at the spindle equator (Figure 10.6).

Thus metaphase is dominated by two events: the orientation of each chromosome's kinetochores toward opposite poles, and the alignment of all chromosomes halfway between the poles. This alignment is crucial for the chromosomal movements that follow. Once it is completed, metaphase is over.

Anaphase

Two events characterize **anaphase**: sister chromatids of each chromosome are split away from each other, and those former partners are moved to opposite poles. Once they do split, they are no longer referred to as chromatids; each is now an independent chromosome.

How are the sister chromatids made to part company? According to one model, remember, a duplicated chromosome may still have a tiny portion of *unduplicated* DNA at its centromere. It may be that duplication is finished at anaphase. This would allow the sister chromatids to break apart, for each would now have a complete DNA double helix. In other words, each would be a completed chromosome:

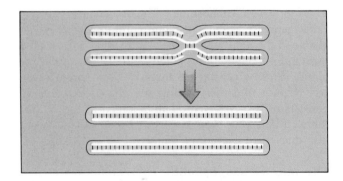

Regardless of what goes on, all the duplicated chromosomes split apart at about the same time and move to opposite poles at the same rate (one micrometer per

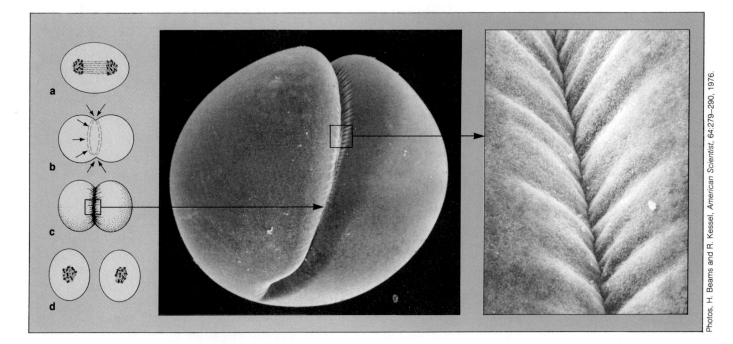

Photos, H. Beams and R. Kessel, *American Scientist*, 64:279–290, 1976.

Figure 10.9 Cytokinesis in an animal cell. The scanning electron micrographs show the furrowing of the plasma membrane caused by the contraction of a microfilament ring just beneath it. (**a**) Nuclear division is complete; the spindle is disassembling. (**b**) Microfilament rings at the former spindle equator contract, like a purse string closing. (**c**) Contractions cause furrowing at the cell surface. (**d**) The cytoplasm is pinched in two.

second). Although metaphase may last a long time, anaphase begins abruptly and is commonly over within a few minutes.

How do the chromosomes actually move away from the equator? At present there is no widely accepted model for the mechanisms involved. One way or another, the microtubules attached to kinetochores move poleward, and they drag the attached chromosomes in their wake (Figure 10.8). Other microtubules extending from both poles to the spindle equator lengthen, and their interactions with parallel microtubules also may contribute to the movement.

Telophase

Telophase begins once the separated chromosomes arrive at opposite spindle poles. During telophase, the chromosomes decondense and become extended in the threadlike form. Small vesicles (the fragments of the old nuclear envelope) fuse together to form patches of membrane alongside the chromosomes. Patch joins with patch, and eventually a new, continuous nuclear envelope separates the hereditary material from the cytoplasm. Once the nucleus is completed, telophase is completed—and so is mitosis.

CYTOKINESIS: DIVIDING UP THE CYTOPLASM

Cytoplasmic division, or **cytokinesis**, usually accompanies nuclear division. In most cases, it coincides with the period from late anaphase through telophase. Exceptions do exist. For example, cytokinesis does not follow mitosis in some insect and plant cells and many fungal cells, so the cell body ends up with more than one nucleus.

For most animal cells, the onset of cytokinesis is marked by the appearance of scattered deposits of material around microtubules at the spindle equator. The deposits accumulate until they form a distinct layer across (typically) the cell midsection. Soon a shallow, ringlike depression appears at the plasma membrane, a ring that corresponds to the periphery of this layer (Figure 10.9). The depression is a **cleavage furrow**.

At a cleavage furrow, a ring of contractile microfilaments attached to the plasma membrane pulls the membrane inward. (The contractile force generated is strong enough to bend a fine glass needle inserted into dividing cells.) The inward movement eventually cuts the cytoplasm in two.

The cells of most land plants have fairly rigid walls that do not lend themselves to the formation of cleavage

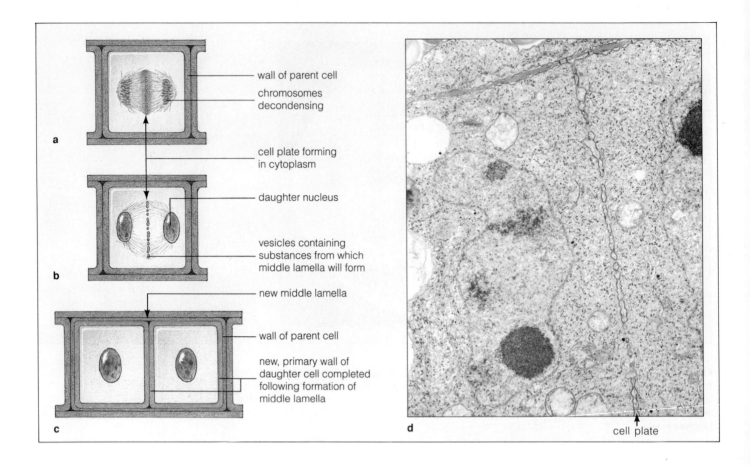

wall of parent cell

chromosomes decondensing

cell plate forming in cytoplasm

daughter nucleus

vesicles containing substances from which middle lamella will form

new middle lamella

wall of parent cell

new, primary wall of daughter cell completed following formation of middle lamella

cell plate

Figure 10.10 Cytokinesis following mitosis in a plant cell. **(a-b)** Vesicles form at the spindle equator and gradually fuse to form a cell plate. As the micrograph **(d)** indicates, the cell plate grows outward until it reaches the parent cell wall. The vesicles contain substances that will form the middle lamella, which will cement together the primary walls of the daughter cells **(c)**; the membrane of the vesicles is used to form the plasma membrane on both sides of the cell plate.

furrows. Hence plant cells use a different mechanism of cytokinesis, called **cell plate formation** (Figure 10.10).

A cell plate is a crosswall that usually forms in conjunction with microtubular remnants of the spindle. Small vesicles filled with wall-building material fuse together at the microtubular array, forming a disklike structure. This structure develops into the cell plate, which marks the boundary between the two new plant cells. Cellulose microfibrils are deposited inside the plate and the new cell wall is completed.

SUMMARY

1. DNA has all the hereditary instructions necessary to construct a new cell of a given species.

2. Before a cell divides, its DNA is duplicated (the process is called DNA replication). This is true of both prokaryotes and eukaryotes.

3. When a cell divides, each new cell receives (a) hereditary instructions encoded in DNA, and (b) cytoplasmic components that serve as "start-up machinery" for carrying out the instructions.

4. Bacterial cells reproduce through prokaryotic fission. The single DNA molecule is duplicated, and one whole molecule ends up in each daughter cell when the parent cell divides at its midsection.

5. Somatic (body) cells of multicelled eukaryotes commonly have a diploid number of chromosomes. That is, they have two of each type of chromosome characteristic of the species. The pairs of "homologous" chromosomes generally are alike in length, shape, and which heritable traits they deal with.

6. Eukaryotic chromosomes are in threadlike form when they are duplicated between cell divisions. Whereas each was one DNA molecule (and associated proteins), now it consists of two, which stay attached for a while as sister chromatids.

7. According to a current model, a small portion of the DNA may not be duplicated in the region where sister chromatids are attached to each other. This region of unduplicated DNA is the centromere. Two kinetochores (groups of proteins) form on the outer faces of the centromere, one for each chromatid.

8. Eukaryotes employ different division mechanisms that serve different functions:

a. *Mitosis* is a type of nuclear division that maintains the parental number of chromosomes in each of two daughter nuclei. (Thus if the parental nucleus is diploid, the daughter nuclei will be diploid, too.)

b. Mitosis is the basis of bodily growth for multicelled eukaryotes. It also is the basis of asexual reproduction for some eukaryotes.

c. *Meiosis* is a type of nuclear division that reduces the parental chromosome number by half (to the haploid number) in each of four daughter nuclei.

d. Meiosis is the basis of gamete formation. It occurs only in germ cells (a cell lineage set aside for sexual reproduction).

e. For most organisms, actual cytoplasmic division, or *cytokinesis*, proceeds late in mitosis or meiosis or at some point afterward.

9. Mitosis is a small part of the eukaryotic cell cycle. A nondividing cell spends about ninety percent of its life in interphase, when it increases in mass, doubles its number of cytoplasmic components, and duplicates its chromosomes.

10. The stages of mitosis are as follows:

a. Prophase: Duplicated, threadlike chromosomes condense; a microtubular spindle forms and establishes the poles for nuclear division. The nuclear envelope starts to break up late in prophase.

b. Metaphase: Spindle microtubules harness the chromosomes and orient the kinetochores of sister chromatids toward opposite spindle poles. All chromosomes become aligned at the spindle equator.

c. Anaphase: Sister chromatids of each chromosome split apart to become independent chromosomes, which move to opposite poles.

d. Telophase: Chromosomes decondense to the threadlike form, and a new nuclear envelope forms around each of the two parcels of chromosomes. Mitosis is completed.

Review Questions

1. Describe the mechanism by which a single-celled prokaryote reproduces following DNA replication. *139*

2. Define the two types of nuclear division mechanisms used by eukaryotes. What is cytokinesis? *139*

3. Table 10.1 indicates that a fern has many, many more chromosomes in its body cells than we do. What does this tell you about the correlation between the amount of DNA and its information content?

4. What is a chromosome? What is it called in its unduplicated state? In its duplicated state (that is, with two sister chromatids)? *140*

5. Define somatic cell and germ cell. In which type of cell does mitosis occur? In which does meiosis occur? *141*

6. Define homologous chromosome. Do pairs of homologous chromosomes interact during mitosis, meiosis, or both? *141*

7. What is a diploid number of chromosomes? Explain how mitosis maintains the diploid number and how meiosis reduces it (to the haploid number). *141–142*

8. Describe a microtubular spindle and its general function in nuclear division processes. *145*

9. Name the four main stages of mitosis, and characterize each stage. *143–147*

10. When does mitosis take place in the cell cycle? *142*

Readings

Burns, M. 1983. *Cells*. Second edition. Philadelphia: Saunders. Clear explanations of current theories and speculations on cell division mechanisms.

Smith-Klein, C., and V. Kish. 1988. *Principles of Cell Biology*. New York: Harper & Row.

Zimmerman, A., and A. Forer (editors). 1981. *Mitosis/Cytokinesis*. New York: Academic Press. Chapters 5–7, 10–12, 16, 17, and 20 are especially worth reading for those who would like to know more about cytological studies of mitosis and cytokinesis.

11

A CLOSER LOOK AT MEIOSIS

ON ASEXUAL AND SEXUAL REPRODUCTION

If you were a newly formed cell, what would be the absolute necessities for your existence in a given environment? You would need (1) a duplicate of the parental DNA and (2) enough cytoplasm pinched from a parent cell to provide you with preexisting enzymes and other metabolic machinery to start up your own operation.

Inheriting the DNA and cytoplasm is fairly straightforward with prokaryotic fission or even with mitosis. Bacteria use fission for asexual reproduction, and many eukaryotes can use mitosis for the same thing (page 139). In **asexual reproduction**, one parent passes on to offspring a duplicate of all of its genes. This means, of course, that the offspring can only be genetically identical copies, or clones, of the parent.

The preceding paragraph assumes you know what genes are. But in case you don't, "genes" are specific stretches of DNA, each being the inherited instructions for producing or influencing a specific trait in offspring.

Inheritance is much more interesting with meiosis, the division mechanism that is the basis of sexual reproduction. A typical case of **sexual reproduction** involves two parent organisms, each with two genes for every trait. Both parents pass on one of each gene to offspring by way of gamete formation and fertilization. Thus the first cell of a new individual (the zygote) inherits two genes for every trait—half from one parent, and half from the other.

If the instructions encoded in every pair of genes were identical down to the last detail, then sexual reproduction would produce clones, also. Just imagine—you, everyone you know, every member of the entire human species would be clones and might all end up looking

Figure 11.1 (**a**) The chin fissure, a heritable trait arising from a rather uncommon form of a gene. Actor Kirk Douglas received a gene that influences this trait from each of his parents. One gene called for a chin fissure and the other didn't, but one is all it takes in this case. The photograph in (**b**) shows what Mr. Douglas' chin might have looked like if he had inherited two ordinary forms of the gene instead.

Through meiosis and fertilization, old gene combinations are broken up and new ones are put together. The immediate consequence is variation in the physical and behavioral traits of offspring. The long-term consequence can be evolutionary change, as we will see in this unit of the book.

a

b

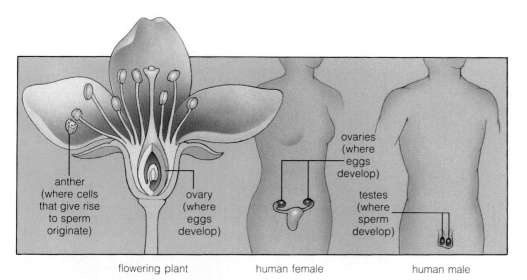

Figure 11.2 Examples of the location of germ cells that give rise to sperm and eggs.

anther
(where cells
that give rise
to sperm
originate)

ovary
(where
eggs
develop)

ovaries
(where
eggs
develop)

testes
(where
sperm
develop)

flowering plant

human female

human male

exactly alike. What has prevented such a boring state of affairs?

It happens that mutations can affect a gene and change a few details about its instructions for a trait. Different mutations can produce various **alleles**—that is, alternative forms of a gene—and each one may cause differences in how a trait is expressed in offspring. (See Figure 11.1, for example.) Going back to our typical sexually reproducing parent, then, there is a chance that the two genes it has for any trait may not "say" the same thing—*and those alternative forms of genes are shuffled during meiosis and combined in novel ways at fertilization.* That is why offspring with novel combinations of traits are the usual outcome of sexual reproduction.

This chapter describes meiosis—the sequence of events by which the parental DNA is divided for distribution to forthcoming gametes. It also starts us thinking about some far-reaching consequences of the gene shufflings that occur during meiosis. Novel gene combinations among offspring lead to variations in their physical and behavioral traits. Such variation, recall, is a testing ground for agents of selection—*and a basis of evolutionary change.*

OVERVIEW OF MEIOSIS

Think "Homologues"

The preceding chapter mentioned a few points about meiosis in multicelled eukaryotes, and now we can put those points together in the following picture.

Meiosis is a nuclear division mechanism that occurs in germ cells, a cell lineage destined to give rise to the gametes (sperm and eggs) used in sexual reproduction. Germ cells develop in a variety of reproductive structures and organs (Figure 11.2). Like the somatic cells of eukaryotes, they commonly have a diploid number of chro-

mosomes (these being structures composed of DNA and numerous proteins tightly attached to it).

Recall that "diploid" means there are two chromosomes of each type, or $2n$, in a cell. The two are called **homologous chromosomes**. Homologues have the same length, the same shape, and the same genes—and they line up with each other at meiosis. Two types of sex chromosomes, designated X and Y, differ in form and in which genes they carry, but they still function as homologues at meiosis.

Meiosis reduces the diploid number by half, to the "haploid" number (n). And not just any half: *each gamete ends up with one of each pair of homologous chromosomes.* To give an example, the diploid number for humans is 46 (that is, 23 + 23 homologues). And a human gamete ends up with one of each type of 23 chromosomes.

Overview of the Two Divisions

We don't have to start from scratch in discussing how meiosis divides the parental number of chromosomes, because this division mechanism bears some resemblances to mitosis, which was covered earlier. While a germ cell is still in interphase, each chromosome is duplicated (by the process called DNA replication). The duplicates remain attached as **sister chromatids** at the centromere:

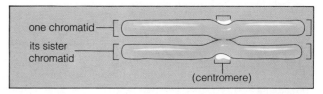

one chromatid

its sister
chromatid

(centromere)

As in mitosis, a kinetochore forms on each chromatid at the centromere. It will serve as an attachment site for microtubules of the spindle apparatus, which move chromosomes during nuclear division (page 145).

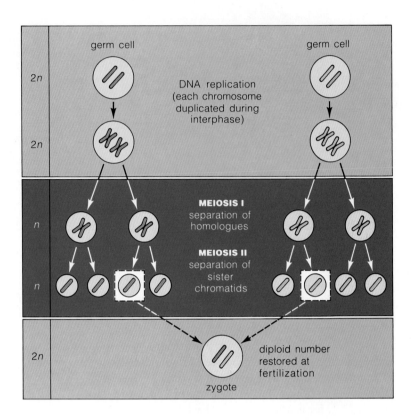

Figure 11.3 Overview of meiosis, showing how a diploid number of chromosomes is reduced by half, then restored following meiosis (at the time of fertilization).

Unlike mitosis, however, there are *two consecutive divisions*, called meiosis I and II:

First Division	Between Divisions	Second Division
prophase I		prophase II
metaphase I	interkinesis	metaphase II
anaphase I	(*no DNA replication*)	anaphase II
telophase I		telophase II

During meiosis I, each duplicated chromosome lines up with its partner, *homologue to homologue*, then the partners are separated from each other. Here we show just one pair of homologous chromosomes, but the same thing happens to all pairs in the nucleus:

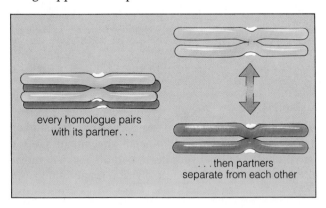

every homologue pairs with its partner. . .

. . .then partners separate from each other

Cytokinesis typically follows. Each daughter cell has a haploid number of chromosomes—but each of those chromosomes is still duplicated.

The transition to the second nuclear division is called interkinesis. There is no DNA replication at this time, as there was at interphase. During meiosis II, *the sister chromatids of each chromosome are separated from each other*:

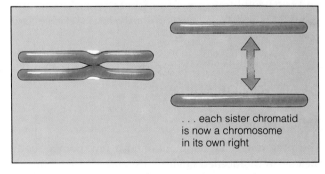

. . . each sister chromatid is now a chromosome in its own right

Cytokinesis typically follows this separation, also. Overall, one DNA replication during interphase, then two nuclear divisions and two cytoplasmic divisions have produced four haploid cells. If we follow the fate of just one pair of homologues in germ cells from two parents, Figure 11.3 shows how fertilization restores the diploid number. Let's now look more closely at meiosis, the stages of which are shown in Figures 11.4 and 11.5.

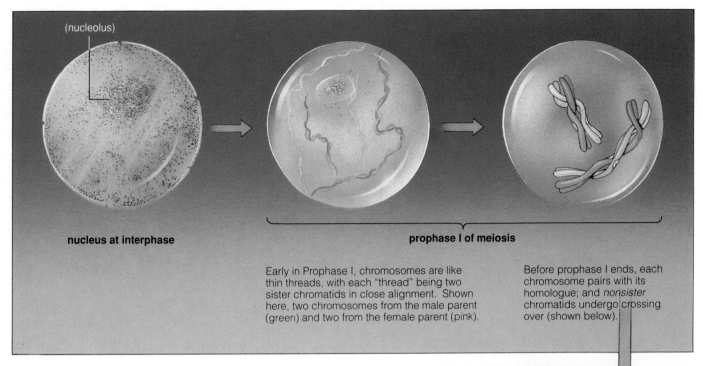

(nucleolus)

nucleus at interphase

prophase I of meiosis

Early in Prophase I, chromosomes are like thin threads, with each "thread" being two sister chromatids in close alignment. Shown here, two chromosomes from the male parent (green) and two from the female parent (pink).

Before prophase I ends, each chromosome pairs with its homologue; and *nonsister* chromatids undergo crossing over (shown below).

STAGES OF MEIOSIS

Prophase I Activities

The first stage of meiosis, **prophase I**, is a time of major gene shufflings between homologous chromosomes. Homologues begin to pair at the onset of prophase I. Each one looks like a very thin thread, for its sister chromatids are closely aligned with each other. The thread-like homologues are drawn together by a process called **synapsis**. It is as if they become stitched point by point along their entire length, with little space between them. (The X and Y chromosomes synapse at one end only.)

Synapsis puts all four chromatids of the homologues into parallel alignment. The arrangement favors **crossing over**, whereby *nonsister* chromatids undergo breakage at one or more sites along their length and exchange corresponding segments at the breakage points. Figure 11.4 shows only a single crossover. On the average, between two and three crossovers are thought to occur between each pair of homologues in human germ cells undergoing meiosis.

Crossing over would be rather pointless if each type of gene never varied from one chromosome to the next. But remember that a gene can come in alternative forms—alleles. And you can safely bet that all the alleles on one chromosome will not necessarily be an identical match to the ones on its homologue. Thus,

Crossing over is an event by which old combinations of alleles in a chromosome are broken up and new ones put together.

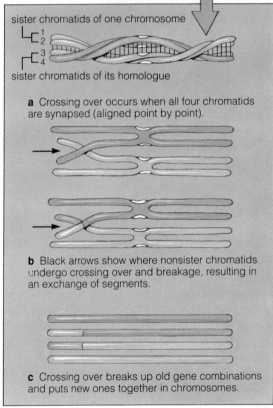

sister chromatids of one chromosome

1
2
3
4

sister chromatids of its homologue

a Crossing over occurs when all four chromatids are synapsed (aligned point by point).

b Black arrows show where nonsister chromatids undergo crossing over and breakage, resulting in an exchange of segments.

c Crossing over breaks up old gene combinations and puts new ones together in chromosomes.

Figure 11.4 Prophase I of meiosis, when crossing over occurs between homologous chromosomes.

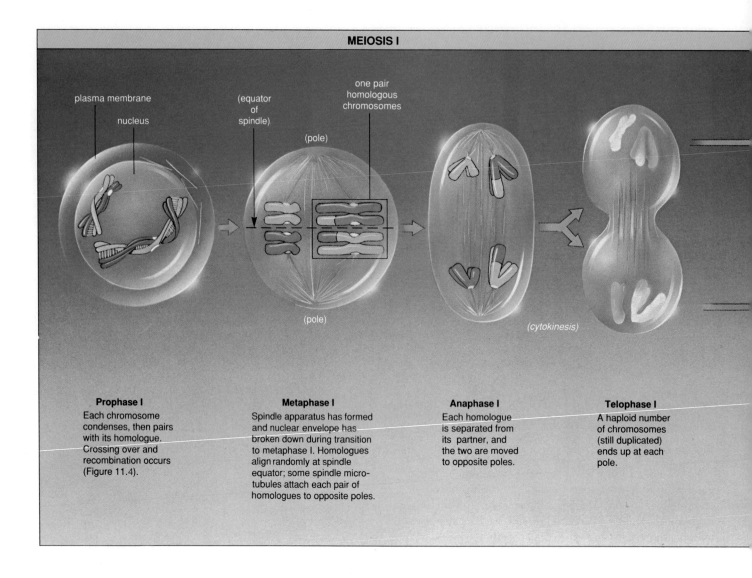

Prophase I

Each chromosome condenses, then pairs with its homologue. Crossing over and recombination occurs (Figure 11.4).

Metaphase I

Spindle apparatus has formed and nuclear envelope has broken down during transition to metaphase I. Homologues align randomly at spindle equator; some spindle microtubules attach each pair of homologues to opposite poles.

Anaphase I

Each homologue is separated from its partner, and the two are moved to opposite poles.

Telophase I

A haploid number of chromosomes (still duplicated) ends up at each pole.

We will look at the mechanism of crossing over in later chapters. For now it is enough to know that crossing over leads to **genetic recombination**, which in turn leads to variation in the traits of offspring.

After segments have been exchanged, all four chromatids can spread apart somewhat—but they remain joined at a few places where nonsister chromatids extend across each other (Figure 11.6). Each crosslike, temporary attachment between two nonsister chromatids is a **chiasma** (plural, chiasmata). Such attachments will play a role in aligning the homologues during metaphase I.

Separating the Homologues

The second stage of meiosis, **metaphase I**, is a time of major shufflings of whole chromosomes, before their distribution into daughter nuclei.

Suppose this stage is proceeding right now in one of your germ cells. We can call its homologous chromosomes "maternal" and "paternal" (that is, one of each

type was inherited from your mother and their homologues were inherited from your father). All the homologues line up at the spindle equator. The spindle itself formed earlier, during the transition from prophase I to metaphase I, and now the homologues become attached by microtubules to the spindle poles (Figure 11.5).

Do all the maternal chromosomes become attached to one pole and their homologues to the other? Maybe, maybe not. The spindle apparatus harnesses each pair at random—*and it doesn't matter which one gets attached to which pole*. Even in a cell with only three pairs of homologues, any of these alignments might occur at metaphase I (here, pink indicates maternal chromosomes and green, paternal):

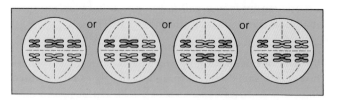

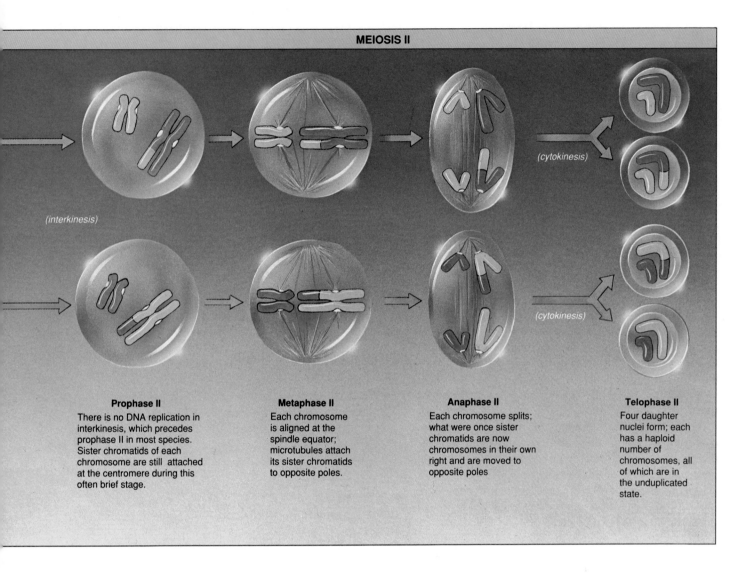

Prophase II
There is no DNA replication in interkinesis, which precedes prophase II in most species. Sister chromatids of each chromosome are still attached at the centromere during this often brief stage.

Metaphase II
Each chromosome is aligned at the spindle equator; microtubules attach its sister chromatids to opposite poles.

Anaphase II
Each chromosome splits; what were once sister chromatids are now chromosomes in their own right and are moved to opposite poles

Telophase II
Four daughter nuclei form; each has a haploid number of chromosomes, all of which are in the unduplicated state.

Figure 11.5 Meiosis: the nuclear division mechanism by which the parental number of chromosomes is reduced by half (to the haploid number) for forthcoming gametes. Only two pairs of homologous chromosomes are shown. The green ones are derived from one parent, and the pink ones are their homologues from the other parent.

During **anaphase I**, the homologues are separated from each other and moved to opposite poles. As you can tell from the sequence in Figure 11.5, the chromosomes at each pole are destined for different gametes. Returning to our three pairs of homologues, we see that 2^3 or 8 combinations of maternal and paternal chromosomes are possible for those gametes:

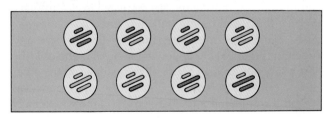

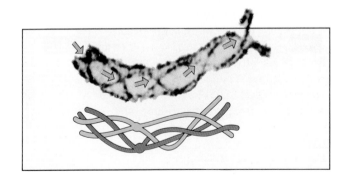

Figure 11.6 Two homologous chromosomes held together by chiasmata (gold arrows). A chiasma is not the same thing as a crossover. Although it indicates crossing over occurred earlier, it is not always located at the site of breakage and exchange. The reason is that chiasmata are not stationary attachments; they are forced toward the chromatid tips as prophase I draws to a close.

Imagine how many chromosome combinations are possible when the 23 homologous pairs of chromosomes in a human germ cell align at metaphase I!

Because homologues align randomly at metaphase I, different gametes will end up with different mixes of maternal and paternal chromosomes.

Typically, anaphase I proceeds to telophase I and **interkinesis**, which often are fleeting stages before the final nuclear division. There is no DNA replication during interkinesis, but remember that each chromosome is still in the duplicated form.

Separating the Sister Chromatids

Meiosis II has one overriding function: separation of the two sister chromatids of each chromosome. During prophase II, nuclear membranes break down and spindle formation proceeds very rapidly. During metaphase II, each duplicated chromosome is moved to the spindle equator. Each one is split at anaphase II; its (formerly) sister chromatids are now chromosomes in their own right.

At each spindle pole, there are only half the number of chromosomes that were present in the parent nucleus. But that haploid number includes one of each homologous pair; there is one of each type of chromosome characteristic of the species. Nuclear membranes form around the cluster of chromosomes at each pole in telophase II. Meiosis is completed.

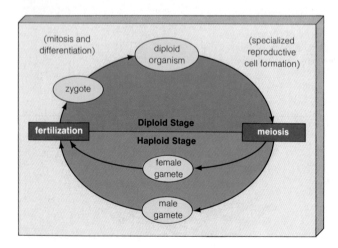

Figure 11.7 Generalized life cycle for animals.

More Gene Shufflings at Fertilization

The diploid number of the parents is restored at **fertilization**, the fusion of nuclei of two gametes in the **zygote** (the first cell of a new multicelled individual). Fertilization, too, contributes to variation in the traits of offspring. Of all the male and female gametes that are produced, *which* two will get together is a matter of chance. Because each gamete has a novel combination of alleles, the sheer number of new combinations that can result at fertilization is staggering. For reasons that will become apparent in the next chapter, even if the parents differ in only ten gene pairs, almost *60,000* different combinations are possible!

The mix of different combinations of alleles from two different gametes at fertilization contributes to variation in the traits of offspring.

MEIOSIS AND THE LIFE CYCLES

Meiosis, gamete formation, fertilization, and mitosis all occur in the life cycles of the plants and animals with which you are probably most familiar. These events vary quite a bit in their details, which are covered in later units. For now, it is enough to keep in mind the following generalized picture.

Animal Life Cycles

Gametogenesis refers to both meiosis and gamete formation in animals. It is followed by fertilization, which marks the formation of the zygote. The zygote grows by mitosis and cytokinesis into a multicelled organism (Figure 11.7).

In *male* animals, meiosis and gamete formation are called **spermatogenesis** (Figure 11.8). Typically a diploid germ cell increases in size and becomes a primary spermatocyte. This large, immature cell undergoes meiosis I and gives rise to two secondary spermatocytes. Both of these cells undergo meiosis II and cytokinesis. The result: four haploid spermatids, which develop into **sperm** (mature male gametes).

In *female* animals, a similar sequence of events is called **oogenesis**. There are two main differences, however. Compared to a primary spermatocyte, many more cytoplasmic components accumulate in a primary oocyte, the female germ cell that undergoes meiosis. Also, the cells formed after meiosis differ in size and function (Figure 11.9). Following meiosis I, one cell (the secondary oocyte) has nearly all the cytoplasm; the other (small) cell is a "polar body." Both cells undergo meiosis II, and the outcome is one large cell and three extremely

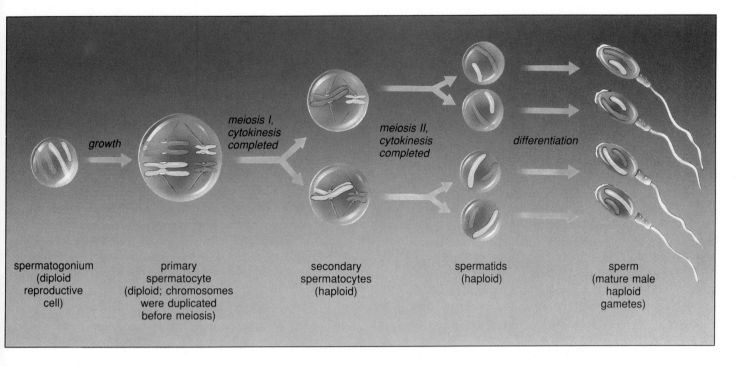

Figure 11.8 Generalized picture of spermatogenesis in male animals. (For the sake of clarity, the nuclear envelopes are not shown in Figures 11.8 and 11.9.)

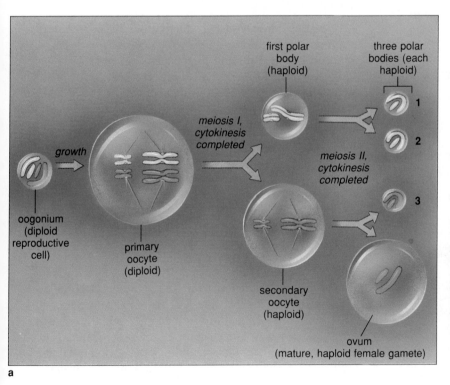

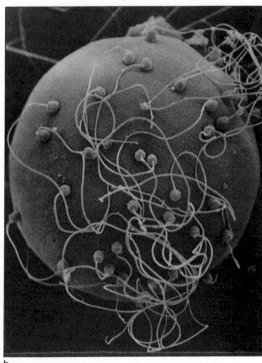

a

b

Figure 11.9 **(a)** Generalized picture of oogenesis in female animals. This sketch is not drawn to the same scale as Figure 11.8. A primary oocyte is *much* larger than a primary spermatocyte, as suggested by the scanning electron micrograph of the egg and sperm of a clam in **(b)**. Also, the polar bodies are extremely small compared to an ovum (see Figure 35.10b, for example).

small ones (that is, three polar bodies). The large cell develops into the mature **egg**, or **ovum**. The polar bodies do not function as gametes. During meiosis they serve as "dumping grounds" for one set of the parental chromosomes, so that the egg will end up with the necessary haploid number (Chapter Thirty-Five).

Plant Life Cycles

Among the familiar land plants, such as pine trees and roses, some special events occur between meiosis and gamete formation. Germ cells within the plant tissues divide by meiosis and produce haploid *meiospores* (Figure 11.10). Later on, mitotic cell divisions transform the meiospore into a multicelled haploid body called a **gametophyte**. (The word means gamete-producing body.)

The female gametophytes of pines are tiny structures perched on scales of some of the pine cones. The pollen grains you may have seen drifting like clouds of dust from tree branches at certain times of the year are the male gametophytes. Sooner or later, some cells from both kinds of gametophytes divide by mitosis and produce cells that will function as eggs and sperm.

After fertilization, the diploid zygote grows into the multicelled diploid body we call "the pine tree." A pine tree is a **sporophyte** (spore-producing plant); it gives rise to the meiospores, in ways that will be described in chapters to come.

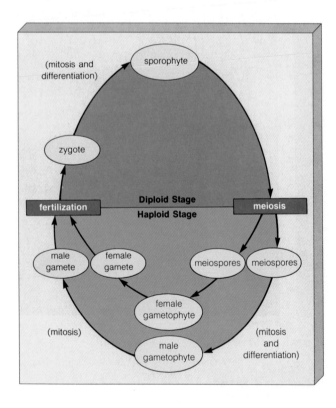

Figure 11.10 Generalized life cycle for complex land plants.

SUMMARY

Key Features of Sexual Reproduction

1. In sexual reproduction, the chromosomes from more than one parent (or from male and female parts of a self-fertilizing organism) are passed on to offspring by events that proceed from meiosis in germ cells to gamete formation, then on to fertilization.

2. Sexually reproducing organisms commonly have a diploid number of chromosomes ($2n$), or two of each type characteristic of the species. The two are "homologous" chromosomes, with the same length, shape, and gene sequence. Homologues pair with each other during meiosis. (The X and Y chromosomes do vary in terms of length, shape, and which genes they carry, but they still have homologous regions that allow them to pair during meiosis.)

3. Commonly, one of each pair of homologues is "paternal" (derived from a male parent) and the other is "maternal" (derived from a female parent).

4. Before meiosis, all chromosomes in a germ cell are duplicated by way of DNA replication. The duplicates remain attached (as sister chromatids) at the centromere.

Stages of Meiosis

1. Meiosis consists of two consecutive divisions of the chromosomes in a germ cell. In meiosis I, each chromosome pairs with and then separates from its homologue. In meiosis II, the sister chromatids of each chromosome separate from each other. In both cases, microtubules of a spindle apparatus organize and move the chromosomes.

2. The following events occur during meiosis I:

   a. Prophase I: Homologues pair with each other. Crossing over occurs (nonsister chromatids break and exchange corresponding segments). Crossing over leads to genetic recombination (new combinations of alleles in chromosomes), which can lead to variation in traits among offspring.

   b. Metaphase I: Each chromosome aligns randomly with its homologue at the spindle equator. This leads to genetic recombination, for the random alignment ensures that different mixes of maternal and paternal chromosomes will end up in different gametes.

   c. Anaphase I: Each chromosome is separated from its homologue and moved to the opposite spindle pole. Nuclear membrane starts to form around the chromosomes at each pole. There is a haploid number of chromosomes at each pole—but each chromosome is still duplicated.

   d. Telophase I: This is a very brief stage that commonly gives way to interkinesis (the transition to meiosis II). There is no DNA replication during interkinesis.

3. The following events occur during meiosis II:

a. Prophase II: Nuclear membranes break down and spindle formation occurs very rapidly.

b. Metaphase II: Microtubules attach to the centromere of each chromosome and move all chromosomes to the spindle equator.

c. Anaphase II: Sister chromatids of each chromosome are separated and move to opposite poles. Once separated, they are chromosomes in their own right.

d. Telophase II: A new nuclear envelope forms around each of the four clusters of separated chromosomes; four haploid nuclei are the result.

4. Following cytokinesis (cytoplasmic division), there are four haploid cells, one or all of which may function as gametes:

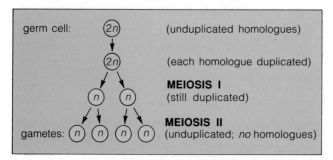

5. For most multicelled animals, four haploid cells are produced through meiosis and gamete formation (gametogenesis). Some or all of those cells may function as male gametes (sperm) or female gametes (eggs). Fusion of a sperm and an egg nucleus at fertilization produces a diploid cell (zygote), which develops into the multicelled form by mitotic cell division.

6. For most land plants, meiosis leads to formation of haploid meiospores. Through mitotic cell divisions, the meiospore gives rise to a multicelled haploid body (gametophyte), which produces gametes. The zygote resulting from fertilization grows into a new sporophyte (which produces meiospores).

Meiosis Compared With Mitosis

In this unit, our main focus has been on mitosis and meiosis—two nuclear division mechanisms used in the reproduction of eukaryotes. Mitosis underlies asexual reproduction of the single-celled forms as well as growth of the multicelled forms. Meiosis occurs only in germ cells used in sexual reproduction. Figure 11.11 summarizes the similarities and differences between the two mechanisms.

The major difference between them is this: Mitotic cell division produces clones (genetically identical copies of the parent cell). Meiosis and fertilization give rise to novel combinations of alleles in offspring which, as a consequence, vary from the parents and one another in

the details of their traits. Three events are responsible for the variation:

1. Crossing over and genetic recombination during prophase I of meiosis.

2. The random alignment of each chromosome with its homologue before their separation during metaphase I, so that forthcoming gametes can end up with mixes of maternal and paternal chromosomes.

3. The chance mix of different combinations of alleles from two different gametes at fertilization.

The variation so produced is a testing ground for agents of selection, hence for the evolution of populations.

Review Questions

1. Define sexual reproduction. How does it differ from asexual reproduction? What is one of its presumed advantages? *150*

2. Is this statement true: Asexual reproductive modes occur only among prokaryotes; eukaryotes rely exclusively on sexual reproduction. *150*

3. The nucleus of diploid cells contains pairs of chromosomes that resemble each other in length, shape, and which genes they carry. What are the pairs called? *151*

4. Refer back to Table 10.1, which gives the diploid number of chromosomes in the body cells of a few organisms. What would be the *haploid* number for the gametes of humans? For the garden pea? *140*

5. Suppose the diploid cells of an organism have four pairs of homologous chromosomes, designated AA, BB, CC, and DD. How would its haploid number of chromosomes be designated? *151*

6. When, and in which type of cells, does meiosis occur? *151*

7. Define meiosis and characterize its main stages. In what respects is meiosis like mitosis? In what respects is it unique? *152, 159*

8. Does crossing over occur during mitosis, meiosis, or both? At what stage of nuclear division does it occur, and what is its significance? *153*

9. Distinguish between: (a) gamete and meiospore, and (b) gametophyte and sporophyte. *156–158*

10. Outline the steps involved in spermatogenesis and oogenesis. *156-158*

Readings

John, B. 1976. "Myths and Mechanisms of Meiosis." *Chromosoma* 54:295–325.

Mitchison, J. 1972. *Biology of the Cell Cycle.* New York: Cambridge University Press.

Prescott, D. 1988. *Cells.* Boston: Jones & Bartlett.

Strickberger, M. 1985. *Genetics.* Third edition. New York: Macmillan. Contains excellent introduction to chromosomes and meiosis.

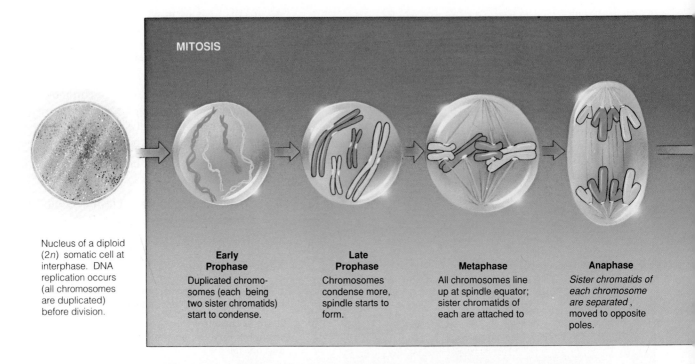

MITOSIS

Nucleus of a diploid (2n) somatic cell at interphase. DNA replication occurs (all chromosomes are duplicated) before division.

Early Prophase
Duplicated chromosomes (each being two sister chromatids) start to condense.

Late Prophase
Chromosomes condense more, spindle starts to form.

Metaphase
All chromosomes line up at spindle equator; sister chromatids of each are attached to

Anaphase
Sister chromatids of each chromosome are separated, moved to opposite poles.

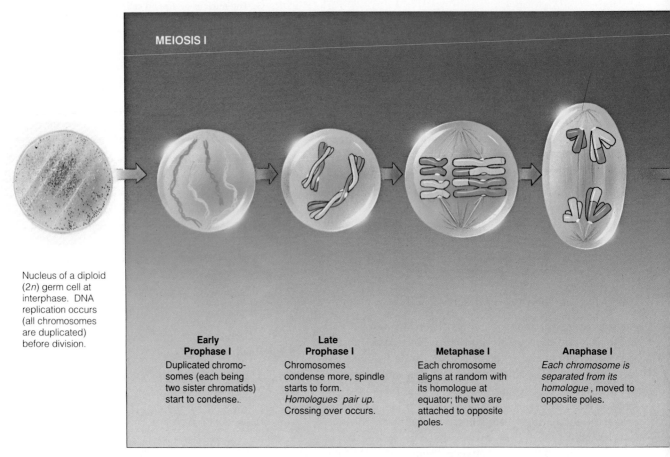

MEIOSIS I

Nucleus of a diploid (2n) germ cell at interphase. DNA replication occurs (all chromosomes are duplicated) before division.

Early Prophase I
Duplicated chromosomes (each being two sister chromatids) start to condense.

Late Prophase I
Chromosomes condense more, spindle starts to form. *Homologues pair up.* Crossing over occurs.

Metaphase I
Each chromosome aligns at random with its homologue at equator; the two are attached to opposite poles.

Anaphase I
Each chromosome is separated from its homologue, moved to opposite poles.

Figure 11.11 Summary of mitosis and meiosis, using a diploid (2n) animal cell as the example. The diagram is arranged to help you compare the similarities and differences between the two division mechanisms. (Chromosomes derived from the male parent are green; their homologues from the female parent are pink.)

Telophase
Each daughter nucleus is *diploid* (2*n*), with pairs of homologous chromosomes. Each chromosome is in the unduplicated state.

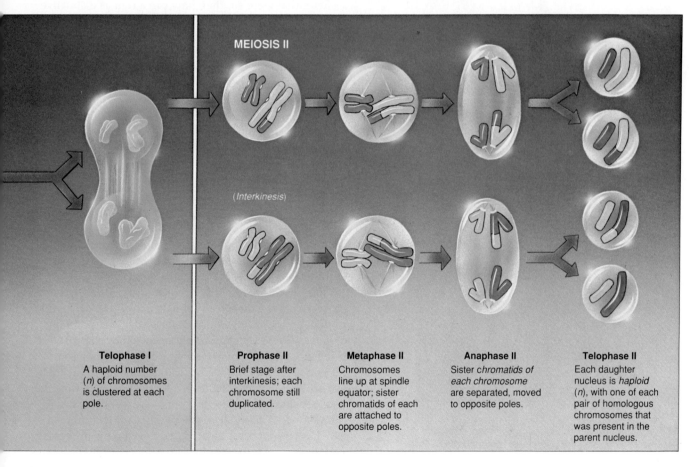

MEIOSIS II

(Interkinesis)

Telophase I
A haploid number (*n*) of chromosomes is clustered at each pole.

Prophase II
Brief stage after interkinesis; each chromosome still duplicated.

Metaphase II
Chromosomes line up at spindle equator; sister chromatids of each are attached to opposite poles.

Anaphase II
Sister *chromatids of each chromosome* are separated, moved to opposite poles.

Telophase II
Each daughter nucleus is *haploid* (*n*), with one of each pair of homologous chromosomes that was present in the parent nucleus.

12

OBSERVABLE PATTERNS OF INHERITANCE

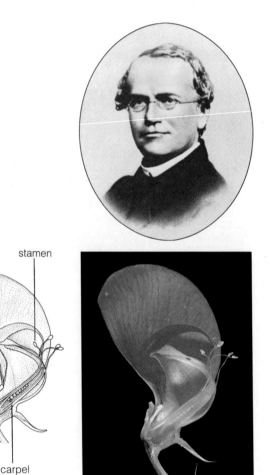

stamen

carpel

Figure 12.1 Gregor Mendel, founder of modern genetics, and the focus of his experiments: the garden pea plant *(Pisum sativum)*. A flower has been sectioned to show the location of stamens (male reproductive organs) and the carpel (the female reproductive organ).

MENDEL'S INSIGHTS INTO THE PATTERNS OF INHERITANCE

Toward the end of the nineteenth century, biology was dominated by talk of Darwin and Wallace's theory of evolution by natural selection. According to the theory, a population could change (evolve) only if its members show variation in heritable traits (such as height, body color, and so on). Variations that improved chances of surviving and reproducing would show up more often in each generation, those that didn't would show up less and might even be eliminated, and in time the character of the population would change. (We considered the reasoning behind this theory in Chapter Two.)

Not everyone accepted the theory. For one thing, it did not fit with a prevailing notion of how males and females pass on their traits to offspring. It was common knowledge that sperm and eggs must carry the instructions about heritable traits—but how were those instructions combined at fertilization? Many biologists thought they blended together, much like cream into coffee.

Yet if that were so, then how can a trait as distinctive as freckles keep turning up among generations of nonfreckled offspring? How come the descendants of a herd of white stallions and black mares are not all uniformly gray? "Blending" scarcely explained the observable fact that distinctive traits aren't diluted out of populations, but it was considered a rule anyway. And Darwin had problems on this account, because uniform populations would present no variation whatsoever for selective agents to act upon. That being the case, "evolution" simply could not occur.

Even before the Darwin-Wallace theory was made public, however, someone was gathering evidence which, in time, would help support its basic premise about the heritability of traits. In a monastery garden in Brünn, northeast of Vienna, a scholarly monk named Gregor Mendel was beginning to identify the rules governing inheritance.

The monastery of St. Thomas was somewhat removed from the European capitals, which were then the centers of scientific inquiry. Yet Mendel was not a man of narrow interests who simply stumbled by chance onto principles of great import. Having been raised on a farm, he was well aware of agricultural principles and their application. He kept abreast of breeding experiments and developments described in the available journals. Mendel was a founder of the regional agricultural society. He won several awards for developing improved varieties of fruits and vegetables. After entering the monastery, he spent two years studying mathematics at the University of Vienna.

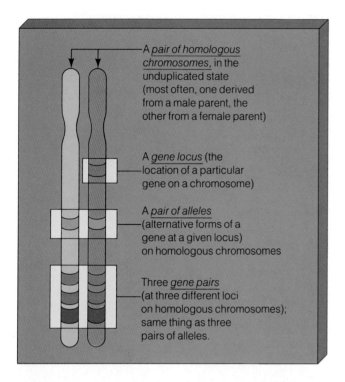

Figure 12.2 A few genetic terms illustrated. In Mendel's time, no one knew about meiosis or chromosomes, but it was clear that offspring received hereditary material from parents by way of sperm and eggs. As we now know, the hereditary material (genes) is packaged in homologous chromosomes (one from the male, one from the female parent). Thus at each gene locus along the chromosomes, one allele has come from the male parent and its partner has come from the female parent.

Shortly after his university training, Mendel began experiments on the nature of plant diversity. Through his combined talents in plant breeding and mathematics, he perceived patterns in the emergence of traits from one generation to the next.

Mendel's Experimental Approach

Mendel experimented with the garden pea plant, *Pisum sativum* (Figure 12.1). This plant fertilizes itself. Different parts of its flowers give rise to what will become male *and* female gametes (sperm and eggs), and fertilization normally occurs between two gametes of the same flower. Some pea plants are **true-breeding**, in that successive generations of offspring are exactly like the parents in one or more traits. (For example, they all may have white flowers.) Of course, when left to their own devices, true-breeding pea plants show a rather monotonous and uninformative pattern of inheritance.

However, pea plants also lend themselves to **cross-fertilization**, whereby sperm from one individual fer-

tilize eggs from another. To prevent a plant from self-fertilizing, Mendel could open a flower bud and remove the stamens. (Stamens are floral structures that bear pollen grains, in which the sperm develop.) If pollen from another plant were brushed on the "castrated" bud, cross-fertilization could follow.

Why did Mendel tinker with plants this way? He wanted cross-fertilization to occur between two true-breeding plants that showed clearly contrasting traits. In one case, for example, he crossed a white-flowered plant with a purple-flowered plant. If their offspring produced white *or* purple flowers, he could easily identify one plant or the other as the source of the hereditary material for that trait. Thus, if there *were* patterns in the way hereditary material is transmitted from parents to offspring, the use of contrasting traits might be a way to identify them.

It will be useful to retrace a few of Mendel's experiments. The conclusions he drew from them have turned out to apply, with some modification, to all sexually reproducing organisms.

Some Terms Currently Used in Genetics

Having read the chapter on meiosis, you already have insight into the actual mechanisms underlying sexual reproduction—which is more than Mendel had. Neither he nor his contemporaries even knew about chromosomes and so could not have known that the parental chromosome number is reduced by half in gametes, then restored at fertilization. Yet Mendel had some remarkably good hunches about what was going on. As we follow his thinking, let's simplify things by substituting a few terms that have since become commonplace in the study of inheritance:

1. Genes are units of instructions for producing or influencing a specific trait in offspring.

2. Each gene has its own **locus**, or particular location, on a chromosome (Figure 12.2).

3. Diploid cells are common in sexually reproducing species. Cells of this type have inherited two genes for each trait, one on each of two homologous chromosomes:

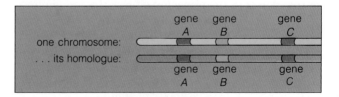

Two genes at homologous loci are called a **gene pair**.

4. Although the two genes of a pair deal with the same trait, they may vary in their information about it. This happens when they differ in form, as when a gene for flower color specifies "red" and a different molecular form of that gene specifies "white." The various molecular forms of a gene are called **alleles**.

5. The gene shufflings during meiosis and fertilization can put together different mixes of alleles in offspring. Thus a pair of alleles at homologous loci might be the same molecular form or different from each other. If they are the same, this is a **homozygous** condition; if different, this is a **heterozygous** condition.

6. Often one allele is "dominant," meaning its effect on a trait actually masks the effect of a "recessive" allele that occupies the homologous locus. We normally use capital letters to indicate dominance and lowercase letters to indicate recessiveness (for example, *A* and *a*).

7. Putting this all together, we say that a **homozygous dominant** individual has two dominant alleles (*AA*) for the trait being studied. A **homozygous recessive** individual has two recessive alleles (*aa*). In a **heterozygous** individual, the two alleles are different (*Aa*).

8. To keep the distinction clear between genes and the traits they specify, we use the word **genotype** for the sum total of an individual's genes (or even for one gene pair at a time). And we use **phenotype** for an individual's traits—that is, the observable aspects of its structure, physiology, and behavior.

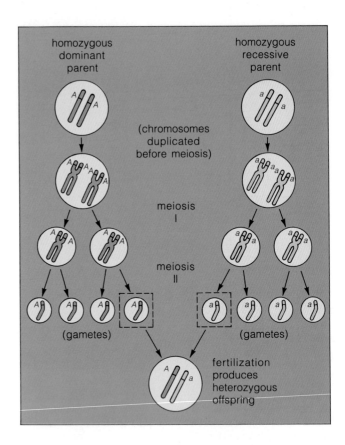

Figure 12.3 Segregation of alleles in a monohybrid cross. Two parents that are true-breeding for contrasting forms of a single trait can give rise only to heterozygous offspring.

The Concept of Segregation

Let's now turn to Mendel's first series of experiments, which we would call "monohybrid crosses." In this type of cross, two parents that are true-breeding for contrasting forms of a single trait give rise to heterozygous offspring. (Remember that the two alleles for a trait are different in heterozygotes.) Mendel studied seven different traits, one at a time (Figures 12.3 and 12.4).

In one monohybrid cross, a purple-flowered plant and a white-flowered plant yielded offspring with purple flowers *only*. What happened to the white-flower trait? Had it disappeared? Maybe not. Mendel allowed the purple-flowered offspring to self-fertilize and produce seeds, rather than crossing them with new plants. (Can you guess why?) Later, some plants grown from the seeds had white flowers! The white-flower trait had not been lost, after all.

Results were much the same for all of Mendel's monohybrid crosses. One form of the trait seemed to disappear in the hybrid offspring of the first-generation plants, only to show up again in some of the second-generation plants.

To explain those results, Mendel assumed that each plant had two "unblended" units of hereditary material for the trait. When it came time to produce sperm (or eggs), the two units somehow were moved apart (segregated) and ended up in separate gametes. Mendel reasoned that one parent had two "dominant" units and the other had two "recessive" units. Why? The hybrid offspring, which received one unit from each parent, had *purple flowers*: the unit specifying purple was being expressed, but the one specifying white was not.

Let's recast Mendel's conclusion in light of what we know about meiosis and fertilization. Pea plants are diploid, with pairs of homologous chromosomes. Assume one parent is homozygous dominant (*AA*) for flower color and the other, homozygous recessive (*aa*). Following meiotic cell division, a chromosome carrying the gene for flower color will be present in each gamete. When gametes from the parents combine at fertilization, only one outcome is possible: *A* + *a* = *Aa*. Figure 12.3 shows why this is so. As you might deduce from this figure, it isn't necessary to depict every single gamete

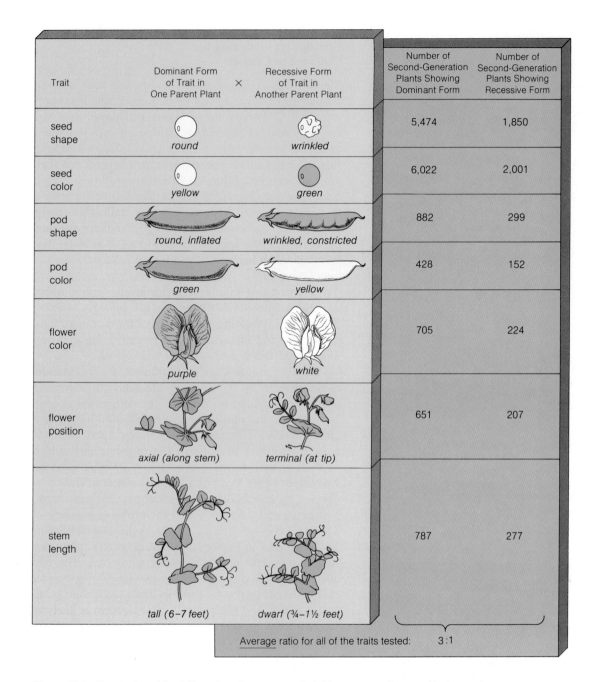

Trait	Dominant Form of Trait in One Parent Plant	×	Recessive Form of Trait in Another Parent Plant	Number of Second-Generation Plants Showing Dominant Form	Number of Second-Generation Plants Showing Recessive Form
seed shape	round		wrinkled	5,474	1,850
seed color	yellow		green	6,022	2,001
pod shape	round, inflated		wrinkled, constricted	882	299
pod color	green		yellow	428	152
flower color	purple		white	705	224
flower position	axial (along stem)		terminal (at tip)	651	207
stem length	tall (6–7 feet)		dwarf (¾–1½ feet)	787	277

Average ratio for all of the traits tested: 3:1

Figure 12.4 Results from Mendel's series of seven monohybrid cross experiments with the garden pea.

that forms. With homozygous parents, all gametes will have the same genotype. With heterozygous parents, they can only have one of two genotypes. Thus we can use a simpler diagram to show the genotypes possible when the two alleles segregate:

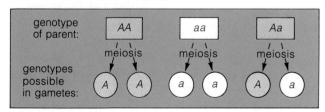

Mendel formulated a principle on the basis of such reasoning. Even though his principle has been modified somewhat over the years, it is still a useful starting point for a discussion of the nature of inheritance. We state it here in modern terms:

Mendelian principle of segregation. Diploid organisms inherit a pair of genes for each trait (on a pair of homologous chromosomes). The two genes segregate from each other during meiosis, so that each gamete formed will end up with one or the other gene, but not both.

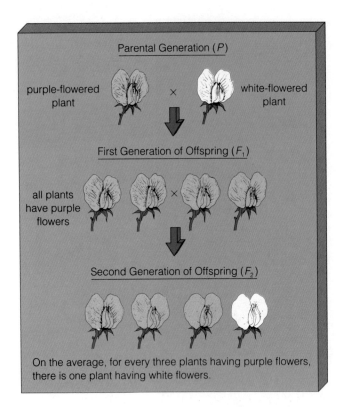

Figure 12.5 Example of a pattern that emerged regularly in Mendel's crosses between true-breeding parents differing only in one contrasting trait (here, flower color). All first-generation plants showed the dominant form of the trait. On the average, though, three of every four second-generation plants showed the dominant form and one showed the recessive.

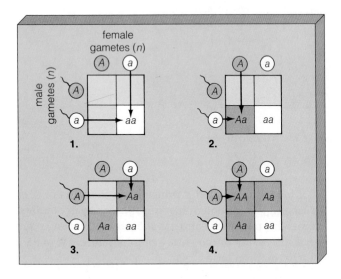

Figure 12.6 Punnett-square method of predicting the probable ratio of traits that will show up in offspring of self-fertilizing individuals known to be heterozygous (*Aa*) for a trait. The circles represent female gametes, or eggs; circles with "tails" represent male gametes, or sperm. The letters inside gametes represent the dominant or recessive form of the gene being tracked. Each square depicts the genotype of one kind of offspring.

Probability: Predicting the Outcome of Crosses

Mendel crossed hundreds of plants and kept track of thousands of offspring, rather than restricting his experiments to a few plants as others had done. Also, he carefully *counted* and *recorded* the number of dominant and recessive offspring for each cross (see, for example, Figure 12.4). Certain numerical ratios emerged, and they strongly suggested that the hereditary material for each trait does not "blend" at fertilization but rather retains its distinct identity. In describing those ratios, let's make use of a few convenient symbols. *P* stands for parental generation, F_1 for first generation (the *F* stands for *filial*, which is taken to mean offspring), and F_2 for the second generation.

An intriguing ratio emerged from the large number of crosses between F_1 plants. On the average, of every four of the F_2 offspring, three showed the dominant form of the trait and one showed the recessive (Figure 12.5). How could this 3:1 phenotypic ratio arise?

Drawing on his knowledge of mathematics, Mendel came up with an explanation. He began by assuming that each particular sperm is not precommitted to combining with one particular egg; fertilization has to be a chance event. This meant his crosses could be interpreted according to certain rules of probability, which apply to chance events. ("Probability" simply means the number of times that one particular outcome will occur divided by the total number of all possible outcomes.)

The easiest way to predict the probable outcome of a cross between two F_1 plants is the *Punnett-square method*, shown in Figure 12.6. Assume each F_1 plant carried a dominant and a recessive allele (*Aa*). The two alleles segregated at meiosis, so two kinds of sperm (or two kinds of eggs) were produced in equal proportions: half were *A*, and half were *a*. If any sperm is likely to fertilize any egg, there were four possible outcomes for each encounter:

Possible Event:	Probable Outcome:	
sperm *A* meets egg *A*	$\frac{1}{4}$ *AA* offspring	
sperm *A* meets egg *a*	$\frac{1}{4}$ *Aa*	or $\frac{1}{2}$ *Aa*
sperm *a* meets egg *A*	$\frac{1}{4}$ *Aa*	
sperm *a* meets egg *a*	$\frac{1}{4}$ *aa*	

As you can see, a new plant had three chances in four of carrying one or both dominant alleles. And it had only one chance in four of carrying two recessive alleles. The probable phenotypic ratio was 3:1. Figure 12.7 illustrates this outcome for one of Mendel's monohybrid crosses, using the Punnett-square method.

Keep in mind that Mendel's ratios weren't *exactly* 3:1. (See, for example, the numerical results in Figure 12.4.) To understand why, flip a coin a few times. We all know that a coin is just as likely to end up heads as tails.

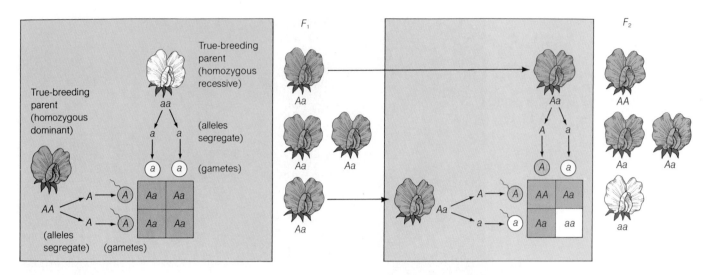

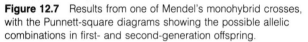

Figure 12.7 Results from one of Mendel's monohybrid crosses, with the Punnett-square diagrams showing the possible allelic combinations in first- and second-generation offspring.

But often it ends up heads, or tails, several times in a row. When you flip the coin only a few times, the actual ratio may differ greatly from the predicted ratio of 1:1. Only when you flip the coin *many* times will you come close to the predicted ratio. Almost certainly, Mendel's reliance on a large number of crosses and his understanding of probability kept him from being confused by minor deviations from the predicted results.

Testcrosses

Mendel gained support for his concept of segregation through the **testcross**, in which first-generation hybrids are crossed to an individual known to be true-breeding for the same recessive trait as the recessive parent.

In one case, purple-flowered F_1 plants were backcrossed with true-breeding, white-flowered plants. If Mendel's idea were not correct (if the recessive unit had lost its identity when combined with the dominant unit), then only the dominant form of the trait would show up in the testcross offspring. If his idea were correct, though, there would have to be about as many recessive as dominant plants in the offspring from the testcross:

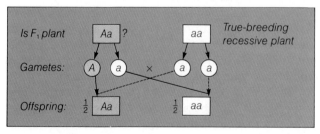

This is exactly what happened in the testcrosses. As predicted (Figure 12.8), about half the testcross offspring were purple-flowered (*Aa*) and half were white-flowered (*aa*).

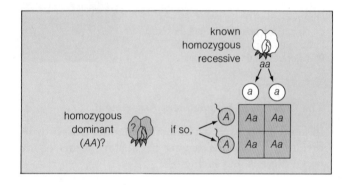

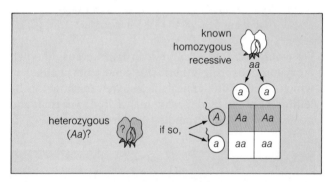

Figure 12.8 Punnett-square method of predicting the outcomes of a testcross between an individual known to be homozygous recessive for a trait (here, white flower color) and an individual that shows the dominant form of the trait. If the individual of unknown genotype is homozygous dominant, all offspring will show the dominant form of the trait. If the individual is heterozygous, about half the offspring will show the recessive form.

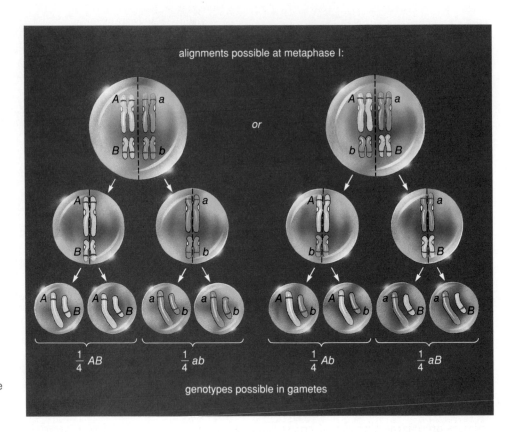

alignments possible at metaphase I:

or

$\frac{1}{4}$ AB $\frac{1}{4}$ ab $\frac{1}{4}$ Ab $\frac{1}{4}$ aB

genotypes possible in gametes

Figure 12.9 Example of independent assortment, showing just two of the pea plant's seven pairs of homologous chromosomes. The different allelic combinations possible in gametes arise through the random alignment of homologues during metaphase I.

The Concept of Independent Assortment

In another series of experiments, Mendel crossed true-breeding pea plants having contrasting forms of *two* traits. Here is one such cross:

purple-flowered,   white-flowered,
tall plant   ×   dwarf plant

We would call this a "*di*hybrid cross," because the heterozygous offspring inherit *two* gene pairs, neither of which consists of identical alleles. We can show this by letting *A* and *B* stand for dominance in flower color and height, with *a* and *b* standing for recessiveness:

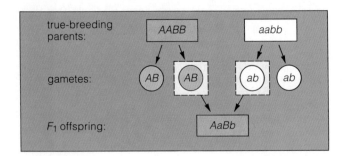

true-breeding parents: AABB   aabb

gametes: AB AB   ab ab

F_1 offspring: AaBb

Mendel anticipated (correctly) that all the F_1 offspring would be purple-flowering and tall. But what would happen when it was time for those *AaBb* offspring to form sperm and eggs of their own? Would a gene for flower color and a gene for height *travel together or independently of each other* into gametes?

Only two combinations were possible if the genes for flower color and height from a parent stayed together during gamete formation (either *AB* or *ab*, just like the parental gametes). But if the two gene pairs segregated independently of each other, then four allelic combinations in gametes were possible:

$$\frac{1}{4}AB \qquad \frac{1}{4}Ab \qquad \frac{1}{4}aB \qquad \frac{1}{4}ab$$

This is indeed what happens in pea plants.

As we now know, a gene pair tends to segregate into gametes independently of other gene pairs—when the others are located on *non*homologous chromosomes. Remember the random alignment of maternal and paternal chromosomes during metaphase I of meiosis? By depicting just two of the pea plant's seven pairs of chromosomes, we can readily see how the four different combinations of alleles can arise (Figure 12.9).

Now think of the variation that can arise at fertilization. Simple multiplication (four kinds of sperm times four kinds of eggs) shows that sixteen allelic combinations are possible in dihybrid F_2 plants. Figure 12.10 shows the possibilities, using the Punnett-square method. When we add up the possible combinations,

Figure 12.10 Results from Mendel's dihybrid cross between true-breeding parent plants differing in two traits (flower color and height). Here, *A* and *a* represent the dominant and recessive alleles for flower color. *B* and *b* represent the dominant and recessive alleles for height. On the average, the phenotypic combinations in the F_2 generation occur in a 9:3:3:1 ratio. Keep in mind that working a Punnett square is really a way to show *probabilities* of certain combinations occurring as a result of allele shufflings during meiosis and fertilization.

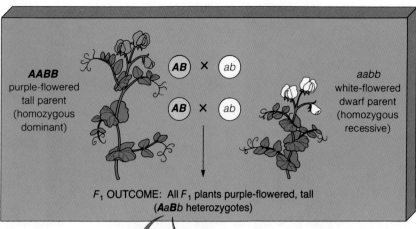

AABB
purple-flowered
tall parent
(homozygous
dominant)

AB × **ab**

AB × **ab**

aabb
white-flowered
dwarf parent
(homozygous
recessive)

F_1 OUTCOME: All F_1 plants purple-flowered, tall
(*AaBb* heterozygotes)

we get $\frac{9}{16}$ tall purple-flowered, $\frac{3}{16}$ dwarf purple-flowered, $\frac{3}{16}$ tall white-flowered, and $\frac{1}{16}$ dwarf white-flowered plants. That is a phenotypic ratio of 9:3:3:1.

Results were close to a 9:3:3:1 ratio in all of Mendel's dihybrid crosses. Results of this sort led to the formulation of another principle:

Mendelian principle of independent assortment. Each gene pair tends to assort into gametes independently of other gene pairs located on nonhomologous chromosomes.

The variety resulting from independent assortment and hybrid crossing is staggering. In a monohybrid cross (involving a single gene pair), only three genotypes are possible (*AA*, *Aa*, and *aa*). We can represent this as 3^n, where *n* is the number of gene pairs. When more gene pairs are involved, the number of possible combinations increases dramatically. Even if parents differ in only ten gene pairs (3^{10}), almost 60,000 genotypes are possible for their offspring. If they differ in twenty gene pairs (3^{20}), the number is close to $3\frac{1}{2}$ billion!

On the basis of his experimental results, Mendel was convinced that hereditary material comes in units that retain their physical identity. In 1865 he reported this idea before the Brünn Society for the Study of Natural Science. His report made no impact whatsoever. The following year his paper was published. Apparently it was read by few and understood by no one. Remember, Mendel was challenging the well-entrenched blending theory of inheritance. His mathematical analysis of traits probably would not have made sense to anybody but mathematicians—who probably would not have had the least bit of interest in pea plants.

In 1871 Mendel became an abbot of the monastery, and his experiments gave way to administrative tasks. He died in 1884, never to know his work would be the starting point for the development of modern genetics.

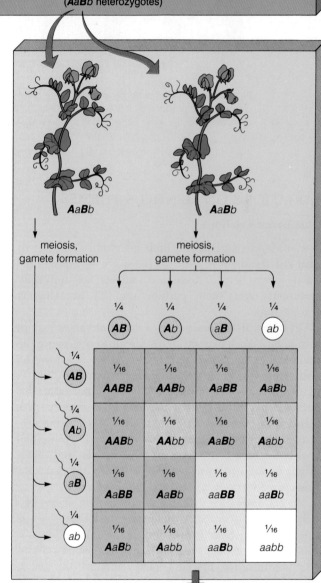

AaBb

AaBb

meiosis,
gamete formation

meiosis,
gamete formation

¼ · ¼ · ¼ · ¼

AB · **Ab** · **aB** · **ab**

¼ **AB**	¹⁄₁₆ **AABB**	¹⁄₁₆ **AABb**	¹⁄₁₆ **AaBB**	¹⁄₁₆ **AaBb**
¼ **Ab**	¹⁄₁₆ **AABb**	¹⁄₁₆ **AAbb**	¹⁄₁₆ **AaBb**	¹⁄₁₆ **Aabb**
¼ **aB**	¹⁄₁₆ **AaBB**	¹⁄₁₆ **AaBb**	¹⁄₁₆ **aaBB**	¹⁄₁₆ **aaBb**
¼ **ab**	¹⁄₁₆ **AaBb**	¹⁄₁₆ **Aabb**	¹⁄₁₆ **aaBb**	¹⁄₁₆ **aabb**

ADDING UP THE F_2 COMBINATIONS POSSIBLE:

◻ ⁹⁄₁₆ or 9 purple-flowered, tall

◻ ³⁄₁₆ or 3 purple-flowered, dwarf

◻ ³⁄₁₆ or 3 white-flowered, tall

◻ ¹⁄₁₆ or 1 white-flowered, dwarf

Figure 12.11 Incomplete dominance at one gene locus. Red-flowering and white-flowering homozygous snapdragons produce pink-flowering plants in the first generation. The red allele (R^1) is only partially dominant over the white allele (R^2) in the heterozygous state.

VARIATIONS ON MENDEL'S THEMES

Dominance Relations

It was Mendel's genius to limit his studies to cases of clear-cut dominance. Not all cases are as straightforward, since some "dominant" alleles are not fully dominant over their partner on the homologous chromosome.

In **incomplete dominance**, a dominant allele cannot completely mask expression of the allele occupying the homologous locus. When homozygous red-flowered and white-flowered snapdragons are crossed, the F_1 plants all have *pink* flowers. Without further tests, this outcome might imply that the hereditary material was blended in offspring. But cross-fertilization between F_1 individuals yields these phenotypes: $\frac{1}{4}$ red, $\frac{1}{2}$ pink, and $\frac{1}{4}$ white (Figure 12.11). Apparently a single "red" allele is not enough to form sufficient pigment to make the flowers appear red, as two red alleles can do.

In **codominance**, the expression of *both* alleles of a pair can be discerned in heterozygotes. Expression of one does not mask expression of the other. Consider that some proteins at the surface of red blood cells act like identification markers. (They identify the cell as being of a certain kind.) A gene coding for one of the markers occurs at the so-called ABO blood group locus. Because there are more than two types of alleles for this locus, we call them a **multiple allele system**.

Two of the alleles are codominant and code for two forms of the marker (designated A and B). The third allele is recessive, and in homozygous recessive people there is an absence of either A or B (a condition called

O). Thus four blood types are possible, depending on which two alleles are present in a person's cells. Using the symbols I^A and I^B to represent the codominant alleles and i to represent the recessive allele, then:

Allelic Combination:	Blood Type Produced:
$I^A I^A$ or $I^A i$	A
$I^B I^B$ or $I^B i$	B
$I^A I^B$	AB
ii	O

In **variable expressivity**, the phenotype associated with a dominant allele varies, by different degrees, from one individual to the next. This condition occurs among humans who show *polydactyly* (the presence of extra fingers, toes, or both). As a human embryo develops, a dominant allele (D) controls how many bone sets will form within the paddlelike appendages destined to become hands and feet (see Figure 34.13). The *Dd* genotype varies in how it is expressed. Some carriers of the dominant allele have five-fingered hands but six-toed feet; others have five-toed feet but six-fingered hands. Still others may have five fingers on one hand and six on the other. Figure 12.12 shows a pedigree of polydactyly. A **pedigree** is simply a chart of the genetic relationships of individuals.

Interactions Between Different Gene Pairs

On thinking about the traits described so far, you might conclude (as Mendel did) that each trait arises from the expression of a single gene pair. Generally, however,

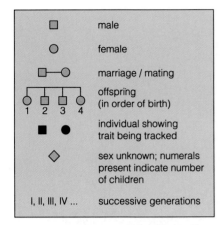

☐	male
○	female
☐—○	marriage / mating
○ ☐ ☐ ○ 1 2 3 4	offspring (in order of birth)
■ ●	individual showing trait being tracked
◇	sex unknown; numerals present indicate number of children
I, II, III, IV ...	successive generations

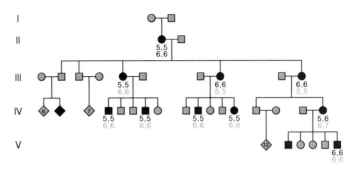

Figure 12.12 Pedigree of polydactyly, showing variable expressivity of the dominant allele for the trait. The phenotype of female I is uncertain, but she probably was polydactylous. The number of digits on each hand is shown in black numerals, the number on each foot is shown in blue. The boxed inset explains some of the symbols used in constructing pedigree diagrams.

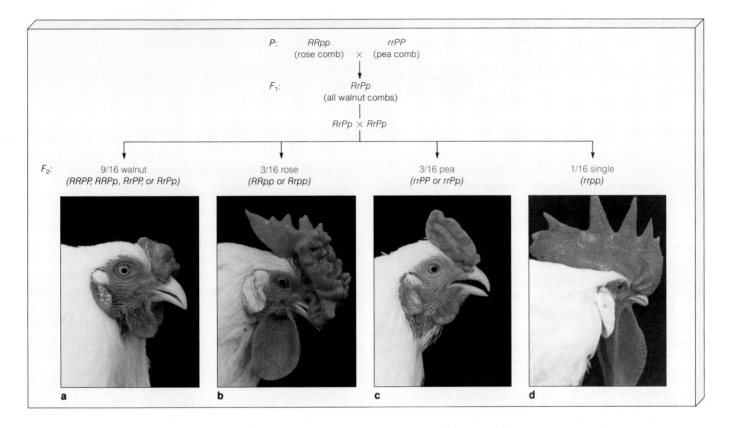

Figure 12.13 Interaction between two genes affecting the same trait in domestic breeds of chickens. The initial cross is between a Wyandotte (with a rose comb on the crest of its head) and a brahma (with a pea comb). With complete dominance at the gene locus for pea comb and at the gene locus for rose comb, the products of these two nonallelic genes interact and give a walnut comb (**a**). With complete recessiveness at both loci, the products interact and give rise to a single comb (**d**).

genes interact with one another to produce some effect on phenotype.

In some interactions, two gene pairs *cooperate* to produce a phenotype that neither can produce alone. W. Bateson and R. Punnett identified two gene pairs that cooperate to produce comb shape in chickens. The allelic combinations *rr* at one locus and *pp* at the other give rise to the most common phenotype, the single comb. Other phenotypes occur when the two dominant alleles, *R* and *P*, are present. Depending on the allelic combinations, the two gene pairs can produce rose, pea, or walnut combs as well as the single comb (Figure 12.13).

In another interaction, called **epistasis**, one gene pair *masks* the expression of another and some expected phenotypes do not appear at all. This is what has happened in a white-haired, pink-eyed rabbit. As with other

Figure 12.14 A rare albino rattlesnake, showing the pink eyes and white coloration characteristic of animals that are unable to produce the brownish-black pigment melanin. In birds and mammals, surface coloration is due almost entirely to the color of feathers and fur. In fishes, amphibians, and reptiles, it is due to color-bearing cells located in the skin. Some of the cells contain melanin. Others contain red to yellow pigments. Still others contain crystals that reflect light and alter the effect of other pigments present.

The mutation affecting melanin production in the snake shown here has no effect on the production of yellow-to-red pigments and light-reflecting crystals. Hence the snake's skin appears iridescent yellow as well as white.

mammals, a rabbit's coat color results from interactions among different genes that help control the type, distribution, and amount of *melanin* (a brownish-black pigment molecule) in a given body region. At one gene locus, the *B* allele for "black coat" is dominant to the *b* allele for "brown coat." Yet a dominant allele (*C*) at another locus influences the actual phenotype.

The *C* allele codes for tyrosinase, the first enzyme needed in a series of reactions that produce melanin-containing granules. If a rabbit or some other mammal is homozygous recessive (*cc*) at this second locus, no melanin can be produced. It makes no difference which combination of *B* and *b* alleles exists at the first gene locus. The animal will be an *albino*, a phenotype arising from the absence of melanin.

Albino mammals have white hair, light skin, and red or pink eyes. (Red light is reflected from blood vessels in the eye when the retina lacks absorptive pigment.) Albinism also occurs among fishes, amphibians, reptiles, and birds (Figure 12.14).

When an albino guinea pig with the genotype *cc BB* is mated to a brown guinea pig with the genotype *CC bb*, the F_1 generation will be *Cc Bb* and will have black coats. What will be the phenotypic ratio of the offspring produced when two of the F_1 individuals are mated?

Many gene pairs modify, interfere with, or prevent the expression of other gene pairs at different loci.

Multiple Effects of Single Genes

Since Mendel's time, studies have also shown that a single gene can exert effects on seemingly unrelated aspects of an individual's phenotype. This aspect of gene expression is known as **pleiotropy**.

One type of allele with pleiotropic effects gives rise to *sickle-cell anemia*, which is actually a group of symptoms of an underlying genetic disorder. The blood of adult humans carries a red oxygen-transporting pigment, hemoglobin A (or HbA). Some people carry a variant form of the pigment called HbS. The HbS molecules can still transport oxygen, which is vital for cellular respiration. After giving up oxygen to tissue regions, however, the molecules interlock with one another. They actually stack up like rigid poles, and the stacking often distorts red blood cells into a "sickle" shape. Figure 12.15c shows a few cells that have been distorted as a result of the abnormal molecular interactions.

The deformed cells clump together in capillaries (blood vessels having tiny diameters) and hamper the flow of oxygen to cells and carbon dioxide wastes from them. The impaired gas flow causes severe damage to many tissues and organs. Because heterozygotes (HbA/HbS) still have one functional allele, they show few symptoms of the disorder even though a portion of their blood cells are sickled. Only homozygous recessives (HbS/HbS) show severe phenotypic consequences (Figure 12.15a).

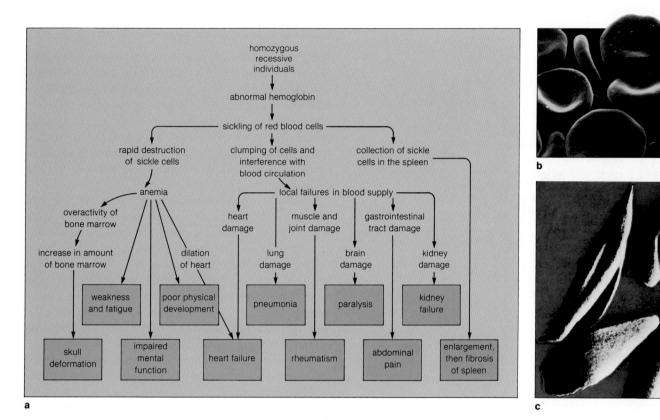

a

b

c

Figure 12.15 (**a**) Pleiotropic effects possible in sickle-cell homozygotes. The color-enhanced scanning electron micrographs show (**b**) normal red blood cells and (**c**) cells characteristic of sickle-cell anemia. Because of their abnormal, asymmetrical shape, sickle cells do not flow smoothly through blood capillaries. They pile up in clumps that impede blood flow. Tissues served by the capillaries become starved for oxygen and nutrients even as they become saturated with waste products.

Environmental Effects on Phenotype

Interactions between genes or gene products and the environment can lead to variations in phenotypes. Environmental temperature, for example, influences coat color in Siamese cats (Figure 12.16). The main pigment molecule in these cats is a brown-black form of melanin, the production of which depends on the enzyme tyrosinase. One allele coding for tyrosinase produces a heat-sensitive form of the enzyme that is less active at warmer temperatures. In cats homozygous for this allele, warmer body parts have light fur and the relatively cool extremities (paws, ears, tail, nose) have dark fur.

The water buttercup (*Ranunculus aquatilis*) provides another example of environmental effects on phenotype. This plant grows in shallow ponds, and some of its leaves develop underwater. The submerged leaves are finely divided, compared with the leaves growing in air. When a leaf-bearing stem is half in and half out of the water, its leaves display both phenotypes (Figure 12.17). The genes responsible for leaf shape produce very different phenotypes under different environmental conditions.

Figure 12.16 Environmental effect on phenotype in the Siamese cat. Fur on the paws and ears is darker than on the rest of the body. The skin temperature in these regions is normally lower than the temperature in the rest of the body. Some cats are homozygous recessive for a gene involved in the formation of the dark pigment melanin. The enzyme produced by this recessive allele is heat-sensitive; it is less active at warmer temperatures. Hence the fur on warmer body parts is lighter in color.

Figure 12.17 Variable expressivity resulting from variation in the external environment. Leaves of the water buttercup (*Ranunculus aquatilis*) show dramatic phenotypic variation, depending on whether they grow underwater or above it. This variation occurs even in the same leaf if it develops half in and half out of water.

Continuous and Discontinuous Variation

The traits Mendel studied would now be called examples of **discontinuous variation**: phenotypes fell into one or another of a few clearly discernible classes. Such differences in phenotype were important in Mendel's work, for they were the only markers he could use to identify and track genotypes. You might say his studies were qualitative in focus, because differences could be established by simple observation, without precise measurements.

Most phenotypic differences in a population are not qualitative. Not all humans can be readily classed as tall *or* short, fat *or* thin, and so forth. For most traits, humans and other organisms show **continuous variation**: small degrees of phenotypic variation occur over a more or less continuous range (Figure 12.18). Measurements of small differences must be *quantitative*, requiring precise measurements of individuals of the population. The term **quantitative inheritance** refers to the transmission of traits showing continuous variation.

H. Nilsson-Ehle, R. Emerson, and E. East developed the idea that quantitative inheritance arises through the additive influence of three or more gene pairs concerned with the same trait. For example, human skin color ranges through hues of blacks, browns, and whites. As for other mammals, a number of different genes help control the type, distribution, and amount of pigment in the skin (hence its color). Their effect is roughly additive, with the intensity of skin pigmentation being determined by the sum total of alleles active at all the different loci.

It is important to understand that quantitative expressions are not a result of "blending," a term that suggests loss of the original identity of hereditary traits. Rather, *all genes retain their physical identity, regardless of the phenotype produced by their combined positive and negative effects.*

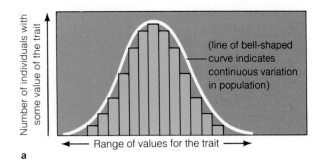

a

Figure 12.18 (**a**) Generalized bell-shaped curve typical of populations showing continuous variation in some trait.

(**b**) Example of continuous variation in a population sample: height distribution in a group of 175 U.S. Army recruits about the turn of the century.

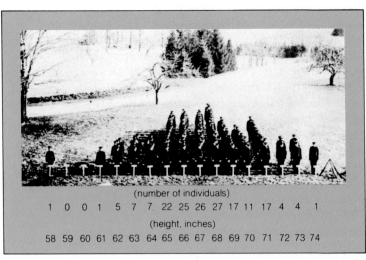

(number of individuals)

| 1 | 0 | 0 | 1 | 5 | 7 | 7 | 22 | 25 | 26 | 27 | 17 | 11 | 17 | 4 | 4 | 1 |

(height, inches)

| 58 | 59 | 60 | 61 | 62 | 63 | 64 | 65 | 66 | 67 | 68 | 69 | 70 | 71 | 72 | 73 | 74 |

b

SUMMARY

1. Mendel established the foundations for genetic analysis (the study of inheritance patterns). His hybridization studies with garden pea plants demonstrated that diploid organisms have two units of hereditary material (genes) for each trait. The genes in a sperm do not "blend" at fertilization with genes in an egg; rather, genes retain their identity when passed on to offspring.

2. Mendel conducted monohybrid crosses (between two true-breeding plants having contrasting forms of a single trait). The crosses indicated that there can be alternative forms (alleles) of the same gene, some of which are dominant over other, recessive forms.

3. A homozygous dominant plant (or any other individual) has two dominant alleles (*AA*) for the trait being studied. A homozygous recessive has two recessive alleles (*aa*); a heterozygote has two different alleles (*Aa*).

4. In Mendel's monohybrid crosses (*AA* × *aa*), all first-generation offspring were *Aa*. Later, crosses between the first-generation plants (*Aa* × *Aa*) produced the following allelic combinations:

sperm:		egg:		
A	×	*A*	=	*AA* (dominant)
A	×	*a*	=	*Aa* (dominant)
a	×	*A*	=	*Aa* (dominant)
a	×	*a*	=	*aa* (recessive)

This produces an expected phenotypic ratio of 3:1 in the second generation. (The term phenotype refers to an individual's observable traits.)

5. Results from such monohybrid crosses support the Mendelian principle of segregation: Diploid organisms have a pair of genes for each trait, on a pair of homologous chromosomes. The two genes segregate from each other during meiosis, such that each gamete formed will end up with one or the other, but not both.

6. Mendel also performed dihybrid crosses (between two true-breeding plants having contrasting forms of two different traits). Results from many experiments were close to a 9:3:3:1 phenotypic ratio:

9 dominant for both traits
3 dominant for *A*, recessive for *b*
3 dominant for *B*, recessive for *a*
1 recessive for both traits

7. On the basis of such dihybrid crosses, the Mendelian principle of independent assortment was formulated: Each gene pair tends to assort into gametes independently of other gene pairs located on *non*homologous chromosomes.

8. We still rely on Mendel's most important insight for analyzing monohybrid and dihybrid crosses—that the two genes of a pair segregate from each other prior to gamete formation.

9. However, Mendel's concept of dominance had to be greatly modified. In any population, most traits are not clearly contrasting (as in the garden pea). They show continuous variation, which does not lend itself readily to simple analysis.

10. Continuous variation arises through interactions between genes and through additive effects of the allelic combinations at different gene loci. Thus,

   a. There may be degrees of dominance between the alleles that occupy homologous loci. The expression of one or both may be fully dominant, or one may be incompletely dominant over the other.

   b. The activity of a single gene may have major or minor effects on more than one trait.

   c. The activity of a gene may be influenced by other genes, with a sum total of their positive and negative actions producing some effect on phenotype.

11. The environment can influence the expression of genes and their contribution to phenotype.

Review Questions

1. State the Mendelian principle of segregation. Does segregation occur during mitosis or meiosis? *164-165*

2. Distinguish between the following terms: (a) gene and allele, (b) dominant trait and recessive trait, (c) homozygote and heterozygote, (d) genotype and phenotype. *164-165*

3. Give an example of a self-fertilizing organism. What is cross-fertilization? *163*

4. Distinguish between monohybrid and dihybrid crosses. What is a testcross, and why is it valuable in genetic analysis? *164, 167, 168*

5. State the Mendelian principle of independent assortment. Does independent assortment occur during mitosis or meiosis? *169*

6. How does quantitative inheritance differ from the notion of "blending" of heritable traits? *174*

7. Contrast continuous and discontinuous variation, and outline the genotypic basis for both. *174*

8. Mendel's concept of dominance was based on observations of inheritance patterns in clearly contrasting traits. How has this concept since been modified? *170*

9. List some of the factors influencing the expression of phenotype. *171-173*

Readings

Dunn, L. 1965. *A Short History of Genetics.* New York: McGraw-Hill.

Mendel, G. 1959. "Experiments in Plant Hybridization." Translation in J. Peters (editor), *Classic Papers in Genetics.* Englewood Cliffs, New Jersey: Prentice-Hall.

Singer, C. 1962. *A History of Biology to About the Year 1900*. New York: Abelard-Schuman.

Stern, C., and E. Sherwood (editors). 1966. *The Origin of Genetics*. San Francisco: Freeman. Includes a modern translation of Mendel's paper and correspondence.

Strickberger, M. 1985. *Genetics*. Third edition. New York: Macmillan. Still the classic introduction to genetics.

Witkop, C., Jr. 1975. "Albinism." *Natural History* 84(8):48-59. Good account of the various forms of albinism—their biochemical and genetic bases and their phenotypic manifestations.

Genetics Problems (Answers appear in Appendix II)

1. One gene has alleles A and a; another gene has alleles B and b. For each of the following genotypes, what type(s) of gametes will be produced? (Independent assortment is expected.)
   a. *AA BB*
   b. *Aa BB*
   c. *Aa bb*
   d. *Aa Bb*

2. Still referring to the preceding problem, what genotypes will be present in the offspring from the following matings? (Indicate the frequencies of each genotype among the offspring.)
   a. *AA BB × aa BB*
   b. *Aa BB × AA Bb*
   c. *Aa Bb × aa bb*
   d. *Aa Bb × Aa Bb*

3. In one experiment, Mendel crossed a true-breeding pea plant having green pods with a true-breeding pea plant having yellow pods. All of the F_1 plants had green pods. Which trait (green or yellow pods) is recessive? Can you explain how you arrived at your conclusion?

4. Being able to curl up the sides of your tongue into a U-shape is under the control of a dominant allele at one gene locus. (When there is a recessive allele at this locus, the tongue cannot be rolled.) Having free earlobes is a trait controlled by a dominant allele at a different gene locus. (When there is a recessive allele at this locus, earlobes are attached at the jawline.) The two genes controlling tongue-rolling and free earlobes assort independently. Suppose a woman who has free earlobes and who can roll her tongue marries someone who has attached earlobes and who cannot roll his tongue. Their first child has attached earlobes and cannot roll the tongue.
   a. What are the genotypes of the mother, the father, and the child?
   b. If this same couple has a second child, what is the probability that it will have free earlobes and be unable to roll the tongue?

5. In addition to the two genes mentioned in Problem 1, assume you now study a third gene having alleles C and c. For each of the following genotypes, indicate what type (or types) of gametes will be produced:
   a. *AA BB CC*
   b. *Aa BB cc*
   c. *Aa BB Cc*
   d. *Aa Bb Cc*

6. A man is homozygous dominant for ten different genes, which assort independently. How many genotypically different types of sperm could he produce? A woman is homozygous recessive for eight of these ten genes, and she is heterozygous for the other two. How many genotypically different types of eggs could she produce? What can you conclude regarding the relationship between the number of different gametes possible and the number of heterozygous and homozygous genes that are present? For a given population, what might be the biological benefits, if any, of possessing a large number of heterozygotes?

7. Recall that Mendel crossed a true-breeding tall, purple-flowered pea plant with a true-breeding dwarf, white-flowered plant. All the F_1 plants were tall and purple-flowered. If an F_1 plant is now self-pollinated, what is the probability of obtaining an F_2 plant heterozygous for the genes controlling height and flower color?

8. Assume that a new gene was recently identified in mice. One allele at this gene locus produces a yellow fur color. A second allele produces a brown fur color. Suppose you are asked to determine the dominance relationship between these two alleles. (Is it one of simple dominance, incomplete dominance, or codominance?) What types of crosses would you make to find the answer? On what types of observations would you base your conclusions?

9. The ABO blood system has often been employed to settle cases of disputed paternity. Suppose, as an expert in genetics, you are called to testify in a case where the mother has type A blood, the child has type O blood, and the alleged father has type B blood. How would you respond to the following statements of the attorneys:
   a. "Since the mother has type A blood, the type O blood of the child must have come from the father, and since my client has type B blood, he obviously could not have fathered this child." (*Made by the attorney of the alleged father*)
   b. "Further tests revealed that this man is heterozygous and therefore he must be the father." (*Made by the mother's attorney*)

10. In mice, at one gene locus, the dominant allele (B) produces a dark-brown pigment; and the recessive allele (b) produces a light-brown, or tan, pigment. An independently assorting gene locus has a dominant allele (C) that permits the production of all pigments. Its recessive allele (c) makes it impossible to produce any pigment at all. The pigmentless condition is called "albino."
   a. A homozygous *bb cc* albino mouse mates with a homozygous *BB CC* brown mouse. In what ratios would the phenotypes and genotypes be expected in the F_1 and F_2 generations?
   b. If an F_1 mouse from part (a) above were backcrossed to its albino parent, what phenotypic and genotypic ratios would be expected?

11. Certain dominant alleles are so important for normal development that the mutant recessive alleles, when homozygous, lead to the death of the organism. However, such recessive alleles can be perpetuated as heterozygotes (Ll), which in many cases are not phenotypically different from homozygous (LL) normals. (In some cases, individuals carrying a recessive lethal allele do have a mutant phenotype.) Consider the mating of two such heterozygotes, $Ll \times Ll$. Among their *surviving* progeny, what fraction will be heterozygous?

12. In corn, a series of three independent pairs of gene loci (A, C, and R) affect the production of pigment that leads to kernel color. If any one of the three pairs is in the homozygous recessive state, then no pigment will form in the kernels. However, if at least one dominant allele of each locus is present, then pigment can form in the kernel. Two corn plants with the following genotypes were crossed:

$$Aa\ cc\ Rr \quad \times \quad aa\ Cc\ Rr$$

What fraction of the progeny kernels will be pigmented? (Note: each kernel represents a separate (potential) individual; it will exhibit the pigment phenotype of the plant that can be grown from it.)

RETURN OF THE PEA PLANT

The year was 1884. Mendel's paper on hybridization of pea plants had been gathering dust in at least a hundred libraries for nearly two decades, and Mendel himself had just passed away. Ironically, the experiments described so carefully in that ignored, forgotten paper were about to be devised all over again, as a way to test ideas emerging from another line of research. Cytology, the study of cell structure and function, was about to converge with genetic analysis.

Improvements in microscopy had rekindled efforts to locate the cell's hereditary material, and cytologists were zeroing in on the nucleus. By 1882, Walther Flemming had observed threadlike bodies—chromosomes—in the nuclei of dividing salamander cells. He called the division process mitosis, after the Greek word for thread. By 1884, questions were beginning to take shape: *Were those threadlike chromosomes being parceled carefully into daughter cells? Could they be the hereditary material?*

Then a key observation was made: Each sperm or egg has half the number chromosomes as a fertilized egg. In 1887, August Weismann proposed that a special division process must reduce the chromosome number by half before gametes form. Sure enough, in that same year meiosis was discovered. Weismann now began to promote his theory of heredity: The chromosome number is halved during meiosis, then restored when sperm and egg combine at fertilization; thus half the hereditary material in offspring is paternal in origin, and half is

13

CHROMOSOMAL THEORY OF INHERITANCE

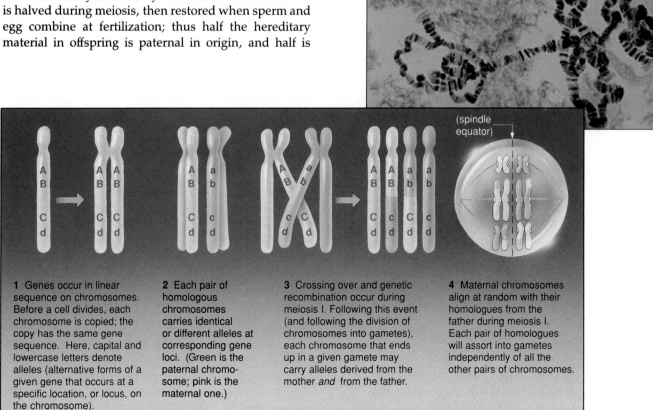

1 Genes occur in linear sequence on chromosomes. Before a cell divides, each chromosome is copied; the copy has the same gene sequence. Here, capital and lowercase letters denote alleles (alternative forms of a given gene that occurs at a specific location, or locus, on the chromosome).

2 Each pair of homologous chromosomes carries identical or different alleles at corresponding gene loci. (Green is the paternal chromosome; pink is the maternal one.)

3 Crossing over and genetic recombination occur during meiosis I. Following this event (and following the division of chromosomes into gametes), each chromosome that ends up in a given gamete may carry alleles derived from the mother *and* from the father.

4 Maternal chromosomes align at random with their homologues from the father during meiosis I. Each pair of homologues will assort into gametes independently of all the other pairs of chromosomes.

Figure 13.1 Overview of some key aspects of chromosome structure and function. The micrograph shows the chromosomes from a salivary gland of a *Drosophila* larva.

maternal. His views were hotly debated, and the debates drove researchers into testing the theory. Throughout Europe there was a flurry of experimental crosses—just like the ones Mendel carried out.

Mendel's pioneering work finally was acknowledged in 1900. Three independent researchers came across his paper while checking for literature related to their own hybridization studies. To their chagrin, they saw that their results merely confirmed what Mendel already had said: Diploid cells have two "units" of instruction for each heritable trait, and the units segregate prior to gamete formation. Later, proof would come that Mendel's "units" are carried on Flemming's "threads."

THE CHROMOSOMAL THEORY

This chapter covers some observations and experiments that unfolded in the decades after the rediscovery of Mendel's work. They lend impressive support to what is now called the **chromosomal theory of inheritance**. From the preceding chapters, we are already familiar with some of the key points of this theory:

1. Chromosomes carry *genes*, the units of instruction for heritable traits. The genes are arranged one after the other along chromosomes (Figure 13.1).

2. Sexual reproduction begins with meiosis and gamete formation and ends at fertilization. By this process, offspring receive the same number and type of chromosomes found in the body cells of their parents.

3. Diploid (2*n*) cells have two chromosomes of each type, called *homologues*. Generally, two homologues have the same length, shape, and gene sequence—and they pair at meiosis.

4. In most animal species, two of the chromosomes in diploid cells are sex chromosomes. Although they are not always the same (as when an X and Y chromosome occur in the same cell), they still pair as homologues at meiosis.

5. After pairing, each chromosome is *segregated* from its homologue. Through segregation, the diploid number is reduced to the haploid number (*n*) for gametes.

6. Chromosomes *assort independently* at meiosis. This means each gamete ends up with one of each pair of homologues—and it doesn't matter which of the two it gets.

7. Genes on the same chromosome tend to stay together during meiosis and end up in the same gamete. But *crossing over* (breakage and exchange of corresponding segments of homologues) can disrupt such linkages and lead to genetic recombination.

8. *Chromosomal aberrations* sometimes occur. A chromosome segment may be deleted, duplicated, inverted left to right, or moved to a new location. Chromosomes also may not separate properly at meiosis, and so gametes end up with an abnormal chromosome number.

9. Independent assortment, crossing over, and chromosomal aberrations play roles in evolution. By changing the genotype (genetic makeup), they lead to variations in phenotype (observable traits) upon which selective agents can act.

CLUES FROM THE INHERITANCE OF SEX

Sex Determination

In the early 1900s, it became apparent that males and females have almost (but not quite) the same assortment of chromosomes. Nearly all of their chromosomes are the *same* in number and kind; these were named **autosomes**. Depending on the species, however, one or two *differ* in number or kind between males and females, and these were named **sex chromosomes** (Figure 13.2).

Although there are notable exceptions (birds being among them), females most often have two identical sex chromosomes designated "X." Males have one X chromosome and another, physically different chromosome designated "Y." For example, female fruit flies (*Drosophila*) are XX and the males are XY. The same is true for humans.

Because a diploid female has two X chromosomes, each gamete she produces will carry an X chromosome. The X and Y chromosomes of a male act as homologues at meiosis, so half of his gametes will carry an X and half will carry a Y chromosome. When a sperm and an egg both carry an X chromosome and combine at fertilization, the zygote develops into an XX female. When the sperm carries a Y chromosome, the zygote develops into an XY male (Figure 13.3).

The most common sex chromosome pattern is XX (female) and XY (male).

Which sex chromosome carries the actual instructions about **gender**—that is, whether a new individual will be male or female? It was not until the end of 1987 that researchers pinpointed the location of the key gene in humans and other mammals. The Y chromosome carries a copy of this gene, which specifies a protein (testis determining factor, or TDF) that dictates gender.

Until the sixth or seventh week of pregnancy, a human fetus is not yet "committed" to being female or male. Then, the presence or absence of TDF gene products dictates whether the fetus will develop ovaries or testes (the primary reproductive organs of females and males). This is the first step in the pathway of sex differentiation. Once the developmental "choice" is made, the ovaries or testes secrete hormones that trigger the development of female or male characteristics.

The picture may be more complicated than this. A TDF gene (or something very much like it) also has been located on the X chromosome. Do the products of two TDF genes cooperate or play off each other to determine sex? Is the TDF gene active only in XY (not XX) individuals? We don't know the answer yet.

Sex-Linked Genes

Even though the Y chromosome carries a copy of the key sex-*determining* gene, it carries very few others (at least in humans, fruit flies, and many other animals). In contrast, the X chromosome carries at least 100 and possibly 200 genes. Among them are genes that deal with nonsexual traits. In fruit flies, for example, one deals with eye color, another with an enzyme for carbohydrate metabolism, and so on.

Strictly speaking, the genes on either sex chromosome can be called **sex-linked**. It is becoming more common to refer to them more precisely as *X-linked* or *Y-linked*. Certainly it is less confusing to do so, given that many of the genes (on the X chromosome especially) have nothing to do with *sexual* characteristics—those specifically associated with maleness or femaleness. (In humans, for example, sexual characteristics include the distribution of hair and body fat, both of which are notably different in males and females.) Autosomes as well as both sex chromosomes carry genes that influence the overall male or female phenotype.

Morgan's Studies of Inheritance

Our understanding of X-linked genes began in the early 1900s, when the embryologist Thomas Hunt Morgan began work to explain the apparent relationship between gender and certain nonsexual traits. For example, the blood's ability to clot is a nonsexual trait, and in hemophilic individuals, blood clotting is abnormal (page 191). For centuries, hemophilia was perceived as being a heritable disorder, one that most often shows up in the males but not females of a family lineage. This phenotypic outcome was not like anything that turned up in Mendel's hybrid crosses between pea plants. In those crosses, remember, one parent plant or the other carried a recessive allele. It made no difference *which* parent carried it; the phenotypic outcome was the same.

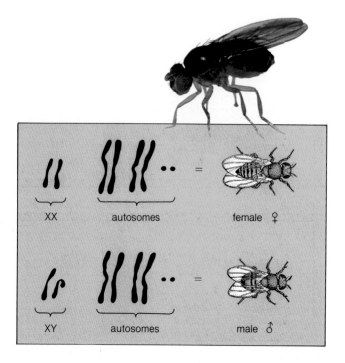

Figure 13.2 Sex chromosomes and autosomes of the fruit fly, *Drosophila melanogaster*. Together they represent a diploid number of eight (four pairs of homologous chromosomes).

In both fruit flies and humans, females are XX and males are XY. The reverse is true for birds, moths, butterflies, and a few other animals (XY for females, XX for males). And still other animals, including grasshoppers, have no Y chromosome at all (females are XX, and males, X0; the 0 indicates the absence of a second sex chromosome).

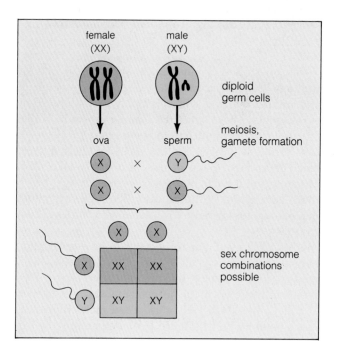

Figure 13.3 Sex determination in humans. This same pattern occurs in many animal species. Only the sex chromosomes, not the autosomes, are shown. Males transmit their Y chromosome to their sons, but not to their daughters. Males receive their X chromosome only from their mother.

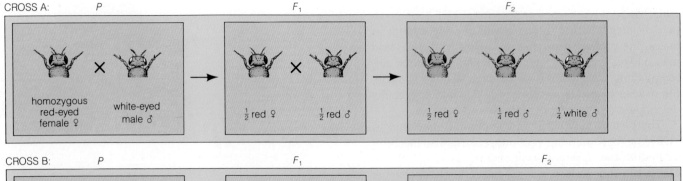

CROSS A:

P — homozygous red-eyed female ♀ × white-eyed male ♂

F₁ — ½ red ♀ × ½ red ♂

F₂ — ½ red ♀ — ¼ red ♂ — ¼ white ♂

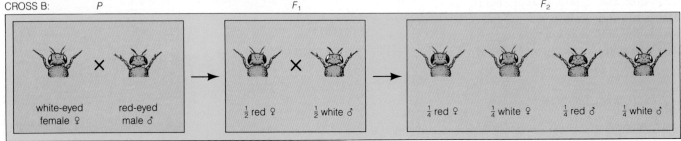

CROSS B:

P — white-eyed female ♀ × red-eyed male ♂

F₁ — ½ red ♀ × ½ white ♂

F₂ — ¼ red ♀ — ¼ white ♀ — ¼ red ♂ — ¼ white ♂

a

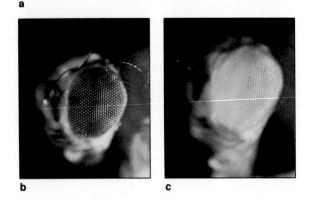

b   c

Figure 13.4 (**a**) Phenotypic results in the F_1 and F_2 generations produced by a reciprocal cross. In the first of these paired crosses, Morgan mated a red-eyed female with a white-eyed male fruit fly. In the second cross, he mated a white-eyed female with a red-eyed male. (**b**) Wild-type eye color for *Drosophila* and (**c**) white-eye phenotype.

For his studies of inheritance, Morgan relied on the fruit fly *Drosophila melanogaster* rather than on the plants and animals that had been commonly used for breeding experiments. These small flies can be grown in bottles on a little bit of cornmeal, molasses, and agar. A female lays hundreds of eggs in a few days, and her offspring reach reproductive age in less than two weeks. Thus Morgan could track hereditary traits through nearly thirty generations of thousands of flies in a year's time! Before long, his laboratory was filled with bottles of busy fruit flies.

Drosophila eye color turned out to be a most informative trait. At first, all the flies Morgan raised were wild-type for eye color; they had brick-red eyes. ("Wild type" simply refers to the normal or most common allele at a given locus.) In 1910, a *white-eyed* male cropped up in a laboratory bottle (Figure 13.4). Apparently the variant form arose through a spontaneous mutation in a gene controlling eye color.

Morgan established true-breeding strains of white-eyed males and females. Then he did a series of **reciprocal crosses**. (In the first of a pair of crosses, one parent displays the trait in question; in the second cross, the other parent displays the trait.) White-eyed males were mated with true-breeding (homozygous) red-eyed females. In the reciprocal cross, white-eyed females were mated with true-breeding red-eyed males. The phenotypic outcomes for the paired crosses were not the same (Figure 13.4).

Clearly the gene controlling eye color was related to gender, and it probably was located on one of the sex chromosomes. But which one? Since females (XX) could be white-eyed, the recessive allele would have to be on one of their X chromosomes. Now suppose white-eyed males (XY) also carry the recessive allele on their X chromosome—*and that there is no corresponding eye-color gene on their Y chromosome*. Those males would have white eyes because the recessive allele would be the only eye-color gene they had!

Figure 13.5 shows the results that can be expected when the idea of an X-linked gene is combined with Mendel's concept of segregation. (Similar examples of X-linked inheritance for humans are included in the next chapter.) By proposing that a specific gene occurs on the X but not the Y chromosome, Morgan was able to explain the seemingly odd outcome of his reciprocal crosses. The results of the experiments matched the predicted outcomes.

LINKAGE

During their *Drosophila* studies, Morgan and his coworkers found evidence that many traits are inherited as a group from one parent or the other. Their studies of an X-linked gene supported the hypothesis that each gene is located on a specific chromosome, and it seemed likely that several genes located on the *same* chromosome might stay physically linked together during meiosis and end up in the same gamete.

White eyes had served as an observable "marker" for tracking the inheritance of one gene, and now other mutations were used in a similar way. (By 1915, Morgan's group has isolated more than eighty types of flies with distinctive mutations.)

For example, a wild-type fly has straight, flat wings and a gray body, and the two genes for those traits were designated *C* and *B*. Some mutants had curved wings (*c*) and a black body (*b*), and they were mated with wild-type flies in dihybrid crosses. In this type of cross, remember, parents true-breeding for two traits of interest produce first-generation offspring that are heterozygous for both traits. All offspring of the *CC BB* × *cc bb* cross were *Cc Bb*.

If the two genes were not linked in some way, the alleles on the two chromosomes that each F_1 offspring had inherited would assort independently, and four second-generation phenotypes would be produced in about a 9:3:3:1 ratio. (See, for example, Figure 12.10.) But that did not happen. There were too many of the parental phenotypes (*CC BB* and *cc bb*) and not enough of the other two. The wing-shape gene and the body-color gene were probably located on the same chromosome.

In time, four groups of apparently linked genes were identified in *D. melanogaster*, and it seemed likely that they corresponded to the haploid number of chromosomes (four) in that organism's gametes. The tendency of genes located on the same chromosome to be transmitted together into gametes eventually came to be called **linkage**.

Two or more genes located on <u>nonhomologous</u> chromosomes tend to assort independently of each other and end up in different gametes.

Two or more genes located on the <u>same</u> chromosome tend to be transmitted together into a gamete.

You may have noticed a little hedging in the two preceding sentences. As it happens, even when genes are linked (located on the same chromosome), they might *not* end up in the same gamete because of an event we have already considered briefly. That event is called crossing over.

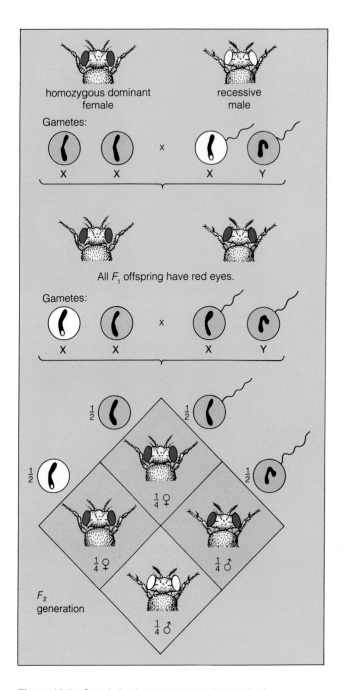

Figure 13.5 Correlation between sex and eye color in *Drosophila*. Given the genetic makeup of the F_2 generation, the recessive allele (depicted here by the white dot) must be carried on the X chromosome only.

This diagram corresponds with Cross A in Figure 13.4. Can you construct a similar diagram to explain the results for Cross B?

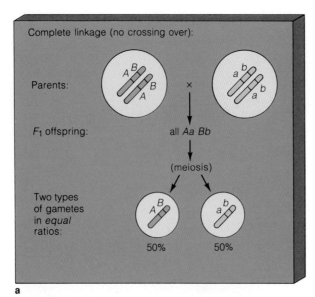

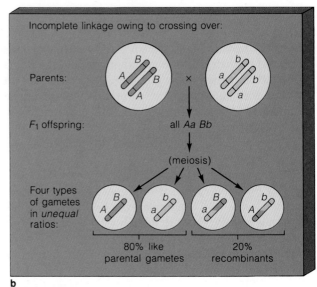

Figure 13.6 Example of how crossing over can affect gene linkage. (Only two of the four chromatids brought together at prophase I of meiosis exchange parts; refer to Figure 11.4.)

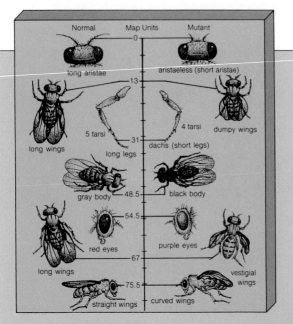

Figure 13.7 Genetic mapping of genes on a segment of chromosome 2 in *D. melanogaster*. Such maps don't show actual physical distances between genes. Rather they show relative distance between gene locations that undergo crossing over and other chromosomal rearrangements. Only if the probability of crossing over were equal along the chromosome's length (which it is not) would it be possible to calculate physical distance exactly.

Here, distances between genes are measured in map units, based on the frequency of recombination between the genes. (One genetic map unit = 1 percent recombination.) Thus, if the frequency turns out to be 10 percent, the genes are said to be separated by 10 map units. The amount of recombination to be expected between "vestigial wings" and "curved wings," for instance, would be 8.5 percent (75.5 − 67).

CROSSING OVER AND LINKAGE MAPPING OF CHROMOSOMES

Crossing over is the breakage and exchange of corresponding segments between homologous chromosomes, and it leads to novel combinations of the parental traits in offspring. This event is shown in Figures 11.4 and 13.1.

Crossing over can disrupt gene linkages at any point along the length of a chromosome. As the geneticist Alfred Sturtevant discovered, however, *the probability of crossing over and recombination occurring at a point somewhere between two genes is proportional to the chromosomal distance separating them.* Suppose that two genes *A* and *B* are twice as far apart as two other genes, *C* and *D*:

We would expect crossing over to disrupt the linkages between the first two much more frequently.

Two genes located physically close together on a chromosome nearly always end up in the same gamete; they are very closely linked (Figure 13.6a). Two genes relatively far apart are more vulnerable to crossing over and recombination, compared to closely linked genes (Figure 13.6b). Two genes very far apart on the same chromosome are affected by crossing over so often that they may appear to assort independently.

For example, Mendel thought he was observing independent assortment for seven different traits of the garden pea, and this plant does have seven chromosomes in its gametes. We now know that the genes for two of the traits Mendel studied are far apart on the *same* chromosome. Crossing over and recombination occurred so frequently between them that it *seemed* as if they were assorting independently!

The farther apart two genes are on a chromosome, the greater will be the frequency of crossing over and recombination between them.

Genes are distributed into gametes in patterns that tell us something about their organization on chromosomes. The patterns are so regular that they can be used to determine the *positions* of genes relative to one another. Plotting the positions of genes on a given chromosome is called **linkage mapping**. Figure 13.7 gives an example of a linkage map.

Of the several thousand known genes in the four chromosomes of *Drosophila* gametes, the positions of about 1,000 have been mapped. There are many more genes in human chromosomes, but the locations of relatively few have been identified. We have a long way to go in mapping the physical basis of heredity. But it is clear that genes are carried linearly, one after another, on human chromosomes—just as they are in fruit flies and all other organisms.

CHANGES IN CHROMOSOME STRUCTURE

Chromosome Banding and Karyotypes

In chromosomes that have been stained for microscopy, clear banding patterns emerge because some regions are more condensed than others and absorb the stain differently. The distinct banding pattern is the same in all chromosomes of a given type.

As early as the 1930s, it was known that the chromosomes in certain insect cells have faint banding patterns that can be enhanced by staining. In these *polytene* chromosomes, the DNA was duplicated repeatedly and the duplicates remained packed in parallel, forming large structures of the sort shown in Figure 13.1. But it was not until the late 1960s that a new method revealed banding patterns in the chromosomes of vertebrates and some plants.

Today, individual chromosomes can be precisely characterized by length, banding patterns at metaphase (when they are in their most condensed form), and so on. Such features are used to create a **karyotype**, a visual representation in which the individual chromosomes of a cell are arranged in order, from largest to smallest (Figure 13.8). In the karyotype of Figure 13.9, the chromosomes have been prepared by a method called G-banding (after the Giemsa stain).

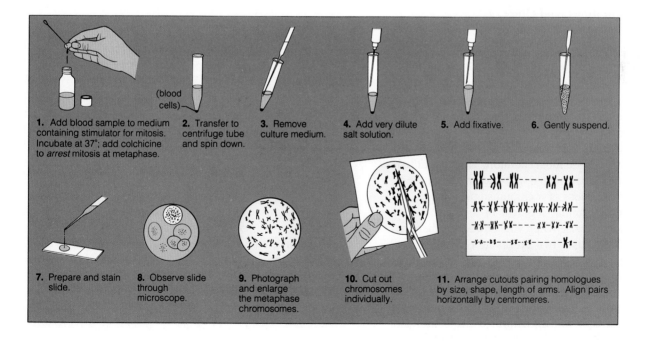

1. Add blood sample to medium containing stimulator for mitosis. Incubate at 37°; add colchicine to *arrest* mitosis at metaphase.
2. Transfer to centrifuge tube and spin down.
3. Remove culture medium.
4. Add very dilute salt solution.
5. Add fixative.
6. Gently suspend.
7. Prepare and stain slide.
8. Observe slide through microscope.
9. Photograph and enlarge the metaphase chromosomes.
10. Cut out chromosomes individually.
11. Arrange cutouts pairing homologues by size, shape, length of arms. Align pairs horizontally by centromeres.

Figure 13.8 Simplified picture of karyotype preparation.

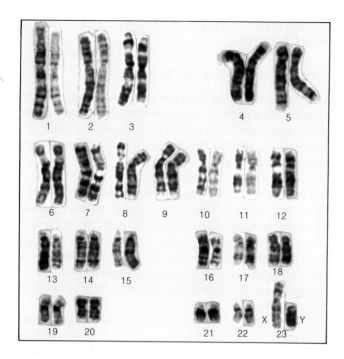

Figure 13.9 Karyotype of a human male. Human cells have a diploid chromosome number of 46. The nucleus contains 22 pairs of autosomes and one pair of sex chromosomes (X and Y). Each chromosome of a given type has already undergone replication. These chromosomes were prepared for microscopy by G-staining, which provides the banded appearance of chromosomes. The bands are used in karyotype analysis.

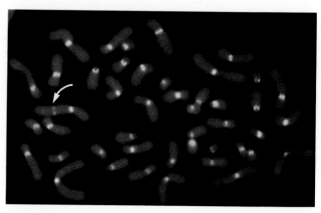

Figure 13.10 Translocation. These metaphase chromosomes were stained with substances that react preferentially with kinetochore proteins, which show up as fluorescent yellow in micrographs. The arrow points to the inactivated kinetochores of chromosome 9—which fused with chromosome 11. (The normal kinetochores of chromosome 11 are visible to the right of the fusion point.)

Deletions, Duplications, and Other Structural Rearrangements

On rare occasions, a chromosome undergoes abnormal structural rearrangements that change its banding pattern. Typically the structural change produces a change in some phenotypic trait that can be correlated with a specific gene product. Thus altered banding patterns can be used to identify specific gene locations.

Of the structural rearrangements that have been identified so far, four are most common. They are called deletions, duplications, inversions, and translocations.

A **deletion** is the loss of a chromosome segment. It can arise when a terminal segment of a chromosome breaks off. It also can arise when viral attack, irradiation, or chemical action causes two breaks in a chromosome region. Enzymes can repair the breaks, but sometimes they accidentally leave out the segment in between. A deletion almost always causes problems, for genes influencing one or more traits may be lost entirely. For example, mental retardation and a malformed larynx are just two of the phenotypic consequences of a rare deletion that affects human chromosome 5. When affected infants cry, the sounds produced are more like meowing—hence the name of this genetic disorder, cri-du-chat (meaning "cat-cry").

In a **duplication**, a chromosome gets a gene sequence in excess of its normal amount. Sometimes the added sequence is adjacent to the one it repeats. Suppose a chromosome has two identical sequences (ABCDABCD). It is possible that homologues will not align properly at this region when they pair at prophase I of meiosis. If crossing over affects the improperly aligned region, one chromosome might end up with three sequences and its partner with only one:

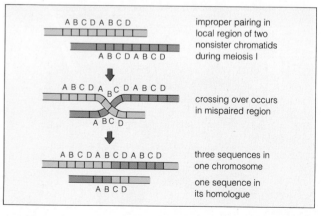

An **inversion** is a chromosome segment that has been excised and rejoined at the same place—but in the reverse order. The reversal alters the position and sequence of its genes:

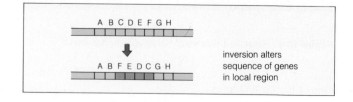

In a **translocation**, part of one chromosome is transferred to a nonhomologous chromosome. In some types of cancerous cells in humans, for example, part of chromosome 8 has become transferred to chromosome 14. Expression of genes on that segment had been precisely regulated by virtue of its normal chromosomal location, but controls are lost at the new location.

Once in awhile, translocation occurs in such a way that an entire chromosome becomes fused to a nonhomologous chromosome—and the two continue to function. This structural rearrangement changes the chromosome number. Figure 13.10 shows the result of such a fusion in a cell from a human female. Her diploid chromosome number is 45 instead of the normal 46.

CHANGES IN CHROMOSOME NUMBER

Translocations that change the chromosome number are rare events. **Nondisjunction**, the failure of one or more chromosomes to separate during meiosis, is a more frequent event that can cause the chromosome number to change. In some cases, a chromosome fails to separate from its homologue at anaphase I; in other cases, sister chromatids of a chromosome fail to separate at anaphase II (Figure 13.11). In *complete* nondisjunction, all of the chromosomes fail to separate. Whatever the case, gametes (and new individuals) end up with an abnormal chromosome number.

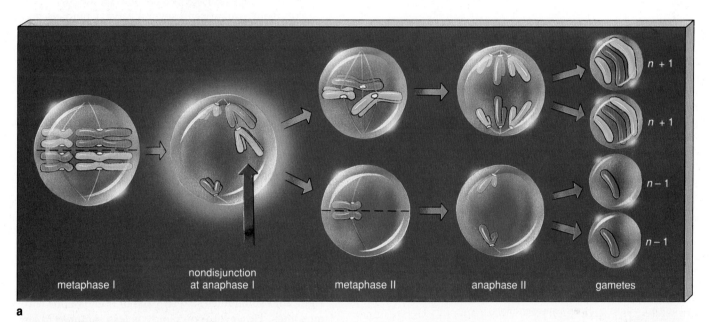

metaphase I   nondisjunction at anaphase I   metaphase II   anaphase II   gametes

a

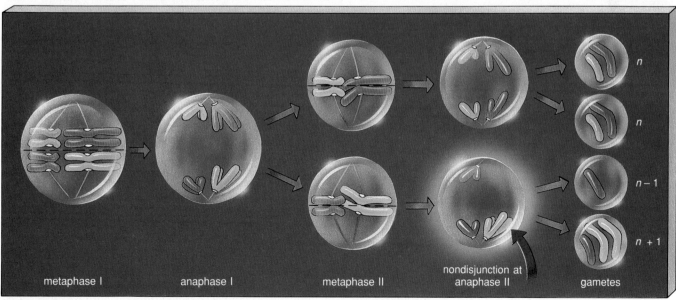

metaphase I   anaphase I   metaphase II   nondisjunction at anaphase II   gametes

b

Figure 13.11 Two examples of nondisjunction, an event that can change the chromosome number in gametes (hence in offspring).

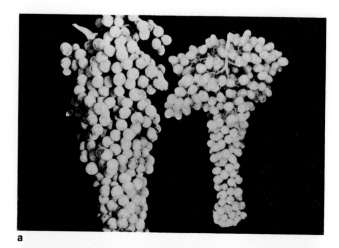

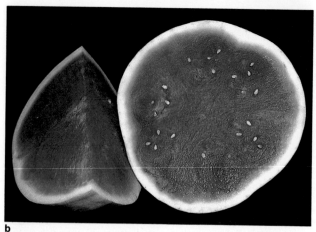

Figure 13.12 Examples of commercial polyploid species. (**a**) Artificially increased cluster size in a colchicine-induced tetraploid (4n = 76) representative of the Sultana grape (left) compared with its diploid counterpart (right).

(**b**) The seedless watermelon: the result of a cross between a tetraploid and a diploid. The triploid fruit matures, but fertility is reduced; hence most of the seeds do not develop normally. The triploid condition is characteristic of commercial bananas, many pears, and apples as well as watermelons.

Missing or Extra Chromosomes

Nondisjunctions of the sort shown in Figure 13.11 can lead to **aneuploidy**, a condition in which offspring have one extra chromosome or one less than the parental number. The next chapter describes the phenotypic consequences for humans; here, we can simply outline how the conditions arise.

Suppose nondisjunction occurs in a diploid germ cell, and a resulting gamete gets an extra chromosome (n + 1). If that gamete combines with a normal one at fertilization, the new individual will have a chromosome number of 2n + 1. This is a case of "trisomy"—the individual has three of one type of chromosome in what would otherwise be diploid cells.

Similarly, if a gamete deprived of a chromosome combines with a normal gamete, the new individual will have a chromosome number of 2n − 1. This is a case of "monosomy"—one chromosome will be without a homologue in what would otherwise be diploid cells. Monosomy is severe; in humans it can be lethal.

Polyploidy

Polyploidy means having a multiple of the parental chromosome number—that is, three or more of each type of chromosome characteristic of the parental stock. This condition is not as rare as you might think; about half of all flowering plant species are polyploid!

Polyploidy can arise in several ways. Complete nondisjunction in a germ cell undergoing meiosis is one way. Suppose the resulting gametes are diploid. If such a gamete combines with a normal (haploid) gamete at fertilization, a "triploid" (3n) individual results. If two diploid gametes combine, a "tetraploid" (4n) individual results.

Triploid plants are usually sterile. About half of all triploid human fetuses abort spontaneously. Generally speaking, any polyploid with an *uneven* chromosome number (3n, 5n, 7n . . .) is likely to have a problem surviving or reproducing, because the extra chromosomes have no homologues with which they can pair at meiosis, so normal gametes cannot be produced.

Polyploidy also can arise when cells duplicate their DNA but do not divide. This sometimes occurs spontaneously in plant cells that go on to produce gametes. It occurs in human liver cells, which can have two, four, eight, and sometimes sixteen pairs of homologous chromosomes. However, polyploidy is less common in animals than in plants. One reason is that polyploidy can upset the balance between autosomes and sex chromosomes, this being crucial for animal development and reproduction (plants generally have no sex chromosomes). Also, most animals cannot self-fertilize, as many plants can do. And even if they are sterile, many plants can still reproduce asexually.

Sexual reproduction normally involves two parents of the *same* species. When the chromosome number increases because one or both gametes from them was diploid (or polyploid), the change is called "autopolyploidy." (The word means self-increase.) There are triploid bananas and winesap apples, and tetraploid potatoes, peanuts, coffee plants, and grapes. All are autopolyploid plants having great commercial value (Figure 13.12). The plants are larger and hardier than their ancestors.

(Plant breeders often induce autopolyploidy with colchicine, a chemical derived from the autumn crocus, *Colchicum autumnale*. Among other things, colchicine keeps microtubular spindles from forming, so chromosomes

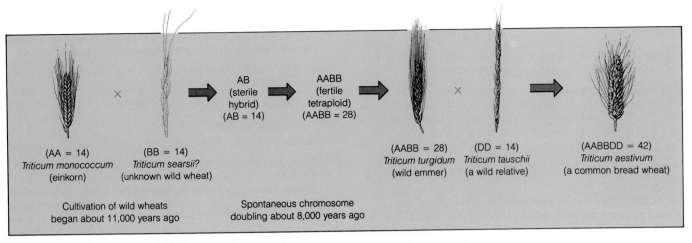

Figure 13.13 Proposed scheme of hybridizations in wheat that led to chromosome doublings and polyploidy.

Wheat grains dating from 11,000 B.C. have been found in the Near East. Several species of wild diploid wheat still grow there. They have 14 chromosomes (two sets of 7, designated AA). Also growing in the region is a wild grass with 14 chromosomes, designated BB. They differ from the A chromosomes, judging from their failure to pair with them at meiosis. One tetraploid wheat species has 28 chromosomes; analysis during meiosis shows that they are AABB. The A chromosomes pair with A's, and the B chromosomes pair with B's. A hexaploid wheat has 42 chromosomes (six sets of seven). Its chromosomes are AABBDD, the last set (DD) coming from *Triticum taushchii*, another wild grass.

cannot separate during mitosis. The chromosome number will keep doubling, depending on the colchicine concentration and length of exposure to it. When the cells are freed from exposure, they divide and give rise to polyploid cells that may go on to produce polyploid gametes.)

Sexual reproduction also occurs *between* members of closely related plant species. If one or both gametes involved in the hybridization are diploid (or polyploid), the resulting increase is called "allopolyploidy." Hundreds of allopolyploids are known, including common bread wheat, one of our major crops (Figure 13.13).

SUMMARY

1. The chromosomal theory of inheritance was formulated between 1900 (following the rediscovery of Mendel's work) and the 1920s. The key points of this theory were summarized earlier (page 178).

2. Most pairs of homologous chromosomes in diploid cells are called autosomes, meaning they are the same in number and kind in males and females. One pair, the sex chromosomes, is different in males and females. The most common sex chromosome pattern is XX (female) and XY (male).

3. Genes that influence the overall male and female phenotype occur on autosomes as well as on the sex chromosomes. (The key sex-*determining* gene apparently is located on the Y chromosome in humans and other mammals.)

4. *X-linked genes* occur on the X chromosome and have no alleles on the Y chromosome. Many of those genes have nothing to do with sexual traits (the eye-color gene of *Drosophila* is an example).

5. Genes located on nonhomologous chromosomes tend to assort independently into gametes. Genes located on the same chromosome tend to remain together in crosses; this is called linkage.

6. Crossing over can affect independent assortment and linkage. The farther apart two genes are on a chromosome, the greater will be the frequency of crossing over and recombination between them.

7. The following abnormal events can alter chromosome structure:
   a. Deletion (loss of a chromosome segment).
   b. Duplication (repeat of a gene sequence).
   c. Inversion (reversal of a gene sequence).
   d. Translocation (transfer of part or all of a chromosome to a nonhomologous chromosome).

8. Nondisjunction is the failure of chromosomes to separate during meiosis. A pair of homologous chromosomes might fail to separate at anaphase I, or sister chromatids of a chromosome might fail to separate at anaphase II. Either event can lead to aneuploidy (one extra or one missing chromosome in offspring).

9. Complete nondisjunction (of all chromosomes) is one route to polyploidy (having a multiple of the parental chromosome number, as when the diploid number $2n$ becomes $3n$ or $4n$, and so on).

Readings

Feldman, M., and E. Sears. January 1981. "The Wild Gene Resources of Wheat." *Scientific American* 244(1):102–112.

Morgan, T., A. Sturtevant, H. Muller, and C. Bridges. 1915. *The Mechanism of Mendelian Heredity.* New York: Holt, Rinehart & Winston. For those who like to browse through original research papers.

Roberts, L. 1988. "Zeroing in on the Sex Switch." *Science* 239:21–23.

Strickberger, M. 1985. *Genetics.* Third edition. New York: McGraw-Hill.

Sturtevant, A. 1965. *A History of Genetics.* New York: Harper & Row. A history of genetics before explosive advances in molecular biology made it almost impossible to continue to do historical surveys of genetics.

Genetics Problems (Answers appear in Appendix II)

1. Recall that human sex chromosomes are XX for females and XY for males.

   a. Does a male child inherit his X chromosome from his mother or father?

   b. With respect to an X-linked gene, how many different types of gametes can a male produce?

   c. If a female is homozygous for an X-linked gene, how many different types of gametes can she produce with respect to this gene?

   d. If a female is heterozygous for an X-linked gene, how many different types of gametes can she produce with respect to this gene?

2. One human gene, which may be Y-linked, controls the length of hair on men's ears. One allele at this gene locus produces non-hairy ears; another allele produces rather long hairs (hairy pinnae).

   a. Why would you *not* expect females to have hairy pinnae?

   b. If a man with hairy pinnae has sons, all of them will have hairy pinnae; if he has daughters, none of them will. Explain this statement.

3. Suppose that you have two linked genes with alleles *A,a* and *B,b* respectively. An individual is heterozygous for both genes, as in the following:

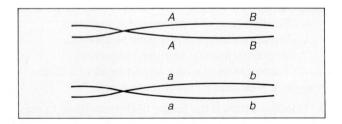

If the crossover frequency between these two genes is 0 percent, what genotypes would be expected among gametes from this individual, and with what frequencies?

4. In *D. melanogaster*, a gene influencing eye color has red (dominant) and purple (recessive) alleles. Linked to this gene is another that determines wing length. A dominant allele at this second gene locus produces long wings; a recessive allele produces vestigial (short) wings. Suppose a completely homozygous dominant female having red eyes and long wings mates with a male having purple eyes and vestigial wings. First-generation females are then crossed with purple-eyed, vestigial-winged males. From this second cross, offspring with the following characteristics are obtained:

   252 red eyes, long wings
   276 purple eyes, vestigial wings
   42 red eyes, vestigial wings
   30 purple eyes, long wings

   600 offspring total

Based on these data, how many map units separate the two genes?

5. Suppose you cross a homozygous dominant long-winged fruit fly with a homozygous recessive vestigial-winged fly. Shortly after mating, the fertilized eggs are exposed to a level of x-rays known to cause mutation and chromosomal deletions. When these fertilized eggs subsequently develop into adults, most of the flies are long-winged and heterozygous. However, a few are vestigial-winged. Provide possible explanations for the unexpected appearance of these vestigial-winged adults.

6. Individuals affected by Down syndrome typically have an extra chromosome 21, so their cells have a total of 47 chromosomes. However, in a few cases of Down syndrome, 46 chromosomes are present. Included in this total are two normal-appearing chromosomes 21 and a longer-than-normal chromosome 14. Interpret this observation and indicate how these few individuals can have a normal chromosome number.

7. Refer to Figure 13.13. *Triticum turgidum* should produce gametes having 14 chromosomes (AB), and *T. tauschii* should produce gametes with 7 chromosomes (D). When these gametes combine, offspring having 7 + 14 = 21 chromosomes should be produced. How, then, did *T. aestivum* originate having 42 chromosomes (AABBDD)? If *T. aestivum* is backcrossed to *T. turgidum*, how many chromosomes would be present in the offspring? Would you expect these offspring to be fertile?

8. The mugwump, a type of tree-dwelling mammal, has a reversed sex-chromosome condition. The male is XX and the female is XY. However, perfectly good sex-linked genes are found to have the same effect as in humans. For example, a recessive allele *c* produces red-green color blindness. If a normal female mugwump mates with a phenotypically normal male mugwump whose mother was color blind, what is the probability that a son from that mating will be color blind? A daughter?

9. Childhood *muscular dystrophy* is a recessive, sex-linked trait in humans. A slowly progressing loss of muscle function leads to death, usually by age twenty or so. Unlike color blindness, this disorder is restricted to males, not ever having been found in a female. Suggest why.

Peas, beans, corn, flies, molds, bacteria—these organisms lend themselves to genetic analysis. They can grow and reproduce rapidly in small quarters, under controlled conditions. Their life spans are far shorter than the life span of geneticists who observe them, so a trait can be tracked through many generations in relatively little time.

Humans are another story. Humans live under variable conditions in diverse environments. Typically they find a mate by chance and reproduce if and when they want to. Human subjects live just as long as the geneticists who study them, so tracking traits through generations is rather tedious. The small size of human families doesn't help much, either. There just aren't enough numbers for meaningful statistical inferences about inheritance patterns.

Despite such obstacles, human genetics is a burgeoning field. Researchers routinely use standardized methods for constructing pedigrees as a way to identify inheritance patterns and to track genetic abnormalities through several generations. (*Pedigrees* are charts of genetic relationships of individuals; Figure 12.12 is an example.) By studying the same trait in many families, researchers increase the numerical base for statistical analysis. They also gather comparative data on genetically distinct populations throughout the world.

Human geneticists are not dabbling in a mere smattering of traits. Already they have identified more than 3,500 abnormalities arising from single-gene mutations alone! Even though some of the abnormalities can be readily diagnosed, none can be permanently cured. There are not even that many **phenotypic treatments**, which involve the use of drugs, restricted diets, or environmental adjustments to compensate for the abnormalities. People so treated may be symptom-free, but they still may bestow the abnormal allele on some or all of their children.

Genotypic cures, in which defective genes are replaced, may eventually be possible for a few disorders. In some preliminary experiments, the normal counterpart of a defective gene was inserted into cells of mammalian embryos—and those cells went on to produce a normal gene product.

This chapter gives examples of the genetic problems we deal with as individuals and as members of society. The examples serve as a framework for considering some practical and ethical aspects of genetic screening, counseling, and treatment programs for genetic disorders.

Keep in mind that "disorder" and "abnormality" are not always interchangeable terms. **Abnormal** means deviation from the average—that is, a rare or less common occurrence, as when a person is born with six toes on each foot instead of five. Whether such a trait is considered to be disfiguring or merely interesting is subjective; there is nothing inherently problematic or life-threatening or even ugly about it. Other genetic

14

HUMAN GENETICS

Table 14.1 Examples of Human Genetic Abnormalities

Disorder	Main Consequences*
Autosomal Recessive Inheritance:	
Albinism	Absence of pigmentation (melanin) (172)
Sickle-cell anemia	Pleiotropic effects leading to severe tissue, organ damage (172)
Galactosemia	Brain, liver, eye damage (190)
Phenylketonuria	Mental retardation (194)
Autosomal Dominant Inheritance:	
Polydactyly	Extra fingers, toes, or both (170)
Achondroplasia	A type of dwarfism (191)
Huntington's disorder	Progressive, irreversible degeneration of nervous system (191)
X-Linked Inheritance:	
Hemophilia A	Deficient blood-clotting (191)
Testicular feminizing syndrome	Absence of male organs, sterility (364)
Chromosomal Aberrations:	
Cri-du-chat	Mental retardation, malformed larynx (184)
Down syndrome	Mental retardation, heart defects (192)
Turner syndrome	Sterility, abnormal development of ovaries and sexual traits (193)
Klinefelter syndrome	Sterility, mental retardation (194)
XYY condition	Mild mental retardation in some cases; no symptoms in others (194)

*Number in parentheses indicates the page on which the disorder is described.

abnormalities cause mild to severe medical problems, and **disorder** is the more appropriate word here.

Also keep in mind that a genetic disorder is not the same thing as a disease, although sometimes the literature is ambiguous on this point. A "disease" is an impairment of the body's structure or functioning that is caused by an infectious agent, such as a parasitic worm, a virus, or a bacterium. Although some genetic factors are known to make a person more susceptible to a disease, the disease itself is not heritable.

DISORDERS ARISING FROM GENE MUTATIONS

Autosomal Recessive Inheritance

The genetic disorder **galactosemia** arises when a breakdown product of lactose (milk sugar) cannot be metabolized. About 1 in 100,000 newborns is galactosemic. Early symptoms include malnutrition, diarrhea, and severe vomiting; often the eyes, liver, and brain become damaged. Without treatment, galactosemics usually die in childhood.

A single gene mutation causes galactosemia. The normal gene product (an enzyme) is part of a metabolic pathway by which lactose is converted first to galactose and ultimately to glucose-1-phosphate. (That phosphorylated compound can be broken down by the reactions of glycolysis, as described in Chapter Nine.) Someone who has two recessive alleles cannot produce molecules

of the enzyme necessary to convert galactose-1-phosphate to glucose-1-phosphate:

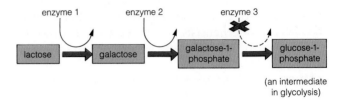

(an intermediate in glycolysis)

The intermediate as well as galactose itself builds up in the blood, and in large concentrations they damage tissues of the brain, liver, and eyes.

Unusually high levels of galactose can be detected in urine samples from homozygous recessive infants. If the abnormality is detected early enough, infants can be put on a diet that includes milk substitutes. They can grow up symptom-free as long as they avoid all sources of galactose.

Galactosemia is an example of *autosomal recessive inheritance*, which has the following characteristics:

1. Both males and females can carry the recessive allele on an autosome (not on a sex chromosome).

2. Heterozygotes of either sex are symptom-free. Homozygotes of both sexes are affected.

3. When both parents are heterozygous, there is a fifty-percent chance that each child born to them will be heterozygous also. There is a twenty-five percent chance that each child will be homozygous recessive.

4. When both parents are homozygous for the recessive allele, all of their children will be affected.

The Punnett square in Figure 14.1 shows an inheritance pattern for an autosomal recessive condition.

Autosomal Dominant Inheritance

Recessive alleles that cause genetic disorders can persist at fairly high frequencies in populations, because heterozygotes may still survive and reproduce. (Their normal allele usually can cover the function of the abnormal one.) But what if a *dominant* allele—which is always expressed to some extent—causes the disorder? If its expression reduces a person's chance of surviving and reproducing, there is less chance that the harmful allele will be passed on, and its frequency in the population will decrease.

Nevertheless, dominant alleles that are responsible for a few pronounced disorders do not disappear from the population. Even if a dominant allele is selected against, new mutations arise and can lead to its continuance in the population. Also, *some dominant alleles do not affect reproduction or they are only expressed after reproductive age.*

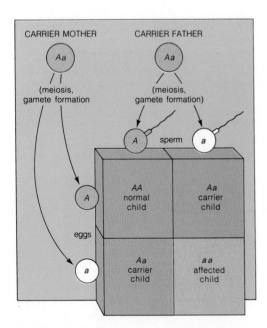

Figure 14.1 Possible phenotypic outcomes for autosomal recessive inheritance when both parents are heterozygous carriers of the recessive allele (shaded red here).

About 1 in 10,000 individuals is affected by **achondroplasia**, a type of dwarfism that is caused by an autosomal dominant allele. When long bones develop in affected children, cartilage forms in ways that lead to disproportionately short arms and legs. Affected persons end up being less than four feet, four inches tall. The dominant allele has no other phenotypic effects in heterozygotes. The heterozygotes normally are fertile; however, homozygous fetuses usually are stillborn.

Huntington's disorder is a less common form of autosomal dominant inheritance. It is characterized by progressive degeneration of the nervous system, and it eventually is lethal. In about half the cases, symptoms are expressed from age forty onward—after most people have already had children. In time, movements are convulsive and brain function deteriorates rapidly.

Although Huntington's disorder is rare worldwide, its frequency has been much higher in South Africa, where about 500 cases have been identified. Researchers discovered that all affected persons were directly or indirectly descended from the same Dutchman who settled there in 1658. The unusually high frequency of the disorder is an example of the founder effect (Chapter Thirty-Six).

Figure 14.2 shows an inheritance pattern for an autosomal dominant condition.

X-Linked Recessive Inheritance

Some genetic disorders are associated with genes located on the X chromosome. (Recall that those genes have no corresponding alleles on the Y chromosome in males.) A few bleeding disorders fall in this category.

Normally, a clotting mechanism quickly comes into play whenever an injury causes bleeding. Some people bleed for an abnormally long time because the mechanism is defective in one way or another. Several different gene products are required in the series of reactions that lead to clot formation. Mutation of any one of the genes coding for those products can give rise to a particular type of bleeding disorder.

A mutated form of one gene, which normally codes for clotting factor VIII, gives rise to **hemophilia A**. Males who inherit a recessive allele on their one X chromosome are always affected. Without medical attention, cuts, bruises, or internal bleeding could lead to death. The blood-clotting time is more or less normal in heterozygous females; the nonmutated gene on their other X chromosome codes for enough of factor VIII to cover the required function. Hemophilia A is an example of *X-linked recessive inheritance*, which has these characteristics:

1. The gene responsible occurs only on the X (not the Y) chromosome. Heterozygous females are phenotypically normal; their one functional allele codes for enough of the required product. Males typically are affected (they have only one allele for the trait, and it is recessive).

2. When the male is normal but the female is heterozygous, there is a fifty percent chance that each daughter born to them will be a carrier and a fifty percent chance that each son will be affected.

3. When the female is homozygous recessive and the male is normal, all daughters will be carriers and all sons will be affected.

Figure 14.3 shows an inheritance pattern for an X-linked recessive condition.

The recessive allele for hemophilia A is rare. Only about 1 in 7,000 human males is affected, and unless hemophilic males marry close relatives, it is not likely that they will have affected (homozygous recessive) daughters. (If a male is hemophilic, the probability that a close relative will carry the recessive allele is greater than if he picked his mate at random from the general population.)

Compared to the general population, the frequency of the recessive allele was much higher among the royal families of nineteenth-century Europe, whose members often intermarried. Queen Victoria of England was a carrier, as were two of her daughters (Figure 14.4). At one time it was calculated that, of her sixty-nine descendants, eighteen were affected males or female carriers.

Crown Prince Alexis of Russia was one of Victoria's hemophilic descendants. His affliction and the cast of

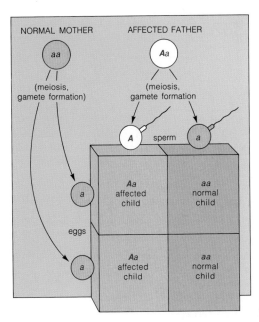

Figure 14.2 Possible phenotypic outcomes for autosomal dominant inheritance, assuming the dominant allele is fully expressed in the carriers. (The dominant allele is shaded red.)

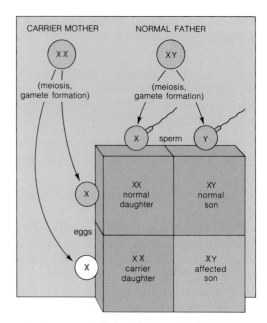

Figure 14.3 Possible phenotypic outcomes for X-linked inheritance when the mother carries a recessive allele on one of her X chromosomes (shaded red here).

Figure 14.4 Descendants of Queen Victoria, showing carriers and affected males that possessed the X-linked gene conferring the disorder hemophilia A. Many individuals of later generations are not shown in this family pedigree.

characters it indirectly drew together—Czar Nicholas II, Czarina Alexandra (a granddaughter of Victoria and a carrier), the power-hungry monk Rasputin who manipulated the aggrieved family to his political advantage—helped catalyze events that brought an end to dynastic rule in the Western world.

DISORDERS ARISING FROM CHANGES IN CHROMOSOME NUMBER

Gene mutations are not the only culprits behind human genetic disorders. Nondisjunction, the failure of a pair of chromosomes to separate normally at meiosis, is a common cause of miscarriage, a natural expulsion of the fetus from the uterus in the first six months of pregnancy. (Two examples of nondisjunction are shown in Figure 13.11.) Moreover, an average of one in every 200 liveborns will be affected by a change in chromosome number that is brought about by nondisjunction.

Down Syndrome

Sometimes a gamete receives two copies of chromosome 21 as a result of nondisjunction. Fusion of the abnormal gamete with a normal one produces an individual with three copies of that chromosome (Figure 14.5a). The condition, called *trisomy 21*, leads to **Down syndrome**. (The word "syndrome" means a set of symptoms that typically occur together and that characterize a particular disorder.) Trisomic 21 embryos are often lost through

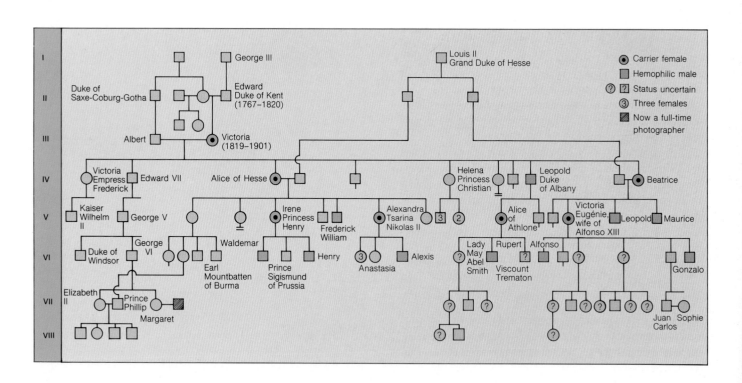

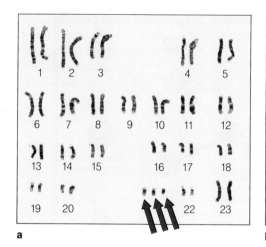

a

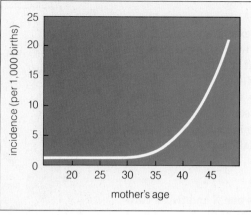

b

Figure 14.5 (**a**) Karyotype of a girl with Down syndrome; red arrows identify the trisomy of chromosome 21. (**b**) Relationship between the frequency of Down syndrome and the mother's age. Results are from a study of 1,119 children with the disorder who were born in Victoria, Australia, between 1942 and 1957.

miscarriage; however, about 1 of every 1,000 liveborns in North America alone is destined to develop the disorder.

Most children with Down syndrome show moderate to severe mental retardation, and about forty percent have serious heart defects. Their skeleton develops slower than normal and muscles are rather slack, so older children are shorter than normal and have certain characteristic facial features (Figure 14.6). For example, most have a small skin fold over the inner corner of the eyelid.

Trisomy 21 is not usually a heritable disorder. (Only about three percent of families with one affected child run a risk of having another owing to a certain chromosomal aberration in one of the parents.) However, as women age, they run a much greater risk of giving birth to a trisomic 21 child than do women in their early twenties (Figure 14.5b). Apparently, nondisjunction of chromosome 21 in female reproductive cells occurs more frequently from age thirty-five onward, according to some geneticists. Recent studies suggest that the risk also increases when the father is past age fifty-five.

Turner Syndrome

About 1 in every 5,000 newborns is destined to have **Turner syndrome**, a *sex chromosome abnormality* that arises through nondisjunction (Figure 14.7). Affected persons have only 45 chromosomes instead of 46; only one sex chromosome (X0) is present in their somatic cells. The frequency of Turner syndrome is lower than for other sex chromosome abnormalities, probably because most affected embryos are miscarried early in pregnancy.

Turner syndrome leads to a female phenotype, but the phenotypic distortions are pronounced. The females are sterile. Their ovaries are nonfunctional, and secondary sexual traits fail to develop at puberty. Often they age prematurely, and they have shortened life expectancies.

a

b

c

d

Figure 14.6 Children with Down syndrome. Facial features associated with the syndrome are not always evident in younger children, as (**a**) and (**b**) indicate.

Klinefelter Syndrome

Nondisjunction leads to other sex chromosome abnormalities, including XXY males who show **Klinefelter syndrome** (Figure 14.7). Affected males are sterile and mentally retarded. Their testes develop to only about a third of normal size, body hair is sparse, and there may be some breast enlargement. Injections of the hormone testosterone can reverse the feminized phenotype but not the sterility or mental retardation.

XYY Condition

About 1 in every 1,000 males has one X and two Y chromosomes, a condition called **XYY**. It is probably inappropriate to apply the term "syndrome" to this chromosome abnormality. XYY males tend to be taller than average and some may show mild mental retardation, but most are phenotypically normal.

At one time, XYY males were thought to be genetically predisposed to become criminals. However, a comprehensive study in Denmark showed that the number who do end up in prison is no more notable than the percentage of other tall men. Compared to normal (XY) males, their rate of conviction was indeed greater (41.7 percent compared to 9.3 percent). But this is not necessarily proof of predisposition to crime. With their moderately impaired mental ability, XYY males simply may be easier to catch.

PROSPECTS AND PROBLEMS IN HUMAN GENETICS

Chances are, you personally know of someone who has a genetic disorder. Of all newborns, possibly one percent will have pronounced problems arising from a chromosomal aberration. Between one and three percent more will have problems because of mutant genes that either cannot code for vital products or code for defective ones. Of all patients in children's hospitals, ten to twenty-five percent are treated for problems arising from genetic disorders.

Geneticists are working to develop diagnostic tools and treatments for heritable disorders. However, the disorders apparently cannot be approached in the same manner as infectious diseases (such as influenza, measles, and polio). Infectious agents are enemies from the environment, so to speak, that attack without much warning. We have had no qualms about mounting counterattacks through immunizations and antibiotics, which either eliminate the agents or at least bring them under control. With genetic disorders, the problem is inherent in the hereditary material of individual human beings.

How do we attack an "enemy" within? Do we institute regional, national, even global programs to identify

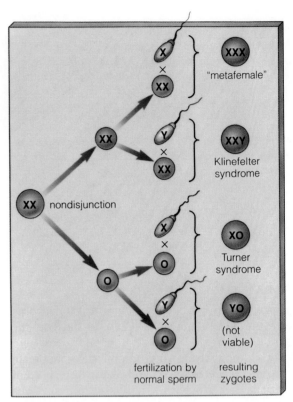

Figure 14.7 Genetic disorders that result from nondisjunction of X chromosomes followed by fertilization involving normal sperm.

affected persons? Do we inform them that they are defective in particular ways, and that they run a risk of bestowing the disorder on their children? Who decides which alleles are "harmful"? Should society bear the cost of treating disorders such as Down syndrome? If so, should society also have some say in whether affected fetuses will be born at all, or aborted? These questions are only the tip of an ethical iceberg, and answers have not been worked out in universally acceptable ways.

The rest of this chapter describes therapeutic and preventive measures currently being used to detect and treat genetic disorders, as listed in Table 14.2.

Treatments for Phenotypic Defects

Although genetic disorders cannot be cured permanently, in some cases their phenotypic consequences can be circumvented. Treatments include diet modifications, environmental adjustments, surgery, and chemotherapy.

Diet Modification. The outward symptoms of several genetic disorders can be suppressed or minimized by controlling the diet. Galactosemia, described earlier, is controlled this way. So is **phenylketonuria**, or PKU. A certain gene normally codes for an enzyme that converts phenylalanine (an amino acid) to tyrosine (another amino acid). However, phenylalanine levels build up in people who are homozygous recessive for the mutant gene. The excess is diverted to other pathways, which may produce compounds in amounts that interfere with normal body functioning. High levels of one compound, phenylpyruvic acid, can damage the nervous system and lead to mental retardation.

The symptoms of PKU can be alleviated when phenylpyruvic acid is detected early enough in urine or blood samples. Affected persons simply adhere to a diet that gives them only as much phenylalanine as they need for protein synthesis; thus their body is not called on to dispose of excess amounts and they can lead normal lives. There is some evidence that homozygous PKU females will give birth to mentally retarded infants unless they are placed on a low-phenylalanine diet well before they become pregnant.

Environmental Adjustments. Some treatments for genetic abnormalities require adjustments to surrounding conditions. True albinos (page 172) avoid direct sunlight. Sickle-cell individuals have two mutant alleles that code for a defective form of hemoglobin, the protein that transports oxygen in the blood (Figure 12.15). They are instructed to avoid strenuous activity when oxygen levels are low, as at high altitudes and in unpressurized aircraft cabins.

At the extreme of environmental adjustments are children whose immune system does not function and who must be raised in sterile environments to avoid infectious agents. One child who was raised this way survived to age twelve, then died following unsuccessful surgery to improve immunological function.

Surgical Correction. Many phenotypic defects are corrected or minimized by surgical reconstructions. For example, one form of **cleft lip** is a genetic abnormality of the upper lip. A vertical fissure is present at the lip midsection and often extends into the palate (the roof of the mouth). Surgery can usually correct the abnormality in terms of appearance and function.

Chemotherapy. Knowledge of the molecular basis of genetic disorders is being used to chemically modify gene products or to compensate for their absence. For example, **Wilson's disorder** arises from an inability to utilize copper, which is essential for the action of several enzymes. Copper deposits build up in tissues and can lead to brain and liver damage. Convulsions and death are the outcome. By avoiding copper-containing foods (such as chocolate) and by taking certain drugs, people who are diagnosed early enough can lead a nearly normal life. One drug (a penicillin derivative) binds with copper in the body, and the complex is flushed from the body by way of the urinary system.

Table 14.2 Some Therapeutic and Preventive Measures Used in Treating Genetic Disorders

Therapeutic Measures:

Diet modification	Providing substances that the body cannot produce for itself, or restricting intake of substances that the body cannot tolerate
Environmental adjustment	Avoiding environmental conditions that the body cannot tolerate (for example, low oxygen levels at high altitudes in the case of sickle-cell anemics)
Surgical correction	Surgically repairing deformities (such as cleft lip, heart defects)
Chemotherapy	Using chemicals (such as drugs) to modify or inhibit gene expression or gene product function

Preventive Measures:

Mutagen reduction	Avoiding or eliminating environmental substances (such as x-rays or radioactive substances) that can induce mutations
Genetic screening	Methodically searching through populations and identifying individuals affected by (or heterozygous for) a particular genetic disorder
Genetic counseling	Conveying information about genetic disorders, as well as the social and medical options available, to affected individuals and their families
Prenatal diagnosis	Detecting chromosomal alterations and metabolic disorders in the fetus
Gene replacement*	Substituting normal alleles for mutant alleles

*Research stage only.

Genetic Screening

The therapeutic measures described so far assist persons who are already affected by a genetic disorder. Preventive measures are also being used to allow early detection and treatment before symptoms can develop. Other measures are used to identify carriers who show no outward symptoms but may give birth to affected children.

Genetic screening usually refers to large-scale programs to detect affected persons or carriers in a population. One extensive program began in the late 1950s and continues today as a means of detecting PKU. Most hospitals in the United States routinely screen all newborns for PKU, so it is becoming less common to see people who show symptoms of the disorder.

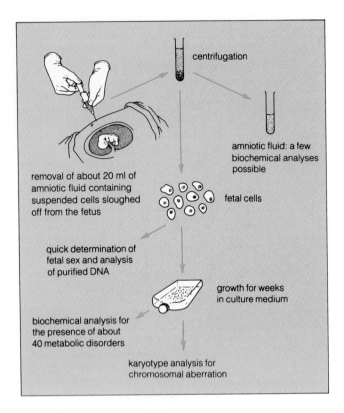

centrifugation

amniotic fluid: a few
biochemical analyses
possible

removal of about 20 ml of
amniotic fluid containing
suspended cells sloughed
off from the fetus

fetal cells

quick determination of
fetal sex and analysis
of purified DNA

growth for weeks
in culture medium

biochemical analysis for
the presence of about
40 metabolic disorders

karyotype analysis for
chromosomal aberration

Figure 14.8 Steps in amniocentesis, a procedure used in prenatal diagnosis of many genetic disorders.

Genetic Counseling and Prenatal Diagnosis

Sometimes prospective parents suspect they are at high risk of producing a severely afflicted child. Either their first child or a close relative shows an abnormality and they now wonder if future children will be affected the same way. How are such people counseled? Typically, several consultants are brought in, including clinical psychologists, geneticists, and social workers who can give emotional support to families with affected children.

Counseling begins with an accurate diagnosis of the parental genotypes, which may reveal the potential for a specific disorder. Biochemical tests are performed to detect many metabolic disorders. Family pedigrees are constructed as completely and accurately as possible to aid in the diagnosis. For disorders that follow simple Mendelian inheritance patterns, it is possible to predict the likelihood of producing affected children—but not all disorders follow Mendelian patterns. Even those that do can be influenced by other factors, some identifiable, others not. Even when the extent of risk has been determined with some confidence, prospective parents are made to understand that the risk is the same for *each* pregnancy. Thus, if there is one chance in four that the child will be born with a genetic disorder, there is one chance in four that the next child will be also.

What happens when a female is already pregnant? For example, suppose a pregnant woman, forty-five years old, wants to know if her child will show Down syndrome. Through prenatal diagnosis, this and more than a hundred other genetic disorders can be detected early in pregnancy.

A major detection procedure is based on **amniocentesis**: a sampling of the fluid surrounding the fetus in the mother's uterus (Figure 14.8). During the fourteenth to sixteenth week of pregnancy, the thin needle of a syringe is carefully inserted through the mother's abdominal wall and into the fluid-filled sac (amnion) containing the fetus. Floating in this fluid are epidermal cells that have been shed from the fetus. Some of the fluid, along with its sample of fetal cells, is withdrawn by the syringe. The cells can be cultured and allowed to undergo mitosis. Abnormalities can be diagnosed through a variety of tests, including karyotype analysis of metaphase chromosomes (page 183). Amniocentesis is also being used to obtain cells that can be tested for biochemical defects, such as the one that is responsible for sickle-cell anemia. A small risk is associated with the procedure; care must be taken not to puncture the fetus or cause infection.

A more recently developed procedure, **chorionic villi sampling** (CVS), uses cells drawn from the chorion (a membranous sac surrounding the amnion). This procedure can be used earlier in pregnancy (by the eighth week). And results often are available in one or two days, not weeks as with amniocentesis. Is CVS more or less safe than amniocentesis? That is still being evaluated.

Unfortunately, there is no known cure for disorders arising from changes in chromosome structure or number. If the embryo is diagnosed as having a severe disorder such as trisomy 21, the parents may decide on an abortion (an induced expulsion of the embryo from the uterus). Such decisions are bound by ethical considerations. The role of the medical community should be to provide information that the prospective parents need in order to make their own choice. That choice in turn must be consistent with their own values, within the broad constraints imposed by society.

RFLPs

As you might imagine, detection and treatment of genetic disorders will become much easier once all twenty-three human chromosomes have been completely mapped (page 242). The mapping is proceeding more rapidly now, owing to recent advances in recombinant DNA technology.

Although recombinant DNA technology is described in a later chapter, let's consider here one of the techniques that has special potential for human genetic analysis and treatment. This technique uses "restriction fragment length polymorphisms," or **RFLPs**, in human chromosomes.

The "restriction fragment length" part of the name refers to the fact that researchers use certain enzymes to make cuts in human DNA, and these so-called restriction enzymes (there are more than a hundred) always produce a characteristic set of fragments. Each makes its cut only at sites having a short, specific sequence of DNA subunits. The resulting DNA fragments can be separated according to their sizes. The separation procedure (electrophoresis, Figure 16.2) spreads out the fragments so that particular ones can be identified.

The "polymorphism" part of the name refers to the fact that each person's DNA has a slightly unique pattern of restriction sites and so produces a different pattern of DNA fragments. (The only exceptions are identical twins, who have identical DNA.)

It is possible to detect differences in the pattern of DNA fragments by using special radioactive probes. And it turns out that mutant alleles responsible for genetic disorders usually have a unique restriction site that is *not* present in the DNA of normal persons. Sickle-cell anemia is an example; a mutant allele associated with a defective hemoglobin chain can be detected in this manner (Figure 14.9).

The use of another kind of RFLP (called hypervariable regions) is expected to revolutionize many criminal investigations. Each person has a unique array of such regions, which are inherited from each parent in a Mendelian pattern. Thus each person has a unique "genetic fingerprint." Already paternity cases have been resolved by comparing the genetic fingerprint of the child with those of the disputed parents.

Murderers also will become much easier to identify if they leave even a few drops of their blood at the scene of the crime. There will be enough DNA in the bloodstain so that the genetic fingerprint can be compared to those of suspects. The hope is that rapists also will be deterred by the new technology. A genetic fingerprint from just a drop or two of semen recovered from a victim can identify the rapist. Courts have already ruled for conviction when a suspect's genetic fingerprint matches that found in semen samples.

SUMMARY

1. Of all humans being born, possibly one percent will be affected by major problems arising from chromosomal aberrations. Between one and three percent will have problems because of mutant genes that code for defective proteins (or for none at all). Of all patients in children's hospitals, between ten and twenty-five percent are being treated for problems arising from genetic abnormalities.

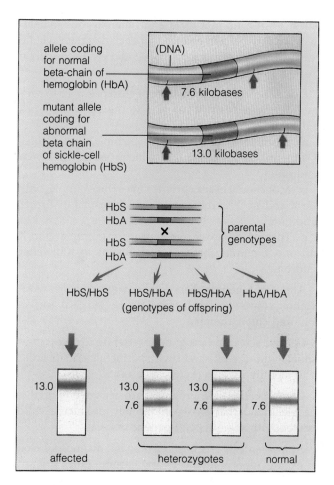

Figure 14.9 Prenatal diagnosis of sickle-cell anemia using RFLPs. The yellow bars represent the electrophoresis results possible for fetal cells that are obtained by chorionic villi sampling.

2. Some genetic disorders show patterns of autosomal recessive inheritance. The recessive allele is expressed in homozygotes but not in heterozygotes of either sex. With two heterozygous parents, there is a fifty percent chance that each child will be heterozygous and a twenty-five percent chance that it will be affected. Two homozygous parents always produce affected children.

3. In genetic disorders showing patterns of autosomal dominant inheritance, the dominant allele is always expressed to some extent. Even though it may be selected against, the dominant allele may remain in the population for three reasons. The dominant allele may be replaced by recurring mutations, it may not preclude reproduction, or it may be expressed after reproductive age.

4. In genetic disorders showing patterns of X-linked recessive inheritance, the recessive allele is masked in heterozygous females and is almost always expressed in

males. With a normal male and heterozygous female, there is a fifty percent chance that each daughter will be a carrier and a fifty percent chance that each son will be affected.

5. Nondisjunction is the failure of chromosomes to separate during meiosis. It can lead to the following types of disorders:

    a. Trisomy of autosomes (for example, three copies of human chromosome 21, which results in Down syndrome).

    b. Monosomy of sex chromosomes (for example, only one X and no Y chromosome, which results in Turner syndrome).

    c. Trisomy of sex chromosome (for example, two X and one Y chromosomes, which results in Klinefelter syndrome).

6. At present, genetic disorders are treated by diet modification, environmental adjustments, surgical correction, and chemotherapy.

7. Genetic screening programs are set up to detect individuals who are affected by a specific genetic disorder or to identify carriers (heterozygotes).

8. Genetic counseling programs are designed to predict the likelihood that two parents may produce a child affected with some type of genetic disorder.

9. Amniocentesis is a form of prenatal diagnosis based on examination of fetal cells and chromosomes prior to birth. It can be used to detect a large number of genetic disorders. Chorionic villi sampling (a more recent diagnostic procedure) can be done earlier in pregnancy.

10. Gene therapy (replacing defective alleles with normal ones) is not yet possible.

11. Restriction fragment length polymorphisms (RFLPs) are being used to screen for genetic disorders. They also are being used to solve certain crimes.

Readings

Edlin, G. 1988. *Genetic Principles: Human and Social Consequences.* Second edition. Portola Valley, California: Jones & Bartlett. Accessible introduction to human genetics.

Fuchs, F. June 1980. "Genetic Amniocentesis." *Scientific American* 242(6):47–53. Summary article of a major diagnostic procedure.

Fuhrmann, W., and F. Vogel. 1986. *Genetic Counseling.* Third edition. New York: Springer-Verlag.

Holden, C. 1987. "The Genetics of Personality." *Science* 237:598–601. For students interested in human behavioral genetics, this is a good article.

Patterson, D. August 1987. "The Causes of Down Syndrome." *Scientific American* 257(2):52–60. Describes research to identify the genes whose products may contribute to this genetic disorder.

Genetics Problems (Answers in Appendix II)

1. In hemophilia A, the body's blood-clotting mechanism is defective. This condition has been traced to a recessive allele of an X-linked gene. Refer now to Figure 14.4. Why are only the females shown as carriers of the recessive allele?

2. Huntington's disorder is due to a dominant autosomal allele. Usually this disorder does not manifest itself until after age thirty-five. Individuals having Huntington's disorder are almost always heterozygous. As a genetic counselor, you are visited by a twenty-year-old woman. Her mother has Huntington's disorder but her father is normal. What is the probability that this woman will develop Huntington's disorder as she grows older? Suppose, at her present age, she marries someone with no family history of the disorder. If they have a child, what is the probability that it will have Huntington's disorder?

3. Color blindness is an X-linked trait. A woman heterozygous for color blindness (*Gg*) marries someone who has normal color vision. What is the probability that their first child will be color blind? Their second child? If they have two children only, what is the probability that both will be color blind?

4. A person affected by Turner syndrome has only a single sex chromosome (X only), yet may survive. In contrast, a person having a single Y chromosome and no X chromosome cannot survive. What does this tell you about the genetic content of the X and Y chromosomes?

5. Fertilization of a normal egg by a sperm that has no sex chromosomes (male nondisjunction) can lead to Turner syndrome. Also, fertilization of an egg that has no sex chromosomes (female nondisjunction) by a sperm carrying one X chromosome can lead to the same disorder. Suppose a hemophilic male and a carrier (heterozygous) female have a child. The child is nonhemophilic and is affected by Turner syndrome. In which parent did nondisjunction occur?

6. The trisomic XXY condition is also called Klinefelter syndrome. How could this syndrome arise if nondisjunction occurred in the female parent of an affected individual? How could it arise if nondisjunction occurred in the male parent?

7. If nondisjunction occurs for the X chromosomes during oogenesis, then some eggs having two X chromosomes and others having no X chromosomes are produced at about equal frequencies. If normal sperm fertilize these two types of eggs, what genotypes are possible?

8. Phenylketonuria (PKU) is an autosomal recessive condition. About 1 of every 50 persons who inherit the gene responsible for the disorder is heterozygous and displays no symptoms of the disorder.

    a. If you select a symptom-free female at random from the population, what is the probability that she will be heterozygous?

    b. If you select a symptom-free male and a symptom-free female at random, what is the probability that both will be heterozygous? What is the probability that they could have a child with PKU?

9. Laws restricting marriage between close relatives (consanguineous matings) are widespread, the rationale being that such marriages generally lead to an increase in the incidence of genetic disorders among offspring. Suppose you are a carrier (heterozygous) for PKU. If you pick a potential mate at random from the population, what is the probability that he or she would also be a PKU carrier? If you marry your first cousin, do you think he or she would have the same probability of being a PKU carrier as your randomly selected mate? Why?

DISCOVERY OF DNA FUNCTION

One might have wondered, in the spring of 1868, why Johann Miescher was collecting cells from the pus of open wounds and, later, from the sperm of a fish. Miescher wanted to identify the chemical composition of the nucleus, and he was interested in those cells because they are composed mostly of nuclear material, with very little cytoplasm. He succeeded in isolating an acidic substance, one with a large amount of phosphorus. Miescher called it "nuclein." He had discovered what came to be known as deoxyribonucleic acid, or DNA.

The discovery caused scarcely a ripple through the scientific community. At the time, no one knew much about the physical basis of inheritance—that is, *which* cellular substance actually encodes the instructions for reproducing parental traits in offspring. Only a few researchers suspected that the nucleus might hold the answer. In fact, seventy-five years passed before DNA was recognized as having profound biological importance.

With this chapter, we turn to investigations that led to our current understanding of DNA structure and function. As you will see, a few turned out to be blind alleys in terms of their intended purpose—yet they are revealing of how ideas are generated in science. *Ideas typically develop as a community effort, with individuals sharing not only what they can explain but also what they do not understand.* There are two points to keep in mind:

1. Even if an experiment "fails," it may turn up information that others can use or lead to questions that others can answer.

2. Unexpected results from an experiment are reported, for they may be clues to something important about the natural world.

A Puzzling Transformation

In 1928, Fred Griffith attempted to develop a vaccine against the bacterium *Streptococcus pneumoniae*, which causes the lung disease pneumonia. (Many vaccines are preparations of killed or weakened bacteria which, when introduced into the body, mobilize the body's defenses prior to a real attack.) He never did create a vaccine, but his experiments unexpectedly opened a door to the molecular world of heredity.

Griffith isolated two strains of the bacterium, which he designated *S* and *R*. (The surface appearance of the bacterial colonies was smooth for one strain and rough for the other.) He used the strains in four different experiments and came up with the results that are summarized in the following list.

15

THE RISE OF MOLECULAR GENETICS

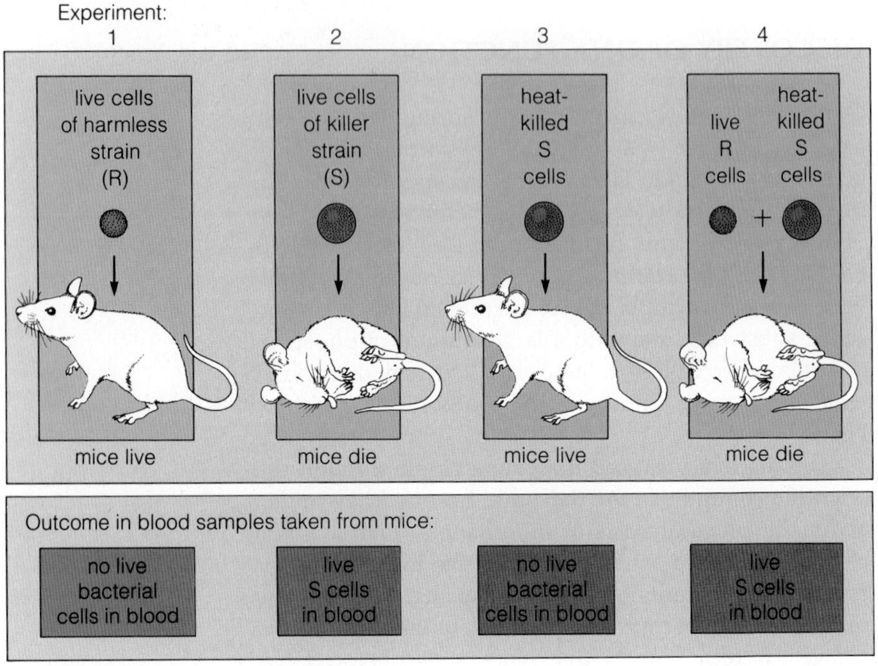

Figure 15.1 Griffith's experiments with harmless (R) strains and disease-causing (S) strains of *Streptococcus pneumoniae*, as described in the text. (You may be wondering why the S form is deadly and the R form harmless. The pathogenic (virulent) S form produces a thick external capsule, which protects it from attack by the host's immune system. The R form has no such capsule; hence the host's defense system destroys the R form before it can cause disease.)

1. Laboratory mice were injected with live R cells. They did not contract pneumonia; the R strain was harmless (Figure 15.1).

2. Mice were injected with live S cells. The mice contracted pneumonia and died, and blood samples taken from them teemed with live S cells. The S strain obviously was pathogenic (disease-causing).

3. S cells were killed by exposing them to high temperature. Mice injected with these heat-killed cells did not develop pneumonia.

4. Live R cells were mixed with heat-killed S cells. Curiously, mice injected with that combination came down with pneumonia and died—and blood samples from the mice teemed with live S cells!

What was going on in the fourth experiment? Maybe the heat-killed pathogens in the mixture were not really dead. But if that were true, the group of mice injected with heat-killed pathogenic cells alone would have contracted the disease, too. Maybe the harmless R cells in the mixture had mutated into the killer S form. But if that were true, the group of mice injected with R cells alone would have died.

The simplest explanation was this: Although heat did kill the pathogenic cells, it did not damage the substance containing their hereditary instructions—including the part that specified "how to cause infection." Somehow the substance had been liberated from those dead cells and it entered living cells of the harmless strain—where it combined with the recipient's hereditary substance.

Further experiments made it clear that the harmless cells had indeed picked up instructions for causing infection and had been permanently transformed into pathogens because of it. Hundreds of generations of bacteria

descended from the transformed cells also caused infections!

A few years later, researchers found that extracts of the killed pathogenic bacteria also could cause hereditary transformation. The microbiologist Oswald Avery and his colleagues began work to purify and experiment with the chemical substances of those extracts. This was a time when most biochemists believed that hereditary instructions were encoded in proteins, not DNA. (After all, heritable traits are spectacularly diverse. The molecules containing the instructions for those traits surely were structurally diverse also, and proteins—with their potentially limitless combinations of amino acid subunits—seemed to be the best candidates.) That prevailing belief was challenged in 1944, when Avery's group reported that DNA probably was the substance of heredity.

Avery's key experiments could not be explained away. He could *block* transformation of harmless bacteria by adding pancreatic deoxyribonuclease to extracts of the pathogenic strain. That enzyme degrades DNA molecules but has no effect on proteins. In contrast, protein-degrading enzymes had no effect at all on the transforming activity.

How were Avery's impressive findings received? Many (if not most) biochemists refused to give up on the proteins. His experimental results, they said, probably applied only to bacteria.

Bacteriophage Studies

While work was going on in Avery's laboratory, Max Delbrück, Alfred Hershey, and Salvador Luria were studying a class of viruses called **bacteriophages**. Dif-

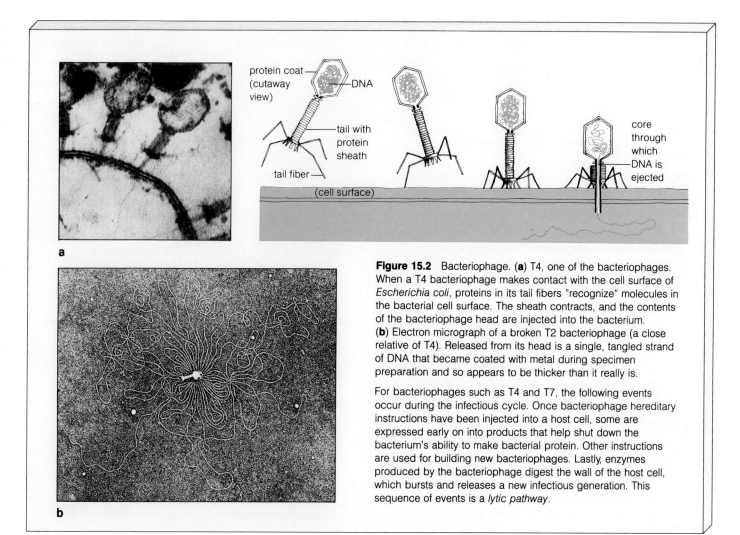

Figure 15.2 Bacteriophage. (**a**) T4, one of the bacteriophages. When a T4 bacteriophage makes contact with the cell surface of *Escherichia coli*, proteins in its tail fibers "recognize" molecules in the bacterial cell surface. The sheath contracts, and the contents of the bacteriophage head are injected into the bacterium. (**b**) Electron micrograph of a broken T2 bacteriophage (a close relative of T4). Released from its head is a single, tangled strand of DNA that became coated with metal during specimen preparation and so appears to be thicker than it really is.

For bacteriophages such as T4 and T7, the following events occur during the infectious cycle. Once bacteriophage hereditary instructions have been injected into a host cell, some are expressed early on into products that help shut down the bacterium's ability to make bacterial protein. Other instructions are used for building new bacteriophages. Lastly, enzymes produced by the bacteriophage digest the wall of the host cell, which bursts and releases a new infectious generation. This sequence of events is a *lytic pathway*.

ferent bacteriophages infect different bacterial cells. The infectious cycle starts when bacteriophages adsorb to a target host cell. In some cases, they inject their contents into the cell (Figure 15.2). Within sixty seconds, the cell starts making the nucleic acids and proteins (including enzymes) necessary to build new bacteriophages. Then lysis occurs: viral enzymes degrade the bacterial cell wall and the cell bursts, thereby liberating a new infectious generation.

Many bacteriophage studies confirmed Avery's conclusion that DNA is the hereditary substance. As an example, consider some experiments conducted in 1952 by Hershey and his colleague Martha Chase. By then it was known that bacteriophages contain only DNA and protein. Hershey and Chase devised a way to track the two substances through the infectious cycle.

Bacteriophage proteins contain sulfur but no phosphorus, and DNA contains phosphorus but no sulfur. Both substances have radioactive isotopes (^{35}S and ^{32}P). Bacterial cells were grown on a culture medium containing amino acids labeled with ^{35}S. When they synthesized

proteins, they took up the isotope and so became radioactively labeled. When the cells were later infected, the new generation of bacteriophages assembled inside them also contained the labeled protein. (Why? They could be built only of materials, including radioactive ones, available from their hosts.) Similarly, cells grown on a culture medium containing nucleotides labeled with ^{32}P ended up with labeled DNA.

As Figure 15.3 shows, labeled bacteriophages were now allowed to infect unlabeled cells. After the cycle of infection ran its course, Hershey and Chase determined that the new bacteriophage generation incorporated labeled DNA—not the labeled protein.

Many different experiments have since been performed on cells of a variety of species from all five kingdoms. All confirm the following principle:

Instructions for producing the heritable traits of single-celled and multicelled organisms are encoded in DNA.

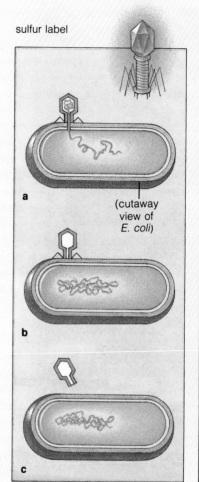

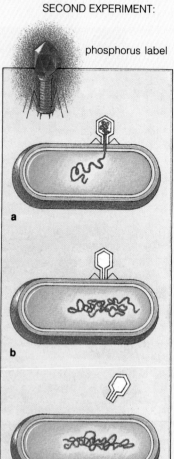

Figure 15.3 Hershey-Chase experiments to determine the paths taken by bacteriophage DNA and protein during the infection of *E. coli*.

(a) In the first experiment, bacteriophages were labeled with a radioactive isotope of sulfur, ^{35}S, to tag the protein (green) making up the bacteriophage body. In the second experiment, bacteriophages were labeled with a radioactive isotope of phosphorus, ^{32}P, to tag their DNA (red).

The labeled bacteriophages were mixed with a suspension of bacterial cells, and they attached to those cells and injected their DNA into them. Following injection, the bacteriophage bodies remained attached to the cells **(b)**.

(c) The suspension of infected cells was agitated in a kitchen blender. The shearing forces cleaved the bacteriophage bodies from the cells, which then were separated from the suspension medium.

When ^{35}S had been used to label protein (the first experiment), radioactivity remained in the suspension medium and was associated with the bacteriophage bodies. When ^{32}P was used to label bacteriophage DNA (second experiment), radioactivity remained with the bacterial cells.

In other samples of sheared bacteria, it was shown that shearing did not prevent subsequent reproduction of new bacteriophages in the infected cells.

Because the bacteriophage DNA had entered the cell and the protein had remained outside, the researchers concluded that the DNA contained the hereditary information required to produce the new bacteriophages.

DNA STRUCTURE

Components of DNA

Long before the studies just described were proceeding, biochemists had shown that DNA contains only four kinds of nucleotides, which are the building blocks of nucleic acids (page 59). All kinds have the same five-carbon sugar (deoxyribose), a phosphate group, and one of the following nitrogen-containing bases:

adenine	guanine	thymine	cytosine
(A)	**(G)**	**(T)**	**(C)**

The component parts of all four nucleotides are bonded together in much the same way (Figure 15.4). But notice, in Figure 15.5, that T and C are smaller, single-ring structures (pyrimidines). A and G are larger, double-ring structures (purines).

By 1949, the biochemist Erwin Chargaff had added three more crucial insights about DNA structure:

1. The four bases in DNA are not present in equal amounts.

2. The four bases differ in relative amounts from one species to the next.

3. However, in any DNA molecule, the amount of adenine is always the same as the amount of thymine, and the amount of guanine is always the same as the amount of cytosine (A = T and G = C).

Now, here was something to think about! Could it be that the arrangement of the four kinds of bases in the DNA molecule represented the hereditary instructions?

Maurice Wilkins, Rosalind Franklin, and others thought they might be able to identify that arrangement through x-ray diffraction methods.

Like other substances, DNA can be crystallized by using chemical agents that dehydrate the molecule. The atoms of a crystallized substance form ordered arrays that will disperse an x-ray beam. If the atoms are arranged regularly in the crystal, the beam will be dispersed in a regular pattern. Such patterns show up as dots and streaks on a piece of film placed behind the crystal. By itself, the pattern on the exposed film does *not* reveal molecular structure. But it can be used to calculate the position of groups of atoms relative to one another in the crystal.

Franklin obtained the best x-ray diffraction images, which provided convincing evidence that DNA had the following features. First, DNA had to be long and thin, with a uniform 2-nanometer diameter. Second, its structure had to be highly repetitive: some part of the molecule was repeated every 0.34 nanometer, and a different part was repeated every 3.4 nanometers. Third, DNA could be helical, with an overall shape like a circular stairway.

Patterns of Base Pairing

In the early 1950s, James Watson (a postdoctoral student from Indiana University) teamed up with Francis Crick (a Cambridge researcher). By piecing together available data, they deduced the structure of DNA.

According to Chargaff's data, A = T and G = C. According to Franklin's data, DNA has a uniform diameter. Watson and Crick reasoned that the *double*-ringed A and G molecules were probably paired with the *single*-ringed T and C molecules along the entire length of DNA. (Otherwise, DNA would bulge where two double rings were linked and narrow down where two single rings were linked.)

Watson and Crick shuffled and reshuffled paper cutouts of the nucleotides. They realized that in certain orientations, A and T could become linked by two hydrogen bonds, and G and C could become linked by three. Suppose there were *two strands* of nucleotides, with their bases facing each other. Then hydrogen bonds could easily bridge the gap between them, like rungs of a ladder.

Scale models were constructed of how the "ladder" might look. The only model that fit all available data had A–T and G–C pairs. And those pairs formed the proper hydrogen bonds only when the sugar-phosphate backbones of the two DNA strands ran in *opposing directions* and were twisted together to form a *double helix* (Figures 15.6 and 15.7).

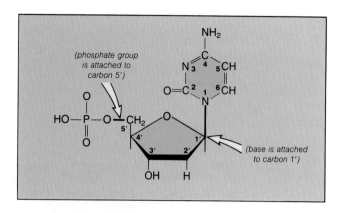

Figure 15.4 Example of one of the four nucleotides of DNA. Each nucleotide has a sugar (deoxyribose), a phosphate group, and one of four bases (shown here, cytosine). The small numerals are used for reference purposes. The carbon atoms of the sugar are numbered 1′, 2′, and so forth simply to distinguish them from the numerals used for the base.

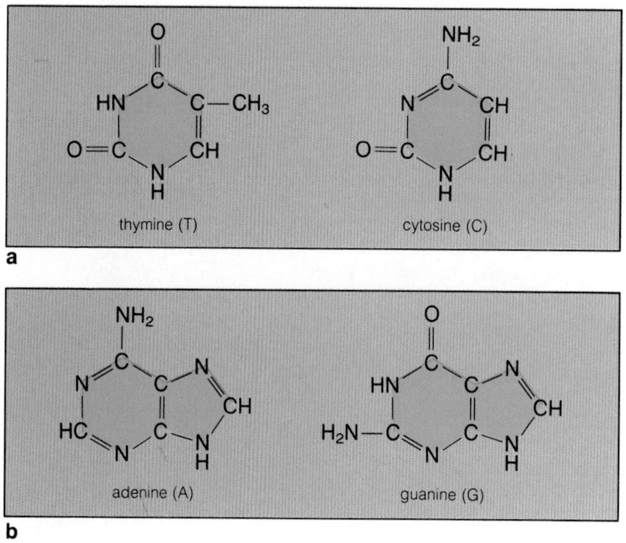

Figure 15.5 Structural formulas for the four nitrogen-containing bases of DNA. (**a**) The single-ring pyrimidines and (**b**) the double-ring purines.

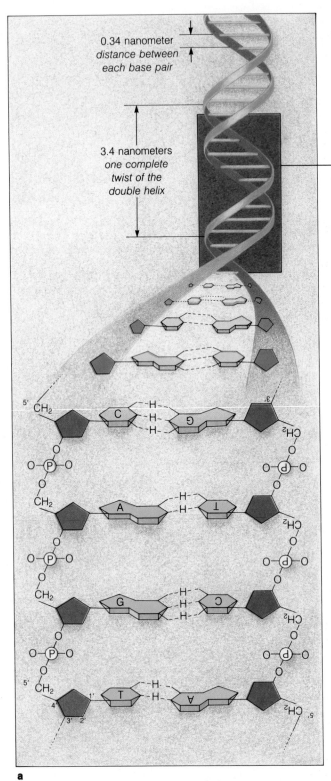

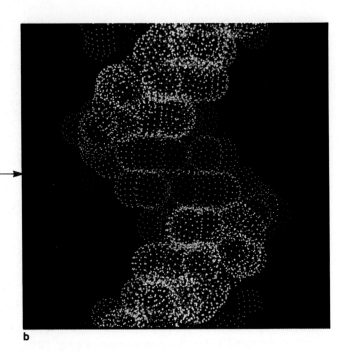

b

Figure 15.6 (**a**) Representation of a DNA double helix. Notice how the two sugar-phosphate backbones run in *opposing directions*. (In other words, the sugar molecules of one strand are upside down. When you compare the numbers used to identify the carbon atoms of a sugar molecule, you can see that one strand runs in the 5′ → 3′ direction and the other, in the 3′ → 5′ direction.) (**b**) Computer-graphics model of a section from a DNA molecule, viewed from the side.

In the Watson-Crick model, A and T are always present in equal amounts, and the same is true of G and C. Yet the amount of A–T can differ relative to the amount of G–C in the DNA of different species because any pair can follow any other in sequence. Even in one small region of a DNA double helix, the sequence might be:

T···A		A···T		C···G		C···G
A···T	or	T···A	or	G···C	or	A···T
T···A		C···G		G···C		C···G

a few base pairs in
the DNA double helix

and so on. The very structure of DNA provides a molecular basis for the unity and diversity of life:

In all species, there is constancy in base pairing (adenine to thymine, guanine to cytosine) between the two strands of the DNA double helix.

The DNA of different species shows variation in which base pair follows the next in a strand.

DNA REPLICATION

How DNA Is Duplicated

The discovery of DNA structure was a turning point in studies of inheritance. Until then, no one had any idea of how the hereditary material is replicated (that is, duplicated) prior to cell division. Now the Watson-Crick model suggested at once how this might be done.

Figure 15.7 James Watson and Francis Crick posing in 1953 by their newly unveiled model of DNA structure.

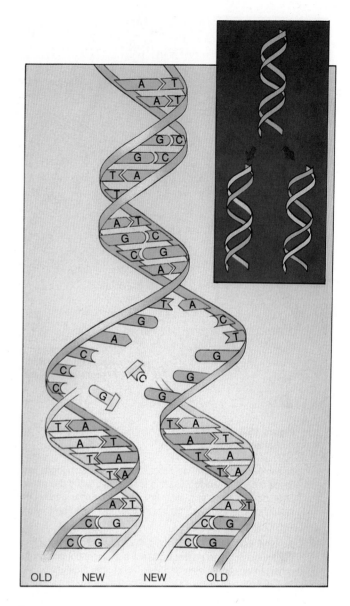

Figure 15.8 Semiconservative nature of DNA replication. The original two-stranded DNA molecule is shown in blue. A new strand (yellow) is assembled on each of the two original strands.

The two strands of the DNA double helix could be unwound from each other, thereby exposing their bases. Then free nucleotides (drawn from cellular pools) could pair with the exposed bases on both unwound strands. As you can see from Figure 15.8, *each parent strand remains intact, and a new companion strand is assembled on each one.* Given the principle of base pairing, the base sequence for the new strand would have to be complementary to the parent (template) strand.

During the replication process, each parent strand is twisted into a double helix with its new, partner strand. In the two double helices that are formed, one "old" DNA strand has been conserved. This so-called semiconservative mode of DNA replication is now known to be nearly universal among organisms.

Prior to cell division, the double-stranded DNA molecule unwinds and is duplicated through semiconservative replication: each parent strand remains intact—it is conserved—and a new, complementary strand is assembled on each one.

Thus semiconservative replication produces two "half-old, half-new" DNA molecules.

A Closer Look at Replication

Origin and Direction of Replication. Where does replication of the DNA molecule actually begin? The two strands of the double helix start to unwind at one or more distinct sites, each being a short, specific base sequence called the "origin." A viral or bacterial DNA molecule usually has only one origin, but a eukaryotic

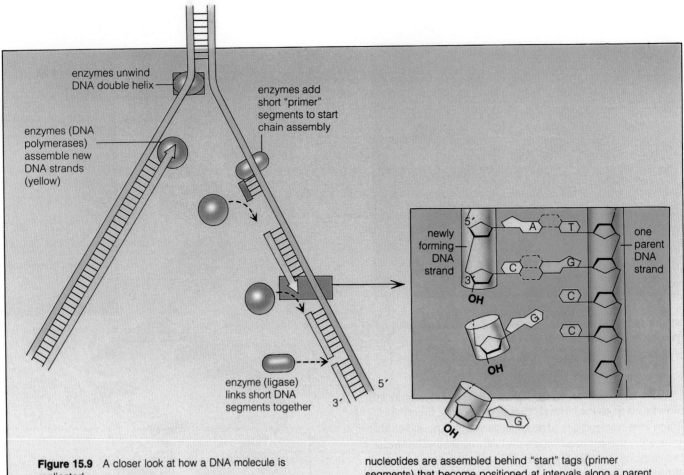

Figure 15.9 A closer look at how a DNA molecule is replicated.

Assembly of new strands proceeds at replication forks. In these limited V-shaped regions, enzymes unwind the parent DNA double helix, and other enzymes assemble a new DNA strand on the exposed regions of each parent strand, which serves as a template.

As discovered by Reiji Okazaki, DNA assembly is usually *continuous* on one parent template but *discontinuous* on the other. In discontinuous synthesis, short stretches of

nucleotides are assembled behind "start" tags (primer segments) that become positioned at intervals along a parent DNA strand. Then enzymes link the short stretches of DNA into a single chain.

Why are there two assembly mechanisms? As the boxed inset suggests, nucleotides can be added to a newly forming DNA chain in the 5′ → 3′ direction only. Bases projecting from a parent template dictate which kind of nucleotide can be added next. But an exposed —OH group must be present on the growing end of a DNA strand if enzymes are to catalyze the addition of more nucleotides to it.

DNA molecule has many of them. Unwinding usually proceeds simultaneously in both directions away from an origin. Strand assembly occurs behind each "fork" that continues to advance as the double helix is being unwound:

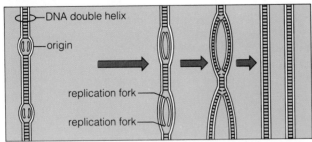

Energy and Enzymes for Replication. A DNA double helix does not unwind all by itself during replication. Batteries of enzymes and other proteins unwind the molecule, keep the two strands separated, and assemble a new DNA strand on each of them. For example, **DNA polymerases** are enzymes that catalyze the assembly of short nucleotide segments on a parent strand (Figure 15.9). Those enzymes also "proofread" the growing DNA double helices for mismatched base pairs, which are replaced with correct bases. The proofreading function is one reason why DNA is replicated with such remarkable accuracy. On the average, for every 100 million nucleotides added to a growing strand, only *one* mistake slips through the proofreading net.

Where does the energy come from to drive replication? It happens that the free nucleotides brought up for strand assembly are not quite in the form shown in Figure 15.4. They are triphosphates (they have three phosphate groups attached, not one). Triphosphates, recall, readily give up energy for cellular reactions (page 105). Energy released when two of the phosphate groups are split away is used by the DNA polymerases for adding nucleotides to a growing strand. The unwinding process runs on energy provided by ATP.

SUMMARY

1. Many different experiments demonstrate that DNA contains the instructions for producing the heritable traits of all single-celled and multicelled organisms.

2. DNA is a macromolecule assembled from small organic molecules called nucleotides. Each nucleotide is composed of a five-carbon sugar (deoxyribose), a phosphate group, and one of the following nitrogen-containing bases:

adenine (A)   guanine (G)
thymine (T)   cytosine (C)

Thymine and cytosine are single-ring structures; adenine and guanine are double-ring structures.

3. The relative amounts of the four nucleotide bases vary in the DNA of different species. (Each species has its own characteristic sequence of base pairs in its DNA.)

4. In the Watson-Crick model, DNA is a long molecule of constant diameter. It consists of two nucleotide strands that are twisted together into a double helix. Hydrogen bonds join the bases of one strand with the bases of the other, and the two strands are oriented in opposing directions. One of the bases in each hydrogen-bonded pair is a single-ring structure and its partner is a double-ring structure.

5. The Watson-Crick model is based on a key principle of genetics—that there is constancy in base pairing between the two strands of the DNA double helix. Adenine always pairs with thymine (A = T), and cytosine always pairs with guanine (G = C).

6. DNA is replicated in a semiconservative manner. The two strands of the DNA double helix unwind from each other and a new strand of complementary sequence is assembled on each one. Two double-stranded molecules result, in which one strand is "old" (it is conserved) and the other is "new."

7. A variety of enzymes and other proteins take part in DNA replication. DNA polymerases are examples; they catalyze strand assembly and also perform proofreading functions.

8. DNA replication usually is bidirectional (it proceeds simultaneously from each origin, which is a specific, short sequence of bases that functions as an initiation site).

Review Questions

1. How did Griffith's use of control groups help him deduce that the transformation of harmless *Streptococcus* strains into deadly ones involved a change in the hereditary material of the harmless forms? *200*

2. What is a bacteriophage? In the Hershey-Chase experiments, how did bacteriophages become labeled with radioactive sulfur and radioactive phosphorus? Why were these particular elements used instead of, say, carbon or nitrogen? *200-202*

3. DNA is composed of only four different kinds of nucleotides. Name the three molecular parts of a nucleotide. Name the four different kinds of nitrogen-containing bases that may occur in the nucleotides of DNA. What kind of bond holds nucleotides together in a single DNA strand? *202-203*

4. What kind of bond holds two DNA chains together in a double helix? Which nucleotide base-pairs with adenine? Which pairs with guanine? Do the two DNA chains run in the same or opposite directions? *203-204*

5. The four bases in DNA may differ greatly in relative amounts from one species to the next—yet the relative amounts are always the *same* among all members of a single species. How does the concept of base pairing explain these twin properties—the unity and diversity—of DNA molecules? *204*

6. When regions of a double helix are unwound during DNA replication, do the two unwound strands join back together again after a new DNA molecule has formed? *204-205*

Readings

Cairns, J., G. Stent, and J. Watson (editors). 1966. *Phage and the Origins of Molecular Biology.* Cold Spring Harbor, New York: Cold Spring Harbor laboratories. Collection of essays by the founders of and converts to molecular genetics. Gives a sense of history in the making—the emergence of insights, the wit, the humility, the personalities of the individuals involved.

Felsenfeld, G. October 1985. "DNA." *Scientific American* 253(4):58–67. Describes how the DNA double helix may change its shape during interactions with regulatory proteins.

Taylor, J. (editor). 1965. *Selected Papers on Molecular Genetics.* New York: Academic Press.

Watson, J. 1978. *The Double Helix.* New York: Atheneum. Highly personal view of scientists and their methods, interwoven into an account of how DNA structure was discovered.

Watson, J., et al. 1987. *Molecular Biology of the Gene.* Fourth Edition. Menlo Park, California: Benjamin/Cummings.

16

PROTEIN SYNTHESIS

DNA is like a book of instructions in each cell. The alphabet used to create the book is simple enough: A, T, G, and C. However, merely knowing the letters does not tell us how they give rise to the language of life, with sentences (genes) evoking precise meanings (proteins). And the letters alone cannot tell us how or why a cell reads certain DNA passages and skips others at any particular moment. The more we learn about DNA, the more we realize it is no simple book. It was put together over a time span of epic proportions, and it is correspondingly complex. It is known to contain passages of great clarity and precision. It contains long stretches of unknown function. It even contains genetic gibberish, arising from the random tamperings of mutations and recombinations.

Even though this evolving, ancient book of inheritance still holds many mysteries, one of the triumphs of our time is that the words themselves have been deciphered! How those words are translated into proteins is the subject of this chapter.

ONE GENE, ONE POLYPEPTIDE

In the early 1900s, the physician Archibald Garrod was tracking metabolic disorders that apparently were heritable (they kept recurring in some families). Blood or urine samples from affected persons contained abnormally high levels of an intermediate known to be produced at one step in a specific metabolic pathway. Most likely, the enzyme that was supposed to use the intermediate as its substrate at the *next* step in the pathway was defective. Unused intermediate molecules accumulated because the pathway was blocked from that step onward. Only one thing distinguished the affected persons from normal ones: they had inherited a metabolic defect. Garrod concluded that specific "units" of inheritance (genes) must function through the synthesis of specific enzymes.

Thirty-three years later, Garrod's hypothesis received support from George Beadle and Edward Tatum, who used the bread mold (*Neurospora crassa*) to study gene function. This organism will grow on a minimal medium containing only sucrose, mineral salts, and biotin (one of the B vitamins). It can synthesize all other substances it requires, including other vitamins, and the steps of those synthesis pathways were known. If an enzyme for one of the steps was defective as a result of a gene mutation, Beadle and Tatum would be able to detect the outcome. Affected strains of *N. crassa* would not be able to synthesize one of the substances that they normally do; they would be "nutritional mutants."

One mutant strain grew only when vitamin B_6 was added to the minimal medium. Another grew only when

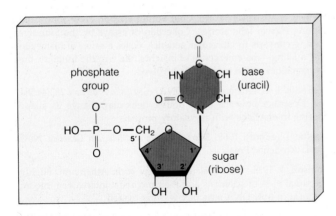

Figure 16.1 By what mechanisms do the hereditary instructions encoded in DNA become translated into proteins? The answer lies with a class of nucleic acids called the RNAs, a major focus of this chapter. Shown here, one of the four nucleotide subunits of the RNA molecule.

vitamin B$_1$ was added, and so on. Analysis of cell extracts of mutants revealed a different defective enzyme in each mutant strain. *Each inherited mutation corresponded to a defective enzyme.* Here was evidence favoring the "one gene, one enzyme" hypothesis.

The hypothesis became refined through studies of sickle-cell anemia. As we have seen, this heritable disorder arises from the presence of abnormal hemoglobin (HbS instead of HbA) in red blood cells. In 1949 Linus Pauling and Harvey Itano subjected HbS and HbA molecules to **electrophoresis**, a technique for measuring the rate and direction of movement of organic molecules in response to an electric field (Figure 16.2). They found that both HbS and HbA molecules move toward the positive pole of the field, but HbS does so more slowly. HbS, it seemed, had fewer negatively charged amino acids.

A few years later, Vernon Ingram pinpointed the difference. Hemoglobin, recall, is a protein with four polypeptide chains, two designated "alpha" and the other two, "beta." As Figure 16.3 shows, one amino acid (valine) has replaced another (glutamate) in each beta chain of HbS. Unlike glutamate, which carries a negative charge, valine has no net charge, and that is why HbS behaved differently in the electrophoresis studies. More importantly, Ingram's discovery suggested that *two* genes code for hemoglobin (one for each kind of polypeptide chain), and that genes code for proteins in general, not just enzymes. And so the more precise "one gene, one polypeptide" hypothesis emerged:

One gene codes for the amino acid sequence of one polypeptide chain—the basic structural unit of proteins.

THE PATH FROM GENES TO PROTEINS

Enter RNA

In the early 1950s, James Watson and Francis Crick ushered in a new era of research when they discovered the structure of the DNA molecule, this being two strands of nucleotides twisted together into a double helix (Figure 15.6). When DNA replicates itself, the two strands unwind and each serves as a structural pattern, or **template**, upon which a new, complementary DNA strand is built.

Were proteins also assembled on the unwound DNA templates? Probably not. Experiments with eukaryotic cells showed that protein synthesis occurs in the cytoplasm, not in the nucleus where the DNA resides. Most likely, some other molecule picked up the protein-building instructions from an exposed DNA template, moved

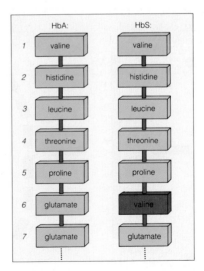

1. Proteins are suspended in a solution that includes a detergent strong enough to keep them from interacting with one another.

2. The mixture is applied to a slot in a slab of gel:

slab of gel

(glass plate)

3. Electrodes are positioned at opposite ends of the slab and voltage is applied. HbS migrates more slowly through the gel, toward the positive electrode; it has fewer negatively charged subunits than HbA:

HbA HbS

Figure 16.2 Gel electrophoresis, used to measure the rate and degree of migration of organic molecules through a gel in response to an electric field. A mixture of different proteins is placed in the gel, and each type will move toward one end of the slab of gel or the other. The rate and direction of movement depend partly on a molecule's net surface charge. (They also depend on its size and shape; for example, larger molecules resist moving through the gel more than smaller ones do.)

HbA:		HbS:
1	valine	valine
2	histidine	histidine
3	leucine	leucine
4	threonine	threonine
5	proline	proline
6	glutamate	valine
7	glutamate	glutamate

Figure 16.3 A substitution of one amino acid for another in a polypeptide chain. In sickle-cell hemoglobin, valine is substituted for glutamate at position 6 of the beta chain (see above). This mutation puts a "sticky" (hydrophobic) patch on the surface of a hemoglobin molecule. When oxygen concentrations in the blood are low, the sticky patches interact and hemoglobin molecules aggregate into rods, distort red blood cells, and cause them to clump in blood vessels (page 172).

into the cytoplasm, and itself served as the template for a polypeptide chain. Attention now turned to another class of nucleic acids, the RNAs.

"RNA" stands for **ribonucleic acid**. An RNA molecule is usually a single strand of nucleotides. Each nucleotide consists of a sugar (ribose), a phosphate group, and one of four kinds of nitrogen-containing bases. Three bases are the same in DNA and RNA (both have adenine, cytosine, and guanine). However, RNA has the base **uracil** instead of thymine (Figure 16.1). Even so, uracil base-pairs with adenine, just as thymine does. The potential was there: *Ribonucleotides could be joined one after another into an RNA molecule by using an unwound region of a DNA double helix as the template.*

Overview of Protein Synthesis

In 1956, Crick summarized the current thinking about the probable path of protein synthesis:

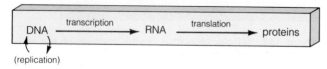

It was a shorthand way of signifying that an exposed DNA strand can serve as the template for assembling an RNA strand (that is, the RNA can be "transcribed off" the DNA). It signified also that RNA can serve as the template upon which genetic information is translated into proteins.

We now know that there are exceptions to this direction of information flow, the retrovirus responsible for AIDS being one of them (page 430). The exceptions are so rare, however, that the summary is still essentially valid. Let's now take a closer look at the path leading from genes to proteins, beginning with the overview in Figure 16.4. The two major steps in this path are called transcription and translation.

During **transcription**, a single strand of RNA is assembled on an unwound portion of a DNA double helix, with one of the DNA strands serving as the template. The ribonucleotides are linked one after another into chains called **RNA transcripts**. The same region of DNA can be transcribed over and over again during the life of a cell. Often, many thousands of RNA transcripts are made of the same gene.

During **translation**, three types of RNA molecules interact to convert the original DNA instructions into a

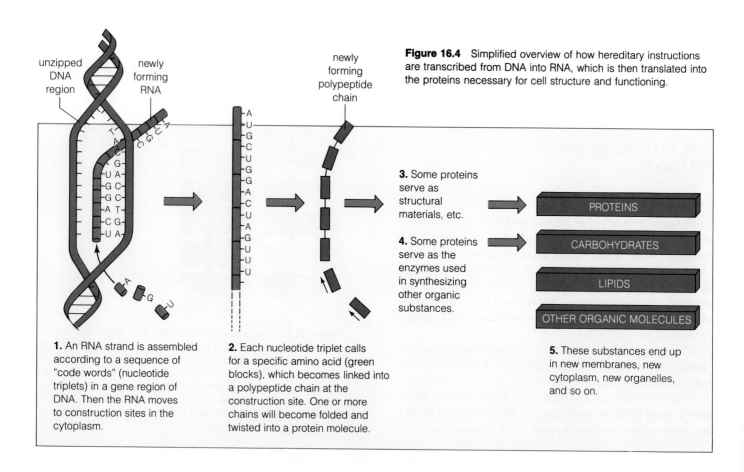

Figure 16.4 Simplified overview of how hereditary instructions are transcribed from DNA into RNA, which is then translated into the proteins necessary for cell structure and functioning.

1. An RNA strand is assembled according to a sequence of "code words" (nucleotide triplets) in a gene region of DNA. Then the RNA moves to construction sites in the cytoplasm.

2. Each nucleotide triplet calls for a specific amino acid (green blocks), which becomes linked into a polypeptide chain at the construction site. One or more chains will become folded and twisted into a protein molecule.

3. Some proteins serve as structural materials, etc.

4. Some proteins serve as the enzymes used in synthesizing other organic substances.

5. These substances end up in new membranes, new cytoplasm, new organelles, and so on.

specific sequence of amino acids. All three types are transcribed from particular DNA regions, but only one serves as the template for translation:

ribosomal RNA (rRNA)
a type of molecule that combines with certain proteins to form the *ribosome* (the structural "workbench" on which a polypeptide chain is assembled)

messenger RNA (mRNA)
the "blueprint" (a linear sequence of specific nucleotides) that is delivered to the ribosome for translation into a polypeptide chain

transfer RNA (tRNA)
an adaptor molecule; it can pick up one type of amino acid *and* pair with an mRNA "code word" (a specific sequence of three bases) calling for that amino acid

TRANSCRIPTION

Synthesis of RNA

The word "transcript" means "copy," although it might help to keep in mind that an mRNA transcript is *complementary to* the DNA template strand on which it is formed, not an identical copy of it.

In two respects, RNA strand assembly is like DNA replication (page 205). Nucleotides are added to a new RNA strand according to the rules of base pairing:

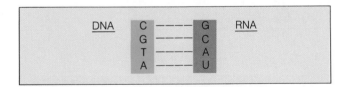

Also, the nucleotides are added to a growing RNA strand only in the 5′ → 3′ direction (here you may wish to refer to Figure 15.9).

RNA strand assembly differs from DNA replication in the following ways. *First*, enzymes called **RNA polymerases** are involved in assembling the transcripts (Figure 16.5). *Second*, several RNA polymerases can move along the same exposed DNA region, one after another, thereby assembling several RNA transcripts of the same gene. Figure 16.6 shows an example of this. *Third*, only one of the two unwound DNA strands is transcribed. (The base sequences of the two strands are complementary, not identical, and would code for different polypeptides.)

Transcription starts at **promoters**, which are specific base sequences at the beginning of genes. The promoter for a given gene is located on one strand or the other, not both. It is the only place where RNA polymerases

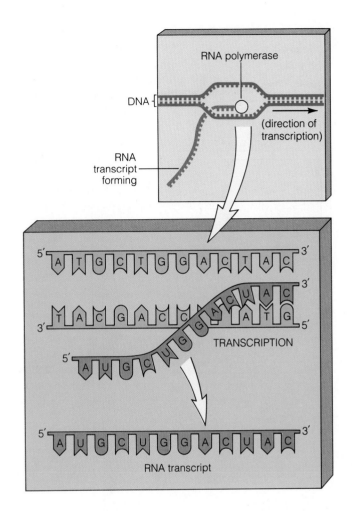

Figure 16.5 Transcription: the synthesis of an RNA molecule on a DNA template. Following transcription, the unwound region of DNA winds up again into a double helix.

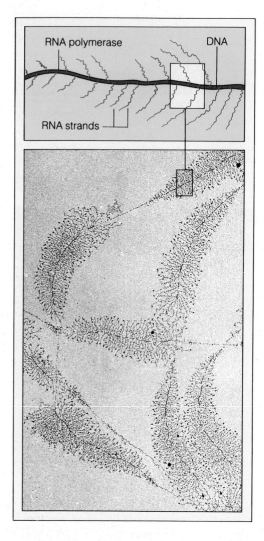

can initiate the correct binding to DNA. After binding occurs, an RNA polymerase moves along the template, joining RNA nucleotides one after another into a transcript. When it reaches the end of the gene region, another signal causes the transcript to be released.

RNA Transcripts

Not all the nucleotides in a eukaryotic gene specify an amino acid sequence. Between the beginning and end of a gene, there can be some number of "noncoding" nucleotide sequences that do not get translated into a polypeptide chain. Those sequences are called *intervening DNA*, or **introns** for short. The parts that are coding sequences and that do get translated are called **exons**.

All introns and exons of a given gene are transcribed into RNA. As a result, a *primary* mRNA transcript contains more than the code for a string of amino acids. However, from the time a transcript peels off a gene until it leaves the nucleus, it is extensively modified. A "cap" is attached to the start of the transcript and a "poly-A tail" is attached to its end. (The cap may be the "start" signal for translation. The function of the tail is not known.) Finally, the introns are snipped out and the exons are spliced together to form the mature mRNA transcript (Figure 16.7). We will consider some details of transcript processing in the next chapter. For now, it is enough to keep in mind that it is the mature mRNA transcript that carries protein-building instructions out of the nucleus and into the cytoplasm.

Figure 16.6 (Above) Micrograph providing evidence of transcription of some DNA isolated from a maturing egg of the spotted newt. The short RNA strands are in early stages of synthesis; the longer strands are nearing completion. The assembly of each one is being catalyzed by an RNA polymerase. Magnification 16,500×.

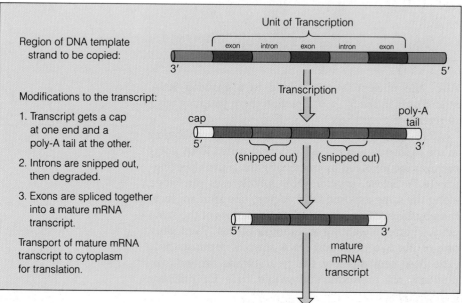

Figure 16.7 Transcript processing in eukaryotes.

TRANSLATION

The Genetic Code

Whatever instructions an mRNA molecule carries away from DNA, they are still in the form of a linear sequence of nucleotides. So we are left with a central question: *What are the protein-building "words" encoded in the linear sequence of nucleotides?*

The building blocks of proteins are twenty common amino acids. The "words" along a DNA or RNA strand must dictate the combination of amino acids that will be joined, one after another, into a given polypeptide chain. But neither DNA nor RNA has more than four kinds of nucleotides. Obviously, each nucleotide doesn't code for only one type of amino acid; that would be only four (4^1) choices. A sequence of two nucleotides would yield only sixteen (4^2) choices, not twenty. A sequence of three nucleotides would yield sixty-four (4^3) choices—more than the twenty required, but logically more appealing than the next possibility (4^4, or 256).

Crick, Sidney Brenner, and their colleagues deduced the nature of the **genetic code**—that is, how the nucleotide sequence in DNA and then mRNA is related to the amino acid sequence in a polypeptide chain. Through their genetic experiments, it became apparent that the nucleotide bases are read sequentially, three at a time, with each triplet signifying an amino acid. A start signal built into the DNA or RNA molecule establishes the correct "reading frame" for blocking out every three nucleotides in the sequence.

Hereditary instructions for building proteins are encoded in the nucleotide sequence of DNA and mRNA.

Every three nucleotides (a base triplet) specifies an amino acid, which becomes linked with other amino acids into a polypeptide chain.

H. Gobind Khorana, Marshall Nirenberg, Severo Ochoa, Robert Holley—these and so many others did the meticulous work to decipher the genetic code. We now know that it consists of sixty-four different nucleotide triplets. Sixty-one actually specify amino acids. The remaining three act like stop signs or termination points in protein synthesis; they signify that no more amino acids are to be added to the polypeptide chain.

Each nucleotide triplet in mRNA that codes for an amino acid or for chain termination is called a **codon**. Only two amino acids are specified by a unique codon (AUG calls only for methionine, and UGG only for tryptophan). As you can see from Figure 16.8, the other eighteen amino acids can be specified by two or more different codons. Notice, for example, that CCU, CCC, CCA, and CCG all specify proline.

First Letter	Second Letter				Third Letter
	U	C	A	G	
U	phenylalanine	serine	tyrosine	cysteine	U
	phenylalanine	serine	tyrosine	cysteine	C
	leucine	serine	stop	stop	A
	leucine	serine	stop	tryptophan	G
C	leucine	proline	histidine	arginine	U
	leucine	proline	histidine	arginine	C
	leucine	proline	glutamine	arginine	A
	leucine	proline	glutamine	arginine	G
A	isoleucine	threonine	asparagine	serine	U
	isoleucine	threonine	asparagine	serine	C
	isoleucine	threonine	lysine	arginine	A
	(start) methionine	threonine	lysine	arginine	G
G	valine	alanine	aspartate	glycine	U
	valine	alanine	aspartate	glycine	C
	valine	alanine	glutamate	glycine	A
	valine	alanine	glutamate	glycine	G

Figure 16.8 The genetic code by which an mRNA molecule, with its linear array of codons, is transcribed from a gene region of DNA. Each codon consists of three nucleotides, and that triplet will call for a specific amino acid during mRNA translation. In this diagram, the first nucleotide of any triplet is given in the left column. The second is given in the middle columns; the third, in the right column. Thus we find (for instance) that tryptophan is coded for by the triplet U G G . Phenylalanine is coded for by both U U U and U U C .

With very few exceptions, the genetic code is the universal language of protein synthesis (see *Commentary*). Codons that specify particular amino acids in bacteria specify the same amino acids in almost all protistans, fungi, plants, and animals.

Codon-Anticodon Interaction

So now we have an mRNA molecule, with its linear array of codons, arriving in the cytoplasm. Here it will interact with its molecular relatives, the tRNAs and rRNAs.

Each kind of tRNA has an **anticodon**, this being a sequence of three nucleotide bases that can pair with a specific mRNA codon. Each tRNA also has a molecular

Mitochondria: Exceptions to the Rule

In the evolutionary view, the genetic code probably dates back to the most ancient cells from which all existing species arose. The basic code is identical for all living species, from bacterial to human. The absence of variation among species suggests that the genetic code arose before the first cells diverged into separate evolutionary lines. Once the code had been established, the rare mutations that did change one or more codons must have been selected against, for reasons given later in the chapter (page 219).

It happens, though, that a slightly altered code exists in mitochondria of several species. A mitochondrion, recall, is an organelle that specializes in ATP formation by way of the aerobic pathway in aerobic energy metabolism. In some respects, it operates somewhat independently of the cell in which it is housed. It has its own DNA, which is replicated independently of the nuclear DNA. Mitochondrial DNA governs the synthesis of some RNA and proteins that the organelle requires for its specialized tasks. Although mitochondrial codes are almost the same as the one used by cells, a *few* codons have different meanings. (For instance, UGA is a termination codon in most genetic messages; for mitochondria, it specifies tryptophan.)

What is the significance of the slightly different mitochondrial code? Lynn Margulis and many other biologists suspect that the original ancestors of mitochondria were free-living cells, similar to some modern bacteria. In some way, those bacteria took up permanent residence inside the cytoplasm of other kinds of cells (Chapter Thirty-Eight). The guest and the host cell became locked in a *symbiotic* relationship, with each gaining some benefit from the activities of the other. Over time, the "guest" lost the means to perform some function that the host cell was performing for it—and so it lost the capacity for an independent existence.

If this is what happened, then the mitochondrial code might be a remnant of a primitive code from an entirely separate species, long since vanished. However, it might also be true that the variant code arose *after* the joint living arrangement became permanent. In other words, maybe mitochondria have a later, more streamlined version of the "universal" code.

Figure 16.9 (**a**) Structural features that all tRNA molecules hold in common. Notice how the ribonucleotide strand folds back on itself into hairpin loops, which are held in place by hydrogen bonds. (**b**) Three-dimensional structure of one tRNA molecule.

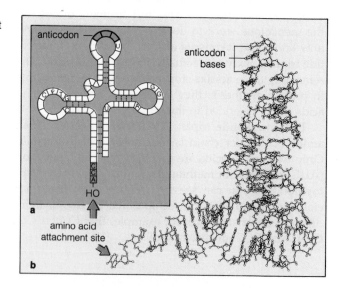

"hook," an attachment site for a particular amino acid (enzymes ensure that the correct amino acid is attached to each kind). Figure 16.9 gives an idea of the structure of tRNA molecules. For the sake of clarity in later illustrations, the structure will be portrayed in this simplified way:

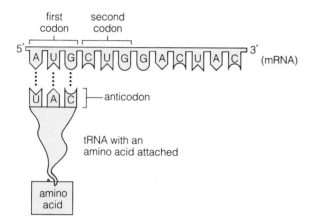

The first two bases of an anticodon must be precisely complementary to an mRNA codon for pairing to occur. However, recognition of the *third* base often is less precise. In yeast cells, for instance, the tRNA carrying an alanine molecule on its "hook" can pair with three different codons—GCU, GCC, or GCA. The freedom in codon-anticodon pairing at the third base is called the "wobble" effect.

A codon (base triplet of mRNA) is a recognition site for an anticodon (complementary base triplet of tRNA).

Ribosome Structure

All codon-anticodon interactions take place on the surface of the **ribosome**, the cytoplasmic structure on which polypeptide chains are assembled. Ten thousand ribosomes can be present in one prokaryotic cell; a eukaryotic cell may contain many tens of thousands. At its widest dimension, a ribosome is only twenty-five nanometers (about a millionth of an inch).

Each ribosome has two parts (Figure 16.10). The small subunit consists of an rRNA molecule and a number of different proteins. The large subunit consists of two different rRNA molecules and a number of different proteins—among them the enzymes that link together amino acids being delivered to the ribosome by tRNAs.

Several ribosomes can translate the same mRNA molecule, one after another, so that several copies of a given polypeptide chain can be formed at the same time. A cluster of ribosomes on the same mRNA is called a **polysome** (Figure 16.11).

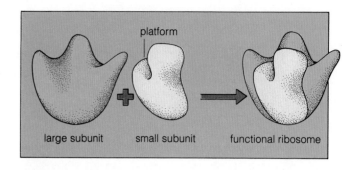

Figure 16.10 A current model of the two-part structure of eukaryotic ribosomes.

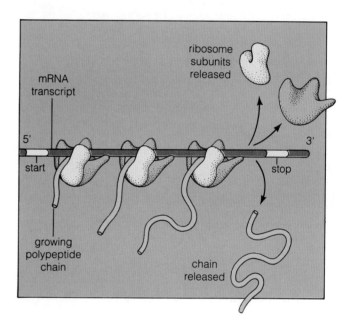

Figure 16.11 A polysome (group of ribosomes) on an mRNA transcript. Each ribosome in the group functions independently, so that several polypeptide chains can be formed at the same time from the same transcript.

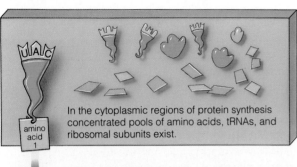

In the cytoplasmic regions of protein synthesis concentrated pools of amino acids, tRNAs, and ribosomal subunits exist.

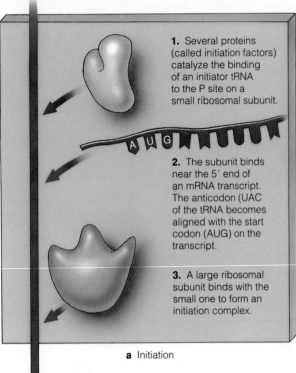

1. Several proteins (called initiation factors) catalyze the binding of an initiator tRNA to the P site on a small ribosomal subunit.

2. The subunit binds near the 5′ end of an mRNA transcript. The anticodon (UAC of the tRNA becomes aligned with the start codon (AUG) on the transcript.

3. A large ribosomal subunit binds with the small one to form an initiation complex.

a Initiation

Stages of Translation

Translation of an mRNA transcript into a polypeptide chain proceeds through three stages: initiation, elongation, and termination.

In *initiation*, a special initiator tRNA binds to a specific site (the so-called P site) on the small ribosomal subunit, which in turn becomes bound near the 5′ end of an mRNA transcript. In eukaryotes, the mRNA moves along the subunit until its initial start codon arrives at the P site. The codon base-pairs with the tRNA's anticodon, and it stays there until chain elongation begins (Figure 16.12b). In the final step of initiation, a large ribosomal subunit joins with the small one. We now have an intact ribosome, an mRNA transcript, and an initiator tRNA occupying the P site. With this "initiation complex," chain elongation can begin.

In *chain elongation*, the start codon on the mRNA defines the reading frame for assembling amino acids in sequence. As Figure 16.12c shows, other tRNAs deliver amino acids to a site on the ribosome that is adjacent to the P site. Once both sites are occupied by two tRNAs, an enzyme catalyzes the formation of a peptide bond between their amino acids. Again, the codons of mRNA specify which amino acids will be joined, one after another, into a polypeptide chain.

With *chain termination*, no more amino acids can be added to the polypeptide chain. The presence of a stop

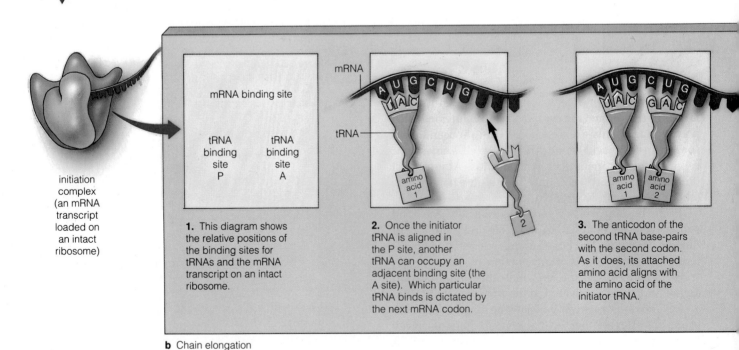

initiation complex (an mRNA transcript loaded on an intact ribosome)

mRNA binding site

tRNA binding site P

tRNA binding site A

1. This diagram shows the relative positions of the binding sites for tRNAs and the mRNA transcript on an intact ribosome.

2. Once the initiator tRNA is aligned in the P site, another tRNA can occupy an adjacent binding site (the A site). Which particular tRNA binds is dictated by the next mRNA codon.

3. The anticodon of the second tRNA base-pairs with the second codon. As it does, its attached amino acid aligns with the amino acid of the initiator tRNA.

b Chain elongation

Figure 16.12 Simplified picture of chain initiation and elongation during the translation stage of protein synthesis.

codon (UAG, UAA, or UGA) in the mRNA transcript triggers events that allow the polypeptide chain to become detached from the ribosome. Detachment requires the participation of specific proteins called release factors; tRNAs in themselves cannot recognize a stop signal because no tRNA anticodon is complementary to a stop codon. In eukaryotes, the chain either joins the pool of free proteins in the cytoplasm or enters the cytomembrane system for further processing, beginning with the endoplasmic reticulum. Here you may wish to refer briefly to page 73, which outlines the final destinations for the newly synthesized proteins.

Figure 16.13 summarizes the flow of information along the path leading from genes to proteins.

MUTATION AND PROTEIN SYNTHESIS

In general, the sequence of base pairs in DNA must be preserved from one generation to the next, otherwise offspring might not be able to synthesize all the proteins necessary for their own growth, development, and reproduction. Yet changes do occur in the DNA, some of which are reversible and some of which are not. The changes can be grouped into three categories: gene mutation, crossing over and recombination, and chromosomal aberration.

At the cytological level, we can detect *crossing over and recombination* during meiosis, these being events that lead to new combinations of alleles at different gene locations in chromosomes. Also detectable are *chromosomal aberrations*, which change the structure or number of chromosomes in a cell (Chapter Thirteen). *Gene mutation* must be described at the molecular level, for this type of heritable alteration may involve no more than one or two nucleotides.

Mutation at the Molecular Level

A **gene mutation** is a change of one to several bases in the nucleotide sequence of DNA. Sometimes one kind of base or base pair replaces another; at other times, one or more base pairs are added or deleted.

Some gene mutations are induced by **mutagens**, which are environmental agents that can interact with a DNA molecule and cause modification of its structure. Mutagens include viruses, ultraviolet radiation, and certain chemicals.

Other gene mutations are spontaneous rather than induced. One type arises mainly from replication errors, as when adenine pairs with cytosine instead of with thymine. This type of mutation results in a "base-pair substitution." Another type of spontaneous change is

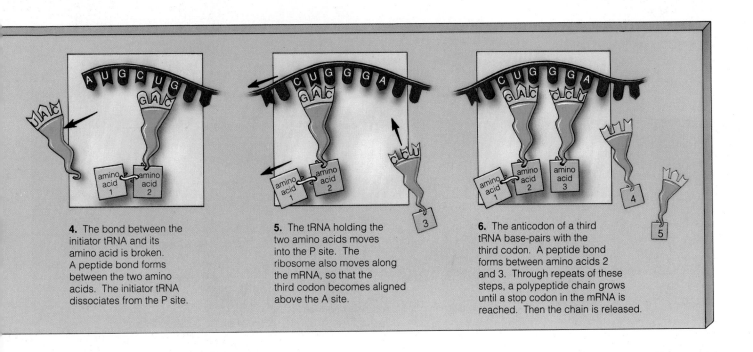

4. The bond between the initiator tRNA and its amino acid is broken. A peptide bond forms between the two amino acids. The initiator tRNA dissociates from the P site.

5. The tRNA holding the two amino acids moves into the P site. The ribosome also moves along the mRNA, so that the third codon becomes aligned above the A site.

6. The anticodon of a third tRNA base-pairs with the third codon. A peptide bond forms between amino acids 2 and 3. Through repeats of these steps, a polypeptide chain grows until a stop codon in the mRNA is reached. Then the chain is released.

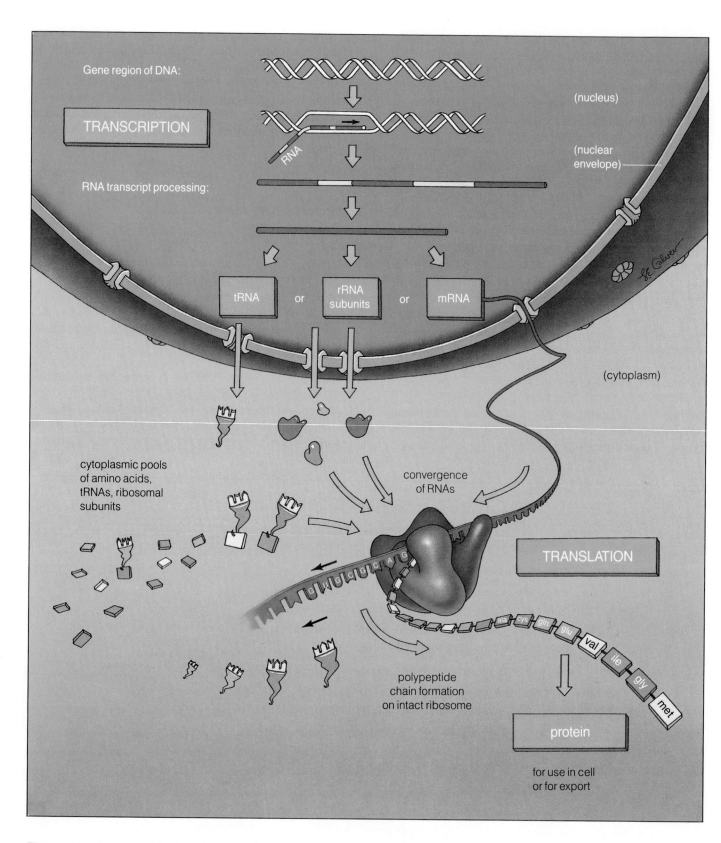

Figure 16.13 Summary of the flow of genetic information in protein synthesis in eukaryotic cells.

the "frameshift mutation." Here, the insertion or deletion of one to several base pairs in a DNA molecule puts the nucleotide sequence out of phase, so that the reading frame shifts during protein synthesis (Figure 16.14). Because genetic instructions are not read correctly, an abnormal protein is produced.

Mutation Rates

The probability that a given gene will mutate during or between DNA replications is called the gene's **mutation rate**. On the average, the mutation rate for a gene is only one in a million (10^6) replications.

Genes mutate independently of one another. To determine the chance of any two mutations occurring in the same cell, we would have to multiply the individual mutation rates for two of its genes. For example, if the rate for the first gene is one in a million per cell generation and the rate for the second is one in a billion (10^9), then there is only one chance in a million billion (10^{15}) that both genes will mutate in the same cell—which is rare, to say the least.

The extremely remote likelihood of two genes mutating at the same time is the reason why two or more antibiotics are used concurrently for treating certain bacterial diseases, such as tuberculosis. If the bacterial invader has a heritable resistance to one antibiotic (say, streptomycin), it is highly unlikely that it will simultaneously be resistant to another.

Mutation and Evolution

In the natural world, gene mutations are rare, chance events. It is impossible to predict exactly when, and in which organism, they will appear. They also are inevitable. As long as sunlight reaches the earth, for example, its ultraviolet component will damage some DNA molecules.

A mutation may turn out to be beneficial or harmful. The outcome depends on how the protein specified by the mutated DNA region is received in the environment, and on how it meshes in the coordinated workings of the entire individual. Because of this, most mutations do not bode well for the individual. No matter what the species, each organism generally inherits a combination of many genes that already are fine-tuned for a given range of operating conditions. A mutant gene is likely to code for a protein that is less functional, not more so, under those conditions.

Over evolutionary time, agents of selection undoubtedly perpetuated those packages of genes having a history of survival value. They must have favored those forms of DNA replication enzymes that could catalyze not only the addition of new nucleotides to growing

DNA strands, but also the removal of mismatched ones. What is clear is that replication enzymes have protected the overall stability of the vulnerable molecules of inheritance that have been replicated through billions of years.

Yet every so often through that immense time span, mutations appeared that provided their bearers with advantages or did them no harm. Selection processes worked to perpetuate mutations having adaptive consequences. Other mutations, it seems, have produced DNA regions with no currently assignable function, but those so-called "rusting hulks" are replicated along with everything else. After more than 5 billion years, molecular descendants of the first strands of DNA are replete with mutations. Each is a patchwork molecule, with variant numbers and kinds of genes stitched into it.

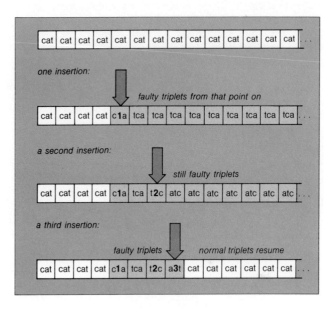

Figure 16.14 Simplified picture of frameshift mutation. Studies of such mutations in bacteriophage led to the discovery of the genetic code. During translation, the base sequence of mRNA is read in blocks of three (that is, as base triplets) according to the genetic code.

The word *cat* is used here to represent every base triplet. Suppose an extra nucleotide ("1") is inserted into a gene. When mRNA is transcribed off that gene, the insertion will put the reading frame out of phase and the wrong amino acids will be called up during translation. A second insertion ("2") would not improve matters. A third insertion ("3") will restore the reading frame, so only part of the resulting protein will be defective. Insertions of extra bases into a DNA molecule are a type of *frameshift mutation*, and they often give rise to mutant phenotypes.

Table 16.1 Sources of Change in the Hereditary Material	
Gene Mutation	
Base substitution	A purine or a pyrimidine replaces another purine or pyrimidine
Frameshift	One to several extra base pairs inserted into or lost from the DNA sequence
Transposition	DNA segments moved to a new location in the same or a different chromosome
Crossing Over and Recombination	
Change at the molecular level	New base-pair sequence in a stretch of DNA
Change at the cytological level	New combinations of alleles at different gene locations in a chromosome
Chromosomal Aberration	
Deletion	Segment lost from a chromosome
Duplication	Chromosome segment repeated
Inversion	Chromosome segment inverted
Translocation	Chromosome segment transferred to a nonhomologous chromosome
Aneuploidy	One of the chromosomes of a given type present only once or more than twice in an otherwise diploid set
Polyploidy	Three or more of each type of chromosome present

PERSPECTIVE

DNA—deoxyribonucleic acid—is the molecule of inheritance in every living cell. It is a helically coiled, double-stranded storehouse of information that can be used to synthesize all the proteins required for cell structure and functioning. Every DNA molecule is composed of only three kinds of substances: a sugar, a phosphate group, and nitrogen-containing bases (adenine, thymine, guanine, and cytosine). Every DNA molecule is replicated according to the same base-pairing rules. When the time comes for new bases from the cellular environment to become paired with old bases on a DNA strand, adenine is paired only with thymine, and guanine only with cytosine.

What this means is that every living thing on earth shares the same chemical heritage with all others. Your DNA is made of the same kinds of substances, and follows the same base-pairing rules, as the DNA of earth-

worms in Missouri and grasses on the Mongolian steppes. Your DNA is replicated in much the same way as theirs, and the same genetic code is followed in translating its messages into proteins. In the evolutionary view, the reason you don't look like an earthworm or a flowering plant is largely a result of selective agents sifting through new genetic information that resulted from mutations and recombinations. Mistakes and shufflings in base-pair sequences made their entrance during the past $3\frac{1}{2}$ billion years and they led, in their unique divergent ways, to the three of you. And so the *sequence* of base pairs along the DNA molecule has come to be different in the three of you.

Dinosaur DNA, too, presumably was assembled from the same chemical stuff as yours. But the mutations and recombinations that gave rise to the unique sequences of base pairs that specified "build dinosaurs" made those creatures unsuitable, when environmental conditions changed, for continuing their journey.

We have, in this unit, introduced three concepts of profound importance. *First, DNA is the source of the unity of life. Second, mutations and other changes in the structure and number of DNA molecules are the source of life's diversity. Finally, the changing environment is the testing ground for the success or failure of the proteins specified by each novel DNA sequence and assortment that appears on the evolutionary scene.*

SUMMARY

1. The path leading from genes to proteins can be summarized this way:

2. Protein-building instructions are encoded in the linear sequence of nucleotides in DNA. A given gene region of DNA is transcribed into RNA, then RNA becomes translated into a linear sequence of amino acids—that is, a polypeptide chain, which is the basic structural unit of proteins.

3. Protein synthesis occurs in two stages. In *transcription*, an exposed region of one strand of the DNA double helix serves as the template for assembling an RNA strand. In *translation*, three classes of RNA molecules interact to convert the message originally encoded in DNA into a polypeptide chain:

   a. rRNA: a structural part of the ribosome on which the polypeptide chain is assembled.

   b. mRNA: the actual blueprint for building a polypeptide; it consists of a sequence of nucleotides that are "read" in blocks of three (base triplets).

c. tRNA: an adaptor molecule that has both an attachment site for an amino acid and three base triplets that can pair with a base triplet in mRNA.

4. The relationship between a sequence of DNA or RNA base triplets and the sequence of amino acids it specifies is called the genetic code. In mRNA, each base triplet is called a codon. In tRNA, the complementary base triplet is called an anticodon.

5. Transcription follows the same base-pairing rule that applies to DNA replication, but in this case uracil (not thymine) is paired with the adenine present in the DNA template strand:

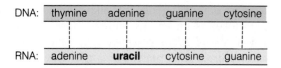

6. In eukaryotes, mRNA transcripts undergo extensive modification before they move into the cytoplasm for translation. To give one example, its introns (nucleotide sequences that do not code for parts of the polypeptide chain) are excised and the exons (coding sequences) are spliced together.

7. Translation consists of three stages:

a. *Initiation*: a small ribosomal subunit binds with an initiator tRNA, then with an mRNA transcript. The small subunit then binds with a large ribosomal subunit to form the initiation complex.

b. *Chain elongation*: tRNAs deliver amino acids one after another to the ribosome, with the tRNA anticodons pairing to different codons that occur one after the other in the mRNA; then peptide bonds form between the amino acids.

c. *Chain termination*: a stop codon triggers events that cause the polypeptide chain to detach from the ribosome.

8. Overall, the protein-building instructions encoded in DNA are preserved through the generations. But the instructions are sometimes altered as a result of three types of events (Table 16.1):

a. *Gene mutation*: for example, insertion or deletion of several bases in the DNA nucleotide sequence (frameshift mutation), or incorrect base pairing during replication (base-pair substitution).

b. *Crossing over and recombination*: introduction of new combinations of alleles at different gene locations in a chromosome (page 178).

c. *Chromosomal aberration* (page 178).

9. Most mutations and other changes in the DNA are harmful; an individual generally inherits a combination of many genes already fine-tuned for a given range of operating conditions. But if conditions change, the product of a mutant form of a gene may be more advantageous than that of the nonmutated form.

10. There is an underlying chemical unity among all organisms. Regardless of the species, DNA is composed of the same substances, follows the same base-pairing rule, and is replicated in much the same way. The genetic code by which its instructions are translated into proteins is nearly universal.

11. Mutations and other changes in the DNA are the source of diversity among organisms. Whether the changes prove to be harmful or beneficial depends on the environment in which the proteins specified by the altered DNA are expressed.

Review Questions

1. Are the products specified by DNA assembled *on* the DNA molecule? If so, state how. If not, tell where they are assembled, and on which molecules. *209–210*

2. Study Figure 16.4, which shows the steps by which hereditary instructions are transcribed from DNA into RNA, which is then translated into proteins. Then, on your own, write a description of this sequence, taking care to define the terms transcription and translation. *210*

3. If sixty-one triplets actually specify amino acids, and if there are only twenty common amino acids, then more than one nucleotide triplet combination must specify some of the amino acids. How do triplets that code for the same thing usually differ? *213*

4. Define "genetic code." Is the same basic genetic code used for protein synthesis in all living organisms? What significance is attached to that fact by most biologists? *213, 220*

5. Define the three types of RNA. What is a codon? An anticodon? Where is each physically located? *211, 213*

6. Define intron and exon. What happens to introns before an mRNA transcript is shipped from the nucleus to the cytoplasm? *212*

7. If genetic information were transmitted precisely from generation to generation, organisms would never change. What are some mutations that give rise to phenotypic diversity? *217, 220*

8. How is your DNA like the DNA of earthworms, grasses, and (presumably) dinosaurs? How is your DNA different? *220*

Readings

Darnell, J. October 1985. "RNA." *Scientific American* 253(4):68–78.

Nomura, M. January 1984. "The Control of Ribosome Synthesis." *Scientific American* 250(1):102–114.

Prescott, D. 1988. *Cells*. Boston: Jones and Bartlett. Chapter 8 of this textbook contains an excellent introduction to protein synthesis.

Watson, J., et al. 1987. *Molecular Biology of the Gene*. Fourth edition, volume 1. Menlo Park, California: Benjamin/Cummings.

17

CONTROL OF GENE EXPRESSION

Figure 17.1 Why is this female calico cat "calico"? Each of her cells contains two X chromosomes. One chromosome carries an allele coding for black coat color and the other carries an allele coding for yellow. When she was an embryo developing in her mother's body, one of the two X chromosomes was inactivated at random in each of the cells that had formed by that time. In all the cellular descendants of each cell, the same chromosome was inactivated, leaving only one functional allele for the coat color trait. And so we see different patches of color, depending on which allele was inactivated in cells making up the tissue in a given region. (The white patches result from interaction with another gene locus—the so-called spotting gene—that determines whether any color appears at all.)

What we are seeing is a pattern arising through the controlled expression of genetic instructions—controls that are the subject of this chapter.

All the different cells of your body carry the same genetic instructions, and they use most of those instructions to synthesize proteins that are basic to any cell's structure and functioning. That is why proteins of the cytoskeleton are the same from one cell to the next, as are ribosomal proteins and many of the enzymes used in metabolism.

Yet each type of cell also uses a small fraction of the genetic instructions in highly specialized ways. Even though they all carry the genes for hemoglobin, only red blood cells activate those genes. Even though they all carry the genes for antibodies (which are protein "weapons" against specific agents of disease), only certain white blood cells activate them. *Escherichia coli* cells living in your gut rapidly transcribe special genes after you drink a glass of milk—genes that allow them to use milk sugar (lactose) as an energy source. *These and all other living cells control which genes are active and which gene products appear, at what times, and in what amounts.*

Cells don't consciously control how their genes are used, of course. Control is exerted through molecules that interact with DNA, with mRNA transcribed from DNA, and with polypeptide chains resulting from mRNA translation. The molecules include enzymes and other proteins that help activate, enhance, or inhibit the transcription of particular genes; they include hormones that influence gene expression.

Why do those molecules interact with one another? They interact in response to chemical changes within the cell or its surroundings. In terms of your own cells, chemical conditions change when you vary your diet or level of activity. Conditions also have been changing in inevitable ways ever since you were a tiny mass of cells growing in your mother's body. Within each responding cell, gene activity has been changing appropriately, either to keep the cell itself functioning or to contribute to your overall growth and development over time.

Controls over gene activity and protein synthesis are exerted at different levels, although we have yet to decipher how most of them work. In this chapter we will focus on the mechanisms that are best understood.

GENE REGULATION IN PROKARYOTES

Mechanisms of Regulation

In prokaryotes, the main controls over gene expression deal with the overall rate of transcription—that is, the number of mRNA molecules transcribed per unit time from a given gene. Recall that transcription starts at promoters (page 211). In prokaryotes, a **promoter** is a base sequence preceding one or more genes and serving as a binding site for RNA polymerase. Some promoters bind RNA polymerase more strongly than others do, depending on the particular bases that occur in the

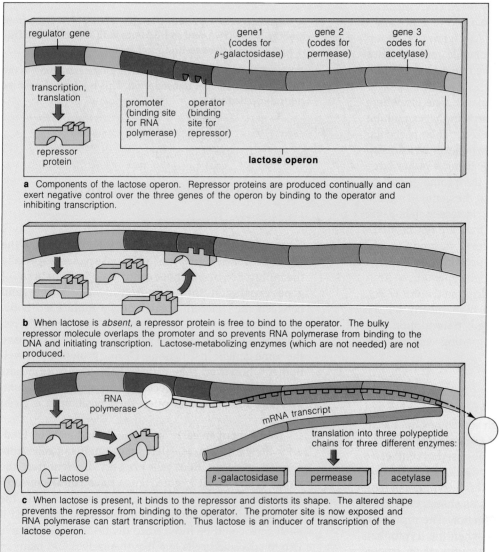

Figure 17.2 Negative control of the lactose operon. The first gene of the operon codes for an enzyme that splits lactose into two subunits (glucose and galactose). The second one codes for an enzyme that transports lactose molecules across the plasma membrane and into the cytoplasm. The third plays a complex role in lactose metabolism.

a Components of the lactose operon. Repressor proteins are produced continually and can exert negative control over the three genes of the operon by binding to the operator and inhibiting transcription.

b When lactose is *absent*, a repressor protein is free to bind to the operator. The bulky repressor molecule overlaps the promoter and so prevents RNA polymerase from binding to the DNA and initiating transcription. Lactose-metabolizing enzymes (which are not needed) are not produced.

c When lactose is present, it binds to the repressor and distorts its shape. The altered shape prevents the repressor from binding to the operator. The promoter site is now exposed and RNA polymerase can start transcription. Thus lactose is an inducer of transcription of the lactose operon.

sequence. Genes with strong promoters are transcribed more often than genes with weak ones—the products of which are required only in small amounts, usually. This type of transcriptional control mechanism therefore is built into the DNA itself.

Another type of control is exerted through regulatory proteins, called repressors or activators, that interact with the DNA to change the rate of transcription. Like RNA polymerase, some regulatory proteins can bind to a promoter. Others can bind to an **operator**, this being a short base sequence located between the promoter and the start of a gene. A bound **repressor protein** interferes with transcription; its inhibitory effect affords *negative control* over genes. A bound **activator protein** enhances transcription; it affords *positive* control over genes.

To get an idea of how regulatory proteins function, let's begin with some examples from *E. coli*, a type of bacterium that lives in mammalian hosts.

Negative Controls

Lactose Operon. For the first few weeks or months following its birth, a young mammal takes in nothing but milk, which is a food source not only for the mammal but also for the *E. coli* cells living in its gut. Once the weaning period is over, most mammalian hosts never take in milk again. (The only exceptions are humans and a few of their pets.) Do subsequent generations of bacterial cells keep on making lactose-degrading enzymes, even though they have no use for them? Not at all. Many years ago, François Jacob and Jacques Monod demonstrated that the bacterial genes coding for lactose-metabolizing enzymes are expressed only when lactose is present (Figure 17.2). When *E. coli* cells were grown on a lactose-free medium, they did not produce the enzymes, but they did so within minutes after being placed on a lactose-enriched medium.

Today we know that a region of *E. coli* DNA includes a promoter, an operator, and three adjacent genes associated with lactose metabolism. Any gene (or group of genes), together with its promoter and operator sequences, is called an **operon**. The one we are describing here is the lactose operon.

As Figure 17.2a shows, a regulator gene elsewhere in the DNA codes for a repressor protein that can inhibit transcription of the lactose operon. This particular repressor binds with the operator whenever lactose concentrations are low (Figure 17.2b). Being a rather large molecule, the repressor overlaps the promoter and so blocks RNA polymerase's access to the genes. Through this negative control mechanism, lactose-metabolizing enzymes are not produced when they are not needed.

When lactose is present, however, it binds to and alters the shape of the repressor protein. In its altered shape, the repressor cannot bind to the operator, so RNA polymerase is free to initiate transcription (Figure 17.2c). When lactose concentrations are high, nearly all the repressor molecules are inactivated; transcription proceeds rapidly and the lactose-degrading enzymes are synthesized.

Tryptophan Operon. Negative controls also govern other operons in *E. coli*, although sometimes the mechanisms are slightly different. For example, genes of the tryptophan operon code for enzymes used to synthesize the amino acid tryptophan. In this case, a repressor protein cannot block transcription unless a tryptophan molecule binds with it first and alters its shape. Only then can the repressor bind to the operator. The repressor-tryptophan complexes form only when the tryptophan concentration is high. In excess amounts, then, tryptophan itself shuts down its further synthesis.

Positive Controls

The positive control mechanisms afforded by activator proteins resemble the negative controls just described—but with opposite results. Unlike repressor proteins, the activator proteins bind with the promoter and, in so doing, enhance the binding of RNA polymerase to it. The stronger binding means that more mRNA molecules will be transcribed in a given period. Transcription is turned down only when something removes the activator protein from the promoter.

In some cases, a specific substance binds with the activator protein and causes it to break away from the promoter. In other cases, binding can occur only when the activator protein is already complexed with another substance. If that substance is not available in high concentrations, binding is prevented and transcription cannot proceed.

A bound repressor protein prevents RNA polymerase from binding to the DNA and so interferes with transcription. This is a type of negative control mechanism.

A bound activator protein helps RNA polymerase bind to the DNA and so enhances transcription. This is a type of positive control mechanism.

GENE REGULATION IN EUKARYOTES

Selective Gene Expression

Compared to prokaryotes, much less is known about the controls over gene expression in eukaryotes, especially the multicelled ones. Those controls underlie patterns of gene expression that vary considerably within and between the different tissues of the multicelled body. Because all the cells are descended from a single cell (such as a zygote produced at fertilization), they all have the same genetic instructions. During growth and development, however, they **differentiate**. The word means that the cells become specialized in appearance, composition, function, and often their position in a given tissue.

Think about an early experiment conducted by Frederick Steward, who isolated small clumps of cells from carrot root tissue. Root cells are very different from the single-celled carrot zygote. Some have hairlike absorptive structures, others have a waxy strip outside the plasma membrane, and so on. You might think that if such cells could be induced to divide, they would give rise to specializd cells just like themselves. Yet many of the root cells isolated by Steward were able to give rise to normal embryos, sometimes even carrot plants. *Clearly the specialized cells had retained all the genes and control elements required to build the different cell types of a complete plant.*

Differentiation arises through *selective* gene expression in different cells. Depending on the cell type and the control agents acting on it, some genes might be turned on only at a particular stage of development, others might be left on all the time, or others might never be activated at all. Still other genes might be switched on and off throughout the individual's life.

In vertebrates especially, hormones and other signaling molecules play crucial roles in selective gene expression. A **signaling molecule** is a chemical secretion from one cell type that alters gene expression and other activities in target cells. (By definition, a target cell is one having receptors to which a specific signaling molecule can bind.) Some hormones affect gene expression in many different cell types. For example, somatotropin (also called growth hormone) is secreted from the pitui-

Figure 17.3 The not-so-simple slime molds: an example of how gene activity is influenced by controls operating within and between cells, and between cells and the environment.

At one stage of the life cycle, *Dictyostelium discoideum* (a slime mold) produces reproductive cells called asexual spores. Each spore gives rise to a single-celled amoeba. The amoebas feed on bacteria present in soil, and they grow and divide as long as bacteria are available. When their food supply dwindles, the amoebas stream toward one another (**a**). Possibly 12 to more than 100,000 cells may gather together into a slug, which crawls about as a coordinated body (**b,c**).

The dwindling food supply is an environmental cue for biochemical and behavioral changes. The cells secrete a chemical signal (cyclic AMP) and become sticky. The cyclic AMP, secreted at intervals of about five minutes, stimulates the dispersed amoebas to move toward central collection points (**a**). The sticky amoebas adhere to one another, and the contact seems to activate receptors at the surface of their plasma membrane. Studies of amoebas raised in isolation suggest that signals from the receptors must control genes involved in differentiation. Unless amoebas receive the signals, they continue to synthesize only those proteins required for growth and division—and the isolated cells never do differentiate.

(**b,c**) Once amoebas form a slug, three distinct cell types give rise to two tissues. *Prestalk* cells (shaded violet in the sketch) make up the slug's anterior tissue. They produce and secrete substances not found in other cells—meaning they transcribe different genes. *Prespore* cells (gold) and *anteriorlike* cells (blue dots) make up the posterior tissue. The latter cells are scattered among the prespore cells and make up about ten percent of the developing cell mass.

For as long as the slug migrates, the anteriorlike cells stay where they are, possibly in response to an inhibitor substance secreted by anterior cells. The slug generally crawls up through the soil, toward light. There is more light at the soil surface, and less moisture. In response to those cues, the slug stops migrating and starts to form a fruiting body—a spore-bearing reproductive structure (**d,e**).

The fruiting body develops as prestalk and prespore cells differentiate into *stalk cells* and *spores* respectively. The anteriorlike cells, now apparently freed from regulation, sort into two groups that become positioned at each end of the posterior tissue. Cyclic AMP secretions from anterior cells may guide their movement. There is speculation that the anteriorlike cells play a role in elevating the nonmotile spores for dispersal from the top of the fruiting body.

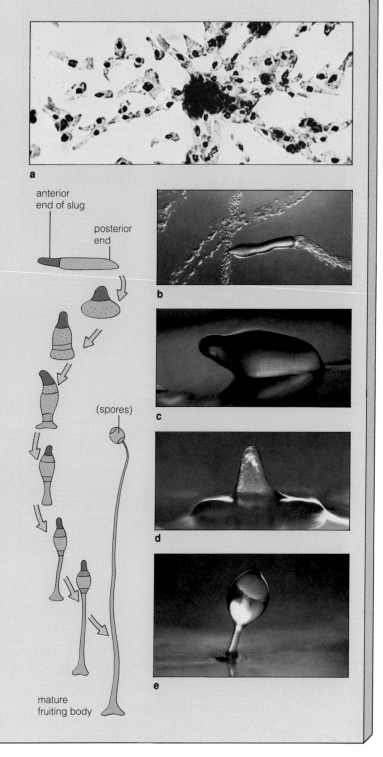

anterior end of slug

posterior end

(spores)

mature fruiting body

a

b

c

d

e

tary gland and induces protein synthesis in most of the body's cells. Other hormones only affect gene expression in certain cells at certain times. Prolactin is like this. Its target cells in mammary glands have exclusive responsibility for milk production, beginning a few days after the mammalian female gives birth.

Explaining hormonal control of gene activity is like explaining a full symphony orchestra to someone who has never seen one or heard it perform. Many separate parts must be defined before their intricate interactions can be understood! We will take up this topic later, in several chapters. For now, Figure 17.3 will give you an

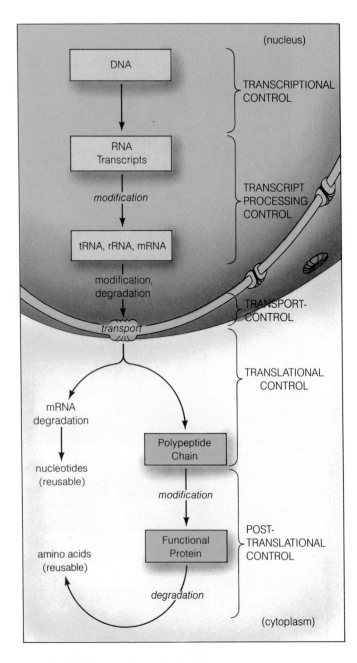

Figure 17.4 Control of eukaryotic gene expression: levels at which regulatory mechanisms can be brought into play. (Here, the steps are superimposed on a sketch of nuclear and cytoplasmic regions of a eukaryotic cell.)

idea of why there is no such thing as a "simple" eukaryote, when it comes to gene regulation. It shows the development of tissues in a slime mold, a very small but very complex protistan.

Cell differentiation occurs in multicelled eukaryotes as a result of selective gene expression. Although all the cells in the body have the same genes, they activate or suppress some fraction of those genes in different ways to produce pronounced differences in cell structure or functioning.

Selective gene expression is controlled by agents that operate within cells, between cells, and between cells and the environment.

Levels of Gene Control

Let's now turn to the levels of control of gene expression in eukaryotes. As Figure 17.4 and the following list indicate, control is exerted at five levels:

1. *Transcriptional controls* influence when and how a particular gene will be transcribed (if at all).

2. *Transcript processing controls* govern modification of the initial mRNA transcripts in the nucleus.

3. *Transport controls* determine which mature mRNA transcripts will be shipped out of the nucleus and into the cytoplasm for translation.

4. *Translational controls* govern which mRNA molecules that reach the cytoplasm will be translated into polypeptide chains at the ribosomes.

5. *Post-translational controls* govern how the polypeptide chains become modified into functional proteins. For example, some chains have specific sugar or phosphate groups attached to them; others are cleaved into smaller, active fragments.

The most common controls occur at the levels of transcription and transcript processing. (Here you may wish to review Figures 16.4 and 16.7.)

Regulatory proteins probably function in transcriptional control, as they do in prokaryotes. However, it seems likely that most are activator rather than repressor proteins, at least in mammalian cells. Why? More than ninety percent of mammalian DNA never is transcribed at all. It hardly seems likely that a cell would make tens of thousands of different kinds of repressor proteins simply to keep all of that DNA silent. In addition to positive control mechanisms, controls over transcription also are afforded by the structural organization of the DNA itself, in ways that will now be described.

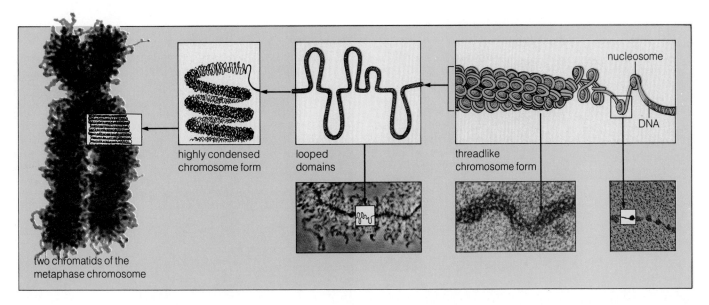

Figure 17.5 A current model of chromosome structure, based on studies of electron micrographs of the sort shown here (bottom row). Each nucleosome consists of DNA wound as a double-stranded thread on an individual spool of histone molecules. The DNA loops around each spool. In an unknown way, further levels of folding lead to the condensed form of the metaphase chromosome.

Chromosome Organization and Gene Activity

Each eukaryotic chromosome contains one very long DNA molecule (Figure 17.5). If you could line up all forty-six human chromosomes end to end, they would extend over a meter. Obviously, all that DNA would become a tangled mess if it were not organized in some way. Throughout the cell cycle, eukaryotic DNA is tightly bound with proteins called **histones** as well as other proteins. Collectively, the histone and nonhistone proteins are about equal in mass to the DNA. Some histones form a series of "spools" for winding up small stretches of DNA. Each histone-DNA spool is a **nucleosome**. The nucleosome packing arrangement is maintained even during interphase, when chromosomes are extended in thin, threadlike form. In electron micrographs of high magnification, the interphase chromosome looks like a beaded chain (Figure 17.6).

Interactions between the histones and DNA cause the eukaryotic chromosome to assume even more organized states. Supercoiling of the "beaded chain" results in a threefold increase in diameter (Figure 17.6). Further folding results in a series of **looped domains**, with each loop containing one or at most a few protein-coding sequences. Figure 17.7 gives an idea of the sheer magnitude of the organization in a condensed human chromosome.

The packing variations just described influence the accessibility and activity of genes in different regions and at different times in the life of the cell, as the following examples will indicate.

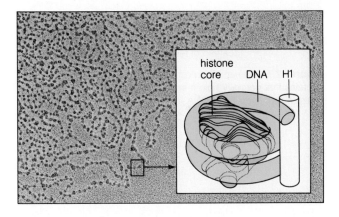

Figure 17.6 Electron micrograph of the nucleosome "beads" on a stretched-out *Drosophila* chromosome. The inset shows one model of the nucleosome. Each nucleosome consists of a double loop of DNA around a core of eight histone molecules. Another histone molecule (H1) stabilizes the arrangement and plays a role in further condensation of the chromosome.

DNA

(scaffold of residual chromosomal proteins)

Figure 17.7 Human chromosome at metaphase, stripped of the proteins that normally hold it in its highly condensed form. The maze of thin lines is the DNA.

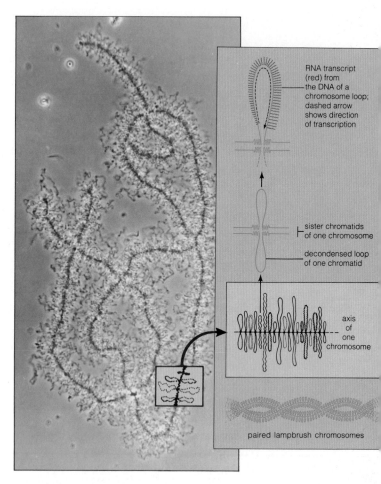

RNA transcript (red) from the DNA of a chromosome loop; dashed arrow shows direction of transcription

sister chromatids of one chromosome

decondensed loop of one chromatid

axis of one chromosome

paired lampbrush chromosomes

Figure 17.8 Electron micrograph of a pair of lampbrush chromosomes at prophase I of meiosis. These homologous chromosomes are from an immature egg of a female newt (*Triturus viridescens*).

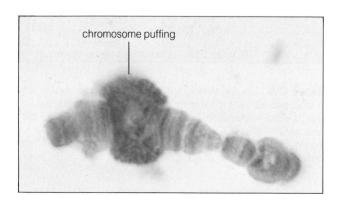

chromosome puffing

Figure 17.9 Chromosome puffing in a polytene chromosome from the salivary gland of a midge (a type of insect). The red-violet stain reveals the regions that are transcriptionally active; the blue stain indicates inactive regions.

Lampbrush Chromosomes. In some cases, transcription can be directly correlated with visible changes in chromosome structure. During prophase I of meiosis, **lampbrush chromosomes** become visible in cells destined to become the eggs of amphibians and other animals. Those chromosomes have decondensed regions where thousands of looped domains have uncoiled. When viewed with the light microscope, the decondensed chromosomes resemble bristle brushes that were once used to clean oil lamps (hence the name).

In a lampbrush chromosome, the DNA has been selectively loosened up by enzymes that are making gene regions accessible for transcription. Regulatory proteins must be involved here, for the chromosome decondenses in regions corresponding to specific gene sequences (Figure 17.8).

Chromosome Puffs. Intense transcription also can be observed in the chromosomes of certain fly larvae. In secretory cells of larval salivary glands, DNA replication occurs repeatedly, producing the polytene chromosomes described on page 183. The hormone ecdysone acts on those cells during different stages of development (Figure 34.17). The hormone binds to a specific regulatory protein, which then binds to the DNA to promote transcription. While transcription proceeds, the chromosomes open up and extend outward, forming **chromosome puffs** (Figure 17.9). The amount of transcription in cells containing polytene chromosomes has been correlated with how large and diffuse these puffs become.

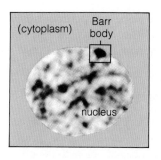

Figure 17.10 Barr body from a cell of a mammalian female.

X Chromosome Inactivation. A dramatic correlation between transcription and chromosome structure comes from studies of cells of mammalian females. Each cell has two X chromosomes. Apparently the gene products of only one X chromosome are necessary for normal cell functioning, and it may even be that a double dose of gene products from two X chromosomes would prove lethal. Whatever the case may be, a control mechanism permanently inactivates one of the two X chromosomes in every cell of mammalian female embryos.

In a female mouse embryo, for example, an X chromosome is shut down in each cell between the third and sixth day of development. Which of the two becomes condensed is entirely a matter of chance. The condensed X chromosome is quite distinct in the interphase nucleus (Figure 17.10). It is called a **Barr body** after its discoverer, Murray Barr.

Although transcription of most of its genes is suppressed, the condensed chromosome is replicated and passed on to all the progeny cells produced by mitotic cell divisions. The progeny cells are clones; that is, they are genetically identical to the parent cell. Because the paternal X chromosome was randomly inactivated in some cells and the maternal X chromosome was inactivated in others, every adult heterozygous female is a "mosaic" of X-linked traits. Typically, there are patches of tissue in which a paternal or maternal allele is being expressed, because most clones of cells tend to remain together in a tissue during growth and development. The mosaic tissue effect arising from random X chromosome inactivation is called **Lyonization** (after its discoverer, Mary Lyon).

The mosaic effect is visible in female calico cats, which are heterozygous for black and yellow coat-color alleles that reside on the X chromosome. Coat color in a given body region depends on which of the two X chromosomes is functioning and on which of the particular alleles is available for transcription (Figure 17.1).

The mosaic effect also is evident in human females affected by *anhidrotic ectodermal dysplasia* (Figure 17.11). Here, a mutant allele on one X chromosome gives rise to a skin disorder characterized by an absence of sweat

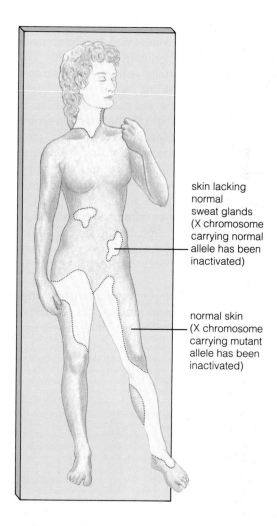

skin lacking normal sweat glands (X chromosome carrying normal allele has been inactivated)

normal skin (X chromosome carrying mutant allele has been inactivated)

Figure 17.11 Pattern of gene expression in a woman affected by anhidrotic ectodermal dysplasia, a disorder in which patches of skin do not have normal sweat glands. The pattern arises through random inactivation of the X chromosome during embryonic development. The mutant allele responsible for the disorder is on one X chromosome, and its corresponding allele on the other chromosome is normal. Depending on which of the two chromosomes is inactivated in an embryonic cell, all of the clonal descendants of that cell will display the same pattern of gene expression.

glands. In patches of defective skin, the X chromosome bearing the normal allele has been inactivated and the X chromosome bearing the mutant allele is functional.

Transcript-Processing Controls

So far, we have considered examples of transcriptional controls. Keep in mind that other levels of gene control may be equally important for normal cell activities; we just don't know as much about them.

For example, we still have a long way to go in identifying the controls over transcript processing (the modification of the initial mRNA transcripts), although researchers are giving us some interesting things to think about. Consider what happened (rather, what didn't happen) to some mRNA molecules that were transcribed from artificially constructed genes in cultured cells. The genes had been put together without any introns. Introns, recall, are noncoding sequences between exons—that is, between the nucleotide sequences in DNA that actually code for specific polypeptide chains (Figure 16.7). Introns are snipped out of an initial transcript, and the coding sequences are spliced together to form the mature transcript. You might think that because the artificial transcripts did not have any introns to begin with, they did not have to be processed and so just zipped on past the nuclear envelope for translation in the cytoplasm. They didn't. Apparently, they could not be recognized by whatever machinery exists for controlling the shipment of specific transcripts at specific times.

As another example, the primary mRNA transcript from a single gene is sometimes processed in alternative ways to produce two or more distinct kinds of proteins.

Although the proteins are very similar, each is unique in a certain region of its amino acid sequence (Figure 17.12). Thus, transcripts from a gene that codes for a contractile protein (troponin-T) are processed differently in different cell types. All the resulting proteins still function in contraction. But they do so in different ways—which may account for subtle variations in the way that different types of muscles in your body function.

Gene Control of Cell Division

This chapter has barely touched on the controls over gene expression in eukaryotes. How is it possible to leave you with a strong impression of how important those controls are? Perhaps with one last example that relates to a pervasive process—cell division. During the life cycle of a multicelled eukaryote, cell divisions underlie the growth, replacement, and repair of the specialized tissues making up different body regions. Transcription and translation of specific genes produce the proteins required for the division process itself.

Every second of the day, for example, millions of cells in different parts of your body divide and replace their worn-out, dead, and dying predecessors. The cells do not all divide at the same rate. Some have long-term roles and are arrested at interphase; nerve cells are like this. Others have short-term roles and divide rapidly, as do cells in the protective epithelium near the stomach wall, which is constantly exposed to the corrosive effects of gastric fluids.

No one knows exactly which genes govern cell growth and division, but controls clearly exist. When most of a rat's liver is surgically removed, protein synthesis and cell divisions accelerate to a phenomenal

Figure 17.12 Alternative processing of a primary mRNA transcript from a single gene. The primary transcript can be processed in different ways to produce distinct mRNA molecules that code for similar but distinct proteins.

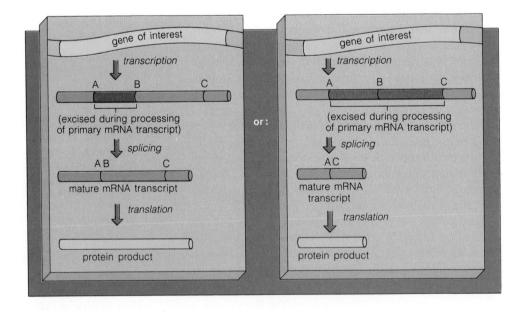

Altered Regulatory Genes and Cancer

Characteristics of Cancer Cells

On rare occasions, genetic controls over cell division become permanently altered. A cell divides again and again, until its offspring begin to crowd surrounding cells and interfere with tissue functioning. The alteration has spawned a **tumor**: a population of cells that are dividing to an abnormal extent. Some of the tumors are "clonal," meaning that all of their cells are genetically identical and derived from a single, genetically altered cell.

The problem is not that tumor cells divide at a horrendous rate; normal cells that replace a surgically removed portion of the liver divide much faster. Rather, tumor cells no longer respond to the controls telling them when to stop. They will not stop as long as conditions for growth remain favorable. We know this from studies of "immortal" tumor cells, which have been dividing for decades under controlled laboratory conditions.

When tumor cells simply divide more than they should but still remain in one place in the body, the tissue mass is a *benign* tumor. When it is surgically removed, its threat to the individual's health ceases. Sometimes, however, a tumor cell can migrate and then grow and divide in other organs of the body. Such tumor cells are said to be *malignant*, or cancerous. By definition, a cancer cell has (at the minimum) the following characteristics:

1. *Profound abnormalities in the plasma membrane.* Membrane transport and permeability are amplified. Some proteins at the surface are lost or altered, and new ones appear.

2. *Profound changes in the cytoplasm.* The cytoskeleton shrinks, becomes disorganized, or both. Enzyme activity shifts (as in a greatly increased reliance on glycolysis, even when oxygen is available).

3. *Abnormal growth and division.* Inhibitors of overcrowding in tissues are lost. Cell populations increase to unusually high densities; proteins are

formed that trigger the abnormal proliferation of blood capillaries that service the growing cell mass.

4. *Diminished capacity for adhesion to substrates.* Secretions needed for adhesion dwindle; cells cannot become properly anchored in the parent tissue.

Normally, recognition proteins at the plasma membrane enable cells of like type to recognize each other and bind together in tissues and organs (page 87). When the genes coding for recognition proteins are altered or suppressed, the cell can leave its proper place and travel by way of blood or lymph to other tissues, where it can form a new growth. This process is called **metastasis**.

Whatever their site of origin, cancer cells of a primary tumor can invade other tissues and multiply into secondary tumors. Medical treatment is very difficult in such cases, for even if a malignant tumor can be removed from one site, another may appear elsewhere.

Oncogenes

Until a decade ago, the possibility of unraveling the secrets of cancer seemed remote, because cells of different cancers differ greatly in form, behavior, metabolic requirements, surface properties, and growth rates. However, *it now appears that a small number of altered regulatory genes may contribute to at least some kinds of cancer.*

Several viruses can cause cancerous transformations in vertebrate cells. They include the Rous sarcoma virus, or RSV (which causes cancer in chickens), papovaviruses (some species cause warts, others cause tumors), adenoviruses (which cause lung infections as well as tumors), and herpes viruses (different species cause fever blisters, chickenpox, genital infections, and cancer). In all cases, the viral genetic instructions become integrated into the DNA of a host cell and are subsequently expressed in all offspring of that cell.

In the early 1970's, researchers identified RSV mutants that could infect host cells but not render them cancerous. Through recombinant DNA methods (page 239), it became evident that the viral mutants were missing a segment of their hereditary material, called the *src* gene. The viral mutants were able to complete their infectious cycles in the absence of an *src* gene—which therefore was not required for infection or replication. The only function they had lost was the ability to induce cancer in chickens. Any gene like the *src* gene that has the potential to induce cancerous transformations is now known as an **oncogene**.

Even with these insights, a genetic theory of cancer seemed elusive. The problem was that most types of cancer arise with no help at all from viruses or viral genes! However, in 1975, J. Michael Bishop and Harold Varmus found that a gene almost identical to the *src* gene exists in the host DNA of chicken cells—*and it codes for a protein that takes part in normal cell activities!*

More than fifty oncogenes have now been isolated from a variety of animals. In each case, an identical gene sequence, called a **proto-oncogene**, has been discovered in the normal DNA of the host animal—and the cells carrying them rarely become cancerous. The cellular counterparts of viral oncogenes are highly conserved in diverse species, ranging from humans to yeasts; apparently, the proto-oncogenes have not changed much in over a billion years.

We can assume that natural selection has not removed these genes from cells because their *normal* expression is vital—even though their abnormal expression may be lethal. *Proto-oncogenes are inherent parts of vertebrate DNA, and they code for proteins necessary in normal cell functioning. They may become cancer-causing genes only on those rare occasions when specific mutations alter their structure or their expression.*

How Oncogene-Encoded Proteins Induce Cancer

Oncogenesis can occur when viral DNA is inserted into a region of cellular DNA and thereby changes the growth properties of the cell. For example, insertions of a viral oncogene adjacent to an active promoter can cause a high level of transcription from the oncogene. Cancer also can arise by the action of a **carcinogen**, which is an agent that causes cancer. Carcinogens include numerous natural and synthetic compounds (such as asbestos and certain components of cigarette smoke), x-rays, gamma rays, and ultraviolet radiation. The carcinogens introduce changes in oncogenes themselves or in the sites that regulate their expression.

Yet cancer seems to be a multistep process, and it may be that several genetic changes are necessary to bring it about. For example, in retinoblastoma (a cancer of the eye), two genetic changes are required. In other cancers, environmental as well as genetic factors are involved.

In normal cells, extracellular signals trigger cell division. At the very least, signaling mechanisms of the pathways controlling division must include the following:

1. Growth factors (molecules that carry growth signals from one cell to another).

2. Signal receptors at the membrane surface of the receiving cell.

3. Molecules that transmit signals from the cell surface to specific targets inside.

No one knows where oncogene proteins intervene in the processes that control cell division. Of the known oncogenes, one group codes for proteins that can enhance transcription of genes that may be central in cell growth and division. Because the protein products of these genes become concentrated in the cell nucleus, there is speculation that they operate by way of transcription. Intriguingly, the normal counterpart of one oncogene protein appears in interphase cells just before the DNA is replicated—*an event that generally foreshadows cell division.*

degree: four *billion* replacement cells are produced in four days. Then brakes are applied and the division rate slows; by the seventh day, most of the missing liver tissue has been replaced.

On rare occasions, controls over cell division are lost when cells undergo cancerous transformation, a process that is described in the *Commentary*. Possibly more than any other example, this transformation brings home the critical extent to which you and all other organisms depend on controls over gene expression.

SUMMARY

1. In both prokaryotes and eukaryotes, changes in gene activity represent a chemical responsiveness to shifts in cellular or extracellular conditions. In multicelled eukaryotes, for example, conditions vary as a result of changes in diet and levels of activity; they vary inevitably during growth and development.

2. Gene expression is controlled by many interacting elements, including control sites built into DNA molecules, regulatory proteins, enzymes, and hormones. Their interactions govern which gene products appear, at what times, and in what amounts.

3. The best understood gene controls are transcriptional. In prokaryotes, operon controls influence transcription rates. In eukaryotes, the timing and rate of transcription are influenced by chromosome organization as well as by control elements.

4. In prokaryotes especially, negative control of transcription is afforded by repressor proteins that can bind to control sites near specific genes. A bound repressor protein inhibits transcription by blocking the access of RNA polymerase to those genes.

5. Positive control of transcription is afforded by activator proteins that also bind to control sites near specific genes. Bound activator proteins promote transcription by helping RNA polymerase bind and start transcription. Positive controls are probably the more common type in eukaryotes.

6. In complex eukaryotes, cells differentiate (they become different in appearance, composition, function, and often position). Differentiation arises through selective gene expression. The term means that different types of cells activate and suppress some fraction of their genes in a variety of ways that lead to pronounced differences in their structure and functioning.

7. Selective gene expression is controlled by agents that operate within cells, between cells, and between cells and the environment.

Review Questions

1. Define these terms: promoter, operator, repressor protein, and activator protein. *222–223*

2. Cells depend on controls over which gene products are synthesized, at what times, at what rates, and in what amounts. Describe one type of control over transcription in *E. coli*, a type of prokaryote. Then list five general kinds of controls involved in eukaryotes. *223–224, 226*

3. Define differentiation. How does it arise? *225, 226*

4. A plant, fungus, or animal is composed of diverse cell types. How might this diversity arise, given that all of the body cells in each organism inherit the *same* set of genetic instructions? *224–225*

5. Somatic cells of human females have two X chromosomes. During what developmental stage are genes on *both* chromosomes active? Explain what happens to one of those chromosomes after that stage. *229*

6. What are the characteristics of cancer cells? Explain the difference between a benign tumor and one exhibiting metastasis. *231*

Readings

Brown, D. 1981. "Gene Expression in Eukaryotes." *Science* 211: 667–674.

Croce, C., and G. Klein. March 1985. "Chromosome Translocations and Human Cancer." *Scientific American* 252(3):54–60.

Goldman, M., G. Holmquist, et al. 1984. "Replication Timing of Genes and Middle Repetitive Sequences." *Science* 224:686–692. Example of the kinds of experiments that are yielding insights into the mechanisms of gene regulation.

Klug, A., et al. 1981. "A Low-Resolution Structure of the Histone Core of the Nucleosome." *Nature* 287:509–515.

Prescott, D. 1988. *Cells*. Boston: Jones and Bartlett. Excellent chapters on prokaryotic and eukaryotic gene regulation.

Sachs, L. January 1986. "Growth, Differentiation, and the Reversal of Malignancy." *Scientific American* 254(1):40–47. Describes the isolation of proteins that induce differentiation and that halt the growth of some cancer cells.

Weinberg, R. September 1988. "Finding the Anti-Oncogene." *Scientific American* 259(3):44–51.

18

RECOMBINANT DNA AND GENETIC ENGINEERING

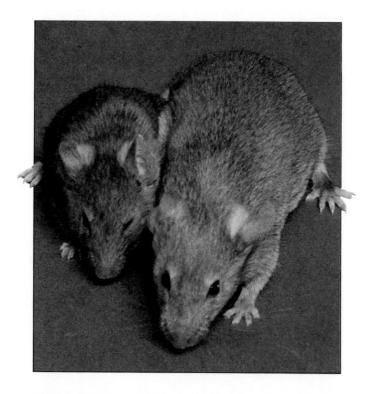

Figure 18.1 Ten-week-old mouse littermates, the one on the left weighing twenty-nine grams, and the one on the right, forty-four grams. The larger mouse grew from a fertilized egg into which the gene for human somatotropin (growth hormone) had been inserted.

For at least $3\frac{1}{2}$ billion years, nature itself has been conducting genetic experiments of one sort or another. Mutation, crossing over and recombination at meiosis, the random gene mixing at fertilization, hybridizations between species—all such events have contributed to the current diversity of life on earth.

We humans have been manipulating the genetic character of different species for more than ten thousand years. One need only compare the modern strains of domesticated wheat and corn, or compare the varied breeds of cattle, poultry, dogs, and cats with their wild ancestral stocks to know that this is so. Artificial selection has produced chickens and turkeys with broader and meatier breasts, larger and sweeter oranges, seedless watermelons, and flamboyant ornamental plants. Hybridizations have given us the tangelo (tangerine crossed with grapefruit) and the mule (donkey crossed with horse). In an indirect way, surgery, transfusions, vaccinations, drug therapy, and other medical practices are forms of genetic manipulation, for they preserve genotypes that might otherwise be selected against and lost from the population.

Today, we have an enhanced capability to change the genetic character of organisms through **recombinant DNA technology**. With this technology, DNA molecules from different species are cut into segments and spliced together to form recombinant DNA, which is propagated in a line of dividing cells. Genes of interest are isolated for analysis; many are modified and reinserted into the same organism (or transplanted into a different one), where they are expressed as functional proteins.

From the historical perspective, our current manipulations are novel not in concept but rather in their magnitude. We are already "engineering" new genetic changes that never would have emerged in nature. Genes of animals are being transferred into plant cells, and genes of plants into animal cells. Synthetic genes have been constructed so that silk can be manufactured without silkworms. The gene for human insulin has been inserted into bacterial cells that serve as factories for producing clinically useful amounts of this protein. The gene for human somatotropin (growth hormone) was inserted into fertilized mouse eggs and was expressed as the eggs developed into young mice; Figure 18.1 shows evidence of its general growth-promoting effects.

Genetic engineering has staggering potential for improving agricultural yields, enhancing our medical arsenal against diseases, diagnosing genetic disorders, and creating diverse new products for home and industry. At the same time, it has triggered public concern and confusion, created problems for governmental regulatory agencies, and raised fears among some individuals that introducing modified organisms into the environment may have unforeseen and disastrous consequences. With this chapter, we consider the conceptual foundations of recombinant DNA technology. We also address some of the ecological, social, and ethical questions related to its applications.

NATURAL RECOMBINATION MECHANISMS

When homologous chromosomes cross over and exchange fairly long segments at meiosis, each typically ends up with a new combination of genes compared to the parental genotypes. You, for example, are chockful of recombinant DNA molecules as a consequence of this mechanism, which we can call **homologous recombination**. It is not that you received more or fewer genes along a chromosome of a given type; alleles at various gene locations were merely swapped (page 153). Homologous recombination is a *reciprocal* exchange, with each chromosome of a pair donating and receiving an equivalent number of genes arranged in exactly the same sequence as before. Other natural recombination mechanisms, described next, are not reciprocal at all.

Transposable Elements

Until a few decades ago, researchers believed that nothing other than rare chromosomal aberrations, such as the inversions described on page 184, could change the number or order of genes on a chromosome. In the late 1940s, however, Barbara McClintock became convinced of the existence of what has come to be called **transposable elements**. The term refers to certain parts of the DNA that frequently "jump" (transpose themselves) to new locations in the same DNA molecule or a different

one. Often they inactivate the genes into which they become inserted and give rise to observable changes in phenotype.

McClintock had been analyzing the phenotypic effects of mutations in genes responsible for the color of individual kernels of maize, or Indian corn. All the cells in a kernel have the same genes, so you might expect the whole kernel to be the same color. Some are indeed fully colored—but some are entirely colorless and others are spotted (Figure 18.2). Early in plant growth, the random insertions of transposable elements cause mutations in individual cells—mutations that affect pigment synthesis in all the daughter cells produced by repeated divisions.

Most geneticists didn't know what to think of McClintock's insight, and for nearly twenty years they didn't think much of it at all. Then, other transposable elements were discovered in *Escherichia coli*; they were given the name **transposons**. A transposon can insert itself at apparently random locations in the bacterial chromosome, often with a variety of disruptive effects on phenotype. Each carries one or more genes that are flanked on both sides by a short sequence of specific nucleotides, called an *insertion sequence* (IS). A simple transposon codes for an enzyme (transposase) that catalyzes its own insertion into the bacterial chromosome, as shown in the following sketch:

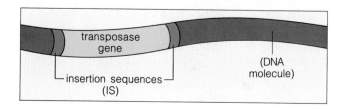

Figure 18.2 Color variation in kernels of maize, or Indian corn. The variation arose because movable genetic elements jumped about in the chromosomes of certain cells of the growing plant. Their insertions into new chromosomal locations produced mutations, some of which inactivated genes that affected pigment synthesis.

Figure 18.3 An *E. coli* cell that has ruptured, thereby releasing a portion of its main DNA molecule and several plasmids (blue arrows).

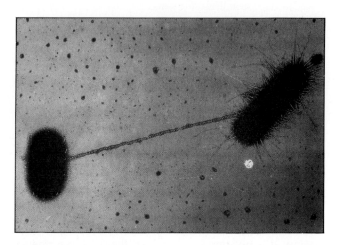

Figure 18.4 Early stage of conjugation between a recipient (F⁻) cell, at left, and a donor (F⁺) cell of *E. coli*. The long appendage joining the two bacteria is an F pilus; it will bring the two participants into close contact so that DNA can be transferred.

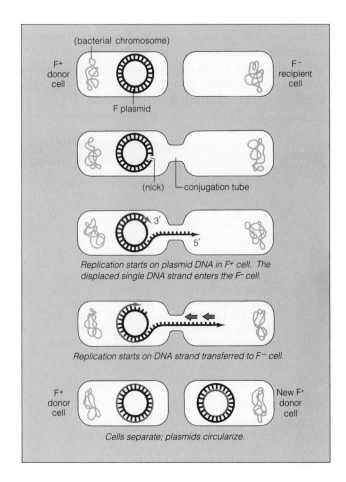

Figure 18.5 Transfer of an F plasmid between two *E. coli* cells during conjugation. For clarity, the bacterial chromosomes are greatly reduced in size; a plasmid actually contains only about $\frac{1}{40}$ as much genetic information as the chromosome.

The complex transposons carry more genetic information. For example, one consists of a pair of insertion sequences and one or more genes that confer antibiotic resistance:

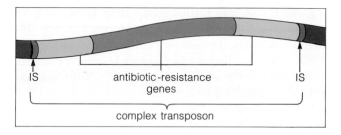

This type of transposon can move out of the chromosome of one bacterium and into the chromosome of another—with clinical consequences, as you will see shortly.

Plasmids

You may have noticed that we are calling the bacterial DNA molecule a chromosome, even though it is not comparable in structure to the eukaryotic chromosome. We do this to distinguish it from **plasmids**, which are much smaller, "extra" DNA molecules in the cells of many bacterial species (Figure 18.3).

Like the bacterial chromosome, a plasmid is a circular molecule equipped with a single origin, a site to which DNA polymerase can bind and initiate replication (page 206). A plasmid also contains insertion sequences. What capabilities do those two structural features confer upon plasmids? *A plasmid can be replicated independently of the bacterial chromosome, and it can on occasion become integrated into the chromosome.*

The bacterial chromosome contains all the genes necessary for normal growth and development. The genes of plasmids are not essential, yet they are still replicated in a given cell. They are transmitted through successive generations; they are even transferred to bacterial cells of different species. What might be the selective advantage of spreading around this seemingly excess genetic baggage? The products of plasmid genes do serve useful functions under some circumstances, as we know from studies of the so-called **F plasmids** and **R plasmids**. (The F stands for *Fertility* and the R, for *Resistance*.)

F Plasmids. Genes carried on an F plasmid code for proteins that promote **bacterial conjugation**, a process by which one bacterial cell transfers DNA to another. Only a cell having an F plasmid can be a donor; such cells are designated F^+. Only a cell lacking an F plasmid (designated F^-) can be a recipient. Sometimes the transfer is said to be a form of sexual reproduction between "male" and "female" bacterial cells, although comparing this to sex among the eukaryotes requires a rather breathtaking leap of the imagination.

Some F plasmid genes code for the proteins necessary to construct an F pilus, a long appendage that can attach to a recipient cell and draw it right up against the donor (Figure 18.4). The attachment apparently activates an enzyme into cutting one strand of the plasmid DNA. The cut strand starts to unwind from the other strand and enters an F^- cell. As Figure 18.5 shows, DNA replication proceeds on the exposed bases of both strands in both cells. Once the transfer and replication are completed, the cells separate. Each is now an F^+ cell.

Occasionally, an F plasmid can become integrated into the chromosome of a recipient cell by recombination (Figure 18.6). According to one model, IS sequences in the chromosome and in the F plasmid are preferred sites for recombination and plasmid integration.

Many strains of *E. coli*, designated Hfr, have an integrated F plasmid. Like F^+ cells, the Hfr cells serve as donors in conjugation, but in this case a recipient usually remains F^-. Why? Conjugation-induced replication starts when a strand cut is made near one end of the integrated plasmid DNA. The cut end peels away from the chromosome, enters the recipient cell, and drags one strand of the chromosomal DNA with it (Figure 18.7). The rest of the plasmid DNA (including the conjugation-promoting genes) is at the tail end of the molecule and rarely makes it into the recipient cell. The fragile cell-to-cell contact nearly always is broken before the transfer (which takes 100 minutes) can be completed.

R Plasmids. Like the F plasmid just described, the R plasmids carry conjugation-promoting genes. They also carry one or more complex transposons, each with genes

that confer resistance to ampicillin or other antibiotics. An **antibiotic** is a normal metabolic by-product of certain microbes, especially the actinomycetes. More than 2,500 of these by-products have been identified. We discovered by accident that they can interfere with gene expression or other functions in microbial pathogens, and we have been using them ever since as wonder drugs to treat infectious diseases.

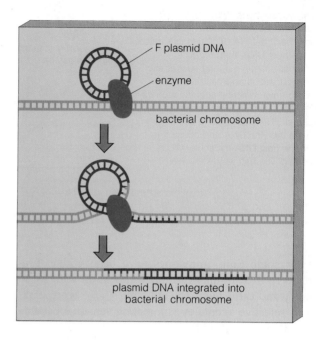

Figure 18.6 Integration of an F plasmid into the bacterial chromosome through recombination. Only a small stretch of the circular bacterial chromosome is shown.

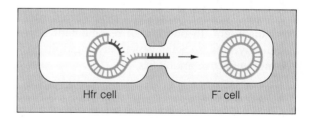

Figure 18.7 Transfer of chromosomal genes from an Hfr cell to an F^- cell. The bacterial chromosomes are shown in blue, and the integrated plasmid DNA in red. Replication proceeds in both cells while the transfer is occurring.

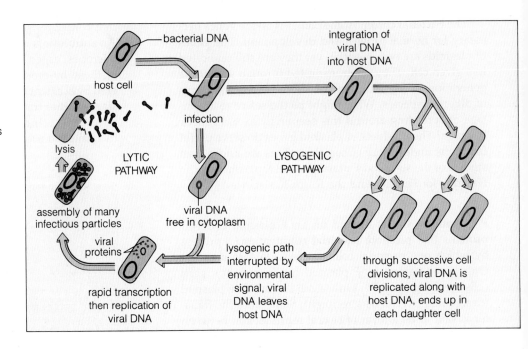

Figure 18.8 Replication cycle of lambda bacteriophage. Depending on environmental factors, infection proceeds by way of either the lytic pathway or the lysogenic pathway.

In the lysogenic pathway, the bacteriophage enters a latent state in which the viral DNA becomes integrated into the host DNA, then remains functionally inactive during successive DNA replications and cell divisions. Specific environmental agents (such as ultraviolet radiation, which damages the bacterial DNA and leads to cell death) activate the viral DNA and cause it to leave the host DNA molecule. When it does, the lytic pathway is followed.

For some time, antibiotics have been widely prescribed to treat human infections and as an additive to poultry and cattle feed. The antibiotics have been working as selective agents by eliminating sensitive bacteria and favoring the evolution of more resistant ones.

The R plasmids being transferred among bacteria of different genera are especially troubling. For example, *Shigella*, a type of bacterium responsible for a serious form of dysentery, is now often resistant to antibiotics. Antibiotic resistance also has been conferred upon the microbes responsible for gonorrhea (a sexually transmitted disease), typhoid, meningitis, and assorted intestinal tract disorders. In one line of recombinant DNA work, actinomycetes are being genetically engineered to produce more effective antibiotics.

Viruses

Certain bacteriophages, such as lambda, also can integrate their DNA into host chromosomes. Lambda bacteriophage can spread through a population of *E. coli* and destroy it through repeated cycles of infection, this being called the lytic pathway (Figure 18.8). On rare occasions, they enter a "lysogenic" pathway, in which the viral DNA is integrated into the DNA of progeny cells. How is this done?

A viral enzyme cuts the bacterial chromosome at a specific site, then the viral DNA is inserted between the cuts and sealed in place. The modified chromosome is replicated and passed on to succeeding cell generations. Later, the viral DNA may move out of the chromosome in much the same way that it moved in, and an infectious cycle begins again.

To date, transfer of genes by recombination has been discovered in many organisms, including a variety of bacteria, bacteriophages, yeasts, fruit flies, and mammals. It appears that gene transfer may be common to most (if not all) organisms. As we turn now to the kind of recombination techniques going on in the laboratory, what is the point to keep in mind? Simply this:

Genes can be transferred from one DNA molecule to another by bacteriophages, plasmids, and transposons.

RECOMBINANT DNA TECHNOLOGY

You might say that recombinant DNA technology grew out of experiments with *E. coli* and the bacteriophages that infect it. In the late 1960s and early 1970s, researchers learned how to use a variety of cutting and splicing enzymes to make DNA fragments and "package" them in plasmids for insertion into host cells. They developed ways to pinpoint the DNA fragments of interest in individual lines of dividing cells. They also started to identify the nucleotide sequences of individual genes and to sequence the **genome** (determine the order of bases in

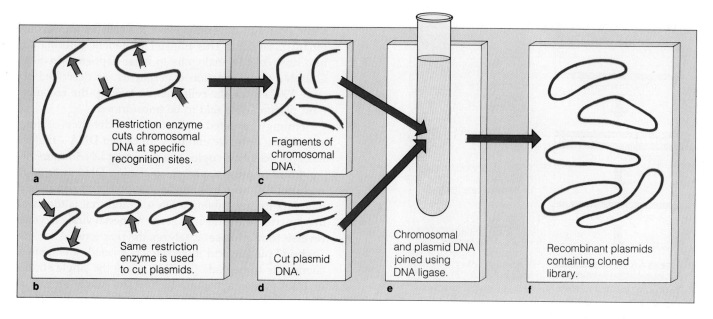

Figure 18.9 Formation of a DNA library.

the DNA in a haploid set of chromosomes) for different species. The following examples will give you a sense of what the new technology entails.

Producing Restriction Fragments

All cells have nucleases, which are enzymes capable of cutting DNA strands during normal repair and degradation reactions. Bacteria also have special nucleases called **restriction enzymes**, the sole function of which is to cut apart foreign DNA that has entered the cell. Several hundred restriction enzymes have been identified. Each makes a cut only within a short "recognition" sequence that reads the same on both strands (that is, in the 5′→3′ direction). For example:

portion of the two strands of a DNA molecule, showing a recognition site for the restriction enzyme EcoRI

Sometimes the two cuts are exactly across from each other and blunt-ended DNA fragments are produced. More often, off-center cuts produce fragments with a short, single-stranded tail:

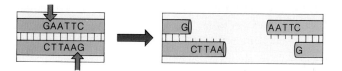

The fragments of any DNA molecule cut by a given restriction enzyme can base-pair with any other DNA molecule that has been cut by the same enzyme.

Preparing and Cloning a DNA Library

When a restriction enzyme cuts an *E. coli* chromosome, thousands of fragments are produced. Suppose the same enzyme is used to cut plasmids. Suppose next that all the chromosomal fragments and the now-linear plasmids are mixed together to allow random base-pairing between them. Their temporary union can be made permanent by exposing them to **DNA ligase**, an enzyme normally involved in DNA replication and repair. Researchers routinely use DNA ligase to seal different DNA molecules together (Figure 18.9).

So now we have plasmids into which DNA fragments are spliced for propagation in host cells. A plasmid or any other self-replicating genetic element used to insert DNA into a host cell for propagation is called a **cloning vector**. Any collection of DNA fragments produced by restriction enzymes and incorporated into cloning vectors is called a **DNA library**. Each library contains DNA fragments from a single species only.

E. coli cells are still being used as hosts for most DNA clonings. How are they induced to take up the recombinant plasmids? Typically the cells are incubated with them in the presence of calcium salts, which make the cell membrane permeable to DNA. (The calcium treat-

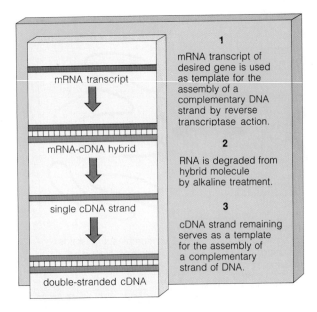

1

mRNA transcript of desired gene is used as template for the assembly of a complementary DNA strand by reverse transcriptase action.

2

RNA is degraded from hybrid molecule by alkaline treatment.

3

cDNA strand remaining serves as a template for the assembly of a complementary strand of DNA.

Figure 18.10 Formation of double-stranded cDNA from an mRNA transcript.

Figure 18.11 (Below) Use of a cDNA probe to identify the colony of transformed bacterial cells that have taken up the DNA of interest.

ment does not adversely affect the cells or the recombinant plasmids.) This transfer of genetic information into host cells is analogous to what happened in Fred Griffith's transformation experiments, as described on page 199. Hence the cells that do take up the recombinant plasmids are said to be transformed.

Through repeated replications and divisions of individual host cells, we end up with **cloned DNA**—that is, multiple, identical copies of a particular DNA fragment.

An alternative approach to cloning DNA is to start with mRNA, which is single-stranded, and use it as a template for assembling a DNA strand that is identical in sequence to the desired gene. A special viral enzyme, **reverse transcriptase**, catalyzes the reactions. After the "hybrid" nucleic acid molecule has formed, it is treated chemically to degrade the RNA. Only the single strand of DNA remains (Figure 18.10). The single strand of DNA is converted to double-stranded form by the action of DNA polymerase.

Any DNA "copied" from an mRNA transcript is called **complementary DNA**, or **cDNA** for short. A cDNA library also can be inserted into plasmid vectors and then cloned.

Identifying the Cloned DNA of Interest

When recombinant plasmids are mixed with potential host cells, not all of the cells take them up. How is it possible to identify the ones that do?

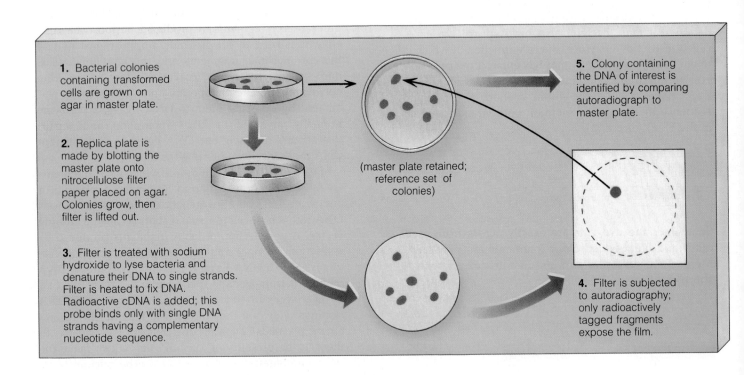

1. Bacterial colonies containing transformed cells are grown on agar in master plate.

2. Replica plate is made by blotting the master plate onto nitrocellulose filter paper placed on agar. Colonies grow, then filter is lifted out.

3. Filter is treated with sodium hydroxide to lyse bacteria and denature their DNA to single strands. Filter is heated to fix DNA. Radioactive cDNA is added; this probe binds only with single DNA strands having a complementary nucleotide sequence.

(master plate retained; reference set of colonies)

5. Colony containing the DNA of interest is identified by comparing autoradiograph to master plate.

4. Filter is subjected to autoradiography; only radioactively tagged fragments expose the film.

In some cases, the plasmids used as cloning vectors have antibiotic-resistance genes. If the bacterial hosts are spread onto a medium that has been supplemented with antibiotics, all the cells *except* the ones carrying the plasmids will be killed or their growth inhibited. The transformed cells will be the only ones that are able to grow into colonies.

In other cases, colonies are analyzed with **nucleic acid hybridization techniques**, such as cDNA probes. A **cDNA probe** is assembled from radioactively labeled subunits; it contains a nucleotide sequence known to be complementary to that of the desired gene.

As Figure 18.11 indicates, colonies to be screened are allowed to grow on a petri plate. A replica of the master plate is made by blotting it onto nitrocellulose filter paper and allowing the transferred cells to grow into new colonies. (These correspond to the locations of the original colonies on the master plate.) The filter paper is treated with sodium hydroxide, which lyses the cells and denatures their DNA into single-stranded form. The radioactive cDNA probe will hybridize only with the DNA having the complementary base sequence and will thereby tag the location of the bacterial colony. Later, the location can be pinpointed by autoradiography, a technique by which an X-ray film is exposed by the presence of a radioactive substance.

Selected Gene Amplification

Cloning DNA fragments is one method of gene amplification (a millionfold or more). An alternative method that does *not* involve cloning, called the polymerase chain reaction, can be used if the nucleotide sequences on both sides of the gene are known.

First, short nucleotide sequences complementary to the ones flanking the desired gene are synthesized. Because they will base-pair with the flanking regions, they can serve as *primers* for DNA polymerase. That enzyme, recall, cannot synthesize a new DNA strand unless a primer is already positioned on the existing one (Figure 15.9). Second, the DNA molecules containing the genes are heated and so converted to single-stranded form. Third, primers are added to the preparation and double-stranded DNA is produced. The preparation is heated again to make the DNA single-stranded, more primers are added to prompt DNA polymerase into action, and so on over and over again (Figure 18.12).

You may be wondering why the DNA polymerase (a protein) is not denatured along with the DNA during the heating phases of this technique. A special DNA polymerase extracted from bacterial residents of hot springs is used; it remains active at the higher temperatures required.

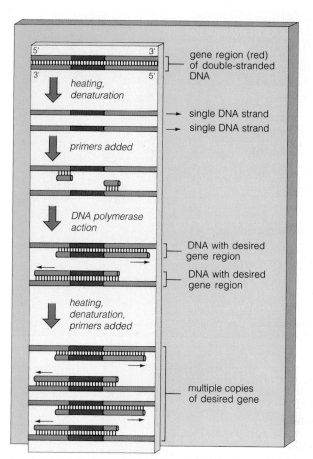

Figure 18.12 Polymerase chain reaction used in gene amplification, as described in the text. Genes amplified in this way can be separated from the rest of the DNA by gel electrophoresis (the gene copies will be the shortest molecules).

Expressing the Cloned Gene

Even when a desired gene has been isolated and amplified, it may not be possible for a host cell to transcribe and translate it into functional protein. As indicated in earlier chapters, human genes contain noncoding regions (introns) as well as coding regions (exons). Those genes cannot be translated unless the introns are spliced out and the exons spliced together into a mature mRNA transcript. Bacterial hosts simply do not have the proper splicing enzymes. The problem can be circumvented by using cDNA—which is synthesized from mature mRNA transcripts.

Using cDNA raises other problems, however. The genes synthesized from mRNA cannot be transcribed unless they occur between a *promoter* (a site to which RNA polymerase can bind and initiate transcription) and a stop signal. An mRNA transcript has neither one. Sometimes a gene can be inserted close to a known promoter in a cloning vector—but that "foreign" promoter may not be recognized in the host cell.

Table 18.1	Examples of Cloned Human Gene Products Approved for Use or Under Development
Protein	**Used in Treating**
Insulin	Diabetes
Somatotropin (growth hormone)	Pituitary dwarfism
Erythropoetin	Anemia
Factor VIII	Hemophilia
Factor IX	Hemophilia
Interleukin-2	Cancer
Tumor necrosis factor	Cancer
Interferons	Some cancers, viral infections
Monoclonal antibodies	Infectious diseases
Atrial natriuretic factor	High blood pressure
Tissue plasminogen factor	Heart attack, stroke

Despite these and other obstacles, several human gene products are currently being mass produced or developed (Table 18.1). Human insulin, a protein hormone that helps regulate glucose metabolism, is one of them. Insulin is absent or deficient in individuals affected by a type of diabetes (page 362), and they can lead relatively normal lives only when insulin is administered daily to them. Before recombinant DNA technology, human insulin could not be obtained. (Diabetics were given insulin extracted from pancreatic tissues of pig cadavers. Insulin from pigs is not exactly the same as insulin from humans, so it caused severe allergic reactions in some diabetics.)

Gene Sequencing

To sequence a cloned gene, DNA is labeled with radioactive phosphorus (^{32}P), then divided into four samples. Recall that there are only four kinds of nucleotides in DNA (abbreviated A, T, C, and G). Each sample of the cloned genes is treated with a different chemical that attacks only one of the four kinds. The treatment is mild—only a single nucleotide is eliminated at a time—so fragments of different lengths are produced. Suppose the labeled genes have this sequence:

Random elimination of one A in different copies of the gene will produce this family of fragments:

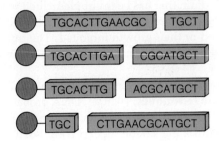

Similar families of DNA fragments are produced in the other samples. Each sample is subjected to gel electrophoresis (page 209). As they move through the gel, the fragments separate by size, which in this case corresponds to length. The final positions of the labeled fragments can be detected by autoradiography, and the nucleotide sequence can be deduced by a "countdown" through the banding pattern produced in all four samples (Figure 18.13).

Gene sequencer machines currently can sequence 10,000 nucleotides of chromosomal DNA on a daily basis. To sequence the estimated 3 billion nucleotides contained in all twenty-four types of human chromosomes, many researchers must collectively devote at least 1,000 years of effort (using the available technology), and at present no computer is big enough to hold all the sequence information that would be generated. Even so, many laboratories are already collaborating in the mapping attempt.

Some individuals question the wisdom of channeling so much research time into such a formidable task. Others say that once we have made such a staggering advance in our knowledge of the human genome, in all likelihood we will put it to use in ways that we cannot yet even dream of. It is also likely that the sequencing technology itself will advance rapidly.

GENETIC ENGINEERING: RISKS AND PROSPECTS

Genetically Engineered Bacteria

Paul Berg and his colleagues were the first to insert foreign DNA into a plasmid. That was in the mid-1970s, and their work made it clear that natural restrictions on the transfer of genes between different species could be bypassed in the laboratory. Could such transfers produce bacteria that would be dangerous to humans or to the environment? Exhaustive investigations carried out since indicate that the risks probably are small. The *E. coli* strains used in many genetic engineering experiments are not pathogenic and have been altered by mutation to prevent their survival outside the laboratory. Yet

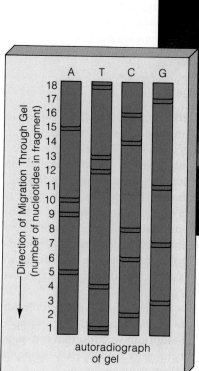

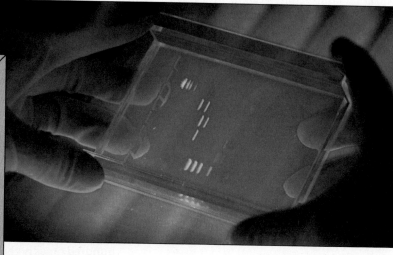

Figure 18.13 One method of gene sequencing. Four separate families of radioactively labeled DNA fragments are produced by chemicals that cleave A nucleotides in the first batch, T in the second, C in the third, and G in the fourth. The batches are put in four parallel lanes cut in the same gel, then subjected to electrophoresis.

In the photograph, DNA fragments have been stained. In the diagram to the left, they have been detected by autoradiography. The nucleotide closest to the start of the gene is the one that has migrated the farthest. Look at level "1" across the four lanes and you see that it is T for this particular gene. Look at level "2" and you see the next nucleotide is G, and so on. The entire sequence, read from bottom to top of the autoradiograph is TGCACTTGAACGCATGCT.

there is ongoing concern about releasing genetically engineered bacteria or other organisms into the environment, as the case of the "ice-minus" bacteria makes clear.

Pseudomonas syringae is a common bacterium that lives on leaves and stems, and it happens to make plants susceptible to frost damage. Proteins on the bacterial cell surface catalyze the formation of ice crystals—even when the air temperature is several degrees above freezing. If it were not for the presence of *P. syringae*, host plants would not freeze even at several degrees below 0°C, and fruits and vegetables would not be damaged by frosts. As it is, the bacterium significantly lowers the yields of many of our food crops.

Not long ago, Steven Lindow of the University of California at Berkeley genetically engineered a strain of *P. syringae* by excising the "ice-forming" gene. Cells of the modified strain were unable to synthesize the ice-forming protein. Lindow and others wanted to spray the so-called **ice-minus bacteria** on strawberry plants in an isolated field just before a frost. Doing so would tell them whether or not the plant cells would now resist freezing.

The proposed experiment only involved an organism from which a harmful gene had been *deleted*, yet it triggered a bitter legal debate over whether to permit the deliberate release of genetically engineered microbes. After several years of litigation, the courts ruled in favor

Figure 18.14 Spraying an experimental strawberry patch in California with "ice-minus" bacteria. (Government regulations required that the sprayer use elaborate protective gear.)

of allowing the experiment to proceed. In April 1987 a small patch of strawberries finally was sprayed (Figure 18.14). As most knowledgeable individuals had predicted, no ecological disaster followed—except a few environmental activists entered the patch at night and pulled up the plants.

The lessons of the ice-minus controversy are important. Rules governing the release of genetically engineered organisms have since been clarified. Environmental impact reports must be filed first. Perhaps most importantly, biotechnology companies have learned that they must communicate openly and effectively with the public about their work.

Genetically Engineered Plants

In the early 1980s, researchers succeeded in inserting DNA fragments into a plasmid from *Agrobacterium tumefaciens*, a bacterium that infects many flowering plants. This so-called Ti plasmid carries genes responsible for the formation of crown gall tumors of the sort shown in Figure 18.15. (The Ti stands for "tumor-inducing.")

The tumor-inducing genes become integrated into the DNA of plant cells that have become infected by the bacteria. When the plasmid is used as a vector, however, investigators first remove the tumor-causing genes and substitute other, desired genes. Plants are then propagated from single cells containing the new genes. In some cases, the foreign genes are expressed normally in the plant tissues and alter the plant's phenotype in some desired manner.

Research along these lines may eventually allow us to genetically engineer crop plants in ways that will help increase global food production. Millions of humans die each year from starvation. In the long run, simply increasing global food production to keep pace with the burgeoning human population is not a solution to the problem, as Chapter Forty-Three will make clear. However, it is one of the few short-term options available to us.

Consider the research interest in developing salt-tolerant crop plants alone. *Halophytes* are plants, including some species of barley and sugar beets, that are adapted to moderately salty environments. These crop plants can grow in soils far saltier than *glycophytes* ("sweet-water" plants) can tolerate. Almost all conventional crops are salt-sensitive. Can the genes of halophytes and glycophytes be recombined to produce salt-resistant, high-yield strains? The question is not trivial. Many of the world's agricultural regions must be irrigated to be productive. Enormous amounts of salt are brought in with the irrigation water, and typically they cannot be flushed from the soil because of the scarcity of rain in those regions. All over the world, croplands are becoming "salted out," which means the number of acres able to support existing crop species is declining rapidly. Thus each day in India alone, 100 hectares (250 acres) are being removed from agriculture.

Genetically Engineered Animals

Ralph Brinster, Richard Palmiter, and their colleagues succeeded in introducing the rat gene for somatotropin into fertilized mouse eggs in 1982. When the mice grew, it was clear that the rat gene had been integrated into the mouse DNA and was being expressed, for the mice were larger than their normal littermates. Their cells had up to thirty-five copies of the gene, and blood concentrations of the growth-promoting hormone were several hundred times higher than normal values. Since the time of that pioneering experiment, the gene for human somatotropin was successfully introduced and expressed in mice (Figure 18.1).

Similar experiments conducted with large, domesticated animals have not been successful. For example, when the gene for somatotropin was inserted into pigs, it was expressed—but the pigs developed a variety of disorders, including arthritis-like symptoms.

Gene modification in animals is extremely difficult. New or modified genes must be inserted either in somatic cells of the target animal or in gametes (that is, a sperm or an egg). They must be incorporated into specific chromosomal locations (for example, between an appropriate promoter and a termination signal). And they must not disrupt the function of other genes.

In one approach, genes are inserted into the sperm nucleus that has just penetrated an egg. Only one fertilized egg can be modified at a time, however, and the eggs are so vulnerable to being poked that the procedure has a high rate of failure. Even when the gene delivery is successful, researchers still cannot control *where* in the genome an inserted gene will become integrated. What happens if its integration activates an oncogene (page 231)? Or causes a mutation? Or alters expression of related genes? It is one thing to lose bacterial colonies to gene insertion mishaps; they can be readily and inexpensively replaced. It is quite another thing to use and lose large laboratory animals.

Human Gene Therapy

Inserting one or more normal genes into the somatic cells of an organism to correct a genetic defect is called **gene therapy**. The idea of doing this to alleviate the suffering of humans afflicted by severe genetic disorders seems to be socially acceptable at present, even though the technology by which gene therapy might be accomplished is still in the research stage.

Figure 18.15 Crown gall tumors on a willow tree, as caused by a tumor-inducing plasmid from *A. tumefaciens*.

By contrast, inserting genes into a normal human individual (or sperm or egg) in order to modify or enhance a particular trait is called many things, including **eugenic engineering**; but mostly it is called a horrifying idea. Who decides which traits are "desirable"? What if prospective parents could pick the gender of their children by way of genetic engineering? (Three-fourths of one recently surveyed group of people said they would choose a boy. What would be the long-term social implications of such a drastic shortage of girls?) If it is okay to engineer lighter skinned or blue-eyed or taller individuals, would it be okay to engineer "superhuman" offspring with exceptional strength or intelligence? Fortunately, perhaps, intelligence and most other traits arise through complex interactions among many genes and environmental factors, and this will put them outside the reach of genetic manipulation for some time.

We have only touched on some of the social and ethical issues raised by recombinant DNA technology and genetic engineering. Some individuals say that the integrity of the DNA of each species is inviolate and should not be tampered with. But as the earlier discussion should make clear, nature itself alters DNA much of the time. The real argument, of course, is whether we as a species have achieved the wisdom to bring about beneficial changes without causing harm to ourselves or to the environment.

When it comes to manipulating the human genome, one is reminded of our very human tendency to leap before we look. W. French Anderson, a biochemist and human geneticist, put it this way: "Our knowledge of how the human body works is still elementary. Our understanding of how the mind, both conscious and subconscious, functions is even more rudimentary. The genetic basis for instinctual behavior is largely unknown. Our disagreements about what constitutes 'humanhood' are notorious. And our insight into what, and to what extent, genetic components might play a role in what we comprehend as our 'spiritual' side is almost nonexistent. We simply should not meddle in areas where we are so ignorant."

Yet, when it comes to the prospect of prohibiting genetic engineering of any sort, one also is reminded of an old saying: "If God had wanted us to fly, he would have given us wings." Certainly when a plane crashes and everyone on board is killed, we wonder what we are doing up in the air in the first place. But something about the human experience gave us the *capacity* to imagine wings of our own making—and that capacity carried us to the frontiers of space.

Where are we going from here with recombinant DNA technology, this new product of our imagination? To gain perspective on the question, spend some time reading the history of our species. It is a history of survival in the face of all manner of threats, of expansions, of bumblings, and sometimes of disasters on a grand scale. It is also a story of increasingly intertwined interactions with the environment and with one another. The questions confronting you today are these: Should we be more cautious, believing that one day the risk takers may go too far? And what do we as a species stand to lose if the risks are *not* taken?

SUMMARY

1. Genetic "experiments" have been occurring in nature for billions of years. Mutations, homologous recombination at meiosis, the novel assortments of alleles brought together at fertilization, and hybridizations between species have all contributed to the current diversity among organisms.

2. Genetic changes also occur naturally through the activity of transposable elements, including transposons, plasmids, and viral DNA. These elements have the means to insert themselves into another DNA molecule (or another location in the same molecule) through recombination.

3. Humans have been manipulating the genetic character of different species for ten thousand years or more. The emergence of recombinant DNA technology in the past few decades has enormously expanded our capacity to cause genetic change. *Recombinant DNA technology* is

founded on procedures by which DNA molecules can be cut into fragments, inserted into cloning vectors (plasmids, bacteriophage DNA), and propagated in microorganisms.

4. Special bacterial enzymes, called *restriction enzymes*, make cuts at recognition sequences, these being short sequences of specific nucleotides that the enzyme identifies as its substrate. (An example is EcoRI, which makes a cut wherever the sequence GAATTC occurs in a DNA molecule.) Another enzyme, *DNA ligase*, can join any two DNA fragments that have been produced by the same restriction enzyme.

5. A *cloning vector* is a plasmid, virus, or any other self-replicating genetic element that can be inserted into a host cell for propagation. A collection of DNA fragments produced by restriction enzymes and incorporated into cloning vectors and amplified in cells is called a *DNA library*.

6. A *DNA clone* is any DNA sequence that has been amplified in dividing cells. DNA sequences also can be amplified in vitro (in test tubes) by the polymerase chain reaction method.

7. Recombinant DNA technology and genetic engineering have enormous potential for research and applications in medicine, agriculture, and home and industry. As with any new technology, the potential benefits must be weighed carefully against the potential risks, including ecological and social disruptions. Although the new technology has not developed to the extent that the human genome can be modified, it seems appropriate that the social, legal, ecological, and ethical questions should be explored in detail before such an application is possible.

Review Questions

1. Explain homologous recombination. Does this mechanism commonly occur in nature? *235, 237*

2. What is a plasmid? Which of its structural features lend themselves to DNA cloning work? *236, 238*

3. What is a restriction enzyme? Do such enzymes occur naturally in organisms? *239*

4. Recombinant DNA technology involves the following:
   a. Producing DNA restriction fragments.
   b. Preparing and cloning a DNA library.
   c. Identifying a desired gene in the cloned DNA.
   d. Determining the nucleotide sequence of particular genes.

Briefly describe one of the methods used in each of these four categories. *239–242*

5. Having read about some of the examples of genetic engineering in this chapter, can you think of some additional potential benefits of this technology? Can you envision other potential problems?

Readings

Anderson, W. F. 1985. "Human Gene Therapy: Scientific and Ethical Considerations." *Journal of Medicine and Philosophy* 10:274–291.

Brill, W. 1985. "Safety Concerns and Genetic Engineering in Agriculture." *Science* 227:381–384.

Primrose, S. 1987. *Modern Biotechnology*. Boston: Blackwell. Clear introduction to genetic engineering. Paperback.

Palmiter, R., et al. 1983. "Metallothionein–Human GH Fusion Genes Stimulate Growth of Mice." *Science* 222:809–814. Report on landmark experiments in mammalian gene transfers.

Watson, J., et al. 1987. *Molecular Biology of the Gene*. Fourth edition, volume 1. Menlo Park, California: Benjamin/Cummings. Advanced topics, but the writing is exceptionally easy to follow.

UNIT FOUR

PLANT SYSTEMS AND THEIR CONTROL

19

PLANT CELLS, TISSUES, AND SYSTEMS

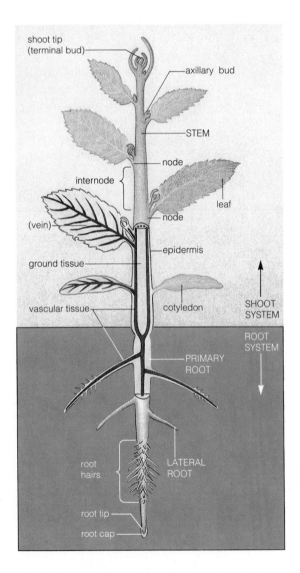

Figure 19.1 Body plan for one type of vascular plant. Some vascular tissues by which water and nutrients move through the plant body are shown in red. Notice how they thread through the ground tissue (shaded yellow), which makes up the bulk of the plant body. Dermal tissue (not shown) forms a covering over all parts of the shoot and root systems.

On a summer morning in 1883, a cataclysmic explosion blew apart the Indonesian island of Krakatoa, and life there ended abruptly. All that remained beneath the hot ashes and pumice was the jagged peak of a volcanic cone, which continued to exude lava. For about a year this smoldering remnant of the island was essentially sterile. Yet even then, winds and water were carrying seeds and spores to it from nearby islands—and half a century after the explosion, a dense forest cloaked the "new" Krakatoa.

In the spring of 1980, Mount St. Helens in southwestern Washington exploded violently, and within minutes, hundreds of thousands of mature trees near the volcano's northern flank were blown down like matchsticks. Thick ashes and pumice turned the previously forested region into a barren sweep of land.

Events of this magnitude dramatize what the world would be like without plants, reminding us that we could no more survive without them here than on the rock-strewn surface of the moon. What characterizes these remarkable organisms, which directly or indirectly nourish all others and make the land habitable? Can we identify patterns of structural organization among them? Do plants, like animals, have intricate systems for circulation, gas exchange, and nutrition? How do plants reproduce? What governs their growth and development? These are questions addressed in this unit.

THE PLANT BODY: AN OVERVIEW

There are more than 275,000 species of plants, and no one species can be used as a "typical example" of their body plans. Plants live in fresh water, in seawater, on land, even high above a forest floor (attached to other plants), and their features are correspondingly diverse. In size alone, plants range from microscopic algae to giant redwoods. Most are **vascular plants**, with well-developed conducting tissues through which water and solutes are transported within the plant body (Figure 19.1). There are fewer than 30,000 species of **nonvascular plants** (the red, brown, and green algae and the bryophytes); either they have no internal transport systems or they have very simple ones.

The most familiar vascular plants are the angiosperms and gymnosperms. **Angiosperms** are flowering plants, such as roses, apple trees, and corn. Besides producing flowers (reproductive structures), angiosperms produce seeds that are completely enclosed in protective tissue layers. **Gymnosperms** are chiefly conifers, such as pine trees and junipers. They produce "naked" seeds that are borne on surfaces of reproductive structures rather than surrounded by tissues. The majority of vascular plants are angiosperms, and they will be our focus here.

Monocots and Dicots

There are two classes of flowering plants, referred to informally as **monocots** and **dicots**. The grasses, lilies, orchids, irises, cattails, and palms are examples of monocots. Nearly all familiar trees and shrubs (other than conifers) are dicots.

Monocots and dicots are similar in their structure and functioning, but they differ in some distinctive ways. For example, monocot seeds have one cotyledon and dicot seeds have two. A "cotyledon" is a leaflike structure originating in the seed, as part of the plant embryo. After the seed germinates (starts to grow), the cotyledons may unfurl somewhere along the length of the tiny seedling. Figure 19.2 shows other differences between monocots and dicots.

Shoot and Root Systems

Flowering plants typically have well-developed shoot and root systems (Figure 19.1). The **shoot system** consists of stems and leaves that usually grow above ground. Its vascular tissues conduct water, minerals, and organic substances between roots, leaves, and other plant parts. The stems serve as frameworks for upright growth (which gives photosynthetic tissues in leaves favorable exposure to light) and for displaying the flowers. Some parts of the system store food.

The **root system** usually consists of parts that are below ground. Its main functions are to absorb water and dissolved minerals from soil and to conduct water and solutes to aerial plant parts. The root system also stores food and anchors (and sometimes structurally supports) the plant.

The tissues of root and shoot systems start to form in the embryo, within the seed. (A plant *tissue* is a group of cells interacting in a specialized task, such as water conduction.) A later chapter describes plant embryonic development. Here, the main point is that when the seed germinates, cells divide and elongate at the tips of the first (primary) root and shoot. The growth originating at root and shoot tips is called **primary growth**, and the parts of the plant body arising from this activity are called "primary" tissues. Many monocots and dicots die after one season of primary growth. Others show primary growth through many seasons; the growth of a "new plant" from an iris bulb each spring is an example.

Many dicots and a *few* monocots (such as palms) also undergo **secondary growth**, which originates at sites other than the root and shoot tips. The resulting "sec-

Figure 19.2 Main differences between monocots and dicots.

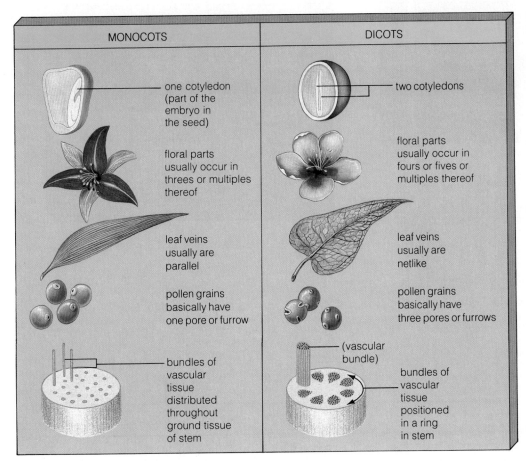

MONOCOTS	DICOTS
one cotyledon (part of the embryo in the seed)	two cotyledons
floral parts usually occur in threes or multiples thereof	floral parts usually occur in fours or fives or multiples thereof
leaf veins usually are parallel	leaf veins usually are netlike
pollen grains basically have one pore or furrow	pollen grains basically have three pores or furrows
bundles of vascular tissue distributed throughout ground tissue of stem	(vascular bundle) bundles of vascular tissue positioned in a ring in stem

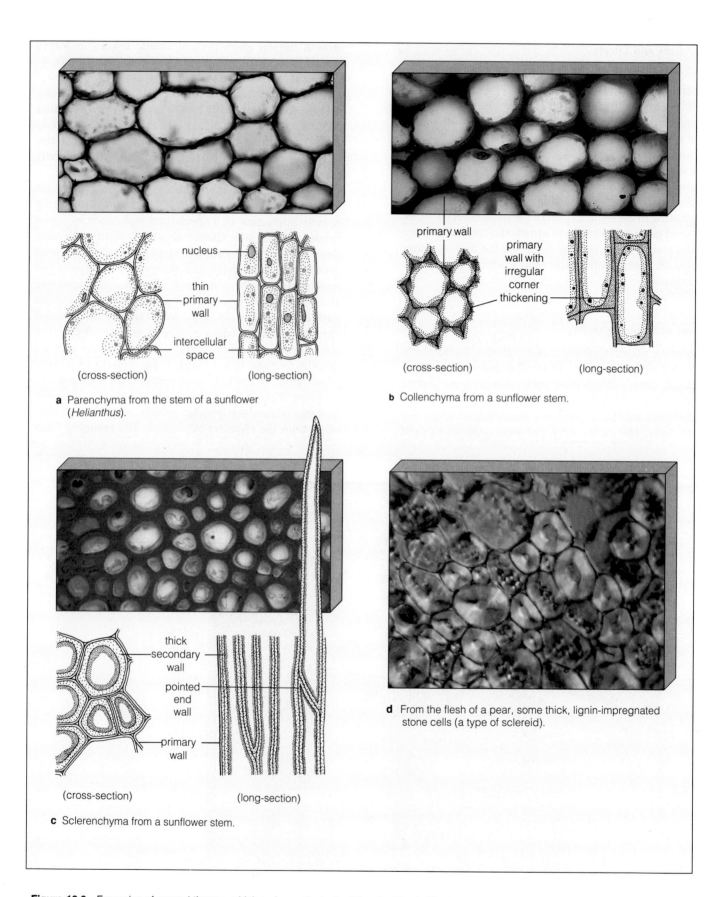

a Parenchyma from the stem of a sunflower (*Helianthus*).

nucleus

thin primary wall

intercellular space

(cross-section)　(long-section)

b Collenchyma from a sunflower stem.

primary wall

primary wall with irregular corner thickening

(cross-section)　(long-section)

thick secondary wall

pointed end wall

primary wall

(cross-section)　(long-section)

c Sclerenchyma from a sunflower stem.

d From the flesh of a pear, some thick, lignin-impregnated stone cells (a type of sclereid).

Figure 19.3 Examples of ground tissues, which make up the bulk of the plant body. The most common types are parenchyma, collenchyma, and sclerenchyma. (The stone cells in **d** are a specialized form of sclerenchyma.)

ondary" tissues increase the diameter (girth) of older roots and stems. Thus each spring, a maple tree puts on primary growth at its root and shoot tips, and secondary growth adds to its woody parts.

Let's turn now to the three types of tissues of flowering plants. **Ground tissues** called parenchyma, collenchyma, and sclerenchyma make up the bulk of the primary plant body. **Vascular tissues** (xylem and phloem) thread through the ground tissue system, and **dermal tissues** form a protective covering for the plant.

PLANT TISSUES

Ground Tissues

Parenchyma is a tissue composed of living, generally thin-walled parenchyma cells (Figure 19.3a). Its cells are the most abundant type of cell in the plant body; they are massed together in stems, roots, leaves, flowers, and the flesh of fruits. Mature parenchyma cells function in healing wounds and often in regenerating plant parts. As you will see, parenchyma cells are involved in photosynthesis, storage, secretion, and other tasks.

Collenchyma, a tissue that helps strengthen the plant, commonly occurs as strands or cylinders beneath the dermal tissue of stem and leaf stalks. (For example, this tissue is a component of the pliable "strings" of a celery stalk.) The cells of collenchyma are alive at maturity. The primary cell walls become thickened with cellulose and pectin, often at their corners (Figure 19.3b). Bonding interactions between the two substances make collenchyma quite pliable.

Sclerenchyma is a tissue that strengthens mature plant parts. Sclerenchyma cells have thick secondary walls that are usually impregnated with lignin, a polymer that helps make the wall rigid. The cells commonly form strands or sheets, but they also may be scattered among other cell types. Some sclerenchyma cells, called *fibers*, are long and tapered (Figure 19.3c). The fibers of hemp and flax are used to produce paper, textiles, thread, and rope. Other sclerenchyma cells, the *sclereids*, are components of seed coats and nut shells; they also give pears their gritty texture (Figure 19.3d).

Vascular Tissues

The two kinds of vascular tissue are called **xylem** and **phloem**. Both kinds contain specialized conducting cells, some fibers, and some parenchyma cells.

Xylem. Xylem conducts water (and dissolved minerals) absorbed from the soil; it also mechanically supports the plant. Its main water-conducting cells are *tracheids* and *vessel members* (Figure 19.4). Tracheids are components

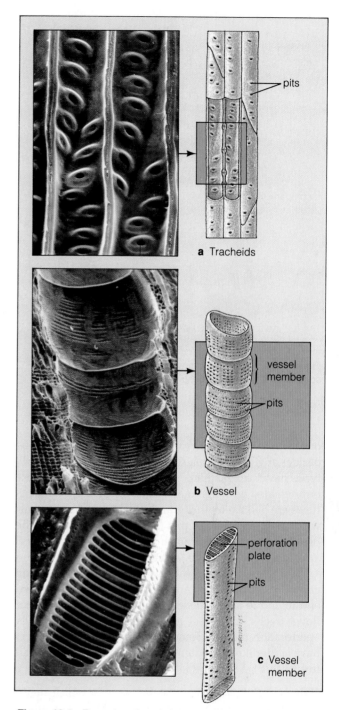

a Tracheids

b Vessel

c Vessel member

Figure 19.4 Examples of tracheids and vessel members, the main cells in xylem that conduct water and dissolved mineral salts through the plant body.

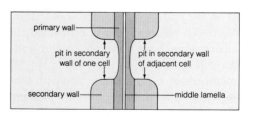

Figure 19.5 A simple pit-pair (long-section).

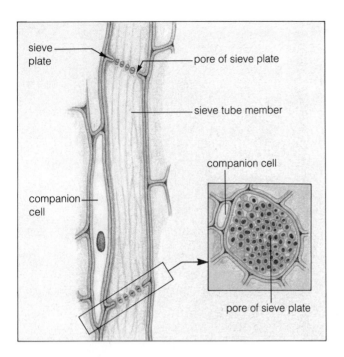

Figure 19.6 Mature sieve tube member and adjacent companion cell. The inset shows one type of sieve plate.

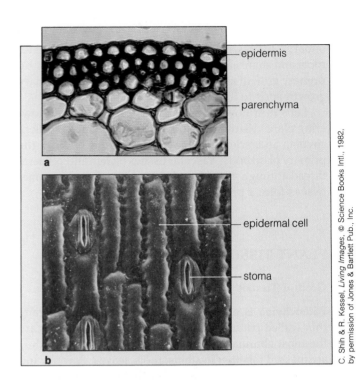

Figure 19.7 Epidermis of corn *(Zea mays)*: transverse section of stem (**a**) and surface of lower leaf (**b**).

of xylem in nearly all vascular plants, but few plants besides angiosperms also have vessel members. Both cell types have strong, thick, multilayered walls of cellulose impregnated with lignin and other substances. Both cell types are dead at maturity, and all that remains are walls with many recesses, or pits (Figure 19.5). Typically, the pits of adjacent walls match up, and these "pit-pairs" are highly permeable to water.

Tracheids are long cells with tapered, overlapping ends. Vessel members are shorter cells that are joined end to end to form a *vessel*, a tube through which water flows freely. All vessel members have open ends, or "perforation plates." Some are completely open; others have ladderlike bars extending across the open end (Figure 19.4c).

Phloem. Phloem is the vascular tissue by which sugars and other solutes are rapidly transported throughout the plant body. Its main conducting cells are alive at maturity. In angiosperms, the conducting cells are *sieve tube members*; in gymnosperms and ferns, they are simply *sieve cells*. The term "sieve" refers to clusters of pores that are located on the end walls and side walls. Here, the cytoplasm between adjacent cells is connected. In sieve tube members alone, large pores on the end walls form "sieve plates" (Figure 19.6).

Sieve tube members are functionally linked with *companion cells*, which are adjacent to them in the phloem. Companion cells have accessory roles in moving sugars

from regions of photosynthesis to other plant parts by way of the phloem. Mature sieve tube members no longer have a functional nucleus, but the companion cell nucleus may direct activities of both cells.

Dermal Tissues

A continuous layer of tightly packed cells, the **epidermis**, covers the primary plant body (Figure 19.7). On above-ground parts, waxes and cutin impregnate the outer walls of epidermal cells to form a noncellular surface coating called a *cuticle*. A cuticle restricts water loss and might confer some resistance to microbial attack.

Epidermis often contains highly specialized cells. For example, *root hair cells* are thin-walled epidermal cells of roots (page 270). They have long protuberances that increase the cell surface and thereby enhance absorption of water and nutrients from the soil. *Guard cells*, to be described shortly, span the epidermis and play roles in governing water loss from leaves or stems and the movement of carbon dioxide into them.

A protective cover replaces the epidermis when roots and stems undergo secondary growth. This cover, the **periderm**, consists of an outermost cork tissue, then a tissue called cork cambium, and an inner tissue of parenchyma cells. The cells of cork tissue are not alive at maturity, but the cell walls are impregnated with su-berin, a waxy secretion that functions in waterproofing.

How Plant Tissues Arise: The Meristems

As in animals, cell division, enlargement, and differentiation give rise to the specialized cells and tissues of the plant body. In animals, however, nearly all cells are committed to being one type of cell only. In contrast, plants have many "meristems," or perpetually young tissue regions where some cells retain the capacity to divide again and again. The shoot tips that give rise to the new blossoms and new leaves of a fruit tree each spring are like this.

Primary growth originates in meristems of shoot and root tips. Each shoot tip has a dome-shaped **apical meristem**. Some of its cells retain the capacity to divide repeatedly. Others produce daughter cells that become committed to producing three kinds of **primary meristematic tissues**, the cells of which divide and differentiate into the primary tissues of the plant body:

protoderm	*gives rise to the epidermis*
ground meristem	*gives rise to the ground tissue*
procambium	*gives rise to primary xylem and phloem*

Secondary growth, which increases the diameter of older stems and roots, originates at two types of **lateral meristems**:

vascular cambium	*gives rise to secondary xylem and phloem after the primary plant body has formed*
cork cambium	*gives rise to periderm (the protective covering that replaces epidermis)*

Figure 19.8 shows the location of meristems in a dicot stem. All of its cells are derived originally from apical meristem at its tip.

The tissue organization of stems, leaves, and roots will be described next. When looking at the photographs accompanying the text, keep in mind the following terms, which identify the way a given tissue specimen was cut from the plant:

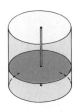
transverse

(cross-section cut perpendicular to long axis of root or stem)

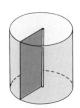
radial

(longitudinal section cut parallel with the radius of root or stem)

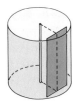
tangential

(longitudinal section cut at right angles to radius of root or stem)

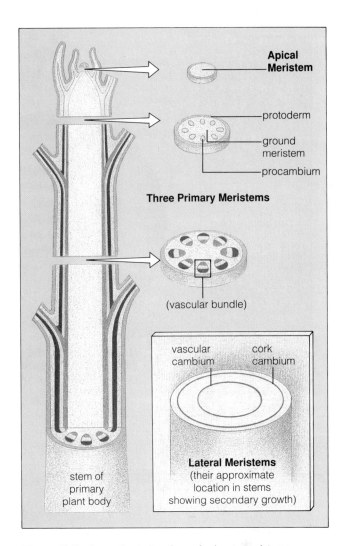

Figure 19.8 Approximate locations of primary meristems (yellow) and lateral meristems (red) in plants that show both primary and secondary growth.

THE PRIMARY SHOOT SYSTEM

Stem Primary Structure

Flowering plants show considerable variation in their shoot systems. Even so, the primary tissues of the stem usually are organized in one of two patterns, based on the distribution of vascular bundles. A **vascular bundle** is a strand of vascular tissue that contains primary xylem and phloem. Commonly, some fibers and parenchyma cells form a sheath around the strand. In most monocots and some dicots, the vascular bundles are distributed throughout the ground tissue (Figure 19.9). In conifers and most dicots, the vascular bundles are arranged as a ring that divides the ground tissue into *cortex* and *pith* (Figure 19.10).

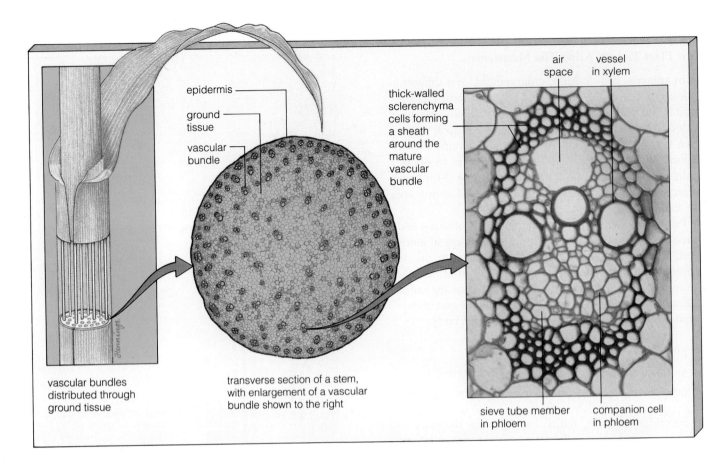

Figure 19.9 Stem structure of corn *(Zea mays)*, a monocot.

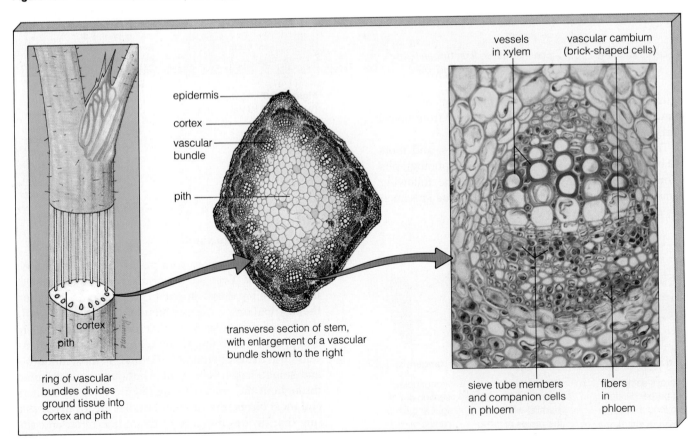

Figure 19.10 Stem structure of alfalfa *(Medicago sativa)*, a dicot.

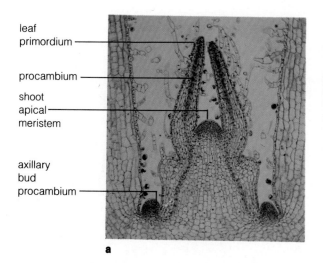

leaf
primordium

procambium

shoot
apical
meristem

axillary
bud
procambium

a

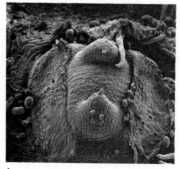

b

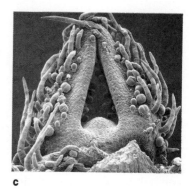

c

Figure 19.11 (**a**) Leaf primordia in the shoot tip of *Coleus*, longitudinal section. (**b,c**) Scanning electron micrographs of leaf development in the same plant.

Formation of Leaves and Buds

Leaves develop from lateral outgrowths of apical meristem that are called **leaf primordia** (singular, primordium). At first, leaf primordia are closely spaced (Figure 19.11), but gradually, cell elongation causes the stem to lengthen between nodes. Each *node* is a point on the stem where one or more leaves are attached. Each stem region between two successive nodes is an *internode* (Figure 19.12).

Meristematic cells also become organized into **axillary bud primordia**, the beginnings of lateral branchings of a stem. As each bud primordia develops, its apical meristem forms either a terminal bud of a branch or a flower. Buds form only at nodes, in the upper angles (axils) where leaves are attached to the stem.

Leaf Structure

Monocot and Dicot Leaves. Each node of a stem may have one or more **leaves**, which usually are sites of photosynthesis. Many dicot leaves have two parts: a broad *blade* and a stalklike *petiole* attached to the stem (Figure 19.13a). Most monocot leaves have no petiole; the base of the blade simply encircles the stem, forming a sheath. Corn is an example (Figure 19.13b).

A "simple" leaf has only one blade, although in some species it is deeply lobed (Figure 19.13c). In a "compound" leaf the blade is divided into smaller units (leaflets), each of which may be attached to the petiole by a small stalk (Figure 19.13d).

There are numerous variations on these basic leaf plans. For example, some leaves have hairs and scales, others have hooks that impale predators. The two-lobed

a

c

b

d

Figure 19.12 Terminal bud of a dogwood tree showing its development into leaves and the dogwood "flower."

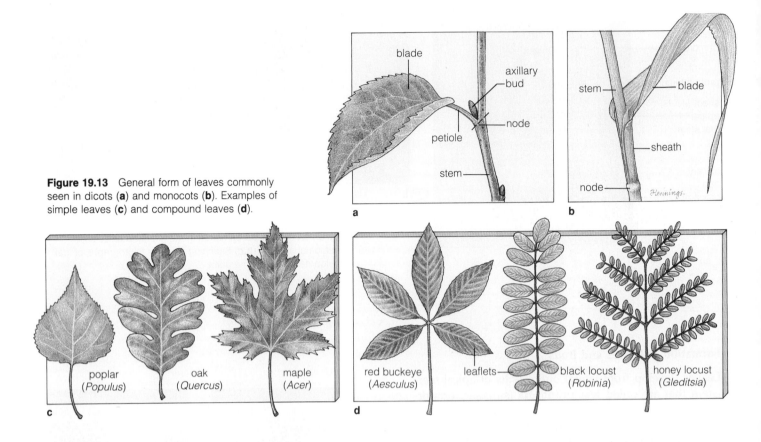

Figure 19.13 General form of leaves commonly seen in dicots (**a**) and monocots (**b**). Examples of simple leaves (**c**) and compound leaves (**d**).

a: blade, axillary bud, node, petiole, stem

b: stem, blade, sheath, node

c: poplar (*Populus*), oak (*Quercus*), maple (*Acer*)

d: red buckeye (*Aesculus*), leaflets, black locust (*Robinia*), honey locust (*Gleditsia*)

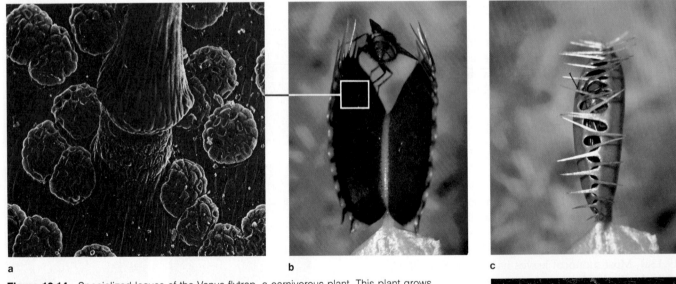

a

b

c

Figure 19.14 Specialized leaves of the Venus flytrap, a carnivorous plant. This plant grows in nitrogen-poor soil. Its two-lobed leaves open and close like a clamshell; the spines fringing the leaf margins intermesh when the lobes close.

Suppose an insect lands on the leaf and moves against one of its long epidermal hairs (the base of one is shown in **a**). The movement triggers cellular changes at the leaf midrib, and the leaf closes (**b, c**). Glandlike epidermal cells (the pincushion-like structures in the micrograph) secrete enzymes that digest proteins of the trapped insect body. Nitrogen released from the proteins is used by the plant in biosynthesis.

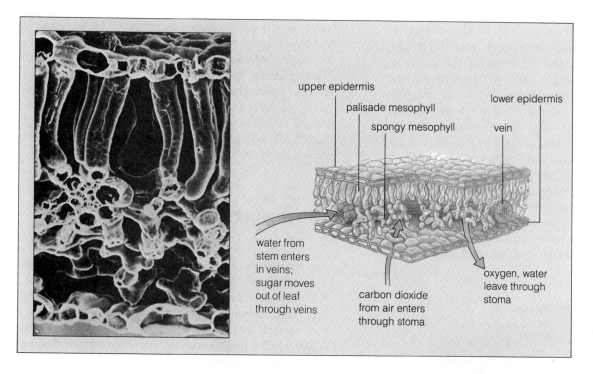

Figure 19.15 Example of leaf internal structure. The scanning electron micrograph shows a mature broadbean leaf, transverse section. The sketch identifies the different leaf cells.

Labels in figure: upper epidermis, palisade mesophyll, spongy mesophyll, lower epidermis, vein; water from stem enters in veins; sugar moves out of leaf through veins; carbon dioxide from air enters through stoma; oxygen, water leave through stoma

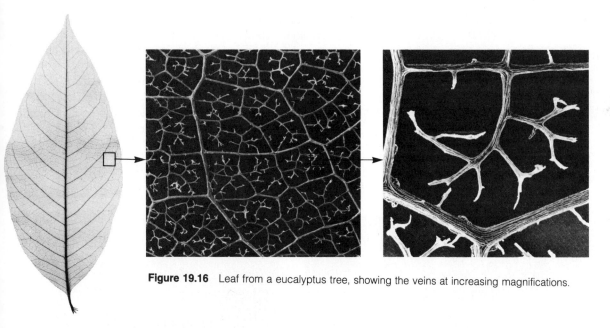

Figure 19.16 Leaf from a eucalyptus tree, showing the veins at increasing magnifications.

leaf of the Venus flytrap even turns the table on plant-eating animals, so to speak (Figure 19.14).

Regardless of the diversity, most leaves are alike in being short-lived. "Deciduous" species such as birches are periodically devoid of leaves, which drop away from the stem as winter approaches. Species such as camellias also drop leaves, but they appear "evergreen" because the leaves do not all drop at the same time.

Leaf Internal Structure. The leaves just described have a large surface area exposed to sunlight and to carbon dioxide in the air. Inside the leaf, the membranes of photosynthetic cells collectively represent a large surface area for sunlight reception and gas exchange. A network of **veins** (vascular bundles) through the leaf moves water and solutes to photosynthetic cells and carries products from them (Figures 19.15 and 19.16).

Figure 19.15 shows the tissue layers common to many leaves. Uppermost is a protective epidermis, with cuticle covering its outer surface. Next comes the **palisade mesophyll**, a loosely packed tissue of parenchyma cells that are capable of photosynthesis. Below the palisade tissue is the **spongy mesophyll**: even more loosely packed parenchyma cells that also are photosynthetic. Between thirty and fifty percent of a leaf consists of air spaces around the spongy mesophyll and around most of the palisade cell walls. Below the spongy mesophyll is another cuticle-covered epidermal layer.

The lower epidermal layer usually contains most of the tiny openings through which water vapor moves out of leaves and carbon dioxide enters them. Each opening is a **stoma** (plural, stomata). It is defined by two **guard cells**, which are illustrated in Figure 19.17. Guard cells can swell under turgor pressure (page 273). When they do, their shape is so distorted that the opening between them widens. When turgor pressure drops, the cells become flaccid (limp); then, there is no opening between them. Stomata open and close in response to environmental conditions, thereby enhancing the movement of carbon dioxide into the plant and limiting water loss during drought or at night.

THE PRIMARY ROOT SYSTEM

Figure 19.1 showed one type of primary root system. Any root system must absorb enough water and dissolved minerals to sustain plant growth and metabolism. If the detailed measurements of one rye plant are any indication, adequate absorption requires a tremendous root surface area. By the time that rye plant was four months old it had developed a root system—including root hairs—with a surface area of 639 square meters (764 square yards), about 130 times more extensive than the surface area of its shoot system!

The root system penetrates downward and spreads out laterally, anchoring the aboveground parts. Most roots also function as storage sites for photosynthetically produced food, some of which is used by root cells and some of which is later transported back to aboveground parts. For example, carrots and sugar beets require two growing seasons to complete the life cycle. During the first growing season, food is stockpiled in roots. During the second, stored food is tapped for the formation of flowers, fruits, and seeds.

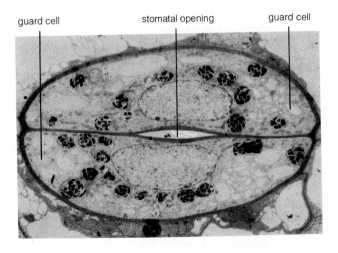

guard cell   stomatal opening   guard cell

Figure 19.17 Paired guard cells defining a stoma on the stem of a beavertail cactus *(Opuntia)*.

a   b

Figure 19.18 (**a**) Taproot system of a dandelion. (**b**) Fibrous root system of a grass plant.

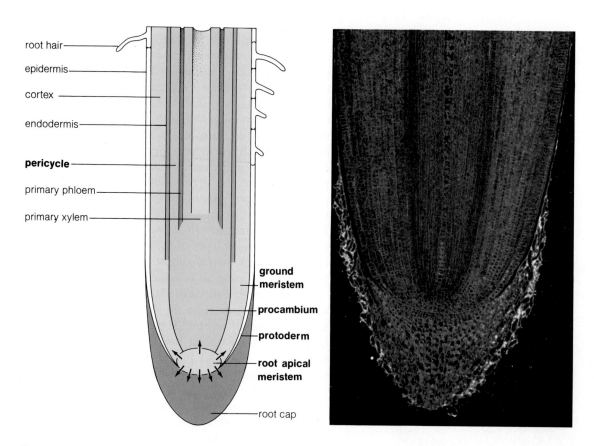

Figure 19.19 Generalized root tip, sliced lengthwise. The micrograph shows a corn root tip.

Taproot and Fibrous Root Systems

The first plant root that emerges upon germination is the **primary root**. In most dicots and gymnosperm seedlings, the primary root increases in diameter and grows downward, and younger branchings called **lateral roots** emerge sideways along its length. The youngest branchings are near the root tip. A primary root and its lateral branchings are a **taproot system**. Figure 19.18a shows an example.

A carrot has a taproot system. So does a pine tree, the roots of which can penetrate the soil to depths of six meters (about six and a half yards) and more. In these plants, the taproot is quite distinct from its branchings. In other species, the branchings enlarge and the primary root stops growing, so in time the roots superficially look alike.

Generally, the primary root is short-lived in monocots such as grasses. In its place, numerous **adventitious roots** arise from the stem of the young plant. (The term "adventitious" refers to any structure arising at an unusual location, such as roots that grow from stems or leaves.) Adventitious roots and their branchings are all somewhat alike in length and diameter, and they form a **fibrous root system** (Figure 19.18b). Generally, fibrous root systems do not penetrate the soil as deeply as do taproots.

Root Primary Structure

Cells of root tips divide, elongate, and mature in different zones. They divide at the apical meristem and just behind it, where differentiation starts. Cells elongate in the next few millimeters of tissue; past that region they may mature further, as when they develop thick walls, but they do not grow longer.

Root Cap. At the root tip is a dome-shaped cell mass, the **root cap** (Figure 19.19). The root apical meristem produces the root cap and in turn is protected by it. As the root elongates, the root cap is pushed forward and some of its cells are torn loose. The slippery remnants lubricate the cap and enhance its penetration of the soil.

Root Epidermis. The epidermis, ground tissue, and vascular tissues form behind the root cap as the root elongates. **Root epidermis** is the absorptive interface with the environment. In the region behind the elongating portion of a root, some epidermal cells send out the long protuberances called root hairs (Figures 19.19 and 20.5). Root hairs greatly increase the surface available for taking up water and solutes. That is why you should never yank a plant out of the ground when transplanting it; doing so would tear off too much of this fragile absorptive surface.

Vascular Column. Most often, the vascular tissues of a root are arranged as a central cylinder, or **vascular column**. The column is surrounded by ground tissue called the root cortex (Figure 19.20). In corn and some other species, the vascular tissues are arranged as a ring that divides the ground tissue into cortex and pith (Figure 19.21). Abundant air spaces in the ground tissue allow oxygen to reach the living root cells, which depend on it for aerobic respiration. Also, cortical cells have many plasmodesmata, the cell junctions that connect the cytoplasm of adjacent plant cells (page 82). Water entering the root can move from cell to cell by way of plasmodesmata or along the cell walls—that is, until it reaches the innermost cell layer of the root cortex. That layer, the **endodermis**, helps control the movement of water and dissolved minerals into the vascular column.

Think of each walled cell of the endodermis as having six sides. One side faces the cortex, the opposite side faces the vascular column (Figure 19.20). The remaining four sides press against neighboring endodermal cell walls—and embedded in the walls of these abutting sides is a thin strip of waxy suberin deposits, called the *Casparian strip*.

The Casparian strip prevents water and minerals from moving indiscriminately into the vascular column, because water cannot penetrate through adjoining walls of the endodermis. To get into the vascular column, water and minerals must pass through the cytoplasm of

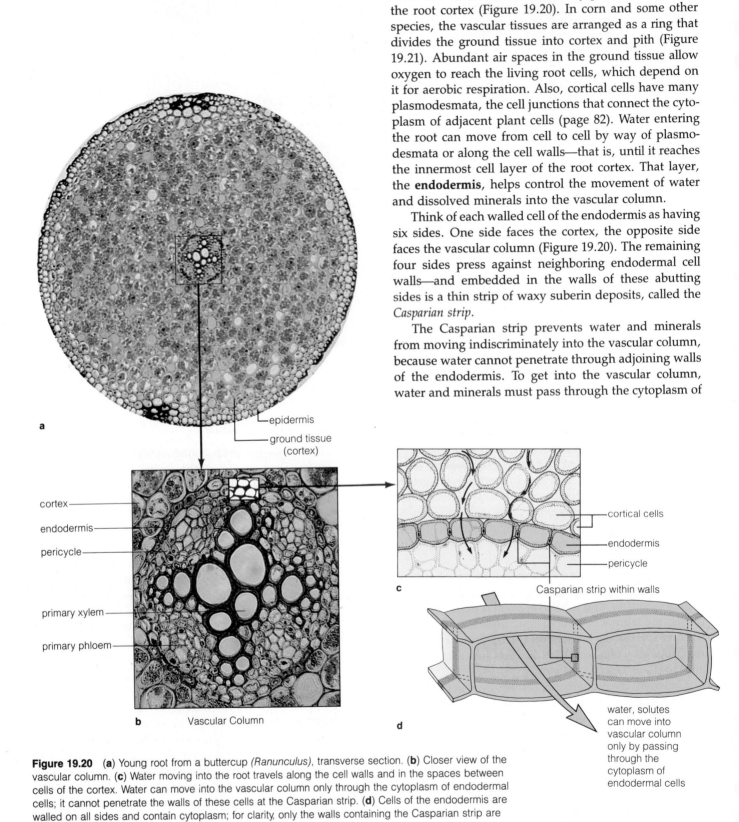

a

epidermis

ground tissue
(cortex)

cortex

endodermis

pericycle

primary xylem

primary phloem

b   Vascular Column

cortical cells

endodermis

pericycle

c

Casparian strip within walls

d

water, solutes can move into vascular column only by passing through the cytoplasm of endodermal cells

Figure 19.20 (**a**) Young root from a buttercup *(Ranunculus)*, transverse section. (**b**) Closer view of the vascular column. (**c**) Water moving into the root travels along the cell walls and in the spaces between cells of the cortex. Water can move into the vascular column only through the cytoplasm of endodermal cells; it cannot penetrate the walls of these cells at the Casparian strip. (**d**) Cells of the endodermis are walled on all sides and contain cytoplasm; for clarity, only the walls containing the Casparian strip are shown here.

endodermal cells (Figure 19.20d). This means they must cross the plasma membrane, with all of its built-in transport mechanisms. *Membrane transport mechanisms enable the endodermis to function as a control point, where different substances can be selectively transported into the vascular column or barred from it.*

Just inside the endodermis is the **pericycle**. This part of the vascular column consists of one or more layers of parenchyma cells that maintain a high degree of meristematic potential. For example, the pericycle gives rise to lateral roots, which grow outward through the cortex and epidermis (Figure 19.22). In species showing secondary growth, the pericycle also contributes to the formation of the vascular and cork cambia, the meristems responsible for increasing the diameter of stems and roots.

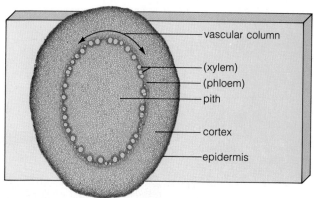

Figure 19.21 Root from a corn plant *(Zea mays)*, transverse section.

SECONDARY GROWTH

Seasonal Growth Cycles

The life cycle of flowering plants extends from seed germination to seed formation, then eventual death. Most monocots and some dicots are called nonwoody, **herbaceous plants** because they show little or no secondary growth during the life cycle. In contrast, many dicots and all gymnosperms are called **woody plants** because they show secondary growth during two or more growing seasons. Herbaceous and woody plants are characterized as follows:

annuals	*life cycle completed in one growing season; little (if any) secondary growth. Examples: snap beans, corn, marigolds.*
biennials	*life cycle completed in two growing seasons (root, stem, leaf formation the first season; flowering, seed formation, death the second). Examples: carrots, foxgloves.*
perennials	*vegetative growth and seed formation continue year after year. Some have secondary tissues, others do not. Examples: the herbaceous cacti, woody shrubs (roses), vines (ivy, grape), and trees (apples, elms, magnolias).*

Vascular Cambium Activity

In species showing secondary growth, stems and roots increase in diameter when lateral meristems (vascular cambium and cork cambium) become active. When fully developed, the vascular cambium is like a cylinder, one or a few cells thick. Its meristematic cells (called "initials") give rise to secondary xylem and phloem that conduct water vertically and horizontally through the enlarging stem or root (Figure 19.23).

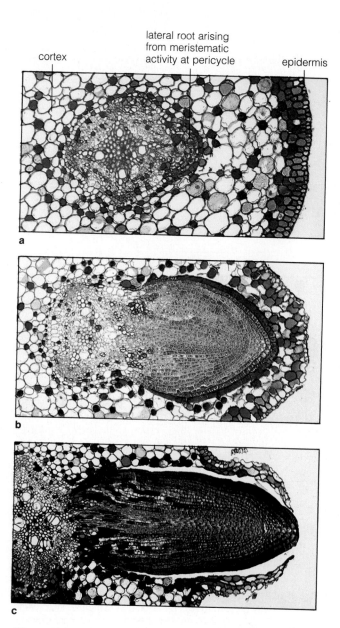

Figure 19.22 Lateral root formation in a willow *(Salix).*

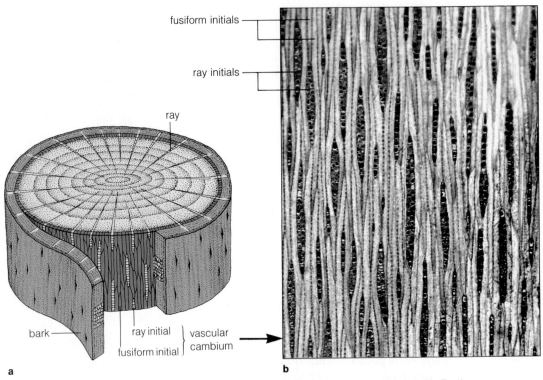

fusiform initials

ray initials

ray

bark

ray initial

fusiform initial

vascular cambium

a

b

Figure 19.23 (**a**) Location of vascular cambium in an older stem showing secondary growth. Fusiform initials (meristematic cells) produce the "vertical system" of secondary xylem and phloem, which conducts water and food up and down the stem. Ray initials (other meristematic cells) produce the "ray system" of mostly parenchyma cells that act as lateral conduits for water and as food storage centers in wood. (**b**) Tangential section through the vascular cambium in the stem of an apple tree *(Malus)*, showing the fusiform initials and ray initials.

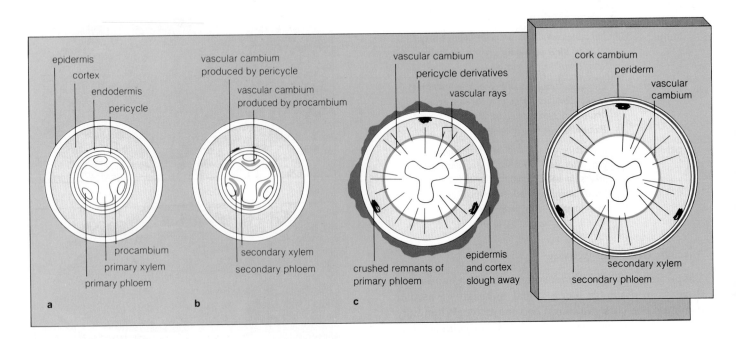

epidermis

cortex

endodermis

pericycle

procambium

primary xylem

primary phloem

a

vascular cambium produced by pericycle

vascular cambium produced by procambium

secondary xylem

secondary phloem

b

vascular cambium

pericycle derivatives

vascular rays

crushed remnants of primary phloem

epidermis and cortex slough away

c

cork cambium

periderm

vascular cambium

secondary xylem

secondary phloem

Figure 19.24 Secondary growth in a dicot root, transverse section. (**a**) Arrangement of tissues at the end of primary growth. (**b–c**) Formation of a complete ring of vascular cambium. The vascular cambium gives rise to secondary xylem and phloem. The root increases in diameter as cell divisions proceed parallel to the vascular cambium. The cortex ruptures as the tissue mass increases. (**d**) Epidermis is replaced by periderm, which arises from cork cambium (see boxed inset).

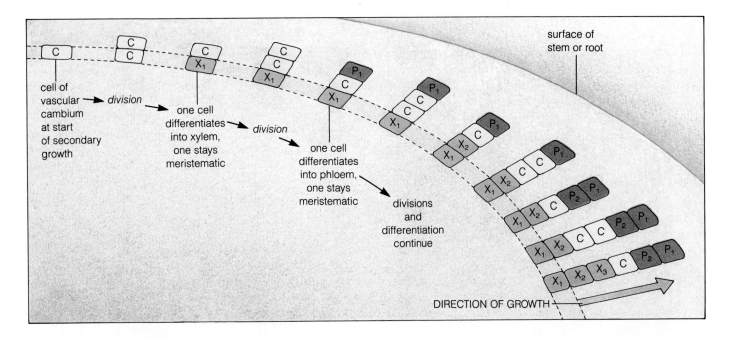

Figure 19.25 Relationship between the vascular cambium and its derivative cells (secondary xylem and phloem). Notice how the ongoing divisions displace the cambial cells, moving them steadily outward even as the core of xylem increases the stem or root thickness.

Xylem forms on the inner face of the vascular cambium, and phloem forms on the outer face (Figures 19.24 and 19.25). As the xylem increases in mass, the vascular cambium is displaced outward. Cells of the vascular cambium do not divide in one direction only, in line with the radius of the root or stem. They also divide in a direction parallel with the root or stem axis, thereby adding new initials that contribute to increases in the circumference of the vascular cambium.

In most species, the number of secondary xylem cells formed is far greater than the number of phloem cells. As the mass of xylem increases season after season, it usually crushes the thin-walled phloem cells from the preceding growth period, leaving only the thick-walled cells. In order to sustain plant growth and development, new rings of phloem cells must be produced each year, outside the growing inner core of xylem.

Cork Cambium Activity

In time, increases in the stem or root diameter cause the cortex and outer phloem to rupture. Parts of the cortex split away and carry epidermis with them (Figure 19.24). But pericycle cells continue to divide. Their derivative cells—including cork cambium—produce periderm, the corky covering that replaces epidermis. ("Cork" is not the same as "bark," a nontechnical term that refers to all living and nonliving tissues between the vascular cambium and the stem or root surface.) In some plants, sec-

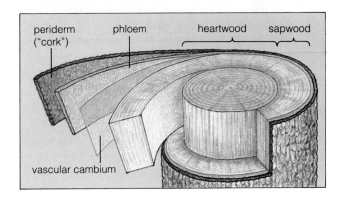

Figure 19.26 Structure of a woody stem showing extensive secondary growth. Heartwood, the core of the mature tree, is devoid of living cells. Sapwood is the cylindrical zone of xylem between the heartwood and vascular cambium; it also contains living parenchyma cells. All the living and nonliving tissues between the vascular cambium and the surface of a woody stem are often called bark.

ondary growth of this sort occurs year after year and produces massive, woody structures.

As you can see from Figure 19.26, living phloem in older trees is confined to a thin zone beneath the periderm. If this narrow band of phloem is stripped all the way around a tree's circumference (an action called "girdling"), the tree will die. The phloem cells are stripped away, so there is no way to transport photosynthetically derived food to the roots, which will die.

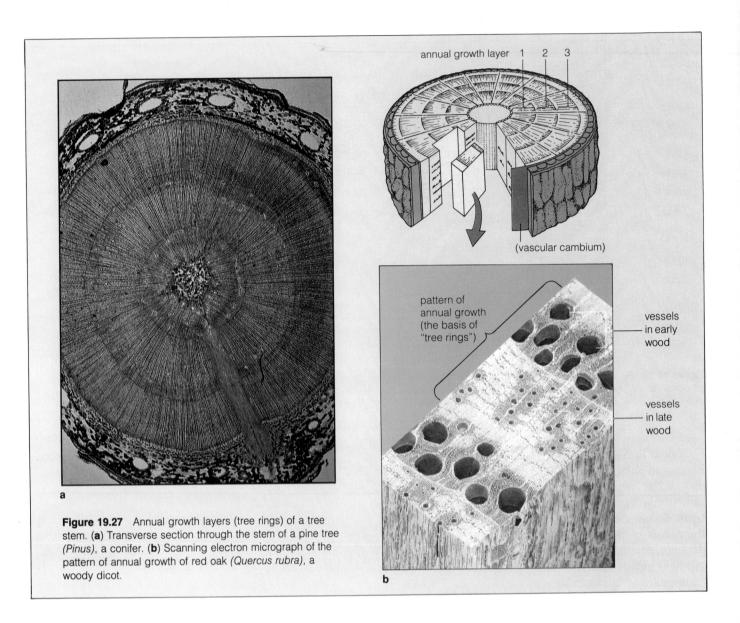

annual growth layer 1 2 3

(vascular cambium)

pattern of annual growth (the basis of "tree rings")

vessels in early wood

vessels in late wood

Figure 19.27 Annual growth layers (tree rings) of a tree stem. (**a**) Transverse section through the stem of a pine tree (*Pinus*), a conifer. (**b**) Scanning electron micrograph of the pattern of annual growth of red oak *(Quercus rubra)*, a woody dicot.

a

b

Early and Late Wood

In regions having prolonged dry spells or cool winters, the vascular cambium of stems and roots becomes inactive during parts of the year. The first xylem cells produced at the start of the growing season tend to have large diameters and thin walls; they represent **early wood** (Figure 19.27b). As the season progresses, the cell diameters become smaller and the walls thicker; these cells represent **late wood**.

The last-formed, small-diameter cells of late wood will end up next to the first-formed, large-diameter cells of the next season's growth. Although we don't see individual cells with the naked eye, there is enough difference in light reflection from a stem cross-section to reveal the alternating light bands (early wood) and dark bands (late wood). These alternating bands represent annual growth layers, which often are called "tree rings."

In some wet, tropical regions, the growing season is continuous. Conditions in other tropical regions allow several spurts of growth during one year. The growth layers of many tropical woody plants are faint, nonexistent, or do not correspond to a single year's growth.

SUMMARY

1. Many plants contain vascular tissues, which are highly specialized for conducting water, dissolved minerals, and dissolved organic compounds such as sucrose through the plant body. The most familiar plants are gymnosperms (chiefly conifers) and angiosperms (flowering plants). This chapter focused on monocots and dicots, the two classes of flowering plants. The root systems and shoot systems (stems, leaves) of monocots and

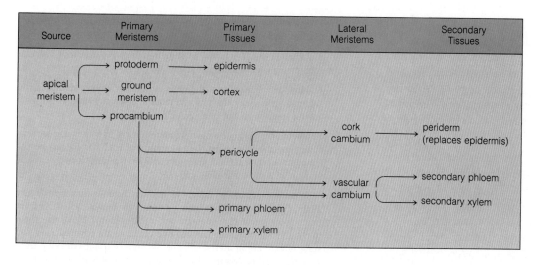

Figure 19.28 Summary of primary and secondary growth during the development of a root from a vascular plant.

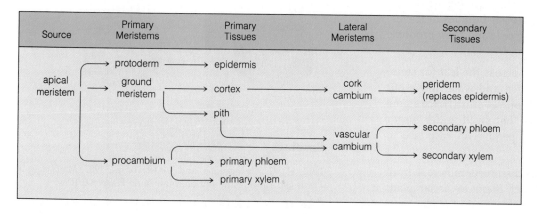

Figure 19.29 Summary of primary and secondary growth during the development of a stem from a vascular plant.

dicots are composed of three kinds of tissues (ground, vascular, and dermal tissues).

2. There are three ground tissues. *Parenchyma* makes up the bulk of fleshy plant parts; its generally thin-walled, living cells function in photosynthesis, storage, and other tasks. *Collenchyma* provides mechanical support for growing plant parts; its elongated, living cells have irregularly thickened, flexible primary walls. *Sclerenchyma* supports and strengthens plant parts that have ceased elongating; its cells have thick, usually lignified secondary walls and often are dead at maturity.

3. Vascular tissues include xylem and phloem. *Xylem* contains cells (actually the walls of cells dead at maturity) for conducting water and dissolved minerals. The main cells are pitted tracheids (in nearly all vascular plants) and pitted, perforated vessel members (in few vascular

plants besides angiosperms). *Phloem*, the food-conducting tissue, contains sieve elements (living cells, joined end to end to form sieve tubes). Companion cells (specialized parenchyma cells) seem to direct the metabolic activity of sieve elements..

4. Dermal tissues include *epidermis* (which covers and protects the surfaces of young plant parts) and *periderm* (which replaces epidermis on plants showing secondary growth).

5. The primary growth of roots and shoots originates at *apical meristems*, the undifferentiated plant tissue at root and shoot tips. Some cells of the apical meristem produce daughter cells that give rise to the *primary meristematic tissues*. These tissues give rise to the primary tissues (Figures 19.28 and 19.29). Herbaceous annual plants, which live one season, show little or no secondary growth.

6. *Lateral meristems* function in secondary growth (increases in girth of stems and roots). One type, the vascular cambium, gives rise to secondary xylem and phloem that conduct water and solutes vertically and horizontally through enlarging stems or roots. Another type, the cork cambium, produces periderm. Perennial plants, which live for many seasons, show secondary growth.

7. Stems are usually upright structures that give photosynthetic tissues favorable exposure to light and that display reproductive structures. Their vascular tissues distribute substances to and from roots, leaves, and other plant parts. Monocot stems have vascular bundles distributed throughout the ground tissue; dicot stems have vascular bundles arrayed as a cylinder, which separates the ground tissue into an outermost cortex and a central pith.

8. In most leaves, the chloroplast membranes of photosynthetic cells collectively represent a huge surface area for intercepting sunlight and exchanging carbon dioxide and oxygen. Veins (vascular bundles in leaves) move water and solutes to these cells and carry products away from them.

9. Stomata are small openings across the leaf (or stem) epidermis through which oxygen and water move out of the plant and carbon dioxide moves into it. Each stoma is flanked by two guard cells that swell under turgor pressure (and so widen the opening) in response to changing conditions.

10. Roots absorb water and dissolved minerals from the surroundings and conduct them to aerial plant parts; they anchor and sometimes support the plant and often have food storage regions. A root cap protects the apical meristem at root tips. Protuberances of certain root epidermal cells (root hairs) greatly enhance water absorption.

11. Vascular tissues form a central column within the root cortex. A single layer of cortical cells (the endodermis) is wrapped around the vascular column. Water and solutes enter the column only by passing through the cytoplasm of those cells (a beltlike Casparian strip seals off the adjoining walls of endodermal cells). Passage of water and solutes is regulated by transport controls built into the cell membranes. A pericycle, a tissue that gives rise to lateral roots and contributes to lateral growth, lies just inside the endodermis.

Review Questions

1. Distinguish between the following tissues, based on their structure and function.
a. parenchyma, collenchyma, and sclerenchyma *251*
b. xylem and phloem *251–252*
c. epidermis and periderm *252*

2. Define meristem. Which meristem regions produce the primary plant body? *253*

3. What is the functional relationship between the root vascular column and the Casparian strip? *260–261*

4. How are annual growth layers formed in woody stems? If you were to strip away the phloem from a tree's circumference, what would happen to the tree? *263–264*

5. Sketch the stem of a monocot and a dicot in transverse section. Can you label the main tissue regions of each, and describe the functions of their cellular components? Do the same for a transverse section of a root. *253–254, 260–261*

6. With a sheet of paper, cover the last three columns in Figure 19.28. Can you now state which primary tissues of roots arise from protoderm, ground meristem, and procambium? Move the sheet to the right so that only the last two columns are covered. Can you state which lateral meristems and secondary tissues develop from various primary tissues? Do the same exercise with Figure 19.29 for the tissues of stems. *265*

7. Label the following stem regions: *254*

Readings

Bold, H., C. Alexopoulos, and T. Delevoryas. 1987. *Morphology of Plants and Fungi.* Fifth edition. New York: Harper & Row.

Core, H., W. Côté, and A. Day. 1979. *Wood Structure and Identification.* Second edition. New York: Syracuse University Press. Stunning scanning electron micrographs of softwood and hardwood structure.

Esau, K. 1977. *Anatomy of Seed Plants.* Second edition. New York: Wiley. A classic reference for seed plant structure.

Raven, P., R. Evert, and S. Eichhorn. 1986. *Biology of Plants.* Fourth edition. New York: Worth. Exquisite color micrographs and illustrations of plant cells and tissues.

Rost, T., et al. 1984. *Botany: An Introduction to Plant Biology.* Second edition. New York: Wiley.

Stern, K. 1988. *Introductory Plant Biology.* Fourth edition. Dubuque, Iowa: W. C. Brown. Beautifully illustrated, accessible introduction to plant structure and function. Paperback.

It took you eighteen years or so to grow to your present height. A corn plant can exceed that in three months! Yet how often do we stop to think that plants do anything impressive? Being endowed with mobility, intelligence, and varied emotions, we tend to be endlessly fascinated with ourselves and somewhat indifferent to the immobile, expressionless plants around us. Besides, from experience and educational biases, most of us simply have acquired more knowledge about animals than we have about plants.

And yet, think about what it must take for plants such as an elm tree or a dandelion to survive. They require only sunlight, water, carbon dioxide, and assorted minerals. But most soils are frequently dry, and what does the plant do then? In a given volume of air, the carbon dioxide concentration averages only about 340 parts per million, so how can enough carbon dioxide be taken up to sustain photosynthesis? Minerals dissolved in soil water are scarce, so how does the plant accumulate them against concentration gradients?

Now think about which aspects of plant structure and function might be responses to the low concentrations of environmental resources. For one thing, central vacuoles increase the volume of most of the living cells (and the overall volume of the plant) and so they increase the surface area for absorbing scarce materials. (A central vacuole is mostly water, and fluid pressure that builds up inside it exerts force on the walls of growing cells and leads to their enlargement.) For another thing,

20

WATER, SOLUTES, AND PLANT FUNCTIONING

Figure 20.1 Sunlight filtering through a grove of California coast redwoods. How are water and essential nutrients transported to the tops of such giant trees? How are photosynthetically derived organic molecules distributed from leaves to all parts of the massive plant body, even down to the roots? These are questions that will be addressed in this chapter.

Table 20.1 Essential Elements for Most Complex Land Plants

Element	Symbol	Form Available to Plants	Percent by Weight in Dry Tissue	
Carbon	C	CO_2	45	
Oxygen	O	O_2, H_2O, CO_2	45	96% of total dry weight
Hydrogen	H	H_2O	6	
Nitrogen	N	NO_3^-, NH_4^+	1.5	
Potassium	K	K^+	1.0	
Calcium	Ca	Ca^{++}	0.5	
Magnesium	Mg	Mg^{++}	0.2	
Phosphorus	P	$H_2PO_4^-$, HPO_4^{--}	0.2	
Sulfur	S	SO_4^{--}	0.1	
Chlorine	Cl	Cl^-	0.010	
Iron	Fe	Fe^{++}, Fe^{+++}	0.010	
Boron	B	H_3BO_3	0.002	
Manganese	Mn	Mn^{++}	0.0050	
Zinc	Z	Zn^{++}	0.0020	
Copper	Cu	Cu^+, Cu^{++}	0.006	
Molybdenum	Mo	MoO_4^-	0.00001	

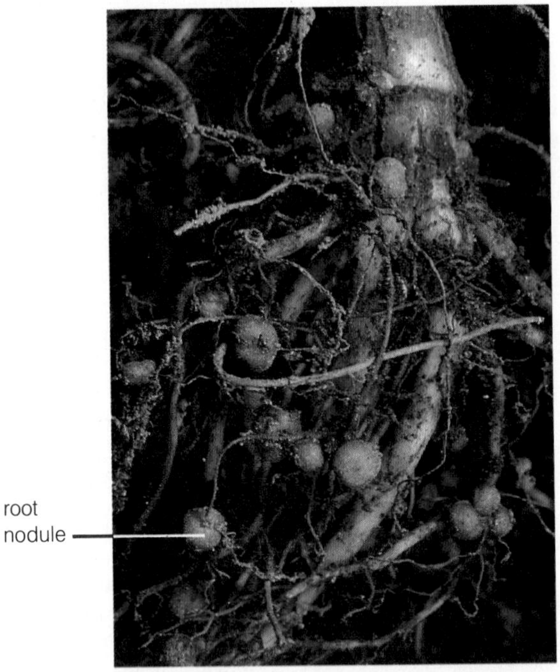

root
nodule —

Figure 20.2 Root nodules, where symbiotic nitrogen-fixing bacteria live.

leaves that are thin and broad represent a large surface area for absorbing sunlight and carbon dioxide.

Obviously you won't ever see broad, leaflike roots (imagine trying to push a narrow strip of paper through soil). But plants have roots shaped like thin cylinders, which collectively represent a large surface area and individually overcome the soil's resistance to penetration. Also, their root systems grow in many directions and expose the root absorptive surface areas to a large volume of moist soil. Finally, even though shoots move farther and farther from roots during growth, plants have vascular systems that conduct water, minerals, and organic molecules from one region to another.

Many aspects of plant structure and function are concerned with exposing a large surface area to a large volume of the environment and with enhancing the uptake of dilute raw materials.

How plant systems function in response to the environment is the focus of *plant physiology*. We have already considered how plants acquire energy to produce food (by photosynthesis), then use their self-produced food (by aerobic respiration). In this chapter, we will see how plants acquire and distribute materials used directly or indirectly in these and other activities.

ESSENTIAL ELEMENTS AND THEIR FUNCTIONS

Oxygen, Carbon, and Hydrogen

As indicated in Table 20.1, plants generally require sixteen essential elements to grow and reproduce, although most studies have been carried out with crop plants only. Oxygen, carbon, and hydrogen are building blocks for all organic compounds (carbohydrates, lipids, proteins, and nucleic acids). Together, these three elements account for about ninety-six percent of the plant's dry weight. The oxygen comes from water, gaseous oxygen (O_2), and carbon dioxide (CO_2) in the air; the carbon dioxide also is the source of carbon. The hydrogen comes from water molecules.

Mineral Elements

Thirteen of the essential elements are minerals. These naturally occurring, inorganic substances become available to plants in ionized form; they are "mineral ions." Table 20.2 lists their main functions, along with the symptoms associated with mineral deficiencies. Six of these minerals are called *macronutrients*; each makes up

Table 20.2 Role of Mineral Elements in Plant Function

Element	Some Known Functions	Deficiency Symptoms
Macronutrients:		
Nitrogen	Component of amino acids, proteins, chlorophyll, nucleic acids, coenzymes	Stunted growth, delayed maturity, light green older leaves; lower leaves turn yellow and die
Potassium	Activates enzymes used in protein, sugar, starch synthesis, helps maintain turgor pressure*	Reduced yields; mottled, spotted or curled older leaves; marginal burning of leaves; weak root system, weak stalks
Calcium	Part of middle lamella (helps cement cell walls together); necessary for spindle formation in mitosis and meiosis	Deformed terminal leaves, reduced root growth. Dead spots in dicot leaves. Terminal buds die
Magnesium	Component of chlorophyll; activates many enzymes used in photosynthesis, respiration, protein synthesis	Plants usually chlorotic (interveinal yellowing of older leaves); leaves may droop
Phosphorus	Component of nucleic acids, ADP and ATP, phospholipids	Purplish veins in older leaves, stems, and branches often turn dark green; reduced yields of seeds and fruits, stunted growth
Sulfur	Component of some amino acids, two vitamins, most proteins	Light green or yellow leaves, including veins; reduced growth. Weak stems. Similar to nitrogen deficiency
Micronutrients:		
Chlorine	Aids in root and shoot growth. Aids in photolysis in noncyclic pathway of photosynthesis	Plants wilt. Chlorotic leaves. Some leaf necrosis. Bronzing in leaves
Iron	Helps synthesize chlorophyll; component of electron transport systems of photosynthesis and respiration	Paling or yellowing of leaves (chlorosis) between veins at first. Grasses develop alternate rows of yellowing and green stripes (veins) in leaves
Boron	Affects flowering, pollen germination, fruiting, cell division, nitrogen metabolism, water relations, hormone movement	Terminal buds die; lateral branches begin to grow, then die. Leaves thicken, curl, and become brittle
Manganese	Chlorophyll synthesis; acts as coenzyme for many enzymes	Network of major green veins on light green background. Leaves later become white and fall off
Zinc	Used in formation of auxins, chloroplasts, and starch; component of several enzymes	Abnormal roots; mottled bronzed or rosetted leaves. Interveinal chlorosis
Copper	Component of enzymes used in carbohydrate and protein metabolism	Terminal leaf buds die. Leaves have chlorotic or dead spots. Stunted growth. Terminal leaves die
Molybdenum	Essential in enzyme-mediated reaction that reduces nitrate; component of enzyme used in nitrogen fixation	Plants may become nitrogen deficient. Pale green, rolled or cupped leaves with yellow spots

Data from Hartmann et al. *Plant Science*, 1981, and others.
*All solutes contribute to the osmotic solute and ion balance.

at least a tenth of a percent of the total dry weight of the plant. The rest are *micronutrients*, or trace elements, which represent only a few parts per million of the plant's dry weight.

Nitrogen is one of the macronutrients. It is not exactly scarce; N_2 molecules make up seventy-eight percent of the air. But plants do not have the metabolic machinery for breaking apart the three covalent bonds ($N\equiv N$) of these molecules. However, some microbes living independently of plants can break the bonds and reduce each nitrogen atom to ammonium ions (NH_4^+); other microbes living in root nodules can do the same thing. **Nodules** are localized swellings on the roots of legumes (for example, peas) and other plants where symbiotic, nitrogen-

fixing bacteria live (Figure 20.2). The nodule residents use up some of the plant's organic molecules. However, the ammonium they produce is used in assembling their own amino acids and nucleic acids—and most of the ammonium and amino acids remain in the plants, which thereby are provided with a nitrogen source.

Most of the minerals listed in Table 20.2 function in activating the enzymes of protein synthesis, starch synthesis, photosynthesis, and aerobic respiration. They also function in setting up solute concentration gradients across plasma membranes. Because of those gradients, water moves by osmosis into cells and cell vacuoles, and turgor pressure increases. (Turgor pressure, recall, is the internal pressure applied to walls as water is

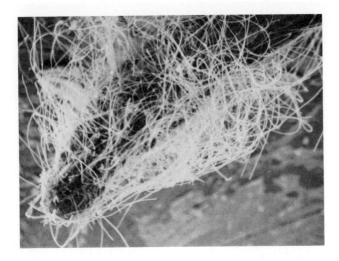

Figure 20.3 Mycorrhiza (fungus-root) of a lodgepole pine tree. White threads are fungal strands.

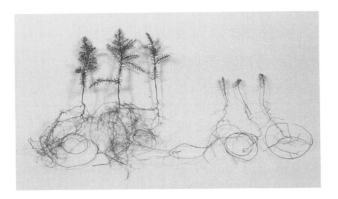

Figure 20.4 Effect of mycorrhizal fungi on plant growth. The six-month-old juniper seedlings on the left were grown in sterilized low-phosphorous soil inoculated with a mycorrhizal fungus. The seedlings on the right were grown without the fungus.

Figure 20.5 Scanning electron micrograph of root hairs, 46×.

absorbed into cells.) This pressure is necessary for cell enlargement during growth. Also, in mature cells with walls that no longer stretch, turgor pressure keeps the plant from wilting (Figure 6.7).

Mineral ions function in many metabolic activities. They also help set up solute concentration gradients; and water necessary for growth and for maintaining plant shape thereby moves by osmosis into cells.

WATER UPTAKE, TRANSPORT, AND LOSS

Water Absorption by Roots

Generally, annual grasses have a highly branched, fibrous root system that spreads out near the soil surface, and most dicots have a taproot system that penetrates more deeply into the soil (Figure 19.18). However, the environment influences the extent to which any root system develops. Millions, sometimes billions of root hairs might develop in a single system. Roots branch out in some locations, then they are replaced by roots that branch into different locations as conditions change. It is not that the roots "explore" the soil in search of minerals or water. Rather, outward root growth (hence growth of the entire plant) is stimulated in regions where water and dissolved mineral ions happen to be more concentrated.

In nearly all vascular plants, absorption of water and dissolved minerals apparently is enhanced by mycorrhizae, or "fungus-roots." A **mycorrhiza** is a mutually beneficial association between a fungus and a young root (Figure 20.3). The symbiotic fungi form an extensive mat of thin filaments (hyphae). Collectively, the hyphae have a tremendous surface area for absorbing mineral ions from a large volume of the surrounding soil. The root uses some of the ions during growth (Figure 20.4); the fungus absorbs some sugars and nitrogen-containing compounds in the root.

Take a moment to look at Figure 20.5, which shows the hair cells of a typical root. Especially in regions of root hairs, water from the soil diffuses inward through epidermal cells. Once inside the root cortex, much of the water apparently is conducted along the porous walls of cortical cells until it reaches the endodermis surrounding the vascular column. There it encounters the waxy Casparian strip, which prevents water from entering the vascular column except by way of the cytoplasm of endodermal cells (Figure 19.20d). Inside the vascular column, water enters the conducting cells of xylem, where its movement is nearly always governed by forces of the sort to be described next.

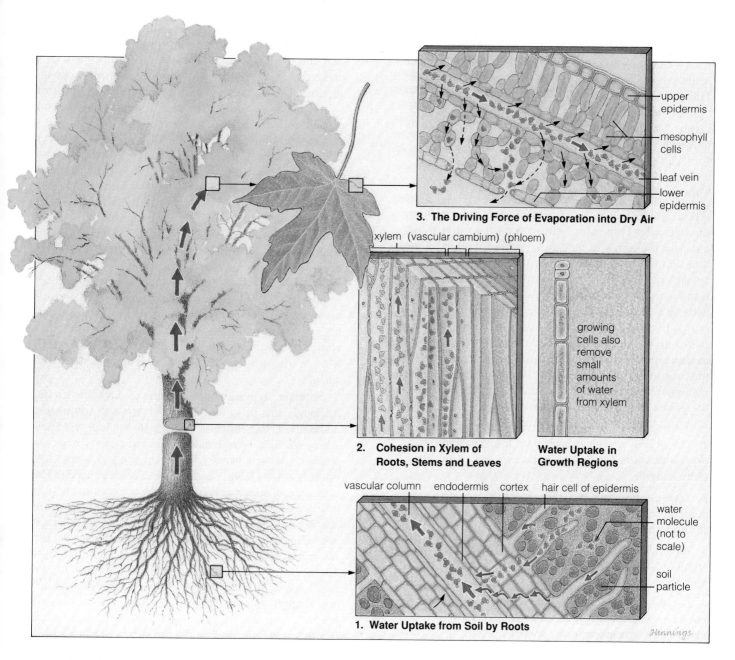

Figure 20.6 Cohesion theory of water transport. Tensions in water in the xylem extend from leaf to root. These tensions are caused mostly by transpiration (the evaporation of water from plant parts). As a result of the tensions, columns of water molecules that are hydrogen-bonded to one another are pulled upward.

Transpiration and Water Conduction

Water moves from roots to stems, then into leaves. A small part is used for growth (through cell enlargement) and in metabolism, but most evaporates into the air. Water evaporation from plant parts (stems and leaves especially) is called **transpiration**. How does water get to those parts which, in the tallest trees, are 100 meters above ground? *Water is pulled upward by continuous negative pressures (tensions) that extend downward from the leaf to the root* (Figure 20.6).

The question becomes this: What causes the tension in the xylem cells? *First*, the air around a plant nearly always causes evaporation from the walls of mesophyll cells inside the leaf. As some water molecules escape, others diffuse out of the cell cytoplasm and into the walls as replacements. When they do, water molecules from the xylem in leaf veins move into the mesophyll cells. *Second*, when water moves out of the veins, replacements are pulled in from the xylem cells leading into the veins from the stem. Because of the pulling action, water inside all of those conducting cells is in a state of tension. *Third*, replacement water molecules move into the root xylem even when the soil is somewhat dry, and more soil water is drawn into the plant, following its osmotic gradient. This inward movement will continue until the soil becomes so dry that an osmotic gradient no longer exists. Figure 20.6 shows how water is pulled uphill, from roots to leaves.

Cohesion Theory of Water Transport

When water moves as a continuous, fluid column through xylem pipelines, why doesn't the "stretching" cause the molecules to snap away from each other? Some time ago the Irish botanist Henry Dixon came up with an explanation, which has since been named the **cohesion theory of water transport**. According to this theory, hydrogen bonds are strong enough to keep water molecules from separating from one another as they are pulled up through the plant body. Dixon had no way of measuring how much tension exists in xylem (and therefore of convincing skeptics that it really exists). Confirmation of his explanation came much later. In any event, the points to remember are these:

1. The drying power of air causes transpiration (evaporation of water from plant parts exposed to air, especially leaves).

2. Transpiration causes a state of tension (negative pressure) in xylem that is continuous from leaves, down through the stems, to roots.

3. As long as water molecules continue to vacate transpiration sites, replacement molecules are pulled under tension.

4. Because of the cumulative strength of hydrogen bonds between water molecules that are confined in the narrow, tubular xylem cells, water is pulled up as continuous columns to transpiration sites.

5. Although hydrogen bonds are strong enough for water molecules to cohere in the xylem, they are not strong enough to prevent them from breaking away from each other during transpiration.

The Dilemma in Water and Carbon Dioxide Movements

Of the water moving into a leaf, more than ninety percent is usually lost through transpiration. About two percent is used in photosynthesis, membrane functions, and other activities. However, when water loss by transpiration exceeds water uptake by roots, the resulting dehydration of plant tissues will interfere with these water-requiring activities.

Even under mild conditions, plants would rapidly wilt and die if it were not for the *cuticle*, the waxy covering that reduces the rate of water loss from aboveground plant parts (page 252). The cuticle does conserve water, but it also limits the rate of diffusion of carbon dioxide into the leaf.

Transpiration and carbon dioxide uptake occur largely at *stomata*, the small openings in the epidermis of leaves and stems (page 258). Two guard cells flank each opening. When the cells are swollen with water, turgor pressure distorts their shape in such a way that they move apart (Figure 20.7). Their separation produces a gap between them (the actual stoma). When the water content of the guard cells dwindles, turgor pressure drops and the stoma closes.

When stomata are open, enough carbon dioxide can be absorbed rapidly for photosynthesis—but when they are open, water nearly always moves out! In itself, carbon dioxide uptake and water loss at the same sites would present something of a dilemma for the plant, if it were not for controls over stomatal opening and closure.

Stomata typically open during daylight hours. Although the processes involved are not completely understood, this much is clear: *A stoma opens and closes mainly according to how much water and carbon dioxide are present in the two guard cells.* When the sun comes up, photosynthesis begins and carbon dioxide is used in the formation of starch and sucrose. As carbon dioxide concentrations dwindle in guard cells, potassium ions are actively pumped into them from surrounding epidermal cells (Figure 20.8). When the ions accumulate inside, water moves in by osmosis from the surrounding cells. The resulting increase in turgor pressure leads to stomatal opening. Transpiration proceeds, and so does carbon dioxide movement into the leaf. Photosynthesis keeps carbon dioxide concentrations low, so the plant continues to lose water and gain carbon dioxide during the day.

Transpiration nearly always occurs when carbon dioxide, which is present only in dilute concentrations in air, is being absorbed. When stomata are open and carbon dioxide is moving into the plant, water tends to move out.

When the sun goes down, photosynthesis stops. Carbon dioxide is no longer used, but it is still being released during aerobic respiration. As a result, carbon dioxide accumulates in all cells. Much of the potassium inside guard cells now moves out, and water follows it osmotically. Turgor pressure decreases, stomata close, transpiration is greatly reduced, and water is conserved.

As long as the soil is moist, stomata can remain open during daylight. When the soil is dry and the air is also

Figure 20.7 Where stomata occur on a typical dicot leaf, and how they function. (**a**) Stomata are found among hairlike structures on a cucumber leaf's lower epidermis. The box identifies one of the stomata. (**b**) Closer look at stomatal structure. Here, we can peer through the gap between the stomatal guard cells and view parts of mesophyll cells inside the leaf. (**c**) Sketches of a closed stoma and one opened as guard cells swell.

dry and hot, the stomata of land plants close or do not open as much, so little water is absorbed and transpired. Although photosynthesis (and growth) slows as a consequence, the plants still survive short drought periods. They can do so repeatedly. Briefly, such stressful conditions trigger the production of a plant hormone called **abscisic acid**. The hormone is synthesized faster when a leaf is water stressed. When abscisic acid accumulates in a leaf, it somehow causes guard cells to give up potassium ions, hence water, so the stomata close.

UPTAKE AND ACCUMULATION OF MINERALS

Active Transport of Mineral Ions

Water uptake depends only on an osmotic gradient. Yet if a cell in a root (or in any other part of the plant) is to retain water, it must maintain a high solute concentration so that an osmotic gradient can be produced. This means that a cell must expend energy to actively accumulate solutes, particularly dissolved mineral ions. Without energy outlays, diffusion would equalize the solute concentrations on both sides of a plasma membrane. Energy from ATP drives the membrane pumps involved in active transport. These pumps are mem-

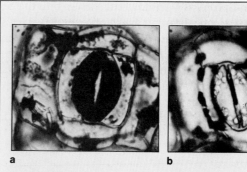

Figure 20.8 Evidence for potassium accumulation in stomatal guard cells undergoing expansion. Strips from the leaf epidermis of a dayflower (*Commelina communis*) were immersed in solutions containing dark-staining substances that bind preferentially with potassium ions. (**a**) In leaf samples having opened stomata, most of the potassium was concentrated in the guard cells. (**b**) In leaf samples having closed stomata, very little potassium was in guard cells; most was present in normal epidermal cells.

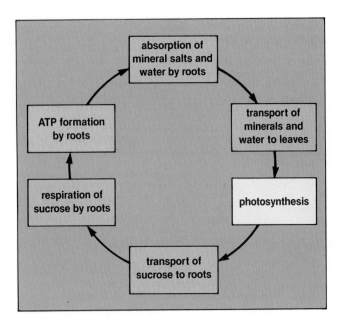

Figure 20.9 Interrelated processes that influence the coordinated growth of roots, stems, and leaves. When one process is rapid, the others also speed up. Any environmental factor limiting one process eventually slows growth of all plant parts.

brane-bound proteins that move substances into the cell even against a concentration gradient (page 93).

In photosynthetic cells, ATP necessary for the membrane pump operation is formed during both photosynthesis and aerobic respiration. What about nonphotosynthetic cells, such as those in roots? How do they get all the ATP necessary for active transport? Here, ATP is formed almost entirely through aerobic respiration in the mitochondria of individual cells.

Controls Over Ion Absorption

Solute absorption and accumulation must be coordinated throughout the plant. For example, root cells receive sugars (commonly sucrose) from leaves, especially when photosynthesis is rapid during the daytime. These cells absorb oxygen from air in the soil (unless the soil is waterlogged). When the soil is moist enough, dissolved ions move rapidly to roots, where some ions are actively transported from cell to cell into the xylem. When the soil is dry, more air is present and more oxygen can be absorbed. But insufficient water limits growth and ion absorption no matter how much oxygen is present. Also, dry soil causes stomata to close partly or completely. Thus, leaves absorb less carbon dioxide when water and ions are in short supply, and photosynthesis and growth slow down. Then, leaf cells cannot send as much sucrose

to roots. Without enough sucrose, aerobic respiration slows down in roots, and so does ion absorption.

Figure 20.9 summarizes part of what is known about coordinated growth between different plant regions.

Mineral ion absorption and accumulation are coordinated throughout the plant body in ways that have profound influences on growth.

TRANSPORT OF ORGANIC SUBSTANCES IN PHLOEM

So far, we have considered some ways in which plants acquire and distribute raw materials. Let's now consider how organic molecules are distributed through the plant body.

Sucrose and related organic compounds formed in leaves are the major building blocks used for biosynthesis in plants. Certain organic molecules not used by the leaf cells are transported to buds, roots, stems, flowers, and fruits, which also require those molecules for growth. Many of the organic molecules are also stockpiled. Starch is the main storage form of carbohydrates in most plants. Fat stores are especially prevalent in many seeds and some fruits, such as the avocado. Proteins are stored in granules in many seeds and grains.

Starch molecules are too large to cross cell membranes; they cannot leave the cells in which they are formed. Even if they could, they are too insoluble for transport in water to other regions of the plant body. Fats are largely insoluble in water, and they cannot be transported out of their storage sites. Storage proteins do not lend themselves to transport, either.

The energy and building blocks inherent in starch, fats, and proteins are made available through different chemical reactions. For example, starch molecules are first hydrolyzed, and in most plants the glucose units released are combined with fructose. The resulting molecule is sucrose, which is soluble and transportable. Similarly, proteins are converted to soluble amino acids and amides.

Transport of organic substances through the plant requires that storage starch, fats, and proteins be converted to smaller subunits that are soluble and transportable.

Translocation

How do soluble organic molecules travel from photosynthesis sites or storage sites to organs that require them? Here we must turn to a process known as **trans-**

location. In botany, the word can mean the relatively long-distance transport of water or solutes, but most often it is used to signify the transport of sucrose and other compounds through phloem. As we have seen, sieve tube members of the phloem are joined into long, interconnecting pipelines (page 252). The pipelines lie side by side in overlapping array within vascular bundles, and they extend from leaf to root. Unlike water-conducting xylem cells, sieve tube members are alive at maturity. Water and organic molecules are transported rapidly through their numerous wall perforations at rates up to 100 centimeters an hour.

Interestingly, the feeding habits of small insects called aphids tell us something about translocation. An aphid feeds on leaves and stems. It forces a mouthpart (stylet) into sieve tube cells, which contain dissolved sugars and other organic compounds. The contents of the cells are under high pressure, often five times as much as in an automobile tire. This pressure seems to force the fluid right through the aphid gut and clear out the other end as "honeydew" (Figure 20.10). Park your car under trees being attacked by aphids and it might get a spattering of sticky honeydew droplets, thanks to sieve-tube pressures.

In some experiments, feeding aphids were anesthetized with carbon dioxide. Then their bodies were severed from their stylets, which were left embedded in the sieve tubes that the aphids had been attacking. Analysis of the fluid being forced out of the tubes verified that sucrose is the main carbohydrate being transported through the plant body in most species.

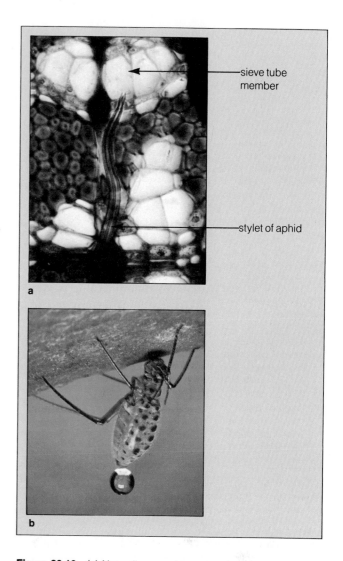

Figure 20.10 (**a**) Naturally occurring research tool: the aphid stylet, here penetrating a sieve tube member. (**b**) Honeydew droplet at the tail end of a well-fed aphid.

Pressure Flow Theory

The movement of organic molecules in phloem follows a "source-to-sink" pattern. The main *source regions* are the sites of photosynthesis in leaves. A *sink region* is any plant part that depends on inputs of organic molecules to meet its nutritional needs or that stockpiles organic molecules for later use. Growing leaves, fruits, seeds, and roots are examples of sink regions. (Although storage tissues are sinks when they receive organic molecules, they also serve as sources when the stores are tapped.)

Obviously, a plant doesn't "know" which way it should be translocating organic molecules through the phloem. But what causes the directional movement? According to the **pressure flow theory**, translocation through the phloem depends on pressure gradients between source and sink regions.

To understand the points of the pressure flow theory, let's begin with what happens when sucrose and other organic molecules move into a sieve tube in a leaf. As solute concentrations increase in this source region, the water potential decreases. Water potential, recall, is the

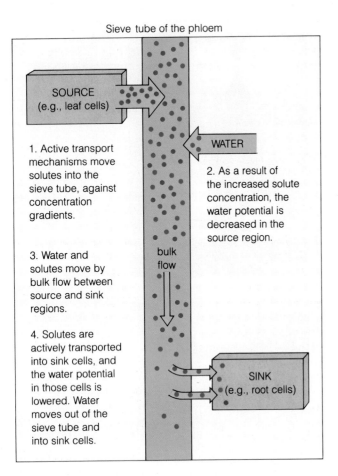

Sieve tube of the phloem

SOURCE (e.g., leaf cells)

WATER

1. Active transport mechanisms move solutes into the sieve tube, against concentration gradients.

2. As a result of the increased solute concentration, the water potential is decreased in the source region.

3. Water and solutes move by bulk flow between source and sink regions.

bulk flow

4. Solutes are actively transported into sink cells, and the water potential in those cells is lowered. Water moves out of the sieve tube and into sink cells.

SINK (e.g., root cells)

Figure 20.11 Proposed mechanism of pressure flow in the phloem of vascular plants.

sum of two opposing forces: water movement into cells by osmosis (when internal solute concentrations increase), and an outward-directed movement of water when turgor pressure increases. The resulting increase in local turgor pressure causes water to flow to regions in the tube system where pressure is lower (Figure 20.11).

Sieve tubes play a passive role in the bulk transport of organic molecules from source to sink regions. These molecules are loaded into the tubes by active transport mechanisms. The energy required for active transport is expended by companion cells alongside the sieve tube members or by neighboring parenchyma cells (page 252).

What maintains the low pressure at sink regions? There, organic molecules are unloaded from the tubes and actively transported into sink cells (which use them in wall building, starch formation, aerobic respiration, and other activities). The water potential in the sink cells

is lowered when the solutes move in, so water also moves in by osmosis.

In growing tissues where cells are still enlarging, the expanding cell walls prevent an increase in turgor pressure (which would counter the inward water movement). As water continues to enter the sink cells, the dilution of solutes allows even more solutes to enter rapidly—and the cells continue to grow.

SUMMARY

1. Plants have a large surface-to-volume ratio that favors the uptake of water and nutrients, which often are present in relatively low concentrations in their surroundings.

a. A central vacuole increases the volume of each cell (hence of the plant), thereby increasing the absorptive surface area.

b. Leaves are generally thin and broad, with a large surface area for sunlight interception and carbon dioxide uptake.

c. Root structure and patterns of growth enhance absorption of water and dissolved minerals from a large volume of soil.

d. Vascular systems conduct water, minerals, and organic molecules through the rather long distances between shoots and roots.

2. Plants build their own organic compounds with oxygen, carbon, and hydrogen (which together make up about ninety-six percent of the plant's dry weight) and with thirteen mineral ions. Water provides oxygen and hydrogen; the air provides gaseous oxygen and carbon dioxide.

3. The mineral ions serve as macronutrients (each being at least 0.1 percent of the total dry weight) or micronutrients (present in trace amounts). They are ionized forms of nitrogen, potassium, calcium, and so on that function in metabolic activities (including enzyme activation) and in establishing solute concentration gradients across the plasma membranes of the plant cells. Such gradients are the basis for water movement into the cells, hence for maintaining cell shape and growth.

4. Water and solute absorption is enhanced by mycorrhizae, mutually beneficial associations between young roots and fungi (which form an extensive mat of thin absorptive filaments around the root). The fungus provides the plant with absorbed mineral ions; the root provides the fungus with some sugars and nitrogen-containing compounds.

5. In the water-conducting system (xylem) of plants, continuous negative pressures (tensions) extend down from leaves to roots. The tension is caused by transpi-

ration (the evaporation of water from leaves and other plant parts exposed to air).

a. When water molecules vacate transpiration sites, replacements are pulled under tension to the site.

b. The cumulative strength of hydrogen bonds between water molecules (which are confined in the narrow, tubular cells of xylem) allows continuous "columns" of water to be pulled up. This concept is the cohesion theory of water transport.

6. A waxy cuticle (which retards water loss) covers most aerial plant parts. Thus transpiration and carbon dioxide uptake occur mostly at stomata, small openings in the epidermis of leaves and stems.

7. Physiological mechanisms cause stomata to open during the day, and this enhances carbon dioxide uptake (hence photosynthesis), although the "cost" is evaporative water loss. However, the mechanisms also cause stomata to close at night, when photosynthesis shuts down, so water loss via transpiration is reduced. The mechanisms involve the movement of potassium ions into and out of the two guard cells flanking each stoma, leading to changes in cell shape (and in the opening between them).

8. Dissolved mineral ions enter cells by active transport, which requires ATP energy. Aerobic respiration in the roots provides the ATP but in turn depends on an adequate supply of sucrose from the leaves.

9. Sucrose and other compounds are translocated to roots and other plant parts through the vascular system called the phloem. These nutrients support growth; some are converted to storage forms (starch, proteins, or fats).

10. Translocation can be explained by the pressure flow theory:

a. Translocation of organic compounds through plants occurs in sieve tube members (living cells in the phloem).

b. Translocation is driven by differences in water pressure between source regions (photosynthesis sites or storage organs) and sink regions (any metabolically active or growing tissue).

c. Organic compounds are actively transported into sieve tube members in source regions. Water potential thereby decreases in those local regions, and the water and solutes move by bulk flow to tube regions of lower pressure.

d. In sink regions, organic compounds are unloaded by active transport mechanisms that move them into individual cells.

e. The movement of organic compounds out of the phloem and into sink cells lowers the water potential in sink cells. Thus water moves out of the phloem and into the sink cells. Cell growth also can lower the pressure

(hence the water potential) in sink cells and cause more water to move out of the sieve tubes. In such ways, low-pressure regions are maintained in sink regions.

Review Questions

1. Give examples of the features that enable land plants to absorb water and nutrients from their surroundings, which have dilute concentrations of these required substances. *267-268*

2. Which three elements make up most of the dry weight of a land plant? Which six elements are considered macronutrients for land plants? *268-269*

3. Describe some of the specific roles that mineral ions play in plant functioning. How do solute absorption and accumulation affect plant growth? *268-270*

4. Define mycorrhiza. Why do you think is it important to include some of the native soil around roots when transplanting a plant from one place to another? *270*

5. Describe transpiration. State how the cohesion theory of water transport helps explain what is going on in this form of water movement. *271-272*

6. Transpiration competes with other water-requiring cell processes. Can you name some of these processes? *272-273*

7. Look at Figure 20.9. Then, on your own, diagram the feedback relations that influence the coordinated growth of stems, roots, and leaves. *274*

8. Sucrose transport from one plant organ to another is called translocation. Can you explain how it works in terms of the four key points of the pressure flow theory? How did aphids help show that sucrose is indeed the main substance being transported through the phloem pipelines? *274-276*

Readings

Apfel, R. 1972. "The Tensile Strength of Liquids." *Scientific American* 227(6):58–71. A fairly simple treatment of theory and experiments providing evidence that liquids can exist under tension as well as pressure.

Epstein, E. 1973. "Roots." *Scientific American* 228(5):48–58.

Galston, A., P. Davies, and R. Satter. 1980. *The Life of a Green Plant.* Englewood Cliffs, New Jersey: Prentice-Hall. A simplified treatment of much of plant physiology.

Hewitt, E., and T. Smith. 1975. *Plant Mineral Nutrition.* New York: Wiley. Techniques and results of studies of plant mineral nutrition.

Mengel, K., and E. Kirkley. 1982. *Principles of Plant Nutrition.* Worblaufer-Bern, Switzerland: International Potash Institute.

Peel, A. 1974. *Transport of Nutrients in Plants.* New York: Wiley. A short, simple treatment of transport processes occurring in xylem and phloem.

Salisbury, F., and C. Ross. 1985. *Plant Physiology.* Third edition. Belmont, California: Wadsworth. Excellent, comprehensive book covering most plant functions.

Torrey, J., and D. Clarkson (editors). 1975. *The Development and Function of Roots.* New York: Academic Press. One of the few modern treatments of root structure and function.

21

PLANT REPRODUCTION AND EMBRYONIC DEVELOPMENT

Although it probably is not something you think about very often, flowering plants engage in sex. They produce sperm and egg cells, as humans do. They, too, have elaborate reproductive systems that protect and nourish sex cells during their formation. As in human females, the female organs of flowering plants house the embryo during its early development. Those exquisite forms called flowers are, in effect, exclusive or open invitations to third parties—pollinators—that function in getting sperm and egg together (Figure 21.1). Long before humans ever thought of it, flowering plants were using tantalizing colors and fragrances in improving the odds for sexual success.

Plants also do something that humans cannot do (at least not yet). They can reproduce asexually. Because asexual reproduction occurs by way of mitosis only, all individuals of the new generation are genetically identical copies of the parent plant. (In contrast, sexual reproduction requires the formation of gametes followed by fertilization. Thus two different sets of genetic instructions, from two gametes, are present in the fertilized egg.) Both sexual and asexual reproduction occur in plant life cycles, and the reproductive details vary from group to group. We will look at these variations in Chapter Forty, which is an evolutionary survey of plant diversity. Here, we will begin with sexual reproduction and embryonic development among angiosperms—the flowering plants.

SEXUAL REPRODUCTION OF FLOWERING PLANTS

Life Cycles of Flowering Plants

What we usually think of as "the plant" is the **sporophyte**, a vegetative body commonly composed of diploid (2n) cells. Radish plants, cactus plants, and elm trees are examples of sporophytes.

At some point in the life cycle of angiosperms, the sporophyte produces **flowers**, which are shoots modified for reproduction (Figure 21.1). Some cells within the floral tissues divide by meiosis to form the haploid (n) cells called meiospores. (As we saw in Chapter Eleven, sporophyte means "spore-producing body.") Each meiospore may divide by mitosis to produce a multicelled haploid gametophyte. A **gametophyte** is a "gamete-producing body." The ones that produce sperm are male gametophytes; the ones that produce eggs are female gametophytes. In short, life cycles of flowering plants show **alternation of generations**, in which multicelled diploid bodies alternate with multicelled haploid bodies (Figure 21.2).

Keep in mind that the female gametophytes are not free-living plants. In most species, they are tiny multi-

Figure 21.1 Example of a flower, a shoot that is specialized for reproduction. Many flowers have coevolved with animals, such as this hummingbird, which play indirect but vital roles in the reproductive process.

celled bodies embedded within floral tissues. In contrast, the male gametophytes are released from the flower (as tiny pollen grains). They are like shipping crates for the sperm-producing cell until they actually land on a female flower part.

Floral Structure

A floral shoot has whorls or spirals of specialized structures called carpels, stamens, petals, and sepals. Most or all of the appendages are attached to the *receptacle*, the modified end of the floral shoot.

A **carpel** is a female structure. In this closed vessel, eggs develop, fertilization takes place, and seeds mature. Some flowers have only one carpel; others have several carpels, either separate or fused together.

Figure 21.3 shows the single carpel of a cherry blossom sliced lengthwise. Notice the *ovary*. In this chamber, the structure called an ovule develops into the cherry seed. (Other ovaries contain more than one ovule. For example, a pea pod is a mature ovary in which several ovules have developed into several "peas," or seeds.) Above the ovary is the *stigma*, a landing platform for pollen. In this plant the ovary and the stigma are separated by a *style*, a region in which the carpel narrows into a slender column.

The **stamens** shown in Figure 21.3 are male reproductive structures. Most commonly, a stamen consists of a *filament* (a slender stalk) capped by an anther. An *anther* contains chambers called pollen sacs, in which pollen grains develop.

The flower just described is a "perfect" flower, having stamens and one or more carpels (it produces both sperm and eggs). In contrast, "imperfect" flowers have stamens *or* carpels, but not both. Often they are called male and female flowers. In some species, such as oaks, male and female flowers appear on the same plant. In other species, such as willows and American holly, they occur on separate plants.

Most flowers also have petals and sepals. **Petals** are attached to the receptacle, below the male and female flower parts, and collectively form a "corolla." Often the color, patterning, and shapes of corollas function in attracting bees and other pollinators. The leaflike, often green **sepals** form the "calyx," the outermost whorl of appendages. Commonly, the calyx encloses all other flower parts (as it does in rosebuds) before a bud opens.

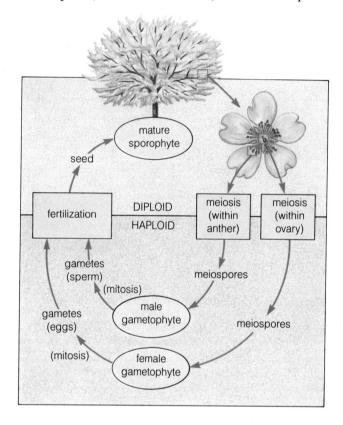

Figure 21.2 Alternation of generations in the life cycle of flowering plants.

Figure 21.3 Common arrangement of floral appendages. Shown here, a cherry (*Prunus*) flower, with a single carpel.

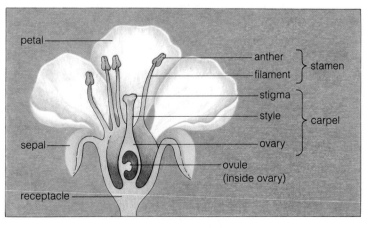

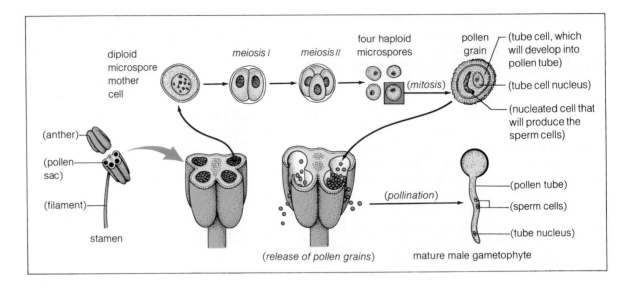

Figure 21.4 Stages in the development of a male gametophyte, beginning with microspore production in the anther. A germinated pollen grain, with its pollen tube and two sperm cells, constitutes the mature male gametophyte.

GAMETE FORMATION

Microspores to Pollen Grains

Let's now turn to reproductive events that typically unfold in a flower. While an anther is growing inside a flower bud, mitotic divisions produce four masses of diploid cells. Each of these so-called mother cells is destined to give rise to four haploid *microspores* (which is the name given to the meiospores produced in anthers). Each cell mass becomes surrounded by several layers of cells that form a **pollen sac**. In this walled chamber, pollen grains develop (Figure 21.4).

Each microspore in a pollen sac divides once, and the two resulting cells stay together as a pollen grain.

One cell will eventually produce the sperm. The other cell will develop into a *pollen tube*, an elongate structure that will grow through carpel tissues and thereby transport sperm to an ovule. Both cells are surrounded by a tough wall (Figure 21.5). This wall will protect the pollen grain from drying out during its journey from the anther to a stigma.

Pollen grains develop from haploid microspores, then later develop into sperm-bearing gametophytes.

Megaspores to Eggs

Meanwhile, in the carpel of the flower, one or more dome-shaped cell masses have been developing on the inner wall of the ovary. In some carpels, only one mass develops; in others, hundreds or thousands form. Each mass is the start of an ovule, which, if all goes well, is destined to become a seed.

As the mass grows, some of its cells form a stalk. The rest develop into an inner *nucellus* and one or two protective layers called *integuments*:

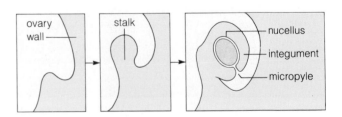

Figure 21.5 Scanning electron micrographs of pollen grains from (**a**) day lily, 260× and (**b**) ragweed, 625×.

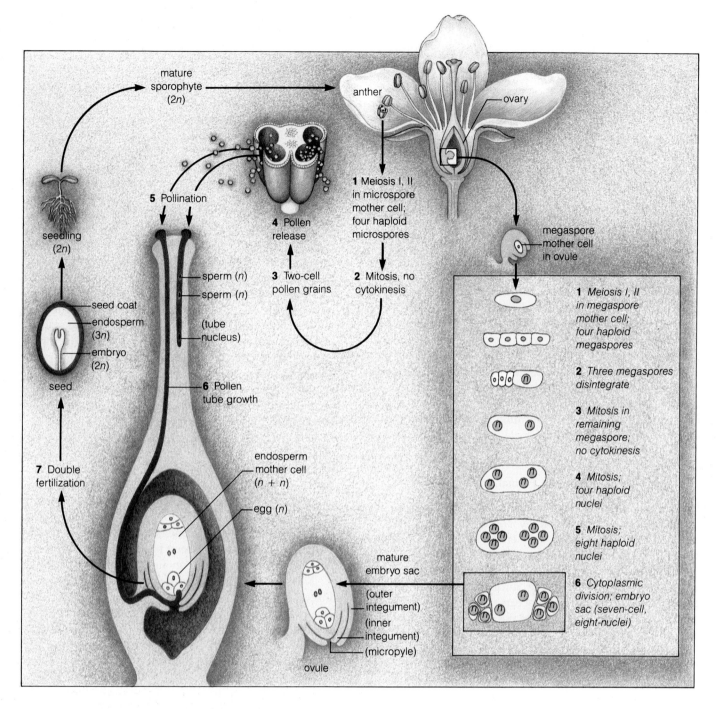

Figure 21.6 Generalized life cycle of a flowering plant, showing details of embryo sac formation within an ovary with one ovule. Compare this illustration with Figure 21.4.

In the nucellus a diploid mother cell gives rise to haploid *megaspores* (which is the name given to the meiospores produced in carpels). Only a tiny part of the nucellus does not become covered with integuments. Most commonly this tiny gap (the micropyle) is where a pollen tube will penetrate into the ovule.

In most flowering plants, three of the four megaspores disintegrate. The one remaining undergoes mito-

sis three times *without* cytoplasmic division, so at first it is a single cell with eight nuclei (Figure 21.6). Each nucleus migrates to a specific location in the cell, then cytoplasmic division occurs. The result is a female gametophyte, or **embryo sac**, that consists of seven cells. One of the seven cells, the "endosperm mother cell," contains two haploid nuclei. It will help give rise to *endosperm*, a nutritive tissue that will surround the forthcoming embryo. Another cell, which ends up near the micropyle, is the egg.

A female gametophyte (embryo sac) is a seven-celled, eight-nucleate body. One cell of the sac is the egg. Another cell has two nuclei and will help give rise to endosperm, a nutritive tissue.

POLLINATION AND FERTILIZATION

Pollination and Pollen Tube Growth

The word **pollination** refers to the transfer of pollen grains to the surface of a stigma (Figure 21.6). A pollen grain germinates (resumes growth) once it has been deposited on the proper stigma. One cell of the pollen grain starts growing as the pollen tube. The other cell divides, and two sperm cells form before or during tube growth. The pollen tube penetrates the stigma, style, and ovarian tissues on its journey to an ovule. The germinated pollen grain, with its tube nucleus and two sperm, constitutes the mature male gametophyte.

Fertilization and Endosperm Formation

Upon reaching the ovarian chambers, a pollen tube grows toward an ovule. Typically it enters the micropyle and penetrates the embryo sac, then the tip of the pollen tube ruptures and the sperm are released.

"Fertilization" generally means the fusion of one sperm nucleus with one egg nucleus. In flowering plants, **double fertilization** occurs. One of the two sperm nuclei fuses with that of the egg, and a *diploid* zygote (2n) forms. This marks the beginning of a new sporophyte. Meanwhile, the other sperm nucleus and both nuclei of the endosperm mother cell all fuse together, forming a *triploid* (3n) nucleus. This marks the formation of the primary endosperm cell, which will give rise to tissues that will nourish the seedling until leaves form and photosynthesis is under way.

In double fertilization, one sperm nucleus fuses with the egg nucleus, and a diploid zygote results. The other sperm nucleus fuses with the two nuclei of the endosperm mother cell, which will give rise to triploid (3n) nutritive tissue.

Before looking at how the sporophyte initially develops, let's digress for a moment and consider the relationship between floral structure and the pollinators necessary for a new generation to begin. It is one of the most intriguing of all evolutionary stories.

Case Study: Coevolution of Flowering Plants and Pollinators

Origins of Pollination Vectors. Diverse flowering plants live almost everywhere on land, from snow-covered flanks of mountains to low deserts. Many species live in freshwater or saltwater environments. How did this diversity and distribution come about, given that plants (unlike animals) cannot just pick up and move off to new places?

For the answer, we must go back 400 million years or so, to the time when plants began their invasion of the land. When the pioneers took hold, mud-dwelling insects that could scavenge on moist, decaying plant parts and tiny spores were probably right behind them. With the evolution of stronger stems and taller plants, some of the insect scavengers underwent structural modifications that allowed them to leave the moist litter on the ground and withstand exposure to air. The initial absence of competition for edible but aerial plant parts would have favored such modifications. Among other things, sucking, piercing, and chewing mouthparts emerged in some insect lineages. And some of the descendants of those insects developed wings.

The first seed-bearing plants probably made their entrance in the warm, moderately humid coastal forests of the Devonian, some 350 million years ago. These plants gave rise to gymnosperms and, presumably, flowering plants (Chapter Thirty-Eight). Ovules and pollen sacs were borne on modified leaves or scales; often the scales were arranged in conelike structures. The mode of pollination must have been passive indeed, with pollen grains simply drifting on air currents to the ovules.

Now, pollen grains are rich sources of protein. If insects began zeroing in on the pollen at its predictable locations in cones, then they would have begun serving as pollination vectors. (A *vector* is an outside agent that acts like a bridge between two reproductive structures or reproductive stages.) Some of the dustlike pollen would be eaten, but some would cling to the insect body and be transported to ovules. To be sure, insects clambering about reproductive cones would assist pollination in an inadvertent way—but the assistance would be more efficient than that provided by air currents alone. Pollen would be delivered right to the door, so to speak.

The tastier the pollen, the more home deliveries, and the more seeds formed. The greater the number of seeds formed, the greater the reproductive success. What we are describing here is an example of **coevolution**: the joint evolution of two (or more) species interacting in close ecological fashion. When one species evolves, the change affects selection pressures operating between the two species and the other also evolves. Thus, in this case, there was natural selection of variant plants able to attract beneficial insects. Concurrently, there was selection of the vector insects which, because of their ability to recognize a particular food and locate it quickly, would outcompete other foraging insects.

Another, perhaps related change should be mentioned here. Existing pollen-eating beetles have strong mouthparts, and many also chew on ovules that they pollinate. We can speculate that chewing behavior was a selective force contributing to the evolution of floral structure. Instead of being borne naked and vulnerable on cone scales, as they are in gymnosperms, the ovules of today's flowering plants are sequestered inside closed carpels—which offer some protection from hungry insects.

Nectar as an Attractant. Nectaries and their specialized visitors are another example of coevolution. A **nectary** is a glandlike organ that secretes *nectar*, a fluid rich in sugars, amino acids, proteins, and lipids. Before flowering plants emerged, insects were indirectly tapping into nectarlike solutions. The aphid, recall, is a tiny insect that inserts its strawlike mouthpart into phloem (Figure

Table 21.1	Correlation Between Characteristics of Flowering Plants and Their Pollinators
Characteristic of Flowering Plant	Related Characteristic of Pollinator
Floral color Color patterning Floral odor	Sensory apparatus (photoreceptors, olfactory receptors, etc.)
Flower size Flower shape Structure and location of reproductive parts	Size of body Shape of body parts (such as feeding apparatus)
Nectar composition	Diet
Timing of flowering	Foraging behavior (by day, by night, by season)

Figure 21.7 Flowers with red and yellow components, colors that attract bird pollinators. (**a**) Flower of the bird-of-paradise, *Strelitzia reginae*. (**b**) Glorybower, a favorite of hummingbirds. (**c**) Saguaro cactus blossoms, a favorite of gila woodpeckers. (The white component of these blossoms also attracts night-flying bats.)

20.10). The sugary solution transported by phloem is under so much pressure that it courses through the insect body, with the excess exuded from the tail end as a "honeydew" droplet. Ants are so fond of honeydew that they tend "herds" of aphids, even to the extent of protecting them from predators such as wasps. At the same time, some ants protect the plant by attacking insects that chew on it.

Between 100 million and 65 million years ago, the diversity among flowering plants increased dramatically. Fossil ants date from that period. Were those ancient ants tending aphids and protecting plants even then? We might well speculate that nectar coevolved with ben-

eficial insects. Whatever the case, between 60 million and 40 million years ago, nectar became varied in composition and attracted a variety of pollinators with different nutritional needs—insects, then birds and bats.

Existing Flowers and Their Pollinators. Today it is possible to correlate many characteristics of flowering plants with those of their pollination vectors (Table 21.1). As an example of these correlations, consider the glory-bower and similar flowers in which some of the petals form a tube (Figure 21.7). The nectary is deep in the flower, at the base of the tube, and typically it contains copious amounts of nectar. The petals themselves are large and brightly colored, often with red components. Typically these flowers have no odor to speak of.

Could such flowers visually attract insects such as beetles and honeybees? Probably not, because these insects cannot detect red wavelengths. Also, although beetles and honeybees are attracted to strong odors, the flowers offer them no olfactory guides. Even if these insects accidentally scrambled over the flower, there would be no easy way for them to reach the nectary, short of navigating the floral tube, and there is a distinct possibility of drowning in some of the larger nectar cups.

If not insects, then what serves as the pollinators? Birds. Birds have a keen sense of vision, and they can detect red wavelengths—and red petals are like bright flags in the distance for these pollinators. Birds also have a poor sense of smell—so floral odors would be wasted on them. Some birds have long beaks that can penetrate the length of the floral tube, and anthers and stigmas are located where the bird head will brush against them while they drink from the nectary (Figure 21.7). Birds

Figure 21.8 Nectar guide of the marsh marigold (*Caltha palustris*), which is visible to bees. Unlike the human eye, the bee eye can detect ultraviolet light. Although the petals of this flower appear solid yellow to us, its distinctive markings become apparent when the flower is experimentally illuminated with ultraviolet light.

a   b   c

Figure 21.9 Coevolutionary match between a flower and its pollinator. The color of Scotch broom attracts bees. Some of the petals serve as a landing platform that corresponds to the size and shape of the bee body. The pollinator's weight on the platform forces petals apart. The pollen-laden stamens, which are positioned to strike against the bee, are thereby released. The pollen brushes against dense hairs covering the bee body.

Periodically, the bee grooms itself, packing pollen for its trip back to the hive. The orange-colored mass of pollen visible in (**b**) has been packed in a pollen basket, formed by stiff hairs on the outer hind leg. Pollen baskets occur on honeybees, bumblebees, and their kin.

require considerable energy to power their flights, and they depend on more than the fluid of one nectar cup to sustain their high metabolic rate. Thus birds visit many plants of the same species—and in so doing they promote cross-pollination.

In contrast, if a tiny beetle were equipped to locate a nectar cup, it probably never would leave the flower, let alone the plant, and cross-pollination would be thwarted. Thus, some flowers pollinated by beetles (and flies) have what we perceive as strong and awful odors, reminiscent of decaying meat, moist dung, and the like. Perhaps olfactory guides of this sort originally resembled the smells of decaying matter in the forest litter, where beetles originally evolved.

Bees, too, are attracted to strong odors, but for them the preferred odors are sweet. Also, bee eyes are sensitive to wavelengths of yellow, blue, purple, and ultraviolet, but not red. Most flowers they visit have bright yellow, blue, purple, or ultraviolet components—but not red (Figure 21.8). Most bees are covered with plumelike hairs that retain pollen from the flowers being visited. Some bee-pollinated flowers have landing platforms that actually position the bee body in the most favorable way for brushing against pollen-laden anthers (Figure 21.9).

What about butterflies and moths? Butterflies forage by day and are commonly attracted to flowers that are sweet-smelling, often red, and upright, with a more or less horizontal landing platform. Most moths forage by night. They pollinate flowers with strong, sweet fragrances and white or pale-colored petals, which are more visible in the dark (Figure 21.10). Butterflies and moths both have long, narrow mouthparts that correspond to narrow floral tubes or spurs. The most extreme example is the Madagascar hawkmoth. When uncoiled, its proboscis is twenty-two centimeters long— the exact same length as the floral tube of the orchid *Angraecum sesquipedale!* Like other hawkmoths, this moth does not require a landing platform; it hovers in front of the floral tube.

Wind alone disperses the pollen of some flowers, such as the flowers of diverse grass species. These flowers have little or no color or odor; they have no nectar at all. Many do not even have petals. Pollen of wind-pollinated plants typically does not get carried much farther than a hundred meters from its release point. These plants are found in temperate regions of North America, Europe, and eastern Asia. Often, many wind-pollinated trees of certain deciduous species are crowded together. Wind through the bare branches is effective for cross-pollination among them.

Wind-pollinated plants are relatively scarce in wet tropical regions. In these regions, many diverse plants are so crowded that individuals of the same species may be few and far between. Here, cross-pollination depends on birds, bees, and the like, which travel considerable distances and actually search for particular plants.

a

b

c

Figure 21.10 Nectar guides for insect pollinators. (**a**) Bees use the close-range guide of the passion flower (*Passiflora caerula*). (**b**) The vibrant red and red-orange center of *Arctotis acaulis*, a composite, attracts butterflies. (**c**) Stephanotis, like other night-flowering plants, has no distinctive color pattern. Its white petals and strong scent attract moth pollinators (white or pale colors are more visible at night).

Figure 21.11 A few stages in the development of *Capsella* (shepherd's purse), a dicot. All micrographs are longitudinal section but not to the same scale. (**a**) The zygote, showing the arrangement of cell parts. This arrangement is inherited and leads to differentiation during the first divisions that produce the multicelled embryo, identified by yellow boxes in (**b**) through (**d**) .

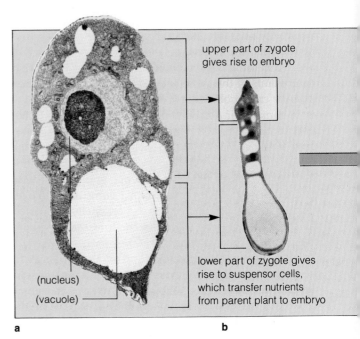

upper part of zygote gives rise to embryo

(nucleus)

(vacuole)

lower part of zygote gives rise to suspensor cells, which transfer nutrients from parent plant to embryo

a

b

a

b

Figure 21.12 Two examples of economically important crop plants. (**a**) The seeds of nutmeg (*Myristica fragrans*) are ground to produce one of the most popular spices throughout the world. The spice called mace comes from the fleshy seed coat, which appears as bands of red tissue in this photograph. (**b**) The pods of cacao (*Theobroma cacao*) contain seeds that are processed into chocolate and cocoa.

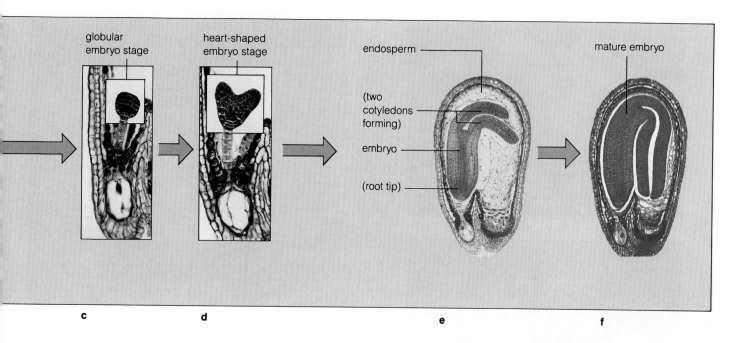

globular embryo stage

heart-shaped embryo stage

endosperm

(two cotyledons forming)

embryo

(root tip)

mature embryo

c   d   e   f

EMBRYONIC DEVELOPMENT

Clearly, the coevolutionary relationships between flowering plants and pollinators are remarkable. As you will see, these plants rely on still other outside agents to assure reproductive success. This part of their story begins with the formation of the zygote, the first cell of the new plant formed through fertilization.

From Zygote to Plant Embryo

When a flowering plant zygote first forms, it is still attached to the parent plant. Even before it begins the mitotic divisions that will produce the multicelled embryo, this single cell already has undergone some development. For example, notice in Figure 21.11 how most organelles, including the nucleus, reside in the top half of a *Capsella* zygote. A vacuole takes up most of the lower half. Once divisions begin, some of the daughter cells give rise only to a simple row of cells (the suspensor) that transfer nutrients from the parent plant to the embryo. Other daughter cells give rise to the mature, multicelled embryo.

Seed and Fruit Formation

The embryo, recall, is housed in an ovule. After double fertilization, the ovule undergoes expansion, integuments harden and thicken, and the endosperm forms. A fully mature ovule is a **seed**; its integuments are the seed coat (Figures 21.12 and 21.13).

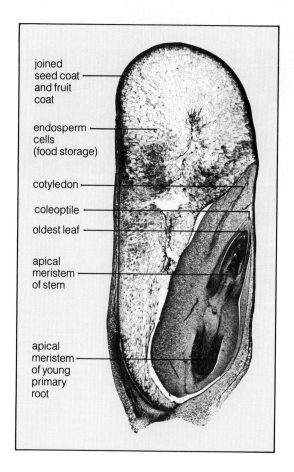

joined seed coat and fruit coat

endosperm cells (food storage)

cotyledon

coleoptile

oldest leaf

apical meristem of stem

apical meristem of young primary root

Figure 21.13 Longitudinal section through a grain of corn (*Zea mays*), a monocot. Most of the volume of this seed is composed of endosperm cells.

a

b

c

Figure 21.14 Fruit formation on an apple tree. (**a**) Petals dropping away from the flower usually signify that fertilization has been successful. After this, the ovary expands (**b**). Sepals and stamens can still be observed on the immature fruit (**c**).

Table 21.2	Kinds of Fruits of Some Flowering Plants	
Type	Characteristics	Some Examples
Simple (formed from single carpel, or two or more united carpels of one flower)	1. Fruit wall *dry; split* at maturity	pea, magnolia, tulip, mustard
	2. Fruit wall *dry; intact* at maturity	sunflower, wheat, rice, maple
	3. Fruit wall *fleshy*	grape, banana, lemon
Aggregate (formed from numerous but separate carpels of single flower)	*Aggregate* (cluster) of matured ovaries (fruits), all attached to common receptacle (modified stem end)	strawberry, blackberry, raspberry
Multiple (formed from carpels of several associated flowers)	*Multiple* matured ovaries, grown together into a mass; may include accessory structures (such as receptacle, sepal, and petal bases)	pineapple, fig, mulberry

In many plants, such as *Capsella*, the embryo develops "seed leaves," or cotyledons, which absorb the endosperm and function in food storage (Figure 21.11). In other plants, the cotyledons remain thin but may produce enzymes that aid in transferring stored food from the endosperm to the germinating seedling. For example, endosperm in corn seeds (Figure 21.13) is filled with proteins, fats, oils, and starch.

While an ovule is developing into a seed, the ovary surrounding it is developing into most or all of the structure called a **fruit**. Sometimes the shriveled sepals and stamens are still present on the mature fruit, as they are on apples (Figure 21.14).

We tend to think of fruits as juicy edible structures, as indeed many are. But fruits are any matured or ripened ovary, with or without accessory parts (Table 21.2). Grains and nuts are dry fruits; the fruit wall of a walnut is dry and intact at maturity. Apples and tomatoes are fleshy fruits. A raspberry is an aggregate of many fruits from one flower. Multiple ovaries remain clustered together in a pineapple (Figure 21.15).

All fruits function in seed protection and dispersal in specific environments. Thus, for example, the light-

floral
remnants   ovary

a

seed (in carpel)

wing

b

receptacle

ovary

c

Figure 21.15 (**a**) Pineapple, a multiple fruit; (**b**) maple, a dry fruit; and (**c**) strawberry, an aggregate fruit.

weight, tiny seeds of the orchid fruit can be widely dispersed on air currents when the ovarian walls rupture. A maple fruit has winglike extensions (Figure 21.15). When the fruit drops, the wings cause it to spin sideways. With such spinnings, seeds can be dispersed to new locations, where they will not have to compete with the parent plant for soil water and minerals. Many fruits have hooks, spines, hairs, or sticky surfaces. By such means they adhere to feathers or fur of animals that brush against them—and thereby can be taxied to new locations.

Fleshy fruits such as those of strawberries and cherries are tasty to many animals and are well adapted for moving through the animal gut, which contains powerful digestive enzymes. The enzymes remove just enough of the hard seed coats to enhance the likelihood of successful germination when the indigestible seeds are expelled from the body.

Fruits are specialized structures that protect the seed and enhance its dispersal in specific environments.

ASEXUAL REPRODUCTION OF FLOWERING PLANTS

The sexual reproductive modes just described are the most prevalent among flowering plants. But plants also can reproduce asexually by several means (Table 21.3). For example, a strawberry plant can send out horizontal aboveground stems, known as *runners*. Along such runners, new roots and shoots develop at every other node. As another example, oranges reproduce every so often by *parthenogenesis*: the development of an embryo from an unfertilized egg. In some plants, parthenogenesis is stimulated when pollen contacts a stigma even though a pollen tube has not grown down the style. Hormones (either formed in the stigma or produced by pollen grains) apparently diffuse down to the unfertilized egg and stimulate embryo formation. The embryos become $2n$ by fusion of products of egg mitosis, or a $2n$ cell outside the gametophyte may be stimulated to form an embryo.

Vegetative reproduction occurs naturally among wounded plants. For example, when a leaf falls or is torn away from a jade plant, a new plant can develop from

Table 21.3 Asexual Reproductive Modes of Flowering Plants		
Mechanism	Representative Species	Characteristics
Reproduction on modified stems:		
1. Runner	Strawberry	*New plants arise at nodes of an aboveground horizontal stem*
2. Rhizome	Bermuda grass	*New plants arise at nodes of underground horizontal stem*
3. Corm	Gladiolus	*New plant arises from axillary bud on short, thick, vertical underground stem*
4. Tuber	Potato	*New shoots arise from axillary buds on tubers (enlarged tips of slender underground rhizomes)*
5. Bulb	Onion, lily	*New bulb arises from axillary bud on short underground stem*
Parthenogenesis	Orange, rose	*Embryo develops without nuclear or cellular fusion (e.g., from unfertilized haploid egg; or adventitiously, from tissue surrounding embryo sac)*
Vegetative propagation	Jade plant, African violet	*New plant develops from tissue or organ (e.g., a leaf) that drops or is separated from plant*
Tissue culture propagation	Carrot, corn, wheat, rice	*New plant arises from cell in parent plant that is not irreversibly differentiated; laboratory technique only*

meristematic tissue adjacent to vascular bundles just inside the wound. (Meristematic cells are cells that retain the potential for continued division.)

Vegetative propagation also occurs with a little help from humans. For example, a whole orchard of individual pear trees might have been grown from cuttings or buds of a parent tree. Obviously, the "offspring" have the same DNA instructions as the parent. Any organism reproduced asexually in such a way that it is genetically identical with its parent is a **clone**. Many of our food crops, including Bartlett pears, MacIntosh apples, and Thompson seedless grapes, are clonal populations.

In some pioneering experiments, Frederick Steward and his coworkers propagated carrot plants by culturing cells from differentiated carrot tissues (page 224). Such induced vegetative propagations are also accomplished with shoot tips or other parts of individual plants. The techniques are particularly useful when an advantageous mutant arises. For example, one such mutant might show resistance to a disease that is particularly crippling to wild-type plants of the same species. Tissue culture propagation can lead to hundreds and even thousands of identical plants from one mutant specimen. This technique is already being used in efforts to improve major food crops, such as corn, wheat, rice, and peas.

SUMMARY

1. In flowering plant life cycles, a multicelled diploid or polyploid stage (the sporophyte, or "spore-producing plant") alternates with a multicelled haploid stage (the gametophyte, or "gamete-producing plant").

2. The sporophyte is a vegetative body with roots, stems, leaves and, at some point, flowers. Certain cells in the floral tissues divide by meiosis and form meiospores (haploid cells that will divide by mitosis to form the gametophytes).

3. A flower consists of one or more carpels (female reproductive structures), stamens (male reproductive structures), petals, and sepals, most or all of which are attached to the receptacle.

   a. Each stamen has a slender stalk (filament) capped by an anther. An anther is a structure containing pollen sacs, or chambers in which pollen grains develop. A pollen grain is a male gametophyte.

   b. Each carpel includes an ovary and a stigma. The carpel is a closed vessel in which eggs develop, fertilization takes place, and seeds mature. The stigma is the landing platform for pollen.

4. "Microspores" are the meiospores produced in pollen sacs of anthers. Each divides once, producing a two-celled pollen grain. One cell produces the sperm; the

other develops into a pollen tube that grows through carpel tissues, transporting the sperm with it.

5. "Megaspores" are the meiospores produced in carpels. They form inside the nucellus, a tissue mass that grows from the inner wall of the ovary and becomes covered with one or two protective layers (integuments). The nucellus, integuments, and the stalk attaching them to the ovarian wall collectively are called an ovule.

6. Of every four megaspores produced by meiosis, all but one usually disintegrate. The one remaining undergoes mitosis three times without cytoplasmic division, the nuclei migrate to specified positions, then cytokinesis produces the female gametophyte (embryo sac). It is a seven-celled, eight-nuclei body, with one cell being the egg. The cell with two nuclei (endosperm mother cell) will help give rise to endosperm, a nutritive tissue.

7. Upon pollination (transfer of pollen grains to an appropriate stigma), the pollen grain germinates, the pollen tube develops and, later, two sperm cells form. The germinated pollen grain (with its two sperm) is the mature male gametophyte.

8. Many flowering plants coevolved with insects, birds, and bats, which serve as pollination vectors (outside agents that act like bridges between male and female reproductive structures of flowers).

   a. Coevolution is the joint evolution of two or more species interacting in close ecological fashion; when one evolves, the change affects selection pressures operating between them and the other species also evolves.

   b. Thus the color, patterning, and odor of flowers can be correlated with the sensory apparatus (such as photoreceptors) of certain pollinators. Floral size and shape, and the structure and location of reproductive parts, can be correlated with the size and shape of the pollinator's body parts (such as mouthparts).

9. At double fertilization, one sperm nucleus fuses with one egg nucleus to form a diploid (2n) zygote. The other sperm nucleus and both nuclei of the endosperm mother cell fuse to form a cell that will give rise to triploid (3n) nutritive tissue in the forthcoming seed.

10. After double fertilization, the ovule expands, the integuments harden and thicken, and endosperm forms. A fully mature ovule is a "seed," with its integuments forming the seed coat.

11. Simultaneously, the ovary undergoes development into the fruit. Mature fruits can be fleshy or dry; they can be simple, aggregate, or multiple structures.

12. Fruits function to protect and disperse the seeds. Fleshy fruits attract animals and other "dispersing" agents; lightweight fruits can be dispersed by winds. Some fruits have hooks and such that attach to animals.

13. The key characteristics of asexual reproduction of flowering plants may be summarized this way:

   a. New plants may arise asexually by mitotic divisions at nodes or buds along modified stems of the parent plant.

   b. Some plants reproduce asexually by parthenogenesis (development of an embryo from an unfertilized egg).

   c. New plants also may arise by vegetative propagation, either natural or induced.

Review Questions

1. Label the floral parts in the following diagram. Explain floral function by relating some floral structures to events in the life cycle of flowering plants: 279

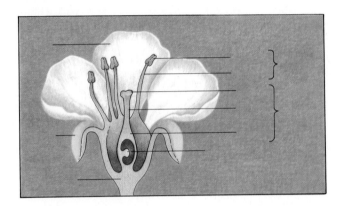

2. Distinguish between these terms:
   a. Megaspore and microspore 280-281
   b. Pollination and fertilization 292
   c. Pollen grain and pollen tube 280
   d. Ovule and female gametophyte 280-282

3. Observe the kinds of flowers growing in the area where you live. On the basis of what you have read about the likely coevolutionary links between flowering plants and their pollinators, can you perceive what kinds of pollinating agents your floral neighbors might depend upon? (Refer to pages 282-285)

4. Describe the steps involved in the formation of an eight-cell, seven-nucleate embryo sac. 280-282

5. Describe what happens to the endosperm mother cell and the egg following fertilization. 282

6. Give some specific examples of adaptations that enhance seed protection and dispersal. 288-289

Readings

Proctor, M., and P. Yeo. 1973. *The Pollination of Flowers*. New York: Taplinger. Beautifully illustrated introduction to pollination.

Raven, P., R. Evert, and H. Curtis. 1986. *Biology of Plants*. Fourth edition. New York: Worth. Outstanding illustrations.

Rost, T. et al. 1984. *Botany: A Brief Introduction to Plant Biology*. Second edition. New York: Wiley.

Salisbury, F., and C. Ross. 1985. *Plant Physiology*. Third edition. Belmont, California: Wadsworth.

22

PLANT GROWTH
AND DEVELOPMENT

In the preceding chapter, we traced the events by which a zygote of a flowering plant becomes transformed into an embryo, housed inside a protective seed coat. Before and after seed dispersal from the parent plant, embryonic growth idles. But at some point following dispersal, **seed germination** occurs: the embryo absorbs water, resumes growth, and breaks through the seed coat. Now the embryo increases in volume and mass; its cells divide and develop into the tissues and organs of the seedling. For many species, leaves unfurl and, as the life cycle progresses, flowers, fruits, and new seeds develop. Then, in autumn, old leaves drop away from the plant. What internal mechanisms govern such developmental events, and what kinds of environmental signals set them in motion? In this chapter, we will consider some answers (and best guesses) to these questions.

SEED GERMINATION

Water, oxygen, temperature, and usually light are major environmental factors that influence seed germination. Most mature seeds do not contain enough water for cell expansion or metabolism. In many parts of the world, water availability is seasonal and germination coincides with the return of spring rains. In a process called **imbibition**, water molecules move into the seed, being especially attracted to hydrophilic groups of the stored proteins. The seed swells as more and more water molecules move inside and the coat finally ruptures.

Once the seed coat splits, oxygen moves in more easily from the surrounding air. Cells of the embryo switch to aerobic pathways and metabolism moves into high gear. The embryo increases in volume, giving rise to the seedling (Figure 22.1).

Figure 22.1 (a) Seedling of corn, a monocot, and (b) seedling of a bean, a dicot.

a   primary root   b

PATTERNS OF GROWTH

Inside the germinating seeds of nearly all species, cells of the **radicle** (the embryonic root) are the first to grow. Continued cell division and cell elongation in the radicle produce the **primary root** (the first root of the plant). When the primary root visibly protrudes from the seed coat, germination is completed.

As the seed germinates, the embryo within it grows, then embryonic cells and their daughter cells grow in ways that lead to roots, stems, and leaves (Figure 22.2). Often the cell divisions are unequal, with one daughter cell ending up with more cytoplasm than the other. The unequal distribution of cytoplasmic substances gives rise to differences in cell composition and structure. In growing tissues, cells divide in different planes and expand in different directions. These genetically controlled differences lead to tissues and organs with diverse shapes.

Keep in mind that division itself does not constitute growth; the two daughter cells have about the same volume as the parent cell (Figure 22.2). However, on the average, about half the daughter cells enlarge—often by twenty times! Thus, cell division increases the *capacity* for overall growth. The remaining daughter cells stay meristematic (they retain the capacity to divide). They grow no larger than the parent cell, then probably divide again.

Water uptake drives cell enlargement. When water moves into new cells, increased turgor pressure (page 92) forces the cell wall to expand. In some ways, the expansion resembles a balloon being blown up with air. Balloons with soft walls are easy to inflate; cells with soft walls grow rapidly under little turgor pressure. A major difference is that a balloon wall gets thinner as it "grows," but a cell wall does not. New polysaccharides are added to the wall during growth, and more cytoplasm forms between the wall and the central vacuole (Figure 22.2).

The primary walls of young cells are somewhat *elastic* and can be stretched, although they do show some tendency to return to their original shape (much as a filled balloon does when the air is let out). But the walls also are *plastic* and tend to retain much of their stretched shape (as does a blown-up bubble of bubble gum when the air has been let out). These two expansive properties encourage **true growth**, as defined by an increase in cell volume.

PLANT HORMONES

Types of Plant Hormones

Like the growth of other multicelled organisms, plant growth and development are influenced in powerful but poorly understood ways by hormones. A **hormone** is a signaling molecule released from one cell and transported to (typically) nonadjacent target cells. A *target cell* is one with receptor sites for a given signaling molecule. In plants, the receptors have not been identified. They may be proteins at the surface of the plasma membrane or proteins within the cell.

Five hormones (or groups of hormones) are known to exist in most plants, and there is evidence of others. The five are auxins, gibberellins, cytokinins, abscisic acid, and ethylene (Table 22.1).

Auxins are best known for their ability to promote elongation of stem cells. The main auxin is indoleacetic

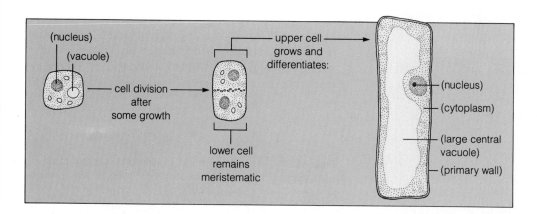

Figure 22.2 The nature and direction of plant growth. Through divisions, meristems provide new cells for growth. At the tip of a young root, small meristematic cells double in size, then divide. The lower cell remains meristematic. The upper cell grows into a mature parenchyma cell of pith or cortex, for example. Tiny vacuoles in young cells absorb water, fuse, and form the central vacuole of mature parenchyma cells.

Table 22.1	Main Plant Hormones and Some Known (or Suspected) Effects
Auxins	*Promote cell elongation in coleoptiles and stems; long thought to be involved in phototropism and gravitropism*
Gibberellins	*Promote stem elongation (especially in dwarf plants); might help break dormancy of seeds and buds*
Cytokinins	*Promote cell division; promote leaf expansion and retard leaf aging*
Abscisic acid	*Promotes stomatal closure; might trigger bud and seed dormancy*
Ethylene	*Promotes fruit ripening; promotes abscission of leaves, flowers, and fruits*
Florigen (?)	*Arbitrary designation for as-yet unidentified hormone (or hormones) thought to cause flowering*

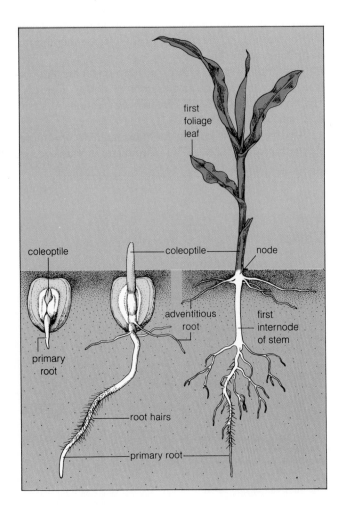

first foliage leaf

coleoptile

coleoptile — node

adventitious root

first internode of stem

primary root

root hairs

primary root

acid (IAA). Several synthetic compounds behave much like natural auxins and are used as *herbicides*, which are selective killers of plant growth. Like other physiologically active compounds, they are toxic when applied in high concentrations to some plant species but not to others. For example, one of the most common herbicides, 2,4-dichlorophenoxyacetic acid (2,4-D), is a potent killer of dicot weeds but has little effect on grasses. It is used to kill weeds in fields of cereal grains.

Gibberellins, like auxins, are best known for their promotion of stem elongation. More than sixty kinds have been identified in plants and fungi. The most familiar kind, *gibberellic acid*, is widely used in growth experiments.

Cytokinins were named for their ability to stimulate cell division (the name refers to "cytokinesis"). Leaf cells also grow larger when exposed to these hormones. The most common and abundant cytokinin seems to be *zeatin*, first isolated from immature seeds of *Zea Mays* (corn).

Abscisic acid (ABA) is important in promoting stomatal closure. It also promotes seed dormancy, bud dormancy, and resistance to water stress in many species. This hormone was once thought to be important in abscission (the loss of flowers, fruits, and leaves from the plant body), but this effect does not seem to be common.

Among other things, **ethylene** (C_2H_4) stimulates most fruits into ripening. The ancient Chinese knew that a bowl of fruit ripens faster when incense is burned near it, although they didn't know that ethylene present in the smoke promotes its ripening.

It is not yet possible to distinguish between hormonal and environmental effects on plant development. One reason is that the environment often controls the amount and distribution of hormones. Even so, let's look at what *is* known by following the development of a monocot (corn) and a dicot (soybean).

Examples of Hormonal Action

In the seedlings of corn and other grasses, a **coleoptile** is a hollow, cylindrical organ that protects the tender young leaves growing within it (Figure 22.3). Without a coleoptile, the young leaves would be torn apart by soil

Figure 22.3 A few stages in the development of a corn plant (*Zea mays*), a monocot. The coleoptile encloses the young leaves and protects them as the shoot grows up through the soil. Adventitious roots develop at the first-formed node at the coleoptile base. When a corn seed is planted deeply, the first internode of the stem elongates as shown here. When the seed is close to the soil surface, light inhibits elongation of the first internode; then the primary and adventitious roots look as if they are all originating at the same region, just below the soil surface.

particles during early growth. While underground, both the coleoptile and the tightly curled oldest leaf grow in a coordinated fashion. Once exposed to sunlight, the coleoptile stops elongating; its task is completed. The leaf now breaks through the coleoptile, uncurls and becomes flattened, engages in photosynthesis, and provides organic molecules for the plant parts below.

The auxin IAA stimulates coleoptile growth (Figure 22.4). IAA moves from the tip of the coleoptile into cells between the tip and the base. Metabolic activities change in those cells and, as a result, cell walls become more plastic—and plasticity encourages elongation.

Gibberellin also can promote growth, especially of stems that never elongate much. For example, dwarf varieties of corn (with short stems) often carry a mutant gene that codes for an enzyme needed in gibberellin production. After application of a gibberellin, the stems of dwarf plants grow longer; yet similar application to stems of a normal variety of corn does not have much of an effect (Figure 22.5).

Most or all hormones required by roots and leaves are synthesized by root and leaf cells. Generally, roots or leaves do not respond much to hormone applications.

Hormones also promote growth of a soybean stem (Figure 22.6). Like all dicots, soybeans have young, growing cells near the tip of the stem; cells at the base are older and larger. An auxin is synthesized near the tip, especially in very young leaves. When the tip is cut off, the dicot stem stops elongating; when an auxin is applied to the stump, growth resumes. This response suggests that an auxin produced at the tip of a dicot stem promotes growth of stem cells below.

Finally, gardeners have long known that pinching off shoot tips encourages increased branching of the stem below and produces bushier plants. An as-yet unknown hormone associated with the shoot tip normally inhibits lateral bud growth, this being called **apical dominance**. As the shoot elongates, the inhibitory effect lessens and lateral bud development proceeds.

PLANT RESPONSES TO THE ENVIRONMENT

Once an embryonic root breaks through the seed coat and begins its downward growth, the plant cannot move on if conditions are not ideal for growth and development. Yet it still has a genetically based capacity to respond to a range of conditions characteristic of the environment in which that type of plant evolved. *Through interactions with the environment, the plant can adjust its patterns of growth.* It can make adjustments to unique circumstances. (For example, it might sprout alongside a highway, a paper bag tossed out of a passing car may come to rest right on top of it, and the shoot

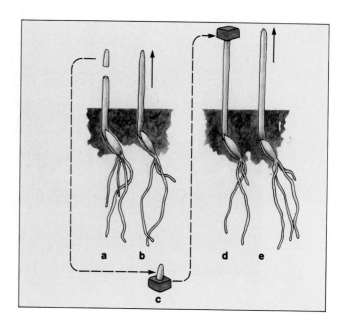

Figure 22.4 Experiment demonstrating that IAA present in the tip of a grass coleoptile promotes elongation of cells below. (**a**) An oat coleoptile tip is cut off. Compared with a normal coleoptile (**b**), the stump does not elongate much. When the excised tip is placed on a tiny block of agar for several hours (**c**), IAA moves into the agar. When the agar is placed on another de-tipped coleoptile (**d**), elongation proceeds about as rapidly as in an intact coleoptile (**e**).

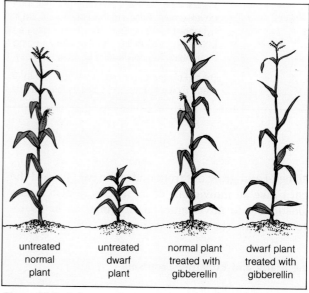

Figure 22.5 Influence of gibberellin on the height of a normal corn plant, compared to its influence on a dwarf corn plant that is mutant for one gene that codes for an enzyme necessary in gibberellin production.

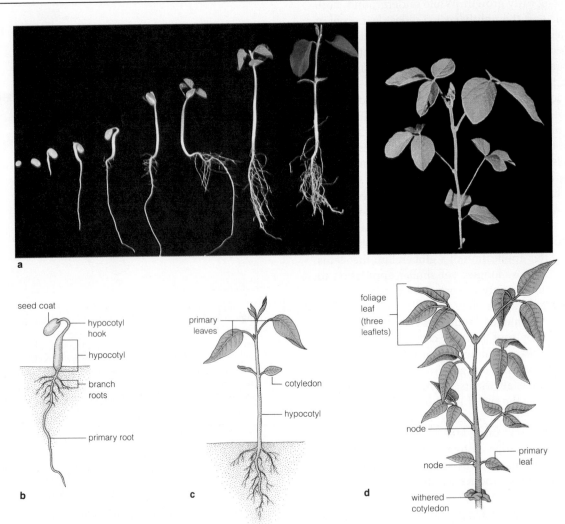

Figure 22.6 Some stages in the development of a soybean plant, a dicot. (**a**) Germination and seedling development. Food-storing cotyledons are lifted aboveground when cells of the hypocotyl elongate. The hypocotyl forms a hook (**b**), which makes a channel through the soil as it grows. The cotyledons can be pulled up through the channel without being torn apart. At the soil surface, light causes the hook to straighten. The cotyledons become photosynthetic for several days, then wither and fall off.

(**c–d**) Foliage leaves, each divided into three leaflets, develop after the first (primary) leaves along the stem. Flowers develop in axillary buds at the four upper nodes shown here.

will bend and grow out from under the bag, toward the light.) Like all individuals of its species, it also makes adjustments to environmental rhythms, as in its responses to the changing seasons.

The Many and Puzzling Tropisms

Often an environmental stimulus is more intense on one side of a stem or some other plant part. The plant responds by making adjustments in the direction and rate of growth of that part. Growth responses of this sort are called "tropisms," and no one knows how they work.

Phototropism. When light strikes one side of a plant more than the other, the stem will curve toward the light (Figure 22.7). Also, the flat surface of its leaf blades will turn until the leaf is perpendicular to the light. This growth response is a form of **phototropism** (growth toward or away from light shining mainly on one side of the organism).

In the late 1800s, Charles Darwin showed that a coleoptile curves more rapidly when light strikes its tip rather than the bending portion below. In the 1920s, Frits Went realized that a growth-promoting substance is present in the tips of coleoptiles. He called the substance

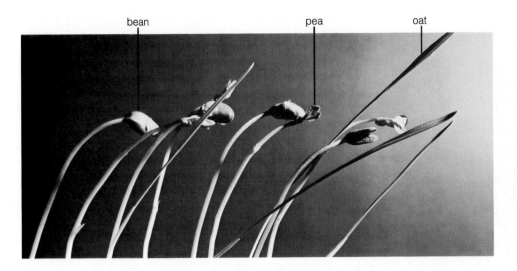

bean   pea   oat

Figure 22.7 Phototropism in bean, pea, and oat seedlings. These plants were grown in darkness, then exposed to light from the right side for a few hours before being photographed.

auxin (after the Greek word meaning "to increase"). Went showed that the substance must move down from the tip, especially into cells that are not as exposed to the light, and then stimulate the cells into elongating.

Today we know that Went's auxin is IAA. We also know that light of blue wavelengths is the main stimulus for phototropism. A large, yellow pigment molecule called *flavoprotein* probably plays a role in phototropic bending, given its capacity to absorb blue wavelengths. We have no idea how that absorption might cause IAA to move horizontally across a coleoptile, to the shaded side.

Gravitropism. A growth response to the earth's gravitational force is called **gravitropism** (or geotropism). You can observe this response after seeds germinate: the primary root always curves down and the coleoptile or stem always curves up. Look at Figure 22.8, which shows what happened when a potted seedling was turned on its side in a dark room. Cell elongation on the upper side of the horizontal stem decreased markedly but increased on the lower side. With the adjusted growth pattern, the stem curved upward, even in the absence of light.

What role, if any, do plant hormones play in gravitropism? At least in bending coleoptiles, some downward movement of IAA has been detected. But the IAA concentration gradient from the tip to the base of the coleoptile just doesn't seem to be great enough or to form fast enough to account for the rapid, extensive differences in growth that lead to bending. Besides, many dicot stems bend in response to gravity even though the IAA gradient seems to be negligible or nonexistent.

Perhaps the main adjustment made by a horizontal stem of a seedling is in the *sensitivity* of cells to IAA already there. Cells on the bottom of the stem, which grow faster, might become more sensitive to IAA; those

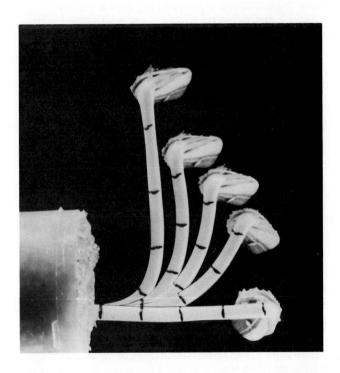

Figure 22.8 Composite time-lapse photograph of gravitropism in a dark-grown sunflower seedling. In this plant, the two cotyledons emerge aboveground, because the stem portion just below the cotyledons elongates. Just before this five-day-old plant was positioned horizontally, it was marked at 0.5-centimeter intervals. After thirty minutes, upward curvature was detectable. The most upright position shown was reached within two hours.

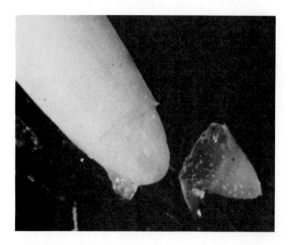

Figure 22.9 Root cap of a young corn plant, shown removed from the root tip. With cap removal, cells just behind the apical meristem elongate faster. With cap replacement, their growth slows. These and other experiments suggest that a growth inhibitor is synthesized in the root cap and is transported to the bottom of a horizontally positioned root. The inhibitor causes the root to curve downward as it continues to grow.

Figure 22.10 Effect of mechanical stress on tomato plants. The plant to the far left was the control; it was grown in a greenhouse, protected from wind and rain. The center plant was mechanically shaken thirty seconds at 280 rpm for twenty-eight consecutive days. The plant to the far right received two such shakings daily for twenty-eight days.

on top might become less sensitive. Perhaps calcium ions or gibberellins figure in the growth response; they, too, are transported from one side to the other of a horizontal coleoptile or stem. These possibilities are just now being investigated, although they were first proposed several decades ago.

What about gravitropism in roots? The root cap plays a role here. A horizontally positioned root will not curve downward if its root cap has been surgically removed, but it will if the cap is put back on (Figure 22.9).

Elongating cells don't stop growing when the root cap is removed; if anything, they grow faster. Could root cap cells contain a growth inhibitor that becomes redistributed in a horizontally oriented root? If gravity somehow causes the inhibitor to move out of the cap and accumulate in cells on the lower side of the horizontally positioned root, those cells would not elongate as much as cells on the upper side—and the root would curve downward. The identity of the growth inhibitor continues to elude us. At one time abscisic acid seemed to be the most likely candidate, but there is now strong evidence that it has nothing to do with root gravitropism.

Thigmotropism. Peas, beans, and many other plants with long, slender stems are climbing vines; they generally do not grow upright without physical support. These plants show **thigmotropism**, or unequal growth resulting from physical contact with solid objects in their surroundings. Suppose one side of the stem grows against a fencepost or the trunk of a tree. Cells stop elongating on the side of the stem making contact, and within minutes the stem starts to curl around the post or trunk. It might do so several times before cells on both sides start elongating at about the same rate once again. How contact affects elongation is a mystery, although auxin and ethylene seem to be involved in the response.

Response to Mechanical Stress

Certain kinds of mechanical stress can inhibit growth of the whole plant, although stem elongation is especially affected. Contact with rain, grazing animals, farm machinery, even winds can inhibit overall growth. Shak-

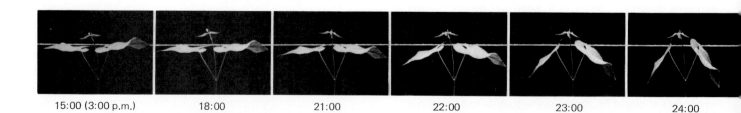

| 15:00 (3:00 p.m.) | 18:00 | 21:00 | 22:00 | 23:00 | 24:00 |

ing some plants daily for a brief period can do the same thing (Figure 22.10). Such responses to mechanical stress help explain differences between plants of the same species when some are grown indoors. Often the plants growing outdoors are shorter, have somewhat thicker stems, and are not easily blown over. These characteristics appear to be responses to wind stress, although the response mechanism is not understood.

Biological Clocks in Plants

Plant growth and development—indeed, nearly all plant activities—are influenced by biological clocks. A **biological clock** is an internal time-measuring mechanism that has a biochemical basis. Such mechanisms have not been identified, although there is speculation that components of the plasma membrane are central to their operation.

There is plenty of indirect evidence of biological clocks. Flowers open in the morning and close at dusk. Some flowers secrete nectar at specific times of day, thereby prompting visits only at those times by bees and other pollinators (which have biological clocks of their own). Leaves of many plants, especially beans and other legumes, are horizontal in the daytime but fold closer to the stem at night (Figure 22.11). Remarkably, even if you keep one of those plants in constant light or darkness for a few days, it will fold its leaves into the "sleep" position in the evening anyway! Clearly the plant has a means of measuring time that is independent of light-on (sunrise) and light-off (sunset) signals.

Activities that occur regularly in cycles of about twenty-four hours are called **circadian rhythms** (from the Latin *circa*, meaning about; and *dies*, meaning day).

And yet, environmental conditions in a twenty-four-hour period are not the same in summer as they are in winter. For example, in North America, winter temperatures are cooler and daylength is shorter. Somehow, biological clocks must be reset, for plants make *seasonal adjustments* in their patterns of growth, development, and reproduction. Let's look first at one time-measuring mechanism before we turn to the question of how that mechanism is continually synchronized with changing conditions.

Photoperiodism

A biological response to a change in the relative length of daylight and darkness in a twenty-four-hour cycle is called **photoperiodism**. What is the biochemical basis of this response? **Phytochrome**, a blue-green pigment molecule, is known to be part of a switching mechanism that promotes or inhibits growth of different plant parts. The switch is turned on when phytochrome absorbs light of red wavelengths; it is turned off (or kept off) when phytochrome absorbs light of far-red wavelengths. After sunrise, red wavelengths dominate the sky. Then, phytochrome is converted to its active form (Pfr). In darkness, phytochrome reverts to its inactive form (Pr), and plant responses are curtailed (Figure 22.12).

Pfr takes part in controls over seed germination, stem elongation, leaf expansion, stem branching, and the formation of flowers, fruits, and seeds. When plants that are adapted to full sunlight are grown in darkness, the effects of Pfr deficiency are readily apparent. The plants put more resources into stem elongation and less into leaf expansion or stem branching (Figure 22.13). Shaded plants show the same sort of responses. (Plants growing above them absorb most of the red wavelengths and use the absorbed energy in photosynthesis, hence the shaded plants are deficient in Pfr.)

The biochemical pathways by which Pfr exerts its effects on plant growth are not understood. Phytochrome molecules probably are embedded in plasma membranes, and conversion of Pr to Pfr may be

Figure 22.11 Leaf movements in bean plants. These so-called sleep movements are among the best-studied rhythms that verify the existence of a biological clock. What is their advantage? Perhaps the sleep position prevents bright moonlight from being absorbed by phytochrome in the leaves (moonlight could otherwise interrupt a dark period necessary to induce flowering). Or perhaps it helps slow heat loss from leaves (folded leaves radiate heat to each other rather than to the cold air around them).

The bean plant shown here was kept in constant darkness, and its leaf movements continued independently of sunrise (6 AM) and sunset (6 PM).

1:00 (a.m.)   2:00   4:00   6:00   9:00   12:00 (noon)

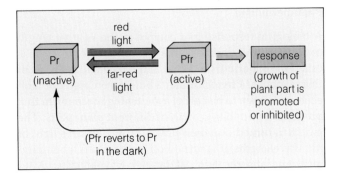

Figure 22.12 Interconversion of phytochrome from the active form (Pfr) to the inactive form (Pr). This pigment is part of a switching mechanism that can promote or inhibit growth of different plant parts.

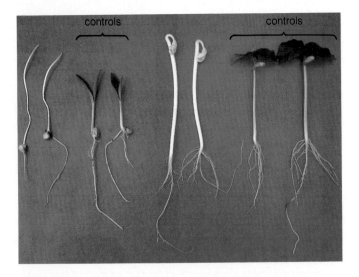

Figure 22.13 Effects of the absence of light on young corn and bean plants. The two plants at the right of each group served as the control; they were grown in a greenhouse. The others were grown in darkness for eight days. Dark-grown plants were yellow: they could form carotenoid pigments but not chlorophyll in darkness. They also had longer stems, smaller leaves, and smaller root systems. (Why do you suppose roots of dark-grown plants grow less, considering that those of light-grown plants aren't exposed to light, either?)

necessary before certain hormones can bind to the membrane or move into the cytoplasm of target cells. Whatever the pathway, there is evidence that some genes but not others are being transcribed when phytochrome is in its active form. This suggests that phytochrome activation helps control which enzymes are being produced in particular cells—and specific enzymes are necessary for specific growth responses.

THE FLOWERING PROCESS

As a flowering plant matures, its activities become directed toward producing flowers, seeds, and fruits. Annuals such as corn, soybeans, and peas begin flowering after only a few months of growth. Perennials such as roses typically flower each successive year or after several years of vegetative growth. Biennials such as carrots and cabbages typically produce only roots, stems, and leaves the first growing season, die back to soil level in autumn, then grow a new flower-forming stem from a bud that remains alive in a protected stem region underground.

The question becomes this: How does a particular plant "know" when to flower? Apparently temperature is one environmental cue. Unless buds of some biennials and perennials are exposed to low winter temperatures, flowers do not form on stems in spring. Low-temperature stimulation of flowering is called **vernalization** (after a Latin term meaning "to make springlike").

In themselves, temperature changes are somewhat unreliable cues about what season is approaching, because weather is often fickle. A better cue is the relative length of day and night, which each year is nearly constant at the same day of the month in a given region. For example, in response to the shorter daylengths of late summer and early fall, many plants form flowers and seeds, develop dormant buds, attain more cold hardiness, and shed their leaves (Figure 22.14). These responses are so predictable that flowering plants can be categorized in terms of daylength:

Long-day plants flower in the spring when daylength becomes longer than some critical value. The critical value can be longer or shorter than twelve hours.

Short-day plants flower in late summer or early autumn when daylength becomes shorter than some critical value. The critical value can be longer or shorter than twelve hours.

Day-neutral plants flower whenever they become mature enough to do so, irrespective of daylength.

To get an idea of the effect of daylength on flowering, look at Figure 22.15, which shows what happens when some long-day plants are grown under short-day conditions and under long-day conditions. These examples illustrate why long-day plants such as spinach never flower and produce seeds in the tropics. In order to flower, spinach needs fourteen hours of light each day for two weeks—and this never happens in the tropics.

The long days of spring promote flowering of many species, including most biennials and winter wheats (which are planted in fall and are able to survive winter conditions). The shorter days and longer nights of late summer promote flowering in other species. At the equator, daylength remains nearly constant, and here you will find day-neutral plants. Many crop plants, including corn, peas, and tomatoes, are also day-neutral. These crop plants have been subject to artificial selection for many generations, and variant forms can flower and produce seeds under a wide range of daylengths.

One short-day plant, the cocklebur, measures time with uncommon sensitivity. Just a single night longer than $8\frac{1}{2}$ hours is enough to promote flowering in most cockleburs. Yet if that dark period is experimentally interrupted with even a minute or two of artificial light, flowering is inhibited! This type of interruption also inhibits flowering in less sensitive short-day plants. Why is this so?

For all short-day plants, red light interrupts the dark period most effectively—and red light is detected by phytochrome. During the day, phytochrome is mostly in the active form (Pfr), which is needed during daylight for flowering to occur. During the night, if red light interrupts the dark period for short-day plants, Pr is then converted to Pfr at night, and flowering is inhibited.

In long-day plants, the extended daylengths provide Pfr for longer periods. As a result, these plants flower more rapidly than do short-day plants, and they produce more flowers.

Again, we do not know how Pfr actually exerts its effects on growth responses such as flowering, but hormonal action is thought to be involved. In anticipation of its discovery, the elusive hormone thought to control flowering was designated **florigen** decades ago. Evidence for the existence of florigen comes from experiments with cocklebur and other plants. Flowers can form only at bud primordia (page 255). Before flowers can form, the bud primordia have to stop producing new cells for leaves and stems. Somehow, meristematic cells in those regions receive signals to start the developmental events that produce flowers.

Experiments suggest that the signal (probably florigen) is produced in leaves and transported to the bud primordia. For example, when all but one leaf is trimmed from a cocklebur plant, and when that leaf is covered with black paper for at least $8\frac{1}{2}$ hours, the plant flowers. But when the leaf is cut off immediately after its dark period, the plant does *not* flower. Apparently some timing mechanism in the leaf detects the length of day and night, and florigen is produced in response. (If the leaf is cut off quickly, the substance remains in the discarded leaf.)

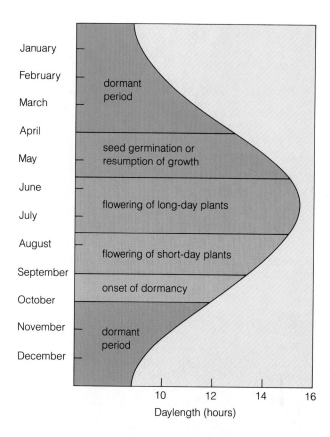

Figure 22.14 Correlation between daylength and plant developmental events in temperate regions of North America. (These regions show seasonal variation in rainfall and temperature.)

Figure 22.15 Effect of daylength on flowering of long-day plants (LDP). In each photograph, the plant on the left was grown under short-day conditions; the plant on the right was grown under long-day conditions. (**a**) *Spinacia* or spinach, (**b,c**) *Sedum spectabile* and *Kalanchoe*.

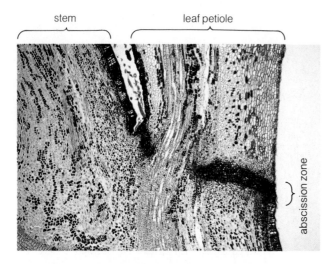

stem | leaf petiole

abscission zone

Figure 22.16 Abscission zone in a maple (*Acer*). This longitudinal section is through the base of the petiole of a leaf.

flowers removed | control group

Figure 22.17 Delay of senescence in soybean plants as a result of the daily removal of flower buds.

SENESCENCE

Plants make a major investment in their reproduction. Nutrients are actually withdrawn from the leaves, roots, and stems and are distributed to newly forming flowers, fruits, and seeds. In older leaves, enzymes synthesized seemingly on demand degrade proteins, chlorophyll, and other organic molecules. The breakdown products then move through the phloem (along with mineral ions) to reproductive structures, where they are converted to fats, starches, and proteins in seeds. Annuals and most perennial plants end up with tan-colored, dead leaves. In deciduous species (which shed leaves at the end of each growing season), nutrients are transported to pa-

renchyma cells in twigs, stems, and roots for storage before the leaves wither and fall off.

The dropping off of leaves (or flowers, fruits, or other plant parts) is called **abscission** (Figure 22.16). Ethylene is formed in cells near the break points and may trigger the process. Studies suggest that ethylene activates enzymes that break down the polysaccharides in cell walls in a narrow zone (the abscission zone).

The sum total of processes leading to the death of a plant or any of its structures is called **senescence**. The drain of nutrients during the growth of reproductive parts might be one stimulus for senescence of leaves, stems, and roots. When the drain of nutrients is stopped (for example, by removing each newly emerging flower from the plant), the leaves and stems stay green and healthy much longer than they otherwise would (Figure 22.17). Gardeners routinely remove flower buds from many plants to maintain vegetative growth.

And yet, other stimuli must be involved in senescence. When a cocklebur is experimentally induced to flower, its leaves turn yellow regardless of whether the nutrient-demanding young flowers are left on or pinched off. It is as if a "death signal" forms during short days and leads to both flowering and senescence. Experiments with soybeans indicate that the signal (whatever it is) counteracts the effects of cytokinins, which delay senescence. Painting the surface of a mature soybean leaf with a cytokinin solution will increase the cytokinin content of the leaf. Leaves painted this way often stay green longer (Figure 22.18).

DORMANCY

As autumn approaches and days grow shorter, growth slows or stops in many evergreen trees, deciduous trees, and nonwoody perennials. Apical meristems in their buds simply stop producing new stem and leaf cells, and they do so even if temperatures are still moderate, the sky is bright, and enough water is available. When a perennial or biennial plant stops growing under conditions that seem (to us) quite suitable for growth, it has entered a state of **dormancy**. Its buds become tolerant of drought and lower temperatures, and ordinarily the buds will not resume growth until early spring.

Short days and long nights represent a strong environmental cue for dormancy. Their effect can be demonstrated with Douglas fir plants (Figure 22.19). When a short period of red light interrupts the long dark period for these plants, the plants respond as if nights are shorter (and the days longer). They continue to grow taller.

In this experiment, conversion of Pr to Pfr by red light during the dark period prevents dormancy. But what happens in nature? Perhaps buds go dormant because less Pfr can form when daylength decreases in late summer. Perhaps also, the decline in Pfr levels causes leaves to synthesize a dormancy-triggering hormone (abscisic acid?) that is transported to buds. Finally, in some unexplained way, cold nights, dry soil, and nitrogen deficiency also promote dormancy.

What breaks dormancy in the spring? Between fall and spring, a major dormancy-breaking process is at work. The process involves exposure to low temperatures for hundreds of hours in winter (Figure 22.20). The actual temperature needed to break dormancy varies greatly among species. For instance, Delicious apples grown in Utah require 1,230 hours near 43°F (6°C); apricots grown there require only 720 hours. Generally, tree varieties growing in the southern United States require less cold exposure than those growing in northern states and Canada. So if you live in Colorado and order a young peach tree from a Georgia nursery, the tree might start spring growth too soon and be killed by a late frost or heavy snow.

Probably both gibberellin and abscisic acid help control dormancy. When gibberellins are applied to dormant buds of several species, the buds grow and dormancy is often broken. Abscisic acid extends dormancy and partially counteracts the effects of gibberellins.

Seeds of most native (noncultivated) species also exhibit dormancy, which provides survival value for them as well as for buds. (In contrast, dormancy is rare in seeds of highly selected agricultural crops.)' The mechanisms by which seed dormancy develops and ends are variable. In some species such as honey locust and alfalfa, hard seed coats are formed. These coats prevent absorption of water and oxygen. Dormancy ends when the seed coat is abraded (perhaps as strong winds and rains drive the seed across sand), when it is chemically digested (by bacteria, by fungi, or in the gut of a bird or mammal), perhaps even when fire burns it away.

For many plants, moistened seeds must be exposed to low temperatures for weeks or months before dormancy ends. Such seeds are shed from the plant in summer or autumn, and built-in controls prevent their germination before spring. Without these controls, the seedlings would be killed by frost. Experiments with many plants indicate that gibberellins and abscisic acid help control seed dormancy.

Finally, many species depend on red wavelengths to break dormancy. Once again the light-activated form of phytochrome is involved. At the very least, reliance on this cue helps assure that seedlings will have enough light for photosynthesis and resumed growth. Thus, in the shade of other plants, many types of seeds never do break dormancy.

Figure 22.18 Effect of local cytokinin applications on senescence in a bean plant. These are the primary leaves formed on the stem. Senescence was delayed in one of the primary leaves by covering its upper surface at four-day intervals with a cytokinin solution.

Figure 22.19 Effect of the relative length of day and night on Douglas fir plant growth. The plant at the left was exposed to twelve-hour light and twelve-hour darkness for a year; its buds became dormant because daylength was too short. The plant at the right was exposed to twenty-hour light and four-hour darkness; buds remained active and growth continued. The middle plant was exposed to twelve-hour light, eleven-hour darkness, and one-hour light in the middle of the dark period. This light interruption of an otherwise long dark period also prevented bud dormancy. Such light causes Pfr formation at an especially sensitive time in the normal day-night cycle.

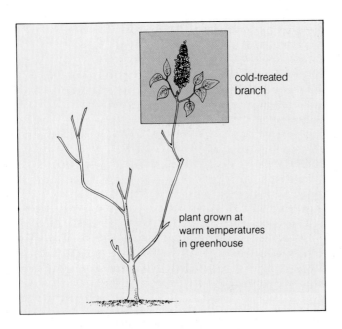

Figure 22.20 Effect of cold temperature on breaking the bud dormancy in many woody plants. In this experiment, one branch (boxed portion) of a lilac plant was positioned so that it protruded from a greenhouse during winter; the rest of the plant remained inside, at warm temperatures. Only the buds on the branch exposed to low outside temperatures grew again in spring. This experiment suggests that low-temperature effects are localized.

cold-treated branch

plant grown at warm temperatures in greenhouse

Case Study: From Embryogenesis to the Mature Oak

Where the ocean breaks along the central California coast, the land rolls inward as steep and rounded hills. Sixty-five million years ago, these sandstone hills had their genesis on the floor of the Pacific. At that time, violent movements in the earth's crust caused parts of the submerged continental shelf to start crumpling upward into a jagged new coastal range. Since then, the rain and winds of countless winters played across the land, softening the stark contours and sending mineral-laden sediments into the canyons. Grasses came to cloak the inland hills, and their organic remains accumulated and enriched the soil. On these hillsides, in these canyons, the coast live oak (*Quercus agrifolia*) began to evolve more than ten million years ago.

Quercus agrifolia is a long-lived giant; some trees are known to be three hundred years old. They can reach heights of a hundred feet; their evergreen branches may spread even wider. In early spring, clusters of male flowers develop and form pendulous, golden catkins among the leafy branches. Wind carries pollen grains from the catkins to female flowers clustered inconspicuously near the branch tips of the same or neighboring trees. After pollination, a sperm-bearing pollen tube grows through the style, toward the ovary where the eggs reside. After a sperm and egg nuclei unite at fertilization, cells of the newly formed zygote divide repeatedly, giving rise to root and shoot tips and to large cotyledons. Integuments of the ovule form the seed coat, and ovary walls develop into a shell. By early fall, the seed reaches maturity and is shed from the tree as a hard-shelled acorn.

Three centuries ago, long before Gaspar de Portola sent landing parties ashore to found colonies throughout Upper California, oaks were shedding the seeds of a new generation. Suppose it was then that a scrubjay, foraging at the foot of a hillside, came across a worm-free acorn. In storing away food for leaner days, the bird used its beak to scrape a small crater in the soil, dropped in the acorn, and covered the acorn-filled crater with decaying leaves. Although a scrubjay might remember many such hiding places most of the time, this particular acorn lay forgotten. The next spring, it germinated.

From the moment of germination, the oak seed embarked on a journey of dynamic, continued growth. Cells divided repeatedly, grew longer, and increased in diameter. Water pressure forced the enlargement— water taken up osmotically as ions accumulated in the newly forming primary root, and as hormones caused a softening of cell walls that otherwise would have been too strong to allow expansion under pressure.

A root cap formed and protected the primary root during its downward growth through the soil. It produced cortex and epidermis, as well as a vascular column through which water and ions would flow. Lateral roots developed, probably under the influence of growth regulators. As new roots continued to elongate, absorptive surfaces increased. When the primary shoot began its upward surge, separate vascular bundles began forming; eventually they would become consolidated into a continuous cylinder of secondary xylem and phloem.

In parenchyma cells of developing roots, stems, and leaves, large central vacuoles formed. As vacuoles enlarged, they pressed the cytoplasm outward so that it became no more than a thin zone against the cell wall. Cytoplasmic interaction with the environment was thereby enhanced; essential ions, gases, and water could be harvested rapidly in spite of their dilute concentrations in the surrounding air and soil. Mycorrhizae, the symbiotic association of roots and fungal strands, enhanced the absorption process. In leaves, stomata developed and regulated carbon dioxide movements and water loss.

At the whim of a scrubjay, the seed had sprouted in a well-drained, sandy basin at the foot of a canyon, in full sunlight. Rainwater accumulated there each winter, keeping the soil moist enough to encourage luxuriant growth during the spring and through the dry summers.

Figure 22.21 From germination to the mature oak (*Quercus agrifolia*).

Out in the open, red wavelengths of sunlight activated phytochrome pigments in the seedling, and hormonal events were triggered that encouraged stem branching and leaf expansion. All the while, delicate hormone-mediated responses were being made to the winds, the sun, the tug of gravity, the changing seasons. Lignin and cellulose strengthened most of the cell walls in secondary xylem. With this strengthening came resistance to the strong winds racing through the canyon. As the oak matured, phytochrome responded to the subtle shifts in daylength throughout the year and the tree responded to the changing seasons. The shorter days of late summer promoted bud dormancy, hence resistance to autumn drought and winter cold.

As the oak seedling matured into the adult form, roots continued to develop and snake through a tremendous volume of the moist soil. Branches continued to spread beneath the sun. Leaves proliferated—leaves where the oak put together its own food from water, carbon dioxide, the few simple inorganic substances it mined from the soil, and the sunlight energy it harnessed to drive the reactions. Continued activity at meristems increased the girth of roots and branches and the height of the tree.

Thus the oak increased in size every season, year after year, century after century. On their way to the gold fields, prospectors of the California Gold Rush rested in the shade of its immense canopy. The great earthquake of 1906 scarcely disturbed the giant, anchored as it was by a root system that spread out eighty feet through the soil. By chance, the brush fires that periodically sweep through California's coastal canyons did not seriously damage the tree. Fungi that could have rotted its roots never took hold; the soil was too well-drained and the water table too deep. Leaf-chewing insects were kept in check by protective chemicals in the leaves and by the predatory birds abounding in the canyon.

In the 1960s, the human population underwent a tremendous upward surge in California. The land outside the cities began to show the effects of population overflow as the native plants gave way to suburban housing. The developer who turned the tractors on the canyon in which the giant oak had grown was impressed enough with its beauty that the tree was not felled. Death came later. How could the new homeowners, just arrived from the east, know of the ancient, delicate relationships between the giant trees and the land that sustained them? Soil was graded between the trunk and the drip line of the overhanging canopy; flower beds were mounded against the trunk; lawns were planted beneath the branches and sprinklers installed. Overwatering in summer created standing water next to the great trunk—and the oak root fungus (*Armillaria*) that had been so successfully resisted until then became established. With its roots rotting away, the oak began to suffer the effects of massive disruption to the feedback relationships among its roots, stems, and leaves. Eventually it had to be cut down. In their fifth winter, in their red brick fireplace, the homeowners began burning three centuries of firewood.

SUMMARY

1. Following dispersal from the parent, the seeds of flowering plants germinate (the embryo absorbs water in a process called imbibition, it resumes growth, and it breaks through the seed coat). Germination is completed when the primary root (which develops from the radicle, or embryonic root) protrudes from the seed coat.

2. Following germination, the plant increases in volume and mass; tissues and organs of the seedling develop; later, flowers, fruits, and new seeds form, then older leaves drop away from the plant. These developmental events are governed by plant hormones: signaling molecules released from one cell and transported to (typically nonadjacent) target cells, which have receptors for the particular hormone.

3. Five hormones have been identified in most flowering plants:

   a. Auxins (especially IAA) promote elongation of coleoptile and stem cells.

   b. Gibberellins promote stem elongation and may help seeds and buds break dormancy.

   c. Cytokinins stimulate cell division, promote leaf expansion, and retard leaf aging.

   d. Abscisic acid promotes stomatal closure and may trigger seed dormancy and bud dormancy.

   e. Ethylene promotes fruit ripening, abscission.

4. An as-yet unidentified hormone (dubbed "florigen") is thought to cause flowering.

5. Plants adjust their patterns of growth in response to environmental rhythms and to unique environmental circumstances. Among these responses are tropisms (differences in the rate and direction of growth on two sides of an organ such as a stem or root).

6. In phototropism, light is more intense on one side of a coleoptile, stem, or leaf, which responds by bending toward the light. Apparently light of blue wavelengths is absorbed by flavoprotein (a yellow pigment), which causes IAA in shoot tips to move into shaded tissue regions, where it promotes the cell elongation that leads to bending.

7. In gravitropism (a growth response to the earth's gravitational force), cells on the lower side of a horizontally positioned stem grow faster than those on the top, causing the stem to curve up; or cells on the upper side of a horizontally positioned root grow faster than those below, causing the root to curve down.

8. Plants have biological clocks (internal time-measuring mechanisms that have a biochemical basis). They also can "reset" the clocks (make seasonal adjustments in their patterns of growth, development, and reproduction).

9. In photoperiodism, plants respond to a change in the relative length of daylight and darkness in a twenty-four hour period. A switching mechanism involving phytochrome (a blue-green pigment) promotes or inhibits germination, stem elongation, leaf expansion, stem branching, and formation of flowers, fruits, and seeds.

10. Long-day plants flower in spring (when daylength is long relative to the night); short-day plants flower when daylength is relatively short. (Flowering of day-neutral plants is not regulated by light.) The flowering response of all may be controlled by phytochrome, which may regulate florigen activity.

    a. When phytochrome absorbs red wavelengths (after sunrise), it is converted from inactive to active form (Pr to Pfr) and helps control the production of enzymes necessary for specific growth responses.

    b. In darkness (at night or in shade), Pfr reverts to Pr and growth responses are curtailed.

11. Senescence is the sum total of processes leading to the death of a plant or plant structure.

12. Dormancy is a state in which a perennial or biennial stops growing even though conditions appear to be suitable for continued growth. A decrease in Pfr levels may trigger dormancy.

Review Questions

1. List the five known plant hormones (or groups of hormones) and the main functions of each. *293-294*

2. Which of the following plant cells or organs synthesize most or all of the hormones required in normal growth and development: coleoptiles, dicot stems, root cells, and leaf cells? Describe one experiment that tells us which ones do this, or which ones don't. *295*

3. Explain how sunlight exposure influences leaf expansion, stem elongation, and stem branching during primary growth. *296*

4. What is phytochrome, and what is its role in plant growth? *299-300*

5. Define plant tropism. *296*

6. Define annual, biennial, and perennial plants. Then describe long-day, short-day, and day-neutral plants. *300*

Readings

Bowley, J. D., and M. Black. 1985. *Seeds: Physiology of Development and Germination*. New York: Plenum Press.

Nickell, L. 1982. *Plant Growth Regulators: Agricultural Uses*. New York: Springer-Verlag. Concise explanations of agricultural practices that include use of growth regulators.

Salisbury, F., and C. Ross. 1985. *Plant Physiology*. Third edition. Belmont, California: Wadsworth. Chapter 20 explains in detail selected examples of time measurement and what we know and don't know about the nature of the biological clock.

Villiers, T. 1975. *Dormancy and the Survival of Plants*. London: Edward Arnold. A short book summarizing major dormancy mechanisms.

Whatley, F. R., and J. M. Whatley. 1980. *Light and Plant Life*. London: Edward Arnold. A short book summarizing effects of light on germination and development with ecological descriptions, too.

UNIT FIVE

ANIMAL SYSTEMS AND THEIR CONTROL

23

ANIMAL CELLS, TISSUES, AND ORGAN SYSTEMS

ANIMAL STRUCTURE AND FUNCTION: AN OVERVIEW

Levels of Organization

With this unit, we turn to the structure of the animal body (its anatomy) and the mechanisms by which the body functions in the environment (its physiology). Like the vascular plants just described, animals are composed of only a few types of tissues. An animal **tissue** is an aggregation of cells and intercellular substances functioning together in a specialized activity. Plants have dermal, vascular, and ground tissues; animals have epithelial, connective, nerve, and muscle tissues. In both cases, the specialized tissues represent a division of labor among cells that contributes to the survival of the organism as a whole.

In animals, the division of labor becomes intricate. Here we see **organs**, the structural units in which specialized tissues are combined in proportions and patterns that allow them to perform a common task. The heart is an organ; it consists of all four types of animal tissues organized as a pump for blood. In animals we see **organ systems**, with two or more organs interacting chemically, physically, or both in the performance of specific tasks. For example, organs of your digestive system (the mouth, pharynx, stomach, pancreas, intestines, and so on) work together to ingest and prepare food for absorption by individual cells and to eliminate food residues.

Not all of the 2 million or so known species of animals have elaborate organ systems. Some do not even have organs. The simplest is *Trichoplax adhaerens*, an organless, pancake-shaped marine animal only two or three milli-

Figure 23.1 With this unit we turn to the diverse ways in which the organs and organ systems of animals function in the environment. Shown here, cigarette smoke swirling down the human windpipe and into the two bronchial routes to the lungs—the consequences of which will be described in chapters to come.

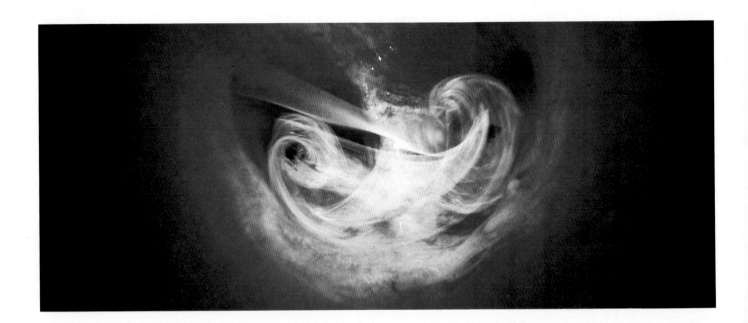

meters across. One of the most complex is the giant squid, which is more than eighteen meters (sixty feet) long; most of its organ systems are as sophisticated as yours. Both species belong to a diverse collection of animals called **invertebrates** (meaning they have no backbone). We mention them only to emphasize that there is no such thing as a "typical" animal. The ones you probably think are typical—the bony fishes, amphibians, reptiles, birds, and mammals—are **vertebrates** (animals with a backbone). Yet fewer than three percent of all species in the animal kingdom are vertebrates!

We humans tend to be interested in vertebrates in general (and humans in particular), and so they will be our focus in this unit. But comparative examples also will be drawn from the invertebrates to keep the structural and functional diversity of the animal kingdom in perspective.

Some Anatomical Terms

Throughout this unit, we will be using some standard terms for describing the location of organs and organ systems in the vertebrate body. Take a moment to study Figure 23.2, which shows the location of some major body cavities in which organs occur. Also study Figure 23.3, which defines some anatomical terms that apply to most animals.

Organ Systems and the Internal Environment

To stay alive, your body's cells must be continually bathed in fluid that supplies them with nutrients and carries away metabolic wastes. In this they are no different from *Paramecium* or any other free-living, single-celled organism (page 91). However, many *trillions* of cells are crowded together in your body—and they all must draw nutrients from and dump wastes into the same fifteen liters (less than sixteen quarts) of fluid.

In all vertebrates, the **extracellular fluid** (that is, the fluid *not* inside cells) is classified as interstitial fluid and plasma. **Interstitial fluid** occupies the spaces between cells and tissues; **plasma** is the fluid portion of blood. As long as the animal is alive, blood constantly exchanges substances with interstitial fluid, which in turn exchanges substances with the cells it bathes.

Extracellular fluid is functionally continuous with the fluid inside cells, so drastic changes in its composition and volume obviously will have drastic effects on cell activities. Its concentrations of hydrogen, potassium, calcium, and other ions are especially important in this regard. If those concentrations are not maintained at levels that are compatible with the survival of the body's individual cells, the organism itself cannot survive.

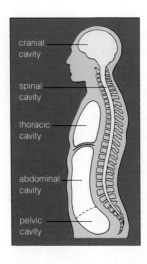

Figure 23.2 Major body cavities in humans. Yellow signifies posterior cavities; white, anterior cavities.

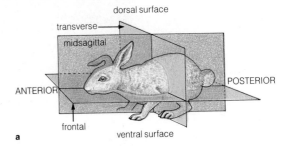

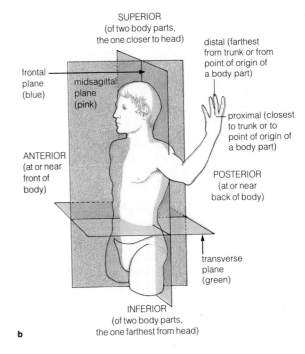

Figure 23.3 (a) Directional terms and planes of symmetry for most kinds of animals. The midsagittal plane divides the body into right and left halves. The transverse plane divides the body into anterior (front) and posterior (back) parts. The frontal plane divides it into dorsal (upper) and ventral (lower) parts. (b) Humans move with the main body axis perpendicular to the earth. "Ventral" corresponds to anterior, and "dorsal" corresponds to posterior.

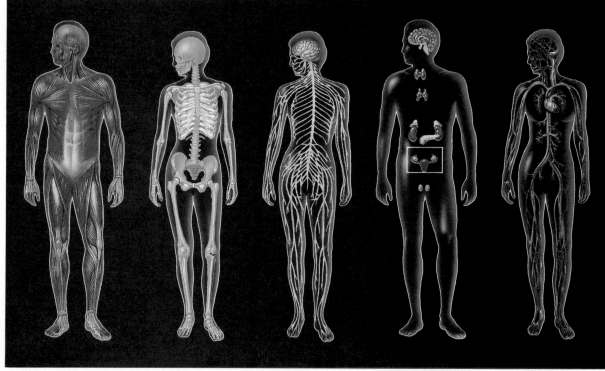

INTEGUMENTARY SYSTEM	MUSCULAR SYSTEM	SKELETAL SYSTEM	NERVOUS SYSTEM	ENDOCRINE SYSTEM	CIRCULATORY SYSTEM
Protection from injury and dehydration; body temperature control; excretion of some wastes; reception of external stimuli; defense against microbes.	Movement of internal body parts; movement of whole body; maintenance of posture; heat production.	Support, protection of body parts; sites for muscle attachment, blood cell production, and calcium and phosphate storage.	Detection of external and internal stimuli; control and coordination of responses to stimuli; integration of activities of all organ systems.	Hormonal control of body functioning; works with nervous system in integrative tasks.	Rapid internal transport of many materials to and from cells; helps stabilize internal temperature and pH.

Figure 23.4 Organ systems of the human body. All vertebrates have the same types of systems, serving similar functions.

It makes no difference whether the animal is simple or complex. *The component parts of any animal work together to maintain the stable fluid environment required by its living cells.* This concept is absolutely central to our understanding of the structure and function of animals, and it may be summarized this way:

1. Each cell of the vertebrate body engages in basic metabolic activities that ensure its own survival.

2. Concurrently, the cells of a given tissue typically perform one or more activities that contribute to the survival of the whole organism.

3. The combined contributions of individual cells, organs, and organ systems help maintain the stable internal environment—that is, the extracellular fluid—required for individual cell survival.

Major Organ Systems

Figure 23.4 is an overview of the kinds of organ systems that occur in all vertebrates. You might think we are stretching things a bit when we say that each of those systems contributes to maintaining the internal environment. After all, what could the body's skeleton and musculature have to do with the composition and volume of extracellular fluid? Yet through their interactions, the skeletal and muscular systems provide a means of moving about—toward sources of nutrients and water, for example. Some of their components help circulate substances through the body, as when contractions of certain leg muscles help move blood in veins back to the heart. Even the reproductive system makes hormonal contributions to the body as a whole.

Let's now take a look at the component tissues of vertebrate organ systems.

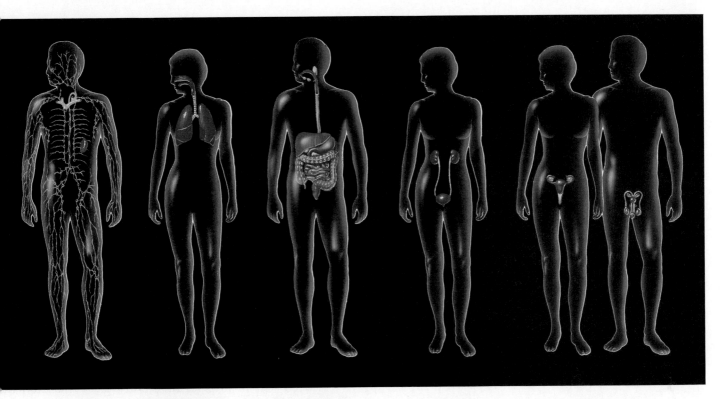

LYMPHATIC SYSTEM	RESPIRATORY SYSTEM	DIGESTIVE SYSTEM	URINARY SYSTEM	REPRODUCTIVE SYSTEM
Return of some extracellular fluid to blood; roles in immunity (defense against specific invaders of the body).	Provisioning of cells with oxygen; removal of carbon dioxide wastes produced by cells; pH regulation.	Ingestion of food, water; preparation of food molecules for absorption; elimination of food residues from the body.	Maintenance of the volume and composition of extracellular fluid.	Male: production and transfer of sperm to the female. Female: production of eggs; provision of a protected, nutritive environment for developing embryo and fetus. Both systems have hormonal influences on other organ systems.

ANIMAL TISSUES

Tissue Formation

The specialized tissues of vertebrates arise through cell differentiation in each new individual (Chapter Seventeen). At some point in the life cycle, meiosis occurs in **germ cells**, which are the cells that give rise to sperm or eggs. Fusion of a sperm and egg results in the formation of a zygote, which undergoes cell divisions that produce the animal **embryo**.

Soon after a vertebrate embryo starts to develop, cells become arranged into three distinct tissues, or **germ layers**, which are called ectoderm, mesoderm, and endoderm (page 497). Cell divisions and differentiation in the three layers lead to all the specialized tissues of the animal body. **Ectoderm** gives rise to the outer layer of skin and to tissues of the nervous system. **Mesoderm** gives rise to muscle; the organs of circulation, reproduction, and excretion; most of the internal skeleton; and connective tissue layers of the gut and body covering.

Endoderm gives rise to the lining of the gut and the major organs derived from it.

Some of the resulting tissues are the "glue" that binds neighboring cells or tissues; others are coverings, skeletal or muscular elements, absorptive surfaces, and so on. All of those body tissues are composed of **somatic cells**. ("Somatic" comes from the Greek *soma*, meaning body.) The only other cells in animals are the germ cells.

Many cell interactions are necessary for tissue formation, as we will see in Chapter Thirty-Four. The point to keep in mind here is this: *Properties of individual cells promote recognition between similar cells and their adhesion to one another in tissues.*

This point was demonstrated in the 1960s, when Tom Humphreys and Aron Moscona experimented with a simple kind of animal, the sponge. A thin tissue of flat cells defines the surface of the sponge body; inside, other cells are loosely positioned in a protein gel. There are no organs; amoebalike cells creeping about in the gel secure food and distribute it through the body (Figure

Figure 23.5 Examples of cell junctions in animal tissues.

(a) *Tight junctions* seal off spaces between adjacent cells in epithelium and so prevent leakage across it. Substances cross the epithelial sheet by moving through the cytoplasm of its individual cells. The sealing strands are composed of proteins, which are embedded in the lipid bilayer of the plasma membranes of adjacent cells.

(b) *Adhering junctions* mechanically link cells and permit them to function as a structural unit. They occur within and between all types of animal tissues. Some buttonlike junctions occur between cells of the outer layer of skin, which is subject to stretching; they help keep the tissue from being pulled apart under stress. Cytoskeletal elements extend from the "buttons" into the cytoplasm of each cell. Other elements extend across the intercellular space and link the two buttons together.

(c) *Communication junctions* are regions where clusters of channels across the plasma membrane of one cell match up with similar channels of an adjacent cell. These junctions influence the passage of ions and small molecules between cells and so help coordinate individual cell activities throughout a tissue. Each channel is composed of protein molecules embedded in the lipid bilayer.

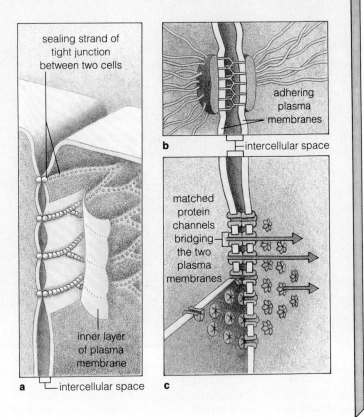

41.7). Humphreys and Moscona forced red-colored sponges and yellow-colored sponges through a fine sieve. This produced single cells and tiny clumps of cells, which were then swirled together in a nutrient medium. As the cells collided, they adhered to one another—but red cells adhered only to red cells, and yellow cells adhered only to yellow. As the clusters became larger, the cells actually arranged themselves in patterns characteristic of the intact sponge body!

Through biochemical studies and electron microscopy, we now know that recognition proteins embedded in the plasma membrane promote cell interactions of this type, which are necessary for tissue formation (pages 87 and 501).

Recognition proteins promote the aggregation of cells into tissues, but what keeps the cells together? In many tissues, specialized regions of the plasma membrane called **cell junctions** take part in sealing, adhesion, and communication between cells. Figure 23.5 shows three types of cell junctions. In addition, the cells of many tissues are embedded in an **extracellular matrix**. An example is the jellylike substance that provides adhesion

between cells in cartilage, a strong yet pliable tissue giving shape to the tip of your nose. The extracellular matrix of cartilage and some other tissues is jellylike, but the matrix also can be more solid or rather fluid.

Now that we have a general idea of what it takes to be a tissue, let's consider some features of the four specific categories of animal tissues—the epithelial, connective, muscular, and nervous tissues.

Epithelial Tissue

Epithelia (singular, epithelium) consist of cells that are linked tightly together by sealing junctions, with little space or extracellular material between them (Figure 23.5a and 23.6). Epithelia are single or multiple layers of cells, but in all cases one surface is free (it does not have an overlying tissue), and the other adheres to an underlying layer called a basement membrane. This type of tissue covers external body surfaces, lines the body's internal cavities and tubes, and forms the secretory portion of glands. It plays many roles, ranging from protection and selective absorption to secretion.

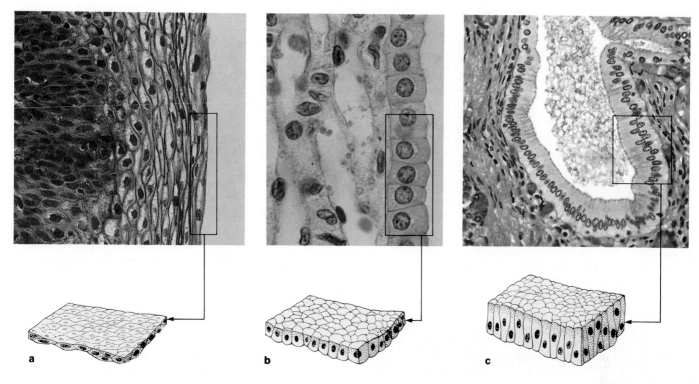

Figure 23.6 Basic cell shapes in epithelial tissues: (**a**) squamous, (**b**) cuboidal, and (**c**) columnar.

Epithelia can be classified by the number of cell layers at the free surface. **Simple epithelium** consists of a single layer of cells that function in diffusion, secretion, absorption, and filtration. For example, gases diffuse readily across the simple epithelium that lines the microscopic air sacs of lungs. **Stratified epithelium**, which consists of cells stacked into several layers, functions in protection. One such tissue forms the outer layer of skin.

Simple and stratified epithelia can be further classified by the predominant shape of their surface cells (Figure 23.6). The cells are shaped like flat paving stones (squamous epithelium), cubes (cuboidal epithelium), or columns (columnar epithelium). Table 23.1 summarizes the main kinds of simple and stratified epithelium.

Epithelium is said to be glandular when it contains single-celled or multicelled secretory structures called **glands**. There are two main kinds of glands: endocrine and exocrine. Generally, an **endocrine gland** secretes hormones into the bloodstream. (Animal hormones are signaling molecules, secreted by endocrine glands, that travel by way of the bloodstream and trigger specific reactions in target cells.) In contrast, an **exocrine gland** secretes substances directly or through ducts to the free epithelial surface (Figure 23.7). Exocrine products include mucus, saliva, wax, oil, milk, and digestive enzymes.

Table 23.1 Classification of Some Epithelial Tissues

Category	Main Functions	Some Common Locations
Simple epithelium:		
Squamous	Filtration, diffusion	Air sacs in lungs, capillary walls, lining of blood and lymph vessels
Cuboidal	Secretion, absorption	Lining of kidney tubules, duct lining for some glands, covering of ovaries
Columnar	Protection, secretion, absorption	Surface layer of lining of stomach, intestines, part of respiratory tract
Stratified epithelium:		
Squamous	Protection	Outer layer of skin, lining of mouth, throat, lining of vagina
Cuboidal	Protection	Lining of ducts of some glands

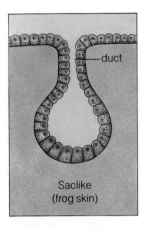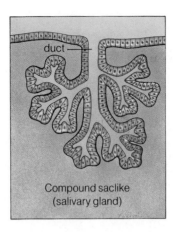

Saclike
(frog skin)

Compound saclike
(salivary gland)

Figure 23.7 Examples of glands found in glandular epithelium, as they would appear in longitudinal section.

Table 23.2	Classification of the Main Types of Connective Tissue	
Category	Main Functions	Some Common Locations
Connective tissue proper:		
Dense connective tissue	Flexible, strong connections between structures; covers some organs	Dermis, tendons around muscles, capsule around kidney, liver, cornea of eye
Loose connective tissue	Support, elasticity	Beneath most epithelia; subcutaneous layer of skin
Adipose tissue	Energy reserve, insulation, padding for some organs	Beneath the skin, around the kidneys, on the surface of the heart
Cartilage	Firm, flexible support; maintenance of shape; shock absorption	Nose, ear regions, tracheal rings, intervertebral disks
Bone	Firm support; rigidity necessary for leverage in movements; protection of internal organs	Bones of vertebrate skeleton
Blood	Transport of varied substances to and from cells; transport medium for infection-fighting proteins and cells; stabilization of pH and temperature throughout body	Within the blood vessels and heart

Connective Tissue

There are four groups of **connective tissue**, called connective tissue proper, cartilage, bone, and blood (Table 23.2). Tissues in the first three groups have an extensive extracellular matrix that contains structural elements called fibers. The cells are widely separated from one another in the matrix. Both the cells and the fibers are surrounded by a watery or jellylike material called the "ground substance."

Connective Tissue Proper. In the tissues of this group, the ground substance is more or less fluid, with many types of cells scattered through it. The predominant cell type, the fibroblast, produces the ground substance and the proteins from which the fibers are constructed.

Dense connective tissue provides strong connections between different tissues. Its collagenous fibers are bundled together irregularly or in parallel array. ("Collagenous" means the fibers are constructed mainly of the protein collagen.) Dense connective tissue occurs in tendons (which attach muscle to bone) and in ligaments (which connect bones at skeletal joints)—two regions in which tension is exerted in only one direction. Their dense connective tissue contains parallel fibers that collectively resist being pulled apart under tension (Figure 23.8a).

Loose connective tissue supports most epithelia and many organs, and it surrounds blood vessels and nerves. Its collagenous fibers, which are tough yet flexible, impart strength to the tissue. Its thinner fibers, which contain the protein elastin, impart elasticity (Figure 23.8b).

In many regions of the body, the connective tissue contains dense clusters of large cells that are specialized for fat storage. This so-called **adipose tissue** is an energy reserve and it also pads some organs. Fat usually accumulates in a single vacuole in each storage cell. As the vacuole increases in volume, it pushes the cytoplasm and nucleus to the cell periphery; thus the fat droplet dominates the cell's appearance in micrographs (Figure 23.8c).

Cartilage. The connective tissue called cartilage cushions body parts and provides a framework for maintaining the shape of some body regions. It consists of a dense network of collagen-containing fibers and elastic fibers, positioned firmly in a jellylike ground substance. The components of cartilage allow it to resist compression while maintaining resiliency.

Cartilage is present on the ends of bones at many joints, in parts of the nose, and in the external ear. One kind of cartilage forms shock pads (such as intervertebral disks of the backbone). Most vertebrate embryos have skeletons of cartilage that are replaced by bone during development.

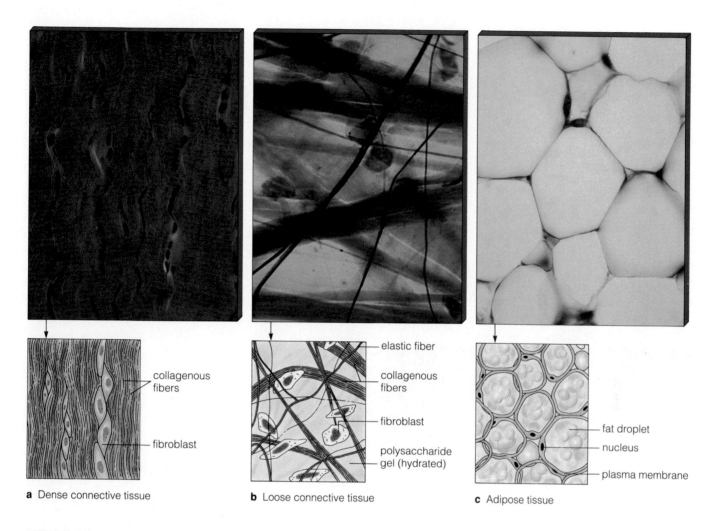

collagenous fibers

fibroblast

elastic fiber

collagenous fibers

fibroblast

polysaccharide gel (hydrated)

fat droplet

nucleus

plasma membrane

a Dense connective tissue

b Loose connective tissue

c Adipose tissue

Figure 23.8 Examples of connective tissues. (**a**) Dense connective tissue, in which many tightly packed collagenous fibers predominate. (**b**) Loose connective tissue, which has many fibroblasts that produce collagenous fibers and elastic fibers. (**c**) The connective tissue called adipose tissue contains clusters of fat-storing cells.

Bone. This connective tissue is also known as osseous tissue. It forms the flat plates, cylinders, and other structures called bones, which are attached to one another at joints. Bones and the cartilage in the joints between them constitute the skeletal system of all vertebrates. The bones support and protect softer tissues and organs. Together with muscles, limb bones form a leverlike system for the body's mechanical movements. Bones also are storage sites for mineral salts, and the marrow of some bones functions in red blood cell production.

Bone has living cells and a network of collagenous fibers distributed through a ground substance. Mineral salts become deposited around the fibers during development, such that the tissue becomes ossified (hardened). Calcium phosphate and calcium carbonate deposits make up sixty-seven percent of the weight of mammalian bones.

Bone is not completely solid. "Canals" of different sizes and lengths extend through the hardened portions, and small spaces called lacunae provide chambers for the living bone cells (osteocytes).

"Spongy" bone tissue is latticelike rather than dense, although it is still firm and strong (Figure 23.9). In many bones (such as the breastbone), red marrow fills the spaces in the spongy tissue. **Red marrow** is a major site of blood cell formation. In adults, a reserve tissue called **yellow marrow** fills the interior cavities of most bones. When blood loss is excessive, as it can be during traumatic injury, this tissue assists the red marrow in producing red blood cell replacements.

"Compact" bone tissue forms a dense layer over spongy bone tissue. It also forms the shaft of long bones and helps them withstand mechanical stress. In mature bone, the matrix is clearly organized as thin concentric

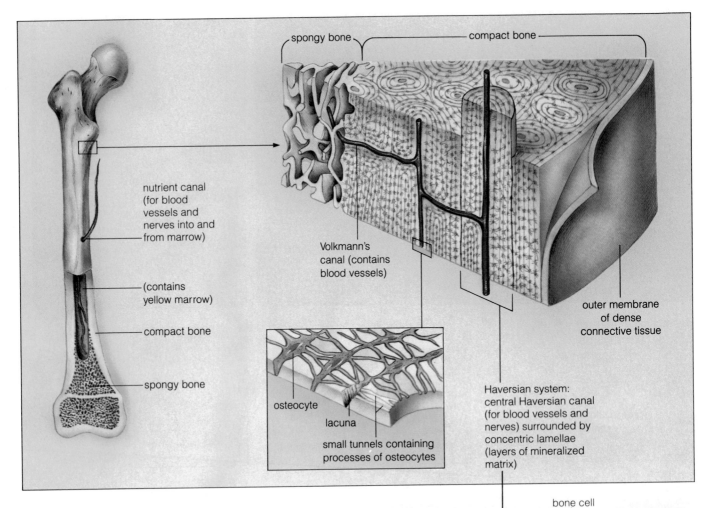

spongy bone

compact bone

nutrient canal (for blood vessels and nerves into and from marrow)

(contains yellow marrow)

compact bone

spongy bone

Volkmann's canal (contains blood vessels)

outer membrane of dense connective tissue

osteocyte

lacuna

small tunnels containing processes of osteocytes

Haversian system: central Haversian canal (for blood vessels and nerves) surrounded by concentric lamellae (layers of mineralized matrix)

layers (lamellae). The lamellae are laid down around Haversian canals, which are small, interconnected channels that run more or less parallel to the bone (Figure 23.9). Blood vessels that transport materials to and from living bone cells extend through the canals. So do nerve fibers, which relay signals that cause changes in the diameter of the blood vessels (and thereby affect the amount of blood flowing to a given region).

Blood. The connective tissue called blood is more viscous than water (this means it is thicker and has an adhesive, or "sticky," quality). It transports oxygen to cells and wastes away from them; it also transports hormones and enzymes. Some of its components protect against blood loss (through clotting mechanisms), and others defend against foreign agents such as microbes. Understanding the nature of this tissue and its complex functions depends on prior understanding of the organ system of which it is part, so we will postpone its description until Chapter Twenty-Nine.

bone cell

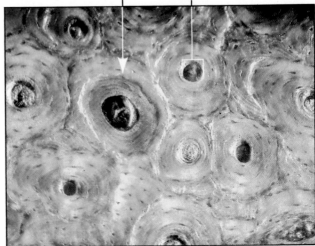

Figure 23.9 Structural organization of the long bones of mammals. The micrograph shows several Haversian systems. Nutrients and hormones reach living bone cells (osteocytes) by way of blood vessels present in these systems. The osteocytes reside in lacunae (spaces in the bone tissue). Small tunnels connect neighboring lacunae.

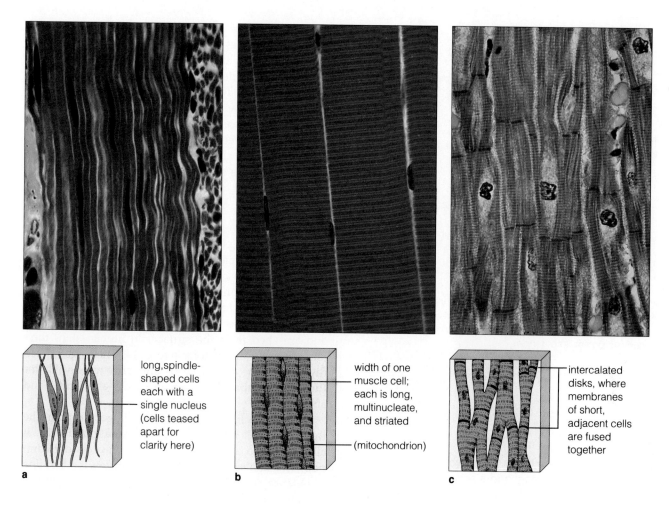

long, spindle-shaped cells each with a single nucleus (cells teased apart for clarity here)

width of one muscle cell; each is long, multinucleate, and striated

(mitochondrion)

intercalated disks, where membranes of short, adjacent cells are fused together

a   b   c

Figure 23.10 Three kinds of muscle tissue: (**a**) smooth, (**b**) skeletal, and (**c**) cardiac.

Muscle Tissue

Muscle tissue specializes in contraction. Tissues containing muscle cells actively contract (shorten) in response to stimulation, then passively lengthen and so return to their resting state. Muscle contraction is the basis of many different movements, such as locomotion of the entire body, adjustments in the orientation of individual body parts, propulsion of food through the gut, and so on. Muscle activity also functions in heat production, which helps maintain overall body temperature.

The three kinds of muscle tissues are called smooth, skeletal, and cardiac. In light micrographs, skeletal and cardiac muscle tissues appear striated (striped), with alternating light and dark bands (Figure 23.10).

Smooth muscle tissue consists of spindle-shaped cells held together by connective tissue. This tissue occurs in the walls of hollow internal structures, such as blood vessels, the stomach, the intestines, and some ducts. In vertebrates, smooth muscle is said to be *invol-*

untary, because the individual usually cannot directly control its contraction.

Skeletal muscle tissue generally attaches by tendons to the vertebrate skeleton and is responsible for movements that can be voluntarily initiated. Skeletal muscle tissue contains many long, cylindrical cells. Some muscle cells in your body are up to thirty centimeters (twelve inches) long. Typically, connective tissue envelops a number of skeletal muscle cells to form a muscle bundle. Several bundles are usually enclosed in a tougher connective tissue sheath to form muscle organs such as the biceps (Figure 28.8).

Cardiac muscle tissue is the contractile tissue of the vertebrate heart. Cardiac muscle cells are shorter than most skeletal muscle cells, and the membranes of adjacent cells are fused at regions called intercalated disks. Because of communication junctions at these fusion points, the cells do not function on an independent basis. Rather, when one muscle cell receives a signal to contract, its neighbors are also stimulated into contracting (as is also the usual case for smooth muscle).

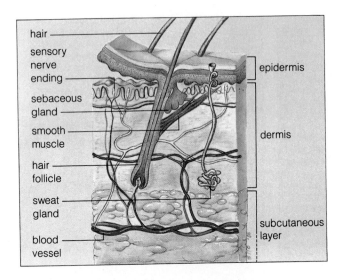

Figure 23.11 A section through vertebrate skin and the underlying hypodermis (or subcutaneous layer). Skin is a two-layered organ, with a thin outer portion (the epidermis) and a thicker, underlying portion (the dermis).

Nervous Tissue

Of all tissues, the nervous tissue of vertebrates is the most intricate in its functioning, and a separate chapter is devoted to explaining its properties. For now, two points should be kept in mind. First, the basic functional units of **nervous tissue** are the *neurons*, or nerve cells, which are organized into lines of communication that extend throughout the body. Second, different types of neurons specialize in detecting environmental change, coordinating the body's immediate and long-term responses to change, and relaying signals to muscles and glands that can carry out those responses.

Case Study: The Tissues of Skin

Having reviewed the basic types of animal tissues, we can start to think about how different tissues are combined in organs. Let's use vertebrate skin as an example of the tissue combinations that exist, even in organs that superficially might seem to be rather unremarkable in comparison to others.

Skin is the largest organ of the vertebrate body. It covers the outer surface and is continuous with the mucous membranes found inside the eyelids, nostrils, and other openings of the body. The skin has two main layers, the **epidermis** and the **dermis** (Figure 23.11). The skin is separated from deeper tissues by what is called a subcutaneous layer (or "hypodermis," meaning below the skin).

Skin structure varies considerably among vertebrates. Some fishes have hard dermal scales; other fishes and amphibians have bare skin covered with mucus. Reptiles have epidermal scales, birds have feathers, and mammals, hairs. Their scales, feathers, and hair, along with beaks, hooves, horns, claws, nails, and quills, are all produced by cell differentiation in epidermal tissue (Figure 23.12). Even the skin that covers the human body is not the same in all regions. For example, the epidermis is thicker in areas subject to friction, such as the palms of the hands and the soles of the feet.

Epidermis. The epidermis of mammalian skin is stratified squamous epithelium. As part of a process called **keratinization**, epidermal cells in the middle layers are transformed into dead bags, so to speak, that contain the protein keratin. Rapid cell divisions in the deepest layers push cells toward the surface of the skin, so that fully keratinized cells pass through the uppermost layer and are shed from the surface. In human skin, their journey takes about forty-five days. The ongoing cell divisions in the deeper layers are partly responsible for the skin's capacity to repair itself after cuts and abrasions.

Keratin is insoluble in water. Together with the toughened cell membrane of fully keratinized cells, it allows skin to act as a barrier that prevents dehydration. The skin's impenetrability also makes it a first line of defense against many toxic substances and disease-causing microbes.

Cells called melanocytes, which produce the brown-black pigment melanin, are present in the deeper layers of epidermis. Melanin affords protection from ultraviolet wavelengths present in sunlight. When light-colored skin "tans" after exposure to sunlight, the melanin concentrations in the epidermis have increased. Carotene, oxyhemoglobin, and other pigments also contribute to skin color.

The surface of human skin is covered with tiny grooves and ridges that form patterns unique to each individual. The patterns are used in the identification procedure called fingerprinting.

Dermis. The dermis is mostly dense connective tissue that cushions the body from everyday stresses and strains. It is tightly connected to the epidermis by an intervening basement membrane. During development, hair follicles, sebaceous glands, and sweat glands grow into the dermis from the epidermis. Each hair grows from a follicle, and the waxy secretions of sebaceous glands associated with the follicles lubricate both the hairs and the skin surface. Smooth muscles are attached to the follicles, and when they contract, they "stand hair on end" (hairs normally lie at oblique angles to the skin surface). Nerves and the sensory endings of nerve cells

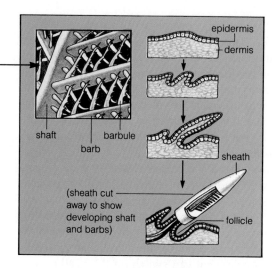

Figure 23.12 Differentiation of a region of epidermal tissue into a feather. The photograph shows a courtship display of a peacock, which relies on spectacularly specialized feathers to capture the attention of a peahen.

are also present in the dermis, where they gather information on touch, pressure, pain, cold, and heat.

The subcutaneous layer beneath the skin is loose connective tissue, and its transition into the dermis is indistinct. It may be thickened by the presence of stored water, adipose tissue, or skeletal muscle tissue. The adipose tissue stores energy and provides insulation against cold; the skeletal muscle tissue allows skin to move somewhat independently of underlying tissues.

Blood vessels extend through the dermis and subcutaneous layer. They are absent from epidermis; oxygen and nutrients from the blood must diffuse through the connective tissues to reach the epidermal cells. Blood circulation to the skin is regulated, and changes in that flow can affect body temperature.

One final point: Skin is both strong and flexible because of the keratin in its epidermis and the collagenous fibers in its dermis. Young skin stretches easily to accommodate movements by the body. As the individual ages (and during some aging disorders), the skin's plasticity is reduced and wrinkles develop.

HOMEOSTASIS AND SYSTEMS CONTROL

Now that we have completed our overview of animal tissues and organs, we can begin to look more closely at the mechanisms by which they interact to maintain a stable internal environment. The importance of those mechanisms was first perceived by the nineteenth-century physiologist Claude Bernard. For example, Bernard discovered that glucose (and many other nutrients) transported by blood is absorbed by liver cells, and that these cells convert the glucose to a storage form (glycogen). When glucose levels fall below a certain point in the blood—hence in extracellular fluid—the liver breaks down glycogen so that glucose can be returned to the blood and circulated to the cells requiring it.

In other studies, Bernard showed that the amount of nutrients and oxygen that blood is supplying to different regions of the body can be regulated by the constriction and dilation of small blood vessels. Controls over the distribution of blood allow it to be diverted to regions that require increased amounts of oxygen and nutrients at a given time.

Much later, the physiologist Walter Cannon extended Bernard's line of thinking. Cannon perceived that maintaining internal conditions is possible only through coordinated **homeostatic mechanisms** that operate to keep physical and chemical aspects of the body within tolerable ranges.

Homeostatic control mechanisms maintain physical and chemical aspects of the internal environment within ranges that are most favorable for cell activities.

Many homeostatic controls work through feedback mechanisms. The word "feedback" refers to a circular situation in which information is fed back into a system. In **negative feedback mechanisms**, which are the most common, an initial condition that causes a change in the internal environment is reversed (Figure 23.13). Negative feedback affects numerous metabolic reactions within cells (page 104); it also affects ecological events in the biosphere.

How does negative feedback work? By analogy, consider how a thermostatically controlled furnace operates. A thermostatic device senses the air temperature and "compares" it to a preset point on a thermometer built into the furnace control system. When the temperature falls below the preset point, the thermostat signals an integrating device that can switch on the heating unit. When the air becomes heated enough to match the prescribed level, the thermostat signals the integrating device, which shuts off the heating unit.

Similar controls are at work in the human body. For example, feedback mechanisms help raise or lower the internal body temperature so that it is maintained near 37°C (98.6°F) even during extremely hot or cold weather.

Homeostatic control mechanisms require three components: receptors, integrators, and effectors (Figure 23.13). Specialized cells or cell parts serve as **sensory receptors** for a stimulus. A **stimulus** is a specific energy change in the environment that the body is able to detect. For example, when someone holds one of your hands and squeezes it, there is a change in pressure, which is a form of mechanical energy. Sensory receptors in the skin of your hand translate the stimulus energy into a signal that can be sent to the **integrator**, a control point where different bits of information are pulled together in the selection of a response. Your brain and spinal cord are integrating centers. They can send signals to your muscles or glands (or both), which are **effectors** that carry out the appropriate response.

Integrating centers receive information that indicates not only how the system *is* operating (the information from sensory receptors), but also how it *should be* operating (information from a "set point," which is sometimes built into the center itself). The difference between these two bits of information influences whether the integrating center will signal the effectors to increase or decrease their activity and thereby bring the system back to its most effective operating range.

Under some circumstances, **positive feedback mechanisms** set in motion a chain of events that *intensify* the original input. Positive feedback is associated with instability in a system. For example, sexual arousal leads to increased stimulation, which leads to more stimulation, and so on until an explosive, climax level is reached (page 519). As another example, during childbirth, pressure

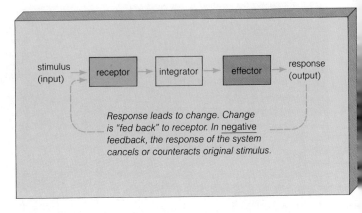

Figure 23.13 Components necessary for negative feedback at the organ level.

of the fetus on the uterine walls stimulates production and secretion of the hormone oxytocin. Oxytocin causes muscles in the walls to contract, this increases pressure on the fetus, and so on until the fetus is expelled from the mother's body.

In addition to negative and positive feedback mechanisms, animals also have **feedforward mechanisms**, which generally use parallel inputs to integrating centers. For example, when numerous sensory cells in human skin detect a drop in air temperature, they send signals to the brain that "anticipate" a change in blood temperature. The brain sends signals to metabolic and muscular systems that can function in raising internal body temperature. With feedforward control, corrective measures can sometimes begin even before an outside change significantly alters the internal environment.

What we have been describing here is a general pattern of monitoring and responding to a constant flow of information about the external and internal environments. During this activity, organ systems operate together in coordinated fashion. Throughout this unit, we will be asking the following questions about their operation:

1. What physical or chemical aspect of the internal environment are organ systems working to maintain as conditions change?

2. By what means are organ systems kept informed of change?

3. By what means do they process incoming information?

4. What mechanisms are deployed in response?

With these questions in mind, we will turn next to the two systems under whose dominion all others must fall—those of neural and endocrine control.

SUMMARY

1. This chapter outlined the structural features (anatomy) of the animal body and introduced some basic concepts about its physiology (the mechanisms by which the body functions in the environment).

2. The cells of most animals interact at three levels of organization: in *tissues*, many of which are combined as *organs*, which are components of *organ systems*.

3. A *tissue* is an aggregation of cells and intercellular substances united in the performance of a specialized activity. Cardiac muscle tissue (which functions in heart contraction) is an example. An *organ* is a structural unit in which tissues are combined in definite proportions and patterns that allow them to perform a common task. The heart (an organ that pumps blood) is composed of all four tissue types. An *organ system* has two or more organs that interact chemically, physically, or both in ways that contribute to the survival of the animal as a whole.

4. Each cell engages in basic metabolic activities that assure its own survival. Concurrently, the cells of a tissue or organ perform activities that contribute to the survival of the animal body as a whole. The combined contributions of individual cells, tissues, organs, and organ systems help maintain the stable internal environment (the *extracellular fluid*) required for individual cell survival. This concept is central to understanding the functioning of any organ system, regardless of its complexity.

5. There are four categories of animal tissues. *Epithelial* tissues cover external body surfaces, line its internal cavities and tubes, and form the secretory portion of glands. *Connective* tissues provide support for other tissues or bind them together; they include connective tissue proper, cartilage, bone, and blood. *Muscle* tissue is the basis of contraction, which underlies mechanical movements of internal and external body parts and locomotion through the environment. *Nervous* tissue detects and coordinates information about change in the internal and external environments, and it controls responses to those changes.

6. This chapter described how tissues of all four categories are combined and arranged in the largest organ system of vertebrates—the integumentary system (skin). The tissue components and organization of the other organ systems are described in chapters to follow.

7. Organ systems work largely through homeostatic control mechanisms that help maintain a stable internal environment. In negative feedback (the most common mechanism), the response to a disturbance in a system *decreases* the original disturbance. In positive feedback, a response *intensifies* the original disturbance. Feedforward mechanisms allow the system to institute corrective measures before a disturbance can occur.

8. Control of the internal environment depends on the body's receptors, integrators, and effectors. Sensory cells and tissues are receptors for stimuli (which are specific environmental changes that the body is able to detect). Integrating centers receive information from receptors about how some aspect of the body *is* operating and compare it to a "set point" about how it *should be* operating. On the basis of this information, signals are relayed to effectors (muscles, glands, or both), which carry out the appropriate response.

Review Questions

1. What is an animal tissue? An organ? An organ system? Can you list the major organ systems of the human body, along with their functions? *308–311*

2. Define extracellular fluid, interstitial fluid, and blood plasma. *309*

3. Perhaps the most important concept in animal physiology relates the functioning of cells, organs, and organ systems to the internal environment. Can you state the three main points of this concept? *310*

4. Epithelial tissue and connective tissue differ from each other in overall structure and function. Can you describe how? *312–314*

5. State the overall functions of (a) muscle tissue and (b) nervous tissue. *317–318*

6. What properties of human skin are associated with the protein keratin? *318*

7. Define homeostasis. What are the three components necessary for homeostatic control over the internal environment? *319–320*

8. What are the differences among negative feedback, positive feedback, and feedforward mechanisms? *320*

Readings

Bloom, W., and D. W. Fawcett. 1986. *A Textbook of Histology*. Eleventh edition. Philadelphia: Saunders. Outstanding reference text.

Kessel, R., and R. Kardon. 1979. *Tissue and Organs: A Text-Atlas of Scanning Electron Microscopy*. San Francisco: Freeman. Outstanding, unique micrographs.

Leeson, C.R., T. Leeson, and A. Paparo. 1985. *Textbook of Histology*. Philadelphia: Saunders.

Ross, M. and E. Keith. 1985. *Histology: A Text and Atlas*. New York: Harper & Row.

Vander, A., J. Sherman, and D. Luciano. 1985. *Human Physiology: The Mechanisms of Body Function*. Fourth edition. New York: McGraw-Hill. Perhaps the clearest, in-depth introduction to human organ systems and their functioning.

24

INFORMATION FLOW AND THE NEURON

From time to time, the human body has been likened to a city, state, or some other social unit composed of separate but interdependent parts. These analogies are wonderfully optimistic about our capacity for social organization. In truth, our cities and states do not begin to approach the degree of integration of any complex animal. Whether the animal is asleep, relaxed, or alert to danger, diverse body parts are being made to work in coordinated ways. Activities of each part are continually monitored and evaluated, not for their sake alone but also for how they are contributing to the whole. In this unit, we will look at neural and endocrine systems that integrate and control different activities in the animal body.

NEURONS: FUNCTIONAL UNITS OF NERVOUS SYSTEMS

The basic unit of communication in all nervous systems is the **neuron**, or nerve cell. Neurons do not act alone; each is part of a local or long-distance circuit of cells in the body. It is a *system* of neurons, not the individual neuron, that coordinates vital functions such as breathing. It is a *system* that senses change, integrates the sensory inputs, then signals different body parts to carry out appropriate responses. Parts of the nervous system store and retrieve information about previous experiences. In mammals especially, some parts contribute to insight and creativity, two of the highest forms of mental activity.

Classes of Neurons

In complex animals, there are three classes of nerve cells, called *sensory neurons*, *interneurons*, and *motor neurons*. We can define each class in terms of its role in a control scheme described earlier, whereby the nervous system monitors and responds to change (page 320). The inputs into the system are **sensory stimuli**, these being different forms of energy in the internal or external environment that can be detected by receptor cells.

Sensory neurons serve as the body's *receptors* (or are activated by receptors) for specific stimuli. For example, some sensory neurons in your eyes respond to light energy; others in the skin of your feet respond to pressure, which is a form of mechanical energy. The stimulus energy causes an electrical disturbance at the receptor's surface, and such disturbances generate messages that are relayed to the *integrators* in our control scheme. The brain and spinal cord are integrators. Here we find the **interneurons** that combine the information arriving on sensory lines and then influence other neurons in turn. **Motor neurons** relay information away from the integrator to muscle or gland cells, these being the body's

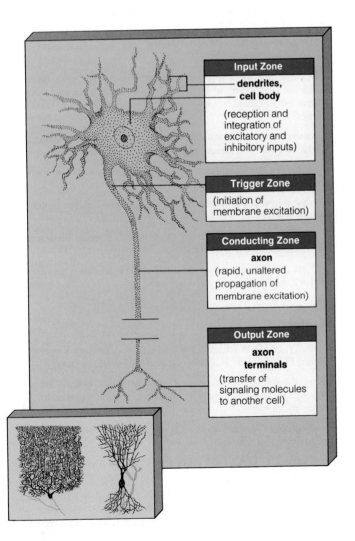

Input Zone

dendrites,
cell body

(reception and integration of excitatory and inhibitory inputs)

Trigger Zone

(initiation of membrane excitation)

Conducting Zone

axon

(rapid, unaltered propagation of membrane excitation)

Output Zone

axon
terminals

(transfer of signaling molecules to another cell)

Figure 24.1 Main information-processing zones of a motor neuron present in vertebrates. In this case, the targets of the signaling molecules released at the output zone are muscle cells. Keep in mind that not all neurons look like motor neurons. Great structural diversity exists among them, as the two examples in the green box suggest.

effectors. Response to change is initiated when signals from motor neurons alter the activity of effector cells:

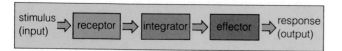

Structure of Neurons

All neurons have a cell body, which contains the nucleus and the metabolic machinery for protein synthesis. Most neurons have "processes," which are slender cytoplasmic extensions of the cell body. The processes differ greatly in number and length (Figure 24.1), so there really is no such thing as a "typical" neuron. The one described most often is the motor neuron that *innervates* (carries neural messages to) striated muscles in vertebrates.

The motor neuron shown in Figure 24.1 has many **dendrites** (short, slender processes) and one **axon** (a long, cylindrical process). Finely branched endings of the axon, called axon terminals, make functional connections with muscle cells. Traditionally, the dendrites (and often the cell body) are viewed as the *input* zone, where the neuron receives and integrates information. The input may cause an electrical signal to be produced at an adjacent, specialized area of membrane called the *trigger* zone. That signal rapidly propagates itself along the *conducting* zone, which extends from the start of the axon to its terminals. The axon terminals are *output* zones, where signaling molecules are sent to another cell. This picture of functional zones is simplified, for a neuron can have many input and output sites. Even so, it is still a good starting point for considering the principles of neural function.

Neuroglia

The three classes of neurons just described make up only about half the volume of vertebrate nervous systems. The rest is **neuroglia** ("nerve glue"), which consists of different cells that provide physical support, metabolic assistance, and protection for the neurons. For example, some neurological cells impart structure to the brain, much as connective tissues do for other body regions. Some segregate groups of neurons. Others wrap around axons like jellyrolls, and they affect how fast a signal travels over long distances.

Nerves and Ganglia

The human nervous system has more than 100 billion neurons in the brain alone. We will consider its organization and functioning in the next chapter. Here we need only define a few of the structures that are composed of either axons or cell bodies of neurons.

Axons of sensory neurons, motor neurons, or both are often packed tightly in bundles within connective tissue to form **nerves** (Figure 24.2). These cordlike communication lines connect the brain and spinal cord with the rest of the body. Within the brain and spinal cord, bundles of axons are called nerve *pathways* or *tracts*.

Typically, the cell bodies of neurons are organized into distinct clusters. Within the brain or spinal cord, the clusters are called *nuclei*; in other body regions they are called *ganglia*.

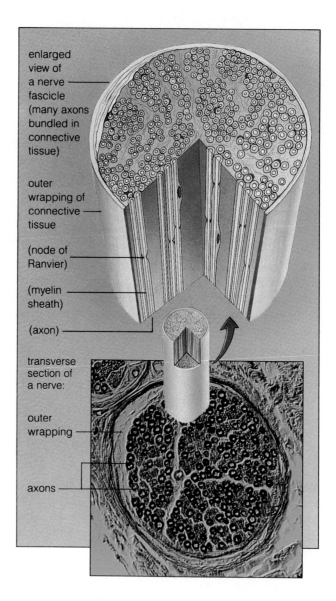

enlarged view of a nerve fascicle (many axons bundled in connective tissue)

outer wrapping of connective tissue

(node of Ranvier)

(myelin sheath)

(axon)

transverse section of a nerve:

outer wrapping

axons

Figure 24.2 Structure of a nerve, the components of which are described in this chapter and the next.

ON MEMBRANE POTENTIALS

Membrane Excitability

To understand how a nervous system works, we can begin with the plasma membrane of the neuron, for it is here that signals are carried. As in other cells, the membrane separates two chemically distinct regions: the cytoplasm and extracellular fluid. The two regions differ in their amounts of potassium ions, sodium ions, and other charged substances, so there can be an electric gradient as well as individual concentration gradients across the membrane (page 89). Overall, the inside is more negatively charged than the outside; there is a *polarity* of charge across the membrane.

When a neuron is stimulated, ions move across the membrane and so create a current (a flow of electric charge). A large current can briefly reverse the polarity (the inside of the neuron becomes more positive than the outside). Such reversals are called **action potentials**. They are so sudden and of such magnitude, they are like an electrical impulse. Neurons and other cells that produce action potentials are said to show *membrane excitability*.

Excitability is influenced by three membrane features, in ways that will be described shortly:

1. Being mostly lipid, a membrane tends to limit the movement of ions across it.

2. Ions *can* cross the membrane through ion channels that span the bilayer (Figure 24.3). Some are **open channels**, and ions "leak" (passively diffuse) through them all the time; others are **gated channels** that open during action potentials.

3. Membrane pumps, including **sodium-potassium pumps**, actively transport ions across the bilayer. Pumping maintains the concentration gradients across the membrane and restores them after an action potential.

The Neuron "At Rest"

Visualize a neuron at rest—that is, when its plasma membrane is not responding to a stimulus. There are far more potassium ions (K^+) inside than outside. Also, there are far fewer sodium ions (Na^+) inside than outside:

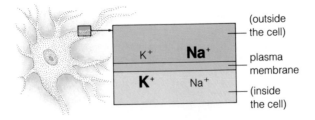

where large letters denote which side of the membrane has the greater concentration. As an example of the magnitude of this difference, consider a motor neuron from a cat. For every 150 potassium ions in a given volume of cytoplasm, there are only 5 in the same volume of fluid outside the cell. For every 15 sodium ions, there are 150 on the outside.

In the resting state, most of the channels that are *open* are ones for potassium ions. These positively charged ions leak out of the neuron, flowing passively down their concentration gradient. With the outward flow of positive charge, the interior becomes more negative—and the flow continues until the increasing negativity attracts some K^+ ions back in. When the inward pull of the electrical force balances the outward-directed diffusive force, there is no more *net* movement of potassium across the membrane.

The inside of the neuron is still more negative than the outside when the two forces acting on potassium are in equilibrium. We say there is a voltage difference across the membrane which, in many cells, holds steady at about seventy millivolts. (A "voltage difference" is simply the amount of energy between two differently charged regions that is potentially available to do work—in this case, to carry an electrical signal.) The steady voltage difference across the plasma membrane is called the **resting membrane potential**.

Sodium ions do not have much to do with maintaining the resulting membrane potential. Most sodium channels are *shut* during the resting state. (A few sodium ions do leak in, but membrane pumps run at a low rate to send just as many back out.) The sodium ions are brought into play during an action potential.

Changes in Membrane Potential

Suppose our neuron "at rest" is a sensory neuron that detects pressure applied to a thumb. When pressure stimulates an input zone (receptive surface) of the neuron, it causes a brief change in membrane potential. The stimulus energy has been converted to an electrical signal, which in sensory neurons is called a **receptor potential** (page 368). As you will see, an electrical signal also can arise at input zones of other neurons and muscle cells; it is called a **synaptic potential**. For now, the point to keep in mind is that both are graded, purely local signals.

"*Graded*" means the signal can vary in magnitude, depending on the intensity and duration of the energy input. (In our example, the stronger the pressure, the greater the electrical disturbance.)

"*Local*" means the signal does not spread far (half a millimeter or less, most often). The reason is that an input zone does not have the type of ion channels needed to propagate a signal farther than this. However,

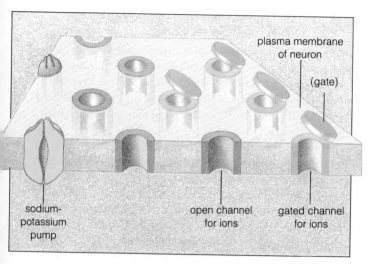

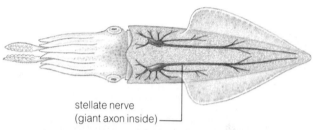

Figure 24.3 Pathways for ions across the plasma membrane of a neuron. These pathways are provided by proteins embedded in the lipid bilayer. Certain ions diffuse through perpetually open protein channels. Other protein channels have gates that only open when the neuron is stimulated; when they are opened, ion movements are more rapid across the membrane. Also, a membrane-bound enzyme system (the sodium-potassium pump) actively transports sodium in one direction and potassium in the opposite direction across the membrane.

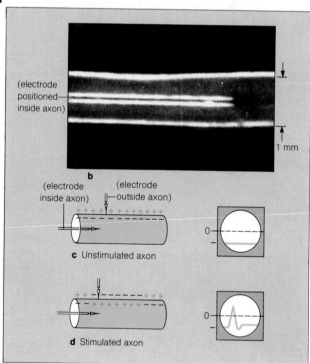

Figure 24.4 (**a**) Approximate location of giant axons that innervate the muscular body wall (the mantle) of the squid *Loligo*. (**b**) The micrograph shows the axon diameter relative to the size of an electrode, a device used in measuring voltage changes. Being large enough to accommodate such devices, the giant axon lent itself to early studies of nerve functioning. (**c**) In a resting neuron, the inside of the axon is negative with respect to the outside, as registered on the screen of an oscilloscope. When the electrodes detect an action potential, a waveform of the sort depicted in (**d**) appears on the screen.

when the stimulation is intense or prolonged, graded signals can reach an adjacent trigger zone of the membrane (Figure 24.1). There, a brief reversal in the polarity of charge—an action potential—can be initiated.

THE ACTION POTENTIAL

The action potential is analogous to a pulse of electrical activity (hence its original name, nerve impulse). Its existence can be demonstrated with the use of so-called giant axons from the squid *Loligo*. These axons, which innervate the muscular body wall of the squid, can be up to a millimeter in diameter. This means that an electrode (a device used to measure voltage changes) can be inserted rather easily into one of them. One fine electrode is inserted into the axon; another is positioned in the fluid outside the axonal membrane. Both are connected to a voltmeter and an oscilloscope. With an oscilloscope, voltage changes show up as deflections of a beam traveling across a fluorescent screen (Figure 24.4).

Measurements of the voltage difference across a membrane before, during, and after an action potential reveal this pattern:

1. In a neuron at rest, the membrane is *polarized* (the inside is more negative with respect to the outside).

2. During an action potential, the membrane is *depolarized* (the inside is more positive than the outside).

3. Following an action potential, the membrane is *repolarized* (resting conditions are restored).

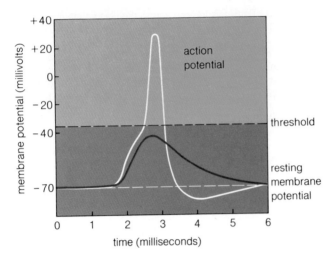

Figure 24.5 Recording of an action potential (white line). The red line is a recording of a graded signal that did not reach threshold.

Figure 24.6 Simplified picture of how action potentials are propagated along the axon of a neuron. The plasma membrane is shown in yellow. The only channels shown are the gated channels for sodium and potassium.

Mechanism of Excitation

For an action potential to occur, the graded signals reaching a trigger zone must depolarize the membrane by a certain minimal amount, this being called the **threshold** level. Depolarization of the membrane to threshold can be brought about by an accelerating flow of sodium ions into the neuron.

Briefly, when a strong graded signal reaches the trigger zone, the ensuing depolarization of that area of membrane causes gated sodium channels to open. As the positively charged ions move into the neuron, the voltage inside becomes less negative. The changing voltage causes more gates to open, which admits more sodium, which increases the positivity inside, and so on until the membrane potential reverses (Figure 24.5). The escalating flow of sodium is an example of *positive feedback*, whereby an original event is increasingly intensified as a result of its own occurrence.

Once threshold is reached, the opening of sodium gates no longer depends on the strength of the stimulus (because the positive-feedback cycle has started). Then, a portion of the energy stored in the concentration and electric gradients for sodium is released automatically. The amount of energy released is *not* related to the strength of the stimulus, any more than the effect of dynamite depends on the size of the match that lit the fuse. All action potentials generated in a given neuron will "spike" to the same voltage level above threshold.

Thus an action potential is an **all-or-nothing event:** If threshold is reached, nothing can stop the maximal spiking, but if threshold is not reached, the electrical disturbance will subside once the stimulus is removed (Figure 24.5).

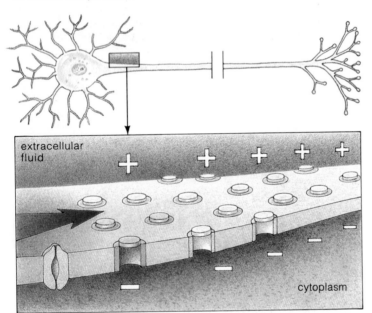

a Membrane at rest (inside negative with respect to the outside). An electrical disturbance (red arrow) spreads from an input zone to an adjacent trigger region of the membrane, which has many gated sodium channels (green).

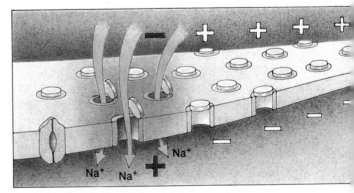

b A strong disturbance (a strong graded signal or a summation of more than one) initiates an action potential. Sodium gates open, the inflow decreases the negativity inside; this causes more gates to open, and so on, until threshold is reached and the voltage difference across the membrane reverses (depolarization).

Duration of Action Potentials

For most neurons, each action potential lasts only a few milliseconds, and several hundred can occur in a single second. Certainly billions of action potentials occur in your nervous system during the time it takes you to read about a single one of them.

Why does an action potential end so abruptly? There are two reasons. First, the gated channels for sodium ions have *another* gate, near the cytoplasmic side. When voltage inside goes from negative to positive during an action potential, this opposing gate slams shut and closes off the sodium flow. Second, gated channels for potassium open when the membrane depolarizes, about halfway through the action potential. Now potassium diffuses out of the neuron, following its concentration gradient. These ion movements restore the voltage across the membrane to its original value.

Even though the voltage is restored, the sodium and potassium gradients are ever so slightly reduced in size at the end of many action potentials. These tiny reductions in concentration gradients must be countered, otherwise the ability of the neuron to respond to stimuli would gradually disappear. The gradients are restored by sodium-potassium pumps. The pumps actively transport potassium ions into the neuron and, at the same time, sodium ions are sent out. This is an example of the cotransport mechanism shown earlier in Figure 6.9.

Propagation of Action Potentials

Once an action potential occurs in a trigger zone of the membrane, it is propagated in undiminished fashion through an adjacent conducting zone, such as the one shown in Figure 24.1. In brief, the electrical disturbance spreads and causes depolarization in adjacent membrane regions, so that sodium channels open there, also. Sodium moves into the neuron and, after a short delay, potassium moves out. The new disturbance causes channels to open in the next adjacent region, and so it goes, away from the original stimulation site (Figure 24.6). These alterations in voltage across the membrane are the "message" that is propagated along the neuron.

The propagation of an action potential can be summarized this way:

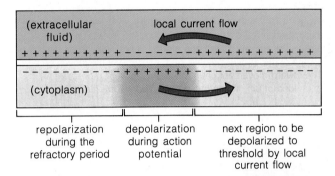

repolarization during the refractory period | depolarization during action potential | next region to be depolarized to threshold by local current flow

Notice how the current flows in a circuit (arrows) between regions that differ in voltage. When positively charged ions move into the neuron at the site of an action potential, a voltage difference is created between the point of entry and adjacent regions of the cytoplasm— so current flows laterally (down the electric gradient, to less negative regions of the cytoplasm). While this is going on, positive charge flows laterally outside the neuron, back toward the more negative region created by the movement of positive ions into the neuron. Such local current flows allow the depolarization to spread to the adjacent membrane, where it opens more sodium channels and thereby propagates the action potential.

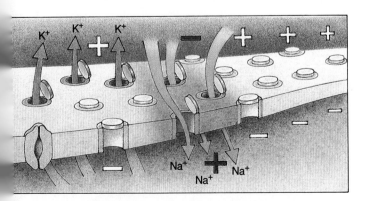

he depolarization causes sodium gates to shut and potassium gates pen at this site. Potassium follows its gradient (out of the neuron) voltage is restored; the membrane is repolarized. Meanwhile, the ctrical disturbance triggers another action potential at the adjacent mbrane site, and so on, away from the point of stimulation.

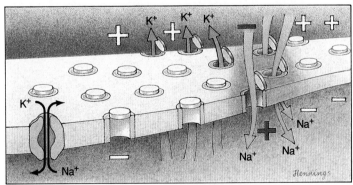

d Repolarization means the inside of the membrane becomes negative again following each action potential, but the sodium and potassium concentration gradients are not yet fully restored. Active transport at sodium-potassium pumps restores the gradients.

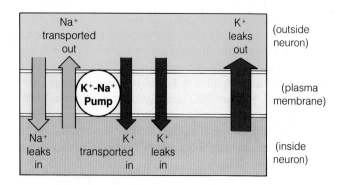

Figure 24.7 Balance between pumping and leaking processes that maintain the distribution of sodium and potassium ions across the plasma membrane of a neuron at rest. The relative widths of the arrows indicate the magnitude of the movements. The inward movement counteracts the outward movement for each kind of ion; hence the ion distributions are maintained.

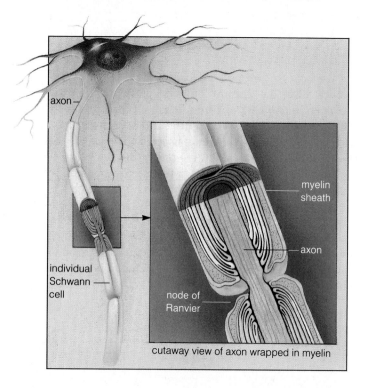

Figure 24.8 Myelinated axon of a motor neuron, formed by wrappings of Schwann cell membranes (shaded in gold).

Figure 24.9 Conduction of action potentials in myelinated neurons. (**a**) The influx of sodium ions at the first node triggers an action potential (green arrows), which generates a local current flow to the second node. (**b**) The current flow depolarizes the next node and triggers a new action potential, which generates a local current flow to the third node.

Notice also how the summary sketch in Figure 24.6 shows action potentials traveling *away* from the stimulation site. They always do this. There is no backflow, largely because each action potential is followed by a **refractory period**. During this period, when sodium gates are shut tight and potassium gates are wide open, the membrane is insensitive to stimulation. Later, after the resting membrane potential has been restored, potassium channels close and sodium channels return to their initial state, ready to be opened when the membrane potential next reaches threshold.

Figure 24.7 summarizes the pumping and leaking processes that keep the membrane in a state of readiness between action potentials.

Saltatory Conduction

Many of the long, thin axons that connect different parts of the vertebrate nervous system are sheathed by neuroglial cells. The sheaths afford protection, and some play a passive but important role in message propagation over long distances.

For example, in peripheral nerves (those in body regions other than the brain and spinal cord), modified neuroglial cells called **Schwann cells** spirally wrap their plasma membrane around some axons (Figure 24.8). The membrane is so tightly wrapped around the axon, it forms a specialized tissue called the *myelin sheath*.

In peripheral nerves, each sheath is separated from the next by a *node of Ranvier*: a small gap where the axon is exposed to the extracellular fluid. Current can flow rapidly across the membrane at the nodes, because each node is loaded with gated channels for sodium—some 12,000 per square micrometer! Only the trigger zone comes close to having this channel density.

Between the nodes, the tight sheaths hinder current flow across the membrane, so the current flowing in at

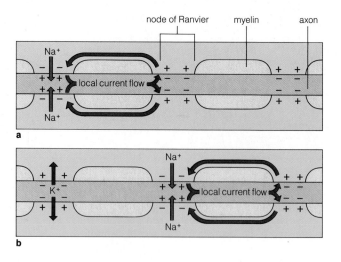

one node spreads laterally through the cytoplasm to the next node in line (Figure 24.9). Because less current is crossing the sheathed portion of the membrane, the electric gradient *between* two nodes does not drop as much as it would in unsheathed axons. Thus the membrane at the next node in line is brought to threshold much faster. In a manner of speaking, the action potentials "jump" from one node to the next and thereby increase the overall speed of propagation. The rapid, node-to-node hopping by lateral current flow is called **saltatory conduction**, after the Latin word meaning "to jump."

Saltatory conduction affords the most rapid signal propagation with the least metabolic effort by the cell. In the largest myelinated axon, signals travel 120 meters per second (270 miles per hour).

SYNAPTIC POTENTIALS

Chemical Synapses

Where does the message go, once it reaches the output zone of a neuron? Here we come to the graded signals called synaptic potentials. These signals arise at a **chemical synapse**, which is a type of cell junction. (Cell junctions, recall, are local regions of the plasma membrane where neighboring cells are linked structurally, functionally, or both.) The simplest chemical synapses are

sites of near-contact between the output terminals of one neuron and the membrane of another neuron. A small extracellular space, the *synaptic cleft*, separates the two cells (Figure 24.10).

Information flows *from* the cell that releases a transmitter substance into the synaptic cleft. A **transmitter substance** is a type of signaling molecule that can cause a rapid change in the membrane potential of a receptive cell at a chemical synapse. An example is acetylcholine (ACh), which has excitatory or inhibitory effects, depending on the properties of the target cell. Another is gamma aminobutyrate (GABA), which has mostly inhibitory effects.

A neuron that releases a transmitter substance into the cleft is called the **presynaptic cell**; the one affected by the substance is the **postsynaptic cell**. Often it is easy to tell which is which in electron micrographs because only the presynaptic cell contains numerous vesicles near the synaptic site. Transmitter substances are known to fill at least some of these vesicles (Figure 24.10).

Given the cleft between presynaptic and postsynaptic cells, an action potential cannot be propagated directly from one to the other. What happens is this: When an action potential arrives at the presynaptic cell membrane facing the cleft, the depolarization causes gated *calcium* channels to open. Calcium ions are more concentrated outside the cell, and when they move inside (down the gradient), they trigger the fusion of synaptic vesicles

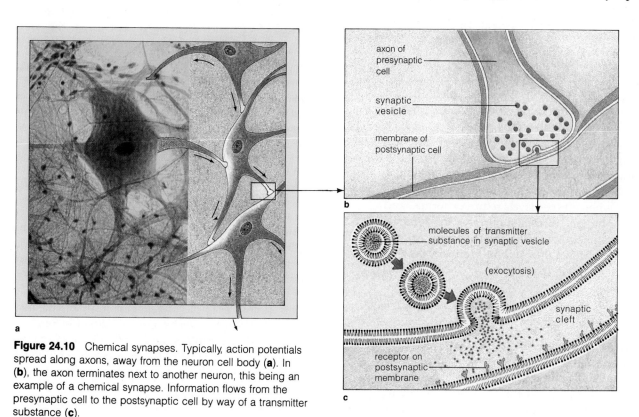

Figure 24.10 Chemical synapses. Typically, action potentials spread along axons, away from the neuron cell body (**a**). In (**b**), the axon terminates next to another neuron, this being an example of a chemical synapse. Information flows from the presynaptic cell to the postsynaptic cell by way of a transmitter substance (**c**).

with the plasma membrane. This is an example of exocytosis (page 94).

Once molecules of transmitter substance are released into the synaptic cleft, diffusion carries them to the postsynaptic cell. The molecules bind briefly to membrane receptors that are coupled to ion channels. With this binding, the channels open and give rise to a synaptic potential. The size of the potential depends on (1) the amount of transmitter substance reaching the membrane receptors and (2) the electrical state of the receiving cell.

Excitatory and Inhibitory Postsynaptic Potentials

Transmitter substances can initiate graded signals in postsynaptic cells. You may be thinking that these signals function only to trigger an action potential. Some do, but others can inhibit it. In other words, a synaptic potential can be excitatory *or* inhibitory:

excitatory postsynaptic potential (EPSP)	*brings the postsynaptic membrane closer to threshold (it is depolarizing)*
inhibitory postsynaptic potential (IPSP)	*drives the postsynaptic membrane away from threshold (it is hyperpolarizing) or maintains it at its resting level*

Especially in the brain or spinal cord, the input of hundreds of excitatory potentials usually must be combined before a large postsynaptic cell will respond to a stimulus with action potentials.

Signaling molecules other than transmitter substances also can act on neurons in ways that do not simply elicit an EPSP or IPSP. These so-called "neuromodulators" regulate the responsiveness of their target neurons to synaptic inputs. They enhance or reduce excitability, and their effects take place over minutes, hours, or days, not milliseconds. An example is cortisol, a hormonal secretion of the adrenal gland that helps raise glucose levels in the blood. Quite a bit of cortisol may be circulating when glucose levels return to normal, and at high concentration, cortisol suppresses the synaptic activity of certain neurons in the brain. Cortisol actually makes them slow down their output of substances that trigger cortisol secretion (page 360).

Synaptic Potentials in Muscle Cells

Much of what is known about chemical synapses came from Bernard Katz and his colleagues, who studied **neuromuscular junctions** (the synapses between motor neurons and muscle cells). Where an axonal branching approaches a muscle cell membrane, it is splayed out in finer branchings. The axon terminals in this region are positioned in troughs in the muscle cell membrane. These troughs are called the *motor end plate* (Figure 24.11).

An action potential traveling along the motor axon spreads through all the terminals, which respond by releasing the transmitter substance ACh. Within a mere fifty microseconds, ACh molecules diffuse across the synaptic cleft to the muscle cell. Their arrival is like a whack on the muscle cell membrane. They give rise to a response that commonly initiates an action potential, which in turn leads to contraction (page 394).

ACh interacts with receptors on the muscle cell membrane—receptors that also happen to be gated channels for ions. By some estimates, each square micrometer of membrane of a motor end plate has 10,000 of those gated channels. While all those gates are being opened and shut, enzyme molecules on the membrane are inactivating ACh. The enzyme (acetylcholinesterase) gives ACh no more than 1/500 of a second to act. Having helped open a gate, the ACh molecule diffuses away from the receptor and is broken down; and so the receptor site is cleared for the next incoming signal.

FROM SYNAPSE TO NEURAL CIRCUIT

Synaptic Integration

At each neuron in a nervous system, excitatory and inhibitory signals compete for control of the membrane potential. As Charles Sherrington perceived, the competing signals are combined in the neuron, in a process called synaptic integration.

Synaptic integration is the moment-by-moment combining of excitatory and inhibitory signals acting on adjacent membrane regions of a neuron.

Suppose, for example, an excitatory synapse and a nearby inhibitory synapse become active at the same time. The white lines in Figure 24.12 show how each resulting change in membrane potential would register on an oscilloscope screen *if it were occurring alone*. The red line shows what happens when they occur simultaneously. The combined effect is that the excitatory potential is pulled away from the threshold of an action potential.

This simple example should not mislead you into thinking that all synapses have equal influence. Remember that synaptic potentials decay with distance from the site of initiation. This means the extent of their influence depends on the direction and magnitude of the current flows that are actually produced. The most critical variable is how much current across the postsynaptic membrane is induced by each synapse. Location also can be

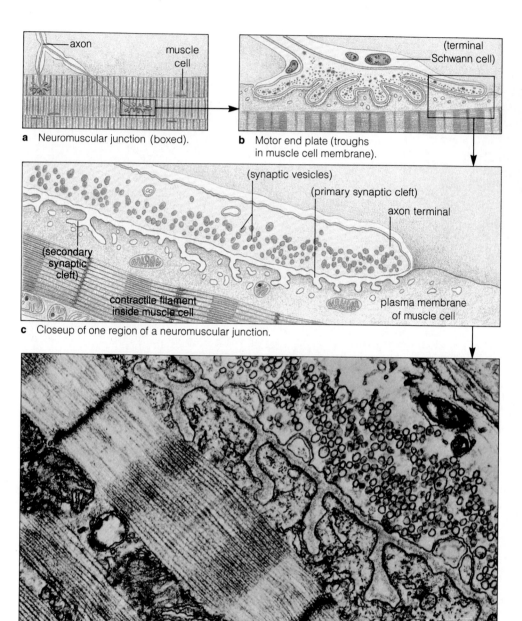

a Neuromuscular junction (boxed).

axon

muscle cell

b Motor end plate (troughs in muscle cell membrane).

(terminal Schwann cell)

(synaptic vesicles)

(primary synaptic cleft)

axon terminal

(secondary synaptic cleft)

contractile filament inside muscle cell

plasma membrane of muscle cell

c Closeup of one region of a neuromuscular junction.

d Transmission electron micrograph of a neuromuscular junction.

Figure 24.11 (**a**) A neuromuscular junction (the region of chemical synapse between a neuron and a muscle cell). (**b**) At this junction, axon terminals act on troughs in the muscle cell membrane (the motor end plate). The myelin sheath of the axon stops at the junction, such that the membranes of the two interacting cells are exposed to each other. (**c-d**) Secondary clefts form deep channels into the muscle cell membrane.

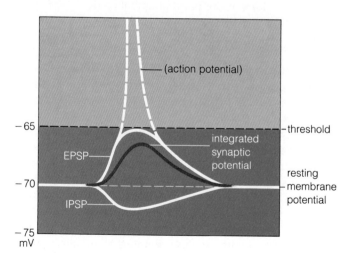

(action potential)

threshold

integrated synaptic potential

EPSP

resting membrane potential

IPSP

−65

−70

−75
mV

Figure 24.12 Synaptic integration. In this example, an excitatory synapse and an inhibitory synapse nearby are activated at the same time. The IPSP reduces the magnitude of the EPSP from what it could have been, pulling it away from threshold. The red line represents the integration of these two synaptic potentials. Threshold is not reached in this case; hence an action potential cannot be initiated.

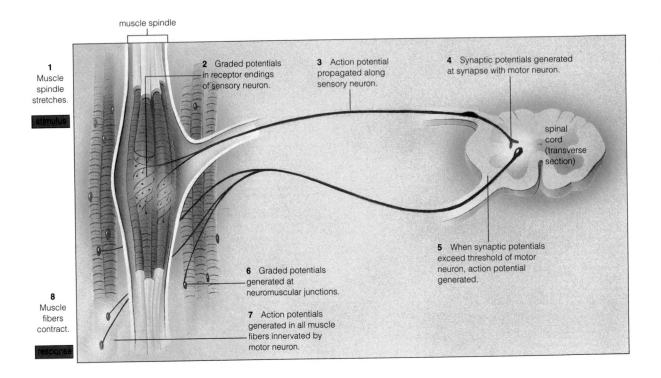

Figure 24.13 Reflex arc governing the stretch reflex in mammals. Sensory axon is shown in purple; motor axon in red. Other inputs and outputs exist, but they are omitted here for the sake of clarity.

important. For example, when synapses are arrayed along dendrites, those closest to the cell's trigger zone will have the greatest influence over the initiation of an action potential. In short, *synaptic potentials can be added together in a given time span (temporal summation) and in a given region (spatial summation).*

When foreign substances in the body interfere with synaptic integration, the consequences can be deadly. For example, on rare occasions the anaerobic bacterium *Clostridium tetani* is carried accidentally into the body's internal tissues. This can occur when the skin is punctured, as from an animal bite or by a soil-covered nail. If the tissues become necrotic (die off, as happens when the injury prevents further delivery of oxygen to them), the bacterium can multiply. One of its metabolic products functions as a neurotoxin in this setting. It interferes with inhibitory synapses on motor neurons in the brain and spinal cord. The unbalanced excitation causes excessive contraction (as in lockjaw). Muscles cannot be released from the contracted state. The result is *tetanus*: prolonged, spastic paralysis of muscles, a condition that can lead to death.

Circuit Organization

Typically, action potentials spread outward along an axon's membrane, away from the cell body. The *direction* in which a given message flows through the nervous system depends on the way neighboring neurons and

their processes are organized relative to one another, as the arrows in Figure 24.10 suggest. With synapses as the connections, neurons form *circuits* for transmitting and processing the electrical and chemical signals. The circuits may involve only a few neurons; they may involve hundreds, thousands, even millions.

When message transfers are confined to a given part of the brain or spinal cord, the interacting neurons form a "local circuit." Local circuits function in processing messages, integrating them with other inputs, and controlling the output to other regions. At a different level of organization are the nerve pathways or tracts. Those parts of circuits are formed by connections between two or more regions in the brain or spinal cord.

The Stretch Reflex

One of the simplest examples of a nerve pathway is the **reflex arc**. A *reflex* is a simple, stereotyped, and repeatable motor action that is elicited by a specific sensory stimulus. It is a basic element of motor behavior.

Many types of reflex arcs are necessary in maintaining and adjusting body posture. One is the **stretch reflex**. This neural mechanism underlies the reflex contraction of a muscle when it is stretched from its resting length. Even when you are not aware of the stretch reflex, it is helping you maintain an upright posture despite small shifts in your balance. This reflex also takes part in voluntary movements, such as helping to maintain arm muscles under tension when you are carrying a heavy package that would otherwise pull your arm down.

As Figure 24.13 shows, stretch-sensitive receptors are located within skeletal muscles. These receptors are part of **muscle spindles**, which are made of small, specialized cells enclosed in a sheath that runs parallel with the muscle itself. The receptors are specialized endings of sensory neurons.

When a muscle in which a spindle resides is stretched, the disturbance is transmitted to and encoded by the receptors, which act in a negative feedback capacity to return the muscle to its original length. Action potentials are generated and conducted rapidly toward the spinal cord, where the axon terminals synapse with (among other things) motor neurons—which have axons leading right back to the muscle that was stretched. Under enough stimulation arriving from sensory neurons, action potentials are generated in the motor neurons. They travel to the axon terminals, where they trigger the release of the excitatory transmitter substance ACh. The ACh activates the muscle cell membrane and initiates contraction. Thus the stretch reflex maintains a degree of tension in the muscle.

In a few reflex arcs, sensory neurons synapse directly on motor neurons. Such direct connections, in which no other neurons intervene to mediate the response, are said to be *monosynaptic pathways*. John Nicholls identified three monosynaptic pathways in the leech. These pathways are involved in the **withdrawal reflex**, a rapid pulling away from an unpleasant or potentially harmful stimulus. By simple reflex action, this invertebrate jerks away from touch, pressure, and noxious chemicals.

In most animals, the withdrawal reflex involves *polysynaptic pathways*. Here, sensory neurons in the pathway make connections with a number of interneurons, whose processes collectively activate or suppress the motor neurons necessary for a coordinated response. If you have ever accidentally touched a hot stove, you know that the withdrawal reflex action can be completed even before you become conscious that it has occurred.

SUMMARY

In this chapter, we have considered several aspects of the neuron. Before we turn to the systems based on this specialized cell, let's put the key points in perspective:

Excitability

1. The neuron, or nerve cell, is the basic unit of communication in all nervous systems. Neurons *collectively* sense environmental change, swiftly integrate sensory inputs, and then activate effectors (such as muscle cells and gland cells) that can carry out responses.

2. Like all cells, the neuron shows a polarity of charge across its plasma membrane: the inside is more negative than the outside. The polarity results from differences in the concentrations of potassium ions, sodium ions, and other charged substances present in cytoplasm and extracellular fluid.

3. Most neurons and some other cells (including skeletal muscle cells) show excitability: in response to stimulation, the polarity of charge can undergo brief but sudden reversals (action potentials); that is, the inside becomes positive with respect to the outside.

4. The neuron receives and integrates stimuli at *input* zones (dendrites, the cell body). The stimulus energy creates an electrical disturbance that may reach a *trigger* zone, where action potentials can be generated. Action potentials travel along *conducting* zones (axons) to the *output* zones (axon terminals), where other kinds of signals are sent to target cells.

5. The neuron's ability to receive and respond to stimuli depends on these membrane properties:

   a. The lipid bilayer is impermeable to ions.

   b. Open and gated channels across the membrane allow ion movements (that is, currents) between the cytoplasm and extracellular fluid.

   c. Pumps (sodium-potassium pumps, especially) spanning the bilayer actively transport ions in ways that help restore and maintain the concentration gradients between action potentials.

Resting Membrane Potential

1. In a neuron at rest, potassium channels are open and sodium channels are shut. For the most part, electrical and diffusive forces acting on *potassium* ions establish the resting membrane potential (a steady voltage difference across the membrane), which commonly is about seventy millivolts.

2. Local, graded signals at input zones can change the membrane potential. Receptor potentials (in sensory neurons) and synaptic potentials (in other neurons and muscle cells) are examples of these signals.

3. Graded signals decay with distance from the point of stimulation. And they vary in magnitude, depending on the intensity and duration of the stimulus.

4. Most graded signals serve as input signals for triggering or suppressing an action potential.

The Action Potential

1. An action potential is triggered when the electrical disturbance created by graded signals causes the voltage difference across the membrane to change by a certain minimum amount (the threshold level). Then, gated sodium channels open in an accelerating way (due to

positive feedback). The resulting current briefly reverses the voltage difference by a dramatic amount, which registers as a "spike" on recording devices.

2. An action potential is an all-or-nothing event. Once threshold is reached, nothing can stop the maximal spiking (because the positive-feedback cycle is now at work). If threshold is not reached and the stimulus is removed, the electrical disturbance will subside and no action potential will occur.

3. An action potential lasts only a few milliseconds. At the membrane region where it occurred, sodium gates shut, potassium gates open, and ion movements across the membrane restore the original voltage difference.

4. Although voltage is restored, the sodium and potassium gradients are slightly reduced. Membrane pumps restore the original gradients.

5. The overall pattern before, during, and after an action potential is this: polarization (inside negative with respect to outside), depolarization (inside becomes positive), and repolarization (inside becomes negative again).

6. An action potential is self-propagating. The electrical disturbance and current flows created by an action potential will trigger the same current flows in the adjacent membrane region, and so on away from the stimulation site.

Synaptic Potentials

1. Chemical synapses are junctions between two neurons, or between a neuron and a muscle cell or gland cell. A small gap, the synaptic cleft, separates them.

2. Usually, only one of the two cells releases a transmitter substance (a type of signaling molecule) into the synaptic cleft. Its action identifies it as the *presynaptic* cell, which relays signals to the other, *post*synaptic cell.

3. Transmitter molecules bind briefly to receptors on the postsynaptic cell membrane. Depending on the receptors that are activated and the electrical state of the membrane, the cell may respond with excitatory or inhibitory potentials.

4. An excitatory postsynaptic potential (EPSP) is depolarizing; it brings the membrane closer to the threshold of an action potential.

5. An inhibitory postsynaptic potential (IPSP) is usually hyperpolarizing; it drives the membrane away from threshold.

6. Most often, many excitatory potentials acting together are needed to depolarize the membrane to threshold and to initiate an action potential, and they must be strong enough to override the effects of inhibitory potentials at nearby synapses.

Synaptic Integration and Neural Circuits

1. Integration is the moment-by-moment combining of all graded signals—excitatory and inhibitory—acting at different synapses on a neuron.

2. Synaptic integration means that signals arriving at a neuron can be reinforced or dampened, sent on or suppressed.

3. The direction of information flow through the body depends on the organization of neurons into interconnecting circuits.

4. In the brain and spinal cord, local circuits are sets of interacting neurons confined to a single region. The parts of circuits called nerve pathways extend from neurons in one region to neurons in different regions.

Review Questions

1. Define sensory neuron, interneuron, and motor neuron. *322*

2. Label the functional zones of a motor neuron on the following diagram *322–323*:

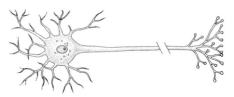

3. What is the difference between a neuron and a nerve? *322–323*

4. Two major concentration gradients exist across a neural membrane. What are they, and how are they maintained? *324*

5. An electric gradient also exists across a neural membrane. Explain what the electric and concentration gradients together represent. *324*

6. Distinguish between an action potential and a graded potential. What is meant by "all-or-nothing" and "self-propagating" messages? *324–328*

7. What is a synapse? Explain the difference between an excitatory and an inhibitory synapse. Define neural integration. *329–330*

8. What is a reflex? Describe the sequence of events in a stretch reflex. *332–333*

Readings

Berne, R., and M. Levy (editors). 1988. *Physiology*. Second edition. St. Louis: Mosby. Section II is an authoritative introduction to neural functioning.

Dunant, Y., and M. Israël, April 1985. "The Release of Acetylcholine." *Scientific American* 252(4):58–83. Experiments showing how this major neurotransmitter functions.

Lent, C., and M. Dickenson. June 1988. "The Neurobiology of Feeding Behavior in Leeches." *Scientific American* 258(6):98–103. Describes the relationship between serotonin (a transmitter substance) and feeding behavior in an invertebrate.

NEURAL PATTERNING:
AN OVERVIEW

Of all organisms on earth, why do animals alone have nervous systems? Plants rely on hormonal integration of activities, and if hormones, why not nerves? Plants also move their leaves in response to environmental stimuli. The Venus flytrap even uses electrical signals in springing its spiny trap—and the response to its insect prey is far more rapid than the response of, say, a sponge to a pinprick. The sponge contracts slowly, in a diffuse sort of way, and the response never extends more than a few millimeters beyond the point of stimulation.

But sponges are just about the simplest members of the animal kingdom. Other animals have specialized means of detecting stimuli and responding swiftly to them. Almost all animals reach out or lunge after food; they pull back, crawl, swim, run, or fly when they are about to become food themselves. And think about what they have to do to find a mate and slow it down or hold its attention. (Think about all the things you have to do.) The more complex the environment and life style, the more elaborate and rapid are the animal modes of sensory reception, signal integration, and response.

There are more than a million known species of animals, so the examples used in this chapter are necessarily limited. Even so, the examples will help reinforce the following points, which together constitute a **theory of neural patterning**:

1. Reflexes are simple, stereotyped movements made in response to sensory stimuli. In the simplest reflex pathways, a sensory neuron directly signals a motor neuron, which acts on muscle cells (page 332).

2. Reflex pathways are the basic operating machinery of nervous systems.

3. Nervous systems evolved through accretion: a layering of additional nervous tissues over reflex pathways of more ancient origin.

4. Nervous systems evolved along with sensory organs (such as eyes) and motor structures (such as legs and wings), and together they provided the foundation for more active and intricate life-styles.

5. The oldest parts of the vertebrate brain deal with reflex coordination of sensory inputs and motor outputs beyond that afforded by the spinal cord alone.

6. Even the most recent layerings of nerve tissue deal partly with reflex coordination. But they also deal with storing, comparing, and using experiences to initiate novel, nonstereotyped action. These regions are the basis of memory, learning, and reasoning.

25

NERVOUS SYSTEMS

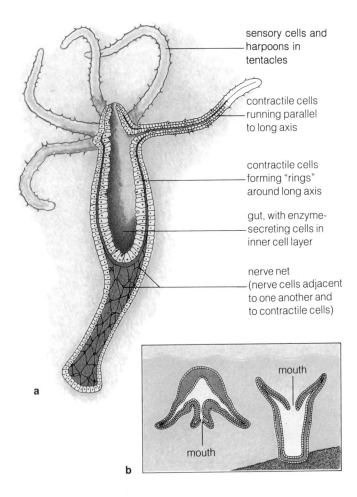

sensory cells and harpoons in tentacles

contractile cells running parallel to long axis

contractile cells forming "rings" around long axis

gut, with enzyme-secreting cells in inner cell layer

nerve net (nerve cells adjacent to one another and to contractile cells)

a

b

mouth

mouth

Figure 25.1 (**a**) Nerve net, sensory cells, and contractile cells in *Hydra*, a type of cnidarian. (**b**) Regardless of whether a cnidarian is free-floating or sedentary, the nerve net is arranged radially about the mouth and gut.

INVERTEBRATE NERVOUS SYSTEMS

Nerve Nets

Of all animals, the cnidarians (sea anemones, hydras, and jellyfishes) have the simplest nervous systems. Some forms of cnidarians are attached to the sea floor, others are free-floating, yet they all face similar challenges. For them, food and danger are likely to appear not on water's surface or on the bottom but anywhere in between (Figure 25.1). The systems by which they sense and respond to the environment show **radial symmetry**, with body parts arranged radially about a central axis, much like the spokes of a bike wheel.

The cnidarian nervous system, called a **nerve net**, is based on reflex pathways between epithelial receptor cells and contractile cells (Figure 25.1). For example, in the pathway concerned with feeding behavior, nerve cells extend from receptors in tentacles to contractile cells around the mouth. In jellyfishes, some pathways are involved in slow swimming movements and in keeping the body right-side up.

Cephalization and Bilateral Symmetry

Complex nervous systems may have evolved from arrangements as simple as nerve nets. Some cnidarian life cycles include a ciliated, cigar-shaped larval stage called a planula. Planulas swim or crawl about, very much in the manner of a more complex animal called the flatworm:

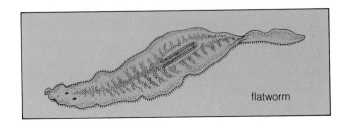

flatworm

Flatworms show **cephalization**: a concentration of sensory structures and coordinating centers at their anterior end (that is, the head end). The selective advantage of cephalization seems clear enough. The forward end of a crawling animal is the first to encounter stimuli, such as food odors, and responses can be more rapid and effective with sensory receptors up front.

Flatworms also show **bilateral symmetry**: they have a body plan with roughly equivalent right and left halves. For example, motor structures for pulling the body forward are developed the same way on both sides of the body, not just one; nerves to control the motor structures are developed the same way on both sides; and so on. What would have been the selective advantage of bilateral symmetry in the early evolution of animals? Perhaps this: *A shift from radial to bilateral symmetry*

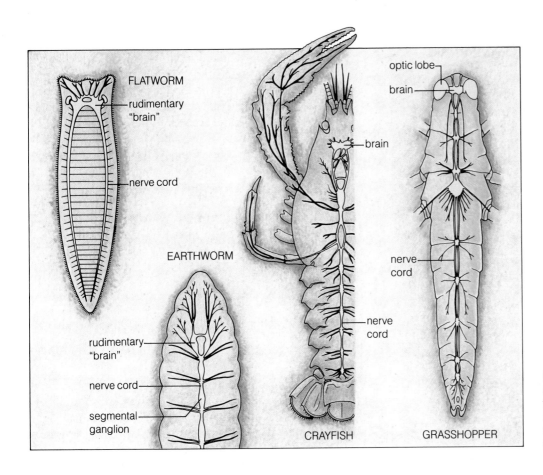

FLATWORM
— rudimentary "brain"
— nerve cord

EARTHWORM
rudimentary "brain"
nerve cord
segmental ganglion

optic lobe
brain
— brain
nerve cord
nerve cord
nerve cord
CRAYFISH
GRASSHOPPER

Figure 25.2 Bilateral symmetry and segmentation evident in the nervous systems of a few invertebrates. The sketches are not to scale relative to one another.

could have led to paired nerves and muscles, paired sensory structures, and paired brain centers.

Intriguingly, simple flatworms have a nerve net, but there are flatworms with *ganglia* (clustered cell bodies of neurons) and with *nerves* and two *nerve cords* (these being bundled-together processes of neurons). The ganglia form a brainlike structure at the head end and coordinate signals from paired sensory organs, including two eyespots. They also provide some control over the nerve cords, which have nerves branching bilaterally from them and which carry signals to such body parts as the muscles used in swimming and crawling (Figure 25.2).

Segmentation

Like flatworms, the annelids (such as earthworms) and arthropods (such as grasshoppers) have bilateral nerves. These animals also show pronounced **segmentation**: the body is composed of repeating units that are more or less the same in structure and function. Each segment has a pair of nerves and a ganglion, which controls muscles in that segment and (sometimes) in its immediate neighbors. A nerve cord extends through the segments, and at its anterior end, nerve cells are fused into a "brain" (Figure 25.2). The nerve branchings and ganglia of your own nervous system echo this pattern of segmentation.

THE VERTEBRATE PLAN

Evolution of Vertebrate Nervous Systems

It appears that the vertebrate nervous system evolved largely through modification of three key characters: bilateral symmetry, a notochord, and a hollow nerve cord.

A **notochord** is a long rod of stiffened tissue, neither cartilage nor bone, that serves as a supporting structure for the body. It first appeared in marine chordates (which, by definition, have a notochord). Although this rod is present in all vertebrate embryos, it is greatly reduced or absent in adults. In almost all species it is replaced during development by hard, bony segments called **vertebrae**, which are serially arranged in a **vertebral column** (the backbone).

The vertebral column proved to be an ideal skeletal axis against which the force of muscle contraction could be applied. And some bony segments of the skeleton proved to have enormous potential for modification into powerful, bony jaws. The evolution of the notochord into this bony column foreshadowed the evolution of fast-moving, predatory animals.

A related key character in vertebrate evolution was a single, hollow **nerve cord** running dorsally above the notochord. Early on, this hollow structure underwent

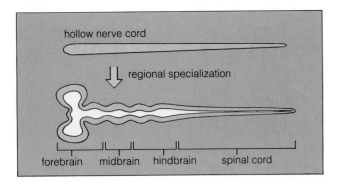

Figure 25.3 Schematic portrayal of the evolution of the anterior end of the vertebrate nerve cord into a spinal cord and specialized brain regions.

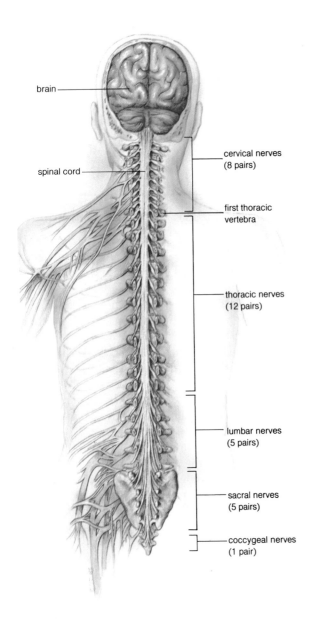

brain

spinal cord

cervical nerves
(8 pairs)

first thoracic
vertebra

thoracic nerves
(12 pairs)

lumbar nerves
(5 pairs)

sacral nerves
(5 pairs)

coccygeal nerves
(1 pair)

expansion and regional modification into a spinal cord and brain. In the changing world of fast-moving vertebrates, sensory receptors became more complex. Paired nasal structures, eyes, and ears gathered information from the outside and fed it into the anterior end of the tube. In time, that end became variably thickened with nervous tissue and functionally divided into three parts: forebrain, midbrain, and hindbrain (Figure 25.3).

Functional Divisions of the Vertebrate Nervous System

The nervous system of all existing vertebrates has two main divisions:

central
nervous system

the spinal cord and brain

peripheral
nervous system

nerves (leading into and from the spinal cord and brain) and ganglia

Figure 25.4 shows the relative locations of the thirty-one pairs of spinal nerves that connect with the central nervous system of humans. There also are cranial nerves (not shown) that connect directly with the brain. The sensory and motor axons contained in these nerves make connections with interneurons, which are confined entirely within the central nervous system. The pathways or tracts running up and down the spinal cord consist primarily of bundled-together long axons of interneurons.

PERIPHERAL NERVOUS SYSTEM

In the peripheral nervous system, the nerves carrying sensory input to the central nervous system are said to be *afferent* (a word meaning "to bring to"). The ones carrying motor output away from the central nervous system to muscles and glands are *efferent* ("to carry outward").

As Figure 25.5 shows, the body's efferent nerves actually form two distinct systems. Efferent nerves leading to skeletal muscles form the **somatic system**. Those leading to the heart, smooth muscles, and glands form the **autonomic system**; they service the "visceral" portion of the body (internal organs such as the heart, lungs, and gut).

Figure 25.4 Human nervous system. Listed are the paired spinal nerves that connect with the central nervous system (brain and spinal cord). Not visible are twelve pairs of cranial nerves that connect with brain centers.

Nerves Serving Autonomic Functions

The nerves of the autonomic system fall into two categories, called **sympathetic** and **parasympathetic nerves**. They differ in where they connect with the central nervous system and in where their ganglia are located (Figures 25.5 and 25.6).

A superficial look at Figure 25.6 might lead you to believe that both sympathetic and parasympathetic motor nerves service all organs and play off each other, with one type of nerve stimulating an organ and the other slowing down its activity. This is true in *most* but by no means all cases. Some organs have a dual nerve supply; others do not. Also, the nerves of both divisions can have excitatory *or* inhibitory effects. (In other words, one division doesn't turn everything off and the other turn everything on.)

The general functions of the two nerve systems become apparent only when we consider the overall state of the animal body relative to its surroundings. When the animal is not receiving much outside stimulation, parasympathetic nerve action tends to slow down overall body activity and divert energy to basic "housekeeping" tasks, such as digestion. During times of heightened awareness, excitement, or danger, sympathetic nerve action tends to slow down housekeeping tasks and, simultaneously, to increase overall body activities that prepare the animal to fight or flee (or frolic intensely, as in play behavior and sexual behavior).

For example, under signals from sympathetic nerves, heart rate increases, blood glucose levels rise, and blood circulates faster, distributing oxygen and packets of quick energy (glucose molecules) through the body. Bronchioles dilate and more air enters the lungs, so cells get more oxygen for increased metabolic output.

At all times, both types of autonomic nerves carry signals that bring about minor adjustments in visceral organs. For example, even while the heart may be receiving low levels of sympathetic signals that cause its rate of beating to increase slightly, low levels of parasympathetic signals are opposing this effect. At any moment, the actual rate is the *net* outcome of opposing signals.

One final point should be made here. The word autonomic means "self-managed." It was coined at a time when the reflexes influencing smooth muscles and glands were thought to be self-governed, without inputs from the central nervous system. Today we know that these reflexes are integrated by commands from the brain and spinal cord, as experiments with biofeedback make clear.

(*Biofeedback* refers to conscious efforts to enhance or suppress autonomic and other physiological responses. For example, electronic devices can be used to detect changes in heart rate, which are registered on display equipment in the form of auditory or visual cues. An individual uses these cues to recognize and reinforce some desired behavior. Thus, with conscious effort, cardiac muscle contractions detected electronically can be slowed down slightly.)

Cranial Nerves

The peripheral nervous system also includes twelve pairs of **cranial nerves**, which connect directly with the brain. Some cranial nerves contain only sensory axons. (The optic nerves, which carry visual signals from the eyes, are like this.) Others contain sensory and motor axons. For example, the vagus nerves have sensory axons leading into the brain as well as motor axons leading out to muscles in the lungs, gut, and heart.

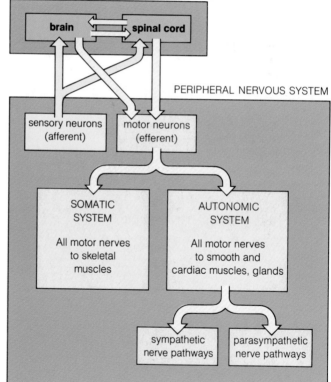

Figure 25.5 Divisions of the vertebrate nervous system.

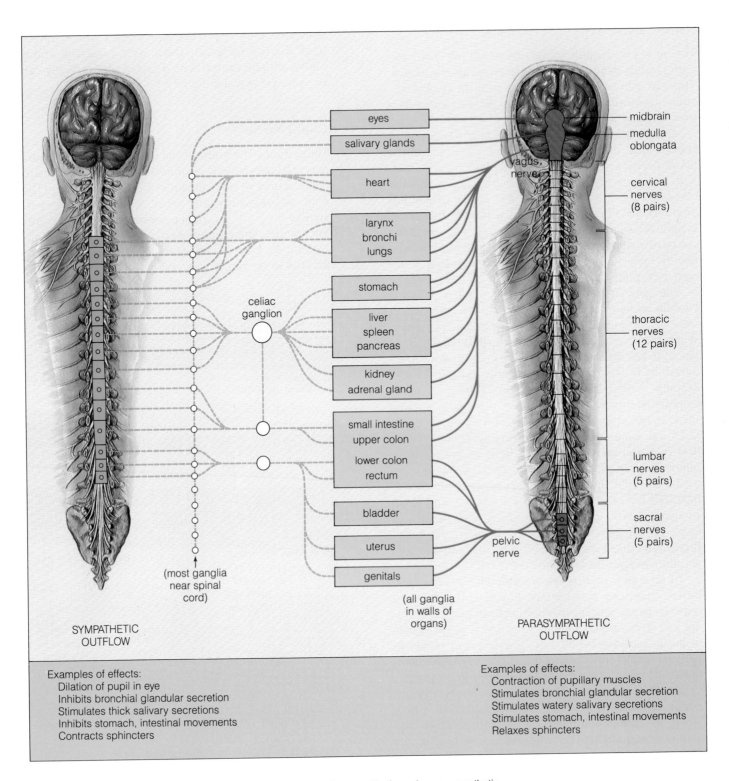

Figure 25.6 Autonomic nervous system. Shown here are the main sympathetic and parasympathetic pathways leading out from the central nervous system to some major organs. As the lists of examples suggest, in some cases the sympathetic and parasympathetic nerves operate antagonistically in their effects on the organ. Keep in mind that both systems have *paired* nerves leading out from the brain and spinal cord.

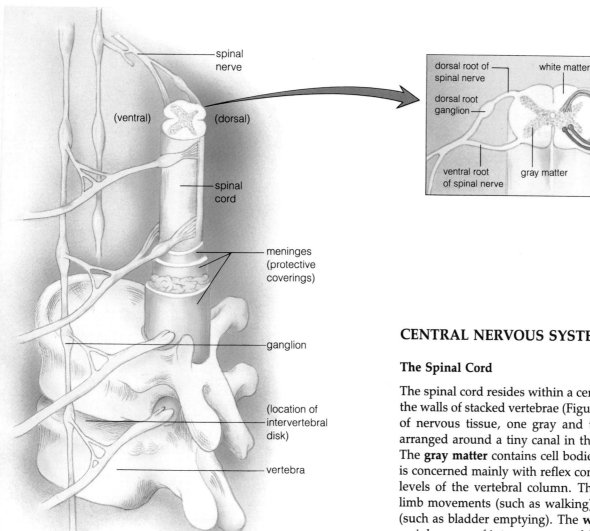

spinal nerve

(ventral) (dorsal)

spinal cord

meninges (protective coverings)

ganglion

(location of intervertebral disk)

vertebra

dorsal root of spinal nerve

dorsal root ganglion

white matter

sensory input lines

ventral root of spinal nerve

gray matter

motor output lines

Figure 25.7 Organization of the spinal cord and its relation to the vertebral column.

CENTRAL NERVOUS SYSTEM

The Spinal Cord

The spinal cord resides within a central canal formed by the walls of stacked vertebrae (Figure 25.7). Two regions of nervous tissue, one gray and the other white, are arranged around a tiny canal in the center of the cord. The **gray matter** contains cell bodies and dendrites and is concerned mainly with reflex connections at different levels of the vertebral column. The reflexes deal with limb movements (such as walking) and visceral events (such as bladder emptying). The **white matter** contains mainly axons of interneurons and is so colored because of the glistening myelin sheaths around them. Many of these axons are bundled into sensory nerve tracts that ascend the cord and end at specific brain centers. Others are bundled into motor tracts that run in the opposite direction.

The functional connections made in the spinal cord may be summarized this way:

1. Direct reflex connections between the sensory and motor neurons controlling the limbs and trunk are made in the spinal cord.

2. Most often, interneurons are interposed between sensory input and motor output.

3. Interneurons connect with one another up, down, and laterally in the gray matter of the spinal cord, providing a degree of control over activities within the cord itself.

4. Major nerve tracts (bundles of myelinated axons in the white matter of the cord) ascend into and descend from specific brain centers that provide more refined control over activities.

Divisions	Main Components	Some Functions
FOREBRAIN	Cerebrum	Two cerebral hemispheres. Most complex coordinating center; intersensory association, memory circuits
	Olfactory lobes	Relaying of sensory input to olfactory structures of cerebrum
	Limbic system	Scattered brain centers. With hypothalamus, coordination of skeletal muscle and internal organ activity underlying emotional expression
	Thalamus	Major coordinating center for sensory and motor signals; relay station for sensory impulses to cerebrum
	Hypothalamus	Neural-endocrine coordination of visceral activities (e.g., solute-water balance, temperature control, carbohydrate metabolism)
	Pituitary gland	"Master" endocrine gland (controlled by hypothalamus). Control of growth, metabolism, etc.
	Pineal gland	Control of some circadian rhythms; role in mammalian reproductive physiology
MIDBRAIN	Tectum and other wall regions	Largely reflex coordination of visual, tactile, auditory input; contains nerve tracts ascending to thalamus, descending from cerebrum
HINDBRAIN	Pons	"Bridge" of transverse nerve tracts from cerebrum to both sides of cerebellum. Also contains longitudinal tracts connecting forebrain and spinal cord
	Cerebellum	Coordination of motor activity underlying refined limb movements, maintaining posture, spatial orientation
	Medulla oblongata	Contains tracts extending between pons and spinal cord; reflex centers involved in respiration, cardiovascular function, gastric secretion, etc.

anterior end of spinal cord

Figure 25.8 Summary of the five regional subdivisions of the vertebrate brain. The drawing is highly simplified and flattened. In the developing embryo, the brain flexes forward and the walls develop in complex ways.

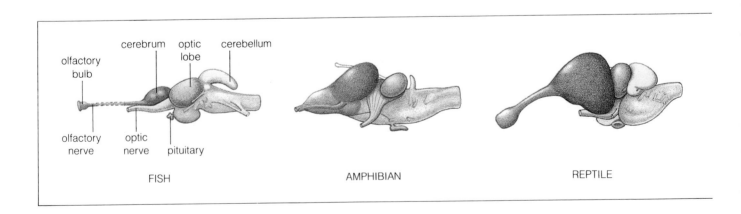

olfactory bulb

cerebrum   optic lobe   cerebellum

olfactory nerve   optic nerve   pituitary

FISH   AMPHIBIAN   REPTILE

Divisions of the Brain

Anatomically, the brain begins as a continuation of the spinal cord. The protective membranes wrapped around the cord extend up and wrap around the brain also. The central canal of the cord also extends upward, but in the brain it is expanded into fluid-filled chambers (ventricles) and channels (aqueducts). The "walls" around these fluid-filled regions are the actual brain tissue. The cranial cavity (a chamber formed by some bones of the skull) houses the mass of brain tissue.

The brain undergoes its regional expansion during embryonic development. A hollow tube of nervous tissue forms and develops into the primordial hindbrain, midbrain, and forebrain. Further development produces five regional subdivisions, which are more elaborate in some species than in others (Figures 25.8 and 25.9).

Hindbrain

The hindbrain, which is continuous with the spinal cord, includes the **medulla oblongata**, **cerebellum**, and **pons** (Figure 25.8). The medulla oblongata contains reflex centers for vital functions, including respiration, blood circulation, and vomiting. These centers are part of the **reticular formation**, a loosely organized net of nerve cell bodies that extends from the top of the spinal cord into the forebrain. The formation functions in motor coordination; it helps activate higher brain centers involved in sleeping, dreaming, and arousal; and it coordinates complex reflexes (such as sneezing and coughing).

The cerebellum contains reflex centers for maintaining posture and for spatial orientation. Especially in mammals, it helps coordinate refined limb movements. In the cerebellum, constant integration of sensory signals from the eyes, muscle spindles, tendons, skin, and inner ear provides information on how the body and limbs are positioned, how much particular muscles are contracted or relaxed, and in which direction the body or limbs happen to be moving. When the cerebellum receives commands from higher brain centers, it uses this information to produce the most effective motor response possible at that moment.

The pons is largely a region through which nerve tracts pass on their way from one brain center to another. The word pons (meaning "bridge") refers to prominent bands of axons that extend into each side of the cerebellum.

Midbrain

The midbrain originally was a center for coordinating reflex responses to visual input, but gradually it took on added functions. The roof of the midbrain, the **tectum**, is a thickened region of gray matter where visual and auditory signals are integrated. In fishes and amphibians, the tectum exerts major control over the body. (You can surgically remove the cerebral hemispheres from a frog brain and the frog can still do just about everything it normally does.) The tectum is also important in reptiles and birds. In mammals, sensory information still converges on the tectum but is rapidly sent on to higher centers for further processing.

Forebrain

For the early vertebrates, chemical odors emanating from food must have been a powerful selective agent, for olfactory structures came to dominate the anterior end of the brain. Like their descendants, those vertebrates had two **olfactory bulbs**: centers that receive input from olfactory nerves. They had a primitive **cerebrum**, a brain center that at first was concerned mainly with integrating olfactory input and selecting motor responses to it.

Below the cerebrum, the **thalamus** served as a center for relaying and coordinating sensory signals. Also, some ascending and descending motor pathways con-

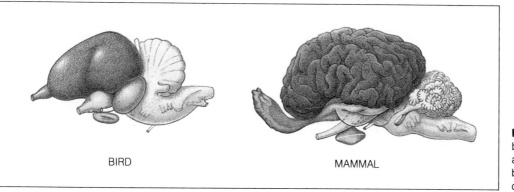

BIRD   MAMMAL

Figure 25.9 Comparison of the brain structure of a fish (codfish), amphibian (frog), reptile (alligator), bird, and mammal (horse). The drawings are not to the same scale.

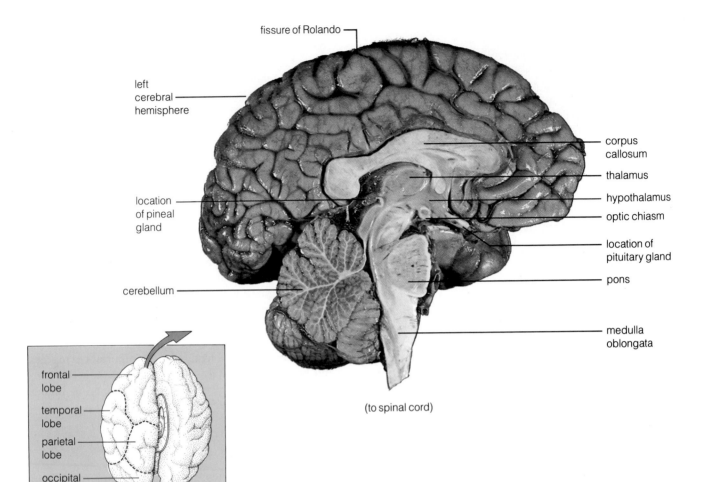

fissure of Rolando

left cerebral hemisphere

location of pineal gland

cerebellum

corpus callosum

thalamus

hypothalamus

optic chiasm

location of pituitary gland

pons

medulla oblongata

(to spinal cord)

frontal lobe

temporal lobe

parietal lobe

occipital lobe

left cerebral hemisphere

right cerebral hemisphere

Figure 25.10 Human brain, sagittal section. The corpus callosum is a major nerve tract that runs transversely, connecting the two cerebral hemispheres. The boxed inset shows the two hemispheres pulled slightly apart; normally they are pressed together, with only a longitudinal fissure separating them.

verged here. The **hypothalamus** monitored visceral activities and influenced forms of behavior related to those activities, such as thirst, hunger, and sex (Chapter Twenty-Six).

Over evolutionary time, some gray matter came to be positioned as a thin layer over each half of the cerebrum. This **cerebral cortex** provided added control over the body's activities. When mammals first appeared, it expanded spectacularly into information-encoding and information-processing centers, which will be described next.

Parts of the forebrain interconnect to form the **limbic system**, which influences learning and emotional behavior. This system includes parts of the cortex, the thalamus, and the hypothalamus, and it has links with many other parts of the central nervous system. Thus information from diverse receptors may converge in the limbic system, leading (for example) to sweating, laughing, crying, and other responses associated with different emotions.

THE HUMAN BRAIN

The Cerebral Hemispheres

Now that we have considered the vertebrate nervous system in general, we can turn to the intricate cerebrum of the human brain. Much of the gray matter of the cerebrum forms a thin surface layer called the cerebral cortex. This layer folds back on itself to a remarkable extent, suggesting that evolution of the mammalian cerebrum outpaced enlargement of the skullbones housing it. A fissure divides the cerebrum into right and left hemispheres. Although there is some variation from one individual to the next, the folds and fissures of each hemisphere follow certain patterns and are regionally divided into lobes (Figure 25.10).

Nerve Tracts in the Hemispheres. The white matter of the cerebral hemispheres consists of major nerve tracts. One, the *corpus callosum*, is a broad strap of white matter that keeps the two hemispheres in communication with each other (Figure 25.10). Many nerve tracts keep the cerebral cortex in communication with the rest of the body. Although these tracts are rather jammed together in the brainstem, they fan out extensively in the hemispheres. Each hemisphere also has its own set of tracts that provide communication among all of its parts.

Functional Regions of the Cortex. Different regions of the cortex have different functions. Some are *motor centers*, where instructions for motor responses are coordinated (Figure 25.11). For example, parts of the motor cortex connect directly with descending motor tracts. Stimulation of different points on the motor cortex triggers contractions of muscle groups in different body parts. A relatively large area of the motor cortex is devoted to muscles that control thumb and tongue movements (Figure 25.12). The size of this area reflects the control required for intricate hand movements and oral expression.

Other regions are *primary receiving centers* for sensory input, including that from the eyes and ears. Just behind the motor cortex is the somatic sensory cortex, a primary receiving center for sensory input from the skin and joints.

Lying outside the motor and somatic sensory regions, but connected to them by neural pathways, are regions of *association centers*. In these regions, information from memory stores is added to the primary sensory information to give it fuller meaning. It is not known which parts of the brain actually give rise to conscious awareness of the world (Figure 25.13).

Memory

Conscious experience is far removed from simple reflex action. It entails *thinking* about things—recalling objects and events encountered in the past, comparing them with newly encountered ones, and making rational connections based on the comparison of perceptions. Thus conscious experience entails a capacity for **memory**: the storage of individual bits of information somewhere in the brain.

The neural representation of information bits is known as a **memory trace**, although no one knows for sure in what form a memory trace occurs, where it resides, or even if it has a specific regional location. So far, experiments strongly suggest that there are at least two stages involved in its formation. One is a *short-term* formative period, lasting only a few minutes or so; then, information becomes spatially and temporally organized in neural pathways. The other is *long-term* storage; then, information is put in a different neural representation that lasts more or less permanently.

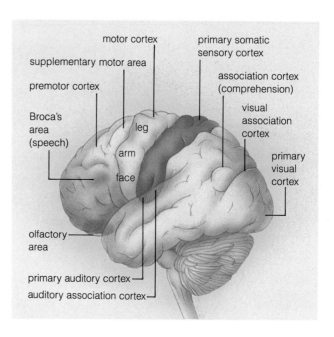

Figure 25.11 Primary receiving and association areas for the human cerebral cortex. Signals from receptors on the body's periphery enter primary cortical areas. Sensory input from different receptors is coordinated and processed in association areas. The text describes the main cortical regions. Also shown here are the *premotor area*, involved in intricate motor activity (as typified by a concert pianist performing); the *supplementary motor area*, which helps coordinate sequential voluntary movements; and *Broca's area*, which coordinates muscles required for speech.

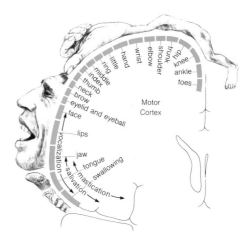

Figure 25.12 Body regions represented in the motor cortex of humans. The size of the body parts in the human figure illustrates the size of the cortical region devoted to controlling that part. The motor cortex runs from the top of the head to just above the ear on the surface of each cerebral hemisphere. The sketch is a cross-section through the right hemisphere of someone facing you.

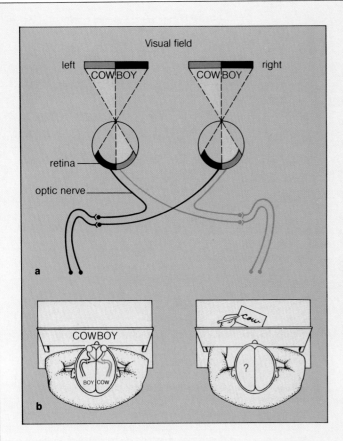

Figure 25.13 Sperry's experiments and the association cortex.

Experiments performed by Roger Sperry and his coworkers demonstrated some intriguing differences in perception between the two halves of the cerebrum. The subjects of the experiments were epileptics. Persons with severe epilepsy are wracked with seizures, sometimes as often as every half hour of their lives. The seizures have a neurological basis, analogous to an electrical storm in the brain. What would happen if the corpus callosum of epileptics were cut? Would the electrical storm be confined to one cerebral hemisphere, leaving at least the other to function normally? Earlier studies of animals and of humans whose corpus callosum had been damaged suggested this might be so.

The surgery was performed. And the electrical storms subsided in frequency and intensity. Apparently, cutting

the neural bridge between the two hemispheres put an end to what must have been positive feedback of ever intensified electrical disturbances between them. Beyond this, the "split-brain" individuals were able to lead what seemed, on the surface, entirely normal lives.

But then Sperry devised some elegant experiments to determine whether the conscious experience of these individuals was indeed "normal." After all, the corpus callosum contains no less than 200 million axons; surely *something* was different. Something was. "The surgery," Sperry later reported, "left these people with two separate minds, that is, two spheres of consciousness. What is experienced in the right hemispheres seems to be entirely outside the realm of awareness of the left."

In Sperry's experiments, the left and right hemispheres of split-brain individuals were presented with different stimuli. It was known at the time that visual connections to and from one hemisphere are mainly concerned with the opposite visual field. (Receptors from the left visual field send signals to the right hemisphere; those from the right visual field send signals to the left hemisphere.) Sperry projected words—say, COWBOY—onto a screen. He did this in such a way that COW fell only on the left visual field, and BOY fell on the right (see sketch). The subject reported seeing the word BOY. The left hemisphere, which received the word, controls language. However, when asked to write the perceived word with the left hand—a hand that was deliberately blocked from the subject's view—the subject wrote COW.

The right hemisphere, which "knew" the other half of the word, had directed the left hand's motor response. But it couldn't tell the left hemisphere what was going on because of the severed corpus callosum. The subject knew that a word was being written, but could not say what it was!

The functioning of the cerebral hemispheres has been the focus of many experiments. Taken together, the results have revealed the following information about this functioning:

1. Each cerebral hemisphere can function separately, but it functions mainly in response to signals from the opposite side of the body. Normally, of course, the corpus callosum is intact, and the two hemispheres work together.

2. The main regions responsible for spoken language skills generally reside in the left hemisphere.

3. The main regions responsible for nonverbal skills (music, mathematics, and other abstract abilities) generally reside in the right hemisphere.

Observations of people suffering from **retrograde amnesia** tell us something about memory. These people can't remember anything that happened during the half hour or so before losing consciousness after a severe head blow. Yet memories of events before that time remain intact! Such disturbances temporarily suppress normal electrical activities in the brain. These observations may mean that whereas short-term memory is a fleeting stage of neural excitation, long-term memory depends on *chemical* or *structural* changes in the brain.

In addition, information seemingly forgotten can be recalled after being unused for decades. This means that individual memory traces must be encoded in a form somewhat immune to degradation. Most molecules and cells in your body are used up, wear out, or age and are constantly being replaced—yet memories can be retrieved in exquisite detail after many years of such wholesale turnovers. Nerve cells, recall, are among the few kinds that are *not* replaced. You are born with billions, and as you grow older some 50,000 die off steadily each day. Those nerve cells formed during embryonic development are the same ones present, whether damaged or otherwise modified, at the time of death.

The part about being "otherwise modified" is tantalizing. *There is evidence that neuron structure is not static, but rather can be modified in several ways.* Most likely, such modifications depend on electrical and chemical interactions with neighboring neurons. Electron micrographs show that some synapses "wither" as a result of disuse. Such regression weakens or breaks connections between neurons. The visual cortex of mice raised without visual stimulation showed such effects of disuse. Similarly, there is some evidence that intensively stimulated synapses may form stronger connections, grow in size, or sprout buds or spines to form more connections! The chemical and physical transformations that underlie changes in synaptic connections may correspond to memory storage.

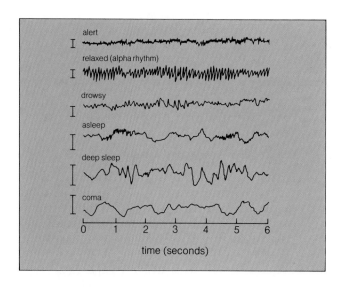

Figure 25.14 Examples of EEG wave patterns. The vertical bars indicate fifty millivolts each, with the irregular horizontal lines indicating the response with time.

States of Consciousness

Between the mindless drift of coma and total alertness are many *levels* of consciousness, known by such names as sleeping, dozing, meditating, and daydreaming. These levels are governed by the central nervous system. They also are subject to alteration by psychoactive drugs (see *Commentary*).

Throughout the spectrum of consciousness, neurons in the brain are constantly chattering among themselves. This neural chatter shows up as wavelike patterns in an **electroencephalogram** (EEG). An EEG is an electrical recording of the frequency and strength of potentials from the brain's surface.

EEG Patterns. Figure 25.14 gives examples of EEG patterns. The prominent wave pattern for someone who is relaxed, with eyes closed, is an *alpha rhythm*. In this relaxed state of wakefulness, EEG waves are recorded in "trains" of one after the other, about ten per second. Alpha waves predominate during the state of meditation. With a transition to sleep, wave trains gradually become larger, slower, and more erratic. This *slow-wave sleep* pattern shows up about eighty percent of the total sleeping time for adults. It occurs when sensory input is low and the mind is more or less idling. Subjects awakened from slow-wave sleep usually report that they were not dreaming. If anything, they seemed to be mulling over recent, ordinary events.

But slow-wave sleep is punctuated by brief spells of *REM sleep*. Rapid Eye Movements accompany this pattern (the eyes jerk beneath closed lids). Also accompanying REM sleep are irregular breathing, faster heartbeat, and twitching fingers. Most people awakened from REM sleep report they were experiencing vivid dreams.

With the transition from sleep (or deep relaxation) into alert wakefulness, EEG recordings show a shift to low-amplitude, higher frequency wave trains. Associated with this accelerated brain activity are increased blood flow and oxygen uptake in the cortex. The transition, called *EEG arousal*, occurs when conscious effort is made to focus on external stimuli or even on one's own thoughts.

The Reticular Activating System. Which brain regions govern changing levels of consciousness? The reticular formation, recall, forms connections with the spinal cord, cerebellum, and cerebrum as well as back with itself. It constantly samples messages flowing through the central nervous system. The flow of signals along

Drug Action on Integration and Control

Classes of Psychoactive Drugs

Broadly speaking, a drug is any chemical substance that has no nutritive value but that is introduced into the body to elicit some effect on physiological processes. A **psychoactive drug** is one that affects those parts of the central nervous system concerned with states of consciousness and behavior.

There are five classes of psychoactive drugs: depressants and hypnotics, stimulants, narcotic analgesics, hallucinogens and psychedelics, and antipsychotics (see Table). Many exert their effects by altering synaptic transmission (for example, by modifying or blocking the synthesis, release, or uptake of transmitter substances).

Depressants, Hypnotics. These drugs lower the activity in certain parts of the brain and nerves, hence they reduce activity throughout the body. Some inhibit synaptic transmission in the reticular activating system and at nerve tracts in the thalamus.

Depending on the dosage, most of these drugs can produce different degrees of general behavioral modification, ranging from relief from anxiety, through sedation, sleep (hypnosis), anesthesia and coma, to death. At low doses, inhibitory synapses are often suppressed slightly more than excitatory synapses, so that initially the person feels excited or euphoric. At increased doses, excitatory synapses are also suppressed and the overall effect is depression. The effects of depressants and hypnotics are additive, in that one will amplify the effect of another. For example, the combined use of alcohol and barbiturates exaggerates behavioral depression.

Alcohol (ethyl alcohol) differs from the drugs just described because it acts not at synapses but directly on the plasma membrane to alter cell function. It is often mistakenly viewed as a stimulant because it produces an initial "high." It is also widely held to be relatively harmless, even though it is one of the most powerful psychoactive drugs and a major cause of death. Short-term use of alcohol in seemingly small doses typically produces disorientation, uncoordinated motor functions, and diminished judgment. Long-term addiction destroys nerve cells and causes permanent brain damage; it can permanently scar and otherwise damage the liver (cirrhosis).

Stimulants. Caffeine, nicotine, amphetamines, and cocaine are examples of stimulants. At first they increase alertness and body activity but inevitably they lead to a period of depression.

Caffeine, one of the most widely used stimulants, is a component of coffee, tea, chocolate, and many soft drinks. Caffeine apparently blocks breakdown of cyclic adenosine monophosphate (cyclic AMP), plays a role in glycogen breakdown into glucose, and influences many receptor functions. At low doses, caffeine stimulates the cerebral cortex first, so its initial effects include increased alertness and restlessness. At higher doses it affects the medulla oblongata and thereby disrupts motor coordination as well as intellectual coherence.

Nicotine, a component of tobacco, has powerful effects on both the central and peripheral nervous systems. It chemically mimics acetylcholine at some receptors and can directly stimulate a number of sensory receptors. Its short-term effects include irritability, increased heart rate and blood pressure, water retention, and gastric upsets. Its long-term effects are serious (pages 408 and 447).

Low doses of amphetamines and cocaine reduce fatigue, elevate mood, and heighten awareness. High doses elicit anxiety, irritability, even psychotic behavior (such as paranoia) and permanent mental abnormalities. Both drugs apparently mimic or enhance the excitatory effect of norepinephrine, which underlies the "fight, flee" responses associated with autonomic nerves. Both drugs can lead to psychological and physiological dependency, with addicts experiencing extreme emotional and physical depression when deprived of them.

Analgesics. When stress leads to physical or emotional pain, the brain produces its own analgesics, or pain relievers. Endorphins and enkephalins are examples. Receptors for brain analgesics have been identified in many parts of the nervous system. When bound to receptors, these naturally occurring substances seem to inhibit neural activity. Endorphins (including enkephalins) have been identified in high concentrations in brain regions concerned with emotions and perception of pain.

The narcotic analgesics sedate the body as well as relieve pain. They include codeine and heroin, both of which are extremely addictive. The body develops a tolerance of these drugs, in that progressively larger doses are needed to produce the same effects. At the same time, the body becomes physically dependent on them. Deprivation following massive doses of heroin leads to cravings for the drug, hyperactivity and anxiety, fever, chills, and painful gastrointestinal disruption (violent vomiting, cramping, and diarrhea).

Psychedelics, Hallucinogens. These drugs alter sensory perception (particularly visual and auditory); hence they have been described as "mind-expanding." Some skew acetylcholine or norepinephrine activity. Others, such as LSD (lysergic acid diethylamide), affect serotonin activity. Even in tiny doses, LSD dramatically warps perceptions.

Marijuana is another hallucinogen. The name refers to the drug made from crushed leaves, flowers, and stems of the plant *Cannabis sativa*. (Hashish, an extract of the same plant, can contain ten times as much of the psychoactive ingredient.) In low doses marijuana is like a depressant; it slows down but does not impair motor activity, it relaxes the body and elicits mild euphoria. Unlike true depressants it cannot, in high doses, lead to coma and death. It can produce disorientation, increased anxiety bordering on panic, delusions (including paranoia), and hallucinations.

Like alcohol, marijuana can affect an individual's ability to perform complex tasks, such as driving a car. In one study, commercial pilots showed a marked deterioration in instrument-flying ability for more than two hours after smoking marijuana. Recent studies point to a link between marijuana smoking and suppression of the immune system.

Antipsychotic Drugs. Emotional states—joy, elation, anxiety, depression, fear, anger—are normal responses to changing conditions in our complex world. Sometimes, through imbalances in the central nervous system, one or another of these states becomes pronounced.

For example, anxiety (a vague sense of uneasiness and tenseness) might give way to emotional instability in the neurotic, who becomes guilt-ridden, anxious, and chronically upset over mundane frustrations. Schizophrenic persons become despairing; they withdraw from society and focus obsessively on themselves. Schizophrenia is an example of psychotic behavior. "Psychosis" means an impairment of a person's ability to communicate and otherwise effectively interact with others, combined with a distorted perception of reality.

Classes of Psychoactive Drugs	
Class	Examples
Depressants, hypnotics	Barbiturates (e.g., Nembutal, Quaalude)
	Antianxiety drugs (e.g., Valium, alcohol)
Stimulants	Caffeine
	Nicotine
	Amphetamines (e.g., Dexedrine)
	Cocaine
Narcotic analgesics	Codeine
	Opium
	Heroin
Psychedelics, hallucinogens	Lysergic acid diethylamide (LSD)
	Cannabis sativa (hashish, marijuana)
Antipsychotics	Chlorpromazine, lithium

Low doses of antipsychotic drugs are tranquilizing, in that they reduce anxiety levels and neurotic behavior. High doses produce tremors, blurred vision, and constipation or diarrhea.

A Biological Perspective on Drug Abuse

For the most part, we do not understand much about how any one drug works. Given the complexity of the brain, it could scarcely be otherwise at this early stage of inquiry. Despite our ignorance of the mechanisms of drug action, one of the major problems in the modern world is drug abuse—the self-destructive use of drugs that alter emotional and behavioral states. The consequences show up in unexpected places: in seven-year-old heroin addicts, in the highway wreckage left by individuals whose perceptions were skewed by alcohol or amphetamines, in victims of addicts who steal and sometimes kill to support the drug habit, in suicides on LSD trips who were deluded into believing that they could fly and who "flew" off buildings and bridges.

Each of us possesses a body of great complexity. Its architecture, its functioning are legacies of millions of years of evolution. It is unique in the living world because of its highly developed nervous system—a system that is capable of processing far more than the experience of the individual. One of its most astonishing products is language, the encoding of *shared* experiences of groups of individuals in time and space. Through the evolution of our nervous system, the sense of history was born, and the sense of destiny. Through this system we can ask how we have come to be what we are, and where we are headed from here. Perhaps the sorriest consequence of drug abuse is its implicit denial of this legacy—the denial of self when we cease to ask, and cease to care.

these circuits—and the inhibitory or excitatory chemical changes accompanying them—has a great deal to do with whether you stay awake or drop off to sleep. In fact, damage to some parts of the reticular formation can lead to unconsciousness and coma.

Within this formation are neurons collectively called the **reticular activating system** (RAS). Excitatory pathways connect the RAS to the thalamus (the forebrain's switching station). Messages routed from the RAS arouse the brain and maintain wakefulness.

Also in the reticular formation are *sleep centers*. One center contains neurons that release serotonin, a transmitter substance with an inhibitory effect on RAS neurons (high serotonin levels are linked to drowsiness and sleep). Another sleep center, in part of the reticular formation within the pons, has been linked to REM sleep. Substances released from the second center counteract the effects of serotonin, perhaps enabling the RAS to maintain the waking state.

SUMMARY

1. Nervous systems provide specialized means of detecting stimuli and responding to them swiftly.

2. Reflexes are simple, stereotyped movements made in response to sensory stimuli. In the simplest reflex, activity in sensory neurons directly signals motor neurons, which act on muscle cells. Reflexes are the basic operating machinery of nervous systems.

3. The oldest parts of the vertebrate brain deal with reflex coordination of sensory inputs and motor outputs beyond that afforded by the spinal cord alone. Nervous systems evolved through the gradual layering of additional nervous tissue over reflex pathways of more ancient origin.

4. Increasingly complicated sensory organs and motor structures evolved along with the changing nervous systems.

5. Even the newest layerings of the brain deal partly with reflex coordination. They also deal with storing, comparing, and using experiences to initiate novel, nonstereotyped action. Thus, these regions are the basis of memory, learning, and reasoning.

6. The simplest nervous systems are nerve nets, such as those of sea anemones, hydra, and jellyfishes.

7. More complex nervous systems involve cephalization and bilateral symmetry with paired nerves, muscles, sensory structures, and brain centers.

8. The central nervous system of all vertebrates consists of a brain and spinal cord; the peripheral nervous system consists of the nerves, which pass between the brain and spinal cord and the innervated structures, and ganglia.

9. The peripheral nervous system is further divided into components that are afferent (conduct information to the central nervous system) and efferent (conduct information from the central nervous system to muscles and glands).

10. Efferent nerves are further divided into the somatic nervous system (which innervates skeletal muscles) and the autonomic nervous system (which innervates the heart, smooth muscles, and glands).

11. The autonomic nervous system contains two divisions. Sympathetic nerves function more during times of activity, excitement, and stress, whereas parasympathetic nerves are more active when the body is at rest.

12. The spinal cord has regions of cell bodies (gray matter) and myelinated cell processes (white matter). The white matter contains major nerve tracts. In the spinal cord, connections (often involving interneurons) are made between sensory and motor neurons for reflexes controlling the limbs and trunk.

13. The hindbrain is the part of the brain that connects with the spinal cord. It includes the medulla oblongata, pons, and cerebellum and contains reflex centers for vital functions and muscle coordination. It also contains nerve tracts (passing between the spinal cord and other parts of the brain) and the reticular formation, which is involved in the coordination of sensory and motor functions and arousal.

14. The midbrain functions in the coordination and relay of visual and auditory information.

15. The forebrain includes the cerebrum, below which are the thalamus (relays sensory information, helps coordinate motor responses) and hypothalamus (monitors visceral activities and coordinates some behavioral, autonomic, and endocrine responses with basic drives such as thirst, hunger, and sex). The limbic system is active in learning and emotional behavior.

16. The cortex, or outer layer of the cerebrum, has regions devoted to specific functions, such as receiving information from the various sense organs, integrating this information with memories of past events, and coordinating motor responses.

17. Memory is thought to occur in two stages: a short-term formative period and long-term storage, which depends on chemical or structural changes in the brain.

18. States of consciousness vary between total alertness and deep coma. The levels are governed by activity in the reticular activating system and sleep centers in the reticular formation. They are subject to the influence of psychoactive drugs. Different EEG patterns are characteristic of the various states of consciousness.

Review Questions

1. Label the parts of the human brain:

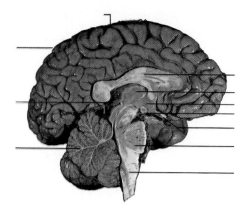

2. Can you list the main points of the theory of neural patterning given at the start of the chapter? *335*

3. Describe three key characters that apparently played central roles in the evolution of the vertebrate nervous system. *336–337*

4. What constitutes the central nervous system? The peripheral nervous system? *338*

5. Can you distinguish among the following:
   a. neurons and nerves *323*
   b. ganglia and nerve pathways (or tracts) *323*
   c. somatic system and autonomic system *338*
   d. parasympathetic and sympathetic nerves *339*

6. Review Figure 25.8. Then, on your own, describe the components of the three main subdivisions of the vertebrate brain. *342*

7. What is a psychoactive drug? Can you describe the effects of one such drug on the central nervous system? *348–349*

Readings

Julien, R. 1985. *A Primer of Drug Action.* Fourth edition. New York: Freeman. Effectively fills the gap between popularized (and often superficial or misleading) accounts of drug action and the upper-division books in pharmacology. Paperback.

Penfield, W., and T. Rasmussen. 1952. *The Cerebral Cortex of Man.* New York: Macmillan. Fascinating account of early mappings of cortical regions of the brain.

Romer, A., and T. Parsons. 1986. *The Vertebrate Body.* Sixth edition. Philadelphia: Saunders. The classic reference book on vertebrate structure and functioning. Marvelous insights into the evolution of vertebrate nervous systems.

Shepherd, G. 1983. *Neurobiology.* New York: Oxford.

Sperry, R. 1970. "Perception in the Absence of the Neocortical Commissures." *Perception and Its Disorders.* Research Publication of the Association for Research in Nervous and Mental Diseases, vol. 48.

Wurtman, R. April 1982. "Nutrients That Modify Brain Function." *Scientific American* 246(4):50–59.

26

INTEGRATION AND CONTROL: ENDOCRINE SYSTEMS

"THE ENDOCRINE SYSTEM"

The word "hormone" dates back to the early 1900s, when W. Bayliss and E. Starling were trying to figure out what triggers the secretion of pancreatic juices that act on food traveling through the canine gut. At the time, it was known that acids mix with food in the stomach, and that the pancreas secretes an alkaline solution when the acidic mixture has passed into the small intestine. Was the nervous system or something else stimulating the pancreatic response?

To find the answer, Bayliss and Starling cut off the nerve supply to the upper small intestine but left its blood vessels intact. When an acid was introduced into the denervated region, the pancreas still was able to make the secretory response. More telling, extracts of cells taken from the epithelial lining of that intestinal region also induced the response. Glandular cells in the lining had to be the source of a substance that stimulated the pancreas into action.

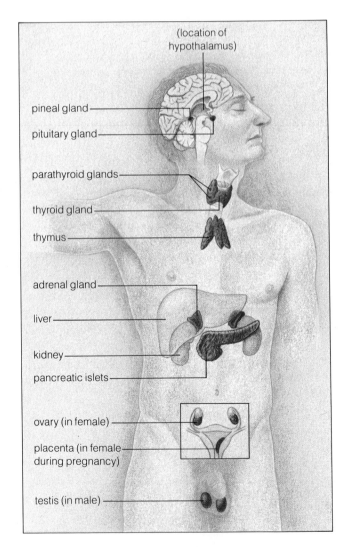

Figure 26.1 Location of endocrine glands in the human body. The liver and kidneys do other things besides secreting hormones, but these organs do have component endocrine cells. Intestinal epithelium also has endocrine cells; so does the heart.

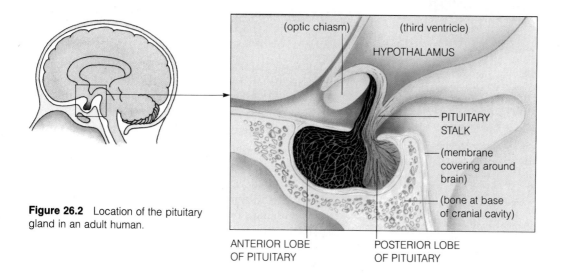

Figure 26.2 Location of the pituitary gland in an adult human.

The substance itself came to be called secretin. Demonstration of its existence and its mode of action was the first confirmation of an idea that had been around for centuries: *Internal secretions released into the bloodstream influence the activities of tissues and organs.* Starling coined the word hormone for such internal glandular secretions (the Greek *hormon* means to set in motion).

Later work led to the discovery of many hormones and their sources (Figure 26.1). The sources include but are not limited to the following array, which is characteristic of most vertebrates:

> Pituitary gland
> Pineal gland
> Thyroid gland
> Parathyroid glands (the number varies; four in humans)
> Adrenal glands (two)
> Gonads (two)
> Thymus gland
> Pancreatic islets (multiple)
> Endocrine cells of gut, liver, kidneys, placenta

These scattered sources of hormones came to be viewed as "the endocrine system." The phrase implied that they formed a separate means of control within the body, apart from the nervous system. (The Greek *endon* means within; *krinein* means to separate.)

In time, however, refined biochemical studies and improvements in electron microscopy revealed that the boundaries between the nervous and endocrine systems are not so tidy. Glands thought to be free of neural control turned out to be well innervated. Some neurons secrete hormones. The hypothalamus (a brain region) exerts major control over the pituitary—the so-called master gland of the endocrine system (Figure 26.2).

We no longer can say that a natural division exists between the nervous and endocrine systems, given their intricate overlapping. For that reason, we will begin with the secretions that serve as agents of integration, regardless of their neural or endocrine origin.

HORMONES AND OTHER SIGNALING MOLECULES

Cells have built-in means of responding to chemical changes in their immediate surroundings, including mechanisms for the uptake and release of different substances. In complex animals, the individual responses of thousands, millions, even many billions of cells must be integrated in ways that benefit the body as a whole.

Integration depends on *signaling molecules*. These are hormones and other chemical secretions that alter the behavior of target cells. (Any cell is a "target" if it has receptors to which specific signaling molecules can bind and elicit a cellular response. A target may or may not be adjacent to the secreting cell.) At least five types of cells and glands produce signaling molecules:

1. Endocrine glands or cells. Endocrine glands or cells secrete **hormones**, which are transported by the bloodstream to *nonadjacent targets*.

2. Neurosecretory cells. Some neurons also secrete hormones, which again are transported by the bloodstream to *nonadjacent targets*. (The secretions are also called neurohormones when the emphasis is on their source.)

3. Synapsing neurons. Most neurons secrete transmitter substances which, recall, act on *immediately adjacent*

Table 26.1 Examples of Signaling Molecules and Their Sources

Source	Type of Secretion	Examples
Endocrine glands or cells	Hormones	Somatotropin, glucagon, testosterone
Neurosecretory cells	Hormones (or neurohormones)	ADH, oxytocin, TRH, dopamine, epinephrine
Synapsing neurons	Transmitter substances	Acetylcholine, dopamine
Local mediator cells	Local signaling molecules	Prostaglandins, histamine, nerve growth factor
Exocrine glands or cells	Pheromones	Bombykol

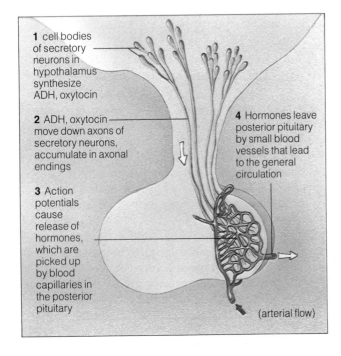

1 cell bodies of secretory neurons in hypothalamus synthesize ADH, oxytocin

2 ADH, oxytocin move down axons of secretory neurons, accumulate in axonal endings

3 Action potentials cause release of hormones, which are picked up by blood capillaries in the posterior pituitary

4 Hormones leave posterior pituitary by small blood vessels that lead to the general circulation

(arterial flow)

Figure 26.3 Functional links between the hypothalamus and the posterior lobe of the pituitary in humans.

target cells. Transmitter substances reach high concentrations in the synaptic cleft, act swiftly, and are rapidly degraded or recycled.

4. Local mediator cells. Many cells secrete local signaling molecules that alter chemical conditions in the *immediate vicinity.* Most of the molecules act swiftly on neighboring cells, then are degraded rapidly; they do not reach notable concentrations in the bloodstream.

5. Exocrine glands. Some exocrine glands secrete **pheromones**, which have *targets outside the body.* Pheromones diffuse through water or air and act on cells of other animals of the same species. They integrate activities between animals by serving as sexual attractants, alarm signals, and so on.

Table 26.1 summarizes the five types of signaling molecules. Let's look at a few examples of how target cells respond to these molecules. Our main focus will be on vertebrates, including humans. As you will see, the degree and duration of their responses are under homeostatic controls.

NEUROENDOCRINE CONTROL CENTER

The Hypothalamus-Pituitary Connection

In vertebrates, the hypothalamus and the pituitary work as a **neuroendocrine control center**. Together they integrate many of the body's activities. The hypothalamus, recall, monitors internal organs (such as the stomach) and influences forms of behavior related to their activities (such as hunger). The pituitary is suspended from a slender stalk that extends downward from the base of the hypothalamus (Figure 26.2). In adult humans, it is about the size of a garden pea.

The pituitary gland has lobes of nervous tissue and glandular tissue. The *posterior lobe* is nervous tissue. It secretes two neurohormones, which are actually produced in the hypothalamus but stored in the lobe. The *anterior lobe* is mostly glandular tissue. It produces and secretes six hormones and controls the release of several more from other endocrine glands. The pituitary of many vertebrates (not humans) also has an *intermediate lobe* of mostly glandular tissue. Often its secretions induce changes in the body's surface coloration.

Posterior Lobe Secretions

Figure 26.3 shows the links between the hypothalamus and the posterior lobe of the pituitary. Notice the neurosecretory cell bodies (in the hypothalamus) and their axons (which extend down the pituitary stalk and into the lobe). **Antidiuretic hormone** (ADH) and **oxytocin** are produced in the cell bodies and stored in the axonal endings, next to a capillary bed. A *capillary bed* has many thin-walled, highly permeable blood vessels called capillaries. Here, hormones as well as nutrients and wastes are exchanged between blood and the interstitial fluid around cells.

When stimulated, the neurosecretory cells release ADH or oxytocin, which diffuses through the interstitial fluid and into the capillaries. The hormones travel the bloodstream until they encounter target cells in different regions (Figure 26.4).

Cells making up the walls of so-called "collecting ducts" in the kidneys are targets for ADH. When the body must conserve water, ADH makes the walls permeable to water, which moves out of the ducts and into interstitial fluid. In this way, water that otherwise would leave the body in urine is conserved. Through its action, ADH helps control the volume and solute levels of the body's extracellular fluid (that is, blood as well as interstitial fluid).

Arterioles (the smallest type of artery) are other targets for ADH. Neural and endocrine signals cause arterioles throughout the body to constrict or dilate in ways that divert the flow of blood to regions of greatest metabolic activity. ADH is one of the signaling molecules. At high levels, it causes smooth muscle cells in arteriole walls to constrict, which helps increase the effective blood volume. (These effects account for ADH's other name: vasopressin.)

Contractile cells also are targets for oxytocin, which plays a role in mammalian reproduction. Oxytocin stimulates contraction of smooth muscles of the uterus during labor as well as contractions associated with milk-secreting glands in the breasts. Its action causes liquid from the glands to move into ducts from which milk is ejected.

Anterior Lobe Secretions

Role of the Hypothalamus. As we have seen, two hypothalamic secretions (ADH and oxytocin) have targets *outside* the pituitary. Other hypothalamic secretions, called **releasing hormones**, act on target cells *within* the anterior lobe of the pituitary, which in turn secrete hormones that act elsewhere. Most releasing hormones stimulate secretion from their pituitary targets; some inhibit it (Table 26.2).

Releasing hormones are produced in neurosecretory cell bodies in the hypothalamus. They are secreted from the axonal endings of those cells, which lie next to a capillary bed in the base of the hypothalamus. The capillaries carry the releasing hormones down the pituitary

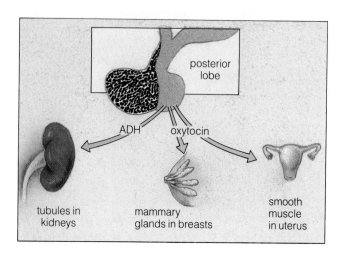

Figure 26.4 Secretions of the posterior lobe of the pituitary and some of their targets.

Table 26.2	Effect of Releasing Hormones on Anterior Pituitary Secretions	
Releasing Hormone	Influences Secretion of:	Effect*
Corticotropin-releasing hormone (CRH)	Corticotropin (ACTH)	+
Thyrotropin-releasing hormone (TRH)	Thyrotropin (TSH)	+
	Prolactin (PRL)	+
Gonadotropin-releasing hormone (GnRH)	Follicle-stimulating hormone (FSH)	+
	Luteinizing hormone (LH)	+
STH-releasing hormone (STHRH)	Somatotropin (STH); also called growth hormone (GH)	+
Somatostatin	Somatotropin, other hormones	−
Dopamine	Prolactin	−

*Stimulatory (+) or inhibitory (−).

Table 26.3 Hormones Released From the Mammalian Pituitary Gland

Pituitary Lobe	Tissue Type	Secretions	Abbreviation	Known Targets	Primary Actions
Posterior	Nervous tissue (extension of hypothalamus)	Antidiuretic hormone (or vasopressin)	ADH	Kidneys, water storage organs, etc.	Induces fluid movements underlying the control of extracellular fluid volume (and, indirectly, solute concentrations)
		Oxytocin		Mammary glands	Induces milk movement into secretory ducts
				Uterus	Induces uterine contractions
Anterior	Mostly glandular tissue	Corticotropin	ACTH	Adrenal cortex	Stimulates release of adrenal steroid hormones involved in stress responses
		Thyrotropin	TSH	Thyroid gland	Stimulates release of thyroid hormones concerned with growth, development, metabolic rate
		Gonadotropins:			
		Follicle-stimulating hormone	FSH	Ovaries, testes	In females, stimulates follicle growth, helps stimulate estrogen secretion and ovulation; in males, promotes spermatogenesis
		Luteinizing hormone	LH	Ovaries, testes	In females, stimulates ovulation, corpus luteum formation; in males, promotes testosterone secretion, sperm release
		Prolactin	PRL	Mammary glands	Stimulates and sustains milk production
		Somatotropin (also called growth hormone)	STH (GH)	Most cells	Has generalized growth-promoting effects in young animals; induces protein synthesis and cell division; stimulates release of somatomedins; plays role in glucose and protein metabolism in adults
Intermediate*	Mostly glandular tissue	Melanocyte-stimulating hormone	MSH	Pigmented cells in integument	Induces color changes in response to external stimuli; affects behavior

*Present in most vertebrates (not adult humans).

stalk to portal vessels that extend into the anterior lobe. (A *portal vessel* connects two capillary beds. In most circulation routes in humans, blood flowing from and back to the heart passes through only one bed, not two.)

Figure 26.5 shows this route between the hypothalamus and the anterior lobe of the pituitary. Inside the anterior lobe, the portal vessels branch into a second capillary bed. There, the releasing hormones leave the blood and bind to receptors on target cells. The binding stimulates or inhibits the release of hormones by specific cell types in the anterior pituitary.

Releasing hormones were identified by monumental research efforts that began in 1955, most notably by Roger Guillemin and Andrew Schally. Over one four-year period, Guillemin's team purchased 500 tons of sheep brains from meat processing plants and extracted 7 tons of hypothalamic tissue from it. They eventually ended up with a single milligram of thyrotropin-releasing hormone.

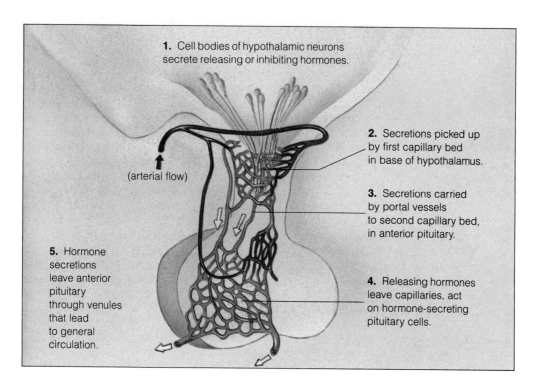

1. Cell bodies of hypothalamic neurons secrete releasing or inhibiting hormones.

(arterial flow)

2. Secretions picked up by first capillary bed in base of hypothalamus.

3. Secretions carried by portal vessels to second capillary bed, in anterior pituitary.

5. Hormone secretions leave anterior pituitary through venules that lead to general circulation.

4. Releasing hormones leave capillaries, act on hormone-secreting pituitary cells.

Figure 26.5 Functional links between the hypothalamus and the anterior lobe of the pituitary.

Anterior Pituitary Hormones. Different cells of the anterior pituitary produce and secrete the following hormones:

Corticotropin-stimulating hormone (ACTH)

Thyrotropin-stimulating hormone (TSH)

Follicle-stimulating hormone (FSH)

Luteinizing hormone (LH)

Prolactin (PRL)

Somatotropin (STH); also called growth hormone (GH)

The first four hormones act on endocrine glands, which in turn produce other hormones (Table 26.3). Later chapters will describe the elegant roles these hormones play in the functioning of specific systems. (For example, the human reproductive system depends on interplays between FSH and LH.) The last two hormones listed, prolactin and somatotropin, have effects on body tissues in general (Figure 26.6).

Prolactin influences a variety of activities among vertebrate species ranging from primitive fishes to humans. One of its functions is to stimulate and sustain milk production in mammary glands during lactation (page 525). Prolactin exerts its effects on mammary glands only when the tissues have been primed by other hormones.

Somatotropin, or growth hormone, profoundly influences overall growth, especially of cartilage and bone. It induces protein synthesis in most cells by stimulating

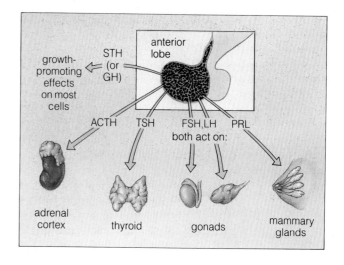

growth-promoting effects on most cells

STH (or GH)

anterior lobe

ACTH   TSH   FSH,LH   PRL

both act on:

adrenal cortex

thyroid

gonads

mammary glands

Figure 26.6 Secretions of the anterior lobe of the pituitary and some of their targets.

Table 26.4 Hormone Sources Other Than the Mammalian Hypothalamus and Pituitary

Source	Its Secretion(s)	Main Targets	Primary Actions
Adrenal cortex	Glucocorticoids (e.g., cortisol)	Most cells	Raise blood sugar level; help control lipid, protein metabolism; mediate responses to stress
	Mineralocorticoids (including aldosterone)	Kidney	Promote sodium reabsorption; control salt, water balance
	Sex hormones (including testosterone)	General	Influence sexual characteristics, general growth
Adrenal medulla	Epinephrine (adrenalin)	Muscle, adipose tissue	Raises blood sugar level by stimulating glucose production; raises blood levels of fatty acids; increases heart rate, force of contraction
	Norepinephrine	Smooth muscle in walls of blood vessels	Promotes vasoconstriction or vasodilation
Thyroid	Triiodothyronine, thyroxine	Most cells	Regulates carbohydrate, lipid metabolism; contributes to growth, development (including brain development, function)
	Calcitonin	Bone	Lowers calcium levels in blood by inhibiting calcium reabsorption from bone
Parathyroids	Parathyroid hormone (or parathormone)	Bone, kidney, gut	Elevates calcium levels in blood by stimulating calcium reabsorption from bone, kidneys, and absorption from gut
Gonads: Testis (in males)	Testosterone (an androgen)	General	Has key roles in spermatogenesis, development of genital tract, and maintenance of accessory sex organs and secondary sex traits; influences growth, development
Ovary (in females)	Estrogens	General	Have key roles in oogenesis; stimulate thickening of uterine lining for pregnancy; other actions same as above
	Progesterone	Uterus, breasts	Prepares, maintains uterine lining for pregnancy; stimulates breast development
Pancreatic islets	Insulin	All cells except most neurons in brain and red blood cells	Lowers blood sugar by stimulating glucose uptake by cells; fat storage; protein synthesis
	Glucagon	Liver	Raises blood sugar by stimulating glucose production
	Somatostatin	Insulin-secreting cells in pancreas	Influences carbohydrate metabolism
Endocrine cells of gastrointestinal epithelium	Gastrin, cholecystokinin, secretin, and others	Stomach, pancreas, gallbladder	Influences activity of stomach, pancreas, liver, gallbladder
Liver	Somatomedins	Most cells	Stimulates overall growth, development
Kidney	Erythropoietin*	Bone marrow	Stimulates red blood cell production
	Angiotensin*	Adrenal cortex, arterioles	Helps control blood pressure and aldosterone secretion
	Vitamin D₃* (an active form)	Bone, gut	Enhances calcium resorption from bone; enhances calcium uptake from gut
Heart	Atrial natriuretic hormone	Kidney, blood vessels	Increases sodium excretion; lowers blood pressure
Thymus	Include thymosin	Lymphocytes, plasma cells	Promote development of infection-fighting abilities and lymphocyte function in immune responses
Pineal	Melatonin	Gonads (indirectly, perhaps via hypothalamus)	Influences daily biorhythms, sexual activity and sexual development

*These hormones are not produced in the kidneys but are formed when *enzymes* produced in kidneys activate specific plasma proteins.

a

b

Figure 26.7 (**a**) Effect of somatotropin (STH) on overall body growth. The person at the center is affected by gigantism, which resulted from excessive STH production during childhood. The person at right displays pituitary dwarfism, which resulted from underproduction of STH during childhood. The person at the left is average in size.

(**b**) Manute Bol, currently with the Golden State Warriors, is 7 feet 6¾ inches tall owing to excessive STH production during childhood.

uptake of amino acids, RNA synthesis, and ribosome activity. It also induces the cell divisions necessary for growth. Somatotropin indirectly promotes these same activities by stimulating the release of somatomedins from the liver and other tissues. Somatomedins are growth factors that travel the bloodstream to cartilage and other target tissues, where they trigger growth-related events.

Figure 26.7 shows what can happen with too little or too much somatotropin. *Pituitary dwarfism* results when not enough somatotropin was produced during childhood. The adult is proportionally similar to a normal person but is much smaller in size. *Gigantism* results when excessive amounts of somatotropin were produced during childhood. The adult is similar in proportion to a normal person but is much larger in size.

When somatotropin secretions are excessive during adulthood (when long bones no longer can lengthen), *acromegaly* follows. Adult bones of the hands, feet, and jaws thicken, as do epithelial tissues of the skin, nose, eyelids, lips, and tongue (Figure 26.8). Skin thickening is pronounced on the forehead and soles of the feet.

Table 26.4 lists hormones from sources other than the hypothalamus and pituitary. Later we will look closely at the effect of these hormones on digestion, circulation, reproduction, and other activities. For now, we will simply preview their sources and consider a few examples of the kinds of controls governing their secretion.

age nine   sixteen

thirty-three   fifty-two

Figure 26.8 Acromegaly, which resulted from excessive production of somatotropin (STH) during adulthood. Before this female reached maturity, she was symptom-free.

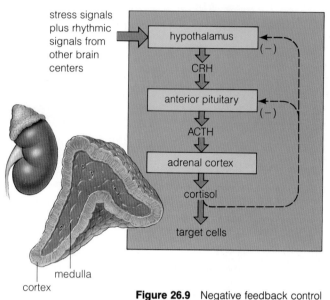

stress signals plus rhythmic signals from other brain centers

hypothalamus (−)

CRH

anterior pituitary (−)

ACTH

adrenal cortex

cortisol

target cells

cortex

medulla

Figure 26.9 Negative feedback control of cortisol secretion.

As Figure 26.9 shows, when the hypothalamus secretes CRH (a releasing hormone), the anterior pituitary is stimulated into secreting corticotropin, or ACTH. (The abbreviation refers to its former name, adrenocorticotropic hormone.) In turn, ACTH stimulates the adrenal cortex to secrete cortisol, which acts on target cells in the liver, muscles, and elsewhere. When the hypothalamus and pituitary detect increased levels of cortisol in the blood, they respond by reducing CRH and ACTH secretions. Cortisol secretion slows down as a result.

The feedback mechanism just described is overridden when the body is abnormally stressed, as from a frightening experience, painful injury, or prolonged illness—any of which may trigger tissue inflammation. Then, increased secretion of cortisol and other signaling molecules is prolonged until conditions return to normal. High levels of cortisol block inflammation. (That is why cortisol derivatives, such as cortisone, are administered to persons suffering from arthritis and other inflammatory disorders.)

ADRENAL GLANDS

Adrenal Cortex

Like other vertebrates, humans have a pair of adrenal glands, one above each kidney. The outer region of each gland is called the **adrenal cortex** (Figure 26.9). Cells of the cortex produce glucocorticoids and mineralocorticoids. They also produce some sex hormones, although the gonads are the primary sources.

Glucocorticoids function in carbohydrate, lipid, and protein metabolism. They also take part in inflammatory responses to tissue injury or infection. Mineralocorticoids, such as aldosterone, help maintain extracellular fluid when food intake and metabolic activity change the amounts of solutes entering the blood. (Aldosterone promotes sodium retention, which affects water retention by the body.)

Cortisol, one of the glucocorticoids, helps maintain the blood concentration of glucose between meals. Among other things, it stimulates liver cells to form glucose from amino acids when glucose levels drop (page 467).

Cortisol secretion is a good example of the **homeostatic feedback loops** that exist between most endocrine glands and the neuroendocrine control center. In loops of this sort, an endocrine gland steps up or slows down its hormone secretion. The hypothalamus or pituitary (or both) detects an ensuing increase or decrease in the concentration of hormone molecules in some body region, and it responds in ways that inhibit or stimulate further secretion.

Adrenal Medulla

Neural and endocrine functions overlap in the **adrenal medulla**, the inner region of the adrenal gland (Figure 26.9). Here we find modified neurons that secrete the hormones epinephrine and norepinephrine. Both hormones help control blood circulation as well as carbohydrate metabolism. Brain centers, including the hypothalamus, govern their secretion by way of sympathetic nerves.

During times of excitement or stress, the adrenal medulla helps mobilize the body by enhancing the effects of the sympathetic nervous system. In response to epinephrine and norepinephrine, the heart beats faster and harder, blood flow increases to heart and muscle cells, airways in the lungs dilate, and more oxygen is delivered to cells throughout the body. These are features of the "fight-flight" response (page 339).

THYROID GLAND

The human **thyroid gland** is positioned at the base of the neck, in front of the trachea (Figure 26.10). Its main secretions, thyroxine and triiodothyronine, influence overall metabolic rates, growth, and development. We can gain insight into the extent of that influence by considering what happens when the thyroid does not function properly.

Insufficient thyroid output leads to *hypothyroidism*. Hypothyroid adults experience lethargy, weight gain, intolerance of cold, and dried-out skin. When hypo-thyroid infants are not detected early enough, the consequences can be mental retardation and dwarfism. Excessive thyroid output can lead to *hyperthyroidism*. Hyperthyroid adults suffer from increased heart rate, elevated blood pressure, profuse sweating, weight loss, and intolerance of heat. Typically they show nervous, agitated behavior and have trouble sleeping.

Thyroid hormones cannot be synthesized without iodine, and this can cause problems in regions where farm soil and drinking water have little or no iodine. In the absence of iodine, thyroid hormone levels in the blood decrease. The anterior pituitary detects the decrease and secretes more TSH—excessively so. Excess TSH overstimulates the thyroid gland and causes it to enlarge. (Since thyroid hormones are not being synthesized, the outcome is hypothyroidism.) The resulting tissue enlargement is a form of *goiter* (Figure 26.11). Goiter caused by iodine deficiency is no longer common in countries where iodized salt is widely used. Elsewhere, hundreds of thousands of people still suffer from the disorder, which is easily preventable.

The thyroid gland also secretes calcitonin. This hormone plays a minor but direct role in controlling extracellular levels of calcium ions (Ca^{++}). When the levels rise, calcitonin promotes calcium deposition into bones. When the levels return to normal, thyroid cells decrease their secretion of calcitonin.

PARATHYROID GLANDS

Parathyroid glands are embedded in the posterior surface tissues of the thyroid gland; humans have four of them (Figure 26.10). These glands secrete PTH (parathyroid hormone), which governs blood calcium levels and thereby influences the availability of calcium ions for gene activation, muscle contraction, and many other diverse tasks.

PTH is a good example of a secretion that is not stimulated by other hormones or nerves. Rather, it is a direct homeostatic response to a change in the internal environment—specifically, to a drop in extracellular concentration of calcium ions.

PTH stimulates calcium (and phosphate) removal from bone and its movement into extracellular fluid. It increases the kidney's reabsorption of calcium. It also helps activate vitamin D, which enhances calcium absorption from food moving through the gut. (Vitamin D deficiency leads to *rickets*, a disorder arising from the lack of enough calcium for proper bone development.) When calcium levels rise in response to PTH-induced events, stimulation of the parathyroid glands decreases as a result.

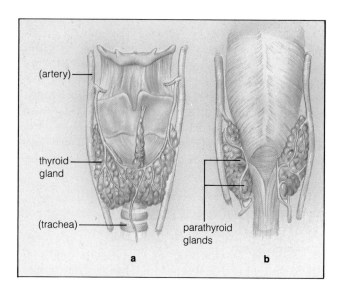

Figure 26.10 (**a**) Anterior view of the human thyroid gland. (**b**) Posterior view, showing the location of the four parathyroid glands embedded in it.

Figure 26.11 A mild case of goiter, as displayed by Maria de Medici in 1625. During the late Renaissance, a rounded neck was considered a sign of beauty; it occurred regularly in parts of the world where iodine supplies were insufficient for normal thyroid function.

GONADS

The **gonads**, the primary reproductive organs, are called testes in males and ovaries in females. Gonads produce gametes; they also secrete sex hormones (estrogens, progesterone, and androgens, including testosterone), which govern reproductive function and secondary sexual traits. The gonads are influenced by the gonadotropins FSH and LH (Table 26.3). Chapter Thirty-Five describes the intricate interplays among these hormones.

OTHER ENDOCRINE ELEMENTS

Pancreatic Islets

The pancreas has a dual function. It has exocrine tissues that secrete digestive enzymes. It also has about 2 million endocrine cell clusters, the **pancreatic islets** (Figure 26.1). Three types of cells in the islets secrete hormones:

1. *Alpha cells* secrete **glucagon**. Between meals, cells use the glucose circulated to them by the bloodstream. The blood glucose level decreases, at which time glucagon secretions cause glycogen (a storage polysaccharide), amino acids, and fatty acid components to be converted to glucose. In such ways, *glucagon raises the glucose level in the blood*.

2. *Beta cells* secrete **insulin**. After meals, when the blood glucose level is high, insulin stimulates uptake of glucose by liver, muscle, and adipose cells especially. It also promotes synthesis of proteins and fats, and inhibits their conversion to glucose. Thus *insulin lowers the glucose level in the blood*.

3. *Delta cells* secrete **somatostatin**, which has a regulatory function (it can inhibit the secretion of insulin and glucagon).

Interplays among these hormones and others help keep the blood glucose level relatively constant, despite great variation in when and how much we eat. We can gain insight into the importance of this function by considering what happens when the body cannot produce enough insulin or when insulin's target cells cannot respond to it. Then, severe disorders in carbohydrate, protein, and fat metabolism arise.

Insulin deficiency can lead to *diabetes mellitus*. Without insulin's contribution to lowering the blood glucose level, the body's water-solute levels become imbalanced, urination becomes excessive, and dehydration occurs. At the same time, protein and fat synthesis decrease even while glucose-starved cells start degrading proteins and fats for energy. The attack on proteins leads to weight loss and an impaired ability to repair damaged tissues. Ketones (normal by-products of fat breakdown) increase in the blood. They, too, upset the body's water-solute balance by promoting excessive water and sodium loss by urination. The extracellular pH decreases; in combination with dehydration, it affects brain function. In extreme cases, diabetic persons may die.

In "type 1 diabetes," insulin levels are nonexistent (or nearly so) because beta cells in the pancreas have been destroyed. This disorder is the least common among diabetics. It appears to arise from a convergence of factors, including a heritable susceptibility to the disorder, viral infection, and an autoimmune response mounted against the body's own insulin-secreting cells (page 429). Type 1 diabetics survive with insulin injections. (Insulin cannot be taken orally because it is a protein; it would be destroyed by protein-degrading enzymes in the gut before it could even be absorbed.)

In "type 2 diabetes," insulin levels are nearly normal or even above normal—but the target cells cannot respond to the hormone. Either the cells have an insufficient number of insulin receptors or the receptors themselves are abnormal. Generally, this condition can be managed by regulating the diet and by using drugs that stimulate insulin production.

Thymus Gland

The lobed **thymus gland** is located behind the breastbone and between the lungs. Lymphocytes (white blood cells) multiply, differentiate, and mature in this gland, which secretes a group of hormones collectively called thymosins. These hormones affect the functioning of lymphocytes that defend the body against disease (Chapter Thirty).

Pineal Gland

Until about 240 million years ago, vertebrates commonly had a third eye, in the middle of the forehead. Lampreys and a few other vertebrates still have one of these photosensitive organs, buried beneath the skin. In mammals, birds, and most reptiles, the organ persists in modified form as an internal **pineal gland** (Figure 25.10). This gland secretes melatonin, a hormone that functions in the development of gonads and in reproductive cycles.

Melatonin is secreted in the absence of light. Thus melatonin levels vary diurnally (from day to night) and seasonally (as when winter days are shorter than summer days). In hamster reproductive cycles, for example, high melatonin levels in winter suppress sexual activity; in summer, when melatonin levels are low, sexual activity peaks.

In humans, decreased melatonin secretion may help trigger the onset of puberty, the age at which reproductive structures start to mature. Nearly a century ago, the physician Otto Heubner performed an autopsy on a boy who had entered puberty prematurely (at age four) and discovered that a brain tumor had destroyed the boy's pineal gland. Normally, melatonin levels are high through age five; then they steadily decrease until, at puberty's end, they are only about one-fourth of the peak levels.

LOCAL CHEMICAL MEDIATORS

Every tissue of the mammalian body has mediator cells that detect changes in the surrounding chemical environment and alter their activity, often in ways that either counteract or amplify the change. These are local responses, confined to the immediate vicinity of change; most of the molecules secreted are taken up so rapidly that not many are left to enter the general circulation. Among the local secretions are prostaglandins and growth factors.

Prostaglandins

More than sixteen different prostaglandins have been identified in tissues throughout the body. They are released continually, but the rate of synthesis often increases in response to local chemical changes. The stepped-up secretion can influence neighboring cells as well as the prostaglandin-releasing cells.

At least two prostaglandins help adjust blood flow through local tissues. When stimulated by epinephrine and norepinephrine, they cause smooth muscle in the walls of blood vessels to constrict or dilate. Prostaglandins have similar effects on smooth muscle of airways in the lungs. Allergic responses to airborne dust and pollen may be aggravated by prostaglandins (page 428).

Prostaglandins have profound effects on some mammalian reproductive events. During the menstrual cycle, many women experience painful cramping and excessive bleeding—both of which have been traced to prostaglandin action. (Anti-prostaglandin drugs block synthesis of this local mediator and alleviate the discomfort.) Prostaglandins also influence the corpus luteum, a glandular structure that develops from cells which earlier surrounded a developing egg in the ovary. If pregnancy does not follow ovulation (release of the egg from the ovary), a corpus luteum self-destructs; it produces copious amounts of prostaglandins that interfere with its own function.

Growth Factors

Some signaling molecules, called growth factors, influence growth by regulating the rate at which certain cells divide. For example, **nerve growth factor** (NGF), dis-

covered by Rita Levi-Montalcini, promotes survival and growth of neurons in the developing embryo. (One experiment demonstrated that certain immature neurons survive indefinitely in tissue culture when NGF is present but die within a few days if it is not.) NGF also may define the direction of growth for these embryonic neurons, laying down a chemical path that leads the elongating processes to target cells. **Epidermal growth factor** (EGF), discovered by Stanley Cohen, influences the growth of many cell types.

SIGNALING MECHANISMS

Hormones and other signaling molecules induce diverse responses in target cells. They can trigger or alter the rate of protein synthesis or they can cause modification in proteins already present in the cell. What dictates the nature of the response?

For one thing, not all cells *can* respond to all types of signals. Consider that nearly all cells have receptors for some hormones, including insulin, and in such cases the signal has widespread effects. Yet only a few cell types have receptors for certain other hormones, in which cases the signal has highly directed effects.

For another thing, different signals activate different cellular mechanisms of response. Let's consider examples of the responses to two main categories of hormones (Table 26.5):

1. *Steroid hormones* (or steroidlike compounds), which are assembled from cholesterol and are largely insoluble in water but soluble in lipids.

2. *Nonsteroid hormones*, which are assembled from amino acids and which are water-soluble.

Table 26.5 Two Main Categories of Hormones	
Type of Hormone	Examples
Steroid	Estrogens, aldosterone, cortisol
Nonsteroid:	
Amines	Norepinephrine, epinephrine
Peptides	ADH, oxytocin, TRH
Proteins	Insulin, somatotropin, prolactin
Glycoproteins	FSH, LH, TSH

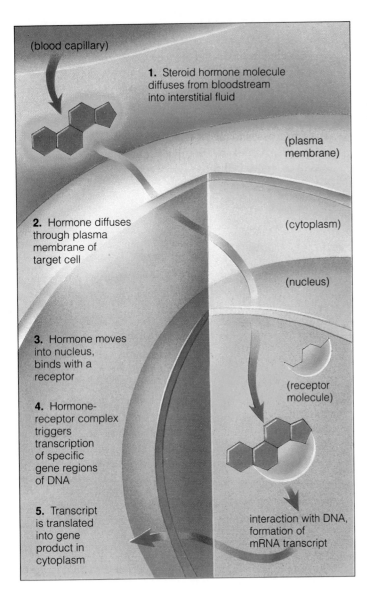

Figure 26.12 Proposed mechanism of steroid hormone action on a target cell. This same type of mechanism is also thought to occur for thyroid hormones.

Labels in figure:

(blood capillary)

1. Steroid hormone molecule diffuses from bloodstream into interstitial fluid

(plasma membrane)

2. Hormone diffuses through plasma membrane of target cell

(cytoplasm)

(nucleus)

3. Hormone moves into nucleus, binds with a receptor

(receptor molecule)

4. Hormone-receptor complex triggers transcription of specific gene regions of DNA

5. Transcript is translated into gene product in cytoplasm

interaction with DNA, formation of mRNA transcript

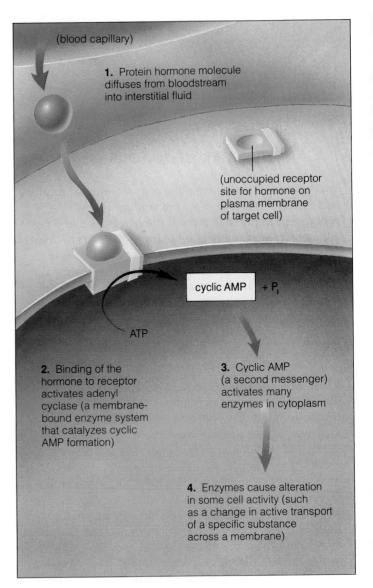

Figure 26.13 Proposed mechanism of protein hormone action on a target cell. The response is mediated by a second messenger inside the cell: in this case, cyclic AMP. Other chemical messengers may be involved, depending on the particular hormone and its particular target cell.

Labels in figure:

(blood capillary)

1. Protein hormone molecule diffuses from bloodstream into interstitial fluid

(unoccupied receptor site for hormone on plasma membrane of target cell)

cyclic AMP + P$_i$

ATP

2. Binding of the hormone to receptor activates adenyl cyclase (a membrane-bound enzyme system that catalyzes cyclic AMP formation)

3. Cyclic AMP (a second messenger) activates many enzymes in cytoplasm

4. Enzymes cause alteration in some cell activity (such as a change in active transport of a specific substance across a membrane)

Steroid Hormone Action

All steroid hormones stimulate the synthesis of new proteins by activating certain genes; they do not alter the activity of already existing proteins.

Being lipid-soluble, steroid hormones can diffuse rather easily across the plasma membrane of target cells (Figure 26.12). Once inside, they move into the nucleus, where they bind with receptors specific for them. The three-dimensional shape of the hormone-receptor complex allows it to associate with chromosomal proteins. The complex activates certain gene regions and leads to the synthesis of particular mRNA transcripts, then to specific proteins (page 218).

One steroid hormone, testosterone, influences the development of male sexual traits. In *testicular feminization syndrome*, the receptor to which testosterone binds is defective. Genetically, the affected individuals are males; they have functional testes that secrete testosterone. However, none of the target cells can respond to the hormone, and the secondary sexual traits that do develop are entirely female or female-like.

Nonsteroid Hormone Action

How do protein hormones and other water-soluble signaling molecules induce responses in target cells, given that they cannot cross the plasma membrane? Some move into the cell by receptor-mediated endocytosis, which was shown in Figure 6.12. Others bind to a receptor and cause ion channels across the membrane to open. Certain ions move inward, and their cytoplasmic concentration changes in ways that affect cell activities.

Most protein hormones, including insulin, activate **second messengers**, which are intracellular mediators of the response to their signal. One second messenger is **cyclic AMP**. (The full name is cyclic adenosine monophosphate.)

The response in which cyclic AMP plays a part begins when a hormone molecule binds to a receptor on the plasma membrane of a target cell. The binding alters the activity of a membrane-bound enzyme system (Figure 26.13). Adenyl cyclase is activated; this enzyme catalyzes the conversion of ATP to cyclic AMP.

Many molecules of adenyl cyclase, not just one, are activated by a hormone-receptor complex. Each of these enzyme molecules increases the rate at which many ATP molecules are converted to cyclic AMP. Each second-messenger molecule so formed then activates many enzymes. Each of the enzymes so activated can convert a very large number of substrate molecules into different enzymes, and so on until the number of molecules representing the final cellular response to the initial signal is enormous. Thus, second messengers are a way of *amplifying* the response to a signaling molecule.

SUMMARY

This chapter concludes our survey of controls over the integration of body activities in multicelled animals. Throughout this unit, we will be looking at examples of these controls, so keep the following concepts in mind:

1. For metabolic activity to proceed smoothly, the chemical environment of a cell must be maintained within fairly narrow limits.

2. In complex animals, thousands to billions of cells continually remove some substances from the extracellular fluid and secrete other substances into it. The nature and amount of the substances can change with the diet or level of activity; they inevitably change during the course of development.

3. It follows that the myriad withdrawals and secretions must be integrated in ways that assure cell survival through the whole body.

4. Integration is accomplished by signaling molecules: chemical secretions by one cell that adjust the behavior of other, target cells. (A target is one having receptors to which specific signaling molecules can bind and elicit a cellular response. It may or may not be adjacent to the signaling cell.)

5. Signaling molecules include hormones (and neurohormones), transmitter substances, local signaling molecules, and pheromones.

6. A neuroendocrine control center integrates many activities for the vertebrate body. This center consists of the hypothalamus and pituitary.

7. Two hypothalamic neurohormones (ADH and oxytocin) are stored and released from the posterior lobe of the pituitary. ADH influences extracellular fluid volume. Oxytocin influences contraction of the uterus and milk release from mammary glands.

8. Six other hypothalamic neurohormones (called releasing hormones) control the secretions by cells of the anterior lobe of the pituitary.

9. The anterior lobe of the pituitary produces and secretes six hormones. Two (prolactin and somatotropin, or growth hormone) have general effects on body tissues. The remainder (ACTH, TSH, FSH, and LH) act on specific endocrine glands.

10. Hormone secretion is controlled by neural signals, hormonal signals, and changes in the extracellular concentration of metabolites (such as calcium). Secretion of many endocrine glands is controlled through homeostatic feedback inhibition of the neuroendocrine control center.

11. Fast-acting hormones such as PTH or insulin generally come into play when the extracellular concentration of a substance must be homeostatically controlled. Slow-acting hormones such as somatotropin have more prolonged, gradual, and often irreversible effects, such as those on development.

12. Cells respond to specific hormones or other signaling molecules only if they have receptors for them. Steroid hormones have receptors in the nucleus of target cells. Nonsteroid hormones (the amines, peptides, proteins, and glycoproteins) have receptors on the plasma membrane of target cells; responses to them are often mediated by a second messenger (such as cyclic AMP) inside the cell.

13. Steroid hormones trigger gene activation and protein synthesis; nonsteroid hormones alter the activity of proteins already present in target cells. These cellular responses contribute in some way to maintaining the internal environment or to the developmental or reproductive program.

Review Questions

1. Name the main endocrine glands and state where each is located in the human body. *352–353*

2. Define the difference between transmitter substances and neurosecretory hormones. *353–354*

3. Define hormone. What functions do hormones serve? How do these functions differ from those of transmitter substances? *353–354*

4. There are two general classes of hormones: steroid and polypeptide. How is each thought to act on a target cell? *364–365*

5. The hypothalamus and pituitary are considered to be a neuroendocrine control center. Can you describe some of the functional links between these two organs? *354*

6. How does the hypothalamus control secretions of the posterior lobe of the pituitary? The anterior lobe? *354–356*

7. Name three endocrine glands and a substance that each one secretes. What are the main consequences of their secretion? *358, 360–361*

8. Which hormone secreted by the anterior pituitary has an effect on most body cells rather than on a specific cell type? What are the clinical consequences of too little or too much secretion of this hormone? *357, 359*

9. Which secretions of the posterior and anterior lobes of the pituitary glands have the targets indicated? (Fill in the blanks; pages *355* and *357*):

Readings

Berridge, M. October 1985. "The Molecular Basis of Communication Within the Cell." *Scientific American* 253(4):142–152.

Cantin, M., and J. Genest. February 1986. "The Heart As an Endocrine Gland." *Scientific American* 254(2):76–81.

Fellman, B. May 1985. "A Clockwork Gland." *Science 85* 6(4):76–81. Describes some of the known functions of the pineal gland.

Franklin, D. 1984. "Growing Up Short." *Science News* 125:92–94. Describes prospects and problems of using synthetic somatotropin to boost height of hypopituitary children.

Hadley, M. 1988. *Endocrinology.* Second edition. Englewood Cliffs, New Jersey: Prentice-Hall.

Norris, D. 1985. *Vertebrate Endocrinology.* Second edition, Philadelphia: Lea & Febiger.

Snyder, S. October 1985. "The Molecular Basis of Communication Between Cells." *Scientific American* 253(4):132–141.

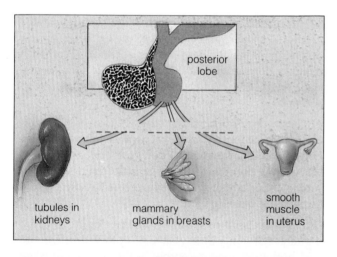

tubules in kidneys   mammary glands in breasts   smooth muscle in uterus

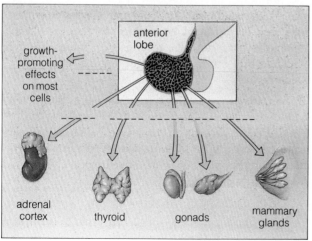

growth-promoting effects on most cells

adrenal cortex   thyroid   gonads   mammary glands

Looking back over the past few chapters, you might be left with the impression that the nervous system is an immense collection of communication lines that are silent until signals are fed into them from the outside, much as telephone lines wait to carry phone calls from all over the country. It is the wrong impression. The crucial feature of a nervous system is not what it is made of but *what it does*—and from its formation in the embryo until death, it never stops doing things, even beyond its responses to signals from the outside.

Many neurons in the brain centers concerned with breathing and other vital functions never do rest. They start firing rhythmically early in development, and although incoming sensory signals may alter the firing frequency, nothing short of damage or death stops their basic activity. Also during development, vast gridworks of neurons are laid down and begin to interact in genetically prescribed patterns. These patterns of activity are inherited **neural programs**. They provide each new individual with a means of responding to situations that members of its species are likely to encounter.

For example, among other things, the python shown in Figure 27.1 eats small mammals that forage at night. At night the temperature difference between its prey and

27

SENSORY SYSTEMS

Figure 27.1 Reticulated python of southern Asia, equipped with infrared-sensitive receptors that enable it to detect warm-blooded prey in the dark. This particular snake has the receptors in thirteen pairs of pits above and below the mouth, as visible here.

the surrounding air is usually large. This snake has receptors for detecting heat (infrared energy), which is used to pinpoint its prey in the dark. Input from the receptors is fed into a brain center where signals about the location of objects are compared against a neural gridwork. The signals are analyzed with great precision, and the snake directs its strike with great accuracy. (At most, the strike may be only a few degrees off center.) Even so, the same snake might slither past a motionless yet edible frog. Frog skin is cool and typically blends with background colors. The snake has neither receptors nor a neural program for responding to it.

SENSORY PATHWAYS

Receptors Defined

A **stimulus** is any form of energy the body is able to detect by means of its receptors. **Receptors** are finely branched peripheral endings of sensory neurons (or specialized cells adjacent to them) that respond to specific stimuli. Often, receptors, epithelial tissues, and connective tissues are organized into **sensory organs** (such as eyes), which amplify or focus the energy of a stimulus. The major categories of receptors may be described in the following way:

Chemoreceptors detect chemical energy (ions or molecules dissolved in body fluids next to the receptor). They include olfactory and taste receptors.

Mechanoreceptors detect mechanical energy associated with changes in pressure, position, or acceleration. They include receptors for touch, stretch, hearing, and equilibrium.

Photoreceptors detect photon energy of visible and ultraviolet light.

Thermoreceptors detect radiant energy associated with temperature. They include infrared receptors.

Table 27.1 summarizes the common forms of stimuli and lists some of the receptors for them. The kind and number of receptors vary among animals. For example, you do not have receptors for ultraviolet light, as honeybees do; and you do not "see" many flowers the way they do (Figure 21.8). Unlike dogs, you have receptors that contribute to color vision, but dogs have far more olfactory receptors in the nose. Hence humans and dogs sample the environment in different ways and have different perceptions of it.

Principles of Receptor Function

A sensory receptor is a *transducer*: it converts one form of energy into another. When the plasma membrane of a receptor cell is stimulated, the outcome is a **receptor potential** (a brief change in voltage across the membrane). The change is a graded response. Like the synaptic potentials described in Chapter Twenty-Four, it can vary in magnitude, depending on the intensity and duration of the stimulus. Graded responses acting together lead to action potentials, the means by which messages travel through the nervous system.

As we have seen, action potentials are all alike; they never vary in magnitude. How, then, do they convey information that gives rise to different sensations such as smell, taste, and color? Part of the answer is that some signals from specific receptors always end up in specific parts of the brain. (In the embryo, sensory nerves destined to carry those signals always develop along genetically ordained routes from one part of the body to another.) Thus, when you accidentally poke your eye in the dark, you "see stars." Action potentials from visual receptors are always interpreted as "light," simply because the brain region in which they end up interprets all signals coming in on a particular pathway as "light."

But now the question becomes this: How do action potentials convey information about variations in stimulus intensities? The same receptors detect energy associated with a throaty whisper or a wild screech, and the same sensory pathways carry information about both to the brain—but how? When the strength of a stimulus increases, the magnitude of receptor potentials also increases, so action potentials are fired more frequently. Figure 27.2 shows some frequency variations corresponding to differences in sustained pressure on human skin.

Also, stronger stimuli "recruit" more and more receptors in a larger area of a given tissue. For instance, when you lightly press a finger into skin on your arm, you activate a small number of receptors. When you push the finger firmly into the same region, more receptors are disturbed. The increased disturbance sets off action potentials in many sensory axons at the same time, and the simultaneous activity reflects the increase in stimulus intensity.

A particular sensation is not triggered by action potentials themselves but by their travel along a particular nerve pathway.

Variations in stimulus intensity are encoded in the frequency of action potentials in a single axon.

Variations in stimulus intensity are also encoded in the number of axons recruited into action in a given tissue.

Table 27.1 Receptors Associated With Different Senses

Category	Stimulus	Examples of Receptors	Main Functions
Chemical senses:			
Internal chemical sense	Chemical energy	Carotid body (page 427), free nerve endings	Signal changes in extracellular concentrations of specific ions, molecules
Taste	Chemical energy	Taste receptor cells	Signal presence of nutritious or noxious substances
Smell	Chemical energy	Olfactory receptor cells	Signal presence of specific airborne ions, molecules
Somatic senses:			
Touch, pressure	Mechanical energy	Mechanoreceptors (free nerve endings, Pacinian corpuscles)	Signal changes at body surface (e.g., mechanical distortion)
Temperature	Thermal energy (heat)	Free nerve endings	Signal changes in temperature
Pain	Varied (e.g., mechanical, thermal, chemical energy)	Several, including free nerve endings	Signal damage or threat of damage
Muscle sense and kinesthesia	Mechanical energy (e.g., stretching, rotation at joints)	Mechanoreceptors in joints, tendons, skin	Signal changes in position, movement; contribute to sensations of effort, force, and weight
Sense of balance (equilibrium)	Mechanical energy, gravitational energy	Hair cells; mechanoreceptors in tendons, joints	Signal changes in body position and movements underlying orientation relative to environment
Acoustical sense (hearing)	Varies (compressional waves or distortions in air, water)	Hair cells in lateral line organs, in mammalian inner ear	Signal sound waves and other physical displacements of environmental medium
Photosensitivity, sense of vision	Light energy	Photoreceptors (e.g., rods and cones)	Varies among species (signal changes in light intensity; motion, form, depth, and color in visual field; polarization of light)

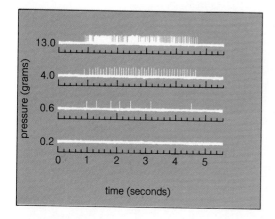

Figure 27.2 Action potentials recorded from a single pressure receptor on the human hand. The recordings correspond to variations in stimulus strength. A thin rod was pressed against the skin with the force indicated. Vertical bars above each thick horizontal line represent individual action potentials. Notice the increases in frequency, which correspond to increases in stimulus strength.

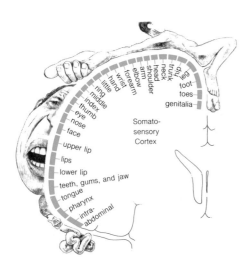

Figure 27.3 Body regions represented in the primary somatic sensory cortex. This region is a strip a little more than an inch wide, running from the top of the head to just above the ear on the surface of *each* cerebral hemisphere. The diagram is a cross-section through the right hemisphere of someone facing you. Compare this with Figure 25.12.

Labels on figure: dig, leg, foot, toes, genitalia, trunk, neck, head, shoulder, arm, elbow, forearm, wrist, hand, little, ring, middle, index, thumb, eye, nose, face, upper lip, lips, lower lip, teeth, gums, and jaw, tongue, pharynx, intra-abdominal, Somato-sensory Cortex

Primary Sensory Cortex

Sensory nerve pathways from different receptors lead to different parts of the cerebral cortex. For example, signals from receptors in the skin and joints travel to the primary somatic sensory cortex. Cells of this region are laid out like a map corresponding to the body surface. Some map regions are larger than others; they represent areas that are functionally more important and that have more receptors. In humans, a large part of the primary somatic sensory cortex responds to receptors located in the fingers, thumb, and lips (Figure 27.3).

Let's now focus on a few kinds of receptors to illustrate how they detect changes in the surrounding environment.

CHEMICAL SENSES

Animals can distinguish between nutritious and noxious substances with their **taste receptors**. These receptors are present in the mouth of most vertebrates. They also are distributed on snail antennae, insect legs, octopus tentacles, and some fish fins. In many vertebrates, taste receptors are components of sensory organs called **taste buds**, which are distributed mostly on the tongue. A taste bud is composed of modified epithelial cells, includ-

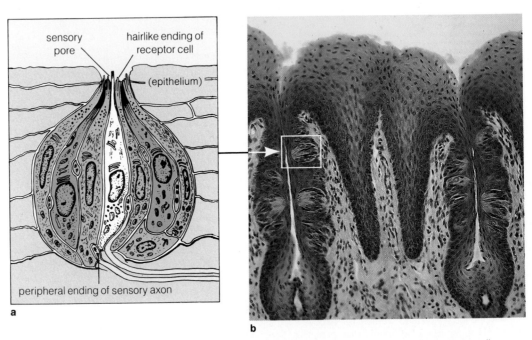

Labels on figure a: sensory pore, hairlike ending of receptor cell, (epithelium), peripheral ending of sensory axon

Figure 27.4 Taste bud (**a**) in the walls of narrow pits in a rabbit tongue (**b**). A mature taste receptor cell, which synapses with a sensory axon, is shown in white. An immature taste receptor cell appears in the lower-right region of the taste bud.

ing the receptor cells (Figure 27.4). Sensory nerves carry signals from the receptors to the brain.

Animals also distinguish among different odors. The detection of odors, or **olfaction**, is one of the most ancient senses. Olfactory receptors still must be important for survival, judging from their sheer abundance. Humans have about 10 million olfactory receptors in the nose; German shepherds have more than 220 million. Olfactory receptors respond to molecules released from food, from predators, and from members of the same species. They all are modified endings of axons that carry signals to the brain.

In many animals, specialized olfactory receptors respond to pheromones (signaling molecules having cellular targets outside the body, in other animals of the same species). Pheromones are social signals. Animals use them to raise an alarm about impending danger, attract a mate, influence sexual behavior, identify individuals of the same social group, disperse a group, or mark territorial boundaries.

Many olfactory receptors are sensitive to pheromones. Those on the antennae of male silk moths (*Bombyx mori*) can detect one bombykol molecule in 10^{15} molecules of air! Each receptor, making contact with one molecule per second, can trigger an action potential. Female silk moths secrete bombykol as a sex attractant, which enables a male to find a female, even in the dark, and even more than a kilometer upwind from him.

SOMATIC SENSES

The term **somatic senses** refers to sensations of touch, pressure, temperature, and pain near the body surface. Two types of somatic receptors are free nerve endings and Pacinian corpuscles (Figure 27.5). Both are activated when their plasma membrane is mechanically deformed or altered in some other way. Free nerve endings in skin contribute to sensations of light pressure, temperature, and pain; Pacinian corpuscles contribute to sensations of deep, rapid pressure and vibration.

"Pain" is the perception of injury to some region of the body. Specific pain receptors have been identified in invertebrates such as the leech, and it seems likely that they exist in mammals as well. Although invertebrates certainly react to injury, it is not known whether they "feel" pain as we do. The perception of pain seems to have an important emotional component, mediated perhaps by the limbic system (page 344). The invertebrate brain does not have a comparable system.

Responses to some types of pain depend on accurately assessing where the stimulus is located. This can be done for most types of somatic pain and some types of visceral (internal) pain. But most sensations of visceral pain apparently are based on erroneous signals, the outcome being that pain is felt in a tissue remote from the original point of stimulation. (For example, as Figure 27.6 suggests, a heart attack may be felt as pain in the skin

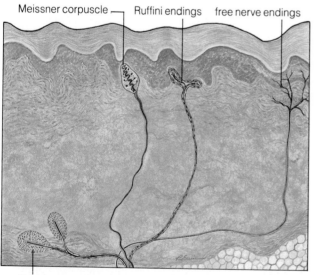

Figure 27.5 Tactile receptors in human skin. Free nerve endings contribute to sensations of temperature, light pressure, and pain. Pacinian corpuscles (see photograph) contribute to sensations of deep, rapid pressure. Meissner corpuscles are stimulated at the onset and end of sustained pressure; the Ruffini endings react continually to ongoing stimuli.

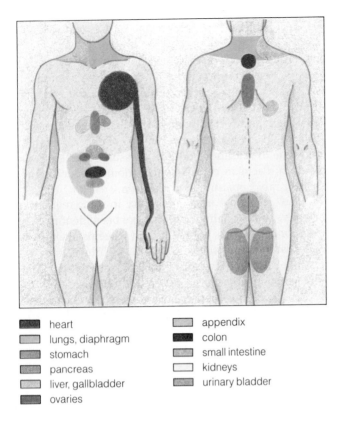

heart	appendix
lungs, diaphragm	colon
stomach	small intestine
pancreas	kidneys
liver, gallbladder	urinary bladder
ovaries	

Figure 27.6 Regions of referred pain. Sensations of certain visceral disorders are erroneously localized to the skin areas indicated.

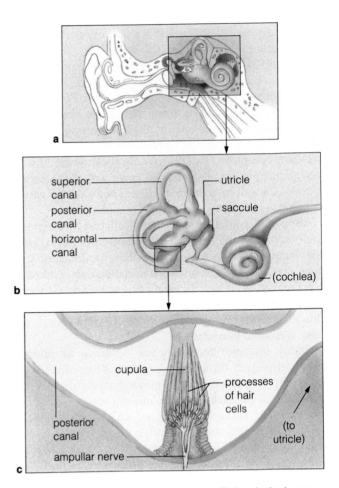

Figure 27.7 (**a**) Location of organs of equilibrium in the human ear. (**b**) The vestibular apparatus, with three semicircular canals and two sacs called the utricle and saccule. (**c**) Closer view of the ampula (bulge) of the posterior canal; the other two canals have similar components. The cupula, a gelatinous membrane positioned inside the canal, bends under fluid pressure. When the body moves, fluid in the canal corresponding to the plane of movement is displaced and pushes against the cupula, so that the hair cells are bent and thereby activated.

The wall of each utricle and saccule contains an otolith organ, in which a patch of gelatinous membrane floats directly over hair cells. The membrane slides in response to gravity, like an unbaked cookie on a greased cookie sheet. When the head tilts, the membrane slides over the hair cells and activates them.

above the heart and along the left shoulder and arm.) This phenomenon is called "referred pain."

What is the basis of referred pain? Nerves from the injured (or malfunctioning) organ and nerves from the skin segment to which the pain is referred usually started out close together when the animal embryo was first developing. By the time they reached their normal position in the adult body, they had become physically separated. Even so, the nerves still enter the same segment of the spinal cord and synapse on the same interneurons. The brain continues to refer the internal pain to the position of the skin segment in the adult.

THE SENSE OF BALANCE

Almost all animals must have a sense of what the "natural" position for their body must be, given the way they return to it after being tilted, turned upside-down, and so on. The baseline against which animals assess displacement from their natural position is called **equilibrium**. At equilibrium, the body (or some part of it, such as a wing) is balanced in relation to gravity, velocity, acceleration, and other forces that influence position and movement.

Organs of equilibrium incorporate mechanoreceptors called **hair cells**. These cells are arranged in an epithelium and have one or more processes ("hairs") projecting from their surface. When the processes bend in response to gravity and other forces, receptor potentials are produced and signals about changing position are sent to the brain.

In vertebrates, hair cells are present in the **vestibular apparatus**, which is a closed system of fluid-filled sacs and canals inside the ear (Figure 27.7). Together with other input from receptors in the eyes, skin, and joints, the vestibular apparatus contributes to the sense of balance. It consists of *semicircular canals* (which detect changing movements) and an *otolith organ* (which detects changes in the head's orientation relative to gravity). Amphibians, birds, and mammals have three semicircular canals, positioned at angles corresponding to three planes of space (Figure 27.7).

Jellyfishes have an organ of equilibrium called a *statocyst* (Figure 27.8). This infolding of epidermis contains hair cells and statoliths, which are dense crystals of sand grains (or mineral salts and organic material). When the body tilts, statoliths also tilt and cause neighboring hair cells to bend. Thus receptor potentials are produced.

THE SENSE OF HEARING

The sense of sound, or **hearing**, also depends on the bending of hair cells under fluid pressure. In this case, vibrations initiate the fluid movements. A "vibration" is a wavelike form of mechanical energy, transmitted outward from a stimulus by a series of compressions and rarefactions of molecules of the surrounding medium. ("Rarefaction" means a low-pressure state.)

For example, clapping your hands produces waves of compressed air. The clapping force drives many molecules together in the air between your hands, creating a high-pressure state in which the molecules collide faster and more often. The collisions send molecules flying outward, where they collide with more distant molecules, and so on away from your hands. Each time molecules are forced outward, a low-pressure state is created in the region they vacated. These pressure vari-

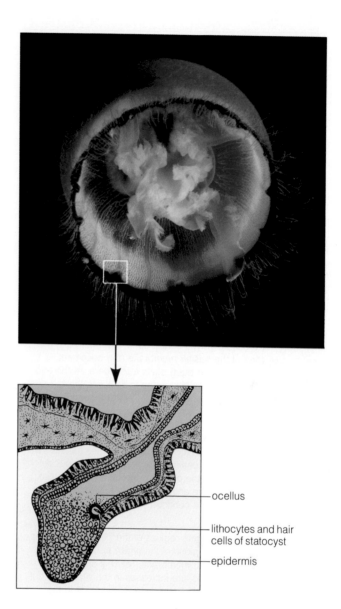

Figure 27.8 Statocyst at the bell margin of the jellyfish *Aurelia*. When the bell tilts, lithocytes (statolith-containing cells) press against neighboring hair cells and activate them. Signals from the hair cells lead to vigorous muscle contractions that correct body position.

ocellus

lithocytes and hair cells of statocyst

epidermis

Figure 27.9 Case study: Receptor activation in the human ear.

When pressure waves reach the outer ear, they are directed inward and funneled through the ear canal. Then they strike the eardrum (*tympanic membrane*), which bows in and out at the same frequency as the waves. The movement activates the middle earbones, a lever system spanning the distance between the eardrum and the *oval window* (an elastic membrane over the entrance to the coiled inner ear; **a** and **b**). The middle earbones transmit the force of pressure waves to the much smaller surface area of the oval window, thereby amplifying the stimulus.

The oval window now bows in and out, producing fluid pressure changes in the inner ear. Pressure waves are propagated through two of the three fluid-filled ducts (the *scala vestibuli* and *scala tympani*). At the end of the second duct, pressure waves reach a membranous *round window*, causing it to bulge. Without this capacity to yield under pressure, fluid would not be able to move inside the inner ear.

How are pressure waves sorted out in the inner ear? The construction of the *basilar membrane* (the basement membrane of the third duct) plays a role here. At the end of the cochlear duct closest to the middle ear, the membrane is narrow and somewhat rigid. It becomes broader and more flexible deep in the coil of the inner ear. High-frequency waves, which carry more energy, cause the greatest displacement of the stiff region. Most of the energy associated with these waves becomes transformed into membrane vibrations here, so the waves die out before traveling deeper into the coil. Low-frequency waves also set up vibrations at the entrance, but the vibrations are lower in amplitude and continue into the more elastic regions.

Displacement of different regions of the basilar membrane stimulates different regions of the *organ of Corti*, which is perched on the basilar membrane. (**c**) This organ contains hair cells. As fluid pressure increases in the scala vestibuli, the vestibular membrane moves down, pressure in the cochlear duct rises, and the basilar membrane is pushed down. When that happens the hair cell processes move in relation to an overhanging flap (the *tectorial membrane*). The movement changes the membrane permeability of the hair cell, leading to receptor potentials that excite associated sensory neurons.

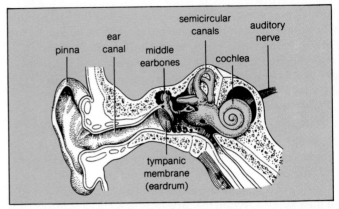

a

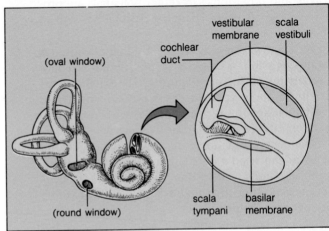

b

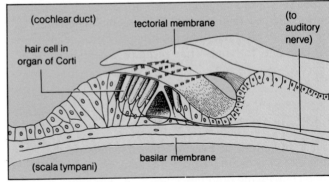

c

hair cell

a

b

c

Figure 27.10 Effect of intense sound on the inner ear. Normal organ of Corti from a human (**a**) and from a guinea pig (**b**), showing three rows of outer hair cells. (**c**) Organ of Corti after twenty-four-hour exposure to noise levels approached by loud rock music (2,000 cycles per second at 120 decibels).

ations can be depicted as a wave form in which the peaks represent compressions and the valleys, rarefactions:

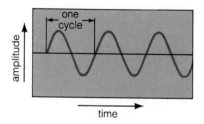

one cycle

amplitude

time

The pressure peaks (amplitude) of sounds are perceived as loudness. The stronger the stimulus, the more compressed the molecules become, and the louder the sound. The frequency of a sound is the number of wave cycles per second. Each "cycle" extends from the start of one peak of the wave to the start of the next. The more cycles per second, the higher the frequency and the higher the perceived pitch of sound.

Which Animals Hear?

Almost all vertebrates have receptors and organs adapted to the higher frequencies associated with hearing. So do many arthropods. (For example, mosquitoes and spiders are attracted to audible, high-frequency vibrations of a tuning fork.) But the most familiar organs of hearing are the paired ears of land-dwelling mammals.

When sound waves spread out through air, they rapidly become weaker with distance. The mammalian **ear** has three parts in which weak signals are received, amplified, and sorted out. The *outer ear* is a system of external flaps (which collect sound waves) and an internal auditory canal (which channels the waves inward to an eardrum). The *middle ear* is a system of small bones (hammer, anvil, and stirrup) that transfers eardrum vibrations to the inner ear. The *inner ear* contains a coiled tube (cochlea) divided into three fluid-filled ducts, where pressure waves are sorted out and where hair cells are stimulated. Figure 27.9 shows how the three systems operate.

Some sounds are barely perceptible to humans, but others are extremely intense and cause structural damage to inner ear regions. Among these intense sounds are amplified music and the thundering of jet planes taking off. Such recent developments exceed the functional range of the evolutionarily ancient hair cells in the ear. Figure 27.10 shows receptors that have been damaged by intense sound.

Echolocation

In a process called **echolocation**, bats, dolphins, and whales emit high-frequency sound waves, and echoes from the waves bounce back to their ears from objects in the surroundings. By perceiving frequency variations in the echoes, these mammals can pinpoint the distance and direction of movement of predators, prey, and the like. Bats can pinpoint prey even though the echoes are only 1/2,000 the amplitude of their cries. The frequency

Figure 27.11 An echolocating bat, listening to echoes of self-produced ultrasonic noises as they bounce back from objects in the environment.

of those cries is almost beyond the range of human ears. Moreover, the cries have been measured electronically at 100 decibels—about the same intensity as thunder cracking overhead or the rumble of a passing freight train. Figure 27.11 shows a bat echolocating.

SENSE OF VISION

In **photoreception**, pigment molecules embedded in the membranes of receptor cells absorb light energy and thereby trigger events that lead to action potentials in neighboring cells. Photoreception is *not* the same as vision. All organisms, whether they see or not, are sensitive to light. Shine a bright light on a single-celled amoeba moving about and it will stop abruptly. Some small invertebrates have neither photoreceptors nor pigments, yet they display **phototaxis**: they orient themselves toward or away from the direction of incoming light.

What we call "vision" depends on highly developed sensitivity to light. A **visual system** includes (1) structures that focus patterns of light energy onto a dense layer of photoreceptors, and (2) a neural gridwork in the brain that can deal with those patterns. The brain analyzes various aspects of a visual stimulus, including its position, shape, brightness, distance, and movement.

A **lens** is present in most visual systems. This spherical or cone-shaped body of transparent protein fibers channels incoming light to photoreceptor cells located behind it. But a lens alone does not lead to vision. Some invertebrates have eyes equipped with lenses, yet they cannot see as we do. Their lenses channel light either in front of or behind their photoreceptors, the result being a very diffuse kind of stimulation. These invertebrates detect a general change in light intensity, as when another animal passes overhead in the water. But they cannot discern the size or shape of objects.

Vision requires precise light focusing onto a photoreceptor cell layer that is dense enough to sample details of the light stimulus, followed by processing of visual information in the brain.

Invertebrate Photoreception

Many invertebrates have ocelli, or **eyespots**. These organs function in photoreception, but vision does not follow. The eyespots are simply clusters of photosensitive cells, usually arranged in a cuplike depression in epidermis (Figure 27.12). The pigmented membrane of these cells is often folded into tiny, fingerlike projections

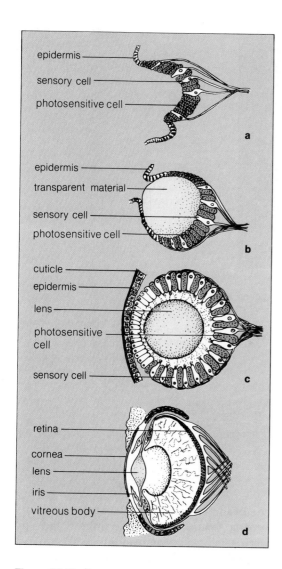

a

b

Figure 27.12 Some molluscan photoreceptors, long section. (**a**) Limpet eyespot, a shallow epidermal depression. (**b**) Abalone eye; the secreted material may serve as a lens. (**c**) Snail eye. (**d**) Octopus eye.

Figure 27.13 Examples of the well-developed eyes of mollusks. (**a**) A red-mouthed conch (*Strombus*), peering about the Great Australian Barrier Reef. (**b**) Slit eye of the octopus.

called microvilli. This pattern also occurs in vertebrate photoreceptors.

Mollusks are the simplest animals with **eyes**: well-developed photoreceptor organs that contribute to some degree of image formation. Some molluscan eyes are closed, fluid-filled vesicles (Figure 27.13). They are equipped with a transparent lens, a **cornea** (transparent cover) and a **retina** (a tissue containing densely packed photoreceptors).

Squids, cuttlefishes, and octopuses are fast-moving, predatory mollusks of dimly lit underwater worlds. All have large, paired eyes capable of forming clear images. Both eyes, which are positioned just behind prey-grabbing tentacles, are used in aligning the tentacles at

the correct striking distance from prey. Muscles control movements of the eyeball and of the **iris**, an adjustable ring of contractile and connective tissues within the eye. The open center of this contractile ring (the *pupil*) can be varied in size to admit more or less light.

Bright light causes the octopus pupil to shrink into a narrow slit. If you happen to bump into a good-sized octopus while snorkeling or diving, its large slit pupils may flare open suddenly, one or both at a time, in response. The octopus uses its startling stare to secure the attention of a potential mate and possibly to warn away potential enemies.

Insects and crustaceans such as crabs have **compound eyes** that contain closely packed photosensitive

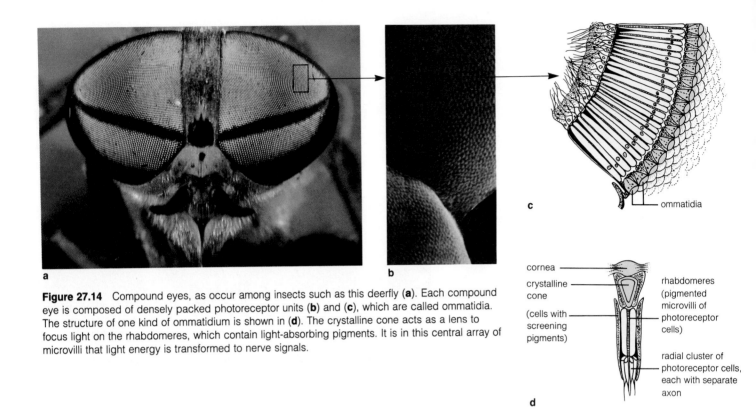

Figure 27.14 Compound eyes, as occur among insects such as this deerfly (**a**). Each compound eye is composed of densely packed photoreceptor units (**b**) and (**c**), which are called ommatidia. The structure of one kind of ommatidium is shown in (**d**). The crystalline cone acts as a lens to focus light on the rhabdomeres, which contain light-absorbing pigments. It is in this central array of microvilli that light energy is transformed to nerve signals.

cornea
crystalline cone
(cells with screening pigments)
rhabdomeres (pigmented microvilli of photoreceptor cells)
radial cluster of photoreceptor cells, each with separate axon

Figure 27.15 An approximation of light reception in the insect eye. This image of a butterfly was actually formed when a photograph was taken through the outer surface of a compound eye that had been detached from an insect. However, it may not be what the insect actually "sees." Integration of signals sent to the brain from photoreceptors in the eye may produce a more crisply defined image. The representation shown here is useful insofar as it suggests how the overall visual field may be *sampled* by separate ommatidia.

units, of the sort shown in Figure 27.14. Some compound eyes have many thousands of these units, which are called *ommatidia* (singular, ommatidium). We do not know how signals from the photoreceptors in each unit are processed in the brain to form visual images. According to the **mosaic theory** of image formation, each ommatidium samples only a small part of the overall visual field, which varies in light intensities. An image is built up according to signals about different light intensities, with each unit contributing a separate bit to a visual mosaic (Figure 27.15).

Vertebrate Photoreception

Almost all vertebrates have eyes capable of forming clear images. As Figure 27.16 and Table 27.2 indicate, the eyeball has a lens, a **sclera** (a tough outer coat), a **choroid** (a dark-pigmented tissue), and a retina densely packed with receptors. A transparent, light-focusing cornea is continuous with the sclera and covers the front of the eye. Choroid tissue extends inward from the front of the eye to form an iris. The iris is richly endowed with light-screening pigments, and it has radial and circular muscle fibers for controlling the amount of incoming light. A clear fluid (aqueous humor) fills the space between the

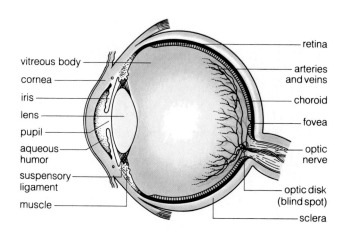

vitreous body
cornea
iris
lens
pupil
aqueous humor
suspensory ligament
muscle

retina
arteries and veins
choroid
fovea
optic nerve
optic disk (blind spot)
sclera

Figure 27.16 Main components of the human eye.

Table 27.2 Components of Eyes of Birds and Mammals

Eye Region	Description	Main Function
Outer layer:		
Sclera	White fibrous tissue	Protection
Cornea	Curved, transparent tissue	Helps focus light rays entering eyes
Middle layer:		
Choroid	1. Dark-pigmented tissue	Helps prevent light scattering
	2. Iris (extension of choroid tissue having adjustable ring of contractile and connective tissue)	Controls amount of incoming light
	3. Pupil (open center of iris)	Entrance for incoming light
Lens	Spherical or cone-shaped body of transparent protein fibers	Adjustable focusing of light rays onto photoreceptors
Aqueous humor	Clear, alkaline solution between cornea and lens	Light transmission, maintenance of pressure within interior of eye
Vitreous body	Jellylike substance filling chamber behind lens	Light transmission, support for lens and wall of eyeball
Inner layer:		
Retina	Tissue containing densely packed photoreceptors (rods and cones)	Light reception and transduction
Fovea	Funnel-shaped depression near center of retina where cones are concentrated	Area of greatest visual acuity
Beginning of optic nerve	Axons of retinal ganglion cells	Carries signals from photoreceptors to brain

cornea and iris. The lens is positioned behind the iris, and a jellylike substance (vitreous body) fills the chamber behind the lens.

Vertebrates can adjust the way in which light from distant or close-up objects converges on their photoreceptors. Light rays entering the curved cornea are bent toward some focal point. (You might want to review the description of light refraction in Figure 5.2.) If the angle of bending is not enough, the focal point will end up behind the retina. If it is too much, the focal point will end up in front. However, movements of the lens can adjust the path of the light rays. The process by which lens adjustments are made to bring about precise focusing of light onto the retina is called **accommodation**.

Fish use eye muscles that move the entire lens forward or back, thereby adjusting its distance from the retina. Increasing the distance moves the focal point forward; decreasing the distance moves it back. Other vertebrates rely on coordinated stretching and relaxation of eye muscles and fibers, which are attached to the lens (Figure 27.17).

In the vertebrate eye, lens adjustments assure that the focal point for light rays will be on the retina.

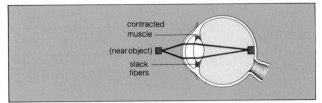

a Accommodation for nearby objects (lens bulges at equator)

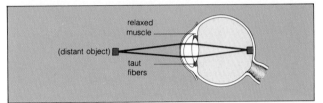

b Accommodation for distant objects (lens flattens out)

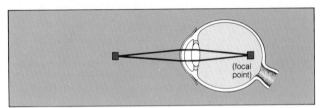

c Focal point in nearsighted vision

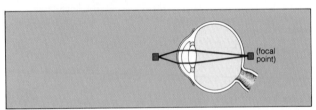

d Focal point in farsighted vision

Figure 27.17 Visual accommodation in the human eye.
(**a**) Close objects are brought into focus when eye muscles contract enough to slacken certain fibers interposed between them and the lens, and this causes the lens to thicken at its equator. (**b**) Distant objects are brought into focus when eye muscles relax, thereby putting tension on the fibers and stretching the lens into a flatter shape. (**c**) In the eyes of *nearsighted* people, the retina is too far behind the lens; light from distant objects is focused in front of the retina. (**d**) In the eyes of *farsighted* people, light from nearby objects is focused behind the retina.

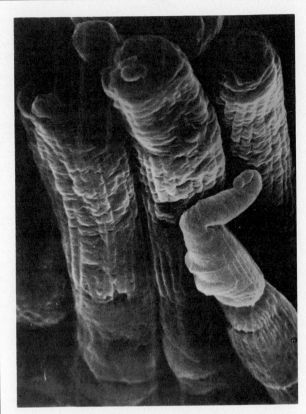

a Scanning electron micrograph of rods and a cone, x 7000

Figure 27.18 Vertebrate rods and cones, and the sensory pathway leading from the retina of the eye to the brain.

Rods and Cones

The retina is well developed in birds and mammals. Its basement layer is composed of pigmented epithelium and covers the choroid. Nerve tissue that contains photoreceptors rests on the basement layer. The pigments of this layer help prevent light scattering and thereby keep the photoreceptors from being stimulated in a diffuse way, which could lead to blurred images.

The photoreceptors are called **rod cells** and **cone cells** because of their shape (Figure 27.18). Rods are sensitive to very dim light; they contribute to coarse perception of changing light intensity caused by movements across the field of vision. Rods are typically abundant in the periphery of the retina.

Cones respond to high-intensity light; they contribute to sharp daytime vision and, usually, color perception. There are three known types of cone cells, which have pigments that are most sensitive to wavelengths corresponding to the colors red, green, and blue. The cones of human eyes are densely packed in the *fovea*, a funnel-shaped depression near the center of the retina,

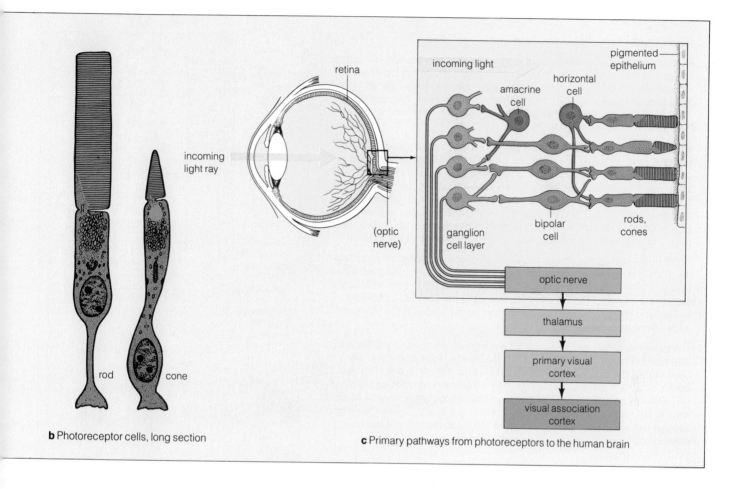

b Photoreceptor cells, long section

c Primary pathways from photoreceptors to the human brain

where nerve tissue is thinner. Although the fovea is only a millimeter across, the cones clustered here contribute the most to visual acuity (precise discrimination between adjacent points in space).

The top of each rod is a dense stack of flattened membrane sacs (Figure 27.18). Embedded in the membranes are **rhodopsin** molecules, each consisting of a protein (opsin) to which a side group (*cis*-retinal) is attached. When the side group absorbs light energy, it is temporarily converted to a slightly different form:

cis-retinal   becomes   *trans*-retinal

In this altered form, the side group breaks away from the protein. The breakdown of rhodopsin leads to a change in the voltage difference across the membrane in which the protein is embedded. In effect, the change

signals the presence of light to neighboring neurons, which then relay signals to ganglion cells (Figure 27.18c). The axons of these cells lead out from the nerve tissue in the eye and converge to form the optic nerve. From there, signals travel to the thalamus, then on to visual processing centers in the cortex.

Processing Visual Information

Visual information undergoes some processing in the retina before it is conveyed to the brain. Signals converge and diverge among neurons of the retina; they spread out and inhibit activity in adjacent cells. As a result, the frequencies at which signals are transmitted to the brain change in response to changes in the size, location, and intensity of light stimuli.

Our understanding of how visual information is processed in the vertebrate brain comes mainly from experiments with cats and monkeys. Figure 27.19 describes some of the pioneer work done in this area.

Figure 27.19 From signaling to visual perception.

Parts of the cerebral cortex concerned with vision contain neurons stacked in columns at right angles to the brain's surface. Connections run between neurons in each column and between different columns. Each column apparently analyzes only one kind of stimulus, received from only one location. The transformations from signaling to visual perception have been traced through several levels of synapses.

What do eight synaptic levels tell us, given the *billions* of connections in the brain? A great deal. The neurons in the brain's surface layer fall into a few categories. In each category, the neurons seem to be tripped into action in the same way. For instance, excitatory signals traveling up through the brain's surface layer or traveling parallel to its surface activate certain neurons, which then send out inhibitory signals to other neurons. The excitatory and inhibitory signals between neurons form narrow bands of electrical activity. In fact, *the pattern of excitation through specific columns of neurons is highly focused*.

David Hubel and Torsten Wiesel implanted electrodes in individual neurons in the brain of an anesthetized cat. After the cat woke up, they positioned it in front of a small screen, then projected images of different shapes (including a bar) onto the screen. The cat observed images of the bar tilted at different angles, and changes in electrical activity that corresponded to the different angles were recorded.

The strongest activity was recorded for one type of neuron when the bar image was positioned vertically (numbered 5 in the sketch). When the image was tilted slightly, the signals were less frequent. When the image was tilted past a certain angle, the signals stopped. In other experiments, another neuron fired only when an image of a block was moved from left to right across the screen; another fired when the image was moved from right to left. These experiments suggest that visual perception is based on the organization and synaptic connections between columns of neurons in the brain.

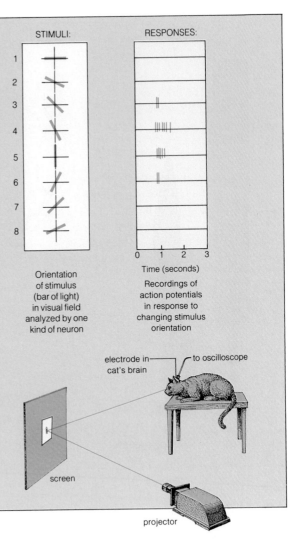

STIMULI:

RESPONSES:

Time (seconds)

Orientation of stimulus (bar of light) in visual field analyzed by one kind of neuron

Recordings of action potentials in response to changing stimulus orientation

electrode in cat's brain — to oscilloscope

screen

projector

SUMMARY

1. Many neurons are active almost continually; their pattern of activity depends on inherited neural programs.

2. A stimulus is any form of energy the body is able to detect by means of its receptors.

3. Receptors are fine peripheral endings of sensory neurons (or specialized cells adjacent to them) that respond to specific stimuli, such as chemical energy, mechanical energy, photon energy, or the radiant energy associated with temperature. Animals can respond to events in the outside world only if they have receptors sensitive to the energy of the particular stimulus.

4. Receptors are transducers which convert one form of energy into another. When they are stimulated, receptors undergo a graded electrical response known as a receptor potential.

5. A particular sensation is not triggered by action potentials themselves but by their travel along a particular nerve pathway. Sensory nerve pathways from different receptors lead to different parts of the cerebral cortex.

6. Variations in stimulus intensity are encoded in the frequency of action potentials in a single neuron and in the number of axons recruited into action in a tissue.

7. Taste receptors are in sensory organs called taste buds. Olfactory receptors respond to molecules released from food and predators and to pheromones released by members of the same species. Taste and olfaction are examples of chemical senses.

8. Somatic senses include the sensations of touch, pressure, temperature, and pain near the body surface.

9. The sense of balance in vertebrates depends on information from hair cell receptors in the vestibular apparatus, which consists of semicircular canals and an otolith organ.

10. Hearing, like the sense of balance, depends on the bending of hair cells by changes in fluid pressure. Echolocation in bats, dolphins, and whales depends on hearing echoes from self-generated, high-frequency sound waves.

11. In vision (photoreception), photopigment molecules are changed by light, the energy of which gives rise to receptor potentials in the rod and cone cells of the eyes. Processing of the visual information begins in the retina and continues in the visual cortex.

Review Questions

1. Label the component parts of the human eye:

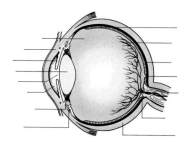

2. What is a stimulus? Receptor cells detect specific kinds of stimuli. When they do, what happens to the stimulus energy? *356*

3. Give some examples of chemoreceptors and mechanoreceptors. What kind of mechanoreceptor occurs repeatedly in sensory organs of different kinds of animals? *369, 373*

4. What is sound? How are amplitude and frequency related to sound? Give some examples of animals that apparently perceive sounds. *375*

5. What is pain? Can you name one of the tactile receptors associated with pain? *371–372*

6. How does vision differ from photoreception? What sensory apparatus does vision require? *376*

7. How does the vertebrate eye focus the light rays of an image? What is meant by nearsighted and farsighted? *379–380*

Readings

Eckert, R., and D. Randall. 1983. *Animal Physiology: Mechanisms and Adaptations*. Second edition. New York: Freeman.

Hubel, D. H., and T. N. Wiesel. September 1979. "Brain Mechanisms of Vision." *Scientific American* 241(3):150–162. Describes studies on information processing in the primary visual cortex.

Hudspeth, A. January 1983. "The Hair Cells of the Inner Ear." *Scientific American* 248(1):54–66.

Jacobs, G. 1983. "Colour Vision in Animals." *Endeavor.* 7(3):137–140.

Kandel, E., and J. Schwartz. 1985. *Principles of Neural Science*. Second edition. New York: Elsevier. Advanced reading, but good coverage of sensory perception.

Newman, E. A., and P. H. Hartline. March 1982. "The Infrared 'Vision' of Snakes." *Scientific American* 246(3):116–127.

Parker, D. November 1980. "The Vestibular Apparatus." *Scientific American* 243(5):118–130.

Stryer, L. July 1987. "The Molecules of Visual Excitation." *Scientific American* 257(1)42–50. Well-written description of the cascade reactions that give rise to nerve signals in the retina.

Wu, C. H. November-December 1984. "Electric Fish and the Discovery of Animal Electricity." *American Scientist* 72(6):598–607.

Young, J. 1978. *Programs of the Brain*. New York: Oxford University Press. An extraordinary book, beautifully written.

28

MOTOR SYSTEMS

The sensory systems described in the preceding chapter sample the surroundings with great precision and keep the nervous system informed of change. In turn, the nervous system sends out commands to a motor system so that the body moves in coordinated ways. In all animals, the operation of a motor system is based on units of contraction that can shorten under stimulation and then relax. As you will see, its operation requires the presence of some medium or structural element against which force can be applied.

We have already looked at the "motor systems" of single cells, many of which use cilia and flagella to move through a liquid medium (page 78). Here we will consider a few examples of the main types of motor systems of animals, as listed in Table 28.1.

INVERTEBRATE MOTOR SYSTEMS

One of the simplest motor systems is found in sea anemones, and it consists mainly of longitudinal and circular muscles in the body wall. In the longitudinal muscles, the muscle fibers are bundled together by connective tissue and run parallel with the body axis. (A *muscle fiber*, recall, is the same thing as a muscle cell.) When the longitudinal muscles contract, the body shortens. In the circular muscles, the muscle fibers are arranged in rings around the body axis, and when they contract, they cause the body to lengthen (Figure 28.1). Together, these muscles work as an **antagonistic system**, in which the action of one motor element opposes the action of the other.

Annelids such as earthworms have a different motor system. Their body cavity is divided into a series of seg-

Figure 28.1 Motor system by which the sea anemone extends its body upward through the water (a feeding behavior) and then relaxes (between feeding periods). Some muscle fibers in the body wall are parallel with the body axis. When they contract, the body shortens (**a**). Other muscle fibers are arranged in rings around the body axis. When they contract, the body is lengthened (**b**).

ments, each with a flexible wall surrounding a fluid-filled chamber. Each segment has a set of longitudinal and circular muscles. When the circular muscles contract, the force of contraction is applied against the fluid-filled interior. Because fluid resists compression, the fluid interior acts as a **hydrostatic skeleton**. (In all hydrostatic systems, body fluids are used to transmit force.) Thus, instead of being compressed, the fluid is squeezed along the axis of the body, much like toothpaste being squeezed down a tube. As a result, the longitudinal muscles stretch and sets of bristles (setae) that project from the body grip the ground, acting like toes for the stretched-out worm. When the longitudinal muscles contract, their force is applied against the bristles and the body is pulled forward.

There are some intriguing variations on the hydrostatic theme. For example, a rapid surge of blood under high pressure can extend the hind leg spines of spiders. Fluid under pressure in tubes is called hydraulic pressure. Figure 28.2 shows what a jumping spider looks like when it uses blood pressure as a hydraulic source for leaping at prey.

Insects, crabs, and other arthropods also have segmented bodies. Each segment has a *cuticle* (a hardened covering of chitin, protein, and sometimes lipid secretions). Besides covering the segments, the cuticle also covers antagonistic muscles that bridge the gap between segments. The cuticle remains pliable at these gaps and acts like a hinge when muscles move it in different directions. Thus the cuticle forms an external skeleton, or **exoskeleton**.

VERTEBRATE MOTOR SYSTEMS

Humans and other existing vertebrates have an internal skeleton, or **endoskeleton**, of bone and cartilage (or cartilage alone). Some fishes have a flexible skeleton composed of an elastic, translucent form of cartilage; it almost looks like glass. The shark skeleton is composed of an opaque form of cartilage that is hardened with calcium deposits (Figure 28.3). However, most vertebrate skeletons are constructed primarily of bone.

Bones have several functions. Bones are mineral reservoirs. Some are sites of red blood cell production. Tough tendons connect bones to skeletal muscles, and ligaments help hold bones together at joints. Interactions between the bones and skeletal muscles are the basis of mechanical movements and body positions, and they provide support and protection for other parts of the body.

Let's focus now on the structure and functioning of the human motor system, starting with its skeletal components.

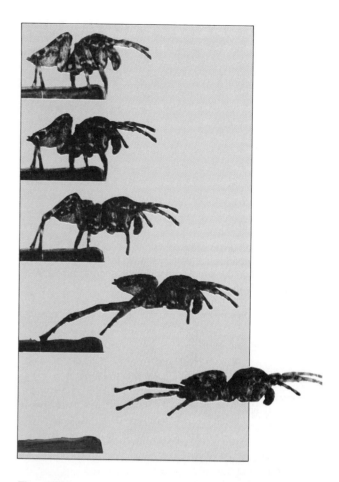

Figure 28.2 Hydraulic leap of the jumping spider (*Sitticus pubescens*), which can soar ten centimeters through the air to pounce on prey. The leap is based on the hydraulic extension of the hind legs when blood surges into them under high pressure.

Table 28.1 Main Categories of Motor Systems

Type	Representative Animals
Muscles alone, no skeleton	Sea anemone
Hydrostatic skeletons:	
1. Body fluid + soft body wall	Earthworm, octopus
2. Body fluid + rigid body part	Spider (hind legs only)
Rigid skeletons:	
1. Exoskeleton	Beetle, grasshopper, crab
2. Endoskeleton	Frog, snake, bird, human

Figure 28.3 Endoskeletons of a shark (left), a generalized early reptile (middle), and a generalized mammal (right).

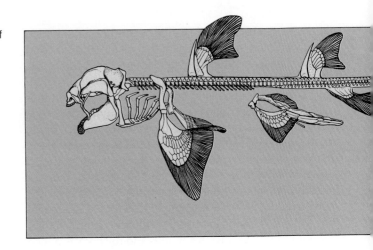

Figure 28.4 The human skeleton, with the axial portion shaded yellow and the appendicular portion shaded tan. Can you identify similar structures in the endoskeletons of the generalized reptile and mammal shown in Figure 28.3?

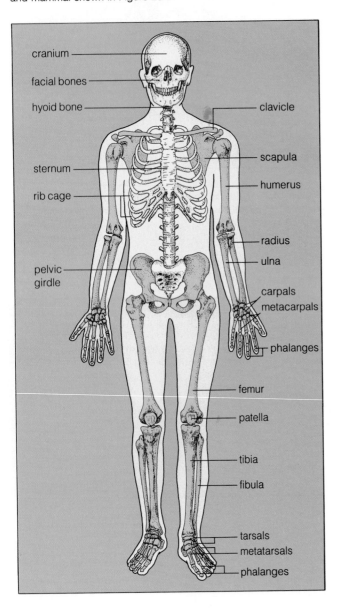

cranium
facial bones
hyoid bone
clavicle
scapula
sternum
humerus
rib cage
radius
ulna
pelvic girdle
carpals
metacarpals
phalanges
femur
patella
tibia
fibula
tarsals
metatarsals
phalanges

SKELETAL STRUCTURE AND FUNCTION

The human skeleton can be viewed as having two divisions, called the axial and appendicular skeletons (Figure 28.4). The *axial skeleton* includes the skull, a series of bones called vertebrae (which are arranged as a backbone), the ribs, and the sternum (breastbone). The *appendicular skeleton* includes the bones of the arms, hands, legs, feet, pelvic girdle (which is at the hips), and the pectoral girdle (at the shoulders).

Axial Skeleton. Of the skull's twenty-eight bones, eight are arranged as the cranium, which protects the brain, eyes, and ears. The rest are facial bones and middle-ear bones. The joint between the skull and the uppermost vertebra of the backbone allows the head to nod back and forth, as when you nod "yes." The joint between this vertebra and the next allows the head to rotate, as when you shake your head "no."

Figure 28.5 shows how vertebrae are arranged in the human backbone. There are twelve pairs of rib bones, and each bone forms a joint with one or two vertebrae and with the breastbone (or with cartilage that connects to it). Together, the ribs, breastbone, and backbone form a cage that protects two vital organs, the heart and lungs. Only one bone of the axial skeleton does not form a joint with any other. That is the hyoid bone; it supports the base of the tongue.

Appendicular Skeleton. The appendicular skeleton is joined to the axial portion at the shoulders and hips. Its pectoral girdle consists of two long, slender collarbones (clavicles) and two large, flat shoulder blades (scapulas). Each clavicle is joined to the breastbone and to a shoulder

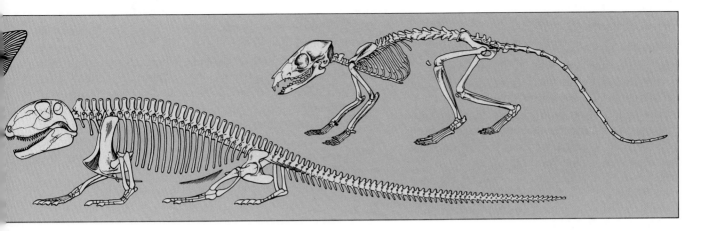

blade. This is not exactly a sturdy arrangement; if you fall on an outstretched arm, the force is transmitted to the body's trunk, and you might end up with a dislocated shoulder or a fractured clavicle. (The clavicle is the most frequently broken bone of the body.)

The scapula meets the upper arm bone (humerus) at the shoulder, and the humerus joins with forearm bones (ulna and radius) at the elbow. The ulna in turn joins with several carpals at the wrist (Figure 28.4). Bones called metacarpals support the palm of the hand; they connect with phalanges (bones of the fingers and thumb).

The pelvic girdle attaches to the backbone and provides strong connections for the two legs that carry the weight of the body. Each upper leg bone (femur) meets the larger of two lower leg bones at the knee joint, which is protected by the patella (knee cap). At the ankle joint, the lower leg bones fit over one of the tarsals at the back of the foot. Five metatarsals support the front of the foot, and phalanges support the toes.

Types of Bones

The bones we have been describing fall into four broad categories. *Long bones* are longer than they are wide, and the two ends usually are larger in diameter than the shaft. A chicken drumstick is like this; so are bones in your legs, arms, hands, and feet. *Short bones* are nearly the same in length and width; they occur at ankles and wrists. *Flat bones*, including those of the skull, usually consist of spongy bone sandwiched between two flat plates of compact bone. *Irregular bones*, including some facial bones and the vertebrae, have complex shapes and differ from one another in the amounts of compact and spongy bone.

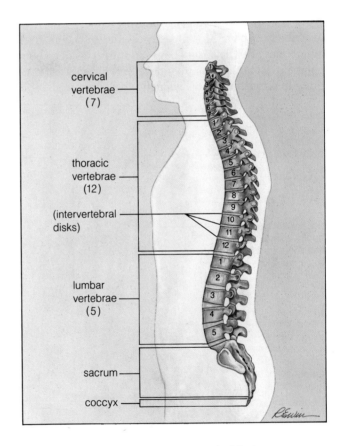

cervical vertebrae (7)

thoracic vertebrae (12)

(intervertebral disks)

lumbar vertebrae (5)

sacrum

coccyx

Figure 28.5 Vertebral column (backbone) of the human skeleton, side view. The cranium balances on the uppermost vertebra of the column.

Development of Bones

How do bones develop? Some, such as the cranial bones, form directly in connective tissue in the embryo. Bone-forming cells called **osteoblasts** secrete material that forms many splinterlike fragments of bone tissue. Over time, the fragments they produce fuse together and, after extensive remodeling, develop into a bone. Mature osteoblasts become trapped by their own secretions; then, they are called **osteocytes** (Figure 23.9).

Most bones form in cartilage, however. A cartilage model for a future long bone, for example, is assembled early in embryonic life. Osteoblasts start to form bone tissue inside the shaft of the cartilage; they receive nutrients and developmental signals by way of a blood vessel that has invaded the model (Figure 28.6). Calcium becomes deposited in the cartilage, the living cartilage cells die, and the intercellular material degenerates so that cavities form in the cartilage model. The cavities grow together to form the marrow cavity. (Marrow, described on page 315, fills the cavities in the shaft of long bones and the spaces between the plates of spongy bone almost as soon as they develop.) Osteoblasts continue to deposit bone tissue along the cavity walls.

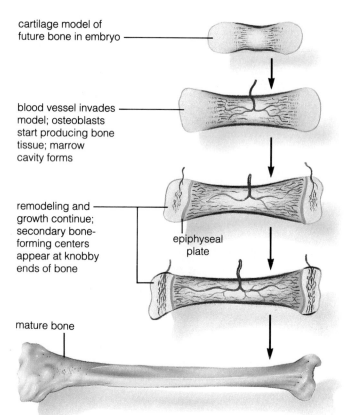

cartilage model of future bone in embryo

blood vessel invades model; osteoblasts start producing bone tissue; marrow cavity forms

remodeling and growth continue; secondary bone-forming centers appear at knobby ends of bone

epiphyseal plate

mature bone

Figure 28.6 Formation of a long bone in mammals.

These activities are repeated in the knobby ends of the long bone. Eventually, the only cartilage left is in the joints at both ends of the shaft, where it forms epiphyseal plates that extend across the width of the bone. A long bone increases in length as calcified cartilage is removed and bony tissue is deposited at the inside surfaces of the epiphyseal plates. It grows wider as new bone tissue is deposited under the dense connective tissue membrane surrounding the bone. When growth finally is over, the epiphyseal plates disappear.

Types of Joints

Soon we will be considering the muscles that interact with bones to bring about the body's mechanical movements and positional changes. Their interactions occur at **joints**, or points of contact (or near-contact) between bones or between cartilage and bones. There are three broad categories of joints: fibrous, cartilaginous, and synovial.

Fibrous joints, or sutures, have no gap between the bones, which hardly move at all. The flat cranial bones of a human fetus are united at sutures, but loosely so. The arrangement allows the bones to slide over each other slightly as the fetus makes its rather arduous trip through the birth canal, and so helps prevent skull fractures during childbirth. The newborn still has "soft spots" (fontanels), which are open areas between cranial bones. Later, during normal growth, bone replaces the fibrous connective tissue at the fontanels.

Cartilaginous joints are bridged by cartilage. Although there is no gap between the bones, some movement is possible. The intervertebral disks of the backbone are an example (Figure 28.5). The disks form strong joints that allow some movement along the backbone. They also help absorb vertical shocks, as when you fall out of a tree and land on your feet (but not a very tall tree).

Synovial joints, the most common type, are freely movable. A cavity separates bones, which nevertheless are held together by a flexible capsule of dense connective tissue and often by accessory ligaments. Cartilage covers the surface of the bone ends in the joint. A membrane that lines the inner wall of the capsule produces synovial fluid, which lubricates the joint. Unfortunately, freely movable joints sometimes move too freely and the structural organization of the joint is disrupted (see *Commentary*).

Freely movable joints are subject to wear and tear over time. In *osteoarthritis*, the cartilage at the bone end simply has worn away. *Rheumatoid arthritis* is a degenerative disorder that has a genetic basis. The synovial membrane thickens and becomes inflamed; consequently, cartilage here degenerates and bone is deposited in the joint.

On Runner's Knee

When you run, one foot and then the other is pounding hard against the ground. Each time a foot hits the ground, the knee joint above it must absorb the full force of your body weight. The knee joint allows us to do many things. It allows the leg bones beneath it to swing and, to some degree, to bend and twist. And the joint can absorb a force nearly seven times the body's weight—but there is no guarantee that it can do so repeatedly. Nearly 5 million of the 15 million joggers and runners in the United States alone suffer from "runner's knee," which refers generally to various disruptions of the bone, cartilage, muscle, tendons, and ligaments at the knee joint.

Like most joints, the knee joint permits considerable movement. The two long bones joined here (the femur and tibia) are actually separated by a cavity. They are held together by ligaments, tendons, and a few fibers that form a capsule around them. A membrane that lines the capsule produces a fluid that lubricates the joint, and where the bone ends meet, they are capped with a cushioning layer of cartilage.

Between the femur and tibia are wedges of cartilage that add stability and act like shock absorbers for the weight placed on the joint. Here also are thirteen fluid-filled sacs (bursae) that help cut down the friction between the parts of the joint, tendons, and skin that have to slide over each other.

When the knee joint is hit hard or twisted too much, its cartilage can be torn. Once cartilage is torn, the body often cannot repair the damage. Orthopedic surgeons usually recommend removing most or all of the torn tissue; otherwise it can cause arthritis (a general term for painful disorders that can render the joint practically immovable). Each year, more than 50,000 pieces of torn cartilage are surgically removed from the knees of football players alone. Football players, tennis players, basketball players, weekend joggers—

all are helping to support the burgeoning field of "sports medicine."

The seven ligaments that strap the femur and tibia together are also vulnerable to injury. A ligament is not meant to be stretched too far, and blows to the knee during collision sports (such as football) can tear it apart. A ligament is composed of many connective tissue fibers. If only some of the fibers are torn, it may heal itself. If the ligament is severed, however, it must be surgically repaired. (Edward Percy likens the surgery to sewing two hairbrushes together.) Severed ligaments must be repaired within ten days. The fluid that lubricates the knee joint happens to contain phagocytic cells that remove the debris resulting from day-to-day wear and tear in the joint. The cells will also go to work indiscriminately on torn ligaments and turn the tissue to mush.

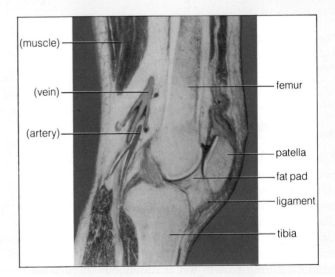

Longitudinal section through the knee joint.

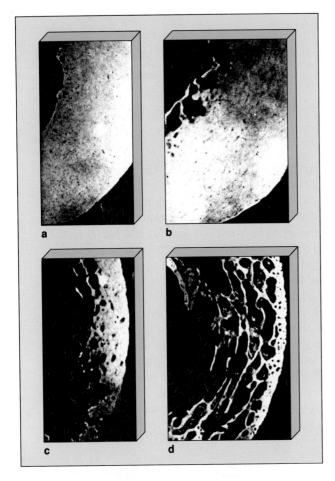

Figure 28.7 Osteoporosis, showing the progressive bone loss from normal density (**a**) to fragile structures at high risk of fracturing (**d**).

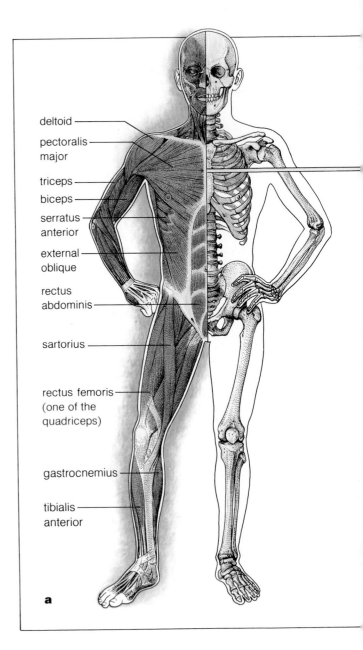

deltoid
pectoralis major
triceps
biceps
serratus anterior
external oblique
rectus abdominis
sartorius
rectus femoris (one of the quadriceps)
gastrocnemius
tibialis anterior

a

Bone Tissue Turnover

Bone is constantly being renewed by way of deposition and resorption activities that serve two functions: (1) remodeling and (2) maintaining calcium levels for the body as a whole. Bone cells called osteocytes and osteoclasts are both important in bone resorption. The **osteoclasts** resorb bone directly at the surfaces during the body's remodeling programs. Their activity is most pronounced during growth but is continuous throughout the individual's life. Osteocytes engage in resorption within the bone tissue; their activity is central to the hormonal control of calcium balance (page 361).

With increasing age, a problem can develop in bone turnover, particularly among older women. The bone mass decreases, especially in the vertebral column, legs, and feet. As a result, the vertebral column can collapse and curve to the extent that ribs drop down and come to rest on the rim of the pelvic girdle. The changing bone positions lead to complications in internal organs as well. This syndrome is called *osteoporosis* (Figure 28.7).

Although the causes of osteoporosis are not known, the suspected factors include decreased activity of the osteoblasts, dwindling levels of sex hormones, calcium deficiency, excessive protein intake, and decreased physical activity as the person grows older.

MUSCLE STRUCTURE AND FUNCTION

Chapter Twenty-Three described three types of muscle tissue (skeletal, smooth, and cardiac). In vertebrates, smooth muscle occurs mostly in the wall of internal organs. Smooth muscle helps propel or regulate the

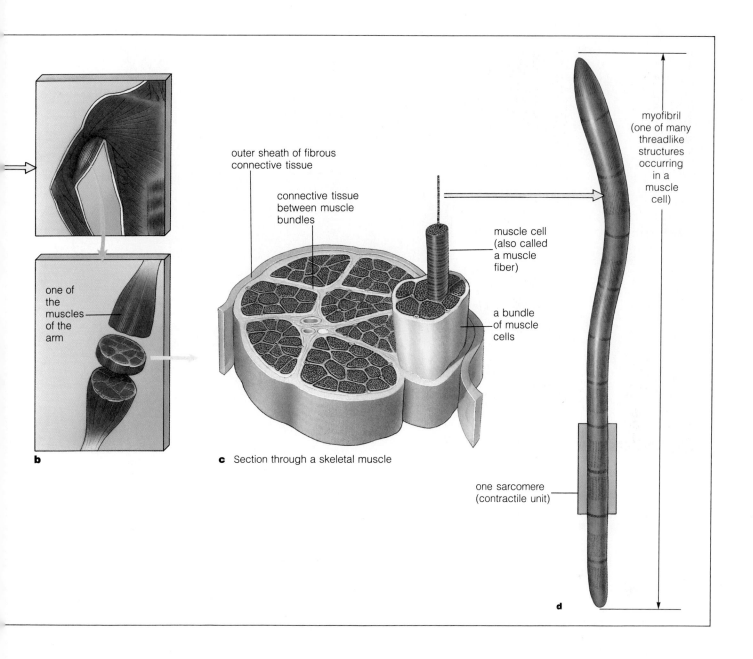

outer sheath of fibrous
connective tissue

connective tissue
between muscle
bundles

muscle cell
(also called
a muscle
fiber)

a bundle
of muscle
cells

myofibril
(one of many
threadlike
structures
occurring
in a
muscle
cell)

one of
the
muscles
of the
arm

b

c Section through a skeletal muscle

one sarcomere
(contractile unit)

d

movement of a variety of substances (see page 454, for example). Cardiac muscle occurs only in the walls of the heart, and its action is described in the next chapter. Here we will focus on skeletal muscle—the only type of muscle concerned with locomotion and positional changes of body parts.

Figure 28.8 shows the main skeletal muscles of the human body. Those muscles are composed of a few hundred to many thousands of individual muscle cells (which also are called muscle fibers). Each muscle cell contains many threadlike structures called **myofibrils** (Figure 28.8). Two types of protein filaments are packed together in each myofibril. One type, the *actin filament*, looks like two beaded chains twisted around each other. The other type, the *myosin filament*, consists of 200 to 400

Figure 28.8 (**a**) Some of the major skeletal muscles of the human skeletal-muscular system. (**b-c**) Closer views of the fine structure of an individual skeletal muscle. (**d**) A myofibril, one of the threadlike structures occurring in a muscle cell. Notice the arrangement of sarcomeres (contractile units) along the myofibril.

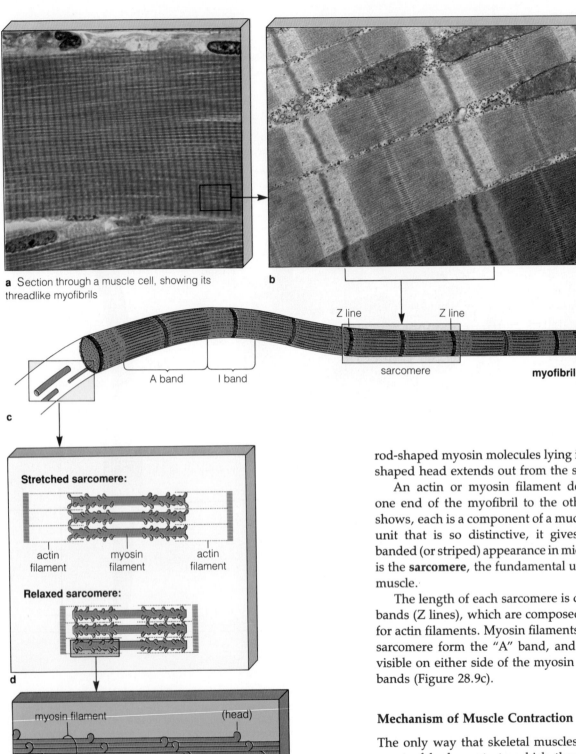

a Section through a muscle cell, showing its threadlike myofibrils

b

Z line   Z line

A band   I band

sarcomere   **myofibril**

c

Stretched sarcomere:

actin   myosin   actin
filament   filament   filament

Relaxed sarcomere:

d

myosin filament   (head)

actin filament

e

Figure 28.9 Fine structure of a muscle cell (**a, b**). Interactions between actin and myosin filaments in each sarcomere (**c-e**) are the basis of skeletal muscle contraction.

rod-shaped myosin molecules lying in parallel. A globe-shaped head extends out from the surface of each rod.

An actin or myosin filament doesn't extend from one end of the myofibril to the other. As Figure 28.9 shows, each is a component of a much smaller structural unit that is so distinctive, it gives skeletal muscle a banded (or striped) appearance in micrographs. This unit is the **sarcomere**, the fundamental unit of contraction in muscle.

The length of each sarcomere is defined by two dark bands (Z lines), which are composed of fibrous anchors for actin filaments. Myosin filaments at the center of the sarcomere form the "A" band, and the actin filaments visible on either side of the myosin region form the "I" bands (Figure 28.9c).

Mechanism of Muscle Contraction

The only way that skeletal muscles bring about movement of body parts to which they are attached is to shorten. When a skeletal muscle shortens, its component muscle cells are shortening. When a muscle cell shortens, its component sarcomeres are shortening. *The combined decreases in length of the individual sarcomeres account for contraction of the whole muscle.*

The question becomes this: How does a sarcomere alternately contract and relax? According to the **sliding-filament model**, actin filaments physically slide over myosin filaments. They move toward the center of the sarcomere during contraction and move away from it when the sarcomere is relaxing.

The yellow box in Figure 28.9c shows that each sarcomere has two sets of actin filaments and a set of myosin filaments between them. For the sliding movement to occur, heads on the myosin filaments must first attach to binding sites on the actin molecules. When attached, the myosin heads are *cross-bridges* between the two types of filaments (Figure 28.10). When cross-bridges are activated, the myosin heads tilt inward in a short power stroke, toward the center of the sarcomere. Because the actin filaments are attached to cross-bridges, they move slightly inward, also. The myosin heads then detach, reattach at another actin binding site in line, and move the actin filaments a little bit more. A single contraction takes a whole series of these power strokes in each sarcomere.

The energy needed to change the angle of attachment of cross-bridges during the sliding movements comes from ATP. When a muscle is relaxed, each myosin head has picked up an ATP molecule and has hydrolyzed it into ADP and phosphate. The energy associated with this hydrolysis is used to activate the myosin head, which is now like a loaded, pulled-back spring of a mousetrap. The energy brings about changes in the shape of the myosin head and thereby leads to the power stroke. The myosin head picks up a new ATP molecule, which causes the cross-bridge to detach, and the cycle begins again.

In the absence of ATP, the cross-bridges never do detach. The muscle becomes rigid, a condition known as *rigor*. Following death, ATP production stops along with other metabolic activities. Cross-bridges remain locked in place and all skeletal muscles in the entire body become rigid. This condition, called *rigor mortis*, lasts up to sixty hours after death.

Clearly, ATP is necessary for muscle contraction. As you might expect, many muscle cells are richly endowed with mitochondria, which produce ATP through aerobic respiration. Such cells can contract for extended periods, when oxygen and the substrates for ATP synthesis are supplied continuously by the circulation system.

ATP *can* be replaced rapidly by the glycolytic breakdown of glycogen stored in the cells. (Glycogen is the primary fuel supply for cells that consume ATP rapidly.) When oxygen concentrations are low in muscle tissue, as they are during strenuous exercise, glycolysis is critical for ATP synthesis. However, this alternate route cannot be followed for long, for muscle cells fatigue quickly when the glycogen is depleted. In a resting muscle, energy is stored in the form of creatine phosphate. This compound readily gives up phosphate to ADP and thereby helps replace ATP used until metabolic activity increases to match ATP consumption following muscle stimulation.

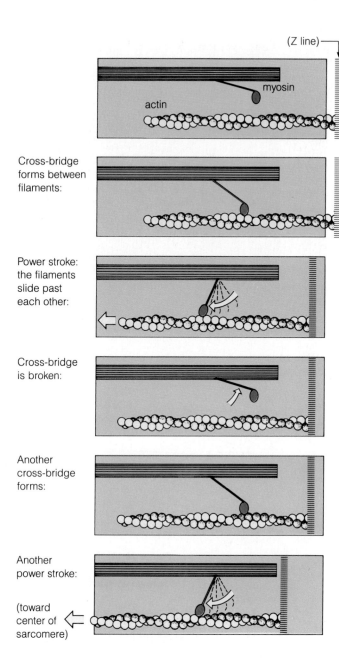

Figure 28.10 Simplified picture of the sliding-filament model, which explains the mechanism of contraction in the sarcomeres of muscle cells.

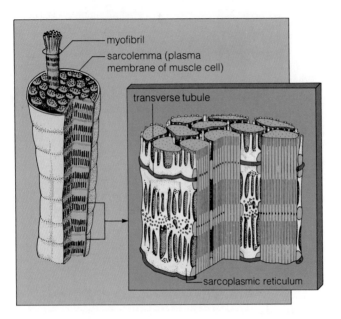

- myofibril
- sarcolemma (plasma membrane of muscle cell)
- transverse tubule
- sarcoplasmic reticulum

Figure 28.11 Location of sarcoplasmic reticulum, the calcium ion storage site within a muscle cell.

Control of Muscle Contraction

To understand how muscle contraction is controlled, we have to consider the connections between three types of membranes:

sarcolemma	*the plasma membrane that surrounds the entire muscle cell*
sarcoplasmic reticulum	*a continuous system of membrane-bound chambers that surround myofibrils within the muscle cell and that store calcium ions*
transverse tubule system	*a system of tubular membranes that extend from the sarcolemma all the way through the muscle cell and that are in intimate contact with the sarcoplasmic reticulum*

Figure 28.11 shows the arrangement of those membranes in a section from a muscle cell.

Signals that initiate contraction begin at neuromuscular junctions between the sarcolemma and motor neurons. A motor neuron releases acetylcholine (page 329). This transmitter substance interacts with receptors on the sarcolemma, leading to changes that produce an action potential. The action potential travels along the sarcolemma and invades the interior of the muscle by way of the transverse tubule system. This signal results in an increase in the permeability of the sarcoplasmic reticulum to calcium ions, which simply diffuse into the cytoplasm. *The calcium ions bind to control sites on the actin filaments and clear the way for the formation of cross-bridges with the myosin filaments.* Figure 28.12 describes the role of calcium ions in the formation of cross-bridges.

When calcium ions are actively taken up after contraction and stored in the sarcoplasmic reticulum, the muscle relaxes. When calcium ions are released from the sarcoplasmic reticulum, the muscle contracts. Now, the

Figure 28.12 Role of calcium in the formation of cross-bridges between actin and myosin.

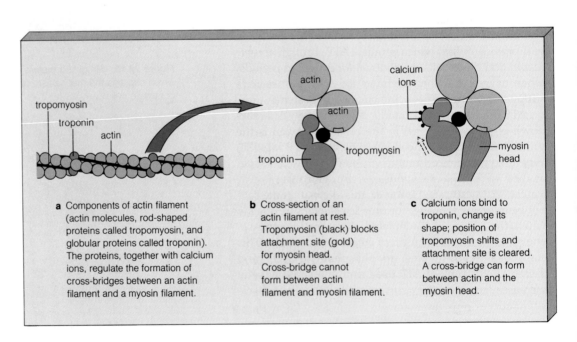

a Components of actin filament (actin molecules, rod-shaped proteins called tropomyosin, and globular proteins called troponin. The proteins, together with calcium ions, regulate the formation of cross-bridges between an actin filament and a myosin filament.

b Cross-section of an actin filament at rest. Tropomyosin (black) blocks attachment site (gold) for myosin head. Cross-bridge cannot form between actin filament and myosin filament.

c Calcium ions bind to troponin, change its shape; position of tropomyosin shifts and attachment site is cleared. A cross-bridge can form between actin and the myosin head.

nervous system dictates which motor neurons will carry action potentials, and at what frequency. By controlling the action potentials that reach the sarcoplasmic reticulum, the nervous system controls calcium ion levels in muscle tissue—and so exerts control over contraction.

SKELETAL-MUSCULAR INTERACTIONS

Until now, we have focused on the organization of the human skeleton and on the basis of contraction of its muscles. Let's put this information together to get a sense of how bones and muscles interact as a system for carrying out motor responses.

The skeleton, together with its muscles, is analogous to a system of levers in which rigid rods (bones) move about at fixed points (the joints). Most attachments are close to their joints. This means that the muscle has to contract only a small distance to produce a corresponding large movement of some body part.

A limb can be extended and rotated around a joint because of arrangements between pairs or groups of muscles. Muscles in limbs are arranged in antagonistic pairs, such as the biceps and triceps shown in Figure 28.13. Notice how those muscles bridge both the elbow and shoulder joints. When one member of this antagonistic pair (the biceps) contracts, the elbow joint flexes (bends). As it relaxes and its partner (the triceps) contracts, the limb extends and straightens. **Reciprocal innervation** in the spinal cord contributes to this coordination. Here, inhibitory signals sent to one set of motor neurons prevent one muscle of a pair from contracting while the other muscle is being stimulated.

Thus, when the biceps contracts, inhibitory neurons are acting at the same time on the motor neurons of its partner, the triceps, which relaxes. And when the triceps contracts, inhibitory signals are acting on the motor neurons of the biceps, which relaxes. (Reciprocal innervation can be overridden. For example, you can contract your biceps and triceps at the same time by holding your arm upright, like a stiff pillar.)

Also contributing to the coordination of skeletal muscle contraction are stretch receptors. Within each skeletal muscle, special muscle cells and stretch-sensitive receptors are enclosed in a sheath to form the muscle spindle (Figure 24.14). The central nervous system uses signals from muscle spindles to produce coordinated muscle contractions.

Before leaving the topic of skeletal muscle contraction, think about how the contraction of muscle cells is related to the contraction of the muscle organ in which they occur. Refer to Figure 24.12, which shows how the branched endings (axon terminals) of a motor neuron innervate several muscle cells. Signals from the brain

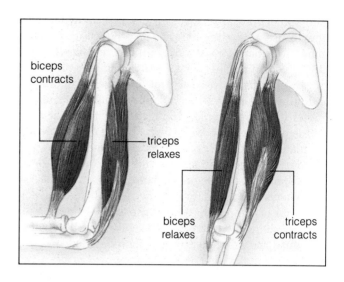

Figure 28.13 Antagonistic muscle pair, showing how two muscles can produce movement in opposite directions.

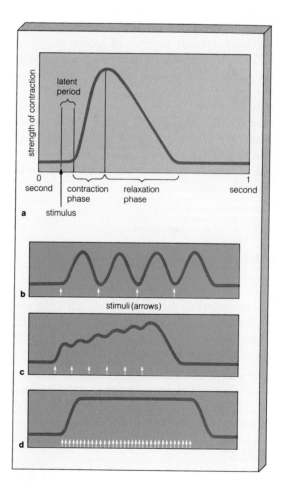

Figure 28.14 Recording of a muscle twitch (**a**). Recordings of a series of muscle twitches caused by about two stimulations per second (**b**); recordings of a summation of twitches resulting from about six stimulations per second (**c**); and a tetanic contraction resulting from about twenty stimulations per second (**d**).

and spinal cord are relayed to the axon terminals and they cause all the muscle cells to contract at the same time. Together, a motor neuron and the muscle cells under its control are called a **motor unit**.

Physiologists use electrical impulses (stimuli) to study motor units. When a single stimulus activates a motor unit, the resulting contraction of its muscle cells is called a **twitch** (Figure 28.14a). There is a brief interval (the latent period) between stimulation and the onset of contraction. The contraction builds up to a peak force, then diminishes during relaxation.

When a motor unit is stimulated again before a twitch response is completed, it contracts again. The strength of the contraction depends on how far the twitch response has proceeded by the time the second signal arrives. A motor unit stimulated repeatedly does not have time to relax; instead, it is maintained in a state of contraction called **tetanus** (Figure 28.14d).

Motor units in the body are either relaxed or contracted, nothing in between. However, a muscle organ contracts gradually, generating a strength of contraction according to need. How is this done? For a weak contraction, only a small number of motor units are activated. For a stronger one, a larger number are activated, and so on. This process is called "motor unit recruitment."

Except during sleep, some motor units are contracted even in a muscle that seems to be resting. What we call "muscle tone" refers to the sustained, partial contraction of a muscle that is important in maintaining posture. The degree of muscle tone is controlled by the brain (specifically, the cerebellum), acting through the muscle spindle.

SUMMARY

1. All motor systems require the presence of some medium or structural element against which force can be applied. This is true of systems based on antagonistic muscles alone; it is also true of systems based on interacting skeletal and muscle elements.

2. Vertebrates have an endoskeleton (internal skeleton) composed of bone and cartilage. In humans, the endoskeleton is divided into an axial portion (the skull, backbone, ribs, and breastbone) and an appendicular portion (bones of the arms, hands, legs, feet, pelvic girdle, and pectoral girdle).

3. In combination with skeletal muscles, the endoskeleton works like a system of levers in which rigid rods (bones) move about at fixed points (the joints). A limb can be extended and rotated around a joint because of the way pairs or groups of muscles are arranged relative to joints.

4. The skeletal muscle cells of vertebrates are composed of many threadlike myofibrils, each of which contains actin and myosin filaments that are partitioned into sarcomeres, the basic units of contraction in muscle. Actin and myosin filaments interact through a sliding mechanism that shortens the sarcomere.

5. All the sarcomeres contract when action potentials arrive from motor neurons and trigger the release of calcium ions from a membrane system (sarcoplasmic reticulum) around the myofibrils. Calcium binding alters the actin filaments so that the heads of adjacent myosin filaments can bind to them.

6. ATP energizes individual myosin heads, causing a short power stroke that makes the actin filament slide past the myosin, toward the center of the sarcomere. The deenergized myosin head picks up another ATP and detaches from the actin. The myosin head becomes energized and attaches to another actin binding site in line. Contraction requires a series of power strokes in each sarcomere.

7. ATP energy drives the power stroke underlying muscle contraction, muscle relaxation, and the accumulation of calcium in the sarcoplasmic reticulum. The nervous system controls the release of calcium and thereby exerts control over contraction itself.

Review Questions

1. Distinguish between the axial and appendicular portions of the human skeleton. How does this endoskeleton differ from an exoskeleton? *385–387*

2. Describe three types of joints. *388*

3. What are the functions of osteoblasts? *388*

4. Look at Figure 28.8. Then, on your own, sketch and label the fine structure of a muscle, down to one of its individual myofibrils. Can you identify the basic unit of contraction in a myofibril? *391–392*

5. How do actin and myosin interact in a sarcomere to bring about muscle contraction? What role does ATP play? *392–393*

Readings

Alexander, R. M. July-August 1984. "Walking and Running." *American Scientist* 72(4):348–354. The biomechanics of traveling on foot.

Eckert, R., and D. Randall. 1983. *Animal Physiology: Mechanisms and Adaptations.* Second edition. New York: Freeman.

Hoyle, G. 1983. *Muscles and Their Neural Control.* New York: Wiley.

Huxley, H. E. December 1965. "The Mechanism of Muscular Contraction." *Scientific American* 213(6):18–27. Old article, great illustrations.

Luttgens, K., and K. Wells. 1982. *Kinesiology: Scientific Basis of Human Motion.* Seventh edition. Philadelphia: Saunders.

CIRCULATION SYSTEMS: AN OVERVIEW

A cell survives by exchanging substances with its surroundings, and most of those substances simply diffuse inward and outward across the plasma membrane. (Diffusion, recall, is the random movement of like molecules down their concentration gradients.) The exchanges are not that complicated for the cells of sponges, jellyfishes, and flatworms. These animals do not have massive bodies, and substances simply diffuse through the tissue fluid around individual cells. However, in most invertebrates and all vertebrates, interior cells are too far from the body surface to exchange substances efficiently with the external environment. These animals have a **circulation system**, which consists of the following components:

blood | *a fluid connective tissue composed of water, diverse solutes, and formed elements (for example, blood cells and platelets)*

heart (or heartlike structure) | *a muscular pump that generates the pressure needed to keep blood flowing throughout the body*

blood vessels | *tubes of varying diameter through which blood is transported*

Most animals have a **closed circulation system**, in which the walls of the heart and the blood vessels are continuously connected (Figure 29.1). As you will see, the vertebrate system includes large-diameter blood vessels that rapidly move blood to and from tissues, and an enormous number of small-diameter blood vessels (capillaries) that function as sites of rather leisurely diffusion.

Not all animals have a closed system of fluid transport. Arthropods (including insects and spiders) and most mollusks (snails, clams, and their kin) have an **open circulation system**. Fluid is pumped from the heart into a set of tubes and then is dumped into a space or cavity in the body tissues. There, it mingles with intercellular fluids before moving into open-ended tubes that lead back to the heart (Figure 29.2). In snails and clams, fluid is pumped into a network of spaces in spongy tissue that provides a large surface area for diffusion. In insects, fluid is pumped into a cavity that cannot expand much under the fluid pressure (because of the rigid exoskeleton of the insect body). The fluid has nowhere to go except back to the heart, which pumps it out again.

Even closed circulation systems are not completely sealed off. A slight amount of fluid is always filtering out of the capillaries, and materials are continually passing between the capillaries and the surrounding tissues. A supplementary network of tubes, the **lymph vascular system**, reclaims the excess fluid and returns it to the circulation.

29
CIRCULATION

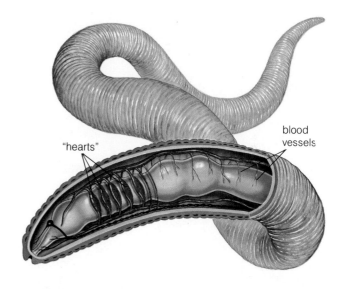

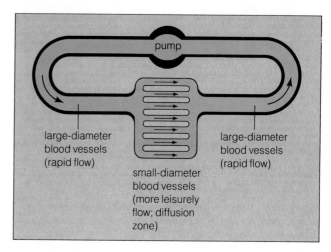

Figure 29.1 Fluid flow through a closed circulation system. The upper sketch shows components of the closed system of the earthworm, which has several muscular "hearts" near its anterior end and several blood vessels running lengthwise through its series of body segments. The earthworm also has pairs of transverse blood vessels in each body segment.

Table 29.1 Components of Blood

	Function	Number per Microliter	Volume Percent
Cellular Portion (40%–50% of total volume)			
1. Red blood cells	Oxygen, carbon dioxide transport	4,500,000–5,500,000	
2. White blood cells:			
Neutrophils	Phagocytosis	3,000–6,750	
Lymphocytes	Central to immune response	1,000–2,700	
Monocytes	Phagocytosis	150–720	
Eosinophils	Phagocytosis	100–360	
Basophils	Source of substances that increase capillary permeability and show anticlotting activity	25–90	
3. Platelets	Source of substances that aid in blood clotting	250,000–300,000	
Plasma Portion (50%–60% of total volume)			
1. Water	Serves as solvent		91–92
2. Plasma proteins	Play diverse roles (infection fighting, blood clotting, lipid transport, etc.)		7–8
3. Other solutes (ions, sugars, lipids, amino acids, hormones, vitamins, dissolved gases)	Play diverse roles (maintaining extracellular pH, fluid volume, etc.)		1–2

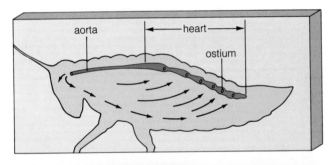

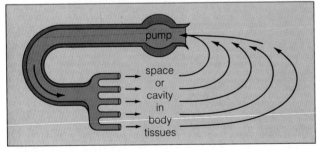

Figure 29.2 Fluid flow through an open circulation system. The sketch above shows the open system of the grasshopper. Like other insects, the grasshopper has a "heart" in the posterior portion of its body that pumps blood through a vessel (aorta) which dumps the blood into tissues at the anterior end of the body. After diffusing through body spaces, blood moves back into the heart through lateral openings (ostia).

CHARACTERISTICS OF BLOOD

Functions of Blood

The simplest animals with a true circulation system are ribbon worms (Figure 41.1). Their blood serves mainly to carry wastes away from cells; it also contains phagocytic cells that engulf foreign particles. In many invertebrates and all vertebrates, blood not only transports products and wastes from cells, but it also transports nutrients and oxygen to them. It contains phagocytic cells that function as scavengers and infection fighters, and it serves as the transport route for hormones. By virtue of its composition, blood helps stabilize internal pH. In birds and mammals, blood helps equalize body temperature by carrying excess heat from regions of high metabolic activity (such as skeletal muscles) to the skin, where it can be dissipated from the body.

Blood is a transport fluid that carries raw materials to cells, carries products and wastes from them, and helps maintain an internal environment that is favorable for cell activities.

Blood Volume and Composition

On the average, an adult human male who weighs 70 kilograms (about 150 pounds) has a blood volume of about 5 liters, or a little more than 5 quarts. The volume varies, depending on the size of the body and also on

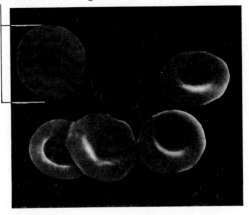

8 micrometer average diameter

b

Figure 29.3 (a) Photomicrograph of red blood cells in capillaries. (b) Scanning electron micrograph showing the biconcave shape of red blood cells.

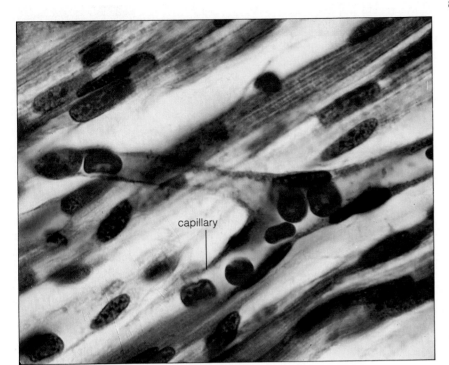

capillary

a

changes in the concentration of water and solutes in the internal environment.

In all vertebrates, blood consists of red blood cells, white blood cells, platelets, and plasma (Table 29.1). Normally, the plasma constitutes about fifty to sixty percent of the total blood volume in an adult human.

Plasma. The portion of blood called **plasma** is mostly water, which functions as a solvent. But this straw-colored liquid also contains hundreds of different *plasma proteins*, including albumin, the globulins, and fibrinogen. The concentrations of plasma proteins influence the distribution of water between blood and interstitial fluid. Albumin is important in this respect, for it represents sixty percent of the total amount of plasma proteins. Some alpha and beta globulins transport lipids and fat-soluble vitamins. Gamma globulins function in immune responses, and fibrinogen serves in blood clotting.

Plasma also contains diverse ions, simple sugars such as glucose, amino acids, vitamins, hormones, and dissolved gases (mostly oxygen, carbon dioxide, and nitrogen). The ions help maintain extracellular pH and fluid volume. The lipids present in plasma include fats, phospholipids, and cholesterol. Lipids that are transported from the liver to different body regions are generally bound with proteins to form lipoproteins.

Red Blood Cells. The task of transporting oxygen to cells is mainly the responsibility of erythrocytes, or **red blood cells**. A mammalian red blood cell is a biconcave

disk, thicker around the rim than in the center (Figure 29.3). Its red color comes from hemoglobin, an iron-containing protein molecule present in its cytoplasm (page 58). When oxygen from the environment first diffuses into the bloodstream, it quickly binds with the iron and forms *oxyhemoglobin*. Blood rich in oxyhemoglobin is bright red. Blood somewhat depleted of oxygen is darker and appears blue when observed through blood vessel walls (hence the "blue" veins that are visible at your wrists).

In addition to oxygen, hemoglobin also transports some of the carbon dioxide wastes of aerobic metabolism. Most of the carbon dioxide is simply dissolved or combined with water to form bicarbonate (HCO_3^-) in the bloodstream, but some binds with hemoglobin to form *carbaminohemoglobin*.

Red blood cells originate in bone marrow (page 315). As each cell matures, its nucleus disappears, but it already has enough enzymes and other proteins to remain functional for its expected life span of about 120 days. The oldest red blood cells are continually removed from the bloodstream, mainly by phagocytic cells in the liver and spleen.

Feedback mechanisms keep the red blood cell count fairly stable. (A *cell count* is the number of cells of a given type in a microliter of blood.) When oxygen levels in tissues are low, the kidneys secrete an enzyme that converts a plasma protein into a key hormone (erythropoietin). The hormone stimulates an increase in red blood cell production in red bone marrow. New oxygen-carrying cells enter the bloodstream, and within a few

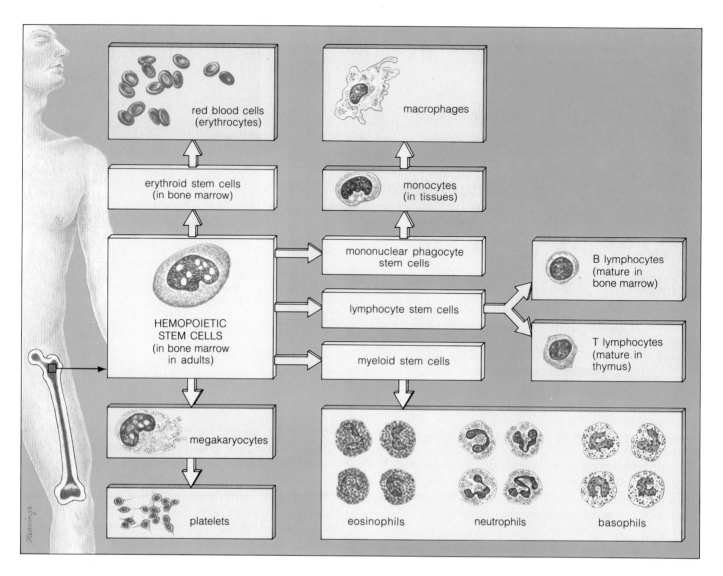

Figure 29.4 Cellular components of vertebrate blood. Stem cells are in yellow boxes; all white blood cells are in beige boxes. There are two classes of white blood cells (leukocytes). The "granular" cells (eosinophils, neutrophils, and basophils) have large granules in the cytoplasm. The "agranular" cells (monocytes, which mature into macrophages, and the lymphocytes) do not.

days there is a rise in oxygen levels in the tissues. Information about the increase is fed back to the kidneys, production of the key hormone dwindles, and the production of red blood cells drops accordingly. With this mechanism, the red blood cell count is maintained, on the average, at 5.4 million in healthy adult males and 4.8 million in healthy females.

White Blood Cells. The day-to-day housekeeping and defense activities that help keep tissues functioning are the responsibility of white blood cells, or **leukocytes**. Some of these cells are scavengers of dead or worn-out cells; others respond to tissue damage and invasion by

bacteria, viruses, and other foreign agents. All are derived from immature cells, called *stem cells*, in bone marrow. Blood is like a reservoir of white blood cells, for most of the housekeeping and defense functions of those cells are expressed after they leave the blood capillaries and enter tissues.

Five types of white blood cells can be distinguished on the basis of size, nuclear shape, and staining traits (Figure 29.4). The functions of these cell types, which are called lymphocytes, neutrophils, monocytes, eosinophils, and basophils, are summarized in Table 29.1. There are two main subcategories of lymphocytes (B cells and T cells). Both are central to immune responses, which are described in the next chapter.

The number of white blood cells in each microliter of human blood varies, depending on whether the body is highly active, in a state of health, or under siege. For example, during bacterial infections, the white blood cell count increases above the levels shown in Table 29.1; during viral infections, it can drop below 5,000.

Platelets. In bone marrow, bits of cytoplasm are pinched off "giant" cells (megakaryocytes). These cell fragments are called **platelets**. As you will see, substances released from platelets aid in blood clotting; hence platelets function in preventing blood loss from damaged blood vessels.

CARDIOVASCULAR SYSTEM OF VERTEBRATES

In all vertebrates, blood pumped out of a muscular heart enters large and then medium-size vessels called arteries. From there it travels into small, muscular arterioles, which branch into tiny vessels called capillaries. Blood travels from capillaries into small vessels called venules. Finally it flows into larger vessels, the veins, which return it to the heart. The entire system is called a **cardiovascular system** (from the Greek *kardia*, meaning heart; and the Latin *vasculum*, meaning vessel). In humans, it is arranged as shown in Figure 29.5.

Before we consider the components of this system, let's step back and think about its overall "design." The blood flows continuously through the system, and the volume flow through each part—arteries, capillaries, or veins—is the same. Yet if blood were to flow as rapidly through a capillary as it does through an artery, there would not be enough time for adequate diffusion of substances to and from cells. As it happens, blood spreads out into **capillary beds**, each of which contains vast numbers of capillaries that run in parallel. The total cross-sectional area of a capillary bed is much larger than that of the transport tubes leading into it. As a result, the rate of flow through each capillary is decreased, the time available for exchange of materials is increased—and the volume flow for the whole system remains constant:

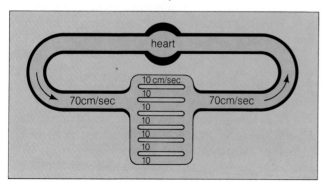

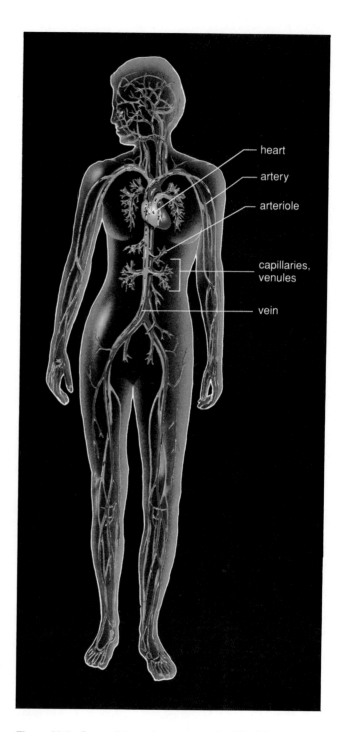

Figure 29.5 Some of the major components of the human cardiovascular system.

heart

artery

arteriole

capillaries, venules

vein

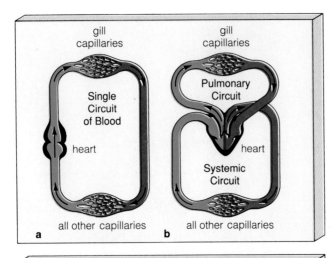

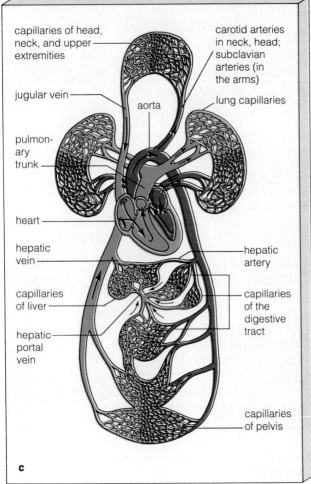

Figure 29.6 Comparison of blood circulation pattern of fishes (**a**) and mammals (**b**). The circulatory routes for the human body are shown in (**c**).

Blood Circulation Routes

Not all circulation routes are alike. In fishes, for example, blood passes through at least two capillary systems during its circuit away from and back to the heart (Figure 29.6a). Blood first encounters the capillary beds of the fish gills, where it picks up oxygen from the surrounding water, then it encounters capillary beds in the rest of the body (page 437). There are two sharp drops in blood pressure along the route, corresponding to the two capillary systems.

In contrast, a given volume of blood being circulated away from and back to the mammalian heart passes through only *one* major capillary bed, in most cases, so there is less of a pressure drop. This efficient circulation pattern is possible because the heart is divided into two pumps. In the **pulmonary circuit**, blood from the right half of the heart is pumped to the lungs, where it picks up oxygen and gives up carbon dioxide; then it flows to the left half of the heart. In the **systemic circuit**, the oxygenated blood is pumped through the rest of the body (where oxygen is used and carbon dioxide is produced), then it flows to the right half of the heart (Figure 29.6b).

There are only a few exceptions to the overall pattern of blood flow. For example, blood leaving the capillary beds of the digestive tract is rich in glucose and other substances that have been absorbed from food. As Figure 29.6c shows, that blood enters the hepatic portal vein, which leads into a capillary bed in the liver—an organ with a key role in nutrition. The decreased flow rate through this capillary bed gives the liver more time to monitor the blood concentrations of absorbed substances.

The Human Heart

Heart Structure. During a seventy-year life span, the human heart beats some $2\frac{1}{2}$ billion times, and it rests only briefly between heartbeats. Its structure (Figure 29.7) reflects its role as a durable pump. The bulk of the human heart, the *myocardium*, is cardiac muscle tissue. It is protected by a tough cover of fibrous connective tissue, the *pericardium*. Connective tissue and endothelium form the *endocardium*, the heart's smooth inner lining. ("Endothelium" is a single layer of epithelial cells lining the heart cavities and the space, or lumen, inside blood vessels.)

The heart's left and right halves are two distinct pumps, each with two chambers. In each half, blood flows first into a thin-walled **atrium** (plural, atria), then into a thick-walled **ventricle**. Between the two chambers in each half are membrane flaps called an **atrioventricular valve** (or AV valve). Another flap, the **semilunar**

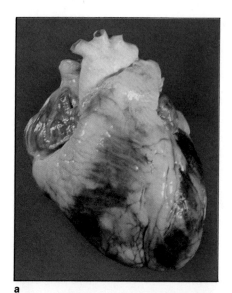
a

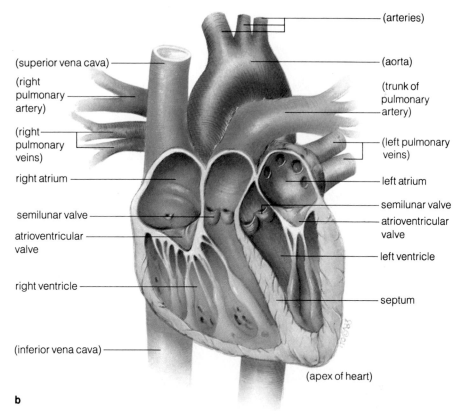

(arteries)

(superior vena cava)

(aorta)

(right pulmonary artery)

(trunk of pulmonary artery)

(right pulmonary veins)

(left pulmonary veins)

right atrium

left atrium

semilunar valve

semilunar valve

atrioventricular valve

atrioventricular valve

left ventricle

right ventricle

septum

(inferior vena cava)

(apex of heart)

b

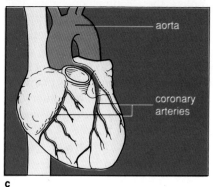

aorta

coronary arteries

c

Figure 29.7 The human heart, external view (**a**) and partial view of the interior (**b**). Location of the coronary arteries (**c**).

valve, spans the exit from each ventricle. The valves open and close in response to fluid pressure changes that are produced when the heart beats. Their passive movements help keep the blood moving in one direction and so prevent backflow.

Heart muscle cells are not serviced by blood in the heart cavity itself. Instead, the heart has its own "coronary circulation" (Figure 29.7c). Oxygen is delivered to heart muscle cells by two coronary arteries that lead into an extensive capillary bed. These arteries are the first to branch off the **aorta**, the major artery that carries oxygenated blood away from the heart. The branches of the coronary arteries are small in diameter, and they can become blocked during cardiovascular disorders, as described in the *Commentary* on page 408.

Cardiac Cycle. Each heartbeat is a sequence of muscle contraction and relaxation called the **cardiac cycle**. In each cycle, the four heart chambers go through phases of contraction (called **systole**) and relaxation (**diastole**). While the atria are relaxed and filling, the ventricles are also relaxed (Figure 29.8). As pressure rises in the atria,

the AV valves are forced open, and the ventricles fill completely when the atria contract. Then the ventricles contract, causing the AV valves to snap shut, and the pressure inside the ventricles rises sharply above the pressure in the vessels leading out from them. One of those vessels is the aorta; the other is the pulmonary artery. With the increased pressure, the semilunar valves open and blood flows out of the heart. After blood has been ejected, the ventricles relax and the cycle starts over.

The blood and heart movements during the cardiac cycle generate vibrations that produce a "lub-dup" sound. The sound can be heard at the chest wall. At each "lub," the AV valves are closing as the ventricles contract. At each "dup," the semilunar valves are closing as the ventricles relax.

What are the key points to remember about the contraction and relaxation phases of the cardiac cycle? First, the atria contract and then begin to relax before blood leaves the ventricles. Second, their contraction simply helps to fill the ventricles; it cannot be the driving force for circulation. In short, *it is the contraction of the ventricles that forces blood through the circulation system.*

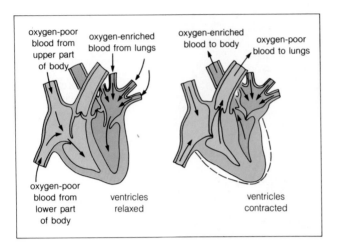

oxygen-poor blood from upper part of body

oxygen-enriched blood from lungs

oxygen-enriched blood to body

oxygen-poor blood to lungs

oxygen-poor blood from lower part of body

ventricles relaxed

ventricles contracted

Figure 29.8 Blood flow through the heart during relaxation (diastole) and contraction (systole) of muscle tissue in the walls of the ventricles.

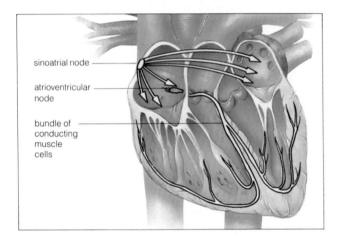

sinoatrial node

atrioventricular node

bundle of conducting muscle cells

Figure 29.9 Location of conducting muscle cells that make up the cardiac conduction system.

Cardiac Conduction System. As you know, skeletal muscle contracts in response to signals from the nervous system, but cardiac muscle is different. Although the nervous system does control the rate and strength of the heartbeat, the heart will keep on beating even if all autonomic nerves leading to the heart are severed! The heart makes itself contract because some of its muscle cells are self-excitatory: they initiate and conduct action potentials. The cells are the basis of the **cardiac conduction system** (Figure 29.9).

In all mammals, the excitation begins in the **sinoatrial node** (or SA node), a region of conducting cells where major veins enter the right atrium of the heart. Here, membrane properties change repeatedly and action

potentials are produced, with no stimulation from the outside. One wave of excitation follows another, seventy or eighty times a minute in the human heart. The rhythmic excitation begins soon after heart muscle cells appear in a developing embryo, and from then on it triggers all cardiac muscle contractions.

Although all cells of the cardiac conduction system are self-excitatory, the SA node fires at the highest frequency and comes to threshold first in each cardiac cycle. Thus the SA node is the *cardiac pacemaker*: its rhythmic firing is the basis for the normal rate of heartbeat.

Each wave of excitation from the pacemaker spreads over both atria, causing them to contract almost at the same time. The wave also passes to the **atrioventricular node** (or AV node), which consists of conducting cells in the floor of the right atrium. Signals are conducted more slowly through this node. The delay gives the atria enough time to complete their contraction before the ventricles start to contract.

A bundle of conducting muscle cells (sometimes called the bundle of His) leads away from the AV node and branches extensively through the inner walls of both ventricles. The bundle is the *only* conduction pathway between the atria and ventricles, which are otherwise separated from each other by nonconducting tissue. In response to signals carried by this bundle of cells, all parts of the ventricle walls contract more or less in unison.

The sinoatrial node is the cardiac pacemaker. Its spontaneous, repetitive excitation spreads along a system of conducting muscle cells that stimulate contractile tissue in the atria, then the ventricles, in a rhythmic cycle.

Heart Muscle Contraction. With each heartbeat, the individual muscle cells of the heart contract in unison, as if they were a single unit. Yet the units of contraction in cardiac muscle are the same as in skeletal muscle. (Individual sarcomeres are organized one after another along the length of each cell, as shown in Figure 28.8.) However, unlike skeletal muscle cells (whose endings are attached to bones), cardiac muscle cells branch and then abut with one another at the branched endings. Here, the plasma membranes are joined together to produce strong cohesion between cells. Each end-to-end region of membrane is an **intercalated disk** (Figure 29.10).

Intercalated disks contain communication junctions, where signals travel rapidly between cells (page 312). Excitatory signals cause ions to flow directly from one cardiac muscle cell to another, changing the membrane properties as they go. The excitation spreads so rapidly that the muscle tissue contracts as a unit.

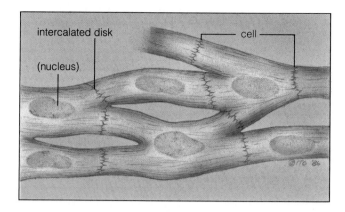

Figure 29.10 Intercalated disks between abutting heart muscle cells. Each cell has its own nucleus, but plasma membranes of adjacent cells are joined together at junctions between them. The membrane-to-membrane junctions are known as intercalated disks.

Blood Pressure in the Vascular System

The fluid pressure that is generated by heart contractions forces blood through the entire circulation system; it is called **blood pressure**. However, the blood pressure level is not the same along the circuits away from and back to the heart. Normally, pressures are high to begin with and they drop along the way (Figure 29.11). As you will see, the pressure drops result from the loss of energy that is used to overcome resistance to flow as blood moves through the circulation.

Table 29.2 summarizes the differences in blood pressure and volume along the systemic route. Let's now consider some of the reasons for these differences.

Arterial Blood Pressure. The heart ejects blood into **arteries**, the transport tubes that conduct oxygen-poor blood to the lungs and oxygenated blood to the rest of the body (Figure 29.12). The thick, impermeable wall of an artery contains smooth muscle and elastic connective tissue that can be distended under surges of fluid pressure. The wall also recoils elastically, and the recoil forces blood onward. Because of these wall properties, the arteries serve as pressure reservoirs that smooth out the changes in blood pressure associated with the cardiac cycle. Pressure is stored here when blood is ejected into the circulation by the contracting ventricles.

Ordinarily, blood pressure is measured at large arteries of the systemic route, such as arteries in your upper arms. First a systolic reading is taken of the highest pressure during a cardiac cycle (generated by the contracting ventricles). In young adults at rest, systolic pressure is about 120 mm Hg. (This means that the measured amount of pressure would make a column of mercury, or Hg, rise a distance of 120 millimeters.) Next, a diastolic reading is taken of the lowest arterial pressure at the end

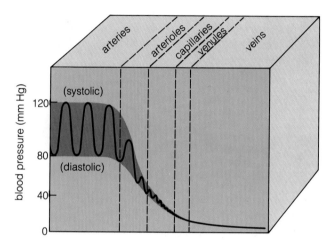

Figure 29.11 Drops in blood pressure in the systemic circulation.

Table 29.2	Blood Volume Distribution Through the Human Cardiovascular System	
	Average Blood Volume (percent of total)	Mean Blood Pressure (mm Hg)
Heart chambers:		
Atria	2	0 (right); 6 (left)
Ventricles	10	0–24 (right); 4–120 (left)
Aorta and other arteries:	10	100
Arterioles:	1	60
Capillaries:	5	30
Veins:	54	10
Pulmonary circulation:	18	15
	100 percent	

Data from Robert C. Little, MD.

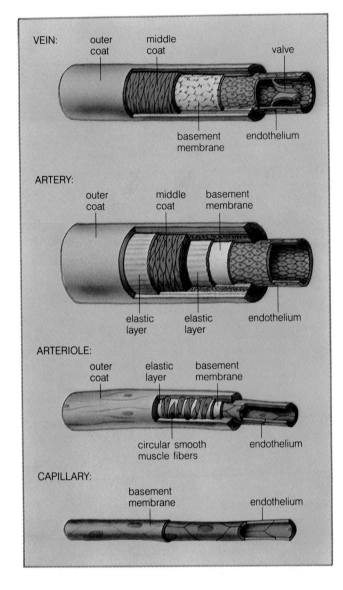

VEIN:
outer coat
middle coat
valve
basement membrane
endothelium

ARTERY:
outer coat
middle coat
basement membrane
elastic layer
elastic layer
endothelium

ARTERIOLE:
outer coat
elastic layer
basement membrane
circular smooth muscle fibers
endothelium

CAPILLARY:
basement membrane
endothelium

Figure 29.12 Structure of blood vessels. The outer coat consists of loose connective tissue. The middle coat contains elastic fibers and smooth muscle. The basement membrane consists of connective tissue elements, including collagenous fibers.

of a cardiac cycle, just before blood is pumped out of the heart again. Diastolic pressure is generally about 80mm Hg. Thus the difference between systolic and diastolic readings (the so-called *pulse pressure*) is 120 − 80, or 40mm Hg (Figure 29.13).

Resistance at Arterioles. Arteries branch into **arterioles**, which are transport tubes of smaller diameter that have rings of smooth muscle cells in their walls (Figure 29.12). Arterioles are the point at which control is exerted over the relative distribution of blood to different parts of the body.

Arterioles can increase or decrease in diameter and thereby offer variations in resistance to flow. As you will see, arteriole diameter is adjusted by neural and endocrine controls that govern blood pressure for the body as a whole. But the diameter also can be adjusted in response to changes in local chemical conditions of a tissue. For example, when metabolic activity increases in skeletal muscle tissue, the concentration of oxygen decreases, and the concentrations of carbon dioxide, hydrogen ions, potassium ions, and other substances increase. Such chemical changes serve as local signals that act on the smooth muscle cells in the arteriole wall. The cells respond by relaxing, the arteriole diameter enlarges, and more blood flows past, delivering more raw materials and carrying away cell products and wastes.

Any enlargement in the diameter of a blood vessel is called **vasodilation**. Any decrease is called **vasoconstriction**. For example, reversal of the local conditions described above could trigger such a decrease. Controlled vasodilation and vasoconstriction of arterioles direct blood to regions of greatest metabolic activity. The more active the cells of a given region, the greater the blood flow to them.

Capillary Function. A **capillary** is a tube about a millimeter long that is specialized for exchanging substances with interstitial fluid. The tube wall is a single layer of flat endothelial cells, separated from each other only by narrow clefts. The capillary diameter is so small that red blood cells squeeze through it single file (Figure 29.3). The small diameter resists flow, but the vast numbers of capillaries afford such a huge cross-sectional area that the overall resistance of a given capillary bed is less than that of the arteries leading into it.

Capillaries thread through nearly every tissue in the body, coming within 0.01 centimeter of every living cell and thereby enhancing diffusion. The density of capillaries in a tissue is directly related to metabolic output. For example, a great deal of blood must be circulated through muscle tissue to provide individual cells with oxygen for aerobic respiration. (Only the aerobic pathway can produce enough ATP energy to drive extensive contraction.)

Most of the solutes exchanged with interstitial fluid, including oxygen and carbon dioxide, simply diffuse across the capillary wall. But some proteins are also exchanged here, probably by endocytosis and exocytosis (page 94). Moreover, certain ions and small, water-soluble molecules probably pass through the clefts between cells. The clefts are wider in some capillary beds than in others, and those beds are more "leaky" to solutes.

Some fluid also moves by bulk flow across capillary walls. The movement plays no significant role in the exchange of solutes. Rather, bulk flow helps maintain the balance of extracellular fluid between the bloodstream and the surrounding tissues. This fluid distribution is important, because blood pressure is maintained only when there is an adequate blood volume. Interstitial fluid is a reservoir that is tapped when blood volume drops to the point where there is a decrease in blood pressure (as during hemorrhage and other events).

As Figure 29.14 indicates, the fluid movements are determined by two opposing forces. One force is the difference between capillary blood pressure and interstitial fluid pressure. Because of the difference, some plasma (but very few plasma proteins) leaves the capillary. This outward fluid movement is called **filtration**.

The other force is the difference in water concentration between plasma and interstitial fluid. (Plasma has

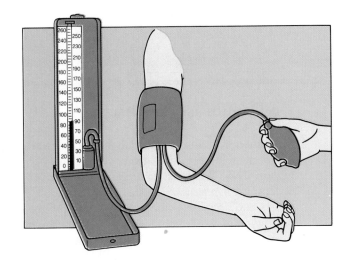

Figure 29.13 A sphygmomanometer, which is used to measure blood pressure. A hollow cuff, attached to a pressure gauge, is wrapped around the upper arm and inflated with air to a pressure above the systolic blood pressure. When that pressure is reached, no sounds can be heard through a stethoscope positioned above the artery, just below the cuff. Now the air in the cuff is slowly released, and soft, tapping sounds can be heard intermittently through the stethoscope. The sounds correspond to the turbulence and vibration caused by blood flowing into the slightly opened artery. The value on the pressure gauge at this time corresponds to the systolic pressure.

The pressure in the cuff is lowered further, and when the cuff pressure reaches the diastolic blood pressure, the sounds become dull and muffled. Just below this point, the blood flow is continuous and the turbulence stops—and so do the tapping sounds. The disappearance of sounds corresponds to the diastolic blood pressure.

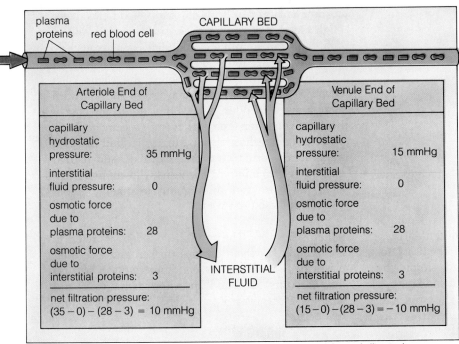

plasma proteins	red blood cell	CAPILLARY BED		

Arteriole End of Capillary Bed		Venule End of Capillary Bed	
capillary hydrostatic pressure:	35 mmHg	capillary hydrostatic pressure:	15 mmHg
interstitial fluid pressure:	0	interstitial fluid pressure:	0
osmotic force due to plasma proteins:	28	osmotic force due to plasma proteins:	28
osmotic force due to interstitial proteins:	3	osmotic force due to interstitial proteins:	3
net filtration pressure: $(35 - 0) - (28 - 3) = 10$ mmHg		net filtration pressure: $(15 - 0) - (28 - 3) = -10$ mmHg	

INTERSTITIAL FLUID

Outward-directed filtration favored

Inward-directed absorption favored

Figure 29.14 Example of fluid movements in an idealized capillary bed. Such movements play no significant role in diffusion, but they are important in maintaining the distribution of extracellular fluid between the bloodstream and interstitial fluid.

On Cardiovascular Disorders

More than 40 million Americans have cardiovascular disorders which, in any given year, claim about a million lives. The most common cardiovascular disorders are *hypertension* (sustained high blood pressure) and *coronary artery disease* (a progressive narrowing of the arterial lumen). They are the major causes of most *heart attacks*—that is, the damage or death of heart muscle due to an interruption of its blood supply. (They also can cause a *stroke*, this being damage to the brain due to an interruption of blood circulation to it.)

Most of the time, a heart attack manifests itself as a "crushing" type of pain behind the breastbone that lasts a half hour or more. Frequently, the pain radiates into the left arm, shoulder, or neck. The pain can be mild but usually is excruciating. Often it is accompanied by sweating, nausea, vomiting, and dizziness or loss of consciousness.

Risk Factors in Cardiovascular Disorders. Cardiovascular disorders are the leading cause of death in the United States. Curiously, many factors associated with those disorders have been identified *and are controllable*—yet millions ignore the information. These are the known risk factors:

1. High level of cholesterol in the blood.

2. High blood pressure.

3. Obesity (page 462).

4. Lack of regular exercise.

5. Smoking (page 447).

6. Diabetes mellitus (page 362).

7. Genetic predisposition to heart failure.

8. Age (the older you get, the greater the risk).

9. Gender (until age fifty, males are at much greater risk than are females).

The last four factors obviously cannot be avoided; but the first five can be. All five have the effect of increasing the risk of cardiovascular disorders. The risk associated with them can be minimized simply by watching your diet, exercising, and not smoking.

For example, the fatter you become, the more your body develops additional blood capillaries to service the increased number of cells, and the harder the heart has to work to pump blood through the increasingly divided vascular circuit. As another example, the nicotine in tobacco stimulates the adrenal glands to secrete epinephrine, which is a powerful vasoconstrictor that triggers an accelerated heartbeat and a rise in blood pressure. The carbon monoxide present in cigarette smoke has a greater affinity for binding sites on hemoglobin than does carbon dioxide—and its action means that the heart has to pump harder to rid the body of carbon dioxide wastes. In short, smoking not only can destroy your lungs; it also can destroy your heart.

The following descriptions will provide a better understanding of the tissue destruction that can result from cardiovascular disorders.

Hypertension. Hypertension arises through a gradual increase in resistance to blood flow through the small arteries; eventually, blood pressure is sustained at elevated levels even when the person is at rest. Heredity may be a factor here (the disorder tends to run in families). Diet also is a factor; for example, high salt intake can raise the blood pressure in persons predisposed to the disorder. High blood pressure makes the heart work harder, and in time it can become enlarged and fail to pump blood effectively. High blood pressure also can cause arterial walls to "harden" and so influence the delivery of oxygen to the brain, heart, and other vital organs.

Hypertension has been called the silent killer because affected persons may show no outward symptoms; they often believe they are in the best of health. Even when their high blood pressure has been detected, some hypertensive persons tend to resist medication, corrective changes in diet, and regular exercise. Of 23 million Americans who are hypertensive, most are not undergoing treatment. About 180,000 will die each year.

Atherosclerosis. The term "arteriosclerosis" refers to a condition in which arteries thicken and lose their elasticity. In *atherosclerosis*, conditions worsen because lipid deposits also build up in the arterial walls and shrink the diameter of the arterial lumen. How does this occur?

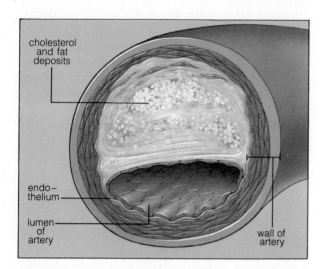

cholesterol
and fat
deposits

endo–
thelium

lumen
of
artery

wall of
artery

a Diagram of an atherosclerotic plaque

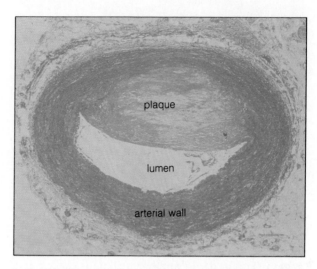

plaque

lumen

arterial wall

b Partially obstructed lumen of artery, cross-section.

plaque (yellow)

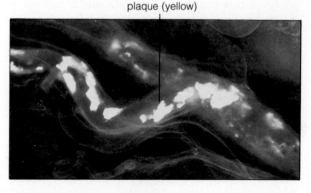

c Plaques in the coronary arteries of a heart patient.

Recall that lipids such as fats and cholesterol are insoluble in water (page 54). Lipids absorbed from the digestive tract are picked up by lymph vessels that empty into the bloodstream. There, the lipids become bound to protein carriers that keep them suspended in the blood plasma. In atherosclerosis, abnormal smooth muscle cells have multiplied and connective tissue components have increased in arterial walls. Lipids have been deposited within cells and extracellular spaces of the wall's endothelial lining. Calcium salts have been deposited on top of the lipids, and a fibrous net has formed over the whole mass. This *atherosclerotic plaque* sticks out into the lumen of the artery (Figures a–c).

Sometimes platelets become caught on the rough edges of plaques and are stimulated into secreting some of their chemicals. When they do, they initiate clot formation. As the clot and plaque grow, the artery can become narrowed or blocked. Blood flow to the tissue that the artery supplies now diminishes or may be blocked entirely. A clot that stays in place is called a *thrombus*; if it becomes dislodged and travels the bloodstream, it is called an *embolus*.

With their narrow diameter, the coronary arteries and their branches (Figure 29.7b) are extremely susceptible to clogging through plaque formation or occlusion by a clot. When such an artery becomes narrowed to one-quarter of its former diameter, the resulting symptoms can range from mild chest pains (angina pectoris) to a full-scale heart attack.

Atherosclerosis can be diagnosed on the basis of several procedures, including stress EKGs (recording the electrical activity of the cardiac cycle while a person is exercising on a treadmill) and *angiography* (injecting a dye that will stain plaques into the blood and then taking x-rays of the arteries; see Figure c). Treatments of serious blockages include *coronary bypass surgery*. During this operation, a section of an artery from the chest is sutured to the aorta and to the coronary artery below the narrowed or blocked

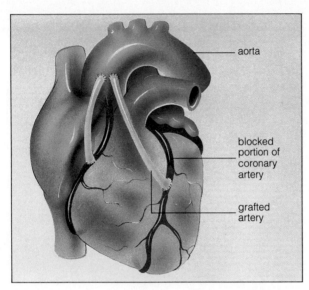

d Two coronary bypasses (green).

region (Figure d). In another technique, called *laser angioplasty*, highly focused laser beams are used to vaporize the atherosclerotic plaques. *Balloon angioplasty* is more common; here, a small balloon is inflated within a blocked artery to break up plaques. All such procedures do not cure the underlying cardiovascular problem; they only buy time for the individual.

Plaque formation is related to cholesterol intake, but other factors are also at work here. For example, when cholesterol is transported through the bloodstream, it is bound to one of two kinds of protein carrier molecules: high-density lipoproteins (HDL) and low-density lipoproteins (LDL). High levels of LDL are related to a tendency toward heart trouble. It appears that LDLs, with their cholesterol cargo, have a penchant for infiltrating arterial walls. In contrast, the HDLs seem to attract cholesterol out of the walls and transport it to the liver, where it can be metabolized. (Atherosclerosis is uncommon in rats; rats have mostly HDLs. It is common in humans, who generally have mostly LDLs.) In addition, it appears that unsaturated fats can reduce the level of LDLs in the blood. Unsaturated fats have one or more double bonds in their fatty acid tails (page 54); they include olive oil and fish oil. The latter contains the so-called omega-3 fatty acids that seem especially beneficial (at least when ingested with other parts of the fish; the effect of omega-3 capsules is still being debated).

Arrhythmia. Arrhythmias are irregular or abnormal heart rhythms. They can be detected by an *electrocardiogram* (EKG), which is a recording of the electrical events of the cardiac cycle (Figure e). Some arrhythmias are normal. For example, the resting cardiac rate of many athletes who are trained for endurance is lower than average, a condition called *bradycardia*. Parasympathetic inhibition of their cardiac pacemaker has increased as an adaptive response to ongoing strenuous exercise. A cardiac rate above 100 beats per minute *(tachycardia)* also occurs normally during exercise or stressful situations.

Serious tachycardia can be triggered by drugs (including caffeine, nicotine, and alcohol), hyperthyroidism, and other factors. Coronary artery disease or syphilis also can cause arrhythmias by blocking the atrioventricular node. A complete block can lead to a very slow ventricular rate, which may cause dizziness, unconsciousness, or convulsions.

A coronary occlusion and certain other disorders also may cause abnormal rhythms that can degenerate rapidly into a dangerous condition called *ventricular fibrillation*. Here, cardiac muscle in different parts of the ventricles contracts haphazardly, at asynchronous rates that pump little or no blood and therefore signify impending death. Sometimes a strong electric shock delivered to the chest can stop the fibrillation and may restore normal cardiac function.

a greater solute concentration, with its protein components, and therefore a lower water concentration.) Because of the difference, some interstitial fluid moves into the capillary. This inward fluid movement is called **absorption**.

As Figure 29.14 suggests, fluid filtration at the arteriole end of a capillary bed tends to be balanced by absorption at the venule end. Normally, there is only a small net filtration of fluid, which is returned to the blood by lymph vessels. Let's turn now to events in the body's venules and veins.

Venous Pressure. Capillaries merge into "little veins," or **venules**. Some diffusion also occurs across the venule wall, which is only a little thicker than that of a capillary. Also, contraction of smooth muscle in the wall can cause

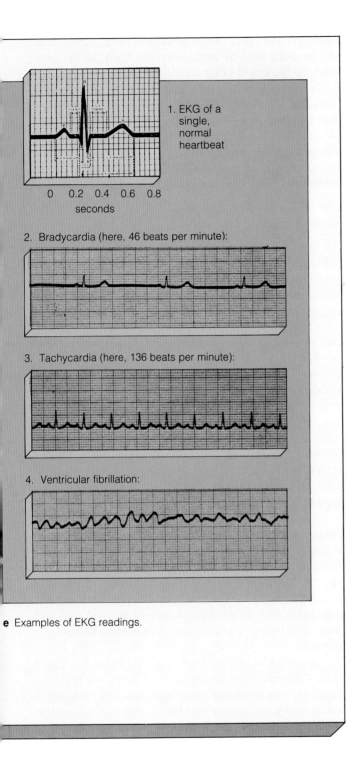

1. EKG of a single, normal heartbeat

0   0.2   0.4   0.6   0.8
seconds

2. Bradycardia (here, 46 beats per minute):

3. Tachycardia (here, 136 beats per minute):

4. Ventricular fibrillation:

e Examples of EKG readings.

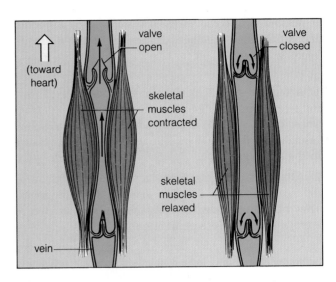

Figure 29.15 Role of skeletal muscle contractions and venous valves in returning blood to the heart.

fluid pressure to increase in the capillaries and thereby affect filtration.

Venules merge into veins, the large-diameter tubes for blood flow back to the heart (Figure 29.12). By the time blood has been circulated into **veins**, fluid pressure has dropped to only 10–15mm Hg. However, fluid pressure in the right atrium of the heart is close to 0mm Hg, and the large diameter of veins presents very low resis-

tance to flow. Thus, even though the pressure gradient between the veins and the heart is rather small, it is enough to allow blood to complete the circuit. Also, many veins have tissue folds into their lumen that serve as valves. When fluid in a vein starts moving backward because of gravity, it pushes the folds into the lumen and so prevents backflow (Figure 29.15).

Veins are more than low-resistance transport tubes; they also are blood volume reservoirs. At any given time, veins contain fifty to sixty percent of the total blood volume for the body. This large reservoir can be tapped to increase cardiac output. How? The vein wall is thinner and can distend much more than an arterial wall. When circulation must be stepped up to meet the demands of increased body activities, the nervous system sends signals to smooth muscle in the vein walls. When the muscle cells contract, the walls become stiffer and cannot distend as much, so pressure in the veins rises. The increased pressure drives more blood from the veins to the heart. With the increase in the cardiac filling pressure, the ventricles contract more forcefully.

The pressure in veins also is increased when limb muscles move (Figure 29.15). When contracting skeletal muscles "bulge" against an adjacent vein, the diameter of the vein is effectively decreased. Internal pressure rises and forces the blood forward. In addition, breathing plays an indirect role in returning blood to the heart. When you inhale, your rib cage expands. Pressure in the chest decreases and the diaphragm pushes down on organs in the abdominal cavity, increasing the pressure there (page 442). The pressure changes help increase the pressure gradient between the heart and the veins that are returning blood from the lower half of the body.

In summary, these are the key points to keep in mind about the differences between the tubes of the cardiovascular system:

1. Arteries are pressure reservoirs that keep blood flowing away from the heart while the ventricles are relaxing. These blood vessels have large diameters that offer low resistance to flow, so there is little drop in blood pressure in the arteries.

2. Arterioles are control points for the distribution of different volumes of blood to different diffusion zones. They offer a great deal of resistance to flow, so there is a major drop in pressure in arterioles.

3. Capillary beds are diffusion zones for exchanges between blood and interstitial fluid. Collectively, they have a greater cross-sectional area than that of the arterioles leading into the beds, so they present less total resistance to flow. There is some drop in pressure here.

4. Venules overlap with capillaries in their function; they afford some control over capillary pressure.

5. Veins are highly distensible blood volume reservoirs and are important in adjusting flow volume back to the heart. They offer only low resistance to flow, and blood pressure is very low in veins.

Regulation of Blood Flow

After a large meal, more blood is diverted to your digestive system, which swings into full gear as other systems more or less idle. When your body is exposed to cold wind or snow for an extended time, blood is diverted away from the skin to deeper tissue regions, so that the metabolically generated heat that warmed the blood in the first place can be conserved.

The rate of flow to different body regions is inversely proportional to resistance. When resistance is decreased, the flow steps up. The question becomes this: How is blood pressure raised and lowered through the cardiovascular system as a whole?

The medulla oblongata (a brain region) contains integrating centers that control blood pressure by coordinating the rate and strength of the heartbeat with changes in the diameter of arterioles (and, to some extent, of veins). This command post integrates information coming in from receptors in certain arteries (such as the carotid arteries in the neck and the aorta) and in cardiac muscle tissue. When blood pressure rises, the receptors signal the medulla, which responds by sending out signals that lead to decreased cardiac output and vasodilation of the arterioles. The combined effect is that

blood pressure falls. When it falls too much, signals from the medulla lead to vasoconstriction and, at the same time, to increased heart activity.

Hormones also play a role in the control of blood pressure. The arterioles in different body regions have different receptors that can be activated by such hormones as epinephrine and angiotensin. Depending on which receptors are present, the epinephrine can cause vasoconstriction or vasodilation. Angiotensin causes widespread vasoconstriction.

Blood Typing

All cells in each individual have certain proteins and other molecules that act like identification flags, or markers; they identify the cell as being of a specific type. Each individual also has proteins called **antibodies**, which can recognize markers on *foreign* cells (page 420). When the blood of two people mixes during transfusions, the antibodies will act against any cells bearing the "wrong" marker. They will do the same thing during pregnancy. (Although the circulation systems of the mother and her unborn child remain separate throughout pregnancy, antibodies diffuse from one system to the other.)

ABO Blood Typing. As we have seen, *ABO blood typing* is based on some of the surface markers on red blood cells (page 170). Type A blood has A markers on those cells, type B blood has B markers, type AB has both, and type O has neither one.

If you are type A, you do not carry antibodies against A markers—but you have antibodies against B markers. If you are type B, you have antibodies against A but not B markers. If you are type AB, you have no antibodies against A or B, and so your body will tolerate donations of type A, B, or AB blood. If you are type O, however, you have antibodies against A and B markers—and those antibodies will act against cells bearing one or both types.

Figure 29.16 shows what happens when the blood from different types of donors and recipients is mixed together. In a response called **agglutination**, antibodies act against the "foreign" red blood cells and cause them to clump together. Such clumps can clog small blood vessels; they may lead to kidney damage and death. In looking at Figure 29.16, can you determine what the responses will be to type AB blood? To type O blood?

Rh Blood Typing. Red blood cells also have other surface markers that can cause agglutination responses. For example, the *Rh blood typing* is based on the presence or absence of an Rh marker (so named because it was first identified in the blood of the *Rhesus* monkey). Rh$^+$ individuals have blood cells with this marker; Rh$^-$ individ-

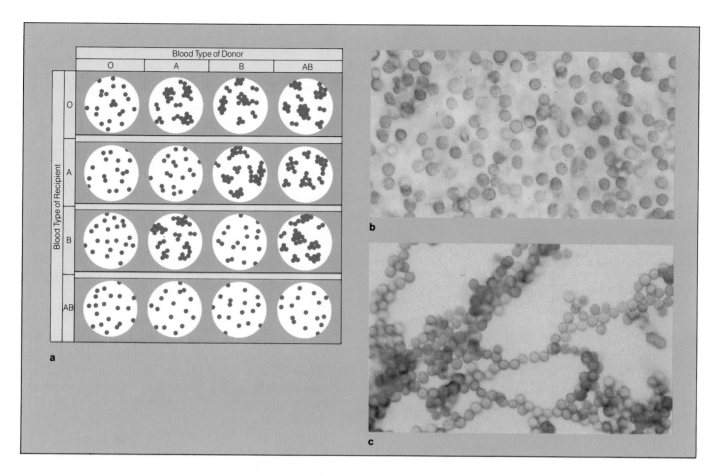

Figure 29.16 (**a**) The agglutination responses in drops of blood of types O, A, B, and AB when mixed with blood samples of the same and different types. (**b**) Micrograph showing the absence of agglutination in a mixture of two different but compatible blood types. (**c**) Micrograph showing agglutination in a mixture of incompatible blood types.

uals do not. Ordinarily, people do not have antibodies that act against Rh markers. However, if someone who is Rh⁻ has been given a transfusion of Rh⁺ blood, antibodies will be produced against it and will continue circulating in the bloodstream.

If an Rh⁻ female becomes pregnant by an Rh⁺ male, there is a chance that the fetus will be Rh⁺. During childbirth, some red blood cells of the fetus may leak from the placenta and enter her bloodstream (page 522). If they do, they will stimulate her body into producing antibodies against the Rh markers (Figure 29.17).

If the woman becomes pregnant again, the Rh antibodies might cross the placenta and enter the bloodstream of the fetus. If this second fetus happens to have Rh⁺ blood, the antibodies will cause a hemolytic response. In this response, red blood cells swell and then rupture, releasing hemoglobin into the bloodstream. In extreme cases of this disorder, called *erythroblastosis fetalis*, so many cells are destroyed that the fetus dies

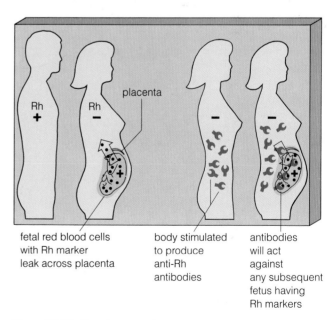

fetal red blood cells with Rh marker leak across placenta

body stimulated to produce anti-Rh antibodies

antibodies will act against any subsequent fetus having Rh markers

Figure 29.17 Development of antibodies in response to Rh⁺ blood.

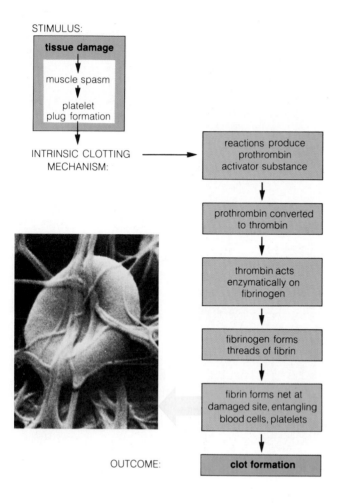

STIMULUS:

tissue damage

muscle spasm

↓

platelet
plug formation

↓

INTRINSIC CLOTTING
MECHANISM: →

reactions produce
prothrombin
activator substance

↓

prothrombin converted
to thrombin

↓

thrombin acts
enzymatically on
fibrinogen

↓

fibrinogen forms
threads of fibrin

↓

fibrin forms net at
damaged site, entangling
blood cells, platelets

↓

OUTCOME:

clot formation

Figure 29.18 Blood coagulation at a cut or at ruptured blood vessel tissue. The micrograph shows a red blood cell trapped in a fibrin net.

before birth. If the child is born alive, all of its blood can be slowly replaced with blood that is free of the Rh antibodies. Currently, a known Rh⁻ female can be treated right after her first pregnancy with a blood product (an anti-Rh antibody) that inactivates any Rh antibodies circulating in her bloodstream, thereby protecting the fetus of the next pregnancy.

Hemostasis

Agglutination is not the same thing as blood clotting, which plays a role in stopping blood loss (as may occur through ruptures or cuts in blood vessels). In many invertebrates and all vertebrates, bleeding can be stopped by blood vessel spasm, platelet plug formation, and blood coagulation. The collective action of these mechanisms, called **hemostasis**, can be described as follows.

First, smooth muscle in the damaged wall contracts in a reflex response called a spasm. The spasm lasts only a few minutes, but it causes the vessel to constrict and stop the flow of blood. Second, platelets clump together, temporarily plugging the rupture. The platelets do this by developing spinelike extensions that adhere to exposed collagenous fibers in the damaged walls. The spiny platelets release substances that attract more platelets. They also release calcium ions (which promote clumping) and a vasoconstrictor (which helps prolong the spasm). Third, blood coagulates (converts to a gel) and forms a clot. Finally, the clot retracts into a compact mass, drawing the ruptured walls of the vessel together.

Blood coagulates mainly through an **intrinsic clotting mechanism** that comes into play when damage exposes the collagenous fibers of blood vessels. A plasma protein becomes activated and triggers reactions that lead to the conversion of another plasma protein into thrombin (Figure 29.18). Thrombin is an enzyme that acts on fibrinogen to form fibrin, a large, rod-shaped protein. Molecules of fibrin become cross-linked to form a net that entangles blood cells and platelets. The entire mass is a blood clot.

Blood also can coagulate through an **extrinsic clotting mechanism**. (The contribution of this mechanism to hemostasis is unclear. However, it is definitely involved in walling off bacteria and preventing the spread of bacterial infection from invaded tissue regions.) "Extrinsic" means that the series of reactions leading to blood clotting is triggered by the release of enzymes and other substances *outside* of the blood itself (that is, from damaged blood vessels or from the surrounding tissues). The substances lead to thrombin formation, and the remaining steps parallel those shown in Figure 29.18. Overall, however, fewer steps are involved than in the intrinsic clotting mechanism, and the reactions occur much more rapidly.

LYMPHATIC SYSTEM

We conclude this chapter with a brief description of the **lymphatic system**, which supplements the circulation system by returning excess tissue fluid to the bloodstream. But think of this description as a bridge to the next chapter, on immunity, for the lymphatic system is also vital to the body's defenses against injury and attack. The lymphatic system consists of transport tubes and lymphoid organs. The tubes constitute the lymph vascular system, which supplements pulmonary and systemic circulation. When tissue fluid has moved into these tubes, it is called **lymph**. The lymphoid organs, which take part in defense responses, are structurally and functionally connected with both the blood and lymph vascular systems (Figure 29.19).

Lymph Vascular System

The **lymph vascular system** includes lymph capillaries, lymph vessels, and ducts that drain the processed fluid back into the circulation system. It serves these functions:

1. Return of excess filtered fluid to the blood.

2. Return of small amounts of proteins that leave the capillaries.

3. Transport of fats absorbed from the digestive tract.

4. Transport of foreign particles and cellular debris to disposal centers (lymph nodes).

At one end of the lymph vascular system are **lymph capillaries**, which are no larger in diameter than blood capillaries. These vessels occur in the tissues of almost all organs, and they seem to be permeable to all substances dissolved in interstitial fluid. Lymph capillaries are "blind-end" endothelial tubes. They have no entrance at the end residing in interstitial regions; the only opening is one that merges with larger lymph vessels (Figure 29.20).

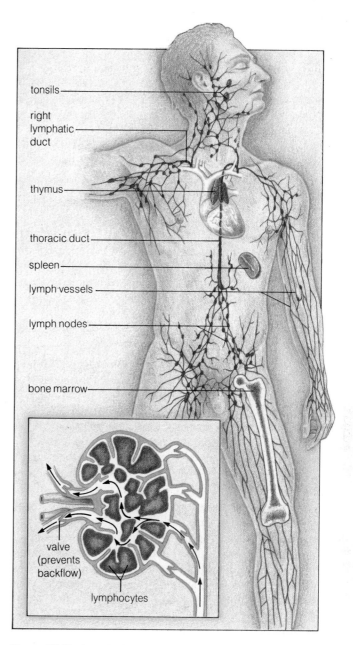

Figure 29.19 Lymphatic system, which includes the lymph vascular network and the lymphoid organs and tissues. Purple dots show some of the major lymph nodes. The inset illustrates the internal structure of a lymph node. Patches of lymphoid tissue in the small intestine and appendix also are part of the system.

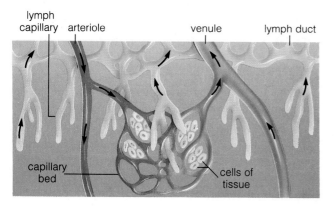

Figure 29.20 Lymph vessels near a capillary bed.

Lymph vessels are structurally similar to veins. They have an outer layer of connective tissue, a midlayer of smooth muscle and elastic fibers, and an inner lining of endothelium. Flaplike valves in these vessels prevent backflow. When you breathe, movements of the rib cage and of skeletal muscle adjacent to the lymph vessels assist in moving fluid through them, just as they do for veins. Lymph vessels converge into collecting ducts, which drain into veins in the region of the lower neck.

Lymphoid Organs

The **lymphoid organs** include the lymph nodes, spleen, and thymus. Other lymphoid structures include the bone marrow, tonsils, and nodules of lymphatic tissue in the digestive and respiratory tracts. These organs and structures function as production centers for infection-fighting cells, including lymphocytes, and as sites for some defense responses.

Like all white blood cells, lymphocytes are derived from stem cells in bone marrow (Figure 29.4). The derivative cells enter the blood and take up residence in lymphoid organs. Upon appropriate stimulation, they divide by mitosis; in fact, most new lymphocytes are produced in those organs, not in the bone marrow.

Lymph nodes are located at intervals along lymph vessels (Figure 29.19). All lymph trickles through at least one of these nodes before being delivered to the bloodstream. Each node is internally partitioned into several chambers, which are packed with lymphocytes and plasma cells. Macrophages in the node help clear the lymph of bacteria, cellular debris, and other substances.

The largest lymphoid organ, the **spleen**, is a filtering station for blood and a holding station for lymphocytes. It is located behind the stomach and beneath the diaphragm. The spleen is also partitioned internally into chambers, which are filled with red and white "pulp." The red pulp contains a large store of red blood cells and macrophages. This region is a production site for red blood cells in developing human embryos.

The **thymus** serves as an endocrine gland, in that it secretes hormones concerned with the activity of lymphocytes. It also is a major organ where certain lymphocytes multiply, differentiate, and mature. The thymus is central to immunity, which is the topic of the chapter to follow.

SUMMARY

1. Cells survive by exchanging substances with the bloodstream. A circulatory system allows rapid movement of these substances from one body region to another. Such systems typically consist of blood, a heart or heartlike structure, and blood vessels.

2. Blood is a transport fluid that carries materials to cells, carries products and wastes from them, and helps maintain an internal environment favorable for cell activities. Blood consists of red and white blood cells, platelets, and plasma.

3. Plasma contains water, ions, nutrients, hormones, vitamins, dissolved gases, and the plasma proteins.

4. Red blood cells transport oxygen (bound to hemoglobin) between cells and the lungs. They are also responsible for the transport of some carbon dioxide.

5. In all vertebrates, some white blood cells are scavengers of dead or worn-out cells or other debris; others serve in the defense of the body against bacteria, viruses, and other foreign agents.

6. In all vertebrates, the heart pumps blood into arteries. From there it flows into arterioles, capillaries, venules, veins, and back to the heart.

7. The mammalian heart is divided into two pumps that drive oxygen-poor blood to the lungs (pulmonary circulation) and oxygen-rich blood to the other tissues of the body (systemic circulation).

8. Heart contractions generate blood pressure. The pressure is high in the arteries but drops as blood flows through the other vessels.

9. The heart has four chambers (two atria and two ventricles). Valves separate the atrium and ventricle on each side of the heart and span the exit from each ventricle.

10. A cardiac cycle consists of one phase of contraction (systole) and relaxation (diastole).

11. The atria contract before the ventricles. Contractions are initiated by the sinoatrial node. Excitation generated there travels from the atria, to the atrioventricular node, and through a conduction system into the ventricles. Contraction of the ventricles forces blood through the circulation system.

12. Arteries are pressure reservoirs that keep blood flowing while the ventricles are relaxing. The blood pressure normally measured is arterial pressure at the upper arm.

13. Arterioles are control points for the distribution of different volumes of blood to different capillary beds.

14. The exchange of substances between the blood, interstitial fluid, and cells occurs in the capillary beds.

15. Veins are highly distensible blood volume reservoirs and are important in adjusting volume flow back to the heart. Valves in veins prevent backflow of blood to the periphery.

16. Venules overlap somewhat with capillaries and veins in their function.

17. Brain centers control blood pressure by varying the rate and strength of the heartbeat and the diameter of arterioles and veins. Hormones also affect blood pressure.

18. Some proteins and other organic molecules at the cell surface act as markers that identify a cell as being of a specific type. Antibodies are other proteins that act against foreign cells (as well as against viruses and microbes) in the body.

   a. Red blood cells have A or B markers, or both A and B, or neither one (type O).

   b. In an agglutination response, antibodies act against foreign blood cells (introduced through an improper blood transfusion, for example) and cause them to clump together. The clumping can clog small blood vessels and lead to other complications.

   c. Red blood cells also may be Rh$^+$ (with an Rh marker) or Rh$^-$ (without this marker).

19. Stopping the loss of blood (hemostasis) includes blood vessel spasm, platelet plug formation, and blood coagulation.

20. The lymph vascular system includes lymph capillaries, lymph vessels, and ducts. This system returns excess, filtered tissue fluid and some proteins to the bloodstream. The fluid being transported in this system is called lymph. It is filtered in lymph nodes and other lymph organs. Lymph vessels transport fat from the digestive tract to the blood.

Review Questions

1. What are some of the functions of blood? *398*

2. Describe the cellular components of blood. Describe the plasma portion of blood. *398–401*

3. Define the functions of the following:
   a. heart *397*
   b. cardiovascular system *401*
   c. lymph vascular system *397, 415*

4. Distinguish between the following:
   a. open and closed circulation *397*
   b. systemic and pulmonary circulation *402*
   c. lymph vascular system and lymphoid organs *415–416*
   d. systole and diastole *403*

5. Describe the cardiac cycle in a four-chambered heart. *403*

6. Can you identify the cardiac muscle cells of the cardiac conduction system and describe how they work? *404*

7. Explain how arteries, arterioles, and capillaries help regulate blood flow to different body regions. *412*

8. State the main function of blood capillaries. What drives solutes out of and into capillaries in capillary beds? *406–407, 412*

9. State the main function of venules and veins. What forces work together in returning venous blood to the heart? *411–412*

10. Label the component parts of the human heart: *403*

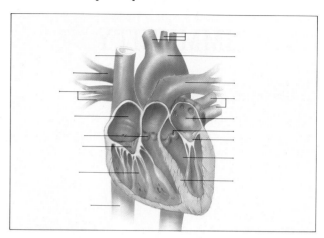

Readings

Berne, R., and M. Levy. 1988. "The Cardiovascular System." In *Physiology*. Second edition. St. Louis: Mosby. Comprehensive introduction to controls over circulatory functions.

Brown, M. S., and J. L. Goldstein. November 1984. "How LDL Receptors Influence Cholesterol and Atherosclerosis." *Scientific American* 251(5):58–66.

Doolittle, R. F. December 1981. "Fibrinogen and Fibrin." *Scientific American* 245(6):126–135. Describes clot formation and clot breakdown.

Eisenberg, M. S., et al. May 1986. "Sudden Cardiac Death." *Scientific American* 254(5):37–43.

Golde, D. W., and J. C. Gasson. July 1988. "Hormones That Stimulate the Growth of Blood Cells." *Scientific American* 259(1):62–70.

Kapff, C. T., and J. H. Jandl. 1981. *Blood: Atlas and Sourcebook of Hematology*. Boston: Little, Brown. Beautiful micrographs of normal and abnormal blood and marrow cells.

Levy, M., and J. Moskowitz. 1982. "Cardiovascular Research: Decades of Progress, a Decade of Promise." *Science* 217:121–126.

Robinson, T. F., et al. June 1986. "The Heart as a Suction Pump." *Scientific American* 254(6):84–91.

Vander, A., J. Sherman, and D. Luciano. 1985. *Human Physiology*. Fourth edition. New York: McGraw-Hill. Clear exposition of the circulatory system.

30

IMMUNITY

Until about a century ago, smallpox swept repeatedly through the world's cities. Some outbreaks were so intense that half or more of the people stricken with that contagious disease died. Those who survived had permanent scars on their face. Intriguingly, the readily identifiable survivors were not vulnerable to subsequent attacks; they were immune to smallpox.

No one knew what caused smallpox, but first in Asia, then in Africa and Europe, there were flurries of inoculations against it. ("Inoculation" means intentionally infecting a healthy person with matter taken from the sores on someone having a mild case of the disease.) If all went well, inoculated individuals also came down with only a mild case of smallpox and thereafter had immunity against it. But sometimes they became seriously ill. And sometimes they infected their family and friends—and triggered another epidemic.

While this immunological version of Russian roulette was going on, Edward Jenner was growing up in the English countryside. At the time, it was known that a rather mild disease called cowpox could be transmitted from cattle to humans. It also was known that people who contracted cowpox never came down with smallpox. In 1776, America declared her independence, and in a less celebrated event an English farmer, Benjamin Jesty, inoculated his family with cowpox matter to protect them against smallpox. But it was not until 1796 that Jenner (by then a physician) scientifically demonstrated the effectiveness of the procedure, and eventually Londoners lined up by the hundreds to be inoculated. The French derided Jenner's procedure, calling it "vaccination" (which translates as "encowment"). Many years later, Louis Pasteur developed similar procedures

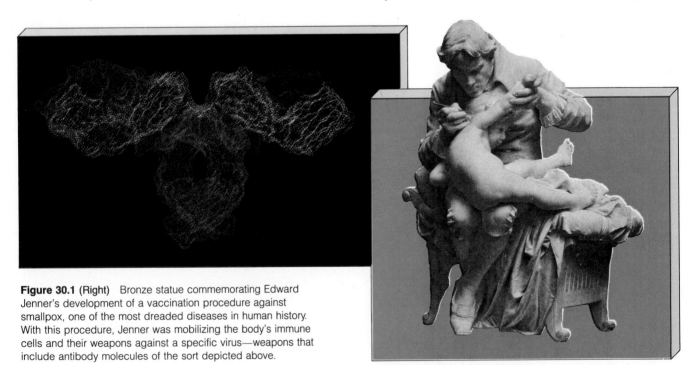

Figure 30.1 (Right) Bronze statue commemorating Edward Jenner's development of a vaccination procedure against smallpox, one of the most dreaded diseases in human history. With this procedure, Jenner was mobilizing the body's immune cells and their weapons against a specific virus—weapons that include antibody molecules of the sort depicted above.

and also called them vaccinations, thereby bestowing respectability on the term.

Today we know that the body is equipped to defend itself against many different *pathogens* (disease-causing agents), including viruses, bacteria, certain fungi, and some protozoans. With his procedure, Jenner was actually mobilizing the body to make an immune response to a specific virus, one of the elegant defenses described in this chapter. Before turning to the specific responses, however, let's start out with the body's generalized, nonspecific defenses against attack.

NONSPECIFIC DEFENSE RESPONSES

Barriers to Invasion

The vertebrate body has an impressive array of physical and chemical barriers against invasion. These barriers typically include the following:

1. Intact skin, which only a few bacteria can penetrate.

2. Ciliated, mucous membranes that line parts of the respiratory tract and that act like sticky brooms to sweep out bacteria and inhaled particles.

3. Secretions from exocrine glands in surface epithelium. For example, mammals secrete lysozyme, an enzyme that helps degrade the cell wall of many bacteria.

4. Gastric fluid in the stomach, the acids of which destroy numerous pathogens.

5. Microbes that normally inhabit the gut and (in females) the vagina. They compete effectively with many types of microbial invaders and thereby help keep them in check.

Phagocytes

What happens when the barriers listed above are breached, as when skin is cut or scraped? Then, an invasion mobilizes the **phagocytes**—cells that engulf and destroy foreign cells and substances by means of endocytosis (Figure 6.11). As you read earlier, the phagocytes in adult vertebrates are derived from stem cells in bone marrow (Figure 29.4). They include neutrophils, eosinophils, and the monocytes that mature into macrophages, the "big eaters."

Phagocytes are strategically distributed cells. Some circulate through the body inside blood vessels, then enter damaged or invaded tissues by squeezing past the endothelial cells that make up capillary walls. Some take up residence in lymph nodes and the spleen. (Here you may wish to review Figure 29.19, which shows tissues and organs of the lymphatic system.) Other phagocytes are stationed in tissues of the liver, kidneys, lungs, and joints. The brain has its own phagocytic glial cells.

Complement System

When certain microbes invade a tissue, about twenty plasma proteins interact as a system—called the **complement system**—to enhance both nonspecific and specific defense responses.

Complement proteins circulate in inactive form. When they encounter any of a variety of bacterial and fungal cells, they are activated one after another in an amplifying cascade of reactions. (In other words, every molecule switched on at the first step switches on many molecules at the second step, each of which switches on many molecules at the third step, and so on through the reaction series.) The reactions have these consequences:

1. Some complement proteins released during the cascade reactions create chemical gradients that attract phagocytes to the scene (this being an example of chemotaxis).

2. Other proteins form a phagocyte-attracting coat on the surface of invading cells.

3. Some proteins promote lysis (gross leakage and subsequent death) of the microbial cell by acting directly on its plasma membrane (Figure 30.2).

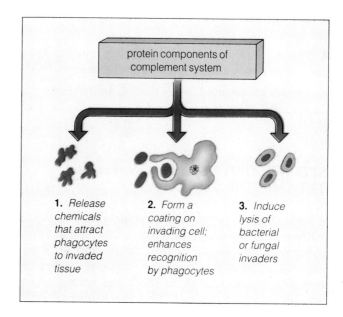

Figure 30.2 Functions of proteins of the complement system. These proteins take part in specific as well as in nonspecific defense responses.

Inflammation

The complement system and other chemical factors take part in the **inflammatory response**: a series of events that destroy invaders and restore tissues and intracellular conditions to normal. The events proceed during non-specific *and* specific defense responses.

Even while complement proteins are being switched on, for example, basophils and mast cells are called into action. Both are derived from stem cells in bone marrow, but basophils circulate freely and mast cells reside in lymphoid organs, in skin, and in connective tissues around blood vessels, nerves, and glands. Both release a potent substance called *histamine*. In humans, histamine dilates fine blood vessels and thereby slows down the flow in them; it also increases permeability of the vessel walls and thereby promotes seepage of fluid into the surrounding tissue. In that fluid are complement proteins and other factors used in the fight against an invasion.

The components of clotting mechanisms also take part in an inflammatory response. As described on page 414, these components help to keep blood vessels intact, and they wall off infected tissues during physical damage or infection.

These and other chemical factors produce the following interrelated events, which characterize an inflammatory response:

1. In damaged or invaded tissues, fine blood vessels dilate and become more permeable. The vasodilation causes localized warmth and redness.

2. Seepage from the blood vessels causes local swelling.

3. Seepage also carries infection-fighting proteins into the tissues.

4. Phagocytes follow chemical gradients to affected tissues, where they engulf invaders and debris.

5. Tissues are repaired, as by clotting mechanisms.

SPECIFIC DEFENSE RESPONSES: THE IMMUNE SYSTEM

As described above, phagocytes summoned to a site of inflammation make a generalized attack response; they indiscriminately engulf cellular debris and anything detected as foreign. Sometimes, however, a general response is not enough to check the spread of an invader, and illness follows. When that happens, highly focused counterattacks are made by three types of white blood cells—macrophages, T lymphocytes, and B lymphocytes. Interactions among these cells are the basis of the **vertebrate immune system**. The hallmarks of this system are *specificity* (its cells zero in on specific invaders) and *memory* (they can mount a rapid attack if the same type of invader returns).

The Defenders: An Overview

Of every 100 cells in your body, one is a white blood cell. The white blood cells and their products that are central to immune responses go by these names:

The Cells

1. **Macrophages:** phagocytic cells that alert helper T cells to the presence of specific foreign agents.

2. **Helper T cells:** master switches of the immune system; they stimulate the rapid division of B cells and killer T cells (which mount counterattacks).

3. **B cells:** lymphocytes that produce potent chemical weapons (antibodies).

4. **Killer T cells and natural killer (NK) cells:** lymphocytes that directly destroy body cells already infected by certain viruses or parasitic fungi; they may also destroy cancer cells.

5. **Suppressor T cells:** lymphocytes with regulatory roles (they slow down or prevent immune responses).

6. **Memory cells:** a portion of B-cell and T-cell populations produced during a first encounter with a specific invader but not used in battle; they circulate through the body and respond rapidly to any subsequent attacks by the same type of invader.

Key Weapons in the Immune Arsenal

1. **Antibodies:** membrane-bound or freely circulating receptor molecules that bind specific foreign targets and thereby tag them for destruction (by phagocytes or by the complement system).

2. **Lymphokines** and **interleukins:** secretions by which white blood cells communicate with each other.

3. **Perforin** and similar proteins, secreted by certain T cells, that kill their targets by punching holes in them.

As you will see, the immune system has two fighting branches, both of which are deployed during most battles. T cells dominate one branch; when they are activated, the events set in motion are called a "cell-mediated" response. B cells dominate the other branch; when they are activated, the associated events are called an "antibody-mediated" response. Figure 30.3 provides an overview of these two immune responses.

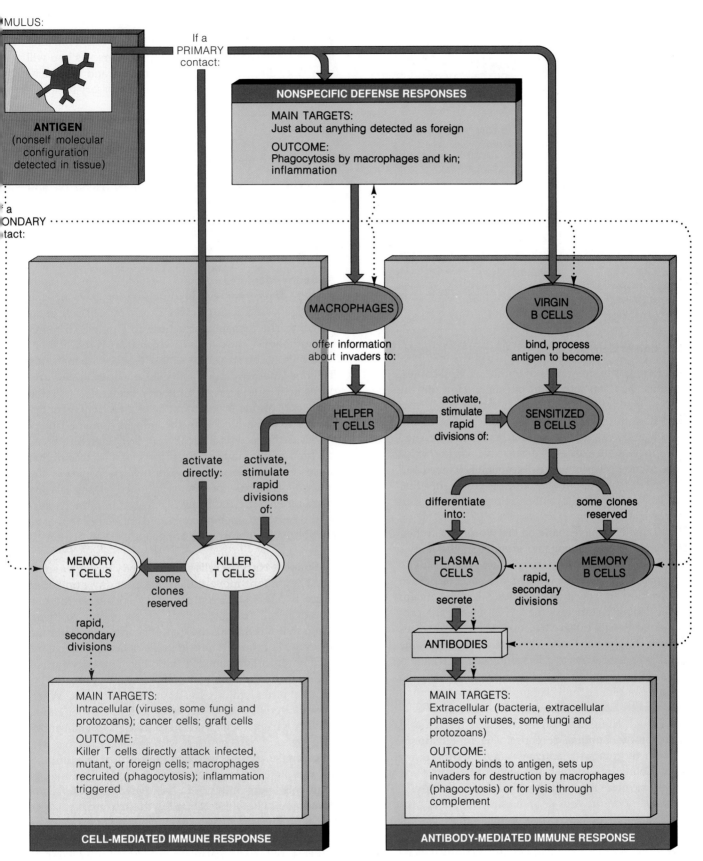

Figure 30.3 Overview of the cell-mediated and antibody-mediated branches of the vertebrate immune system. Green arrows indicate a "primary" response, which follows a first-time encounter with a specific antigen. Dashed arrows indicate a "secondary" response to a subsequent encounter with the same kind of antigen. This illustration can be used as a road map as you make your way through the descriptions in the text. The details of the vertebrate immune system are astonishingly complex; even here, many events have been omitted so the main sequences can be seen clearly.

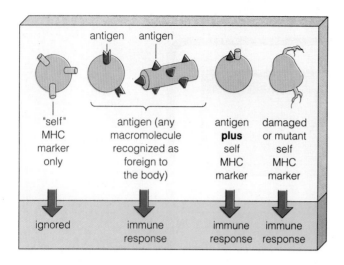

Figure 30.4 Molecular cues that stimulate lymphocytes to make immune responses.

Recognition of Self and Nonself

Before getting into the details of immunological battles, let's think about an important question. How do the defenders distinguish *self* (the body's own cells) from *nonself* (harmful foreign agents)? Such recognition is absolutely essential, for the immune reactions unleashed by lymphocytes are extremely destructive. We know this because on rare occasions, the distinction is blurred and lymphocytes make an autoimmune response. In other words, they turn on the body itself, sometimes with lethal consequences.

Recall that cells have many different proteins in their plasma membrane. Among the proteins of the body's own cells are "self" markers that project from the membrane surface. Usually the markers are parts of protein molecules that are specified by genes called the "major histocompatibility complex" (or MHC). Hence the proteins themselves are called **MHC markers**. Lymphocytes ignore any cells having "self" MHC markers alone. Some respond in specific ways to cells having self MHC markers *plus* antigen, others to antigen that is circulating freely in extracellular fluid (Figure 30.4).

An **antigen** is a macromolecule (protein or polysaccharide, most often) that is recognized as foreign and that triggers an immune response. Different lymphocytes have receptors for antigens—and binding of an antigen stimulates a cellular response. Certain proteins at the surface of a virus particle or bacterium, "nonself" MHC markers on the cells of an organ transplant, proteins in ragweed pollen or bee venom—these are all examples of antigens.

Macrophages that have engulfed foreign substances or invaders have self MHC markers *and* antigen at their surface; so do infected cells. (Antigens or antigen fragments that end up in the lysosomes of those cells are processed and then moved to the plasma membrane,

where they are "displayed.") Similarly, B cells bind antigen and display it with self MHC markers. As you will now see, the combination of antigen and self-MHC markers stimulates the cell divisions necessary for a full-blown immune response.

Primary Immune Responses

The actions taken by macrophages, lymphocytes, and their products in response to a *first-time* encounter with an antigen represent a **primary immune response**. Here we will consider an antibody-mediated response to a first-time encounter, then a cell-mediated response.

Antibody-Mediated Immune Response. An **antibody** is a type of receptor with binding sites for a specific kind of antigen, and it is produced only by a B cell or its progeny. The Y-shaped structures in Figure 30.5 represent antibodies, the details of which are covered later on.

B cells synthesize antibodies while they are maturing in bone marrow. Each of the immature cells produces just one kind of antibody molecule. (All the antibodies of one B cell might be of a molecular configuration that can bind with the antigen characteristic of the virus that causes mumps; those of a different B cell can bind with antigen at the surface of a bacterium that causes ear infections, and so on.) Many copies of the antibody molecule become positioned at the plasma membrane before the B cell is released into the circulation. B cells with membrane-bound antibodies that have not yet made contact with antigen are called "virgin" B cells.

Suppose bacterial cells enter the body through a small cut. They might move undetected past any number of virgin B cells—but eventually they encounter the one type of B cell "precommitted" to responding to them. When that B cell binds antigen, it is primed to enter the cell cycle. For the first time in its life, it becomes sensitive to communication signals from other lymphocytes that will drive it into mitosis.

What other lymphocytes are we talking about? When those bacteria invaded the body, a few were engulfed by macrophages. The invaders became enclosed in endocytic vesicles, which fused with lysosomes (page 94). Lysosomal enzymes degraded the bacterial cells but did not completely break down the antigens. The antigen fragments were moved to the macrophage's plasma membrane, where they bound with MHC markers (Figure 30.5). The combination of antigen and MHC markers was recognized by helper T cells. With its coupling to a helper T cell, each macrophage secreted interleukin-1, which stimulated helper T cells to secrete lymphokines.

Lymphokines stimulate the rapid growth and division of helper T cells—and they stimulate the division of B cells. Thus, *the binding of antigen plus signals from*

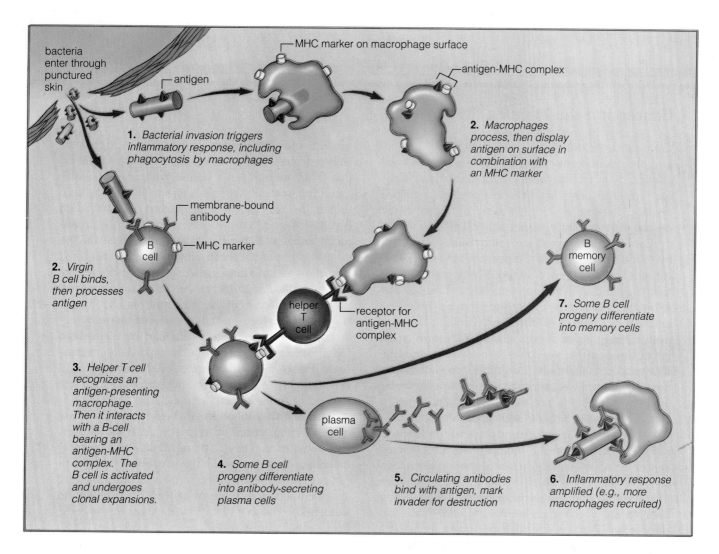

Figure 30.5 Amplification of the inflammatory response by specific immune reactions. This example is of an *antibody-mediated* response to a bacterial invasion. Plasma cells (the progeny of activated B cells) release antibodies, which circulate and mark invaders for destruction by other defense agents, including more macrophages recruited to the battle scene.

other immune fighters stimulate a sensitized B cell to enter the cell cycle. The divisions give rise to a clone—a population of B cells that will all produce the same kind of antibody. In due time, some of the progeny of the dividing B cells become sensitive to other signals that cause them to differentiate into plasma cells, each of which will secrete the same kind of antibody against the particular antigen that triggered the response in the first place.

Plasma cells are weapons factories: they secrete about 2,000 antibody molecules per second into the extracellular fluid! They devote nearly all of their metabolic machinery to sustaining that rate of antibody production and secretion, and they die in less than a week. The antibodies they secrete cannot destroy a target directly; rather, they bind to it and mark it for disposal by other agents.

Different classes of antibodies, called the **immunoglobulins** (Ig), enlist the aid of different cells and substances. The IgM and IgG antibodies activate macrophages and the complement system. The IgE antibodies stimulate mast cells into releasing histamine. IgA secretions are present in saliva, tears, the lungs, and the intestinal lumen, where they form first lines of defense against microbes. The function of IgD antibodies, which occur on B cell surfaces, is not known.

The main targets of an antibody-mediated response are bacteria as well as *extracellular* phases of viruses and certain parasitic fungi and protozoans. In other words, antibodies can't act against their target if it is already sequestered in the cytoplasm of a host cell; it has to be circulating in extracellular fluid or attached to the surface of host cells.

Cancer and the Immune System

Cancer cells might arise in your body at any time as a result of mutations induced by events such as viral attacks, chemical bombardment, or irradiation. "Cancer" refers to cells that have lost controls over cell division. Through their berserk divisions, they destroy surrounding tissues (page 231).

Cancer cells can be destroyed by killer T cells and by natural killer cells. Sometimes, though, cancer cells simply are not detected. Maybe the mutation that triggered the cancerous transformation did not affect the cell surface markers; or maybe the markers became chemically disguised. Perhaps they are even released from the cell surface and begin circulating through the bloodstream to lead the immune fighters down false trails. Sometimes, too, individuals are not genetically equipped to respond to a particular antigen. A person's age and overall state of health also seem to play a role in resistance to cancer.

At present, surgery, drug treatment (chemotherapy), and irradiation are the only weapons against cancer. Surgery works when a tumor is fully accessible and has not spread, but it offers little hope when cancer cells have begun wandering (page 231). When used by themselves, chemotherapy and irradiation destroy good cells as well as bad. *Immune therapy* is a promising prospect. The idea here is to mobilize the immune system by deliberately introducing agents that will set off the immune alarm.

Interferons, a group of small proteins, were early candidates for immune therapy. Most cells produce and release interferon following a viral attack. The interferon binds to the plasma membrane of other cells in the body and induces resistance to a wide range of viruses. So far, however, interferon has been useful against only some rare forms of cancer.

Monoclonal antibodies hold promise for immune therapy. It is difficult to get normal, antibody-secreting B cells to grow indefinitely and thereby mass-produce pure antibody in useful amounts. But Cesar Milstein and Georges Kohler discovered a way to do this. They immunized a mouse with a specific antigen. (The point of doing this was to allow lymphatic tissues in the mouse—the spleen especially—to become enriched with B cells specific for the immunizing antigen.) Later, B cells were extracted from the mouse spleen and were fused with a malignant B cell that showed indefinite growth. Some of the hybrid cells multiplied as rapidly as the malignant parent and produced quantities of the same type of antibodies as the parent B cells from the immunized mouse.

Clones of such hybrid cells can be maintained indefinitely and they continue to make the same antibody. Hence the name "monoclonal antibodies." All the antibody molecules are identical, and all are derived from the same parent cell.

Monoclonal antibodies are being studied for use in passive immunization against malaria, flu viruses, and hepatitis B. They also are candidates for cancer imaging. By using scanning machines along with radioactively labeled monoclonal antibodies that are specific for certain types of cancer, it is possible to home in on the exact location of cancer in the body. Such scans indicate whether cancer is present, where a tumor is located, and its size.

Monoclonal antibodies might also help overcome one of the major drawbacks to drug treatment of cancer. Such treatments have severe side effects because the drugs used are highly toxic and cannot discriminate between normal cells and cancerous ones. A current goal is to hook up drug molecules with a monoclonal antibody. As Milstein and Kohler speculated, "Once again the antibodies might be expected to home in on the cancer cells—only this time they would be dragging along with them a depth charge of monumental proportions." Such is the prospect of targeted drug therapy.

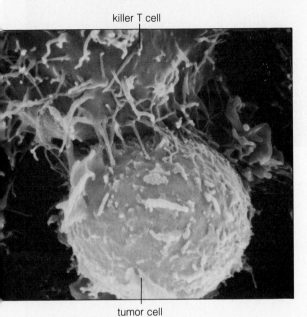

killer T cell

tumor cell

) A killer T cell recognizes and binds tightly to a tumor cell,
en secretes pore-forming proteins that will destroy the
tegrity of the target cell membrane.

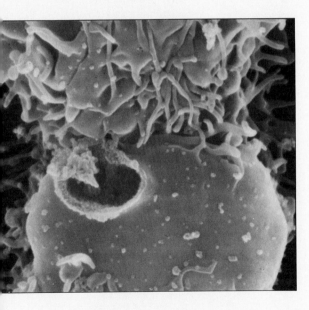

) The target cell has become grossly leaky and has
allooned under an influx of the surrounding fluid; soon there
ill be nothing left of it.

Cell-Mediated Immune Response. As you have seen, one branch of the immune system depends on an arsenal of antibodies. The other depends on T cells. The T cells start their development in bone marrow, but then they move into the thymus gland. There, each one assembles many copies of an antigen-binding receptor. (These receptors are very similar to antibodies.)

T cells do not bind with circulating antigens. Rather, the ones with receptors for particular antigens are stimulated into dividing by helper T cell secretions. They also will directly destroy any of the body's own cells that have the target antigen *combined with* an MHC marker—in other words, cells already infected by a specific virus or some other pathogen.

When a killer T cell locks tightly onto one of its targets, it secretes perforin and other chemicals that punch holes in the plasma membrane. Thus infected cells are destroyed before the infectious agent can reproduce inside them.

The main targets of killer T cells are virus-infected cells of the body. Another kind of lymphocyte, the natural killer (NK) cell, also destroys virus-infected cells. And both kinds may attack mutant and cancerous cells. (However, NK cells are less specialized; they destroy tumor cells of many sorts. The killer T cells are pickier about their targets.) The figures in the *Commentary* show how they act.

The killer cells may also be the reason why organ transplants often are rejected by the body. Normally the body tolerates transplants between identical twins (who have identical sets of DNA, hence identical MHC markers). Transplants between unrelated individuals are risky, because molecular differences between their MHC markers are recognized as foreign by killer T cells. Organ recipients take drugs that destroy killer T cells, but this compromises the body's ability to mount immune responses to other invaders. (That is why the kind of pneumonia caused by the bacterium *Streptococcus pneumoniae* is one of the leading causes of death among transplant recipients.)

Control of Immune Responses

Although it is beyond the scope of this book to discuss the mechanisms involved, keep in mind that cell-mediated and antibody-mediated immune responses are regulated events. To give just one brief example, antibody production itself is subject to feedback inhibition. When the tide of battle turns, antibody molecules "saturate" the binding sites on the viruses or foreign cells that have not yet been disposed of. With fewer exposed antigens, fewer lymphocytes become activated, so less antibody is secreted. Also, secretions from suppressor T cells call off the counterattack by the other lymphocytes. In such ways, immune reactions are kept from spiraling out of control.

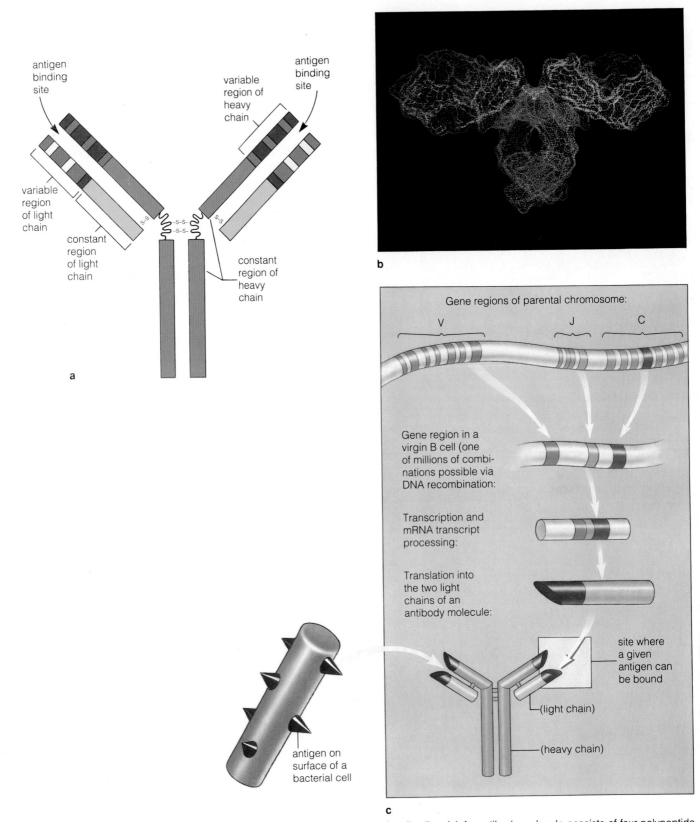

Figure 30.6 Structure and synthesis of antibodies. (**a**) An antibody molecule consists of four polypeptide chains joined by disulfide and noncovalent bonds into a Y-shaped structure. (**b**) Computer-generated model of an antibody molecule. The end of each arm of the Y-shaped molecule contains a site where the surface structure can vary, depending on which of an assortment of gene segments are expressed. One antigen can become bound to this site on each arm. Only the protein-coding regions for the light chain are shown in (**c**); different regions code for the heavy chain.

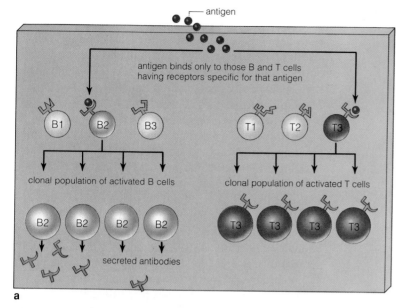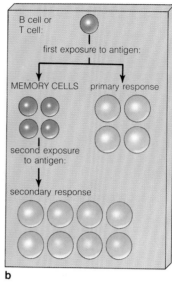

Figure 30.7 **(a)** Clonal selection of lymphocytes having receptors for specific antigens. The proteins from which the receptors are constructed are produced through random shufflings of DNA segments while lymphocytes are maturing. Only antigen-specific lymphocytes will become activated and will give rise to a population of immunologically identical clones. **(b)** Immunological memory. Not all cells of the activated lymphocyte populations are used in the primary immune response to an antigen. Many continue to circulate as memory lymphocytes, which become activated during a secondary immune response.

Antibody Diversity and the Clonal Selection Theory

The human body can be invaded by an enormous variety of viruses and microbes, each with many different proteins at its surface. To complicate matters, those proteins are subject to mutation. Even though mutations are rare in any given generation, microbial populations reproduce so rapidly and in such vast numbers that new, mutated proteins emerge more often than you might think. The point is, the human body may be exposed to millions of different antigens. How do lymphocytes produce the millions of different receptors required to detect all the potential threats?

For the answer, consider the Y-shaped antibody molecule, which consists of four polypeptide chains (Figure 30.6). The tail of the Y is the same (or nearly so) in all antibody molecules, but the end of each arm has a variable region, and this is where antigen binds. Each kind of antibody has a unique antigen-binding site. Here, the two polypeptide chains of each arm are folded to form a groove or cavity that is complementary to the contour and electric charge of the antigen.

The diversity in antigen-binding sites arises through genetic recombination that occurs in each B cell while it is maturing in bone marrow. As Figure 30.6c shows, recombination occurs at certain gene regions of the parental chromosome. It gives rise to a unique amino-

acid sequence in the polypeptide chains for antibodies that are constructed in each B cell—hence to unique contours and electric charges at the antigen-binding sites. *Through genetic recombination, each B cell becomes pre-committed to responding to the molecular configuration of one of millions of antigens.*

Thus, it is not that you or any other individual inherited a limited war chest from your ancestors, useful only against types of invaders that were successfully fought off in the past. Even if you encounter an entirely new antigen (as might occur when an influenza virus has mutated), it may be that gene recombinations in one of your maturing B cells produced the exact molecular configuration that can lock onto the invader. By happy accident, that lymphocyte has the precise weapon needed.

According to the **clonal selection theory**, first proposed by Macfarlane Burnet in 1955, a lymphocyte that has been activated when antigen combines with its receptor will multiply rapidly, and all of its descendants will retain specificity against that antigen only. They constitute a *clone* of cells that are immunologically identical for the antigen that "selected" them (Figure 30.7a).

Secondary Immune Response

The clonal selection theory also explains how a person has "immunological memory," which is the basis of a more rapid response to a subsequent invasion by the

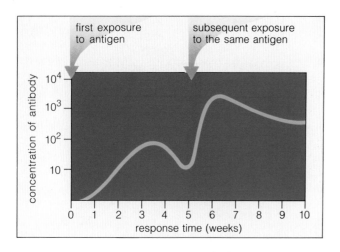

Figure 30.8 Differences in magnitude and duration between a primary and a secondary immune response to the same antigen. (The secondary response starts at week 5.)

same antigen. A *primary* immune response, recall, is made to a first-time encounter with an antigen. Full mobilization of the body's defenses usually takes five or six days. In contrast, a **secondary immune response** to the same antigen can occur in two or three days, and it is greater and of longer duration (Figure 30.8).

Why is this so? A primary response to an antigen does not use up all of the clonal populations of lymphocytes. A portion continues to circulate for years (decades, in some cases), as patrolling battalions of **memory lymphocytes**. Any one of those cells will be stimulated to divide at once if it encounters the same type of antigen (Figure 30.7b). Thus a large clone of active lymphocytes can be produced—and it can be produced in a matter of days.

IMMUNIZATION

Jenner didn't know why his cowpox vaccine provided immunity against smallpox. Today we know that the viruses causing the two diseases are related, and they bear cross-reacting (similar) antigen at their surface. Let's express what goes on in modern terms.

When an antigen is deliberately introduced into the body, the body may counterattack and then produce memory lymphocytes. Deliberately provoking the production of memory lymphocytes is known as **immunization**. Preparations used to stimulate their appearance, called **vaccines**, are injected into the body or taken orally. An initial injection elicits a primary immune

response. A second injection (the "booster shot") elicits a secondary response, which stimulates the production of additional antibodies and memory cells to provide long-lasting protection against the disease.

Many vaccines are prepared from killed or weakened bacteria or viruses. (Sabin polio vaccine, for example, is a preparation of weakened polio virus.) Other vaccines are prepared from the toxic but inactivated by-products of dangerous organisms, such as the bacteria that cause tetanus.

Recently, selected antigen-encoding genes from pathogens were incorporated into the vaccinia virus. The virus was then used successfully to immunize laboratory animals against hepatitis B, influenza, rabies, and other serious diseases. A genetically engineered virus is not as potentially dangerous as a weakened but still-intact pathogen (which might revert to virulent form).

For people already exposed to diphtheria, tetanus, botulism, and some other bacterial diseases, antibodies are injected directly to confer **passive immunity**. The term indicates that the effects are not lasting (the person's own lymphocytes are not producing the antibodies). However, the antibodies may help counter the immediate attack.

ABNORMAL OR DEFICIENT IMMUNE RESPONSES

Allergies

Sometimes the immune system can damage the body instead of protecting it. An **allergy** is a type of secondary immune response to a normally harmless substance, and it can cause tissue damage. Some allergic reactions occur explosively within minutes; others are delayed. About fifteen percent of the human population has a genetic predisposition to become sensitized to dust, pollen, insect venom or secretions, drugs, certain foods, and other seemingly innocuous substances. Exposure triggers the production of IgE antibodies, which give rise to many of the symptoms of immediate allergic reactions.

Emotional state, changes in air temperature or pressure, and infections can trigger or complicate reactions to dust and other substances that the body perceives as antigens. With each new exposure to the antigen, IgE antibody molecules are produced and become attached to cells that release chemicals such as histamine and prostaglandins. Histamine, which acts on exocrine glands as well as blood vessels, causes mucous secretions and increased capillary permeability. Prostaglandins constrict smooth muscle in assorted organs and contribute to platelet clumping. Such chemical factors initiate a local inflammatory response. In *asthma* and *hay-*

fever, the resulting symptoms include a drippy nose, sneezing, congestion, and labored breathing.

In a few hypersensitive people, secretions are so copious that an inflammatory response can be explosive and life-threatening. People who are hypersensitive to, say, wasp or bee venom can die within minutes following a single sting. Air passages leading to their lungs undergo massive constriction. Plasma escapes too rapidly from grossly permeable capillaries; blood pressure plummets and can lead to circulatory shock.

Some allergy sufferers can be stimulated to manufacture "blocking antibodies" when the allergy-producing substance has been identified by tests. Increasingly large injections of the antigen over a period of time provoke the production of IgG antibodies. Circulating IgG antibodies can bind with and mask the substance before it interacts with IgE to produce the inflammatory response.

Autoimmune Disorders

In an **autoimmune response**, the destructive action of lymphocytes is turned against the body's own cells. Some cases of type 1 diabetes, described on page 362, are autoimmune disorders. So is *rheumatoid arthritis*, characterized by prolonged and sustained inflammation of body tissues. No one knows the cause of this crippling disorder. An antigen produced either by the body itself or by a foreign agent probably serves as the trigger. Its continued presence may incessantly stimulate the body's immune system.

Many afflicted persons have high levels of rheumatoid factor, an antibody that unfortunately locks onto the body's own IgG molecules as if they were antigens. Most of the antigen-antibody complexes that form seem to be deposited on the synovial membrane of joints (page 388), which thus become the prime target for chemical factors released in an inflammatory response. The complement system is activated and signals macrophages to migrate into the joint. Other factors dilate blood vessels, and escaping fluid accumulates in the joint cavity. The fluid lifts the membrane from underlying tissues, and cells of the membrane respond by proliferating. The joint thickens. The clotting mechanism produces fibrin deposits, which cause more swelling. These and other events continue in cycles of inflammation that do not end until the joint is totally destroyed.

Deficient Immune Responses

On rare occasions, the body's natural, cell-mediated immunity is weakened. When that happens, the individual becomes extremely vulnerable to infections that might not otherwise be life-threatening. This is what happens in *acquired immune deficiency syndrome*, or AIDS. The virus responsible for AIDS is called "human immunodeficiency virus," or HIV. It is one of the most deadly of all infectious agents.

The *Commentary* on page 430 describes some immunological aspects of HIV infection, how the virus replicates itself inside a human host, and what the prospects are for treating or curing infected individuals. The social implications of the growing, worldwide pool of infection are described in Chapter Thirty-Five, in the *Commentary* on sexually transmitted diseases.

Case Study: The Silent, Unseen Struggles

Let us conclude this chapter with a case study of how the immune system helps *you* survive attack. Suppose on a warm spring day you are walking barefoot to class. Abruptly you stop: a tack that had been lying point-upward punctured one of your toes. You pull out the tack right away, but the next morning the punctured area is red, tender, and swollen. A few days later, though, your foot is back to normal and you have forgotten the incident.

All that time, your body had been struggling against an unseen enemy. Your walk took you across a spot where a bird died the night before and had lain until a scavenging animal carried it off. Your toe encountered some bacteria that had parted company from their feathered host—and when the tack penetrated your skin, it carried several thousand bacterial cells with it.

Within your tissues the bacteria found conditions suitable for growth. They soon doubled in number and were on their way to doubling again. At the same time, they were releasing metabolic products that interfered with your own cell functioning. Left alone, they would have threatened your life.

Yet even as the tack penetrated your skin, your body's defenses were being mobilized. Blood from ruptured capillaries began to pool and clot around the wound. Basophils and mast cells released histamine and other potent chemicals that triggered vasodilation and increased permeability to plasma proteins, including those of the complement system.

Now phagocytes crawled through clefts that had opened in the capillary walls. Like bloodhounds on the trail, they moved in the direction of higher concentration of complement proteins and began engulfing any particle not having the proper markers that meant, "I'm *self*." Dirt, rust, bits of broken host cells, and bacteria were engulfed indiscriminately.

If bacteria had not entered your body or if they were unable to multiply rapidly in your tissues, then phago-

Acquired Immune Deficiency Syndrome (AIDS)

AIDS is a constellation of disorders that follow infection by the human immunodeficiency virus, or HIV. Two types are known: HIV-1 (first identified in 1981) and HIV-2 (isolated in 1985). Successful invasion by the virus weakens the immune system, thereby leaving the body dangerously susceptible to opportunistic infections and some otherwise rare forms of cancer. At this time there is no vaccine against HIV, and there is no cure for those already infected.

From 1981 to early 1988, there were more than 50,000 cases of AIDS and an estimated 1 million or more HIV-1 carriers in the United States alone. HIV-2 is found mainly in West African countries. Although it may be less virulent than HIV-1, it has been associated with diseases there and in other countries. The World Health Organization (WHO) believes there already may be 5 to 10 million HIV carriers worldwide. No one can say with certainty how many people will be infected by HIV in the next decade; the number could be as high as 50 or 100 million.

HIV Replication Cycle. HIV cripples the immune system by attacking helper T cells (also called T4 lymphocytes) as well as macrophages. Sometimes the virus directly attacks the central nervous system, the outcome being severe mental impairment and loss of motor function. HIV causes other symptoms directly by attacking intestinal cells, particularly in the rectal portion of the colon.

To understand what goes on in an infected cell, we can begin with the fact that HIV is a *retrovirus*; that is, its genetic material is RNA rather than DNA. Each virus particle consists of a protein core that surrounds the viral RNA and several copies of an enzyme (reverse transcriptase). The core itself is wrapped in a lipid envelope derived from the plasma membrane of a host T4 cell. Once inside a host, the enzyme uses the viral RNA as a template for making DNA, which is then inserted into a host chromosome.

Soon after infection, antibodies to several HIV proteins can be detected in the body. But the antibodies do not eliminate the virus-infected cells, so a host cell can harbor the viral DNA for months, even years. However, when the body is called upon to make a secondary immune response, the infected cell may be activated. It transcribes portions of its DNA—including the foreign insert. Transcription yields copies of the viral RNA, which are translated into viral proteins. The RNA and proteins are assembled into new virus particles, which bud from the host membrane or are released when the cell membrane undergoes lysis (see Figures a-e). And so it goes, with more and more T4 cells destroyed through new rounds of infection.

In time, the T4 cell population is depleted and the body loses its ability to mount immune responses. The initial symptoms are not clear-cut, but they include unexplained weight loss, flu-like symptoms of fatigue and malaise, fevers and bed-drenching night sweats, as well as enlarged lymph nodes. If such symptoms persist for more than three months and cannot be attributed to any other disease, then they themselves may be diagnostic of AIDS. Opportunistic infections, including a form of pneumonia caused by a protozoan (*Pneumocystis carinii*), may superimpose

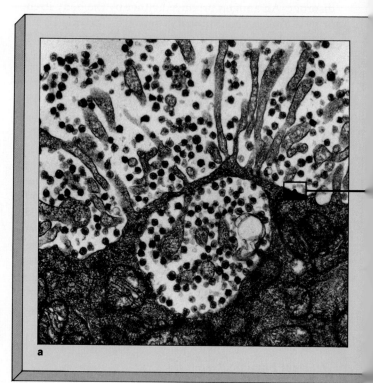

a

their own set of symptoms on the initial ones. Primarily in homosexual patients, blue-violet or brown-colored spots may appear on the skin of the legs especially, the latter being associated with Kaposi's sarcoma (a deadly form of skin cancer).

Modes of Transmission. Like any human virus, HIV requires a medium by which it can leave the body of its host, and enter a susceptible cell that can support its replication.

HIV is transmitted when bodily fluids of an infected person enter another person's tissues. In the United States, transmission has occurred most often among male homosexuals and intravenous drug abusers who share needles. Heterosexual transmission predominates in Africa and seems to be on the rise in Latin America, the Caribbean, and the United States. In New York City, AIDS is currently the leading cause of death in men 25–44 years old and women who are 25–34 years old. (Here you may wish to refer

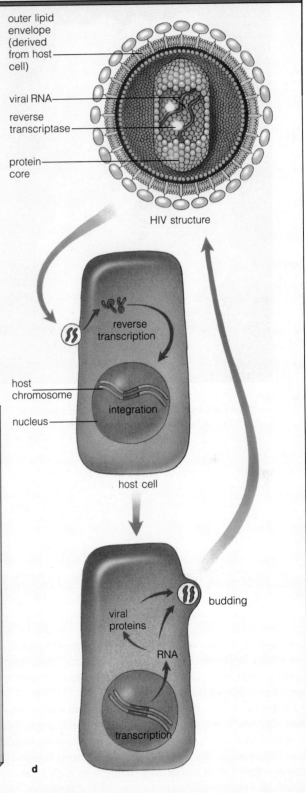

HIV structure

host cell

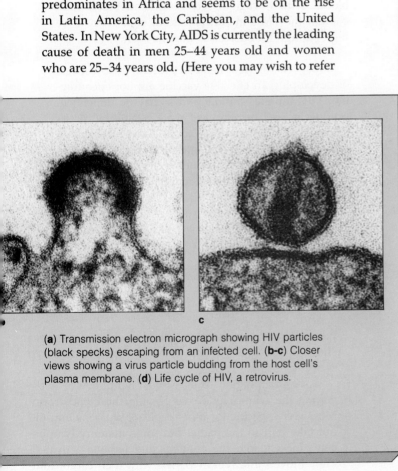

(**a**) Transmission electron micrograph showing HIV particles (black specks) escaping from an infected cell. (**b-c**) Closer views showing a virus particle budding from the host cell's plasma membrane. (**d**) Life cycle of HIV, a retrovirus.

(e) Color-enhanced scanning electron micrograph of a helper T cell being attacked by HIV particles (blue).

to the *Commentary* on sexually transmitted diseases, page 532.) HIV also has been transmitted from infected mothers to their infants during pregnancy, birth, and breast-feeding. In the United States, contaminated blood supplies accounted for a few cases before screening for HIV was implemented in 1985. In several developing countries, HIV has spread through contaminated blood transfusions and through reuse of unsterile needles by health care providers.

HIV cannot survive for more than about one or two hours outside the human body. Individual virus particles on needles and other objects are readily destroyed by disinfectants, including household bleach. At this time, there is no evidence that HIV can be effectively transmitted by way of food, air or water, or casual contact. The virus *has* been isolated from blood, semen, vaginal secretions, saliva, tears, breast milk, cerebrospinal fluid, amniotic fluid, and urine; it is likely to be found in other bodily fluids, secretions, and excretions. However, only infected blood, semen, vaginal secretions, and breast milk contain the virus in concentrations that seem to be high enough for successful transmission.

Prospects for Treatment. At present, the prospects of developing an effective vaccine are not encouraging. HIV mutates rapidly and it may be difficult to produce a vaccine that will work against all of its mutated forms. Even if a vaccine can be developed that could coax the body into producing antibodies to HIV, the antibodies may not protect against AIDS. There is evidence that antibodies do not neutralize the virus.

Recently, researchers successfully cloned the genes coding for several growth factors that influence

cytosis and inflammation would have cleaned things up. As it was, bacterial divisions were outpacing those nonspecific defenses. The battle had to be turned over to lymphocytes and their products.

If this had been your first encounter with the bacterial species, few lymphocytes would have been around to respond to the call. A primary immune response would have been mounted, and it would have been five or six days before the lymphocytes divided enough times to produce enough antibody to control the invasion. But when you were a child, your body did fight off the same type of bacterium and it still carries vestiges of the strug-

gle—memory lymphocytes. When the bacterial species showed up again, it encountered a lymphocyte trap ready to spring.

As inflammation progressed, lymphocytes were among the white blood cells leaving the bloodstream. Most were specific for other types of antigens and did not take part in the battle. But some memory lymphocytes locked onto the antigens and became activated. They moved into lymph vessels with their cargo, tumbling along until they reached a lymph node and were filtered from the fluid. For the next few days, memory cells steadily accumulated in the node. They divided blood cell production. Among the factors are GM-CSF (granulocyte macrophage colony-stimulating factor) and interleukin 3. Other researchers isolated the stem cells in bone marrow that actually give rise to blood cells, including those of the immune system. Perhaps work of this sort will benefit AIDS patients; it might be a way to increase dramatically the cell count of macrophages and other white blood cells that can fight infections. One problem that must be overcome here is that the macrophages themselves can become infected with HIV and actually spread the virus.

The drug AZT (azidothymidine) is being used to prolong the life of AIDS patients, although it cannot cure them. At this writing, researchers are attempting to develop useful drugs for those who are already infected; until they do, checking the spread of HIV depends absolutely on implementing behavioral controls through education on a massive scale. We return to this topic in Chapter Thirty-Five.

several times a day, so the number of lymphocytes and their products increased rapidly.

For the first two days the bacteria appeared to be winning; they were reproducing faster than phagocytes, antibody, and complement were destroying them. By the third day, antibody production was in full swing and the tide of battle turned. For two weeks or more, antibody production will continue until the invaders are wiped out. After the response draws to a close, memory lymphocytes will go on circulating, prepared for some future struggle.

SUMMARY

1. The vertebrate body is equipped for the following tasks:

   a. *Defense* against pathogens (disease-causing agents), including many viruses and bacteria, certain fungi, and some protozoans.

   b. Possible *defense* against mutant or cancerous cells.

   c. *Extracellular housekeeping* that helps maintain the tissue environment by eliminating dead or damaged body cells and structures.

2. "External" lines of defense against invasion include intact skin; lysozyme and other secretions from exocrine glands in surface epithelium; gastric fluid, microbial inhabitants of the gut (which compete effectively against many invaders); and ciliated, mucous membranes that trap and lead to the expulsion of invaders entering the respiratory tract.

3. The initial "internal" lines of defense include phagocytic cells (which engulf foreign agents) and the complement system. Phagocytes are distributed through the body; many are massed in lymph nodes and the spleen. Complement proteins, which circulate in inactive form, act against bacteria and certain fungi in three ways:

   a. Complement proteins generated during cascade reactions attract phagocytes (by chemotaxis).

   b. Some proteins coat the surface of an invading cell, enhancing its recognition by phagocytes.

   c. Some proteins cause the invading cell to lyse.

4. In the *inflammatory response*, phagocytes and chemical factors such as complement are mobilized to destroy invaders, then tissues and conditions are restored to normal. The coordinated activities occur during nonspecific *and* specific defense responses.

5. The vertebrate immune system has *specificity* (a response is made to a particular invader, rather than being a generalized attack response), and *memory* (a much more rapid attack can be mounted against subsequent encounters with the same type of invader).

6. Immune responses are carried out by interactions among macrophages and various lymphocytes (helper T cells, antibody-producing B cells, killer T cells, suppressor T cells, and memory cells); see Table 30.1.

7. Cells of the immune system communicate with one another by cell-to-cell contact and by chemical secretions (notably lymphokines), which stimulate rapid growth and division of B cells, helper T cells, and killer T cells into large armies against particular invaders.

8. MHC markers on the surface of the body's own cells allow lymphocytes to distinguish "self" from "nonself" (foreign agents).

9. Lymphocytes are precommitted to recognizing a particular antigen. An *antigen* is a macromolecule that is

Table 30.1 Summary of White Blood Cells and Their Roles in Defense

Cell Type	Take Part In	Main Characteristics
Lymphocytes:		
1. Killer T cell	Cell-mediated immune response	Each type equipped with membrane receptors specific for one type of antigen; each can directly destroy virus-infected cells (and possibly cancer cells) by punching holes in them
2. Helper T cell	Cell-mediated and antibody-mediated immune responses	Master switch of immune system; stimulates rapid proliferation of killer T cell and B cell populations
3. Suppressor T cell	Same as above	Modulates degree of immune response (slows down or prevents activity by other lymphocytes)
4. Virgin B cell	Antibody-mediated immune response	Inactivated lymphocyte with *membrane-bound* antibodies (serving as antigen-specific receptors at plasma membrane)
5. Plasma cell	Same as above	*Antibody-secreting* descendant of an activated B cell
6. Memory cell	Cell-mediated or antibody-mediated immune responses	One of a clonal population of T cells or B cells set aside during a primary immune response and that can make a rapid, secondary immune response to another encounter with the same type of invader
Macrophages	Inflammatory, cell-mediated, and antibody-mediated immune responses	Phagocytic (engulfs foreign agents and infected, damaged, or aged cells); develop from circulating monocytes and take up stations in tissues; present antigens to immune cells; secretions trigger T cell and B cell proliferation
Neutrophil	Inflammatory response	Phagocytic; most abundant type of white blood cell; dominates early stage of inflammation
Eosinophil	Inflammatory response	Phagocytic (engulfs antigen-antibody complexes, certain parasites); combats effect of histamine in allergic reactions
Basophil and mast cell	Inflammatory response	Release histamine and other substances that contribute to vasodilation and a rapid inflammatory response
Natural killer cell (NK)	Cell-mediated immune response	Directly destroy tumor cells, some virus-infected cells; distinct from both T and B cells

foreign to the body and that elicits an immune response).

10. An *antibody-mediated immune response* proceeds as follows against bacteria and extracellular phases of viruses, fungi, or protozoans:

a. Macrophages and B cells become sensitized to a particular antigen (they display antigen-MHC complexes at their cell surface).

b. Helper T cells detect antigen-MHC complexes on the macrophages and B cells, then stimulate the B cells to grow, divide, and develop into plasma cells, which secrete antibodies.

c. Antibodies (proteins of the class also called immunoglobulins, or Ig) bind to specific antigens and thereby mark invaders for disposal (by macrophages or by complement proteins).

11. A *cell-mediated immune response* proceeds as follows against cells already infected (and possibly against cancerous or mutant cells):

a. T cells do not react to circulating antigens; rather, they recognize antigen-bearing cells. These include macrophages and infected cells (which have processed and displayed antigen together with MHC markers at the cell surface), as well as B cells with bound antigen.

b. Helper T cells are activated by the combination of antigen and self MHC markers; they secrete chemical signals that activate killer T cells precommitted to act against the antigen. The killer T cells are stimulated to grow and divide rapidly.

c. Killer T cells (and natural killer cells) directly attack and destroy infected cells by "punching" holes in them with perforin and other secretions.

12. Following a primary (first-time) immune response, portions of the B and T cell populations produced continue to circulate as memory lymphocytes. They are available for a rapid, amplified response to subsequent encounters with the same antigen (a secondary immune response).

Review Questions

1. The vertebrate body has physical and chemical barriers against invading pathogens. Can you name five such barriers? *419*

2. Phagocytes are cells that engulf and destroy foreign cells and substances by means of endocytosis. Where in the body are phagocytes located? Are they deployed during nonspecific defense responses or immune responses only? *419, 434*

3. What are the five events that characterize an inflammatory response? What are some of the chemical factors associated with this response? *420*

4. The vertebrate immune system is characterized by *specificity* and *memory*. Can you describe what these two terms mean? *420*

5. Define the following types of white blood cells, which are central to immune responses: macrophages, helper T cells, B cells, killer T cells, suppressor T cells, and memory cells. *420*

6. Antibodies, lymphokines, and interleukins also are central to immune responses. Can you define them? *420*

7. What is an MHC marker? An antigen? An antibody? How do interactions among these three types of macromolecules allow the body to distinguish self from nonself? *422*

8. A secondary immune response to a particular antigen involves macrophages, lymphocytes, and their products. How does it differ from a primary immune response—which involves the same things? *422, 427–428*

9. Two fighting branches of the immune system are deployed during most primary and secondary immune responses. One is the cell-mediated response, the other is the antibody-mediated response. Can you give an example of what goes on during each kind? *420–425*

10. Refer to Figure 30.7. Then on your own, diagram the cell divisions involved in a clonal selection of an antigen-specific B cell or T cell. *427*

11. What is an immunization? A vaccine? *428*

12. What is the difference between an allergy and an autoimmune response? Is AIDS an autoimmune response? *428–429*

Readings

Buisseret, P. August 1982. "Allergy." *Scientific American* 247 (2):86–95.

Cooper, E. 1982. *General Immunology*. New York: Permagon Press. Contains a discussion of the evolution of immune responses.

Edelson, R. and J. Fink. June 1985. "The Immunologic Function of Skin." *Scientific American* 252(6):46–53.

Fauci, A. February 5, 1988. "The Human Immunodeficiency Virus: Infectivity and Mechanisms of Pathogenesis." *Science* 239(4840): 617–622.

Francis, D. and J. Chin. March 13, 1987. "The Prevention of Acquired Immunodeficiency Syndrome in the United States." *Journal of the American Medical Association* 257(10): 1357–1366.

Golub, E. 1987. *Immunology: A Synthesis*. Second edition. Sunderland, Massachusetts: Sinauer Associates. Readable account of cellular immunology.

Kimball, J. 1986. *Introduction to Immunology*. Second edition. Advanced reading, but highly recommended for serious students of immunology.

Leder, P. May 1982. "The Genetics of Antibody Diversity." *Scientific American* 246(5):102–115. Describes how a few hundred DNA segments can be shuffled and recombined to make billions of different antibodies.

Marx, G. October 30, 1987. "Structure of MHC Protein Solved." *Research News* 238:613–614.

Roitt, I., J. Brostoff, and D. Male. 1985. *Immunology*. St. Louis: Mosby. Lavishly illustrated introduction to basic and clinical immunology.

Scientific American. October 1988, volume 259, number 4. Entire issue is devoted to articles on AIDS.

Tizard, I. 1984. *Immunology: An Introduction*. Philadelphia: Saunders.

Tonegawa, S. October 1985. "The Molecules of the Immune System." *Scientific American* 253(4):122–131. Well-written survey article that incidentally illustrates how rapidly the field of immunology has advanced over the past few years.

Young, J. and Z. Cohn. January 1988. "How Killer Cells Kill." *Scientific American* 258(1):38–44.

31

RESPIRATION

In most animals, energy stored in food becomes available through aerobic metabolism, which requires oxygen. Most aquatic animals get oxygen from the water, land-dwelling animals get it from the air. In both cases, oxygen first crosses the body surface before it reaches individual cells. Carbon dioxide wastes of aerobic metabolism also cross the body surface and are dissipated in the surroundings.

Acquiring oxygen and eliminating carbon dioxide from the body as a whole is called **respiration**. Although the systems of respiration are diverse, all are alike in their reliance on diffusion. To be sure, in many systems the gas molecules are swept along by bulk flow, as when they are sucked into and exhaled from the body or when they hitch rides on the blood circulation highways. But recall that bulk flow has one overriding function in the body: it enhances the *rate of diffusion* between the environment and living cells.

SOME PROPERTIES OF GASES

Diffusion of any substance depends largely on differences in its concentration between two regions. For gases, the differences are expressed in terms of pressure. The more concentrated the molecules of a gas are outside the body, the higher the pressure and the greater the force available to drive individual molecules inside, and vice versa.

Air is a mixture of gases. At sea level, the gases are present in these amounts:

nitrogen	78 percent by volume
oxygen	21
argon, other gases	0.97
carbon dioxide	0.03

Figure 31.1 Countercurrent flow in most fish gills. (**a**) Location of gills beneath the operculum (a bony cover), which has been removed for this sketch. (**b,c**) Each gill filament has an incoming and an outgoing blood vessel. Capillary beds, arranged in thin membrane folds (lamellae), connect the two. (**d**) Blood flow from one vessel to the other in these beds runs counter to the direction of water flow.

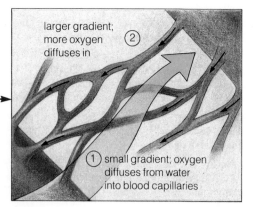

water in

water out

a

gill arch

blood vessels in gill filament

b

oxygen-poor blood

oxygen-enriched blood

direction of water flow past lamellae

c

larger gradient; more oxygen diffuses in ②

① small gradient; oxygen diffuses from water into blood capillaries

d

Because individual gas molecules are so spread out in air, we can think of each gas and the pressure it exerts as being independent of all the others. **Partial pressure** is the pressure exerted by one gas in a mixture of gases. Partial pressures can simply be added to give us the total pressure for the gas mixture.

At sea level, atmospheric pressure is about 760mm Hg, as measured by a mercury barometer:

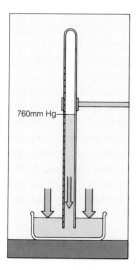

760mm Hg

Atmospheric pressure as measured by a mercury barometer. At sea level, the level of mercury (Hg) in a glass column is about 760 millimeters (29.91 inches). At this level, the pressure exerted by the column of mercury equals atmospheric pressure outside the column.

Oxygen represents about 21 percent of the total atmospheric pressure, so the partial pressure of oxygen is (760 × 21/100), or about 160mm Hg. The partial pressure of carbon dioxide in air is about 0.3mm Hg.

Any gas will diffuse from a region of higher partial pressure to a region of lower partial pressure. For example, oxygen diffuses from the air into surface waters of lakes and seas (where its partial pressure is only about 0.5 percent), and thereby replenishes oxygen supplies in aquatic habitats. Diffusion also figures in the movement of dissolved gases into and out of cells, extracellular fluid, and the bloodstream of animals.

RESPIRATORY SURFACES

In the simplest respiratory system, gases are exchanged across the body surface as a whole. In others, they are exchanged at specialized respiratory surfaces called gills, tracheas, and lungs (Figures 31.1 through 31.4).

Integumentary Exchange

Many animals do not have massive bodies or high metabolic rates. Their demands for gas exchange are not great, and the outer layer of the body itself provides enough of a respiratory surface. These animals rely on **integumentary exchange**, in which gases diffuse through a thin, vascularized layer of moist epidermis at the body surface. In aquatic animals, the layer is kept moist by the surrounding water; in land dwellers, it is kept moist by body secretions.

For example, earthworms secrete mucus that helps moisten the integument, which is a single layer of epidermal cells. Oxygen molecules between soil particles dissolve in the mucus and diffuse across the integument. From there, oxygen diffuses into blood capillaries that project, fingerlike, between the epidermal cells. Pressure generated by muscular contractions of the body wall and by the pumping action of tiny "hearts" causes blood to circulate in the narrow, tubelike body. The bulk flow enhances the diffusion rates for individual cells.

Most annelids, some small arthropods, and mollusks called nudibranchs rely on integumentary exchange. To a large extent, so do amphibians. But the integument of other animals is too thick, is too hardened, or has too few blood vessels to serve as an efficient respiratory surface. Moreover, their integument alone cannot provide enough surface area for adequate gas exchange, and specialized respiratory organs are required.

Gills

The respiratory organ called a **gill** typically has a moist, thin, vascularized layer of epidermis that functions in gas exchange. Larval forms of a few fishes, amphibians, and some insects have *external* gills projecting from the main body mass. Adult fishes have *internal* gills: a series of slits or pockets that originate in the pharynx (a muscular tube at the back of the mouth) and extend to the body surface. Water enters the mouth, moves down the pharynx, and flows out across the gills (Figure 31.1).

Not much oxygen is dissolved in water compared to air, and oxygen takes about 300,000 times longer to diffuse through water, which is more dense and viscous than air. Fish gills can take up adequate oxygen under these conditions. An extensive network of blood vessels services the epidermal surface of the gills. Because of the way gills are oriented, water flows over them in one direction and blood is circulated in the opposite direction. Such movement of fluids in opposing directions is called **countercurrent flow**.

Water passing over a fish gill first encounters the domain of a vessel that transports oxygen-rich blood *into* the body. The partial pressure of oxygen in this region is lower than it is in the water, so some oxygen diffuses in at this time. Then, just before the water moves completely past the gill, it passes over the domain of a vessel carrying blood *from* deep body regions—and this blood has less oxygen than the (by now) oxygen-poor water. With this even greater pressure difference, more oxygen diffuses inward (Figure 31.1d).

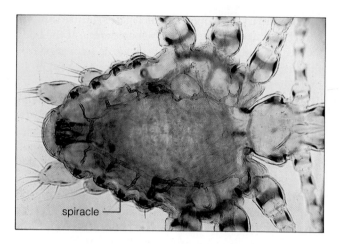

Figure 31.2 Tracheal system of a louse. Notice the spiracles leading inward from the body surface.

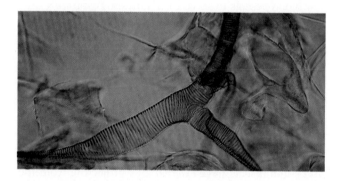

Figure 31.3 Closer look at the air-conducting tubes in the louse body.

Figure 31.4 Evolution of vertebrate lungs and swim bladders. The esophagus (a tube leading to the stomach) is shaded gold; the respiratory tissues, pink.

Lungs originated as pockets off the anterior part of the gut; they increased the surface area for gas exchange in oxygen-poor habitats. In some lineages, lung sacs became modified into swim bladders: buoyancy devices that help keep the fish from sinking. Adjusting gas volume in the bladders allows fishes to remain suspended at different depths.

Trout and other less specialized fishes have a duct between the swim bladder and esophagus; they replenish air in the bladder by surfacing and gulping air. Most bony fishes have no such duct; gases in the blood must diffuse into the swim bladder. Their swim bladder has a dense mesh of blood vessels (rete mirabile) in which arteries and veins run in opposite directions. Countercurrent flow through these vessels greatly increases gas concentrations in the bladder. Another region of the bladder allows reabsorption of gases by the body tissues.

Tracheas

Insects, centipedes, some mites, and some spiders use air-conducting tubes called **tracheas** for gas exchange. In most land-dwelling insects, these chitin-reinforced tubes branch finely through the body and provide a rather self-contained system of gas conduction and exchange; assistance by a circulatory system is not required.

Two main tracheal trunks run the length of the insect body (Figures 31.2 and 31.3). In most insects, smaller air-conducting tubes extend from the trunk lines to openings, called spiracles, at the body surface. Many species have a lidlike structure that spans each opening and prevents water loss through evaporation.

Some of the larger and more active insects move their body in ways that enhance air movement through the tracheal system. Have you ever noticed that foraging bees stop every so often and pump the segments of their abdomen back and forth? The segments extend and retract like a telescope, forcing air into and out of the spiracles. The stepped-up oxygen intake and carbon dioxide removal help support the high rate of metabolism required for insect flight.

Lungs

A **lung** is an internal respiratory surface that is generally in the shape of a cavity or sac. Scorpions, trap-door spiders, tarantulas, and a few other invertebrates have lunglike cavities. Fishes that lived more than 450 million

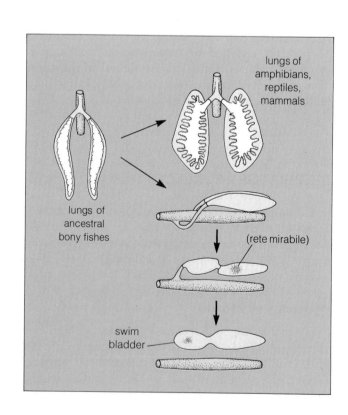

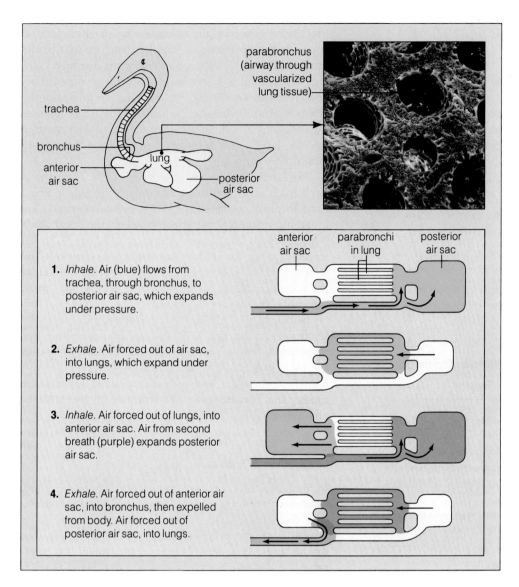

parabronchus (airway through vascularized lung tissue)

trachea

bronchus

anterior air sac

lung

posterior air sac

1. *Inhale.* Air (blue) flows from trachea, through bronchus, to posterior air sac, which expands under pressure.

2. *Exhale.* Air forced out of air sac, into lungs, which expand under pressure.

3. *Inhale.* Air forced out of lungs, into anterior air sac. Air from second breath (purple) expands posterior air sac.

4. *Exhale.* Air forced out of anterior air sac, into bronchus, then expelled from body. Air forced out of posterior air sac, into lungs.

anterior air sac   parabronchi in lung   posterior air sac

Figure 31.5 Respiratory system of birds. High metabolic rates and efficient gas exchange sustain flight and other activities. The rates are possible because of a unique ventilating system.

Typically, four air sacs are attached to each bird lung, which is somewhat small and inelastic. The sacs are not respiratory surfaces; they are more like bellows.

Inhaled air travels from bronchi into large posterior air sacs. When the bird exhales, this air is forced into bronchial branches that lead to parabronchi: small tubes (open at both ends) present in vascularized lung tissue. This is the respiratory surface (see photograph).

Air is not merely drawn into bird lungs: it is drawn *through* them. Air sacs and intricate lung airways make possible a *continuous* flow of air across the respiratory surface.

years ago had simple lungs that apparently played a supplementary role in respiration (Figure 31.4). In some of their descendents, the lungs developed into moist, thin-walled swim bladders. (A swim bladder is an organ where gas volume is adjusted to maintain body position and balance; some oxygen is also exchanged with blood and the surrounding tissues.) In other descendents, the lungs became complex respiratory organs.

The evolution of lungs may be reflected in the respiration systems of existing vertebrates. African lungfish have gills, but they also use lungs to supplement respiration. In fact, they will drown if they are kept from gulping air at the water's surface. Integumentary exchange still predominates in amphibians, but they, too, supplement respiration with a pair of small lungs. In all reptiles, birds, and mammals, paired lungs are the major respiratory surfaces.

OVERVIEW OF VERTEBRATE RESPIRATION

The respiratory surface is a boundary layer between the external and internal environments. In all animals with lungs, *airways* carry gas to and from one side of the respiratory surface of the lungs, and *blood vessels* carry gas to and from the other side. Respiration itself involves the following events:

1. Air moves by bulk flow into and out of the lungs, and new air is delivered to the respiratory surface. These two events are called *pulmonary ventilation*.

2. Gases diffuse across the respiratory surface of the lungs.

3. Pulmonary circulation (the bulk flow of blood to and from the lung tissues) enhances the diffusion of dissolved gases into and out of lung capillaries.

4. Gases diffuse between blood and interstitial fluid.

5. Gases diffuse between interstitial fluid and individual cells.

Pulmonary ventilation occurs within a series of air-conducting tubes that branch to become shorter, smaller in diameter, and more numerous along the route from the body surface into the lungs. Numerous, thin-walled outpouchings of the smallest air-conducting tubes are the actual sites of gas exchange.

Let's focus now on the human respiratory system, for the principles governing its operation apply to almost all vertebrates. The major exception is the respiratory system of birds, which is shown in Figure 31.5.

HUMAN RESPIRATORY SYSTEM

Air-Conducting Portion

The human respiratory system is shown in Figure 31.6. Normally, air enters and leaves through two narrow channels, called **nasal cavities**, in the nose. (Some air also enters and leaves through the mouth.) Dust and other foreign particles are filtered from the air by hairs at the entrance of the channels and by cilia along the epithelial lining of the channels. Blood vessels embedded in the lining warm the incoming air, and mucous secretions moisten the air before it flows into the lungs.

From the nasal cavities, air moves into the throat, or **pharynx**, which is the entrance to both the respiratory tract and the digestive tract. The throat cavity connects with the **larynx** (which leads to the lungs) and with the esophagus (which leads to the stomach). The larynx consists of muscles and cartilages bound in elastic connective tissue. Part of the cartilage is attached to and supports the **epiglottis**, a flaplike structure that points upward and allows air to enter the larynx during breathing. When food is being swallowed, muscular contractions force the food back into the pharynx and raise the larynx against the base of the tongue. This action presses the epiglottis down, so that it partly covers the opening into the larynx and helps prevent the food from going down the wrong tube.

The larynx contains two **true vocal cords**. These thickened folds of the larynx wall contain the elastic fibers used to produce the sounds of speech. When you are breathing normally, the space between the vocal cords remains open. This space is the **glottis** (Figure 31.7). Air forced through the glottis gives rise to sound waves. The stronger the air pressure on the vocal cords, the louder the sound produced. The greater the muscle tension on the cords, the higher the sound.

During inhalation, air from the larynx moves into the windpipe, or **trachea**. The trachea branches into two main airways called **bronchi** (singular, bronchus). After the main bronchi enter the lungs, they divide into two branches, then the branches divide again many more times. The branchings continue down to the smallest airways, the terminal bronchioles.

Gas Exchange Portion

In the lungs, terminal bronchioles divide into *respiratory bronchioles*, which can have a few cup-shaped outpouchings from their walls. These outpouchings are called **alveoli** (singular, alveolus), and gas exchange between the air and blood occurs across their walls. The respiratory bronchioles lead into *alveolar ducts*. Here, the outpouchings typically are clustered together to form a larger pouch called an **alveolar sac** (Figure 31.6). Thus the gas exchange portion of the respiratory system extends from the respiratory bronchioles to the alveolar sacs.

Lungs and the Pleural Sac

Human lungs are a pair of elastic, cone-shaped organs separated from each other by the heart and other structures (Figure 31.6). They are positioned in the rib cage above the **diaphragm**, which is a muscular partition between the chest cavity (or thoracic cavity) and the abdominal cavity.

A pair of lungs contains more than 300 million alveoli surrounded by a dense mesh of blood capillaries. Collectively, the alveoli provide a tremendous surface area (about 50 to 100 square meters) for exchanging gases with the bloodstream. If the alveolar epithelium were stretched out as a single layer, it would cover the floor of a racquet ball court!

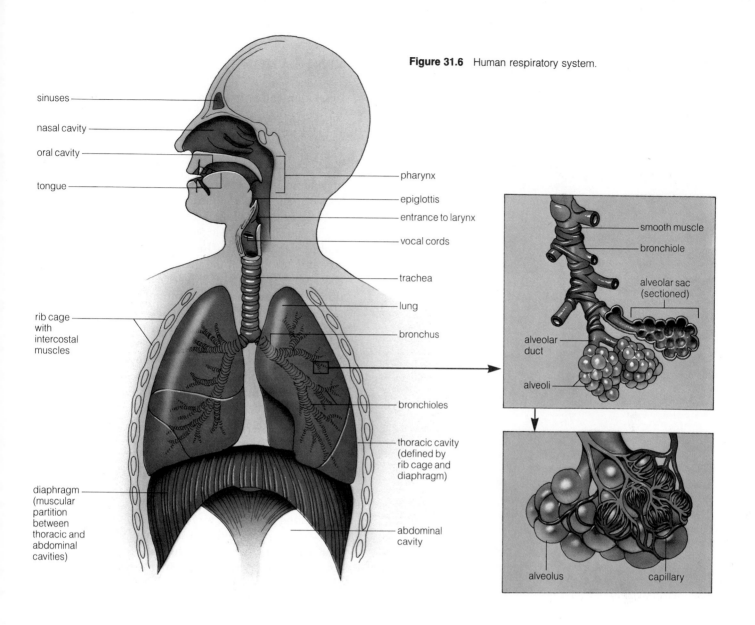

Figure 31.6 Human respiratory system.

sinuses

nasal cavity

oral cavity

tongue

pharynx

epiglottis

entrance to larynx

vocal cords

trachea

lung

bronchus

bronchioles

thoracic cavity (defined by rib cage and diaphragm)

abdominal cavity

rib cage with intercostal muscles

diaphragm (muscular partition between thoracic and abdominal cavities)

smooth muscle

bronchiole

alveolar sac (sectioned)

alveolar duct

alveoli

alveolus

capillary

Lungs are not attached directly to the wall of the chest cavity. Each lung is positioned within a **pleural sac** (a thin membrane of epithelium and loose connective tissue). By analogy, imagine pushing a closed fist into a fluid-filled balloon (Figure 31.8). A lung occupies the same kind of position as your fist; and the pleural membrane folds back on itself, as does the balloon. Only an extremely narrow *intrapleural space* separates the two facing surfaces of the membrane.

Part of the pleural membrane (the parietal pleura) adheres to the wall of the chest cavity. The other part (the pulmonary pleura) is firmly attached to the lungs. The facing surfaces of the membrane are coated with a thin film of lubricating fluid that prevents friction between them while you breathe. (When the pleural membrane becomes inflamed and swollen, friction does occur and breathing can be painful. This condition is called *pleurisy*.)

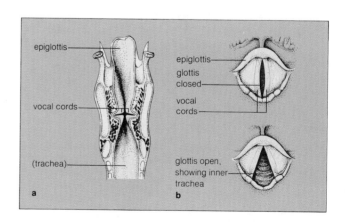

epiglottis

vocal cords

(trachea)

epiglottis

glottis closed

vocal cords

glottis open, showing inner trachea

a

b

Figure 31.7 Where the sounds necessary for speech originate. (a) Front view of the larynx, showing the location of the vocal cords. (b) The two vocal cords as viewed from above when the glottis (the space between them) is closed or opened.

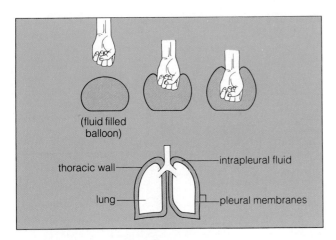

Figure 31.8 Position of the lungs and pleural sac relative to the chest (thoracic) cavity. By analogy, when you push a closed fist into a fluid-filled balloon, the balloon completely surrounds the fist except at your arm. A lung is analogous to the fist; the balloon, to the pleural sac. Here, intrapleural fluid volume is enormously exaggerated for clarity.

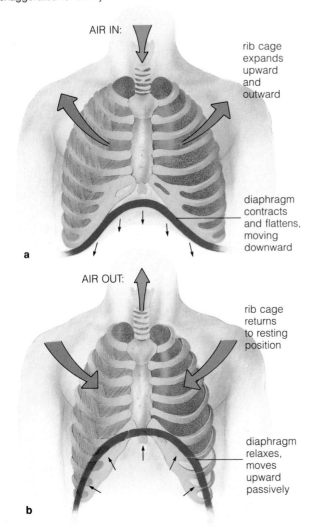

Figure 31.9 Changes in the size of the chest cavity during breathing. Blue line indicates the position of the diaphragm during inhalation (**a**) and exhalation (**b**).

VENTILATION

Inhalation and Exhalation

When you breathe, air is inhaled (drawn into the air-conducting tubes), then exhaled (expelled from them). These air movements result from rhythmic increases and decreases in the volume of the chest cavity. The changes in volume lead to reversals in the pressure gradients between the lungs and the atmosphere, and gases in the respiratory tract follow those gradients.

As you start to inhale, your dome-shaped diaphragm contracts and flattens. The action of this muscular sheet accounts for most of the increase in the volume of the chest cavity (Figure 31.9). Muscles that move the rib cage upward and outward also contribute to the expansion.

When the chest cavity expands, the rib cage moves away very slightly from the lung surface, and pressure drops in the intrapleural space. Even between breaths the intrapleural pressure is lower than atmospheric pressure in the lungs, but now the pressure difference becomes large enough to push the lungs outward. Thus incoming air does not expand the lungs; the lungs are *already* expanded. Now fresh atmospheric air flows down the air-conducting tubes, almost to the terminal bronchioles.

As you start to exhale, the muscular contractions that brought about expansion of the chest cavity (and the lungs) have ceased. The muscles now relax, and the elastic lung tissue that was stretched during inspiration now recoils passively toward its resting volume. As a result, the volume of the chest cavity decreases and thereby compresses the air in the alveoli. With this compression, the alveolar pressure becomes greater than the atmospheric pressure, and air follows the gradient and moves out from the lungs.

Lung Volumes

When you are resting, about 500 milliliters of air enter or leave your lungs in a normal breath. This is called the "tidal volume." The maximum volume of air that can move into and out of your lungs in a single, deep breath is called the "vital capacity." You rarely use more than half the total vital capacity, even when you breathe deeply during strenuous exercise. (To do so would exhaust the muscles used in respiration.) Even at the end of your deepest breath and exhalation, your lungs still would not be completely emptied of air; about 1,000 milliliters (the "residual volume") would remain.

How much of the 500 milliliters of inhaled air is actually available for gas exchange? About 150 milliliters remain in the air-conducting tubes between breaths. Thus only (500 − 150), or 350 milliliters of fresh air reach the alveoli with each inhalation. When you breathe, say, ten times a minute, you are supplying your alveoli with (350 × 10) or 3,500 milliliters of fresh air per minute.

GAS EXCHANGE AND TRANSPORT

Gas Exchange in Alveoli

Each alveolus is like a tiny, empty bowl, its rim continuous with the walls of an alveolar duct. Each "bowl" is no more than a single layer of epithelial cells, surrounded by a thin basement membrane. Thus, at most, the gas in alveoli is separated from the blood in pulmonary capillaries by only a thin film of interstitial fluid and the thin capillary and alveolar walls (Figure 31.10). Under normal conditions, gas can diffuse rapidly across this narrow space.

Figure 31.11 shows the partial pressure gradients for oxygen and carbon dioxide through the human respiratory system. Passive diffusion alone is enough to move oxygen across the respiratory surface and into the bloodstream. And it is enough to move carbon dioxide in the reverse direction (from the bloodstream into the spaces inside the alveolar sacs). Carbon dioxide is about twenty times more soluble than oxygen, and it diffuses more rapidly across the respiratory surface.

Driven by its partial pressure gradient, oxygen diffuses from alveolar air spaces, through interstitial fluid, and into the blood capillaries.

Carbon dioxide, driven by its partial pressure gradient, diffuses in the reverse direction.

Gas Transport Between Lungs and Tissues

Diffusion of oxygen and carbon dioxide down their gradients is enough to move adequate amounts of these gases across the respiratory surface in the lungs. But the amounts that can be carried by blood in dissolved form is not enough to meet the needs of the body, and their transport must be enhanced. Oxygen-carrying hemoglobin in red blood cells increases the oxygen transport by seventy times. Carbon dioxide transport is increased seventeen times by a series of reversible reactions, which will be described shortly.

Oxygen Transport. With each new breath, air that is low in carbon dioxide and rich in oxygen enters the lungs and flows down into the alveoli. Within the lung capillaries adjacent to the alveoli, the blood is low in oxygen and rich in carbon dioxide. Oxygen diffuses into the

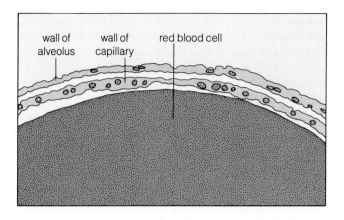

Figure 31.10 Diagram of a section through an alveolus and an adjacent blood capillary. By comparison to the diameter of the red blood cell, the diffusion distance across the capillary wall, the interstitial fluid, and the alveolar wall is exceedingly small.

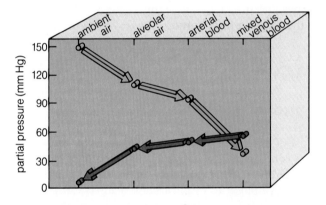

Figure 31.11 Summary of partial pressures for oxygen (blue) and carbon dioxide (red) in different body regions. The graph shows that both gases diffuse down gradients of decreasing partial pressure.

blood plasma, then into red blood cells. When it does, it rapidly forms a weak, reversible bond with hemoglobin. Recall that each hemoglobin molecule can bind four oxygen molecules at the same time. Oxygen combined with hemoglobin is called **oxyhemoglobin** (HbO_2). The amount of oxygen that does combine with hemoglobin depends on the partial pressure of the gas. The higher the pressure, the more oxygen will be picked up (until the hemoglobin binding sites are saturated).

Hemoglobin holds onto oxygen rather weakly and tends to give it up in regions where the partial pressure of oxygen is less than it is in the lungs. It also lets go of oxygen in regions where the blood is warmer and shows a decrease in pH. A decrease in the partial pressure of oxygen, increased carbon dioxide production, declining pH values, and local temperature increases are all characteristics of tissues that are showing greater metabolic activity. That is why more oxygen is released from oxyhemoglobin in highly active tissues (such as muscles during periods of exercise).

---

1. Oxyhemoglobin gives up oxygen when oxygen partial pressure is low, carbon dioxide production is high, temperature is elevated, and pH values are low.

2. These conditions are characteristic of tissues showing increased metabolic activity.

Carbon Dioxide Transport.   The partial pressure of carbon dioxide in blood flowing through the capillaries is lower than it is in the surrounding tissues. Thus carbon dioxide diffuses down its pressure gradient, from the tissues into the capillaries. From there, it is transported to the lungs in three forms:

CO_2 (carbon dioxide dissolved in blood plasma)

$HbCO_2$ (carbon dioxide combined with hemoglobin, forming *carbaminohemoglobin*)

HCO_3^- (bicarbonate ions)

Most of the carbon dioxide produced in the body's tissues is transported to the lungs in the form of bicarbonate. Carbon dioxide combines with water in the plasma portion of blood to form carbonic acid (H_2CO_3), which then dissociates into bicarbonate and hydrogen ions:

$$CO_2 + H_2O \rightleftharpoons H_2CO_3 \rightleftharpoons HCO_3^- + H^+$$

Although the reaction proceeds slowly in plasma, much of the carbon dioxide diffuses into red blood cells, which contain the enzyme carbonic anhydrase. With this enzyme, the reaction rate increases by 250 times and the concentration of free carbon dioxide in the blood drops rapidly. Its action helps maintain the concentration gradient that keeps carbon dioxide diffusing from interstitial fluid into the bloodstream.

Hemoglobin acts as a buffer for the hydrogen ions that are produced by the dissociation of carbonic acid. (A buffer, recall, is a molecule that combines with or releases hydrogen ions in response to changes in cellular pH.) The bicarbonate ions tend to diffuse out of the red blood cells into the blood plasma. Typically, about 70 percent of the carbon dioxide in blood is transported as bicarbonate. Only about 7 percent remains dissolved in plasma (as carbon dioxide). The remaining 23 percent is transported as carbaminohemoglobin.

In the lungs, where the partial pressure of carbon dioxide in the alveoli is lower than it is in the capillaries, the reactions proceed in the reverse direction. Carbonic acid dissociates to form water and carbon dioxide, which diffuses down its concentration gradient and is exhaled from the body. The blood is now ready for another round trip through the systemic circulation.

---

1. Hemoglobin is central to the transport of oxygen from the lungs to tissues throughout the body. It combines with or releases oxygen in response to shifts in the partial pressure of oxygen, carbon dioxide concentration, pH, and temperature.

2. Hemoglobin helps maintain the partial pressure gradient of oxygen between the lungs and body tissues. It does this by taking up oxygen from the blood plasma in the lungs and releasing oxygen in metabolically active tissues.

3. Carbon dioxide transport is enhanced by reactions that rapidly convert the gas to bicarbonate and hydrogen ions in tissues, then back to carbon dioxide in the lungs. These conversions help maintain the concentration gradient necessary for the gas to diffuse into and out of the bloodstream.

MATCHING AIR FLOW AND BLOOD FLOW DURING VENTILATION

Gas exchange in the alveoli is most efficient when the rate of air flow is matched with the rate of blood flow. Both rates can be adjusted locally (in the lungs) and throughout the body as a whole.

Neural Control Mechanisms

The nervous system controls oxygen and carbon dioxide levels in arterial blood for the entire body. It does this through homeostatic mechanisms that influence the rate and depth of breathing. In general, these are the main elements in respiratory control:

Respiratory centers in the brain

Chemoreceptors in the brain and in the walls of arteries

Respiratory muscles (effectors of change)

Contraction of the diaphragm and muscles that move the rib cage is under the control of cells of the reticular formation (page 343). One cluster of cells, in the medulla oblongata of the brain, is concerned mostly with coordinating the contractions associated with inhalation. Another is concerned with coordinating the signals for

Figure 31.12 The atmosphere contains the same percentage of oxygen at high and low altitudes. But the atmospheric pressure is lower at high altitudes, so the partial pressure exerted by its oxygen component is not as great. Hence less oxygen is available to move from the air into the body. People who are not adapted to high altitudes can experience hypoxia, or cellular oxygen deficiency.

exhalation. Respiratory centers in other parts of the brain can stimulate or inhibit these cell clusters; they work to fine-tune the rhythmic contractions of the muscles concerned with respiration.

Moveover, chemoreceptors in the medulla respond indirectly to rising carbon dioxide levels in the blood. (Such increases affect the level of H$^+$ in the fluid in the brain, and the shift in pH stimulates the receptors.) Other chemoreceptors include the *carotid bodies* (which are located at the branching of the carotid arteries to the brain) and the *aortic bodies* (which are specialized structures in arterial walls near the heart). These receptors detect decreases in the partial pressure of oxygen dissolved in arterial blood, and they can trigger an increase in the rate and depth of respiration (hence more oxygen delivery to tissues).

Local Control Mechanisms

In the lungs themselves, changes in the diameter of bronchioles help control the proportion of air and blood flow to alveoli. When there is not enough air flow and too much blood flow, local levels of carbon dioxide rise. Smooth muscles in the bronchiole walls are sensitive to such increases and they dilate in response—and thereby increase the local air flow. Similarly, a decrease in carbon dioxide levels causes the bronchiole walls to constrict—and thereby decrease the air flow.

Local changes also occur in the diameter of blood vessels that supply different lung regions. If air flow is too large relative to the blood flow, oxygen levels rise in these regions. The increase directly affects smooth muscle in the blood vessel walls, which undergo vasodila-

tion—and thereby increase blood flow to the region. Similarly, if air flow is too small, vasoconstriction leads to a decrease in blood flow (page 412).

Hypoxia

When tissues are not supplied with enough oxygen, *hypoxia*, or cellular oxygen deficiency, is the result. For example, hypoxia can occur at high altitudes, where the partial pressure of oxygen is lower than it is at sea level (Figure 31.12). At an altitude of 2,400 meters (about 8,000 feet), a person attempts to compensate for the oxygen deficiency by hyperventilating (breathing much faster and more deeply than would otherwise be done at a given level of activity). At 3,650 meters (about 12,000 feet), the partial pressure of oxygen is only 100mm Hg, and oxygen deprivation can cause headaches, nausea, and lethargy. At 7,000 meters (23,000 feet), hypoxia can lead to loss of consciousness and death.

Hypoxia also occurs when the partial pressure of oxygen in arterial blood falls because of *carbon monoxide poisoning.* Carbon monoxide (CO) is a colorless, odorless gas that is present in the exhaust fumes from gasoline-powered vehicles, in tobacco smoke, and in the smoke from coal or wood burning. It combines with hemoglobin at least 200 times faster than oxygen does. Even at a partial pressure of only 0.5mm Hg, carbon monoxide will tie up *half* the hemoglobin in the body, forming carboxyhemoglobin (HbCO). When that happens, HbCO levels are high and HbO$_2$ levels are low. This means that the oxygen content of arterial blood is low even though the partial pressure for oxygen remains normal. Thus the chemoreceptors that are supposed to

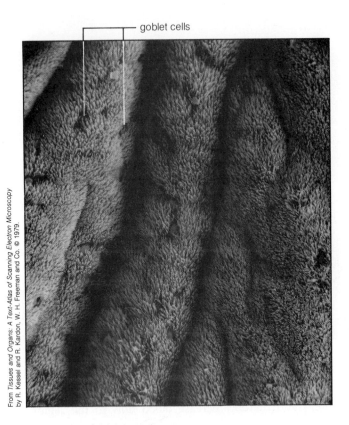

goblet cells

From *Tissues and Organs: A Text-Atlas of Scanning Electron Microscopy* by R. Kessel and R. Kardon, W. H. Freeman and Co. © 1979.

Figure 31.13 Ciliated epithelium of the human trachea, 830×.

detect decreases in oxygen levels are not stimulated, the body does not step up ventilation, and carbon monoxide poisoning occurs. This condition can be reversed in some cases by administering pure oxygen at a partial pressure of 600mm Hg to slowly replace the HbCO.

HOUSEKEEPING AND DEFENSE IN THE RESPIRATORY TRACT

The respiratory surface of the lungs is exposed to what can be a harsh environment. Vapors from acids, airborne particles of dust (which may be carrying bacteria), smoke, ashes, soot, oil, asbestos particles, and other substances may be inhaled daily.

Large airborne particles are filtered out at the nose, and smaller particles are filtered out at the air-conducting tubes, which are lined with ciliated epithelium. Scattered between the hairlike cilia are mucus-secreting goblet cells (Figure 31.13). The small particles stick in the mucus of the epithelium, and the rhythmically upward-beating cilia sweep the debris-laden mucus toward the mouth. The mucus is then swallowed or expelled. In the alveoli, migrating macrophages engulf what they can, and leukocytes eliminate some of the foreign material.

Table 31.1 To Smoke or Not to Smoke: Some Comparisons*

Risks Associated With Smoking	Benefits of Quitting
Shortened Life Expectancy: Nonsmokers live an average of 8.3 years longer than those in mid-twenties who smoke two packs of cigarettes a day	Cumulative reduction of risk; after 10–15 years, the life expectancy of ex-smokers approaches that of nonsmokers
Chronic Bronchitis, Emphysema: Smokers have 4–25 times more risk of dying from these diseases than do nonsmokers	Greater chance of improving lung function and slowing down rate of deterioration
Lung Cancer: Cigarette smoking the major cause of lung cancer	After 10–15 years, risk approaches that of nonsmokers
Cancer of Mouth: 3–10 times greater risk among smokers	After 10–15 years, risk is reduced to that of nonsmokers
Cancer of Larynx: 2.9–17.7 times more frequent among smokers	After 10 years, risk is reduced to that of nonsmokers
Cancer of Esophagus: 2–9 times greater risk of dying from this form of cancer	Risk is proportional to amount smoked, so quitting should reduce risk
Cancer of Pancreas: 2–5 times greater risk of dying from pancreatic cancer	Risk is proportional to amount smoked, so quitting should reduce risk
Cancer of Bladder: 7–10 times greater risk for smokers	Risk decreases gradually over 7 years to that of nonsmokers
Coronary Heart Disease: Cigarette smoking a major contributing factor	Risk drops sharply after a year; after 10 years, risk reduced to that of nonsmokers
Effects on Offspring: Women who smoke during pregnancy have more stillbirths, and weight of liveborns averages less (hence babies are more vulnerable to disease, death)	When smoking stops before fourth month of pregnancy, risk of stillbirth and lower birthweight eliminated
Impaired Immune System Function: Increase in allergic responses, destruction of macrophages in respiratory tract	Avoidable by not smoking

*Based on data published in 1980 by the American Cancer Society, Inc.

a

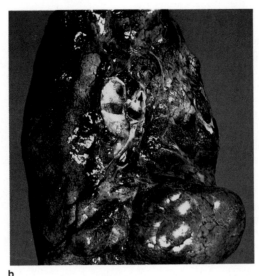
b

Figure 31.14 (a) Normal appearance of human lung tissue and (b) appearance of a lung taken from a person who suffered emphysema.

When Defenses Break Down

In urban environments, in certain occupations, even near a cigarette smoker, airborne particles and certain gases are present in abnormal amounts, and they put an extra workload on the respiratory tract. For example, ciliated epithelium in the air-conducting tubes is extremely sensitive to cigarette smoke, probably because of the chemical nature of the concentrated particles.

Bronchitis. Consider the conditions that lead to a lung disorder called bronchitis. Smoking and other forms of bronchial irritants increase the secretion of mucus while interfering with ciliary action in the air-conducting tubes of the lungs. Mucus and the particles it traps—including bacteria—begin to accumulate in the trachea and bronchi.

Now coughing sets in as the body attempts to clear away the mucus. If the irritation continues, the coughing reflex persists. Coughing aggravates the condition because it further irritates the bronchial walls, which become inflamed and infected. Cilia diminish in numbers, and mucus-secreting cells multiply as the body works to fight against the accumulating debris. With all of this aggravation, fibrous scar tissue forms that can obstruct parts of the respiratory tract.

Emphysema. A person suffering an acute attack of bronchitis who is otherwise in good health responds to medical treatment. But what happens if the irritation persists, particularly in a person who is prone to develop lung infections? As fibrous scar tissue builds up in the respiratory tract, the bronchi become progressively clogged with more mucus. Air becomes trapped in alveoli, and the alveolar walls break down. The remaining alveoli enlarge and the balance between air flow and blood flow is altered. The outcome is *emphysema*, in which the lungs are distended and gas exchange efficiency is compromised (Figure 31.14). Running, walking, and even exhaling can be difficult.

For some people, early environmental conditions—poor diet, chronic colds, other respiratory ailments—apparently can create a predisposition to emphysema later in life. Also, some who suffer from emphysema have a hereditary deficiency in their ability to form antitrypsin, a substance that inhibits tissue-destroying enzymes produced by bacteria. These individuals may be at a disadvantage in fighting off respiratory infections when they do strike.

Emphysema is insidious. It can develop slowly, over twenty or thirty years. By the time it is detected, the damage to lung tissue is irreparable. About 1,300,000 individuals in the United States alone now suffer from the disorder.

Effects of Cigarette Smoke. Table 31.1 lists some effects of cigarette smoke on the airways. The noxious particles in smoke from one cigarette can prevent the cilia from beating for several hours. The particles also stimulate excessive secretions of mucus, which can eventually clog the airways. "Smoker's cough" is one of the least serious consequences; the coughing can contribute to the development of bronchitis and emphysema, as described above. Cigarette smoke can also destroy phagocytic cells that defend the respiratory epithelium. Extensive lung damage has also been documented in marijuana smokers.

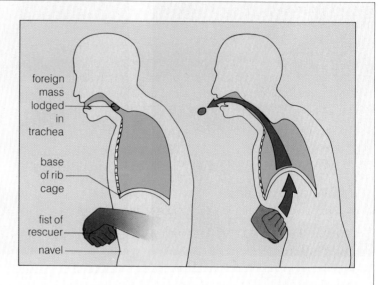

Figure 31.15 The Heimlich maneuver, an emergency procedure used to dislodge foreign matter blocking the respiratory tract.

First, stand behind the victim, make a fist with one hand, then press it thumb-side in against the victim's abdomen. The fist must be slightly above the navel and well below the rib cage.

Second, press the fist into the abdomen with a sudden upward thrust. Repeat the thrust several times if needed. The maneuver can be performed on a victim who is standing, sitting, or lying down.

Once the foreign matter is dislodged, be sure the victim is seen at once by a physician, for an inexperienced rescuer can inadvertently cause internal injuries or crack a rib. It could be argued that the risk is well worth taking, given that the alternative is death.

foreign mass lodged in trachea

base of rib cage

fist of rescuer

navel

Cigarette smoke also contains compounds that can lead to lung cancer. These compounds, including methylcholanthrene, are found in coal tar and cigarette smoke. It appears that they become chemically modified in the body through the action of natural substances, thereby turning into highly reactive intermediates that are the real carcinogens. The carcinogens act on gene expression in the cells of lung tissues in ways that lead to uncontrolled cell division.

Susceptibility to lung cancer is related to the number of cigarettes smoked each day as well as to the extent and depth of inhalation. At least eighty percent of all lung cancer deaths are the legacy of cigarette smoking. In its terminal stage, lung cancer is agonizing. It is a disorder that only ten out of a hundred afflicted individuals will survive, with varying degrees of tissue damage and malfunctioning.

The Heimlich Maneuver

Sometimes large chunks of food become lodged in the respiratory tract. Each year, several thousand people die from strangulation when food enters the trachea instead of the esophagus. Strangulation can occur when the air flow is blocked for as little as four or five minutes.

Such misdirected chunks often can be dislodged by an emergency procedure called the *Heimlich maneuver*. With this procedure, the diaphragm is forcibly elevated, causing a sharp decrease in the volume of the chest cavity and a sudden increase in alveolar pressure. The increased pressure forces air up the trachea, and it is often enough to dislodge the obstruction (Figure 31.15).

SUMMARY

1. Aerobic metabolism requires oxygen and produces carbon dioxide. The process of acquiring oxygen and eliminating carbon dioxide is called respiration.

2. Diffusion of substances, including gases, depends on differences in concentration between two regions. The concentration of gases is measured in terms of partial pressure (the pressure exerted by one gas in a mixture of gases).

3. Respiratory systems make use of the diffusive properties of oxygen and carbon dioxide, which tend to move from a region of higher partial pressure to a region of lower partial pressure.

4. In some respiratory systems, gases are exchanged across the whole body surface (integumentary exchange); in others, they are exchanged at specialized respiratory surfaces known as gills, tracheas, and lungs. All these systems have a moist, thin vascularized layer of epidermis across which gases are exchanged.

5. In vertebrates, airways carry gas to and from one side of the respiratory surface of lungs, and blood vessels carry gas to and from the other side.

6. Air moves by bulk flow into and out of the lungs, and thus new air is delivered to the respiratory surface. This air movement is called pulmonary ventilation. Gases diffuse across the respiratory surface into the blood of the pulmonary circulation.

7. The air-conducting portion of the human respiratory system consists of the nasal cavities, pharynx, larynx, trachea, main bronchi, and bronchioles. Gas exchange

occurs only in the alveoli, which are located at the end of the air-conducting system.

8. Human lungs are paired structures in the thoracic cavity, which is separated from the abdominal cavity by the diaphragm. Each lung is enclosed in a pleural sac and is separated from the thoracic wall by the inter pleural space.

9. During inhalation, the chest cavity becomes larger because of contraction of the diaphragm and some of the chest-wall muscles; the pressure in the lungs falls below atmospheric pressure, and air moves by bulk flow down the pressure gradient into the lungs. During normal, quiet exhalation these processes are reversed.

10. There is still air left in the lungs (the residual volume) at the end of each exhalation.

11. Driven by its partial pressure gradient, oxygen diffuses from alveolar air spaces into the blood capillaries. Carbon dioxide diffuses in the reverse direction.

12. Once in the blood, oxygen diffuses into the red blood cells where it combines loosely with hemoglobin to form oxyhemoglobin. In the tissue capillaries, oxyhemoglobin gives up oxygen, which diffuses out of the capillaries and into nearby cells.

13. In the tissues, carbon dioxide diffuses into the capillaries. It is transported in the blood as a dissolved gas, in combination with proteins (carbamino compounds), and as bicarbonate ions. In the lungs, the reactions that occurred in the tissue capillaries are reversed, and the carbon dioxide diffuses from the blood out into the air spaces of the alveoli.

14. Hemoglobin is central to the transport of oxygen from the lungs to the tissues throughout the body. It combines with or releases oxygen in response to shifts in oxygen levels, pH, and temperature.

15. Respiratory control centers in the reticular formation of the brainstem alter the rate and depth of respiration to maintain the levels of oxygen, carbon dioxide, and hydrogen ions within ranges compatible with life.

Review Questions

1. What is the main requirement for gas exchange in animals? What types of systems are used for gas exchange in (a) water-dwelling animals and (b) land-dwelling animals? *436, 437–439*

2. Explain how a countercurrent flow mechanism works in a fish gill. *437*

3. Define respiration. What five events occur during human respiration? *436, 440*

4. By what mechanisms do carbon dioxide move out of your body and oxygen into it through the *same* system of branched tubes? *442*

5. What governs the rate and depth of breathing? *444–445*

6. What force drives oxygen from alveolar air spaces, through interstitial fluid, and across capillary epithelium? What force drives carbon dioxide in the reverse direction? *443*

7. How does hemoglobin help maintain the oxygen partial pressure gradient during gas transport in the body? What reactions enhance the transport of carbon dioxide through the body? *444*

8. Label the component parts of the human respiratory system: *441*

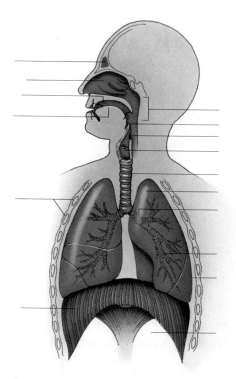

Readings

American Cancer Society. 1980. *Dangers of Smoking; Benefits of Quitting and Relative Risks of Reduced Exposure.* Revised edition. New York: American Cancer Society, Inc.

Baker, P. 1969. "Human Adaptation to High Altitude." *Science* 163:1149.

Ganong, W. 1979. *Review of Medical Physiology.* Ninth edition. Los Altos, California: Lange Publications. Advanced reading, but one of the most authoritative books on respiration.

Hickman, C., et al. 1983. *Integrated Principles of Zoology.* Seventh edition. St. Louis: Mosby.

Vander, A., J. Sherman, and D. Luciano. 1985. *Human Physiology: The Mechanisms of Body Function.* Fourth edition. New York: McGraw-Hill. Clear introduction to the respiration system and its functioning.

West, J. 1985. *Respiratory Physiology: The Essentials.* Third edition. Baltimore: Williams & Wilkins. Excellent, brief introduction to respiratory functions. Paperback.

Wyman, R. 1977. "Neural Generation of the Breathing Mechanism." *Annual Review of Physiology* 39:417.

32

DIGESTION AND ORGANIC METABOLISM

Nutrition—here is a word that has to do with all those processes by which the body takes in, digests, absorbs, and uses food. The word signals that you are about to begin one more educational trek through the animal gut. This time around, however, you will move beyond passive memorization of names for specialized tissues and organs. This time your main concern will be with *systems integration*—with how systems interact in meeting the metabolic needs of the entire body.

Consider the female bear in Figure 32.1 and the destination of that salmon in her mouth. Is it enough, really, to assume the nutritional picture begins and ends in her gut? If nutrients are to reach all living cells, they must cross the gut lining and enter the bloodstream. If cells are to use nutrients, a respiratory system must supply them with oxygen (for aerobic metabolism) and relieve them of carbon dioxide. Because bears do not eat around the clock, there must be ways to store and release nutrients at different times. So "nutrition" in complex animals requires more than food digestion and absorption. It also requires systems of distributing, storing, and using particular substances in controlled ways (Figure 32.2).

This chapter describes the components of digestive systems and their functions. Then it describes how the activities of three systems—digestive, circulatory, and respiratory—are integrated. The main examples are from an organism with which you are already more or less acquainted: yourself.

TYPES OF DIGESTIVE SYSTEMS AND THEIR FUNCTIONS

A **digestive system** is some form of body cavity or tube in which food is first reduced to particles, then to molecules small enough to move into the internal environment. (The internal environment, recall, is the body's *extracellular fluid*. It consists of interstitial fluid that bathes living cells and, in animals with circulatory systems, the plasma portion of blood.) A layer of cells serves as a lining for the body cavity or tube, and nutrients enter the internal environment by crossing that lining.

An **incomplete digestive system** has only one opening, and what goes in but cannot be digested goes out the same way. A flatworm called a planarian has this type of system. A muscular organ (pharynx) opens into a highly branched cavity that serves both digestive and circulatory functions (Figure 32.3). Food is partly digested and transported to cells even while residues are being sent back out through the pharynx. Because of the two-way traffic, this cavity cannot be subdivided into specialized regions for food transport, processing, and storage.

Figure 32.1 Digestion includes those interrelated processes by which food is ingested, prepared for absorption, and moved into the internal environment.

Annelids, mollusks, arthropods, echinoderms, and chordates have a **complete digestive system**. They have an internal tube with an opening at one end for taking in food and an opening at the other end for eliminating unabsorbed residues (Figure 32.3). Between the two openings, food generally moves in one direction through the *lumen* (the space inside the tube). And the tube itself is subdivided into specialized regions for food transport, processing, and storage. For example, one part of the digestive tube of earthworms and birds is modified into a crop (a food storage organ). Another part is modified into a gizzard (a muscular organ in which food is ground into smaller bits).

The specialized regions of complete digestive systems can be correlated with the animal's feeding behavior. Animals with *discontinuous* feeding habits may eat large amounts of food when it is available, then go for long periods without eating at all. Certain organs in these animals store food that is taken in faster than it can be digested and absorbed. Other organs help maintain an adequate distribution of nutrients when food is not being ingested.

Many grazing animals eat almost continuously and face other digestion tasks. For example, they have to digest cellulose, which requires specific enzymes and a rather long processing time for the enzymes to work. **Ruminants**, such as deer, cattle, and goats, have multiple stomach chambers in which the tough cellulose

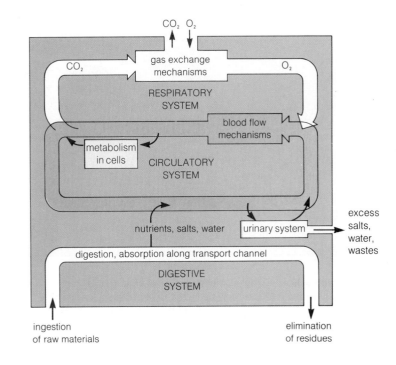

Figure 32.2 Interconnected systems for moving food into the internal environment and for assuring that cells can utilize the energy stored in food once it reaches them. The connections represented here are characteristic of most complex animals. Sensory receptors in each system channel information to the nervous system, which coordinates the interrelated activities. Hormonal controls are also at work.

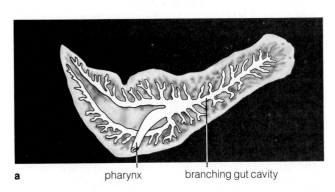

a   pharynx   branching gut cavity

b   anus   intestine   crop   mouth

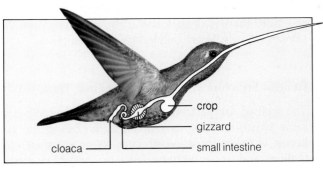

c

Figure 32.3 (a) Incomplete digestive system of the planarian, a flatworm. The pharynx of this animal is a muscular tube that protrudes from the body during feeding. Complete digestive systems of animals ranging from earthworms (b) to birds (c) and humans have special food-processing regions and a one-way movement of material.

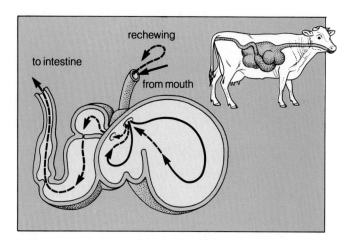

Figure 32.4 Complete digestive system of ruminants, such as cattle. Ruminants swallow partially chewed plant material, which moves into two stomachlike chambers. Then they regurgitate the material, chew it more, and swallow it again. The double chewing time mechanically breaks apart the plant material, which contains tough cellulose fibers. Symbiotic bacteria present in the digestive tract produce enzymes that can digest cellulose. The double chewing gives the enzymes more time and more surface area upon which to act. Altogether, food is processed in four stomachlike chambers before being sent on to the small intestine for absorption.

fibers are processed gradually (Figure 32.4). The first two chambers house vast microbial populations. Among the microbes are symbiotic bacteria that produce the enzymes capable of degrading cellulose (and other nutrients). Ruminants regurgitate food from the first two chambers and rechew it before swallowing again. Thus the plant material is mixed and pummeled more than once, and more cellulose fibers are exposed to agents of digestion before continuing through the digestive tract.

No matter how regionally specialized they have become, the digestive systems of most animals have four main functions:

Motility: mechanical breakdown and mixing of ingested nutrients, passage of nutrients through the digestive tract, and elimination of undigested or unabsorbed residues from the body.

Secretion: release of enzymes, hormones, and other substances that take part in digestion.

Digestion: chemical reduction of ingested nutrients into particles, then into molecules small enough to cross the lining of the gut and reach the internal environment.

Absorption: passage of digested nutrients from the gut lumen into the blood or lymph, which distributes them through the body.

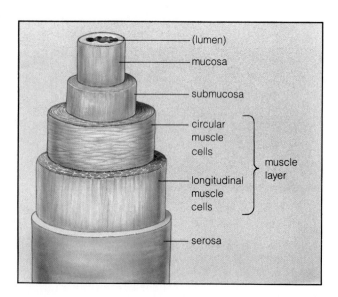

Figure 32.5 Generalized sketch of the wall of the gastrointestinal tract.

HUMAN DIGESTIVE SYSTEM: AN OVERVIEW

Components of the Digestive System

Humans have discontinuous feeding habits, and they dine on various and diverse foodstuffs. From this you might deduce (correctly) that the human digestive system is a tube with many regional specializations. The length of the tube, or **gastrointestinal tract** (Figure 32.5), is between 6.5 and 9 meters (21 to 30 feet) in adults. Its regional subdivisions are the mouth, pharynx, esophagus, stomach, small intestine, large intestine (or colon), rectum, and anus (Figure 32.6). Glandular organs having accessory roles in digestion and absorption include the salivary glands, liver, gallbladder, and pancreas.

General Structure of the Gastrointestinal Tract

The wall of the gastrointestinal tract has basically the same structure along its entire length. Facing the gut lumen is the **mucosa**, which consists of a surface epithelium and an underlying layer of connective tissue (Figure 32.5). The mucosa is surrounded by the **submucosa**, a connective tissue layer in which blood and

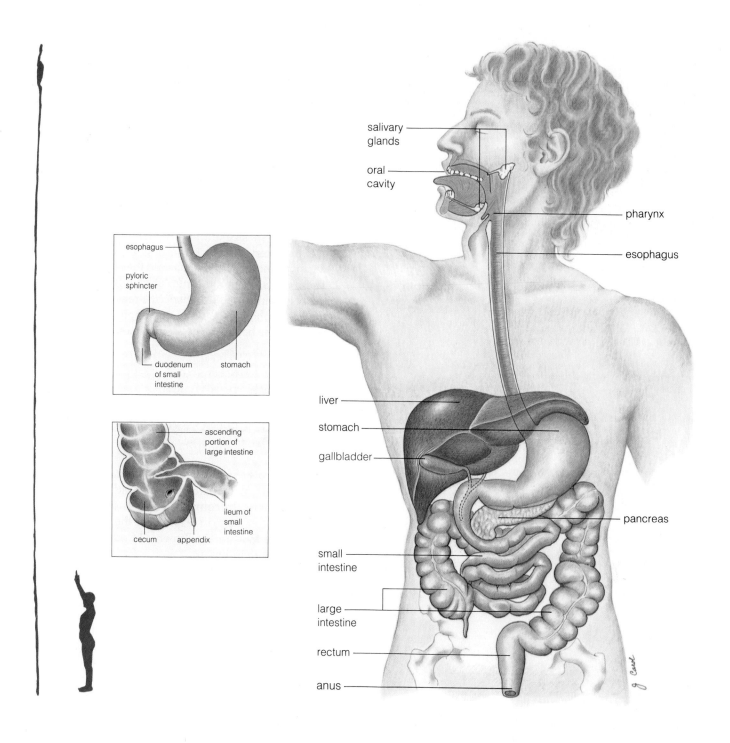

Figure 32.6 Simplified picture of the human digestive system. (Far left: If you have ever wondered how far a stretched-out gastrointestinal tube extends, now you know.)

lymph vessels are meshed. Next is a **muscle layer**, with two sublayers of smooth muscle arranged in longitudinal and circular directions relative to the tube axis. The thin outermost layer of connective tissue is the **serosa**.

Gastrointestinal Motility

Coordinated contractions in the muscle layers of the gastrointestinal tract mix food material with secretions and move it forward. Two common types of movement in the tract are peristalsis and segmentation.

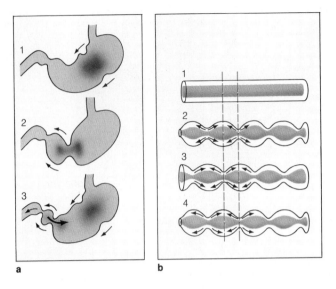

Figure 32.7 (**a**) Peristaltic wave, as it occurs in the stomach. (**b**) Segmentation, or oscillating movement, in the intestines.

During **peristalsis**, a mass of food material advances through the tract when rings of circular muscles contract behind it and relax in front of it. As the mass moves, it expands the tube wall, the expansion stimulates peristalsis, and so on (Figure 32.7). For example, peristaltic waves move down the stomach walls about three times a minute.

Segmentation occurs only in the intestines. Rings of smooth muscle in the intestinal wall repeatedly contract and relax, creating an oscillating (back-and-forth) movement in the same place. This movement constantly mixes the contents of the lumen and forces them against the absorptive surface of the intestinal wall (Figure 32.7).

Sphincters influence the flow of material from one region to another in the tract, and they prevent backflow. These rings of smooth or striated muscle are located at the beginning and end of specific regions. For example, there is a sphincter between the esophagus and stomach, and another between the stomach and small intestine.

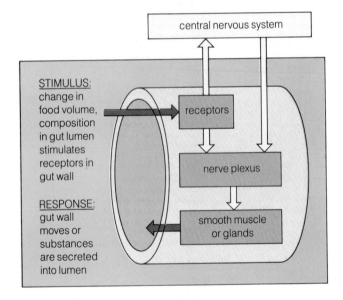

Figure 32.8 Local and long-distance reflex pathways called into action when food is in the digestive tract.

CONTROL OF GASTROINTESTINAL ACTIVITY

Most controls over body activity operate in response to some aspect of the extracellular fluid (such as oxygen concentration). In contrast, controls over the gastrointestinal tract operate in response to the volume and composition of food in the lumen—in other words, to the "external" environment.

Ingested food distends the gut wall, and it varies in solute concentrations and in acidity. During the digestion of carbohydrates, proteins, and fats, the breakdown products as well as certain secretions accumulate in the lumen. Receptors embedded in the gut wall trigger reflex responses to these various stimuli. As Figure 32.8 indicates, a short reflex pathway operates independently of the central nervous system. Signals from the receptors travel through nerve plexuses in the gut wall that can directly influence the wall contractions and secretions into the gut lumen. (A nerve plexus, recall, is a network of nerves outside the central nervous system.) A long-distance reflex pathway connects the receptors and effectors with the central nervous system. One or both pathways can act to control activities of the gastrointestinal tract.

Moreover, several hormones secreted by endocrine cells of the tract help regulate digestion and absorption. These hormones include gastrin, secretin, cholecystokinin (CCK), and gastric-inhibitory peptide (GIP). Table 32.1 lists their sources and functions.

Table 32.1 Functions of Primary and Accessory Organs of Digestion

Organ	Secretions	Main Functions
Mouth	—	Mechanically breaks down food
Salivary glands (accessory organs)	Water	Moistens food
	Mucus	Lubricates and binds food into bolus
	Salivary amylase	Starts breakdown of starch, glycogen
	Bicarbonate	Buffering action neutralizes acidic food in mouth
Stomach	—	Stores, mixes, dissolves food; regulates emptying of chyme into small intestine
Secretory cells in stomach mucosa	Hydrochloric acid	Dissolves food particles; kills many microorganisms
	Pepsinogens	Activated forms (pepsins) split apart peptide bonds in protein chains
	Mucus	Lubricates and protects stomach lining
	Gastrin	Stimulates hydrochloric acid secretion
Small intestine	—	Digestion and absorption of most nutrients; mixes and propels chyme forward
Secretory cells in intestinal mucosa*	Assorted enzymes	Break down major food molecules
	Mucus	Lubricates chyme
	Secretin	Stimulates pancreatic bicarbonate secretion, inhibits gastric acid secretion
	Cholecystokinin	Stimulates gallbladder contraction, pancreatic enzyme secretions; inhibits stomach emptying
	Gastric-inhibitory peptide	Inhibits stomach acid secretion and motility
Pancreas (accessory organ)	Assorted enzymes (e.g., lipase)	Break down all major food molecules
	Bicarbonate	Buffering action neutralizes hydrochloric acid entering small intestine from stomach
Liver (accessory organ)	Bile salts	Hydration of emulsified fat droplets
	Bicarbonate	Buffering action neutralizes hydrochloric acid entering small intestine from stomach
Gallbladder (accessory organ)	—	Stores and concentrates bile from liver
Large intestine (colon)	—	Stores, concentrates undigested matter by absorbing water and salts; mixes and propels material forward
Secretory cells in intestinal mucosa	Mucus	Lubricates undigested residues
Rectum	—	Distension triggers defecation reflex that rids body of undigested and unabsorbed residues

*Most enzymes are embedded in plasma membrane facing the lumen; some are released into lumen when cells are shed and disintegrate.

STRUCTURE AND FUNCTION OF GASTROINTESTINAL ORGANS

Mouth and Salivary Glands

Mechanical reduction of food begins in the **mouth** (oral cavity), as does the digestion of polysaccharides. Most animals have a mouth, but only humans and other mammals *chew* food in the mouth. Adult humans normally have thirty-two teeth (sixteen in the upper jawbone and sixteen in the lower). Each **tooth** consists of an enamel coat (hardened calcium deposits), dentine (a thick bone-like layer), and an inner pulp (which houses nerves and blood vessels). The teeth in the back of the mouth are flat-surfaced *molars*, which grind food. Teeth in the front are chisel-shaped *incisors*, useful in biting off chunks of food. In between are cone-shaped *cuspids*, for grasping and tearing food (see also page 692).

While the teeth and tongue are mechanically reducing food in the mouth, they are also mixing it with *saliva*, a fluid secreted from several **salivary glands**. Ducts of these exocrine glands empty into different parts of the oral cavity.

Salivary amylase, an enzyme that takes part in the initial breakdown of starch, is a component of saliva. So are bicarbonate ions (HCO_3^-), which act as buffers in keeping salivary pH between 6.5 and 7.5 even when acidic foods are in the mouth. Other components are the *mucins*, glycoproteins that bind bits of food together into a softened, lubricated ball called a bolus.

Pharynx and Esophagus

Once food has been processed into a bolus, voluntary muscle contractions move the tongue toward the roof of the mouth. The movement forces the bolus into the pharynx, where it stimulates mechanoreceptors that call for contraction of the walls of the pharynx and esophagus. That contraction initiates swallowing. A swallow can be *initiated* voluntarily, but thereafter it is a reflex controlled by a brain region.

In humans, the **pharynx** is a muscular tube continuous with the **esophagus**, which leads to the stomach. The pharynx is also continuous with the trachea, which leads to the lungs. The swallowing reflex opens a sphincter at the start of the esophagus, and normally it closes off the trachea (hence prevents choking) while food is moving into the esophagus. Neither the pharynx nor the esophagus contributes to digestion. Their peristaltic movements simply propel food to the stomach.

Stomach

The **stomach** is a muscular, distensible sac having three main functions. First, it stores and mixes food received from the esophagus. Second, it secretes substances that help dissolve and degrade food. Third, it helps control the rate at which food moves into the small intestine.

Components of Gastric Fluid. Among the cells scattered throughout the stomach mucosa are exocrine cells that release hydrochloric acid (HCl), pepsinogens, mucus, and other substances directly into the stomach lumen. Each day they secrete as much as two liters of these substances, which together constitute the gastric fluid. Also, endocrine cells in one part of the stomach mucosa release hormones (such as gastrin), which travel by way of the bloodstream to target cells. The secretions of their target cells also contribute to the gastric fluid.

The HCl secreted into the stomach dissolves bits of food, producing a solution called chyme. It also kills most of the microbes entering the body in food. The H^+ concentration in the lumen increases when the HCl molecules dissociate into H^+ and Cl^-. At times the H^+ level in the stomach becomes three million times higher than it is in arterial blood!

An increase in stomach acidity contributes to protein digestion. First, it alters the ionized groups of proteins and exposes some of their peptide bonds. Second, it converts inactive forms of protein-degrading enzymes (called *pepsinogens*) to active forms (called *pepsins*). These enzymes break certain peptide bonds of proteins and thereby produce peptide fragments.

HCl secretion can be stepped up in several ways. Receptors in the stomach wall trigger reflexes that stimulate secretion whenever food distends the wall. Also,

From *Tissues and Organs: A Text-Atlas of Scanning Electron Microscopy.* by R. Kessel and R. Kardon, W.H. Freeman Co. © 1979.

a

b   intestinal villi

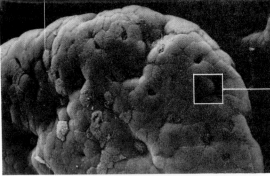

one epithelial cell of villus

c

the peptide fragments of protein breakdown stimulate the release of gastrin, which in turn stimulates the HCl-secreting cells. Other substances, such as the caffeine in coffee, tea, chocolate, and cola drinks also cause an increase in the rate of secretion. Finally, reflexes initiated by food in the mouth as well as by stress can also stimulate HCl secretion by way of the vagus nerve to the stomach (Figure 25.6).

Sometimes part of the stomach mucosa becomes damaged by the digestive action of the gastric fluid. This leads to formation of a *peptic ulcer*. In some way, normal control mechanisms that protect the mucosa are deficient. When the surface of the stomach breaks down, hydrogen ions diffuse into the mucosa and thus trigger the release of histamine. Histamine in turn triggers vasodilation and increased capillary permeability, and it stimulates HCl secretion. Thus a positive feedback loop is set up and leads to tissue damage, which may result in bleeding into the stomach and abdomen.

Stomach Emptying. Peristaltic waves in the stomach mix the chyme. The waves build up force as they approach the pyloric sphincter between the stomach and small intestine (Figure 32.7). Normally the sphincter is relaxed, but with the arrival of a strong peristaltic contraction, it closes. Most of the chyme is squeezed back; only a small amount moves into the duodenum, the first part of the small intestine.

Three factors control how fast the stomach empties. *First*, stomach distension following a meal activates mechanoreceptors in the stomach wall. The larger the meal, the more these mechanoreceptors trigger reflexes that increase the force of contraction (hence stomach emptying). *Second*, increases in acidity, osmotic pressure, and fat content stimulate receptors in the duodenum.

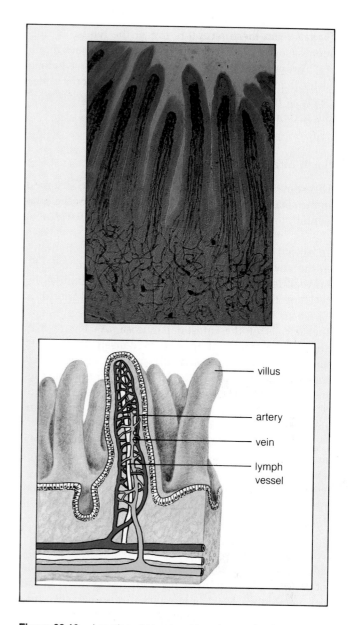

Figure 32.10 Location of blood and lymph vessels within intestinal villi.

microvillus at cell surface

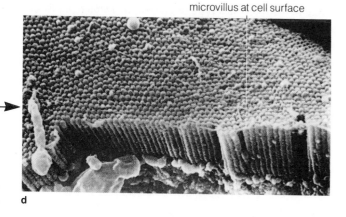

d

Figure 32.9 (**a,b**) Location of villi in the mammalian intestine. (**c**) Tip of a single villus, 825x. (**d**) The dense crown of microvilli at the surface of an epithelial cell. Microvilli face the lumen of the intestine and enhance the absorptive and secretory processes.

Signals from these receptors trigger the release of hormones (such as CCK and GIP) that inhibit stomach motility. Through such slowdowns, food is not moved along faster than it can be processed. *Third*, emotional states (such as fear and depression) can trigger signals from the nervous system that also inhibit motility.

Small Intestine

Digestion is completed and most nutrients are absorbed in the **small intestine**, which has three regions: the duodenum, jejunum, and ileum.

Intestinal Villi.   Most proteins, fats, and carbohydrates in the chyme have been broken down to amino acids, fatty acids, monoglycerides, and monosaccharides by the time they are halfway through the small intestine. These small organic molecules readily move across epithelial cells of the intestinal mucosa. As Figure 32.9 shows, the intestinal mucosa is densely folded into absorptive structures called **villi** (singular, villus), which look like tiny tongues. Beneath its one-cell-thick epithelium, each villus houses blood and lymph vessels (Figure 32.10). Contraction of the villus promotes fluid flow through these vessels.

In themselves, villi greatly increase the surface area available for interactions with the chyme. Moreover, epithelial cells of each villus have a surface crown of **microvilli**: threadlike projections of plasma membrane that further increase the surface area available for absorption (page 80 and Figure 32.9).

Absorption Mechanisms.   The movement of glucose and certain other monosaccharides into epithelial cells of the villi depends on active transport systems in the cell membranes. Most amino acids are also actively transported; others move passively into the epithelial cells. Free fatty acids and monoglycerides can diffuse across the cells because they are soluble in the lipid bilayer of the plasma membrane.

Once these small molecules have moved across the epithelial cells, they travel one of two routes. Glucose and amino acids diffuse into blood capillaries. Absorbed fatty acids and monoglycerides recombine into triglycerides inside the epithelial cells, and the triglycerides and other fats aggregate into small droplets (called chylomicrons). These droplets leave the cells by exocytosis and then enter lymph vessels, which drain into the circulation system.

Besides absorbing organic molecules, the small intestine absorbs water and dissolved mineral ions (such as Na^+ and Cl^-). Each day, about nine liters of fluid enter the small intestine from the stomach, liver, and pancreas. Of that, all but five percent is absorbed across the intestinal mucosa.

Role of the Pancreas in Digestion.   Ducts leading from the pancreas and liver join to form a common duct that empties into the duodenum. Exocrine cells in the **pancreas** secrete enzymes into this duct in response to hormonal and neural signals. The enzymes digest carbohydrates, fats, proteins, and nucleic acids. For example, like pepsin in the stomach, the pancreatic enzymes *trypsin* and *chymotrypsin* digest proteins into peptide fragments. The fragments are then degraded to free amino acids by *carboxypeptidase* (from the pancreas) and by *aminopeptidase* (present on the surface of the intestinal mucosa).

The pancreas also secretes bicarbonate ions, which help neutralize the HCl arriving from the stomach. If there were no such neutralization, pancreatic enzymes could not function.

Other pancreatic secretions do not function in digestion, but they still play a role in nutrition. Certain patches of endocrine cells in the pancreas (the islets of Langerhans) secrete the hormones *insulin* and *glucagon* into the blood. As you will see shortly, these two hormones are important in the feedback control of metabolism.

Role of the Liver in Digestion.   The **liver** is the largest glandular organ in the vertebrate body. One of its functions is the secretion of *bile*, a solution containing bile salts, bile pigments, cholesterol, and lecithin. Between meals, bile is stored and concentrated in the **gallbladder**, a small sac that branches off the common duct leading to the duodenum.

Part of a bile salt molecule (its cholesterol part) repels water, but another part dissolves in water. Because of these properties, bile salts enhance the breakdown and absorption of fats. Fats, recall, are insoluble in water. Most of the fats we eat are in the form of triglycerides, which clump together as large fat globules. In the small intestine, fat globules in the presence of bile salts are mechanically broken into smaller droplets by segmentation movements. Once that happens, the water-repelling parts of bile salts dissolve on the surface of the fat droplets. But other parts stick out from the surface and interact with water—which prevents the fat molecules from clumping together again. This suspension of small fat droplets is a type of emulsion. *Through the emulsifying effects of bile salts, pancreatic lipase has access to more triglycerides—hence fat digestion is enhanced.*

Bile salts also combine with the breakdown products of fat digestion (mostly monoglycerides and free fatty acids) to form micelles. A **micelle** is an aggregate of lipid molecules with a surface coat of bile salts. Since they are only three to ten nanometers across, micelles are small enough to move among microvilli of the intestinal epithelium. The monoglycerides and fatty acids diffuse down their concentration gradient—from micelles into the cells of these absorptive structures.

Human Nutrition and Gastrointestinal Disorders

The United States harbors one of the best-fed populations in the world. Yet digestive disorders among its individuals are on the increase.

Along with affluence, it appears we have picked up some bad eating habits. We skip meals, eat too much and too fast when we do sit down at the table, and generally give our gastrointestinal tracts erratic workouts. Worse yet, our diet tends to be rich in sugar, cholesterol, and salt—and low in bulk. (Here, bulk means the volume of fiber and other undigested food materials that cannot be decreased by absorption.) The problem with too little bulk in the diet comes from the longer transit time of feces through the colon. This material has irritating and even potentially carcinogenic effects. The longer the material is in contact with the colon walls, the more damage it can do. Thus, the more steadily the contents of the colon are cleared out by natural processes, the better. Increased bulk produces increased pressure on the colon walls, which stimulates expulsion of the material from the body.

Disorders such as appendicitis and cancer of the colon are practically nonexistent in rural Africa and India, where the inhabitants cannot afford to eat much more than whole grains. Whole grains happen to be high in fiber content. When individuals from those rural areas move to urban centers of the more affluent nations, they tend to become more susceptible to appendicitis and colon cancer. This suggests that diet is a key factor here. In addition, what we eat is known to affect the distribution and diversity of bacterial populations living in the gut. Do these changes somehow contribute to gastrointestinal disorders? That is not known.

Certainly the emotional stress associated with living in complex societies seems to compound the nutritional problem. Urban populations seem to be more susceptible to the irritable colon syndrome (once called colitis). Its symptoms include abdominal pain, diarrhea (excretion of watery feces), and constipation. Diarrhea can be brought on by emotional stress. There seems to be a genetic predisposition to some kinds of ulcers—inflammations of the stomach, the lower end of the esophagus, and the duodenum. But emotional stress apparently is a contributing factor in the development of some ulcers.

Where does this leave us? Short of surgery, there may not be much we can do about many inherited structural disorders of the gastrointestinal tract. Learning to handle stress is one way that we can ease up on the tract, though, and certainly learning how to eat properly is another.

Yet what is "eating properly"? In 1979 the United States Surgeon General released a report representing a medical consensus on how to promote health and avoid such afflictions as high blood pressure, heart disorders, cancer of the colon, and bad teeth. The report advised us to eat "less saturated fat and cholesterol; less salt; less sugar, relatively more complex carbohydrates such as whole grains, cereals, fruits, and vegetables; and relatively more fish, poultry, legumes (for example, peas, beans, and peanuts); and less red meat."

The controversies over what constitutes proper nutrition rage on. In the meantime, it might not be a bad idea to think about your own eating habits and how moderation in some things might help you hedge your bets. Put the question to yourself: Do you look upon a bowl of bran cereal with the same passion as you look upon, say, french fries and ice cream, prime rib, and chocolate mousse? Now put the same question to your colon.

Large Intestine

Each day, contractions force the 500 milliliters or so of chyme remaining in the small intestine into the **large intestine**, or **colon**. The colon functions mainly in storing and concentrating *feces*, a mixture of undigested and unabsorbed material, water, and bacteria. This material becomes concentrated when (1) epithelial cells of the colon actively transport sodium ions into the internal environment and (2) water follows passively as a result.

The colon is about 1.2 meters long. The "ascending" part of the colon extends upward on the right side of the abdominal cavity, the "transverse" part cuts across to the other side, then the "descending" part extends down the left side. There it ends in an "S" (sigmoid) shape. The sigmoid colon is continuous with a small tube called the **rectum**. Distension of the rectal wall triggers reflex actions by which material is expelled from the body. This reflex can be overridden by nerves under voluntary control. Those nerves cause contraction of a striated muscle sphincter of the **anus**, the terminal opening of the gut.

The small intestine does not lead into the start of the ascending colon. Because of the location of a sphincter between the two organs, the ascending colon begins as a blind pouch (the cecum). The **appendix** is a small, narrow projection from the cecum (Figure 32.6). Although the appendix has no known digestive functions, it does contain lymphatic tissue, which suggests a role in body defense. The appendix can become infected, a condition called *appendicitis*. If ignored, an infected appendix can rupture. Then, bacteria normally inhabiting the colon can spread into the abdominal cavity and cause serious infections.

Enzymes of Digestion: A Summary

Now that you have completed this tour of the gastrointestinal tract, you may find it helpful to scan Table 32.2. This table summarizes the locations of carbohydrate, protein, fat, and nucleic acid digestion, the enzymes responsible, and the breakdown products at each stage. It should be readily apparent, from this summary, that the small intestine is the major site of digestion and absorption of all foodstuffs.

HUMAN NUTRITIONAL NEEDS

It now appears that the earliest members of the human lineage lived on fruits, seeds, and other plant material. From this nutritional beginning, humans in many parts of the world have moved to diets rich in fats, sugars, and salts—and low in fiber. Some of the suspected consequences of this long-term shift in diet include a predisposition to colon cancer, breast cancer, cardiovascular disorders, kidney stones, as well as obesity (see *Commentary* on page 408).

The body grows and maintains itself in working order when it is kept supplied with energy and materials from foods of certain types, in certain amounts. Let's take a brief look at the energy requirements and nutrients essential for health.

Energy Needs

In nutritional studies, energy is measured in units called "calories," which unfortunately is taken to be equivalent to "kilocalories." (Both units are supposed to mean 1,000 calories of energy.) To avoid confusion, in this book we will express energy requirements in terms of kilocalories. In order to maintain an acceptable weight and keep the body functioning normally, caloric intake must be balanced with energy output. The output varies from one individual to the next, depending on such factors as the extent of physical activity, basic rate of metabolism, age, sex, hormone activity (especially epinephrine and thyroid hormone secretions), and emotional state. Some of these factors have a genetic basis; others are influenced by the social environment of the individual.

In most adults, the energy input and output are so balanced that the body weight remains remarkably constant over long periods. As any dieter knows, it is as if the body has a set point for weight and works to counteract deviations from that set point.

What amount of calories should be taken in daily to maintain what you consider to be an acceptable body weight? You can calculate the amount in two steps. First, multiply the acceptable weight (in pounds) by 10 if you are not very active physically, by 15 if you are moderately active, and by 20 if you are quite active. Then, depending on your age, subtract the following amount from the value obtained from the first step:

Age:	Subtract:
25–34	0
35–44	100
45–54	200
55–64	300
Over 65	400

Thus, for example, if you want to weigh 120 pounds and if you are highly active, $120 \times 20 = 2,400$ kilocalories. And if you are thirty-five years old, then you should take in a total of $(2,400 - 100)$, or 2,300 kilocalories a day.

Table 32.2 Major Enzymes of Digestion

Enzyme	Source	Where Active	Substrate	Main Breakdown Products*
Carbohydrate digestion:				
Salivary amylase	salivary glands	mouth	polysaccharides	disaccharides
Pancreatic amylase	pancreas	small intestine	polysaccharides	disaccharides
Disaccharidases	small intestine	small intestine	disaccharides	monosaccharides (e.g., glucose)
Protein digestion:				
Pepsins	stomach mucosa	stomach	proteins	peptide fragments
Trypsin and chymotrypsin	pancreas	small intestine	proteins and polypeptides	peptide fragments
Carboxypeptidase	pancreas	small intestine	peptide fragments	amino acids
Aminopeptidase	intestinal mucosa	small intestine	peptide fragments	amino acids
Fat digestion:				
Lipase	pancreas	small intestine	triglycerides	free fatty acids, monoglycerides
Nucleic acid digestion:				
Pancreatic nucleases	pancreas	small intestine	DNA, RNA	nucleotides
Intestinal nucleases	intestinal mucosa	small intestine	nucleotides	nucleotide bases, monosaccharides

*Yellow parts of table identify breakdown products that can be absorbed into the internal environment.

Carbohydrates

The body's main sources of energy are complex carbohydrates. As we have seen, these carbohydrates can be readily broken down to provide the body with glucose—the primary energy source for the brain, muscles, and other body tissues. Many nutritionists recommend that fleshy fruits, cereal grains, legumes, and other fibrous carbohydrates should make up at least fifty-five to fifty-eight percent of the daily caloric intake.

The average American takes in as much as 128 pounds of refined sugar (sucrose) per year. Sucrose adds calories to the diet but does so without the fiber afforded by complex carbohydrates.

Fats

Currently, fats constitute forty percent of the average diet in the United States, and most of the medical community agrees that it should be less than thirty percent. The body can manufacture most fats, including cholesterol, from protein and carbohydrates. (That is exactly what it does when you eat too much protein and carbohydrate.) However, your body needs to be supplied with about one tablespoon a day of polyunsaturated fat (such as corn or safflower oil), which serves as a source of certain essential fatty acids that the body cannot synthesize from other nutrients. Butter and other animal fats are forms of saturated fats, which tend to raise the blood levels of cholesterol. Too much cholesterol is believed to promote atherosclerosis (page 408).

Proteins

In the United States, the typical daily intake of proteins is about 125 grams. The amino acids released through protein digestion are absorbed and used to build the body's own proteins. Of the twenty common amino acids, eight are **essential amino acids**. Our cells cannot build these molecules; they must be obtained from the diet. Those amino acids are phenylalanine (and/or tyrosine), isoleucine, leucine, lysine, threonine, tryptophan, cysteine (and/or methionine), and valine.

Most animal protein contains all of the essential amino acids; plant proteins do not. To get enough protein from plant foods, different plants must be eaten in combination. To compare proteins from different sources, nutritionists use a measure called **net protein utilization** (NPU). NPU values range from 100 (all essential amino acids present in ideal proportions) to 0 (one or more absent; the protein is useless when eaten alone).

Table 32.3 Efficiency of Some Single Protein Sources in Meeting Minimum Daily Requirements

Source	Protein Content (%)	Net Protein Utilization (NPU)	Amount Needed to Satisfy Minimum Daily Requirement (grams)	Amount Needed to Satisfy Minimum Daily Requirement (ounces)
Eggs	11	97	403	14.1
Milk	4	82	1,311	45.9***
Fish*	22	80	244	8.5
Cheese*	27	70	227	7.2
Meat*	25	68	253	8.8
Soybean flour	45	60	158**	5.5**
Soybeans	34	60	210**	7.3**
Kidney beans	23	40	468**	16.4**
Corn	10	50	860**	30.0**

*Average values.
**Dry weight values.
***Equivalent of 6 cups. The figure is somewhat misleading, for most of the volume of milk is water. Milk is actually a rich source of high-quality protein.

Balancing the diet with different proteins can make up for deficiencies.

For much of the world, cereal grains are the main foods. As Table 32.3 suggests, cereal grains such as corn are low in protein content and NPU value. In contrast, beans are high in protein. Although NPU values for beans are no higher than those for cereal grains, beans are deficient in *different* amino acids. When beans are eaten *with* grain, the overall NPU value is raised.

Given the pervasive role of proteins in body structure and function, it is easy to see that protein deficiency has serious consequences. Protein deficiency is most damaging among the young, for rapid brain growth and development occur early in life. Unless enough protein is taken in just before and just after birth, irreversible mental retardation occurs. Even mild protein deprivation can retard growth and affect mental and physical performance.

Vitamins and Minerals

Normal metabolic activity depends on very small amounts of more than a dozen organic substances called **vitamins**. Most plant cells can synthesize all of these substances. In general, animal cells have lost the ability to do so; hence animals must obtain vitamins from food. Human cells need at least fourteen different vitamins (Table 32.4).

In addition to vitamins, cells require inorganic materials known as **minerals**. (Some minerals are called *trace elements* because they are needed in extremely small amounts.) Most cells require calcium and magnesium in a host of enzyme-mediated reactions. All cells need potassium during protein synthesis, for maintaining osmotic balances, and for muscle and nerve function. All cells require iron in building cytochromes, which are components of electron transport chains (Chapter Seven). Iron is also needed to produce the hemoglobin present in red blood cells (Table 32.5).

The sensible way to supply cells with essential vitamins and minerals is to eat a well-balanced selection of carbohydrates, fats, and proteins. (About 32–42 grams of protein, 250–500 grams of carbohydrates, and 66–83 grams of fat should do the trick.) In recent years, there have been claims that massive doses of certain vitamins and minerals are spectacularly beneficial. To date, there is no clear evidence that vitamin intake exceeding the recommended daily allowance leads to better health. To the contrary, excessive vitamin doses are often merely wasted or even harmful.

For example, the body simply will not hold more vitamin C than it needs for normal functioning. Vitamin C is not fat-soluble and the excess is excreted. Direct chemical analysis shows that any amount above the recommended daily allowance ends up in the urine almost immediately after it is absorbed from the gut. Abnormal intake of at least two other vitamins (A and D) can cause serious disorders. The reason is that, like all fat-soluble vitamins, vitamins A and D can accumulate in the fat deposits of the body (Table 32.4). *Severe shortages and massive excess of vitamins and minerals both can disturb the delicate balances that promote physiological health.*

Objective and Subjective Views of Obesity

By definition, **obesity** is an excess of fat in the body's adipose tissues, caused by imbalances between caloric intake and energy output. Clearly obese persons run a greater risk of developing high blood pressure, atherosclerosis, and diabetes, among other problems. But extremely underweight persons are also risking their health.

What constitutes the "ideal weight" for a person? Many charts have been developed, primarily by insurance companies who want to identify individuals who are overweight and therefore are an insurance risk. Figure 32.11 is an example of this sort of chart. There is widespread agreement that persons who are twenty-five percent heavier than the "ideal" are obese. Some mortality studies suggest that the "ideal" should be ten to fifteen pounds heavier than the charts indicate; adherents to various nutritional programs are convinced the values given in these charts should be less.

Table 32.4 Vitamins Necessary for Normal Cell Functioning

Vitamin	RDA* (milligrams)	Dietary Sources	Major Body Functions	Possible Outcomes of Deficiency	Possible Outcomes of Excess
Water-Soluble					
Vitamin B$_1$ (thiamine)	1.5	Pork, organ meats, whole grains, legumes	Coenzyme (thiamine pyrophosphate) in the removal of carbon dioxide	Beriberi (peripheral nerve changes, edema, heart failure)	None reported
Vitamin B$_2$ (riboflavin)	1.8	Widely distributed in foods	Constituent of two flavin nucleotide coenzymes involved in energy metabolism (FAD and FMN)	Reddened lips, cracks at corner of mouth (cheilosis), lesions of eye	None reported
Niacin	20	Liver, lean meats, grains, legumes (can be formed from tryptophan)	Constituent of two coenzymes involved in oxidation-reduction reactions (NAD$^+$ and NADP$^+$)	Pellagra (skin and gastrointestinal lesions, nervous, mental disorders)	Flushing, burning and tingling around neck, face, and hands
Vitamin B$_6$ (pyridoxine)	2	Meats, vegetables, whole grain cereals	Coenzyme (pyridoxal phosphate) involved in amino acid metabolism	Irritability, convulsions, muscular twitching, kidney stones	None reported
Pantothenic acid	5–10	Widely distributed in foods	Constituent of coenzyme A, which plays a central role in energy metabolism	Fatigue, sleep disturbances, impaired coordination, nausea (rare in humans)	None reported
Folic acid	0.4	Legumes, green vegetables, whole wheat products	Coenzyme (reduced form) in carbon transfer in nucleic acid and amino acid metabolism	Anemia, gastrointestinal disturbances, diarrhea, red tongue	None reported
Vitamin B$_{12}$	0.003	Muscle meats, eggs, dairy products	Coenzyme in carbon transfer in nucleic acid metabolism	Pernicious anemia, neurological disorders	None reported
Biotin	Not established. Usual diet provides 0.15–0.3	Legumes, vegetables, meats	Coenzyme in fat synthesis, amino acid metabolism, glycogen formation	Fatigue, depression, nausea, dermatitis, muscular pains	None reported
Choline	Not established. Usual diet provides 500–900	All foods containing phospholipids (egg yolk, liver, grains, legumes)	Constituent of phospholipids. Precursor of neurotransmitter acetylcholine	None reported for humans	None reported
Vitamin C (ascorbic acid)	45	Citrus fruits, tomatoes, green peppers, salad greens	Maintains intercellular matrix of cartilage, bone, and dentine. Important in collagen synthesis	Scurvy (degeneration of skin, teeth, blood vessels, epithelial hemorrhages)	Relatively nontoxic. Possibility of kidney stones
Fat-Soluble					
Vitamin A (retinol)	1	Provitamin A in green vegetables. Retinol in milk, butter, cheese, margarine	Constituent of rhodopsin (visual pigment). Maintenance of epithelial tissues	Xerophthalmia (keratinization of ocular tissue), night blindness, permanent blindness	Headache, vomiting, peeling of skin, anorexia, swelling of long bones, liver damage
Vitamin D	0.01	Cod liver oil, eggs, dairy products, margarine	Promotes bone growth, mineralization. Increases calcium absorption	Rickets (bone deformities) in children. Osteomalacia in adults	Vomiting, diarrhea, weight loss, kidney damage
Vitamin E (tocopherol)	15	Seeds, green leafy vegetables, margarine	Functions as an antioxidant to prevent cell membrane damage	Possibly anemia; never observed in humans	Relatively nontoxic
Vitamin K (phylloquinone)	0.03	Green leafy vegetables. Small amount in cereals, fruits, and meats	Important in blood clotting (involved in formation of active prothrombin)	Deficiencies associated with severe bleeding, internal hemorrhages	Synthetic forms at high doses may cause jaundice

*Recommended daily allowance, for an adult male in good health.
From "The Requirements of Human Nutrition," by Nevin S. Scrimshaw and Vernon R. Young.

Table 32.5 Minerals Necessary for Normal Cell Functioning

Mineral	Amount in Adult Body (grams)	RDA* (milligrams)	Dietary Sources	Major Body Functions	Possible Outcomes of Deficiency	Possible Outcomes of Excess
Calcium	1,500	800	Milk, cheese, dark-green vegetables, dried legumes	Bone and tooth formation Blood clotting Nerve transmission	Stunted growth Rickets, osteoporosis Convulsions	Not reported for humans
Phosphorus	860	800	Milk, cheese, meat, poultry, grains	Bone and tooth formation Acid-base balance, ATP formation, etc.	Weakness, demineralization of bone, loss of calcium	Erosion of jaw (fossy jaw)
Sulfur	300	(Provided by sulfur amino acids)	Sulfur amino acids (methionine and cystine) in dietary proteins	Constituent of active tissue compounds, cartilage and tendon	Related to intake and deficiency of sulfur amino acids	Excess sulfur amino acid intake leads to poor growth
Potassium	180	2,500	Meats, milk, many fruits	Acid-base balance Body water balance Nerve function	Muscular weakness Paralysis	Muscular weakness Death
Chlorine	74	2,000	Common salt	Formation of gastric juice Acid-base balance	Muscle cramps Mental apathy Reduced appetite	Vomiting
Sodium	64	2,500	Common salt	Acid-base balance Body water balance Nerve function	Muscle cramps Mental apathy Reduced appetite	High blood pressure
Magnesium	25	350	Whole grains, green leafy vegetables	Activates enzymes. Involved in protein synthesis	Growth failure. Behavioral disturbances Weakness, spasms	Diarrhea
Iron	4.5	10	Eggs, lean meats, legumes, whole grains, green leafy vegetables	Constituent of hemoglobin and enzymes involved in energy metabolism	Iron-deficiency anemia (weakness, reduced resistance to infection)	Siderosis (iron deposition in tissues) Cirrhosis of liver
Fluorine	2.6	2	Drinking water, tea, seafood	May be important in maintenance of bone structure	Higher frequency of tooth decay	Mottling of teeth. Increased bone density. Neurological disturbances
Zinc	2	15	Widely distributed in foods	Constituent of enzymes involved in digestion	Growth failure Small sex glands	Fever, nausea, vomiting, diarrhea
Copper	0.1	2	Meats, drinking water	Constituent of enzymes associated with iron metabolism	Anemia, bone changes (rare in humans)	
Iodine	0.011	0.14	Marine fish and shellfish, dairy products	Constituent of thyroid hormones	Goiter (enlarged thyroid)	Very high intakes depress thyroid activity
Cobalt	0.0015	(Required as vitamin B_{12})	Organ and muscle meats, milk	Constituent of vitamin B_{12}	None reported for humans	Industrial exposure: dermatitis and diseases of red blood cells

*Recommended daily allowance, for an adult male in good health.
From "The Requirements of Human Nutrition," by Nevin S. Scrimshaw and Vernon R. Young.

Man's Height	Size of Frame		
	Small	Medium	Large
5' 2"	128–134	131–141	138–150
5' 3"	130–136	133–143	140–153
5' 4"	132–138	135–145	142–156
5' 5"	134–140	137–148	144–160
5' 6"	136–142	139–151	146–164
5' 7"	138–145	142–154	149–168
5' 8"	140–148	145–157	152–172
5' 9"	142–151	148–160	155–176
5'10"	144–154	151–163	158–180
5'11"	146–157	154–166	161–184
6' 0"	149–160	157–170	164–188
6' 1"	152–164	160–174	168–192
6' 2"	155–168	164–178	172–197
6' 3"	158–172	167–182	176–202
6' 4"	162–176	171–187	181–207

a

Woman's Height	Size of Frame		
	Small	Medium	Large
4'10"	102–111	109–121	118–131
4'11"	103–113	111–123	120–134
5' 0"	104–115	113–126	122–137
5' 1"	106–118	115–129	125–140
5' 2"	108–121	118–132	128–143
5' 3"	111–124	121–135	131–147
5' 4"	114–127	124–138	134–151
5' 5"	117–130	127–141	137–155
5' 6"	120–133	130–144	140–159
5' 7"	123–136	133–147	143–163
5' 8"	126–139	136–150	146–167
5' 9"	129–142	139–153	149–170
5'10"	132–145	142–158	152–173
5'11"	135–148	145–159	155–176
6' 0"	138–151	148–162	158–179

b

Figure 32.11 The "ideal weights" for adult men (**a**) and adult women (**b**), according to one insurance company in 1983. The values shown are for persons twenty-five to fifty-nine years old wearing shoes with one-inch heels and five pounds of clothing (men) or three pounds of clothing (women).

Dieting has become nearly epidemic in the United States especially, where starvation is not the problem that it is in much of the world. Millions are dieting every day. Unfortunately, in a growing number of cases dieting becomes an obsession that leads to a potentially fatal eating disorder called *anorexia nervosa*. The disorder occurs primarily in women in their teens and in their early twenties.

Disturbances in the hypothalamus may trigger some of the weight loss and lead to the skewed perception of body weight that characterizes anorexia nervosa. But it appears that emotional factors contribute more to the disorder. (For example, some individuals fear growing up in general and maturing sexually in particular; others have irrational expectations of what they can accomplish.) Severe cases require psychiatric treatment.

Another eating disorder on the rise is *bulimia* ("an oxlike appetite"). Some surveys show that at least twenty percent of college-age women are now suffering to varying degrees from this disorder. The bulimic person goes on eating binges, taking in enormous amounts of food, then vomits or purges the body with laxatives after each binge. In some cases, the person does this once a month; others go through the binge-purge routine several times a day. Some women start doing this because it seems like a simple way to lose weight. But with repeated vomiting, the stomach acids brought into the mouth can erode teeth to stubs; repeated purgings can severely damage the digestive tract. Psychiatric treatment is also used for severe cases of this disorder.

ORGANIC METABOLISM

So far, we have looked at the routes by which food molecules enter the internal environment. We have also looked at the types and proportions of organic molecules necessary for proper nutrition. Once those molecules are inside the body, some are used as building blocks for structural components of cells. Others are funneled into ATP-producing pathways. Figure 32.12 summarizes the main routes by which organic molecules enter and leave the body.

Figure 32.12 also shows the main routes by which organic molecules are shuffled and reshuffled once they are inside. With few exceptions (such as DNA), most of these molecules are continually being broken down, with some of their component parts picked up and used again in new molecules. At the molecular level, your body undergoes massive and sometimes rapid turnovers.

The Vertebrate Liver

The liver is central to the storage and interconversion of absorbed carbohydrates, fats, and proteins (Figure 32.12). It also helps regulate the concentrations of organic components of blood and removes many toxic substances from blood. Most hormones are inactivated in the liver, then sent to the kidneys for excretion from the body.

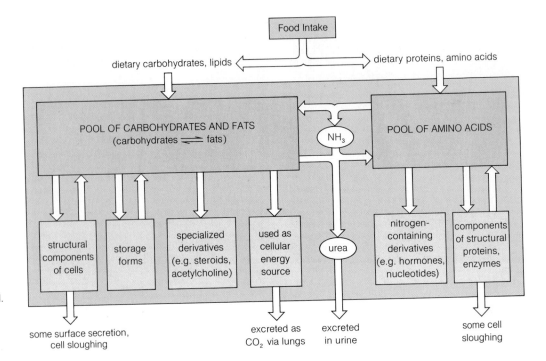

Figure 32.12 Summary of major pathways of organic metabolism. Urea formation occurs primarily in the liver. Carbohydrates, fats, and proteins are continually being broken down and resynthesized.

Food Intake

dietary carbohydrates, lipids ← → dietary proteins, amino acids

POOL OF CARBOHYDRATES AND FATS
(carbohydrates ⇌ fats)

NH₃

POOL OF AMINO ACIDS

structural components of cells

storage forms

specialized derivatives (e.g. steroids, acetylcholine)

used as cellular energy source

urea

nitrogen-containing derivatives (e.g. hormones, nucleotides)

components of structural proteins, enzymes

some surface secretion, cell sloughing

excreted as CO₂ via lungs

excreted in urine

some cell sloughing

Table 32.6 Some Activities That Depend on Liver Functioning

1. Carbohydrate metabolism
2. Control over some aspects of plasma protein synthesis
3. Assembly and disassembly of certain proteins
4. Urea formation from nitrogen-containing wastes
5. Assembly and storage of some fats
6. Fat digestion (bile is formed by the liver)
7. Inactivation of many chemicals (such as hormones and some drugs)
8. Detoxification of many poisons
9. Degradation of worn-out red blood cells
10. Immune response (removal of some foreign particles)
11. Red blood cell formation (liver absorbs, stores factors needed for red blood cell maturation)

Glucose and amino acids absorbed across the intestinal wall are transported directly by capillaries to the hepatic portal vein, which leads to the liver capillary bed (Figure 29.6). In the liver, excess glucose is stored as glycogen or converted to fat. Here, too, excess amino acids are converted to forms that can be sent through the Krebs cycle (as an alternate energy source) or converted to fat.

When cells degrade amino acids, ammonia (NH₃) is a reaction product. Ammonia is potentially toxic to cells. However, it travels through the blood to the liver, where it is converted to urea (a much less toxic waste product). Urea is excreted from the body by way of the kidneys.

Table 32.6 summarizes the main functions of the liver.

Absorptive and Post-Absorptive States

In terms of the total nutritional picture, there are two functional states of organic metabolism:

absorptive state	*ingested organic molecules enter the bloodstream from the gastrointestinal tract*
post-absorptive state	*gastrointestinal tract is not supplying nutrients; the body draws from its internal pools of organic molecules*

During the absorptive state, the body builds up its pools of organic molecules. Excess carbohydrates and other dietary molecules are transformed mostly into fats, which are stored in adipose tissue. Some also are

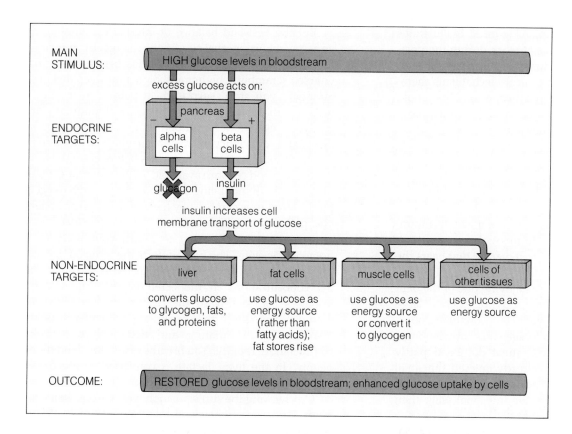

MAIN
STIMULUS: HIGH glucose levels in bloodstream

excess glucose acts on:

pancreas

ENDOCRINE
TARGETS: — alpha cells beta cells +

glucagon insulin

insulin increases cell
membrane transport of glucose

NON-ENDOCRINE
TARGETS: liver fat cells muscle cells cells of other tissues

converts glucose to glycogen, fats, and proteins use glucose as energy source (rather than fatty acids); fat stores rise use glucose as energy source or convert it to glycogen use glucose as energy source

OUTCOME: RESTORED glucose levels in bloodstream; enhanced glucose uptake by cells

Figure 32.13 Main metabolic routes and the endocrine commands during the absorptive state.

converted to glycogen in the liver and in muscle tissue. Most cells use glucose as the primary energy source at this time; there is no net breakdown of protein in muscle or other tissues.

During the post-absorptive state, there is a notable shift in the type of food molecules used to support cell activities. A key factor in this shift is the need to provide brain cells with glucose, the major nutrient they use for energy.

When glucose is being absorbed, its concentrations in the bloodstream are readily maintained. But how does the body maintain blood glucose concentrations during the post-absorptive state? *First,* glycogen stores (particularly in the liver) are rapidly broken down to glucose, which is released into blood. *Second,* body proteins are broken down to provide amino acids, which are transported by the blood to the liver. There they are converted to glucose, which can be released into blood.

Most cells use fats as the major energy source during the post-absorptive state. Fats stored in adipose tissue are broken down into glycerol and fatty acids, which are released into blood. The glycerol can be converted to glucose in the liver; the circulating fatty acids are used by most cells in ATP production.

During the absorptive state, glucose moves into cells, where it can be used for energy and where the excess can be stored.

During the post-absorptive state, most cells use fat as the main energy source; stored fats are mobilized. Brain cells are kept supplied with glucose mainly by the conversion of amino acids into glucose by the liver.

Controls Over Organic Metabolism

Both endocrine and neural controls govern metabolism during the absorptive and post-absorptive states. The most important control agents are hormones secreted by clusters of endocrine cells in the pancreas (the islets of Langerhans). These clusters include alpha and beta cell types, which function antagonistically. *Beta cells* secrete **insulin**, a hormone that enhances glucose uptake, storage, and use by cells. *Alpha cells* secrete **glucagon**, a hormone that prods liver cells into converting glycogen into glucose and that inhibits glycogen synthesis.

Figure 32.13 illustrates the endocrine controls over organic metabolism during the absorptive state, which we might call times of "feasting." Figure 32.14 illustrates endocrine and neural controls at work during the post-absorptive state, including times of "fasting" (or starvation). A discussion of the control mechanisms themselves would be beyond the scope of this book. However, the following case study will give you a sense of the marvelous nature of their interactions.

Case Study: Feasting, Fasting, and Systems Integration

Suppose, this morning, you are vacationing in the mountains and decide on impulse to follow a forested trail. You fail to notice the wooden trail marker that bears the intriguing name, "Fat Man's Misery." As you walk down the tree-lined corridor, you are enjoying one of the benefits of discontinuous feeding. Having eaten a large breakfast, you have assured your cells of ongoing nourishment; you do not have to forage constantly amongst the ferns as, say, a nematode must do. Food partly digested in the stomach has already entered the small intestine. Right now, amino acids, simple sugars, and fatty acids are moving across the intestinal wall, then into the bloodstream.

With the surge of nutrients, glucose molecules are entering the bloodstream faster than your cells can use them. The level of blood glucose begins to rise slightly. However, your body has a homeostatic program for converting glucose into storage form when it is flooding in, then releasing some of the stores when it is scarce.

With the rise in blood glucose, pancreatic beta cells are called upon to secrete insulin. Blood concentrations of insulin rise—and the hormonal targets (liver, fat, and muscle cells) quickly begin using or storing the glucose molecules (Figure 32.13). At the same time, alpha cells are prevented from secreting glucagon—which slows the liver's conversion of stored glycogen into glucose.

What is the outcome? High levels of glucose that have entered the circulation from your gut move out of the blood and into cells, where it can be burned as fuel or stored for later use.

Even though you are no longer feeding your body, your brain cells have not lessened their high demands for glucose. Neither have your muscle cells, which are getting a strenuous workout. Little by little, blood glucose levels drop. Now endocrine activities shift in the pancreas. With less glucose binding to them, beta cells decrease their insulin output. With less glucose to inhibit them, alpha cells increase their glucagon output. When glucagon reaches your liver, it causes the conversion of glycogen back to glucose—which is returned to your blood. This prevents blood glucose from falling below levels required to maintain brain function.

But the best-laid balance of internal conditions can go astray when external conditions change. In your case, the "miserable" part of the trail has begun. You find yourself scrambling higher and higher on steep inclines. Suddenly you stop, surprised, in pain. You forgot to reckon with the lower oxygen pressure of mountain air, and your leg muscles cramped. Your body has already detected its deficiency of oxygen-carrying red blood cells at this altitude, but it will take days before enough additional red blood cells are available. In the meantime, your muscle cells are not being supplied with enough oxygen for the strenuous climb. They have switched to an anaerobic pathway in which lactate is the end product.

Again, systems interact to return your body to a homeostatic state. The body detects the reduced oxygen pressure and an accompanying increase in hydrogen ion concentrations in the cerebrospinal fluid. Nerve impulses course toward the respiratory center in the medulla. The result: the diaphragm and other muscles associated with inflating and deflating your lungs contract more rapidly. You breathe faster now, and more deeply. In the liver, lactate is converted to glucose—which is returned to the blood.

On checking the sun's position, you see it is well past noon. And guess what: you forgot about lunch. When you start the long walk back, the drop in blood glucose levels triggers new homeostatic mechanisms. Under hypothalamic commands, your adrenal medulla begins secreting epinephrine and norepinephrine. Its main targets: the liver, adipose tissue, and muscles. In the liver, glycogen synthesis stops. In body tissues generally, glucose uptake is blocked. In fat cells, fats are converted to fatty acids, which are routed to the liver, muscles, and other tissues as alternative energy sources (Figure 32.14). For every fatty acid molecule sent down metabolic pathways in those tissues, several glucose molecules are held in reserve for the brain.

You do get back to the start of the trail by sundown. However, your body had enough stored energy to sustain you for many more days, so the situation was never really desperate. The balance of blood sugar and fat is constantly monitored and controlled by the liver and hormones. Glucose levels only drop beyond the set point to stimulate glycogen conversion and fat conversion, and vice versa. It takes several days of fasting before blood sugar levels are markedly reduced.

Even after several days of fasting, your energy supplies would not have run out. Another hypothalamic command would have prodded your anterior pituitary into secreting ACTH (page 360). The ACTH would have signaled adrenal cortex cells to secrete glucocorticoid hormones, which have a potent effect on the synthesis of carbohydrates from proteins and on the further breakdown of fat. Slowly, in muscles and other tissues,

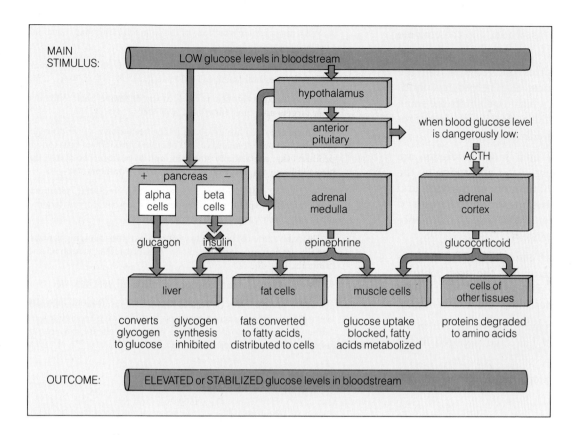

MAIN
STIMULUS: **LOW glucose levels in bloodstream**

hypothalamus

anterior pituitary

when blood glucose level is dangerously low:

ACTH

pancreas

+ alpha cells   − beta cells

adrenal medulla

adrenal cortex

glucagon   insulin   epinephrine   glucocorticoid

liver   fat cells   muscle cells   cells of other tissues

converts glycogen to glucose | glycogen synthesis inhibited | fats converted to fatty acids, distributed to cells | glucose uptake blocked, fatty acids metabolized | proteins degraded to amino acids

OUTCOME: **ELEVATED or STABILIZED glucose levels in bloodstream**

Figure 32.14 Main metabolic routes and the endocrine and neural commands during the post-absorptive state.

your body's proteins would have been disassembled. Amino acids from these structural tissues would have been used in the liver to build new glucose molecules—and once more your brain would have been kept active.

As extreme as this last pathway might be, it would be a small price to pay for keeping your brain functional enough to figure out how to take in more nutrients and bring you back to a homeostatic state.

SUMMARY

1. Nutrition has to do with all the processes by which a body takes in, digests, absorbs, and uses food.

2. A digestive system is a body cavity or tube which has four main functions:

a. Motility: mechanical breakdown of ingested nutrients and elimination of unabsorbed residues from the body.

b. Secretion: release of enzymes, hormones, and other substances that take part in digestion.

c. Digestion: chemical reduction of ingested nutrients into particles, then into molecules small enough to cross the lining of the gut and thereby to reach the internal environment.

d. Absorption: passage of digested nutrients from the gut lumen into the blood or lymph, which distributes them through the body.

3. The human digestive tract has regional subdivisions: the mouth, pharynx, esophagus, stomach, small intestine, large intestine (colon), rectum, and anus. Glands associated with digestion are the salivary glands, liver, gallbladder, and pancreas.

4. Coordinated contractions of the muscle layer of the gastrointestinal tract mix food with secretions (segmentation) and move it forward (peristalsis). Sphincters control the flow of contents from one region to another.

5. Controls over gastrointestinal activity operate in response to the volume and composition of food in the lumen of the tract. The response can be a change in muscle activity, a change in the secretory rate of hormones or enzymes, or both.

6. Saliva contains mucins, ions, water, and the enzyme salivary amylase.

7. The stomach stores and mixes food received from the esophagus, secretes substances that help digest food, and helps control the rate at which food enters the small intestine. The secretions of the stomach include hydrochloric acid and protein-degrading enzymes.

8. Digestion is completed and most nutrients are absorbed in the small intestine. After absorption by the intestinal mucosa, glucose, other monosaccharides, and most amino acids pass into the intestinal capillaries and then go directly to the liver. Fatty acids and monoglycerides enter the lymph system.

9. The pancreas secretes into the small intestine bicarbonate ions (which help neutralize the acid contents arriving from the stomach) and enzymes (which help digest proteins, carbohydrates, nucleic acids, and fats).

10. The liver secretes bile, a substance essential for the breakdown and digestion of fats. Bile is stored in the gallbladder between meals.

11. The large intestine functions mainly in storing and concentrating feces.

12. Nutritional energy is measured in kilocalories. To maintain weight and health, caloric intake must balance energy output.

13. Complex carbohydrates, which can be broken down to produce glucose, are the body's main energy source. Fats are produced by the body as the storage form of protein and carbohydrate. Dietary protein provides the amino acids needed for the body's synthesis of protein. Eight essential amino acids and a few essential fatty acids must be provided by the diet.

14. Vitamins and inorganic minerals must be supplied by the diet.

15. During the absorptive state, nutrients are being absorbed into the bloodstream by the digestive tract. Secretion of insulin by the pancreas increases and glucose moves into cells, where it can be used for energy and where the excess can be stored.

16. During the post-absorptive state, the gastrointestinal tract is not supplying nutrients and the body draws from its internal pools of organic molecules. Secretion of glucagon by the pancreas increases.

Review Questions

1. Study Figure 32.2. Then, on your own, diagram the connections between metabolism and the digestive, circulatory, and respiratory systems. *451*

2. Explain the difference between digestion and absorption. *452*

3. What are the main functions of the stomach? The small intestine? The large intestine? *456*

4. In what ways are food materials mixed and propelled through the gastrointestinal tract? *453–454*

5. Name three hormones at work in the digestive tract. What are their targets and their functions? *454–455*

6. Which enzymes are involved in the breakdown of (a) polysaccharides, (b) proteins, and (c) fats? Name four kinds of breakdown products that are actually small enough to be absorbed across the intestinal mucosa and into the internal environment. *460–461*

7. A glass of milk contains lactose, protein, butterfat, vitamins, and minerals. Explain what happens to each component when it passes through your digestive tract. *455–461*

8. Describe some of the reasons why each of the following is nutritionally important: carbohydrates, fats, proteins, vitamins, and minerals. *460–462*

9. Describe some of the functions of the liver. *458, 465–466*

10. What are the roles of insulin and glucagon in organic metabolism? When blood glucose levels are high, glucagon secretions are (enhanced/inhibited). When blood glucose levels are low, insulin secretions are (enhanced/inhibited). *362, 467*

Readings

Clemente, C. 1987. *Anatomy: A Regional Atlas of the Human Body.* Third edition. Baltimore: Urban and Schwartzenberg. Stunning, detailed illustrations of human anatomy. Drawings of the gastrointestinal tract are among the best available.

Hamilton, W. 1985. *Nutrition: Concepts and Controversies.* Third edition. Menlo Park, California: West. Information on digestion, nutrition, diet, and health; evaluates fads and erroneous ideas about nutrition in light of current research.

Kessel, R. G., and R. H. Kardon. 1979. *Tissues and Organs: A Text-Atlas of Scanning Electron Microscopy.* San Francisco: Freeman. Outstanding, unique micrographs, accompanied by well-written descriptions of major tissues and organs.

Krause, M., and L. Mahan. 1984. *Food, Nutrition, and Diet Therapy.* Seventh edition. Philadelphia: Saunders.

Kretchmer, N., and W. van B. Robertson. 1978. *Human Nutrition.* San Francisco: Freeman. Excellent collection of articles from *Scientific American* that consider nutrition at the cellular level and the global level. Paperback.

Vander, A., J. Sherman, and D. Luciano. 1985. *Human Physiology: The Mechanisms of Body Function.* Fourth edition. New York: McGraw-Hill.

Judging from the fossil record, animals first evolved in the shallow waters of ancient seas. From the beginning, then, their tissues and organ systems were geared to operating in a salty fluid where temperatures were relatively stable.

About 375 million years ago, some animals began invading the land. They were able to do so partly because the physical and chemical aspects of their "internal environment" approximated conditions in the seas they left behind. But there were new challenges on land, where winds and radiant energy from the sun could dehydrate the animal body, where water might not be available to replenish body fluids, and where most water was fresh, not salty. Over time, the body plans, physiology, and behavior of land-dwelling animals became modified in response to those threats to the stability of the internal environment.

Today, as then, animals make responses to physical conditions over which they often have little or no control. For example, nonmigratory birds of the arctic tundra are obliged to lose heat to their cold surroundings in winter. Yet many responses can be adjusted, as when those same birds fluff their feathers and so expand their "insulation" for holding in body heat. *To maintain a hospitable internal environment, all animals make controlled adjustments as well as obligatory exchanges with the external environment.* In this chapter, we turn to the exchanges and adjustments that affect the body's temperature and its water and solute balances.

33

TEMPERATURE CONTROL AND FLUID REGULATION

Figure 33.1 After a cold night in their Kalahari Desert dens, meerkats stand seemingly at attention, allowing a large surface area of their bodies to absorb the warm rays of the morning sun. Like many other animals, meerkats rely on such behavioral adaptations to help maintain their body temperature even though the outside temperature changes.

CONTROL OF BODY TEMPERATURE

Temperatures Suitable for Life

In a manner of speaking, the animal body runs on enzymes—and enzyme activity is affected by temperature. The enzymes of most animals commonly remain functional within the 0°–40°C range (Table 33.1). Above 40°C or so, they do not function as well because denaturation occurs. (Denaturation disrupts the chemical interactions holding a molecule in its required three-dimensional shape.) Also, the rate of enzyme activity generally decreases by at least half every time the temperature drops by ten degrees. Clearly, then, metabolism can be upset if body temperatures exceed or fall below the proper range.

The question becomes this: How do animals keep their body temperature fairly constant? They do this by balancing heat gains and heat losses.

Heat Gains and Heat Losses

Enzyme-mediated reactions proceed simultaneously in the millions or billions of cells of a large-bodied animal. Heat is produced as a result of all that metabolic activity, and it contributes to the body's heat content. So does heat absorbed from the environment. At the same time, the body loses heat to the environment. A simple equation summarizes these heat gains and losses:

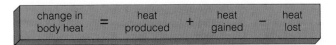

$$\text{change in body heat} = \text{heat produced} + \text{heat gained} - \text{heat lost}$$

The body gains or loses heat through four processes, called radiation, conduction, convection, and evaporation.

Radiation is a process by which the sun, animals, and all other objects emit energy in the form of infrared and other wavelengths. Animals lose heat by infrared radiation when their surroundings are cooler than they are; they gain heat from objects that are warmer than they are.

Conduction is the direct transfer of heat from one object to another when they are in direct contact, because of collisions between their atoms and molecules. Because heat moves down thermal gradients, we lose heat by conduction when (for example) we sit on cold ground, and we gain heat when we sit on warm sand at the beach.

Convection is the transfer of heat by way of moving fluid, such as air or water currents. The process involves conduction (heat moves down the thermal gradient between the body surface and the air or water next to it). It also involves mass transfer, with currents carrying heat away from or toward the body. When your skin temperature is higher than the air temperature, you lose heat by convection. Even when there is no breeze, your body loses heat by creating its own convective current. (Air becomes less dense as it is heated and rises away from the body.)

Evaporation is a process whereby a substance changes from a liquid to a gaseous state. Humans and some other mammals have sweat glands that move water and specific solutes through pores to the skin surface. Water at the surface can absorb some of the skin's thermal energy. Individual water molecules thereby become more energetic, and the hydrogen bonds holding them to their neighbors can break. They leave the skin's surface, and some body heat is dissipated.

Keep in mind that sweat dripping from the skin does *not* dissipate body heat by evaporation. You may have noticed this if you have exercised strenuously on hot, humid days. Your body sweats profusely as it attempts to balance the large heat gain resulting from your muscular activity and the hot environment—yet there is so much moisture in the air, complete evaporation is not possible.

Animals gain heat by way of metabolism, and they gain or lose heat by radiation, conduction, convection, and evaporation.

Classification of Animals Based on Temperature

Ectotherms. We humans have high metabolic rates that sustain our active way of life. But most animals have low metabolic rates, and on top of that, they are poorly insulated. This means they rapidly absorb and gain heat, especially when they are small-bodied animals. They maintain their body temperature mostly by heat gains from the environment, not from metabolism. Such animals are **ectotherms** (meaning "heat from outside"). Lizards and other reptiles are examples.

Ectotherms are not entirely at the mercy of their environment, however, for they can make behavioral adjustments to changing external temperatures. We call this **behavioral temperature regulation**.

Lizards move about, putting themselves in places where they minimize heat or cold stress. To warm up, they move out of shade and keep orienting their body to expose the maximum surface area to the sun's infrared radiation. They gain heat by conduction when they bask on rocks that absorbed heat from the sun earlier in the day. In such ways, the lizard body can warm up as fast as 1°C per minute.

Lizards lose heat just as rapidly when the sun goes down and temperatures drop. Then, metabolic activity decreases and they become almost immobilized. Before that happens, they usually crawl into crevices or under rocks, where heat loss is not as great and where they are not as vulnerable to predators.

Endotherms and Heterotherms. Like most other mammals, we are **endotherms** ("heat from within"). So are most birds. Our body temperature is controlled mainly by (1) metabolic activity and (2) controls over heat conservation and dissipation. We also make behavioral adjustments that supplement the physiological controls.

Most endotherms have an active life style, made possible by high metabolic rates. It is a costly adaptation. A foraging mouse uses up to thirty times more energy than a foraging lizard of the same weight. Yet such energy outlays are advantageous, for they are the main reason why endotherms can be active under a wide range of temperatures. Cold nights or cold seasons don't stop them from foraging, for example, or escaping from predators, or digging a burrow.

Endotherms have a variety of adaptations for conserving or dissipating the heat associated with high metabolic rates. Think about the ways their bodies are shaped and insulated. Some mammalian species in cold regions (such as arctic hares) are more massive than closely related species in warmer regions (such as jackrabbits). Compared to lightweight or thin-limbed bodies, a massive body has a greater volume of cells for generating heat and less surface area for losing it. Fur, feathers, and layers of fat also help reduce heat loss.

Like ectotherms, endotherms also adjust behaviorally to heat stress. During the day, some desert mammals of north temperate regions have body temperatures that are much lower than temperatures of the air and the ground surface. Outside temperatures often are lower during the night as well as during winter. However, the soil well below the surface never heats up much, and it is here that most desert rodents and other mammals find refuge from the daytime heat. Typically those animals forage by night and spend the hottest part of the day in burrows or in the shade of bushes or rock outcroppings.

Some birds and mammals fall between the ectothermic and endothermic categories. Part of the time, these so-called **heterotherms** allow their body temperature to fluctuate as ectotherms do—and at other times they control heat exchanges as endotherms do. Hummingbirds have very high metabolic rates for their size, and they devote much of the day to locating and sipping nectar as an energy source for metabolism. Because hummingbirds do not forage at night, they could rapidly run out of energy unless their metabolic rates decreased considerably. At night, they may enter a sleeplike state and their body temperature may approach that of their cooler surroundings.

Table 33.1	Temperatures Favorable for Metabolism, Compared With Environmental Temperatures	
Temperatures generally favorable for metabolism:	0°C to 40°C (32°F to 104°F)
Air temperatures above land surfaces:	−70°C to +85°C (−94°F to +185°F)
Surface temperatures of open oceans:	−2°C to +30°C (+28.4°F to +86°F)

Advantages of Ectothermy Versus Endothermy. In general, ectotherms are at an advantage in the warm, humid tropics. They do not have to expend much energy to maintain body temperature, and more energy can be devoted to other tasks, including reproduction. Indeed, reptiles far exceed mammals in numbers and species diversity in the tropics. However, endotherms have the advantage and are more abundant in moderate to cold settings. High metabolic rates allow some endotherms to occupy even the polar regions, where you would never find a lizard.

Temperature Regulation in Mammals

Environmental temperatures can change quickly, and exercise or other activities can alter the metabolic rate for the mammalian body. Such changes trigger slight increases or decreases in the normal **core temperature**. "Core" refers to the body's internal temperature, as opposed to temperatures of the tissues near its surface. Here we will consider some of the physiological and behavioral responses that can rapidly restore the normal temperature.

Responses to Cold Stress. Among mammals, **peripheral vasoconstriction** is a normal response to a drop in outside temperature. The response is initiated when thermoreceptors at the body surface signal the hypothalamus about the temperature change (Figure 33.2). The hypothalamus sends out commands to smooth muscles in the walls of blood vessels in the skin. When the muscles contract, vasoconstriction occurs—and the bloodstream's convective delivery of heat to the body's surface is reduced. How effective is the response? To give an example, when your fingers or toes become cold, all but one percent of the blood that would otherwise flow to their skin is curtailed.

In another response to a drop in outside temperature, smooth muscle controlling the erection of hair or feathers

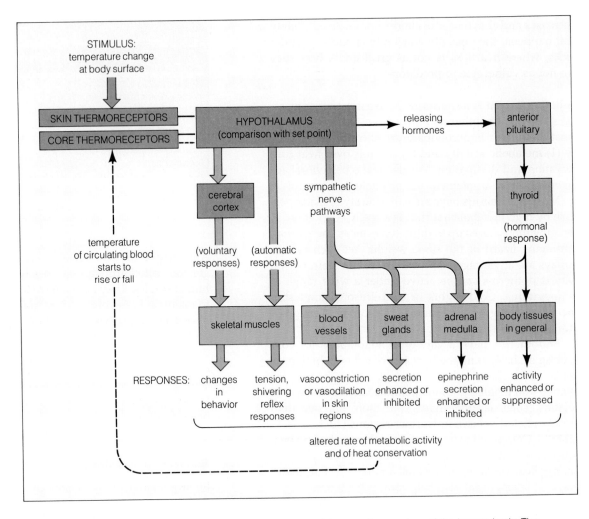

Figure 33.2 Overview of the feedback relationships that control the core temperature of the human body. The dashed line shows how the feedback loop is completed. The purple arrows indicate the main control pathways.

is stimulated to contract. This is called a **pilomotor response**. The plumage or pelt fluffs up, and this creates a layer of still air that reduces convective and radiative heat losses from the body. Further reductions in heat loss are possible through behavioral responses, as when cats curl up into a ball or when you hold both arms tightly against your body.

When other responses are not enough to counter cold stress, the hypothalamus calls for an increase in skeletal muscle activity that leads to **shivering**. (The word refers to rhythmic tremors in which the muscles contract about ten to twenty times per second.) Within a short time, heat production throughout the body increases several times over. Shivering comes at a high energy cost and is not effective for very long.

Prolonged or severe cold exposure also can lead to **nonshivering heat production**, which results when hormone secretions trigger an increase in metabolic rates (Figure 33.2). The response occurs mainly in the brown adipose tissue of newborns and animals that become

acclimatized to cold. Human infants have this specialized tissue; adults do not.

When defenses against cold are not adequate, the result is *hypothermia*, a condition in which the core temperature falls below normal. In humans, a drop of only a few degrees affects brain function and leads to confusion; further cooling can lead to coma and death (see *Commentary*). Many animals can recover from profound hypothermia. However, cells that become frozen may be destroyed unless thawing is precisely controlled (this sometimes can be done in hospitals). Tissue destruction through localized freezing is called *frostbite*.

Responses to Heat Stress. The hypothalamus also plays a role in responses to increases in core temperature. In a major response called **peripheral vasodilation**, hypothalamic signals cause peripheral blood vessels to dilate. More blood flows from deeper body regions to skin regions, where the excess heat it carries is dissipated (Figure 33.3).

Falling Overboard and the Odds for Survival

In 1912, the ocean liner *Titanic* set out from Europe on her maiden voyage to America. In that same year, a huge chunk of the leading edge of a Greenland glacier broke off and began floating out to sea. Late at night on April 14, off the coast of Newfoundland, the iceberg and the *Titanic* made their ill-fated rendezvous. Lifeboats and survival drills had been neglected, and only about a fourth of the 2,000 people on board managed to scramble into the lifeboats that could be launched. What happened to the rest of the passengers? Within two hours, rescue ships were on the scene—yet 1,513 bodies were recovered from a calm sea. All were wearing life jackets. None had drowned. Probably every one of those individuals had died from hypothermia—from a drop in body temperature below tolerance levels. The following are responses at the body temperatures indicated:

Body Temperature	
36°–34°C (about 95°F)	Shivering response, increase in respiration. Increase in metabolic heat output. Constriction of peripheral blood vessels; blood is routed to deeper regions. Dizziness and nausea set in.
33°–32°C (about 91°F)	Shivering response stops. Metabolic heat output drops.
31°–30°C (about 86°F)	Capacity for voluntary motion is lost. Eye and tendon reflexes inhibited. Consciousness is lost. Cardiac muscle action becomes irregular.
26°–24°C (about 77°F)	Ventricular fibrillation sets in (page 410). Death follows.

Evaporative heat loss is another response that can be influenced by the hypothalamus, which can activate sweat glands. Your skin has $2\frac{1}{2}$ million or more sweat glands, and considerable heat is dissipated when the water they give up to the skin surface evaporates. (With extreme sweating, as might occur in a marathon race, the body loses an important salt—sodium chloride—as well as copious amounts of water. Such losses may change the character of the internal environment to the extent that the runner may collapse and faint.)

What about mammals that sweat very little or not at all? Some of them, including dogs, pant. "Panting" refers to shallow, rapid breathing that increases evaporative water loss from the respiratory tract. Cooling occurs when the water evaporates from the nasal cavity, mouth, and tongue.

Sometimes peripheral blood flow and evaporative heat loss are not enough to counter heat stress, and *hyperthermia* results. This is a condition in which the core temperature increases above normal. For humans and other endotherms, an increase of only a few degrees above normal can be dangerous.

Figure 33.3 A jackrabbit (*Lepus californicus*) cooling off on a hot summer day in the mountains of Arizona. Notice the dilated blood vessels in its large ears. Both the large surface area of the ears and the extensive vascularization are useful for dissipating heat (by way of convection and radiation).

Fever. During a *fever*, the hypothalamus actually resets the body's "thermostat" that dictates what the core temperature should be. The normal response mechanisms are brought into play, but they are carried out to maintain a higher temperature! At the onset of fever, heat loss decreases and heat production increases. At the time the person feels chilled. When the fever "breaks," peripheral vasodilation and sweating increase as the body attempts to reduce the core temperature to normal; then, the person feels warm.

Fever may be an important defense mechanism against infections, and perhaps against cancer. Following infection, monocytes, macrophages, and other cells secrete signaling molecules (including interleukin-1, interferons, and other factors). Their secretions stimulate the hypothalamus into causing an increase in temperature. The increase seems to enhance the effectiveness of the body's immune response. Given this possibility, questions are being raised about the widespread practice of administering aspirin and other drugs to suppress modest fevers. (There is no question that such drugs must be used in cases of extreme fevers.)

Table 33.2 summarizes the controlled responses to cold stress and heat stress in mammals.

CONTROL OF EXTRACELLULAR FLUID

Water Gains and Losses

Just as the mammalian body exchanges heat with its surroundings in obligatory and controlled ways, so also does it exchange water and solutes. Ordinarily, water losses are balanced precisely by water gains (Table 33.3). The body *gains* water through two processes:

1. Absorption of water from liquids and solid foods in the gastrointestinal tract.

2. Metabolism (specifically, the breakdown of carbohydrates and other organic molecules in reactions that yield water as a by-product).

Thirst behavior influences the gain of water. This behavior is under the control of the hypothalamus, as will be described later in the chapter.

The body *loses* water by several processes, the most important of which are these:

1. Excretion by way of the urinary system.

2. Evaporation from the respiratory surface.

3. Evaporation from cells of skin.

4. Sweating.

5. Elimination by way of the gastrointestinal tract.

For mammals, the process of greatest importance in controlling water loss is **excretion**: the elimination of excess water and excess (or harmful) solutes from the internal environment by way of organs called the kidneys. The evaporative processes listed are called "insensible water losses" because the individual is not aware that they are taking place. As we have seen, sweating occurs through temperature-regulating mechanisms, and normally very little water leaves the body in feces.

Table 33.2	Summary of Mammalian Responses to Cold Stress and to Heat Stress	
Environmental Stimulus	Main Responses	Outcome
Drop in temperature	Vasoconstriction of blood vessels in skin; changes in behavior (e.g., curling up the body to reduce surface area exposed to the environment)	Heat is conserved
	Increased muscle activity; shivering; nonshivering heat production	Heat production increases
Rise in temperature	Vasodilation of blood vessels in skin; sweating; changes in behavior; panting	Heat is dissipated from body
	Decreased muscle activity	Heat production decreases

Table 33.3	Normal Balance Between Water Gain and Water Loss in Humans and in Kangaroo Rats			
Organism	Water Gain (milliliters)		Water Loss (milliliters)	
Adult human (measured on daily basis)	Ingested in solids:	850	Urine:	1,500
	Ingested as liquids:	1,400	Feces:	200
	Metabolically derived:	350	Evaporation:	900
		2,600		2,600
Kangaroo rat (measured over four weeks)	Ingested in solids:	6.0	Urine:	13.5
	Ingested as liquids:	0	Feces:	2.6
	Metabolically derived:	54.0	Evaporation:	43.9
		60.0		60.0

Solute Gains and Losses

Aside from oxygen (which is absorbed at the respiratory surface), diverse solutes are added to the internal environment by three processes:

1. Absorption from the gastrointestinal tract. The absorbed substances include *nutrients* such as glucose (used as energy sources and in biosynthesis reactions), as well as drugs and food additives. They also include *mineral ions*, such as sodium and potassium ions.

2. Secretion (such as hormone secretions).

3. Metabolism (including *waste products* of degradative reactions).

Carbon dioxide is the most abundant waste product of metabolism, and it is eliminated from the body at the respiratory surface. Aside from carbon dioxide, the main metabolic wastes eliminated from the body are these:

1. *Ammonia,* formed in "deamination" reactions whereby amino groups are stripped from amino acids. If allowed to accumulate in the body, ammonia can be highly toxic.

2. *Urea,* produced in the liver in reactions that link two ammonia molecules to carbon dioxide. Urea is the main nitrogen-containing waste product of protein breakdown and is relatively harmless.

3. *Uric acid,* formed in reactions that degrade nucleic acids. If allowed to accumulate, uric acid can crystallize (and sometimes collect in the joints).

Other metabolic wastes, including phosphoric acid and sulfuric acid, are usually produced in small amounts during protein breakdown. Some of those compounds are highly toxic and may be responsible for many of the symptoms associated with kidney failure.

What we have, then, are ongoing inputs of water, nutrients, and ions and the production of metabolic wastes. Let's take a look at the major system for eliminating excess amounts of those substances so that stable conditions prevail in the internal environment.

Urinary System of Mammals

Urine is a fluid composed of water and mineral ions, organic wastes, and other substances that are filtered from the blood. It forms in the **kidneys,** a pair of fist-sized organs located behind the back wall of the abdominal cavity, on either side of the aorta (Figure 33.4). The water and solutes leaving the kidneys as urine are only

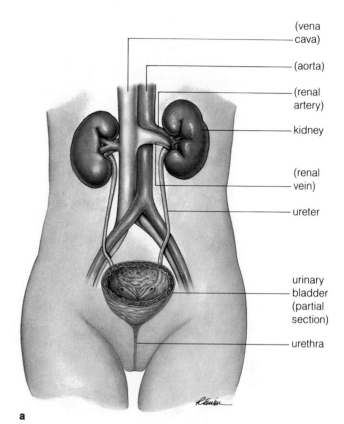

(vena cava)

(aorta)

(renal artery)

kidney

(renal vein)

ureter

urinary bladder (partial section)

urethra

a

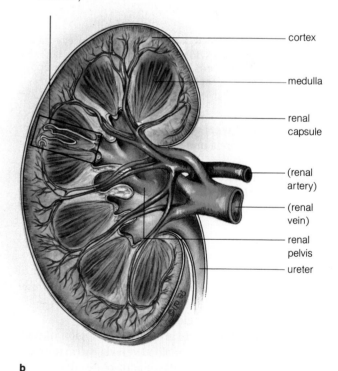

nephron (greatly exaggerated in diameter)

cortex

medulla

renal capsule

(renal artery)

(renal vein)

renal pelvis

ureter

b

Figure 33.4 (a) Components of the human urinary system. (b) Closer look at the kidney.

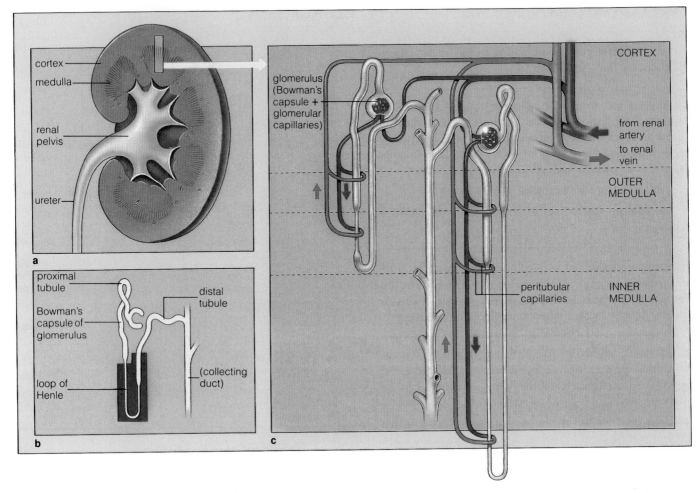

Figure 33.5 Anatomy of the kidney. (**a**) A human kidney, cutaway view. (**b**) Diagram of a nephron, the functional unit of the kidney. (**c**) Blood vessels associated with the nephron. The glomerular capillaries are shown in detail in Figure 33.10. The peritubular capillaries thread profusely around the tubular parts of the nephron.

Labels in figure:

(a) cortex, medulla, renal pelvis, ureter

(b) proximal tubule, Bowman's capsule of glomerulus, loop of Henle, distal tubule, (collecting duct)

(c) glomerulus (Bowman's capsule + glomerular capillaries), peritubular capillaries, CORTEX, from renal artery, to renal vein, OUTER MEDULLA, INNER MEDULLA

a tiny portion of what goes in, since the kidneys quickly return all but about one percent to the blood. But the composition of the fluid that *is* returned has been adjusted in numerous and absolutely vital ways. *Through their action, kidneys regulate not only the volume but also the concentrations of many individual components of the internal sea.*

Each kidney has a coat of connective tissue, called the renal capsule (from the Latin *renes*, meaning kidney). Its *cortex* extends from the capsule toward the center of the kidney and surrounds a number of lobes. The inner portion of each renal lobe is called a *medulla*; the outer portion is its cortical region (Figure 33.5).

Urine formation starts at the cortical region of each lobe, which is composed of blood vessels and many slender tubes called *nephrons*. Water and solutes filter out of the blood and enter the tubes. Most of the filtrate is reabsorbed, but some moves into tubelike *collecting ducts*. Fluid flows through the collecting ducts and into the

kidney's central cavity, or *renal pelvis*; this is the urine.

The renal pelvis is continuous with the **ureter**, a tube that carries urine to a storage organ called the **urinary bladder**. Urine arriving from two ureters (one from each kidney) accumulates here. It leaves the bladder through a single tube, the **urethra**, which opens to the body surface at the end of the penis (in males) or just above the vagina's entrance (in females). The two kidneys, two ureters, urinary bladder, and urethra constitute the **urinary system** of mammals (Figure 33.4).

You may have heard about "kidney stones," these being deposits of uric acid, calcium salts, and other substances that have precipitated out of urine and have collected in the renal pelvis. At times the stones become lodged in the urethra, where they interfere with urine flow and intermittently cause a great deal of pain. Kidney stones usually pass naturally from the body; if they do not, they can be eliminated by a variety of medical or surgical procedures.

Urination, or urine flow from the body, is a reflex response. As a urinary bladder fills, tension increases in its strong, smooth-muscled walls. The increased tension causes muscles that prevent the flow of urine into the urethra to relax; at the same time, the bladder walls contract and force fluid through the urethra. The reflex response is basically involuntary, but it can be consciously inhibited.

Nephrons. In each kidney, urine forms in more than a million long, slender tubes—the **nephrons**. The tube wall is only a single layer of epithelial cells, but the cells and junctions between them differ in structure and function along the nephron's length. Most importantly, some wall regions are highly permeable to water and solutes, yet others bar the passage of solutes except at active transport systems built into the cell membranes.

The start of each nephron is a blood-filtering unit called the **glomerulus**, which is located in the kidney cortex (Figure 33.5). Here, the nephron wall balloons, cuplike, around a compact cluster of small-diameter blood vessels, the *glomerular capillaries*. The cuplike wall region is called *Bowman's capsule*; it is a receptacle for water and solutes being filtered from blood. From here, the filtrate flows into the **proximal tubule**, a tubular section of the nephron that precedes the **loop of Henle**. This hairpin-shaped section descends toward or plunges into the medulla, forms a sharp turn, then thickens again (Figure 33.5b). The nephron's last section is the **distal tubule** (the part most distant from Bowman's capsule).

The distal portions of neighboring nephrons merge to form one of the preliminary collecting ducts for urine. Different collecting ducts merge with one another to form larger ducts as they approach the renal pelvis.

Blood Vessels at the Nephron. Blood is carried to the kidneys by the renal arteries, which branch off the main artery from the heart (the aorta). The renal artery branches into smaller arteries, then into a series of arterioles. Each arteriole is said to be "afferent," for it carries blood *into* a glomerulus. Inside, it branches into glomerular capillaries. A rather unusual vascular connection is made here, for those capillaries do not merge into veins. Instead they converge to form an "efferent" (outgoing) arteriole, which divides into *another* set of capillaries. Those **peritubular capillaries** thread profusely around the proximal and distal sections of the nephron (Figure 33.5c). Eventually they merge to form veins, which carry blood out of the kidney.

Overview of Urine Formation

Figure 33.6 and the following list give us a simple picture of the urine-forming processes in the nephron:

1. Filtration occurs at the glomerulus. Blood pressure (generated by heart contractions) causes the *bulk flow* of water and solutes out of the glomerular capillaries and into Bowman's capsule. The blood is said to be filtered because blood cells, proteins, and other large solutes are left behind as smaller solutes (such as glucose, sodium, and urea) are forced out. The filtrate passes into the proximal tubule.

2. Reabsorption occurs along the tubular sections of the nephron, which are in intimate contact with the peritubular capillaries. Water and solutes move *out* of the nephron by diffusion or active transport, then back into the capillaries. By this process, the body reclaims most of the water and usable solutes.

3. Secretion occurs along the peritubular capillaries. In this highly regulated process, excess amounts of hydrogen ions, potassium ions, and a few other substances move *out* of the capillaries and into adjacent tubular sections of the nephron. Secretion also rids the body of foreign substances (such as penicillin), uric acid, the products of hemoglobin breakdown, and other wastes.

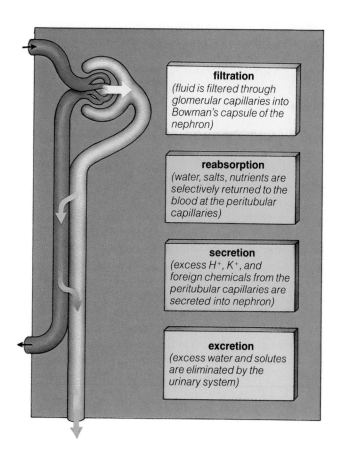

filtration
(fluid is filtered through glomerular capillaries into Bowman's capsule of the nephron)

reabsorption
(water, salts, nutrients are selectively returned to the blood at the peritubular capillaries)

secretion
(excess H⁺, K⁺, and foreign chemicals from the peritubular capillaries are secreted into nephron)

excretion
(excess water and solutes are eliminated by the urinary system)

Figure 33.6 Overview of the processes involved in the formation and excretion of urine in mammals.

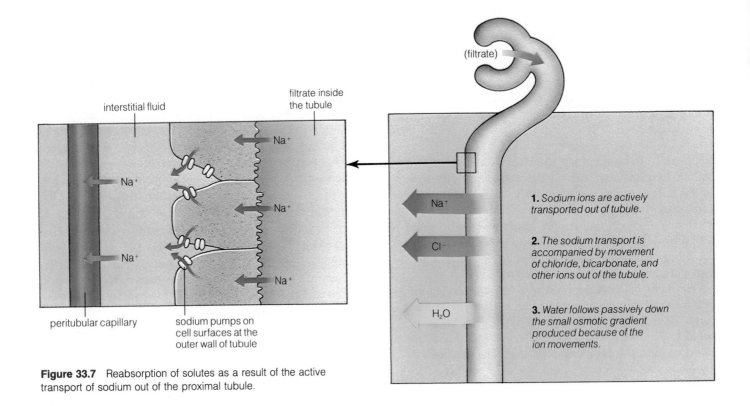

interstitial fluid

filtrate inside the tubule

(filtrate)

Na⁺

Na⁺

Na⁺

Na⁺

Na⁺

peritubular capillary

sodium pumps on cell surfaces at the outer wall of tubule

Na⁺

Cl⁻

H₂O

1. *Sodium ions are actively transported out of tubule.*

2. *The sodium transport is accompanied by movement of chloride, bicarbonate, and other ions out of the tubule.*

3. *Water follows passively down the small osmotic gradient produced because of the ion movements.*

Figure 33.7 Reabsorption of solutes as a result of the active transport of sodium out of the proximal tubule.

Through these three processes, a concentrated urine is formed. The urine contains waste products as well as water and solutes in excess of the amounts necessary to maintain the extracellular fluid.

A Closer Look at Filtration

Rate of Filtration.   Each day, more blood flows through the kidneys than through any other organ except the lungs. Nearly one-fourth of the cardiac output (about 1½ quarts per minute) goes to the kidneys. How can those two fist-sized organs handle blood flowing through on such a massive scale? There are two reasons.

First, the arterioles delivering blood to a glomerulus have a wider diameter—and less resistance to flow—than most arterioles do. As a result, the hydrostatic pressure (caused by heart contractions) does not drop as much when blood flows through them. *Second*, the glomerular capillaries are highly permeable. They do not allow blood cells or platelets to escape, but they are 10 to 100 times more permeable to water and small solutes than other capillaries in the body. Because of the higher hydrostatic pressure and the greater capillary permeability, the kidneys can easily filter an average of about 45 gallons (180 liters) per day!

Factors Influencing Filtration.   Filtration rates are set mainly by the rate at which water is reabsorbed; and as you will see shortly, reabsorption is under hormonal control. Filtration rates also are influenced by neural controls over blood flow through the body as a whole. When you exercise, for example, more blood than usual must be diverted to your heart and skeletal muscles to sustain the increased activity. Blood is diverted from the kidneys when neural signals stimulate smooth muscle in the walls of afferent arterioles to contract. The resulting vasoconstriction reduces the flow volume to the kidneys and so lowers the filtration rate.

Local chemical signals also influence filtration rates. When arterial blood pressure decreases, locally produced chemicals stimulate the afferent arterioles into dilating, so more blood flows into the kidneys. When blood pressure rises, arterioles are stimulated into constricting, so flow through the kidneys decreases.

A Closer Look at Reabsorption

Each day, a minimum volume of about 400 milliliters of urine is required to rid the body of metabolic wastes. We call that amount the "obligatory water loss." Urine increases in volume and is less concentrated when water

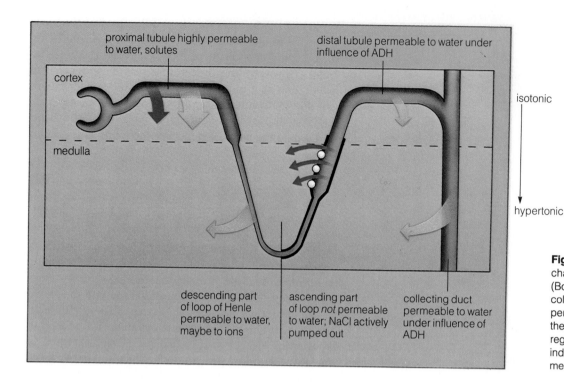

proximal tubule highly permeable to water, solutes

distal tubule permeable to water under influence of ADH

cortex

medulla

isotonic

hypertonic

descending part of loop of Henle permeable to water, maybe to ions

ascending part of loop *not* permeable to water; NaCl actively pumped out

collecting duct permeable to water under influence of ADH

Figure 33.8 Permeability characteristics of the nephron. (Both the distal tubule and the collecting duct have very limited permeability to solutes; most of the solute movements in these regions are related directly or indirectly to active transport mechanisms.)

intake rises; it decreases in volume and is more concentrated when water intake is reduced.

Adjustments in urine composition and volume are made during the reabsorption process. As Table 33.4 indicates, reabsorption involves surprisingly large quantities of water and solutes. If for some reason the reabsorption process stopped, all the water of the bloodstream would be urinated away in less than a half hour!

The question becomes this: How are most of the water and usable solutes that have entered the nephron sent back to the bloodstream in precisely regulated amounts? Their controlled reclamation depends on solute concentration gradients that are maintained between the nephron and the interstitial fluid surrounding it. We can get an idea of how that gradient is established and maintained by considering the role played by sodium ions (Na$^+$), one of the key solutes in extracellular fluid.

Transport Processes at the Proximal Tubule. Sodium ions are reabsorbed at the proximal tubule of the nephron. Sodium pumps (a type of active transport system) exist in the plasma membrane of epithelial cells making up the tubule's outer wall (Figure 33.7). They all pump sodium in one direction: out of the filtrate and into the surrounding interstitial fluid.

An outward movement of chloride, bicarbonate, and other ions accompanies the active transport of sodium. The resulting drop in solute concentrations inside the

Table 33.4	Average Daily Reabsorption Values for a Few Substances		
	Filtered	Excreted	Proportion Reabsorbed
Water	180 liters	1.8 liters	99%
Glucose	180 grams	None, normally	100%
Sodium ions	630 grams	3.2 grams	99.5%
Urea	54 grams	30 grams	44%

tubule is slight, but it affects the osmotic gradient for water between the tubule and the interstitial fluid. Water passively follows the gradient out of the tubule, then is reabsorbed by diffusing into the peritubular capillaries.

Countercurrent Multiplication. The body can reclaim water only as long as the solute concentration gradient is maintained between the nephron and interstitial fluid. The gradient starts at the boundary between the cortex and medulla, and it increases with increasing depth into the medulla. It is established partly through *countercurrent multiplication*. "Countercurrent" refers to flow in opposing directions through a pair of interacting tubes. Such a flow takes place at the loop of Henle.

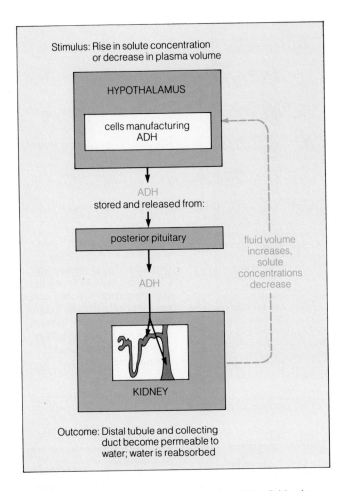

Figure 33.9 Homeostatic control over extracellular fluid volume. A rise in extracellular solute concentration or a decrease in volume stimulates the release of ADH, which acts on distal tubules and collecting ducts and makes them permeable to water. Water is thus returned to the body and conserved, and only a small volume of highly concentrated urine is excreted.

ADH secretion is inhibited when blood volume increases or its solute concentration decreases (dashed line), as might occur after water ingestion.

When fluid enters the descending part of the loop, it has the same solute concentration as blood (this being an outcome of transport processes in the proximal tubule). However, the descending part is permeable to water, and as it plunges into the medulla, water moves out and the solute concentration inside increases.

What happens after the fluid travels around the bend and enters the ascending part of the loop? This section of the nephron is not permeable to water, but a salt— sodium chloride, or NaCl—is actively pumped out here (Figure 33.8). As the fluid in the ascending part continues on its way, the pumping causes a progressive decrease in its solute concentration. More salts are pumped out, the surroundings become saltier, and more water is drawn out of the descending part. A positive feedback mechanism is at work here, "multiplying" the solute concentration of the surrounding tissue as the sequence of events just described is repeated over and over.

We have only considered the role of sodium in reabsorption. It should be pointed out that urea and other substances also have their effects. (This is evident in individuals on low-protein diets. Their ability to form a concentrated urine is impaired unless they are administered urea, a product of protein metabolism.) Basically, when the urea concentration is high in urine, a portion moves passively from the collecting ducts into the medulla. To the extent that it raises the solute concentration there, urea contributes to water reabsorption from the loop of Henle and to the urine concentrating mechanism.

Control of Water Reabsorption. When the body must conserve water, **antidiuretic hormone**, or ADH, is called into action. ADH triggers increased permeability in cells making up the walls of distal tubules and collecting ducts. When ADH secretion is low, the walls are impermeable to water, so the urine remains dilute. When ADH secretion is stepped up, the rate of water reabsorption increases, solutes remain behind, and the urine that leaves the collecting ducts is more concentrated.

The day-to-day regulation of ADH secretion is mainly under hypothalamic control. The hypothalamus has receptors that are sensitive to changes in the total solute concentration of extracellular fluid. Detection of such changes accounts for the delicate balance between variations in water intake and urine output. For example, when you drink a lot of water, the solute concentration of extracellular fluid is diluted slightly. ADH secretion is inhibited, and you excrete a large volume of urine. With water deprivation, solute levels rise in extracellular fluid, ADH secretion steps up, and the kidney is stimulated to conserve water (Figure 33.9).

Thirst Mechanism. When the body loses more water than it takes in, a thirst mechanism is activated. The hypothalamus contains a cluster of nerve cells that are

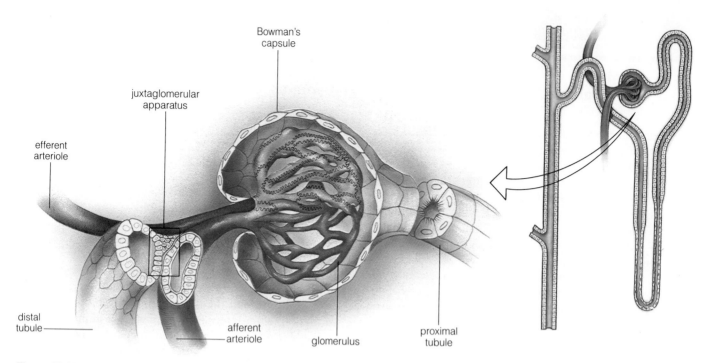

Figure 33.10 Juxtaglomerular apparatus, which plays a role in sodium reabsorption.

Labels on figure:
- Bowman's capsule
- juxtaglomerular apparatus
- efferent arteriole
- distal tubule
- afferent arteriole
- glomerulus
- proximal tubule

also stimulated when ADH secretion is being stepped up. When receptors detect a rise in solute levels (a drop in water volume), thirst center cells are stimulated into sending signals that initiate water-seeking behavior.

Control of Sodium Reabsorption. Another hormone, called **aldosterone**, plays a major role in reabsorbing sodium ions, which are important components of the internal environment.

Consider what happens when the body loses more sodium than it takes in. Then, the volume of all extracellular fluid is reduced. Stretch or pressure receptors in the heart chambers, in blood vessels in the chest cavity, and possibly elsewhere detect the reduction, and now mechanisms are switched on that call certain cells in the kidney into action. Those cells are part of the "juxtaglomerular apparatus," a region of contact between the afferent arteriole and the distal tubule of the nephron (Figure 33.10). The cells are stimulated to secrete an enzyme, called renin.

Renin's substrate is a protein produced in the liver and released in inactive form into the bloodstream, where it is always circulating. Renin catalyzes the removal of a polypeptide from the protein, and the fragment (angiotensin I) is converted to a hormone (angiotensin II). The targets of angiotensin II are endocrine cells in the adrenal cortex. When stimulated, those cells secrete the hormone aldosterone. Aldosterone, which acts on cells making up the walls of distal tubules and collecting ducts, stimulates sodium reabsorption.

Sodium retention is accompanied by water retention, of course, and so the volume of extracellular fluid is raised. Receptors detect the ensuing rise in blood pressure, and the signaling mechanism that led to aldosterone secretion is switched off.

High blood pressure, or *hypertension*, can adversely affect kidney function. (The "tension" part of the name refers to the muscle tone of walls of arteries and arterioles.) The wear and tear on the kidneys as well as the vascular system and brain may proceed undetected for many years, with severe consequences (page 408). One of the ways in which blood pressure is controlled is to restrict salt intake.

Acid-Base Balance

So far, we have focused on the kidney's primary function—that is, how it controls water and sodium reabsorption, and so influences the total volume and distribution of body fluids. Yet the kidney has another function that has profound impact on the health of the individual. *Together with the respiratory system and other*

Kidney Failure, Bypass Measures, and Transplants

Sometimes the kidneys can no longer perform their filtration, reabsorption, and secretion tasks. For example, diabetes or immunological reactions can damage the glomerulus and so reduce or stop urine formation. The leading cause of kidney failure is *glomerulonephritis*, an inflammatory disorder in which the glomeruli are damaged. In the United States alone, an estimated 13 million individuals in all age groups suffer from some kidney disorder.

When the kidneys malfunction, solute concentrations in the blood are not regulated properly. Substances such as potassium ions as well as various toxic products of protein metabolism can accumulate in the bloodstream. The buildup may lead to nausea, fatigue, loss of memory and, in advanced cases, death. A *kidney dialysis machine* can be employed to restore the proper solute balances. The machine is often called an artificial kidney, not because it resembles the natural organ but because the end result is the same: concentrations of substances are regulated by their selective removal from (and addition to) the bloodstream.

The artificial kidney is based on dialysis: the separation of substances across a membrane between solutions of differing concentrations. In *hemodialysis*, a patient is connected to the machine by tubes leading from an artery or a vein. Blood is then pumped through narrow tubes located in a warm-water bath. The bath contains a precisely balanced mix of salts, glucose, acetate or bicarbonate, amino acids, and other substances that set up the proper gradients with the blood flowing through the tubes. Thus, sub-stances at too high a concentration in the patient's blood will diffuse into the dialysis fluid.

A similar effect can be obtained in some patients by *peritoneal dialysis*, a process in which a fluid of appropriate composition is instilled into the abdominal cavity and, after a suitable interval, drained out. In this process, the lining of the cavity (the peritoneum) serves as the dialysis "membrane."

On the average, hemodialysis takes about four to five hours. The machine does not approach the kidney's efficiency, and blood must circulate over and over again through the tubes before the solute concentrations of the internal environment are improved. Afflicted persons must be treated three times a week.

For temporary disorders, the artificial kidney is used as a bypass measure until normal kidney function resumes. For chronic kidney disorders, it must be used for the remainder of the patient's life or until a functional kidney is transplanted. With treatment and with controlled diet, many individuals are able to resume fairly normal activity.

Transplant surgery is expensive, but it is far less than the cost of hemodialysis year after year (which is about $25,000 to $30,000 annually). More than half the kidneys donated come from individuals who have just died as a result of accidents. Imagine yourself in the ethical position of being severely afflicted with a kidney disorder, waiting and half-hoping for someone else's death. Besides, about a fourth of all kidney transplants induce an immune response and are rejected.

Figure 33.11 A kangaroo rat, master of water conservation in the deserts.

organ systems, the kidneys help keep the extracellular environment from becoming too acidic or too basic.

The overall acid-base balance is maintained by controls over ion concentrations, especially hydrogen ions (H^+). Those controls are exerted through (1) buffer systems, (2) respiration, and (3) excretion by way of the kidneys.

Normally, the extracellular pH for the human body is between 7.35 and 7.45. Maintaining that value means neutralizing or eliminating the various acidic and basic substances entering the blood from the digestive tract and from normal cell metabolism. (Acids lower the pH and bases raise it.) In individuals on an ordinary diet, an excess of acids results from normal cell metabolism. The acids dissociate to release H^+ and so lower the pH. The fall in pH is minimized when the excess H^+ ions react with various buffers. The bicarbonate–carbon dioxide buffer system is an important example. The reactions can be summarized this way:

$$H^+ + HCO_3^- \rightleftharpoons H_2CO_3 \rightleftharpoons H_2O + CO_2$$

Here, the H^+ is neutralized and the carbon dioxide that forms is excreted by the lungs (page 444).

Keep in mind that this buffer system and others do not *eliminate* acid; they only neutralize H^+ temporarily. *Only the urinary system can eliminate excess amounts of H^+ and restore the body's buffers.*

Notice, in the preceding equation, that the reaction arrows also run in reverse. The reverse reactions occur in cells making up the tubular wall of the nephron. *First,* the HCO_3^- that forms in those cells is moved into the interstitial fluid of the kidney, then into the peritubular capillaries. The capillaries deliver the HCO_3^- to the general circulation, where it helps buffer excess acid. *Second,* the H^+ that forms in the cells is secreted into the fluid inside the nephron. There, it can combine with bicarbonate ions to form CO_2, which is returned to the blood and excreted by the lungs. It also can combine with phosphate salts or ammonia (NH_3), which are excreted in urine. In such ways, hydrogen ions can be permanently removed from extracellular fluid.

Case Study: On Fish, Frogs, and Kangaroo Rats

Let's conclude this chapter with a look at how a few vertebrates maintain their water and solute levels in some entirely different settings—in the seas, in fresh water, and on land. Figure 33.11 shows a kangaroo rat, which will be our main focus.

Compared to the tissues of herring, snapper, and other bony fishes living in it, seawater has about three times more solutes. These fishes continually lose water (by osmosis) to their hypertonic environment, and continual drinking brings in replacements. (If the fishes are experimentally prevented from drinking, they die from dehydration in a few days.) Ingested solutes are excreted against their concentration gradients. Although kidneys are present, they are too small to excrete much water or solutes. Instead, most of the excess solutes are pumped out through the membranes of fish gills, the cells of which actively transport sodium ions out of the blood (and potassium ions into it).

In fresh water (a hypotonic medium), lake trout and other bony fishes tend to gain water and lose solutes. The same is true of amphibians. These animals do not drink water; rather, water moves by osmosis into the body, through the thin gill membranes (or, in the case of adult amphibians such as frogs, through the skin). Excess water leaves by way of well-developed kidneys, which excrete a large volume of dilute urine. Some solutes also are excreted, but the losses are balanced by solutes gained from food and by the active transport of sodium ions across the gills, into the body.

For the desert-dwelling kangaroo rats, water is exceedingly scarce (Figure 33.11). Moreover, the air is dry and temperatures can approach 45°C, so water losses could be devastating without behavioral and physiological adaptations for conserving water. Like many other desert animals, kangaroo rats spend the day in deep burrows, where the air temperature seldom exceeds 30°C. They forage during the cooler hours of the night. Although some desert rodents eat moist plant parts, the kangaroo rats eat primarily dry seeds, which contain very little water. They do not drink any water at all. They gain most of their water through the metabolic oxidation of carbohydrates, fats, and proteins in the seeds.

Land-dwelling vertebrates in general lose water by way of the skin, respiratory tract, urine, and (to some extent) feces. Kangaroo rats and other desert rodents reduce all such losses. Their skin has no sweat glands, and it is thick and dry. Their nose is small, with very narrow and convoluted air passages. When these rodents inhale, dry air passing over the moist nasal tissues becomes warmed and saturated with water vapor. Evaporation from the nasal epithelium cools the tissues well below the body temperature. When the rodents exhale, the warm, humid air from the lungs is cooled as it passes over the cooled tissues. Hence water condenses on the nasal epithelium, like it does on the outside of a glass of ice water on a warm day. As a result, considerable respiratory water is recovered, not lost to the environment.

Kangaroo rats are also parsimonious when it comes to giving up water in feces (the water loss is five times less than it is from laboratory rats). Their urine can be twice as concentrated as that of laboratory rats and three times that of humans. This remarkable ability to conserve water loss from the urinary tract is attributed to particularly long loops of Henle in the kidneys. The loops are so long that the renal capsule containing them extends through the renal pelvis, down into the ureter. Through countercurrent multiplication at these long loops, the solute concentration in the surrounding interstitial fluid becomes very high. Thus the osmotic gradient between the interstitial fluid and the urine is so steep that most of the water reaching the (equally long) collecting ducts is reabsorbed; only a small volume of concentrated urine leaves the kangaroo rat body.

SUMMARY

All animals make regulated as well as obligatory exchanges with the external environment and so maintain a hospitable internal environment.

Control of Body Temperature

1. Temperatures between 0°C and 40°C are generally favorable for life.

2. The body temperature of animals is determined by the balance between heat produced, heat absorbed from the environment, and heat lost to the environment. Animals gain heat by way of metabolism, and they exchange heat with the environment by these processes:

   a. *Radiation* (emission of energy in the form of infrared and other wavelengths which, following absorption by the animal body or some other object, is converted to heat energy).

   b. *Conduction* (the direct transfer of heat energy from one object to an immediately adjacent object).

   c. *Convection* (heat transfer by air or water currents; involves conduction and mass transfer of heat-bearing currents away from or toward the animal body).

   d. *Evaporation* (dissipation of heat energy during the conversion of water from the liquid to the gaseous state).

3. The body temperature of different animals is determined by the rate of metabolic activity and by anatomical, behavioral, and physiological adaptations.

   a. Ectotherms are animals whose body temperature is determined more by heat exchange with the environment than by metabolic heat.

   b. The body temperature of endotherms is determined largely by metabolic activity and by precise controls over heat produced and heat lost.

   c. Heterotherms allow their body temperature to fluctuate at some times, and at other times they control heat balance.

Control of Extracellular Fluid

1. Water and solutes are exchanged with the environment in obligatory and controlled ways.

2. The body gains water through absorption from the gastrointestinal tract and from metabolism. It loses water

by evaporation from the respiratory surface and cells of the skin, by sweating, by elimination from the gastrointestinal tract (in feces), and by excretion of urine.

3. Water gain is controlled through thirst. Water loss is regulated mainly by varying the composition and volume of urine.

4. Solutes (nutrients and ions) are gained by absorption from the gastrointestinal tract and by production within the body.

5. The urinary system of mammals includes two kidneys, two ureters, a urinary bladder, and a urethra.

6. Each kidney is composed of a renal capsule, cortex (which represents more than half of its mass), and medulla. Numerous individual tubes (nephrons) produce the urine. The urine drains from the nephrons into collecting ducts.

7. Each nephron consists of a cup-shaped portion (Bowman's capsule), proximal tubule, loop of Henle, and distal tubule.

8. Blood vessels associated with the nephron are an afferent arteriole, glomerular capillaries, an efferent arteriole, and peritubular capillaries.

9. The composition and volume of urine are controlled by the selective reabsorption or secretion of substances. The following processes are involved:

   a. *Filtration* of plasma (minus proteins and other large solutes through the glomerular capillaries into Bowman's capsule); blood pressure provides the force for filtration.

   b. *Reabsorption*, or the movement of substances from the lumen of the nephron, then into the peritubular capillaries.

   c. *Secretion*, or the movement of substances from the peritubular capillaries into the lumen of the nephron.

10. Each day, massive volumes of fluid are filtered at the glomeruli. Most of the water and solutes filtered are reabsorbed.

11. The interstitial fluid of the medulla becomes highly concentrated by a countercurrent multiplier system in the loops of Henle.

12. Antidiuretic hormone (ADH) is produced in the hypothalamus, then stored in and secreted from the posterior pituitary. It acts on cells in the walls of the distal tubule and collecting ducts, making them permeable to water. ADH is secreted when water must be conserved; its secretion is inhibited when excess water must be excreted.

13. Sodium reabsorption in the distal tubule and collecting ducts is under the control of the hormone aldosterone.

Review Questions

1. Define ectotherm and endotherm. In endotherms, which controls help balance the amount of heat lost and heat gained? *473*

2. In your own body, where are thermoreceptors located? Where is the main center of temperature control located? *473–474*

3. All animals have mechanisms for maintaining body fluid concentration and composition. In your own body, which organs cooperate in these tasks? *477–478*

4. Describe what happens during (a) filtration, (b) reabsorption, and (c) secretion in the kidney's nephron/capillary unit. What do these three processes influence? *479–480*

5. Which hormone is involved in the control of water reabsorption? Which hormone plays a major role in sodium reabsorption? Describe how they work. *482–483*

6. The kidneys contribute to maintaining the body's acid-base balance. Which type of ion is especially important in the control process? *483–485*

Readings

Bartholemew, G. 1977. "Body Temperature and Energy Metabolism." In *Animal Physiology: Principles and Adaptations* by M. Gordon, et al. Third edition. New York: Macmillan. Good introduction to thermal regulation in animals.

Gottschalk, C., and W. Lassiter. 1980. "The Kidney and Body Fluids." In *Medical Physiology* (V. Mountcastle, editor). Fourteenth edition. Volume 2, part 10. St. Louis: Mosby.

Smith, H. 1961. *From Fish to Philosopher*. New York: Doubleday. Available in paperback.

Valtin, H. 1983. *Renal Function: Mechanisms Preserving Fluid and Solute Balance in Health*. Second edition. Boston: Little, Brown. Paperback.

Vander, A., J. Sherman, and D. Luciano. 1985. *Human Physiology: Mechanisms of Body Function*. Fourth edition. New York: McGraw-Hill. Contains an excellent introduction to renal functioning.

Vaughan, T. 1986. *Mammology*. Third edition. New York: Saunders. Chapter 22 is an excellent introduction to temperature regulation in mammals.

34

PRINCIPLES OF REPRODUCTION AND DEVELOPMENT

With a full-throated croak that only a female of its kind could find seductive, a male frog proclaims the onset of warm spring rains, of ponds, of sex in the night. By August the summer sun will have parched the earth, and his pond dominion will be gone. But tonight is the hour of the frog! Through the dark, a female moves toward the vocal male. They meet, they dally; he clamps his forelegs about her swollen abdomen and gives it a prolonged squeeze. Out streams a ribbon of hundreds of eggs. As the eggs are being released, the male expels a milky cloud of swimming sperm. Each egg joins with a sperm, and soon afterward, their nuclei fuse. With this fusion, fertilization is completed; a zygote has formed.

For the leopard frog *Rana pipiens*, a drama now begins to unfold that has been reenacted each spring, with only minor variations, for many millions of years. Within a few hours after fertilization, the single-celled zygote begins dividing into two, then four, then eight cells, and many more to produce the early embryo. The cells become smaller and smaller with each successive division until there is a ball of tiny cells, no larger than the zygote. It has formed in fewer than twenty hours.

And now the cells embark on a course of migrations, changes in shape, and cell interactions. A dimple forms on the embryo's surface as some cells sink inward to become the forerunners of internal tissue layers. Other cells lengthen and still others flatten out, causing a groove to form at the surface—a groove destined to become the nervous system. Through interactions between surface cells and interior cells, eyes start to develop. Within the embryo, a heart is forming and will soon beat rhythmically. Fins take shape; a mouth forms. These developments, appearing one after another, are signs of a process going on in *all* the cells that were so recently developed from a single zygote. The cells are becoming different from one another in both appearance and function!

Within twelve days after fertilization, the embryo has become a tadpole, a larval stage that can swim and feed

Figure 34.1 Development of the leopard frog, *Rana pipiens*. (**a**) A male clasping the female in a behavior called amplexus. When the female releases eggs into the water, the male releases sperm over the eggs. (**b**) Frog embryos. (**c**) A tadpole. (**d**) Transitional form between the tadpole and the young adult (**e**).

a

b

c

on its own (Figure 34.1). For several months the larva grows until, in response to hormonal cues, its body starts to change into the adult form. Legs begin to grow from the body; the tail becomes shorter and shorter, then disappears. The small mouth, once suitable for feeding on algae, develops jaws that accommodate insects and worms. Eventually a full-fledged frog leaves the water for life on land. If it is lucky it will avoid predators, disease, and other threats through the months ahead. In time it may even find a pond swollen with the waters of a new season's rains, and the cycle will begin again.

How did that single-celled frog zygote become transformed into all the specialized cells and structures of the adult? With this question we turn to one of life's greatest mysteries— to the incredible orchestration of the processes underlying development, from the time of reproduction to the emergence of the adult form.

THE BEGINNING: REPRODUCTIVE MODES

We have already looked at the cellular basis of **sexual reproduction**, which requires the fusion of nuclei from two (or more) gametes to produce a new individual. We have also looked at the cellular basis of **asexual reproduction**, whereby gametes are not required in the production of offspring. Here we will consider some structural, behavioral, and ecological adaptations that are associated with both reproductive modes. Figure 34.2 hints at the diversity of these adaptations.

Asexual Reproduction

Some of the structurally simple invertebrates can reproduce asexually, most often by fission or budding. In animal **fission**, the entire body of the parent divides into two roughly equivalent parts, with each part then growing into a whole individual. When certain flatworms reproduce this way, the body may be divided in half in either the transverse or longitudinal direction. In **budding**, the new individual develops as an outgrowth of the parent body. When the bud develops a full set of the parental body structures, it breaks away. The buds of one cnidarian (*Hydra*) protrude from the parent body. The "buds" (gemmules) of sponges are produced internally, and each develops into a separate individual when the parent body disintegrates.

Neither fission nor budding promotes genetic variability; the offspring are *clones*, or genetically identical copies of the parents. Cloning is advantageous only so long as the parents are well adapted to the surroundings, and only so long as the surroundings remain stable. When stability prevails, asexual reproduction is biologically inexpensive, so to speak, because progeny can be produced without much energy outlay. However, most animals live in changing and unpredictable environments. Not surprisingly, most depend primarily on sexual reproduction.

Sexual Reproduction

Usually, sexual reproduction involves the fusion of gametes from a male parent and a female parent. However, some sexually reproducing animals occasionally depart from the basic reproductive mode. **Parthenogenesis** refers to the cleavage and subsequent differentiation of an *unfertilized* egg into an adult. Natural parthenogenesis has been observed in arthropods and in all groups of vertebrates except mammals. For example, sometimes beetles and aphids produce fatherless offspring. Changes in pH, temperature, or salinity, and even mechanical stimulation of the egg can trigger parthenogenesis. Such changes can induce parthenogenesis not only in insects but also in frogs, salamanders, turkeys, and lizards.

d

e

liveborn snake in egg sac

a

b

c

Figure 34.2 Examples of where the animal embryo develops, how it is nourished, and how (if at all) it is protected. In most mammals, the fertilized egg is retained inside the mother's body and nourished by her tissues until the time of birth. Such animals are called *viviparous* (*viva-* , alive; *-parous*, to produce).

Some fishes, lizards, and many snakes are *ovoviviparous*. Their fertilized eggs develop within the mother's body. Such eggs are not nourished by the mother's tissues; they are sustained by yolk reserves. (**a**) A copperhead is one of the ovoviviparous snakes. Her liveborn are still contained in the relics of egg sacs.

(**b**) Snails are *oviparous* (egg-producing) but are not doting parents; their fertilized eggs are left unprotected. (**c**) Monotremes, including the duck-billed platypus, are oviparous, yet they also secrete milk to nourish the juveniles. (**d**) Birds are oviparous animals. Their fertilized eggs, which have large yolk reserves, develop and hatch outside the mother's body.

The opossum (**e**) and kangaroo (**f**), both marsupials, are viviparous. But their young emerge in somewhat unfinished form and undergo further fetal development in a pouch on the ventral surface of the mother's body, where they are nourished from mammary glands. (**g**) Young kangaroo in its mother's pouch.

Earthworms and parasitic tapeworms also are not bound to straightforward sexual reproduction. These animals are **hermaphrodites** (the individual has both male and female reproductive organs). Two earthworms can cross-fertilize each other when they lie head to tail and exchange sperm. Parasitic tapeworms can fertilize themselves. One need not look askance at tapeworms, when one realizes that these animals often live by themselves somewhere in the host animal. Given the bountiful resources and the protected habitat, hermaphroditism does have advantages for the lone parasite.

The reproductive modes of animals include asexual processes, the development of unfertilized eggs into adults, self-fertilization, and (most commonly) sexual reproduction between separate male and female forms.

Some Strategic Problems in Having Separate Sexes

Complete separation into male and female sexes imposes biologically expensive demands. After all, how is fertilization to be accomplished? Specialized body structures and behavioral activities are necessary.

In behavioral terms, male and female reproductive cycles usually are *synchronous*, with mature gametes being released at the same time. Synchrony requires hormonal control mechanisms and sensory receptors that detect environmental cues, such as changes in daylength from one season to the next. Males and females must be able to recognize each other as being of the same species, so energy outlays are necessary for constructing structural signals (such as bright feathers in birds) and for performing courtship routines.

f

g

In structural terms, getting sperm and egg together is not too complicated among certain water-dwelling animals, including sea urchins and most bony fishes. Their gametes are simply released into the same region of water, and the motile sperm swim to an egg. Such *external* fertilization would be chancy if only one sperm and one egg were released each season. So energy outlays by each parent are required for the production of large numbers of sperm and eggs.

In contrast, nearly all land-dwelling animals and even some aquatic forms depend on *internal* fertilization. (Sperm released on dry land obviously do not stand much chance of swimming over to an egg.) The males have reproductive organs called **testes**, in which sperm are formed. Many male animals have a **penis**, a copulatory organ by which sperm are deposited into a specialized duct in the female. Many female animals have

a **vagina**, a duct between the outside world and the **ovary** (an organ where the immature eggs grow). Thus the reproductive structures of reptiles, birds, and mammals protect sperm and eggs from harsh external conditions and enhance the probability of successful internal fertilization.

Another problem to consider: How is the embryo to be nourished? Almost all animal eggs contain **yolk** (proteins, lipids, and other nutritive substances), but some eggs contain more yolk than others. Following fertilization, sea urchin eggs develop into feeding-stage larvae within forty hours. Because these eggs are produced in large numbers, the biochemical investment in yolk for each one is limited. Hence there is a developmental premium on rapid development, with the self-feeding larval stage being reached in the shortest possible time. Bird eggs also are released from the mother's body, but these eggs are mostly a ball of yolk (the nonyolky cytoplasm and nucleus are positioned as a thin cap on the surface). The large yolk reserves nourish the embryo through a longer period of development, which proceeds inside a hard eggshell; unlike sea urchins, birds cannot feed on their own when they are hatched (Figure 34.2). Human eggs have very little yolk—but the developing embryo becomes attached to the mother's body and receives nourishment, and respiratory and excretory support, by physical exchanges with her tissues.

The point of these few examples is that tremendous diversity exists in reproductive and developmental strategies. However, some patterns are widespread in the animal kingdom, and these patterns will serve as the framework for discussions to follow.

STAGES OF DEVELOPMENT

Animal development commonly proceeds through the stages listed in Figure 34.3, beginning with **gametogenesis**, or gamete formation. During this first stage, sperm or eggs form and mature within the parental reproductive system. **Fertilization**, the second stage, starts when a sperm penetrates an egg and it is completed when the sperm nucleus fuses with the egg nucleus. Next comes **cleavage**, a series of mitotic cell divisions that subdivide the fertilized egg into many smaller cells. Cleavage produces the early multicelled embryo (the blastula).

During **gastrulation**, the organizational framework for the whole animal is laid out as cells become arranged into two or more simple tissue layers. Interactions between cells in those layers permit the selective expression of specific genes—and this leads to organ formation, or **organogenesis**. Finally, organs increase in volume and acquire their specialized structural and chemical properties during **growth and tissue specialization**, which continue into the post-embryonic period of the life cycle.

Adult animals of different species obviously do not all look alike. Neither do the embryonic forms, even though they commonly pass through all the stages just listed. Figure 34.4 shows the appearance of the types of embryos we will now consider, with amphibian embryos being our main focus.

Gametogenesis

A sperm cell produced during spermatogenesis is little more than paternal DNA, packaged with a few components that are necessary for moving the DNA to an egg (page 511). The immature egg cell, or **oocyte**, is

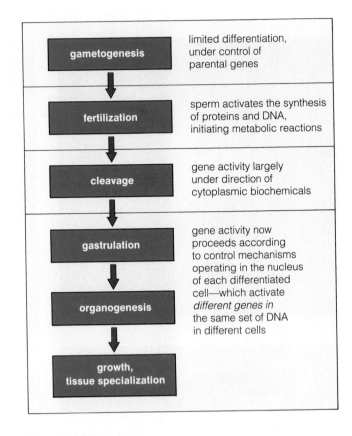

Figure 34.3 Generalized overview of the stages of animal development described in this chapter.

another story. An oocyte increases in volume—by many thousands of times, in the ovary of a female frog. Specialized proteins, including important enzymes, mRNA transcripts, and other molecules, accumulate in its cytoplasm; they will direct the development of the early embryo.

The components of the cytoplasm are not randomly distributed. Even the nucleus has a characteristic position in a frog egg and so imparts polarity to it (in other words, a frog egg has two identifiable poles). The *animal pole* is the one closest to the nucleus; opposite is the *vegetal pole*, where substances such as yolk accumulate (Figure 34.5). All animal eggs show some degree of polarity, which will influence the structural patterning of the embryo.

Fertilization

When a sperm penetrates an egg, it triggers structural reorganization in the egg cytoplasm. You can observe indirect signs of this reorganization in frog eggs, which contain granules of dark pigment in their cortex. (The "cortex" is the plasma membrane and the cytoplasm just

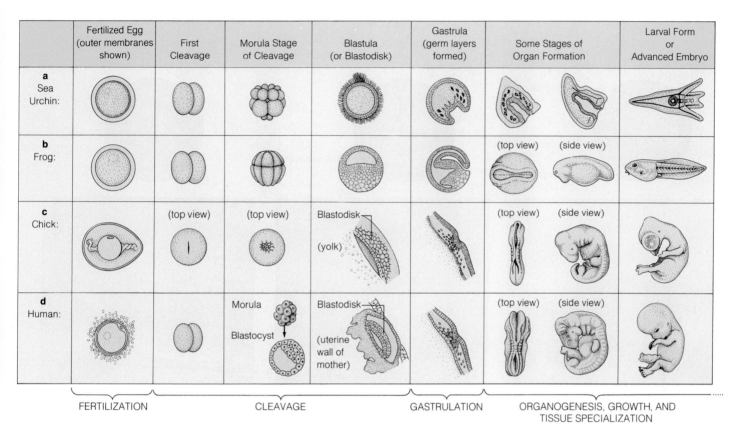

	Fertilized Egg (outer membranes shown)	First Cleavage	Morula Stage of Cleavage	Blastula (or Blastodisk)	Gastrula (germ layers formed)	Some Stages of Organ Formation	Larval Form or Advanced Embryo
a Sea Urchin:							
b Frog:						(top view) (side view)	
c Chick:		(top view)	(top view)	Blastodisk (yolk)		(top view) (side view)	
d Human:			Morula Blastocyst	Blastodisk (uterine wall of mother)		(top view) (side view)	

FERTILIZATION — CLEAVAGE — GASTRULATION — ORGANOGENESIS, GROWTH, AND TISSUE SPECIALIZATION

Figure 34.4 Comparison of embryonic development in four different animals. The drawings are not to the same scale; however, they show the developmental patterns that are common to all four types. For clarity, the membranes surrounding the embryo are not shown from cleavage onward. Blastula and gastrula stages are shown in cross-section; they are described in more detail in this chapter and the next for frogs, birds, and humans.

beneath it.) The animal pole has more granules and is darker than the vegetal pole. Sperm penetration causes the granules and then the cortex itself to shift position. The outcome is a **gray crescent**, an area of intermediate pigmentation near the equator, on the side of the egg *opposite* the penetration site (Figure 34.5). In this area, the body axis of the frog embryo will become established and gastrulation will begin.

Cleavage

Fertilization is followed by mitotic cell divisions that convert the zygote into the early multicelled embryo. A cleavage furrow appears after each mitotic division, defining the plane where the cytoplasm will be pinched in two (page 147). Usually the initial cleavages of the cytoplasm are not accompanied by any increase in the size of the embryo, so all the daughter cells collectively occupy the same volume as did the zygote. (The sketches in the top row of Figure 34.6 show this.) However, the cells differ in size, shape, and activity. Some become

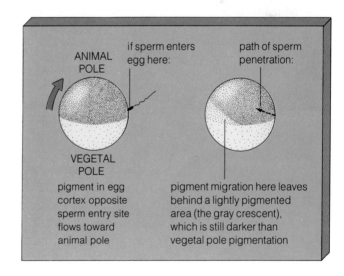

Figure 34.5 Formation of the gray crescent in the eggs of some amphibians, including frogs.

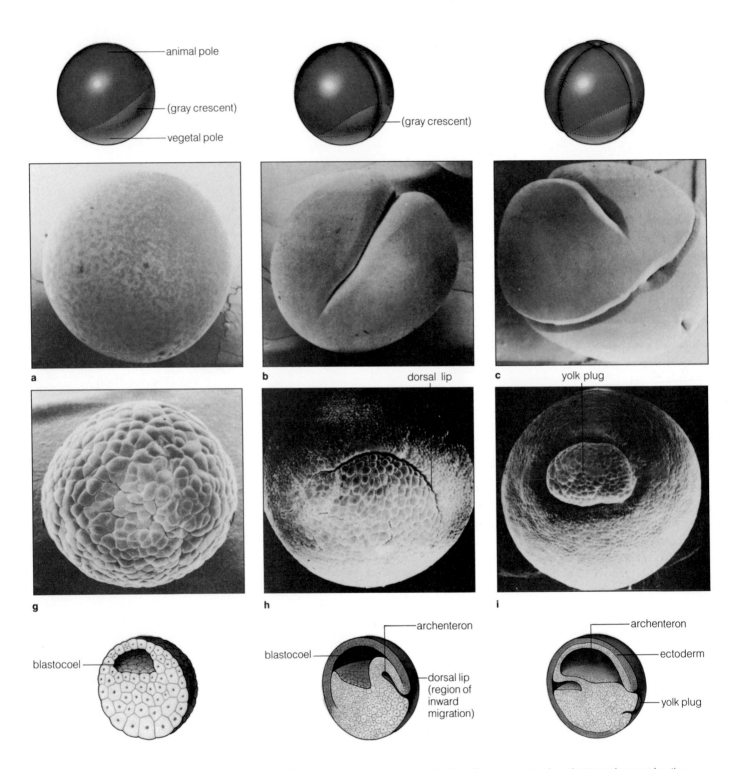

Figure 34.6 Early embryonic development of a frog. For these scanning electron micrographs, the jellylike layer surrounding the egg has been removed. (**a**) Within about an hour after fertilization, a region of differentiated surface cytoplasm (gray crescent) appears opposite the site where the sperm penetrated the egg. (**b-g**) Cleavage leads to a blastula, a ball of cells in which a cavity (blastocoel) has appeared.

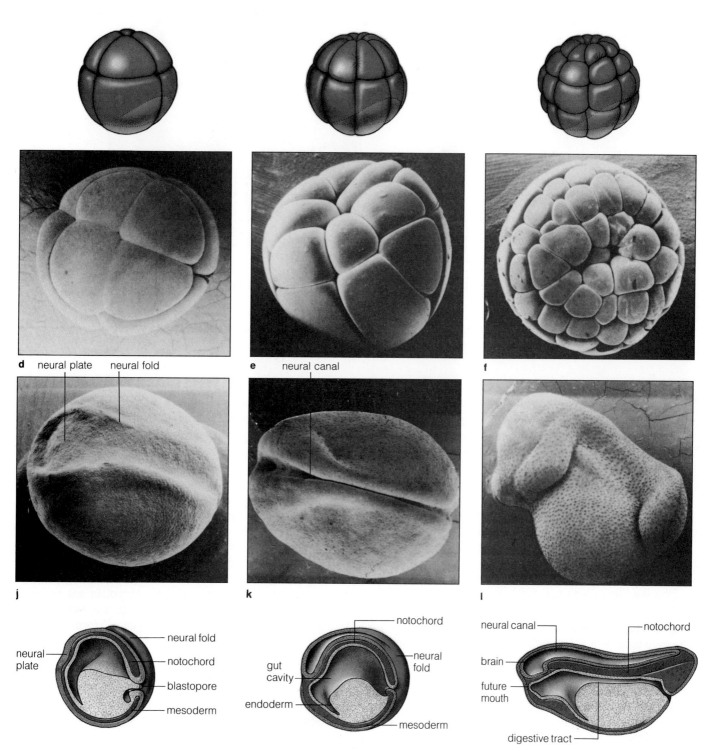

d | **e** | **f**

neural plate   neural fold   neural canal

j | **k** | **l**

neural fold

neural plate — notochord — blastopore — mesoderm

notochord — neural fold — gut cavity — endoderm — mesoderm

neural canal — notochord — brain — future mouth — digestive tract

(**h,i**) Major cell movements and rearrangements occur during gastrulation. Tissue layers form; a primitive gut cavity (archenteron) develops. (**j,k**) Neural developments now take place, and the fluid-filled body cavity in which vital organs will be suspended appears. (**l**) Differentiation proceeds, moving the embryo on its way to becoming a functional larval form. (Photos from R. Kessel and C. Shih, *SEM in Biology*, Springer-Verlag, New York, 1976.)

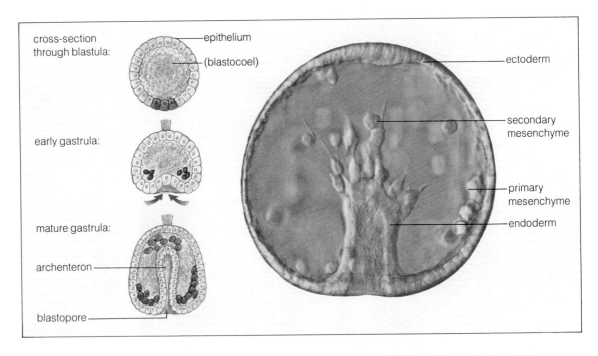

Figure 34.7 Inward migration of surface cells during gastrulation in a sea urchin. The photograph shows the embryo at the mid-gastrula stage. Some of the cells (designated mesenchyme cells) will ultimately develop into a third germ layer, the mesoderm.

flattened, others are rearranged, and physical connections are established that permit intercellular communication. Cells develop microvilli and ridges, the number and shape of which depend on where the cell is located in the embryo.

Successive cleavages produce the early embryonic stage called a **blastula**. In sea urchins, the blastula is a hollow sphere. The hollow portion, the **blastocoel**, is surrounded by a simple epithelium (Figure 34.7). In amphibian eggs, the concentrated yolk impedes cleavage near the vegetal pole, so the amphibian blastula is not entirely hollow. Instead, the blastocoel forms near the animal pole (Figure 34.6g). Eggs of reptiles, birds, and most fishes have so much yolk that cleavage is restricted to a tiny, caplike region at the animal pole. In this case, cleavage leads to a **blastodisk**: two flattened layers with a thin blastocoel between them, perched on the yolk surface (Figure 34.4c).

Early in cleavage, the mammalian embryo is a solid mass of cells, and the embryo proper will form from the "inner cell mass," this being a small group of cells at the interior. The outermost cells eventually form a surface layer that will give rise to **extraembryonic membranes**. Membranes of this type attach the embryo to the mother's uterus and take part in the transfer of nutrients derived from her tissues. By late cleavage, a cavity forms in the cell mass. It becomes filled with fluid that lifts the inner cell mass away from the surface layer on all but one side (Figure 34.4d). The mammalian embryo has become a **blastocyst**. The next chapter describes the fate of the blastocyst during human development.

Gastrulation

As cleavage draws to a close, the pace of cell division slackens. The cells begin to move about and change their positions relative to one another. This stage of cell rearrangements is called gastrulation.

Although there is little (if any) increase in size during gastrulation, the cell rearrangements dramatically change the appearance of the embryo. In sea urchins, for example, a major inward migration of surface cells occurs. Some cells that end up inside will form the lining of the **archenteron**, an internal cavity destined to become the gastrointestinal tract (Figure 34.7). In other species, cell rearrangements establish a long axis for further development following gastrulation. (For example, a **neural tube** will eventually form along this axis in vertebrate embryos—a tube that is the forerunner of the brain and spinal cord.)

This last example brings us to the significance of gastrulation. Think about the structural organization of animals. With few exceptions, animals have an internal region of cells, tissues, and organs that function in digestion and absorption of nutrients. They have a surface tissue region that protects internal parts and is equipped

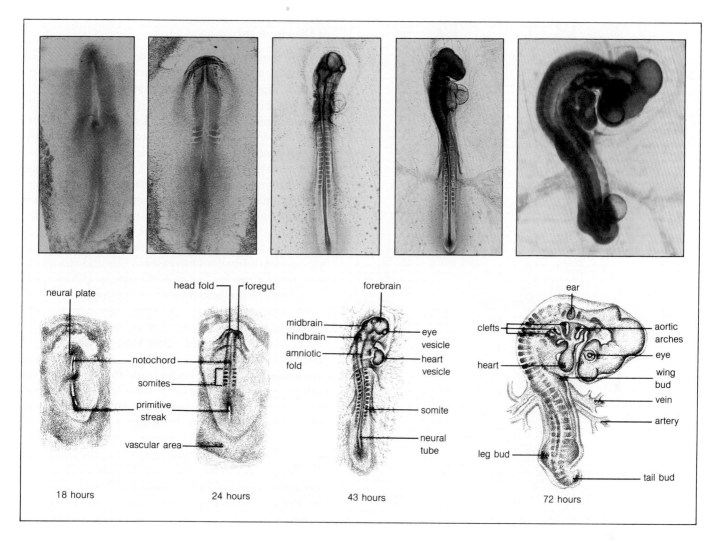

Figure 34.8 Onset of organ formation in a chick embryo during the first 72 hours of incubation. ("Incubate" means to induce development, as by application of heat.) The heart begins to beat at some time between 30 and 36 hours.

with sensory receptors for detecting changes in the outside world. Most animals have an intermediate region of tissues organized into many internal organs, such as those concerned with movement, support, and blood circulation. The three regions develop from three "primordial" tissue layers, or **germ layers**:

endoderm	*inner layer; gives rise to inner lining of gut and organs derived from it*
mesoderm	*intermediate layer; gives rise to muscle, the organs of circulation, reproduction, and excretion, most of the internal skeleton, and connective tissue layers of the gut and body covering*
ectoderm	*surface layer; gives rise to tissues of nervous system and outer layer of body covering*

This three-layered organization, which is typical of most animals, depends on gastrulation.

Organogenesis

Following gastrulation, the simple tissue layers of the gastrula become split into subpopulations of cells that are forerunners of specific organs. We see for the first time the emergence of different cell types, each specialized in ways that will contribute to the functioning of the animal as a whole.

Some of the cell lineages that form are quite complex, with their cells destined to give rise to whole systems of organs, such as the nervous system. The rate of organ formation can be remarkable, as the time line in Figure 34.8 suggests.

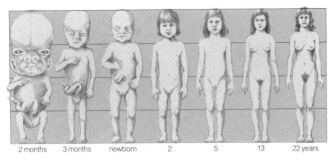

Figure 34.9 Diagram of changes in the proportions of the human body during prenatal and postnatal growth.

| 2 months | 3 months | newborn | 2 | 5 | 13 | 22 years |

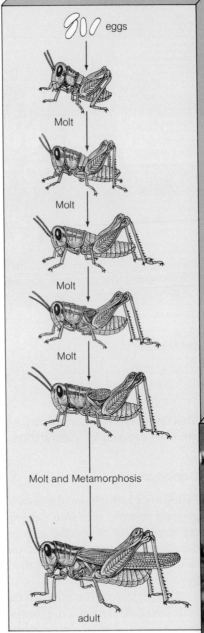

eggs

Molt

Molt

Molt

Molt

Molt

Molt and Metamorphosis

adult

Figure 34.10 Metamorphosis leading to the adult grasshopper.

Post-Embryonic Pathways of Development

Once all the organs necessary for feeding and other vital activities have formed and are functioning, the new individual is ready to lead a more or less independent existence. Embryonic development is over, and now the young animal follows a prescribed course of further growth and development that leads to the **adult**, the sexually mature form of the species.

In nematodes and a few other animals, the transition from embryo to adult is straightforward. The young simply are miniatures of what is to come; all it takes to reach adulthood are increases in size and maturation of the gonads.

The post-embryonic development of reptiles, birds, and mammals involves increases in size and changes in body proportions, as Figure 34.9 suggests. Other organs in addition to the gonads may not be fully developed. Newly hatched birds, for example, have only partially developed primary feathers just under the skin.

For insects and some other kinds of animals, the course of development is "indirect," for a larval stage intervenes between the embryo and the adult. First the embryo grows into a **larva** (a sexually immature, free-living and free-feeding animal), then the larva itself grows and changes into the sexually mature form. In some species, the transformation is gradual, with the immature form simply growing in size to the adult form (see Figure 34.10, for example). In other species, the transformation involves massive tissue reorganization and drastic remodeling into the adult form, as in frogs (Figure 34.1). The reactivated growth and transformation of a larva into the sexually mature adult is called **metamorphosis**.

Patterns of indirect development vary among aquatic and terrestrial animals. Extended larval stages are typical of animals that release small and relatively yolkless eggs into water. Short larval stages are common among animals that lay large, yolky eggs in water. Metamorphosis in insects and some other animals proceeds through a sequence of discrete stages in which the larvae grow in size, then molt (shed their too-small skin or cuticle), then grow again (Figure 34.10).

A different kind of reactivated growth occurs during **regeneration**: the replacement of body parts that have been lost by accident. For example, if a predator grasps one of the legs of a crab, the crab can give it up (the better to make an escape), and then grow a replacement.

MECHANISMS OF DEVELOPMENT

Now that we have completed our survey of the stages of animal development, we can begin to think about the mechanisms responsible for the orderly sequence of developmental events. The mechanisms are complex, and it will help to keep the following conceptual overview in mind when reading about them.

As a general rule, DNA replication and mitotic division assure that all cells in the new individual are *genetically equivalent*. In other words, each new cell receives the same number and kind of genes that were present in the zygote. At the blastula stage, the cells are smaller but they are still all alike in their appearance or behavior. Later, after tissue layers have formed in the gastrula, cells start to exhibit unique structures and functions. This is a reflection of different patterns of gene expression in different cell lineages.

As you will now see, two processes that are absolutely central to development have their foundations in (1) the unequal distribution of cytoplasmic substances to daughter cells at cleavage and (2) cell interactions. In the first process, **cell differentiation**, initial populations of genetically equivalent cells give rise to subpopulations of phenotypically different cell types. In the second process, **morphogenesis**, different cell types become organized into all the specialized tissues and organs of the animal body.

Developmental Information in the Egg

Let's start with the nature of those substances in the egg cytoplasm. To a large extent, the destinies of cell lineages are mapped out within the maturing oocyte—that is, before fertilization. Quantities of mRNA molecules are transcribed from the maternal DNA. Many are translated at once into enzymes and other proteins, such as histones, that will be used for chromosome replications in the early embryo. Ribosomal subunits and other cytoplasmic components that the embryo will require for protein synthesis are stockpiled. Also, numerous mRNA transcripts accumulate in different cytoplasmic regions. They will be allocated to different daughter cells during cleavage and then activated to direct the synthesis of specific sets of proteins.

Also present in the growing oocyte are microtubules and other cytoskeletal elements, oriented in specific directions that will influence the first cell divisions in the embryo. Whenever a cell divides in two, it does so at a prescribed angle relative to the adjacent cells, based on the orientation of microtubules of the mitotic spindle (page 145). Look at the cleavage planes indicated in the top row of sketches for Figure 34.6; this is the division pattern for every normal frog embryo. Microtubules also take part in the movement of pigment granules and cortex that produce the gray crescent of a fertilized frog egg—and this is the site where the long axis and bilateral symmetry of the body are established. Thus the organization of the egg's cytoskeleton represents critical information for the forthcoming embryo.

Finally, the amount and distribution of yolk within the egg cytoplasm profoundly influences the pattern and consequences of cleavage. As we have seen, these factors affect the formation of cleavage planes—hence the size and spatial positions of the resulting cells. In turn, the size and position of a given cell in the embryo affect its interactions with other cells during the gastrula stage. In one experiment, cells from the roof of an amphibian blastocoel were placed next to the yolky vegetal cells at the blastocoel floor. Instead of forming ectodermal tissue, as they normally would have done, the cells differentiated as mesoderm.

Cell Differentiation

All cells in the animal body start out with the same set of genes, so they all have the same developmental potential. During development, however, restrictions are placed on which genes will be expressed in a given cell type. Through controls over gene expression, some cells are kept in an undifferentiated or "uncommitted" state, while others proceed down a given developmental road to become a specialized cell type. Once committed, each cell type is able to use a small fraction of the genome that no other type of cell uses, and so it proceeds to a uniquely differentiated state.

An adult animal has dozens or hundreds of fully differentiated cell types, such as neurons and muscle cells. It also has populations of **stem cells**, which can self-replicate, differentiate, or both. Populations of stem cells are the basis of ongoing replacements of worn-out or dead cells in skin, blood, and the gut mucosa.

With only rare exceptions, the fully differentiated state of a given cell type is reached without any loss of genetic information. *A differentiated cell still has the same*

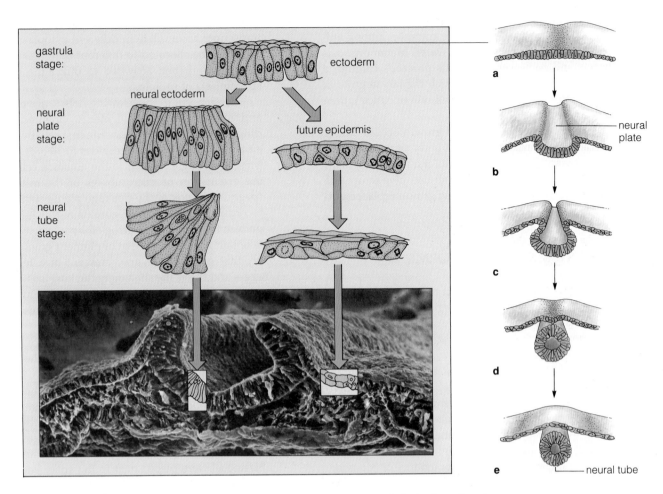

Figure 34.11 Example of morphogenesis—the changes in cell shape that underlie the formation of a neural tube (the forerunner of the brain and spinal cord). As gastrulation draws to a close, the ectoderm is a uniform sheet of cells (**a**). Some ectodermal cells elongate, forming a neural plate, then they constrict at one end to become wedge-shaped. The changes in cell shape cause the ectodermal sheet to fold over the neural plate to form the neural tube (**b-e**). Other ectodermal cells flatten while these changes are occurring; they will become part of the epidermis. The scanning electron micrograph shows the neural tube forming in a chick embryo.

number and kind of genes that were present in the zygote. How do we know this? Consider John Gurdon's experiments with the South African clawed frog, *Xenopus laevis.* Gurdon and his coworkers removed or inactivated the nucleus of an unfertilized frog egg. They also isolated intestinal cells from *Xenopus* tadpoles and carefully ruptured the plasma membrane, leaving the nucleus and most of the cytoplasm intact. Then they inserted the nucleus of the ruptured tadpole cell into the enucleated egg. In some cases, the transplanted nucleus—which was from a highly differentiated cell—directed the developmental program leading to a whole frog! Clearly the intestinal cell nucleus contained the same genes as the zygote nucleus.

As another example, consider what happens when a human embryo spontaneously splits during the first cleavage into two separate cells. The result is not two half-embryos but **identical twins**, or two complete, normal individuals having the same genetic makeup. (In contrast, nonidentical twins occur when two different eggs are fertilized at the same time by two different sperm.)

Twinning is actually the normal pattern of development for armadillos (Figure 2.6). For these animals, the embryo splits at the four-cell stage to produce quadruplets, every time.

Mechanisms Underlying Morphogenesis

For each animal, tissues and organs become organized with great precision into patterns characteristic of the species. What kinds of mechanisms give rise to that organization? Considerable research in embryology has shown that the main ones are these:

cell division
cell migrations
changes in cell size and shape
localized growth
controlled cell death

Let's consider a few examples of the morphogenetic changes that these mechanisms can bring about.

Cell Migrations. During morphogenesis, cells and entire tissues migrate from one site to another. In *active cell migration*, for example, cells move about by means of pseudopods. (Pseudopods are temporary projections from the main cell body, some rather bulbous, others fingerlike.) The cells migrate over prescribed pathways, reaching a prescribed destination and establishing contact with cells already there. These movements must be extremely accurate. Even before you were born, for example, precursors of nerve cells migrated and established the many billions of precise connections that enable your nervous system to function.

How do the cells know where to move? In part, they move in response to chemical gradients, a behavior called **chemotaxis**. The gradients are probably created when specific substances are released from target tissues. This type of movement was shown in Figure 17.3, which described the development of the slime mold *Dictyostelium discoideum*.

Cells also move in response to **adhesive cues**, provided by recognition proteins at the surface of other cells and by other molecules in the extracellular matrix. In vertebrate embryos, pigment cells are following such cues when they move along blood vessels but not along the axons of neurons. Similarly, Schwann cells are following adhesive cues when they migrate along the axons of neurons but not along blood vessels. It seems likely that the synthesis, secretion, deposition, and removal of specific extracellular substances helps coordinate the active migration of cells in a developing embryo.

How do the migrating cells know when to stop moving? Again, adhesive cues appear to be involved. Migrating cells respond to adhesive cues by moving to locations that permit stronger adhesive interactions. Once they become arranged in a manner that maximizes their adhesion, further migration is impeded.

Another morphogenetic movement is the inward or outward *folding of sheets of cells*, this resulting from coordinated changes in cell shapes. Think about what happens after the three germ tissue layers form in the embryos of amphibians, reptiles, birds, and mammals. As Figure 34.11 shows, ectodermal cells at the midline of the embryo elongate and form a **neural plate**, an intermediate structure in the development of the brain and

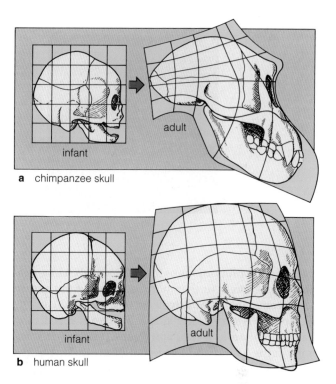

Figure 34.12 Comparison of the proportional changes in a chimpanzee skull and a human skull, both of which are remarkably similar in infants. The deformation in the grid superimposed over the adult skulls shows that the proportions change much more in the chimpanzee.

spinal cord. The change in each cell's shape arises through the elongation of microtubules in its cytoplasm. Next, cells near the middle become wedge-shaped (a ring of microfilaments in the cytoplasm constricts one end of each elongated cell). Collectively, the changes in cell shape cause the neural plate to fold over and meet at the embryo's midline to form the neural tube. Meanwhile, other ectodermal cells have flattened; they will become the epidermis above the tube.

Localized Growth and Cell Death. Morphogenesis depends on localized growth, which contributes to changes in the sizes, shapes, and proportions of body parts. How localized growth occurs in some tissues more than others is not fully understood, but regulatory genes are almost certainly involved.

Consider that adult chimpanzees and humans differ quite a bit in the proportions of their body parts, yet there are very few differences between chimpanzee and human DNA. Intriguingly, the fetuses of both animals develop at about the same rate, in remarkably parallel ways. For example, the infant skull is strikingly similar in both species (Figure 34.12). From infancy onward,

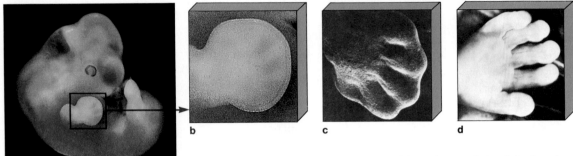

Figure 34.13 Formation of the human hand through controlled cell death and other morphogenetic events. (The hand has been turned palm upward in **d**.)

however, the proportions of the chimp skull change dramatically, whereas further morphogenesis of the human skull seems almost arrested. Because the DNA of the two species is very nearly identical, there is speculation that a mutation in a regulatory gene occurred at some point in the evolution of ancestral forms leading to the first humans. That mutant gene might inhibit major proportional changes in the human skull.

Morphogenesis also depends on **controlled cell death**: the elimination of tissues and cells that are used for only short periods in the embryo or the adult. Controlled cell death is genetically programmed, as the following examples suggest.

Perhaps you have noticed that all kittens and puppies are born with their eyes sealed shut. The eyelids form as an unbroken layer of skin. Just after birth, cells stretching in a thin line across each eyelid die on cue. As the dead cells degenerate, a slit forms in the skin, and the upper and lower lids part company.

Cell death also helps transform the four paddlelike appendages in human embryos into hands and feet. Skin cells between the lobes of the "paddle" die on cue, leaving separate toes and fingers (Figure 34.13). Duck embryos also have paddlelike appendages, but cell death normally does not occur in them; that is why ducks have webbed feet instead of separated toes. In some mice and some humans, a gene mutation blocks cell death in the paddles, and hands and feet remain webbed.

Pattern Formation

For morphogenesis to proceed smoothly in a developing embryo, cells must sense their position relative to one another and use the information to produce ordered, spatial arrangements of differentiated tissues. The term **pattern formation** refers to the mechanisms responsible for the differentiation of tissues and their positioning in space. Here we will consider two of those mechanisms, called ooplasmic localization and embryonic induction. Through **ooplasmic localization**, cells differentiate according to which cytoplasmic substances they inher-

ited during cleavage. Through **embryonic induction**, one body part differentiates in response to signals from an adjacent body part.

As an example of ooplasmic localization, consider how the cytoplasm is divided at the first cleavage of certain molluscan zygotes. The cytoplasm of one daughter cell has microfilaments that constrict part of it to form a protruding lobe; but no such lobe forms in the other daughter cell:

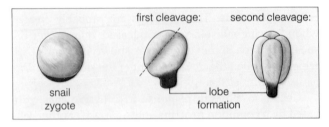

If the two daughter cells are experimentally separated, the one with the lobe can give rise to a normal molluscan larva, complete with a tuft of cilia at one end (this being necessary for swimming). The other cell develops into a stunted embryo that lacks (among other things) the tuft. Similarly, if the lobe is cut off, the whole embryo will develop abnormally and will be missing the ciliary tuft and other organs.

The importance of embryonic induction is illustrated by the precise positioning of bones that occurs when a chick wing develops. Through cell divisions in its mesoderm and the ectodermal covering, the wing bud grows out from the body. When certain groups of ectodermal cells are removed from the tip of a half-grown wing bud, terminal wing bones never develop (Figure 34.14c). Apparently, as new cells are added to the outwardly growing tip, they "assess" which bones were formed previously, then they become the next bones in line. If the ectodermal ridge is surgically removed, there will be no new mesodermal cells to form the remaining bones.

As another example of embryonic induction, consider how the vertebrate eye develops. The retina of an eye originates from the forebrain but its lens, which focuses light onto the retina, originates from the epidermis (Fig-

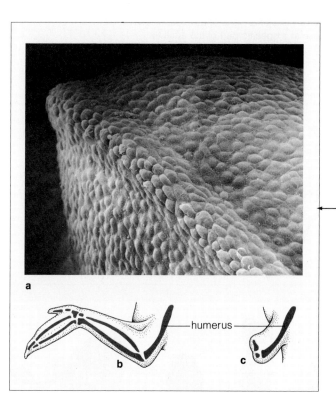

a

humerus

b   c

Figure 34.14 (**a**) Scanning electron micrograph of the normal ectodermal ridge of a wing bud in a chick embryo. (**b**) Normal pattern of bone formation. (**c**) Pattern of bone formation when ectodermal cells at the ridge are surgically removed before the wing bud is fully grown. All new development ceases. Only the bones that have been determined by the time of the surgery will differentiate; notice the deficiencies in the wing skeleton.

Figure 34.15 (Below) Eye formation in a frog embryo. The retina develops as an outgrowth of the brain; the lens as an ingrowth of the ectoderm. (**a**) The optic vesicle grows out of the side of the brain. When it contacts the head ectoderm, it induces the elongation and inward folding of the ectodermal cells to form a lens vesicle (**b**). Meanwhile, the optic vesicle is induced to sink inward, forming the optic cup. (**c**) The cup's inner layer will form the retina. (**d**) Scanning electron micrograph of an optic cup and lens in a chick embryo.

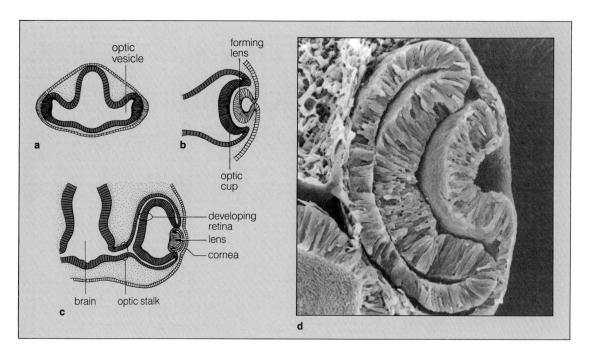

optic vesicle

forming lens

a   b

optic cup

developing retina

lens

cornea

brain   optic stalk

c

d

ure 34.15). Hans Spemann experimented with a salamander embryo in which optic cups had already begun to grow out of the forebrain. He surgically removed one of the optic cups and inserted it under the ectoderm of the belly region. Belly epidermal cells that had come in contact with the transplanted optic cup were induced to form a lens—which fit perfectly into the transplanted optic cup!

What acts as an inducer during pattern formation? How is it transmitted from the inducer tissue to the responding tissue? Several experiments indicate that chemical signals diffuse from one tissue to the other. For example, when the two tissues destined to become the epithelium and connective tissue of a pancreas are surgically separated from each other in a mouse embryo, the future epithelial tissue does not differentiate prop-

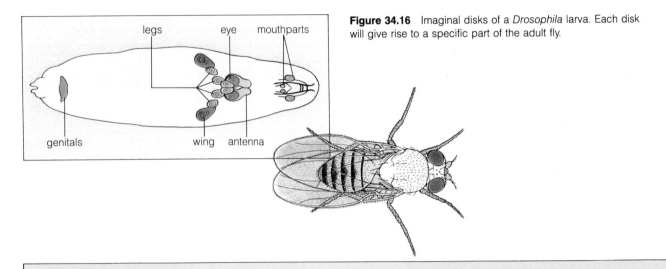

Figure 34.16 Imaginal disks of a *Drosophila* larva. Each disk will give rise to a specific part of the adult fly.

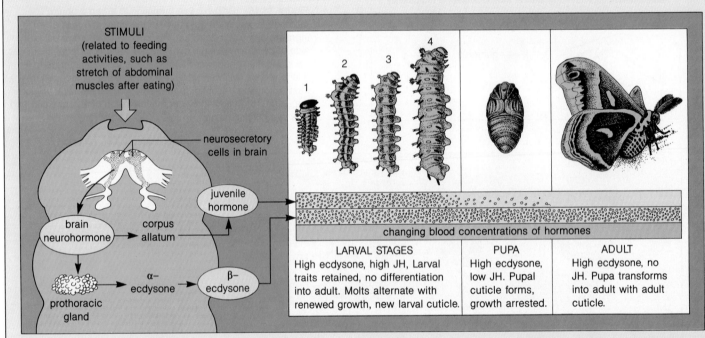

STIMULI
(related to feeding activities, such as stretch of abdominal muscles after eating)

neurosecretory cells in brain

juvenile hormone

brain neurohormone → corpus allatum

prothoracic gland → α–ecdysone → β–ecdysone

changing blood concentrations of hormones

LARVAL STAGES	PUPA	ADULT
High ecdysone, high JH, Larval traits retained, no differentiation into adult. Molts alternate with renewed growth, new larval cuticle.	High ecdysone, low JH. Pupal cuticle forms, growth arrested.	High ecdysone, no JH. Pupa transforms into adult with adult cuticle.

a

Figure 34.17 Neuroendocrine control of the development of a silkworm moth, *Platysamia cecropia*.

A hatched larva eats until it completes five near-doublings in size. Rapid epidermal cell divisions and cell enlargements sustain the massive growth—but the cells also secrete chitin to the epidermal surface. Chitin forms a tough, flexible casing (the cuticle) from which the larval legs protrude.

The cuticle sets an upper limit on increases in mass. When the limit is reached, cell division idles and the larva molts (sheds its cuticle). Cell divisions resume, more chitin is secreted, and a larger cuticle is produced. The larva grows and molts repeatedly. Then chitin is deposited over the whole insect, legs and all; this is the pupal stage.

The insect spins a cocoon around itself for three days, then the body undergoes massive cell destruction and tissue reorganization. The contents of degraded cells form a nutrient-rich soup that sustains the growth of imaginal disks. The disks had been growing slowly, but now they rapidly give rise to what will become adult tissues and organs. The pupa lasts eight winter months. In spring, cell death and tissue changes transform the pupa into the adult.

Transformation of a larva into a moth is an example of metamorphosis. The developmental line leading from the egg to the adult is not direct; it proceeds through immature forms that are radically remodeled. The adult "plan" is already laid out in immature tissues, but neural and endocrine controls dictate when the plan will be fulfilled.

In the larval brain, neurosecretory cells release *brain neurohormone*, which acts on paired prothoracic glands. The glands produce and secrete precursors of a hormone, *β-ecdysone*, that activates gene expression and stimulates molting.

erly. However, suppose the two tissues are grown in a culture medium, separated only by a filter that is permeable to large molecules (but not to cells or cell structures). Then, the responding tissue differentiates properly. Although physical contact with the inducing tissue had been prohibited, signals still reached it.

Some inducer substances (hormones, perhaps) act like green traffic lights; they signal already committed target cells to continue with their differentiation. Other inducer substances are like flagmen at a road construction site; they instruct target cells on which course to follow.

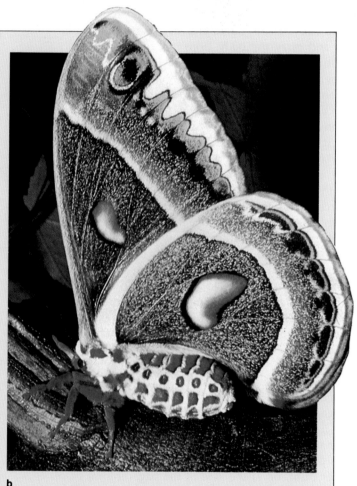

b

Neurohormones also act on two glands (the corpora allata) located behind the brain. The glands secrete *juvenile hormone* (JH) which, at high levels, prolongs the juvenile state. The *absence* of JH triggers cell differentiation into the adult.

High concentrations of both JH and ecdysone promote larval growth and development. But the amount of JH declines steadily and disappears by the fifth larval stage. The pupa forms when ecdysone levels are high and JH levels are low. Then, the developmental restraints are lifted from cells of the imaginal disks.

Experiments with mutant strains of *Drosophila* tell us something about the "instructive" inducers. In *Drosophila* larvae, some cells are arranged in clusters, each of which will give rise to specific body parts of the adult (Figure 34.16). The clusters are called **imaginal disks**. Normally, the fate of the cells in each disk is sealed early in development. However, when an "antenna" disk is surgically removed and exposed to abnormal tissue environments before reinsertion into a larva undergoing metamorphosis, it may later differentiate into a leg.

The same thing happens as a result of homeotic mutations. Apparently, a **homeotic mutation** affects a regulatory gene that activates sets of genes concerned with development. These single-gene mutations lead either to alterations in the inducer substance itself or to abnormal responses to it. One homeotic mutation activates the wrong set of genes in the cells of an "antenna" disk in *Drosophila*, causing those cells to differentiate into a leg.

This completes our discussion of the types of mechanisms underlying specific steps along the road leading from the embryo to the adult form. We conclude with the example in Figure 34.17, which shows how separate mechanisms can be integrated and controlled to bring about the transformation. This figure describes how controls govern the dramatic metamorphosis of a silkworm larva into a moth.

AGING AND DEATH

Following growth and differentiation, the cells of all complex animals gradually deteriorate. Paralleling the deterioration are structural changes and a gradual loss of efficiency in bodily functions, as well as increased sensitivity to environmentally induced stress. Progressive cellular and bodily deterioration is built into the life cycle of all organisms in which differentiated cells show extensive specialization; the process is called **aging**.

Aging in humans leads to structural changes such as loss of hair and teeth, increased skin wrinkling and fat deposition, and decreased muscle mass. Less obvious are gradual physiological changes. For example, metabolic rates decline in kidney cells, so the body cannot respond as effectively as it once did to changes in extracellular fluid volume and composition. As another example, the collagen fibers being produced are structurally changed. Collagen is present in the extracellular spaces of nearly all tissues; it may represent as much as forty percent of the body's proteins. Thus any changes in its structure are bound to have widespread physical effects.

No one knows what causes aging, although researchers have given us some interesting things to think about. More than two decades ago, Paul Moorhead and Leonard Hayflick cultured normal embryonic cells from

Death in the Open

Lewis Thomas

(Printed by permission from the author and the New England Journal of Medicine, *January 11, 1973, 288:92–93)*

Everything in the world dies, but we only know about it as a kind of abstraction. If you stand in a meadow, at the edge of a hillside, and look around carefully, almost everything you can catch sight of is in the process of dying, and most things will be dead long before you are. If it were not for the constant renewal and replacement going on before your eyes, the whole place would turn to stone and sand under your feet.

There are some creatures that do not seem to die at all; they simply vanish totally into their own progeny. Single cells do this. The cell becomes two, then four, and so on, and after a while the last trace is gone. It cannot be seen as death; barring mutation, the descendants are simply the first cell, living all over again. The cycles of the slime mold have episodes that seem as conclusive as death, but the withered slug, with its stalk and fruiting body, is plainly the transient tissue of a developing organism; the free-swimming amoebocytes use this mode collectively in order to produce more of themselves.

There are said to be a billion billion insects on the earth at any moment, most of them with very short life expectancies by our standards. Someone has estimated that there are 25 million assorted insects hanging in the air over every temperate square mile, in a column extending upward for thousands of feet, drifting through the layers of atmosphere like plankton. They are dying steadily, some by being eaten, some just dropping in their tracks, tons of them around the earth, disintegrating as they die, invisibly.

Who ever sees dead birds, in anything like the huge numbers stipulated by the certainty of the death of all birds? A dead bird is an incongruity, more startling than an unexpected live bird, sure evidence to the human mind that something has gone wrong. Birds do their dying off somewhere, behind things, under things, never on the wing.

Animals seem to have an instinct for performing death alone, hidden. Even the largest, most conspicuous ones find ways to conceal themselves in time. If an elephant missteps and dies in an open place, the herd will not leave him there; the others will pick him up and carry the body from place to place, finally putting it down in some inexplicably suitable location. When elephants encounter the skeleton of an elephant in the open, they methodically take up each of the bones and distribute them, in a ponderous ceremony, over neighboring acres.

It is a natural marvel. All of the life on earth dies, all of the time, in the same volume as the new life that dazzles us each morning, each spring. All we see of this is the odd stump, the fly struggling on the porch floor of the summer house in October, the fragment on the highway. I have lived all my life with an embarrassment of squirrels in my backyard, they are all over the place, all year long, and I have never seen, anywhere, a dead squirrel.

I suppose that is just as well. If the earth were otherwise, and all the dying were done in the open, with the dead there to be looked at, we would never have it out of our minds. We can forget about it much of the time, or think of it as an accident to be avoided, somehow. But it does make the process of dying seem more exceptional than it really is, and harder to engage in at the times when we must ourselves engage.

In our way, we conform as best we can to the rest of nature. The obituary pages tell us of the news that we are dying away, while birth announcements in finer print, off at the side of the page, inform us of our replacements, but we get no grasp from this of the enormity of the scale. There are 4 billion of us on the earth, and all 4 billion must be dead, on a schedule, within this lifetime. The vast mortality, involving something over 50 million each year, takes place in relative secrecy. We can only really know of the deaths in our households, among our friends. These, detached in our minds from all the rest, we take to be unnatural events, anomalies, outrages. We speak of our own dead in low voices; struck down, we say, as though visible death can occur only for cause, by disease or violence, avoidably. We send off for flowers, grieve, make ceremonies, scatter bones, unaware of the rest of the 4 billion on the same schedule. All of that immense mass of flesh and bone and conscious-ness will disappear by absorption into the earth, with-out recognition by the transient survivors.

Less than half a century from now, our replace-ments will have more than doubled in numbers. It is hard to see how we can continue to keep the secret, with such multitudes doing the dying. We will have to give up the notion that death is a catastrophe, or detestable, or avoidable, or even strange. We will need to learn more about the cycling of life in the rest of the system, and about our connection in the process. Everything that comes alive seems to be in trade for everything that dies, cell for cell. There might be some comfort in the recognition of synchrony, in the infor-mation that we all go down together, in the best of company.

humans. They discovered that all of the cell lines pro-ceeded to divide about fifty times, then the entire pop-ulation died off. Hayflick took some of the cultured cells and froze them for a period of years. Afterward, he allowed them to thaw and placed them in a culture medium. The cells proceeded to complete the cycle of fifty doublings—whereupon they all died on schedule.

Such experiments suggest that normal cell types have a **limited division potential**, whereby mitosis is genet-ically programmed to decline at a particular stage of the life cycle. But does the change in mitosis *cause* aging or is it a *result* of the aging process? Consider that neurons, which do not divide at all after an early developmental stage, still deteriorate gradually during the life of an animal. Their predictable deterioration might indicate that similar changes may be occurring in dividing cells throughout the body.

Some researchers believe that cells gradually lose the capacity for DNA self-repair, perhaps as a result of an accumulation of environmental insults. Over time, DNA mutations certainly could thwart the production of enzymes and other proteins required for proper cell functioning. Consider that cells depend on smooth exchanges of materials between the cytoplasm and the extracellular fluid. And collagen, recall, is present in extracellular spaces throughout the body. If deteriorating regions of DNA code for defective collagen molecules, it is conceivable that the movement of oxygen, nutrients, hormones, and so forth to and from cells could be ham-pered, with repercussions extending through the entire body.

Finally, consider what might happen as a result of deterioration of genes coding for membrane proteins that serve as self-markers (page 422). If these identifi-cation markers change, does the immune system then perceive the body's own cells as "foreign" and mount an attack on them? According to one theory, such autoimmune responses might intensify over time, thereby producing the increased vulnerability to disease and stress associated with aging.

SUMMARY

1. Asexual reproduction in animals occurs most often by fission or budding. Sexual reproduction requires the fusion of nuclei of two gametes.

2. Animal development commonly proceeds through six stages:

a. *Gametogenesis*, during which the egg and sperm mature within the reproductive tracts of the parents.

b. *Fertilization*, which begins when a sperm pene-trates an egg and is completed when the sperm and egg nuclei fuse.

c. *Cleavage,* when the fertilized egg (zygote) undergoes mitotic cell divisions that form the early multicelled embryo (the blastula). Most genes are inactive during cleavage, and the destiny of cell lineages is established in part by the sector of cytoplasm inherited at this time.

d. *Gastrulation,* when the structural organization of the whole animal is laid out. Cell migrations lead to the formation of primary germ layers. In most animals, these layers are the endoderm, mesoderm, and ectoderm.

e. *Organogenesis,* the onset of organ formation. Cell differentiation and morphogenesis begin at this stage.

f. *Growth and tissue specialization,* when organs acquire their specialized chemical and physical properties. The maturation of tissues and organs continues into post-embryonic stages.

3. Cell differentiation and morphogenesis have their foundations in (a) the distribution of cytoplasmic molecules during cleavage and (b) cell interactions that begin during organogenesis.

4. *Cell differentiation* is a process whereby initial populations of genetically equivalent cells give rise to subpopulations of phenotypically different types of cells.

5. Each cell of an animal starts out with the same number and kind of genes as all others, so each starts out with the *potential* to develop in the same way as any other. However, beginning at organogenesis, restrictions are placed on gene expression in different cell types. As a result, some cells remain in an undifferentiated state, and others proceed through differentiation and develop into specialized cell types.

6. *Morphogenesis* is a process whereby different cell types become organized into all the specialized tissues and organs of the body. It involves the growth, shaping, and arrangement of body parts according to predefined patterns, and it is brought about by cell divisions, cell migrations, changes in cell shapes, localized growth, and controlled cell death.

7. Embryonic cells sense their position relative to each other and use this information to produce ordered, spatial arrangements of differentiated tissues. This kind of developmental activity is called *pattern formation.*

8. Following embryonic development, some animals grow directly into the adult form. Other animals proceed through indirect development, in which a larval stage is interposed between the embryo and the adult. Reactivated growth and development of the larva into the sexually mature adult form is called *metamorphosis.*

9. Cells of all complex animals gradually show changes in structure and lose efficiency (aging). Although aging is part of the life cycle of all animals having extensively specialized cell types, its cause is unknown.

Review Questions

1. What are some of the differences between asexual and sexual reproduction? Describe some forms of asexual reproduction. What are some adaptations associated with external and internal fertilization of sexually reproducing animals? *489–492*

2. Define and describe the main features of the following developmental stages: fertilization, cleavage, gastrulation, and organogenesis. *492–497*

3. All cells in an animal are genetically equivalent. Can you state what this means? *499*

4. Cell differentiation and morphogenesis are two processes that are critical for development. Can you define them? Can you state briefly which two kinds of developmental events serve as the basic foundation for cell differentiation and morphogenesis? *499*

5. What specific types of mechanisms bring about morphogenetic changes? *500–501*

6. Which of your organs formed from endoderm? From mesoderm? From ectoderm? *497*

7. During cleavage, the embryonic cells exhibit few regional differences, but in later stages, the rudimentary tissues and organs are discrete and the different cell groups do not merge. Suggest mechanisms that may be responsible for their discrete character. *499–505*

8. Experimentally, it is possible to divide an amphibian egg so that the gray crescent is wholly within one of the two cells formed. If the two cells are separated from each other, only the cell with the gray crescent will form an embryo with a long axis, notochord, nerve cord, and back musculature. The other cells form a shapeless mass of immature gut and blood cells. What do you think is the explanation of these outcomes? *502*

9. What kind of evidence suggests that aging is a *programmed* event rather than the result of random mutations? *505–507*

Readings

Balinsky, B. 1981. *An Introduction to Embryology.* Fifth edition. Philadelphia: Saunders.

Browder, L. 1984. *Developmental Biology.* Second edition. Philadelphia: Saunders. Good discussions of egg formation and gene expression.

Carlson, B. 1988. *Patten's Foundations of Embryology.* Fifth edition. New York: McGraw-Hill.

Gilbert, S. 1988. *Developmental Biology.* Second edition. Sunderland, Massachusetts: Sinauer. Excellent introduction to animal development; clear illustrations.

Gordon, R., and G. Browland. 1987. "The Cytoskeletal Mechanics of Brain Morphogenesis." *Cell Biophysics* 11:177–234.

Lewis, J. 1983. "Cellular Mechanisms of Development." In *Molecular Biology of the Cell* (B. Alberts et al.,). New York: Garland.

Raff, R., and T. Kaufman. 1983. *Embryos, Genes, and Evolution.* New York: Macmillan.

Saunders, J. 1982. *Developmental Biology: Patterns, Problems, Principles.* New York: Macmillan. Excellent introduction to embryology.

In the preceding chapter, we looked at some general principles of animal reproduction and development. Here, we will focus on humans as a way of presenting an integrated picture of the structure and function of reproductive organs, gametogenesis, and the stages of development from fertilization to birth. As part of this picture, we will also consider some of the mechanisms controlling reproduction.

PRIMARY REPRODUCTIVE ORGANS

The human reproductive system consists of a pair of primary reproductive organs, or **gonads**, and accessory glands and ducts. The male gonads are the **testes** (singular, testis); the female gonads are the **ovaries**. Testes and ovaries have the same general functions: they produce gametes and they secrete sex hormones, which influence reproductive functions and the development of secondary sexual traits. The male gametes are **sperm**; the female gametes are "eggs," or (more accurately) **secondary oocytes**. Among the secondary sexual traits are the amount and distribution of body fat, hair, and skeletal muscle. Those traits are associated with maleness and femaleness but play no direct role in reproduction.

Until the seventh week of development, the gonads look the same in all human embryos. Then, activation of genes on the sex chromosomes triggers the differentiation of the early gonads into testes *or* ovaries. Although gonads and accessory organs are already formed at birth, they do not reach their full size and become reproductively functional until twelve to sixteen years later.

35

HUMAN REPRODUCTION AND DEVELOPMENT

Figure 35.1 Reproductive system of the human male, longitudinal section. There are two each of the following components: testis, epididymis, vas deferens, seminal vesicle, ejaculatory duct, and bulbourethral gland.

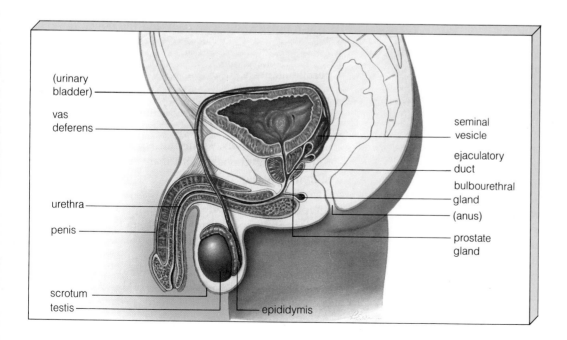

(urinary bladder)

vas deferens

urethra

penis

scrotum

testis

epididymis

seminal vesicle

ejaculatory duct

bulbourethral gland

(anus)

prostate gland

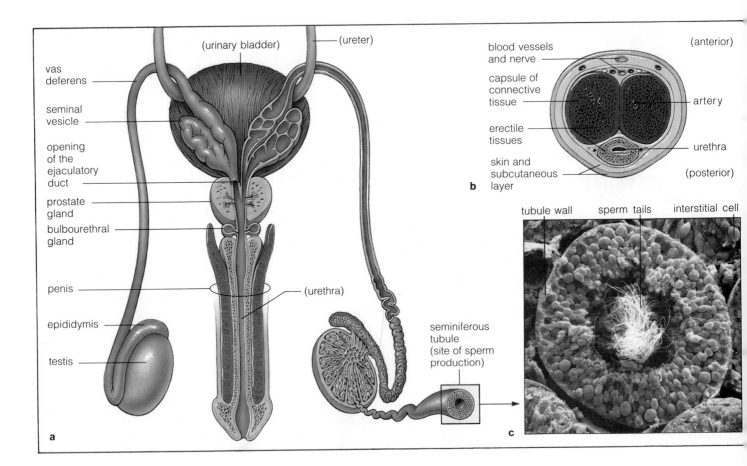

Figure 35.2 (**a**) Posterior view of the male reproductive system. (**b**) Cross-section of the penis. (**c**) Scanning electron micrograph showing the cells inside a seminiferous tubule.

MALE REPRODUCTIVE SYSTEM

Figure 35.1 shows the male reproductive system. The testes themselves are divided internally into 250 to 300 wedge-shaped lobes (Figure 35.2). Each lobe contains two to three **seminiferous tubules** in which sperm form ("seminiferous" means seed-bearing). Although each testis is only about 5 centimeters long, about 125 meters (137 yards) of these highly coiled tubes are packed into it! The connective tissue between the tubes contains *interstitial cells*, a type of endocrine cell that secretes sex hormones, including testosterone.

If sperm are to develop properly, the testes must be kept a few degrees cooler than the body's core temperature. The testes are suspended in an outpouching of skin (the *scrotum*) below the pelvic region. (Although they initially form internally against the back wall of the abdominal cavity, they descend into the scrotum before birth.) Contractions of muscles in the tissues of the scrotum adjust the location of the scrotum relative to the body wall and so help maintain the temperature at about

95°F inside the pouch. When it is cold outside, contractions draw the pouch closer to the warm body; when it is warm outside, muscles relax and the pouch lowers.

Spermatogenesis

Mammals generally reproduce on a seasonal basis, and spermatogenesis (sperm production) usually coincides with the reproductive season. Humans and other primates are exceptions. From puberty onward, sperm are produced continuously, with many millions of cells in different stages of development on any given day.

Figure 35.3 shows how sperm are produced in a seminiferous tubule. Inside, undifferentiated diploid cells called spermatogonia are closest to the tubule wall. These cells are forced away from the wall by ongoing mitotic cell divisions and are gradually transformed into primary spermatocytes. They undergo meiosis I, the result being secondary spermatocytes. Although the cells are now haploid, keep in mind that each chromosome is still in the duplicated state; it consists of two sister chromatids (page 140). The sister chromatids of each chromosome are separated from each other during meiosis II. The resulting cells are haploid spermatids, which gradually develop into mature sperm. The entire process takes about nine to ten weeks. All the while, the developing cells receive nourishment and chemical signals from adjacent *Sertoli cells*, which are the only other type of cell present in the tubule.

A mature sperm consists of a head, midpiece, and tail (Figure 35.4). The head contains the nucleus, which is packed with DNA. The *acrosome*, a cap over most of the head, contains lytic enzymes that will assist the sperm in penetrating the outer layer surrounding an oocyte. The tail contains an array of microtubules, which bend in coordinated fashion to produce whiplike movements. Numerous mitochondria in the midpiece supply the energy necessary for these movements.

Sperm Movement Through the Reproductive Tract

After their formation in a testis, sperm move into a long, coiled duct called the **epididymis**. Between times of male sexual arousal, most of the sperm are stored in the region of the epididymis closest to a thick-walled tube called the **vas deferens**. The vas deferens functions in rapid sperm transport. It merges with a duct from a gland called the **seminal vesicle** (Figure 35.2). From there, sperm move into a short duct that runs through the tissues of the **prostate gland** and empties into the urethra. (The urethra terminates at the tip of the penis. It is the channel that serves in the ejaculation of sperm as well as in the excretion of urine, as described on page 478.)

As the sperm pass through the reproductive tract, they become mixed with fluids secreted from the prostate gland, seminal vesicles, and bulbourethral glands. This sperm-bearing mixture of fluids is called **semen**. The seminal vesicles secrete a fluid containing fructose

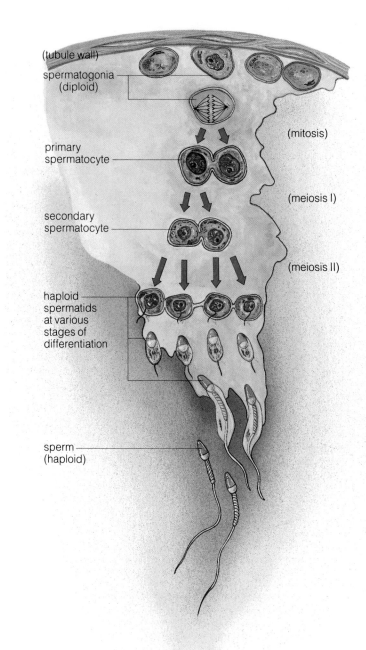

Figure 35.3 Spermatogenesis inside a seminiferous tubule.

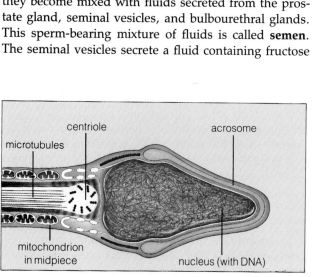

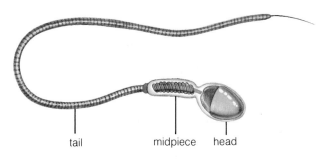

Figure 35.4 Human sperm. The boxed sketch shows the head region of this single cell, longitudinal section.

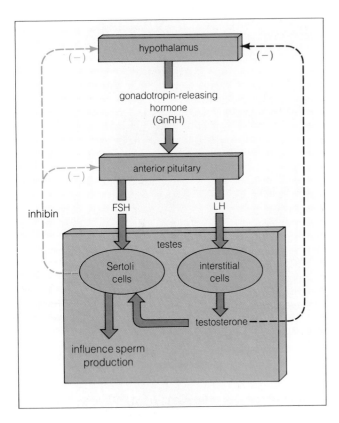

Figure 35.5 Hormonal control of reproductive function in human males. The black dashed line indicates that increased testosterone secretion inhibits LH secretions through its negative effect on hypothalamic GnRH. The blue dashed line indicates that an inhibitory signal from Sertoli cells influences GnRH and FSH secretions. The "signal" is the hormone inhibin.

Table 35.1	Organs and Accessory Glands of the Male Reproductive Tract
Organs:	
Testis (two)	Sperm production, sex hormone production
Epididymis (two)	Sperm maturation site, sperm storage
Vas deferens (two)	Rapid transport of sperm
Ejaculatory duct (two)	Conduction of sperm
Penis	Organ of sexual intercourse
Accessory Glands:	
Seminal vesicle (two)	Secretions make up large portion of semen*
Prostate gland	Same as above
Bulbourethral gland (two)	Same as above

*Semen enhances sperm motility in various ways.

(which nourishes sperm) and prostaglandins (which trigger contractions in the female reproductive tract and thereby may assist sperm in their movement). The prostate gland secretes a thin, alkaline fluid that probably helps neutralize acidic secretions in the vagina. (The pH in the vagina is about 3.5–4, but sperm motility and fertility are enhanced when it increases to about 6.) The bulbourethral glands secrete a small amount of mucus-rich fluid into the urethra during sexual arousal. This fluid lubricates the penis, assisting its penetration into the vagina; it also enhances sperm movement.

Hormonal Control of Male Reproductive Functions

Three major hormones directly or indirectly control the reproductive functions of males. One is **testosterone**, produced by interstitial cells in the testes. The other two are **follicle-stimulating hormone** (FSH) and **luteinizing hormone** (LH). As we have seen, these two gonadotropins are produced by the anterior pituitary. (FSH and LH were first named for their effects in females, but their molecular structure is exactly the same in males.)

Testosterone has multiple functions. First, this hormone stimulates spermatogenesis. Second, it is necessary for the growth, form, and function of all parts of the male reproductive tract. Third, testosterone promotes the normal development and maintenance of sexual behavior. (Aggressive behavior also is somewhat dependent on testosterone.) Fourth, secondary sexual traits such as lowering of the voice, changes in muscle tone, beard growth, and pubic hair growth are testosterone-dependent. Fifth, testosterone has growth-promoting effects on general body tissues.

Testosterone secretion is governed by negative feedback loops involving the hypothalamus, the anterior pituitary, and the testes (Figure 35.5). The hypothalamus secretes **GnRH**, or gonadotropin-releasing hormone. When GnRH reaches the anterior pituitary, it causes the secretion of LH, which is circulated to the testes. There, LH stimulates interstitial cells to increase their secretion of testosterone, which acts on the Sertoli cells in ways that stimulate sperm production. The GnRH also causes the pituitary to secrete FSH, which acts *directly* on the Sertoli cells in the testes. As the testosterone concentration increases, it exerts an inhibitory effect on the hypothalamus to reduce GnRH secretion.

FSH function in a mature testis is not well understood. It is known to be essential for initiating spermatogenesis at puberty. Currently, research is focusing on local controls over the testis. For example, Sertoli cells produce peptide hormones that act on nearby cells to modulate the overall control provided by pituitary hormones, but their exact roles are unclear.

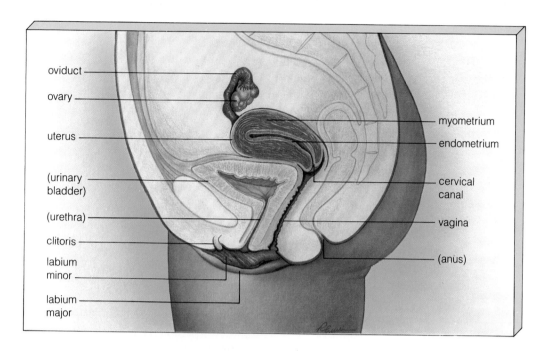

Figure 35.6 Reproductive system of the human female, longitudinal section. There are two ovaries and two oviducts (which lead into the uterus).

Labels (top to bottom, left):
oviduct
ovary
uterus
(urinary bladder)
(urethra)
clitoris
labium minor
labium major

Labels (right):
myometrium
endometrium
cervical canal
vagina
(anus)

FEMALE REPRODUCTIVE SYSTEM

Figure 35.6 shows the female reproductive system. The two ovaries produce and release oocytes on a monthly basis in women of reproductive age, and they secrete the sex hormones **estrogen** and **progesterone**. Adjacent to each ovary but not connected to it is an **oviduct** (or Fallopian tube), which channels oocytes into the uterus. Ciliated, fingerlike projections from the oviduct extend over one end of the ovary, and their sweeping action moves the oocytes into the entrance of the duct.

The **uterus** houses the developing embryo during pregnancy. This hollow organ, which resembles an upside-down pear, is positioned near the floor of the abdominal cavity, above the urinary bladder. The bulk of the uterus is a thick layer of smooth muscle called the *myometrium*. The interior lining of the uterus, the *endometrium*, consists of connective tissue, glands, and blood vessels. The lower portion of the uterus (the narrow part of the "pear") is the *cervix*. The **vagina** is a muscular tube that connects with the uterus (by way of the cervix) to the outside of the body. It is the organ in which sperm from the male are deposited, and it functions as part of the birth canal (Table 35.2).

At the body surface are the external genitalia (or *vulva*). The outermost structures are a pair of skin folds (the labia majora), which contain adipose tissue. Within the cleft formed by these folds are the labia minora—a smaller pair of skin folds that are highly vascularized but have no fatty tissue. At the anterior end of the vulva, these two interior skin folds partly enclose the *clitoris*, a

Table 35.2	Female Reproductive Organs
Ovaries	Oocyte production, sex hormone production
Oviducts	Conduction of oocyte from ovary to uterus
Uterus	Chamber in which new individual develops
Cervix	Secretes mucus that enhances sperm movement into uterus and (after fertilization) reduces the embryo's risk of bacterial infection
Vagina	Organ of sexual intercourse; birth canal

small organ sensitive to sexual stimulation. The opening of the urethra is about midway between the clitoris and the vaginal opening (Figure 35.6).

Oogenesis

Every oocyte that develops in a woman's body was already present, in immature form, before she was born. When she herself was a fetus, certain diploid cells divided by mitosis and enlarged to form primary oocytes, each of which became surrounded by a single layer of *granulosa cells*. A primary oocyte together with its surrounding cell layer is a **primary follicle** (Figure 35.7).

Before the woman was born, her primary oocytes already started meiosis I; that is, the chromosomes in each nucleus paired with their homologues as a prelude

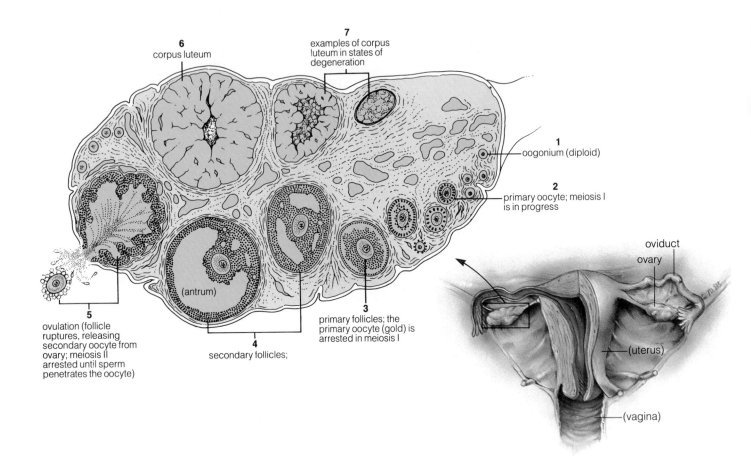

6
corpus luteum

7
examples of corpus
luteum in states of
degeneration

1 oogonium (diploid)

2
primary oocyte; meiosis I
is in progress

5
ovulation (follicle
ruptures, releasing
secondary oocyte from
ovary; meiosis II
arrested until sperm
penetrates the oocyte)

(antrum)

4
secondary follicles;

3
primary follicles; the
primary oocyte (gold) is
arrested in meiosis I

oviduct

ovary

(uterus)

(vagina)

Figure 35.7 A human ovary, drawn as if sliced lengthwise through its midsection. Events in the ovarian cycle proceed from the growth and maturation of primordial follicles, through ovulation (rupturing of a mature follicle with a concurrent release of a secondary oocyte), through the formation and maintenance (or degeneration) of an endocrine element called the corpus luteum. The positions of these structures in the ovary are varied for illustrative purposes only. The maturation of an oocyte occurs at the *same* site, from the beginning of the cycle to ovulation.

to their distribution into haploid gametes (see, for example, Figure 11.9). However, meiosis I was arrested and her primary oocytes remained dormant in her ovaries until puberty.

Of the perhaps 2 million primary oocytes that were present in her ovaries at the time of birth, fewer than 300,000 remain at puberty. Of those, about 400 will mature and be expelled from her body, one at a time on a monthly basis, over the next three decades or so. The longer it takes for the cells to be released from that prolonged meiosis, the greater the risk of chromosome abnormalities (page 193).

When a primary oocyte matures, the granulosa cells deposit a layer of material (the *zona pellucida*) around it. Less than thirty-six hours before ovulation, it completes meiosis I and two cells form: a haploid secondary oocyte (which ends up with almost all the cytoplasm) and a tiny, first polar body. The tiny cell functions only as a "dumping ground" for half of the diploid number of chromosomes, so that the gamete will be haploid.

At **ovulation**, a secondary oocyte is released from the ovary and enters meiosis II. However, this second division is *not* completed unless fertilization occurs. When meiosis II is completed, one cell again retains most of the cytoplasm; it is the mature oocyte, or **ovum**. The other cell is a tiny, secondary polar body. Polar bodies may eventually complete meiosis like the oocyte, but they degenerate soon afterward.

Menstrual Cycle: An Overview

Unlike males, human females have a cyclic and intermittent reproductive capacity. In most mammalian females, the reproductive pattern is called the **estrous cycle**. During estrus, the female comes into "heat" (becomes sexually receptive to the male) only at specific times of year. Also during estrus, hormones stimulate the maturation and release of oocytes, and they prime the endometrium for implantation.

In humans and other primates, the cyclic reproductive capacity in females is called the **menstrual cycle**. Here, too, the reproductive tract and the oocytes maturing within it are primed for fertilization. Unlike estrus, however, there is no correspondence between heat and the time of fertility. All female primates can become physically and behaviorally receptive to the male's overtures at any time. Also unlike estrus, the hormone-primed uterine lining is sloughed off at the end of each cycle. Menstrual cycles begin at about age thirteen in humans and continue until menopause (in the late forties or early fifties).

On the average, it takes about twenty-eight days to complete one menstrual cycle, although the range may be twenty-two to thirty-six days. During this time span, changes take place in the ovary and in the endometrium (Table 35.3). Feedback loops involving the hypothalamus, the anterior pituitary, and the ovaries control these changes. In females, as in males, GnRH secreted from the hypothalamus causes the release of LH and FSH from the anterior pituitary. But in females, those pituitary hormones trigger estrogen and progesterone secretions by cells in the ovary. Estrogen is secreted by cells of the follicles and, after ovulation, by the residual follicle—which becomes a glandular body called the **corpus luteum**. Progesterone is secreted in very small amounts by the granulosa cells, but the major secretion is from the corpus luteum. Let's look first at how these controls influence ovarian function. Then we will look at the concurrent changes in the endometrial lining of the uterus.

Control of Ovarian Function

Follicular Phase. The follicular phase of each menstrual cycle lasts about thirteen days (Table 35.3). During this phase, several follicles grow but ordinarily only one reaches maturity; the others degenerate. As the follicle develops, estrogen-containing secretions from its granulosa cells begin to accumulate and form a fluid-filled space (the *antrum*) inside the follicle. The antrum eventually becomes so large that the mature follicle balloons outward from the ovary's surface. At ovulation, the "balloon" ruptures and fluid escapes, carrying the secondary oocyte with it (Figure 35.7).

At the start of each menstrual cycle, hormone levels are low (Figure 35.8). Then, FSH and LH levels rise and stimulate the follicle into producing and secreting estrogen, which in turn stimulates the mitotic divisions that underlie follicle growth. The rising blood level of estrogen early in the follicular phase has a dampening effect on LH and FSH secretions. About midcycle, however, the estrogen level increases rapidly, and now estrogen has the opposite effect: it causes a brief outpouring of LH from the anterior pituitary. Within thirty-six to forty-eight hours, this **midcycle surge of LH** has triggered the following events:

1. Within the follicle, meiosis I resumes in the primary oocyte, and a secondary oocyte forms. (Thus the parental diploid number of chromosomes has been reduced to the haploid number for the forthcoming gamete.)

2. Estrogen secretion (by cells of the follicle) slows down as the follicle is disrupted.

3. The follicle ruptures, possibly because LH stimulates the production of lytic enzymes that act on the follicle where it balloons out from the ovary. Thus, *high levels of LH trigger ovulation*.

4. The parts of the follicle left behind are transformed into a corpus luteum, which starts secreting quantities of progesterone and estrogen.

Table 35.3	Overview of Events of the Menstrual Cycle	
Phase	Events	Days of the Cycle*
Follicular phase	Menstruation; endometrium breaks down	1–5
	Follicle matures in ovary; endometrium rebuilds	6–13
Ovulation	Secondary oocyte released from ovary	14
Luteal phase	Corpus luteum forms; endometrium thickens and develops	15–28

*Assuming a 28-day cycle.

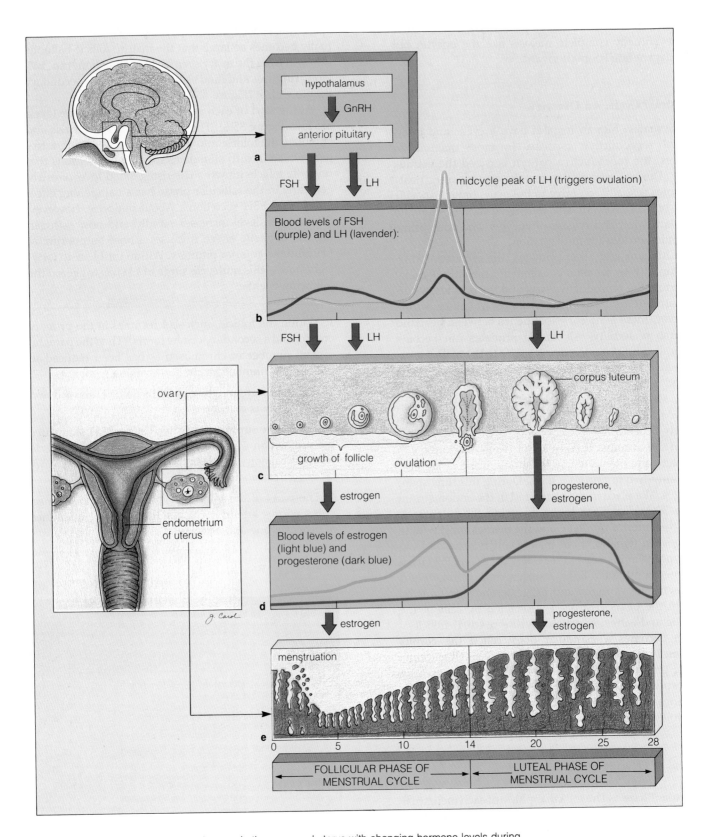

Figure 35.8 Correlation between changes in the ovary and uterus with changing hormone levels during the menstrual cycle. Green arrows indicate which hormones dominate the follicular phase or the luteal phase of the cycle. The hypothalamus (**a**) controls the release of gonadotropins (FSH and LH) from the pituitary. The FSH and LH promote changes in ovarian structure and function (**b, c**), then estrogen and progesterone from the ovary promote changes in the endometrium (**d, e**).

Early in the follicular phase, then, estrogen has a *negative* feedback effect on FSH and LH secretion, but later on it has a *positive* feedback effect on LH secretion. These feedback loops are shown in Figure 35.9.

Luteal Phase. The luteal phase of the menstrual cycle follows ovulation. Under the continued influence of LH, the remaining cells of the ruptured follicle become the corpus luteum, which secretes considerable amounts of progesterone and some estrogen. A corpus luteum can persist for about twelve days if fertilization does not follow ovulation.

During that twelve-day span, the high blood concentrations of progesterone and estrogen have a feedback effect on FSH and LH secretions. The FSH and LH levels fall; hence a new follicle is prevented from developing until the menstrual cycle is completed.

When fertilization does not occur, the corpus luteum starts to degenerate during the last days of the menstrual cycle. Apparently it self-destructs by secreting large amounts of prostaglandins, which interfere with its functions. Once the corpus luteum stops functioning, estrogen and progesterone levels fall rapidly, so their inhibitory effect on the hypothalamus and pituitary is canceled. FSH and LH secretions rise, other follicles in the ovary are stimulated into developing—and the cycle begins anew.

Control of Uterine Function

The changing estrogen and progesterone levels just described cause profound changes in the uterus. Estrogen stimulates the growth of endometrium and uterine smooth muscle. Under the influence of progesterone, blood vessels proliferate in the thickened endometrium; glycogen and enzymes are stockpiled in endometrial glands and connective tissues. In such ways, the uterus is prepared for the possibility of pregnancy.

At the time of ovulation, estrogen causes the cervix to secrete a mucus that is thin, clear, and abundant—an ideal medium through which sperm can travel. (Just after ovulation, progesterone secretions from the corpus luteum act on the cervix. The mucus becomes thick and sticky, and it serves as a physical barrier against bacteria that might enter the vagina and endanger a newly conceived zygote.)

In the absence of fertilization, and with the disintegration of the corpus luteum, blood levels of progesterone and estrogen fall. Without hormonal support, the highly developed endometrium starts disintegrating also. Blood vessels in the endometrium constrict and the tissues (deprived of oxygen and nutrients) die off. Arterioles now dilate, and the increased blood flow ruptures the walls of weakened capillaries. Menstrual flow consists of blood and sloughed endometrial tissues. The flow moves out of the body through the vagina, and its appearance marks the first day of a new menstrual cycle. It continues for three to six days, until rising estrogen levels stimulate the repair and growth of the endometrium.

In a given year, between 4 and 10 million American women are affected by *endometriosis*: the spread and growth of endometrial tissue in areas outside the uterus. According to one theory, endometriosis arises through retrograde menstruation. That is, instead of being discharged during the menstrual cycle, some blood backs up through the oviducts and spills into the pelvic cavity. Normally it is resorbed and so is harmless; in other cases, endometrial cells become implanted at sites throughout the pelvic region. According to another theory, mispositioned embryonic cells have been in the abdominal cavity since birth, and hormones activated at puberty induce their growth and proliferation.

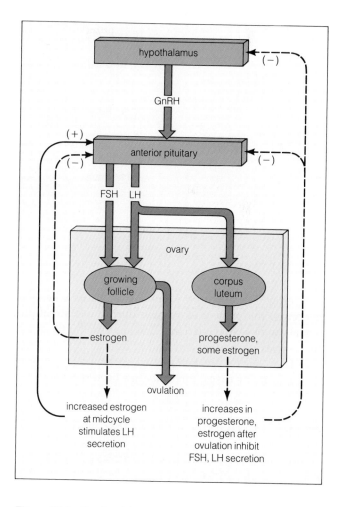

Figure 35.9 Feedback loops among the hypothalamus, anterior pituitary, and ovaries during the menstrual cycle.

Table 35.4 Stages of Human Development: A Summary

Prenatal Period:

1.	**Zygote**	Single cell resulting from the fusion of a sperm nucleus with the nucleus of a secondary oocyte at fertilization; zygote usually forms in the oviduct.
2.	**Morula**	Solid ball of about sixteen cells (called blastomeres), produced through cleavages of the zygote.
3.	**Blastocyst**	Ball of cells with a surface layer (the trophoblast) and an inner cell mass. Produced after the morula enters the uterus; then, fluid enters the ball and lifts some cells to form a cavity.
4.	**Embryo**	Embryo refers to all developmental stages from two weeks after fertilization (at which time the inner cell mass forms a two-layer embryonic disk) until the end of the eighth week. All major body structures begin forming during the embryonic period.
5.	**Fetus**	Fetus refers to all developmental stages from the ninth week until birth, about thirty-eight weeks after fertilization. The rate of overall growth and structural elaborations increases dramatically during the fetal period.

Postnatal Period:

6.	**Newborn**	Individual during the first two weeks after birth (the neonatal period). The physiological transition from life in the uterus to life in the external world requires many gradual changes, as in respiration.
7.	**Infant**	Individual from two weeks to about the first fifteen months after birth. During infancy, body weight triples and height doubles, on the average.
8.	**Child**	Individual from infancy to about twelve or thirteen years of age. Bone formation and overall growth are extensive early in childhood, then slow down; just before puberty, the rate of growth steps up rapidly.
9.	**Pubescent**	Girls between twelve and fifteen years old, and boys between thirteen and sixteen years old; secondary sexual characteristics develop now.
10.	**Adolescent**	Individual from puberty until about three or four years later; a time of physical, mental, and emotional maturation.
11.	**Adult**	During early adulthood (between eighteen and twenty-five years of age), bone formation and growth are finally completed. The maximum human life span is probably 100 years or so, but after early adulthood, developmental changes proceed very slowly.

Estrogen stimulates the endometrial tissue wherever it occurs; this may lead to pain during menstruation, urination, or sexual relations. Endometrial scar tissue on the ovaries or oviduct can lead to infertility. Currently, tiny laser beams are being used to vaporize the scar tissue; in many cases, infertility has been reversed.

SEXUAL UNION AND FERTILIZATION

A secondary oocyte encounters sperm as a result of sexual union, or **coitus**. Sperm are ejaculated from the penis, which contains three cylinders of spongy tissue arranged around the urethra (Figure 35.2b). At the tip of the penis, the ventral cylinder terminates as a mushroom-shaped structure (the glans penis), which contains abundant mechanoreceptors that are stimulated by friction. During normal circumstances, blood vessels leading into the three cylinders are constricted and the penis is limp. During initial sexual excitation, blood flows into the cylinders faster than it flows out. Blood collects in the spongy tissue, and the organ lengthens and hardens. These changes can occur within seconds of sexual excitation, and they facilitate penetration into the vagina.

During coitus, pelvic thrusts stimulate the penis as well as the vaginal walls and clitoral region of the female body. Mechanical stimulation of the penis triggers rhythmic, muscular contractions that force the contents of seminal vesicles and the prostate into the male urethra. Involuntary contractions of other muscles expel semen into the vagina. (During ejaculation, a sphincter

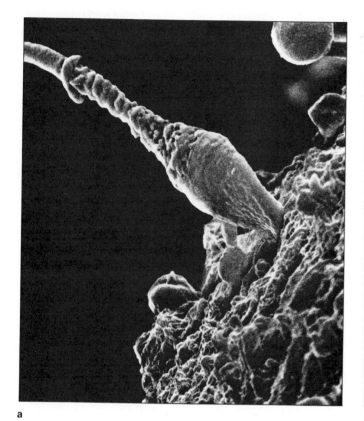
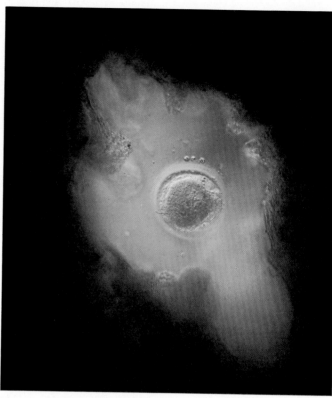

a

b

Figure 35.10 (a) Human sperm about to penetrate the zona pellucida around a secondary oocyte. (b) A mature ovum. Notice the three polar bodies above it; these products of meiosis will degenerate shortly.

closes off the bladder. As a result, sperm cannot enter the bladder and urine cannot be excreted from it.) The involuntary muscular contractions, ejaculation, and associated sensations of release, warmth, and subsequent relaxation constitute the event called **orgasm**.

Female orgasm involves similar physical events, including an intense vaginal awareness, a series of involuntary uterine and vaginal contractions, and sensations of relaxation and warmth. Even if the female does not reach this state of excitation during coitus, she can still become pregnant.

If sperm ejaculation into the vagina coincides with ovulation (if it occurs within a range of about three days before ovulation and three days afterward), pregnancy can result. Within thirty to sixty minutes after ejaculation from the penis, sperm have moved into and up through the uterus, to the oviducts.

Once a sperm reaches its surface, the secondary oocyte is stimulated into completing meiosis II. Only at this stage is the oocyte referred to as a **mature ovum** (Figure 35.10). The sperm nucleus enters the egg cytoplasm and then fuses with the egg nucleus. The paternal and maternal chromosomes intermingle, restoring the diploid number for the new individual.

On the average, between 150 million and 350 million sperm enter the vagina during ejaculation, but only a few hundred reach the upper region of the oviduct where fertilization usually occurs. Before a sperm can penetrate the secondary oocyte, however, it must undergo a physiological change called **capacitation**, which occurs during its passage through the female reproductive tract. The change somehow primes the sperm so that its outer acrosomal membrane and its plasma membrane fuse locally, producing channels through which acrosomal enzymes are released. The action of these digestive enzymes clears a path for the sperm through the zona pellucida. Although several sperm can penetrate the zona pellucida, usually only one enters the secondary oocyte.

PRENATAL DEVELOPMENT

Table 35.4 summarizes the developmental stages that follow fertilization. Let's briefly consider some key events of the prenatal stages ("prenatal" means before birth).

First Week of Development

For the first three or four days after fertilization, the zygote travels down the oviduct, picking up nutrients from maternal secretions and undergoing its first mitotic cell divisions. By the time it reaches the uterus, the cleavages have transformed it into a **morula** (a cluster of about 16 cells). Later, a cavity forms and becomes filled with fluid, so that the interior cells are lifted away from the surface cells except in one region. The human embryo now is called a **blastocyst**; it consists of an inner cell mass and a surface cell layer called the *trophoblast* (Figure 35.11).

Cells of the trophoblast secrete **chorionic gonadotropin**, a hormone that maintains the corpus luteum. Thus the corpus luteum continues to grow and to secrete estrogens and progesterone, which help maintain the uterine lining. (The presence of chorionic gonadotropin in the mother's urine is an early indicator of pregnancy.)

Before the first week ends, the blastocyst contacts and adheres to the uterine lining. Cell divisions in the trophoblast produce two layers, the outer one consisting of a mass of cells that send out fingerlike projections into the uterine lining. Through this invasion, the blastocyst becomes implanted in the mother's tissues (Figure 35.12).

While the invasion proceeds, endoderm starts to form on the inner cell mass of the blastocyst; this will become the first germ layer of the embryo (page 497). Eventually there is a two-layered, **embryonic disk** (Figure 35.12). The disk is usually shown in side view, cut down the middle; but keep in mind that it is shaped rather like an oval pancake.

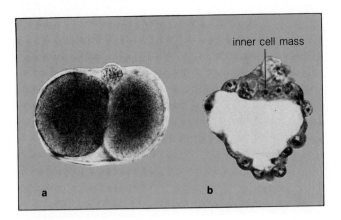

Figure 35.11 (**a**) Two-cell stage of cleavage, about thirty-six hours after fertilization. (**b**) Blastocyst in cross-section after four to six days. The layer of surface cells of the blastocyst is called the trophoblast.

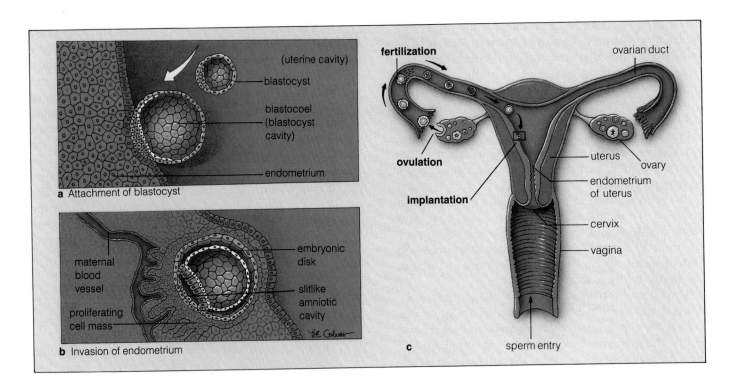

Figure 35.12 Implantation of the blastocyst in the endometrium (the lining of the uterus).

Extraembryonic Membranes

To gain perspective on the events that follow implantation, think about a key character in vertebrate evolution—the shelled egg. Shelled eggs allowed embryonic development to take place on land rather than in the water, where vertebrates first evolved. Inside the shelled eggs of reptiles and birds are several membranes that protect the embryo and function in its nutrition, respiration, and excretion. These membranes are called the **yolk sac**, the **allantois**, the **amnion**, and the **chorion**.

As we have seen, bird eggs have a large store of yolk—but how does the embryo draw nutrients from it? At first, endoderm grows almost completely around the yolk, forming a sac. Then mesoderm spreads over the sac and develops into vessels that convey nutrients to the embryo. Eventually the connection between the yolk sac and the embryo is constricted into a "stalk." Although a human embryo is not housed within a shelled egg, it is still provided with a yolk *sac* (but no yolk) as part of its vertebrate heritage. A membranous sac forms below the embryonic disk *as if* the yolk were still there; parts of this membrane give rise to the digestive tube of the embryo (Figure 35.13).

As we have seen also, nitrogenous waste products of protein metabolism can be toxic in high concentrations—but how are wastes disposed of in a shelled egg? In reptiles and birds, they are stored inside the allantois,

a bladderlike sac that develops as an extension of the primitive gut. Later, a network of blood vessels develops on the surface of the allantois and helps supply the embryo with oxygen. In humans, the allantois no longer functions in handling nitrogenous wastes but its blood vessels still function in oxygen transport.

In all land vertebrates, the amnion is a membranous, fluid-filled sac, a tiny replica of the aquatic environment in which embryos of ancient vertebrates developed. In humans, the amnion forms as a slitlike opening in the inner cell mass of the blastocyst about two weeks after fertilization. Eventually it completely surrounds the embryo and keeps it from drying out. The amniotic fluid also acts like a shock absorber. (The "water" that flows freely from the vagina just before childbirth is amniotic fluid, released when the amnion ruptures.)

As Figure 35.13 shows, the growing human embryo gradually becomes separated from the extraembryonic membranes just described until it is connected to them only by the **umbilical cord**. This cord contains the "stalks" of the yolk sac and allantois, surrounded by the outer layer of the amnion. It is well endowed with blood vessels.

Another membrane, the chorion, develops as a protective membrane around the embryo and the other membrane structures. In humans, it is derived from the trophoblast and reinforcing mesodermal tissues. The chorion becomes a primary component of the placenta.

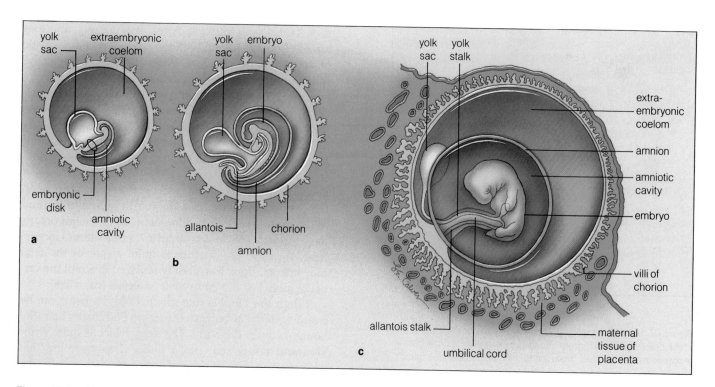

Figure 35.13 Formation of the extraembryonic membranes that protect the embryo and function in its development.

The Placenta

Three weeks after fertilization, almost a fourth of the inner surface of the uterus has developed into a spongy tissue that is composed of endometrium *and* extraembryonic membranes, the chorion especially. This tissue, the **placenta**, is the means by which the embryo receives nutrients and oxygen from the mother and sends out wastes in return, which are disposed of through the mother's lungs and kidneys.

The tiny, fingerlike projections that were sent out from the blastocyst during implantation develop into numerous chorionic villi, which contain embryonic blood vessels (Figure 35.14). The embryonic blood vessels do not "merge" with those of the mother; the two bloodstreams remain separate throughout pregnancy. Small solutes diffuse both ways across the thin endothelial walls of the two sets of blood vessels, moving to and from the maternal lacunae (which are blood-filled spaces in the uterine lining).

The placenta secretes several hormones, including the chorionic gonadotropin that maintains the corpus luteum. Within about two months after fertilization, it also secretes estrogen and progesterone in such large quantities that pregnancy can be maintained even if the corpus luteum is surgically removed.

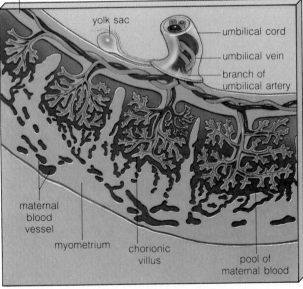

Figure 35.14 Relationship between fetal and maternal tissues in the placenta. The diagram shows how chorionic villi become progressively developed (from left to right across the illustration).

Embryonic and Fetal Development

The nine months of human pregnancy and prenatal development are divided into three trimesters, each consisting of three months.

First Trimester. Once the embryonic disk has formed in the blastocyst, human development proceeds along the same general course described in the preceding chapter for all complex animals.

Gastrulation, which begins during the second week, leads to the formation of endoderm, mesoderm, and ectoderm—the three germ layers. Some surface cells of the gastrula migrate inward to form the mesoderm. When gastrulation is over, the remaining surface tissue (the ectoderm) will give rise to the nervous system and other organs. Inside, the endoderm will form the inner lining of the respiratory and digestive systems; the mesoderm will develop into the heart, muscles, bones, and many other internal organs. The tubelike forerunner of the heart will be beating after the third week.

By the end of the fourth week, the embryo has grown to 500 times its original size (Figure 35.15). Its rapid growth gives way to four more weeks in which the main organs develop rather slowly. The nerve cord and the four chambers of the heart form; the respiratory organs form but are not yet functional. The segmentation characteristic of all bilateral animals is now clearly evident. Some of the paired segments (somites) that are lateral to the main body axis are visible in Figure 35.15. The tissues of these segments develop into connective tissues, bones, and muscles. Arms, legs, fingers, and toes are developing—along with the tail that is characteristic of all vertebrate embryos. After the eighth week, the human tail gradually disappears.

From the start of the ninth week until birth, the developing individual is called a **fetus** (Figure 35.16). Between the ninth and the twelfth week, the fetus begins moving its arms and legs. Movements of facial muscles produce frowns and squints; the sucking reflex also is evident.

Second Trimester. At the start of the **second trimester**, the period extending from the fourth month to the end of the sixth month of development, all the major organs have formed. The fetus is about 5.7 centimeters (3 inches) long. By now, the mother is quite aware of the fetal movements. Bone has already replaced much of the cartilage model of the embryonic skeleton (page 388).

When the fetus is five months old, its heart can be heard with the aid of a stethoscope. Soft, fuzzy hair (the lanugo) covers the body. The skin, which is quite wrinkled and rather red, has a thick, cheesy coating that protects it against abrasion. During the sixth month, the upper and lower eyelids separate and eyelashes form; during the seventh month, the eyes open.

gill arches   somites

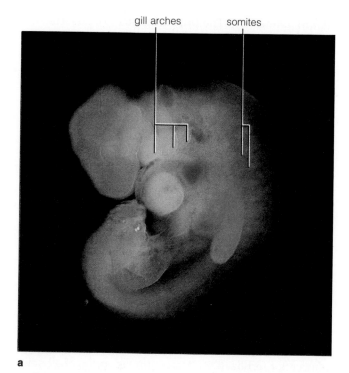

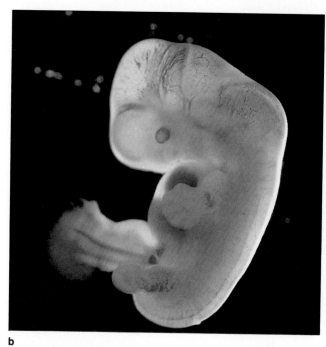

a

b

Figure 35.15 (**a**) Embryo at four weeks, about seven millimeters (0.3 inch) long. Notice the tail and the gill arches, which vaguely resemble a double chin. These features emerge during embryonic development of all vertebrates. Arm and leg buds are also visible now. (**b**) Embryo at the end of five weeks, about twelve millimeters long. The head starts to enlarge and the trunk starts to straighten. Finger rays appear in the paddlelike forelimbs. The umbilical cord has developed.

umbilical cord   amniotic sac

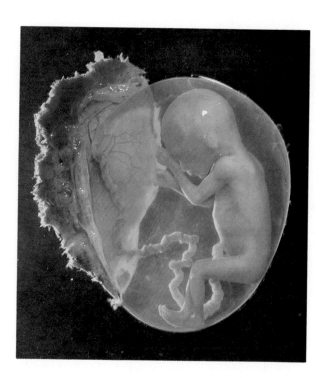

Figure 35.16 The fetus at nine weeks, floating in fluid inside the amniotic sac. (Here, the chorion, which covers the amniotic sac, has been opened and pulled aside.) Notice the blood vessels in the umbilical cord.

Figure 35.17 The fetus at sixteen weeks (sixteen centimeters, or 6.4 inches, long).

Figure 35.18 Fetus at eighteen weeks, about eighteen centimeters (a little more than seven inches) long. The sucking reflex begins during the earliest fetal stage, as soon as nerves establish functional connections with developing muscles. Legs kick, arms wave, fingers make grasping motions—all reflexes that will be vital skills in the world outside the uterus.

Table 35.5	Tissues and Organs Derived from the Three Germ Layers in Human Embryos
Germ Layer	Main Derivatives in the Adult
Endoderm	Various epithelia, as in the gut, respiratory tract, urinary bladder and urethra, and parts of the inner ear; also portions of the tonsils, thyroid and parathyroid glands, thymus, liver, and pancreas
Mesoderm	Cartilage, bone, muscle, and various connective tissues; gives rise to cardiovascular system (including blood), lymphatic system, spleen, and adrenal cortex
Ectoderm	Central and peripheral nervous systems; sensory epithelia of the eyes, ears, and nose; epidermis and its derivatives (including hair and nails), mammary glands, pituitary gland, subcutaneous glands, tooth enamel, and adrenal medulla

Data from Keith Moore, *The Developing Human.*

Third Trimester. The **third trimester** extends from the seventh month until birth. Not until the middle of the third trimester is the fetus developed enough to survive on its own if born prematurely or if removed surgically from the uterus. By the seventh month, fetal development appears to be relatively complete, but fewer than ten percent of infants born at this stage survive, even with the best medical care. In most cases they are not yet able to breathe normally or maintain a normal core temperature, because the air sacs in the lungs and the parts of the nervous system that govern breathing are not completely developed. By the start of the ninth month, survival chances increase to about ninety-five percent.

Table 35.5 summarizes the tissues and organs that developed from the three germ layers during the three trimesters just described.

Birth

On the average, the date of birth is approximately 280 days after fertilization. Almost three-fourths of all newborns have arrived within two weeks of that day. The birth process begins with the onset of contractions of muscles in the uterine wall. The contractions build in strength and increase in frequency over a period that usually lasts anywhere from two to eighteen hours. During that period, the cervical canal of the uterus becomes fully dilated and the amniotic sac usually ruptures. In most cases, the fetus is expelled within less than an hour after full dilation. Immediately afterward, uterine contractions cause the expulsion of fluid, blood, and the placenta (Figure 35.19). The umbilical cord—the lifeline to the mother—is now severed, and the newborn embarks on its nurtured existence in the outside world.

Lactation

During pregnancy, estrogen and progesterone levels are high in the bloodstream. Among other things, the hormones stimulate the growth of mammary glands and ducts in the woman's breasts. There are about twenty lobes of glandular tissue in each breast, and a tiny duct leads from each one to the surface of the nipple (Figure 35.20).

For the first few days after giving birth, the woman's mammary glands produce a fluid (colostrum) that is rich in proteins and lactose. Then prolactin from the pituitary stimulates milk production. When the newborn suckles, nerve endings in the areola of the breast are stimulated. Action potentials are relayed to the hypothalamus, which stimulates the posterior pituitary to release oxytocin into the circulation. When oxytocin arrives in the breast tissues, it stimulates contraction of the milk

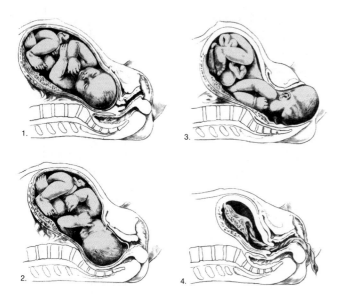

Figure 35.19 Movement of a full-term fetus from the uterus during childbirth. The last frame shows expulsion of the afterbirth (placenta).

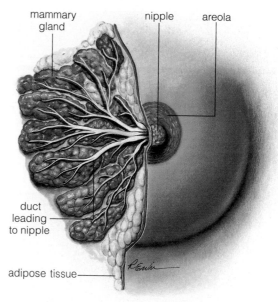

Figure 35.20 Breast of a lactating female, showing the mammary glands and ducts. The *Commentary* on the following pages describes the tissue disruptions that occur during breast cancer.

glands, so milk flows into the ducts. Oxytocin also stimulates uterine contractions that "shrink" the uterus back to its normal size.

Case Study: Mother as Protector, Provider, Potential Threat

Many safeguards are built into the female reproductive system. The placenta, for example, is a highly selective filter that prevents many noxious substances in the mother's bloodstream from gaining access to the embryo or fetus. Even so, from fertilization to birth, the developing individual is at the mercy of the mother's diet, health habits, and lifestyle.

Some Nutritional Considerations. During pregnancy, a balanced diet usually provides enough vitamins and minerals for normal development. The mother's vitamin needs are definitely increased, but the developing fetus is more resistant than she is to vitamin and mineral deficiencies (the placenta preferentially absorbs vitamins and minerals from her blood).

A few years ago, it was accepted medical practice for a pregnant woman to keep her total weight gain to ten or fifteen pounds. It is now clear that if the woman restricts her food intake too severely, especially during the last trimester, fetal development will be affected and the newborn will be underweight. Significantly underweight infants face more post-delivery complications than do infants of normal weight; in fact, they represent nearly half of all newborn deaths. They also will suffer a much higher incidence of mental retardation and other handicaps later in life. In most cases, a woman should gain somewhere between twenty and twenty-five pounds during pregnancy.

As birth approaches, the growing fetus demands more and more nutrients from the mother's body. During this last phase of pregnancy, the mother's diet profoundly influences the course of development. Poor nutrition damages most fetal organs—particularly the brain.

Risk of Infections. During pregnancy, antibodies transferred across the placenta protect the developing individual from all but the most severe bacterial infections. However, certain viral diseases can have damaging effects if they are contracted during the first six weeks after fertilization, the critical time of organ formation. For example, if the woman contracts German measles (rubella) during this period, there is a fifty percent chance that her embryo will become malformed. If she contracts the measles virus when the embryo's ears are forming, her newborn may be deaf. (German measles can be avoided by vaccination *before* pregnancy.) The likelihood of damage to the embryo diminishes after the first six weeks. The same disease, contracted during the fourth month or thereafter, has no discernible effect on the development of the fetus.

Cancer in the Human Reproductive System

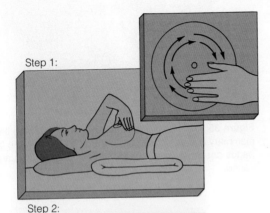

Step 1:

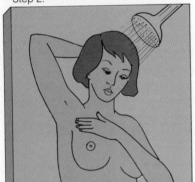

Step 2:

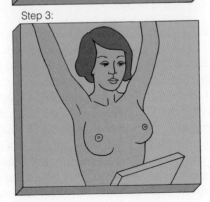

Step 3:

a Steps in breast self-examination

For both men and women, reproductive function depends absolutely on hormonal controls of the sort described in this chapter. Hormonal imbalances contribute to a variety of disorders of the reproductive system, including many forms of cancer. Unless it is eradicated from the body, cancer kills—and it is not often eradicated easily. This makes cancer one of the most feared disorders of modern times. Here we will focus on two types: cancer of the breast and of the testis. Both types often can be detected through routine self-examination, a habit that may end up saving your life.

Breast Cancer

Of all cancers in women, breast cancer currently is second only to lung cancer in having the highest mortality. Despite intensive medical research, that rate has not been lowered by much over the past fifty years. Each year in the United States, well over 100,000 women develop breast cancer; more than a third die from it. Obesity, high blood cholesterol, and excessively high levels of estrogen and perhaps other hormones play roles in the development of cancer, but how they do this is not clear.

The chances for cure are excellent if breast cancer is detected early and treated promptly. That is why a woman should examine herself once a month, about a week after each menstrual period. The following steps are recommended by the American Cancer Society:

1. Lie down and put a folded towel or pillow under the right shoulder, then put your right hand behind your head. With the left hand (fingers flat), begin the examination by following the outer circle of arrows shown in the pink box of Figure **a**. All the while, gently press the fingers in small, circular motions to check for any lump, hard knot, or thickening. Next, follow the inner circle of arrows. Continue doing this for at least three more circles, one of which should include the nipple. Then repeat the procedure for the left breast.

2. For a complete examination, repeat the procedure of step 1 while standing in a shower or tub (hands glide more easily over wet skin).

3. Stand before a mirror, lift your arms over your head, and look for any unusual changes in the contour of your breasts, such as a swelling, dimpling, or retraction (inward sinking) of the nipple. Also check for any unusual discharge from the nipple.

If you discover a lump or any other change during a breast self-examination, it's important to see a physician at once. Most changes are not cancerous, but let the physician make the diagnosis.

Currently, mammography (which uses low doses of x-rays) is the only imaging procedure with a proven record for detecting small cancers in the breast; it is eighty percent reliable. Figure **b** shows a mammogram that revealed a tumor (arrow), which a biopsy indicated to be cancerous. (The white patches at the front of the breast are milk ducts and fibrous tissue.)

Most often, cancerous tumors are removed through modified radical mastectomy. All the breast tissue, the overlying skin, and the axillary lymph nodes (Figure **c**) are removed, but muscles of the chest wall are left intact to permit more normal shoulder motions following surgery. In another procedure (lumpectomy), which is followed by radiation therapy, some of the breast tissue is left in place. In both cases, the removed lymph nodes are examined to determine the need for further treatment and to predict the prospects of a cure. If tumor cells are present in the nodes, there is a high risk of metastasis (cancer spread). Treatment then may include hormone therapy (aimed at shrinking any tumor masses) and radiation therapy. Treatment is most promising when cancer is detected at an early stage.

Cancer of the Testis

You may be surprised to learn that cancer of the testis is a frequent cause of death in young men. About 5,000 cases are diagnosed in a given year in the United States alone. In its early stages, testicular cancer is painless. If not detected in time, however, it can spread to lymph nodes in the abdomen, chest, neck and, eventually, the lungs. Once it has metastasized, the cancer kills as many as half of its victims.

Once a month, from high school onward, men should examine each testis separately after a warm bath or shower (when the scrotum is relaxed). The testis should be rolled gently between the thumb and forefinger to check for any type of lump, enlargement, or hardening. Changes of that sort may or may not cause discomfort—but they must be reported to a physician, who can make a complete examination. Treatment of testicular cancer has one of the highest rates of success—when the cancer is caught before it can spread.

normal tissue   tumor

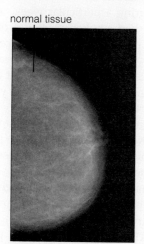
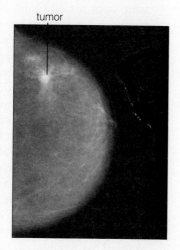

b

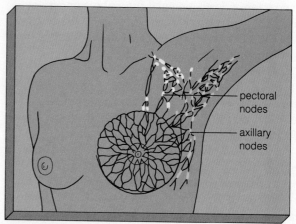

pectoral nodes

axillary nodes

c

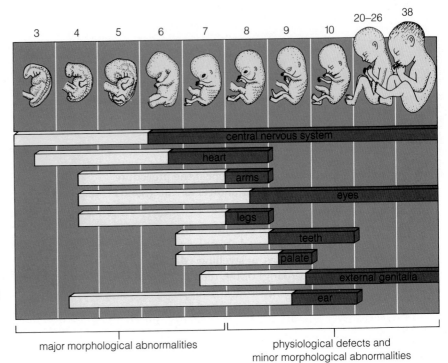

Figure 35.21 Critical periods of human embryonic development. The periods during which embryonic organs are most sensitive to damage by environmental factors (which include cigarette smoke, alcohol, viral infections, and so on) are shown in red. Numbers signify the week of embryonic development.

major morphological abnormalities

physiological defects and minor morphological abnormalities

Effects of Prescription Drugs. During the first trimester, the embryo is highly sensitive to drugs. A shocking example of drug effects came during the first two years after *thalidomide* was introduced in Europe. Women using this prescription tranquilizer during the first trimester gave birth to infants with missing or severely deformed arms and legs. Once the deformities were traced to thalidomide, the drug was withdrawn from the market. However, there is evidence that other tranquilizers (and sedatives and barbiturates) might cause similar, although less severe, damage. Even the drug Accutane, used for treating acne, increases the risk of facial and cranial deformities. Tetracycline, a commonly prescribed antibiotic, causes yellowed teeth; streptomycin causes hearing problems and may affect the nervous system.

At no stage of development is the embryo impervious to drugs in the maternal bloodstream. Clearly, the woman should take no drugs at all during pregnancy unless prescribed by a knowledgeable physician.

Effects of Alcohol. As the fetus matures, its physiology becomes increasingly like that of the mother's. Alcohol passes freely across the placenta and has the same kind of effect on the fetus as on the woman who drinks it. *Fetal alcohol syndrome* (FAS) is a constellation of deformities that are thought to result from excessive use of alcohol by the mother during pregnancy. FAS is the third most common cause of mental retardation in the United States. It also is characterized by facial deformities, poor coordination and, sometimes, heart defects. Between sixty and seventy percent of alcoholic women give birth

to infants with FAS; some researchers now suspect that even two drinks a day during pregnancy may be dangerous for the fetus. Increasingly, physicians are urging total or near-abstention during pregnancy.

Effects of Smoking. Cigarette smoking has an adverse effect on fetal growth and development. Newborns of women who have smoked every day throughout pregnancy have a low birth weight. That is true even when the woman's weight, nutritional status, and all other relevant variables are identical with those of pregnant women who do not smoke. Smoking has other effects as well (Figure 35.21).

For example, for seven years in Great Britain, records were kept for all births during a particular week. The newborns of women who had smoked were not only smaller, they had a thirty percent greater incidence of death shortly after delivery and a fifty percent greater incidence of heart abnormalities. More startling, at age seven, their average "reading age" was nearly half a year behind that of children born to nonsmokers.

In this last study, the critical period was shown to be the last half of pregnancy. Newborns of women who had stopped smoking by the middle of the second trimester were indistinguishable from those born to women who had never smoked. Although the mechanisms by which smoking exerts its effects on the fetus are not known, its demonstrated effects are further evidence that the placenta—marvelous structure that it is—cannot prevent all the assaults on the fetus that the human mind can dream up.

CONTROL OF HUMAN FERTILITY

Some Ethical Considerations

The remarkable transformation of a single-celled zygote into an intricately detailed adult raises profound questions. *When does development begin?* As we have seen, many key aspects of development emerge even before fertilization. *When does life begin?* During her lifetime, a human female can produce as many as four hundred secondary oocytes, all of which are alive. During one ejaculation, a human male can produce a quarter of a billion sperm, which also are alive. Even before sperm and oocyte merge by chance and establish the genetic constitution of a new individual, they are as much alive as any other form of life. It is scarcely tenable, then, to suggest that "life begins" when they fuse. *Life began billions of years ago; and each gamete, each zygote, and each mature individual is only a fleeting stage in the continuation of that beginning.* This fact cannot diminish the meaning of conception, for it is no small thing to entrust a new individual with the gift of life, wrapped in the unique evolutionary threads of our species and handed down through an immense sweep of time.

Yet how can we reconcile the marvel of individual birth with our growing awareness of the astounding birth rate for the human species as a whole? At the time this book is being written, 9,900 infants are being born each hour. By the time you go to bed tonight, there will be 238,000 more people on earth than there were last night at that hour. Within a week, the number will reach 1,700,000 more. *Within one week.* Our worldwide population growth has outstripped our resources, and each year millions face the horrors of starvation. Living as we do on one of the most productive continents on earth, few of us can know what it means to give birth to a child, to give it the gift of life, and have no food to keep it alive.

And how can we reconcile the marvel of birth with the confusion surrounding unwanted pregnancies? Even in highly developed countries there are not enough educational programs concerning fertility control, and often there is reluctance to exercise control. Each year in the United States alone there are about 100,000 "shotgun" marriages, about 200,000 unwed teenage mothers, and perhaps 1,500,000 abortions. On the one hand, many parents encourage boy-girl relationships at early ages. On the other hand, they ignore the possibility of premarital intercourse and unplanned pregnancy. Advice is often condensed to a terse, "Don't do it. But if you do it, be careful!"

The motivation to engage in sex has been evolving for more than 500 million years. Moral sanctions calling for the suppression of that motivation have not prevented unwanted pregnancies, and complex social factors have contributed to a population growth rate that is out of control. How will we reconcile our biological past with the need for a stabilized cultural present?

Whether human fertility is to be controlled—and how it is to be controlled—is one of the most volatile issues of our time. We will return to this issue in Chapter Forty-Three, in the context of principles governing the growth and stability of populations. Here, we can briefly consider some possible control options.

Possible Means of Birth Control

At present, approaches to birth control fall into three categories:

1. *Fertility* control (physical, chemical, surgical, or behavioral interventions that disrupt reproductive function or affect gamete survival).

2. *Implantation* control (physical or chemical interference with the blastocyst's ability to invade the uterine wall).

3. *Abortion* (prevention of embryonic development after implantation has occurred).

Let's consider some behavioral interventions first. The most effective method of preventing conception is complete **abstinence**: no sexual intercourse whatsoever. It is unrealistic to expect many people to practice this method. A modified form of abstinence is the **rhythm method**, in which intercourse is avoided during the woman's fertile period. The fertile period begins a few days before ovulation and ends a few days after. It is identified and tracked either by keeping records of the length of the woman's menstrual cycle or by taking her temperature each morning when she wakes up. The core temperature decreases slightly just before ovulation, then it may rise by one-half to one degree above normal within a day or two. The likelihood of detecting the small change is improved when a special, expanded-scale thermometer is used. However, ovulation can be irregular, and body temperature can fluctuate as a result of infections, emotional disturbances, use of electric blankets, and so forth. Moreover, sperm deposited in the vaginal tract a few days prior to ovulation may survive until ovulation. The method *is* inexpensive (it costs nothing after you buy the thermometer) and it does not require fittings and periodic checkups by a doctor. But its practitioners do run a substantial risk of becoming pregnant (Figure 35.22).

Withdrawal is the removal of the penis from the vagina prior to ejaculation. This contraceptive method dates back at least to biblical times. But withdrawal requires an unusual degree of willpower. In any case,

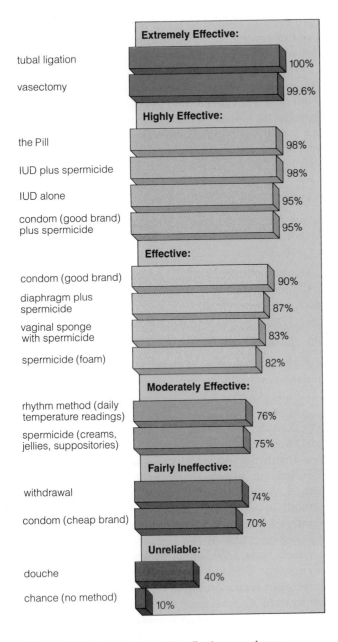

Figure 35.22 Comparison of the effectiveness of some contraceptive methods.

into the vagina just before intercourse. These products are toxic to sperm, yet they are not always reliable unless used with another device, such as a diaphragm or condom. A **diaphragm** is a flexible, dome-shaped device that is inserted into the vagina and positioned over the cervix before intercourse. A diaphragm is relatively effective (1) when it is used with foam or jelly, (2) when it has been fitted by a doctor, (3) when it is inserted correctly with each use, and (4) when foam or jelly is reapplied with each sexual contact.

Condoms are thin, tight-fitting sheaths of rubber or animal skin that are worn over the penis during intercourse. They are about eighty-five to ninety-three percent reliable, and they help prevent the spread of sexually transmitted diseases (see *Commentary* on page 532). However, condoms can tear and leak, in which case they are rendered useless.

You may have heard about the intrauterine device, or **IUD**. This small plastic or metal device, which must be inserted by a trained specialist, can be left in the uterus for years to prevent implantation. Although the IUD is an effective means of fertility control, controversy has surrounded its use. Women who use the device run a greater risk of pelvic inflammatory disease (infection in the upper genital tract) and infertility. The risk of infection seems to be greatest during the first four months after IUD insertion and among women exposed to sexually transmitted diseases. A new generation of presumably safer IUDs is about to become available in most countries. Even then, they will not be recommended for everyone, and they will require insertion and periodic checkups by a competent health care provider.

Another method of fertility control is hormonal intervention in the reproductive cycle. Most widely used is **the Pill**, an oral contraceptive of synthetic estrogens and progesterones taken daily by the female. By suppressing the normal release of gonadotropins from the pituitary, these synthetic hormones prevent the cyclic maturation and release of ova. The Pill is a prescription drug. Formulations differ and are selected to match the individual patient's needs. That is why it is not wise for a woman to borrow the Pill from someone else.

If the woman does not forget to take her daily dosage, the Pill is one of the most reliable methods of controlling fertility. It does not interrupt sexual intercourse, and the method is easy to follow. Often the Pill corrects erratic menstrual cycles and decreases menstrual cramping. However, the Pill must be taken for one month (a full cycle) before it can be considered fully effective. In other words, a woman cannot take one pill the day before intercourse and expect it to work. Also the Pill does have known side effects for a small number of users. In the first month or so of use, it may cause nausea, weight gain, tissue swelling, and minor headaches. Its contin-

the method may fail: fluid released from the penis just before ejaculation may contain viable sperm.

The practice of **douching**, or rinsing out the vagina with a chemical right after intercourse, is next to useless. Sperm can move past the cervix and out of reach of the douche within ninety seconds after ejaculation.

Other methods involve physical or chemical barriers that prevent sperm from entering the uterus and moving to the oviducts. **Spermicidal foam** or **spermicidal jelly** is placed in an applicator, which is inserted and emptied ued use may lead to blood clotting in the veins of a few women (3 out of 10,000) predisposed to this disorder. There have been some cases of elevated blood pressure among the Pill users, and also some abnormalities in fat metabolism that might be linked to a growing number of gallbladder disorders. Possible longer term effects have been difficult to assess.

Hormonal control of male fertility is a more complicated matter. The hormonal methods for suppressing female fertility are based on the cyclic nature of the woman's reproductive capacity. But sperm production in males is not cyclic, and it is under a more diffuse kind of hormonal control. Although various medications have been developed, they are still experimental. It will probably be several years before they become available, and their effectiveness is not yet assured.

Surgical intervention in the reproductive tract is becoming common. In **vasectomy**, a tiny incision is made in the scrotum and each vas deferens is severed and tied off. The simple operation can be performed in twenty minutes in a doctor's office, with only a local anesthetic. After vasectomy, sperm cannot leave the epididymides and therefore will not be present in seminal fluid. (However, sperm present in duct regions below the surgical cuts can be present in the ejaculate for several weeks after the operation.) So far there is no firm evidence that vasectomy disrupts the male hormone system, and surveys suggest there is no noticeable difference in sexual activity. Although vasectomies can be surgically reversed, half of the men who have undergone surgery will develop antibodies against sperm and some of those men may not be able to regain fertility.

For females, surgical intervention includes **tubal ligation**, in which the oviducts are cauterized or cut and tied off. Tubal ligation is more complex than vasectomy and is usually performed in a hospital. A small number of women who have had the operation suffer recurring bouts of pain and inflammation of tissues in the pelvic region where the surgery was performed. The operation can be reversed, although major surgery is required and success is not always assured.

Once conception and implantation have occurred, the only way to terminate a pregnancy is **abortion**, in which the implanted embryo is dislodged and removed from the uterus. (In *miscarriages*, or spontaneous abortion, the embryo is dislodged and expelled spontaneously.)

Until recently, abortions were generally forbidden by law in the United States unless the pregnancy endangered the mother's life. Supreme Court rulings have held that the government does not have the power to forbid abortions during the early stages of pregnancy (typically up to six months). Abortion is now legal in this country. Moving the large number of illegal, unsupervised operations to modern medical facilities has reduced the frequency of dangerous, traumatic, and often fatal attempts to abort embryos, either by pregnant women themselves or by quacks. Newer methods have made abortion relatively rapid, painless, and free of complications when performed during the first trimester. RU-486, the so-called "morning-after Pill" that can induce the termination of pregnancy rather painlessly, is becoming available in Europe at this writing.

In the United States, abortions in the second and third trimesters will probably remain extremely controversial unless the mother's life is clearly threatened. For both medical and humanitarian reasons, however, it is generally agreed in this country that a preferable route to birth control is not through abortion but through control of conception in the first place.

In Vitro Fertilization

In the United States, about fifteen percent of all couples are unable to conceive a child, owing to sterility or infertility. For example, hormonal imbalances may prevent ovulation in females; or the male's sperm count may be too low to assure successful fertilization. With **in vitro fertilization**, external conception is possible, provided that the sperm and secondary oocyte obtained from the couple are normal.

A hormone administered to the female prepares her ovaries for ovulation within thirty-three to thirty-four hours. Then a laparoscope allows the physician to locate and remove the preovulatory oocyte with a suction device. (A laparoscope is a long, metal tube containing a light and an optical system; it can be used to identify the ballooning follicle at the ovarian surface.)

Before the oocyte is removed, a sample of the male's sperm is put in a saline solution that simulates the fluid in oviducts. When the suctioned oocyte is placed in the solution with the sperm, fertilization may occur a few hours later. About twelve hours later, the newly dividing embryo is transferred to a solution that will support further development, and about two to four days after that, it is transferred to the female's uterus. If all goes well, implantation may occur.

Since the first successful childbirth following in vitro fertilization in 1978, clinics specializing in the procedure have opened in several countries. There appears to be no greater risk associated with the procedure than with normally occurring pregnancies. Implantation occurs in only about ten or twenty percent of the attempts, and each attempt costs several thousand dollars. Even so, there are many childless couples who are attempting to have children in this manner.

Sexually Transmitted Diseases

(This Commentary is based on information from the Centers for Disease Control, Atlanta)

Sexually transmitted diseases (STDs) have reached epidemic proportions, even in countries with the highest medical standards. The disease agents are mostly bacteria and viruses, and they usually are transmitted from infected to uninfected persons during sexual intercourse. In the United States alone, 10 million young adults have reported that they have some form of STD; no one can estimate the number of unreported cases.

The economics of this health problem are staggering. By conservative estimates, the cost of treatment is exceeding $2 billion a year—and this does not include the accelerating cost of treating AIDS patients. In Africa and many other developing countries, AIDS alone threatens to overwhelm health-care delivery systems and to unravel decades of economic progress.

The social consequences are sobering. Of every twenty babies born in the United States, one will start out in the world with a chlamydial infection. Of every 10,000 newborns, as many as three will contract systemic herpes. Half of those newborns may die early, and a fourth of the survivors will have serious neurological defects. Each year in the United States, 1 million adolescent and adult females are stricken with pelvic inflammatory disease, usually as a complication of gonorrhea and other STDs. Of those females, about 200,000 are hospitalized, more than 100,000 undergo pelvic surgery that results in permanent sterility, and 900 cannot recover and die. The examples just given only hint at the alarming complications of many sexually transmitted diseases.

AIDS

Acquired immune deficiency syndrome (AIDS) is a set of chronic disorders that can follow infection by the human immunodeficiency virus (HIV). The virus cripples the immune system, in the manner described in Chapter Thirty, and the body becomes highly vulnerable to illnesses, many of which would not otherwise be life-threatening. (Hence the description, "opportunistic" infections.)

AIDS is mainly a sexually transmitted disease, with most infections occurring through the transfer of bodily fluids during vaginal or anal intercourse. (Such fluids include blood, semen, urine, and vaginal secretions.) The virus enters the body through cuts or abrasions in the penis, the vagina, or the rectum. Mucous membranes in the mouth may be another point of entry. Once inside the body, the virus locks onto cells that are capable of sustaining its replication (page 430). Helper T cells (the T4 lymphocytes), macrophages, brain cells, and epithelial cells of the colon and cervix are known targets.

Unlike some other sexually transmitted diseases, AIDS cannot be effectively treated at this time and there are no vaccines against the causative agent. *There is no cure.* Infected persons may be symptom-free at first, but as many as half develop AIDS within five to ten years. Others develop symptoms that are milder than the ones characterizing AIDS (page 430). The milder symptoms are called the AIDS-related complex, or ARC. Will most or all of those infected eventually develop AIDS? We do not know enough about the natural history of the virus and the progression rates of the disease to discount that possibility.

Categories of AIDS Patients*	
Sexually active homosexual and bisexual men (or any man who has had sex with another man since 1977):	65%
Current or past abusers of intravenous drugs:	17%
Homosexual and bisexual men who also are intravenous drug abusers:	8%
Heterosexuals who have sexual relations with someone having AIDS or at risk for AIDS:	4%
Recipients of transfusions of blood or blood products:	2%
Persons with blood-clotting disorders (e.g., hemophilia):	1%
Infants born to infected mothers:	1%

*About 2% of patients do not fall in these groups, but transmission is thought to have occurred in similar ways. (For example, some patients died before complete histories could be taken.)

It is the symptom-free carriers of HIV who unwittingly have been fueling a worldwide AIDS epidemic. *During the next decade, as many as 50 million to 100 million could be infected worldwide.*

HIV apparently has been present in localized regions of Central Africa for at least several decades. However, in the 1970s and early 1980s, it spread to different countries and has since reached epidemic proportions. (AIDS was not even identified until 1981.) Although in Africa the virus is transmitted primarily through heterosexual contact, the initial victims in the United States were male homosexuals. In the developed countries, HIV is still transmitted mainly through homosexual contact as well as through needle-sharing by intravenous drug abusers (see Table). Unfortunately, the use of illicit drugs in conjunction with sexual practices is widespread, not only among homosexuals but also in the heterosexual community at large. Today, in both rural and urban areas, more and more women are infected through needle-sharing and relations with bisexual men.

Who, then, is *not* at risk? Only individuals who fall in these categories:

1. Individuals who are not drug abusers or who do not share unsterile needles or syringes.

2. Individuals who abstain from sexual relations.

3. Mutually monogamous couples who have had no other sexual partners since the 1970s (when AIDS began to spread dramatically).

4. Couples who are shown to be free of infection and who refrain from sexual relations with anyone else.

Free or low-cost, confidential testing for AIDS is available through public health facilities and many physicians' offices. Keep in mind that there may be a time lag from a few weeks to six months or longer before detectable antibodies form in response to infection. The presence of antibodies indicates exposure to the virus, but this in itself does not mean that AIDS will develop. Even so, anyone who tests positive should be considered capable of spreading the virus.

Beyond this, there is confusion about what constitutes "safe" sex. Proper use of high-quality, latex condoms, together with a spermicide that contains nonoxynol-9, is assumed to be highly effective in stopping transmission—but there is still a small risk of irreversible infection. Open-mouthed, intimate kissing with a person who tests positive for the virus should be avoided. Caressing carries no risk—*if* there are no lesions or cuts through which the virus can enter the body. Such lesions commonly accompany other sexually transmitted diseases, and they apparently are correlated with increased susceptibility to HIV infection.

In sum, AIDS has reached epidemic proportions mainly for three reasons. First, we did not know that the virus is transmitted by semen, blood, and vaginal fluid and that *behavioral* controls can limit its spread. Second, we did not have tests that could be used to identify symptom-free carriers who could unwittingly infect others; we do now. Third, we thought AIDS was a threat associated only with homosexual behavior. The medical, social, and economic consequences of its rapid spread throughout the world make it everyone's problem.

Gonorrhea

Unlike AIDS, gonorrhea is a sexually transmitted disease that can be cured by prompt diagnosis and treatment. Gonorrhea ranks first among the reported communicable diseases in the United States, with 1 million new cases reported each year. (There may be anywhere from 3 to 10 million unreported cases.)

Gonorrhea is caused by *Neisseria gonorrhoeae*. This bacterium can infect epithelial cells of the genital tract, eye membranes, and the throat. Since 1960, its incidence in the population has been rising at an alarming rate. The increase has coincided with the use of birth control pills and increased sexual permissivity. (The pills alter vaginal pH, which normally would offer some protection against *N. gonorrhoeae* and other acid-sensitive bacteria.)

Males have a greater chance than females do of detecting the disease in early stages. Within a week, yellow pus is discharged from the urethra. Urination becomes more frequent and painful because the uri-

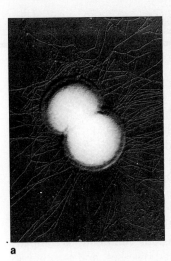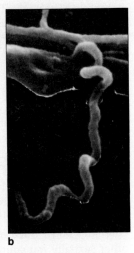

(a) *Neisseria gonorrhoeae*, a spherical bacterium that typically is seen as paired cells, as shown here. The pili evident in this electron micrograph help the bacterium attach to its host, upon which it bestows gonorrhea. (b) *Treponema pallidum*, a spiral bacterium that causes syphilis. This scanning electron micrograph shows a bacterium adhering to a testis cell from a rabbit.

nary tract becomes inflamed. Females may or may not experience a burning sensation while urinating. They may or may not have a slight vaginal discharge; even if they do, the discharge may not be perceived as abnormal. Thus, in the absence of worrisome symptoms, gonorrhea often goes untreated. The bacteria may spread into the oviducts, the eventual outcome being violent cramps, fever, vomiting and, in many cases, sterility due to scarring and blocking.

Complications arising from gonorrheal infection can be avoided with prompt treatment. As a preventive measure, males who have multiple sexual partners can wear condoms to help prevent the spread of infection. Part of the problem is that the initial stages of the disease are so uneventful that the dangers are masked. Also, many infected persons wrongly believe that once cured of gonorrhea, they are safe from reinfection—which simply is not true. Multiple reinfections can and do occur.

Syphilis

Syphilis is caused by a motile bacterium, *Treponema pallidum* (a spirochete). As many as 90,000 humans may become infected in a given year in the United States, but only about 30,000 are reported. The bacterium is transmitted by sexual contact. After it has penetrated exposed tissues, it produces a chancre

(that is, a localized ulcer) that teems with treponeme progeny. The chancre, which is a symptom of the primary stage of syphilis, becomes visible between one to eight weeks following infection. By then, treponemes have already moved into the lymph vascular system and bloodstream.

During the second stage of infection, lesions can occur in mucous membranes, the eyes, bones, and the central nervous system. Afterward, the infection enters a latent stage that has no outward symptoms. Syphilis can be detected only by serology tests during the latent stage, which can last many years. All the while, the immune system works against the bacterium. Sometimes the body does cure itself, but this is not the usual outcome.

If untreated, syphilis in its tertiary stage can produce lesions of the skin and internal organs, including the liver, bones, and aorta. Scars form; the walls of the aorta can weaken. Treponemes also damage the brain and spinal cord in ways that lead to various forms of insanity and paralysis. Women who have been infected typically have miscarriages, stillbirths, or sickly and syphilitic infants.

Chlamydial Infections

There are other infections of the genital tract besides gonorrhea and syphilis; they are called the "nonspecific" sexually transmitted diseases. The most chronic are the chlamydial infections. Each year, anywhere from 3 to 10 million Americans—college students particularly—are affected.

An obligate, intracellular parasite called *Chlamydia trachomatis* is the culprit behind a variety of diseases. Among other things, it infects cells of the genitals and urinary tract. Following infection, the parasites migrate to regional lymph nodes, which become enlarged and tender. The enlargement can impair lymph drainage and lead to pronounced tissue swelling. Chlamydial infections can be treated effectively with tetracycline and sulfonamides. Most of the infections have no long-term complications. However, in some females the infection leads to pelvic inflammatory disease.

Pelvic Inflammatory Disease

A condition called pelvic inflammatory disease (PID) affects about $1\frac{3}{4}$ million women each year. It is one of the serious complications of gonorrhea, chlamydial infections, and other STDs, but it also can arise when normal vaginal microbes ascend into the pelvic region and when intrauterine devices (IUDs) malfunction and cause an inflammatory response in the uterine lining. Most often, the uterus, oviducts, and ovaries are affected. The pain may be so severe, infected women often think they are having an attack of acute appendicitis. The oviducts may become scarred, and this can lead to abnormal (ectopic) pregnancies as well as to sterility.

Genital Herpes

Genital herpes is an extremely contagious viral infection of the genitals. It is transmitted when any part of a person's body comes into direct contact with active *Herpes* viruses or sores that contain them. Mucous membranes (particularly of the mouth or genital area) are susceptible to invasion, as is broken or damaged skin. Transmission seems to require intimate sexual contact; the virus does not survive for long outside the human body.

There are an estimated 5 to 20 million persons with genital herpes in the United States alone. From 1965 to 1979, the number of reported cases increased by 830 percent; 200,000 to 500,000 cases are still being reported annually. Newborns of infected mothers are among those cases. Contact with the mother's active lesions during normal vaginal delivery can lead to a form of herpes that is often fatal. Lesions arising in the infant's eyes can cause blindness. Chronic herpes infection of the cervix is linked statistically to an increased risk of cervical cancer.

There are many strains of *Herpes* viruses, which are classed as types I and II. The type I strains infect mainly the lips, tongue, mouth, and eyes. Type II strains cause most of the genital infections. Disease symptoms occur two to ten days after exposure to the virus, although sometimes symptoms are mild or absent. Among infected women, small, painful blisters appear on the vulva, cervix, urethra, or anal tissues. Among men, the blisters occur on the penis and anal tissues. Within three weeks, the sores crust over and heal without leaving scars.

After the first sores disappear, sporadic reactivation of the virus can produce new, painful sores at or near the original site of infection. Recurrent infections may be triggered by sexual intercourse, emotional stress, menstruation, or other infections. At present there is no cure for genital herpes. Acyclovir, an antiviral drug, decreases the healing time and often decreases the pain and viral shedding.

SUMMARY

1. The human reproductive system consists of a pair of primary reproductive organs (gonads) and accessory ducts and glands. The male gonads are the testes, the female gonads are the ovaries. The gonads produce gametes (sperm in the male, secondary oocytes or "eggs" in the female) and hormones that influence reproductive functions and the secondary sexual traits.

2. Sperm form in seminiferous tubules within the testes from undifferentiated cells (spermatogonia). Interstitial cells between the tubules secrete sex hormones.

3. Sperm move from the testis to the epididymis, where their capacity for motility is partly developed. (Full motility occurs only in the female reproductive tract.) Sperm pass through the male's duct system and finally into the urethra. The sperm, together with secretions from glands along the reproductive tract, form the semen.

4. Testosterone, accompanied by FSH and LH, stimulates spermatogenesis. It also stimulates the development of the male reproductive tract, promotes male sexual behavior, and supports development of secondary sexual traits.

5. In the female, the ovaries release oocytes and secrete the sex hormones estrogen and progesterone. An oviduct next to each ovary transports the oocyte to the uterus. The uterus opens by way of the cervix into the vagina, which opens to the exterior.

6. The oocytes develop before birth into primary oocytes which, together with their surrounding cell layer, are called primary follicles. They do not start to mature until puberty, when one at a time on a monthly basis, they develop into secondary oocytes.

7. In the human female, the reproductive periods are menstrual cycles, during which the interplay of FSH, LH, estrogen, and progesterone causes (1) development of the oocyte and (2) preparation of the lining of the uterus to receive the embryo. If fertilization does not occur, the uterine lining is shed (the menstrual flow), the corpus luteum degenerates, and a new cycle begins.

8. At ovulation (caused by a sudden secretory burst of the hormone LH) a secondary oocyte is released from the ovary and begins its final meiotic division, which is not completed until fertilization; then, the secondary oocyte becomes a mature ovum. The remaining cells of the follicle develop into a corpus luteum (a glandular structure that secretes progesterone and some estrogen, both of which maintain the endometrium).

9. During sexual intercourse, sperm are ejaculated into the vagina, where they then undergo further maturation. Fertilization normally occurs in an oviduct.

10. Embryonic development depends on the formation of extraembryonic membranes that function in the following ways:

   a. *Yolk sac:* parts give rise to the embryo's digestive tube.

   b. *Allantois:* its blood vessels function in oxygen transport.

   c. *Amnion:* forms a fluid-filled sac that surrounds and protects the embryo from mechanical shocks and keeps it from drying out.

   d. *Chorion:* forms a protective membrane around the embryo and the other membranes; a primary component of the placenta.

11. The embryo and the mother exchange substances by way of the placenta (a spongy tissue of endometrium and extraembryonic membranes).

12. Human development proceeds through the formation of three germ layers (endoderm, ectoderm, and mesoderm). By the beginning of the second trimester of pregnancy, all major organs have formed.

13. At delivery, contractions of the uterus dilate the cervical canal and expel the fetus and afterbirth.

14. Estrogen and progesterone stimulate growth of the mammary glands. After delivery, nursing causes the release of hormones that stimulate milk production and release.

15. During intrauterine life, the placental barrier provides some protection for the fetus, but it may suffer harmful effects from the mother's nutritional deficiencies, infections, intake of prescription drugs, alcohol, and smoking.

16. Control of human fertility raises many important ethical questions. These questions extend to the physical, chemical, surgical, or behavioral interventions used in the control of unwanted pregnancies.

Review Questions

1. Study Table 35.1. Then list the main organs of the human male reproductive tract and identify their functions. *512*

2. What are the accessory glands of the male reproductive tract? What are their functions? *511–512*

3. Describe spermatogenesis and the route by which sperm leave the seminiferous tubules, then the body. *510–512, 518*

4. Which hormones have profound influence over male reproductive functioning? Diagram how feedback mechanisms link the hypothalamus, anterior pituitary, and interstitial cells in controlling this functioning. *512*

5. Trace the events leading to the formation of the secondary oocyte. *513–514*

6. Label the component parts of the female reproductive tract: *513–514*

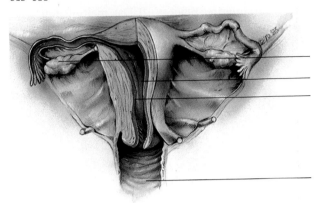

7. What is the menstrual cycle? Which four hormones have profound influence on this cycle? Diagram the feedback loops among the hypothalamus, anterior pituitary, and ovary that govern these hormonal secretions. *515–517*

8. List the four events that are triggered by the surge of LH at the midpoint of the menstrual cycle. *515*

9. What changes occur in the endometrium during the menstrual cycle? *516–517*

10. Describe these extraembryonic membranes: *521*
   a. amnion
   b. yolk sac
   c. allantois
   d. chorion

11. From what tissues and membranes does the placenta form? *522*

12. With a sheet of paper, cover the summary definitions in Table 35.4. Can you state the definition for each of the stages of human development listed? *518*

Readings

Carlson, B. 1988. *Patten's Foundations of Embryology.* Fifth edition. New York: McGraw-Hill.

Gilbert, S. 1988. *Developmental Biology.* Sunderland, Massachusetts: Sinauer. Second edition. Acclaimed reference text.

Nilsson, L. et al. 1986. *A Child Is Born.* New York: Delacorte Press/Seymour Lawrence. Extraordinary photographs of embryonic development.

Saunders, J. W. 1982. *Developmental Biology: Patterns, Problems, Principles.* New York: Macmillan.

Schatten, G. 1983. "Motility During Fertilization." *Endeavor* 7(4):173–182.

Zack, B. 1981. "Abortion and the Limitations of Science." *Science* 213(4505).

UNIT SIX

EVOLUTION

36

POPULATION GENETICS, NATURAL SELECTION, AND SPECIATION

Where would the mallard duck be now if its destiny had been placed in the hands of that eighteenth-century cataloger of life, Carl von Linné? There, awaiting classification, was a bird with emerald-green head feathers and with wings displaying metallic blue patches. There, in the same ponds and marshes, was a drab little brown-feathered duck bearing no obvious resemblance to the more resplendent waterfowl (Figure 36.1a). Thus did von Linné, on the basis of outward appearance alone, pronounce the male and female mallard duck as separate species. (It goes without saying that the male and female duck, paying no attention whatsoever to his pronouncement, continued to produce more ducks.)

You may be thinking it is unfair to von Linné to dredge up one of his mistakes. Yet anyone might have made the same mistake if he or she had attempted to classify the ducks according to a rigid notion of species, as von Linné had done. Recall, from Chapter Two, that studies of organismic diversity were once funneled through the concept that species do not change. The approach was *typological*: an individual was selected as being the perfect standard, or type, for the species based on a formulation of what the "perfect" physical features were. Then newly encountered individuals were compared against the standard to judge whether they belonged to that species. Small variations among similar individuals were viewed as imperfect renditions of a particular species plan. (Thus, for example, there would be no mistaking any of the individuals shown in Figure

a

b

36.1b as being anything other than a snow goose.) Dramatic variations in physical appearance were often viewed as evidence in itself of different species—hence von Linné's mistake.

Of course, it took only a little field observation during the breeding season to discover that male and female mallard ducks are simply an extreme case of sexual dimorphism (phenotypic variation between sexes). But adherents of the typological approach fell into other, more serious conceptual traps. *What could be done about the vexing number of obviously similar organisms that nevertheless showed extreme variation in certain traits?* For example, certain snails living in the Caribbean show hundreds of variations in the color and banding pattern of their shells (Figure 36.1c). With the typological approach, each distinctly shelled snail would have to be assigned a separate "species" status—as, indeed, they were—making that taxonomic category rather useless.

The phenotypic variation illustrated in Figure 36.1 only hints at the immense variation among individuals of all the different species, past and present, on earth. How can this variation be explained? How does it arise, and what is its meaning? In this chapter we will cover some possible answers to these questions, using the following premise as our starting point:

Variation is a fundamental attribute of the individuals of a species and the raw material for evolution.

First we will review the sources of variation among the individuals of a species. Then we will consider how variation is maintained and how it can be modified over time.

POPULATION GENETICS

Sources of Variation

As we have seen, variation arises as a result of events at the molecular and cytological levels (Table 16.1). These events in some way alter an individual's *genotype* (genetic makeup), and the alteration may affect the *phenotype* (the individual's structural, physiological, or behavioral traits).

For example, a single amino acid substitution at one gene locus on a human chromosome produces a recessive allele that codes for a variant form of the hemoglobin molecule, designated HbS. (*Alleles*, recall, are alternative forms of a gene at a given locus on a chromosome.) Individuals who are homozygous recessive at this locus will suffer the phenotypic consequences of sickle-cell anemia. As another example, two copies of chromosome 21 may end up in a human gamete as a result of nondisjunction at meiosis (page 185). Fusion of this gamete with a normal one can lead to a variant individual exhibiting Down syndrome. Independent assortment of homologous chromosomes at meiosis is another source

c

Figure 36.1 Examples of phenotypic variation (and the seeming lack of it) between individuals of the same species. (**a**) Male and female mallard duck, an example of sexual dimorphism (phenotypic variation between sexes). (**b**) Population of snow geese at a wintering ground in New Mexico. Why do all these birds superficially look the same, and why do certain snails—the shells of which are shown in (**c**)—look so different? In more general terms, *what is the meaning of individual variation, and how does it arise*? These are questions addressed in this chapter.

Hardy-Weinberg Principle:

The genotypic frequencies for a population in equilibrium will fit the formula

$$\underset{p^2}{AA} + \underset{2pq}{Aa} + \underset{q^2}{aa}$$

where p = the frequency of allele A and q = the frequency of allele a.

The allele frequencies and the genotypic frequencies will be stable from generation to generation if the following assumptions are true:

There is no mutation, the population is infinitely large and is isolated from other populations, mating is random, and all genotypes are equally viable and fertile.

Figure 36.2 Hardy-Weinberg equilibrium. To prove the validity of the Hardy-Weinberg rule stated above, let's follow the course of two alleles, A and a, through succeeding generations.

For all members of the population, the gene locus must be occupied by either A or a. In mathematical terms, the frequencies of A and a must add up to 1. For example, if A occupies half of all the gene loci and a occupies the other half, then $0.5 + 0.5 = 1$. If A occupies ninety percent of all the gene loci, then a must occupy the remaining ten percent $(0.9 + 0.1 = 1)$. No matter what the proportions of alleles A and a,

$$p + q = 1$$

You know that during sexual reproduction of diploid organisms, the two alleles at a gene locus segregate and end up in separate gametes. Thus p is also the proportion of gametes carrying the A allele, and q the proportion carrying the a allele. To find the expected frequencies of the three possible genotypes (AA, Aa, and aa) in the next generation, we can construct a Punnett square:

	p Ⓐ	q ⓐ
p Ⓐ	AA (p^2)	Aa (pq)
q ⓐ	Aa (pq)	aa (q^2)

Because the frequency of genotypes must add up to 1,

$$p^2 + 2pq + q^2 = 1$$

To see how these calculations can be applied, let's follow the allele frequencies for a population of 1,000 diploid individuals made up of the following genotypes:

$$
\begin{array}{l}
450\ AA \\
500\ Aa \\
\underline{\ 50\ aa} \\
1,000\ \text{individuals (or 2,000 alleles)}
\end{array}
$$

Theoretically, of every 1,000 gametes produced, the frequency of A will be $450 + \frac{1}{2}(500) = 700$, or $p = 0.7$. The frequency of a will be $\frac{1}{2}(500) + 50 = 300$, or $q = 0.3$. Notice that

$$p + q = 0.7 + 0.3 = 1$$

After one round of random mating, the frequencies of the three genotypes possible in the next generation will be as follows:

$$
\begin{array}{l}
AA = p^2 = 0.7 \times 0.7 = 0.49 \\
Aa = 2pq = 2 \times 0.7 \times 0.3 = 0.42 \\
aa = q^2 = 0.3 \times 0.3 = 0.09
\end{array}
$$

and

$$p^2 + 2pq + q^2 = 0.49 + 0.42 + 0.09 = 1$$

Notice that the allele frequencies have not changed:

$$A = \frac{2 \times 490 + 420}{2,000\ \text{alleles}} = \frac{1,400}{2,000} = 0.7 = p$$

$$a = \frac{2 \times 90 + 420}{2,000\ \text{alleles}} = \frac{600}{2,000} = 0.3 = q$$

The genotypic frequencies have changed initially. However, given that the distribution of genotypes fits the equation $p^2 + 2pq + q^2$, the genotypic frequencies will be stable over succeeding generations. You can verify this by calculating the most probable allele frequencies for gametes produced by the second-generation individuals:

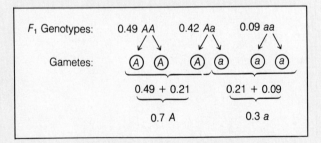

which is back where we started from. Because the allele frequencies are exactly the same as those of the original gametes, they will yield the same frequencies of genotypes as in the second generation.

You could go on with the calculations until you ran out of paper, or patience. As long as the population adheres to the conditions stated in the boxed inset for the Hardy-Weinberg principle, you would end up with the same results. When the frequencies of different alleles and different genotypes remain constant through successive generations, the population is in Hardy-Weinberg equilibrium: it is not evolving.

of phenotypic variation, and so is crossing over. These and other sources of phenotypic variation fall into five broad categories:

1. Gene mutation
2. Chromosomal aberrations
3. Independent assortment of chromosomes at meiosis
4. Crossing over at meiosis
5. Fertilization between genetically varied gametes

Only mutations *create* new alleles. Although individual mutations at any one gene locus are rare events (page 219), the total number of mutations that have occurred in the history of a species has provided the potential for incredible variation. Moreover, the other events listed above shuffle *existing* alleles into new combinations in new individuals. The number of different genotypes made possible by all these events is simply extraordinary.

According to one estimate, more than 10^{600} genetically different human gametes are possible. By comparison, there are fewer than 10^{10} humans alive today, and the estimated number of atoms in the universe is on the order of 10^{70}! Unless you have an identical twin, it is extremely unlikely that another person with your exact genotype has ever lived, or ever will. In short, *far more genetic variation is possible than can ever be expressed in the individuals alive at any one time.*

The Hardy-Weinberg Baseline for Measuring Change

Although it may seem obvious, it is important to keep in mind that individuals don't evolve; *populations* do. Here our focus will be on populations of sexually reproducing, diploid organisms, as defined in the following way:

A population is a group of individuals of the same species for which there are no restrictions to random mating among its members.

The sum total of all the genes of a given population has traditionally been called a gene pool. For sexually reproducing, diploid individuals, it is more instructive to think of it as a pool of alleles. (Such individuals have two sets of chromosomes, hence two alleles for each gene locus.) When you take the whole population into account, you find that for any one locus, some alleles occur more often than others. Thus it is possible to think

of variation in terms of **allele frequencies**: the relative abundance of different alleles carried by the individuals in that population.

Early in this century, the mathematician G. Hardy and the physician W. Weinberg independently discovered a principle that can be used as a baseline against which changes in allele frequencies can be measured. They observed the following:

In the absence of disturbing factors, the frequencies of different genotypes in a population will reach an equilibrium and will remain stable from generation to generation.

This observation is based on a mathematical formula called the **Hardy-Weinberg principle**, which is given in Figure 36.2. The principle applies to an idealized population of sexually reproducing organisms that is not evolving. The deduction is that evolution cannot occur under the following conditions:

1. No mutation
2. Infinitely large population
3. Isolation from any other populations of the same species
4. Equal viability, fertility, and mating ability of all genotypes (no selection)
5. Random mating

Figure 36.2 gives an example of the genetic results when all of these conditions are met. After one generation of mating, the relative frequencies of the three possible genotypes (*AA*, *Aa*, and *aa*) reach an equilibrium described by the formula $p^2 + 2pq + q^2$. Those frequencies, and the frequencies p and q of the two alleles (*A* and *a*), stay the same over succeeding generations. Such stability of allelic and genotypic ratios is called **genetic equilibrium**.

As it happens, a natural population never meets all of these conditions, hence never is able to reach genetic equilibrium. Why, then, do we mention the Hardy-Weinberg rule at all? The reason is that genetic equilibrium is useful as a reference point, signifying zero evolution. *Hence the degree to which deviations are measured from this reference point can serve as a measure of the rates of evolutionary change.*

Factors Bringing About Change

When a population is evolving, its allele frequencies are changing through successive generations as a result of one or more of the following factors:

1. **Mutation**. A heritable change in the kind, structure, sequence, or number of the component parts of DNA.

2. **Genetic drift**. A random fluctuation in allele frequencies over time, due to chance occurrences alone. Genetic drift has its greatest effect when population size is small.

3. **Gene flow**. A change in allele frequencies due to immigration (new individuals enter the population), emigration (some individuals leave), or both.

4. **Natural selection**. Differential survival and reproduction of genotypes within a population.

Each of these factors can be important in some circumstances. Overall, genetic drift and natural selection seem to be the most important mechanisms bringing about evolutionary change, with gene flow not far behind in many groups of organisms. Mutations are important in changing the character of a population through long spans of time, but they generally have little effect on a generation-to-generation basis.

As you probably noticed, genetic recombination is not listed as a factor that causes evolution. Recombination does contribute to phenotypic variation, and other forces (such as selective agents) can act on that variation. However, recombination in itself only reshuffles the combinations of existing alleles; it does not change the *frequencies* of those alleles.

Let's now take a look at the effects of mutation, genetic drift, and gene flow before exploring in detail the effects of natural selection.

MUTATION

Mutation is the original source of genetic variation, and mutations have been accumulating for billions of years. In most populations, mutations are random in terms of when they will happen, which locus will be affected and, most importantly, whether they will be harmful or beneficial to the individual at the time.

The effects of a mutation depend on how it changes the structure, function, or behavior of the individual in the context of prevailing conditions. For example, a certain gene codes for a protein necessary in cartilage formation. Cartilage is a key component of many pathways in normal animal development. A mutation at this gene locus may lead to such deformities as blocked nostrils, narrowed tracheal passageways, thickened ribs, and the loss of elasticity in lung tissue. Here the new allele produced by mutation is expressed in the context of an intricate developmental program, and it may have lethally disruptive effects.

As another example, suppose an ectothermic animal (which has no physiological controls to maintain body temperature) bears a mutant gene coding for an enzyme that functions at a higher temperature, compared to its normal counterpart. If the animal moves into an environment where temperatures are higher than those typically encountered by its species, enzyme function would not be diminished. Here the mutation might be advantageous in the new environment but not in the old one.

The new alleles that appear by mutation in a given population are neither harmful nor beneficial in themselves; their effects depend on the environment in which their gene products are expressed.

Even though mutations are neither good nor bad in themselves, most mutations with large effects are harmful, even lethal. The reason is that a new allele represents a departure from alleles that have withstood the test of time. Overall, the structural, functional, and behavioral traits of an organism already allow it to function well in a particular environmental context. Any drastic change is more likely to derange things than to enhance them.

Of course, just because a new allele is not beneficial does not mean that it cannot be passed on from one generation to the next. First, its effects might be masked by its allelic partner. Second, the mutation may make no difference whatsoever in a given environmental context, in which case there would be no selection for *or* against the mutation. Third, the new allele might be closely linked in the chromosome to highly adaptive genes. Thus it could hitchhike with the adaptive genes through meiosis, linkage, sexual fusion, and environmental tests.

FACTORS RELATED TO POPULATION SIZE

Earlier, we defined a population as a group of individuals of the same species, for which there are no restrictions on random mating among its members. Thus, in theory, any male in a population could select any female as a mate (or vice versa), while encounters between males and females of different populations are less likely. (For example, some populations of the same species are separated by areas of unsuitable habitat and they never do make contact.)

The size of a population (its number of individuals) and the area it occupies vary from one species to the next, and both factors influence the mechanisms and rates of evolutionary change. As an example of a truly spread-out population, mallard ducks move around so much from the time of birth to the breeding season, and even between breeding seasons, that all of the individuals in North America could be considered a single population! Toward the other extreme is the fringe-toed sand lizard, with small populations confined to their own isolated sand dunes in California deserts.

Genetic Drift

Population size influences the rate of genetic drift (the chance increase or decrease in the relative abundance of different alleles). For example, suppose that over several generations in a small population, none of the bearers of an allele designated A reproduces simply by chance. Whether you call it chance or bad luck, they fail to mate, or fall ill, or accidentally die early. Assuming that genetic drift is the only evolutionary factor at work at the time, the A allele would disappear from the population (see Figure 36.3). Such a run of bad luck is much less likely in a large population, where hundreds of individuals might be carrying the A allele; thus there would be a greater chance that at least some of them would reproduce and thereby help maintain the A allele in the population.

By analogy, there are two outcomes when you toss a coin, each with equal probability of occurring. When the coin is tossed ten times, it could by chance end up tails eight times, or two times, or one, rather than the expected 1:1 ratio (half heads, half tails). However, the laws of probability tell you that when the coin is tossed, say, a thousand times, the expected ratio will be more closely approached.

In general, then, genetic drift is the random change in allele frequencies through successive generations of a population, regardless of its size. It can be more rapid in a small population, but even small shifts in each generation of a large population can add up to significant change over time.

One more point can be made here. Some alleles in a population may not be the "best" possible alternatives in contributing to survival and reproduction. Through genetic drift, they may have become established just because other alternatives were lost by chance. Suppose that in a small population of green parakeets and blue parakeets, the allele governing the dominant feather color (green) is lost as a result of genetic drift, and all parakeets of the next generation are blue. This change in feather color has arisen not because the blue color is more advantageous, but because chance events led to a change in genotypes (hence phenotypes).

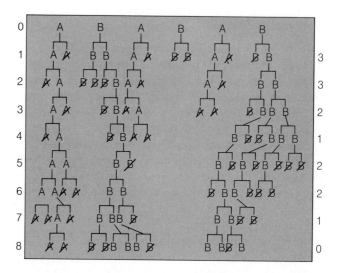

Figure 36.3 Illustration of genetic drift. Each individual who carries an allele designated A produces two identical offspring in each generation, and so does each individual who carries an allele designated B. Half the offspring die before reproductive age (population size remains constant) but which ones die is random. The relative abundance of the two types of individuals fluctuates until A no longer is represented in the population and B becomes fixed in the population

Numbers to the left of the diagram signify the sequential generations; numbers to the right signify the number of individuals bearing the allele A who survived in each generation. (Which individuals were to die in this example was determined by tossing a coin.)

Founder Effect

Sometimes a small number of dispersed individuals manage to establish a new population. Simply by chance, the allele frequencies at many gene loci are likely to be different in these individuals from what they were in the original population, and the new assortment will dictate the genotypic character of the new population. This extreme case of genetic drift is called the **founder effect**.

The founder effect is important in the colonization of oceanic islands and other isolated locations (such as land-locked, glacier-fed lakes that are "seeded" with a few trout for fishermen). It is also important when a few individuals of a species are accidentally or intentionally introduced into new environments. The American starling is a descendant of a small flock from Europe that was released in New York City in the 1800s. As a result of the limited allelic variation in that small flock of birds, the American and European populations are genetically different—even though there now may be more starlings in America than in all of Europe.

Bottlenecks

Genetic drift can also occur when populations go through **bottlenecks**, whereby a normally large population is drastically reduced in size because of unfavorable conditions. Even though the population may eventually recover, genetic drift during the bottleneck can alter the relative abundances of alleles.

For example, just before the turn of the century, hunters destroyed the large population of northern elephant seals. Only about twenty seals survived. Since that time, the population has increased to more than 30,000. Interestingly, there is no allelic variation whatsoever at twenty-four gene loci that have been studied. The lack of variation is unique, compared to other seal species and populations that have not gone through comparable bottlenecks. It suggests that a number of alleles were lost during the bottleneck.

It may be that cheetahs underwent a similarly severe bottleneck. These cats are so genetically uniform that they will accept skin grafts from unrelated individuals, something that is rarely possible to accomplish even among littermates in other species of mammals. Such extreme genetic uniformity is of concern to conservationists. Without genetic variation the whole population, no matter how large, is extremely susceptible to diseases or environmental changes. Species such as the cheetah and the northern elephant seal could suddenly become extinct.

Founder effects and bottlenecks occur when a population originates or is rebuilt from very few individuals. In both cases, the amount of genetic variation in the population may be severely limited.

GENE FLOW

Probably very few populations are completely isolated from other populations of the same species. Individuals migrate between populations, seeds and pollen grains drift or are transported from one place to another, and ocean currents carry larvae of marine organisms for many hundreds of kilometers.

As a specific example, baboons in Africa generally live in troops. Each troop represents a separate allele pool. Commonly, some of the males wander off or are driven from one troop. They may join another troop some distance away. Assuming that migrant males encounter receptive females in the new population, and assuming they have offspring, the pool of alleles changes, just as it changes in the troop the baboons left behind.

For many plant and animal species, most of the offspring, spores, or gametes end up close to the parent. For example, in 1951, Robert Colwell released and then tracked radioactively labeled pollen grains from individual pine trees. Most of the pollen was dispersed downwind only about three to six meters from the tree that produced it. The amount of pollen then dropped off sharply at a greater distance from the source.

When such a dispersal pattern continues over generations, populations with somewhat distinct genotypes can become established in the same regions. At the same time, however, winds, insects, birds, and other dispersal agents may carry some offspring, spores, or gametes beyond the typical dispersal range for the neighboring populations. A small but probably significant amount of gene flow thereby occurs, the result being a homogenizing effect on the genotypic character of neighboring populations of the same species. Thus gene flow is thought to be a main factor in decreasing the variation between populations—variation that might arise through mutation, genetic drift, and selection.

Gene flow is the physical flow of alleles into and out of populations. It tends to decrease the genetic variation between populations that arises through other evolutionary factors.

NATURAL SELECTION

The simple fact that individuals vary in genotype and phenotype means that they are likely to be equipped in different ways to deal with environmental challenges. Some will be more suitable, in certain contexts, for surviving and reproducing. If their phenotypic differences are genetically based, then alleles associated with the advantageous phenotypes will tend to increase in frequency in the population.

The formal statement of this process is called the principle of natural selection. It was formulated in the nineteenth century by Charles Darwin and Alfred Wallace, as you read in Chapter Two. Here we will take a closer look at this principle.

Darwin, recall, was fascinated with the variation that exists in populations. Much of the variation had to be heritable, for it could be passed on from one generation to the next. At the same time, Darwin knew from his studies of artificial selection that the variation was not necessarily static. For example, selective breeding could lead to diverse breeds of pigeons, as shown in Figure 2.9. In the case of pigeons, at least, humans obviously were doing the selecting—but what could be the basis of selection in *natural* populations?

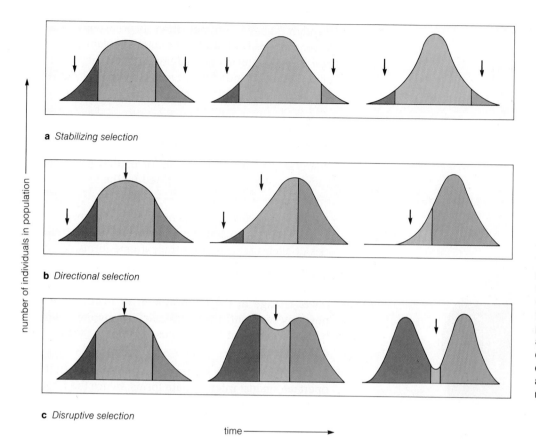

a *Stabilizing selection*

b *Directional selection*

c *Disruptive selection*

number of individuals in population

time ⟶

Figure 36.4 Three modes of natural selection. The blue-shaded and brown-shaded portions of the range of variation encompass individuals that are extreme phenotypes: the tan-shaded region is the range of the most common forms. (**a**) In stabilizing selection, conditions favor the most common forms. The downward-pointing arrows signify that variants are being selected against. (**b**) In directional selection, the phenotypic character of the population shifts as a whole in a consistent direction. (**c**) In disruptive selection, two or more extreme variants are favored and become increasingly represented in the population.

In arriving at the answer, Darwin correlated his observations of inheritance with certain observable characteristics of populations and the environment. First, he knew all populations have enormous reproductive potential. If all the individuals that are born were themselves to reproduce, then population size would burgeon. (The reason is that, with each increase in the number of reproducing members, the potential reproductive base enlarges.) Yet Darwin also knew populations generally do not behave in this fashion. Finally, he knew food supplies and other resources do not increase explosively; in most environments, resources more or less remain within certain limits. Therefore, when the population outstrips the resources necessary to sustain it, there must be *competition* for the resources that are available.

From these and other observations, Darwin put together a concept of natural selection. As noted earlier, observations and experimental work that accumulated since Darwin's time represent such an overwhelming body of evidence in favor of his concept that it is now recognized as one of the most important of all principles.

Because of its importance, the principle of natural selection is restated here:

1. In any population, more offspring are usually produced than can survive to reproductive age.

2. Members of the population vary in form and behavior, and much of this variation is heritable.

3. Some heritable traits are more **adaptive** than others—that is, they improve an individual's chances of surviving and reproducing under prevailing environmental conditions.

4. Because the bearers of adaptive traits have a greater chance of reproducing, their offspring tend to make up an increasingly greater proportion of the reproductive base for each new generation. This tendency is called **differential reproduction**.

5. *Natural selection is the result of the differential reproduction.* Adaptive traits increase in frequency in a population because their bearers contribute proportionally more offspring to succeeding generations.

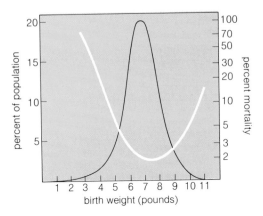

Figure 36.5 Birth weight distribution for 13,730 infants (black curve). The white line is the curve of early mortality in relation to birth weight. The optimum birth weight is between 7½ and 8 pounds.

a

b

Figure 36.6 Two hundred and fifty million years of stabilizing selection? (**a**) Body imprints of a horseshoe crab made that long ago—imprints that could well be made by a modern-day horseshoe crab (**b**), here shown mating and perpetuating the general species form.

It is important to understand that evolution by natural selection is said to be a principle because whenever the conditions necessary for its occurrence (points 1 through 3 above) are met, then its occurrence is inevitable. Hundreds of studies of complex plants, animals, and microorganisms have demonstrated the validity of all the points listed above. Here we will consider just a few examples of natural selection in specific populations.

Modes of Natural Selection

There are three major modes of natural selection, as shown in Figure 36.4 and defined by the following list:

1. *Stabilizing selection* favors intermediate forms of a trait and operates against extreme forms; hence the frequencies of alleles representing the extreme forms decrease.

2. *Directional selection* shifts the phenotypic character of the population as a whole, either in response to a directional change in the environment or in response to a new environment; hence the allelic frequencies underlying the range of phenotypes move in a steady, consistent direction.

3. *Disruptive selection* favors extreme forms of a trait and operates against intermediate forms; hence the frequencies of alleles representing the extreme forms increase.

Stabilizing selection tends to counter the effects of mutation, genetic drift, and gene flow by favoring the most common phenotype. In contrast, directional selection tends to favor one phenotype at either extreme of the range of variation. Disruptive selection tends to foster an increase in the frequency of phenotypes at both extremes. The outcome of disruptive selection is **polymorphism** (two or more distinct phenotypes).

Stabilizing Selection

Human Birth Weight. On the average, human newborns weigh about 7 pounds (3.2 kilograms). Significantly higher or lower birth weight greatly reduces the newborn's chances of survival. The shape of the survival curve in Figure 36.5 suggests that stabilizing selection favors individuals with a birth weight between 7½ and 8 pounds and works against individuals at either extreme. Studies in widely different populations have yielded remarkably similar correlations between birth weight and mortality.

Horseshoe Crabs. The distinctive horseshoe crab (Figure 36.6) of nearshore North Atlantic waters is not a crab at all. Rather it is representative of a very ancient arthropod lineage. Apparently these animals are very well adapted to scavenging on sandy and muddy bottoms in shallow waters, since the fossil record reveals little morphological change among some members of this group for more than 250 million years. This could be cited as an extreme example of stabilizing selection, in which a single body plan has been conserved in a series of species through time.

Directional Selection

The Peppered Moth. Directional selection has been documented in populations of the peppered moth (*Biston betularia*), which is widely distributed in England. Before the mid-1800s, a speckled light-gray form of this moth was prevalent (Figure 36.7). A dark-gray form also existed but was extremely rare, making up less than 1 percent of the population. Between 1848 and 1898, however, the dark form increased in frequency. Near one industrial city, it came to represent about ninety-eight percent of the population.

Today, two gene loci that influence the wing color and body color of these moths have been identified. Thus the trait has a heritable basis and is subject to selection.

In the 1930s, the geneticist E. B. Ford suggested that natural selection was causing the increased frequency of the dark form in industrial areas. He knew that before the industrial revolution, light-gray speckled lichens grew profusely on tree trunks. He also knew that the peppered moth is active at night but rests during the day on tree trunks, where it is vulnerable to bird predators. Light-gray speckled moths resting on the lichens would be well camouflaged from birds, whereas dark moths would be highly conspicuous (Figure 36.7a).

Ford reasoned that with the spread of factories, soot and other pollutants from smokestacks began to kill the lichens and darken the tree trunks. In this new environmental context, the rare dark form blended in with the blackened trees and the light-gray form was no longer camouflaged. Hence the dark form began to survive and reproduce at a greater rate than its light-colored kin, thereby altering allele frequencies.

In the 1950s, H. Kettlewell used the **mark-release-recapture** method to test Ford's hypothesis. He bred both forms of the moth in captivity, then marked hundreds of them so that they could be identified. He released the moths in two areas, one near the heavily industrialized area around Birmingham and the other in the unpolluted area of Dorset. After a time he recaptured as many moths as he could. Table 36.1 shows the results. He recaptured more dark moths in the polluted area—and more light moths in the pollution-free area. By stationing watchers in blinds near groups of moths tethered

a

b

Figure 36.7 An example of variation that is subject to directional selection in changing environments. (**a**) The light- and dark-colored forms of the peppered moth are resting on a lichen-covered tree trunk. (**b**) This is how they appear on a soot-covered tree trunk, which was darkened by industrial air pollution.

Table 36.1	Marked *Biston betularia* Moths Recaptured in a Polluted Area and a Nonpolluted Area			
	Light-Gray Moths		Dark-Gray Moths	
Area	Number Released	Number Recaptured	Number Released	Number Recaptured
Near Birmingham (pollution high)	64	16 (25%)	154	82 (53%)
Near Dorset (pollution low)	393	54 (13.7%)	406	19 (4.7%)

Data after H. B. Kettlewell.

male phenotype
Papilio dardanus

Danaus niavius dominicanus

female morph
(*P. dardanus*)

Amauris jacksoni

female morph
(*P. dardanus*)

Bematistes poggeii

female morph
(*P. dardanus*)

Inedible Models   **Mimics**

to tree trunks, Kettlewell also found by direct observations that birds captured more light moths around Birmingham and more dark moths around Dorset.

About a hundred different species of moths underwent the same kind of directional selection in response to pollution in British industrial regions. However, as a result of strict pollution controls (which went into effect in 1952), lichens have become reestablished and tree trunks are largely free of soot. As predicted, the frequency of the dark forms is now declining in the moth populations.

Pesticide Resistance. Insect populations that develop resistance to insecticides provide other examples of directional selection. The first application of an insecticide kills most of the insects. Some individuals may survive, however, because of physiological, behavioral, or morphological differences that enable them to resist the chemical effects of the insecticide. If the resistance has a genetic basis, it can be passed on. As a result of differential survival and reproduction, the next generation will contain more of the resistant individuals.

Then, as the number of resistant forms of the insects increases, heavier and more frequent applications of the insecticide act as a selective agent that favors them even more. The genotypic structure of the insect population shifts rapidly in a consistent direction—toward more resistant phenotypes. Crop damage from insects is now greater than it was before the widespread use of insecticides.

Disruptive Selection

Disruptive selection has been noted in populations of the African swallowtail butterfly (*Papilio dardanus*). Here, selection is associated with **mimicry**, in which one species is deceptively similar in color, form, or behavior to another species that has a selective advantage. The first species is the mimic, the second is the model. For example, if a model has warning coloration that identifies it as inedible to potential predators, then a tasty species that mimics its coloration (as does *P. dardanus*) will have a better chance of being left alone.

The males in all populations of *P. dardanus* have yellow and black wings with "tails" at the tips. In most of tropical Africa, the females of *P. dardanus* are conspicuously different in appearance from the males—and from

Figure 36.8 Effect of disruptive selection on females of African swallowtail butterfly populations. Disruptive selection is apparently favoring several different female forms.

a

b

c

Figure 36.9 Examples of sexual dimorphism in (**a**) the northern fur seal, (**b**) the sage grouse of North America, and (**c**) the lion of Africa.

one another. The wing patterns and coloration of each form of the female mimic those of an inedible species present in the same region (Figure 36.8). In regions where models are absent, there are no mimicking females. Bird predators are apparently the selective agents promoting variation in these populations.

Unlike the females, males of *P. dardanus* are not mimics. Being recognized by females probably has a greater advantage for the males than does the avoidance of predation by mimicry. The fact that all the female variants continue to recognize and accept the same males apparently counterbalances the effect of disruptive selection and keeps the populations from diverging into separate species. Here, sexual selection, which will now be described, may be stronger than other mechanisms of natural selection in guiding the evolution of the male phenotype.

Sexual Selection

Some of the most dramatic examples of selection are found among animals that show **sexual dimorphism**—that is, differences in appearance and behavior between males and females. Sexual dimorphism is especially striking among birds and mammals (Figure 36.9).

Generally, the males are larger, more conspicuous in color and patterning, and more aggressive than females. Of course, such traits make the males of prey species more visible to predators, and the males of predatory species more visible to their prey. So in some respects, the traits are probably selected against. Nevertheless, the traits are used by females to select mates and thus they are associated directly with reproductive success. Such **sexual selection** may be based on any trait that gives the individual a preferential advantage in mating and in producing offspring (Chapter Forty-Eight).

For example, northern sea lions mate only on small islets and rocky beaches. Males that are large enough to command the rocks enjoy mating privileges with about ten to twenty females. The males that cannot secure a territory do not mate at all, hence they contribute nothing to the allele pool of the next generation. As an evolutionary outcome of sexual selection, males weigh as much as 1,000 kilograms, about twice the weight of the females (Figure 36.9).

As another example, the males of several species of grouse are more strikingly colored than the females, and they engage in behavioral displays. The males congregate in small mating territories, called leks, where they call and display. The drab females, attracted by the fracas, tend to mate with the most dazzling and aggressive males.

Selection and Balanced Polymorphism

Sometimes selection favors a particular allele so strongly that it supplants all other alleles at the gene locus. Often, however, **balanced polymorphism** occurs. Here, two or more alleles of a single gene locus persist at a frequency too high to be maintained by mutation alone; and that frequency, if changed, will return to its former value over several generations.

Sickle-cell polymorphism in humans is one of the best-studied examples of this condition. In West and Central Africa, the HbS allele associated with sickle-cell anemia is maintained at a high frequency relative to the normal HbA allele. Within a given population, HbS/HbS homozygotes make up about 2.5 percent and heterozygotes (HbS/HbA) nearly 30 percent of all genotypes at birth.

HbS/HbS homozygotes often die in their early teens or early twenties, generally before age forty-five. It is the abnormal hemoglobin arising from the mutant allele that puts them at a disadvantage. Whenever the blood oxygen level falls below normal (as it may during overexertion or respiratory illness), HbS molecules crystallize and distort red blood cells into sickle shapes. The sickled cells cannot transport normal amounts of oxygen and tend to clump in the capillaries. As we have seen, the consequences are severe (page 173).

Why doesn't selection remove the HbS allele from the populations? The reason is this: *The survival value of alleles must be weighed in the context of the environment in which they are being expressed.* Wherever the sickle-cell polymorphism persists, malaria is also rampant. About half of all cases of malaria are caused by *Plasmodium falciparum*, a parasite that uses only one kind of mosquito to transmit its sporozoites to the human bloodstream (Chapter Thirty-Nine). Since these mosquitoes live only in the tropics and subtropics, malaria and the sickle-cell polymorphism occur mostly in those parts of the world.

Many studies suggest that the sickle-cell polymorphism is maintained through (1) *differential mortality* and (2) *differential fertility* between the normal members and the heterozygous members of the populations. (The frequency of the HbS/HbS homozygote among adults is so low that its effect can be ignored here.) Let's look first at the survival advantage of the HbS/HbA heterozygote.

In the absence of malarial infection, the HbA/HbA homozygote has an eighty-five percent greater probability of surviving to reproductive age than does the heterozygote. The presence of malaria alters this probability. In 1954, A. Allison reported that heterozygotes have greater resistance to malaria and are more likely to survive severe infections. After infecting fifteen HbA/HbA and fifteen HbS/HbA volunteers with *P. falciparum*, Allison later found malarial parasites in the blood of fourteen of the normal homozygotes and in only two of the heterozygotes. In another study, severe or fatal infections were found to be twice as high in HbA/HbA children as in HbS/HbA individuals.

Thus, the persistence of an extremely deleterious trait (the HbS allele) becomes a matter of relative evils. On an absolute scale, the survival and reproductive rate of HbS/HbA individuals is poor, but the rate is better than it is for HbA homozygotes in regions where malaria is prevalent. As part of the "cost" of maintaining the more adaptive heterozygote, half the infants born will be either HbA/HbA or HbS/HbS homozygotes—neither of which has much chance of surviving to reproductive age.

EVOLUTION OF SPECIES

All the processes described so far operate at the level of populations. What is commonly overlooked is their potential *reversibility* at this level. Alleles lost through genetic drift or through selection may be introduced again by mutation or gene flow, and changes in allele frequency mediated by selection may be reversed if the environment changes in an appropriate direction. Yet if such reversibility is possible, *then how did the diversity we observe in nature arise in the first place?* What prevents populations from simply cycling through the same events again and again? The answer is that an *irreversible step*, acting like a ratchet, causes genetically isolated populations to branch in different evolutionary directions. This step, "evolution's ratchet," is part of the process by which species originate—a process called **speciation**. We will now consider some characteristics of this process, for which we will need the following definition as our point of reference:

For sexually reproducing organisms, a species is one or more populations whose members actually (or potentially) interbreed under natural conditions, produce fertile offspring, and are reproductively isolated from other populations.

Divergence

Within a given population, shifts in allele frequencies either may move the whole population in one direction or another or may produce a balanced polymorphism.

No matter how diverse the individuals become, however, they will remain members of the same species as long as they continue to interbreed successfully and share a common pool of alleles.

Sometimes, however, barriers arise between parts of the population, creating local breeding units. As a result, two or more pools of alleles may arise where there had been only one before. Over time, the absence of gene flow between populations and the action of selection, genetic drift, and mutation can lead to divergence. The term **divergence** refers to a buildup of differences in allele frequencies between reproductively isolated populations. When divergence becomes so great that successful interbreeding is no longer possible under natural conditions, the populations are said to constitute separate species.

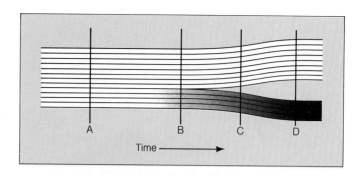

Figure 36.10 Divergence leading to speciation. Because evolution is gradual here, we cannot say at any one point in time that there are now two species rather than one. Each vertical line represents a different population. In A, there is only one species. In D, there are two. In B and C, the divergence has begun but is far from complete.

When Does Speciation Occur?

Except in some unusual cases, which will be described shortly, it is impossible to identify the exact moment at which speciation occurs. As in any gradual process (such as the development of an infant into an adult), species usually do not spring forth at a single moment in time. Figure 36.10 shows the gradual nature of speciation.

Determining whether speciation has occurred is not always a simple matter. Divergence between reproductively isolated populations must have gone so far that interbreeding no longer occurs, even when the opportunity arises. But if the populations are geographically isolated, how can we know for sure that divergence has proceeded to that point? For example, lions and tigers are reproductively isolated in the wild. They certainly look different, and we call them separate species. Yet when they are brought together in captivity, they sometimes interbreed and produce hybrid offspring. Are they, in fact, separate species?

Moreover, if reproductive isolation is one of the criteria used to define a species, how can fossils be assigned to one species or another? We have no idea of whether a now-fossilized individual was once able to interbreed with individuals of similar morphology living in the same region. (Thus, for example, we have different interpretations of what was really going on among the earliest members of the human family, some of which lived in the same regions and were morphologically varied.)

Reproductive Isolating Mechanisms

A **reproductive isolating mechanism** is any aspect of structure, function, or behavior that prevents interbreeding. These mechanisms fall into two categories. First, some take effect before or during fertilization, thereby preventing the formation of hybrid zygotes. Second, some take effect during development of the embryo, resulting in early death, sterility, or even the failure of the hybrid animal to be recognized as an acceptable mate by members of either parental species. All reproductive isolating mechanisms have the same effect: *they prevent the exchange of alleles between populations.*

A reproductive isolating mechanism is any aspect of structure, function, or behavior that prevents successful interbreeding (hence gene flow) between populations.

Mechanical Isolation. Differences in the structure or function of the reproductive organs may prevent individuals of different populations from producing hybrid zygotes. For example, the stigma of one plant species may be so different in size, shape, or length that the pollen tube of another plant species cannot reach its ovary. As another example, two species of sage plants in Southern California differ in the size and arrangement of their floral parts, and their pollinators are so different in size that the pollinator of one species cannot come into contact with the pollen of the other (Figure 36.11).

Isolation of Gametes. Incompatibities between the sperm of one species and the egg (or the female reproductive system) of another may prevent fertilization. This is the case with certain marine animals that rely on external fertilization. For example, even when the eggs and sperm of two species of sea urchin are released at the same time and in the same place, the gametes are rarely attracted to each other. Probably the female does not release the biochemical substances required to attract

 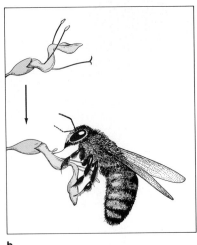

a   b

Figure 36.11 Mechanical isolation between two species of sage (*Salvia mellifera* and *S. apiana*). The first species (**a**) has a small floral landing platform for its small or medium-size pollinators. The second species (**b**) has a large landing platform and long stamens, which extend some distance away from the nectary. Even though small bees can land on this larger platform, they can do so without brushing against the pollen-bearing stamens. It takes larger pollinators to do this. Hence the small pollinators of *S. mellifera* are mostly incapable of spreading pollen to flowers of *S. apiana*; and the large pollinator of *S. apiana* cannot land on and cross-pollinate *S. mellifera*. The plants and their pollinators are all drawn to the same scale.

the sperm to her. Similarly, incompatibilities that exist between the stigma of one plant species and the pollen grains of other species are enough to prevent fertilization.

Isolation in Time. The timing of reproduction may also serve as an isolating mechanism. For most animals and plants, mating or pollination is a seasonal event of relatively short duration, sometimes less than a day. Even closely related species may be reproductively isolated simply because their times of reproduction do not coincide. Two closely related species of cicadas provide an extreme example. In any one location, one of these insects emerges and mates every thirteen years, the other every seventeen years. The possibility of their meeting arises only once every 221 years!

Behavioral Isolation. Behavioral isolation is one of the strongest mechanisms at work among related species in the same territory. For example, complex courtship rituals often must precede mating. The song, head-bobbing, wing-spreading, and dancing by a male bird of one species may stimulate a female of the same species—but the female of a related species probably would

not even recognize such behavior as a sexual overture. Table 36.2 reveals the effectiveness of behavioral isolation in some closely related species of *Drosophila*.

Hybrid Inviability and Infertility. Even when fertilization occurs between the gametes of two species, there may be incompatibilities between the developing embryo and the mother. If the incompatibilities are severe, the embryo will die. Hybrids that do live at first are commonly weak in structure, physiology, or behavior, and their chance of surviving is not good. In a few cases, the hybrid offspring are vigorous but sterile. Thus a cross between a female horse and a male donkey produces a mule, a hybrid that is fully functional *except* in its reproductive capacity.

Finally, even if a first-generation hybrid manages to survive, it may be unable to reproduce. The unusual appearance or behavior of a hybrid animal may preclude its recognition by either species when the time comes for it to mate. Even when such hybrids manage to reproduce, their offspring are typically not vigorous. Crosses between two species of evening primrose produce a partially fertile first generation, but the second-generation plants are slow-growing sterile dwarfs, susceptible to disease.

Modes of Speciation

Allopatric Speciation. Typically, the populations of a given species are not strung out continuously in space, with one merging into the others and thus open to gene flow. Most often the populations are geographically isolated from one another to varying degrees, with gene flow being more of an intermittent trickle than a continuous stream.

On occasion, even the trickles can be shut off completely through the formation of impassable geographic barriers. Sometimes the barriers can form quickly, as when a flood changes the course of a river and isolates populations of, say, an insect species that cannot swim or fly. Sometimes the barriers form slowly, as when the climate changes over time and causes the breakup of formerly continuous environments. (Thus an extensive forest might give way to grasslands with isolated stands of trees as a result of long-term shifts in the pattern of rainfall.) On a more immense time scale, oceans or mountain ranges can form between populations as a result of exceedingly slow but massive realignments of the continents themselves (Chapter Thirty-Eight).

Once geographic separation is absolute, genetic drift or adaptation to different environments may then bring about divergence. If divergence proceeds far enough, the descendants of the separated populations may become so different that they will not interbreed even if they expand their distribution and make contact at some later time. When new species form in this manner, as a result of geographic isolation, the process is called **allopatric speciation** (from *allos*, meaning "different," and *patria*, "native land").

Allopatric speciation is generally considered to be the major route to diversity. It has been well documented in virtually all groups of sexually reproducing organisms in all types of habitats. Until rather recently some biologists believed it to be the *only* route, but with our increased understanding of the genetics of speciation, that view has become outdated.

Parapatric Speciation. It now appears that speciation also can occur without the formation of absolute geographic barriers. **Parapatric speciation** occurs in populations that live adjacent to one another (*para* means "side by side").

For example, suppose the populations of a species extend across different major environments, as from high mountains into an adjacent desert. Quite probably, selection will favor different phenotypes in the two settings. If the transition between the two environments is abrupt, individuals with intermediate phenotypes may be selected against, for they may not be well adapted to either set of operating conditions. Selection will begin to favor the individuals that choose members of their own population type as mates. In time, reproductive isolating mechanisms will become established and will lead to speciation.

Sympatric Speciation. The word sympatric means "same native land." **Sympatric speciation** is the origin of species as a result of ecological, behavioral, or genetic barriers that arise *within* the boundaries of a single population. Suppose a mutation leading to a shift in food preference or to a shift in the timing of reproduction begins to spread in an insect population. In such cases, individuals with that mutation will be in a position to breed only with one another. Similarly, various types of chromosomal aberrations (such as duplications of some or all of the chromosomes, or changes in chromosome numbers by fusion or fission) can result in instantaneous speciation. The bearers of the altered chromosome complement successfully reproduce only with one another, not with other members of the population. (Offspring that are heterozygous for different numbers of chromosomes die early in development, or are sterile if they do survive.)

Sympatric speciation can occur through *polyploidy*, whereby the original chromosome number is multiplied in a particular zygote. This mechanism has been important in plant evolution; about forty percent of all flowering plants are polyploid. The reason for its widespread occurrence is that plants often are self-fertilizing, and most can reproduce asexually by vegetative propagation. Because the polyploid individual need not wait for a sexual partner with a comparably multiplied chromosome number, speciation is instantaneous. That one individual can give rise to a whole population just like itself.

Polyploidy combined with successful *hybridization* can also produce a new species. Usually, interspecific hybrids are sterile because they inherit chromosomes that differ in number or kind from their different parents; hence the "homologues" probably cannot pair during meiosis. However, if the interspecific hybrid undergoes polyploidy, then the *duplicates* can pair with the originals, meiosis can proceed, and viable gametes can form. Wheat is an example of a successful polyploid hybrid (page 186).

Sympatric speciation is relatively rare in animals. It has been documented in certain flies and grasshoppers, and it may also occur in some mammals.

Table 36.2	Behavioral Isolation Between Three Closely Related *Drosophila* Species				
Contact Limited to the Following Combinations:		Number of Females	Number of Matings	Percent Matings
Females	Males			
D. serrata	*D. serrata*	3,841	3,466	90.2
D. serrata	*D. birchii*	1,246	9	0.7
D. serrata	*D. dominicana*	395	5	1.3
D. birchii	*D. birchii*	2,458	1,891	76.9
D. birchii	*D. serrata*	699	7	1.0
D. birchii	*D. dominicana*	250	1	0.4
D. dominicana	*D. dominicana*	43	40	93.0
D. dominicana	*D. serrata*	163	0	0.0
D. dominicana	*D. birchii*	537	20	3.7

Data from F. Ayala. 1965. "Evolution of Fitness in Experimental Populations of *Drosophila serrata*." *Science* 150:903–905.

SUMMARY

1. Within a population, individuals show phenotypic variations which are expressions of gene mutation, chromosomal aberrations, and genetic recombination.

2. Within a population, the frequencies of different alleles change as a result of mutation, genetic drift, gene flow, and natural selection, including sexual selection.

3. A mutation is a heritable change in the kind, structure, sequence, or number of the component parts of DNA. The new alleles produced by mutation arise randomly with respect to timing and location, and without reference to their desirability for the population at the time.

4. Genetic drift is an increase or decrease in the relative abundance of different alleles through successive generations, simply by chance. Its effects are more rapid in small populations.

5. Gene flow is a change in allele frequencies due to immigration or emigration.

6. Natural selection is the result of differential reproduction of genotypes within a population.

7. *Stabilizing* selection tends to counter the effects of mutation, genetic drift, and gene flow by favoring the most common forms of a trait (which are already well adapted to prevailing conditions).

8. *Directional* selection tends to favor phenotypes at one extreme or the other in the range of variation. The frequency distribution of alleles tends to move in a consistent direction, shifting the phenotypic character of the population as a whole.

9. *Disruptive* selection tends to favor an increase in phenotypes at both extremes in the range of variation. The outcome may be polymorphism (in which two or more forms of a trait are common in the population).

10. For sexually reproducing organisms, a species is one or more populations whose members are able to interbreed under natural conditions and produce fertile offspring, and are reproductively isolated from other populations.

11. Over time, one or more populations (or parts within a single population) of a species may become geographically isolated from others. The absence of gene flow between them and the action of selection, genetic drift, and mutation may lead to divergence.

12. Divergence is a buildup of differences in allele frequencies between reproductively isolated populations. Speciation occurs when divergence is so great that successful interbreeding no longer occurs under natural conditions.

13. Any aspect of structure, functioning, or behavior that reproductively isolates different populations may lead to speciation.

14. *Allopatric* speciation may occur when geographic separation prevents gene flow. It is the most common speciation process.

15. *Parapatric* speciation may occur when adjoining populations undergo divergence despite some gene flow.

16. *Sympatric* speciation occurs when populations diverge within the same geographic range. For example, polyploidy has led to sympatric speciation in flowering plants.

Review Questions

1. What is the typological approach to categorizing species? In what fundamental way does the population concept differ from this approach? *538–539*

2. Define these terms: diploid individual, population, and species. *541, 550*

3. What is genotypic variation? Phenotypic variation? Describe evolution in terms of frequency distributions for a given trait, and in terms of the underlying allele frequencies. *539, 541*

4. What is the Hardy-Weinberg baseline against which changes in allele frequencies may be measured? What is Hardy-Weinberg equilibrium? *540, 541*

5. Changes in allele frequencies may be brought about by mutation, genetic drift, gene flow, and selection pressure. Define these occurrences, then describe the way each one can send allele frequencies out of equilibrium. *542–546*

6. What implications might the effect of genetic drift hold for an earlier concept of "survival of the fittest"? As part of your answer, define polymorphism and explain how different phenotypes can persist indefinitely in the *same* population. *550*

7. Natural selection is no longer in the realm of pure theory. Can you recount the ongoing sagas of the peppered moth and the pesticide-resistant pests to explain why it is now considered an operating principle of biology? *547–548*

8. Define stabilizing, directional, and disruptive forms of selection and give a brief example of each. *546–549*

9. Before labeling a particular genotype as being advantageous or disadvantageous, what must you first consider? *550*

10. Give two examples of reproductive isolating mechanisms, and outline what they accomplish. *551–552*

11. What is the difference between allopatric and sympatric specialization? Which do you suppose occurs most often in plants? *552–553*

Readings

Ayala, F. J., and J. W. Valentine. 1979. *Evolving*. Menlo Park, California: Benjamin/Cummings. Short introduction to evolutionary theory.

Cavalli-Sforza, L., and W. Bodmer. 1971. *The Genetics of Human Populations*. New York: Freeman. Objective look at problems of racial differentiation; clear discussion of sickle-cell polymorphism.

Futuyma, D. 1987. *Evolutionary Biology*. Second edition. Sunderland, Massachusetts: Sinauer. Excellent synthesis of modern evolutionary thought.

Grant, V. 1981. *Plant Speciation*. Second edition. New York: Columbia University Press. Good discussion of speciation in plants.

White, M. J. D. 1978. *Modes of Speciation*. San Francisco: Freeman. Excellent review of speciation mechanisms.

The great triumph of evolutionary theory is that it provides a framework for explaining why living organisms are all so similar in their biochemistry and molecular biology yet so different in form and function. Evolution does not create new organisms out of thin air; rather, it proceeds by modifications of the genetic makeup of existing organisms, and this means there is an underlying continuity of relationship among them. Typically this relationship by way of descent is said to resemble a tree, with the living species forming an immense shell-like canopy of leaves that are related by collections of twigs, branches, and major limbs leading to a common ancestral trunk. If we could look back through time, we would see how the failures and successes of vast numbers of previous leaves have uniquely shaped this tree, pruning and nourishing it into its present form. Yet we can directly observe no more than the present leaves and the ones preserved in the fossil record. The rest of the tree must be reconstructed by indirect methods.

It is important to understand that only the *species* are real entities in the tree of descent. The "twigs, branches, and limbs" are not real; they are *categories of relationship* that we superimpose on the species themselves. These categories, called the **higher taxa** (singular, taxon), are listed in Figure 37.1. Obviously, then, we cannot talk about "the evolution" of genera, families, and so on except in terms of the developments that occurred during the history of all the species contained in those increasingly inclusive categories. As we will see, these developments include the origins, persistence, radiations, and extinctions of different groups of species. What we call **macroevolution** refers to the large-scale patterns, trends, and rates of change among groups of *species* since the beginning of life.

37

PHYLOGENY AND MACROEVOLUTION

A QUESTION OF PHYLOGENY

Classification Schemes

Why are species assigned membership in one genus and not another? Why are different genera assigned membership in one family rather than another, and so on? These are questions you should ask when studying any classification system—that is, a system for grouping taxa.

In general, classification schemes are constructed on the basis of phenotypic similarities, relationships through descent, or both. **Phenotype** refers to the morphological, physiological, and behavioral traits of an individual, all of which may be observed or measured. **Phylogeny** refers to evolutionary relationships, starting with the most ancestral species and including all the branchings leading to all of its descendants.

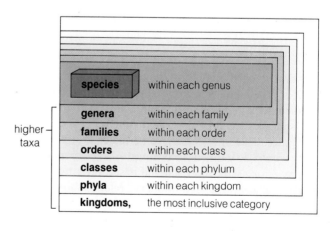

Figure 37.1 The higher taxa—categories used in describing the large-scale patterns, trends, and rates of change among species since the beginning of life.

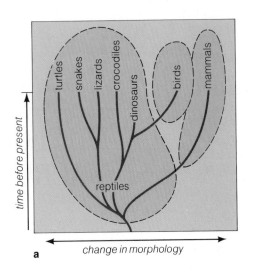

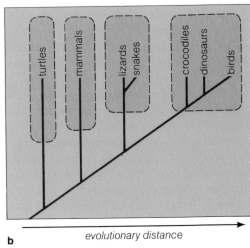

Figure 37.2 How a few major groups of vertebrates are ranked according to two different systems of classification. In evolutionary taxonomy (**a**), the groups are defined by such traditionally accepted characteristics as scales, feathers, or fur. A cladistic classification (**b**) links together the organisms having shared ancestries.

time before present

reptiles

change in morphology

a

evolutionary distance

b

If you were to construct a tree of descent by taking into account both phenotypic and phylogenetic relationships, each twig or branch would be a single line of descent, or **lineage**. The branch points would represent speciation events, and the angle of each branching would indicate the rate of morphological change in a given time period. Horizontal branchings would mean abrupt change and narrow angles would mean gradual change from the ancestral species form. For example,

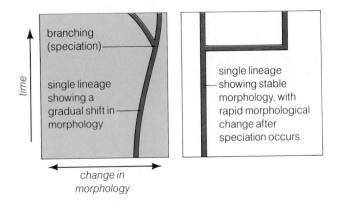

time

branching (speciation)

single lineage showing a gradual shift in morphology

single lineage showing stable morphology, with rapid morphological change after speciation occurs

change in morphology

The time of each branching (how high or low it is in the tree) is now being determined by radioactive dating of rocks in which fossils are found and by biochemical yardsticks that will be described shortly.

Figure 37.2a shows how some major groups of animals would be ranked according to "evolutionary taxonomy," a classification scheme based on a mixture of morphological and evolutionary relationships. Such schemes may be misleading at times, for who is to decide which morphological traits are most important for assigning particular types of animals to one category or another? Here, for example, the dinosaurs are grouped with the reptiles, which are ranked as a class equal to the classes of birds and mammals.

Yet consider how these animals might be ranked in one alternative scheme, called "cladistics." With this approach, relatedness is depicted only in terms of branch points in the lines of descent. (As we will see, branch points are inferred by combining information from the fossil record, comparative morphology, and comparative biochemistry.) What does such a scheme tell us? Notice that there is no class of "reptiles" and that the birds are grouped with their closest relatives, the dinosaurs and crocodiles. With this scheme, we may infer that the dinosaurs may have been much more "birdlike" than "reptilelike" in their lives! Like crocodiles and birds, they may have had a four-chambered heart, territorial behavior and the use of "songs" by males, and complex nesting behavior and parental care of the young. None of these traits characterizes any "reptile" except the crocodiles. The point to keep in mind is this:

The way taxa are categorized affects the ways we think about them and the questions we are led to ask.

Speciation and Morphological Change

Let's now consider a crucial feature of evolutionary trees—specifically, the manner in which they portray the rate of morphological change in an evolving lineage.

Everyday experience tells us there is some order in the diversity among organisms. For example, you know there are several species of cats—lions, tigers, leopards, pumas, bobcats, and so forth—and you could probably identify most of them as cats on sight. You could also identify such members of the dog family as foxes, coyotes, and wolves, and it is unlikely that you would incorrectly identify any member of the dog family as a type of cat. The point is this: *Morphological "gaps" separate even these closely related species, just as increasingly larger "gaps" separate genera, families, and other taxa.*

How does the pattern of clusters and gaps arise—in other words, why are there so few intermediate forms? Let's first address this question at the level of species, then return to it later with reference to higher taxa.

Figure 37.3 shows two ways of interpreting rates of morphological change in an evolving lineage. According to the traditional model, termed **gradualism**, most morphological change occurs *within* species as a result of genetic drift, directional selection, and other processes by which allele frequencies change (Chapter Thirty-Six). For example, continuous layers of deep-sea sediments contain fossils of foraminiferans (a type of shelled protistan) that clearly show gradual changes in form within the same lineage. This fossil sequence, however, is one of only a few well-documented cases of gradualism. Many cases, including the much-repeated example of the evolution of the modern horse, have not held up under close examination.

According to an alternative model, termed **punctuation**, most morphological change takes place rapidly *during* speciation, with relatively little change occurring once a particular species is established. Compared to the average span of existence for a species (often on the order of 2 million to 6 million years), the hundreds or thousands of years that may be required for speciation is a short period. Yet founder effects, bottlenecks, and strong directional selection could accomplish a great deal of genetic change in only a thousand years, and stabilizing selection thereafter could maintain the traits of a well-adapted species within relatively narrow limits.

The punctuational model is consistent with the observation that there are few intermediate forms between closely related species. Moreover, the fossil record provides indirect evidence against gradualism. Although there are cases of gradual change in form, they are greatly outnumbered by cases in which new species were morphologically distinct when they appeared, then changed relatively little thereafter.

RECONSTRUCTING THE PAST

The Fossil Record

Most evidence of extinct species comes from fossilized skeletons, shells, leaves, seeds, and tracks. To be preserved as a fossil, body parts or impressions must be buried before they deteriorate, and the resulting rock layers must remain relatively undisturbed by geologic forces. Some types of organisms and environments are more likely than others to yield fossils. For example, animals with hard, durable shells or skeletons are better represented than are more fragile species, and wholly

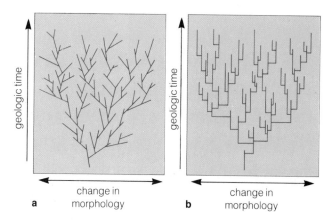

Figure 37.3 Two interpretations of how morphological diversity develops in a lineage. Each vertical line represents a single species. In a *gradual* model (**a**), changes occur more or less steadily but at different rates among species. In a *punctuational* model (**b**), rapid morphological change is associated with speciation, as indicated by the horizontal lines that signify branch points, and the morphologies of established species remain constant through time.

soft-bodied animals such as jellyfishes or worms are preserved only in exceptional environments. The seafloor, flood plains, swamps, and natural traps such as caves and tar pits typically favor fossilization; land surfaces that are undergoing erosion usually do not.

The completeness of the fossil record thus varies as a function of the kinds of organisms involved, the places where they lived, and the geological stability of the region between the time of fossilization and today. Some parts of the record are gone forever, and most of the rest is incomplete. The fossils described so far represent about 250,000 species that lived over the past 600 million years—a very small number when compared to the $4\frac{1}{2}$ million species estimated to be alive today. Still, this is much better than nothing at all. Also, in some cases the fossil record is thought to be nearly complete. For example, in certain shallow marine environments where sediments were deposited continuously, the numbers of shelled invertebrate species known from fossils are closely comparable to the numbers of such species living in the same types of environments today.

The quality of specimens preserved as fossils also varies widely. Because the average fossil is broken, incomplete, and often crushed or deformed (Figure 37.4), painstaking preparation and careful study are needed to gain useful information about the organism it represents. In contrast, a few fossils are spectacular in their completeness and preservation (Figure 37.4), and they tell us a great deal about the structure, function, and even behavior of extinct organisms.

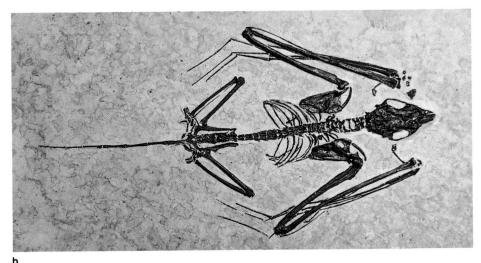

b

Figure 37.4 Examples of vertebrate fossils. (**a**) This photograph shows a good find—parts of the fossilized skeletons of several individuals of an early ducklike bird. Many hours of very careful preparation and study will be required to establish the identity of the organism and to piece together its morphological characteristics. (**b**) A paleontologist's dream, the complete skeleton of a bat that lived 50 million years ago. Fossils of this quality are extremely rare.

a

Comparative Morphology

Most hypotheses about evolutionary relationships and most classification schemes are based on **comparative morphology**, the comparison of body form in major groups of animals. The comparative approach can be powerful if certain limitations are recognized, and misleading if they are not.

For example, comparisons show that the early stages of embryonic development have been highly conserved through vertebrate history. Thus gill slits form in all vertebrate embryos, even though adult fish and many amphibian larvae are the only vertebrates with gills. The striking similarities are just one of the reasons why many animals that otherwise might appear to be quite dissimilar are all classified as vertebrates. It is not that, say, a mammalian embryo "passes through a fishlike stage"; evolution is not repeated all over again every time an embryo develops. Rather, the early stages are fundamental steps in major developmental programs in all the lineages. Most mutations affecting these early stages were probably *selected against* during vertebrate history, for they would have been lethal in terms of further development.

Given that development tends to be a conservative process, what accounts for the morphological variation among vertebrates? Most likely, mutations affected certain genes that control the *rate* of growth of organs and other structures.

For example, the series of bones in the vertebral column develop from the tissues of a series of body seg-

ments (somites) in the embryo. A mutation can increase or decrease the rate at which the segments grow and develop. The change can affect the numbers of vertebrae and segmental muscles, which in turn affects the structure of blood vessels and nerves. This sort of modification may help explain how a snake could evolve from a lizard, or a long-necked swan from a short-necked goose. In human embryos, the developmental period in which brain cells rapidly proliferate is longer than in related primates. Many of the differences in skull and facial structure between humans and other primates are the outcome of a developmental program that produces our larger brain.

Divergence and Convergence in Form

Because morphological features often have clearly functional roles, they must be subject to natural selection. However, a structure often serves several different functions, and the outcome of selection may represent a compromise between the most suitable form for one of its roles and the least disturbance to its other roles. Moreover, evolution can only proceed by the modification of *existing* structures, so the results of selection for a particular role may be quite different from the original function. The power of natural selection to bring about change despite these constraints is evident in examples of morphological divergence.

In **morphological divergence**, selection leads to departures from the ancestral form. Figure 37.5 shows

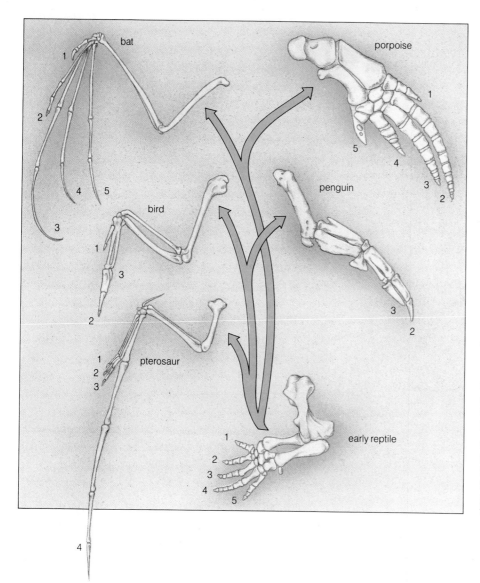

Figure 37.5 Morphological divergence in a structure with a conserved developmental plan—the vertebrate forelimb. From the generalized form of the reptilian limb, structures as diverse as the wings of pterosaurs, birds, and bats, the flippers of penguins and porpoises, and many others have evolved while preserving many similarities in the number and position of bones.

Convergence. The penguin wing has converged on the form of the porpoise flipper, even though the ancestors of penguins had wings used for flight.

examples of divergence in the forelimbs of vertebrates. In each of the three groups of flying vertebrates (pterosaurs, birds, and bats), a different type of wing developed, quite independently, from the five-toed limb characteristic of most terrestrial vertebrates. Even so, all three have the same component parts, as do the flippers of porpoises. We could expand this example to include the long, one-toed limbs of horses, the stout and powerful limbs of moles and other burrowing mammals, and the thick, columnlike limbs of elephants. As another example of morphological divergence, snakes lost all traces of the forelimbs present in their ancestors.

In **convergence**, two or more species from dissimilar and only distantly related lineages adopt a similar way of life and come to resemble one another rather closely. For example, penguins evolved from flying birds but now use their wings in swimming. The structure of their wing bones diverged from the ancestral form and converged on the limb form of porpoises (Figure 37.5).

Convergence often results when the physical requirements of a particular life-style are unusually strict. A classic case of convergence in external morphology occurred among sharks, ichthyosaurs, and porpoises (Figure 37.6). The three lineages are among the most distantly related vertebrates, and the ichthyosaurs and porpoises were derived from fully terrestrial stocks. Yet the three are functionally similar in being adapted for rapid swimming, with a common streamlined shape, similar placement and proportions of stabilizing fins, and a powerful tail.

Comparative Biochemistry

All organisms are fundamentally the same in many aspects of their biochemistry and molecular biology; over ninety percent of the known gene products of human cells have readily identifiable counterparts in yeast cells.

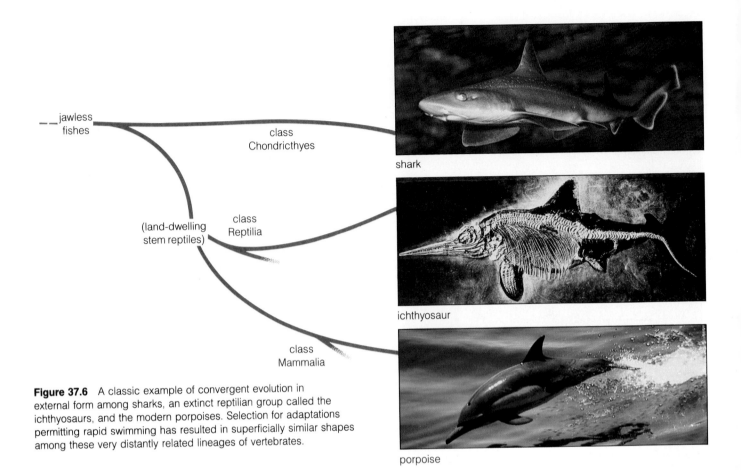

Figure 37.6 A classic example of convergent evolution in external form among sharks, an extinct reptilian group called the ichthyosaurs, and the modern porpoises. Selection for adaptations permitting rapid swimming has resulted in superficially similar shapes among these very distantly related lineages of vertebrates.

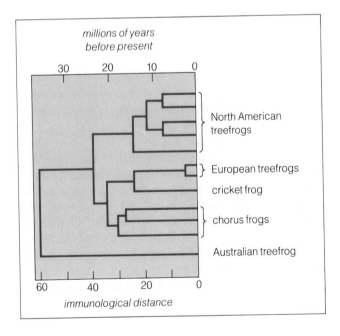

Figure 37.7 Relationships among groups of tree frogs as indicated by immunological data. The immunological distance is simply the number of amino acid substitutions that have accumulated since the two taxa shared a common ancestor.

Yet the genes in the two cell types have been evolutionarily isolated from one another for many hundreds of millions of years.

It is not that the molecular structure of DNA has never been modified in different ways in different lineages. However, we now believe that a significant number of the gene mutations are *neutral* (or nearly so), which means they do not measurably affect the function of the protein being specified. Because neutral mutations have no effect on phenotype, they are not subject to natural selection, and they tend to accumulate through random processes such as genetic drift.

Given the inevitable occurrence of mutations, the probability of neutral mutations becoming permanently incorporated into the DNA of a species increases with time. Through the use of statistical methods, it can be argued that neutral mutations do this at regular rates for major classes of proteins. If that is so, they can be used as a **molecular clock** for dating the divergence of two species from a common ancestor. (Thus, if the sequences of amino acids in a particular protein are quite similar in two species, it can be argued that they shared a recent common ancestor; but if the proteins are quite dissimilar

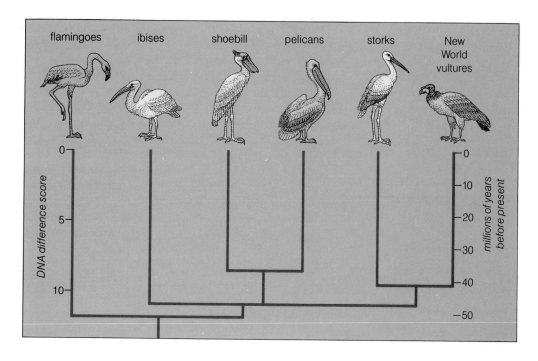

Figure 37.8 Relationships among some New World vultures, storks, and other birds, as indicated by DNA hybridization techniques.

as a result of accumulated neutral mutations, the last common ancestor they shared lived in the very remote past.) When such a clock is calibrated against the branching sequences indicated in the fossil record, it may be possible to determine the actual time of each divergence.

Neutral mutations can be estimated by a variety of methods, two of which are described below. In general, it appears that they represent a reliable molecular clock; they have helped to answer some thorny questions about evolutionary relationships.

1. Molecular approaches to phylogenetic reconstruction rely on the premise that neutral mutations accumulate at a regular rate, and thus that distantly related species will differ at more sites than will close relatives.

2. The methods employed can be used to construct a branching sequence and this can be calibrated against the fossil record to give the ages of divergence points.

The fossil record, comparative morphology, and comparative biochemistry each provide information about phylogenetic relationships. Each has its advantages and limitations, and we would be most confident if the evidence from each approach gave the same result. All three approaches have been used for relatively few such comparisons, but in general, the results have been in agreement.

Immunological Comparisons. One method of estimating the differences in the DNA of vertebrate species is based on the antibody-mediated immune response (page 422). As we have seen, the immune system produces antibodies, a class of proteins that recognize and bind with specific foreign substances in the body and mark them for destruction. The behavior of antibodies can be put to use in reconstructing phylogenies. When purified protein obtained from, say, a species of frog is injected into a rabbit, the rabbit will produce antibodies to the foreign protein. The antibodies can be recovered and mixed in a test tube that contains (1) the same protein from a different species of frog and (2) carefully measured quantities of mammalian red blood cells and complement (page 419).

If the antibodies react with the foreign protein, they "turn on" complement, which causes the blood cells to lyse and release hemoglobin. How much hemoglobin is released depends on the amount of complement activated, which in turn depends on the amount of antibodies that recognize the protein as being foreign. (Recognition is less likely when the protein being compared differs from the one initially injected.) The amount of hemoglobin released is a measure of the overall similarity of the two proteins of the two species being compared. The test is repeated for the same protein from many pairs of species, and the resulting data are used to construct a phylogeny such as that shown in Figure 37.7.

Era	Period	Epoch		Age (millions of years)
Cenozoic	Quaternary	Recent		0.01
Cenozoic	Quaternary	Pleistocene		2.5
Cenozoic	Tertiary	Pliocene		7
Cenozoic	Tertiary	Miocene		25
Cenozoic	Tertiary	Oligocene		38
Cenozoic	Tertiary	Eocene		54
Cenozoic	Tertiary	Paleocene		65
Mesozoic	Cretaceous	Late		100
Mesozoic	Cretaceous	Early		135
Mesozoic	Jurassic			195
Mesozoic	Triassic			240
Paleozoic	Permian			285
Paleozoic	Carboniferous			375
Paleozoic	Devonian			420
Paleozoic	Silurian			450
Paleozoic	Ordovician			520
Paleozoic	Cambrian			570
Proterozoic	oxygen (O_2) abundant			2,000
Proterozoic				2,500
Archean	oldest fossils known			3,500
Archean	oldest dated rocks			3,800
Archean	approximate origin of the earth			4,600

Figure 37.9 Absolute geologic time scale, with dates based on radioactive isotopes from rocks of each interval.

DNA Hybridization Studies. Another approach to constructing phylogenies makes use of DNA hybridization techniques (page 241). Here an investigator isolates short segments of DNA from two species, converts them to single-stranded form, and allows them to recombine to form hybrid double helices. The extent of hybridization is directly related to the similarity of base-pair sequences in the DNA of the two species. The similarity can be determined by carefully heating the hybrid DNA until the two strands dissociate. The higher the temperature required for dissociation, the more complete was the match between the two sequences. Data gathered for many pairwise combinations of species can be used to construct a phylogeny such as that shown in Figure 37.8.

GEOLOGIC TIME

Until fairly recently, there was no way to determine the age of the earth or to develop an actual time scale for the events in geologic and evolutionary history. But well over a century ago, geologists and paleontologists established a *relative* time scale. They had observed fossils of different taxa in various rock layers (or strata), which typically lie on top of one another in an orderly sequence. In some cases the fossils and strata could be found in the same association over huge geographic areas, even on different continents.

Because sedimentary rocks form by the gradual deposition of erosion products and by the steady "rain" of skeletons of microorganisms in the sea, it was logical to assume that the now-deepest sedimentary layers were formed first, followed by those overlying them. Layers might be missing (due to erosion or nondeposition in some places), deformed, or even reversed in sequence by geologic upheavals, but the overall pattern through time was clear. By Darwin's time, this pattern was being used as the basis of a relative time scale, which included a series of major and minor subdivisions. These divisions remain in use today.

In what way are these divisions relevant to a discussion of macroevolution? The short answer is that they correspond in time to major evolutionary events. The geologic time scale is in fact a biological one, as evidenced by the names of the largest divisions, or *eras*. From oldest to youngest these are the Proterozoic ("first life"), Paleozoic ("ancient life"), Mesozoic ("middle life"), and the Cenozoic ("modern life"). An even earlier era, the Archean ("ancient times") has now been partitioned away from the Proterozoic, as shown in Figure 37.9.

These divisions of geologic time were initially recognized on the basis of abrupt transitions in the fossil

Figure 37.10 Radioactive dating. For many years, scientists tried to measure the ages of rocks by assuming that erosion, mountain-building, and other geologic processes occurred at a constant rate. Such attempts failed, because there is no constancy. Episodes of uplift and erosion occur irregularly and proceed at highly variable rates.

More recently, radioactivity has proved to be an accurate timekeeper. An atom of a radioactive element has an unstable combination of protons and neutrons in its nucleus, and it breaks down spontaneously. As it decays, radiation is released. Some combinations are inherently less stable than are others, and thus each radioactive isotope has its own characteristic rate of decay. This rate cannot be modified by temperature, pressure, chemical reactions, or any other environmental influence. *The rate of radioactive decay is constant, and differs among elements and their isotopes.*

Each radioactive element has a half-life, which is the time required for half of the atoms in a sample to decay into a lighter element. We can measure the proportions of the original element and its decay product to determine the age of a particular rock. For example, half of the atoms of a radioactive isotope of potassium will decay into the inert gas argon in 1.25 billion years. If a rock contains this isotope of potassium, all of the argon now present in the rock has been derived by radioactive decay. If the ratio of potassium to argon is 1:1, we know that the rock solidified 1.25 billion

Main Radioactive Elements Used in Dating			
Radioactive Isotope	Half-life	Stable Product	Useful Range (years)
Rubidium 87	49 billion years	Strontium 87	>100 million
Thorium 232	14 billion years	Lead 208	>200 million
Uranium 238	4.5 billion years	Lead 206	>100 million
Uranium 235	704 million years	Lead 207	>100 million
Potassium 40	1.25 billion years	Argon 40	>100,000
Carbon 14	5,730 years	Nitrogen 14	0—60,000

years ago. And if the ratio is 1:3, the rock solidified 2.5 billion years ago.

The radioactive dating method is powerful, but not perfect. For example, we must use an isotope whose decay product is otherwise absent in the rock, is chemically inert, and cannot leak away over time. Also, if a rock is melted again, the decay product may escape, resetting the clock to zero. Finally, there is a small percentage error in the measurement of the ratios, such that the more ancient is the date, the greater is its uncertainty. We can often test the dates given by two or more radioactive elements in the same rock sample to minimize the chances of large errors.

sequence, which we now know were the result of mass extinctions. The great Permian extinction brought the Paleozoic Era to a close; the Cretaceous extinction event ushered out many of the Mesozoic forms. Periods and epochs were similarly defined, at least initially, with reference to less dramatic turnovers. We will consider these events shortly.

The actual ages of the divisions of geologic time were established through radioactive dating methods (Figure 37.10), which demonstrate that the history of life began more than 3.5 billion years ago. We are accustomed to thinking of a year as a long time, and few of us will appreciate the duration of a century from personal experience. Yet a million years is ten thousand centuries end-to-end; and this is still only 1/3500, or 0.028 percent, of the history of life. Think about that for a minute.

MACROEVOLUTIONARY PATTERNS

How do lineages come to thrive, become extinct, or undergo sustained trends in their overall pattern of characteristics? Are *microevolutionary* events (changes in the allele frequencies within populations or species) enough to explain what is going on? The answer is probably yes.

However, the term *macroevolution* still can serve as a useful label for certain trends and processes for which a detailed characterization of each species is unnecessary. In this broader sense, the species become *units of evolution*, in the same way that trees are units of a forest. It is a matter of being able to look at the forest without being concerned with the individuality of every tree.

Evolutionary Trends

Evolutionary trends at the level of higher taxa are produced by variations in the *relative rates* of morphological change, speciation, and extinction. Differences in these rates can lead to highly divergent patterns as the history of a lineage develops, as the following examples will illustrate.

Adaptive Radiation

Certain lineages may dribble along, producing a species here, a species there, for tens of millions of years. Others succeed spectacularly, filling the environment with bursts of evolutionary activity in what are termed **adaptive radiations**. Figure 37.11 gives an example of a large adaptive radiation involving mammals.

If we know an adaptive radiation when we see one, can we then specify the conditions under which an adaptive radiation will occur? The answer is no; we can identify conditions under which a radiation *may* occur, but we cannot predict macroevolutionary events.

Nearly half a century ago, George Simpson developed the concept of adaptive zones to help explain the phenomenon of adaptive radiation. An **adaptive zone** is most simply characterized as a way of life, such as "burrowing in the seafloor," or "catching insects in the air at night." It is usually defined in retrospect, with reference to one or more lineages that have succeeded in occupying it, but there is no reason that an unfilled adaptive zone cannot exist as a concept.

According to Simpson, a lineage must have physical, ecological, and evolutionary access to an adaptive zone in order to become a successful occupant. Physical access is a straightforward criterion. For example, it might involve colonization of a new habitat or a new geographic area, such as by the arrival of the ancestor of Darwin's finches on the Galápagos Islands (page 29). Ecological access may be achieved if there are no species already occupying an adaptive zone or if the colonist is competitively superior to the resident species. Thus Darwin's finches reached an unoccupied adaptive zone as well as an unoccupied physical habitat.

The criterion of evolutionary access is less easy to explain. Often it is presented in terms of the acquisition of a **key innovation**, an adaptation that significantly improves some function or allows its possessors to exploit the environment in a novel way. For example, the evolution of wings opened new adaptive zones to the ancestors of insects, pterosaurs, birds, and bats, just as the evolution of limbs opened the way for the radiations of land-dwelling vertebrates.

Adaptive radiations are common features during the first few millions of years following major extinction events. For example, as shown in Figure 37.11, most of the living orders of mammals made their appearance in the first twelve million years of the Cenozoic Era. These included body forms as distinct as those of whales and bats. The mammalian lineage originated more than 140 million years earlier, but mammals were never diverse or numerous during the Mesozoic. It is often suggested that the extinction of the dinosaurs at the close of the Mesozoic opened up an array of ecological opportunities, which the mammals then exploited.

Origin of Higher Taxa

The fossil record repeatedly refuses to show us a gradual and steady development of organic diversity. Instead, organisms as diverse as bats and whales appear quickly and, for the most part, without a succession of intermediate forms. Virtually all phyla whose members have durable shells or skeletons appear "suddenly," in the span of about 30 million years in the early Cambrian. Flowering plants began their adaptive radiation and achieved much of their present diversity in a ten-million-year interval in the early Cretaceous. Darwin termed these events "inexplicable" and "an abominable mystery," and they continue to be cited by opponents of evolution today.

The fossil record is good enough for us to rule out the possibility of long, unrecorded histories for many of these groups. Biologists now face squarely the evidence that group after group originated and reached most of its morphological diversity in less than ten to twelve million years' time. How do we explain the origins of higher taxa and the major changes in form and function that they represent?

For example, the average duration of a mammalian species is comparatively short (about one million years). A lineage of mammals might have made twelve speciational "steps" in a particular direction between the end of the Cretaceous and the early Eocene. Is there a mechanism that could produce a bat, or a whale, from a generalized mammalian ancestor in only twelve steps? No one knows the answer, but mutations that affect developmental pathways are strongly suspected. Probably such mutations are nearly always lethal, but even if one in a million years were not, it would be enough to accomplish the sort of rapid changes in form evident in the fossil record.

Events of this sort are so rare, we cannot reasonably expect to see them happening except by enormous good luck. Suppose an individual able to survive and reproduce despite a major mutation in its developmental program does indeed appear once every million years per species, on the average. This means that somewhere in the world, four or five such mutants occur each year. Would we be likely to notice those mutants among all the individuals making up the estimated $4\frac{1}{2}$ million species, about two-thirds of which have not even been described yet? Probably not.

Still, it is possible to find examples of regulatory gene mutations that result in less pronounced changes in morphology. Some examples were described in the previous chapter. Also, artificial selection has produced the phenotypic diversity among domestic dogs in an "instant" of geologic time, chiefly by inbreeding the lineages in which regulatory gene mutations have occurred. The differences among breeds such as great danes, greyhounds, chihuahuas, and pekingese were produced in a few hundred years at most. Admittedly, all are still dogs—but would they be if the process continued for a million years?

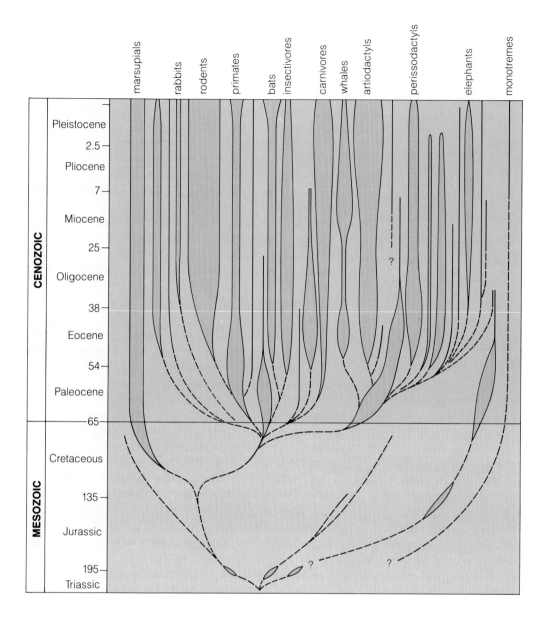

Figure 37.11 Geologic range and phylogenetic relationships of the principal living orders of mammals. In graphs of this sort, the widths of the lines signifying the different lineages indicate the numbers of species through geologic time. Notice the persistent low diversity throughout the Mesozoic, followed by an explosive adaptive radiation of placental mammals in the first 10–12 million years of the Cenozoic Era. Notice that neither the marsupials nor the monotremes (pouched and egg-laying mammals, respectively) participated in this radiation, which is thought to have resulted from the opportunistic invasion of ecological space vacated by the extinction of the dinosaurs in the late Cretaceous.

Extinction and Replacement

Most species that have ever lived are now extinct. Entire lineages that were once dominant have now disappeared or have dwindled in numbers as other radiations have flourished. Often, the extinction of one group has been followed closely by the adaptive radiation of another into the vacated ecological space. In general, it is not that members of the new radiation actively replaced those of the old through competition. Lineages usually declined to extinction millions of years before the radiations of the groups that replaced them began in earnest. Thus, large-scale turnovers are largely opportunistic events. *Radiations may follow extinctions but are rarely the cause of extinctions.*

Table 37.1 Examples of the Estimated Average Durations of Species

Group	Duration (millions of years)
Protistans:	
foraminiferans	20–30
diatoms	25
Plants:	
bryophytes	20+
higher plants	8–20+
Animals:	
bivalves	11–14
gastropods	10–13.5
graptolites	2–3
ammonites	1–2, 6–15
trilobites	1+
beetles	2+
freshwater fishes	3
snakes	2+
mammals	1–2+

From Stanley, 1985.

Background Extinction. If competitive interactions have not influenced extinctions at the level of higher taxa, how are such events explained? Careful study of the fossil record in the past decade has begun to provide some answers. Two separate processes, called background extinction and mass extinction, seem to have been at work. **Background extinction** is defined as the steady rate of species turnover that characterizes lineages through most of their histories. The causes of such extinctions seem to be many and varied ecological factors acting at the level of individual species. There is considerable variation in the rate of background extinction from group to group, as shown in Table 37.1. In general, larger and more complex organisms tend to have higher extinction rates. This means, for example, that the rate of species turnover is higher for trilobites, snakes, or mammals than it is for diatoms or clams.

David Raup and John Sepkoski studied background extinction rates for families of marine animals to provide a basis for comparison with mass extinction events. They found that the average extinction rate has declined from about *five* families per million years in the Cambrian to about *two* families per million years in the Tertiary. Standing far above these values are five events in which extinction rates rose to 10–20 families per million years. These global events will now be described.

Mass Extinction. An abrupt increase in extinction rates affecting several higher taxa simultaneously is called a **mass extinction**. Figure 37.12 shows the timing of five major mass extinctions and their effects on marine animals. Notice in Figure 37.12 that diversity rose rapidly in the early Paleozoic and leveled off, then was cut in half at the Paleozoic-Mesozoic boundary, followed by a recovery phase that continues to the present day. The five major mass extinctions appear as abrupt dips in diversity. Four of these each resulted in the extinction of 11–14 percent of marine animal families; the great Permian extinction carried off more than half of the families. According to Raup's calculations, as many as *96 percent of all marine species became extinct* at this time, making the Permian extinction the greatest crisis in the history of life. The Cretaceous extinction, which was substantial for terrestrial organisms, did not affect marine animals as extensively; it appears as only a small drop in Figure 37.12.

These mass extinctions and other, smaller extinction events have had profound effects on the history and diversity of life. Which taxa survived and which did not have repeatedly redefined the course of evolution and affected all aspects of the biological world. What if the ancestors of the dinosaurs had not survived the Permian extinction? Would the mammals have begun their adaptive radiation 100 million years earlier? What if the ancestors of the mammals had not survived? There are no answers to such questions.

Obviously, mass extinctions were global events of catastrophic proportions. In most cases the explanations of their causes remain only partially formulated theories. The explanations advanced for the Permian and Cretaceous mass extinctions are now being tested and are briefly described below.

In the past quarter century, it has become clear that continents move, separate, and collide, and that ocean basins form and close in the course of geologic time. During the latter half of the Paleozoic, widely separated continents began to converge to form a single "world continent" in the late Permian, and sea levels fell drastically. Together, the drainage of shallow continental shelves, loss of coastlines due to the clumping of continents, and substantial climatic changes resulting from altered oceanic current patterns may have caused the late Permian mass extinction. The shallow marine habitats so important to marine organisms virtually disappeared, resulting in the extinction of a large number of groups.

A different mechanism is suspected as the cause of the late Cretaceous mass extinction. In 1980, Luis Alvarez, Walter Alvarez, and other scientists proposed that a comet or an asteroid had collided with the earth. Because the resulting global dust cloud blocked the sun

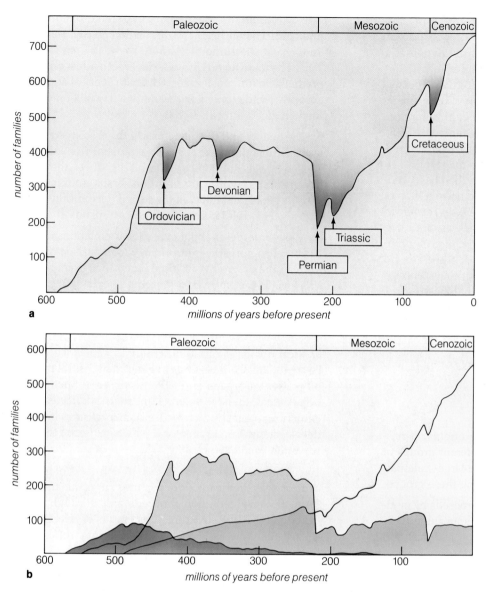

Figure 37.12 Graphs of the family-level diversity of marine animals over the past 600 million years, showing several macroevolutionary patterns. (**a**) All families combined, showing increases in diversity in the early Paleozoic and through Mesozoic and Cenozoic times, separated by a period of stability in the middle and late Paleozoic. Five mass extinction events appear as abrupt decreases in the number of families. The Permian mass extinction was the most severe.

(**b**) The same data as above, partitioned to show the histories of three major assemblages of marine animals. The Cambrian assemblage (pink shading) arose first and declined early, with very few representatives still living today. The Paleozoic assemblage (blue) was catastrophically affected by the Permian mass extinction and has never recovered. In contrast, the Modern assemblage (yellow) suffered little loss at the Permian event and has been dominant ever since.

for a period of months or even years, it drastically reduced temperatures and contributed to or caused widespread extinctions. The evidence for this hypothesis lies in a thin stratum of rock formed precisely at the boundary between the Cretaceous and Tertiary epochs. This stratum contains high concentrations of the element iridium, which is otherwise rare on the earth's surface. Iridium is more common in asteroids, and the thin, worldwide layer of iridium-rich rock may have formed from the dust that settled following the collision.

Such catastrophic, extraterrestrial "bombs" were common in earth's early history, as the ancient craters of the moon testify. Today, a few asteroids remain in earth-crossing orbits, and perhaps there was one more of them in the Cretaceous than there is now. Other theories suggest other physical forces at work in and around the solar system, for which there is little evidence at present. There is even some evidence that mass extinctions may occur at regular intervals of 26-30 million years, for reasons yet unknown.

SUMMARY

1. The history of life encompasses the origins, persistence, radiations, and extinctions of different groups of species. "Macroevolution" refers to the large-scale patterns, trends, and rates of change among these groups.

2. The premise is that since evolution proceeds by modifications of existing organisms, there is an underlying continuity of relationship among all species, past and present. Evolutionary classification schemes are ways of depicting this relatedness through descent.

3. Classification schemes place species in a hierarchy of increasingly inclusive categories: genera, families, orders, classes, phyla, and kingdoms. Categories above the generic level are often called the higher taxa.

4. In evolutionary taxonomy, the prevailing approach to classification, the higher taxa are ranked on the basis of morphological and evolutionary relationships. Because some morphological traits are given more weight than others, the resulting "family trees" may not accurately reflect relatedness.

5. In cladistics, a newer approach to classification, relatedness is depicted only in terms of branch points in lines of descent. The branch points are inferred by combining information from the fossil record, comparative morphology, and comparative biochemistry.

6. Rates of morphological change in an evolving lineage are interpreted in two ways:

    a. According to the gradualistic model, most morphological change occurs gradually *within* a species.

    b. According to the punctuational model, most morphological change occurs *during the time of speciation*, with little change accruing thereafter.

    c. The fossil record provides indirect evidence against gradualism; in most cases, distinct species appeared abruptly (within only a thousand or a few million years), then changed little after that.

7. The quality and completeness of the fossil record varies as a function of the types of organisms (for example, shelled versus softbodied), the places where they died (natural traps favoring fossilization versus eroding soils), and the geologic stability of the region (tectonically active zones versus undisturbed sedimentary plains).

8. Most hypotheses about evolutionary relationships are based on comparative morphology (comparison of body form in major groups of organisms). For example, gill slits form in the embryos of vertebrates even though they are not present in all adult forms; this is one of many striking similarities used to group diverse organisms together as "vertebrates."

9. Morphological variation within a major lineage may be attributable to mutations in genes that control the rate of growth of organs and other structures. Such mutations would not have been retained if they affected early embryonic stages, which are fundamental steps in major developmental programs for the lineage (hence the conservation of gill slits).

10. In morphological divergence, the process of genetic drift or selection leads to departures from the ancestral form. Because evolution proceeds only through modification of existing structures, the results of selection for a particular role for a given structure may differ from the original role.

11. In convergence, two or more species from dissimilar and only distantly related lineages adopt a similar way of life and come to resemble one another closely. Convergence often results when the physical requirements for a particular life style (for example, the rapid swimming required of certain predatory marine vertebrates) are strict.

12. Molecular approaches to establishing relatedness rely on the premise that neutral mutations (which do not significantly alter gene function and hence have little measurable effect on phenotype) accumulate at a regular rate. Thus distantly related species will differ at more gene loci than will close relatives.

13. Molecular methods (including immunological comparisons and DNA hybridization studies) can be used to construct a branching sequence for a lineage. The sequence can be calibrated against the fossil record to give the ages of divergence points.

14. The geologic time scale is based on succession of major groups of organisms; its eras and most of its

periods are bounded by extinction events that affected numerous taxa simultaneously. Lesser divisions of geologic time are recognized on the basis of smaller "turnovers." The ages of geologic events have been established through radioactive dating.

15. Evolutionary trends result from variations in the relative rates of morphological change, speciation, and extinction of species within a lineage.

16. In one evolutionary trend, called adaptive radiation, certain lineages rapidly fill the environment with bursts of evolutionary activity. Those radiations occur when there are unfilled adaptive zones (ways of life, such as "catching insects in the air at night") to which a lineage has physical, ecological, and evolutionary access.

   a. Physical access may involve colonization of a new habitat or a new geographic area for the species.

   b. Ecological access can occur if an adaptive zone is not yet filled or if the invading species can outcompete the resident species.

   c. Evolutionary access may result when a key innovation appears in a species. A key innovation is an adaptation that significantly improves some function or allows the organism to exploit the environment in a novel way (the chloroplast is an example).

17. Most higher taxa, such as bats and whales, appeared rapidly in evolutionary time and usually without an observed succession of intermediate forms. Mutations that affected developmental pathways are strongly suspected as the basis for their appearance, as defined by major changes in form and function.

18. Extinction is one of the most pervasive trends in the history of life. Through most of their history, lineages show a steady rate of species turnover, probably as a result of ecological factors operating at the species level. This process is called background extinction.

19. Mass extinction refers to an abrupt increase in extinction rates above the background level. These are catastrophic, global events in which several higher taxa disappear simultaneously.

20. In the great Permian extinction, about ninety-six percent of all marine species disappeared. The late Cretaceous extinction may have been triggered by the collision of a comet or asteroid with the earth.

Review Questions

1. Why are species considered to be natural units, but higher taxa not? *555*

2. How does classification affect our thinking about macroevolution? *555–556*

3. Does morphological change usually accompany speciation or follow it? *556–557*

4. What factors influence the completeness of the fossil record? On what basis do we subdivide geologic time, and how are actual dates estimated? *557*

5. How do morphological divergence and convergence illustrate constraints inherent in evolutionary processes? *558–559*

6. What is the basic assumption in employing comparative biochemical techniques for reconstructing phylogenies? How are DNA hybridization experiments performed to reveal phylogenies? *559–561*

7. How is an adaptive radiation defined, and what criteria must be met before a radiation can occur? Do extinctions lead to radiations, or do radiations lead to extinctions? Why? *563–565*

8. What is the biological problem posed by the rapid origin of higher taxa, and how are novel body plans thought to arise in short spans of geologic time? *564*

9. Why is an understanding of background extinction rates necessary for the study of mass extinctions? How have the five major mass extinctions reset evolutionary history? *565–568*

Readings

Ayala, F. J., and J. W. Valentine. 1979. *Evolving.* Menlo Park, California: Benjamin/Cummings. Short introduction to evolutionary theory.

Gould, S. J. 1982. "Free to Be Extinct." *Natural History* 91:12–16.

Lewin, R. 12 July 1985. "Pattern and Process in Life History." *Science* 229:151–153.

Minkoff, E. 1983. *Evolutionary Biology.* Reading, Massachusetts: Addison-Wesley. Comprehensive introduction to topics covered in this chapter, accessible reading level.

NASA. 1985. *The Evolution of Complex and Higher Organisms.* Washington, D.C. National Aeronautics and Space Administration Publication SP-478. Outstanding, readable review of current topics in macroevolution.

Raff, R. and T. Kaufman. 1983. *Embryos, Genes, and Evolution: The Developmental Basis of Evolutionary Change.* New York: MacMillan.

Raup, D., and J. Sepkoski, Jr. 19 March 1982. "Mass Extinctions in the Marine Record." *Science* 215:1501–1502. Landmark report on background and mass extinctions over geologic time.

Stanley, S. M. 1979. *Macroevolution, Pattern and Process.* San Francisco. Freeman. Comprehensive introduction to macroevolutionary theory; well written.

38

ORIGINS AND THE EVOLUTION OF LIFE

By the late 1860s, "the fixity of species" was crumbling as a scientific concept, for the accumulated evidence that species evolve, or change through time, had become overwhelming. By the middle of the twentieth century, "the fixity of continents" met a similar fate. Geologists already suspected that the continents have been buckling upward and eroding downward through time. But now it became clear that continents have formed, split apart, wandered about, and collided with one another, opening and closing ocean basins in the process!

This disconcerting discovery was one more piece of evidence that the earth itself has been evolving on a massive scale. The geologic record is one of stability and instability, involving mountain-building, glaciations, and widespread shifts in climate and in sea level. Although changes in the fossil record were known to be associated with episodes of geologic change, the meaning of the association is only now becoming understood.

Today it is clear that *the evolution of life has been linked, from its origin to the present, to the physical and chemical evolution of the earth.* As yet, there is no comprehensive theory drawing together all of the separate lines of evidence, and interpretations continue to be modified and refined as new evidence comes in. We encourage you at the outset to explore and evaluate this evidence for yourself by continuing with the readings listed at the chap-

Figure 38.1 Representation of the primordial earth, about 4 billion years ago. Within another 500 million years, living cells would be present on the surface. (During its formation, the moon presumably was closer to the earth. Here it looms on the horizon.)

ter's end, for all we can do in these few pages is highlight the discoveries and sketch out some prevailing theories, beginning with our current understanding of how the earth was formed.

ORIGIN OF LIFE

The Early Earth and Its Atmosphere

Billions of years ago stellar explosions ripped through our galaxy, leaving behind a dense cloud of dust and gas that extended trillions of kilometers in space. In time the immense cloud cooled, and countless bits of matter gravitated toward one another. By about 4.6 billion years ago, the cloud had contracted back on itself and was now a flattened disk rotating slowly through space. Temperatures at the dense center of the disk became extremely high, driving the fusion between the nuclei of colliding atoms of hydrogen and, to a lesser extent, helium. Thus thermonuclear chain reactions began that would feed themselves for the next 10 billion years; the sun was born a luminous star.

Farther out from the center of the disk, the earth and other planets were also forming under gravitational forces. At first the earth may have been a cold, homogeneous mass. However, contraction and radioactive

heating would have made its core increasingly dense—and hot. By 3.8 billion years ago, the earth was hurtling through space as a thin-crusted inferno (Figure 38.1).

Long before life appeared in that forbidding place, gases trapped beneath the thin crust or formed during reactions in the earth's molten interior were being forced to the outside. Those emissions were the start of an early atmosphere. Although more than twenty percent of the air we breathe today is oxygen, there probably was very little of this gas on the early earth. (More likely, the early atmosphere resembled those of Venus and Mars. Planetary probes have shown that oxygen accounts for less than one percent of the atmospheres of our two sister planets.) The near-absence of free oxygen was probably a prerequisite for the origin of life. Many of the molecules on which life is based (including amino acids and nucleotides) cannot assemble spontaneously in the presence of free oxygen, which attaches to and chemically alters them.

Another prerequisite for the origin of life was the presence of liquid water; all of the activities associated with life cannot proceed for long without it. Certainly a tremendous amount of water vapor would have been released from the breakdown of rocks during volcanic eruptions. But as fast as water vapor condensed in the early atmosphere, it would have evaporated in the intense heat blanketing the rumbling crust. Most likely liquid water accumulated only after the surface cooled enough for it to rain. When the rain began, so began the stripping of mineral salts from the parched rocks, and salt-laden waters collected where basins had formed in the crust.

If the early earth had settled into an orbit closer to the sun, its surface would have remained so hot that water vapor never would have condensed in liquid form. If its orbit had been more distant, the surface would have become so cold that any water formed would have been locked up as ice. And if the earth had not become as large as it did, it would not have had enough gravitational mass to retain the atmosphere with its water vapor. *Because of its size and its distance from the sun, the earth could retain liquid water on its surface.* Without liquid water, life as we know it never would have originated.

Yet exactly where and when did life originate? There is no fossil record of the event. Most of the rocks from that period have been melted, solidified, and remelted many times over because of ongoing movements in the earth. Some are buried far below more recently formed rocks, where they have been subjected to heat and compression. Thus any clues they might have held about the event would be altered beyond recognition.

Even though there is no direct record of the origin of life on earth, it is still possible to gain insight into the manner in which it may have occurred. To see how this can be done, we can start with four questions:

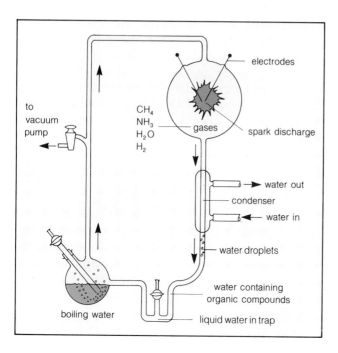

Figure 38.2 Stanley Miller's apparatus used in studying the synthesis of organic compounds under conditions believed to have been present on the early earth. (The condenser cools the circulating steam and causes water to condense into droplets.)

1. What were physical and chemical conditions like at the time of origin?

2. Based on known physical, chemical, and evolutionary principles, could life have originated spontaneously under those conditions?

3. Can we postulate a sequence of events by which the first living systems developed?

4. Can we devise experiments to test whether that sequence could indeed have taken place?

Spontaneous Assembly of Organic Compounds

To begin with, rock samples from meteorites, the earth's moon, and Mars are known to be between 4.5 and 4.6 billion years old. Because meteorites, the moon, and Mars as well as the earth all condensed from the same primordial dust cloud, we can safely assume the rock samples are representative of the earth's early chemical composition. *They all contain the chemical elements found in biological molecules.*

What about energy sources to drive the synthesis reactions? Lightning, hot volcanic ash, even shock waves have enough energy to drive the assembly of carbon-containing compounds. We know this from the pioneering experiments of Stanley Miller and Harold Urey, as

well as from the later work of many other chemists. In 1953, for example, Miller set up a reaction chamber containing a mixture of hydrogen, methane, ammonia, and water (Figure 38.2). For a week he kept the mixture recirculating. All the while, he bombarded it with a continuous spark discharge to simulate lightning as an energy source. By the week's end, he found organic molecules had formed, including many kinds of amino acids.

Such experiments have been repeated many times, with variations in elements, gas mixtures, and energy sources. The results show that all the building blocks in living systems—including lipids, sugars, amino acids, and nucleotides—can form under abiotic conditions. ("Abiotic" means not involving or produced by organisms.) In addition, when inorganic phosphate is present in the starting mixture, ATP will form. That molecule is involved in energy transfers in all living systems.

Speculations on the First Self-Replicating Systems

Given the 300 million years available for it to happen, large quantities of organic material probably accumulated in the shallow waters of the earth. What sequence of events led from this organic "soup" to self-replicating systems? In living cells, these systems include DNA (which contains instructions for building proteins) and RNA (which translates the DNA instructions into actual proteins). The proteins themselves are essential for self-replication. Many are important structural materials, and others are *enzymes*, which greatly increase the rate of chemical reactions. All the reactions needed to maintain and reproduce even the simplest cell could not proceed fast enough without enzymes. So we can rephrase our earlier question: How did the "DNA→RNA→protein" system originate? We do not yet have the answer to this question, although researchers are giving us some interesting things to think about.

Templates for Protein Synthesis. To give you a sense of the kind of work going on, let's look briefly at an idea that the first templates or "patterns" for protein synthesis existed at the bottom of the primordial soup, in the mud of tidal flats and estuaries. G. Cairns-Smith has analyzed microscopic crystals of different clays in which metal ions (such as those of iron and zinc) are embedded. Such crystals can grow by attracting molecules to themselves. When they do, their latticelike organization is repeated again and again, right down to the small imperfections caused by the random inclusions of different metal ions. Now, if you have ever baked meat in a clay or metal pot, you know that bits of protein stick tenaciously to both kinds of heated surfaces. According to Cairns-Smith and J. Bernal, naturally occurring clay crystals might have served as the original templates for assembling free amino acids into protein chains. Without such a template, amino acids might still combine through random collisions in the water, but the template would allow longer chains to form in a shorter period of time and help to prevent them from being broken apart.

Suppose some amino acids brought together on a clay template formed proteins that had weak enzyme activity. Perhaps some of those primordial enzymes hastened the linkages between amino acids. In such ways, some clay templates would have selective advantages over others. *There would have been chemical competition between different types of clay templates for the amino acids available—and selection for the first molecules characteristic of living cells.*

Suppose nucleotides were also attracted by the clay surface or by amino acids stuck to it. Intriguingly, most existing enzymes are assisted by small molecules called coenzymes—*some of which resemble or are identical in structure to RNA nucleotides.*

At this point we run up against an unsolved puzzle. We can show that enzyme-coenzyme systems involving RNA might have developed, and that the more successful combinations would have increased in abundance. We can even hypothesize that in some cases the RNA molecule itself might have replaced the clay crystals as a template for protein assembly. In most living systems, however, RNA plays a secondary role to DNA in mediating protein synthesis. We do not understand how the simple coenzyme relationship between RNA and proteins could have been transformed into a relationship involving DNA as the primary template, coding for RNA as the secondary template, in turn leading to protein synthesis. The problem is that DNA does not work without a suite of enzymes to transcribe it into RNA. It is not at all clear how such enzymes evolved, since both components must be present in order for the system to function. This is the main unsolved problem in devising a scenario for the origin of self-replicating systems, as depicted in Figure 38.3.

Whatever the exact sequence might have been, we should not think of it as a series of purely random chemical reactions. Rather, *there are good reasons to believe the physical and chemical conditions of the early earth made some reactions much more probable than others.* For example, amino acids can exist in two forms that are like mirror images of two hands. Thus one form is called lefthanded, the other righthanded. In almost all living things, amino acids are in the lefthanded form. Was this chemical choice a result of chance? Some recent work by James Lawless suggests it was not. Lawless began by studying meteorites. Samples from meteorites contain disorganized arrays of both lefthanded and righthanded amino acids. Yet Lawless discovered that when such disorganized mixtures are exposed to clay crystals, the clay attracts *only* the lefthanded forms.

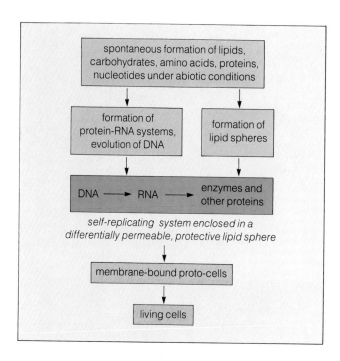

Figure 38.3 One hypothetical sequence of events that might have led to the first self-replicating systems and, later, to the first living cells.

Models for the First Plasma Membranes. All living cells have a surface membrane, or plasma membrane, that is differentially permeable—that is, it lets some substances pass freely into and out of the cell but restricts or controls the passage of other substances. These membranes are composed of lipids and proteins of various sorts. We do not know when membranes emerged in the sequence leading from chemical to organic evolution. What we do know is that metabolism means chemical control—and chemical control is possible only with chemical isolation from the ebb and flow of materials in the environment. If we assert that chemical evolution led to a metabolic system, then we must show that molecular boundaries for a protein-nucleic acid system can arise spontaneously. Here again, experiments suggest some chemical probabilities.

For example, Sidney Fox and his coworkers heated amino acids under dry conditions, which produced a number of long protein chains. The chains were placed in hot water and then allowed to cool. Upon cooling, the proteins assembled into small, stable spheres, or **microspheres** (Figure 38.4a). The microspheres selectively accumulated certain substances in greater concentrations inside themselves than were found on the outside. Moreover, the microspheres tended to pick up lipids from the water—and the outcome was the formation of a lipid-protein film around each sphere.

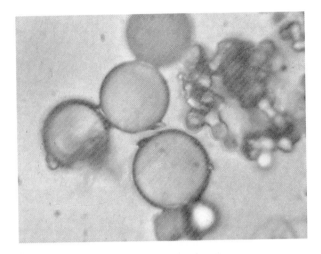

a

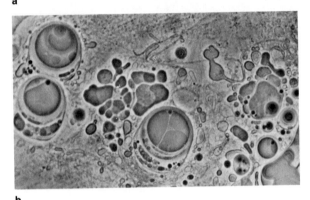

b

Figure 38.4 (a) Microspheres of protein chains, as they appear in water solutions. 11,000×. (b) Liposomes with multiple internal compartments made from simple, straight-chain lipids.

William Hargreaves and David Deamer have proposed that lipids must have been important in the origin and evolution of biological membranes. Although existing membrane lipids are structurally complex, these workers have experimentally produced membranes from simple lipids having only single hydrocarbon tails and water-soluble heads (pages 85–86). These lipids formed **liposomes** (microscopic, closed sacs filled with water) with the following properties:

1. Self-assembly into spheres
2. Formation of aqueous compartments within the spheres
3. Ion permeability, water permeability
4. Fluidity and elasticity (self-repair)

All of these properties are characteristic of lipids that contribute to cell membrane function.

These and other experiments suggest that membranes as well as the self-replicating systems they enclose could have arisen through spontaneous but inevitable chemical events. If lipids were present in the primordial soup, they may well have formed vesicles, for that is their thermodynamically favored state. Early self-replicating systems could have produced more complex lipids and thereby would have increased membrane stability. Whatever the sequence of chemical events, the first living cells that did appear on earth were probably little more than membrane-bound sacs containing the nucleic acids that served as templates for protein synthesis.

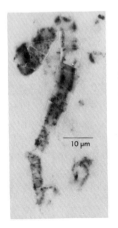

a   b   c

Figure 38.5 Some of the oldest known fossils. (a) From Western Australia, a filamentous form with cross-walls between cells; it is thought to be 3.5 billion years old. (b) Section through limestone stromatolite from Rhodesia, about 3.1 billion years old. The stacked organic (dark) and inorganic (light) layering is almost identical with that laid down by modern communities of photosynthetic microbes. (c) These stromatolites from Western Australia formed between 1,000 and 2,000 years ago in shallow seawater. Calcium carbonate deposits preserved their structure.

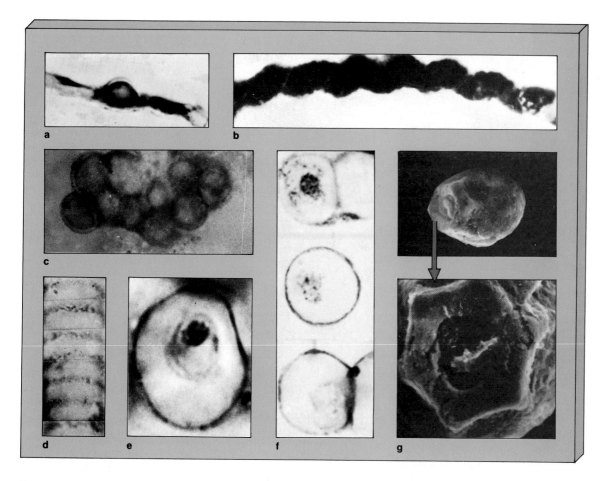

Figure 38.6 (**a**) Fossil cell from South Africa, about 2.25 billion years old; it is similar to modern cyanobacteria. (**b**) From Central Australia, a spiral form 900 million years old. (**c**) Colonial cells found in stromatolites in the USSR, about 650 million years old. (**d**) Filamentous form, about 650 million years old. (**e**) Presumed eukaryotic cell from Central Australia, about 900 million years old; the granules and spots may be remnants of organelles. (**f**) Fossil eukaryotes containing cytoplasmic remnants, about 900 million years old. (**g**) Notice the well-developed mouth of these eukaryotes, 750 million years old.

THE AGE OF PROKARYOTES

Until about 3.7 billion years ago, the earth's crust was highly unstable. Yet rocks that survive from this period contain fossilized cells, which were probably deposited in tidal mud flats. The rocks are dated at about 3.5 billion years, and the cells they contain are already well developed in structure (Figures 38.5 and 38.6).

Prokaryotic Metabolism and a Changing Atmosphere

The first known cells were prokaryotes (cells that do not have a nucleus or other membranous organelles). In their morphology they were like the simple anaerobic bacteria that now live in mud flats, bogs, and pond bottoms. (Anaerobic cells, recall, either cannot use oxygen during metabolism or they die in its presence.) Probably those early cells were heterotrophs, obtaining energy by consuming organic molecules that had accumulated spontaneously in the environment. The nucleotides such as ATP would have been the most accessible energy source. Most likely, metabolic systems became more and more efficient at using the existing ATP, then later developed the means of producing ATP themselves.

Today, the most common anaerobic pathway for producing ATP is fermentation, and it probably was the first to evolve. (In fermentation, recall, hydrogen atoms and electrons are juggled among intermediates of the reactions but are not transferred anywhere else.)

Possibly in one group, electrons started to be transferred away from fermentation products to inorganic electron acceptors in the environment. Such transfers would generate ATP by anaerobic electron transport (page 125). Today, *Desulfovibrio* and related

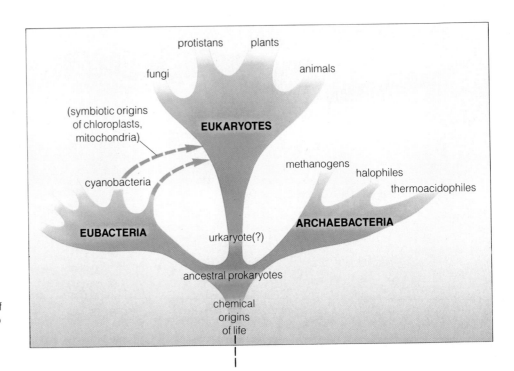

Figure 38.7 Diagram of the proposed relationships among prokaryotic and eukaryotic organisms. The eukaryotes are thought to have gained some of their organelles by entering into a symbiotic relationship with certain prokaryotic eubacteria.

bacteria use this type of pathway. *Desulfovibrio* produces ATP by fermenting sugars—*and* by transferring electrons to sulfate in the surroundings.

Electron transport chains used in such transfers may have been a key character in the evolution of autotrophs ("self-feeders")—including the first photosynthetic cells. In such cells, the transfers leading to ATP formation are driven by a virtually unlimited energy source—sunlight energy (Chapter Eight).

When populations of photosynthetic cells expanded —and the fossil record attests to their success—they had to be releasing oxygen as a by-product of their activities. Over time, the accumulation of oxygen in the atmosphere had two irreversible consequences for the origin and evolution of life. First, an oxygen-rich atmosphere meant that life no longer could arise spontaneously; oxygen now prevented further nonbiological synthesis of organic compounds. Second, the oxygen-rich atmosphere was a challenge of global dimensions. Some anaerobic prokaryotes continued to thrive in oxygen-free habitats, and many other lines became extinct. But some cells were able to adapt metabolically in ways that rendered the oxygen harmless. In one such adaptation, oxygen was used as a dump for electrons from transport chains. Hydrogen joined with the oxygen to form water, the toxic element was removed—and aerobic respiration emerged.

Aerobic respiration proved to be highly efficient at generating ATP (Chapter Nine). It was a key character in the evolution of larger and structurally complex cells; it was pivotal in the evolution of eukaryotes.

Divergence Into Three Primordial Lineages

In reconstructing the past, biologists also search through the "records" built into living cells. For example, they compare the similarities and differences in the nucleotide sequences of DNA and RNA from diverse cell types. In so doing, they have discovered differences suggesting that the forerunners of eukaryotes—cells that have a nucleus—arose simultaneously with the forerunners of existing prokaryotes. Today, research is pointing to the existence of three primordial lineages that arose from a common prokaryotic ancestor:

eubacteria	*true bacteria, including photosynthesizers such as cyanobacteria*
archaebacteria	*strictly anaerobic, including the methane-producing bacteria*
urkaryotes	*long-since-vanished ancestors postulated for eukaryotes*

The proposed relationship among these lineages is shown in Figure 38.7.

Archaebacteria still exist in many settings, and eubacteria are among the most successful groups of organisms (Chapter Thirty-Nine)—but there is no direct evidence of the urkaryotes. Their supposed descendants—the eukaryotic cells—do not appear in the fossil record until more than 2 billion years *after* the oldest known prokaryotes.

THE RISE OF EUKARYOTES

As Figure 38.7 shows, the protistans, plants, fungi, and animals are all eukaryotes. In structural terms, the simplest ones are the soft-bodied protistans known as amoebas. Because these forms do not have hard parts, they usually disappear quickly when they die. If the ancestors of eukaryotes were no more complex than the simplest of those forms, it is not likely that fossil evidence of them will ever be found. Yet there must have been an interesting history of evolution leading to their descendants.

The hallmark of all eukaryotes is the presence of organelles—especially the nucleus—these being membrane-bound compartments for specialized metabolic activities within the cell body. Given the integrated, coordinated functioning between the nuclear and cytoplasmic machinery, the nucleus probably evolved from infoldings of the plasma membrane around the DNA.

We have an example of such infoldings in some existing bacteria. In most bacteria, replicated DNA simply becomes attached to the plasma membrane, which grows and thereby separates the two molecules during cell division (page 138). In some bacteria, however, the plasma membrane folds elaborately around the replicated DNA in structures called **mesosomes** (Figure 38.8). Mesosomes retain their attachment to the plasma membrane. But perhaps structures similar to mesosomes separated permanently from the outer membrane and became the forerunners of the nuclear envelope.

But how did the eukaryotes acquire such organelles as mitochondria and chloroplasts? Lynn Margulis argues persuasively that these organelles had symbiotic origins. The word **symbiosis**, which means "living together," refers to mutually beneficial interactions in which one species serves as host to another species (the guest).

Origin of Mitochondria

Suppose urkaryotes included predatory, amoebalike forms that were only weakly tolerant of free oxygen. Suppose some of the aerobic bacteria they ingested happened to resist digestion. Oxygen released by their respiring prey could have poisoned most of the predators, yet some of the predators might have been more tolerant of the gas. If that were the case, the two species could benefit from continued association. The guest species would find itself in a protected environment, endowed with energy-rich remnants of the anaerobic host's fermentation activities. The guest cells would divide independently and come to be distributed in the cytoplasm of the host daughter cells. Over time, the guest species might lose its genetic independence, for the host could perform many functions for it. And the host species would develop increased metabolic efficiency as a result of the guest's capacity for aerobic respiration. The extra energy provided to the host could be channeled into

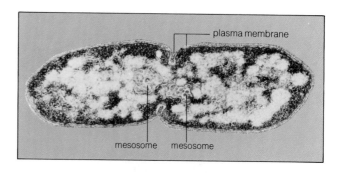

Figure 38.8 Mesosomes in a bacterium that lives in the gut of a rat. These two mesosomes, each an infolding of the plasma membrane that surrounds the DNA, have formed prior to cell division.

growth in size, increased activity, and the construction of more structures, such as hard body parts. The vestiges of symbiotic bacteria within the increasingly complex host cells would become mitochondria, which are now the basis of aerobic metabolism in eukaryotes.

The scenario is not as farfetched as it might seem. Such partnerships can be observed today between cells of entirely different species, as Figure 38.9 suggests. More to the point, aerobic bacteria that serve as models for such a scenario are uncannily like mitochondria in their size, structure, and biochemistry. Moreover, a mitochondrion has its own set of DNA, separate from the cell in which it is located, as well as a somewhat independent ability to replicate it!

Origin of Chloroplasts

After acquiring a capacity for aerobic metabolism, some urkaryotes may have ingested photosynthetic bacteria, which became a new kind of symbiont. Such guests could have been the ancestors of chloroplasts, foreshadowing the evolution of photosynthetic protistans and plants.

In their metabolism and overall DNA sequences, chloroplasts are like cyanobacteria. In their DNA they are also like eubacteria. Like the mitochondrion, a chloroplast has its own complement of DNA, and can replicate it in partial independence of the nuclear division of the cell.

If symbiosis was the route by which some urkaryotes acquired the capacity for photosynthesis, the route was followed by a number of ancestral lines. Today, the simplest photosynthetic eukaryotes resemble one another only in the fact that they are photosynthetic. They resemble different species of bacteria that have different body forms and different life-styles. Their chloroplasts are not the same. Their accessory pigments also vary, so they also differ in color.

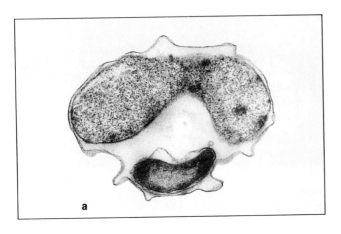

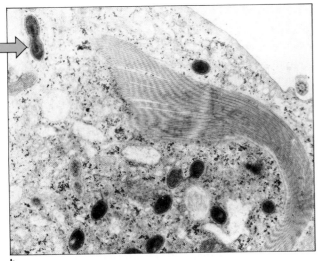

Figure 38.9 Symbiosis in existing prokaryotic and eukaryotic cells. (**a**) A predatory bacterium (small, dark oval) called *Bdellovibrio*, which has ended up living in the space between the cell wall and plasma membrane of a larger bacterium. (**b**) Symbiotic bacteria (small, dark ovals) living in *Pyrsonympha*, a protistan that dwells in the termite hindgut. The bacteria are the same size as mitochondria, they are distributed in the cytoplasm as mitochondria are, and they probably use the breakdown products of sugar molecules as mitochondria do. The arrow points to a bacterium that is dividing inside the cytoplasm (as mitochondria do).

Beginnings of Multicellularity

The evolution of multicelled organisms surely is one of the most important events in the history of life, and it is closely linked to the genetic specializations seen in eukaryotic cells. Unlike prokaryotes, eukaryotes contain organelles such as chloroplasts and mitochondria, and they have genes that regulate organelle activity. Such regulatory genes working *within* cells apparently became adapted to work *between* cells; they began to control the coordinated activities of cells clustered into tissues and, later, organs. In the same way that mitochondria or chloroplasts became dependent on their "host" cell, so did the increasingly specialized cell types of multicelled organisms become dependent on one another to function as a unit.

Multicellularity has many advantages, including larger size and the greater metabolic efficiency that results from the development of specialized organs. Each organ can perform a single function very well, while depending on other tissues or organs to provide for the remainder of its metabolic needs. Thus multicelled organisms benefit from a division of labor, something that is not possible for single cells to do.

While it is easy to picture how multicellularity evolved and how it is advantageous, the fossil record shows that this step took a long time. The first eukaryotes appeared well over a billion years ago. But it is not until about 700 million years ago that we see evidence of multicelled animals (metazoans)—the tracks and burrows on the ocean floor made by soft-bodied animals. Such fossils became widespread by 600 million years ago and were followed quickly by the origin of hard shells and skeletons. This event, 570 million years before the present, marks the beginning of the Paleozoic Era. From that point on, the history of life consists largely of the evolution of multicelled organisms.

FURTHER EVOLUTION ON A SHIFTING GEOLOGIC STAGE

The events described so far took place in the Archean and Proterozoic eras (Figure 37.9). We have seen how the physical and chemical evolution of the earth influenced the origin of life, and how the evolution of photosynthetic pathways led to an oxygen-rich atmosphere. Here we turn to major geologic events during the three subsequent eras—the Paleozoic, Mesozoic, and Cenozoic—and to their effects on the evolution of life.

The continental masses are known to have changed in their position and orientation over the past 540 million years. The *plate tectonic theory*, which is outlined in Figure 38.10, accounts for these changes.

During Paleozoic times, an early continent called **Gondwana** drifted southward from the tropics, across the south polar region, then northward. Other drifting land masses collided to form a tropical continent that is now called **Laurasia**. Then, near the end of the Paleozoic, Gondwana and Laurasia became massed together to form a single world continent, called **Pangea**, that extended from pole to pole. An immense world ocean covered the rest of the earth's surface; to the east of Pangea, it curved around to form the equivalent of a giant tropical bay, called the **Tethys Sea**. Pangea began to break up in the Mesozoic, and the drifting and collisions of the fragmented land masses continue today.

Figure 38.10 Plate tectonic theory. **(a)** The earth's surface is broken up into rigid plates. Today these plates are drifting together or apart, or shifting past one another, in the directions of the arrows. Boundaries between plates are marked by recurring earthquakes and volcanic activity.

(b) Seafloor spreading and continental drift. Plate tectonics is based on observations that the seafloor is slowly spreading away from sites called oceanic ridges, and on measured displacements of the continents relative to those ridges. Thermal convection in the mantle is proposed as the mechanism underlying these movements. More heat is being generated deep beneath oceanic ridges than elsewhere. The hotter material slowly wells up and spreads out laterally beneath the crust (much as hot air rises from a stove and spreads out at the ceiling). Oceanic ridges are places where the material has ruptured the crust. As the cooler material moves away from the ridges, it acts like a conveyor belt, carrying older oceanic crust along with it.

Thus plates grow and spread away at the oceanic ridges. As the plates push against a continental margin, they are thrust beneath it. The thrusting causes the crumpling and upheavals that have formed most major mountain ranges (see, for example, Figure 38.11).

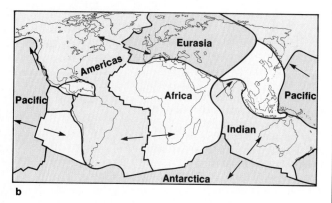

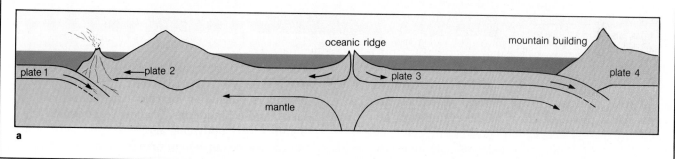

Figure 38.11 The view northward across Crater Lake in Oregon, a collapsed volcanic cone aligned with Cascade volcanoes reaching into Washington—and all paralleling the Pacific Coast. These formations are testimony to the violent upheavals that began in Jurassic times.

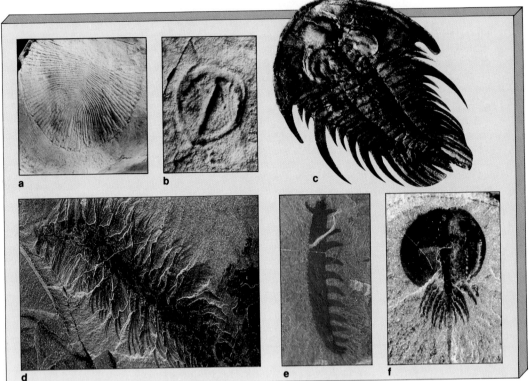

Figure 38.12 (**a, b**) Metazoan fossils, 600 million years old, from South Australia. (**c**) Fossil trilobite, about 570 million years old. Soft-bodied animals (**d, e**) and an arthropod (**f**) from the Burgess Shale in British Columbia. The soft-bodied animals are remarkably well preserved; they died in relatively still, poorly oxygenated water and were buried gently in a layer of fine silt.

All such changes in land masses, shorelines, and oceans had a profound effect on the evolution of life. When land masses were widely spread apart, speciation and adaptive radiations proceeded separately, giving rise to diverse arrays of organisms on land and along the shores. When land masses collided, habitats were lost and the overall diversity tended to decline. Thus the mass extinction 240 million years ago at the Paleozoic-Mesozoic boundary reduced the number of species of marine animals by ninety-six percent (page 567). Finally, with each major shift in land masses, the climatic patterns and the direction of warm or cold ocean currents changed, and once again there were evolutionary repercussions for organisms on land and in the seas. The remainder of this chapter can only hint at the outcomes of these and other interactions between organisms and their changing environment.

Life During the Paleozoic

The Paleozoic Era consisted of six periods: the Cambrian, Ordovician, Silurian, Devonian, Carboniferous, and Permian.

The Cambrian. Most Cambrian organisms apparently lived on or just beneath the surface of the seafloor, feeding on organic debris or particles suspended in water. Among the limited array of species were the trilobites—crawling and burrowing animals (Figure 38.12c). There was plenty of sand and mud for the trilobites and their neighbors, including brachiopods (lampshells), mollusks, and echinoderms. Shallow seas covered the low margins of Gondwana and Laurentia and probably afforded a benign environment; both continents were positioned at tropical latitudes during the Cambrian.

The Ordovician. Over the next 60 million years, Gondwana drifted southward until it eventually straddled the South Pole. The seas made further transgressions of the land; by the late Ordovician, the inundations were widespread. As vast marine environments opened up, there were extensive adaptive radiations. During the mid-Ordovician, cephalopods ancestral to squids and octopuses underwent major diversification. The eyes and brain regions of those animals became highly developed, which reflects a trend toward the active life-styles typical of predators.

What about the vertebrates? Even before this time, the jawless ancestors of modern vertebrates must have been evolving. From the few fossilized armored plates they left behind, we suspect they resembled the jawless fishes of later times (Chapter Forty-One).

The Silurian. During the Silurian, Gondwana drifted northward and lesser land masses moved about in the mid-latitudes. This was a time of major evolution for the armor-plated fishes, for by now the vertebrate jaw had developed. That feeding structure, so useful in biting and chewing food, foreshadowed an explosive proliferation of predatory forms.

Figure 38.13 Fossils of early plants. (**a**) *Cooksonia*, the oldest known plant, from the late Silurian. Its stem was less than seven centimeters tall. (**b**) *Psilophyton*, one of the trimerophytes of the Devonian that may be the earliest ancestors of seed-bearing plants. (**c**) *Archaeopteris*, one of the progymnosperms that were probably on the evolutionary road leading to both gymnosperms and angiosperms. Some were more than twenty meters tall.

Even while animal groups were diversifying, a relatively inconspicuous event was taking place that foreshadowed a major evolutionary trend. In the late Silurian, small stalked plants were evolving in wet mud at the margin of the land (Figure 38.13a).

The Devonian. As the Silurian gave way to the Devonian, several land masses converged and formed Laurasia (Figure 38.14). The surface area of dry land increased dramatically, and plants adapted to surviving near the land's edge now began their forays onto the land itself. Some plants had leafless aerial stems, horizontal root-like structures, and tubelike water-conducting tissue. Lycopods (simple plants with true roots, stems, and leaves) also appeared in tropical land environments during the Devonian, as did the ancestors of seed-bearing plants (Figure 38.13).

In the early Devonian, the major groups of fishes came to dominate the seas and freshwater habitats, as their descendants have done ever since. Toward the end of the Devonian, one of the lineages of lobe-finned fishes succeeded in moving onto land (Figure 38.15). The fins of these animals, already thick and muscular, were perhaps adapted for moving slowly through shallow water and thick aquatic vegetation; their ancestors had long ago begun to rely on air-breathing, as do many fishes living in stagnant waters today. Although the colonization of wet shorelines and swamps required no major structural or physiological changes, it was a major event foreshadowing the evolution of amphibians and fully terrestrial vertebrates.

The Carboniferous. From the beginning to the end of the Carboniferous, major land masses were submerged and drained fifty times, and there were changing sea levels in between. Immense swamp forests with large, scaly-barked trees became established. Over time, the seas moved in and buried them in sediments and debris. The organic mess left behind has since compacted into extensive coal deposits, as in the Appalachians of North America.

Some groups survived—indeed they flourished, for adaptations had appeared among them that allowed movement onto higher (and drier) land. The gymnosperms were one such group. Those plants were not restricted to the water's edge; they could complete their reproductive cycles without free-standing water. The reptiles were another group of survivors. The reptiles, which arose from primitive amphibians, could break away from an aquatic existence because of shelled eggs and internal fertilization. Their adaptations meant that embryonic development could proceed within the eggshell, a moist and (compared to the outside) dependable setting. The Carboniferous also saw the first of the great

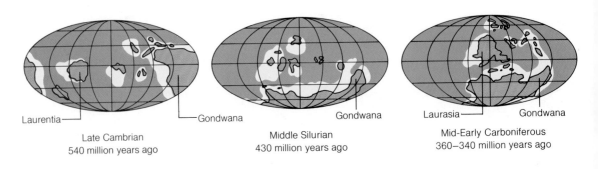

Laurentia — — Gondwana

Late Cambrian
540 million years ago

Gondwana

Middle Silurian
430 million years ago

Laurasia — — Gondwana

Mid-Early Carboniferous
360–340 million years ago

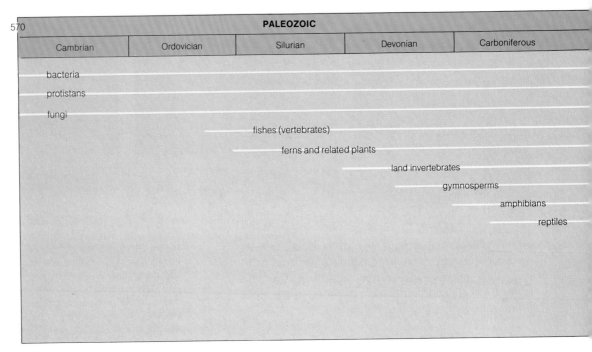

570

PALEOZOIC				
Cambrian	Ordovician	Silurian	Devonian	Carboniferous

bacteria

protistans

fungi

fishes (vertebrates)

ferns and related plants

land invertebrates

gymnosperms

amphibians

reptiles

Figure 38.14 (Above) Correlation between geologic evolution and the rise of major groups of organisms. The light-blue areas on the maps represent known regions of shallow seas. Notice the changing positions of the seas relative to the equator.

Figure 38.15 Reconstruction of lobe-finned fishes venturing onto a Devonian shoreline. Animals such as these were the ancestors of amphibians, reptiles, birds, and mammals.

radiations of insects, which apparently coincided with the evolution of the gymnosperms and, later, the angiosperms, or flowering plants (page 282).

The Permian. As the Carboniferous gave way to the Permian, all land masses had formed one vast continent called Pangea, leaving one immense world ocean. During this transitional period in geologic history, changes in the distribution of water, land area, and land elevation brought pronounced differentiation in world temperature and climate. About ninety percent of all known marine species disappeared at the boundary between the end of the Paleozoic and the beginning of the Mesozoic.

Of the three main lineages of late Permian reptiles, one gave rise to dinosaurs and birds, another to snakes and lizards. The third lineage dominated the Permian animal life. In this lineage were the synapsids, including such animals as the diverse therapsids, or mammal-like reptiles (Figure 38.16), and the sail-backed pelycosaurs (Figure 38.17).

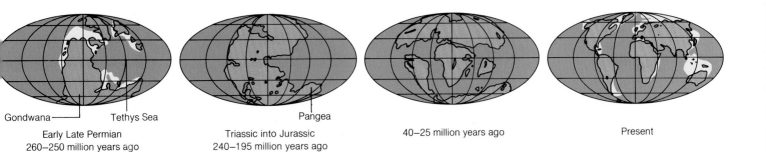

| Gondwana ——— Tethys Sea | Pangea | | |
| Early Late Permian 260–250 million years ago | Triassic into Jurassic 240–195 million years ago | 40–25 million years ago | Present |

	240	**MESOZOIC**		65	**CENOZOIC**	
Permian	Triassic	Jurassic	Cretaceous	Tertiary	Quaternary

marine reptiles

mammals

birds

flowering plants

marine mammals

Life During the Mesozoic

Once the supercontinent Pangea had formed, the patterns of diversity in the seas changed forever. The corals, brachiopods, cephalopods, fishes, and other lineages comprising the previously stable marine communities had disappeared or were reduced to a few species by the dawn of the Mesozoic. Other lineages, including the bivalves, gastropods, crustaceans, and freshwater fishes, had not been so severely tested. On land, the gymnosperms were becoming the dominant plants, and the therapsid reptiles dwindled to near extinction.

Mesozoic Seas. In the words of James Valentine, the late Paleozoic marine communities were "showy," whereas those of the early Triassic were "grubby." In effect, the most ecologically specialized lineages had been hit hardest during the Permian extinction, particularly tropical reef communities that had been as "showy" as the existing one illustrated in Figure 1.5.

Adaptive radiations of new lineages began in the Mesozoic seas, but this time the increase in diversity did not level off, as it did in the Paleozoic; it has continued to the present day.

Mammals, Dinosaurs, and Flowering Plants. Among the few remaining therapsids were the ancestors of mammals, which made their entrance during the Triassic (Figure 38.16). Even so, mammals remained small in size and few in number for the next 140 million years, when other vertebrates dominated the land.

As it happened, the reptiles called thecodonts had already begun a modest radiation. Some would give rise to the archosaurs ("ruling reptiles"). Later on in the Triassic, the first vertebrate experiment in flight—the pterosaurs—was under way, and several reptilian lineages had reinvaded the seas (Figure 38.17). On land, two major orders of dinosaurs diversified through the late Jurassic and early Cretaceous, producing the material for fabulous scenes such as shown in Figure 38.17d.

Figure 38.16 Reconstruction of a therapsid. Such reptiles are thought to have been the forerunners of mammals.

a

d

Most dinosaurs *were* large, many spectacularly so. By reason of their size alone, their body temperatures were likely to have been high and relatively constant. Recent work on fossil trackways and dinosaur skeletal anatomy indicates that many were capable of rapid locomotion and that some of the smaller bipedal carnivores would make a respectable match for modern horses or ostriches in speed. Moreover, at least some of the small, carnivorous forms may have been endotherms, with an insulating coat of feathers. One of the earliest known birds is *Archaeopteryx* (Figure 2.11) from the upper Jurassic. It was so similar to those smaller dinosaurs that only the chance fossilization of feather impressions and the presence of a collarbone distinguish it.

Recently, fossil bones of a crow-sized form that lived 225 million years ago (75 million years before *Archaeopteryx*) were discovered in western Texas. With its bony tail, strong hind legs, and pelvis adapted for running, this animal (designated *Protoavis*) was like the small dinosaurs. But the remains indicate that *Protoavis* had ears and eye sockets like birds, as well as small nodes on the bones to which feathers might have been attached. *Protoavis* may be among the earliest transitional forms between dinosaurs and birds, which began their diversification about 100 million years ago, although this idea is currently highly speculative.

The dinosaurs disappeared abruptly at the Cretaceous-Tertiary boundary. What happened? Currently, the leading hypothesis is that an asteroid collided with the earth, setting in motion a series of events that led to mass extinctions (page 566).

Dinosaurs so dominate our thinking about this period of evolutionary history that we often overlook one of the most important events of all. In the early Cretaceous, flowering plants began their ongoing radiation. In a span of about 10 million years, they surpassed the already declining gymnosperms in terms of their diversity and distribution.

The Cenozoic: The Past 65 Million Years

If you were to be transported 65 million years back in time to the beginning of the Cenozoic Era, you would probably not see anything that would strike you as being unusual. Marine organisms and the flowering plants would not appear that different from their existing descendants, and the conspicuous insects, reptiles, and birds might evoke only a little surprise as unfamiliar species. On reflection, you might conclude that what you *did not* see was the more unusual. For example, where were all the mammals? If all you could find were some unassuming rodentlike and opossumlike creatures in the shrubbery, you would have done well, because that is all there was.

b

c

Over the next 10-12 million years, however, most of the orders of mammals now living appeared (Figure 38.18). The magnitude of this adaptive radiation into bats, whales, ungulates (hoofed, herbivorous mammals), and so forth is remarkable in comparison to the earlier, rather undistinguished history of the group. For example, in the 140 million years since their origin, the Mesozoic mammals comprised only about 100 genera; in the first 60 million years of the Cenozoic, there were more than 3,000. (Figure 38.19 shows examples.)

The Cenozoic is known as the Age of Mammals, but if sheer *numbers* were the criterion it could be termed the Age of Birds. If the criterion were to be the most *rapid* adaptive radiation, the latter half of the Cenozoic would have to be called the Age of Snakes!

The point is, we live in a period of unparalleled biological diversity. Coinciding with that diversity is a far-ranging dispersal of land masses, which are intersecting the tropics and confining the polar regions. The vast island arcs of the tropical Pacific—remnants and omens

Figure 38.17 Reconstructions of Permian and Mesozoic reptiles and associated plants. (**a**) Triassic thecodont reptiles, ancestral to dinosaurs and birds. (**b**) A pterosaur, representing the first group of winged vertebrates. (**c**) Reptiles that returned to the sea: two ichthyosaurs and a plesiosaur. Some plesiosaurs were more than 15 meters (50 feet) long.

(**d**) A mural depicting four periods of the late Paleozoic and the Mesozoic. On the left, sail-backed synapsid reptiles of the Permian and early vascular plants; at left center, Triassic dinosaurs amid gymnosperms. The radiation of dinosaurs flourished in the Jurassic (right center), with some herbivores exceeding 25 meters (90 feet) in length. Dinosaurs of the late Cretaceous (right side) coexisted with early flowering plants. They included the largest of the carnivorous dinosaurs, such as *Tyrannosaurus*, standing nearly 6 meters (19 feet) tall.

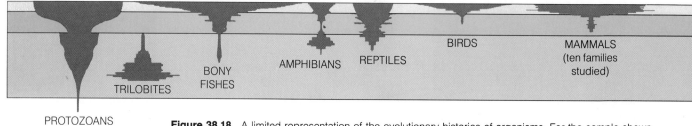

Figure 38.18 A limited representation of the evolutionary histories of organisms. For the sample shown here, notice the diverse patterns of radiation and extinction. Notice also the spectacular current success of the insects. (Here, the yellow band represents the Cenozoic to the present; blue, the Mesozoic; and green, the Paleozoic. The widths of the lineages, all shown in red, represent the approximate numbers of *families* in the case of plants and animals.)

a

b

Figure 38.19 Reconstructions of North American mammals of the Eocene (**a**) and Pliocene (**b**).

of plate tectonic movements—may well be the richest ecosystems ever to appear on the planet, and the extensive tropical forests of South America, Africa, and Southeast Asia are probably not far behind.

PERSPECTIVE

All living things are composed of molecules made of the same chemical elements—primarily carbon, hydrogen, nitrogen, and oxygen. These elements were present in one form or another in the crust and atmosphere of the primordial earth. It has been demonstrated that these elements can combine spontaneously into carbohydrates, lipids, proteins, nucleic acids, even ATP—into the stuff of life—given the right chemical environment and a source of activation energy. It has been demonstrated that, given a particular set of physical and chemical conditions, some of those large molecules can form differentially permeable structures much like simple cell membranes. Even now, research is shedding light on how such structures might have evolved chemically into the first organized, self-reproducing systems—into the first cells.

Whatever the details of such chemical evolution might have been, we suspect that it flowed to the threshold of biological evolution and, eventually, to the explosive diversification of life.

Today you share the earth with perhaps $4\frac{1}{2}$ million species. They, and you, share allegiance to the same principles of energy flow and chemical interactions, and the same molecular and cellular heritage. All these things speak of the underlying unity of life, but they also speak eloquently of its subsequent diversity. For if the environments of the first living things had never changed, if there never had been different abiotic and biotic horizons waiting the first inadvertent explorers, perhaps the world today would hold little more than prokaryotic cells

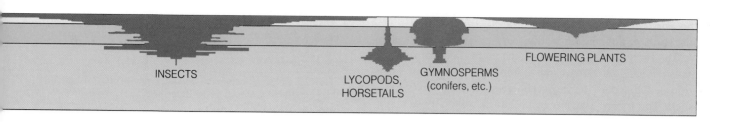

INSECTS

LYCOPODS, HORSETAILS

GYMNOSPERMS (conifers, etc.)

FLOWERING PLANTS

	Millions of Years Before Present	Major Events

Cenozoic Era	Quaternary (1.65–Present)
	  Recent
	  Pleistocene
		North and South America join. Several glacial cycles; extinctions of large mammals; origin of modern humans.

	Tertiary (65–1.65)
	  Pliocene
	  Miocene
	  Oligocene
	  Eocene
	  Paleocene
		Laurasia divides; India, Australia move northward; Alps, Himalayas, Andes, Rockies form. Major radiations of flowering plants; insects, snakes, birds, and mammals. Grasslands arise; succession of mammalian forms; origin of humans.

Mesozoic Era	Cretaceous (135–65)	Pangea continues to separate; broad inland seas; rich marine communities. Major radiation of ray-finned fishes, further radiations of insects and dinosaurs; origin of angiosperms and snakes. Mass extinction removes 11 percent of marine families, and all dinosaurs.
	Jurassic (181–135)	Pangea begins to break up. Marine communities become rich and diverse. Major radiations of dinosaurs; origin of birds. Gymnosperms dominant.
	Triassic (230–181)	Pangea stable. Marine life recovering; gastropods, crustaceans, fishes radiating. Early radiations of thecodont reptiles and dinosaurs; amphibians decline; origin of mammals. Gymnosperms dominant. Mass extinction removes 12 percent of marine families.
Paleozoic Era	Permian (280–230)	Pangea forms, squeezing out shallow tropical seas; now a worldwide ocean. Major radiations of synapsid, therapsid reptiles, and gymnosperms. Mass extinction removes 52 percent of marine families, up to 96 percent of marine species, with lesser effects on land.
	Carboniferous (345–280)	Gondwana moves toward Laurasia, enclosing Tethys Sea. Major radiations of insects and amphibians; origin of reptiles. Spore-bearing plants dominant, gymnosperms present.
	Devonian (405–345)	Gondwana moves north, equatorial land masses converge to form Laurasia. Radiation of fishes continues; origin of amphibians; development of swamp plants. Mass extinction removes 14 percent of known families in sea.
	Silurian (425–405)	Gondwana crosses south pole. Major radiations of jawed fishes; origin of vascular land plants.
	Ordovician (500–425)	Gondwana moves south. Major radiations of corals, crinoids, brachiopods, mollusks, early fishes; swimming arthropods, cephalopods. Marine communities diverse. Mass extinction removes 12 percent of known families.
	Cambrian (570–500)	Continents dispersed near equator. Origin of metazoans with hard parts; 50–100 phyla present; simple marine communities dominated by algae, brachiopods, trilobites.
Archean and Proterozoic Eras	(2,500–570)	Continental drift in progress. Oxygen becomes abundant in atmosphere. Origin of aerobic pathways of metabolism; origin of eukaryotes and multicellular organisms.
	(3,800–2,500)	Formation of continents. Chemical evolution, origin of life; prokaryotes and anaerobic pathways of metabolism.
	(4,600–3,800)	Origin of planets. Cooling, meteorite bombardment; formation of crust, oceans, and atmosphere.

Figure 38.20 Summary of the major features of the evolution of the earth and its life.

of the sort preserved in ancient rocks—cells matted against rocks or suspended in the seas.

Yet the record of earth history tells us that environments *have* changed. Organisms either have been lucky or equipped to adapt to those changes, or they have perished. The record also suggests that the diversity of life has been more than a product of evolution. It also has been an evolutionary force of the first magnitude. "Diversity" not only means adaptations to some combination of temperature, chemical balance, available water, light, dark, and living space. "Diversity" also means adaptations to different kinds of predators, different prey, different competitors after the same resource, and different forms of behavior, coloration, and patterning that help assure reproductive success. *Thus all existing species can be viewed as the evolutionary products of interactions with the environment and one another.*

And therein lies the story of evolution, the story of chemical competition and cooperation leading to the first self-reproducing forms of life, of dinosaurs, of continents on the move. Therein lies the story of simple strategies unchanged since the dawn of life, and of the complex human strategy—as yet unresolved—that can hold a world together or rip it apart. Yet must we predict gloomily that such unresolved activity on our part will end this magnificent story for all time? We doubt it. If the record of earth history tells us anything at all, it is that life in one form or another has survived disruptions of the most cataclysmic sort. That life can evolve tenaciously through tests of flood and fire suggests it has every chance of evolving around and past our transgressions too. *Viva Vida!*

SUMMARY

This chapter has presented a broad historical framework for the evolution of the earth and its organisms. The key points are highlighted in Figure 38.20; details of the evolutionary pathways taken in all five kingdoms are topics of chapters in the unit that follows.

Review Questions

1. Describe the chemical and physical characteristics of the earth 4 billion years ago. How do we know what it was like? *572*

2. Describe the experimental evidence for the spontaneous origin of large organic molecules, the self-assembly of proteins, and the formation of organic membranes and spheres, under conditions similar to those of the early earth. *572–574*

3. Which steps leading to the origin of living cells remain unexplained, and why are these important? *573*

4. Distinguish between heterotrophic and autotrophic prokaryotes, and describe the origins of photosynthesis and respiration. What are the main biological and environmental consequences of photosynthesis? *575–576*

5. How might symbiosis have played a role in the evolution of eukaryotes? How do eukaryotes differ from prokaryotes? *576–577*

6. How does continental drift occur, and in what ways does this process influence changes in biological communities? *579, 580*

7. When did plants, insects, and vertebrates invade the land? *581*

8. The Atlantic Ocean is widening, and the Pacific and Indian oceans are closing. Write a short essay on the possible biological consequences of the forthcoming formation of a second Pangea. (*This will require reflection on major concepts in all three chapters of this unit.*)

Readings

Bambach, R., C. Scotese, and A. Ziegler. 1980. "Before Pangea: The Geographies of the Paleozoic World." *American Scientist* 68(1):26–38. Excellent summary article, complete with color-coded maps of the changing configurations of continents.

Dott, R., and R. Batten. 1980. *Evolution of the Earth*. Third edition. New York: McGraw-Hill. Findings from diverse lines of research are distilled into a stunning picture of earth and life history. We recommend this book for your personal library.

Eigen, M., W. Gardiner, P. Schuster, and R. Winkler-Oswatitch. April 1981. "The Origin of Genetic Information." *Scientific American* 244(4):88–118. Describes experiments that indicate the chemical nature of the first self-replicating protein-RNA template systems.

Gensel, P. and H. Andrews. September-October 1987. "The Evolution of Early Land Plants." *American Scientist* 75(5)484–489.

Lambert, D. 1983. *A Field Guide to Dinosaurs*. New York: Avon Books. Comprehensive review of the diversity of dinosaurs; well-written and capably illustrated.

Margulis, L. 1982. *Early Life*. Boston: Science Books International. Easy to read introduction to the origin and evolution of prokaryotes and eukaryotes. Paperback.

Schopf, J. 1975. "The Age of Microscopic Life." *Endeavour* 34(122):51–58.

Stebbins, G. 1982. *Darwin to DNA, Molecules to Humanity*. New York: Freeman. Excellent overview of evolutionary theory and of experiments at reconstructing the physical and chemical conditions leading to the origin of life. Paperback.

Valentine, J. (ed.) 1985. *Phanerozoic Diversity Patterns: Profiles in Macroevolution*. Princeton, New Jersey: Princeton University Press. Multi-authored volume on major features of plant and animal histories; technical, but clearly written.

Valentine, J., and C. Campbell. 1975. "Genetic Regulation and the Fossil Record." *American Scientist* 63:673–680. Excellent discussion of the origin of multicellular animals and their rapid adaptive radiation.

Woese, C. R. June 1981. "Archaebacteria." *Scientific American* 244(6):98–125. Presents evidence of three primordial lineages of bacteria.

UNIT SEVEN

EVOLUTION AND DIVERSITY

39

VIRUSES, MONERANS, AND PROTISTANS

In the preceding unit, we considered the observations and methods being used to discern the evolutionary relationships among organisms. We also traced the history of life, from its origin some 3.8 billion years ago to the present. If one overriding trend is evident in that story, it is the evolution of single-celled forms into increasingly complex, multicelled organisms.

In this unit we turn to the existing spectrum of life, with the two kingdoms of single-celled forms—the monerans and protistans—being the main focus of this chapter. Most members of these kingdoms are microorganisms, or **microbes**; with few exceptions, they are too small to be seen without a microscope (see, for example, Figure 39.1). Before we begin, however, let's consider the viruses, which also are microscopically small. Although viruses are *not* alive, they do have impact on living organisms in all five kingdoms.

VIRUSES

General Characteristics of Viruses

In ancient Rome, *virus* meant "poison" or "venomous secretion." In the late 1800s, this rather nasty word was bestowed on members of a newly discovered class of infectious agents, smaller than the microscopic pathogens being studied by Louis Pasteur and others of that era. Many viruses deserve their name. Between 1918 and 1920 alone, a Spanish influenza virus killed more than

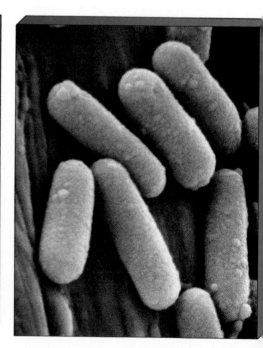

Figure 39.1 How small are bacteria? Shown here, *Bacillus* cells on the tip of a pin at increasing magnification: 85×, 440×, and 11,000×.

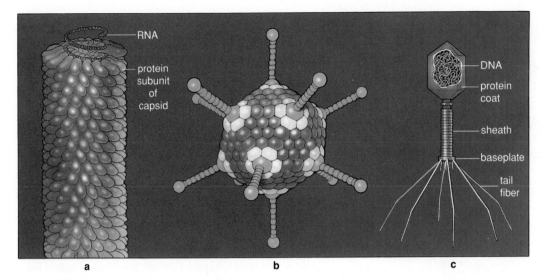

Figure 39.2 Examples of viral capsids. The rod-shaped capsid of a tobacco mosaic virus (**a**) and the hexagonal capsid of an adenovirus (**b**), which causes acute infections of the respiratory tract. (**c**) Component parts of a T4 bacteriophage.

20 million people. Specific viruses have devastating effects on crop plants, cattle, cats, dogs (Table 39.1 lists a few), insects, coconut palms, fungi, bacteria, and even other viruses. You name it, and there probably are one or more kinds of viruses that can infect it.

Today we define a **virus** as a noncellular infectious agent having two characteristics. *First*, a virus consists of a nucleic acid core surrounded by a protective protein coat. Sometimes a lipid envelope encloses its coat, but that is the extent of its structural complexity. *Second*, a virus can replicate only after its genetic material enters a specific host cell and subverts the cell's biosynthetic machinery, causing it to produce the materials necessary to assemble new viruses. When outside of a host, a virus is no more alive than a chromosome is alive.

The genetic material of a given virus is DNA or RNA (but not both), and it may be a single- or double-stranded molecule, depending on the type of virus. The viral coat, which is called a **capsid**, surrounds the nucleic acid. In rod-shaped capsids, protein subunits are coiled helically around the nucleic acid core. The tobacco mosaic virus has 2,220 identical protein subunits arranged that way (Figure 39.2a). In other capsids, protein subunits form a geometric structure to which filaments and other components may be attached (Figures 39.2b and c).

In some viruses, the capsid is enclosed in an envelope derived from the lipid bilayer of a cell membrane. The human immune deficiency virus that causes AIDS has such an envelope (page 430). A portion of the bilayer surrounds a new virus particle as it leaves an infected host cell. A new host cell becomes infected when the viral envelope fuses with the plasma membrane or when pinocytosis transports the virus into the cytoplasm.

Table 39.1 Classifications of Animal Viruses

Category	Some Diseases Produced
DNA Viruses:	
Adenoviruses	*Acute respiratory tract infections; under some circumstances can cause malignant tumors in hamsters*
Parvoviruses	*Some types of gastroenteritis (diarrhea, vomiting); implicated in hepatitis A in humans*
Papovaviruses	*Warts in humans, rabbits, dogs; some cancers in mice, hamsters*
Herpes viruses	*Fever blisters; chickenpox; shingles; certain genital infections with neuro-logical effects; some induce cancers; one (Epstein-Barr virus) implicated in infectious mononucleosis*
Poxviruses	*Smallpox; cowpox; formation of fibromas (nodules or benign tumors)*
RNA Viruses:	
Enteroviruses	*Diarrhea; polio; aseptic meningitis*
Rhinoviruses	*Common colds*
Togaviruses	*Yellow fever; German measles (rubella); equine encephalitis*
Influenza viruses	*Influenzas*
Paramyxoviruses	*Mild respiratory disorders: Newcastle disease; measles; mumps*
Rhabdoviruses	*Rabies*
Arenaviruses	*Meningitis; hemorrhagic fevers*
Coronaviruses	*Upper respiratory disease*
Retroviruses	*Certain tumors (sarcomas); leukemia; AIDS*
Reoviruses	*Mild respiratory disorders; severe diarrhea in humans, cattle, mice*

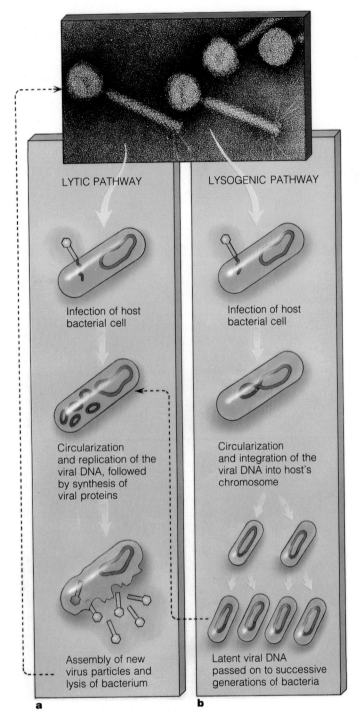

LYTIC PATHWAY

Infection of host
bacterial cell

Circularization
and replication of the
viral DNA, followed
by synthesis of
viral proteins

Assembly of new
virus particles and
lysis of bacterium

a

LYSOGENIC PATHWAY

Infection of host
bacterial cell

Circularization
and integration of the
viral DNA into host's
chromosome

Latent viral DNA
passed on to successive
generations of bacteria

b

Figure 39.3 Replication cycle of lambda bacteriophage, by way of the lytic pathway (**a**) or the lysogenic pathway (**b**).

Viral Infectious Cycles

A virus can be replicated only within a suitable host cell. By definition, its "host" is a cell having specific molecular groups (recognition sites) at its surface to which the virus can selectively bind. Once bound, the whole virus or its nucleic acid alone enters the cell's cytoplasm, where it characteristically escapes degradation. At some point thereafter, instructions encoded in the viral nucleic acid are used to direct the host cell's mechanisms for gene transcription and translation—*viral* genes, in this case. Then the newly synthesized viral proteins and nucleic acids are assembled into new virus particles. Usually, many new particles are assembled and released together, either by budding from the host cell or through a rupturing of its plasma membrane (lysis).

As we saw in Chapter Fifteen, the viruses called bacteriophage (which infect specific types of bacteria) can follow a lytic or lysogenic pathway of infection (Figure 39.3). In the **lytic pathway**, the host bacterial cell ruptures following viral replication. In the **lysogenic pathway**, the infection does not kill the host cell. Instead, the viral DNA becomes integrated into the host DNA, where it remains in a latent state. (In the case of RNA viruses, a DNA "transcript" of the RNA is synthesized and then integrated, as described on page 430.) When the time comes for the bacterium to reproduce, it replicates its own DNA. The viral DNA is replicated along with it and so is passed on to subsequent bacterial generations. Later, the viral DNA may spontaneously leave the host DNA molecule (or be induced to leave it), whereupon the lytic pathway is followed.

Animal Viruses

Many viruses infect only vertebrates; others infect specific insects and other invertebrates. Table 39.1 lists the ones that are troublesome for humans. Among them are the *Herpes* viruses (Figure 39.4a), different strains of which cause mild to serious diseases of the sort described on page 535. Also among the animal viruses are the influenza viruses (Figure 39.4b), which are responsible for **pandemics** (worldwide epidemics) of the common winter flu, Asian flu, Hong Kong flu, and Spanish flu. These serious diseases recur in ten- to forty-year cycles, with local epidemics in intervening years. Infected persons experience chilling, then fever and muscle aches. The virus weakens tissues of the upper respiratory tract. Often the weakened tissues are susceptible to agents of pneumonia, a complication that accounts for a large number of the deaths associated with flu epidemics.

New viral strains arise periodically, and the World Health Organization attempts to identify them as early as possible through its monitoring stations around the world. Once a new strain is identified, researchers work rapidly to develop a modified vaccine. Their task is not easy. There is evidence that flu pandemics are caused by novel—and not readily decipherable—genetic recombinants between human influenza viruses and viruses that infect other animals.

Plant Viruses

More than a thousand plant diseases are known to be caused by viruses. Typically, sucking insects (such as aphids), nematodes, or other invertebrates transmit the virus from an infected plant to another plant. Some viruses replicate in cells of these animal vectors as well as in cells of the host plant. Viral diseases can significantly reduce the yields of a variety of crops, including potatoes, tomatoes, cauliflowers, cucumbers, turnips, and barley.

Figure 39.5 shows the visible effects of one type of viral infection. It is caused by a type of "mosaic virus" that attacks pigment-storing parenchyma cells. Other plant viruses circulate through the phloem tissue and eventually may kill the component cells, with serious consequences for the plant.

Viroids and Prions

Are viruses the most stripped down of all disease agents? It appears not. **Viroids** are linear or closed circles of single-stranded RNA, with no protein coat at all. Under natural conditions, the RNA molecule base-pairs back on itself like a hairpin, so that viroids look like tiny rods. Apparently, viroids depend on a host cell for their replication.

Viroids are mere snippets of genes, smaller than the smallest virus. Yet they can wipe out huge fields of seed potatoes and groves of citrus or avocados. Thirty years ago, viroids almost destroyed the chrysanthemum industry in the United States. At one time, a type of viroid destroyed 12 million coconut palms on Philippine plantations, with major economic consequences. There are suspicions that viroids may cause some forms of human cancer, as they are known to do in hamsters.

Prions are small proteins that do not seem to be associated with any nucleic acid at all—and yet they can be replicated in host cells. This is puzzling; without genetic information, how do they take over the host cell's machinery? One idea is that they might activate a prion-coding gene that exists in the cellular DNA, but no one knows for sure. Medical researchers are interested in finding the answer. Prions are thought to cause some slow, fatal diseases of the central nervous system, such as *scrapie* in sheep and *kuru* and *Creutzfeldt-Jakob* disease in humans.

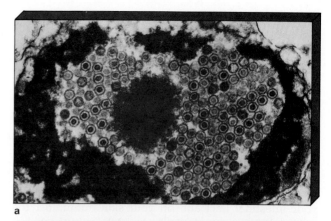

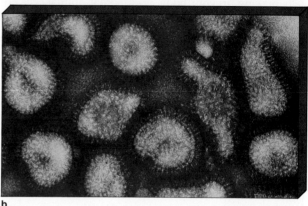

Figure 39.4 (**a**) Particles of *Herpes* virus in an infected cell, cross-section. (**b**) Particles of an influenza virus, each a package of viral RNA enclosed in a protein capsid. 250,000×.

Figure 39.5 Effect of a viral infection on Rembrandt tulips. These and other types of flowers are variegated as a result of a relatively benign viral infection that affects pigment formation in different tissue regions. Virus particles are passed on to new generations of plants, which also produce variegated flowers.

MONERANS

Bacteria, the sole members of the kingdom Monera, are the most abundant of all microbial cells. Some are smaller than the large viruses, but regardless of size, they all are capable of living and reproducing as an independent unit. And they live nearly everywhere! You can find bacteria in boiling mud and hot springs, snow, deserts, "pristine" lakes, and the deepest oceans. A handful of rich soil contains tens of billions of living bacterial cells—several times the number of people on earth. Many bacteria live in or on other organisms. The bacterial cells in your intestinal tract and on your skin outnumber the cells making up your body. Fortunately, animal cells are much larger than bacterial cells, so humans are only a few percent "bacterial" by weight.

Characteristics of Bacteria

Bacterial cells are prokaryotic, meaning they have no nucleus or other membrane-bound organelles for carrying out metabolic activities (see page 67 and Table 39.2). This does not mean their metabolism is simple. ATP production and other complex reactions occur at the plasma membrane—the bacterial equivalent of organelle membranes. Protein synthesis itself is efficient and rapid, being accomplished at numerous ribosomes that are distributed through the cytoplasm or attached to the cytoplasmic side of the plasma membrane.

Some bacterial species are **photosynthetic autotrophs**; they produce their own organic compounds from simple inorganic substances, using sunlight as an energy source. A few species are **chemosynthetic autotrophs**; they, too, produce their own organic compounds, using simple inorganic substances as the energy source. But the vast majority of bacteria are **heterotrophs** of one sort or another. They must use organic compounds produced by other organisms for food energy.

Most species of bacteria reproduce through **binary fission**, a division process whereby a single cell divides into two genetically identical daughter cells following DNA replication (Figure 10.1). In addition to the bacterial chromosome, which is a circular DNA molecule, many bacterial species have "extra" DNA in the form of plasmids, which were described on page 236.

Nearly all bacteria have a semirigid cell wall. Except for the archaebacteria, the wall is a tough mesh of **peptidoglycan**, a molecule in which polysaccharide strands are crosslinked to one another through short polypeptides. In addition, proteins, glycoproteins, or both are abundant at the wall's outer surface. In some species, they are firmly attached to the wall and form a sticky **capsule** (Figure 39.6). In other species, the molecules are only loosely attached to the wall and form a **slime layer**.

With its capsule or slime layer, a bacterium can adhere to a variety of surfaces, including rocks, plant parts, and teeth. The same can be said for **pili** (singular, pilus). As we saw in Chapter Eighteen, these filamentous appendages help bacteria attach to one another during conjugation. But they also help a bacterial cell adhere to a substrate and thereby resist being washed away—as from the urinary or intestinal tract of a human host. Finally, most motile bacteria have one or more **bacterial flagella** anchored to the cell wall and the plasma membrane (Figure 39.6).

The cell wall imparts a characteristic shape to a bacterium. The following terms are used to designate the most commonly occurring shapes:

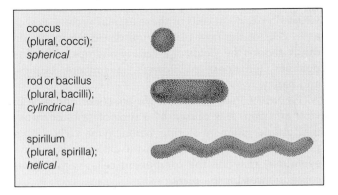

coccus (plural, cocci); *spherical*

rod or bacillus (plural, bacilli); *cylindrical*

spirillum (plural, spirilla); *helical*

Both cocci and rods (and spirilla, much less often) may stick together in pairs, in clusters, or as chains of cells when the daughter cells do not separate completely following cell division.

Classification of Bacteria

Traditionally, bacteria have been characterized on the basis of their overall form, growth patterns, tolerance of environmental conditions (including pH and temperature), and wall characteristics. For example, some bacteria are said to be **Gram-negative** and others, **Gram-positive**. The two terms refer to differences in the way different types of bacteria retain a deep purple stain (devised by Hans Christian Gram). The Gram-positive cells remain colored even when washed with organic solvents; the Gram-negative cells lose the color easily (Figure 39.7). The staining properties depend on the structure and composition of the cell wall.

But "characterizing" a bacterium is not the same thing as "classifying" it. How are the many thousands of known bacterial species related to one another? Bacteria are not well represented in the fossil record, and until recently, they kept their evolutionary secrets to themselves. Advances in molecular biology, including nucleic acid sequencing techniques, are yielding insights that are dramatically changing our concepts about the classification of these organisms. For example, through the work of Carl Woese and others, we now recognize that there are two fundamentally different types of prokaryotic cells: the archaebacteria and the eubacteria.

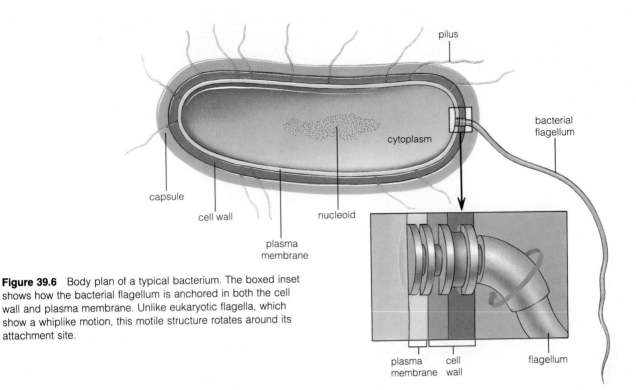

Figure 39.6 Body plan of a typical bacterium. The boxed inset shows how the bacterial flagellum is anchored in both the cell wall and plasma membrane. Unlike eukaryotic flagella, which show a whiplike motion, this motile structure rotates around its attachment site.

Archaebacteria

Only three types of bacteria—the methanogens, extreme halophiles, and thermoacidophiles—are grouped in the subkingdom **Archaebacteria**. They are distinctive in several respects. First, one or another species lives in boiling mud, hot springs, acidic soils, extremely salty water, and other harsh settings reminiscent of the ancient environments that were the stage for the origin of life. (Hence the name, "archaebacteria.") Second, their ribosomes and membrane lipids are structurally unlike those of any other organism. Third, DNA and RNA sequencing studies show that the diverse types of archaebacteria are closely related to one another but not to any other bacteria. Finally, unlike eubacteria, the archaebacteria have no peptidoglycan in their cell wall.

The **methanogens** inhabit swamps, marine and freshwater sediments, and the animal gut—especially the gut of elephants, cows, and other ruminants. They are strict anaerobes that make ATP by converting carbon dioxide and hydrogen gases to methane (CH_4). Much natural gas and the "marsh gas" of swamps and sewage treatment facilities result from their metabolic activities. Collectively, methanogens play a role in the global cycling of carbon (Chapter Forty-Five). They produce about 2 billion tons of methane gas each year and so influence the atmospheric concentration of carbon dioxide. (When methane reacts with oxygen, carbon dioxide forms.)

Some archaebacteria are **extreme halophiles**, which are organisms that tolerate or require high salt concentrations in their surroundings. Species of *Halobacterium*, for example, thrive in salt lakes, salt ponds, and brackish

Table 39.2 Characteristics of Bacterial Cells

1. Bacterial cells are prokaryotic (they have no nucleus or other membrane-bound organelles in the cytoplasm).

2. Bacterial cells have a single chromosome (a circularized DNA molecule); many species also have plasmids.

3. Most bacteria have a cell wall composed of peptidoglycan.

4. Most bacteria reproduce by binary fission.

5. Collectively, bacteria show great diversity in their modes of metabolism.

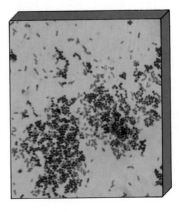

Figure 39.7 Example of Gram staining. Gram-positive *Staphylococcus aureus* remains purple when washed with organic solvents. Gram-negative *Escherichia coli* loses color easily. *E.coli* cells shown here appear red because they have been treated with a light-red dye (safranin) after being washed. Without this "counter-stain," they would be colorless.

Table 39.3 Some Major Groups of Bacteria

Group	Main Habitats	Cell Wall	Other Characteristics	Representatives
Archaebacteria				
Methanogens	Anaerobic sediments of lakes, swamps; also animal gut	Peptidoglycan absent	Chemosynthetic, methane by-products; used in sewage treatment facilities	*Methanobacterium*
Halophiles	Brines (extremely salty water)	Peptidoglycan absent	Chemosynthetic; also have photosynthetic machinery of a unique sort	*Halobacterium*
Thermoacidophiles	Acidic soil, hot springs, hydrothermal vents on seafloor	Peptidoglycan absent	Chemosynthetic; use inorganic substances such as sulfur as a source of electrons for ATP formation	*Sulfolobus, Thermoplasma*
Photosynthetic eubacteria				
Cyanobacteria	Mostly lakes, ponds; some marine, terrestrial	Gram-negative	In photosynthesis, water is electron donor, oxygen a by-product; some fix nitrogen	*Anabaena, Nostoc*
Prochlorobacteria	Live in tissues of marine invertebrates	Gram-negative	In photosynthesis, water is electron donor, oxygen a by-product	*Prochloron*
Purple or green bacteria	Generally anaerobic sediments of lakes, ponds	Gram-negative	In photosynthesis, H_2, H_2S, or S is electron donor, oxygen *not* a by-product	*Rhodospirillum, Chlorobium*
Chemosynthetic eubacteria				
Nitrifying bacteria	Soil, freshwater, marine habitats	Gram-negative	Major ecological role (nitrogen cycle)	*Nitrosomonas, Nitrobacter*
Heterotrophic eubacteria				
Spirochetes	Aquatic habitats; parasites of animals	Gram-negative	Helically coiled, motile; free-living and parasitic species; some major pathogens	*Spirochaeta, Treponema, Borrelia*
Gram-negative, aerobic rods and cocci	Soil, aquatic habitats; parasites of animals and plants		Some major pathogens; some (e.g., *Rhizobium*) fix nitrogen	*Pseudomonas, Neisseria, Azotobacter, Rhizobium, Agrobacterium*
Gram-negative, facultative anaerobic rods	Soil, plants, animal gut		Many are major pathogens; one (*Photobacterium*) is bioluminescent	*Salmonella, Shigella, Proteus, Escherichia, Photobacterium*
Sulfur-, sulfate-reducing bacteria	Anaerobic muds, sediments (as in bogs, marshes)	Gram-negative	Use sulfur or sulfur compounds as final electron acceptor in ATP formation	*Desulfovibrio*
Rickettsias and Chlamydias	Host cells of insects, other animals	Gram-negative	Obligate intracellular parasites; many pathogens	*Rickettsia, Chlamydia*
Myxobacteria	Decaying plant, animal matter, bark of living trees	Gram-negative	Gliding, rod-shaped; aggregate to form fruiting bodies with sporelike structures	*Myxococcus, Chondromyces*
Mycoplasmas	Parasites of plants, animals	Wall absent	Some pathogens (cause some forms of pneumonia and other diseases)	*Mycoplasma*
Gram-positive cocci	Soil; skin and mucous membranes of animals		Some major pathogens	*Staphylococcus, Streptococcus*
Endospore-forming rods and cocci	Soil; animal gut	Gram-positive	Some major pathogens	*Bacillus, Clostridium*
Gram-positive nonsporulating rods	Fermenting plant, animal material; human oral cavity, gut, vaginal tract		Some important in dairy industry, others serious contaminators of milk, cheese	*Lactobacillus, Listeria*
Actinomycetes	Soil; some aquatic habitats	Gram-positive	Include anaerobes and strict aerobes that form branching filaments; major producers of antibiotics	*Actinomyces, Streptomyces*

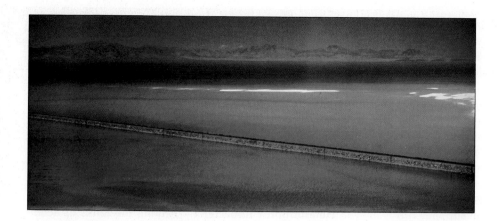

Figure 39.8 Great Salt Lake, Utah. Halophilic bacteria and algae growing in this vast, saline lake impart a pink cast to the water. The diagonal strip across the photograph is a raised bed for a railroad track.

seas. They also can cause spoilage of salted fish and salted animal hides, and they can contaminate "sea salt" (produced commercially from seawater evaporation). Large colonies of halobacteria appear pink, red, or red-orange, owing to carotenoid pigments that are present in individual cells. At times, these colonies and certain algal populations turn the water pink in Great Salt Lake, Utah (Figure 39.8). Apparently, the carotenoids help protect them against the strong sunlight characteristics of their habitats.

Most halophilic bacteria are heterotrophs that use aerobic pathways for ATP formation. When oxygen levels are low, however, some strains can use sunlight for photosynthesis. Patches of a special pigment (bacteriorhodopsin) form in the plasma membrane. When the pigments absorb light energy, H^+ ions are pumped out of the cell. Then the ions reenter the cell through ATP synthase systems, and ATP forms (compare page 129).

Thermoacidophiles can be found in hot springs, highly acidic soils, even near volcanic vents at the ocean floor. One member of the group, the heterotroph *Thermoplasma*, has only one known habitat—the waste piles of coal mines! Since this is a habitat of recent origin (only a few hundred years old), *Thermoplasma* must have evolved in habitats we don't know about.

Eubacteria

We are only beginning to work out the hierarchical relationships among the thousands of diverse **eubacteria**. Here we will consider a few kinds of eubacteria, using their basic modes of nutrition (photosynthetic, chemosynthetic, and heterotrophic) as our conceptual framework.

Photosynthetic Eubacteria. Like plants, some photosynthetic eubacteria use sunlight as an energy source and water as a donor of electrons for ATP formation. **Cyanobacteria**, formerly called blue-green algae, are like this (Figure 39.9). Cyanobacteria live mostly in fresh-

a

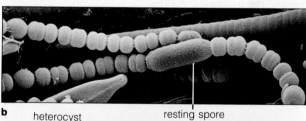

b

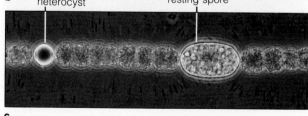

heterocyst   resting spore

c

Figure 39.9 (**a**) A population of cyanobacteria floating near the surface of a pond enriched with nutrients. The resting spores shown in the scanning electron micrograph (**b**) and phase-contrast micrograph (**c**) form when conditions do not favor growth. When favorable conditions return, a resting spore germinates and gives rise to a new chain of cells. A heterocyst, the type of differentiated cell that is specialized for nitrogen fixation, is shown in (**c**).

water ponds, where many species grow as chains of cells that become surrounded by a mucous sheath of their own secretions. Often, chains of cyanobacteria are so intertwined that they look like dense, slimy mats near the water's surface.

Anabaena and other species of cyanobacteria are capable of nitrogen fixation as well as photosynthesis, even though the two sets of reactions are incompatible. (The oxygen produced in photosynthesis inactivates nitrogenase, the enzyme required for nitrogen fixation.) When the environment provides *Anabaena* with plenty of nitrogenous compounds, photosynthesis prevails and all cells in the chain look alike. When the supplies run low, however, some of the cells develop into **heterocysts**, a type of cell in which nitrogenase is synthesized. In the entire chain, heterocysts are the *only* cells in which DNA recombination occurs to produce the exact gene sequence coding for nitrogenase.

Do the heterocysts keep the fixed nitrogen to themselves? Not at all. Junctions connect the cytoplasm of adjacent cells within each chain. Through these junctions, nitrogen compounds move from the heterocysts to the photosynthetic cells, and carbohydrates move from the photosynthetic cells to the heterocysts. It is a simple division of labor—but one that has echoes in more complex organisms.

Unlike *Anabaena*, the **green bacteria** and **purple bacteria** do not produce oxygen as a by-product of their photosynthetic activities because they do not split water to obtain the required electrons. The green bacteria, for example, strip electrons from hydrogen sulfide (H_2S) or hydrogen gas (H_2). In this respect, they may resemble the ancient anaerobic bacteria in which photosynthesis first emerged.

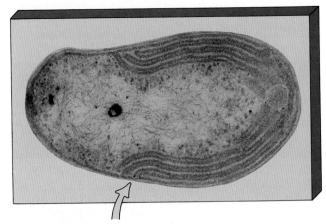

Figure 39.10 *Nitrobacter*, one of the nitrifying bacteria. The arrow points to a site where the plasma membrane folds into the cytoplasm. The membrane infoldings greatly increase the membrane surface area available for metabolic reactions.

Chemosynthetic Eubacteria. The metabolic activities of the **nitrifying bacteria** are vital to the global cycling of nitrogen, an essential element in all nucleic acids and proteins. Nitrifying bacteria use ammonia or nitrite as a source of the electrons required for ATP formation (Chapter Forty-Five). They obtain these nitrogenous substances from the surrounding soil or water. In some species, the reactions are carried out across extensive infoldings of the plasma membrane (Figure 39.10). Other chemosynthetic eubacteria can oxidize ferrous iron (Fe^{++}), sulfur compounds, or hydrogen as an energy source.

Heterotrophic Eubacteria. Among the ranks of heterotrophic eubacteria are some major pathogens of humans and their crops, farm animals, and pets. Here we find the causative agents of syphilis, gonorrhea, and chlamydial infections in humans (page 534) and crown gall tumors in plants (page 245). Here also we find pathogens that are transmitted by ticks and insects to human hosts.

For example, humans can be stricken with typhus, Rocky Mountain spotted fever, and other diseases that cause severe circulatory disorders when they are bitten by ticks or insects that are harboring certain rickettsias. As another example, one spirochete (*Borrelia burgdorferi*) that is transmitted from deer and other wild animals to humans causes **Lyme disease**. The disease was named after a small town in Connecticut where it was first diagnosed. Infected persons develop a rash at the site of the tick bite, then severe headaches, backaches, chills, and fatigue. Without prompt treatment, they can develop conditions that resemble rheumatoid arthritis as well as cardiac disorders and problems with the central nervous system. Penicillin, erythromycin, and the tetracyclines are all highy effective against this pathogen.

Or consider the endospore-forming heterotrophs, of the sort shown in Figure 39.11. An **endospore** is a dormant cell formed by a special division process that puts one daughter cell inside the other. The inner cell becomes the endospore and the outer cell eventually disintegrates. Endospores are resistant to heat and cannot be readily destroyed, even by boiling.

Clostridium botulinum, an endospore-forming bacterium, produces a deadly toxin. Each year, many cattle and birds die after eating fermented grains that contain the toxin. *C. botulinum* also can grow and produce toxin in food that has been improperly sterilized and packaged in an anaerobic environment (as in cans and jars). Eating the contaminated food can lead to **botulism**, a form of poisoning in which the toxin interferes with normal muscle activity; death can follow as a result of respiratory failure. Individuals suspected of having botulism can be given antitoxins; in serious cases, an artificial respirator can be used to prevent breathing failure.

Finally, consider *Escherichia coli*, a normal resident of the mammalian gut. *E. coli* produces compounds that enhance the digestion of fats, and it manufactures vitamin K that can be used by its host in blood-clotting reactions. Its activities also help create chemical conditions that impede colonization of the gut by a variety of pathogens.

Despite its benign and beneficial roles in healthy people, some strains of *E. coli* can be serious pathogens. Some *E. coli* cells carry the genes for a potent toxin and for pili that allow them to colonize the small intestine. When the pathogenic cells are inadvertently ingested, they produce a serious diarrhea. The consequences for adults are mild compared to what they are for infants and children. Tragically, *E. coli* diarrhea is the leading cause of infant mortality for most of the human population.

Lest we give all heterotrophic eubacteria a bad name, we should point out that some of our best defenses against the pathogenic forms are provided by some of their heterotrophic relatives—the actinomycetes, which are major producers of antibiotics (page 238). Other beneficial heterotrophs are *Azotobacter* and *Rhizobium*; Chapter Forty-Five will describe how these nitrogen-fixers contribute to soil fertility and plant growth. Still another is *Lactobacillus*, the first colonizer of the mammalian gut. By breaking down milk sugar, *Lactobacillus* helps newborns digest milk. Some *Lactobacillus* species are important in the production of cheese, sour cream, yogurt, and other fermented milk products.

A Final Word on the "Simple" Bacteria

So far, we have sampled only a small bit of the diversity that exists in the Kingdom Monera, but it is enough to bring us to an important point. Bacteria are small, *but they are not simple.* We can reinforce this point by concluding with a look at bacterial behavior.

Like multicelled organisms, bacteria have sophisticated ways of sensing and responding to their environment. Photosynthetic bacteria sense the intensity of light and move toward its source (or away if it is too bright for them). Heterotrophic bacteria sense and move toward higher concentrations of nutrients. Most bacteria can sense the presence of oxygen and move appropriately (toward for aerobes, away for anaerobes), and many also can detect several toxic chemical compounds and move away.

Bacteria detect the presence of a stimulus through membrane receptors that change shape upon absorbing light or upon binding to a chemical compound. When a bacterium moves in different directions, variations are introduced in the activity of membrane receptors. The variations are the basis of a fleeting biochemical "memory," allowing the cell to compare present conditions against the immediate past.

Many bacteria reverse direction when they sense that conditions are turning bad. Others swim for a period of time, stop and tumble for a few seconds, then swim again. The tumbles orient them in random ways. But a bacterium rarely tumbles when it is moving toward favorable conditions and so tends to keep moving in that direction. A bacterium moving away from favorable conditions tumbles frequently and so keeps changing direction until it senses that it is going the "right" way. **Magnetotactic bacteria** have a chain of magnetite particles in their cytoplasm that serves as a tiny compass (Figure 39.12). The compass helps them sense which way is north (or south in the Southern Hemisphere) and also down. Thus they swim toward the bottom of a body of water, where lower oxygen concentrations are more suitable for their growth.

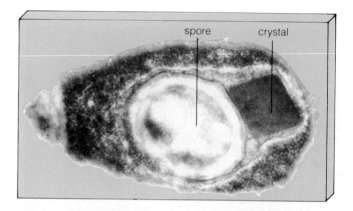

Figure 39.11 Transmission electron micrograph of an endospore-forming bacterium, *Bacillus thuringiensis*, thin section. In this species and others, crystals form inside the endospore. The crystals from this type of bacterium are widely used as an insecticide against the caterpillars of gypsy moths, cabbage worms, and other destructive insects.

Some endospore-forming bacteria cause serious diseases in humans. Because endospores are heat resistant, hospitals sterilize instruments in *autoclaves*, a kind of pressure cooker in which the elevated pressure raises the boiling point of water to a temperature that can destroy the bacterial endospores.

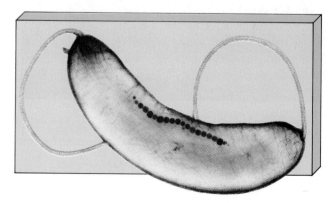

Figure 39.12 Transmission electron micrograph of a magnetotactic bacterium, showing the chain of magnetite particles that act like a tiny compass.

Figure 39.13 Predatory behavior of *Myxococcus xanthus*, shown in a series of frames on a video monitor. A latex bead, five micrometers in diameter, can be seen at the bottom of the screen. The cells sense and turn toward the bead. Thirty-three minutes later they reach it, sense that it is inedible, and move on.

Figure 39.14 Fruiting body of *Chondromyces crocatus*, a myxobacterium in which cells differentiate into specialized types.

Finally, the **myxobacteria** and many other bacterial species show collective behavior. They are not multicellular, but they give a good imitation of it. *Myxococcus xanthus*, for example, behaves like a predatory animal. Millions of cells gather together into a spherical colony to which cyanobacteria and other microbes become stuck. The cells secrete enzymes that degrade the trapped "prey," the contents of which are then absorbed by cells of the colony. *M. xanthus* cells migrating along a substrate also will change direction and *move as a single unit* toward an object that may be a food source, then they will move away as a unit if the object turns out not to be chemically "tasty" (Figure 39.13).

Many species of myxobacteria also form **fruiting bodies**, or spore-bearing structures (Figure 39.14). Under appropriate conditions, some cells in the colony differentiate and produce a slime stalk, other cells form branches, and still others form clusters of single-celled spores. When the clusters burst open, the spores are dispersed; each may form a new colony. As you will see, such structures also form during the life cycles of many eukaryotes.

PROTISTANS

The boundaries of the Kingdom Protista are poorly defined. Several decades ago, Robert Whittaker argued persuasively that all eukaryotes that are single celled should be placed in their own kingdom. We more or less follow his scheme here, as it is the most widely accepted (Table 39.4). But keep in mind that the boundaries defining this kingdom (and others) are artificial, in that we must decide where to impose them on a *continuum of diversity*. The boundaries probably cut across several protistan lineages that spill over into the kingdoms of plants, fungi, and animals. Figure 39.15 illustrates this evolutionary continuum.

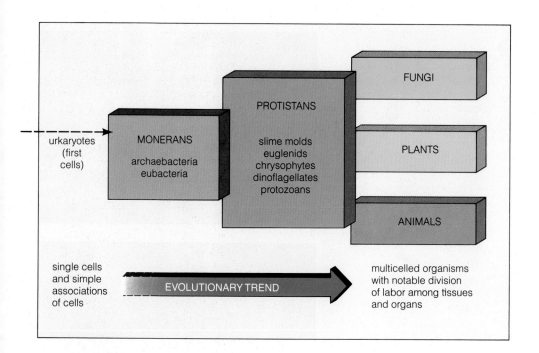

Figure 39.15 Evolutionary trend from single-celled to increasingly complex multicelled states, as reflected by the existing members of the five kingdoms of life.

Slime Molds

Among the problematic protistans are the slime molds (phylum Gymnomycota). Like myxobacteria, the cells of some species aggregate to form cytoplasmic masses that differentiate into spore-bearing structures. Like fungi, some slime molds form thick-walled spores that may be dispersed by air currents. Like animals, some slime molds creep about and engulf food, which for them includes bacteria, spores, and organic remains.

Slime molds are classified in two groups, the **cellular slime molds** and **plasmodial slime molds**. Both are alike in having a phagocytic phase during the life cycle. The organism moves along decaying logs, twigs, and leaves, engulfing food as it goes. But the body of a *cellular* slime mold is an aggregation of distinct, amoebalike cells; the body of a *plasmodial* slime mold is a multinucleate blob of cytoplasm which, when migrating, leaves a slimy track (hence the name).

Dictyostelium discoideum is the best known of the cellular slime molds. Details of its life cycle were presented earlier, in Figure 17.3. Recall that the independent amoebalike cells aggregate to form a slug, which differentiates into a fruiting body. Similar spore-bearing structures form during the life cycle of a plasmodial slime mold. Under suitable conditions, a plasmodium may grow until it spreads out over several square feet. Upon dehydration and exposure to light, the mass gives rise to a fruiting body of the sort shown in Figure 39.16. This is the structure in which spores form by way of meiosis. The spores are walled structures which, upon germination, will give rise to gametes. Union of two gametes gives rise to a new plasmodium.

Table 39.4 Classification of Protistans	
Phylum*	Common Name
Gymnomycota	
Class Myxomycota	Plasmodial slime molds
Class Acrasiomycota	Cellular slime molds
Euglenophyta	Euglenids
Chrysophyta	Yellow-green algae
	Golden algae
	Diatoms
Pyrrophyta	Dinoflagellates
Sarcomastigophora	
Subphylum Mastigophora	Flagellated protozoans
Subphylum Sarcodina	Amoeboid protozoans:
	Amoebas
	Foraminiferans
	Heliozoans
	Radiolarians
Apicomplexa	Sporozoans and kin
Ciliophora	Ciliated protozoans

*Also called Division in some classification schemes; the terms are equivalent.

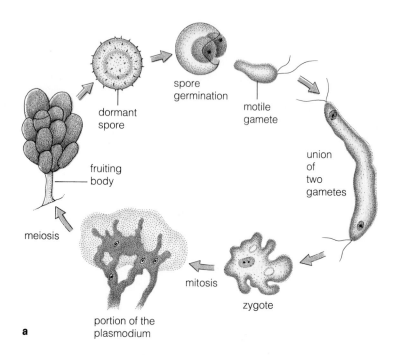

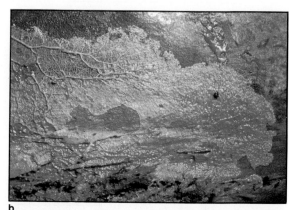

a

meiosis

fruiting body

dormant spore

spore germination

motile gamete

union of two gametes

zygote

mitosis

portion of the plasmodium

b

c

Figure 39.16 Plasmodial slime molds. (**a**) Life cycle of a typical plasmodial slime mold. Compare this with Figure 17.3, which shows the life cycle of a cellular slime mold, *Dictyostelium discoideum*. (**b**) The plasmodium of *Physarum*. (**c**) The spore-bearing structures of *Stemonitis splendens*.

a

eyespot granules   nucleus   mitochondrion   chloroplast   nonrigid pellicle

(b) Photo, C. Shih & R. Kessel, *Living Images*. © Science Books Intl., 1982, by permission of Jones & Bartlett Pub., Inc.

Figure 39.17 Anatomy of a protistan, *Euglena*. In (**a**), the profusion of internal organelles is evident. 5,200×. In (**b**), notice the flagellum; *Euglena* is highly motile. 1,700×.

b

Euglenids

The ponds and lakes of the world—especially the stagnant ones—are home to more than 800 species of flagellated, photosynthetic cells called **euglenids**. Euglenids show great structural complexity (Figure 39.17). They have a **pellicle**, which is a thin, translucent body covering composed mainly of proteins that the cell itself secretes. They also have an **eyespot**—in this case, a light-sensitive region where carotenoid pigments are concentrated in granules. The eyespot is part of a sensory-motor system that is used to position the cell body in regions where light intensity is best for photosynthesis.

Some species of *Euglena* can survive in darkness (that is, without photosynthesizing) as long as they are provided with nutrients. In fact, given sunlight, favorable temperatures, and abundant nutrients, some strains of *Euglena* reproduce faster than they duplicate their chloroplasts and end up as nonphotosynthetic cells. From then on, they are heterotrophs!

Like most other flagellated protistans, the euglenids reproduce by longitudinal fission. The cell grows in circumference while all organelles are being duplicated, then the cell divides along its long axis.

Chrysophytes

The photosynthetic **chrysophytes** are abundant in fresh water and marine water, especially in temperate regions. The members of this group include 450 species of "yellow-green algae," about 500 species of "golden algae," and more than 5,000 species of diatoms. All contain accessory pigments (including xanthophylls and beta-carotene) that mask the color of their chlorophylls.

Most golden algae are flagellated and many have silica scales or a silica skeleton. The diatoms are not flagellated and their walls are thin, double shells of silica that fit together, one on top of the other. Each shell is perforated with many holes, arranged in a pattern characteristic of the species (Figure 39.18). The holes permit the cell's plasma membrane to maintain close contact with the environment. Some diatoms also have longitudinal slots in the wall, through which the cell can make contact with a solid surface and crawl about.

The silica shells of diatoms began to accumulate 100 million years ago, and extensive deposits of diatomaceous earth are found in different parts of the world. Many abrasives, filtering materials, and insulating materials are made of this fine, crumbly substance. More than 270,000 metric tons of it are extracted annually from a quarry near Lompoc, California.

Dinoflagellates

Nearly all of the several thousand known species of **dinoflagellates** are members of marine plankton. ("Plankton" is derived from the Greek *planktos*, meaning to wander. It refers to communities of aquatic protistans and animal larvae, mostly microscopic, that drift or swim weakly through the water.)

Among the dinoflagellates are forms encased in stiff cellulose plates that have grooves between them. One groove circles the cell body and defines a channel for the movement of a ribbonlike flagellum. Another groove runs perpendicular to it and is a channel for another flagellum (Figure 39.19). When the two flagella beat in their channels, they make the cell spin like a top.

a

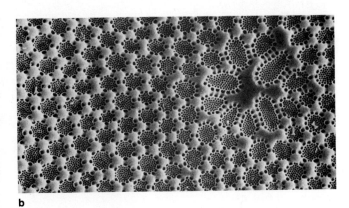

b

Figure 39.18 (**a**) Scanning electron micrograph of the magnificent silica shell of a diatom. 580×. (**b**) Closer view of the shell, showing the intricate perforations. 2,340×.

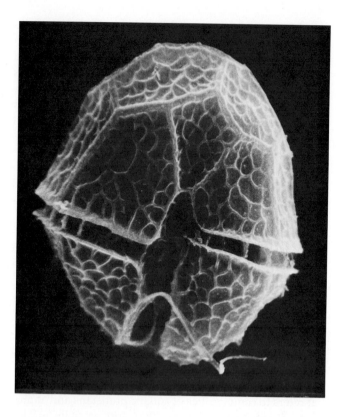

Figure 39.19 One of the "whirling whips"—dinoflagellates with two flagella that beat in opposing grooves in the armor-plated body. One flagellum is visible at the base, another beats in the groove running from left to right of this scanning electron micrograph.

Although a few dinoflagellates are heterotrophs, most are photosynthetic forms that contain chlorophylls. Depending on their array of accessory pigments, they may appear yellow-green, brown, or red. Periodically, populations of the red dinoflagellates, including *Gonyaulax*, increase dramatically in numbers and actually color the seas red or brown. The so-called **red tides** can be extremely devastating, for some dinoflagellates produce a powerful neurotoxin. Fish that feed on the plankton are poisoned; sometimes hundreds of thousands are killed and wash up along the coasts (Figure 39.20). Mollusks such as clams, oysters, and mussels are not affected by the neurotoxin. The poison builds up in their tissues, however, and humans who eat the contaminated mollusks can die from poisoning.

Protozoans

Because of their predatory or parasitic habits, some protistans were once viewed as primitive relatives of multicelled animals. Hence the name **protozoans**, meaning "first animals." We now know that many protozoans are at the pinnacle of single-celled complexity, and their evolutionary relationship to animals is not clear.

More than 65,000 species of protozoans have been identified, and they are classified as four groups: the flagellated, amoeboid, and ciliated protozoans, and the sporozoans (Table 39.4). Fewer than two dozen species cause diseases in humans, yet their influence is staggering. In any given year, hundreds of millions of people suffer protozoan infections!

a

Figure 39.20 (a) Scanning electron micrograph of *Ptychodiscus brevis*, the dinoflagellate responsible for the outbreaks of red tides along the Florida coast. (b) Portion of a fish kill that resulted from a dinoflagellate "bloom."

b

Flagellated Protozoans. The flagellated protozoans include free-living and parasitic forms. They reproduce by longitudinal fission, in the manner described for *Euglena*.

In this group are the **trypanosomes**, some of which are harbored in the salivary glands of biting insects that can infect new hosts. *Trypanosoma brucei*, which causes African sleeping sickness, is transmitted by the tsetse fly. Other trypanosomes develop in the hindgut of insect vectors, leave the insect by way of feces, and enter a new host through broken skin or mucous membranes.

Also in this group are **trichomonads** of the sort shown in Figure 39.21. Among humans, *Trichomonas vaginalis* is a worldwide nuisance. It is transferred to new hosts through sexual intercourse. Trichomonal infection can severely damage vaginal membranes if untreated. Among men, infection can occur in the urinary and reproductive tracts.

By some estimates, about ten percent of the human population in the United States is infected with another flagellated protozoan, *Giardia intestinalis*. Infection leads to mild intestinal disturbances, including diarrhea, but it has severe and sometimes fatal consequences in a few susceptible people. *G. intestinalis* travels from one host to another in cysts. Infection can occur by ingesting food or water that has been contaminated with feces, but hikers should know that cysts have been found even in remote mountain streams. The cysts can be destroyed by boiling.

Amoeboid Protozoans. There are four groups of amoeboid protozoans: the amoebas, foraminiferans, heliozoans, and radiolarians. In all four groups, the adult forms have pseudopods that are used for locomotion, capturing prey, or both. Except for the few parasitic species, food for the amoeboids consists of bacteria, algae, diatoms, other protozoans, even small animals (such as nematodes).

The **amoebas** may or may not have shells. The "naked" species, such as *Amoeba proteus* of biology laboratory fame, constantly change their shape (Figure 5.1). These protozoans can be found in fresh water, sea water, and soil. Although most are microscopic, some species are several millimeters long. Shelled amoebas, which inhabit fresh water, damp soil, and mosses, can extend pseudopods or much of the cell body through a large opening in the shell. Like many other protistans, the freshwater species use organelles called contractile vacuoles, as illustrated in Figure 6.6, to expel excess water from the body.

Some amoeboid protozoans are parasites of humans. *Entamoeba histolytica*, which parasitizes the human gut, causes amoebic dysentery. Infected individuals suffer fever, abdominal cramps, and severe diarrhea. *E. histolytica* proliferates in the gut during part of its life cycle, then it develops into a thick-walled resting cell (a cyst). The cysts are excreted with feces. Without proper sewage treatment facilities, the cysts can contaminate food and water.

The elaborately shelled **foraminiferans** live mostly in the seas. Their hardened shells are often peppered with hundreds of thousands of tiny holes through which sticky, threadlike pseudopods extend. Figure 39.22 shows some foraminiferan shells. Often the shells bear spines, which in some species are long enough that the shell can be seen with the naked eye.

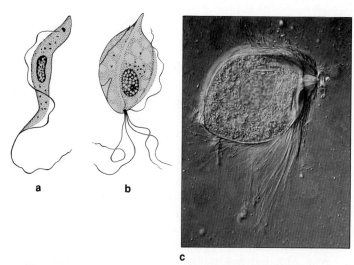

a   b

c

Figure 39.21 A few flagellated protozoans. (**a**) A leaf-shaped trypanosome, with its undulating membrane. (**b**) A trichomonad. These (generally) anaerobic parasites live in vertebrates and invertebrates. (**c**) *Trichonympha*, one of the trichonomads, this one from a termite gut.

Figure 39.22 Shells of some foraminiferans (the word means, loosely, "bearers of windows").

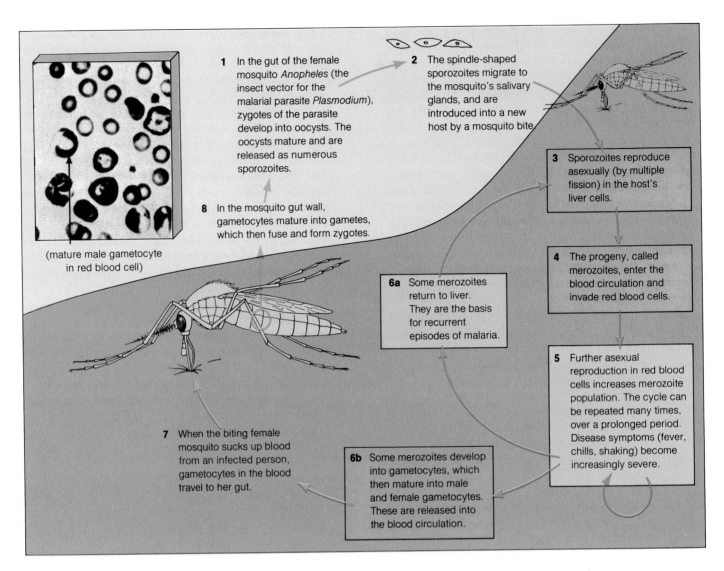

Figure 39.23 Life cycle of the sporozoan *Plasmodium*, which causes the disease malaria. The life cycle unfolds in the human body and in an insect vector (the female *Anopheles* mosquito), which transfers the sporozoan to new hosts during bites. The events described in the boxes occur in the human body.

Within the figure:

1 In the gut of the female mosquito *Anopheles* (the insect vector for the malarial parasite *Plasmodium*), zygotes of the parasite develop into oocysts. The oocysts mature and are released as numerous sporozoites.

2 The spindle-shaped sporozoites migrate to the mosquito's salivary glands, and are introduced into a new host by a mosquito bite.

3 Sporozoites reproduce asexually (by multiple fission) in the host's liver cells.

4 The progeny, called merozoites, enter the blood circulation and invade red blood cells.

5 Further asexual reproduction in red blood cells increases merozoite population. The cycle can be repeated many times, over a prolonged period. Disease symptoms (fever, chills, shaking) become increasingly severe.

6a Some merozoites return to liver. They are the basis for recurrent episodes of malaria.

6b Some merozoites develop into gametocytes, which then mature into male and female gametocytes. These are released into the blood circulation.

7 When the biting female mosquito sucks up blood from an infected person, gametocytes in the blood travel to her gut.

8 In the mosquito gut wall, gametocytes mature into gametes, which then fuse and form zygotes.

(mature male gametocyte in red blood cell)

The **heliozoans**, or "sun animals," have fine, needle-like pseudopods that radiate from the body like sun rays. These largely freshwater protozoans are generally floaters or bottom-dwellers. Part of the cytoplasm forms an outer sphere around a core composed of denser cytoplasm and the bases of microtubular rods.

Perhaps the most beautiful of all protozoans are the **radiolarians**, which are found mostly in marine plankton. In cytoplasmic structure they resemble the heliozoans, but most also have a skeleton of silica. In addition, there are colonial radiolarians in which many individuals are cemented together. Accumulated shells of radiolarians and foraminiferans are key components of many ocean sediments.

Sporozoans. The **sporozoans** and their relatives are a diverse group of parasites that are alike in having an infective, sporelike stage (sporozoites) during the life cycle. Often, insects transmit sporozoans from one host to another. Probably the most notorious sporozoans are the fifty or so species of *Plasmodium*, four of which cause the disease malaria in humans (Figure 39.23). *Plasmodium* requires an insect vector (mosquitoes) and an animal host (birds, mostly, and humans). On a worldwide basis, about 150 million people are stricken with malaria each year. Although malaria has largely been eliminated from Europe and North America, it is still the most serious infectious disease in tropical and subtropical regions of the world.

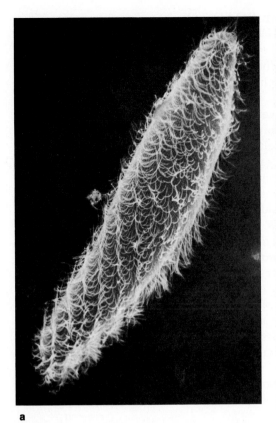

a

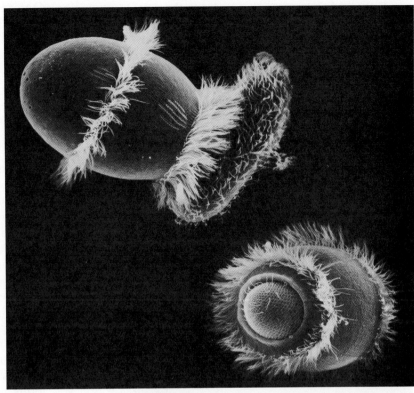

b

Figure 39.24 (a) Anatomy of *Paramecium*, a fast-swimming, predatory ciliate. (b) Mealtime for *Didinium*, a ciliate with a big mouth. Dinner in this case is another ciliate, the cucumber-shaped *Paramecium* poised at the mouth (upper left) and swallowed (lower right). (c) A living ciliate protozoan, as viewed with Nomarski optics, a special microscope technique that does not damage the cell. The nucleus (near the center of the cell) is surrounded by food-filled vacuoles in which filamentous algae and diatoms are visible. The "sunken" spots are contractile vacuoles.

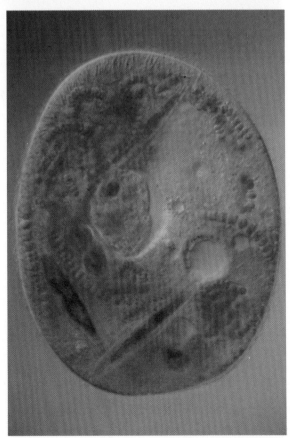

c

Malarial infections persist over time. They are rarely fatal except when *Plasmodium falciparum* is the causative agent. Work is under way to develop a vaccine through genetic engineering. Until a vaccine becomes available, individuals traveling through countries with high rates of malaria are advised to use antimalarial drugs such as chloroquine.

Ciliated Protozoans. Perhaps more than any other organism, the **ciliated protozoans** illustrate the structural complexity that is possible in a single cell. Cilia are present at the cell surface and are synchronized for swimming; in some species they number in the thousands. Often, hundreds of poison-charged, harpoonlike weapons are arrayed at the cell surface—weapons that are fired at other predators and prey.

The 8,000 species of ciliated protozoans inhabit fresh water and marine water. *Paramecium* is one of them (Figure 39.24). Like most ciliated protozoans, *Paramecium* has a gullet, which is a cavity that opens to the outside at the cell surface. Rows of specialized cilia beat bacteria and food particles into the gullet. Once inside, the particles become enclosed in enzyme-filled membranous vesicles and are digested. Wastes are moved to a region known as the anal pore and are eliminated. Like the amoeboid protozoans, *Paramecium* relies on contractile vacuoles to rid the cell of excess water (page 91).

Although some ciliates crawl about or stay put, *Paramecium* is built for speed. Between 10,000 and 14,000 cilia project like rows of tiny, flexible oars from the cell surface. So efficient is the coordinated beating of the rows that some species of *Paramecium* are propelled through their surroundings at a remarkable 1,000 micrometers per second. Even so, *Paramecium* itself is often outmaneuvered by another free-swimming ciliate, *Didinium* (Figure 39.24).

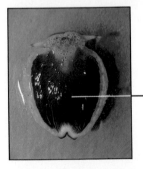

cerata (numerous projections of the dorsal surface) are pushed back here, revealing the extent to which functional chloroplasts become incorporated in tissues

Figure 39.25 One of nature's experiments — *Plakobranchus*, a marine mollusk that feeds on algae, the chloroplasts of which become incorporated in its tissues and continue functioning, providing the animal with oxygen.

Figure 39.26 Another experiment— *Volvox*, a colonial organism that resembles protistans *and* plants.

ON THE ROAD TO MULTICELLULARITY

Among the protistans we see associations of cells that form when their walls remain attached after division. We also see colonies cemented together in a glasslike matrix that the cells themselves have secreted. When we think about complex organisms such as redwoods and whales, it seems almost incomprehensible that such simple associations gave rise to multicelled forms of life. Yet the problem may be one of overlooking what must have been an immense evolutionary parade of intermediate forms, some dead-ends, others not.

Earlier, we described a possible symbiotic relationship that may have figured in the origin of mitochondria and chloroplasts. The idea is that some ancient bacteria were ingested by phagocytic cells, escaped digestion, and took up residence in their "hosts" (page 214). An example of this kind of escape and subsequent residency occurs in an algae-eating mollusk called *Plakobranchus*. Chloroplasts of ingested algae become incorporated into this animal's tissues and continue to engage in photosynthesis (Figure 39.25). If such jarring examples are observable today, is it too farfetched to assume that novel forms made their way onto the evolutionary stage in the past?

In its biochemistry, the photosynthetic bacterium *Prochloron* (Table 39.3) resembles the chloroplasts of green algae. The simplest green alga, *Chlamydomonas*, has one chloroplast, one eyespot, and two flagella. Cells resembling *Chlamydomonas* occur among the volvocines—colonial organisms that resemble both protistans and plants.

The peak of volvocine complexity, *Volvox*, gives us some interesting things to think about. Depending on the species, *Volvox* is a single-layered, hollow sphere of 500 to *600,000* cells. The flagella of all those cells beat in

unison to propel the sphere forward. Reproduction is assigned to a few cells, which divide and give rise to daughter colonies that float inside the parent sphere (Figure 39.26). The daughter cells are released after they synthesize and secrete certain enzymes that dissolve the parental jellylike matrix. Some *Volvox* species also show sexual differentiation (certain cells produce eggs, sperm, or both). Like the myxobacteria and some other organisms described in this chapter, *Volvox* may be one of nature's experiments in **multicellularity**, a word that signifies interdependence and division of labor among specialized cell types.

Consider, now, a tiny marine animal called *Trichoplax adhaerens*, which is little more than a flattened ball of ciliated cells half a millimeter across (Figure 39.27). It has no right side, left side, front, or back; it simply moves in any direction, amoebalike. It has a few more cilia on its top than on its bottom, but there is no other cell differentiation to speak of. *Trichoplax* may be reminiscent of simple multicelled animals that made their entrance during Proterozoic times.

Still other living organisms provide us with ideas about the first experiments in multicellularity. Consider the multicelled forms of green algae—straight or branched filaments merely one cell thick, and simple sheets of cells. As you read in the preceding chapter, the tiny multicelled plants that first invaded the land were not much more complex than this.

Consider, finally, just one of your own multicelled systems—the calcium-containing bones of your endoskeleton. Controls over calcium intake must have existed in the earliest protistans. Calcium, after all, is required in microtubule assembly—and microtubules are required in such essential structures as pseudopods and mitotic spindles. The proteins of existing organisms preferentially bind and release calcium ions. In ancient protistans, such proteins could have latched onto calcium entering the cell from the environment; they could have been calcium storage centers. Such proteins could have become incorporated into cell surface layers—and into internal structures. In the increasingly elaborate shells and internal skeletons of protistans, we have hints of the origins of bones—of our own skeletal system, and our own calcium reservoirs.

SUMMARY

Viruses

1. A virus is a noncellular infectious agent that consists only of a nucleic acid core and a protein coat (capsid), which sometimes is surrounded by a lipid envelope. A virus is not alive; it can replicate only after its genetic material enters a host cell and directs the cellular machinery into synthesizing the proteins and nucleic acids necessary to produce new virus particles.

2. A given virus has a single- or double-stranded molecule of DNA or RNA (but not both). Viral RNA is often replicated by way of reverse transcription (page 430).

3. For most viruses, the replication cycle proceeds by way of a lytic pathway whereby the host cell ruptures following replication of the viral DNA or RNA. In the lysogenic pathway, the infection does not kill the host cell. Instead, the viral DNA (or a DNA transcript from viral RNA) becomes integrated into the host chromosome, and the viral genes are passed on in a latent state to new cell generations. Later, the viral DNA may leave the host chromosome; then the lytic pathway is taken.

4. Viroids are infectious particles, but they are linear or closed circles of single-stranded RNA without a protein coat. Prions are infectious particles that do not encode any genetic information; they consist only of protein.

Monerans

1. Bacteria are the sole members of the Kingdom Monera. All are prokaryotic (they have no nucleus or other membrane-bound organelles for conducting specialized metabolic reactions).

2. Bacterial cells have a single chromosome, which is a circularized DNA molecule. Many species also have "extra," smaller circles of DNA called plasmids.

3. Most bacteria reproduce by binary fission (Figure 10.1). The myxobacteria are spore formers.

4. Great diversty exists in bacterial metabolism; there are many different types of photosynthesizers, chemosynthesizers, and heterotrophs. This is not true of eukaryotes; their pathways of photosynthesis, fermentation, and aerobic respiration are much the same from one species to the next.

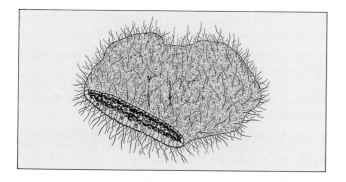

Figure 39.27 *Trichoplax adhaerens*, one of the simplest of all multicelled animals, being little more than a flattened ball of ciliated cells. This tiny animal was discovered crawling about in a seawater aquarium.

5. Recent molecular studies show that there are two different types of prokaryotic cells:

a. Archaebacteria (the methanogens, extreme halophiles, and thermoacidophiles), all of which live in extreme environments thought to be similar to the environments in which life originated. None has peptidoglycan in the cell wall.

b. Eubacteria (all other bacteria), which include photosynthetic autotrophs, chemosynthetic autotrophs, and heterotrophs. Eubacteria have peptidoglycan in the cell wall.

6. Bacteria are small, but they are not simple. They show great variation in their behavior as well as in their metabolism.

Protistans

1. The Kingdom Protista encompasses the single-celled eukaryotes. The boundaries of the kingdom are poorly defined, in that several protistan lineages apparently extend into the kingdoms of plants, fungi, and animals.

2. The cellular and plasmodial slime molds are alike in having a motile, phagocytic phase and a spore-forming phase during the life cycle. The plasmodial slime molds are a multinucleate mass of cytoplasm, and they reproduce sexually.

3. The euglenids are flagellated, photosynthetic cells having a pellicle (a thin, translucent covering) and a light-sensitive eyespot. The eyespot is part of a sensory-motor apparatus that is used to position the cell where light intensity is most favorable for photosynthesis.

4. The photosynthetic chrysophytes include yellow-green algae, golden algae, and diatoms, all of which have accessory pigments (including beta-carotene and xanthophylls) that mask the color of the chlorophylls. Some species have flagella and silica scales, skeletons, or shells.

5. Dinoflagellates are mostly photosynthetic cells encased in stiff celluose plates. Some red forms produce a neurotoxin that is responsible for fish kills associated with "red tides."

6. Protozoans are mostly motile predators or parasites. They include the flagellated protozoans, amoeboid protozoans (which move by pseudopods), sporozoans (which are gliding or nonmotile), and ciliates. Of more than 65,000 known species, only a few infect humans—but in any year, hundreds of millions of individuals are infected with one protozoan or another.

a. The flagellated protozoans include trypanosomes that cause African sleeping sickness and trichomonads that can infect the human reproductive tract.

b. Amoeboid protozoans include shelled or un-shelled amoebas, foraminiferans, heliozoans, and radio-larians, all of which use pseudopods for locomotion or capturing prey.

c. Sporozoans have an infective sporelike stage (sporozoites) during the life cycle. Often insects transmit the parasite from one host animal to another. The sporozoan *Plasmodium* causes malaria.

d. The ciliated protozoans are predators that use numerous cilia for locomotion and feeding behavior.

Review Questions

1. Define a virus. Why is a virus considered to be no more alive than a chromosome is alive? *591*

2. Outline the replication cycle of a virus that enters the lytic pathway, then one that enters the lysogenic pathway. *592*

3. What is a viroid and a prion, and how do they differ from viruses? *593*

4. Describe the key characteristics of a bacterium. What are some differences between archaebacteria and eubacteria? *594–595*

5. Name a few photosynthetic, chemosynthetic, and heterotrophic eubacteria. Describe some that are likely to give you the most trouble recreationally (if you enjoy water sports), medically, and ecologically (if you are worried about the greenhouse effect, for example; see page 758). *597–599*

6. What is an endospore? Are all endospore-forming bacteria dangerous? *598, 599*

7. Name the main categories of protistans. Think about where most of them live. Can you draw a few correlations between biotic and abiotic conditions in their environment and their structural characteristics? *Table 39.4, 601–608*

8. What kinds of protistans are photosynthetic? Predatory? Parasitic? Which move by means of pseudopodia? *601–608*

Readings

Atlas, R. 1988. *Microbiology*. Second edition. New York: Macmillan.

Bold, C., and J. LaClaire. 1987. *The Plant Kingdom*. Fifth edition. Englewood Cliffs, New Jersey: Prentice-Hall. Includes a brief, authoritative account of slime molds and photosynthetic protistans.

Brock, T., and M. Madigan. 1988. *Biology of Microorganisms*. Fifth edition. Englewood Cliffs, New Jersey: Prentice-Hall.

Frankel-Conrat, H., P. Kimball, and J. Levy. 1988. *Virology*. Second edition. Englewood Cliffs, New Jersey: Prentice-Hall.

Frazier, W., and D. Westoff. 1988. *Food Microbiology*. Fourth edition. New York: McGraw-Hill. Good reference on the microbes that have major effects on our food supplies.

Jurand, A., and G. Selman. 1969. *The Anatomy of Paramecium aurelia*. New York: St. Martin's Press. A tribute to the remarkable complexity of this single cell.

Margulis, L., and K. Schwartz. 1988. *Five Kingdoms*. New York: Freeman. An illustrated guide to the diversity of life. Paperback.

Shapiro, J. June 1988. "Bacteria as Multicellular Organisms." *Scientific American* 258(6):82–89.

Woese, C. 1981. "Archaebacteria." *Scientific American* 244(6):98–125. Early report on the molecular studies that pointed to separate origins for archaebacteria and eubacteria.

FUNGI AND PLANTS

How often, if we think of them at all, do we think of plants and fungi as being merely part of the scenery—a stationary backdrop for the riveting activities of animals? To be sure, fungi and plants don't crawl, run, leap, lurch, or fly about. And yet, many millions of years ago, simple aquatic plants managed to make the transition to life on land—and their descendants came to cloak the earth, inch by inch, continent by continent. Their evolution, while not smacking of the drama of, say, the rise and fall of dinosaurs, was nevertheless remarkable.

What kinds of adaptations allowed those organisms to spread through a world vastly different from the lagoons and pools they left behind? Here was a world of dry winds, of more pronounced seasonal variations in temperature and rainfall. Here, those early pioneers were no longer bathed perpetually in water that could provide them with dissolved nutrients and serve as a medium for dispersing their new generations.

Increasingly elaborate stalks, vascular tissues, spores, sporophytes, seeds—these were some of the passports to life on ever higher and drier land. Be glad that ancient plants made the journey. Without those plants, without their interactions with fungi that accompanied them onto land, we humans and all other terrestrial animals never would have evolved.

With this thought in mind, let's begin thinking about the adaptations that characterize the fungi and plants. Like other eukaryotes, their cells contain numerous organelles—the legacy, perhaps, of symbiotic interactions among their prokaryotic ancestors (page 576). But well before the dawn of the Cambrian, some 570 million years ago, the forerunners of fungi and plants were already traveling different evolutionary roads. Like those ancestral forms, all existing fungi are adapted for a heterotrophic way of life; they cannot produce their own food. In contrast, nearly all existing plants are photosynthetic autotrophs. The difference in their mode of nutrition is one of the key reasons for placing these organisms in two entirely separate kingdoms, which we will now describe.

PART I. KINGDOM OF FUNGI

The Fungal Way of Life

Fungi are decomposers of the first rank. They can break down just about anything that has organic components—nature's garbage (such as dead plants and animal wastes), your groceries (bread, meat, fruit, vegetables, cheese), and many of your possessions (clothing, paper products, photographic film, shoe leather, paint). Some fungi even grow in jet fuel and cause trouble by clogging the fuel lines.

Fungi living on the materials just mentioned are **saprobes**: they obtain nutrients from *nonliving* organic mat-

Figure 40.1 *Pilobolus*, one example of the numerous strategies for dispersal by which fungi and plants manage to spread themselves about through land environments. *Pilobolus* is a fungus that grows on animal feces. The dark sacs at the end of each stalked structure contain spores, which are reproductive units that can give rise to a new fungal body. The stalk grows in such a way that incoming rays of sunlight converge at the base of the swollen portion of the stalk (that is, just below the dark sacs). Turgor pressure inside a vacuole in the swollen portion becomes so great that the spore sac can be blasted two meters away—a remarkable feat, considering that the stalk is only five to ten millimeters tall!

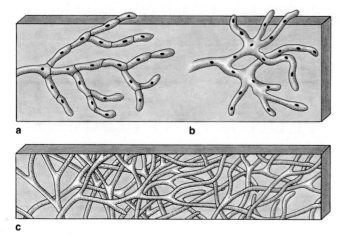

Figure 40.2 Fungal hyphae with cross-walls (**a**) and without cross-walls (**b**). A mycelium (**c**) is a complex mesh of hyphal filaments.

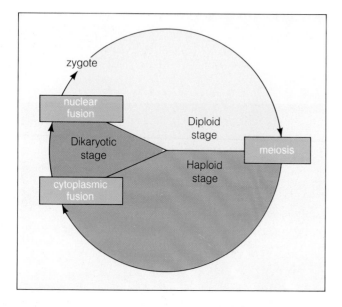

Figure 40.3 Generalized life cycle for most sexually reproducing fungi. The relative length of each stage varies, depending on the species. For example, the dikaryotic stage is extended among species of club fungi but extremely brief among the species of other fungal groups.

ter. Most fungi are saprobes, but many are **parasites**: they obtain nutrients directly from organic matter that is still part of a living host. In both cases, fungal cells secrete enzymes that promote digestion *outside* themselves. Large organic molecules are broken down by the digestive enzymes into smaller components that fungal cells can absorb.

Fungi alter their food source as they feed on it, and this happens to be one of the vital activities in the world of life. Think about the annual volume of wastes and remains of insects, birds, and other animals as well as the debris of plants. Each season, one deciduous elm tree alone may shed 400 pounds of leaves! Without the activity of saprophytic fungi and other decomposers, natural communities gradually would be buried in their own garbage, nutrients could not be recycled, and life could not go on.

Fungal Body Plans

Although there are some single-celled forms, such as the yeasts, the vast majority of fungi are multicelled. In either case, the body plan is adapted for two central tasks: absorbing nutrients and reproducing.

Part of the fungal body grows over or in the material from which nutrients are absorbed—organic matter in soil, dead plant tissues, and so on. For most multicelled fungi, the food-absorbing part of the body is the **mycelium** (plural, mycelia): a mesh of microscopic filaments that branch in different directions (Figure 40.2).

Each filament of a mycelium is a **hypha** (plural, hyphae). Most hyphae are composed of elongated cells that have chitin-reinforced walls. Sometimes the wall is absent between abutting regions of adjacent cells in the filament, and cytoplasm can move unimpeded from one cell to the next. More often, crosswalls occur between the adjacent cells, but they are perforated (some more so than others), so cytoplasm still can flow from cell to cell. The cytoplasmic movement is important, for it allows nutrients to be transported from food-absorbing parts of the body to other, nonabsorptive parts, such as the reproductive structures.

In many multicelled fungi, hyphae become interwoven and modified to form distinct reproductive structures, some with stalks and dense tissue masses. One type of structure is the **gametangium** (plural, gametangia), which produces male or female gametes for sexual reproduction. Another type of structure is the **sporangium** (plural, sporangia), which produces spores.

A **fungal spore** is a walled reproductive unit, derived from either mitosis or meiosis, that contains one or more nuclei and that can be dispersed from the parent body. (In single-celled fungi, spores simply form inside the saclike parent cell.) When the spore of a multicelled fungus germinates, it gives rise to a tubelike structure that grows and differentiates into a new mycelium.

Overview of Reproductive Modes

Depending on the species and on prevailing conditions, a fungus may reproduce asexually, sexually, or both. The *asexual* modes include the following:

1. Spore formation.

2. Binary fission (the parent body divides into two equivalent parts) or budding (part of the parent body pinches off and grows into a new individual).

3. Growth of hyphal fragments (broken away from the parent body) into separate fungal bodies.

What about *sexual* reproduction? As you know, fertilization usually involves the cytoplasmic fusion of two gametes, followed immediately by nuclear fusion to form the zygote. Many fungal species stretch out this event. They have a **dikaryotic stage** in which cells contain two separate haploid nuclei, each derived from a separate parental gamete (Figure 40.3). Often the dikaryotic stage continues for many cell generations before nuclear fusion occurs.

Major Groups of Fungi

Fungi probably arose during the Proterozoic era, long before the first multicelled animals appeared (Figure 38.14). The existing fungi have these features in common: (1) they are eukaryotic, (2) their mode of nutrition is extracellular digestion and absorption, and (3) their vegetative (nonreproductive) cells usually have chitin-reinforced walls.

About 50,000 species of fungi have been identified, although the actual number may be four times as high. Table 40.1 lists the major classes of fungi, which are known as the chytrids, water molds, zygospore-forming fungi, sac fungi, and club fungi. Also listed are the "imperfect" fungi, a group that does not have formal taxonomic status, for reasons that will soon be apparent.

The chytrids and water molds are members of the division Mastigomycota, all of which can reproduce asexually by the production and germination of motile, flagellated spores. The other fungi listed are members of the division Amastigomycota, all of which form nonmotile spores that must be dispersed by air currents or other means.

Chytrids

The chytrids are mostly saprobes in muddy or aquatic habitats. They feed on the cells of decaying plants, although some parasitize living plants, animals, and even other fungi. Some chytrids, such as the one shown in Figure 40.4, are single-celled. They begin life as a motile spore that settles onto a host cell, germinates, then grows into a globe-shaped cell having **rhizoids**—

Table 40.1	Classification of Fungi	
Group	Common Name	Typical Habitats
Chytridiomycetes	Chytrids	Aquatic (mud, decaying plants or animals); some parasitic
Oomycetes	Water molds and related forms	Aquatic; some parasitic
Zygomycetes	Zygospore-forming fungi	Soil, decaying plant parts; some parasitic
Ascomycetes	Sac fungi	Soil, decaying plant parts; many pathogens of plants
Basidiomycetes	Club fungi	Soil, decaying plant parts; many pathogens
Deuteromycetes*	Imperfect fungi	Diverse (e.g., soil, grains, human body)

*These groups of fungi are of undetermined affiliation; the name has no formal taxonomic status.

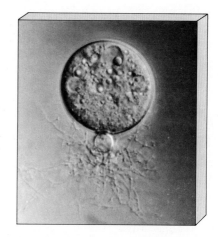

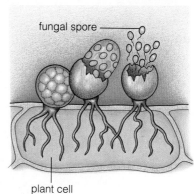

Figure 40.4 A chytrid, one of the microscopically small aquatic fungi. The sketch shows how motile spores are released from one type of chytrid that has parasitized a plant cell.

Figure 40.5 Powdery mildew of grapes, the outcome of an attack by *Plasmopora viticola*, one of the oomycetes.

Figure 40.6 Life cycle of *Rhizopus stolonifer* (the bread mold). As for other zygospore-forming fungi, the haploid stage dominates. Asexual reproduction is common, but different mating strains (+ and −) also can reproduce sexually. In both cases, haploid spores form that give rise to new mycelia.

Chemical attraction between + hypha and − hypha causes them to fuse. Two gametangia form, each with several haploid nuclei inside. Later their nuclei fuse to form a zygote. The zygote develops a thick coat, thereby becoming a zygospore, and may remain dormant for several months. Meiosis occurs during germination of the zygospore, and new spores are produced.

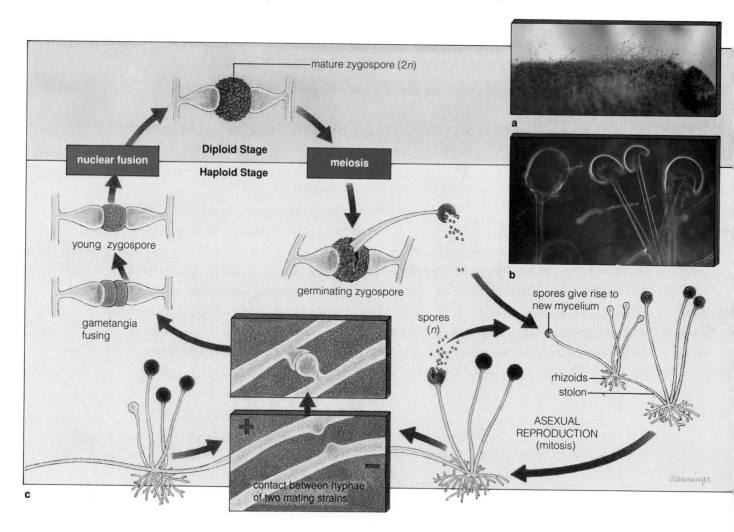

mature zygospore (2n)

nuclear fusion

Diploid Stage

Haploid Stage

meiosis

young zygospore

germinating zygospore

gametangia fusing

spores give rise to new mycelium

spores (n)

rhizoids
stolon

ASEXUAL
REPRODUCTION
(mitosis)

contact between hyphae of two mating strains

a

b

c

Hennings.

rootlike appendages that function in absorption. At maturity, the cell is transformed into a sporangium from which new spores are released.

Water Molds

The water molds (Oomycetes) are important decomposers in aquatic ecosystems. Some parasitize land plants (Figure 40.5), but most are saprobes or parasites on algae or simple animals (such as rotifers) that live in lakes, streams, and other bodies of water. The cottony growths you may have seen on pet goldfish or tropical fish are the well-developed mycelia of *Saprolegnia*, a parasite.

Water molds generally have an extensive mycelium, and some of the hyphae become modified to form gametangia. Fusion of a male and female gamete produces a diploid zygote, which develops a thick wall and thereby becomes a **resting spore**. Spore germination is followed by the development of a new mycelium.

Zygospore-Forming Fungi

Many of the zygospore-forming fungi are saprobes that live in soil, on decaying plant matter, or on food that is being stored. Others are parasites on insects, including the common housefly. All are alike in that each zygote produced during sexual reproduction develops a distinctively thick wall. This thick-walled structure, which can protect the zygote through a period of dormancy, is called a **zygospore**.

Figure 40.6 shows the zygospore of the bread mold (*Rhizopus stolonifer*), which commonly forms on stale baked products. When the zygote breaks dormancy, it gives rise to a sporangium in which haploid spores are produced. Each spore can be the start of an extensively branched mycelium. Sometimes so many sporan-

gia are produced that moldy bread looks black. The spores themselves are lightweight, dry, and readily dispersed by air currents. Figure 40.1 describes how another zygospore-former, *Pilobolus*, disperses its spores.

Sac Fungi

There are more than 30,000 species of sac fungi, which produce haploid spores in pouchlike cells called **asci** (singular, ascus). The simplest members are the yeasts, which are single-celled. Yeasts occur naturally in the nectar of flowers and on fruits and leaves. One commercially important yeast, *Saccharomyces cerevisiae*, produces the carbon dioxide that leavens bread and the ethanol in wine, beer, and other alcoholic beverages (Figure 40.7). Yeasts reproduce asexually by budding or fission; they also reproduce sexually, most often by the fusion of two gametes that resemble the vegetative cells.

Multicelled sac fungi are far more numerous than the yeasts. They all produce **ascocarps**, which are complex reproductive structures that bear or contain the asci. Some ascocarps look like globes, others like flasks or open dishes (Figures 40.8 and 40.9). Their mycelia grow through soil, wood, and other substrates.

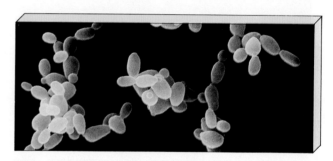

Figure 40.7 Budding cells of *Saccharomyces cerevisiae*, the yeast that makes bread dough rise.

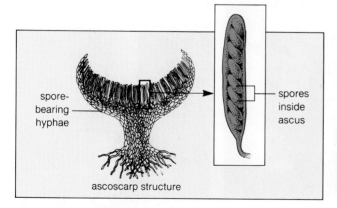

Figure 40.8 Photograph of scarlet cup fungi (*Sarcoscypha*), one of the sac fungi. The sketch shows the structure of a cup-shaped ascocarp, composed of tightly interwoven hyphae. Saclike structures (asci) that bear spores occur within ascocarps.

spore-bearing hyphae

spores inside ascus

ascoscarp structure

a

b

Figure 40.9 (a) Morels (*Morchella*), all of which are edible and delicious sac fungi. (b) Another sac fungus (*Venturia*) causes apple scab.

Some multicelled sac fungi are pathogenic. One species of *Neurospora* is a nuisance in bakeries, but another (*N. crassa*) is an important organism in genetic research. Also included in this class of fungi are the edible truffles and morels. Trained pigs and dogs are used to snuffle out truffles, which grow underground. No one has been able to grow truffles commercially; until this happens, these fungi will remain one of the most expensive luxury foods (they are priced by the gram, not by the pound).

Club Fungi

You may be familiar with some of the 25,000 or so species of club fungi—the mushrooms, shelf fungi, bird's nest fungi, stinkhorns, and puffballs—a few of which are shown in Figures 40.10 and 40.11. Most club fungi are saprobes that are important decomposers of plant debris. Many, including the rusts and smuts, cause serious plant diseases. Some species are edible; in fact, cultivation of the common mushroom (*Agaricus bisporus*) is a multimillion-dollar business.

Amanita muscaria, the fly agaric mushroom shown in Figure 40.10a, causes hallucinations when it is ingested (it contains intoxicants as well as toxins). It was used ritualistically in ancient societies in Central America, Russia, and India. Related species occur in the United States, but they induce nausea and vomiting rather than hallucinations. Others of this genus, including the death cap mushroom (*A. phalloides*, Figure 40.10b) produce deadly toxins. Within eight to twenty-four hours of ingesting even as little as five milligrams of the toxin, vomiting and diarrhea begin. Later, kidney and liver

a   b

Figure 40.10 Club fungi with hallucinogenic or deadly effects. (a) *Amanita muscaria*, the fly agaric mushroom, which contains intoxicants as well as toxins. (b) *A. phalloides*, the death cap mushroom which, if ingested, is usually fatal. (c) A related species (*A. ocreata*) known only in California also has caused fatalities.

c

b

c

d

Figure 40.11 Diversity in form among the club fungi. A coral fungus (**a**), shelf fungi growing outward from a tree trunk (**b**), bird's nest fungi (**c**), and a stinkhorn fungus (**d**).

cells start to degenerate; death can follow within a few days.

The spore-producing structures (basidia) of club fungi are usually club-shaped, and they always bear the spores on their outer surface. These structures typically appear on a short-lived reproductive structure, the basidiocarp. The part of the fungus that is visible above ground or on the surface of a log is the basidiocarp; the mycelium is buried in the soil or decaying wood. What we call "the mushroom" is one type of basidiocarp; it consists of a stalk and a cap. Its spore-producing structures occur on mushroom gills, which are sheets of tissue hanging from the lower surface of the cap (Figures 40.12 and 40.13).

When a basidiospore is dispersed to a suitable site, it germinates and gives rise to a haploid mycelium. Often, mycelia of two compatible mating types grow next to each other and cytoplasmic fusion occurs between them. Nuclear fusion does not follow immediately, so the resulting hypha is dikaryotic; its cells contain one nucleus of each mating type (Figure 40.12). The hypha grows into an extensive mycelium, and under favorable conditions, basidiocarps will form. At first, each spore-bearing structure of a basidiocarp is dikaryotic, but then its two nuclei fuse to form a zygote. The zygote undergoes meiosis, and haploid spores are produced, then dispersed by air currents.

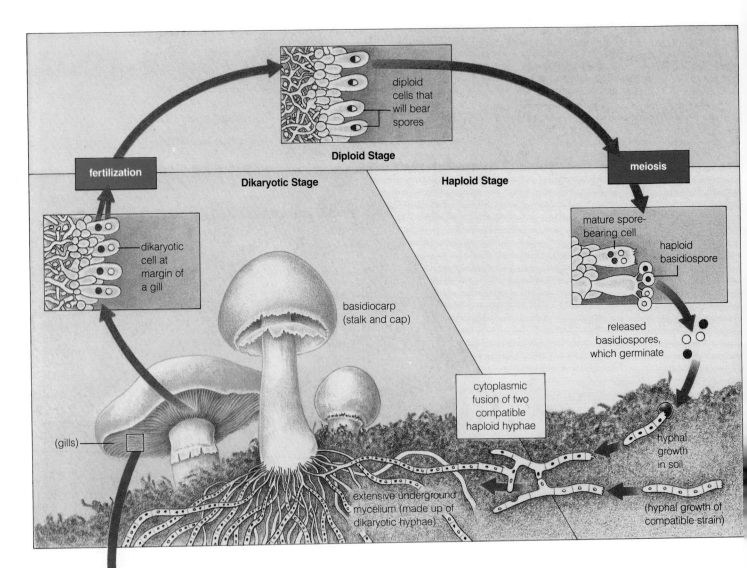

Figure 40.12 Life cycle of a typical mushroom. Notice the dikaryotic stage, in which hyphal cells contain two distinct nuclei.

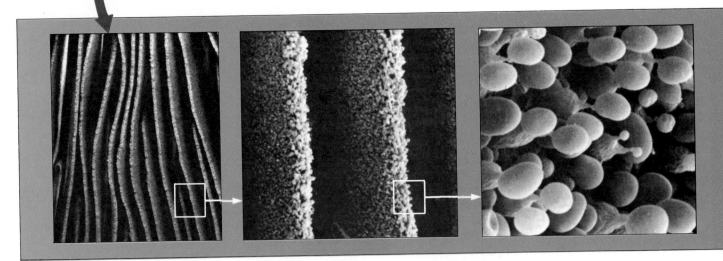

Figure 40.13 A close look at the gills of a basidiocarp from the common field mushroom seen at three magnifications. The highest magnification (far right) shows the stalked, club-shaped structures that bear the basidiospores.

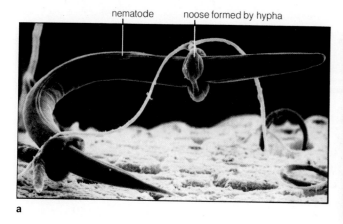

nematode   noose formed by hypha

a

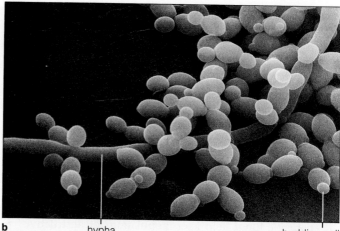

b

hypha   budding cell

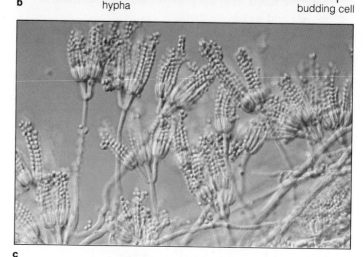

c

Figure 40.14 Representative imperfect fungi.

(**a**) *Arthrobotrys dactyloides*, a predatory fungus. The fungal hyphae form nooselike rings that can swell rapidly with incoming water after the hyphal cell walls are stimulated (as when a nematode brushes past). Within a tenth of a second, the increased turgor pressure shrinks the "hole" in the noose and captures the nematode. Hyphae grow into the animal's body and release digestive enzymes.

(**b**) *Candida albicans*, which causes "yeast infections" of the mucous membranes of the mouth and vaginal tract. A long hypha and budding cells are shown; 1,850×.

(**c**) *Penicillium*. This phase-contrast micrograph shows its distinct type of spore, called the conidium. Spores of this type are common among fungi. They usually form at the end or along the sides of the spore-producing cells and are easily shed at maturity.

Imperfect Fungi

In about 15,000 of the known fungal species, a sexual phase is absent or undetected. These fungi are sometimes called the deuteromycetes and sometimes the "imperfect fungi." They have no place in the formal classification scheme for fungi, which is based mainly on the mode of sexual reproduction. Frequently, researchers discover a sexual phase for one of the species and assign it to a recognized class (the sac fungi or club fungi, most often).

Many of these species are plant and animal pathogens. *Candida albicans* and others cause many of the fungal infections of humans (see Figure 40.14 and the *Commentary*). Some saprophytic members grow in stored grain that has become too damp, and their toxic metabolic wastes can cause cancer in humans who consume the poisoned grain. One type even shows predatory behavior by ensnaring nematodes (Figure 40.14).

Other imperfect fungi are commercially important. *Aspergillus* is used to produce the citric acid that imparts a lemon flavor to candies and soft drinks, and it is used in making soy sauce (the fungus ferments soybeans).

Species of *Penicillium* (Figure 40.14) produce the aroma and distinctive flavors of Camembert and Roquefort cheeses. The antibiotic penicillin, which is effective against many pathogenic bacteria, is also derived from *Penicillium*.

Mycorrhizae

Many land-dwelling plants and fungi enter into symbiotic relationships, in which both species benefit from their intimate association (Chapter Forty-Four). For example, the hyphae of certain club fungi form dense sheaths around the roots of forest trees, orchids, and other plants; they also penetrate between the cells of the root cortex. The combination of plant structures and fungal hyphae is a **mycorrhiza**. Figure 20.4 compared junipers grown with and without the benefit of mycorrhizal mats; the effects on plant growth are significant. Orchid seeds will not even germinate unless a mycorrhizal fungus invades them.

In this symbiotic relationship, the fungus regulates the flow of dissolved mineral ions into the plant; it con-

COMMENTARY

A Few Fungi We Would Rather Do Without

You know you are a serious student of biology when you can view organisms objectively in terms of their roles in nature, not in terms of the impact they have on humans generally and yourself in particular. As a student you can indeed respect that saprophytic fungi are vital decomposers in the web of life, and that many parasitic fungi help keep populations of destructive insects and weeds in check.

The true test is when you raid the refrigerator for a bowl of exorbitantly priced raspberries and discover that a fungal species beat you to them, or when another has begun feeding on the warm, damp tissues between your toes.

Who among us can praise the fungal species that cause athlete's foot and similar diseases that make the skin redden, crack, or turn scaly? Who can wax poetic about the fungi that can destroy wheat, apples, peaches, and nearly all other crop plants? Find the home gardeners who take pride in the presence of black spot or powdery mildew on their roses and other ornamental plants. Find the farmers who happily lose millions of dollars annually following invasions of their crops by rusts or smuts.

Among the pathogens listed in the Table are a few that have had a pronounced effect on the course of human history. *Phytophthora infestans* causes *late blight*, a disease that rots potato and tomato plants. More than a century ago, Irish peasants cultivated potatoes around their cottages and depended upon them as a major food source. Between 1845 and 1860, the growing seasons were cool and damp, year after year. *P. infestans* produced more spores in the cool conditions, and spore dispersal through the watery film on the plants went unimpeded. Thus the fungus spread rapidly and caused widespread destruction of the potato plants. During this fifteen-year period, a third of Ireland's population starved to death, died in the outbreak of typhoid fever that followed as a secondary effect, or fled to the United States and other countries.

And what about *Claviceps purpurea*, one of the sac fungi? *C. purpurea* is parasitic on rye and other grains. The alkaloids that are by-products of its activities can stimulate smooth muscle and block sympathetic nerve pathways in animals that eat the infected grain. In small amounts, the alkaloids have some medical uses. For example, they are used to treat migraine headaches; they also are widely used to return the uterus to prepregnancy size and to prevent hemorrhaging following childbirth. However, when the alkaloids are eaten in large amounts, as might occur in contaminated rye flour, a disease called *ergotism* develops. Symptoms include hysteria, hallucination, and convulsions as well as vomiting, diarrhea, and dehydration. Lesions may develop on the body's limbs and gangrene may occur. Severe cases are fatal.

Ergotism epidemics were common in Europe in the Middle Ages, when rye was a principal crop. Ergotism is known to have thwarted Peter the Great, the Russian czar who was obsessed with conquering ports along the Black Sea for his vast (and nearly landlocked) empire. The soldiers laying siege to the ports ate mostly rye bread and fed rye to their horses; the former went into convulsions, and the latter, into "blind staggers." There is even speculation that outbreaks of ergotism were an excuse to launch the Salem witchhunts in colonial Massachusetts.

Fungal Pathogens—Some Examples	
Oomycetes:	
Phytophthora infestans	Late blight of potato, tomato
Plasmopora viticola	Downy mildew of grapes
Zygomycetes:	
Rhizopus (various species)	Food spoilage; occasionally infect cattle, other animals
Ascomycetes:	
Ophiostoma ulmi	Dutch elm disease
Endothia parasitica	Chestnut blight
Venturia inequalis	Apple scab
Claviceps purpurea	Ergot of rye; ergotism in humans, sheep, cattle
Monilinia fructicola	Brown rot of stone fruits
Basidiomycetes:	
Puccinia graminis	Black stem wheat rust
Uromyces maydis	Smut of corn
Amanita (some)	Severe or fatal food poisoning
Deuteromycetes:	
Verticillium	Plant wilt
Microsporum, *Trichophyton,* *Epidermophyton*	Various species cause ringworms, including athlete's foot
Candida albicans	Infection of mucous membranes (e.g., moniliasis, a common vaginal infection)

serves ions when they are plentiful and makes them available to the plant when they are scarce in the soil. The fungus benefits from the partnership by absorbing carbohydrates from the cells of its host. Many plants seem to have difficulty absorbing phosphorus and other nutrients in the absence of mycorrhizae—which happen to be highly susceptible to acid rain (Chapter Forty-Seven). This susceptibility may adversely affect coniferous forests throughout the world.

Lichens

Some sac fungi and a few club fungi enter into what has been called a symbiotic relationship with a photosynthetic partner—typically a cyanobacterium or a green alga. Together they form a composite organism, the **lichen** (Figures 40.15 and 40.16). Lichens live in seemingly inhospitable places, including bare rock and wind-whipped tree trunks. Their secretions help degrade rock and convert it to soil that can support larger plants. Lichens growing in the arctic tundra, where large plants are scarce, are an important food source for reindeer and other animals. Sometimes air pollution is monitored by observing lichens near cities; lichens cannot grow in heavily polluted air.

The relationship often begins when a fungal mycelium contacts a free-living cyanobacterium, an algal cell, or both. The fungus parasitizes a photosynthetic cell, sometimes killing it in the process. If the photosynthetic cell can survive, however, it multiplies in association with the fungal hyphae. The fungus penetrates cells of its partner with short, specialized hyphae that can absorb as much as eighty percent of the food produced from photosynthesis. The fungus depends entirely on its cap-

tive cells for food, but the drain on nutrients has an adverse affect on the photosynthesizer's growth and reproductive capacity.

It is difficult to see what the photosynthetic partner gets from its enslavement. Of course, by entering into the association, photosynthetic cells can live on bare rock and other places that otherwise would be unavailable to them (the lichen can absorb and retain water, and the fungal hyphae provide some shade in intense sunlight). Only a few highly specialized green algae (*Trebouxia*) truly benefit from the partnership. They grow very slowly and cannot compete successfully with other organisms on their own, but they thrive in the shelter provided by a lichen.

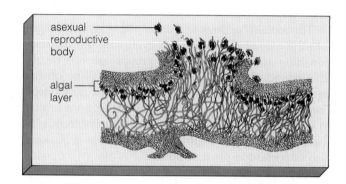

Figure 40.15 Cross-section through a complex lichen, *Lobaria verrucosa*. Fungal hyphae with a gelatinlike coat form the upper, protective layer. Just below is a layer of algal cells, which are functionally connected with loosely interwoven, thin-walled hyphae. The lower fungal layer attaches the lichen to the substrate. Notice the fragments containing both hyphae and algal cells; lichens reproduce asexually by such fragmentation.

b

Figure 40.16 Two lichens: (**a**) *Usnea*, known as old man's beard, as it appears on tree limbs, and (**b**) *Cladonia rangiferina*, sometimes called reindeer "moss."

a

PART II. KINGDOM OF PLANTS

Evolutionary Trends Among Plants

We turn now to a kingdom dominated by multicelled, photosynthetic autotrophs—to the plants that directly or indirectly nourish the vast majority of organisms in other kingdoms.

Plants have a long evolutionary history. We know from the fossil record that multicelled green algae were thriving by 700 million years ago. By 400 million years ago, simple green shoots were established on land. They had simple water-conducting cells within their photosynthetic stems; they absorbed water and nutrients through mycorrhizal fungi associated with their underground parts. By 345 million years ago, there were forests of seed-bearing plants with tall, woody stems and large, fernlike fronds.

The remains of these and other plants give us insight into the overall trends in plant evolution, but the relationships among the different groups are not clearly understood. For that reason, our focus will be on the trends themselves, using existing species as our models. Foremost among these was the trend from nonvascular to vascular body plans. Certain reproductive trends, of the sort outlined in Figure 40.18 and Table 40.2, were also important in plant evolution.

From Nonvascular to Vascular Plants. Plants require water, dissolved mineral ions, and sunlight for photosynthesis. These essential requirements surely influenced the changes in form that occurred during plant evolution. Consider the changes that were required for the transition to life on land. Aquatic plants obviously have no problem with obtaining water; and where upright growth does occur (as among the tall kelps), it occurs because the surrounding water supports much of the plant's weight.

Successful invasion of the land occurred through the evolution of spreading root systems that provided the plant with water and nutrients. It occurred through the evolution of upright shoot systems that exploited an abundant resource—sunlight energy. The evolution of expanded root and shoot systems could not have occurred without the development of vascular tissues.

Here we are talking about **xylem**, which transports water and dissolved mineral ions throughout the plant body; and **phloem**, which transports sugars and other photosynthetic products. Accompanying these developments were the evolution of sclerenchyma and other supporting tissues, which probably proved adaptive in allowing some plants to grow taller and escape the shade created by their neighbors. Similarly, a cuticle and numerous stomata in the epidermis of aboveground parts proved adaptive in limiting water loss in dry environments. These tissue specializations are characteristic of nearly all existing land plants (Chapter Nineteen).

Isogamy to Oogamy. The gametes produced by some aquatic green algae are all motile and of the same size, a condition called **isogamy**. The gametes produced by others are differentiated into motile sperm and immotile eggs, a condition called **oogamy**. In both cases, the gametes are released into the surrounding water, and fertilization occurs when gametes drift together or when one swims to the other.

How did the gametes of the first land plants get together in the absence of water? The simplest land

Figure 40.17 Plants at the boundary between two different worlds, one aquatic, one terrestrial. More than 400 million years ago, the ancestors of existing land plants crossed the boundary and made their successful bids for life on land. Shown here, one of the brown algae (sea palms) of the intertidal zone. Of all the aquatic algae, it appears that only the descendants of certain green algae made it onto dry land.

plants—the mosses and other bryophytes—provide us with a clue. They are all confined to moist environments; they cannot reproduce elsewhere. But they all depend on oogamy, with motile sperm swimming through a film of water to the immotile eggs—which typically remain with the parent body. Similarly, the more complex land plants are oogamous. They have sperm that are adapted for dispersal—and they have increasingly complex ways of holding onto and protecting the developing egg, as you will now see.

The Evolution of Seeds. During their life cycle, plants produce spores by the vegetative body called the sporophyte (page 158). **Plant spores** are reproductive units that form following meiosis, and they give rise to the gametophytes (gamete-producing bodies). The spores produced by some algae and simple vascular plants are alike; these plants are "homosporous." Two types of spores are produced by many vascular plants, which are said to be "heterosporous." They produce **megaspores** (which develop into female gametophytes) and **microspores** (which develop into male gametophytes).

Heterospory was the prelude to the evolution of seeds. In terms of sheer numbers and diversity, the seed-bearers are the most successful plants. All are heterosporous. The female gametophytes that develop from megaspores are surrounded by protective tissue, and they mature while still attached to the parent plant.

What is so important about that? The female gametophytes are nourished by the sporophyte—the stage that is well adapted for obtaining nutrients and water on dry land. The immature male gametophytes, or **pollen grains**, are released from the parent plant and travel to the female gametophytes by air currents, insects,

birds, or mammals. Perhaps more than any other factor, the evolution of pollen grains that carry sperm *without* requiring liquid water allowed seed-bearing plants to radiate into diverse land environments.

Fertilization occurs within the female gametophyte, and the zygote develops into a multicelled embryo sporophyte. The tissues surrounding the female gametophyte develop into a protective outer coat around the embryo and its food reserves. Together, the tissues and the embryo are the **seed** that is shed from the parental body. The seed is an ideal package for traveling through

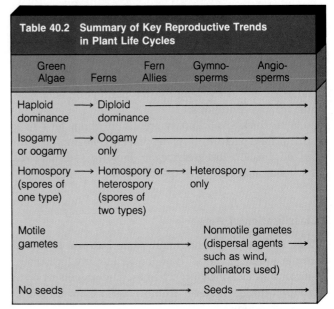

Table 40.2 Summary of Key Reproductive Trends in Plant Life Cycles

	Green Algae	Ferns	Fern Allies	Gymnosperms	Angiosperms
	Haploid dominance →	Diploid dominance	→		
	Isogamy or oogamy →	Oogamy only	→		
	Homospory (spores of one type) →	Homospory or heterospory (spores of two types) →	Heterospory only	→	
	Motile gametes ——————→			Nonmotile gametes (dispersal agents such as wind, pollinators used) →	
	No seeds ——————→			Seeds →	

From James R. Estes, University of Oklahoma.

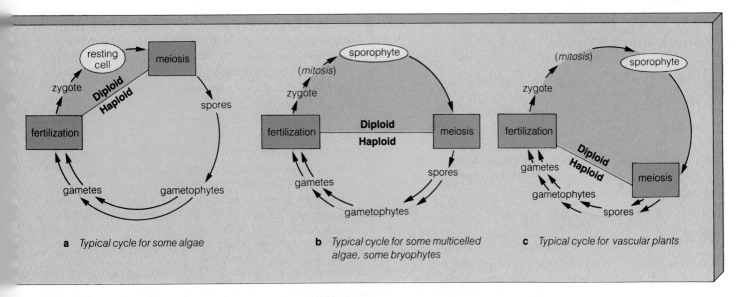

a *Typical cycle for some algae*

b *Typical cycle for some multicelled algae, some bryophytes*

c *Typical cycle for vascular plants*

Figure 40.18 Generalized plant life cycles, indicating evolutionary trends from haploid to diploid dominance that occurred among land plants.

Table 40.3 Overview of Adaptations of the Algae

Division (Phylum)	Some Representatives	Characteristic Habitat	Kinds of Tissue Differentiation
Rhodophyta (red algae)	*Porphyra, Nemalion*	Few freshwater; most marine; more abundant in warmer and coastal tropical seas	Single-celled to branched filaments; some with filaments massed into stemlike and leaflike structures; no vascular (internal transport) tissue
Phaeophyta (brown algae)	*Fucus* (rockweed), *Laminaria* and other kelps, *Sargassum*	Almost all marine, coastal waters especially; most abundant in colder seas	Branched filaments to complex parenchyma bodies; some with leaflike, stemlike (stipe), and anchoring (holdfast) structures; some with ducts for transporting photosynthetic products to different plant regions
Chlorophyta (green algae)	*Ulva* (sea lettuce), *Ulothrix, Spirogyra, Chlamydomonas, Acetabularia*	Most freshwater; moist soils; many in coastal tropical seas	Single-celled to filamentous, and simple colonial to simple sheetlike forms; no conducting tissue

and surviving often lengthy dry periods or otherwise unfavorable conditions.

With these evolutionary trends in mind, let's now turn to the spectrum of plant diversity, beginning with the red, brown, and green algae. Table 40.3 summarizes the major adaptations of these plants.

Classification of Algae

Algae is a term that originally came into use to define simple aquatic "plants." It no longer has any formal significance in classification schemes. The organisms once grouped under the term are now assigned to three different kingdoms:

Kingdom	**Division**
Plantae:	Rhodophyta (red algae)
	Phaeophyta (brown algae)
	Chlorophyta (green algae)
Protista:	Chrysophyta (golden algae, diatoms)
	Euglenophyta (euglenids)
	Pyrrophyta (dinoflagellates)
Monera:	Cyanobacteria (blue-green algae)

As you can see, the red, brown, and green algae are still assigned to the plant kingdom. Most aquatic plants belong to one of these three divisions.

Red Algae

Red algae (Rhodophyta) are important producer organisms of many aquatic communities, and some have commercial uses. Several species are used to make **agar**, a gelatinous substance used as a culture medium in laboratories throughout the world. Agar also is used as a moisture-retaining agent in baked goods and in gelatin desserts (to promote rapid setting). In Japan especially, a red alga ("nori") is commercially grown and harvested for food.

The red algae share three characteristics. First, they have no motile, flagellated cells. Second, their photosynthetic pigments include phycobilins. Third, the sexually reproducing red algae are oogamous, with immotile sperm and specialized, egglike structures.

Except for a few freshwater forms, the 4,000 or so species of red algae are marine organisms. Some live in shallow intertidal zones; others live 175 meters below the surface when the water is clear enough for light penetration and photosynthesis. As Figures 1.6f and 40.19b suggest, most are attached to rocks or to other substrates; only a few are free-floating.

"Red" algae actually can be green, red, purple, or greenish-black, depending on the types and relative amounts of their phycobilins. The phycobilins absorb blue-green wavelengths, which predominate in deeper waters. (The chlorophylls are more efficient at absorbing light of red and blue wavelengths, which do not penetrate far below the water's surface.)

Red algae show complex patterns of sexual reproduction. Most often the zygote develops by mitosis into a small, diploid sporophyte that is "parasitic" on the female gametophyte, deriving nourishment from its tissues. Its diploid spores develop into a larger, independent sporophyte in which meiosis occurs and haploid spores (which will form the gametophyte generation) are produced.

Brown Algae

The brown algae (Phaeophyta) include about 1,500 olive-green, golden, and dark-brown species, many of which are large seaweeds (Figure 40.20b). Abundant accessory

Light-Harvesting Structures	Main Reproductive Strategies
Chloroplasts; dominant phycobilin pigments (trap blue-green light in deep water) plus chlorophylls; some species with radial or bilateral branches (nonoverlapping exposure to sunlight)	Alternation of fertilization and meiosis; sexual reproduction based on oogamy; also some vegetative reproduction from plant body fragments or asexual spores
Chloroplasts; dominant xanthophyll pigments plus chlorophylls; leaflike structures	Most with alternation of multicelled generations; sexual reproduction based on isogamy or oogamy; vegetative reproduction from plant body fragments or asexual spores
Well-organized arrays of chloroplasts with membrane stacks; chlorophylls dominant	Alternation of fertilization and meiosis (some have resting spores); sexual reproduction based on isogamy to oogamy; also vegetative reproduction by fragments or spores

b

Figure 40.19 Two red algae, one showing the more common pattern of growth, which is branched and filamentous (**a**), and the other showing sheetlike growth (**b**).

Figure 40.20 Brown algae. (**a**) *Postelsia palmaeformis*, the sea palm, grows in the intertidal zone, where it is alternately submerged and exposed to air. (**b**) *Laminaria*, one of the kelps that usually stay submerged, even at low tide. Only the uppermost portions are visible here, at the water's surface; the long stalks are anchored to rocks below.

a

b

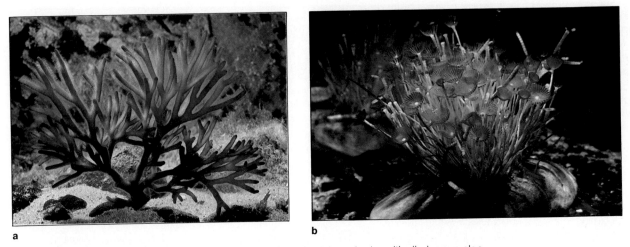

Figure 40.21 Representative green algae of marine habitats. (**a**) A branched, multicelled green alga (*Codium*) and (**b**) a group of "mermaid's wineglass," or *Acetabularia*. Each plant contains a multinucleate mass of cytoplasm (not distinct cells) that has a rootlike structure, a stalk, and a cap in which gametes form.

Figure 40.22 Life cycle of *Chlamydomonas*, a single-celled green alga. Asexual reproduction is most common, but sexual reproduction also can occur between spores of different mating strains.

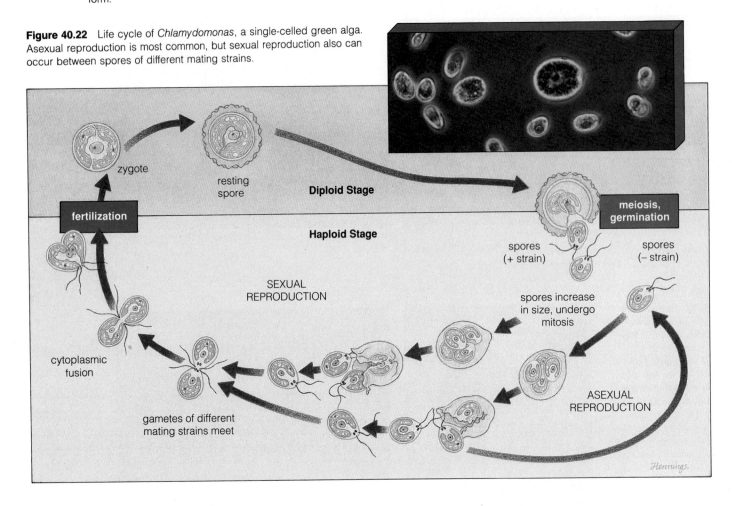

pigments, particularly the xanthophylls, give them their distinct coloration.

Most brown algae have leaflike blades and a **stipe** (a stemlike structure). Sometimes hollow, gas-filled stipe regions called "floats" occur at regular intervals and help hold the plant upright in water. Some species have tube-like strands of cells similar to those found in phloem.

Like many red and green algae, most brown algae show alternation of multicelled generations. Among simpler, filamentous species, the gametophyte and sporophyte may be similar in outward appearance. In more complex species, the gametophyte is extremely small and a large sporophyte dominates the life cycle. The kelps shown in Figure 40.20 are sporophytes; the gametophytes consist of microscopic filaments.

When you consume commercial ice cream, pudding, salad dressing, canned and frozen foods, jellybeans, or beer, when you use cough syrup, toothpaste, a variety of cosmetics, paper, textiles, ceramics, or floor polish, thank the brown algae. Some of these plants produce **algin**. This substance is used as a thickening, emulsifying, or suspension agent; it can control water penetration into porous surfaces and it can control the development of ice crystals.

Many brown algae live offshore or in the intertidal zone (Figure 40.17). They attach to submerged rocks and other substrates by **holdfasts**, which are rootlike structures at the base of the plant. The giant kelps sometimes grow more than fifty meters long, forming underwater forests that sway with the currents. The brown alga *Sargassum* is unusual in that it floats as tangled masses through the vast Sargasso Sea, which lies between the Azores and the Bahamas.

Green Algae

There are at least 7,000 species of green algae. The majority of green algae live in freshwater habitats; a few grow on snow, soil, and tree trunks or in marine habitats. Figure 40.21 shows two representatives.

Green algae do not have widespread commercial application, compared with the brown and red algae, but this picture may change with efforts to develop new food sources. Some Asian countries already harvest *Ulva* (sea lettuce) to a limited extent. *Chlorella*, a single-celled alga, is being investigated as a source of oxygen for submarines and space stations.

Green algae are like complex land plants in terms of their photosynthetic pigments; all have chlorophylls *a* and *b*, carotenoids, and xanthophylls. Also like land plants, their cells store carbohydrates as starch and have cellulose walls.

The life cycles of green algae show diverse patterns of sexual and asexual reproduction. Some species of *Chlamydomonas* are isogamous. Haploid cells of different mating types function as gametes, although they are identical in size and structure (Figure 40.22). There is no multicelled stage, only an alternation between haploid and diploid cells. Other species of *Chlamydomonas* are oogamous, with gametes differing in size and motility.

Spirogyra (watersilk) is a filamentous green alga; Figure 40.23 shows one of its reproductive modes. Spores produced by the leaflike *Ulva* give rise to "male" and "female" gametophytes that look alike. They also look just like the sporophyte (Figure 40.24).

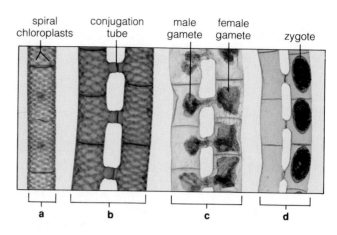

Figure 40.23 One mode of sexual reproduction in watersilk (*Spirogyra*), a filamentous green alga having the structure shown in (**a**). Ribbonlike, spiral chloroplasts occur in each cell; the central dark structure in each cell is the nucleus. (**b**) A conjugation tube can form between two cells of adjacent filaments of different mating strains. (**c**) The contents of one cell function as the male gamete, and the contents of the other, as the female gamete. Fusion of the two gametes produces a zygote, which forms a resting cell. The resting cell later germinates and produces a new filament.

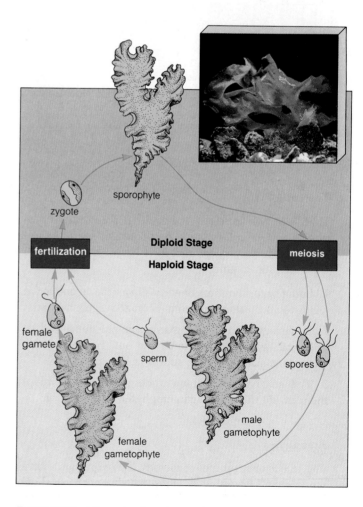

Figure 40.24 Life cycle of sea lettuce (*Ulva*), a green alga.

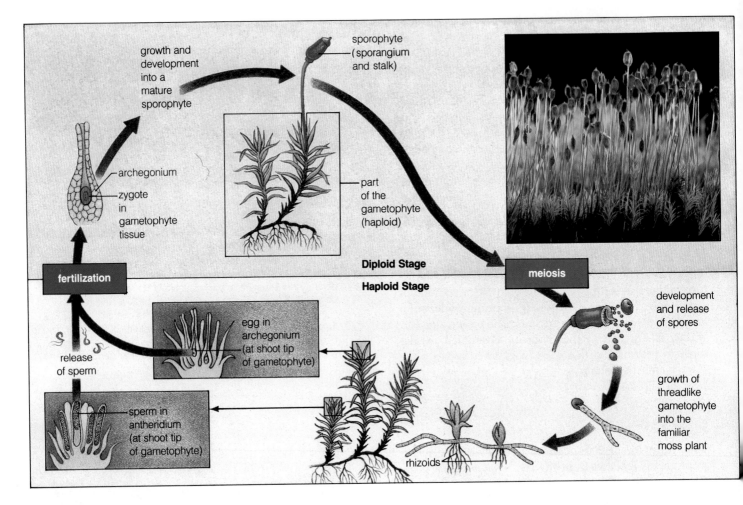

Figure 40.25 Life cycle of a moss, a representative bryophyte.

The Land Plants

There are two lineages of terrestrial plants: the bryophytes and vascular plants. Both lineages share the following features, which must have been adaptive during the transition to land:

1. Some bryophytes and nearly all vascular plants have a cuticle and numerous stomata on aboveground parts.

2. A protective cell layer surrounds the reproductive cells and keeps them from drying out. For example, a *sporangium* surrounds the spores in the sporophyte.

3. The embryo sporophyte undergoes early development *within* the female gametophyte.

Bryophytes

The 16,000 or so species of mosses, liverworts, and hornworts are all bryophytes. Figures 40.25 and 40.26 show examples. These are small plants, generally less than twenty centimeters (less than eight inches) long. Most grow in regions that are moist for at least part of the year. Unless there is a film of water on plant parts, the sperm cannot swim to the eggs and sexual reproduction will not occur.

Bryophytes have leaflike, stemlike, and rootlike organs. (Most species have rhizoids, these being long cells or filaments that attach the gametophyte to the substrate and absorb water and mineral ions.) Because these organs do not have xylem and phloem, they are not regarded as true leaves, stems, and roots.

Mosses are the most common bryophytes. When a moss spore germinates, it produces a green, threadlike gametophyte that resembles a filamentous green alga. The filament grows by way of mitosis into the familar moss plant (Figure 40.25). Generally, each haploid gametophyte produces antheridia, archegonia, or both at the shoot or branch tips. *Antheridia* (singular, antheridium) are protective cell layers that surround sperm in male gametophytes; *archegonia* (singular archegonium) are

Figure 40.26 Sexual and asexual reproductive structures of *Marchantia*, a type of bryophyte called a liverwort. Unlike other liverworts, *Marchantia* produces male *and* female structures on separate gametophytes. Part of the male gametophyte looks like an umbrella (**a**) and the female gametophyte, like an abnormally armed sea star (**b**). Asexual reproduction involves tiny cups that develop on the thallus of this plant. (A "thallus" is a simple plant body, devoid of roots, stems, and leaves.) The cups (**c**) hold vegetative bodies (gemmae) that can grow into new plants after air currents carry them to suitable substrates.

protective cell layers around the eggs. These structures usually occur on separate plants growing near each other, and sperm must swim through a film of water to reach the eggs.

After fertilization, the zygote develops into a mature sporophyte, which consists of a stalk and a sporangium. At first, the sporophyte may be photosynthetic, but it eventually comes to depend on the gametophyte for nutrients and water. The bryophytes are distinct from all other land plants in having an independent gametophyte and a dependent sporophyte.

Lycopods, Horsetails, and Ferns

Lycopods, horsetails, and ferns once were the dominant forms of terrestrial plants (Figure 40.27). Today, most of their descendants are miniature versions of those exotic forms. They all differ from the bryophytes in having an independent, free-living sporophyte that is the most conspicuous stage of the life cycle.

Lycopods. The lycopods (Lycophyta) were highly diverse 350 million years ago, when some forms were even tree-sized. Among the existing species are the tiny club mosses (members of the genus *Lycopodium*) that typically grow on forest floors. The sporophyte of a club moss has true roots, stems, and small leaves with a single

Figure 40.27 Reconstructions of vascular plants of the Carboniferous. Most of the tall, treelike forms shown here became extinct. The descendants of such forms include the modern-day lycopods, horsetails, and ferns.

Figure 40.28 (a) *Lycopodium* sporophytes, showing the conelike strobili in which spores are produced. (b) Horsetails (*Equisetum*) showing the strobili borne on stems. (c) Detail of horsetail strobilus showing clusters of sporangia.

Figure 40.29 Ferns in a temperate rain forest of Tasmania.

strand of vascular tissue. The sporophyte often has **strobili** (singular, strobilus), which are cone-shaped clusters of leaves bearing the sporangia (Figure 40.28). Spores are dispersed from the sporangia and germinate to form small, free-living gametophytes. Although species of *Lycopodium* are homosporous, those of other genera are heterosporous. All require ample water for sperm to swim to the eggs.

Horsetails. The ancient relatives of horsetails (Sphenophyta) included treelike forms about fifteen meters tall. A single genus, *Equisetum*, has survived to the present (Figure 40.28).

Horsetails characteristically grow in moist soil along streams and in disturbed habitats, such as vacant lots, roadsides, and the beds of railroad tracks. Their sporophytes typically have **rhizomes**, these being underground stems, with roots and aerial branches arising at nodes. Scalelike leaves are arranged in whorls about a hollow, photosynthetic stem. Pioneers of the American West used horsetails to scrub their cooking pots; the walls of stem cells contain silica, giving the stems a sandpaper quality.

All horsetails are homosporous. The spores form in strobili at the shoot tips and then are dispersed by air currents. They must germinate within a few days to produce gametophytes, which are independent, free-living green plants about the size of a small pea. Here again, sperm must travel through free water (rain or dewdrops) to reach the eggs.

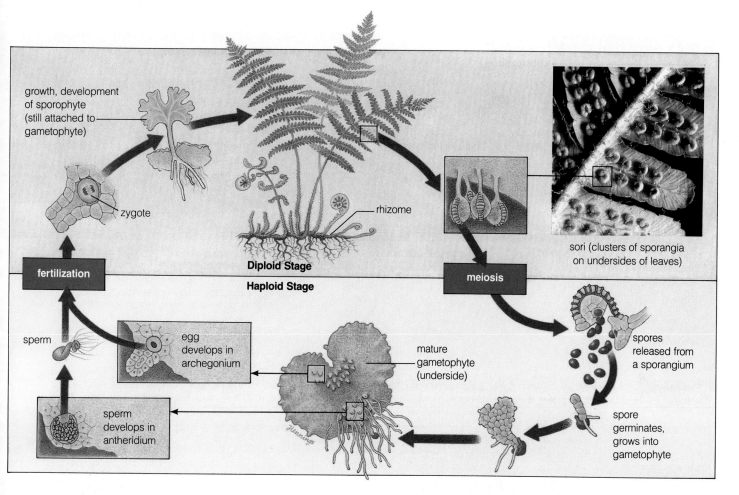

growth, development
of sporophyte
(still attached to
gametophyte)

zygote

rhizome

Diploid Stage

sori (clusters of sporangia
on undersides of leaves)

fertilization

Haploid Stage

meiosis

sperm

egg
develops in
archegonium

mature
gametophyte
(underside)

spores
released from
a sporangium

sperm
develops in
antheridium

spore
germinates,
grows into
gametophyte

Henning

Figure 40.30 Fern life cycle.

Ferns. There are about 12,000 species of ferns (Pterophyta), and they include some of the most popular of all nursery plants. Most ferns are native to tropical and temperate regions (Figure 40.29). Some floating species are less than 1 centimeter (0.4 inch) across; some tropical tree ferns are 25 meters (82 feet) tall. Except for tropical tree ferns, the fern stems are mostly underground. Fern leaves, or **fronds**, are usually featherlike, with the blades divided into segments, but extraordinary diversity exists on the basic plan.

Most fern species are homosporous. You may have noticed rust-colored patches on the lower surface of the fronds of many ferns. These are clusters of sporangia. Each cluster is called a **sorus** (plural, sori); some are protected with flaps of tissue. Generally, each sporangium is microscopic and looks rather like a baby rattle (Figures 40.30 and 40.31). When the time comes for spores to be dispersed, a row of thick-walled cells at the sporangium surface snaps open, causing the spores to be catapulted through the air. When a spore lands at a suitable site, it germinates and develops into a small gametophyte, the most common being the green, heart-shaped type shown in Figure 40.30.

As is the case for lycopods and horsetails, the sporophytes of ferns have well-developed vascular systems.

Figure 40.31 Scanning electron micrograph of sporangia from a fern leaf.

Although all three kinds of plants are adapted for life on land, they are confined largely to wet, humid regions because their short-lived gametophytes have no vascular tissues for water transport. Moreover, the male gametes must have water to reach the eggs. Thus the lycopods, horsetails, and ferns are the "amphibians" of the plant kingdom: *for part of the life cycle they are, in essence, aquatic.* The few species that are adapted for survival in extreme environments such as deserts can reproduce sexually only when adequate water is available.

Table 40.4 Overview of Adaptations of Existing Land Plants

Division (Phylum)	Some Representatives	Characteristic Environment	Kinds of Tissue Differentiation
Bryophytes (non-seed-bearing; depend on free water for fertilization)			
Bryophyta (bryophytes)	mosses, liverworts, hornworts	Most land; moist, humid sites; some arid sites; a few submerged sites	Threadlike anchoring structures (rhizoids); leaflike and stemlike structures, often branched; some with simple water- and food-conducting tissues; some have guard cells in epidermis
Ferns and Fern Allies (non-seed-bearing; depend on free water for fertilization)			
Lycophyta (lycopods or club mosses)	*Lycopodium, Selaginella*	Land; wet, shaded sites in temperate zones, tropics, and subtropics; some arctic or desert	Sporophyte with true vascular system; roots, stems, leaves; stomata with guard cells in one or both leaf surfaces; vascular system makes possible a higher volume-to-surface ratio than in nonvascular plants. *These features also occur in all plant divisions listed hereafter.*
Sphenophyta (sphenopsids)	*Equisetum* (horsetails), only living genus	Land; sand dunes, swamps, moist woodlands, lake margins, railroad embankments	Extensive underground stem (rhizome) with roots and aerial shoots at nodes; aerial stems jointed, hollow, branched or unbranched
Pterophyta (ferns)	Sword ferns, lady ferns, bracken ferns, tree ferns	Land; some epiphytes (attached to but nonparasitic on other plants); most wet, humid sites	True vascular system; conspicuous leaves; most with creeping rhizomes; columnlike stem in tree ferns
Gymnosperms ("naked-seed-bearing"; not dependent on free water for fertilization)			
Cycadophyta (cycads)	*Zamia*	Limited tropical, subtropical land regions	Stems short and bulbous, or columns; palmlike appearance; fernlike leaves; massive cones (reproductive structures bearing pollen or seeds). *Well-developed vascular system; this is also true of all divisions listed hereafter.*
Ginkgophyta (ginkgos)	*Ginkgo biloba* (only existing species)	Land; temperate regions	Tall, woody stem (trunk) with long side branches; fan-shaped, deciduous leaves
Gnetophyta (gnetophytes)	*Gnetum, Ephedra, Welwitschia*	Land; warm-temperate to tropical regions; desert and mountain sand or rocky soil	Some shrubby, branched, with whorled leaves; *Welwitschia* short, bowl-shaped stem, two large, strap-shaped leaves, cones
Coniferophyta (conifers)	Pine, spruce, fir, larch, juniper, hemlock, cypress, redwood	Land; widespread through Northern and Southern hemispheres	Some shrubby, others impressively tall, thick-trunked trees; most evergreen; cones
Angiosperms (flowering, seed-bearing; depend on pollinating agents; not dependent on free water for fertilization)			
Anthophyta Monocotyledonae (monocots)	Grasses, palms, lilies, orchids, onions, pineapple	Almost all land zones; some aquatic	Floral structures; one seed leaf (cotyledon); floral parts generally in threes or multiples of threes; parallel-veined leaves common*
Dicotyledonae (dicots)	Most temperate-zone fruit trees; roses, cabbages, melons, beans, potatoes	Almost all land zones; some aquatic	Two seed leaves; floral parts generally in fours, fives, or multiples of these; net-veined leaves common*

*See Figure 19.2.

Light-Harvesting Structures	Main Reproductive Strategies
Well-organized arrays of chloroplasts with membrane stacks; leaflike structures show radial and bilateral symmetry	Alternation of generations (this is true of all divisions listed hereafter); diploid plant body produces homospores; gametophyte is dominant generation with dependent sporophyte; vegetative reproduction mostly by asexual reproductive bodies (gemmae)
Well-organized arrays of chloroplasts with membrane stacks; most with spirally arranged leaves; some with palisade layer of photosynthetic cells in leaves	Complex, dominant sporophyte independent of gametophyte; some homosporous, others heterosporous
Well-organized chloroplast arrays; upright growth habit; whorled, scalelike leaves but most photosynthesis in stems	Complex, dominant sporophyte free-living (independent of gametophyte); homosporous
Well-organized chloroplast arrays; most photosynthesis in stems	Sporophyte free-living; most homosporous
Well-organized chloroplast arrays; complex leaves, arranged in pattern that allows good sunlight exposure	Dominant, woody sporophyte body, heterosporous; well-developed seed- and pollen-bearing cones
Well-organized chloroplast arrays; complex leaves, arranged in pattern that allows good sunlight exposure	Dominant, woody sporophyte body; heterosporous; well-developed seeds not borne in cones; separate male and female sporophytes
Well-organized chloroplast arrays; in *Ephedra*, most photosynthesis in stems and branches (almost leafless)	Dominant, woody sporophyte body; heterosporous; seed- and pollen-bearing cones; separate male and female sporophytes
Well-organized chlorcplast arrays; most with needlelike or scalelike leaves	Dominant, woody sporophyte body; heterosporous; well-developed seed- and pollen-bearing cones
Well-organized chloroplast arrays; complex leaves, arranged in patterns that allow good sunlight exposure	Dominant woody or herbaceous sporophyte body; diverse floral structures adapted to pollinating agents; heterosporous; seeds enclosed within ovary; fruits (ripened ovaries) aid in seed dispersal by wind, water, animals

Figure 40.32 Two seed-bearing plants; the cycad (**a**) and the ginkgo (**b**). Cycad cones are shown in (**c**); the seeds of the ginkgo are shown in (**d**).

Existing Seed Plants

The ancestors of existing seed-bearing plants first appeared during Devonian times. Fossil evidence of one representative, a "progymnosperm," was shown earlier in Figure 38.13. Apparently those plants were on the evolutionary road leading to the **gymnosperms** and **angiosperms**.

The word gymnosperm is derived from the Greek *gymnos* (meaning naked), and *sperma* (which is taken to mean seed). As the name implies, gymnosperm seeds are perched at the surface of reproductive structures and are rather unprotected. The word angiosperm is derived from the Greek *angeion* (meaning vessel) and *sperma*. The "vessel" part of the name refers to the carpels within which seeds develop (page 279). In other words, the young seeds of angiosperms are surrounded by protective tissue layers.

The Gymnosperms

The sporophytes of nearly all gymnosperms are conspicuous trees or shrubs. The small gametophytes are not free-living but rather are enclosed in seeds by sporophyte tissues. The most familiar gymnosperms are the conifers (Coniferophyta). Others are the cycads (Cycadophyta), ginkgo (Ginkgophyta), and the gnetophytes (Gnetophyta).

Cycads. During the Mesozoic era, the cycads flourished along with the dinosaurs. About 100 species have survived to the present, but they are confined to the tropics and subtropics. At first glance you might mistake a cycad for a small palm tree (Figure 40.32a). Despite having similar leaves and stems, the palms and cycads are not closely related (palms are angiosperms). Cycads have massive, cone-shaped strobili that bear either pollen or ovules (Figure 40.32c). Pollen from the "male" plants is transferred by air currents or crawling insects to the developing seeds on "female" plants. In tropical Asia, cycad seeds and a starchy flower made from cycad trunks are used as food sources.

Ginkgos. The ginkgos are even more restricted in native distribution than the cycads. Only a single species has survived to the present, despite the diversity of the group during the Mesozoic. It seems that several thousand years ago, ginkgo trees were planted extensively in cultivated grounds around Buddhist temples in China. Since then, the natural populations from which the domesticated trees were derived must have become extinct. The near-extinction of this living fossil from the age of dinosaurs is puzzling, for ginkgos seem to be hardier than many other trees. They are planted in cities because of their attractive, fan-shaped leaves (Figure 40.32) and their resistance to insects, disease, and air pollutants. Nevertheless, they have survived only through human protection.

Gnetophytes. There are about seventy species of gnetophytes, which are divided into three genera: *Gnetum*, *Ephedra*, and *Welwitschia*. Of all the gymnosperms, *Welwitschia* is the most unusual. Most of this seed-producing plant is a deep-reaching taproot; the only exposed part is a woody disk-shaped stem that bears cone-shaped strobili and leaves. It never produces more than two strap-shaped leaves, which split lengthwise repeatedly as the plant grows older, the outcome being a rather straggly pile (Figure 40.33).

Conifers. Pine, spruce, fir, hemlock, juniper, cypress, and redwood are all conifers. The strobili of all these plants are distinctively cone-shaped (hence the name conifer, which means cone-bearing). Most conifers are woody trees and shrubs with needlelike or scalelike leaves. Most are evergreens; although they shed old leaves throughout the year, they retain enough leaves to distinguish them from deciduous species.

Conifer life cycles vary in the duration of events, but we can use the pine as a general example (Figure 40.34). As with all conifers, pine seeds develop on the shelflike scales of a female cone. Each immature seed, or ovule, consists of a tissue layer (the megasporangium) surrounded by protective layers that will become the seed coat. Inside the megasporangium, a reproductive cell undergoes meiosis, and one of the resulting haploid spores (the megaspore) survives to form a many-celled female gametophyte.

The male gametophytes originate within pollen sacs (microsporangia), which are borne on the scales of male cones. Inside the sacs, reproductive cells undergo meiosis and microspores are produced. The microspores develop into pollen grains, which contain the immature male gametophytes (Figure 21.4).

Each spring, a pine tree produces millions of pollen grains. Each pollen grain floats through the air, having been equipped with two tiny air sacs that may impart buoyancy to it. With such extravagant discharges from the male cones, at least some pollen grains land on the female cones, where they become trapped by a sticky substance at the tip of the ovule. When the sticky substance dries out and shrinks, it pulls pollen grains inside the ovule.

At least a year passes between the time of pollination and fertilization. During that time, the pollen grain produces a pollen tube (a tubelike projection) that starts to grow toward the developing female gametophyte. Eventually, one cell within each pollen grain divides by mitosis and forms two sperm nuclei (Figure 40.34).

Also during this period, the female gametophyte gradually matures. It contains two or more archegonia, each with a single egg (Figure 40.34). When the pollen tube reaches an archegonium, the stage is set for fertilization, zygote formation, and early development of the embryo. The whole seed, including the embryo, female gametophyte, and seed coat, is shed from the pine cone. The seed coat protects both the embryo and the gametophyte from drying out and injury. The female gamet-

ophyte tissue serves as a food reserve for the embryo during germination.

With their mechanisms of producing, protecting, and dispersing seeds, the conifers became the dominant land plants during the Mesozoic and radiated into many land environments. Since then, the angiosperms have become more successful, although gymnosperms are still dominant in many regions, especially to the north and at high altitudes. Today they are major sources of lumber, paper, and many other commercial and industrial products. Chapter Forty-Six describes some of the world's coniferous forests.

a

b

Figure 40.33 (**a**) *Welwitschia*, a seed-producing gnetophyte. (**b**) Close-up showing the female strobili.

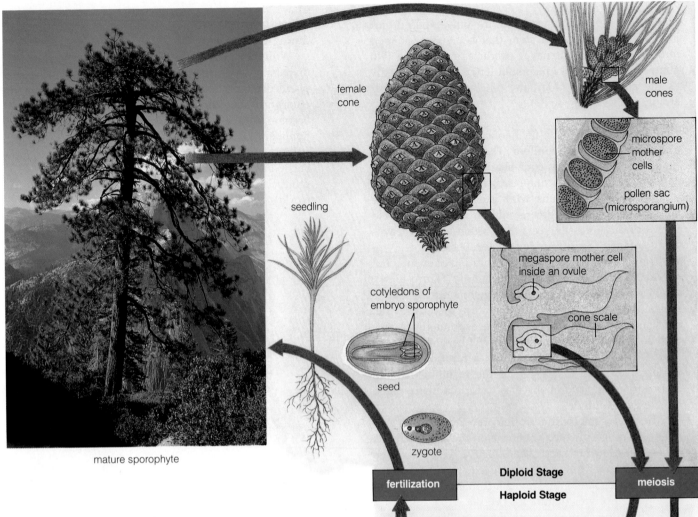

female cone

male cones

microspore mother cells

pollen sac (microsporangium)

seedling

cotyledons of embryo sporophyte

megaspore mother cell inside an ovule

cone scale

seed

zygote

Angiosperms—The Flowering Plants

Of all the divisions of plants, the angiosperms (Anthophyta) are the most successful. These are the "flowering plants" we studied earlier, in Unit Four. There are about 250,000 known species (far more than are found in other plant divisions), and new species are being discovered almost daily in previously unexplored regions of the tropics.

Angiosperms are also the most diverse types of plants. They range in size from tiny duckweeds (about a millimeter long) to *Eucalyptus* trees more than 100 meters (roughly 327 feet) tall. Most are free-living and photosynthetic, but some are saprophytic or parasitic. Diverse species are found on dry land and in wetlands, fresh water, and seawater.

As we have seen, angiosperms are grouped into two classes: the monocots and dicots. Figure 21.6 showed the life cycle of a typical dicot (*Prunus*); to round out the picture, here we show the life cycle of a monocot (*Lilium*, Figure 40.35).

There are nearly 200,000 species of dicots. They include most of the familiar shrubs and trees other than conifers, most herbaceous plants, the cacti, and the

fertilization

Diploid Stage

Haploid Stage

meiosis

egg

(megasporangium)

microspores

megaspore

female gametophyte

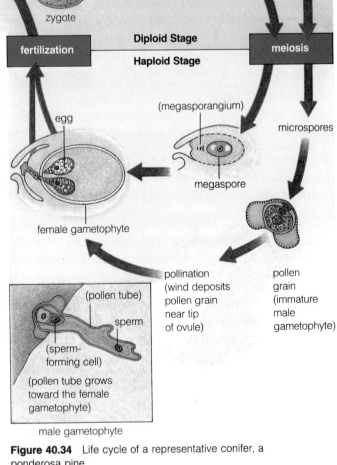

pollination (wind deposits pollen grain near tip of ovule)

pollen grain (immature male gametophyte)

(pollen tube)

sperm

(sperm-forming cell)

(pollen tube grows toward the female gametophyte)

male gametophyte

Figure 40.34 Life cycle of a representative conifer, a ponderosa pine.

mature sporophyte

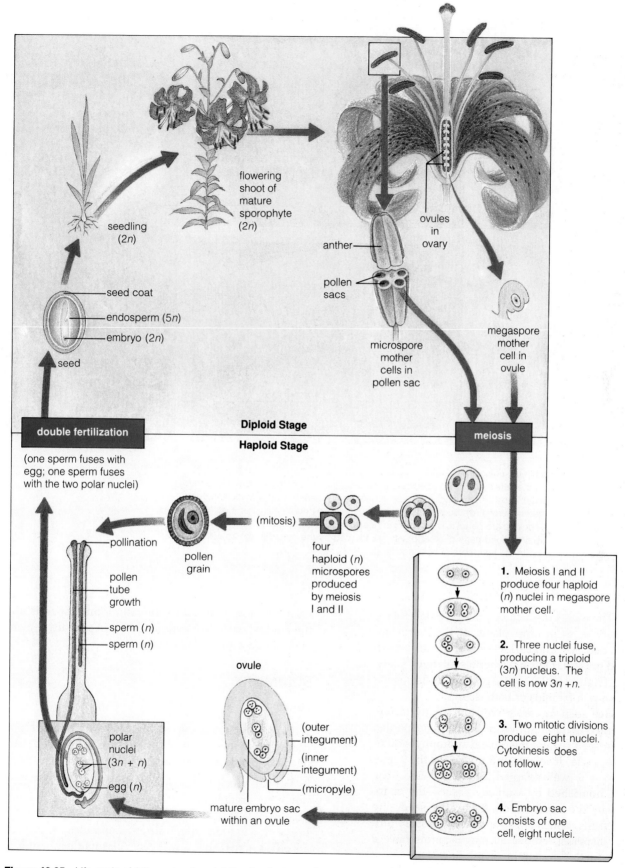

Figure 40.35 Life cycle of *Lilium*, showing details of embryo sac formation. Although about seventy percent of the known flowering plant species follow the pattern shown in Figure 21.6, microscope slides abound of embryo sac formation in lilies, and you are likely to encounter them in the classroom. Like about thirty percent of the flowering plants, lilies do not produce 3*n* endosperm. Depending on the species, the number may be 2*n*, 5*n*, 9*n*, or even 15*n*.

a   b

c   d   e

Figure 40.36 Floral structure typical of the wild iris (**a**), a typical monocot; and of *Hypericum* (**b**), a dicot. Among the far-flung relatives of these fairly common species are wind-pollinated birches of north temperate zones (**c**) and *Rafflesia* of tropical forests, this being a leafless, foul-smelling plant having a fly-pollinated flower that can extend three meters across (**d**). Sugarcane (**e**) is propagated by humans in many parts of the world. Diverse flowering plant species are beautifully adapted to nearly all environments, ranging from the snowline of mountains (**f**), to aboveground parts of other plants (as favored by the orchid in (**g**), to water (the habitat of water lilies, (**h**) and on to the poor soils of hot deserts (**i**).

water lilies. The monocots include grasses, palms, lilies, and orchids. The main crop plants of the world are wheat, corn, rice, rye, and barley—all monocots, and all domesticated grasses. Only a few representatives can be shown here (Figure 40.36), but Appendix II lists some of the well-known families of both classes.

As we have seen in earlier chapters, many factors have contributed to the adaptive success of angiosperms. As with other seed plants, the diploid sporophyte dominates the life cycle. It retains and nourishes the gametophyte, and it is well adapted for life on land. The embryos are nourished by a unique tissue—the endosperm—within the seed. Additionally, the seeds are packaged in fruits, which are structures that assure protection and dispersal. Above all, angiosperms have a unique reproductive system involving flower formation (page 279). Many diverse floral structures have co-evolved with pollination vectors—insects, bats, birds, and rodents. Although some flowering plants are wind-pollinated, it may have been the effectiveness of polli-

nation vectors that led to the rise of angiosperms and the gradual decline of gymnosperms in so many regions over the past 100 million years.

SUMMARY

Fungi

1. Fungi are heterotrophs that are important decomposers for the world of life. Many also are serious pathogens of plants and animals.

2. Fungal species are either saprophytic (feeding on nonliving organic matter) or parasitic (feeding on living organisms). Fungal cells secrete digestive enzymes that break down food into nutrient molecules, which are absorbed across the fungal cell's plasma membrane.

3. Most fungi are multicelled, with haploid and diploid stages alternating in the life cycle. The multicelled body (mycelium) is composed of microscopic filaments called hyphae.

f

g

h

i

4. Fungi can reproduce asexually by spore formation or by binary fission, budding, or fragmentation of the parent body. Many sexually reproducing fungi have a dikaryotic stage that intervenes between cytoplasmic fusion and nuclear fusion of two gametes.

5. There are five main groups of fungi: the chytrids, water molds, zygospore-forming fungi, sac fungi, and club fungi. The "imperfect fungi" include species of as-yet unknown affiliation; this group has no formal taxonomic status.

Plants

1. The plant kingdom includes aquatic and terrestrial forms, nearly all of which are photosynthetic autotrophs. Terrestrial plants are believed to have evolved from multicelled green algae over 400 million years ago.

2. The major trends in the evolution of terrestrial plants were:

   a. Cellular and metabolic adaptations to dry periods (water stress).

   b. In bryophytes, a reliance on the gametophyte for rapid propagation by way of gametes; in vascular plants, a reliance on the sporophyte for rapid propagation by way of spores.

   c. A transition from homospory to heterospory.

   d. A transition from unprotected zygotes to the formation of seeds.

3. The algae include the red, brown, and green algae. The existing "nonvascular" plants (those without well-developed xylem and phloem) include the bryophytes (mosses, liverworts, and hornworts). Existing vascular plants include the lycopods, horsetails, ferns, gymnosperms, and angiosperms (flowering plants).

4. Red and brown algae are the most common seaweeds. Their chlorophylls are masked by accessory pigments. Like land plants, green algae have chlorophylls *a* and *b* within plastids, they store carbohydrates as starch, and they have cellulose-reinforced walls.

5. Land plants typically have a cuticle and stomata that help control water loss. They have archegonia and antheridia (protective cell layers that surround the female and male gametophytes, respectively) and sporangia (protective cell layers around the spores in the sporophytes). In all species, the sporophyte initially develops *within* gametophyte tissues as an embryo.

6. Bryophytes, including mosses and liverworts, live in moist soils; they generally have no vascular tissues. Water is required for fertilization. Bryophytes are unlike other land plants in that their sporophytes remain attached to the parental gametophyte body and cannot live independently of it.

7. The simplest vascular plants are the lycopods, horsetails, and ferns, all of which are characterized by the size, arrangement, and complexity of their leaves and reproductive structures. All show alternation between independent gametophyte and sporophyte generations during the life cycle.

8. The most complex vascular plants are gymnosperms and angiosperms, both of which are seed producers. Gymnosperms include cycads, ginkgos, gnetophytes, and conifers. Angiosperms (flowering plants) are divided into monocots and dicots. In all seed plants, the gametophyte is wholly dependent on the sporophyte.

9. Sexual reproduction may take two years or more in some conifers such as pines. Sexual reproduction in many flowering plants may be completed in weeks. Efficient reproductive structures of flowering plants (including pollinator-attracting flowers as well as fruits) have contributed to the widespread distribution of this group.

Review Questions

1. Name and describe a representative of each of the formal taxonomic groups of fungi. Name and describe one of the deuteromycetes ("imperfect fungi"). 613–619

2. Describe the stage in the life cycle of many fungi that intervenes between cytoplasmic union and nuclear fusion. 613

3. What is the difference between a mycorrhiza and a lichen? 619–621

4. Distinguish among the megaspores, microspores, and gametes of land plants. As part of your answer, define sporophyte and gametophyte. Which is haploid? Which is diploid? 623

5. Describe the major evolutionary trends that figured in the invasion of land by plants. 622–624

6. What are some similarities and differences between algae and bryophytes? Between bryophytes and lycopods? 622–623, 628–631

7. If both gymnosperms and angiosperms are seed-bearing, how do they differ in their reproductive modes? 635–638

8. Choose a garden plant, crop plant, or weed that grows in your neighborhood. Make a diagram of it, labeling its component parts. Can you correlate some of those parts with seasonal variations in temperature, moisture, and other local environmental conditions?

Readings

Bold, H., and J. LaClaire. 1987. *The Plant Kingdom*. Fifth edition. Englewood Cliffs, New Jersey: Prentice-Hall. Paperback.

Moore-Landecker, E. 1982. *Fundamentals of the Fungi*. Second edition. Englewood Cliffs, New Jersey: Prentice-Hall. Well-written introduction to the kingdom of fungi.

Raven, P., R. Evert, and S. Eichhorn. 1986. *Biology of Plants*. Fourth edition. New York: Worth. Lavishly illustrated introduction to botany.

Stern, K. 1988. *Introductory Plant Biology*. Fourth edition. Dubuque, Iowa: Brown. Nice illustrations and packed with examples that students should find interesting. Paperback.

Stewart, W. 1983. *Paleobotany and the Evolution of Plants*. New York: Cambridge University Press. Beautiful synthesis of what is known about ancestral plants.

With this chapter we arrive at the kingdom of multicelled animals, or **metazoans**. Sponges are the simplest members of this kingdom, and **vertebrates** (the only animals with a "backbone," these being fishes, amphibians, reptiles, birds, and mammals) are the most complex. In between are more than 2 million spectacularly diverse but lesser known species, including the ribbon worm shown in Figure 41.1.

Among the **invertebrates** (the ones without a backbone) are species that have not changed much since Cambrian times. (This is not to say they are "primitive," or evolutionarily stunted; they are as exquisitely adapted to certain environments as you are to yours.) Other invertebrates show greater structural and behavioral complexity. In certain respects, some even bear strong resemblances to the vertebrates.

If we accept that (1) all animals are related and (2) evolution occurs only through genetic modification of existing species, then the range of complexity among animals alive today may echo a progression of changes that occurred in the distant past. That apparent progression will be our focus here. In the next chapter, we will look more closely at mammals, then primates, and finally the human species. From invertebrates to vertebrates, including humans and other mammals—this sequence will give insight into the traits we share with other members of the animal kingdom.

41

ANIMAL DIVERSITY

Figure 41.1 One of the lesser known invertebrates: a ribbon worm, a member of the phylum Nemertea. These slender, predatory animals live mostly in marine habitats, and some grow several meters in length.

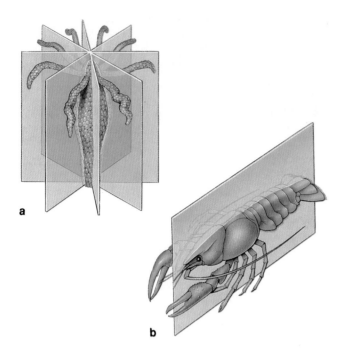

Figure 41.2 Planes of radial symmetry in a hydra (**a**) and of bilateral symmetry in a crayfish (**b**).

GENERAL CHARACTERISTICS OF ANIMALS

Nearly all animals have the following characteristics in common:

1. Animals are multicellular, and (except for sponges) their cells form tissues. The tissues usually are arranged into organs and organ systems.

2. Animals are heterotrophs; they cannot produce their own food and so must eat other organisms or absorb soluble substances produced by other organisms.

3. Animals are diploid organisms that reproduce sexually and, in many cases, asexually.

4. Animal life cycles include a period of embyronic development. In brief, cleavages transform the zygote into an early multicelled embryo. Then the embryonic cells become arranged into germ tissue layers (*ectoderm*, *endoderm* and, in most species, *mesoderm*), and these give rise to all tissues and organs of the adult (page 497).

5. Most animals are motile, at least during part of the life cycle.

Figure 41.3 Body plans of bilateral animals. (**a**) Acoelomate, or without a body cavity, as in flatworms. (**b**) Pseudocoelomate, without a continuous peritoneal lining; various organs, especially of the reproductive system, occupy the pseudocoel. (**c-d**) Coelomate, a plan typical of several invertebrate groups (including annelids and echinoderms) as well as vertebrates. In some coelomate invertebrates, such as insects, the coelom is greatly reduced. To a considerable extent, its function is taken over by a hemocoel, an aggregate of interconnected, blood-filled spaces.

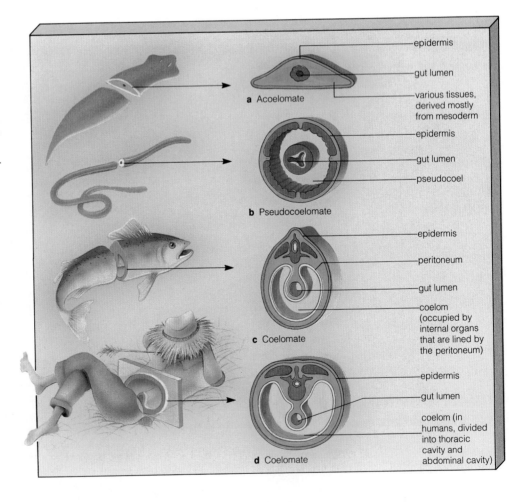

Body Plans

As we track the increasing complexity among different animal groups, we will be looking at the presence (or absence) and degree of elaboration of five body features: *symmetry, type of gut, type of body cavity, segmentation,* and *cephalization.*

Body Symmetry. Except for sponges, animals show radial or bilateral symmetry. Hydras, for example, are radially symmetrical (Figure 41.2). If you slice a hydra lengthwise, you will divide it into equal halves; another slice at right angles to the first will divide it into equal quarters. You cannot get the same result with an earthworm, a crayfish, or a rabbit. These and most other animals have a definite head and are bilaterally symmetrical, with right and left halves that are mirror-images of each other. Even if some body parts are positioned to one side of the midline, as is often the case, this does not violate the basic bilateral plan.

Type of Gut. Almost all animals have a **gut**: a cavity or tube into which food is taken and within which digestion proceeds. The simplest kind of gut is basically a sac with one opening (the mouth). The sac may be branched or subdivided, but no matter how elaborate it is, undigested residues must be eliminated through the mouth. Most animals have a "complete" gut which, by definition, has a mouth and an anus (page 451). This is a more efficient arrangement, partly because the mouth region functions only to take in food. A one-way gut also offers greater possibilities for regional specialization in such tasks as grinding, digestion, and absorption.

Body Cavities. Although a gut lumen is an internal space, we are concerned here with body cavities of a different kind, located between the gut and the body wall. Many invertebrates and all vertebrates have a **coelom**, a type of cavity that is lined by a sheet of epithelial cells called a peritoneum (Figure 41.3). Besides covering the cavity wall, the peritoneum also covers organs that occupy the coelom and helps hold them in place. In your own body, the coelom is partitioned into thoracic and abdominal cavities that house the heart, lungs, liver, bladder, and other organs.

The coelom is reduced in insects and some other animals. These animals have another kind of body cavity, called a **hemocoel**. A hemocoel actually is an aggregate of interconnected, blood-filled spaces that bathe the surrounding tissues. Animals with hemocoels are said to have open circulation systems (page 397).

Flatworms and other animals that do not have a body cavity are said to be **acoelomate**. Nematodes, rotifers, and several other kinds of animals have a body cavity that does not develop the same way as a true coelom does, and it is not lined with peritoneum. This type of cavity is called a **pseudocoelom** ("false coelom").

Segmentation. Earthworms, insects, and other animals are said to be segmented because they consist of a series of body units. Most segments of an earthworm are externally similar to one another, and many are similar on the inside. In insects, individual segments are more specialized and are grouped into body divisions called tagmata (singular, tagma). These divisions are the insect head (with eyes, antennae, and mouthparts), thorax (with legs and usually wings), and abdomen (with reproductive openings and organs for copulation).

Cephalization. Sponges, jellyfishes, and sea stars are *uncephalized* animals. The word means they do not have a head or anything that deserves to be called a brain. Flatworms have a definite head end equipped with sen-

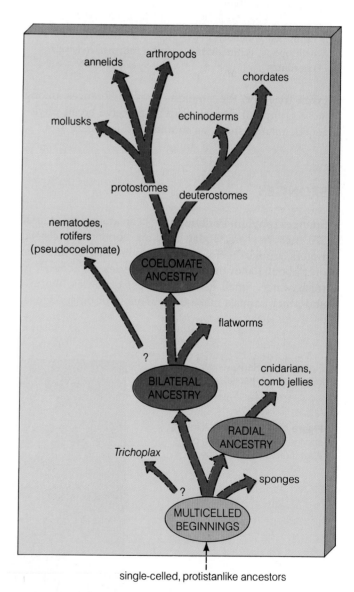

Figure 41.4 Broad phylogenetic framework for considering the relationships among major animal groups. All of the groups shown were established by 570 million years ago.

sory structures and a rudimentary brain. They are among the simplest of the *cephalized* bilateral animals. Even so, when a flatworm moves, the head end generally leads, and sensory receptors located in that region are the first to detect food or stimuli that indicate danger ahead.

Representative Animal Phyla

It is possible to follow several trends in animal evolution without considering every single phylum (there are more than thirty). These are the groups we have chosen to illustrate the major body plans:

1. Sponges — *Asymmetrical, no gut*

2. Cnidarians, comb jellies — *Radial, saclike gut*

3. Flatworms — *Bilateral, saclike gut*

4. Nematodes, rotifers — *Bilateral, pseudocoelomate, gut usually complete*

5. Mollusks, annelids, arthropods, echinoderms, chordates — *Bilateral (echinoderms also show radial patterns), coelomate, gut usually complete*

Taken together, the similarities and differences among these groups of animals help us perceive broad phylogenetic relationships, as summarized in Figure 41.4.

SPONGES

Sponges (phylum Porifera) have been around for at least 570 million years. Today they are common in coastal waters, tropical reefs, and many other marine habitats; only about 100 of the 8,000 or so species are adapted to fresh water. Small fishes, shrimps, barnacles, worms, and other animals make their home in or on sponges.

A sponge has no organs and no nervous system. Although many of its thousands or millions of cells are organized into layers that resemble epithelia, the layers are not comparable to the tissues of other animals. (The cells do not secrete a basement membrane and they are not often linked by tight junctions.)

Sponges vary greatly in size and shape. Some are as small as a fingernail and a few are large enough for a diver to sit in. Many are flat and sprawling; others are compact, lobed, tubular, cuplike, or vaselike (Figures 41.5 and 41.6). Regardless of its shape, the sponge body is organized as a system of pores, canals, and chambers through which water is circulated. Even when the body appears to have symmetry (as in the tubular types), it really has none, because the microscopic chambers and passageways running through it do not conform to a regular pattern.

As Figure 41.7 shows, water moves *into* the sponge body through thousands or millions of small incurrent pores and *out* through one or more outcurrent openings called oscula (singular, osculum). The water movements are caused by the collective activity of flagellated **collar cells**, so named for their "collar" of microvilli around the base of the flagellum. These cells line the central cavity, side pockets, or small chambers that are joined by canals to the central cavity and to pores at the body surface. When water is being drawn in through the pores, the bacteria and other food particles suspended in it become trapped on the collars. The collar cells may digest food completely or transfer it to **amoeboid cells** that further digest food, store it, and distribute it to the other cells.

Flattened cells line the sponge exterior and the internal cavities where there are no collar cells. Between the two linings is a semifluid matrix through which the amoeboid cells crawl and within which skeletal elements are embedded. The skeletal elements are composed of

Figure 41.5 Representative sponges in their marine habitat.

Figure 41.6 (Far right) A tubular sponge (*Verongia archeri*) of the West Indies, releasing sperm.

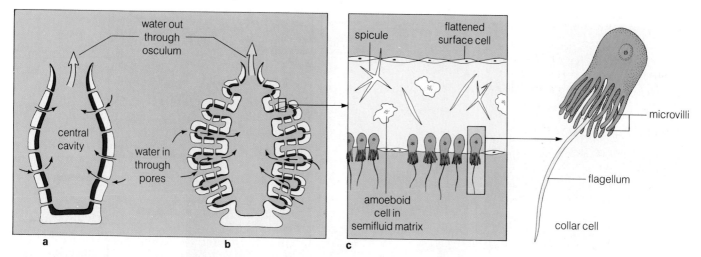

Figure 41.7 Morphology of a simple sponge (**a**) and a more complex form (**b**). Purple shading indicates the location of collar cells, which trap bacteria and other small food particles on their "collar" of microvilli. Black arrows indicate the direction of water flow through the sponge, as caused by the collective beating of the flagella of collar cells. (**c**) Small section through a sponge body wall.

substances secreted by the amoeboid cells. Some are needles (spicules) of silica or calcium carbonate; others are fibers of a flexible protein. (Fibrous skeletons give natural bath sponges their soft, squeezable texture.) Many spicules project to the outside, forming a protective barrier of sharp points, but most function to support the sponge body.

Sponges reproduce sexually, with eggs and sperm being derived from collar cells and perhaps from amoeboid cells. Sperm are released into the surrounding water (Figure 41.6). After fertilization, the eggs may be released or they may spend some time in the canals or chambers of the parent body. The young sponges pass through a swimming larval stage before they grow into adults.

Some sponges reproduce asexually when small fragments break away, settle on a substrate, and develop into new sponges. Most freshwater sponges can produce **gemmules**: aggregations of sponge cells, some of which form a hard covering (with spicules) around the others. The living cells in a gemmule can resist extreme cold or drying out. With the return of favorable conditions, a gemmule germinates and founds a new colony.

CNIDARIANS

The phylum Cnidaria, which consists of about 11,000 species, includes the hydras, jellyfishes, sea anemones, and corals. The oceans are home to most of these animals. Fewer than fifty species (including the hydras) are adapted to freshwater lakes and ponds. Figure 41.8 shows a representative of the phylum.

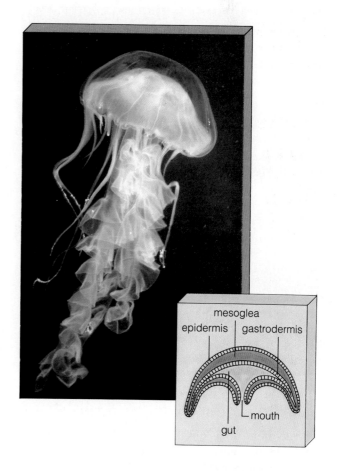

Figure 41.8 Sea nettle (*Chrysaora*), one of the large jellyfishes. The frilled structures are oral arms that assist in capturing and ingesting prey. Jellyfishes of this type usually have an abundance of nematocysts which, in the sea nettle and some other species, give painful stings. The boxed inset shows the basic body form of a jellyfish. This form is called a medusa.

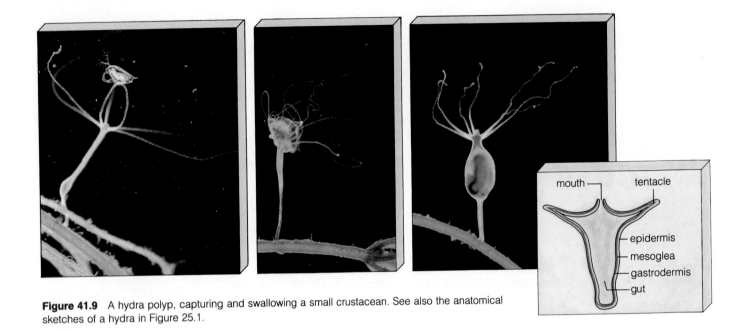

Figure 41.9 A hydra polyp, capturing and swallowing a small crustacean. See also the anatomical sketches of a hydra in Figure 25.1.

mouth — tentacle
— epidermis
— mesoglea
— gastrodermis
— gut

b   mouth of polyp

Figure 41.10 Representative corals from the Great Barrier Reef, Australia. (**a**) *Telesto*, a "soft coral." It lacks an external skeleton but has spicules of calcium carbonate in the mesoglea. Individual polyps are small. (**b**) *Tubastraea*, a type of "hard coral," with an external calcareous skeleton. In this genus the polyps are solitary, not colonial, but they often aggregate into large masses. Only one polyp in the mass shown here is extended.

Most of the massive, reef-forming hard corals are colonial; the many polyps of which they are composed are interconnected. Other examples of corals are shown in Figures 1.5 and 1.7.

a

With more than 6,000 species, the sea anemones, reef-building corals, and their relatives are the largest class of cnidarians. Most corals have external skeletons, reinforced with calcium carbonate, that collectively serve as the main building material for reefs (page 9). Their most extravagant accomplishment is the Great Barrier Reef, which parallels the northeast coast of Australia for about 2,000 kilometers. The northernmost reefs are in Bermuda, where the Gulf Stream waters are relatively warm.

Cnidarians have two main body forms, called the **medusa** (plural, medusae) and the **polyp**. A medusa looks like a bell or an upside-down saucer (Figure 41.8). The mouth is centered under the bell, and the bell's edge has tentacles extending from it. Sometimes "oral arms" extend from the region of the mouth and assist in prey

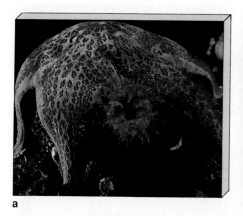
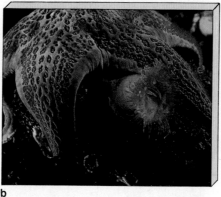

a   b   c

Figure 41.11 A sea anemone escaping from a predatory sea star. The mouth of the sea anemone closes so that the gut can function as a hydrostatic skeleton. Muscle contractions allow the animal to detach from its substrate and make thrashing movements. (Only a few sea anemones can do this.)

capture and in feeding. Polyps are tubelike, and they have tentacles encircling the mouth (Figure 41.9). Typically they are sedentary, being attached by the end opposite the mouth end. Often the polyps are organized into colonies in which the tissues and gut of one member are continuous with those of its neighbors (Figure 41.10a).

Both medusae and polyps show radial symmetry, and both are composed of true epithelial tissues. Epidermis (derived from embryonic ectoderm) covers the body surface, and gastrodermis (derived from endoderm) lines the saclike gut. These tissues contain nerve cells, arranged to form a type of nervous system called a **nerve net** (page 336). In some jellyfishes, sensory cells are organized into structures that detect changes in orientation (page 373). The tissues also contain contractile cells that function in locomotion, shape changes, feeding, and other activities. (These "epitheliomuscular" cells are not the same thing as muscle cells of other animals, which are derived from mesoderm.) Between the two tissues is a layer of secreted material called the **mesoglea** ("middle jelly").

Jellyfishes have a large amount of mesoglea that imparts buoyancy and serves as a deformable skeleton against which the contractile cells can act. This is important for swimming; each contraction that narrows the bell and squirts out a jet of water from the underside is followed by quick recovery to the original shape. Polyps generally have little mesoglea, and they are more likely to use their fluid-filled gut as a hydrostatic skeleton against which contractile cells can act. A dramatic example of this is the response of certain sea anemones to predators (Figure 41.11).

Medusae and polyps are equipped with **nematocysts**: capsules that discharge threads, the function of which is to immobilize prey (Figure 41.12). Only cnidarians produce nematocysts, and the toxin-laden types produced by some species can give humans a painful sting (hence the phylum name Cnidaria, which is derived

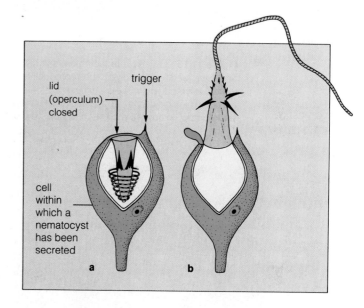

lid (operculum) closed   trigger

cell within which a nematocyst has been secreted

a   b

Figure 41.12 Example of a nematocyst. It has been secreted within a certain type of cell; and it consists of a capsule with an inverted tubular thread inside (**a**). A nematocyst-containing cell may have a modified flagellum that serves as a bristlelike trigger. When a prey organism (or possibly a predator) touches the trigger, or when the whole complex receives an appropriate chemical stimulus, the permeability of the capsule increases and water diffuses into it. This increases the pressure within the capsule, and the thread is forced to turn inside-out (**b**). The tip of the thread may penetrate the prey organism, releasing a toxin in the process.

In other nematocysts, the thread adheres to prey by a sticky secretion or it may simply become entangled with hairs or other structures of the target and so act like a lasso.

Once a nematocyst has been fired, its career is over. But a cnidarian continually produces new nematocysts, so a large supply is always available.

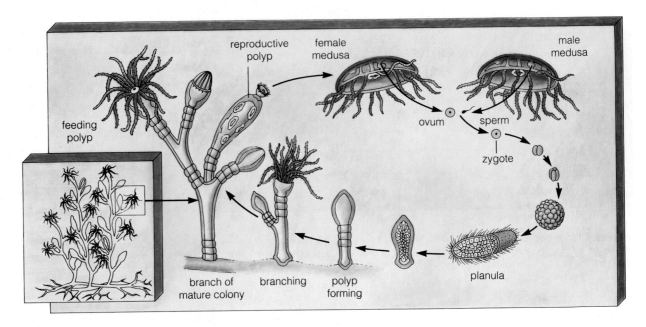

Figure 41.13 Life cycle of the marine hydrozoan *Obelia*. The inset shows a colony, actual size. On its branches are feeding polyps and reproductive polyps. The medusa stage, produced asexually by a reproductive polyp, is free-swimming. Fusion of gametes from male and female medusae leads to a zygote, which develops into a planula. The swimming or crawling planula develops into a polyp, which starts a new colony. As growth continues, new polyps are formed.

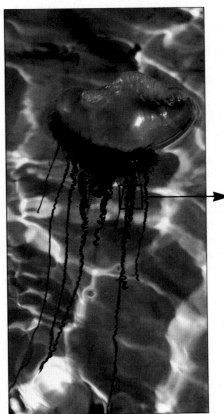

Figure 41.14 The Portuguese man-of-war (*Physalia*), which is a floating colonial hydrozoan. Numerous modified polyps and medusae occur in the colony.

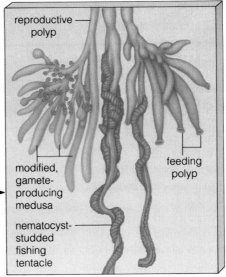

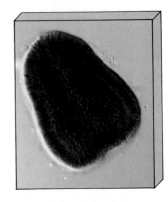

Figure 41.15 Example of a planula, this from the jellyfish *Aurelia*. Magnification 275×.

from the Greek word for "nettle"). A single cnidarian tentacle may have hundreds or thousands of nematocysts embedded in the epidermis. Nematocysts may also occur in the gastrodermis. Figure 41.12 describes how these stinging capsules work.

Many cnidarians have only a polyp stage or a medusa stage, either of which can reproduce sexually or asexually. Others have both polyp and medusa stages in the life cycle, the medusa being the sexual form. An example is *Obelia*, a colonial hydrozoan (Figure 41.13). Modified

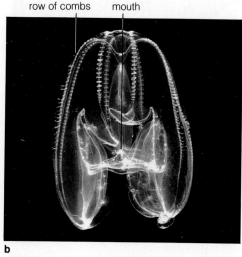

row of combs   mouth

Figure 41.16 Two very different comb jellies: (**a**) *Pleurobrachia*, with long, sticky tentacles, and (**b**) *Bolinopsis*, which has only short tentacles close to the mouth.

The comb jelly shown in (**b**) is giving a good display of bioluminescence, or luminescent flashes. Bioluminescence occurs in about half of the animal phyla and is especially common among comb jellies, cnidarians, squids, and fishes. Most often, interactions between a heat-stable compound (luciferin) and a heat-sensitive enzyme (luciferase) cause the flashes. In comb jellies, these substances are part of a protein complex that is activated by calcium. Most luminous displays are responses to mechanical disturbances and are under neural control.

medusae and polyps also occur together in the colonial hydrozoan *Physalia* (the Portuguese man-of-war), which has to be one of the most remarkable cnidarians known (Figure 41.14). The members of these colonies are specialized for different functions, such as feeding, defense, and reproduction.

Regardless of whether it is the polyp or medusa that reproduces sexually, the resulting zygote nearly always develops into a swimming or creeping larva called a **planula** (Figure 41.15). With few exceptions, planulas have ciliated epidermal cells. Eventually a mouth opens up at its body surface and gradually the larva is transformed into a polyp or medusa, and so the cycle begins anew. Many biologists think that planula-like organisms may have given rise to bilateral animals.

COMB JELLIES

The comb jellies belong to the phylum Ctenophora (meaning "comb-bearing"). These predatory marine animals have eight rows of comblike structures. The combs consist of thick, fused cilia, and all the combs in each row beat in waves to propel the animal forward, mouth first. Comb jellies show modified radial symmetry. (You can slice them lengthwise into two equal halves, and even into quarters—but two of the quarters will be mirror-images of the other two.) Some comb jellies have two long, muscular tentacles, equipped with sticky cells, that serve as an efficient fishing net (Figure 41.16). Others have sticky lips for capturing prey, which may include

other comb jellies or jellyfishes. Many comb jellies are luminescent; they glow in the dark. Together with some jellyfishes, dinoflagellates, and other organisms, they make the water glow at night when a boat or a swimmer cuts through it.

Comb jellies do not produce nematocysts (although sometimes they opportunistically save and use the ones of jellyfish they have ingested). More importantly for our story, they have cells with multiple cilia—a trait characteristic of flatworms and many of the more complex animals. And they have embryonic tissue comparable to the mesoderm of flatworms and more complex animals. This tissue—not ectoderm or endoderm—gives rise to most of their muscles.

FLATWORMS

The 12,700 species of flatworms (phylum Platyhelminthes) include turbellarians, which are mostly free-living, as well as the parasitic flukes and tapeworms. Most of these animals have more or less flattened bodies (hence the name, flatworm).

Unlike cnidarians, the flatworms are bilateral, cephalized animals, and three germ tissue layers form in the embryo. The mesoderm, which gives rise to muscles and reproductive structures, was central in the development of complex animals. Besides enabling contractile cells to develop independently of epidermis and gastrodermis, it eventually became the embryonic source of blood, bones, kidneys, and other internal tissues and organs.

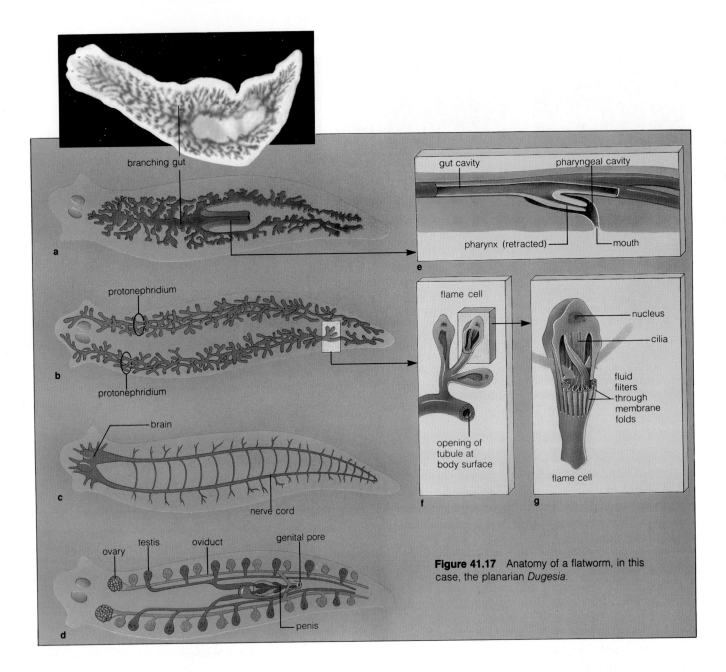

branching gut

a

protonephridium

protonephridium

b

brain

nerve cord

c

ovary
testis
oviduct
genital pore
penis

d

gut cavity
pharyngeal cavity

pharynx (retracted)
mouth

e

flame cell

opening of
tubule at
body surface

flame cell

f

nucleus
cilia
fluid
filters
through
membrane
folds

g

Figure 41.17 Anatomy of a flatworm, in this case, the planarian *Dugesia*.

The flatworm gut is saclike, without an anus. Usually, food enters it by way of a muscular pharynx (Figure 41.17). The flatworm pharynx qualifies as an organ; besides having an epithelial lining, it has muscle tissue and, generally, connective tissue and glands. Other flatworm structures are also at the organ level of complexity.

Turbellarians

Most turbellarians are carnivorous; except for a few species that consume diatoms and other small protistans, they feed on whole small animals or suck tissues from dead or wounded ones. Most live in the seas; some, including the planarians, live in lakes, streams, and ponds. A few can be found in moist places on land.

The freshwater turbellarians are notable for having an organ system that regulates the volume and salt con-

centrations of their body fluid. The system is based on one or more units called **protonephridia** (singular, protonephridium). Each unit consists of branched tubules that extend from a pore at the body surface to many cup-shaped **flame cells** in the body tissues (Figure 41.17). Flame cells get their name from a tuft of cilia that "flickers" within them. When excess tissue fluid moves into the cups formed by these cells, the flickering drives it down the tubule system to the outside.

In cnidarians, the gonads are simple and remain closely associated with the epidermis or gastrodermis; they just rupture and release gametes. In turbellarians and other flatworms, the gonads are nearly always part of a complex **hermaphroditic system** (that is, with male and female gonads in the same individual). Usually there is an organ that functions as a penis and a structure for storing sperm received during copulation; and there are

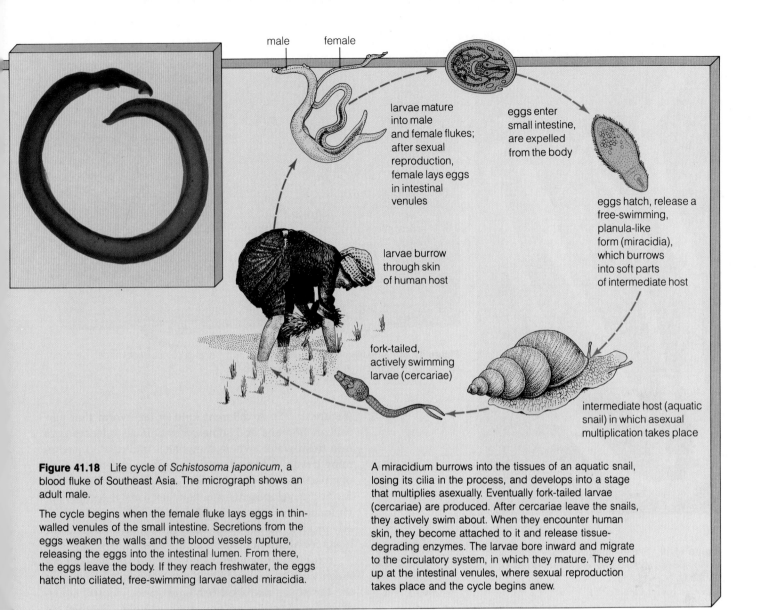

male   female

larvae mature
into male
and female flukes;
after sexual
reproduction,
female lays eggs
in intestinal
venules

eggs enter
small intestine,
are expelled
from the body

eggs hatch, release a
free-swimming,
planula-like
form (miracidia),
which burrows
into soft parts
of intermediate host

larvae burrow
through skin
of human host

fork-tailed,
actively swimming
larvae (cercariae)

intermediate host (aquatic
snail) in which asexual
multiplication takes place

Figure 41.18 Life cycle of *Schistosoma japonicum*, a blood fluke of Southeast Asia. The micrograph shows an adult male.

The cycle begins when the female fluke lays eggs in thin-walled venules of the small intestine. Secretions from the eggs weaken the walls and the blood vessels rupture, releasing the eggs into the intestinal lumen. From there, the eggs leave the body. If they reach freshwater, the eggs hatch into ciliated, free-swimming larvae called miracidia.

A miracidium burrows into the tissues of an aquatic snail, losing its cilia in the process, and develops into a stage that multiplies asexually. Eventually fork-tailed larvae (cercariae) are produced. After cercariae leave the snails, they actively swim about. When they encounter human skin, they become attached to it and release tissue-degrading enzymes. The larvae bore inward and migrate to the circulatory system, in which they mature. They end up at the intestinal venules, where sexual reproduction takes place and the cycle begins anew.

glands concerned with producing a protective capsule around one or more eggs after these have been fertilized. Mating between two individuals usually results in the reciprocal transfer of sperm.

Asexual reproduction is common among the turbellarians. Freshwater planarians, for instance, often divide transversely. Both halves must regenerate the parts they lack. This mode of reproduction is so successful for some planarians that they do not even have a functional system for sexual reproduction.

Trematodes

Adult trematodes, or flukes, are parasites of vertebrates. Some live on the body surface of fishes and amphibians, where they feed on cells or tissue fluids. The vast majority of the internal parasites live in the gut, liver, lungs, bladder, or blood vessels. They have complex life cycles involving at least one intermediate host as well as asexual multiplication by larval stages. Figure 41.18 describes the stages in the life cycle of *Schistosoma japonicum*, one of the blood flukes that cause **schistosomiasis**, a disease of humans and other animals. Although flatworms as a rule are hermaphroditic, the species of *Schistosoma* have separate sexes.

In any given year, millions of people are afflicted with schistosomiasis; this is especially true of parts of Africa and Asia. An infected host typically makes a cellular immune response to fluke eggs being amassed within its body. Leukocytes and other cell types infiltrate and produce granular masses in tissues. The disease leads to deterioration and malfunctioning of the liver, heart, spleen, bladder, and kidneys.

Cestodes

Adult cestodes, or tapeworms, are intestinal parasites of vertebrates. Ancestral tapeworms apparently became so good at letting their hosts provide for them that their own gut became superfluous and eventually disap-

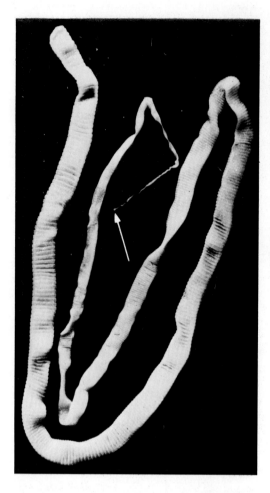

Figure 41.19 Sheep tapeworm, showing the size of the scolex (white arrow) relative to the rest of the body.

Figure 41.20 Scanning electron micrograph of the scolex of *Acanthocirrus retrirostris*, a tapeworm that parasitizes shorebirds.

peared from the body. Their living descendants do no mechanical digestion at all; they simply absorb soluble nutrients from their hosts.

A tapeworm attaches to the intestinal wall by a **scolex**, a structure that usually has suckers, hooks, or both (Figures 41.19 and 41.20). Just behind the scolex is a proliferative region from which units called **proglottids** are budded. Each proglottid is almost like an individual, for it has a complete hermaphroditic reproductive system. Proglottids mate and transfer sperm, and the fertilized eggs may accumulate in the older proglottids (the ones farthest from the scolex). Sooner or later, the oldest proglottids separate from the chain. They may or may not disintegrate before leaving the body with fecal material, but in either case the eggs reach the outside and so become available to infect a new host.

Generally, at least one intermediate host is used in the life cycle of a tapeworm. Pigs are intermediate hosts for one common tapeworm that infects humans; cattle are intermediate hosts for another. Humans become infected when they eat insufficiently cooked pork or beef

(Figure 41.21). A different kind of tapeworm that parasitizes humans and some other animals releases eggs into fresh water, where they hatch into ciliated, swimming larvae. When a larva is eaten by a copepod (a type of crustacean), it develops into a stage that will undergo further development in a fish that eats the copepod. Humans who eat raw, insufficiently cooked, or improperly pickled fish of certain freshwater species can become hosts for adult tapeworms of the species.

You have probably noticed that the life cycles of the flukes and tapeworms just described take advantage of the life styles and food habits of various hosts. Sometimes the life cycles are amazingly complex, but they get the parasite through the sequence of hosts in which it reproduces asexually and then sexually.

The simplest turbellarians and the larval stages of flukes and tapeworms happen to resemble the planula of cnidarian life cycles. The resemblance has inspired the idea that ancient bilateral animals evolved from ancestors that were much like planulas. This could have occurred through increased cephalization and the emergence of tissues derived from mesoderm. In this way, planula-like ancestors may have given rise to most groups of complex animals. This idea is speculative, but it has much to recommend it.

NEMERTEANS

It seems likely that ribbon worms (phylum Nemertea) of the sort shown in Figure 41.1 are closely related to the turbellarians just described. They, too, are ciliated externally and their tissue organization is similar to that of flatworms in general. Unlike flatworms, however,

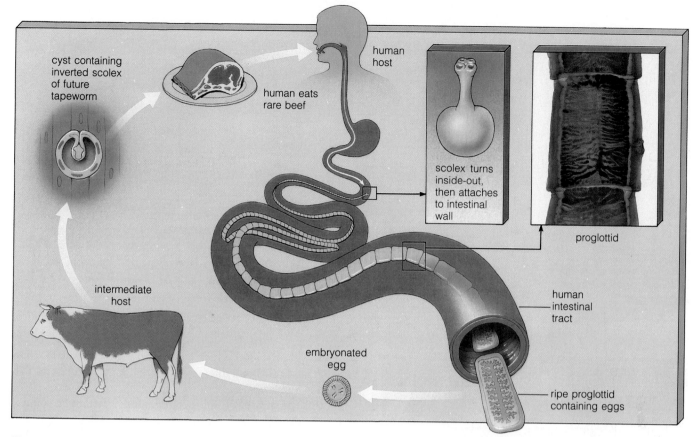

Figure 41.21 Life cycle of a beef tapeworm, *Taenia saginata*.

they have a complete gut and a circulation system, and the sexes are nearly always separate. Also, ribbon worms have a unique proboscis that is used for capturing prey. All of these features are significant departures from the flatworm body plan.

The proboscis lies within a fluid-filled cavity. When circular muscles around the cavity contract, hydrostatic pressure inside is raised, such that the proboscis is forced to turn inside-out and emerge from the mouth or from a separate opening at the anterior end. The proboscis is equipped with glands that secrete a paralytic venom and, often, with a stylet. The stylet is a penetrating device that is plunged into prey (another kind of worm, perhaps, or a small mollusk or crustacean). The prey may be swallowed whole or some of its tissues may be sucked out.

NEMATODES

Being cylindrical, the nematodes (phylum Nematoda) are often called roundworms. The slender nematode body follows a bilateral plan (Figure 41.22). It is covered by a type of cuticle that is tough and not easily stretched, yet flexible. Its gut is complete (it has a mouth and an anus). Between the gut and body wall is a pseudocoel. Reproductive organs may fill nearly all of the space

within the pseudocoel, but the fluid in the cavity distributes nutrients and thus serves as a kind of circulatory system. It also serves as a hydrostatic skeleton.

About 12,000 species of nematodes have been described, but there probably are several times that many. They live in places as diverse as snowfields, deserts, and hot springs. A cupful of rich soil will have thousands of them, and a dead earthworm or a fruit rotting on the ground will almost certainly have an interesting variety of scavenging types. Nematodes also are immensely successful parasites of plants and animals, and therefore are important in agriculture and medicine. About thirty species infect humans alone.

The pinworm (*Enterobius vermicularis*) is perhaps the most common parasitic nematode that infects humans, children especially, in temperate regions. Pinworms live in the large intestine. At night, the females (which are about a centimeter long) migrate to the anal region to lay their eggs. Their presence at the body surface causes itching, and scratchings made in response will transfer the eggs to other objects. When they are laid, the eggs contain embryos—but within a few hours they are juveniles that are ready to hatch if the eggs are inadvertently ingested by another host.

Hookworms, belonging to the genera *Necator* and *Ancylostoma*, can be a problem for humans in moist tropical regions. They live in the small intestine, where they

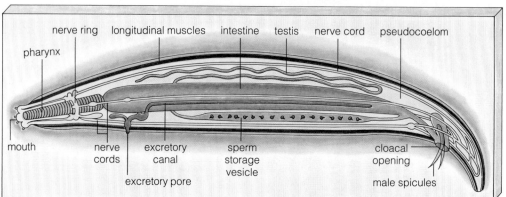

Figure 41.22 Anatomy of a nematode, which is specialized for feeding and reproducing, and little else.

Labels on figure: nerve ring, longitudinal muscles, intestine, testis, nerve cord, pseudocoelom, pharynx, mouth, nerve cords, excretory pore, excretory canal, sperm storage vesicle, cloacal opening, male spicules

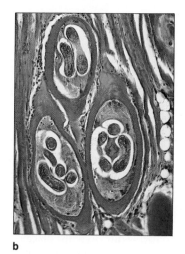

a   b

Figure 41.23 (a) Juveniles of *Trichinella spiralis* in muscle tissue. The adults of this nematode live in the lining of the small intestine of pigs, rats, humans, and some other animals. Female worms release juveniles, and these work their way into blood vessels and travel to muscles, where they become encysted (b). The presence of adults and juveniles in humans causes a variety of painful and sometimes fatal symptoms. The usual source of infective juveniles for humans is pork or wild game that has not been cooked sufficiently.

feed on blood and other tissues. Teeth or sharp ridges bordering the hookworm mouth are used to cut into the intestinal wall. Females, which are about a centimeter long, release approximately 1,000 eggs every day. These leave the host by way of feces, and they hatch into juveniles that feed on the fecal matter or other organic debris. A juvenile molts twice, then it is transformed into a nonfeeding infective stage that can penetrate the skin of a barefoot person. Once inside a host, the infective stage travels the bloodstream to the lungs, where it works its way into the air spaces. Then it goes up the trachea to

the glottis, after which it is swallowed with mucus. Soon it is in the small intestine, where it may mature and live for a few years.

The adults of another nematode, *Trichinella spiralis*, live in the small intestine of humans. After mating, the females produce active young that migrate through the intestinal wall and enter the lymph, then the bloodstream. The young finally encyst in striated muscle (Figure 41.23), including muscles of the diaphragm. The presence of juvenile forms of *Trichinella* in muscles may cause serious medical problems. Humans are not the only hosts for these parasites; pigs, rats, bears, and other vertebrates keep the cycle going in nature. Humans usually become infected by eating insufficiently cooked meat from pigs or certain game animals. The presence of encysted juveniles cannot easily be detected when fresh meat is examined, even in a slaughterhouse.

The nematode *Wuchereria bancrofti* has a more complex life cycle. Adult worms live in the lymph nodes of humans. The females produce active young that travel through the peripheral circulation at night. If blood containing the juveniles is sucked by the right kind of mosquito, the worms go through the wall of the insect's stomach and penetrate the muscles of its thorax. After some growth, they move to the mosquito's proboscis (a sucking device). The proboscis can serve as the vehicle by which they penetrate the skin of another human. Once inside, they enter the lymphatic system and migrate to lymph nodes, where they mature. The adult worms may obstruct the flow of lymph through the nodes. The obstruction causes fluid to accumulate in the legs or (in males) in the scrotum. These body parts undergo grotesque enlargement as a result, a condition called **elephantiasis.**

In spite of their success in so many diverse habitats, nematodes do not show much diversity in structure. It

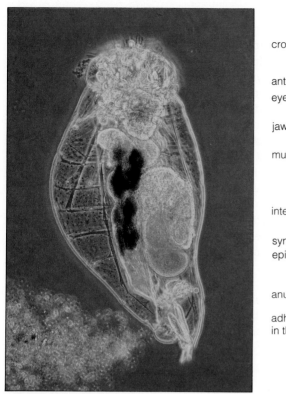
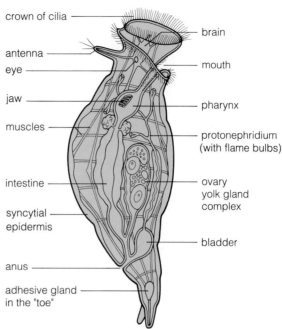

crown of cilia

antenna

eye

jaw

muscles

intestine

syncytial epidermis

anus

adhesive gland in the "toe"

brain

mouth

pharynx

protonephridium (with flame bulbs)

ovary yolk gland complex

bladder

Figure 41.24 Lateral view of a rotifer, which is busily laying eggs. In many rotifer species, males are unknown; females produce diploid eggs that develop into more diploid females. In other species, females can produce diploid eggs that develop into females *or* they can produce haploid eggs that develop into haploid males. If the haploid eggs happen to get fertilized by a male, they develop into females. The male rotifers appear only occasionally, and they are dwarfed and short-lived.

is as if they had a body plan that worked well under many different conditions and just stayed with it. The ancestors of nematodes are not known, but it is not likely that they gave rise to any other groups of living invertebrates.

ROTIFERS

Of the rotifers and a few other pseudocoelomates, we might say this: Seldom has so much been packed into so little space.

Most of the 1,800 or so species of rotifers are less than a millimeter long. Some rotifers float about in aquatic environments (they are planktonic); others crawl over aquatic plants, wet moss, and other substrates. They feed mainly on bacteria and single-celled algae, and other animals in turn feed on them. They are often abundant and so are probably important components of food webs.

Early microscopists called these animals rotifers (meaning "wheel-bearers"), because most kinds have a crown of cilia, the activity of which suggests that a wheel

is turning (Figure 41.24). The cilia function in swimming and in creating currents that direct food to the mouth.

Behind their ciliary crown, rotifers have salivary glands, a pharynx with jaws, an esophagus, digestive glands, a saclike stomach from which most nutrients are absorbed, and usually an intestine and anus. Protonephridia similar to those of flatworms remove excess water from the body. Nerve cells are clustered at the head end, and there often are photosensitive eyespots. Many species have a pair of flexible "toes." Sticky substances exude from the tips of these structures and they enable rotifers to attach themselves to a substrate while feeding. Some rotifers are covered with a fiber-reinforced epidermis that thins down at intervals to form body "joints." Nearly all rotifers have bands of circular and longitudinal muscles in the body wall. As muscles contract and relax, the cuticle flexes and bends to allow the animal to make looping and telescoping movements.

As Figure 41.4 indicated, rotifers are probably one of the side roads of animal evolution, although they do show the structural diversity that is possible with the pseudocoelomate body plan. It is only when we turn to the coelomate animals that we see diversity on a spectacular scale.

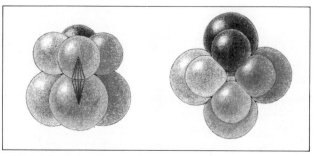

a Radial cleavage, side view (left) and top view

b Spiral cleavage, side view (left) and top view

Figure 41.25 Two major patterns of cleavage in early animal embryos: (**a**) radial, which is characteristic of deuterostomes, and (**b**) spiral, which is characteristic of protostomes.

TWO MAIN EVOLUTIONARY ROADS

The bilateral, coelomate animals that arose during Cambrian times must have been terrific grist for the evolutionary mill. Not long after they started to diversify, they gave rise to two distinct lineages, called protostomes and deuterostomes. Nearly all of the complex coelomate animals belong to one or the other of these lineages:

> **Protostomes:**   *mollusks, annelids, and arthropods*
>
> **Deuterostomes:**   *echinoderms and chordates*

In general, the animals of both lineages have a complete gut and a coelom. However, each follows its own patterns of early embryonic development. Recall that the embryonic stage called a gastrula has a single opening that forms at its surface (page 496). In protostomes, that first opening becomes the mouth; the anus forms elsewhere. (Protostome means "first mouth.") In deuterostomes, the first opening becomes the anus and a second opening, which appears later, becomes the mouth.

For most protostomes, the destiny of embryonic cells is fixed at the first cleavage; after that, each cell is committed to a developmental pathway that will lead to specific cell types. For most deuterostomes, the embryonic cells generally do not become committed until a little later in development. (For example, if the cells of a sea

star embryo are separated at the four-cell stage, each cell can develop into a complete larva.)

Also, the cleavages that produce the early embryo follow different patterns in the two lineages. Cleavage in the embryos of frogs and other deuterostomes occurs parallel and at right angles to the axis established by the animal and vegetal poles (see, for example, Figure 34.6). This pattern, called **radial cleavage**, produces an array of cells each located directly above or below one another (Figure 41.25a). In contrast, early cleavages in the embryos of nearly all protostomes are oblique. After the second division, for instance, each of the four cells of the upper tier lies on the furrow between two cells of the lower tier (Figure 41.25b). The outcome is a tiered array of cells, with each cell in a tier snuggled against two cells of adjacent tiers. This pattern is called **spiral cleavage**.

Finally, the coelom of protostomes arises in mesoderm on both sides of the embryonic gut. A cavity opens up and subsequently enlarges in each of the two masses of mesoderm:

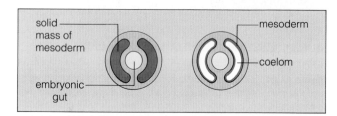

In deuterostomes, pouches form in the walls of the developing gut cavity, not in mesoderm. The pouches separate from the gut wall, and they go on to form the coelom:

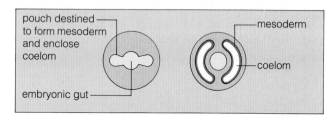

Embryonic differences of this sort are examples of the clues that are used to solve the puzzle of evolutionary relationships among the groups of more complex animals, which we will now consider.

MOLLUSKS

There are probably about 100,000 living species in the phylum Mollusca (the name means "soft-bodied animals"). Snails, slugs, clams, squids, and octopuses are familiar examples of mollusks. Although they all share the same basic body plan, there is enormous diversity in their individual structures and life styles.

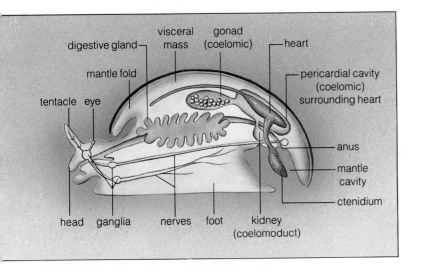

Figure 41.26 Generalized body plan of a mollusk.

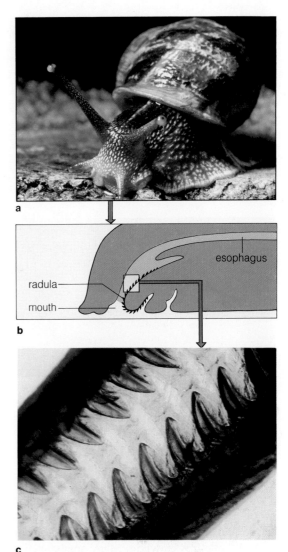

Figure 41.27 (**a**) A land snail. Part of the mantle cavity of this gastropod is modified for air-breathing. (**b**) Diagram of a section through a snail head, showing the radula (a rasping, tonguelike organ). (**c**) Close-up of a radula, showing the teeth that can rasp off pieces of food. The conveyer-belt motion of the radula moves food into the pharynx.

All mollusks have a head, foot, and visceral mass (Figure 41.26). They usually have a shell that consists mostly of calcium carbonate. They have a **mantle**: a fold of tissue that hangs down like a skirt around some or all of the body. The space between this tissue fold and the rest of the body is called the mantle cavity.

Mollusks with a well-developed head have obvious sensory structures, such as eyes and tentacles. Most mollusks other than clams and their relatives have a **radula**, a tonguelike organ with hard teeth (Figure 41.27). The gut represents a large part of the visceral mass. Also in the visceral mass are reproductive organs, kidneys, and a heart. Besides being excretory organs, the kidneys of some mollusks also function as ducts through which gametes are released from the body. Both the anus and the kidney openings are usually in the mantle cavity.

Except for the squids and their relatives, mollusks have an open circulation system, the blood spaces of which form a hemocoel. Their respiration system often consists of a pair of gills, of a type called **ctenidia**, within the mantle cavity. These outgrowths of the mantle wall probably appeared early in molluscan evolution. One or both gills are absent in some mollusks; for them, other outgrowths of the body surface may function as secondary gills.

Let's look briefly at a few mollusks to get an idea of how the basic body plan has been modified in different groups.

Chitons

The marine mollusks called chitons are rather elongated animals, most of which use their radula to graze on algae, hydrozoans, and other low-growing organisms. Chitons have a large, broad foot that enables them to creep over or cling tenaciously to rocks and other hard surfaces.

If a chiton is dislodged, the animal can roll up into a ball and protect itself until it can safely unroll and become reattached elsewhere. It is able to roll up because its dorsal shell is divided into a series of eight plates (Figure 41.28). When a chiton is disturbed or when it is exposed by a receding tide, the muscles in its foot can pull the animal down tightly and the edge of the mantle (which partly or completely covers the shell plates) functions like the rim of a suction cup. Then, the animal is difficult to detach from its substrate.

Figure 41.28 Two chitons from the same habitat: the intertidal zone of Monterey Bay, California. Members of this class of mollusks show beautiful variations in the color and patterns of their dorsal shell.

Figure 41.29 Two of the soft-bodied gastropods of the type called nudibranchs.

Gastropods

Snails and slugs of varied sorts make up the largest group of mollusks, the gastropods ("belly foots"). They are so named because the foot is spread out under the animal while it is crawling. Many snails have a spirally coiled or conical shell. Coiling is a way of compacting the visceral organs into a mass that can be balanced above the rest of the body, much as you would balance a backpack full of books.

Early in their development, most gastropods undergo a strange internal realignment by a process called **torsion**. In this process, contraction of certain muscles as well as differential growth cause the posterior mantle cavity to twist around to the right side and then twist even farther forward, until it is above the head:

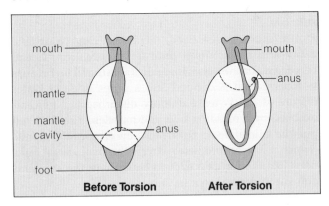

The twisting does provide the head with a cavity into which it can withdraw in times of danger. But it also brings the gills, anus, and kidney openings above the head. As you might imagine, this realignment could create something of a sanitation problem, what with wastes being dumped near the respiratory structures and the mouth. Typically, most gastropods use cilia to create currents that sweep the wastes away from the mouth and from the gills.

Sea slugs (nudibranchs) and some other gastropods apparently underwent detorsion in the course of their evolution. The mantle cavity has largely disappeared. They have lost the ctenidia, but most have other outgrowths that function as secondary gills. Many nudibranchs are striking in their coloration and their array of outgrowths (Figures 41.29 and 41.30).

Bivalves

Clams, scallops, oysters, mussels, and shipworms are bivalves (animals with a "two-valved shell"). Some bivalves are only 1 or 2 millimeters across. The largest, the giant clams of the South Pacific, are more than a meter across and may weigh 225 kilograms (close to 500 pounds).

The head region of a bivalve is not much to speak of, but the foot is usually large and specialized for bur-

Figure 41.30 *Aplysia*, a sea hare, widely used in experimental studies of behavior and physiology. The two flaps over the dorsal surface are extensions of the foot; they undulate and help ventilate the underlying mantle cavity.

Like many other gastropods, *Aplysia* is hermaphroditic. It may act as a male, a female, or both during encounters with others like itself. Neurobiologists are much taken with this organism and have tracked several of its nerve pathways.

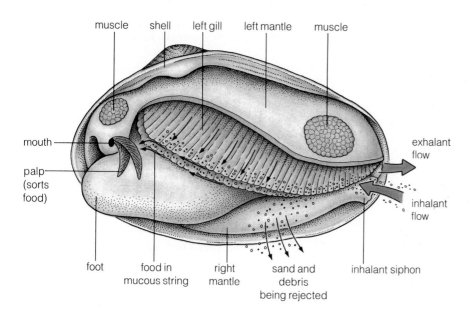

muscle · shell · left gill · left mantle · muscle

mouth

palp (sorts food)

foot · food in mucous string · right mantle · sand and debris being rejected · inhalant siphon

exhalant flow

inhalant flow

Figure 41.31 Anatomy of a clam. The left valve and much of the left mantle have been removed in this diagram. Food trapped in mucus on the gills is sorted by the palps, and suitable particles are carried by cilia to the mouth.

rowing in mud or sand. (Small-footed bivalves, including oysters and mussels, typically attach themselves to a firm substrate or nestle in holes, crevices, and empty shells.) Nearly all bivalves have a pair of well-developed gills. The action of cilia in the gills draws water into the mantle cavity, circulates it over the gills and other body surfaces, and then drives it out.

With few exceptions, the gills also are used in feeding activities. Microscopic food particles are suspended in the water that is circulating through the mantle cavity, and they become trapped in mucus on the gills. Tracts of cilia on the gills move the mixture of food and mucus to a pair of flaps (called palps) next to the mouth. There, acceptable food particles are delivered to the mouth and the rest are rejected (Figure 41.31).

Most bivalves that burrow or nestle have a pair of siphons at their posterior end. The siphons of the giant geoduck (pronounced "gooey duck") of the Pacific Northwest may be more than a meter long. Bivalve siphons are extensions of the mantle edges, fused into tubes. Water is drawn into the mantle cavity through the inhalant siphon. Water leaves through the exhalant siphon and carries wastes from the anus and kidneys along with it.

Scallops and a few other bivalves can swim by clapping their valves together, thus producing a localized jet of water that propels the animal (Figure 41.32). This fast, jerky type of locomotion is advantageous for scallops, which are preyed upon by sea stars, octopuses, and other animals.

Figure 41.32 A scallop swimming away from a predator (a sea star). Scallops and a few other bivalves can clap their two valves together rapidly enough to force a strong jet of water from the mantle cavity.

Figure 41.33 An octopus, illustrating the well-developed eyes and arrangement of the eight arms characteristic of this type of cephalopod.

Cephalopods

The squids, cuttlefishes, octopuses, and nautiluses are examples of cephalopods. Most are fast-swimming predators of the seas. Some are only a few centimeters long, but the giant squid (*Architeuthis*) may reach eighteen meters (about sixty feet) in length. It is the largest invertebrate known. Figures 41.33 and 41.34 show two representative species.

Cephalopods have flexible, muscular arms or tentacles that usually are equipped with suction pads (Figures 27.13b and 41.35). After a prey animal has been captured by the tentacles, it is bitten or crushed by a pair of jaws that resembles an upside-down parrot beak. If venom is secreted, this accelerates the death of the prey, which is drawn into the mouth by a radula.

Cephalopods have a siphon associated with the mantle cavity. Squids and cuttlefishes are able to swim swiftly by a type of jet propulsion, in which a stream of water is forced out of the mantle cavity through the siphon. The mantle wall has both circular and radial muscles. When the radial muscles contract, the cavity enlarges and water is drawn into it. When the circular muscles contract, water is squeezed out. If the free edge of the mantle is simultaneously closed down tightly on the head and siphon, a jet of water shoots out through the siphon. By manipulating its siphon, the animal can partly control the direction in which it propels itself.

Being as active as they are, you might expect the cephalopods to have great demands for oxygen. They do; and they are the only gastropods with a closed circulation system. Blood is pumped from the main heart to the two gills. Each gill has an accessory, "booster" heart at its base to speed the flow and therefore the

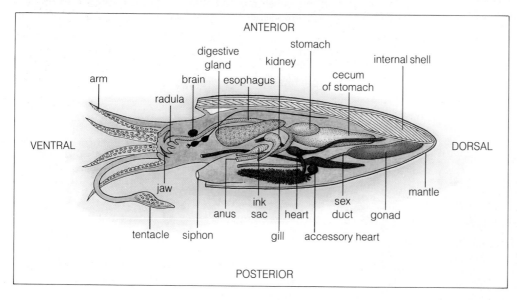

ANTERIOR

Figure labels for the squid/cuttlefish anatomy diagram:

ANTERIOR

digestive gland · brain · esophagus · kidney · stomach · internal shell · cecum of stomach

arm · radula

VENTRAL · DORSAL

jaw

tentacle · siphon · anus · ink sac · heart · gill · accessory heart · sex duct · gonad · mantle

POSTERIOR

Figure 41.34 (Above) A female and a male cuttlefish (*Sepia*) mating, head end to head end.

Figure 41.35 General anatomy of a squid or a cuttlefish. The tentacles are similar to the arms, but they are more slender and specialized for capturing prey. The shell of a cuttlefish is calcified and layered, as shown; squids have only a vestigial shell and it is not calcified.

uptake of oxygen and elimination of carbon dioxide. Even when a cephalopod is resting, it rhythmically draws water into its mantle cavity and then expels it. The movements enhance gas exchange and also help to eliminate wastes that the excretory organs and anus discharge into the mantle cavity.

Cephalopods have well-developed nervous systems, and their brain is larger in proportion to body size, compared to other mollusks. Giant nerve fibers connect the brain with the muscles responsible for jet propulsion, making it possible for a cephalopod to respond quickly to prey or to a dangerous situation. The eyes resemble those of vertebrates, although they develop in a different way. Finally, cephalopods are capable of learning. For example, if an octopus is given a mild electric shock after being shown an object with a distinctive shape, it will avoid that object in the future. In terms of their capacity for learning and memory, the cephalopods are the world's most complex invertebrates.

ANNELIDS

Like the mollusks, the 10,000 or so species of segmented worms (phylum Annelida) are protostomes. The best-known annelids are the earthworms and leeches. Less familiar are the enormously diversified group of (mostly) marine worms called polychaetes.

Figure 41.36 Terrestrial leech attached to a human host, shown before feeding (**a**) and afterward, when it is engorged with blood (**b**). In the past, especially in the early nineteenth century, doctors sometimes were unrestrained in using leeches to "cure" disorders ranging from nosebleeds to obesity; they often applied fifty leeches at a time to patients. Nowadays, leeches are used more selectively, as in relieving congested tissue grafts.

Earthworms

Earthworms are the usual textbook example of annelids. They belong to a group called oligochaetes, some of which are land-dwelling and others, aquatic. Earthworms are scavengers that burrow in moist soil, mud, and silt. They feed largely on decomposing plant material and other organic matter, much of which is ingested during the burrowing process. In soft soil, an earthworm plunges forward and pushes the soil aside. In compacted soil, it literally eats dirt in order to burrow. Every twenty-four hours, an earthworm can ingest its own weight in soil. Earthworms not only aerate soil (which is beneficial to some plants), they also carry subsoil to the surface and so make nutrients available to other organisms.

Leeches

Leeches live in fresh water, the sea, and moist terrestrial habitats in the tropics. Most are predators that swallow small invertebrates or kill them and suck out their juices; the leeches most people have heard about feed on vertebrate blood (Figure 41.36).

Leeches use their sharp jaws to make an incision in their prey, then they plunge a sucking apparatus (a modified pharynx) into the incision. The leech gut has many side branches for storing food taken in during a big meal,

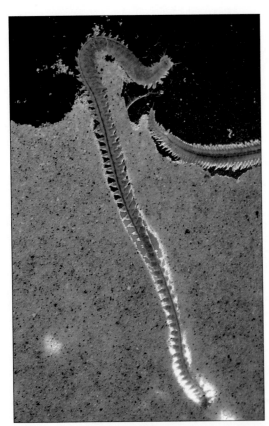

Figure 41.37 Polychaetes. (**a**) A marine polychaete inside its burrow. Notice the setae (chitin-reinforced bristles) along the length of the body. (**b**) A tube-dwelling polychaete, with featherlike structures coated with mucus. The mucus traps small food particles suspended in the water.

this being handy for an animal that may have long waits between meals. The leech body is quite muscular and has a posterior sucker for attachment as well as a sucker around the mouth. Using its suckers, a leech can inch forward over such substrates as the surface of an animal about to provide it with dinner.

Polychaetes

The annelid worms called polychaetes include burrowing, crawling, free-swimming, and attached species. Some excavate vertical or U-shaped burrows in soft mud or sand; others live in tubes made of calcium carbonate, organic secretions, or sand and bits of shell cemented together by secretions. Many polychaetes feed on small invertebrates, some feed on algae, and others use organic matter present in sediments. The predatory forms have a muscular pharynx equipped with jaws that can be everted to capture prey; herbivorous types use their jaws to grasp and tear algae. Most of the tube-dwelling species have ciliated tentacles or featherlike structures originating near the mouth (Figure 41.37). Their mucous secretions trap bacteria, diatoms, and other small particles. The action of cilia carries the mixture of food and mucus to the mouth.

Annelid Adaptations

As a group, the annelids are distinct in having **setae**: chitin-reinforced bristles that occur in pairs or clusters at the body surface and that can be protracted or retracted by special muscles. (Setae are absent only in leeches and a few other forms.) Setae provide the traction required for crawling and burrowing through soil; in swimming species they often are broadened into paddles. Although earthworms have only a few setae per body segment, the marine polychaete worms typically have many of them, often concentrated in fleshy lobes called **parapodia**. The parapodia function in respiration and locomotion.

Nearly all annelids show pronounced segmentation. Internally, the segments are usually separated from one another by partitions, so that there is a series of coelomic chambers. The gut extends continuously through all the chambers, from mouth to anus (Figure 41.38). A "brain" at the head end integrates sensory input for the whole worm. Leading away from the brain is a double nerve cord that extends the length of the body. In each segment, the nerve cord broadens into a ganglion that controls local activity (Figure 25.2).

The coelomic chambers serve as a hydrostatic skeleton against which muscles can operate. Circular muscles are located mostly within the body wall of each segment, and longitudinal ones span several segments. When the longitudinal muscles contract and the circular

ones relax, segments shorten and fatten; they lengthen when the pattern reverses. An earthworm moves forward when its first few segments elongate while segments just behind them hold their position (by protruding their setae). Then the first segments contract and protrude their setae—and the ones behind retract their setae and are pulled forward. Alternating contractions and elongations proceed along the length of the body to move the worm forward.

Annelids generally have a closed circulation system. Contractions of muscularized blood vessels ("hearts")

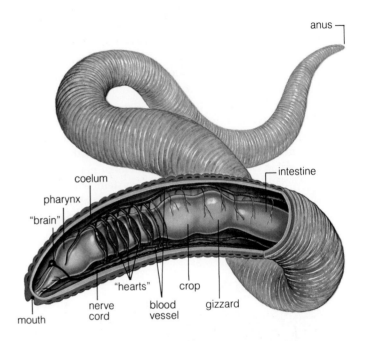

Figure 41.38 Arrangement of nerves, blood vessels, and the digestive system of an earthworm, all of which extend longitudinally through one coelomic chamber after another.

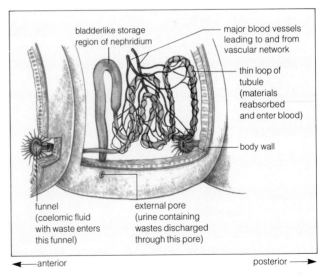

Figure 41.39 Earthworm system of excretion and salt-water balance. The functional unit of the system is the nephridium.

keep blood circulating in one direction. Smaller blood vessels lead to and from the gut, nerve cord, and body wall (Figure 41.38). Functionally linked with the circulation is a system of **nephridia** (singular, nephridium) for regulating the volume and composition of body fluids. Often the functional units of this system have flame cells and so resemble protonephridia, implying an evolutionary link between flatworms and annelids. More commonly, however, the beginning of each nephridium is a funnel-like structure that collects fluid from a coelomic chamber. The funnel leads into a tubular portion of the nephridium that carries the fluid to a pore at the body surface (Figure 41.39). The funnel is located in one chamber but its terminal pore is located in the body wall of the next chamber in line.

In short, the annelids tell us about some major evolutionary trends. Their coelomate and segmented ancestors were an early branch on the phylogenetic tree, foreshadowing a trend toward increases in size and greater complexity of internal organs. Annelids have a rudimentary brain and segmental ganglia—that is, centralized as well as segmented regions of neural control. They have a complete digestive system with regional specializations; they have a closed circulatory system functionally linked with nephridia—the annelid equivalent of the vertebrate kidneys.

ARTHROPODS

Of all the invertebrates, the arthropods show the greatest diversity; there are more than a million known species. Evidently, ancestral annelids or animals like them gave rise to four different arthropod groups. (Figure 41.40 shows an existing animal that bears striking resemblances to both annelids and arthropods.) The groups of arthropods are so distinct that each is now considered to be a separate subphylum:

Subphylum	Representatives
Trilobita	*trilobites (described in Chapter Thirty-Eight); now extinct*
Chelicerata	*horseshoe crabs, spiders, scorpions, ticks, mites*
Crustacea	*crustaceans (including copepods, crabs, lobsters, shrimps, barnacles)*
Uniramia	*centipedes, millipedes, insects*

Arthropod Adaptations

The arthropods are more widely distributed through more habitats than any other animal group. Many are herbivores, feeding on algae or land plants; others are carnivores, omnivores, or parasites. What adaptations have contributed to the success of these animals? Perhaps the most important are these:

1. Hardened exoskeleton
2. Specialization of segments
3. Jointed appendages
4. Specialized respiratory systems
5. Efficient nervous system and sensory organs

The Arthropod Exoskeleton. Annelids, recall, have a cuticle that is thin and flexible enough to permit bending as well as integumentary gas exchange. It also is thin enough to be pierced by predators. From the fossil record, we know that some early annelids developed thicker, hardened cuticles—a pattern that continued among the arthropods.

The arthropod cuticle is an **exoskeleton** (external skeleton). Its outer layers contain protein and chitin, a combination that is flexible, lightweight, and protective. In some species, the cuticle is hardened by calcium deposits that reduce flexibility but help create "armor" plates. At first, exoskeletons were probably useful as a defense against predators. They became useful in other ways when arthropods radiated into land environments. *The arthropod exoskeleton is a superb barrier to evaporative water loss, and it provides support for an animal body deprived of the buoyancy it would have in water.*

An exoskeleton does place restrictions on increases in size, but arthropods enlarge in spurts. Periodically, as part of the molting process, they shed the exoskeleton and grow a new one. If the arthropod lives in water or wet habitats, it swells up a little with water after shedding the exoskeleton and enlarges before the new exoskeleton hardens. If it lives on land, it enlarges first by swelling up with air. Just before molting, some arthropods go into hiding to protect themselves until their new armor hardens.

Specialization of Segments. As arthropods evolved, body segments became more specialized, reduced in number, and grouped or fused together in a variety of ways. In some lineages, for example, several segments became organized to form the head, others as a thorax and abdomen. In other lineages, a few segments formed the forebody, and the rest, a hindbody. Typically the appendages for each body region became highly specialized.

As the arthropod exoskeleton became increasingly firm and unyielding, muscles became concentrated in bundles that could bring about movements of adjacent segments and their specialized appendages. The muscles run from the exoskeleton of one segment to that of the adjacent segments. This arrangement limits the extent to which adjacent segments can be bent or rotated.

Figure 41.40 A velvety "walking worm" of the small phylum Onychophora, the members of which resemble both the annelids and the arthropods. These worms live mostly in humid forests, especially of the tropics. They hide by day under rotting logs and in leaf litter and hunt for prey at night. The largest is about fifteen centimeters (nearly six inches) long. Depending on the species, there are between fourteen and forty-three pairs of stumpy, unjointed legs with claws at the tip (onychophoran means claw-bearer).

Like annelids, these worms have ciliated nephridia. Like arthropods, they molt and they have an open circulation system, with blood in the hemocoel reentering the heart through lateral pores. Moreover, they have tracheal tubes similar to those of millipedes, centipedes, and insects. Such similarities do not mean that onychophorans are "missing links" between annelids and arthropods; they are probably one of many invertebrates that evolved from annelid ancestors and have survived to the present.

Jointed Appendages. Arthropods have jointed appendages (the word arthropod means "jointed foot"). The appendages are specialized for such diverse tasks as walking or swimming, feeding, sensory reception, transfer of sperm, and spinning out silk for a web. The origins of jointed appendages are obscure, but it seems likely that they were derived from fleshy, segmental outgrowths similar to the parapodia of some annelids.

Respiratory Systems. Arthropods have a variety of respiratory structures, but the ones that contributed most to the diversity of this group are the tubes called tracheas, especially among the insects. Tracheas begin at pores on the body surface and branch in such a way that they supply oxygen directly to most tissues (Figure 31.2); the circulatory system has little to do with transporting oxygen. The tracheal system provides adequate support for high metabolic rates in some small-bodied arthropods; and it is the foundation for such energy-consuming activities as insect flight.

Specialized Sensory Structures. The specialized sensory organs of arthropods, described in Unit Five, contribute immensely to their success. Think about the compound eye, which is typical of insects, millipedes, centipedes, and most crustaceans. Most of these arthropods have a wide angle of vision and they can process visual stimuli from many directions. The compound eyes of dragonflies are nearly hemispherical and contain more than 30,000 photoreceptor units (ommatidia); they have much to do with the dragonfly's success in capturing other insects in midair.

With this overview of arthropod adaptations in mind, let's turn now to the specializations among the living members of this phylum: the chelicerates, crustaceans, and insects and their kin.

Chelicerates

The Chelicerates originated in the shallow seas of the early Paleozoic. Except for some mites, the only truly marine survivors are the so-called horseshoe crabs (page 546) and sea spiders. Among the more familiar land-dwelling members of this group are ticks, mites, spiders, and scorpions.

Ticks are blood-sucking external parasites of many land vertebrates; many are vectors for pathogenic microbes, including the ones that cause Rocky Mountain spotted fever and Lyme disease (Chapter Thirty-Nine). Some mites are free-living scavengers; others are parasites that live in such interesting places as the vertebrate ear. Spiders and scorpions are predators.

The spider or scorpion body is divided into a forebody and hindbody (Figure 41.41a). The forebody has six pairs of appendages. Four pairs are walking legs; the other two pairs (chelicerae and pedipalps) function in subduing prey and food handling. The chelicerae inflict wounds and discharge venom. The spider or scorpion pumps digestive enzymes from its gut into the wounds, and then it sucks up the liquefied remains of its prey. Male spiders also use the pedipalps to transfer sperm to the female, and scorpions use their pincerlike pedipalps to grasp their partner in the mating ritual. The forebody has several eyes, used to detect movement of prey.

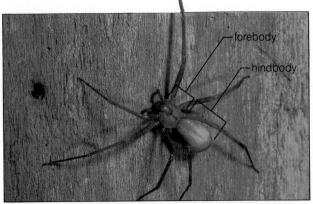

a

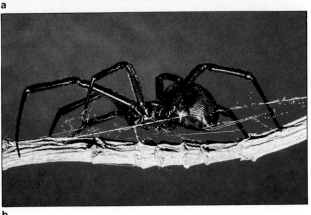

b

c

Figure 41.41 Representative spiders. (**a**) The brown recluse, the bite of which can be severe to fatal for humans. This North American spider lives under bark and rocks, and in and around buildings. It can be identified by the violin-shaped mark on its forebody. (**b**) A female black widow, the bite of which can be painful and sometimes dangerous. (**c**) Wolf spider. Like most spiders, it is harmless to humans and plays a major role in nature in keeping insect pests in check.

Appendages on the hindbody, if any, are extremely modified. Spiders use small posterior appendages to spin out threads of silk for webs and egg cases. The silk threads of most spider webs are nets for capturing prey; they also serve as lines of vibration that inform the spider of a disturbance in the web (as when a prey animal attempts to free itself). One spider spins a vertical thread from which a ball of sticky material is suspended, then uses one of its legs to swing the ball at passing insects!

Spiders and scorpions have respiratory organs called **book lungs**, these being deep pockets on the underside of the body (Figure 41.42). The pockets have extensively folded walls that resemble the pages of a book, and they greatly increase the surface area available for gas exchange. The folds are moist, and blood passing through them picks up oxygen from the surrounding air.

Crustaceans

There are about 40,000 species of crustaceans, including shrimps, crayfishes, lobsters, crabs, barnacles, isopods (such as pillbugs), copepods, and water fleas. Many are predators or herbivores; others are scavengers or parasites. Most live in the seas, many live in fresh water, and numerous species also may be found on land. All are important components of food webs.

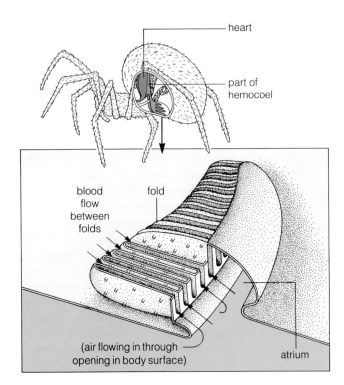

Figure 41.42 A book lung, a type of respiratory organ that occurs among spiders and scorpions.

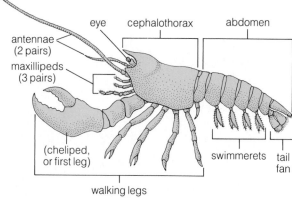

Figure 41.43 Paired appendages of a male American lobster. This and some other crustaceans have a body organized into a cephalothorax (head plus thorax) and an abdomen. The appendage shown on the first segment of the abdomen is used in transferring sperm to the female.

The simplest crustaceans have rather uniform, unspecialized body segments and appendages; they may resemble the ancient arthropods that gave rise to the group. However, most crustaceans show great diversity in the number and organization of body segments and in the specializations of paired appendages. For example, a crab has strong claws for tearing up seaweed, collecting organic debris, or intimidating or attacking another animal; a barnacle uses featherlike appendages for combing microscopic food from the water.

Crustaceans commonly have between sixteen and twenty segments, although some have more than sixty. The segmentation is not readily apparent in crabs, lobsters, and some other types because the dorsal cuticle of the head extends backward, covering some or all segments with a **carapace** (a shieldlike cover). The head appendages are two pairs of antennae, a pair of mandibles, and two pairs of maxillae. The **antennae** are mostly sensory, although they are sometimes used for swimming or other functions. **Mandibles** are jawlike appendages; **maxillae** are food handling and food sorting appendages. Figure 41.43 shows how some of these body parts are arranged in a lobster.

The 7,500 species of copepods are among the most abundant aquatic animals. They are distinctive in having a single eye in the middle of their head (Figure 41.44). Most copepods are weakly swimming, planktonic forms less than two millimeters long. They dine on microscopic algae and other food particles and in turn are eaten by larger aquatic animals. About 1,500 species are parasitic on fishes and various invertebrates.

Barnacles are unusual in having a strong, calcareous shell (actually a modified exoskeleton) that helps them withstand predators and the force of waves or currents. Some barnacles attach to rocks and other surfaces by an

Figure 41.44 A marine copepod, one of the smaller crustaceans (typically no more than a few millimeters long). Some copepods are parasitic, but the ones of greatest ecological importance are the free-swimming herbivores of aquatic communities; they form an important part of the diet of fishes and other carnivores.

Figure 41.45 Stalked barnacles with the body in a shell that is hardened with calcium deposits. This crustacean uses featherlike appendages to comb the seawater for microscopic food.

Figure 41.46 Specialized appendages of a land crab (a crustacean).

elongated stalk (Figure 41.45). Others are stalkless forms that attach directly to rocks, wharf pilings, ship hulls, and other surfaces by their shells; a few species attach only to the skin of whales!

The shrimps, crayfish, lobsters, crabs, and their relatives are called decapods because they have ten legs (deca- means ten). Many are commercially important. Most live in the seas, although there are a few freshwater crabs and shrimps as well as a few crabs and crayfishes that burrow on land. Lobsters walk about on the seafloor, preying on snails, clams, crabs, and small fishes. Although crabs are rather broad and flat (Figure 41.46), their weight is distributed so evenly over their legs that they are remarkably agile walkers.

Insects and Their Kin

Millipedes and Centipedes. Millipedes and centipedes are close relatives of the insects. Both are notable for their numerous paired legs along the trunk, this being a long, segmented region behind the head. Millipedes have a rather cylindrical trunk (Figure 41.47a), ranging from about two millimeters to nearly thirty centimeters in length. All millipedes are slow-moving, nonaggressive scavengers of decaying plant material.

Unlike millipedes, the centipedes have a flattened trunk and they are fast-moving, aggressive carnivores, complete with fangs and venom glands. Centipedes generally live in damp places under rocks, logs, or forest litter. They prey mostly on insects, earthworms, and snails, although some tropical species can subdue small lizards and toads (Figure 41.47b). One long-legged centipede (*Scutigera coleoptrata*) eats insect pests, is harmless to humans, but its unexpected appearance in a house

occasionally startles people. A bite from various species around the world can be quite painful, however.

Insects. Insects are the most successful invertebrates on land. Many also live in fresh water during at least some part of the life cycle; relatively few are adapted for life in the sea or in brackish water. A tracheal system of respiration and an exoskeleton that retards evaporative water loss were key characters that led to their success on land; so did the development of **insect wings**, these being lateral folds of the exoskeleton. Winged insects disperse themselves rapidly, move into areas where food may be abundant, swoop down on prey or capture other insects in midair, and escape from predators.

The insect body consists of a head, thorax, and abdomen. The head has a pair of sensory antennae and three pairs of mouthparts, which are modified for biting, chewing, puncturing, and sucking, depending on the species (Figure 41.48). The thorax has three pairs of legs, used for walking and clinging; in most species the thorax also has two pairs of wings. The abdomen has eleven segments at most, although some may be fused together or reduced. In most insects, the only abdominal appendages are reproductive structures. For example, some female insects have an ovipositor (egg-laying device).

The insect gut is divided into three parts (foregut, midgut, and hindgut). Digestion occurs mostly in the midgut, and water is reabsorbed in the hindgut. The excretory organs of insects are called **Malpighian tubules**, these being named after an early anatomist. The tubules, which are connected to the hindgut, are suspended in the hemocoel, and soluble nitrogenous wastes enter them by diffusion. There, the wastes are converted into harmless crystals of uric acid that enter the hindgut and are eliminated with feces. This excretory system

a

b

Figure 41.47 (a) One of the mild-mannered millipedes, which scavenge on decaying plant parts. (b) One of the not-so-mild-mannered centipedes of Southeast Asia, an aggressive predator that can bring down small frogs and lizards.

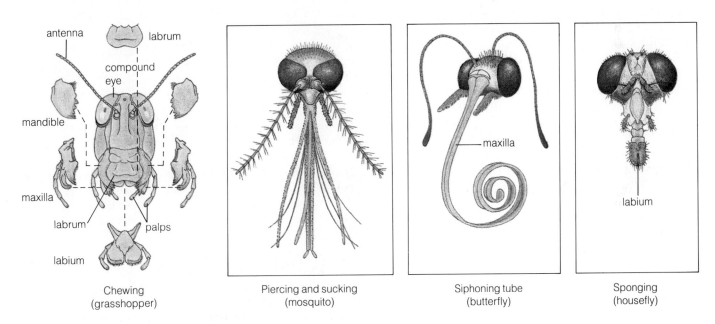

Chewing
(grasshopper)

Piercing and sucking
(mosquito)

Siphoning tube
(butterfly)

Sponging
(housefly)

Figure 41.48 Examples of insect headparts, adapted for feeding in specialized ways.

allows land-dwelling insects to rid themselves of nitrogenous wastes without losing precious water.

Some insects have a staggering reproductive capacity. Perhaps you have seen hundreds of maggots developing from the eggs laid by a single female fly. It has been estimated that if all of a fly's progeny were to survive and reproduce through six more generations, she would have more than 5 trillion descendants!

Insects also are our most aggressive competitors. They eat our vegetable crops and stored food; they destroy wool, paper, and timber. They bite and suck blood from us and our pets, and transmit diseases. On the bright side, insects are important pollinators of some of our major crop plants, and many beneficial insects attack or parasitize insects we would rather do without. Figure 41.49 describes some representatives.

a

b

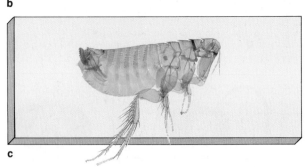

c

d

e

f

Figure 41.49 Representatives of some orders of insects.

(**a**) Mediterranean fruit fly (order Diptera). The larval forms destroy citrus fruits and other crops. (**b**) A duck louse (order Mallophaga), which dines on particles of feathers and bits of skin. (**c**) A flea (order Siphonaptera) with big strong legs, excellent for jumping.

(**d**) The robberfly (order Diptera) looks beelike but is a true fly. It is a predator of honeybees (order Hymenoptera) and other insects. Its relatives include mosquitoes, fruit flies, tsetse flies, and other insects we would mostly rather do without.

(**e**) A scarab beetle, member of the largest order (Coleoptera) in the animal kingdom. There are more than 300,000 known species. (**f**) The flower mantis (order Mantodea). Its segments are almost indistinguishable from the flower petals.

(**g**) Ladybird beetles (order Coleoptera) swarming. These beetles are raised commercially; they serve as natural biological controls of aphids and other insect pests.
(**h**) Stinkbugs (order Hemiptera), newly hatched. (**i**) The carrion-eating insects called fire ants (order Hymenoptera), which can inflict painful bites.

(**j**) Luna moth (order Lepidoptera), one of the most beautiful flying insects of North America. Like most other moths and butterflies, its wings and body are covered by microscopic scales. (**k**) Adult dragonfly (order Odonata), a remarkable aerialist that can capture *and* eat other insects in midflight.
(**l**) European earwig (order Dermaptera), a common household pest.

a

b

c

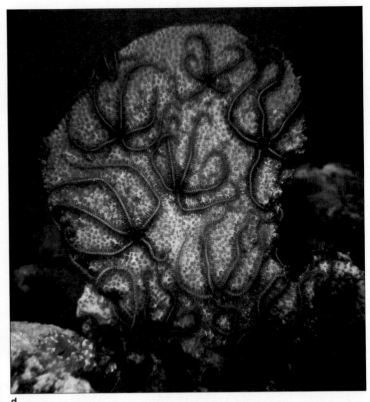

d

Figure 41.50 Representative echinoderms. (**a**) Feather star, one of the crinoids. A feather star clings to pebbles, shells, and other objects, but it can detach itself and swim gracefully by fluttering movements of its arms. (**b**) Cobalt sea star. This colorful species has six rays; most sea stars have five. (**c**) Sea cucumber, with elongated body and lengthwise rows of tube feet. The branching tentacles of the anterior end are modified tube feet, used in collecting food. (**d**) Brittle stars, which move by rapid, snakelike action of their rays.

ECHINODERMS

So far, we have focused on only one of the two major lineages of coelomate animals. We turn now to the deuterostomes, beginning with the members of the phylum Echinodermata. The name means "spiny-skinned" and alludes to the calcareous spines, spicules, or plates in the body wall.

Among the 6,000 or so existing species of echinoderms are sea stars, sea urchins, sand dollars, brittle stars, sea cucumbers, and crinoids (Figures 41.50 and 41.51). All of the adult forms show radial symmetry, although this is complicated by bilateral features, especially in sand dollars, sea urchins, and sea cucumbers.

The nervous system of echinoderms is not like that of clearly bilateral animals; it is decentralized and there is no brain or "head end." The decentralized arrangement is favorable for animals with radial symmetry, for it enables them to respond to stimuli coming from var-

a

b

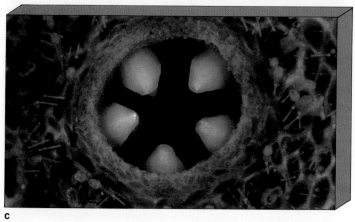

c

tube feet   spine

Figure 41.51 **(a, b)** Two representative sea urchins, another type of echinoderm. Sea urchins do not have rays, and the calcium-containing ossicles in the body wall are flattened and positioned so tightly together that they form a nearly solid shell. Most sea urchins are nearly globular or tomato-shaped. **(c)** Ossicles in the mouth region are organized into a complicated apparatus for scraping and chewing up food, which consists mostly of seaweed.

Sea urchins use their spines and tube feet for locomotion. The spines offer considerable protection; they also are equipped with a venom gland in some species. With or without the venom, a spine that punctures and breaks off beneath human skin can induce an inflammatory response.

ious directions. In most sea stars, any ray ("arm") may become the leader, and the animal moves in that direction.

If you turn a sea star over, you will see **tube feet**: fluid-filled, muscular structures that are commonly tipped with suckerlike adhesive disks. Tube feet are typical of most echinoderms. They can be used for walking, burrowing in sand or mud, clinging to a rock, or gripping the shell of a snail or clam that is to be eaten. The tube feet are part of a **water-vascular system** that is unique to echinoderms (Figure 41.52).

In this system, a main canal runs down each ray, and each is connected to a ring canal that encircles the mouth region. In most sea stars, a perforated plate at the body surface connects with the ring canal and perhaps allows water to enter and mix with the fluid inside. Short side canals extend from each main canal and deliver water to the tube feet. Each tube foot has an **ampulla**, a muscular structure something like the rubber bulb on a medicine dropper. When an ampulla contracts, the fluid inside is forced into the tube foot, causing it to lengthen. Tube feet continually change their shape because fluid

within the water-vascular system is constantly being redistributed by muscle action.

Hundreds of tube feet may move at a time. After being released, each one swings forward, reattaches to the substrate, then swings backward and is released before swinging forward again.

Figure 41.52 also shows the two-part stomach of sea stars. Some species can distend their mouth enough to get a whole prey animal inside the lower chamber of the stomach. Most extrude that part of the stomach around a prey and initiate digestion even before swallowing it. Coarse, undigested residues are eliminated through the mouth. Most sea stars do have a small anus, but this is of no help in getting rid of empty clam or snail shells.

How did echinoderms originate? We do not know, even though they have left a rich fossil record. Some of the early echinoderms were bilaterally symmetrical, others appear to have been asymmetrical. When the life cycle proceeds through a larval stage, the larva is bilateral, suggesting that some group of ancient bilateral invertebrates gave rise to the ancestral stock of the phylum.

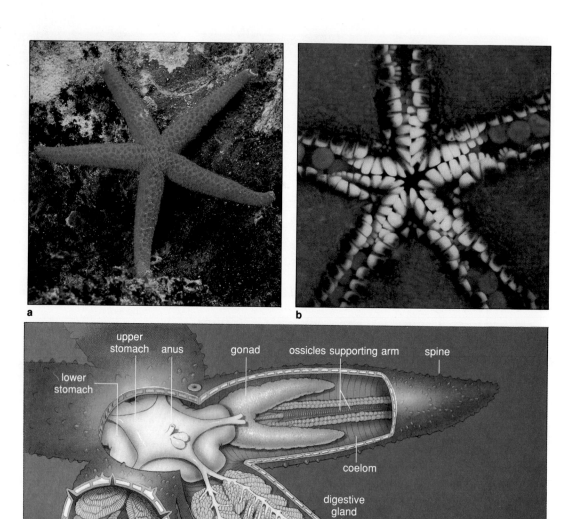

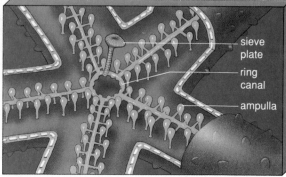

Figure 41.52 Anatomy of a sea star (**a**). External view of the mouth (**b**), and the internal arrangement of major organs (**c**). The water-vascular system, based on tube feet, is shown in (**d**) and (**e**).

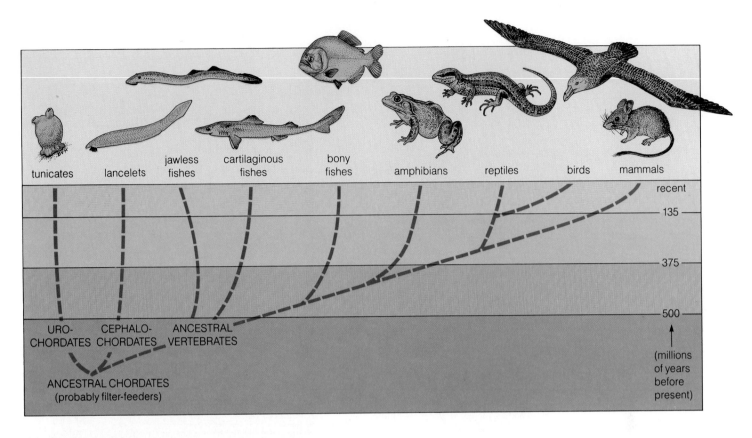

Figure 41.53 One proposed family tree for the vertebrates and their relatives, the urochordates and the cephalochordates.

CHARACTERISTICS OF THE CHORDATES

Figure 41.53 shows representatives of the phylum Chordata. Most chordates are vertebrates—animals whose brain and nerve cord are protected by a column of cartilage or bone. The chordates without such a "backbone" have the following features that give them membership in this phylum:

1. Perforations in the wall of the pharynx, or at least pouches that suggest perforations.

2. A nerve cord dorsal to the gut. It is also tubular, because of the way it originates during embryonic development—as a trough of ectoderm that becomes roofed over and separated from the skin (Figure 34.11).

3. A supporting structure called the *notochord*, located just beneath the nerve cord (page 337).

The three features may not persist throughout the life of a chordate, but they are usually evident during the early stages of development. For example, you have a tubular dorsal nerve cord, but your notochord and gill pouches disappeared when you were still an embryo (Chapter Thirty-Five).

Let's now look briefly at the two groups of invertebrate chordates, called **urochordates** and **cephalochordates**. All of the members of both groups are confined to the seas, whereas vertebrates are found in marine, freshwater, and terrestrial habitats.

THE INVERTEBRATE CHORDATES

Urochordates

The urochordates are commonly called tunicates, a name alluding to a gelatinous or leathery "tunic" these animals secrete around themselves. They also are called seasquirts, because some expel water through their excurrent siphon when they are irritated. They are the smallest of our chordate relatives, being microscopic or, at most, a few centimeters tall.

These animals do not have a coelom, as all other chordates do. When a notochord is present, it is confined to the tail region of an animal that resembles a tadpole (Figure 41.54a). Usually the "tadpole" form is a larval stage, but sometimes the adults have this form. The notochord is a series of large cells that are more or less wrapped around a fluid-filled channel. Being turgid yet flexible, it functions like a torsion bar. After it has been bent by contraction of muscles on one side or the other

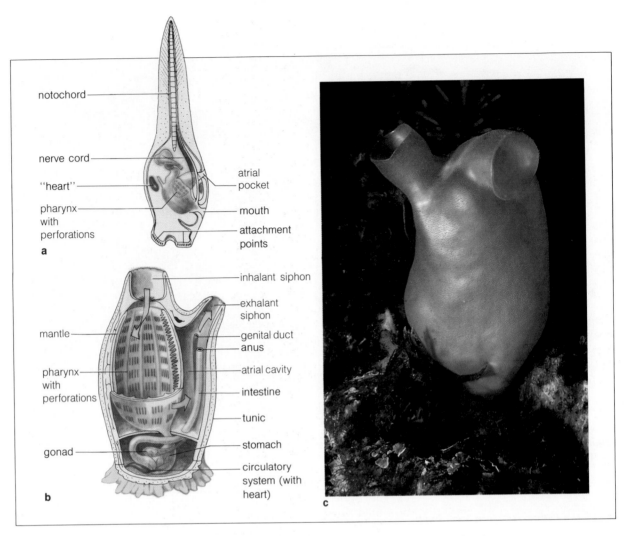

Figure 41.54 One of the tunicates, a "sea-squirt." (**a**) Tadpole larva and (**b**) adult form. These sketches are not to the same scale; the larva is much smaller than the adult. (**c**) A typical tunicate, the sea peach.

of the tail, relaxation of these muscles allows the notochord to straighten. As it springs back, it gives the tail a strong propulsive force.

Most adult tunicates are attached to solid objects, such as rocks, shells, and wooden pilings. The tadpole larva, shown in Figure 41.54a, is a bilateral animal. Its gut is incomplete and nonfunctional, but the pharynx has a few perforations that open into right and left pockets that are open to the outside.

After attaching to a firm substrate, the tadpole undergoes drastic metamorphosis. It resorbs its tail, loses the notochord, and secretes a tunic around itself. The body becomes completely reorganized. The pharynx (into which the mouth opens) enlarges and its perforations become subdivided into many small openings.

The pharynx now resembles a sieve. Its epithelial lining is ciliated and is continually being coated with mucus. The action of the cilia causes water to be drawn into the pharynx through the inhalant siphon and to be moved through perforations into the atrial cavity (Figure 41.54b). The water leaves the body through the exhalant siphon. Small food particles, such as diatoms, become trapped in the mucus, and special tracts of cilia move them to the stomach for digestion.

Besides collecting and concentrating food, the pharynx is an efficient respiratory organ; it is thin-walled and has a good blood supply. There is no real excretory system in these animals, but relatively insoluble nitrogenous wastes are commonly stored in certain tissues. There is a heart beneath the pharynx. Curiously, it pumps blood for a time in one direction, then in the opposite direction. Blood leaving the heart is at first enclosed by vessels, but it soon escapes into spaces comparable to those forming the hemocoel of a mollusk or arthropod.

Are urochordates in the line of evolution of vertebrates? Probably not. The absence of a coelom, the peculiar kind of notochord, and the fact that the perforations of the pharynx are not in a single, linear series suggest there is no direct connection between urochordates and vertebrates. Nevertheless, it is possible that an animal somewhat similar to the swimming larva of a tunicate could have figured in the early evolution of vertebrates. The animals called acorn worms are also interesting in this respect (Figure 41.55).

Cephalochordates

There are only a few species of cephalochordates, which are fishlike animals rarely more than five centimeters (about two inches) long. They are called lancelets because of the way the body is sharply tapered at both ends. The name "amphioxus," often applied to these animals, also refers to this feature (Figure 41.56).

With its well-developed musculature, a lancelet can bury itself quickly in sand or mud. The muscles concerned with the movements are arranged in series on both sides of the notochord, giving the animal a segmented appearance. The notochord itself extends for nearly the full length of the body and consists mostly of disklike muscles. It is decidedly different from the tunicate notochord, as well as from the notochord of living fishes of ancient ancestry.

When a lancelet is buried and in the feeding position, its mouth is at the surface of the sediments. Ciliary activity that draws water into the pharynx also drives it out through numerous gill slits arranged in series on both sides. On passing through the gill slits, the water enters a spacious atrial cavity, then leaves the body by way of a ventral pore. Food trapped in mucus within the pharynx is delivered to the rest of the gut, where digestion, absorption, and compaction of fecal material take place.

Lancelets are almost certainly closer to vertebrates than urochordates are. The larval stage of a lamprey (a jawless fish) is similar to a lancelet, not only in structure but in the method of feeding. The similarity is probably the result of parallel evolution.

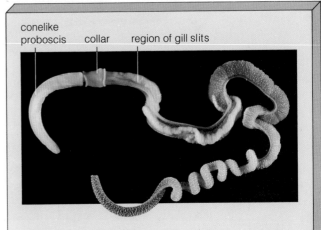

Figure 41.55 An acorn worm, one of about 120 species of the phylum Hemichordata. Acorn worms are sometimes common in tidal and mud flats, where they live in U-shaped burrows. One species was recently discovered on the edge of a deep-sea vent in the Galápagos Rift (Figure 46.35).

Many of these animals are filter-feeders; the mouth region is adapted for sorting and collecting food particles suspended in water. Their gill slits open from the pharynx and resemble the gill slits of chordates in their functioning. The larvae often resemble those of echinoderms. Did the early chordates evolve from animals something like acorn worms? It is interesting to think about.

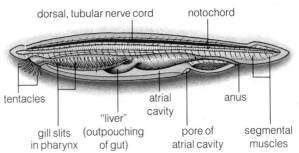

Figure 41.56 Photograph and cutaway view of a lancelet, showing the position of its nerve cord and flexible notochord.

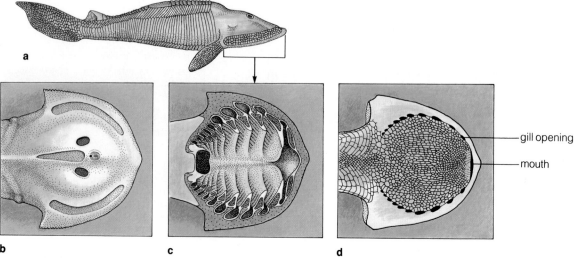

Figure 41.57 (**a**) Reconstruction of an ostracoderm and its skull. This one lived about 100 million years after the oldest known ostracoderms. The dorsal view of the skull in (**b**) shows the openings for three eyes (blue) and a single nostril (red). The ventral view (**c**) shows the small mouth and relatively large, food-straining pharynx; the flattened bones covering all but the gill openings and a slitlike mouth have been removed, but are shown in place in (**d**).

THE VERTEBRATES

From Jawless to Jawed Fishes

The earliest known vertebrates were fishlike animals called **ostracoderms**, a name referring to their external body plates of a material similar to bone (Figure 41.57). The brain was enclosed in a skull of bone or cartilage, and the notochord apparently functioned as a supporting structure. Ostracoderms did not have jaws. In other early evolutionary lines of jawless vertebrates, there were distinct vertebral units of cartilage or bone. This arrangement would be much like that in a lamprey or hagfish. In these existing jawless fishes, the notochord is somewhat gelatinlike, but it is surrounded by a sheath of connective tissue and supplemented by cartilage. Some cartilaginous elements also support the tissue bridges that separate successive gill slits (Figure 41.58); others, above the notochord, bear resemblances to the vertebrae of complex vertebrates. Lampreys and hagfishes also have a simple skull.

Eventually, well-developed vertebrae evolved in fishes. What are the advantages of having a series of these cartilaginous or bony structures? They fit together in such a way that they allow the body to flex, and they provide attachment sites for muscles concerned with this type of movement. Moreover, they completely surround and protect the nerve cord.

Jaws, which are characteristic of all living vertebrates other than lampreys and hagfishes, evolved from the most forward gill supports (Figure 41.58). It may have taken a long time for this new structural development to become functional in biting and chewing, but it certainly opened up many possibilities for feeding by mech-

anisms other than sucking, rasping or nibbling, which would be the only way jawless fishes could feed. It must also have intensified competition for prey and increased the prospect that one fish could be eaten by another. Thus there would have been selective advantages for a fish to be able to recognize food or a predator from a distance. The evolution of more sensitive eyes, olfactory receptors, and sensors that could detect vibrations followed; the brain and motor pathways also became more complex (Chapter Twenty-Five). The evolutionary trends toward more efficient sense organs and nervous systems in the jawed fishes continued in the vertebrates that invaded the land. Thus, *some of the features that evolved first in the fishes became the legacy for amphibians, reptiles, birds, and mammals.*

Key Developments in the Transition to Land

Evolution of Gills, Lungs, and the Heart. It seems that the ancestors of vertebrates were filter-feeders, and in this respect they were similar to urochordates and cephalochordates. The perforations in their pharynx would have been used for eliminating water, and in this way the food particles that had been retained within the pharynx would have become concentrated. The perforations may have enhanced respiration, too, but this would have been a secondary role. By the time the first jawless fishes evolved, their degree of activity and larger size would have required more efficient ways to absorb and distribute oxygen. When rasping and sucking replaced filter feeding, the gill slits and tissues around them became important in respiration.

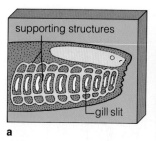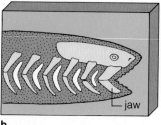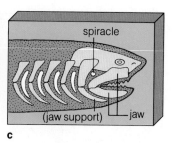

a   b   c

Figure 41.58 Proposed evolution of gill-supporting structures, as found in jawless fishes (**a**), into the hinged vertebrate jaw (**c**). The first gill opening of the mud-dwelling jawless fishes was converted into a spiracle through which water could be drawn. The first gill bars in the series, no longer required as supporting structures, became enlarged and equipped with teeth. The embryonic development of sharks reflects such a sequence.

In existing lampreys and other fishes, the ventricle of a two-chambered heart initially pumps blood to the gills. Then the oxygenated blood is distributed through the rest of the body by a closed circulation system, finally returning to the atrium of the heart (Figure 41.59a). Gills are unsuitable respiratory devices for land animals. If its gills and body surface could be kept moist, and if its gill branches did not become so badly stuck together that they could not function, a fish might be able to live temporarily out of water, as it would have to do in order to get from a pond that has dried up to one that has water. (A few existing fishes are able to do this.) The first land-venturing animals, which lived during the Devonian period, were fishes that had developed lungs. These originated as outpocketings of the pharynx. Lungs continue to function in a few fishes living today in Africa, Australia, and South America, but in most fishes they have been converted into a swim bladder (Figure 31.4). Some of the tissue that lines a swim bladder controls the quantity and proportions of gases within it, and this enables a fish to maintain buoyancy and its position at a particular depth. The swim bladder is still connected to the pharynx in some fishes (including goldfishes and carps), which are able to gulp air.

An important development related to the appearance of lungs was the extension of the nostrils to the mouth, as is typical of lung-breathing amphibians, and all reptiles, birds, and mammals. This makes it possible for an animal to breathe while its mouth is closed.

In addition, the atrium of the heart became divided, so the heart was three-chambered. This is the situation in modern amphibians (Figure 41.59b). In a frog, for instance, blood that has been oxygenated in the lungs is returned to the left atrium, whereas blood from the systemic circulation is returned to the right atrium. From both atria, blood is pumped into the ventricle, where some mixing occurs. Nevertheless, when the ventricle contracts, poorly oxygenated blood, mostly in the anterior part of the ventricle, tends to be forced into the vessels that lead to the lungs. The remainder of the blood, which is moderately well oygenated, goes to the

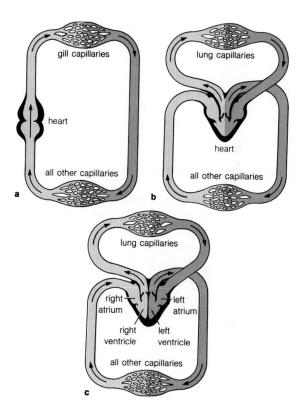

Figure 41.59 Relationship between the heart and blood vessels in the closed circulation system of (**a**) fishes, (**b**) amphibians, and (**c**) mammals.

systemic circulation. The arrangement is adequate for amphibians because they do not have a high rate of metabolism and because the skin and mouth cavity are also generally important in exchange of oxygen and carbon dioxide.

Most reptiles have conserved the amphibian system of circulation, although the ventricle is partly divided into right and left chambers. This reduces the extent to which blood from the right and left atria is mixed. In crocodiles and alligators, in fact, the separation into right and left ventricles is complete, so the pattern of circu-

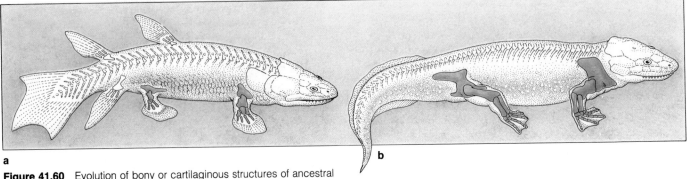

Figure 41.60 Evolution of bony or cartilaginous structures of ancestral lobe-finned fishes (**a**) into the limbs of early amphibians (**b**).

lation is comparable to that of birds and mammals. Blood from the right ventricle goes to the lungs; that from the left ventricle goes to the systemic circulation. This arrangement is absolutely essential for birds and mammals (Figure 41.59c), which maintain a warm internal body temperature, and which generally have high rates of metabolism requiring a large supply of oxygen.

Evolution of Limbs. The legs, arms, and other limbs of the terrestrial vertebrates can be traced back to fins of certain fishes (Figure 41.60). Paired fins developed in the pectoral and pelvic regions of the body, paving the way for the evolution of *lobed* fins, with internal supports of bone or cartilage. Appendages of this type usually permitted more effective locomotion on land, and further evolution led to the four legs of amphibians and eventually to arms and legs of humans, flippers of seals, and wings and legs of bats and birds. Among the few living fishes that have descended from early lobe-finned fishes are the coelacanths (Figure 41.61). They were long thought to be extinct, but since 1938, many specimens have been collected in the deep sea between Africa and Madagascar. Lungfishes are also survivors of the same general group, although their fins are so much modified that they scarcely resemble those of their ancestors.

Evolving Neural and Sensory Structures. By the time that amphibians were established on land, they were encountering stimuli to which their aquatic ancestors had not been exposed. The brain underwent dramatic evolution, particularly in the portions concerned with processing signals related to vision and hearing.

Eyes already had become well developed in fishes, and they were modified for vision in air when amphibians began their evolution on land. The ear, as a complex of structures concerned with balance, also originated in fishes. These structures—including the semicircular canals and statocysts associated with them—have persisted throughout the diversification of vertebrates.

The cerebral cortex of the brain began to develop markedly during the evolution of reptiles. At first the layer of gray matter was thin, but it must have had connections that permitted more complex integration than had been possible in amphibians. It was not until the rise of mammals that the cerebral cortex began to reveal its remarkable potential, gradually expanding to a large mass of information-encoding and information-processing cells (Chapter Twenty-Five). Like other evolutionary developments described in this chapter, the vertebrate brain is the result of many changes taking place over many millions of years.

Today, there are seven classes of vertebrates:

Agnatha	*jawless fishes*
Chondrichthyes	*cartilaginous fishes*
Osteichthyes	*bony fishes*
Amphibia	*amphibians*
Reptilia	*reptiles*
Aves	*birds*
Mammalia	*mammals*

Details of the anatomy, physiology and life styles of these diverse animals are covered in other units of the book, and the major orders are listed in Appendix I. Here we will simply summarize their important characteristics.

Modern Fishes

Of all living vertebrates, the fishes are the most numerous and diverse. Nearly 22,000 species of these mostly predatory, aquatic animals have been identified. About fifty-eight percent of the species live in marine habitats, including deep-sea trenches, the open oceans, coastal waters, and estuaries; forty-one percent live in freshwater ponds, lakes, rivers, and streams. The remaining one percent include salmon and other types that move regularly between freshwater and marine habitats.

The overall body form of most fishes tells us something about the aquatic medium in which they live. Being

Figure 41.61 A coelacanth (*Latimeria*), a "living fossil" that resembles the early lobe-finned fishes.

Figure 41.62 A sea lamprey, here pressing its toothed oral disk to the glass wall of an aquarium. Like other members of class Agnatha, it has a slender, rounded, eel-like body with skin that is soft and lacks scales. The skeleton is cartilaginous.

about 800 times denser than air, water resists movement through it. Fishes typically have streamlined bodies with a large proportion of skeletal muscle tissue organized for forward motion. (Much less muscle tissue is devoted to counteracting the effects of gravity, which are less pronounced in water than on land or in the air.) Most modern fishes also have an array of **fins**, these being appendages that help propel, stabilize, and guide the body through the water.

Many fishes have modified bony plates called **scales**, which vary in type, number, and size among different groups. The gars, sturgeons, and other slow-moving bottom dwellers typically have large scales that serve as armor. What you probably think are "typical" scales cover the body of most free-swimming fishes, including sardines; they are small and apparently provide protection without weighing the fish down. Keep in mind that a surprisingly large number of species have only a few modified scales or none at all. These include eels and other fishes that hide in nooks or crannies (Figure 1.6l) and swordfish and some other fast predators of the open oceans. Tunas have scales, but they are deeply embedded in the body wall.

Jawless Fishes. The jawless fishes flourished between 500 and 350 million years ago. The existing representatives are limited to about seventy species of lampreys and hagfishes. All existing jawless fishes have cylindrical, eel-like bodies, a notochord, and a cartilaginous skeleton.

Lampreys have a suckerlike oral disk with horny, toothlike plates of keratin that can rasp the flesh from prey (Figure 41.62). The predatory lampreys typically spawn in minor coastal streams, and their larval stage is adapted to feeding on algae and detritus in the muddy backwaters. Some species have given the whole group a rather poor reputation; they have a habit of latching onto salmon, trout, and other commercially valuable fishes and then sucking out body juices and tissues.

About the turn of the century, sea lampreys began entering the Great Lakes of North America. The populations of lake trout, whitefishes, and other large fishes collapsed with the invasion of each lake. (Only Lake Erie escaped the lampreys; its waters were too warm and it did not have enough suitable spawning streams.) A concerted research effort culminated in the development of a "lampricide" that poisons the larvae of lampreys. Most of the native fish populations are recovering, but the battle is still going on.

Hagfishes look rather like large worms with specialized "feelers" (tactile organs) around the mouth. The inconspicuous mouth is equipped with four rows of keratinized, toothlike plates that are used to tear flesh from dead fishes. Hagfishes are not on the list of favorites of fishermen. Not only do they burrow into fish trapped by set-lines or nets, the ones that end up on deck secrete copious amounts of sticky, slimy mucus. This is probably why hagfishes are never found in the stomachs of predatory fishes, which seem to have learned that hagfishes can rapidly drench themselves in slime.

Cartilaginous Fishes. The cartilaginous fishes include sharks, skates, sawfishes, rays, and chimaeras. Most are highly specialized predators with a streamlined body; their fins enhance stability and maneuverability. All of these fishes have endoskeletons of cartilage, and most

caudal fin   dorsal fin

pelvic fin (paired)   pectoral fin (paired)

Figure 41.63 Representative cartilaginous fishes (Chondrichthyes): a blue-spotted reef ray (**a**) and a shark (**b**). Representative bony fishes (Osteichthyes): a soldier fish (**c**) and a sea horse (**d**).

types have five to seven gill slits on both sides of the pharynx (Figure 41.63b). A swim bladder is absent. All have scales of one sort or another, although skates and rays typically have no more than a few rows on the back that sometimes are modified into spines. Sharks have numerous conelike scales that give the skin a sandpaper quality.

Some sharks are fifteen meters long; they are among the largest living vertebrates. Sharks that feed on large fishes and marine mammals use their sharp, triangular teeth to capture and rip off chunks of a prey animal. Skates and rays, which are mostly bottom dwellers, have flattened teeth (much like paving stones) that are used for crushing hard-shelled invertebrates. In all species, the teeth are modified scales that are shed and replaced continually.

Skates and rays are distinctive in having enlarged pectoral fins that extend onto the side of the head (Figure 41.63a). Among the rays are species with electric organs in the head. These electric rays can stun prey fishes with up to 200 volts of electricity. The tail of stingrays is equipped with a spine (actually a modified scale) that has a venom gland at its base. Stingrays eat invertebrates, mostly, so the spine is probably used in defense against predators. The manta rays are the largest of the rays; some weigh 1,360 kilograms and measure over six meters from the tip of one pectoral fin to the other.

The chimaeras are commonly called ratfishes; with their bulky body and long, slender tail, they do look rather like a rat. All of the thirty or so species are bottom dwellers. They, too, have a venom gland; it is associated with a spine in front of the dorsal fin.

a

"red eft" stage

aquatic
larva ← — — — — — eggs released — — — — — — — aquatic
                    in water   adult

b

c

Figure 41.64 Representative amphibians. (**a**) Cave salamander, (**b**) terrestrial stage in the life cycle of a red-spotted salamander, and (**c**) American toad.

Bony Fishes. Early in their evolution, the bony fishes diverged into at least four lines: the lungfishes, crossopterygians, bichirs, and ray-finned fishes. Today there are only 6 lungfish species and 1 crossopterygian species (the coelacanth shown in Figure 41.61). The 11 species of bichirs are predators limited to the swamps and rivers of tropical Africa. Except for two highly specialized groups (sturgeons and paddlefishes), most of the 21,000 known species of bony fishes are ray-finned fishes.

Teleosts, the most successful of the ray-finned fishes, include eels, salmon, minnows, catfish, rockfish, perch, tuna, and deep-sea luminescent fishes. They generally have highly maneuverable fins and so are capable of rapid, complex movements. The respiratory system delivers plenty of oxygen to metabolically active tissues. Moreover, teleost scales are light and flexible. The bony vertebrae are reduced in number, lighter, and stronger than in ancestral forms. There are exceptions to this general list of characteristics in forms such as the sea horse (*Hippocampus*), shown in Figure 41.63.

Amphibians

There are three orders of living amphibians: the salamanders, frogs and toads, and apodans. All have a largely bony endoskeleton and, most often, four legs (a few have none). In the strictly aquatic types, respiration occurs through gills and the skin; in terrestrial types, it occurs through lungs, skin, and the lining of the pharynx. Amphibians have a three-chambered heart. With few exceptions, their eggs must be laid in water or damp places.

All adult amphibians are eclectic carnivores; they eat just about anything they can catch. The size of the head dictates the upper limit of the size of the prey. One type of frog (*Ceratophrys*) has such a large head that it is known as the walking mouth.

Amphibian skin generally has no protective scales; it also is thin and vulnerable to drying out. The skin does contain numerous glands, some of which produce toxins. The most noxious species are flamboyantly colored, the better to inform predators of their inedibility. The skin of the South African clawed frog (*Xenopus laevis*) contains peptides that serve in chemical defense against microbes, which flourish in the frog's habitat. Some secretions are neurotoxins that can help a captured frog escape from predatory snakes; the secretions trigger uncontrollable gaping of the snake's mouth.

The three orders of amphibians are distinctive mainly in terms of their mode of locomotion. All salamanders have an elongated body, and when they walk, the body bends from side to side, much as a fish body does when it moves through water. The first quadrupedal (four-legged) animals probably walked this way. Salamanders also are notable in that the adults retain many larval features, including arrested tooth and bone development, and (in mudpuppies and other species) the retention of external gills. Some salamanders have made their home in caves, where they dine on small invertebrates (Figure 41.64).

a

b

c

Figure 41.65 Representative reptiles. (**a**) Green sea turtle off the coast of Hawaii. (**b**) Galápagos tortoise. (**c**) Crocodile. (**d**) Green python, which first suffocates and then eats its prey. (**e**) Coral snake, one of the most deadly venomous snakes. (**f**) Tuatara, a "living fossil." Its body plan has not changed much since the age of dinosaurs, about 150 million years ago. (**g**) Defensive behavioral display of a chameleon (*Chamaeleo gracilis*) caught crossing open ground in Kenya. (**h**) Diagram of an amniotic egg—a key character in the vertebrate invasion of land.

In contrast to salamanders, the frogs and toads have a stiffened vertebral column, strong pelvic bones, and long hindlimbs to which powerful muscles are attached. Together, the bones and skeletal muscles form a leverlike system that can catapult the body into the air or propel it forcefully through the water. Most frogs and toads also have the tongue attached at the *front* of the mouth. The sticky-tipped tongue is flipped out to capture prey.

The apodans (commonly called caecilians) are the least familiar amphibians. They are legless, tiny-eyed forms that tunnel through moist soil or live in aquatic habitats in the tropics.

Reptiles

In the Late Carboniferous, insects began their adaptive radiation into the lushly foliaged habitats on land, and this event probably triggered the evolution of reptiles from amphibious vertebrates. Like their modern descendants, those amphibious forms almost certainly were carnivores; there is little in the fossil record to suggest

otherwise. And almost certainly, the abrupt expansion in the quantity and choices of insects represented a major food source, ready for exploitation. The fossil record tells us that the jaw began evolving in ways that were adapted specifically to feeding on insects; limb bones evolved in ways that permitted more efficient movements on land.

Adaptations to life on land also included an increased reliance on internal fertilization and the development of the **amniotic egg**. This type of egg is characteristic of reptiles, birds, and (in one form or another) mammals—but not amphibians. Its leathery or calcified shell mechanically protects the developing embryo yet also permits the passage of oxygen, carbon dioxide, and water vapor (Figure 41.65h). The amniotic egg departs from the eggs of fishes and amphibians in this important respect: it contains extraembryonic membranes that protect the embryo and provide it with metabolic support. These membranes—the chorion, amnion, and allantois—are described on page 521.

Reptiles underwent an adaptive radiation in the Mesozoic; this was the time of origin for the dinosaurs and related forms that would dominate the land for the

d

e

f

g

allantois   embryo   amnion   chorion

albumin

air space

yolk sac

shell

h

next 125 million years. Their domination ended abruptly when the Cretaceous drew to a close (page 566). Today, the reptiles include about 6,000 species of turtles, crocodilians, and snakes and lizards. The present diversity is greatest in the tropics, although some species of lizards and snakes live as far north as the edge of the Arctic Circle and as far south as the tip of South America.

All living reptiles have a well-developed bony endoskeleton and skin that resists drying out. Keratinized scales are nearly always present. Reptiles depend entirely on lungs, with air being sucked in (not forced in by mouth muscles, as in amphibians). Most reptiles rely on internal fertilization and most lay leathery eggs. (There are exceptions; garter snakes, rattlesnakes, cop-

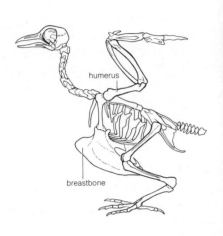

Figure 41.66 An owl (class Aves) on the wing. The diagram shows a generalized skeletal system of birds.

perheads, and some others give birth to fully formed young, as shown in Figure 34.2.)

The turtles are highly specialized for life in a mobile home, so to speak. The ones on land have a rigid, domed shell that they cart around on sturdy legs; the sea turtles have flipperlike legs protruding from a more streamlined shell (Figures 41.65a and b).

The crocodilian body plan is similar throughout the group. Variations occur mainly in the shape of the snout, these being correlated with specialized diets. Until recent times, adult salt-water crocodiles may have grown up to seven meters in length. They grow very slowly, however, and crocodiles have been hunted with such a vengeance over the past two centuries that few ever grow that large today.

Like other reptiles, the crocodilians are ectotherms. They show elaborate behavioral and physiological responses for regulating body temperature (page 472). They also show complex social behavior, as when male and female parents assist hatchlings in their move out of the egg and into the water. Young alligators of the Gulf Coast swamps remain with the mother for two years after hatching.

More than ninety percent of the living reptilian species are snakes and lizards, and the majority of those are snakes. Most lizards are small and weigh no more than twenty grams or so. The largest, the Komodo monitor lizard, may weigh seventy-five kilograms; it can bring down deer and even water buffalo. The tuatara, now classified with the lizards and snakes, has changed little since Mesozoic times (Figure 41.65f).

Snakes lost their legs during the course of evolution (refer to page 25). Also, their skull is constructed in such a way that they can swallow prey animals larger in diameter than the snake's head. Figure 48.7b is a good example of this.

Birds

Birds apparently descended from crocodilian-like reptiles that ran around on two legs during the Jurassic, some 160 million years ago. The oldest known bird (*Archaeopteryx*, shown in Figure 2.11) resembled those reptiles, especially in its limb bones. Birds still resemble reptiles in many of their internal structures, their horny beaks and scaly legs, and their habit of laying eggs. So close is the resemblance that Thomas Huxley once called birds "glorified reptiles" and placed both groups of vertebrates in the same class. (Here you may wish to refer to page 556, which mentions some of the problems in classifying birds and reptiles.)

There are more than 8,900 species of birds. They vary tremendously in size, body proportions, coloration, and capacity for flight, and their songs and other social behaviors are remarkably diverse, as you will see in later chapters. The largest bird (the ostrich) weighs about 150 kilograms (330 pounds); one of the smallest hummingbirds barely tips the scales at 2.25 grams (0.08 ounce). Yet in basic body plan, nearly all birds meet the two key requisites of flight: low weight and high power.

The **bird wing** is a forelimb constructed of feathers, powerful muscles, and lightweight bones. Contraction

Figure 41.67 Arctic fox in winter coat, a representative of the class Mammalia, a topic of the next chapter.

of muscles attached to the humerus (Figure 41.66) is the basis of the powerful downstroke required for flight. Bird bones are strong and yet weigh very little because of air cavities in the bony tissue. For example, the endoskeleton of a frigate bird, which has a seven-foot wingspan, weighs only four ounces. That is less than the feathers weigh! All birds have a large, strong, four-chambered heart and an efficient system for oxygen delivery. Figure 23.12 shows the structure of one type of feather and how it develops from differentiated epidermal cells; Figure 31.5 describes the unique respiratory system of birds.

Mammals

There are three major groups of living mammals: the egg-laying mammals (Monotremata), pouched mammals (Metatheria), and placental mammals (Eutheria). All are descended from reptiles, as described on page 584. Fertilization is internal. Embryonic development is also internal in all groups except for a few egg-laying types, the most famous of which is the duck-billed platypus (Figure 34.2c shows one at its nest).

In what is clearly the most distinctive mammalian character, the young are nourished by milk-secreting glands of the adult. Most mammals are covered with hair (Figure 41.67); whales are not, and reptilian-like scales persist on the tail of beavers, rats, and other rodents. Adults typically have two sets of teeth (reptiles have successive sets). Mammals have lungs, a four-chambered heart, and a well-developed cerebral cortex. They will occupy our attention in the chapter to follow.

SUMMARY

1. Multicelled animals, ranging from simple sponges to vertebrates, are assigned to about thirty phyla. The members of each phylum share certain characteristics and are therefore believed to have descended from a single ancestral stock, or at least from closely related stocks. Table 41.1 summarizes the major characteristics of the main phyla covered in this chapter.

2. In assigning animals to phyla, many aspects of structure, development, physiology, and the fossil record must be considered. Some of these are mentioned below.

3. Sponges (phylum Porifera) are structurally simple multicelled animals. They have neither a gut nor a nervous system; although they have several kinds of cells, the cells are not organized into tissues. Certain cells trap microscopic food particles from water moving through a system of pores, canals, and chambers.

4. Jellyfishes, sea anemones, and their relatives, which belong to the phylum Cnidaria, have radial symmetry and their cells form tissues. One tissue layer (epidermis) covers the outside of the body; the gastrodermis lines the gut, which is basically saclike because it has no anus. The epidermis and gastrodermis are derived from embryonic ectoderm and endoderm, respectively. They are separated by secreted material (mesoglea), and this may make up much of the total mass of the animal, as it does in a jellyfish.

5. Cnidarians have nerve cells, usually organized into networks closely associated with the epidermis and gastrodermis. A distinctive feature of the phylum is the presence of nematocysts, or stinging capsules, which assist in capturing prey and may also serve in protection.

6. Most animals more complex than cnidarians are bilaterally symmetrical, and during their embryonic development a third germ layer (mesoderm) is formed. Mesoderm gives rise to muscles and various other tissues between the epidermis and gastrodermis. The gut may be saclike as in flatworms (phylum Platyhelminthes), but generally it is complete, with an anus as well as a mouth.

7. Most animals more complex than flatworms have either a coelom or a pseudocoel. These cavities separate the gut from the body wall. A true coelom is lined by a peritoneum; a pseudocoel is not lined. Some invertebrates (notably of the phyla Mollusca and Arthropoda) have a coelom but it is much reduced. The blood that bathes their tissues, however, flows through the interconnected spaces collectively called a hemocoel. This blood-filled cavity, like a coelom or pseudocoel, may function as a hydrostatic skeleton as well as in distributing oxygen and nutrients.

8. Many invertebrates, mostly of the phyla Annelida and Arthropoda, are segmented (their bodies are parti-

Table 41.1 Summary of Characteristics of Animal Phyla Described in This Chapter

Phylum (and some representatives)	Typical Environment	Typical Life Style of Adult Form	Integration	Support and Movement
Porifera (8,000)* sponges	Most marine, some fresh water	Attached filter feeders	No nervous system (cell-to-cell transmission)	Support by spicules, protein fibers, or both; contractile cells change openings at body surfaces
Cnidaria (11,000) hydras, sea anemones, jellyfishes, corals	Most marine, some fresh water	Attached, creeping, swimming, or floating carnivores	Nerve net, cell-to-cell transmission	Hydrostatic support (by fluid in gut), by secreted jellylike mesoglea, or by calcareous or hard proteinaceous skeletal elements; contractile fibers in epithelial cells
Ctenophora (100) comb jellies	Marine	Mostly planktonic; few attached or creeping	Nerve net	Support by mesoglea; muscles separate from epithelia; locomotion by cilia in eight rows of comblike structures; sometimes by muscular activity
Platyhelminthes (12,700) flatworms, flukes, tapeworms	Marine, fresh water, some terrestrial in moist places; many parasitic in or on other animals	Herbivores, carnivores, scavengers, parasites	Brain, nerve cords	Hydrostatic support (no secreted skeleton); well-developed muscle tissue
Nemertea (700) ribbon worms	Most marine; few fresh water, terrestrial	Mostly carnivores	Brain, nerve cords	Hydrostatic support; well-developed muscle tissue
Nematoda (12,000) roundworms	Marine, fresh water, terrestrial; many parasitic in animals, plants	Scavengers, carnivores, parasites	Nerve ring, nerve cords	Hydrostatic support (by pseudocoel); tough cuticle; longitudinal muscle in body wall
Rotifera (1,800) rotifers	Marine, fresh water, moisture on mosses	Mostly filter feeders, capturing bacteria, unicellular algae	Brain, nerve cords	Locomotion by cilia; muscles for shape changes
Mollusca (100,000) snails, slugs, clams; squids, octopuses	Marine, fresh water, terrestrial	Herbivores, carnivores, scavengers, detritus or filter feeders; mostly free-moving, some attached	Brain, nerve cords, major ganglia other than brain	Hydrostatic skeleton (using hemocoel) in most; well-developed musculature in foot, mantle, other structures
Annelida (10,000) earthworms, leeches, polychaetes	Marine, fresh water, terrestrial in moist places	Herbivores, carnivores, scavengers, detritus or filter feeders; mostly free-moving	Brain, double ventral nerve cord	Hydrostatic skeleton (using coelom); well-developed musculature in body wall
Arthropoda (1,000,000) crustaceans, spiders, insects	Marine, fresh water terrestrial	Herbivores, carnivores, scavengers, detritus or filter feeders, parasites; mostly free-moving	Brain, double ventral nerve cord	Exoskeleton (of cuticle); jointed appendages; muscles mostly in bundles
Echinodermata (6,000) sea stars, brittle stars, sea urchins, sea lilies, sea cucumbers	Strictly marine	Mostly carnivores, detritus feeders, few herbivores; most free-moving, some attached	Radially arranged nervous system	Endoskeleton (of spines, etc.); muscles for body movement, tube feet often used in locomotion
Chordata (39,000) tunicates, lancelets, jawless fishes, jawed fishes, amphibians, reptiles, birds, mammals	Marine, fresh water, terrestrial	Herbivores, carnivores, scavengers, filter feeders; generally free-moving (most tunicates attached as adults)	Well-developed brain, dorsal and tubular nerve cord in most	Notochord or a bony or cartilaginous endoskeleton; well-developed musculature in most

*Number in parentheses indicates approximate number of species.

Digestive System	Respiratory System	Circulatory System	Mode of Reproduction
No gut; microscopic food particles secured by individual cells	None; respiration by individual cells	None	Sexual (certain cells become or produce gametes); some asexual budding; production of resistant bodies
Saclike gut (may be branched)	None; respiration by individual cells; gut may distribute oxygen	None, other than via gut	Sexual (usually separate sexes; gonads discharge gametes into gut or to exterior); asexual
Saclike, but branched	None; respiration by individual cells; gut may distribute oxygen	None, other than via gut	Sexual (usually hermaphroditic); gonads closely associated with gut
Saclike gut (may be branched)	None; gas exchange across body surface	None	Sexual (usually hermaphroditic, with complex reproductive system); some asexual
Complete gut	None; gas exchange across body surface	Closed system	Sexual (sexes usually separate; reproductive system simple)
Complete gut	None; gas exchange across body surface	Pseudocoel	Sexual (sexes separate; reproductive system fairly complex)
Usually complete gut (sometimes saclike)	None; gas exchange across body surface	Pseudocoel	Mostly parthenogenetic (eggs develop without fertilization); males appear only occasionally
Complete gut	Ctenidia, other gills; mantle can be modified as lung; gas exchange across body surface	Usually open, forming a hemocoel (closed system in cephalopods)	Sexual (hermaphroditic or separate sexes; reproductive system usually complex)
Complete gut	Gas exchange across body surface; varied outgrowths of surface in many	Usually closed; coelom also may function in distribution	Sexual (hermaphroditic or sexes separate; reproductive system simple or complex); asexual also in many
Complete gut	Gills; tracheal tubes; book lungs; general body surface in some	Open system, forming a hemocoel	Sexual (usually separate sexes; reproductive system fairly complex)
Usually complete gut (sometimes saclike)	Gas exchange across general body surface or surface of outgrowths of it (such as tube feet)	Coelom around viscera, also water-vascular coelom	Sexual (reproductive system simple; gonads usually discharge gametes directly to exterior); some asexual
Complete gut	Lungs in most vertebrates other than fishes; perforated pharynx; gills; gas exchange across body surface	Closed system in most (open system in most tunicates); lymphatic system in many vertebrates	Sexual (sexes usually separate, except in most tunicates); asexual in some tunicates

tioned into a series of units). In annelids, many of the segments may be similar, at least externally, but the extent to which the segments are specialized varies greatly. In arthropods, segments tend to be decidedly specialized, and groups of successive segments form distinct body regions such as the head, thorax, and abdomen of an insect. Arthropods also have an exoskeleton and jointed appendages. Both features have contributed to the success of arthropods, of which there are about a million species.

9. Flatworms, the simplest bilaterally symmetrical and cephalized animals, have a brain and two or more longitudinal nerve cords. In annelids and arthropods, the brain is more elaborate and there is a double ventral nerve cord. In mollusks, the usual components are ganglia that form a brain and some other ganglia that are linked with it by major nerves. In vertebrates and other chordates, the nerve cord is dorsal.

10. Of the animals having a true coelom, the echinoderms are the only ones that are uncephalized and radially symmetrical. The nervous system is decentralized, there being no brain. This arrangement is favorable for animals with radial symmetry (such as sea stars) because the sensory structures and nerve tracts of any ray can initiate activity in response to a stimulus.

11. Early in the evolution of vertebrates, the brain and dorsal nerve cord became protected by a skeleton of cartilage or bone. Other skeletal elements, such as jaws, ribs, and supports for the limbs, also developed.

12. The vertebrae around the spinal cord allow for considerable flexibility, and nearly all skeletal structures are sites to which muscles concerned with chewing, locomotion, and some other functions are attached.

13. During vertebrate evolution, certain fishes became equipped with lungs and fins of a type that enabled them to walk on land, leading to the evolution of amphibians.

14. Reptiles, derived from amphibians, became even more successful on land because of their protective scaly skins, physiological adaptations, and ability to reproduce without going back to water (which most amphibians must do). Fertilization is generally internal, and eggs are protected by a leathery covering. These features, together with the appearance of the amnion and other extraembryonic membranes, were important adaptations to life on land.

15. Birds and mammals evolved from separate groups of reptiles. Their success is related in part to their constant and warm body temperature and to the insulation provided by feathers or hair. These characteristics enable birds and mammals to live in many environments. Good vision, a well-developed brain, and appendages specialized for flight, swimming, running, grasping, and other activities are a few more features underlying the success of birds and mammals.

Review Questions

1. Sponges are simple multicelled organisms, but in some situations they colonize much of the available living space. Can you give a few reasons for their success? *645*

2. Taking spiders and insects as examples, explain why certain groups of arthropods have become so successful in terrestrial habitats. *664–665*

3. How do the structural and developmental characteristics of protostomes differ from those of deuterostomes? *656*

4. Explain the difference between a coelom, pseudocoel, and hemocoel. Using nematodes, mollusks, annelids, and arthropods as examples, discuss the importance of all three of these types of fluid-filled cavities. *643, 653, 657, 663*

5. Why was the appearance of segmentation such an important step in the evolution of certain complex invertebrates? In developing your answer, think about the segmentation of an earthworm or polychaete annelid and the segmentation and presence of jointed appendages of an insect and crustacean. *664–665*

6. Compare the exoskeleton of arthropods with the endoskeletons of echinoderms and vertebrates. How do these types of skeletons differ, and what advantages do they confer upon the animals that have them? *664, 672, 678, 680*

7. What were some of the significant steps in the evolution of birds and mammals from early vertebrates that were somewhat comparable to modern lampreys? Be sure to deal with jaws, gill slits, fins and other limbs, lungs, skins, body temperature, and development of eggs and embryos on land versus development in water. *678–680*

Readings

Barnes, R. 1988. *Invertebrate Zoology*. Fifth edition. Philadelphia: Saunders. Good introduction to the invertebrates.

Hickman, C. P., and L. S. Roberts. 1988. *Integrated Principles of Zoology*. Eighth edition. St. Louis: Mosby.

Hildebrand, M. 1985. *Analysis of Vertebrate Structure*. Third edition. New York: Wiley.

Mitchell, L., J. Mutchmor, and W. Dolphin. 1988. *Zoology*. Menlo Park, California; Benjamin/Cummings. Clearly written, beautifully illustrated survey of the animal kingdom.

Romer, A. S., and T. S. Parsons. 1986. *The Vertebrate Body*. Sixth edition. Philadelphia: Saunders. Chapter 16 contains detailed pictures of vertebrate nervous systems. Excellent reference book.

In the preceding chapter, we traced a tremendous span of animal evolution, beginning with the origin of bilateral animals, then segmented ones. We saw how the bilateral, segmented plan evolved in truly spectacular directions. Think back on a few of the adaptations that emerged—paired sensory structures and a complex brain (for refined perception and response to environmental challenges). Think of the blood circulation systems (for transporting substances to and from all the living cells of massive bodies). And think of the pectoral and pelvic fins that gave rise to lobed fins—the forerunners of legs and other limbs (for locomotion).

These and other adaptations led us to the vertebrate branch of the animal phylogenetic tree (Figure 41.53). They represent a common heritage for more than 40,000 existing species of fishes, amphibians, reptiles, birds, and mammals. When we turn to the evolution of any one of those species—as we do here—it helps to keep the common heritage in mind. It reminds us that complex systems were already in place and functioning at each crossroad leading to a new species—*and that new traits have emerged only through modification of what went before.* The history of the human species speaks eloquently of this evolutionary principle.

THE MAMMALIAN HERITAGE

We are members of the class Mammalia, one of the seven classes of vertebrates. Like all other vertebrates, mammals have an axial endoskeleton with two key features: (1) a nerve cord threading through a vertebral column and (2) a skull that houses sense organs and a "three-

42

HUMAN ORIGINS AND EVOLUTION

Figure 42.1 Simplified family tree for primates. The two highest taxa (suborders Prosimii and Anthropoidea) are shown in green. Major groups of living primates are indicated by the gold boxes; representative members are indicated. All monkeys, apes, and humans are anthropoids. Apes, humans, and recent human ancestors are further classified as hominoids; their evolutionary relationships are the focus of this chapter.

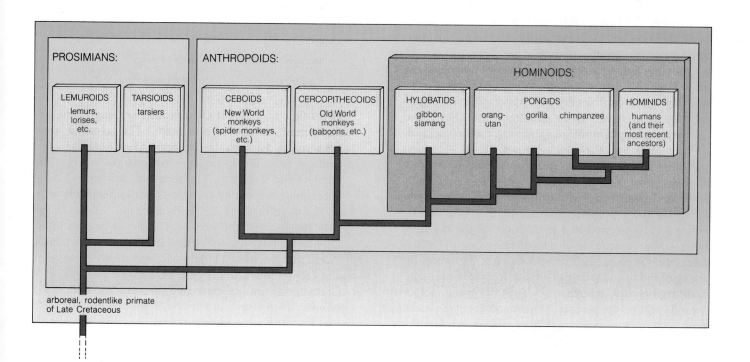

part" brain (the hindbrain, midbrain, and forebrain). These are the traits that best characterize vertebrates; no other animal has them.

Of the various traits that characterize mammals, two are important for our story of human evolution. The first is their **dentition**—that is, the type, number, and size of teeth in their two opposing jaws. Unlike their reptilian ancestors (which had peglike, unmatched teeth), mammals have upper and lower teeth that match up and work together to crush or cut food. The teeth also are differentiated into four types, as this generalized sketch shows:

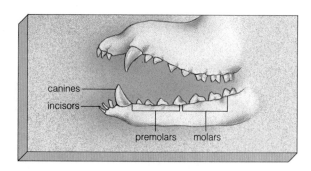

canines
incisors
premolars   molars

Incisors are shaped like flat chisels or cones and are used to nip or cut through food. *Canines* are typically pointed and long enough to project beyond the other teeth; they are used in biting and piercing. The *premolars* and *molars* (the "cheek teeth") are used in grinding or shearing.

Dentition varies in different mammals. Carnivores have pronounced canines, which can hook into prey. Some monkeys and apes also have pointed canines that can split open tough plant parts (such as bamboo stalks). Plant eaters generally have flat-surfaced cheek teeth that form a platform upon which coarse, fibrous plant parts are chewed. Cheek teeth often have *cusps* (surface bumps). Rounded cusps are useful in grinding; pointed ones are useful for shearing meat off bones.

As you can see, teeth tell us quite a bit about what an animal eats, hence something of its life style. Teeth also happen to fossilize very well. Many inferences about the diet and life styles of our earliest ancestors have been drawn from fragments of jaws and teeth they left behind.

Mammals are also characterized by an extended period of **infant dependency** and **learning**. Recall that almost all mammals are placental, with the embryo nourished and maintained within the mother's uterus until birth. (In the few living marsupials, the "unfinished" embryo is transferred to a pouch outside the mother's body, where its early development continues. In the still fewer monotremes, the mother lays eggs, then cares for the eggs and her offspring.) The young of placental mammals are helpless at birth. They depend on the mother for milk and on the parents, other members of the social group, or both for protection. They also depend on the adults as models for behavior.

A great deal of learning goes on during the period of dependency. By observing and imitating what the adults are doing, the young learn a basic set of behaviors that allow them to recognize appropriate food, respond to danger, and so on. However, both young and adult mammals also show *flexibility* in the responses they make to novel situations (as when deer adapt to life in the suburban yards that supplant their natural habitat). As we have seen, such flexibility is possible in mammals having a brain with a large capacity for memory and learning. And it has reached its fullest expression among the primates.

PRIMATE CLASSIFICATION

Among the twenty-one orders of placental mammals we find the order **Primates**, which includes lemurs, lorises, tarsiers, monkeys, apes, and ourselves (Figure 42.1). Most primates live in tropical or subtropical forests, mixed woodlands, and savannas. Like their ancient forerunners, the vast majority are **arboreal**, or tree-dwellers.

Lemurs, tarsiers, and related forms are **prosimians**. The smallest are the size of a mouse; the largest are about as big as a house cat. Most forage at night and have large, forward-directed eyes specialized for nocturnal vision (Figure 42.2). As their name suggests, prosimians are the oldest primate lineage (*pro*, before, *simian*, ape). For millions of years, before the first monkeys or apes appeared, the prosimians dominated the trees in North America, Europe, and Asia. In time, the more agile and larger brained monkeys and apes replaced them in all but a few restricted forests of Africa and Asia.

Monkeys, apes, and humans are **anthropoids** and are further classified this way:

ceboids:	*New World monkeys*
cercopithecoids:	*Old World monkeys*
hominoids:	*apes and humans*

According to a recent classification scheme, there are three main groups of hominoids. The so-called great apes (pongids) include the orangutans, gorillas, and chimpanzees; the lesser apes (hylobatids) include the gibbons and siamangs. Humans and their extinct close ancestors are grouped separately as the **hominids**.

If you go through the primate section of a zoo, compare the faces of the prosimians, monkeys, and apes. Prosimians have a moist nose at the end of a snout. New World monkeys have a flat nose, with nostrils far apart and facing outward. (They are said to be platyrrhine, a Greek term meaning flatnosed.) In contrast, Old World monkeys, apes, and humans have close-together nostrils that face forward and downward. (They are catarrhine, after the Greek word for hooknosed.)

Figure 42.2 Representative primates. Gibbons (**a**) have limbs and a body adapted for brachiation (swinging arm over arm through the trees). Monkeys are quadrupedal (four-legged) climbers, leapers, and runners, as the spider monkey in (**b**) demonstrates. Tarsiers (**c**) are vertical clingers and leapers.

TRENDS IN PRIMATE EVOLUTION

As a group, primates are difficult to define, because no one feature sets them apart from other mammals. Perhaps primates are better defined in terms of *trends in their evolution*, many of which were shaped by their arboreal ancestry:

1. Change in overall skeletal structure and mode of locomotion.

2. Modification of the hands, this leading to increased dexterity and manipulation.

3. Less reliance on the sense of smell and more reliance on daytime vision, including color and depth perception.

4. Change in dentition, toward fewer and smaller, less specialized teeth.

5. Brain expansion and elaboration.

6. Behavioral evolution.

Not all of the trends listed occurred in every primate lineage. (Prosimians have retained many "primitive" features, for example, including a snout.) And where change did occur, it did not proceed at the same time or rate in different lineages. The trends simply reflect the adaptive potential that has been expressed within the primates *as a group.* Let's consider a few key aspects of these trends before we follow their historical development.

From Quadrupeds to Bipedal Walkers

By comparing existing monkeys, apes, and humans, we can gain insight into the skeletal reorganizations that occurred during primate evolution. In certain lineages, structural and functional changes in the shoulders, spinal column, hip bones, legs, and feet took place during a major transition from **quadrupedal** (four-legged) to **bipedal** (two-legged) locomotion.

Figure 42.3 Comparison of the skeletal organization and stance of monkeys, apes (the gorilla is shown here), and humans. Modifications of the basic mammalian plan have allowed three distinct modes of locomotion. The quadrupedal monkeys climb and leap, and apes climb and swing by their forelimbs. Both modes of locomotion are well suited for life in the trees. Humans are habitual two-legged walkers.

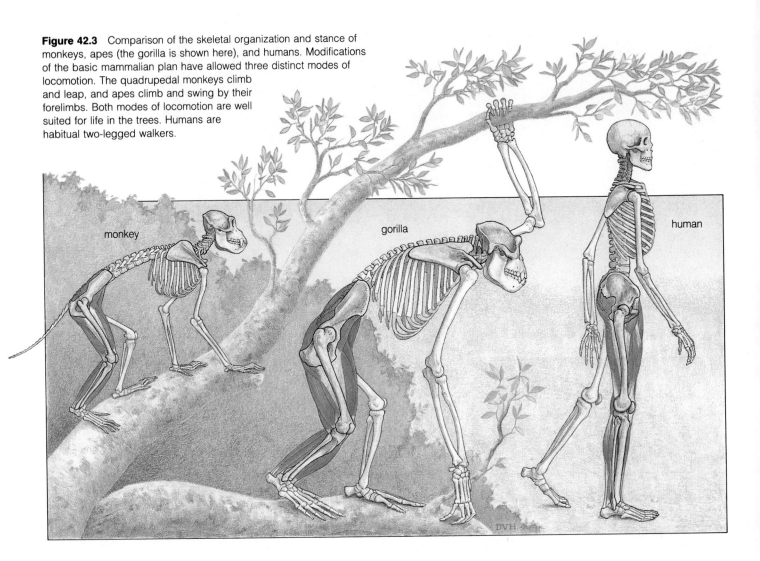

The monkey skeleton is adapted for an arboreal way of life; it permits rapid climbing, leaping, and running along branches of different sizes (Figures 42.2 and 42.3). Monkeys have a long, flexible vertebral column and a narrow pelvic girdle. The angle of the pelvic girdle and shoulder blades is appropriate for a quadrupedal stance. So are the arms, which, being somewhat shorter than the legs, allow monkeys to walk with palms down. (Try doing this yourself for any length of time.)

The ape skeleton is adapted for climbing and swinging by the forelimbs. (This adaptation is most pronounced in gibbons; they brachiate, or swing arm over arm, through the trees.) The vertebral column is short and somewhat rigid. A long pelvis, an enlarged and flared pelvic girdle, and the attachment sites they provide for muscles can support a heavy torso. The position of the shoulder blades allows the upper arms to swivel freely above the head (monkeys cannot do this).

Overall, the ape skeleton permits semierect or erect stances. Also, body weight can be carried not only by the feet but by arms hanging from overhead branches

(Figure 42.3). We can speculate that forelimbs adapted for climbing and swinging allowed early apes to reach previously unreachable fruit and other food on slender branches that would not have supported their body weight. Some apes do this today. Although chimpanzees and gorillas spend much or most of the time on the ground, in skeletal structure they are still vertical climbers and arm swingers.

The human skeleton is adapted for bipedalism. The vertebral column is shorter than in monkeys, S-shaped, and moderately flexible. Together with the position and shape of the shoulder blades and pelvic girdle, it allows a fully upright stance. Of all the vertebrates, only humans are habitually erect, two-legged striders.

The skeletal reorganization that led from apelike to humanlike (bipedal) forms is not difficult to envision; the original selective advantage is another matter. (It may be that our early ancestors did not have much choice in leaving behind the advantages of swinging through the trees. As you will see, owing to changes in climate, the trees left them.)

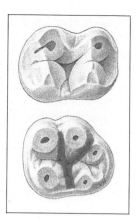

Figure 42.4 A key molar pattern used to identify fossil hominoids. Some molars of Old World monkeys have four cusps, one at each corner of the tooth's crown (above). For more than 20 million years, all hominoids have had some five-cusped molars (below). The arms of the Y embrace the fifth cusp, as shaded in orange.

Modification of Hands

Among certain primate lineages, clawed paws evolved into hands. Think about the four basic movements that you can make with your own hand:

1. *Divergent* movements that spread fingers apart. This ability emerged among the first (four-legged) mammals, which spread their toes apart to help support body weight during walking and running.

2. *Convergent* movements to cup the hands. This trait is characteristic of many mammals that have paws or hands; squirrels, for example, use cupping motions to bring food to the mouth.

3. *Prehensile* movements to wrap fingers around objects. Arboreal mammals generally are prehensile. Such movements developed to a high degree among the branch-clinging, leaping, and brachiating primates of the distant past. The evolution of dermal ridges (fingerprints) and fingernails (instead of claws) was correlated with prehensile movements.

4. *Opposable* movements that allow the thumb to touch the tip of each finger. Such movements were refined in ancestral primates that lived in the trees.

Among the ancestral forms that gave rise to humans, modification of the hand bones led to refinements in all four movements. And they made possible the precision grip and the power grip:

These hand positions underlie the toolmaking and tool-using capacity of humans—which in turn underlie a unique technology and cultural development.

Enhanced Daytime Vision

The early primates had an eye socket on each side of their skull. Later ones had forward-directed eyes, which means they had an overlapping visual field and depth perception. Over time, they became increasingly good at discerning shape, movement, and variations in color and in light intensity (dim to bright). As you will see, these visual stimuli are typical of life in the trees.

Changes in Dentition

Monkeys have longer canines than apes do, and their jaws are rather rectangular. Human teeth are smaller and more uniform in length, and the human jaw is bow-shaped. Like all other hominoids, humans also have a distinct dental feature, the Y-5 molar pattern (Figure 42.4). These and other aspects of dentition are useful in determining the phylogenetic status of different fossils. By analyzing the entire fossil record for primates, we see that the jaws and dental patterns became less specialized during the evolution of forms leading to humans. There was a shift from insect eating, to dining on fruit and leaves, and on to a more omnivorous diet.

Brain Expansion and Elaboration

During primate evolution, the brain (especially the cerebral cortex) increased in mass and complexity. The *largest* brain size in living apes is 650 cubic centimeters (for the gorilla). The *smallest* known brain size in humans is 855 cubic centimeters (the average is 1,350). Using this range as a guide, scientists originally proposed that a cranial volume of 750 cubic centimeters was the "cerebral Rubicon"—the boundary that had to be crossed before the ancestors of humans became fully human. More recently, brain size relative to body size is considered to be a more useful indicator.

Behavioral Evolution

Existing primate groups reflect a trend toward longer life spans, longer periods between pregnancies, single births rather than litters, and longer periods of infant dependency (Figure 42.5). The increased parental investment in fewer offspring is associated with strong social bonds between parents and offspring, intense maternal care, and longer periods of learning.

Like mammals generally, primates rely on a limited set of social behaviors that can be learned and repeated by the young. And they show a capacity to add quickly to their behavioral repertoire. (To give one example, when researchers in Japan introduced a group of macaque monkeys to the sweet potato, one adolescent female monkey carefully dipped it in water before eating

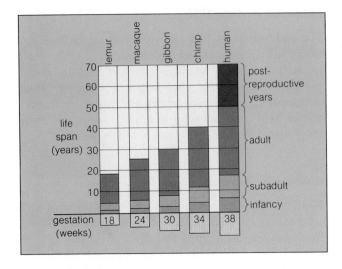

Figure 42.5 Trend toward longer life spans and longer periods of infant dependency among existing primates.

it. Her peers imitated her, and after $3\frac{1}{2}$ years, potato-dipping had become a stereotyped behavior among the new generation.)

The capacity for learning was extended during human evolution, probably because our ancestors were confronted with so much more to learn when their environment changed. It takes a large, complex brain to handle a complex way of life. Modifications of the brain and of behavioral complexity were interlocked, with new developments in one stimulating development of the other. Nowhere is this more evident than in the parallel evolution of human culture and the dramatic expansion of the human brain.

Here we define **culture** as the sum total of behavior patterns of a social group, passed between generations by learning and by symbolic behavior, especially language. It is the culmination of a long evolutionary history.

PRIMATE ORIGINS

The first known primates lived during the Paleocene (65 million to 54 million years ago). In form they resembled small rodents or the tree shrews of Southeast Asia (Figures 42.6 and 42.7).

Tree shrews are nighttime omnivores—highly active, and hungry (they eat the equivalent of their own weight nightly). Like shrews, the first primates probably preferred insects, seeds, buds, and eggs in the underbrush of the forest floor. Like shrews, they had a long snout and a well-developed olfactory sense, useful in snuffling out food and catching the scent of predators, which once included some of the smaller dinosaurs. They were prehensile to a degree; they could turn their ankle bones

and clawed feet toward each other and clamber upward, although not with olympian speed or agility.

Many of the early, ground-dwelling forms died out. Some of their descendants apparently went underground, evolving into moles. Some took to the water and air; their modern descendants are water shrews and bats. And some took to the trees, giving rise to the prosimian forms of the Eocene (54 to 38 million years ago).

The early Eocene climate was somewhat warmer than the Paleocene, and tropical rain forests flourished. The forest canopy was a promising adaptive zone, with its abundance of fruits and insects and its safety from ground-dwelling predators. Yet with all of its advantages, life in the trees also was demanding. It must have been a visually complex world then, as it is now, with dense leaves, dappled sunlight, boughs swaying in the wind, colorful fruit tucked among the leaves, darting insects and other prey, perhaps predatory birds. The mammalian snout would not have been much use in such a world, where air currents disperse the scent of food and predators. However, eyes that could discern color, shape, and movement in a three-dimensional field would have been enormously useful while the animal was leaping and moving among the branches.

At the same time, an arboreal way of life called for a nervous system that could integrate diverse stimuli, then coordinate precise, rapid, yet flexible movements in response. Before walking, swinging, and leaping (especially!) could be undertaken high up in the forest canopy, the brain had to assess many factors—distance, body weight, wind, the suitability of the destination, how to compensate quickly for miscalculations. Not surprisingly, fossils of the Eocene prosimians give evidence of increased brain size, a shorter snout (and enhanced daytime vision), and hand bones capable of refined grasping movements.

The divergences that would lead to modern prosimians took place during the early Eocene. By the time that epoch drew to a close, anthropoids had evolved from one of the prosimian stocks—but we know practically nothing about the original and early evolution of those ancestral forms. Anthropoids were well established by the early Oligocene (38 to 25 million years ago). Our knowledge of them is drawn mostly from fossils found in the Fayum Depression, which is now a small desert southwest of Cairo, Egypt.

About 35 million years ago, the Fayum Depression was cloaked with the lush vegetation of a tropical rain forest. Humid swamps lined the many rivers that ran through the forest and emptied into the Mediterranean Sea. The early anthropoids were adapted for life in the trees. Given the altogether nightmarish predatory reptiles that inhabited the swamps at the time, it is perhaps understandable why they rarely left the branches for the ground—and why it became imperative to think fast and grip strongly. (Slip-ups were always possible; a surprising number of primates still fall out of the trees.)

Aegyptopithecus (Figure 42.7) was one of the many and varied Fayum anthropoids. This quadrupedal, gibbon-sized form lived 30 million years ago and probably predates the divergence leading to Old World monkeys and apes. Compared to the early prosimian forms, it had a slightly increased cranial capacity, better grasping abilities, and more refined visual abilities. Compared to modern monkeys, its snout was somewhat longer and its cranium not as large—yet its teeth had features that persist in modern apes and humans.

The earliest known hominoids were apelike forms that appeared between 23 million and 20 million years ago. While these forms were evolving, major land masses were on the move as crustal plates began to assume their most recent positions (page 578 and Figure 38.14). At the start of the Miocene, some 25 million years ago, the redistribution of land masses led to pronounced shifts in climate. In what is now Africa, Europe, and elsewhere, there began a major cooling trend and a decline in rainfall that would culminate in the Pleistocene. Over time, the tropical and subtropical forests would gradually give way to grasslands. In the interim, there was a mosaic of forests and savannas.

All of this geologic evolution had profound effects on the evolution of hominoids. Early populations of forest-dwelling apes spread from Africa across a newly created land bridge that linked the continent with Eurasia. As the climate continued to become cooler and drier, the belts of grasslands spread, and different subpopulations of apes became reproductively isolated within pockets of forests. Adaptations to those local conditions led to divergences in form. In short, a great adaptive radiation took place, resulting in a widespread distribution of many varieties of hominoids. A lineage established during this time span apparently gave rise to the lesser apes, including gibbons.

By the middle Miocene, a large group of apes, collectively called the **dryopiths**, ranged through Africa into Europe and Eurasia (Figure 42.7). The dryopiths varied in form, size, and apparent life style. In general, they may have differed from earlier anthropoids in food preference; it seems they depended more on fruit than on leaves. They were also larger brained and may have been more behaviorally flexible.

By the late Miocene (10 to 6 million years ago), the Old World monkeys dominated the trees, and hominoids such as the dryopiths were less common. But some dryopiths or other forms must have given rise to the forerunners of modern gorillas, chimpanzees—and humans. The fossil record suggests that divergences marking the origin of those lineages occurred during this epoch. So do many comparative biochemical and immunological studies (pages 559 and 560).

Figure 42.6 One of the nocturnal tree shrews, found on certain islands of Indonesia.

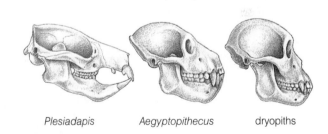

Plesiadapis   Aegyptopithecus   dryopiths

Figure 42.7 Comparison of skull shape and dentition of some extinct primates. *Plesiadapis* was a Paleocene primate with rodentlike teeth. *Aegyptopithecus*, an Oligocene anthropoid, probably predates the divergence leading to Old World monkeys and the apes. The dryopiths had the Y-5 molar pattern characteristic of living apes and humans.

THE HOMINIDS

As Figure 42.1 indicates, humans and their extinct close ancestors are called hominids. The first hominids probably emerged between 6 and 4 million years ago, during the late Miocene or early Pliocene. The earliest known fossils (from 4 to 2 million years old) have been discovered in southern and eastern Africa. They suggest there were a variety of forms, which nevertheless had three features in common:

> Bipedalism
> Omnivorous feeding behavior
> Further brain expansion and reorganization

According to one view, those features developed as responses to a major shift in climate. The cooling and drying trend that began in the Miocene continued into the Pliocene, and vast savannas replaced much of the African rain forests. The hominids were forced to make the transition from forest dwellers to life in mixed woodlands and open grasslands.

Rather than speculate on how large-brained, omnivorous, and bipedal forms actually evolved during the transition, think instead about the plasticity that those three features represent. In this context, **plasticity** means the ability to be flexible and to adapt to a wide range of demands. The first hominids were faced with a new, complex, and highly unpredictable world. Yet they had the brains, the freedom from dependence on only one type of food, and the freedom that bipedalism gave them to use their hands in ways that would promote their survival. Thus their emergence during the course of evolution was a modification of patterns observed among the other primates; *it was based on the primate heritage.*

Australopiths

The earliest known hominids, the **australopiths**, can be grouped into two broad categories:

1. Gracile forms (slightly built), currently going by the names *Australopithecus afarensis* and *A. africanus.*

2. Robust forms (muscular, heavily built), including *A. boisei, A. robustus,* and the puzzling "WT 17000."

Figure 42.8 compares the different forms. We really have little idea of how they were related. The problem is that this apparently was a "bushy" period of evolution, and inferences must be drawn from a limited number of diverse fossil fragments. As one scientist put it, deciphering the course of evolution from those fragments is like trying to follow the story of *War and Peace* from twelve pages torn randomly from the book.

One thing we do know: the early hominids were bipedal. Legbone fragments 4 million years old show muscle insertions typical of two-legged walkers. We also have an unusually "complete" skeleton of a gracile australopith (a female dubbed Lucy), who lived nearly 3 million years ago. The angle of Lucy's thigh bones allowed her weight to be centered directly beneath her pelvis (Figure 42.9); the angle is a hallmark of bipedalism. (Ape thigh bones splay out from the pelvis, and apes have a waddling, quadrupedal gait.) Most convincingly, the australopiths left footprints (Figure 42.10).

Australopiths that lived between 4 million and 3 million years ago are *A. afarensis* in one scheme and part of the *A. africanus* continuum in another (Figure 42.11). Their features are transitional between the Miocene apes and the later hominids. Bipedalism had freed their hands—and fossil handbones suggest they were as dex-

Figure 42.8 Comparison of the size and stature of australopiths with modern humans.

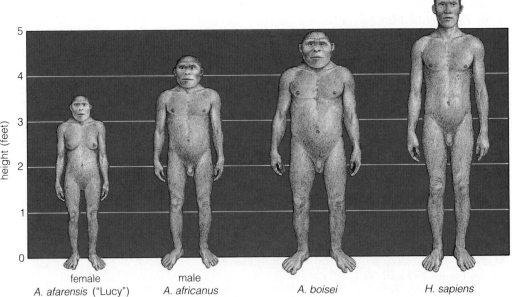

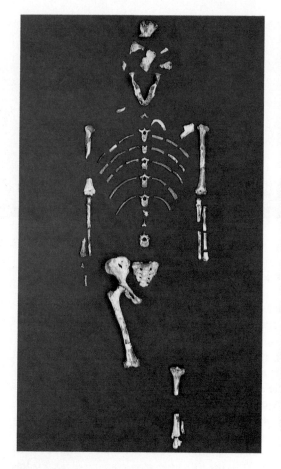

Figure 42.9 Fossil remains of Lucy, one of the earliest known australopiths. Lucy was only 1.1 meters (3 feet 8 inches) tall and weighed 30 kilograms (about 65 pounds). The density of her limb bones is indicative of very strong muscles. Although her face (with its large, jutting jaws) was apelike, her cranium indicates that the brains of these early hominids were already undergoing some reorganization to the humanlike condition.

Figure 42.10 Footprints made in soft, damp, volcanic ash 3.7 million years ago at Laetoli, Tanzania, as discovered by Mary Leakey. The arch, big toe, and heel marks are those of upright early hominids.

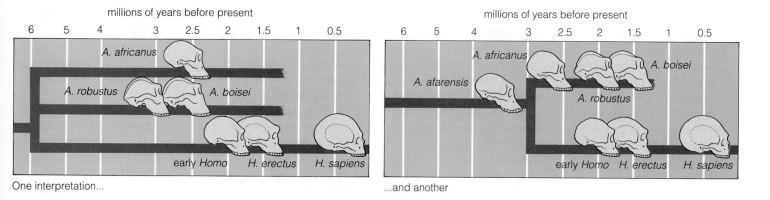

One interpretation...   ...and another

Figure 42.11 Two of several proposed phylogenetic trees for the hominids. There is general agreement that the hominids emerged during the late Miocene, between 6 and 4 million years ago. The *Homo* lineage is well documented; the australopith connections are not yet understood.

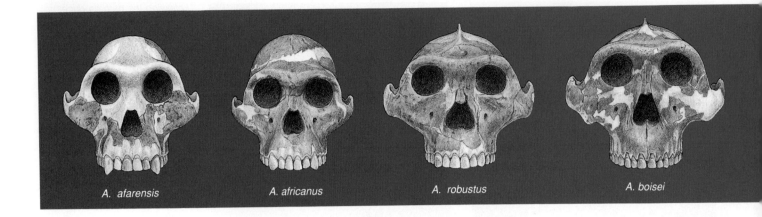

A. afarensis   A. africanus   A. robustus   A. boisei

terous as we are. Like small apes, their cranial capacity was about 400 cubic centimeters. Unlike apes (which have a U-shaped jaw), their jaw was slightly bow-shaped. Finally, the generalized pattern of their molars suggests they were omnivores.

A. africanus fossils, mainly from southern Africa, date from about 3 million to 2.5 million years ago. This form was 1.4 meters (4 feet 6 inches) tall, on the average, and weighed about 27.2 kilograms (60 pounds). The average cranial capacity was 450 cubic centimeters. Its cheek teeth formed a platform that could grind plant matter, but its incisors were relatively large; most likely, A. africanus was an omnivore.

A. robustus fossils date from about 2.5 million to 1 million years ago. On the average, this form was about 1.5 meters (5 feet tall), weighed about 45.4 kilograms (100 pounds), and had a cranial capacity of 530 cubic centimeters. It had a massive skull, deep jaws, and a built-up ridge of cranial bone (crest) to which strong jaw muscles were attached (Figure 42.12). Its cheek teeth, which formed a grinding platform, suggest it was a vegetarian.

A. robustus fossils have been found mainly in southern Africa; A. boisei fossils for the same time period come from the east. The eastern form was even more muscular and powerful, although both had about the same cranial capacity. A. boisei had a massive cranium with very prominent crests. Its molars were quite large, and those farthest back were heavily cusped. The dental pattern suggests it was specialized to the extreme for chewing tough plant matter, including seeds and nuts.

Specimen WT 17000 is more ancient yet more robust than A. robustus. It is like A. afarensis and A. boisei in its crest and other aspects of the skull (and unlike A. africanus, which has no crest). In face and teeth (as large as any hominid teeth ever found) it is like the robust forms, the most specialized australopiths known, yet its brain is no larger than Lucy's—who lived a million years earlier. No one quite knows where WT 17000 fits in the family tree.

Stone Tools and Early *Homo*

By at least 4 million years ago, hominids were walking upright, their hands free to develop new manipulative skills. It appears that more than 2 million years passed before they began making stone tools. During the interim, they may have used branches, sticks, and other perishable materials as tools, much as apes do today.

Who were the first toolmakers? The likely candidates were the east African forms called "early *Homo*" in one scheme (and "late australopith" or *H. habilis* in others). This larger brained, omnivorous hominid overlapped in time and space with its vegetarian cousin, *A. boisei*. Its remains have been found with the bones of frogs, pigs, chameleons, birds, fish, and small antelopes. It may have been a scavenger, a hunter, or both. And it apparently was ancestral to modern humans.

Early *Homo's* skull and teeth were notably different from the australopiths. The cranial capacity (650 cubic centimeters) was about thirty percent larger. (It was still small compared to modern humans.) The face was smaller in proportion to the cranium. A form designated ER 1470, which lived nearly 2 million years ago, had a thin face, high skull, and a cranial capacity of 775 cubic centimeters. Compared to the later australopiths, early *Homo* had a less specialized dental pattern.

Was it chance perception that put early *Homo* on a unique evolutionary road? At some point, those hominids perceived that rocks could be used to crack open animal bones to get at the soft marrow. (Plenty of battered marrow bones are at the same sites as early *Homo* remains.) At some point, too, they must have perceived that they could use sharp-edged flakes, split accidentally (and not too often) from rocks tumbling through a river or down a hill. Such flakes could be used to scrape flesh from animal bones (and plenty of bones with telltale parallel scrapes are found at the same sites). And they must have perceived the advantage of *duplicating* the rare, natural accident to give them a dependable supply of cutting tools.

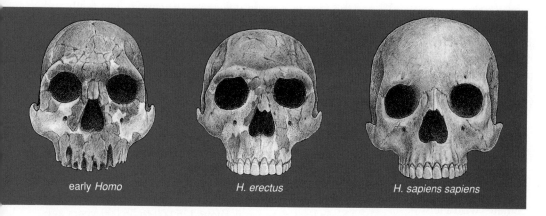

Figure 42.12 Comparison of skull shapes of the early hominids relative to modern humans (*Homo sapiens sapiens*). In general, australopiths had a small brain and large face, compared with the larger brained, smaller faced forms of the genus *Homo*. The drawings are not all to the same scale.

early *Homo* H. erectus H. sapiens sapiens

Figure 42.13 Representatives of the more than 37,000 stone tools recovered from Olduvai Gorge. *Upper row:* A crude chopper and more advanced forms having a joint as well as a sharp edge. The stone ball may represent a transition from passive to aggressive tool use. It resembles the Argentine bolas, which are strung together on lengths of hide and thrown at animals to entangle the legs and bring them down. *Lower row:* A hand ax and a cleaver.

The earliest "manufactured" tools were crudely chipped pebbles, called choppers, which were first discovered by Mary Leakey at Olduvai Gorge in eastern Africa. This gorge cuts through a layercake sequence of geologic deposits. A level dated at 1.9 million years yielded the choppers; ever higher layers yielded ever more sophisticated tools. Together, this tool sequence and later ones give insight into the increasing skills that underlie the technological foundations of modern culture. They give insight also into the increasingly refined exploitation of a major food source—the game that flourished in the open grasslands (Figure 42.13).

Olduvai gives testimony to our early social and cultural development. Nearly 1.9 million years ago, our hunting or scavenging ancestors were making shelters—*they had a concept of "home."* Remains at several sites indicate that hominids were gathering in the shelters to eat captured or scavenged animals and discarding the inedible parts outside what may have been a brushwood windbreak. At one site, a loosely arranged circle of small rocks apparently supported the pile of brushwood (Figure 42.14).

This is an extraordinary development among primates. Other primates forage together, stay together overnight, then move on—but they don't bring back food to a home base. And think about this: early *Homo* traveled several miles from home and back for special types of rocks (lava, quartzite) that were best for making specific types of tools. *Early on, the human brain was assessing distant horizons.*

On reflection, we can say that the emergence of *Homo* was based on *plasticity*, not on specializations that would restrict the genus to a particular habitat and life style. It is a reason why *Homo* continued to survive, adapt, and expand into a wide range of habitats.

Homo erectus

The adaptive potential of the early humans was put to the test between 1.5 million and 300,000 years ago. This was a time of extreme changes in climate, with Pleistocene Ice Ages alternating with interglacial warmings. When vast glacial sheets formed, they tied up tremendous amounts of water and sea levels fell; when the glaciers melted, water was released and sea levels rose—and land bridges as well as coastal regions were alternately submerged and exposed. Large tracts of land were slowly buried under ice when glaciers advanced southward from the poles, with landscape features destroyed and new ones carved out in the process.

During the interglacials, a larger brained human species called *Homo erectus* apparently migrated out of Africa and into Southeast Asia, China, and Europe.

Although *H. erectus* populations showed regional variations in form, they all had certain features in common. Compared with modern humans, the skull was thick-walled and primitive, with a "pentagonal" shape when viewed from behind. The cranial capacity ranged from 775 to 1,225 cubic centimeters (the average being 1,000). Thus the difference in brain size between a large *H. erectus* and a small *H. sapiens* was practically nil.

The increase in brain size correlates with the spectacular travels of *H. erectus* into diverse environments; no other hominids had ranged so far from their original environment. The increase also correlates with refinements in technological skills. *H. erectus* fossils are associated with advanced stone tools. Some sites show evidence of the controlled use of fire—and how advantageous campfires would have been, in the moves northward.

Homo sapiens

Somewhere between 300,000 and 200,000 years ago, *H. erectus* apparently gave rise to a new species, *H. sapiens*. It seems there was a gradual transition in form, often with a variety of types existing at the same time. Then *H. erectus* gradually disappeared. Many Old World sites have yielded transitional forms (with larger but not fully modern cranial capacity) together with fairly sophisticated tools.

Compared with their predecessors, the early members of *H. sapiens* had a rounder, more vaulted skull, and the face was more delicately structured. The teeth and jaw were relatively small; many forms had a chin (its predecessors did not). Eventually the cranial capacity averaged 1,350 cubic centimeters—the same as modern humans. From that point on, cultural evolution far outstripped biological change.

It was not until about 40,000 years ago that anatomically modern humans, *H. sapiens sapiens*, emerged. Before then, a number of archaic groups lived in the Old World. Some groups closely resembled modern populations; others, including the Neandertals, showed distinctive traits.

Neandertal fossils from Europe, China, and the Near East date from about 125,000 to 35,000 years ago. The Neandertals varied in form, but contrary to common belief, they really didn't look that much different from modern humans—who also vary in form. Facial bones were heavy; often there were large browridges. The skull was not as rounded or vaulted as the average modern skull. The "contents" were similar; if anything, Neandertals had larger brains than we do—ranging between 1,300 and 1,750 cubic centimeters.

With their sophisticated tools, Neanderthals survived as proficient hunters and gatherers in diverse environments. They lived in caves, rock shelters, and open-air camps, in varied climates. Yet they disappeared suddenly from the fossil record 35,000 to 40,000 years ago. Did a plague wipe them out? Were they slaughtered by invaders? Or were they simply assimilated into a population of other groups moving into Europe from other parts of the Old World? No one knows; but by 40,000 years ago, human forms were clearly modern.

Since then, human evolution has been almost entirely cultural rather than biological, and so we leave the story. From the biological perspective, however, we can make these concluding remarks: Humans have spread throughout the world by rapidly devising the cultural means to deal with a broad range of environmental conditions. Compared with their predecessors, modern humans have developed spectacularly rich and varied cultures, moving from "stone-age" technology to the age of "high tech." Yet hunters and gatherers persist in parts of the world, attesting to the great plasticity and depth of human adaptations.

SUMMARY

1. The primates include prosimians (lemurs, tarsiers, and related forms) and anthropoids (monkeys, apes, and humans). They are descended from small rodentlike or insectivorous mammals that evolved about 65 million years ago.

2. The anthropoids are further classified as ceboids (New World monkeys), cercopithecoids (Old World monkeys), and *hominoids* (apes and humans).

Figure 42.14 Artist's reconstruction of a campsite of the early hominids, revealed by excavations at Olduvai Gorge. Like other campsites, it was built at the margin of a freshwater lake. The hominids apparently brought back game to their shelters, which may have resembled the brushwood windbreaks still constructed by nomadic tribes.

3. Hominoids include the apes (such as the orangutan, chimpanzee, and gorilla), and the *hominids* (these being humans and their most recent ancestors).

4. The following evolutionary trends have occurred among the primates as a whole and are related largely to an arboreal (tree-dwelling) ancestry:

a. From quadrupedalism (four-legged gait) to bipedalism (habitual free-striding, two-legged gait). This trend involved changes in the shoulders, vertebral column, pelvic girdle, and footbones and legbones. It proceeded from ancestral ground-dwelling forms to climbers and leapers, then forelimb swingers, then upright walkers. The changes developed first in the trees. There, primates were largely free of the selective pressures imposed by ground-dwelling predators but were challenged by a visually complex habitat.

b. Increased manipulative skills owing to modification of the hands, which began to be freed from their load-bearing function among arboreal primates. The forms able to stand or sit upright had an advantage in reaching for fruit and other objects, holding onto their infants, and not falling out of trees and becoming prey to ground-dwelling predators.

c. Less reliance on the sense of smell, more reliance on enhanced daytime vision, including color vision and depth perception.

d. From specialized to omnivorous feeding behavior. This trend was pronounced during the Miocene, when tropical forests were giving way to mixed woodlands and savannas.

e. Brain expansion and reorganization. This occurred among mammals generally but became pronounced during hominid evolution. Changes in hominid cranial capacity are associated with ever more sophisticated technology (from simple to refined tools) and with social development.

f. All of these trends were the foundation for the remarkable *plasticity* displayed by the hominids—that is, the ability to remain flexible and to adapt to a wide range of demands imposed by an unpredictable, complex environment.

5. The oldest known primates date from the Paleocene (65 to 54 million years ago). The first prosimians date from the Eocene (54 to 38 million years ago). The ancestral anthropoids had emerged by the Oligocene (38 to 25 million years ago).

6. An adaptive radiation occurred during the Miocene, when the climate became cooler and drier. Forests shrank, grasslands spread, and different subpopulations of anthropoids became reproductively isolated and adapted to local conditions. Divergences led to the great apes and the first hominids.

7. The first hominids, the australopiths, apparently emerged between 6 and 4 million years ago. All were bipedal, with a larger brain than their predecessors. Some forms were omnivorous and gracile (slightly built); others were vegetarians and quite robust (muscular, taller, more heavily built). The fossils are there, like pieces of a puzzle; the phylogenetic relationships among them have not been worked out.

8. Fossils of early *Homo*, the first known representative of the human lineage, date from 2 million years ago. Early *Homo* was omnivorous and larger brained than the early hominids, and it used simple tools (such as pebble choppers). There is evidence of social development (including a brushwood shelter used as a home base).

9. *Homo erectus* fossils throughout the Old World date from about 1.5 million to 300,000 years ago. Diverse and abundant cultural artifacts are associated with the fossils. This form was larger brained than its predecessors, its cultural evolution was more pronounced, and it was adapted to a wide range of habitats.

10. *H. sapiens* emerged between 300,000 and 200,000 years ago; this larger brained form may have evolved from a *H. erectus* population. By 40,000 years ago, fully modern forms (*H. sapiens sapiens*) had evolved. From that point on, cultural evolution has greatly outstripped biological evolution of the human form.

Review Questions

1. What are the general evolutionary trends that occurred among the primates as a group? What way of life apparently was the foundation for these trends? *693–696*

2. What conditions seem to have been responsible for the great adaptive radiation of apelike forms during the Miocene? *697*

3. What is the difference between "hominoid" and "hominid"? Are we hominoids, hominids, or both? *692*

4. Describe the key characters of hominid evolution. How do they relate to the concept of plasticity? *698*

5. What are the major differences between the australopiths and the genus *Homo*? *700*

Readings

Reader, J. 1981. *Missing Links.* Boston: Little, Brown. Readable, exquisitely illustrated historical account of discoveries concerning human evolution—and penetrating insights into the failings and triumphs of the discoverers.

Rensberger, R. 1981. "Facing the Past." *Science 81* 2(8):41–50. Intriguing look at how artists' reconstructions can bias our perceptions of what the early hominids looked like.

Shipman, P. 1987. "Baffling Limb on the Family Tree." *Discover* 7(9):86–93.

Weaver, K. 1985. "The Search for Our Early Ancestors." *National Geographic* 168(5):560–623. Stunning photographs, reconstructions.

Weiss, M., and A. Mann. 1985. *Human Biology and Behavior.* Fourth edition. Boston: Little, Brown. Excellent perspective on origins and the course of human evolution.

UNIT EIGHT

ECOLOGY AND BEHAVIOR

43
POPULATION ECOLOGY

The optimist proclaims that we live in the best of all possible worlds; and the pessimist fears this is true—James Branch Cabell

Picture a crowded outdoor marketplace in India, with colorful heaps of fruits and vegetables, a dozen kinds of dried beans and lentils in great baskets, and an endless variety of fragrant herbs and spices. The marketplace links the teeming city with the farms and orchards of the countryside, where nearly every usable plot of land is already in use for producing food. Yet the population of India and many other developing countries throughout the world continues to increase at an alarming rate.

Meanwhile, populations in the developed countries are beginning to stabilize, yet they use resources from around the world at a far greater rate than do the people of less developed nations. Some of these resources, like petroleum, cannot be replaced. Others, like tropical forests, which are often cleared to raise beef and other products for export to the developed nations, will take centuries to replace, if they can be replaced at all. On a global scale, the demand for food and other resources by the increasing human population appears to be on a collision course with a supply of resources that, at best, is scarcely holding its own.

In the natural world, the effects of resource scarcity are often felt more quickly. Shift the scene from the Indian marketplace to a quiet hardwood forest in summer. In an oak tree overhead, a warbler forages for insects among the leaves. Each year, the number of baby warblers hatched depends in part on the supply of insects available. Nearby, a squirrel scuffles noisily in the decaying leaves on the forest floor, adding an acorn to its cache of winter rations. In years of poor acorn crops, many squirrels may not survive the winter.

Nearby, a butterfly alights on a wildflower, briefly takes in nectar, and flies off. When butterflies are scarce, the wildflowers they pollinate will produce few seeds.

Whether we are talking about humans or any other kind of organism, certain principles govern the growth and stability of their populations over time. Over the long term, we must reckon with these principles, which influence the patterns of relationships of organisms with one another and their environment. These patterns, in *all* their varied forms, are the focus of ecology. *As a science, ecology seeks to treat the world of nature—including its human component—with a single set of concepts and principles.*

ECOLOGY DEFINED

The word **ecology** was coined in the last century from the Greek *oikos* (meaning "house") to designate the study of organisms in their natural homes. Specifically, it means the study of the interactions of organisms with one another and with the physical and chemical environment. Although it includes the study of environmental problems such as pollution, the science of ecology also encompasses research on the natural world from many viewpoints, using many techniques. Modern ecology relies heavily on experiments, both in the laboratory and in field settings, and on mathematical models. These techniques have proven helpful in testing ecological theories and in arriving at practical decisions in the management of natural resources.

In this unit of the book, we will be considering ecological interactions at several levels of biological organization, which may be defined in the following way:

1. The **population**: a group of individuals of the same species occupying a given area. The place where a population (or an individual) lives is called its **habitat**: tree squirrels live in a "forest habitat," muskrats in a "streambank habitat," dandelions in "disturbed habitats."

2. The **community**: the populations of *all* species that occupy a habitat. Ecologists also use the term to refer to certain groups of organisms in a habitat—the "bird community" or the "plant community," for example. Species in a community play different roles:

a. *Producers*: The "self-feeding" (autotrophic) organisms, which include most plants and some microorganisms. The producers synthesize their own organic compounds from simple inorganic substances with the aid of energy from the sun (photosynthetic autotrophs) or from the inorganic substances themselves (chemosynthetic autotrophs).

b. *Consumers*: All organisms that are "not self-feeding" (heterotrophic) and that ingest other (usually) living organisms in whole or in part to obtain organic nutrients. Thus the consumers called *herbivores* eat plants, *carnivores* eat animals, and *parasites* take in blood, sap, and other tissues from living hosts.

c. *Decomposers*: Mostly heterotrophic bacteria and fungi that obtain organic nutrients by breaking down the remains or products of organisms. The activity of decomposers allows simple compounds to be recycled back to the autotrophs.

d. *Detritivores*: Earthworms, nematodes, crabs, and other heterotrophs that feed on particles of organic matter, such as would be produced by the partial decomposition of plant and animal tissues.

3. The **ecosystem**: a community and its physical and chemical environment. An ecosystem has living (*biotic*) and nonliving (*abiotic*) components. Soils, temperatures, rainfall, even organic matter are examples of the abiotic component. The classes of organisms detailed above form the biotic component.

4. The **biosphere**: all living organisms on earth which, together with their interactions with the global physical and chemical environment, maintain a system of energy use and materials cycling. This system runs on energy flowing into it (from the sun) and it gives up energy (primarily as low-grade heat) to space.

Let's focus first on the ecological relationships that influence the size, structure, and distribution of populations. In subsequent chapters we will consider the relationships at the levels of communities, ecosystems and, finally, the biosphere.

POPULATION DENSITY AND DISTRIBUTION

If you have ever collected fossils, or butterflies, or kept a list of backyard birds over a period of time, you will know that for every common species you find, there are several rare ones. In other words, **population density** (the number of individuals per unit area) generally varies widely among species living in the same locality.

Ecological relationships influence the density of a population. To study these influences, we must determine what the density is to begin with so that we can chart any changes in space or over time. For sparse populations, a simple "head count" (or tree count, and so on) can be made in a defined area; for dense populations, counts can be made at random in small sampling areas to arrive at an estimate of overall density.

Distribution in Space

Even when the density of a species in a particular habitat is known, we still don't know how the individuals are distributed (arranged) in space. They might be clumped together in certain parts of the habitat, distributed randomly throughout, or spread out rather uniformly (Figure 43.2).

Clumped Distribution. Most commonly the members of a population are distributed in clumps through their habitat for three reasons. First, physical and chemical conditions suitable for growth typically are "patchy" rather than uniform. In a pasture, for example, certain kinds of plants grow in soil rich in nitrates where individual cowpats fell weeks or months before. Second, some parts of the habitat offer more protection to prey organisms (or, conversely, better hunting to predators). Third, dispersal of seeds, larvae, and other representative forms of each new generation is often limited. For example, the offspring of some intertidal organisms settle near their parents and one another; gypsy moths and other insect pests often spread slowly from the original point of infestation.

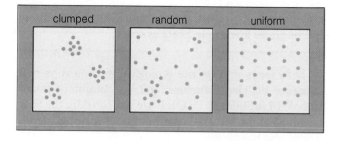

Figure 43.2 Generalized patterns of spatial distribution of individuals in a population.

Random Distribution. Assuming that environmental conditions are the same throughout a habitat and that members of the population are neither attracting nor repelling each other during a given time period, then distribution may be random. Spiders on a forest floor may be an example of random spacing among individuals. In population studies, a random distribution is the theoretical standard from which clumped or uniform distributions depart in opposite directions (Figure 43.2).

Uniform Distribution. Nearly uniform distribution (even spacing), as typified by orchards, is exceedingly rare in nature. For the few cases in which uniformity is approached, competition among individuals for a limited resource seems to be the cause. An example is the nearly uniform spacing of creosote bushes (*Larrea*) in dry scrub deserts of the sort occurring in the American Southwest. Large, mature plants deplete the soil around them of available water, and seed-eating ants and rodents concentrate their activities near the plants. As a result, seeds and seedlings are unable to survive near mature plants, so clumps do not form (Figure 43.3).

Distribution Over Time

Population distribution also varies with time, often in response to environmental rhythms. Few environments are so productive that they yield abundant resources all year long. For example, temperatures drop and food supplies dwindle during the winter in north temperate regions (such as parts of Canada and Europe), where many migratory birds reproduce. Whole populations move southward to winter feeding grounds in (for example) South America and Africa. With the return of spring, when their breeding grounds are warmer and have

Figure 43.3 Nearly uniform spacing among creosote bushes (*Larrea*) near Death Valley, California.

abundant new growth of plants and insect populations, they make the return trip north.

Animals may also move locally between habitats with the seasons. In deciduous tropical forests, trees lose their leaves during the dry season—except for broadleaf evergreen species growing as narrow "gallery forests" along watercourses. Many birds and mammals crowd into these narrow habitats during the dry season, greatly increasing the animal density compared to the wet season.

POPULATION DYNAMICS

Variables Affecting Population Size

To understand how populations change in size, we will begin with a simple concept. The size of any population increases or decreases with change in one or more of the following variables:

natality	*births*
mortality	*deaths*
immigration	*individuals from other populations of the same species join the population*
emigration	*individuals leave the population*

Together, these four variables dictate the rate of change in the number of individuals in the population over a given period of time:

$$\begin{pmatrix}\text{population}\\\text{growth}\\\text{rate}\end{pmatrix} = \begin{pmatrix}\text{births}\\+\\\text{immigration}\end{pmatrix} - \begin{pmatrix}\text{deaths}\\+\\\text{emigration}\end{pmatrix}$$

When the birth rate plus immigration is balanced over the long term by the death rate and emigration, population size is stabilized; there is said to be **zero population growth**.

In the next sections, we will assume that there is no immigration or emigration, to simplify things.

Exponential Growth

The variables that affect population size can be measured either for the entire population, as we did in the equation above, or per individual. For example, if 1,000 baby mice per month are born to a population of 2,000 mice living in a cornfield, then the birth rate per individual is 1,000/2,000 = 0.5. Death rates per individual are computed in the same way; if 800 of the original 2,000 mice in the cornfield die during the same month, the death rate per

individual is 800/2,000 = 0.4. Now we can rewrite the relationship for population growth as:

$$\begin{pmatrix}\text{population}\\\text{growth}\\\text{rate}\end{pmatrix} = \begin{pmatrix}\text{births}\\\text{per}\\\text{individual}\end{pmatrix} - \begin{pmatrix}\text{deaths}\\\text{per}\\\text{individual}\end{pmatrix} \times \begin{pmatrix}\text{number}\\\text{of}\\\text{individuals}\end{pmatrix}$$

Suppose, for the sake of argument, that the birth and death rates remain constant, regardless of how much or how little the population grows. Then we can lump them together as a single variable, called *r*, the *net reproduction per individual*. (In the mouse example, this would be 0.5 − 0.4 = 0.1, the birth rate minus the death rate.) This simplifies the formula for population growth rate to:

$$\begin{pmatrix}\text{population}\\\text{growth}\\\text{rate}\end{pmatrix} = \begin{pmatrix}\text{net}\\\text{reproduction}\\\text{per individual}\end{pmatrix} \times \begin{pmatrix}\text{number}\\\text{of}\\\text{individuals}\end{pmatrix}$$

In symbols, the same relationship can be represented as:

$$G = rN$$

This equation tells us that, when *r* is held constant, any population will show **exponential growth**. The number of its individuals increases in *doubling* increments—from 2 to 4, then 8, 16, 32, 64 and so on. (This is unlike an arithmetic increase from 1 to 2, 3, and so on.)

We can observe a limited period of exponential growth in the laboratory by putting a single bacterium in a culture flask with a complete supply of nutrients. In thirty minutes the bacterium divides in two; thirty minutes later the two divide into four. If no cells die between divisions, the number will double every thirty minutes. The larger the population base becomes, the more bacteria there are to divide. After only 9-1/2 hours (nineteen doublings), the population size will exceed 500,000; and after 10 hours (twenty doublings), it will soar past 1 million. When size increases are plotted against time, the result is a **J-shaped curve**, which is characteristic of populations undergoing unrestricted, exponential growth (curve *a* in Figure 43.4).

So far, we have assumed none of the bacteria die. To test the effect of mortality on growth rate, let's start over with our bacterium. This time, assume that twenty-five percent of the population dies between each doubling time. This death rate does slow things down a bit, because now it takes almost two hours instead of thirty minutes to double population size. *But only the time scale changes.* It now takes thirty hours instead of ten to arrive at a million bacteria—but we still have a J-shaped, or exponential, curve (curve *b* in Figure 43.4).

time (hours)	number of individuals for curve a
10	1,048,576
9½	524,288
9	262,144
8½	131,072
8	65,536
7½	32,768
7	16,384
6½	8,912
6	4,096
5½	2,048
5	1,024
4½	512
4	256
3½	128
3	64
2½	32
2	16
1½	8
1	4
½	2
0	1

Figure 43.4 (a) Exponential growth for a bacterial population that is dividing by fission every half hour. (b) Exponential growth of the population when division occurs every half hour, but when twenty-five percent dies between divisions. Although deaths slow the rate of increase, in themselves they are not enough to stop exponential growth.

As long as the birth rate remains even slightly above the death rate, a population will grow. If the rates remain constant, the population will grow exponentially.

Limits on Population Growth

Clearly there must be factors in nature that limit population growth. (Thus you will never be trampled to death by a billion rabbits when you walk through the woods.) However, those factors are often difficult to determine in nature because of the complex interactions among various populations. For that reason, let's go back to that single bacterium in its culture flask, where we can control the variables.

Assume we enrich the culture medium with glucose and essential elements required for bacterial growth, then allow the bacteria to reproduce for many generations. At first the population grows almost exponentially, then levels off and briefly remains rather stable. Then the population starts to decline rapidly and soon dies out.

What happened? For these bacteria, glucose meant food and energy—but the culture dish held only so much glucose. As the population expanded faster and faster, the glucose was being used up faster and faster. When supplies began to dwindle, so did the basis for growth. *When any essential resource is in short supply, it becomes a limiting factor for population growth.*

Even if we were to supply the bacterial population with all necessary nutrients, it would crash following its initial exponential growth. The increased numbers of bacterial cells would produce increased metabolic wastes which, in high enough concentrations, would drastically alter the environment. Unless we removed the tainted medium every so often from the culture flask and substituted a fresh supply, the bacteria would poison themselves to death.

In sum, the number of limiting factors can be large and their relative effects variable. *For natural populations, the factors that limit population growth can fluctuate considerably over time, with first one and then another setting the upper bound.*

Carrying Capacity and Logistic Growth

As a population increases in size, the same resources must be shared by a greater and greater number of individuals. The decreasing supply of resources may lower the birth rate, increase the death rate, or both—until births and deaths are in balance. At that point, and as long as the resource supply remains constant, the population should stabilize at some equilibrium size, called the **carrying capacity** (K) of the environment. *A sustainable supply of resources (including nutrients, energy, and living space) defines the carrying capacity for a particular population in a particular environment.*

The effect of a limited carrying capacity on population growth is expressed by the **logistic growth equation**:

$$\begin{pmatrix} \text{population} \\ \text{growth} \\ \text{rate} \end{pmatrix} = \begin{pmatrix} \text{maximum net} \\ \text{reproduction} \\ \text{per individual} \end{pmatrix} \times \begin{pmatrix} \text{number} \\ \text{of indi-} \\ \text{viduals} \end{pmatrix} \times \begin{pmatrix} \text{portion of} \\ \text{unexploited} \\ \text{resources} \end{pmatrix}$$

In symbols, this equation becomes:

$$G = r_{max} N\left(\frac{K - N}{K}\right)$$

When a low-density population grows logistically, the growth pattern is slow at first but then steadily accelerates. However, its growth slows more and more after it has passed the halfway mark of the carrying capacity; then it levels off once the capacity is reached. A plot of this growth pattern gives us a sigmoid, or **S-shaped curve**, as shown in Figure 43.5. (Keep in mind that the slope of such curves reflects the rate of population *change*, not population *size*.) Moreover, if a population exceeds the carrying capacity, it declines until it reaches its equilibrium size.

When a low-density population shows a logistic growth pattern, its rate of increase is slow at first, then accelerates and slows again, then levels off as the carrying capacity of the environment is reached.

The logistic growth model is only a baseline for studying population growth. Carrying capacities fluctuate with the seasons and from one year to the next. Moreover, the birth and death rates are not necessarily clearcut (even starving animals can continue to consume resources and bear young). For example, in 1910, four male and twenty-two female reindeer were introduced on one of the Pribilof Islands of Alaska. Within thirty years, the population increased to 2,000—greatly "overshooting" the carrying capacity. The vegetation on which the reindeer grazed almost disappeared, and in 1950 the herd size plummeted to eight members (Figure

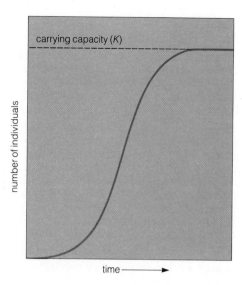

Figure 43.5 S-shaped curve characteristic of logistic growth. Following a rapid growth phase, growth slows and the curve flattens out as the carrying capacity of the environment is approached.

Figure 43.6 Rise and fall of a reindeer herd introduced on one of the Pribilof Islands, Alaska. Rapid population growth led to an overshooting of the carrying capacity of the environment. Growth stopped abruptly, and the population size crashed to eight reindeer—eighteen fewer than were present in the starting population.

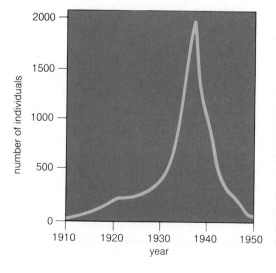

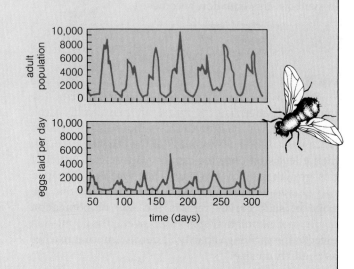

Figure 43.7 Effect of resource availability, time lags, and competition on population density for sheep blowflies (*Lucilia cuprina*). In one laboratory experiment, A. Nicholson fed limited amounts of beef liver to fly larvae. He also fed unlimited amounts of sugar and water to adults that were part of the same experimental group.

When the adult population density was high, so many eggs were laid that the resulting larvae devoured all the food before completing their development, and they all died. Through natural mortality, the size of the existing adult population dwindled and fewer eggs were laid. Fewer eggs meant fewer larvae—and less competition among them. Now some larvae were able to mature into adults. For a while, population density declined because of the time lag between the survival of larvae and the development of egg-laying adults. The delay permitted an increasing number of larvae to survive—and the adult population soared again.

The outcome of this process was a drastic overshoot-undershoot oscillation in population size over time.

43.6). A similar pattern of "overshoot and crashes" is shown in Figure 43.7 for a laboratory culture of flies. These cycles were also caused by time lags in responses to density.

LIFE HISTORY PATTERNS

So far, we have treated populations as if each were made up of identical members. Yet phenotypes obviously differ; and all but the simplest kinds of organisms go through many developmental stages, with each new change in morphology and behavior having its own perils and rewards. In short, *the patterns of birth, death, migrations, and reproduction vary through the life span characteristic of the species.* Let's take a look at a few examples of these **age-specific patterns** before considering some of the environmental variables that might have helped shape them.

Life Tables

Although each species has a characteristic life span, few individuals reach the maximum. The probability of death usually changes with age. Moreover, the reproductive rate and the probability of migration to another population of the same species vary with age in different ways from one species to the next. The study of such age-specific patterns is the subject of *demography*. Although originally developed by the life insurance and health

insurance companies, demographic methods are now widely applied by ecologists to populations of plants and animals.

For example, **life tables** are created to summarize the age-specific patterns of birth and death for a particular population in a particular environment. Typically ecologists follow the fate of a group of newborn individuals until the last one dies. (Such a group is called a *cohort*.) Besides recording the age at death of each individual, they keep track of how many offspring (if any) each surviving individual produces during each age interval of its life. The death rate and birth rate for each age interval are then easily computed, producing birth and death "schedules" for the cohort.

The death schedule is usually transformed to its more cheerful reflection, called **survivorship**. A survivorship schedule is produced by writing down, for each age x, the number of individuals of the original cohort that *survive* to age x.

Table 43.1 presents a life table for a cohort of 996 individuals of an annual plant (*Phlox drummondii*) in Texas. The curve for the death rate, the corresponding survivorship curve, and the "birth" (seed production) pattern are shown in Figure 43.8.

Patterns of Survivorship and Reproduction

Life history patterns are almost bewildering in their diversity, but we can get an idea of the *range* of possibilities by looking at a few extremes.

Table 43.1 Life Table for a Cohort of Annual Plants (*Phlox drummondii*)

Age Interval (days)	Survivorship (number surviving at start of interval)	Number Dying During Interval	Death Rate per Individual During Interval	"Birth" Rate (number of seeds produced per individual) During Interval
0–63	996	328	0.329	0
63–124	668	373	0.558	0
124–184	295	105	0.356	0
184–215	190	14	0.074	0
215–264	176	4	0.023	0
264–278	172	5	0.029	0
278–292	167	8	0.048	0
292–306	159	5	0.031	0.33
306–320	154	7	0.045	3.13
320–334	147	42	0.286	5.42
334–348	105	83	0.790	9.26
348–362	22	22	1.000	4.31
362–	0	0	0	0
		996		

Data from W. J. Leverich and D. A. Levin, *American Naturalist* 1979, 113:881–903.

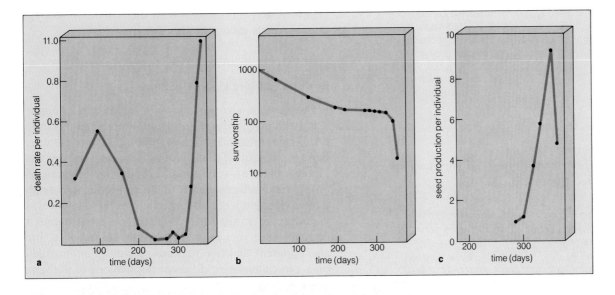

Figure 43.8 Death rate (**a**), corresponding survivorship curve (**b**), and the "birth," or seed production pattern (**c**) for a cohort of 996 individual phlox plants (*Phlox drummondii*), a representative of which is shown in (**d**).

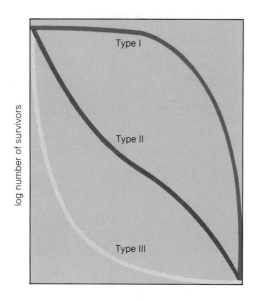

age (percent of lifespan)

Figure 43.9 Three generalized types of survivorship curves. For Type I populations, there is high survivorship until some age, then high mortality. Type II populations show a fairly constant death rate at all ages. For Type III populations, there is low survivorship early in life.

Survivorship Curves. Three "types" of survivorship curves are often noted (types I, II, and III):

I	*survivorship high throughout most of life; most deaths in a short time span after reproduction*
II	*death rate fairly constant at all ages*
III	*high death rate early in life, decreasing with age*

Figure 43.9 provides a generalized picture of these survivorship curves. (Notice that the curve for phlox in Figure 43.8b fits none of them exactly.)

A Type I curve is more or less characteristic of humans in communities with good health care services. Historically, and where health care is poor today, infant mortality puts a sharp drop at the beginning of the curve, followed by a plateau—rather like the phlox curve in Figure 43.8b. In general, large mammals that show extended parental care of offspring approach Type I curves. Survivorship curves approaching a Type II pattern have been reported for some songbirds, lizards, and small mammals, and probably they apply to destruction of seeds prior to germination for many plants. Survivorship patterns approaching the Type III curve are common among marine invertebrates, most insects, and many fishes, amphibians, and reptiles.

Timing of Reproduction. Organisms also vary widely in terms of the age at which they reproduce and in how often they do. Some reproduce once late in life, then die. Mayflies (which spend a single day as adults), annual plants such as the phlox, and certain long-lived plants (such as many bamboos) belong in this category. In contrast, most vertebrates, some insects, and most perennial plants (such as apple trees and roses) reproduce repeatedly throughout maturity.

Number of Offspring. Organisms characterized by a Type I survivorship curve tend to produce only a few offspring, and the offspring are large in relation to the size of the adult. Female elephants, for example, produce only four or five large calves in a lifetime, devoting several years of parental care to each. In contrast, species with a Type III survivorship curve produce vast numbers of offspring that are usually small in relation to the adult size. (Or, to put it more accurately, any species with a high juvenile mortality that did *not* produce large numbers of young is no longer with us.) The small seeds of annual weeds exemplify this pattern.

Evolution of Life History Patterns

For some time, ecologists have been working to understand and predict the relationships among natural selection, environmental variables, and life history patterns. Early on, it seemed that selection processes favored two kinds of patterns—either rapid production of many relatively small offspring early in life, or production of only a few relatively large offspring late in life. However, it is now apparent that these two patterns are extremes at opposite ends of a range of possible life histories. It is also apparent that *both* patterns as well as intermediate ones can characterize different populations of the same species!

Recently, David Reznick and John Endler conducted some elegant studies in Trinidad and, later, in the laboratory to identify which environmental variables were influencing the life history patterns of guppies. These small, live-bearing fish are often purchased for home aquariums (Figure 43.10). Male guppies are smaller than the females, which they attract with brightly colored patterns on the body and with complex courtship displays. Males stop growing once they are sexually mature, but the drab-colored females continue to grow larger as they reproduce.

In the mountains of the island of Trinidad, guppies living in different streams—and even in different parts of the same streams—are subject to different dangers. In some of the streams, a small killifish preys heavily on immature guppies but cannot handle the (larger) adults. In other streams, pike-cichlid and other dangerous pred-

Figure 43.10 Guppies (*Poecilia reticulata*) and their predators in the mountain streams of Trinidad. (**a**) Guppies devote considerable time to courting and avoiding predators and must be adapted to do both. Depending on where they live, guppies might meet up with weak predators (such as the killifish *Rivulus hartii*, about to attack a guppy in (**a**); with *Hemibrycon dentatum* (**b**), a characin that eats guppies often; or with *Crenicichla alta* (**c**) a pike-cichlid that is very dangerous to guppies.

In places where the more dangerous predators live, guppies are smaller, more streamlined, and duller in color patterning (**d**). In places where predators eat guppies only occasionally, the guppies are larger, less streamlined, and more brightly colored (**e**). The differences in morphology are accompanied by pronounced differences in life history patterns (see text).

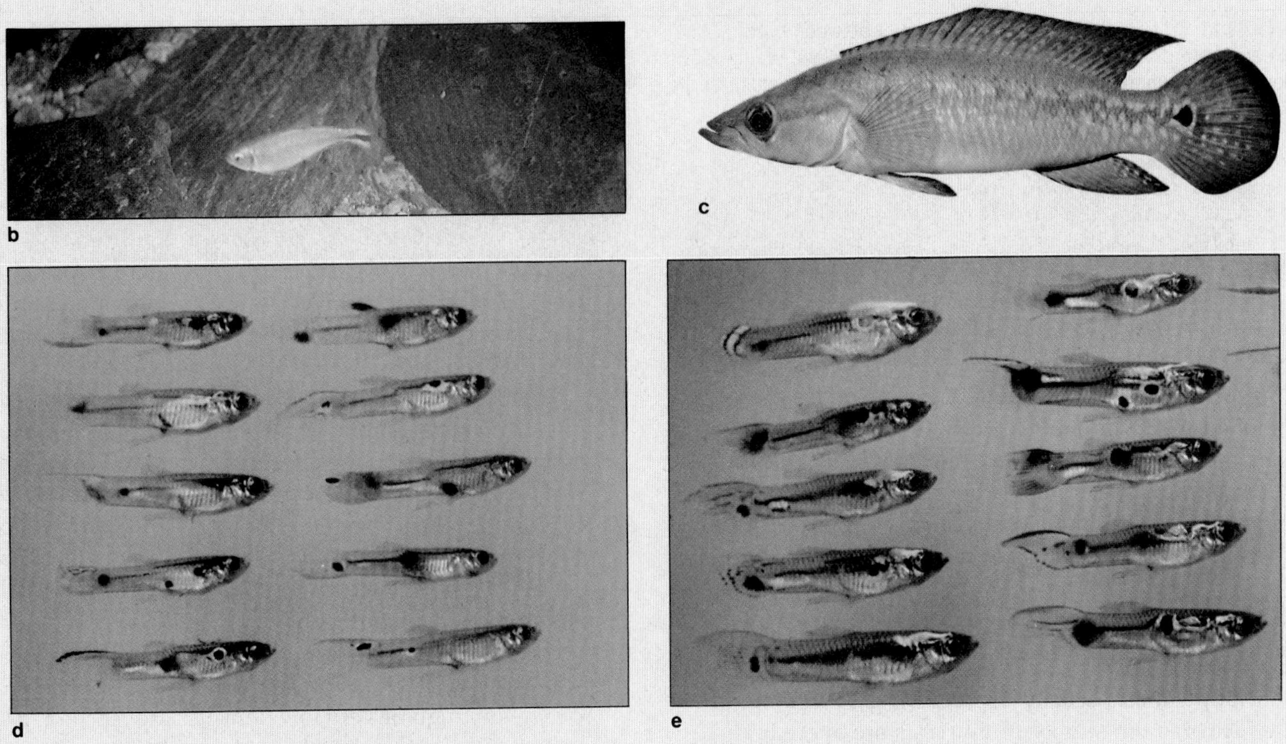

ators prefer mature (bigger) guppies and tend not to waste time hunting the small ones.

As might be predicted from our understanding of selection processes, the individuals of guppy populations confronted with pike-cichlids and comparable predators (which favor large-bodied dinners) mature sooner, are smaller at maturity, and reproduce at a younger age. Moreover, they produce far more offspring and do so more often than their counterparts in killifish streams (Figure 43.11).

As a check on the possibility that the differences were due to some other, unknown differences between the streams, Reznick and Endler raised guppies from each kind of stream in the laboratory for two generations under identical conditions (with no predation). They discovered that the life-history differences were maintained, indicating a genetic (hence a heritable) basis for the differences.

As a check on the role of predators in the evolution of size differences seen in the field, guppies were grown for many generations in the laboratory—some alone, some with killifish, and some with pike-cichlids. As predicted, the guppy lineage subjected to predation over time by killifish became larger at maturity, whereas the lineage living with pike-cichlids showed a trend toward earlier maturity.

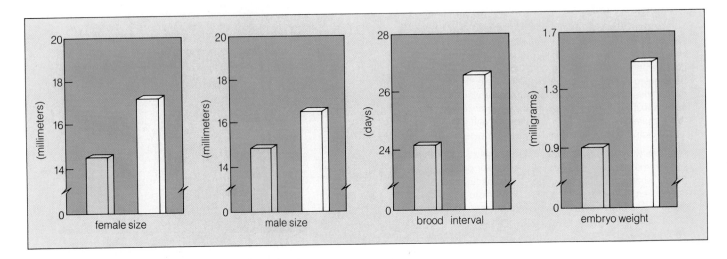

Figure 43.11 Differences in body size, interval between broods, and embryo weights for guppies living in streams with pike-cichlids (lavender) or killifish (yellow) on the island of Trinidad. Pike-cichlids prey on larger guppies, whereas killifish prey on smaller ones, selecting for the differences shown here.

LIMITS ON POPULATION GROWTH

So far in this chapter, we have considered some of the variables shaping the size, structure, distribution, and life histories of populations. Along the way, we have mentioned in passing a few of the factors that act like brakes on population growth (nutrient availability for our flask-bound bacterium, for example). Let's now briefly review the main categories in which these growth-limiting factors are grouped.

Density-Dependent Factors

When population density increases, growth-limiting factors such as the supply of nutrients are altered, and they collectively lead to a reduced birth rate and an increased rate of death, emigration, or both. Such density-dependent factors have a self-adjusting effect on population growth. Once the density decreases, the pressures ease and the net reproductive rate may increase once more.

Density-Independent Factors

Other checks on population growth act more or less independently of population density. Consider that severe changes in temperature or rainfall, wave action in the intertidal zone, and catastrophic events such as hurricanes and volcanic eruptions tend to increase the death rate per individual, *but they do so more or less to the same degree regardless of population density*. A hard freeze in the late spring is just as likely to kill an individual annual plant seedling regardless of whether it is in the middle of a clump of plants or growing alone nearby.

Density-dependent and density-independent factors interact in such complex ways, however, that the effects of one often amplify the effects of the other. For instance, whether rabbits tolerate a sudden freeze depends in part on whether they have enough food and burrows. Food and burrow availability are density-dependent factors, yet here they interact with a change in temperature that would likely kill some rabbits at any density.

For many populations, there seems to be a broad range over which density-dependence is weak or absent. When very high or very low densities are reached, however, density-dependent forces come into play, decreasing or increasing the rate of population growth, respectively. The increase in the net reproductive rate at lower densities is especially important, for if a population were limited only by density-independent factors, there would be nothing to prevent its extinction over the long term.

Competition for Resources

The amount of water, nutrients, and other resources available for each individual often declines as a population increases in density. The decline may temporarily hamper reproductive processes or reproductive behavior without affecting the individual's health over the long term. (For example, in some kangaroos, embryonic development is arrested until there is enough plant food to sustain the mother *and* assure development of the embryo.) However, a decline in resources also can impair health to the extent that the reproductive rate is lowered; this happens to annual plants (such as corn) confronted with drought years.

When individuals of the same species compete for resources, the interaction is called **intraspecific competition**. In general, these interactions take two forms:

exploitation competition	*all individuals have equal access to the resource; only the rate or the efficiency of exploiting it differs*
interference competition	*certain individuals control access to the resource, limiting or preventing use of the resource by others*

The difference between exploitation competition and interference competition becomes clear with a simple analogy. Suppose you and a greedy friend are sharing a milkshake, each with your own straw. If your friend sucks harder than you do, he or she is the superior exploitation competitor. However, if you reach over and pinch your friend's straw, you are the superior interference competitor.

In nature, the most common forms of interference competition among animals are territoriality, social dominance, or simple defense of food sources (Chapter Forty-Nine). Among land plants, larger and taller individuals may shade smaller ones and suppress their growth.

Predation, Parasitism, and Disease

When prey or host populations become increasingly dense, their members face a proportionally greater risk of being dispatched by predators, colonized by parasites, or infected by contagious disease. As the predator, parasite, or pathogen population increases, the abundance of prey or host decreases, and vice versa.

Herbivores—"plant predators"—play an important role in limiting plant populations. For example, foraging rabbits often keep the density of plant populations at depressed levels. Their impact was dramatically illustrated in the English countryside in the 1950s, when the rabbit population was decimated by the disease myxomatosis (itself a good example of population regulation). Shortly afterward, the forage plants grew spectacularly, and formerly rare species became quite common.

Predation, parasitism, and disease are among the most powerful factors shaping the growth of populations. We will look closely at these factors in chapters to follow.

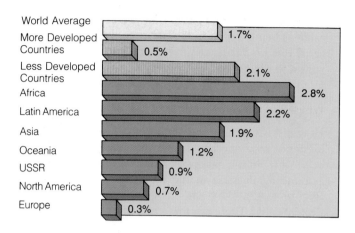

Figure 43.12 Average annual population growth rate in various groups of countries in 1987.

HUMAN POPULATION GROWTH

In 1986 the human population reached 5 billion, having passed 4 billion only eleven years earlier. Even if that number represented a stabilized population level, we would have to contend with monumental problems. In a given year, between 5 to 20 million people now die of starvation and malnutrition-related diseases. What makes future prospects especially troubling is that our population size is by no means static.

In 1986, the annual population growth rate was higher than 4 percent in some parts of the world—Kenya is an example. Continuing at that rate, Kenya's population would double in seventeen years. In contrast, the population growth rate in some highly developed countries has slowed to nearly zero; the doubling rate for the United Kingdom would now be around 460 years.

In 1987, the base population was 5.1 billion and the annual growth rate was 1.7 percent (Figure 43.12). Also, about 86.7 million individuals were added to the world's population in 1988 alone! That amounted to an average of 1.7 million more people per week, 238,000 per day, or 9,900 per hour. Yet the incredible growth of the human population is occurring when at least one in six is hungry or malnourished and without adequate housing, one in four lacks clean drinking water, and one in three lacks access to adequate sewage facilities or effective health-care delivery systems.

With the most intensive effort, we might be able to double food production to keep pace with growth. However, we would succeed in doing little more than maintaining marginal living conditions for most of the world. Under such conditions, deaths from starvation could be 10 to 40 million a year. For a while, it would be like the Red Queen's garden in Lewis Carroll's *Through the Looking Glass*, where one is forced to run as fast as one can to remain in the same place. Can you brush this picture aside as being too far in the future to warrant your concern? It is no farther removed from you than your own sons and daughters.

Where We Began Sidestepping Controls

How did we get into this predicament? For most of our existence as a species, human population growth has been slow. In the past two centuries, there has been an astounding increase in the rate of population growth (Figure 43.13). Why has our growth rate increased so dramatically? There are three possible reasons:

1. We steadily developed the capacity to expand into new habitats and new climatic zones.

2. The carrying capacities of the environments we already occupied were increased.

3. A series of limiting factors was removed.

Let's consider the first possibility. By 50,000 years ago, the human species had radiated through much of the world. For most animal species, such extensive radiation would not have occurred as rapidly. Humans were able to do so with the application of learning and memory—how to build fires, assemble shelters, create clothing and tools, plan a community hunt. Learned experiences were not confined to individuals but spread quickly from one human group to another because of language—our ability for cultural communication. (It took less than seven decades from the time we first ventured into the air until we landed on the moon.) Thus the human population expanded into new environments, and it did so in an extremely short time span compared with the radiations of other organisms.

What about the second possibility? About 10,000 years ago, people began to shift from the hunting and gathering way of life to agriculture—from risky, demanding moves following the game herds to a settled, more dependable basis for existence in more favorable settings. Even in its simplest form, agricultural management of food supplies bypassed the natural carrying capacity of the environment for the old hunter-gatherer way of life. Through the development of irrigation, metallurgy, social stratification (which provided a labor base) and, later, use of fertilizers and pesticides, the

limits were expanded and were met again with a resurgence of human population growth. Thus, with the domestication of plants and animals, the carrying capacity has indeed risen abruptly for human populations.

What about the third possibility—the removal of limiting factors? The potential for growth inherent in the development of agriculture began to be realized with the suppression of contagious diseases. Until about 300 years ago, malnutrition, contagious diseases, and poor hygiene kept the death rate relatively high (especially among infants), balancing the birth rate. Contagious diseases are density-dependent factors, and they spread rapidly through crowded settlements and cities. Without proper hygiene and sewage disposal methods, and plagued with such disease carriers as fleas and rats, population size increased only slowly at first. Then plumbing and sewage treatment methods appeared. Bacteria and viruses were recognized as disease agents. Vaccines, antitoxins, and drugs such as antibiotics were developed.

Thus the effect of one after another major limiting factor on human population growth has been reduced. Plague, diphtheria, cholera, measles, malaria—many diseases have been brought under control in developed countries. Smallpox, once a deadly scourge, has apparently been eliminated everywhere. (Some diseases, including malaria, schistosomiasis, and dysentery, are still prevalent in less developed countries and are again on the rise in some areas for lack of adequate funds for treatment and control measures.)

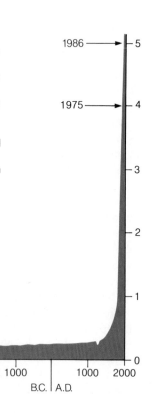

Figure 43.13 The curve of human population growth. The vertical axis of the graph represents world population in billions. The slight dip between 1347 and 1351 shows the time when 75 million people died in Europe as a result of the bubonic plague, or Black Death—a virulent disease spread by fleas that thrived on the rat populations in cities. At present growth rates, another billion will be added over the next twelve-year period, between 1987 and 1999.

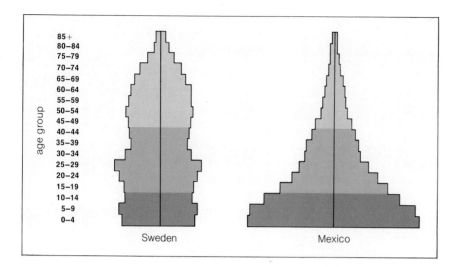

Figure 43.14 Age structure diagrams for two countries in 1977. Dark green indicates the pre-reproductive base. Dark blue indicates reproductive years; light blue, the post-reproductive years. The portion of the population to the left of the vertical axis in each diagram represents males; the portion to the right represents females. Mexico has a very rapid rate of increase. In 1980, Sweden showed zero population growth.

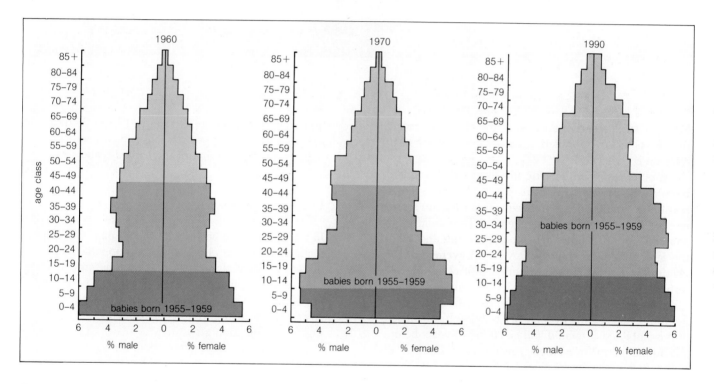

Figure 43.15 Age structure of the U.S. population in 1960, 1970, and 1990 (projected). The population bulge of babies between 1955 and 1959 will slowly move up.

Age Structure and Reproductive Rates

Two important factors influence the degree to which the population growth rate can be slowed. The first is called the **age structure** for a population, this being the relative numbers of its individuals of various ages. Figures 43.14 and 43.15 show examples of age structure diagrams for three human populations growing at different rates. The individuals are assigned to three categories—those before, during, and after reproductive age. (The ages 15 to 44 are used as the average range of childbearing years.)

These diagrams readily show the effects of age structure on population growth. The age structure pyramid for a rapidly growing population has a broad base, as Figure 43.14 suggests. It is filled not only with reproductive-age men and women but with a large number of children who will move into that category during the next fifteen years. As Figure 43.16 indicates, *more than a third of the world population now falls in the broad reproductive*

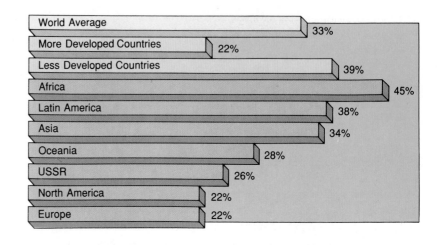

Figure 43.16 Percentage of individuals under age fifteen in various regions in 1987.

Figure 43.17 Projected growth in the number of individuals age sixty-five and older in the United States between 1985 and 2080.

base. Such figures give an idea of the challenge facing the efforts to slow population growth by reducing birth rates on a global scale.

Another factor influencing the short-term picture of population growth is the birth rate itself—the average number of infants born to each woman during her reproductive years. Today the average number of children in a family is 2 in the more developed countries and 4.2 in the less developed countries. A world average of 2.5 children per family is the estimated "replacement level" fertility rate that would bring us to zero population growth, given current mortality patterns. But even assuming that a world average of 2.5 children is achieved and maintained, it would still be 70 to 100 years before the human population stops growing. Why? Because an immense number of existing children are yet to move into the reproductive age category.

One way to slow things down is to encourage delayed reproduction—childbearing in the early thirties as opposed to the mid-teens or early twenties. Later childbearing slows population growth by lengthening the generation time and by lowering the average number of children in each family. In Ireland, women customarily marry later in life. In China, the government has raised the age at which marriage is allowed (age twenty for women, twenty-two for men). Moreover, the Chinese government provides financial, educational, housing, and retirement inducements for couples to have only one child, and free birth control measures. It also imposes penalties for having more than two children. The annual birth rate in China dropped from 32 per

thousand women in 1970 to only 17 per thousand in 1986; the percentage growth rate is now about 1.0 percent, comparable to that of the United States.

PERSPECTIVE

In this chapter, we began with the premise that all populations have the potential for exponential growth—growth at increasing rates to enormous numbers. For the human population, as for all others, the biological implications of exponential growth are staggering. Yet so are the social implications of achieving and maintaining zero population growth.

For instance, most members of an actively growing population fall in younger age brackets. Under conditions of constant growth, the age distribution means that there is a large work force. A large work force is capable of supporting older, less productive individuals with various programs, such as social security, low-cost housing, and health care. With zero population growth, far more people will fall in the older age brackets (see, for example, Figure 43.17). How, then, can goods and services be provided for less productive members if more productive ones are asked to carry a greater and greater share of the burden? These are not abstract questions. Put them to yourself. How much are you willing to bear for the sake of your parents, your grandparents? How much will your children be able to bear for you?

We have arrived at a major turning point, not only in our biological evolution but also in our social evolution. The decisions awaiting us are among the most difficult we will ever have to make, yet it is clear that they must be made, and soon.

SUMMARY

1. Ecology is the study of all organisms, including ourselves, in relation to other species and to the environment. Studies are made of populations of single species, interacting communities of many species, the movement of materials and energy through communities in ecosystems, and global patterns in the biosphere.

2. The number of individuals of a population in a given area determines the population density, and their arrangement in space defines their distribution (which may be clumped, more-or-less random, or fairly uniform).

3. The growth rate of a population (G) depends on the birth rate, the death rate, and the rates of immigration and emigration. If the birth rate per individual exceeds the death rate per individual by a constant amount, the population will grow exponentially (assuming immigration and emigration are zero).

4. If resources are limited, a growing population will level off at a density determined by the carrying capacity of the environment, tracing a pattern of logistic growth.

5. The rates of birth and death among individuals of different ages can be recorded in a life table. Age-specific survivorship and the age-specific pattern of births differ greatly among species, and the variations form one focus of the study of life histories.

6. Some factors that limit population growth in nature (such as starvation) have greater effect with increasing population density; they are density-dependent factors. Other factors (such as freezing weather for plants) tend to limit population growth to the same degree (constant death rate per individual regardless of population density); they are density-independent factors.

7. Currently, human population growth varies from zero in some developed countries to more than 4 percent per year in some developing countries; the world average is about 1.7 percent per year, down from the historic high of 2 percent, but still very high.

8. The present world population of 5 billion is expected to double, at least, before it levels off. The large proportion of humans currently in younger age groups will slow the approach to zero population growth.

9. Rapid growth of the human population during the past 200 years has been made possible primarily by reducing the death rate and expanding our carrying capacity through technology.

Review Questions

1. The following terms will be used repeatedly in this unit. Can you define them? *707*
   a. population, community, ecosystem, biosphere.
   b. producer, consumer, decomposer, detritivore.
   c. herbivore, carnivore, parasite.

2. Why do populations that are not restricted in some way grow exponentially? *709*

3. If the birth rate equals the death rate, what happens to the growth rate of a population? If the birth rate remains slightly higher than the death rate, what happens? *709*

4. Explain the difference between density-dependent and density-independent factors that influence population growth? *716*

5. What defines the carrying capacity for a particular environment? Can you describe what happens when a low-density population shows a logistic growth pattern? *710–712*

6. At present growth rates, how many years will elapse before another billion individuals are added to the human population? (*See Figure 43.13*)

7. How have human populations developed the means to expand steadily into new environments? How have humans increased the carrying capacity of their environments? How have they avoided some of the limiting factors on population growth?—or is the avoidance illusory? *718*

8. If a third of the world population is now below age fifteen, what effect will this age distribution have on the growth rate of the human population? What sorts of humane recommendations would you make that would encourage this age group to limit the number of children they plan to have? *719–720*

Readings

Begon, M., J. Harper, and C. Townsend. 1986. *Ecology: Individuals, Populations, and Communities.* Sunderland, Massachusetts: Sinauer Associates

Bouvier, L. 1984. "Planet Earth 1984–2034: A Demographic Vision." *Population Bulletin.* Vol. 39, no. 1, 1–39.

Krebs, C. 1985. *Ecology.* Third Edition. New York: Harper & Row. Chapters 10 through 12 are more advanced treatments of topics covered in this chapter.

Miller, J. T. 1988. *Living in the Environment.* Fifth edition. Belmont, California: Wadsworth. This author consistently pulls together information on human population growth into a coherent picture.

Pianka, E. 1983. *Population Ecology.* Third Edition. New York: Harper & Row.

Polgar, S. 1972. "Population History and Population Policies From an Anthropological Perspective." *Current Anthropology* 13(2)203–241. Analyzes often-ignored cultural barriers to programs for population control.

Scientific American. 1974. *The Human Population.* San Francisco: Freeman. Entire issue devoted to world population problems.

44

COMMUNITY INTERACTIONS

Flying through the dense rain forests of New Guinea is an extraordinary pigeon the natives call *gara*. It is about the size of a turkey and has cobalt blue feathers and plumes on its head (Figure 44.1). It flaps so slowly and noisily that its flight has been likened to the sound of an idling truck. Like eight smaller species of pigeon living in the same forest, the *gara* perches on tree branches to eat fruit. Why are there nine variously sized species of fruit-eating pigeons in this forest, and not just one species of some optimal size? Wouldn't competition for fruit eventually leave one species the winner?

In fact, each of these related fruit-eating species has its own role in the forest. Members of the larger species must perch on heavier branches when they feed, and

they eat larger fruit. Members of the smaller species eat fruit hanging from branches too thin to support the weight of a turkey-size pigeon, and they have bills too small to open the larger fruits. The species of trees in the New Guinea forest vary in terms of fruit size and the thickness of fruit-bearing branches, so the bird species end up foraging on different trees. Hence parts of the food supply that are used less by one species are used more by others in the same forest.

All of the pigeons benefit the different tree species by dispersing their seeds. The trees benefit from other animals as well—insects, bats, and other birds that serve as pollinators—and from associations with fungi (such partnerships being the mycorrhizae described on page

Figure 44.1 Tropical rain forest of New Guinea—habitat of many diverse species, including the turkey-size *gara* (also known as the Victoria crowned pigeon) and eight species of smaller pigeons.

Within this habitat each species has its own *niche*, an everchanging mosaic of abiotic and biotic conditions under which it lives and reproduces.

270). Meanwhile, many thousands of insect species feed on the leaves, flowers, bark, and roots of those same trees—or eat other insects, decaying vegetation or carcasses, or the blood or droppings of other animals.

This example reminds us that organisms interact as part of **communities**, which are all those populations of all species living together in a given area. In this chapter we turn to what is known about how communities are organized, how they function, and how they change over time.

BASIC CONCEPTS IN COMMUNITY ECOLOGY

Habitat and Niche

As we have seen, a **habitat** is the place where a population (or an individual) of a given species lives—its mailing address, so to speak. The word sometimes refers to a vegetation zone, such as a tropical rain forest, or to a more specific place, such as the height above ground in a forest canopy (some organisms live nowhere else), or a particular depth in the mud at the bottom of a pond (the only place where certain organisms ever will be found).

Each habitat has a characteristic range of physical and chemical conditions, such as the amount of light, typical temperatures, pH of the water, and so on. And each species is adapted to those conditions in terms of its morphology, physiology, and behavior. Moreover, it is adapted to the range of conditions imposed, directly or indirectly, by other organisms in the habitat. (For example, snakes have various adaptations for locating and capturing prey of certain sizes.)

The full range of abiotic and biotic conditions under which a species can live and reproduce is called its **niche**. Within this range, different conditions shift in large and small ways, creating an ever-changing mosaic to which that species responds. Thus a niche is a property of a particular species in a particular setting, including the role it plays in relations with other species.

With enough effort, we could probably identify all the niche characteristics for any species, but we would end up with an unmanageably long list. In attempting to understand how communities work, it is more informative to focus on important *differences* between the niches of species living in the same or similar habitats. For example, given their differences in body size and bill size, the nine species of New Guinea pigeons can eat fruit of different sizes, hence they disperse seeds of different sorts, the differences in dispersal affect where different trees grow—and the tree distribution influences how the entire community is organized.

Table 44.1 Types of Interactions Between Two Species

Type of Interaction	Direct Effect of Interaction*	
	Species 1	Species 2
Neutral	0	0
Commensalism	+	0
Mutualism	+	+
Interspecific competition	−	−
Predation	+	−
Parasitism	+	−

*0 indicates no direct effect on population growth,
+ indicates positive effect, − indicates negative effect.

Types of Species Interactions

Even simple communities consist of dozens or hundreds of species, which interact in a bewildering number of ways. To make progress in disentangling this complexity, let's initially focus on the types of interactions between any two species in a community (Table 44.1). Later, we will explore a few of the interactions that affect the structure and function of entire communities.

Most of the interactions between any two species in a community are **neutral**—that is, neither species *directly* affects the other (even though the two may be linked indirectly through a series of interactions with other species). For example, eagles certainly have a neutral effect on any given species of grass. Because rabbits eat grass and eagles eat rabbits, an eagle indirectly benefits the grass by helping to control the rabbit population—and the grass indirectly benefits the eagle by fattening up its prey. Still, by convention, the eagle-grass interaction is regarded as neutral.

Sometimes one species of a pair benefits significantly from the interaction, while the other is *neither helped nor harmed* to any great degree. This kind of interaction is called **commensalism**. For example, by providing many kinds of birds with roosting or nesting sites, a tall tree affords them protection from many kinds of predators. Usually the tree gets little or nothing in return from the birds that do nothing else but roost or nest (and do not, for example, eat insects that damage the tree).

The most important types of community interactions are known as mutualism, competition, predation, and parasitism. In **mutualism**, both members of a pair of species clearly and directly benefit from the interaction.

(Mutually beneficial relationships involving continuous, intimate contact between species are also called *symbiotic* relationships.) In **interspecific competition**, both species are harmed by the interaction. Finally, in **predation** and **parasitism**, one species (the predator or the parasite) benefits directly from the interaction while the other species (the prey or the host) is directly harmed by it. Let's now take a close look at each of these interactions in turn.

MUTUALISM

Mutually beneficial relationships between interacting species were once regarded as quaint biological curiosities. We now realize that the functioning of natural communities depends critically on several main categories of directly mutualistic interactions. One such interaction is the mycorrhiza, an intimate association between a fungus and the young roots of nearly all vascular plants. As we have seen, the extensive fungal filaments absorb mineral ions from the soil, some of which are used by the plant; and the fungus absorbs some sugars and nitrogen-containing compounds from the root (Figure 20.3). The dominant plants of forests and grasslands rely on this interaction.

Natural communities also depend on the mutualistic interactions between flowering plants and their pollinators as well as their agents of seed dispersal. Many of these remarkable interactions are described in Chapter Twenty-One.

These and other examples of mutually beneficial relationships between species are either facultative or obligatory. In **facultative mutualism**, both species benefit from the interaction but each can live without the other, if necessary. Thus some kinds of plants produce seeds by self-pollination if they fail to be visited by their usual pollinators, and the pollinators may do just as well with nectar from other sources.

In **obligate mutualism**, neither one of two interacting species can survive for long without the other. A classic example is the relationship between the yucca moth and the yucca plant (Figure 44.2). This moth obtains pollen only from the yucca plant; even its larval form dines only on yucca seeds. The yucca plant depends exclusively on this one moth pollinator. Hence the moth's private energy source, available throughout its lifetime, helps assure reproductive success. At the same time, the moth helps assure reproductive success for the plant: the pollen is carried exactly where it must go instead of being randomly spread about by a less picky pollinator that visits several plant species.

Figure 44.2 Mutualism in the high desert of Colorado. There are several species of yucca plants (**a**), but each is pollinated exclusively by only one kind of yucca moth species (**b**). The adult stage of the moth life cycle coincides with the blossoming of yucca flowers. Using mouthparts that have become modified for the task, the female moth gathers up the somewhat sticky pollen and rolls it into a ball. Then she flies to another flower and, after piercing the ovary wall, lays her eggs among the ovules. She crawls out the style and shoves the ball of pollen into the opening of the stigma. When the larvae emerge (**c**), they eat a small portion of the yucca seeds. Then they gnaw their way out of the ovary to continue the life cycle. The seeds remaining are enough to give rise to a new yucca generation.

So refined is this mutual dependency that the moth and larva can obtain food from no other plant, and the flower can be pollinated by no other agent.

INTERSPECIFIC COMPETITION

When different species in a community have some requirements or activity in common, their niches potentially overlap. Some resources (such as the oxygen used in aerobic respiration) are unlimited, so the niche overlaps present no problems. In other cases, however, supplies can be limited and the species may end up competing for them. The greater the potential for niche overlap, the greater will be the potential for competition.

Interspecific competition (that is, between species) usually is not as intense as *intraspecific* competition (within a population of a species). Why is this so? Each individual using a resource may deplete or limit it to some extent, so its activity affects the ability of others to use the resource. Because requirements are much the same for all members of a given population, intraspecific competition can be fierce when resources are in short supply. Although requirements for different species might be similar, they commonly differ more than the

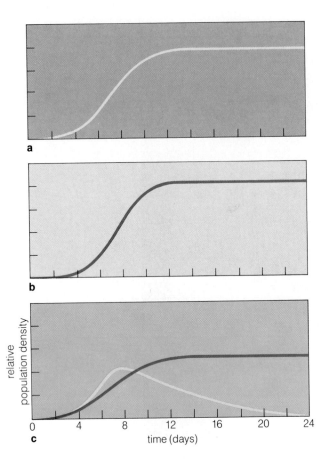

a

b

relative
population density

0   4   8   12   16   20   24

c

time (days)

Figure 44.3 Competition between two species of *Paramecium*. When grown separately, *P. caudatum* (**a**) and *P. aurelia* (**b**) established stable populations. (**c**) When grown together, *P. aurelia* (red curve) drove the other species (gold curve) toward extinction.

dant, exploitation competition cannot occur. (Thus, if you and a friend were sharing a milkshake, each drinking from your own straw, you wouldn't care how fast your friend drank if it were a ten-gallon milkshake, replaced daily.)

With interference activities, the picture becomes more complicated, for one species may limit another's access to resources, whatever the level of resource supply. (You could still pinch your friend's straw, even with a ten-gallon milkshake.)

Exploitation Competition

When two species compete for resources in the same habitat, the simplest of all possible outcomes are these: either they will continue to coexist or one will exclude the other from the habitat. Models developed in the 1920s indicated that if the resources required by both species are similar enough, the population growth rate of one species will depress the growth rate of the other and lead to its exclusion from the habitat. (The difference in the rate of increase would be the outcome of some structural, physiological, or behavioral difference between the two species, a difference that would give one the competitive edge.) However, the two species might continue to coexist in spite of some overlap in their use of scarce resources as long as there was enough of a difference in how they use those resources.

In the 1930s, G. F. Gause tested this idea by growing two species of *Paramecium* separately and then together. Because the two species require similar food, Gause reasoned that there would be strong competition between them. His experiments, summarized in Figure 44.3, suggested that the earlier models were correct. Gause reported evidence that complete competitors cannot coexist indefinitely, a concept now called **competitive exclusion**.

In other experiments, Gause used two species of *Paramecium* that did not overlap as much in their use of resources. (When grown together, one species tended to feed on bacteria suspended in the liquid medium in the culture tube, the other tended to feed on yeast cells at the bottom of the tube.) In this case, the growth rate was lowered for both populations—but not enough for either population to exclude the other; they continued to coexist.

The more similar two species are in their use of scarce resources in the same habitat, the less likely they are to coexist.

---

requirements for individuals of the same species, so competition is usually less intense.

There are two categories of competition, regardless of whether we are talking about individuals of the same or different species. In **exploitation competition**, all individuals have equal access to the resource in question but they differ in how fast or how efficiently they can exploit it. In **interference competition**, certain individuals limit or prevent others from using the resource and thereby control access to it.

Whether competition actually occurs between two or more species that use the same resource depends on two things—the resource supply and the activities of the organisms. In pure exploitation competition, one species detracts from the growth, reproduction, or survival of another only indirectly—by reducing the common supply of resources. Therefore, if shared resources are abun-

alpine
chipmunk

lodgepole
chipmunk

yellow pine
chipmunk

least
chipmunk

Figure 44.4 In the foreground, alpine tundra, with a stand of lodgepole pines in the distance. These are two of the vegetational zones of the eastern slopes of the Sierra Nevada in California. Each zone is the habitat of a different species of chipmunk (*Eutamias*).

Today, ecologists tend to use field experiments to evaluate the effects of exploitation competition. For example, when they suspect competition between two species, they remove one of the species from some test plots, remove the other species from other test plots—and leave still other plots untouched (to serve as controls).

In one field experiment, N. Hairston studied competition between two species of salamanders in the Great Smoky Mountains and in the Balsam Mountains. Although *Plethodon glutinosus* lives at lower elevations than its relative, *P. jordani*, the ranges overlap in certain areas. Hairston did a series of removal experiments in the overlap areas and monitored the results for five years. In the control plots, nothing had changed at the end of the test period—the two species continued to coexist. In plots from which *P. jordani* had been removed, *P. glutinosus* increased in abundance. In the plots from which *P. glutinosus* had been removed, there was no clear change in the abundance of *P. jordani* but the proportion of individuals in younger age classes had increased—a pattern suggesting a growing population.

In another field experiment in Britain, A. Tansley showed that exploitation competition influenced the distribution of two species of bedstraw (plants of the genus *Galium*). One species normally grows on acidic soils and the other, on basic soils. When one species was transplanted to the different type of soil, it could grow—as long as the other species was not there with it! When the two were grown together on either an acidic or a basic soil, the surviving species was the one that normally grew there.

In sum, the niches of two or more species living in the same habitat may overlap in several respects—but the mere existence of overlap is not a sure indicator of competition among them. Birds overlap in their need for oxygen, but they *do not* compete for it. Hairston's salamanders overlap in habitat use, and they *do* compete. Tansley's bedstraws also compete, but they show *no* overlap in habitat use. Finally, of course, most species that share *no* resource do *not* compete (grasses and eagles).

Interference Competition

Nature provides many spectacular examples of interference competition between species. Corals kill neighboring corals of other species by poisoning and growing over them. Hummingbirds chase hummingbirds of other species, even bees, away from defended clumps of flowers. Some limpets slowly but surely "plow" competing species out of defended territories along the seashore.

A strangler fig tree surrounds a victim tree and eventually kills it while growing its own massive canopy of leaves.

In most cases of interference competition, one species displaces another, which survives in or on whatever is left. On the eastern slope of the Sierra Nevada in California, for example, each of four chipmunk species occupies a different habitat (Figure 44.4). The alpine habitat is at the highest elevation, followed by the lodgepole pine habitat, then the yellow pine habitat, and finally, at the base of the mountains, the sagebrush habitat. The so-called "least" chipmunk of the sagebrush habitat can actually live at higher elevations. In the White Mountains, only a short distance to the east, it ranges all the way up into the alpine habitat. But when the least chipmunk was experimentally introduced into higher habitats in the Sierra Nevada, the aggressive behavior it provoked from the yellow pine chipmunk showed why the least chipmunk is restricted to the sagebrush. On the other hand, because of its dietary and water requirements, the yellow pine chipmunk could not long survive in sagebrush habitat.

PREDATION

On "Predator" Versus "Parasite"

Of all community interactions, predation is perhaps the most riveting of our attention (as well as the prey's). Dramatic examples abound, including the confrontation between the leopard and the baboon shown in Figure 44.5. A goat pulling up a thistle plant for breakfast, although less dramatic, is no less a case of predation— the prey is a living organism killed by the predator. But what about grasshoppers or horses, which graze on plants without killing them; or mosquitoes, which take blood from your arm and then fly off? What about ticks and fleas, which live on the host and take blood for long periods but get off to lay their eggs elsewhere? What about internal parasites such as tapeworms, external ones such as lice, or the parasitic plants called mistletoe, which spend their entire lives with the host?

The terminology available for these diverse interactions is extensive and sometimes contradictory. In this section and the next, we will cover all consumer/victim interactions with two broad definitions of *predators* and *parasites*. (To use narrower definitions, we would have to set up half a dozen other categories as well.) **Predators** get their food from other living organisms (their *prey*), but they do not live on or in the prey and may or may not kill it. **Parasites** also get their food from other living organisms (their *hosts*), but they live on or in the host for a good part of their life cycle and may or may not kill it.

Figure 44.5 (Right) Confrontation between predator and prey— the most arresting of all community interactions. As a last resort, the baboon has turned back to face the leopard with threat behavior. The effect has momentarily stopped the predator's advance; under other circumstances it might have meant the difference between capture and escape. Here, in the open, the baboon has run out of alternatives.

Dynamics of Predator-Prey Interactions

Mathematical models predict a great variety of patterns between interacting populations of predator and prey. The patterns range from stable coexistence at steady population levels for both species, to repeated oscillations or cycles of abundance, erratic oscillations, or the extinction of prey through predation. Which outcome is predicted depends on several features of the two populations, including:

---

1. The carrying capacity of the environment for the prey, in the absence of predation. (The carrying capacity, recall, is the equilibrium size of a population as defined by its sustainable supply of resources.)

2. The reproductive rate of the prey.

3. The **functional response** of predators (the capacity of *individual* predators to respond to increases in the density of prey by eating more prey).

4. The **numerical response** of the predator (its reproductive rate).

Stable coexistence for both populations is likely when predation keeps the prey population in check (it prevents the prey population from overshooting its carrying capacity). Predators can do this when they reproduce quickly relative to the prey and are capable of eating more when there are more prey organisms around.

Oscillations are likely when predators reproduce more slowly than their prey, when they eat only so many prey organisms at a time regardless of how many are around, and when the carrying capacity of the environment is high for their prey.

An idealized cyclic oscillation of predator and prey abundance appears in Figure 44.6. The cycling is caused by time lags in the predator's response to changes in the abundance of prey. (A "time lag" is an interval of time between two related events.) Cyclic oscillations of this sort have been documented in a number of predator-prey systems. For example, one long-term study points to a correlation between population density of the Canadian lynx and the snowshoe hare (Figure 44.7). There is

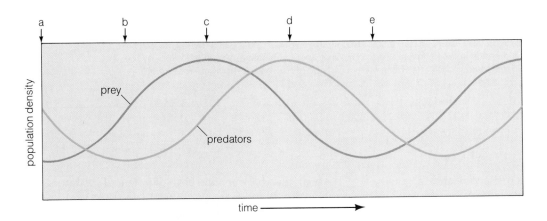

Figure 44.6 Idealized cycles of predator and prey abundance. The cycling is caused by time lags in the predator's response to changes in prey abundance. Starting at time *a*, the prey population is at a low density, so the predator population is hungry and declining. The prey population responds to the decline by increasing, but the predators continue to decline for a while until increased predator reproduction gets under way (time *b*).

Both populations increase until the increasing predation halts further growth of the prey population (time *c*), and it begins to decline. The prey suffer further impact from predation as the predators continue their increase, which slows as they feel the effects of starvation from lower prey density (time *d*). At time (*e*), we are back where we started and a new cycle begins. (The scale on this figure exaggerates predator density—in fact, predators are usually less common than their prey at all points of the cycle.)

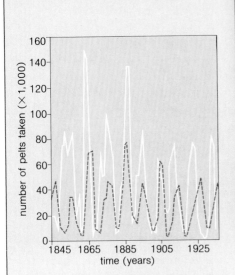

Figure 44.7 Relation between the lynx population (dashed line) and the snowshoe hare population (solid line) in Canada over a ninety-year period. These data are derived not from field observations but from counts of the pelts that trappers sold to Hudson's Bay Co. The curves were long regarded as a general example of the way predation can control the populations of both predator and prey.

This figure is a good test of how willing you are to accept conclusions without questioning their scientific basis. (Remember the discussion of scientific methods in Chapter Two?) For example, what other abiotic and biotic control factors could have been influencing the population levels? Were there also fluctuations in climate over this time span? Were some winters more rigorous, thereby imposing a higher death rate on one or both populations? Although this is called a simple predator-prey system, weren't the hares preying on the vegetation, which may have been overbrowsed in some years but not in others? What about owls, martens, and foxes—which also prey on hares? What if some years there were fewer trappers because of such variables as Indian uprisings? What about fluctuating demands for furs by the fashion industry? What if some years there was more lynx trapping than hare trapping, or vice versa?

indeed a correspondence between the rise and fall of these predator and prey populations, but the assumption that predation alone causes the oscillation is too simplistic.

Time lags certainly contribute to the lynx-hare cycles. When prey population density is at its peak, predators are not usually numerous enough to bring about an immediate decline of the huge hordes of hares. It takes time for the lynx population to increase by way of reproduction and by migration of lynx into areas where prey are abundant. More abundant food can promote survival, more rapid development, even fertility, but there

is a lag time between the birth of potential predators and the point at which they are mature enough to take prey.

In this example, increased predation can make a major contribution to the decline of the prey population, but it probably cannot trigger the decline. What does?

There is some evidence that recurring wildfires, floods, and insect outbreaks have indirect bearing on the hare-lynx cycle. Such environmental disturbances can destroy mature forest canopies that thereby create favorable conditions for different plant species that are adapted to moving in and becoming established in exposed settings. These species flourish until the species typical of the mature forest system become reestablished and gradually succeed them, in ways that will be described shortly.

Now, the so-called early successional species provide ideal forage for the hares; in fact, the peaks in hare population densities correspond to the occurrence of this type of vegetation cover. However, the early successional shrubs and trees include alders, poplar, and birch species—all of which produce toxins in especially high concentrations in new shoots. During winter, the hares prefer to feed on other plant parts and suffer no ill effects. When population density is high, though, the hares destroy the harmless parts of their winter forage, and they are forced to begin feeding on the toxic shoots. Experiments show that hares ingesting high concentrations of the plant toxins rapidly lose weight and become severely stressed.

Thus the cyclic fluctuations in population density may be a consequence of hares preying on plants and are simply amplified by lynx preying on hares. The built-in chemical defenses of certain plants prevent even more devastating increases in the numbers of hares (Figure 44.8).

PREY DEFENSES

When you stop to think about it, the fact that the plants described above defend themselves against being eaten by snowshoe hares is rather remarkable. How do such defenses arise? From what is known about evolutionary processes, it is clear that predators and prey exert continual selection pressure on each other. When some new, heritable means of defense spreads through the prey population, only the predators equipped to counter the defense of their prey survive to reproduce. Thus, when the prey evolves, the predator also evolves to some extent because the change affects selection pressures operating between the predator and prey. Again, this type of joint evolution of two (or more) species that are interacting in close ecological fashion is called **coevolution**. Let's take a look at some of the outcomes of predator-prey interactions.

Camouflage

One evolutionary outcome of predation is the capacity of many species of prey (and of predators) to "hide" in the open. **Camouflage** is an adaptation in form, patterning, color, or behavior that enables an organism to blend with its surroundings, the better to escape detection.

Figure 44.8 Snowshoe hare browsing on an early successional shrub. At the peak of the cyclic fluctuation in hare population density, such plants are overbrowsed. The new shoots they put out have high concentrations of toxins, which severely stress the hares. These conditions may last long enough to trigger a decline in the hare population.

Figure 44.9 Camouflage among the rocks. Find the plants (*Lithops*) that have the form, patterning, and color of stones.

Figure 44.10 The fine art of camouflage, as developed in predator-prey interactions. (**a**) One katydid species "hides" in the open from bird predators by looking like leaves—even down to blemishes and chewed-up parts. (**b**) Caterpillars of some moth species tend to look like bird droppings because of their coloration and the body positions that they assume. (**c**) By lurking motionless against its like-colored background, the yellow crab spider is essentially invisible to prey. (**d**) What bird??? With the approach of a potential predator, the least bittern stretches its reed-colored neck and thrusts its beak upward—and even sways gently, like the surrounding reeds in a soft wind.

For example, a desert plant (*Lithops*) resembles small rocks in shape and color (Figure 44.9). Only during the brief rainy season, when other vegetation and water are more plentiful for herbivores, do these "living rocks" put forth brightly colored flowers and draw the attention of pollinators. Figure 44.10 shows examples of camouflage among prey animals. (Camouflage, of course, is not the exclusive domain of prey. Predators that rely on stealth also blend well with the background. Think about polar bears against snow, tigers against tall-stalked and golden grasses, and pastel spiders against pastel flower petals as in Figure 44.10c.)

Moment-of-Truth Defenses

When cornered, some prey species defend themselves with display behavior that may startle or intimidate a predator (Figures 44.5 and 44.11a). Such behavior can create momentary confusion, and a moment may be all it takes for the prey to escape. When attacked, the bombardier beetle raises its abdomen and sprays a noxious chemical at its predator. (It is an effective adaptation in some cases, but grasshopper mice have a behavioral counteradaptation. These mice pick up the beetle, shove its tail end into the earth, and munch the head end.)

The bombardier beetle is one of many animals as well as plant species that release chemicals to repel or otherwise deter a potential attacker. Chemicals serve as warning odors, repellants, alarm substances, and outright poisons. Earwigs, grasshoppers, and skunks can produce awful odors. The foliage and seeds of some plants contain tannins, which taste bitter and which decrease the digestibility of the plant material; the tissues of other plants incorporate terpenes, which can be toxic. Nibbling on a buttercup (*Ranunculus*) leads to highly irritated mucous membranes in the mouth.

Some pheromones serve as alarm substances among prey fish that travel in schools that have neither spines nor other built-in defenses. (Unlike hormones, which have internal targets, a *pheromone* is an exocrine gland secretion with some target outside the animal body. Pheromones can trigger behavioral changes in other animals of the same species by acting as alarm substances, trail markers, sex attractants, and the like.) The pheromones, released when the skin is broken (as during an attack) trigger an escape response in neighboring fish.

Warning Coloration and Mimicry

One consequence of predation is the existence of many prey species that are bad-tasting, toxic, or able to sting or otherwise inflict pain on their attackers. Often, toxic prey species are decked out with conspicuous colors and bold patterns that serve warning to potential predators.

a

b

c

Figure 44.11 Moment-of-truth defensive behavior. (**a**) A cornered short-eared owl spreads its wings in a startling display that must have worked against some of its predators some of the time; it is part of the behavioral repertoire of the species. (**b**) As a last resort, some beetles spray noxious chemicals at their attackers, which works some of the time but not all of the time. (**c**) Grasshopper mice plunge the chemical-secreting tail end into the ground and feast on the head end.

Inexperienced predators might attack a black-and-white striped skunk, a bright orange monarch butterfly, or a yellow-banded wasp. As a result of the experience, they quickly learn to associate the colors and patterning with pain or digestive upsets.

Other prey species *not* equipped with such defenses can still sport warning colors or patterns that *resemble* those of distasteful, toxic, or dangerous species. The resemblance of an edible species to a relatively inedible one is a form of **mimicry** (Figure 44.12).

PARASITISM

True Parasites

Like predator-prey interactions, the interactions between true parasites and their hosts are remarkably diverse. Like predators, a parasite also takes sustenance from other living organisms—but it doesn't eat them outright. Parasites live on or in their host during some part of their life cycle, and the host may or may not die as a consequence of the association. (For example, the parasite can weaken the host so that the host succumbs to secondary infections. The parasitic blood flukes, described on page 651, can do this.)

Generally, parasites cause death only when they infect a host population with no coevolved defenses against them. After all, a host able to live longer may transmit parasites to more new individuals than a vulnerable, rapidly dying host ever could do. Thus, parasite populations also tend to coevolve with their hosts, producing an intermediate level of negative effects on the host itself.

Parasitoids

In contrast to parasites, some insect larvae are **parasitoids**, which kill their host by completely consuming its soft tissues before the host can grow into an adult. Parasitoids sound horrendous, but fortunately their target hosts are not humans but the larvae of other insect species.

Parasitoids are natural controls over other insect populations in a community. Many have been successfully raised and released as an alternative to the use of chemical pesticides. (Many of the worst pest outbreaks have *followed* applications of pesticides. The pest population soon recovers, but parasitoids formerly keeping the pest population under control were also killed and their populations usually fail to recover as quickly.)

What sort of defenses work against attacks of parasitoids? One example was described by Peter Price, who

a A dangerous species that serves as a model…

Figure 44.12 Mimicry. Many animals—especially those bite-sized morsels, the insects—avoid being eaten by having a bad taste, obnoxious secretion, or painful bite or sting. Among predators, knowledge of these traits is usually not inherited. Each young predator learns about them the hard way, by often unpleasant trials.

Among many prey species, dangerous or unpalatable individuals are easily recognized and remembered. If this were not the case, many individuals would be lost as inexperienced predators learned their lessons. Thus repugnant species tend to have distinctive, memorable appearances—bright colors (such as red, which predatory birds see so well), and bold markings (such as stripes, bands, and spots). These flamboyant species make no effort to conceal themselves. Sometimes they even deliberately flash colors with an uplift of the body or the wings. Their coloration and patterning are called "aposematic" (*apo-*, meaning "away," and *sematic*, meaning "signal").

In the natural world, each of the hundreds of dangerous or unpalatable species does not have a distinct warning signal, for too many signals would tax the learning capacity of predators. Instead there are whole groups of related species having almost identical aposematic appearances. Thus, many species benefit from a single taste trial. In turn, many less related and even totally unrelated species avoid predation by mimicking the appearance and behavior of the repugnant or dangerous model species.

Some mimics are as unpalatable as their models; they are called Müllerian mimics (after Fritz Müller, who named the phenomenon). Others may be quite edible yet are still avoided; they are called Batesian mimics (after Henry Bates, who discovered this class of mimics). Most mimicry series comprise both of these types.

There are other types of mimicry. In aggressive mimicry, parasites or predators bear resemblances to their hosts or prey. In speed mimicry, sluggish, easy-to-catch prey species resemble fast-running or fast-flying species that predators have given up trying to catch. (This might well be termed "frustration" mimicry.)

(a) Numerous and pugnacious yellowjackets (here, *Vespula arenaria*) are models for extensive mimicry series. (b) This

b

c

d

…and three of its mimics (above)

e

A speedy model…

f

…and a slow-moving mimic

g An unpalatable model…

h …and a palatable mimic

masarid wasp is probably a Müllerian mimic of *Vespula*. Other insects, such as beetles (**c**) and flies (**d**) may be Batesian mimics. (**e**) Flesh flies have gray and black bodies, red eyes, and red tail ends. Birds soon give up trying to catch these fast-flying insects. In the American tropics, many sluggish insects, such as the weevil *Zygops rufitorquis* (**f**), closely resemble flesh flies and thus reduce the likelihood of being eaten. This is an example of speed mimicry.

(**g**) Ithomiid butterflies of the New World tropics are frequent models for mimicry. They come in two basic types of coloration: orange with black stripes, and transparent with a white forewing band. Only a specialist can distinguish the many look-alike species (such as *Dismorphia* shown in **h**) and their mimics. In this case, the mimics may be either Müllerian or Batesian. (From Edward S. Ross, California Academy of Sciences)

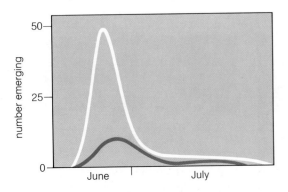

Figure 44.13 Effect of parasitoid wasp activity on sawfly emergence. The white line represents emergence when attack by parasitoids is prevented; the red line represents emergence after parasitoid attack. Notice that most attacks occurred on the would-be early emergers, which did not burrow as deeply into the forest litter.

studied the effects of a parasitoid wasp that lays its eggs on jackpine sawfly cocoons. The sawfly larvae emerge in trees and then drop to the forest floor, where they burrow into the leaf litter and spin their cocoons. The female wasp tends to lay eggs on sawfly cocoons nearest the surface of the litter. From these shallowly buried cocoons, the first adult members of the new generation of sawflies emerge; later in the season, adult sawflies emerge from more deeply buried cocoons (Figure 44.13).

The reproductive activities of the wasps represent a strong selection pressure on the sawfly population by favoring the late-emerging (and deep-burrowing) sawflies. The sawflies, in turn, exert selection pressure on the wasps, with the advantage going to those individual wasps having the capacity to find sawfly cocoons at ever-increasing depths in the litter. As Price points out, the host stays ahead in this coevolutionary contest. Each time the sawfly larvae burrow deeper, the female wasps have to spend more time searching for them. Consequently, fewer wasp eggs are laid and the wasp population is held in check.

Social Parasitism

Not all parasites feed directly on the tissues of a host. In **social parasitism**, one species depends on the social behavior of another to complete its life cycle. (This case stretches even our broad definition of parasitism, but it has basic parallels with the behavior of true parasites and parasitoids.)

For example, the North American brown-headed cowbird never builds a nest, incubates its eggs, or cares for its offspring. Instead, it removes one egg from the nest of a bird of another species and lays one of its own eggs as a "replacement." Some birds recognize the foreign egg and shove it out of the nest. Others, including the Kirtland warbler, do not. The warblers hatch the egg and raise the young cowbird. Since it is larger and more aggressive than the warbler nestlings, the young cowbird usually pushes them out of the nest or takes most of the food, leaving the warbler nestlings to starve to death.

Increasingly, ecologists have come to appreciate the role of parasites and parasitoids in regulating host populations in nature. The individual insect or marine invertebrate with no parasites at all is rare, and an individual bird or a mammal with neither external nor internal parasites is truly exceptional. Only humans who enjoy the benefits of modern health care are normally free of parasites.

COMMUNITY ORGANIZATION, DEVELOPMENT, AND DIVERSITY

Whatever organization and stability we see in communities is the result of antagonistic forces that have come into balance—sometimes an uneasy balance. The growth rate of a population depends on a balance between births and deaths. The coexistence of predators and prey depends on neither species winning the ecological and evolutionary footrace. Competitors have no sense of fair play whatsoever—straw pinching is common. Even mutualists are really antagonists—the flower gives as little nectar as necessary to attract a pollinator, while the pollinator seeks to get the most reward possible for the least effort. The alga and the fungus in a lichen have different gene pools under different and perhaps conflicting selection pressures. What kinds of community patterns and processes do these conflicting forces produce? Let's take a look.

Resource Partitioning

Through numerous studies, we know that groups of functionally similar species living together in a community do different things. If we assume there is indeed a limit to the degree of ecological similarity that will permit coexistence, then such a community pattern could arise in two ways. First, only species that are dissimilar enough from established ones will succeed in joining an existing community. Second, natural selection and other evolutionary processes may act on the competing populations already established in the community, increasing the differences among them.

However the pattern of coexistence arises, the outcome is **resource partitioning**, whereby groups of functionally similar species share the same resource in different ways, in different areas, or at different times.

N. Weland and F. Bazzaz studied resource partitioning among the plants growing in a field that had been abandoned a year after it was plowed. Three species of annual plants were common in the field. Like all land plants, each required sunlight, water, and mineral ions from the soil. The investigators proposed that each species met those requirements without interfering with the others by exploiting different areas of the habitat space (Figure 44.14).

Bristly foxtail grasses became established in soil where the amount of water available varied greatly from day to day. With their shallow, fibrous root systems, the foxtails could rapidly absorb water after the rains and could recover rapidly from droughts. Mallow plants occupied different areas of the soil. With their taproot systems, those plants could exploit soil depths that were moist early in the growing season but less moist later on. The third species, smartweed, had a taproot system that branched in the topsoil and in the continuously moist soil below the rooting zone of the other two species.

Is interspecific competition usually responsible for such differences among functionally similar species, or are the differences more often due to historical accident? There is no conclusive answer at present.

Effects of Predation on Competition

One day Charles Darwin instructed his gardener to stop cutting a small patch of his lawn. English lawns (unlike most American ones) are composed of many low-growing plant species. If cutting was discontinued, Darwin speculated, then competition among the species might intensify. He was correct. Of the twenty species originally growing in the plot, nine were eventually eliminated by competition with the remaining species once the cutting had stopped.

If we liken the lawnmower to a predator, then predation was minimizing the intensity of competition among the "prey" (plant) species. Many experiments show that predation has this effect in natural communities as well.

For example, for several years, Robert Paine kept predatory sea stars out of experimental plots in a rocky intertidal zone. (He left the sea stars with their fifteen species of invertebrate prey in control plots.) In the absence of the predator, the number of prey species declined to eight—the rest were crowded out by mussels, the main prey of sea stars and yet the strongest competitors in their absence.

Species Introductions

In their field experiments, ecologists are extremely careful about minimizing the disturbance to natural communities (hence Paine's limit on a few experimental plots). In contrast to these planned experiments, great numbers of species have been shuffled among different geographic regions, sometimes intentionally and sometimes inadvertently. Most introductions of species from one continent to another probably fail to become established—although few records are kept of such non-events. What we know the most about are two classes of introductions—agriculturally useful species and pest species.

As an example of a pest, in the 1880s the water hyacinth from South America was put on display at the New Orleans Cotton Exposition. Flower fanciers from Florida and Louisiana carried home clippings of the blue-

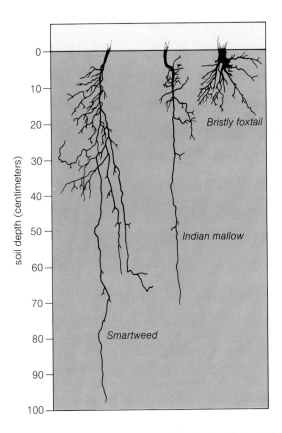

Figure 44.14 Partitioning of a resource (soil, with its nutrients and water) by three annual plant species that became established in a field plowed one year before.

Table 44.2   Effects of Introducing a Few Species into the United States

Species Introduced	Origin	Mode of Introduction	Outcome
Water hyacinth	South America	Intentionally introduced (1884)	Clogged waterways; shading out of other vegetation
Dutch elm disease The fungus *Ophiostoma ulmi* (the disease agent)	Europe	Accidentally imported on infected elm timber used for veneers (1930)	Destruction of millions of elms; great disruption of forest ecology
Bark beetle (the disease carrier)		Accidentally imported on unbarked elm timber (1909)	
Chestnut blight fungus	Asia	Accidentally imported on nursery plants (1900)	Destruction of nearly all eastern American chestnuts; disruption of forest ecology
Argentine fire ant	Argentina	In coffee shipments from Brazil? (1891)	Crop damage; destruction of native ant communities
Camphor scale insect	Japan	Accidentally imported on nursery stock (1920s)	Damage to nearly 200 species of plants in Louisiana, Texas, and Alabama
Japanese beetle	Japan	Accidentally imported on irises or azaleas (1911)	Defoliation of more than 250 species of trees and other plants, including commercially important species such as citrus
Carp	Germany	Intentionally released (1887)	Displacement of native fish; uprooting of water plants with loss of waterfowl populations
Sea lamprey	North Atlantic Ocean	Through Welland Canal (1829)	Destruction of lake trout, lake whitefish, and suckers in Great Lakes
European starling	Europe	Released intentionally in New York City (1890)	Competition with native songbirds; crop damage; transmission of swine diseases; airport runway interference; noisy and messy in large flocks
House sparrow	England	Released intentionally (1853)	Crop damage; displacement of native songbirds; transmission of some diseases
European wild boar	Russia	Intentionally imported (1912); escaped captivity	Destruction of habitat by rooting; crop damage
Nutria (large rodent)	Argentina	Intentionally imported (1940); escaped captivity	Alteration of marsh ecology; damage to earth dams and levees; crop destruction

From David W. Ehrenfeld, *Biological Conservation*, 1970, Holt, Rinehart and Winston and *Conserving Life on Earth*, 1972, Oxford University Press.

flowered plants and set them out for ornamental display in ponds and streams. Unchecked by their natural predators and nourished by the nutrient-rich waters of the region, the fast-growing hyacinths spread rapidly, displacing many native species along the way. Eventually they choked off entire ponds and streams, then rivers and canals. They are still thriving—now as far west as San Francisco—and they are still bringing river traffic to a halt in many areas.

Species successfully introduced into established communities do not always lead to such wholesale disasters, but few (if any) are without ecological consequences. Honeybees, mosquitofish, and ring-necked pheasant are among the species that have been absorbed into existing communities in the United States. But the honeybees have displaced native bees in many natural areas. Table 44.2 lists other introduced species and their effects on established communities.

Succession

The potential repercussions of newly introduced species in a community raise an interesting question. How do communities come to exist in the first place? Obviously they do not arise full-blown in the environment. Whether we are talking about an environment initially devoid of life (such as a newly forming volcanic island) or a disturbed patch of a previously inhabited environment (such as an abandoned pasture), *a rather predictable and repeatable pattern of change takes place.*

The first species become established, flourish, and then decline while others are on the rise. Unless some disturbance halts or sets back the process, the changes continue until a more or less constant array of species is reached, called a **climax community**. This relatively self-sustaining assemblage of species tends to be perpetuated until disturbed by some force such as fire, epidemic, or severe weather.

The changes in species composition that lead to a climax community are collectively called **succession**. When these orderly changes occur in an area previously devoid of life, they constitute *primary succession*. When they proceed in a disturbed area that was previously inhabited, the sequential changes are known as *secondary succession*.

Primary Succession. Primary succession begins with colonization of an uninhabited site by pioneer species, which aid in the development of soil. Pioneer species are able to grow in exposed areas with intense sunlight, wide variations in air temperature, and nutrient-poor soil. Typically the pioneers are small and low-growing. They have short life cycles, and each year they produce abundant small seeds that are quickly dispersed from the parent plants.

At higher latitudes, early successional species must start from scratch with each new season, sprouting from seeds or sending out new shoots from the base of withered stalks. This puts them at a disadvantage when the later successional species begin to appear. Many of the later species are perennials; they live for more than one growing season. Thus they have a head start once they are established and can begin seasonal growth before the pioneer plants get started. They store more biomass and produce relatively few seeds—but the seeds are well endowed with the nutrients necessary for early, secondary growth. Eventually the later species crowd out the pioneers, whose seeds travel as fugitives on the wind or water—destined, perhaps, for a new but equally temporary habitat.

Figure 44.15 shows how primary succession proceeds in a community in the Glacier Bay region of Alaska.

Secondary Succession. Secondary succession is characteristic of abandoned fields and parts of an established forest where falling trees or other disturbances have opened the canopy of leaves, letting sunlight reach the forest floor. In many respects the pattern of change is similar to all but the first stages of primary succession. In secondary succession, however, many plants grow from seeds or even seedlings that are already present when the process begins.

Different mechanisms account for the rise and fall of species during succession. In some cases, early species apparently set the stage for their own replacement by altering the environment in ways that favor later species. Soil-forming pioneer species in primary successions probably fall in this category. Alternatively, both early and late species may do just as well under existing conditions but the late species simply grow more slowly, eventually excluding the early species through competition. This process seems to be common in secondary successions in communities on land. Finally, early species might actually inhibit the growth of later species, which prevail only if some disturbance removes the early species. Algal succession in the intertidal zone seems to proceed in this way.

Cyclic Replacement. Even with its array of well-adapted species, a climax community does not perpetuate itself indefinitely. Natural communities that seem most stable are actually a mosaic of successional patches as a result of major and minor disturbances. Winds, fires, insect infestations, and overgrazing all modify and shape the direction of succession by encouraging the proliferation of some species and eliminating others from the community.

One response to such disturbances is **cyclic replacement**. For example, the Sierra Nevada of California supports isolated groves of sequoia trees. Some of the giant trees of this climax community are more than 4,000 years old, which certainly implies long-term stability in the region. Although brush fires sweep through the community every so often, the disturbance they create actually helps maintain the climax configuration. Sequoia seeds germinate only in the absence of smaller, shade-tolerant plant species. If there is extensive litter on the forest floor, no sequoia seeds will germinate. Modest fires eliminate the species of trees and shrubs that compete with the sequoias, but do not damage the sequoias themselves. Mature sequoias have extraordinarily thick bark, which burns poorly and insulates them against heat damage.

Many sequoia groves are protected in national and state parks. Protection traditionally has meant minimizing the incidence of fires—not just accidental fires from

a

b

c

d

Figure 44.15 Primary succession in the Glacier Bay region of Alaska (**a**), where changes in newly deglaciated regions have been carefully documented. A comparison of maps from 1794 onward shows that ice has been retreating at annual rates ranging from 3 meters (at the glacier's sides) to a phenomenal 600 meters at its tip over bays. (**b**) When a glacier retreats, the constant flow of meltwater tends to leach the newly exposed soil of minerals, including nitrogen. Less than ten years ago, the soil here was still buried below ice. (**c, d**) The first invaders of these nutrient-poor sites are the feathery seeds of mountain avens (*Dryas*), drifting over on the winds. Mountain avens is a pioneer species that benefits from the nitrogen-fixing activities of symbiotic microbes. It grows and spreads rapidly over glacial till.

(**e**) Within twenty years, young alders take hold. These deciduous shrubs also are symbiotic with nitrogen-fixing microbes. Young cottonwood and willows also emerge (**f**) . Eventually the alders form dense thickets (**g**). As the thickets mature, cottonwood and hemlock trees grow rapidly, as do a few evergreen spruce trees. (**h**) By eighty years, the spruce crowd out the mature alders. (**i**) In areas deglaciated for more than a century, dense forests of Sitka spruce and western hemlock dominate. By this time, nitrogen reserves are depleted, and much of the biomass is tied up in peat: excessively moist, compressed organic matter that resists decomposition and that forms a thick mat on the forest floor.

campsites and discarded cigarettes, but also natural fires touched off by lightning. When small, periodic fires are prevented, litter builds up, and other species take hold that are susceptible to fire. Even though these species do not displace the mature sequoias, they prevent the sequoia seeds from germinating and from growing into the replacement trees needed to maintain the climax community.

Moreover, the litter and the undergrowth constitute so much potential fuel that when fires do occur, they are hotter than they otherwise would be—hot enough to damage the giants. Thus fire prevention efforts aimed at preserving the climax community actually have the opposite effect. In recent years, "prescribed burns," set intentionally under carefully controlled conditions, eliminate the underbrush in sequoia groves.

Community stability often requires episodes of instability, which permits cyclic replacements of equilibrium species and thereby maintains the climax community over time.

e

f

g

h

i

Species Diversity: Island Patterns

One day in 1965, about forty kilometers southwest of Iceland, a volcanic eruption began beneath the sea. When it eventually stopped, a new island had been formed and primary succession was under way. Within six months of the island's emergence, the first signs of life had appeared—bacteria, fungi, seeds of several kinds of beach plants, a fly species, and some seabirds. Two years after the eruption, the first vascular plant appeared and, two years later, the first moss. As primary succession and soil formation gradually progressed, the number of plant species continued to increase and is still doing so (Figure 44.16). All these species are colonists from Iceland—none evolved on the island, which is now named Surtsey.

The number of plant and animal species on Surtsey will not increase indefinitely. What will stop it? Detailed studies of community patterns on islands around the world provide us with two ideas. First, the farther an island is from a source of potential colonists, the fewer species it supports (Figure 44.17). This generalization is called the **distance effect**. Islands distant from source areas simply receive fewer colonizing species, at a slower rate, and the ones that do arrive are well adapted for long-distance dispersal. Second, larger islands tend to support more species than smaller islands at equivalent distances from source areas (Figure 44.18). This generalization is called the **area effect**.

The cause of the area effect is not clear-cut. In part, larger islands are physically more complex and often rise higher above sea level, so more kinds of habitat are available. Since many species live only in certain habitats, this effect increases species diversity. To some degree, larger islands probably "intercept" more colonists, being a larger target. However, extinctions may amplify the effect of area on the number of island species. Established colonists on small islands have small populations, which are more vulnerable to extinction from storms, volcanic eruptions, and biotic causes (epidemics, predation, parasitism) than larger populations.

Mainland and Marine Patterns of Species Diversity

In some ways, patterns of species diversity on islands are reflected in communities of the continents and the oceans. The "area effect" also holds for these communities—the larger an area sampled, the more species in the sample. As with islands, the diversity of habitats here is greater in large areas than in small areas. As we enlarge the sampling area, however, we also include more and more rare species with similar habitat requirements yet with only local distributions.

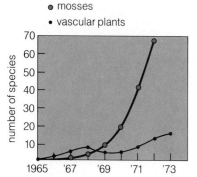

Figure 44.16 The number of species of mosses and vascular plants recorded on the new island of Surtsey from 1965 to 1973.

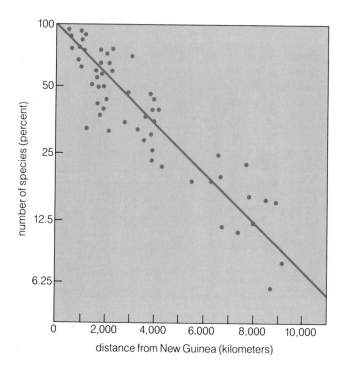

number of species (percent)

distance from New Guinea (kilometers)

Figure 44.17 The "distance effect." The number of species on islands of a given size declines with increasing distance from the source of colonizing species. Here, each point represents the number of species of land birds that live in lowland areas of islands in the South Pacific. The very large island of New Guinea is the source of colonists for these islands, each of which is at least 500 kilometers from this source. To correct for differences in island size (area), the actual number of bird species on each of these distant islands has been expressed as a percentage of the number of bird species on an island of equivalent size close to New Guinea.

The most striking patterns of species diversity correlate with latitude. For most groups, the number of coexisting species rises sharply from the arctic latitudes to temperate zones to the tropics (Figure 44.19). The most obvious difference between tropical regions and higher latitudes is in the annual pattern of sunlight (the tropics receive more direct sunlight spread more evenly throughout the year). Also, where rainfall is high on land in the tropics, communities are extraordinarily productive. In tropical seas, coral reef communities are remarkably productive and diverse.

There are several reasons for the increased diversity at tropical latitudes. First, availability of some resources is more constant than at higher latitudes. For example, although different tree species show different seasonal patterns, the trees as a whole provide new leaves, flowers, and fruit all year long in wet tropical forests and thereby support many resident species of specialized herbivores, nectar foragers, and fruit consumers. In temperate and arctic regions, such specializations would be suicide.

Second, species diversity is self-reinforcing. As soon as more species of plants live together, more species of herbivores emerge, partly because no one herbivore can overcome the chemical defenses of all the plants. (Many wet tropical forests support hundreds of tree species per acre, at least ten times more than temperate forests.) In

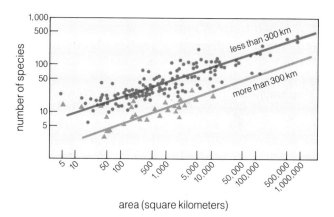

number of species

area (square kilometers)

Figure 44.18 The "area effect." Among islands similarly distant from the source of colonizing species, the larger islands support larger numbers of species. The points in this figure represent the number of bird species on tropical and subtropical islands. The solid circles and the upper line are for islands less than 300 kilometers from their source of colonists, while the triangles and the lower line represent islands more than 300 kilometers from source areas. Thus this figure shows not only the "area effect," but the "distance effect" as well.

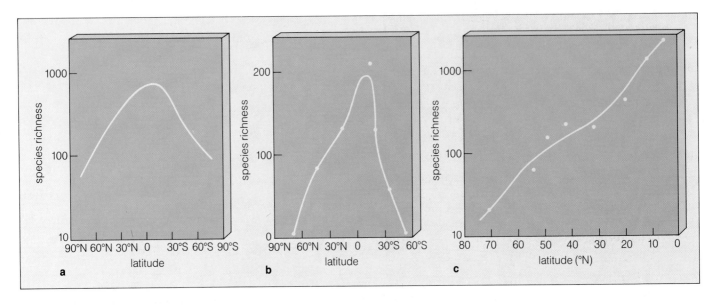

Figure 44.19 Patterns of species diversity corresponding to latitude. (**a**) Marine bivalve mollusks, (**b**) ants, and (**c**) breeding birds of North and Central America.

turn, more kinds of predators and parasites evolve in response to the diversity of prey and hosts. Similarly, the diversity of coral species on a tropical reef supports a pyramid of diverse groups representing many phyla. The reasons for the astonishing diversity of trees in tropical forests and of corals on tropical reefs are still unclear. One hypothesis is that storms and other disturbances, or possibly predation or parasitism, alter community structure in an intermediate way. In other words, they might prevent any one species from becoming dominant, leaving space and resources for many very similar species while causing few extinctions.

Finally, lowland tropical areas have never been glaciated. It may be that the great age of tropical forests itself has contributed to the accumulation of species. In this view, the overall rate of speciation in the tropics has long exceeded the rate of extinction (from natural causes), whereas periodic mass extinctions at higher latitudes have helped to keep diversity low there (Chapter Thirty-Seven).

The next decade will decide the fate of millions of species of organisms in tropical forests, of which ten square miles have been cleared during the past hour. With determination, we may still keep the rate of extinction in the tropical forests low enough to learn the answers to the intriguing questions raised by the patterns of species diversity.

SUMMARY

1. Each component species of a biological community responds to physical factors in its environment as well as to other species in the community. The habitat of the species is defined by the characteristics of the place where it lives. The niche of the species is defined by the conditions it requires for life and by its relations with other species in the community.

2. Many pairs of species in a community derive mutual benefits through their direct interactions, although each partner in such a mutualism has evolved to give as little as necessary to get as much as possible from its partner. Important classes of mutualism include plants and their pollinators, seed dispersal by animals, mycorrhizae, and normal microbial inhabitants of the animal gut.

3. When a resource is in short supply, two species that both require it are likely to compete, either by using up the resource as rapidly or as efficiently as possible (exploitation), or by preventing the other from using it efficiently (interference).

4. Competing species are more likely to coexist in the same community if their niches are rather different than if they are quite similar.

5. Although the population of a predator and the population of its prey may be stable under some conditions,

time lags in the response of the predator population to changes in the prey population may lead to cyclic fluctuations in the abundance of both.

6. The evolutionary race between predators and their prey has produced a rich array of special adaptations for both capture and escape.

7. Like predators, parasites may regulate the population of their hosts. Coevolution tends to favor resistant hosts and only moderately harmful parasites.

8. Related species in a community generally differ in their use of resources. Interspecific competition plays a role in producing this "partitioning" of resources, but the prevalence of competition in producing such patterns is still unclear.

9. Predation can reduce population densities among prey species, reducing competition among them and promoting their coexistence.

10. The introduction of non-native species to communities sometimes produces unexpected and undesirable results.

11. When an unoccupied habitat is colonized, species tend to replace one another in an orderly sequence (succession), eventually leading to a fairly stable "climax" community.

12. The number of species that occupy an area is affected by the size of the area, the rate of colonization by new species, and the rate of extinction of existing species.

Review Questions

1. Define habitat. Why do you suppose it might be difficult to define "the human habitat"? *723*

2. What is the difference between the habitat and the niche of a species? *723*

3. Define mutualism and give a few examples of its occurrence in nature. *724–725*

4. Define interspecific competition. Explain how this form of behavior is incorporated into the concept of competitive exclusion. *725–726*

5. Why do you suppose it is more difficult to observe competitive exclusion in the natural environment, compared to observations of simple two-species interactions (such as bacteria) in a laboratory? (*Refer to the Commentary on page 21 as well as to pages 726–727.*)

6. How might two species that compete for the same resource coexist? Can you think of some possible examples besides the ones used in the chapter? *727–728*

7. What effect does predation have on interspecific competition among prey species? *728–731*

8. Can you explain why predation alone may not account for the long-term oscillations in population growth of the Canadian lynx and snowshoe hare? *728–731*

9. Define coevolution. How might two species coevolve to the extent that they enter a mutualistic relationship? *731–736*

10. What is the difference between camouflage and mimicry? Can you give some examples of mimicry among insects? Camouflage among insects? *731–735*

11. What is the difference between a parasite and a parasitoid? What sort of coevolutionary defenses do host species have against parasitoids? *734–736*

12. Define primary and secondary succession. What is a climax community? *739*

Readings

Begon, M., J. Harper, and C. Townsend. 1986. *Ecology: Individuals, Populations, and Communities*. Sunderland, Massachusetts: Sinauer Associates. Excellent, current introduction to topics covered in this chapter.

Price, P. 1975. *Insect Ecology*. New York: Wiley. Advanced reading, but excellent examples of ecological interactions.

Smith, R. 1980. *Ecology and Field Biology*. Third Edition. New York: Harper & Row. Particularly good descriptions of succession.

West, D., H. Shugart, and B. Botkin, eds. 1981. *Forest Succession: Concepts and Application*. New York: Springer-Verlag. An excellent overview of the modern approach to succession.

45

ECOSYSTEMS

The earth's surface is remarkably diverse. In climate, soils, vegetation, and animal life, its deserts differ from hardwood forests, which differ from prairies, which differ from tropical rain forests and arctic tundra. Oceanic provinces, inland seas, lakes, ponds, and rivers differ in physical and chemical properties as well as in their arrays of organisms. *Yet despite the differences, each region functions as a system in much the same way as the others.*

With few exceptions, each system runs on energy from the sun. Plants and other **photosynthetic autotrophs**, which are the most common "self-feeders," capture sunlight energy and convert it to forms they can use to synthesize organic compounds from simple inorganic substances (Chapter Eight). By doing so, *autotrophs secure the energy used by the entire system.*

Energy stored in their self-assembled organic compounds is transferred through a number of heterotrophs before being lost as heat to the surroundings. The **heterotrophs**, recall, include consumers (herbivores, carnivores, scavengers, and parasites), decomposers (bacteria and fungi that degrade organic remains or products), and detritivores (invertebrates that feed on partially decomposed particles of organic matter).

sedges, mosses, grasses. . . .

snowy owl

shooting star . . .

. . . *Dryas*, and *Silene* (pink)

Figure 45.1 Examples of the primary producers and consumers of the arctic tundra—sedges, mosses, and brightly flowered plants, along with the ptarmigan and lemming (eaters of plant parts) and the snowy owl (eater of lemmings). As in all ecosystems, these diverse organisms interact with one another and with their physical environment through (1) a flow of energy and (2) a cycling of materials, in ways that will be described in this chapter.

In addition, *autotrophs secure nutrients for the entire system*. During growth, they take up water and carbon dioxide (as sources of oxygen, carbon, and hydrogen) along with dissolved minerals, including ionized forms of nitrogen and phosphorus. Such nutrients are essential for the synthesis of carbohydrates, lipids, proteins, and nucleic acids—which autotrophs as well as heterotrophs use in biosynthesis and other cell activities. When decomposers and detritivores get their turn at this organic matter, they break it down completely to inorganic bits. If those bits are not washed away or otherwise removed from the system, they can be used again as nutrients by the autotrophs.

What we have just described in broad outline is an **ecosystem**, which includes all those organisms in a defined region as well as their physical environment. In this type of system, each organism interacts with others and with the environment through (1) a flow of energy and (2) a cycling of materials—both of which have consequences for community structure *and* the environment.

It is important to understand that ecosystems are *open* systems, and so are not self-sustaining. They require an *energy input* (as from the sun) and often *nutrient inputs* (as from minerals carried by erosion into a lake). Because energy cannot be recycled, all ecosystems have an *energy output*. Most of the energy originally fixed by autotrophs eventually is lost to the environment as metabolically generated heat. Although nutrients are typically recycled, some loss occurs from the system (as through soil leaching), so there also is a *nutrient output*.

In this chapter, we consider the inputs, internal transfers, and outputs of different ecosystems. This "ecosystem approach" gives insight into the collective effects of all the organisms in a region and so allows simplification of the immense complexity in nature. It is through this approach that we have come to see how the processes of energy flow and nutrient cycling underlie certain broad similarities among ecosystems.

STRUCTURE OF ECOSYSTEMS

Trophic Levels

For all ecosystems, the feeding relationships among its component members are structured in much the same way. Each member fits into a hierarchy of energy transfers called **trophic levels** (from *troph*, meaning nourishment). "Who eats whom?", we can ask. When organism B eats organism A, energy is transferred from A to B. All organisms that are the same number of transfer steps away from the energy input into an ecosystem are said to be at the same trophic level.

In nearly all ecosystems, photosynthetic autotrophs are the organisms closest to the primary energy source (the sun), and so they make up the first trophic level. An example would be the photosynthetic microbes and aquatic plants of an estuary. Snails and other herbivorous animals that feed directly on plants are at another trophic level; certain birds and other primary carnivores that eat the herbivores are at still another level.

Many kinds of organisms obtain energy from more than one source. Decomposers are like this; so are humans. They really are "trophic groups" that cannot be assigned to a particular trophic level. Even so, the categories in Table 45.1 are useful as a starting point in describing the feeding relationships in an ecosystem.

Food Webs

The general sequence of who eats whom is sometimes called a **food chain**. However, the term implies a simple, isolated relationship that seldom occurs in ecosystems. More typically, the same food resource is part of more than one chain, especially when the resource is at one of the low trophic levels. Such interconnected feeding

ptarmigan

lemming

Table 45.1 Examples of Trophic Levels

Members of Each Level	Energy Source	Representative Organisms
Primary producers:		
photosynthetic autotrophs	sunlight energy	photosynthetic bacteria, plants
chemosynthetic autotrophs	oxidation of inorganic substances	nitrifying bacteria
Primary consumers:		
herbivores	primary producers	grasshoppers, deer, snails
Secondary consumers:		
primary carnivores	herbivores	spiders, small squids
Tertiary consumers:		
secondary carnivores	primary carnivores	Emperor penguins

relationships take the form of **food webs**, of the sort shown in Figure 45.2.

To gain insight into the difference between a food web and a food chain, think about a fisherman netting some fish that were feeding on algae near the ocean's surface. Come lunchtime, he cooks some of his catch. Should he later slip on the deck and fall into the water, where other carnivores lurk, the "chain" might be portrayed as:

algae → fish → fisherman → shark

This chain would be an oversimplification of feeding relationships, however, for it excludes a number of alternatives. Most likely, crustaceans also were grazing on the algae. Small squids and assorted medium-sized fishes might have been feeding on the crustaceans; some larger fishes were probably feeding on smaller ones. Sharks may have been moving in to feed on the large and medium-sized fishes. The fisherman might have cooked his fish in wine and herbs, shifting back and forth between herbivore and carnivore. He would have been even more omnivorous in consuming the alcoholic product of decomposers (yeasts whose fermentation activities yield wine from crushed grapes).

A food web is a network of crossing, interlinked food chains, encompassing primary producers and an array of consumers and decomposers.

ENERGY FLOW THROUGH ECOSYSTEMS

Primary Productivity

The food webs of different ecosystems vary in the number, kind, and life span of their participants, so there will always be exceptions to any generalization about the way energy flows through them. Even so, it is possible to get an idea of how energy flow is studied by thinking about just one type—terrestrial ecosystems, for which multicelled plants are the primary producers.

In terrestrial ecosystems, plants capture solar energy and convert it to the chemical energy of organic compounds. The *rate* at which those primary producers capture and store a given amount of energy in a given interval is called the ecosystem's **primary productivity**.

How much energy actually gets stored in the plants? That depends on the balance between photosynthesis and metabolic activities that use up energy (aerobic respiration, most often). It also depends on how many plants escape the attention of consumers and decomposers in the specified interval.

1. The *gross primary productivity* is the total rate of photosynthesis for the ecosystem during a specified interval.

2. The *net primary productivity* is the rate of energy storage in plant tissues in excess of the rate of respiration by the plants themselves.

3. The *net community productivity* is the rate of energy storage in excess of heterotrophic consumption.

Besides the grazings of herbivores, several other factors affect the net primary productivity of terrestrial ecosystems. How much sunlight is available? Are there enough of all kinds of essential nutrients? What will be the temperatures and rainfall during the growing season? How long is the growing season? Will fire sweep through the community? Such factors influence more than the amount of net primary production. They also influence its seasonal patterns, its spatial distribution, and the size and shape of the primary producers themselves.

For example, the proportion of biomass that a plant allocates to roots versus shoots depends partly on the availability of water and nutrients. The harsher the environment in those respects, the greater the ratio of roots to shoots and the lower the productivity. The more favorable the environment, the more growth is put into shoots.

The age of an ecosystem also affects the rate of energy storage. When a forest is young and trees are small, the net community productivity is high. As the trees get older and larger, more of the energy captured by pho-

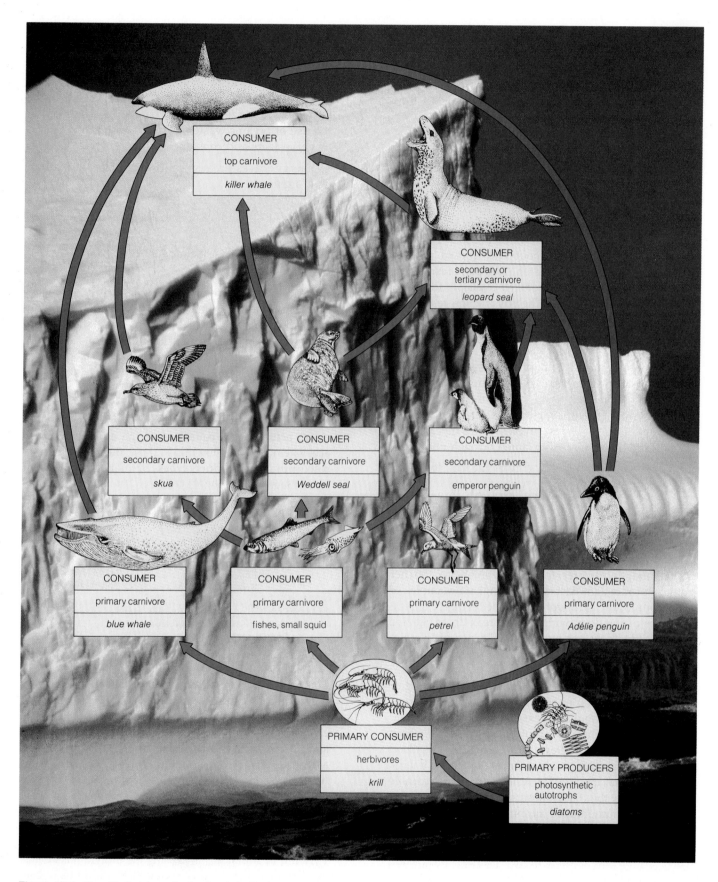

Figure 45.2 Simplified picture of a food web in the Antarctic; there are many more participants, including an array of decomposer organisms.

tosynthesis is used to maintain existing plant parts and more is lost to consumers and decomposers. These examples underscore an important concept:

Net primary productivity varies in space and over time. This is the variable energy base available for consumers in the ecosystem.

Major Pathways of Energy Flow

Figure 45.3 shows the general direction of energy flow through terrestrial ecosystems. Of all the solar radiation reaching the earth, only about one or two percent becomes stored in organic matter. Then plant respiration reduces (often by half) the energy available for other members of the ecosystem. The tiny fraction remaining becomes available in two forms: as living tissues of the producers and as **detritus** (organic wastes, dead tissues, and partly decayed organic matter).

All organisms produce heat as a result of metabolic activities, especially aerobic respiration. At the same time, they radiate heat to the environment (page 472). The pervasive heat losses represent one-way flow of energy out of the ecosystem. In ecosystems that are near equilibrium, the energy being stored through photosynthesis is about equal to that being lost through community-wide metabolism.

This one-way flow of energy proceeds through two interconnected types of food webs. In **grazing food webs**, herbivores consume living plant tissues and in turn are consumed by some array of carnivores (see Figure 45.2, for example). In **detrital food webs**, decomposers and detritivores (not herbivores) use the net primary production. Bacteria and fungi are examples of the decomposers. Earthworms, millipedes, and crabs are examples of the detritivores. They break down organic waste products or remains to simple inorganic molecules, many of which are then picked up by plants and recycled within the ecosystem.

The fraction of energy that flows through grazing or detrital food webs varies from one ecosystem to the next and often throughout the year in the same ecosystem. When cattle feed heavily on plants in a pasture, about half the stored energy flows through the grazing food web. However, cattle don't assimilate all of the energy stored in plants; undigested residues become available (in feces) for the decomposers and detritivores. In marshes, consumption of all but about ten percent of the stored energy is delayed until plant parts die and become available for detrital food webs. In most ecosystems, the largest portion of net primary production passes through detrital food webs.

1. Energy flows into ecosystems from an outside source (the sun, in most cases).

2. Energy flows through ecosystems by way of grazing food webs (based on the consumption of living tissues of photosynthesizers) and detrital food webs (based on the use of organic waste products and remains of photosynthesizers and consumers).

3. Energy leaves ecosystems through heat losses from each organism.

Pyramids Representing the Trophic Structure

The trophic structure of an ecosystem is influenced by the different numbers and sizes of its component organisms, their different metabolic rates, and the amount of energy lost at each transfer. Often the trophic structure is diagrammed as one or another type of "ecological pyramid," in which producers form a base for successive tiers of consumers above them.

Pyramid of Numbers. A pyramid of numbers is based on an actual "head count" of all the individuals in an ecosystem. In the following example, the lowest tier represents the food production base for higher trophic levels of a bluegrass field:

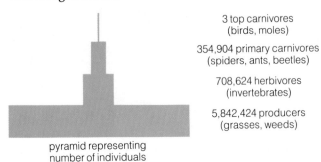

3 top carnivores (birds, moles)

354,904 primary carnivores (spiders, ants, beetles)

708,624 herbivores (invertebrates)

5,842,424 producers (grasses, weeds)

pyramid representing number of individuals

This type of pyramid doesn't always adequately portray the trophic structure, for it doesn't take into account that the sizes of organisms being counted in each trophic level can vary. A count in a redwood forest would yield a small number of large producers (the trees) that support a large number of mostly small herbivores and carnivores (insects)—and so the pyramid of numbers would be "upside-down." Moreover, if the decomposers were counted also, they could outnumber the members of other trophic groups by more than a billion.

Pyramid of Biomass. Weighing the individuals at each trophic level instead of counting them allows us to con-

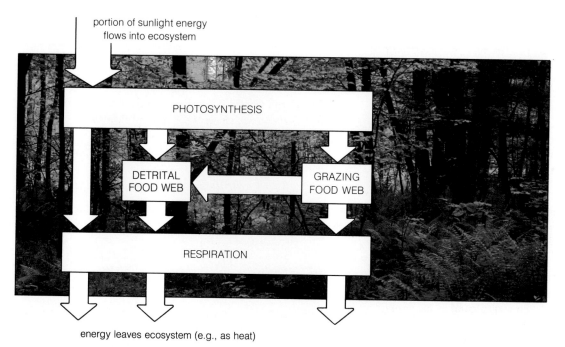

portion of sunlight energy
flows into ecosystem

PHOTOSYNTHESIS

DETRITAL
FOOD WEB

GRAZING
FOOD WEB

RESPIRATION

energy leaves ecosystem (e.g., as heat)

Figure 45.3 Generalized model of the main pathways of energy flow through an ecosystem. Photosynthesizers capture only a small fraction of solar energy reaching the earth. They use it in assembling organic matter, some of which is incorporated in plant tissues. In a grazing food web, herbivores feed on living plant parts, and carnivores feed on herbivores as well as on each other. In a detrital food web, decomposers and detritivores feed on organic wastes, dead tissues, and partially decayed organic matter. All organisms of the ecosystem release energy to the surroundings (typically as heat) as a result of energy metabolism. Thus there is a one-way flow of energy into and then out of the ecosystem.

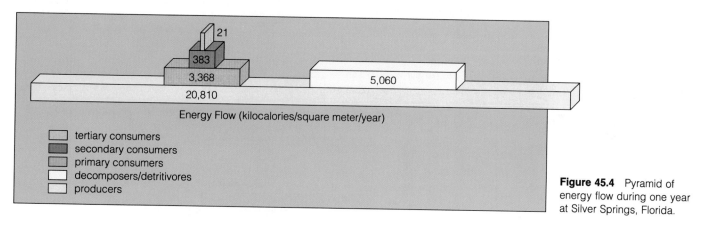

21

383

3,368

5,060

20,810

Energy Flow (kilocalories/square meter/year)

☐ tertiary consumers
■ secondary consumers
▨ primary consumers
☐ decomposers/detritivores
☐ producers

Figure 45.4 Pyramid of energy flow during one year at Silver Springs, Florida.

struct a pyramid of biomass (the total dry weight of all organisms at a given level). For most ecosystems on land, such pyramids have a large base of primary production, with smaller and smaller trophic levels perched on top. In contrast, the producers of some aquatic ecosystems are tiny phytoplankton that grow and reproduce rapidly. Here, the pyramid of biomass can be upside-down, with the consumer biomass at any instant actually exceeding the producer biomass. The phytoplankton are consumed or die about as fast as they reproduce; it is just that the survivors (few as they may be) are reproducing at a phenomenal rate.

Pyramid of Energy. A pyramid of energy represents energy losses at each transfer to another trophic level in an ecosystem. It is probably the most informative way to portray the functional roles of trophic levels. The pyramid shape is not skewed by overemphasis on differences in the size and metabolic rates of the organisms. And it more accurately reflects the laws of thermodynamics; the pyramid is always "right-side up," with a large energy base at the bottom (Figure 45.4).

Constructing an energy pyramid means measuring the actual amounts of energy that individuals take in, how much they burn up during metabolism, how much

Figure 45.5 Annual energy flow (measured in kilocalories per square meter per year) for an aquatic ecosystem in Silver Springs, Florida.

The producers are mostly green aquatic plants. The carnivores are insects and small fish, and the top carnivores are larger fish. The energy source (sunlight) is available all year long.

Only 1.2 percent of the incoming solar energy are actually trapped in photosynthesis to generate new plant biomass. And more than 63 percent of the photosynthetic products are metabolized by the plants themselves to meet their own energy needs. Only 16 percent are harvested by herbivores, and the remainder is eventually decomposed by bacteria and fungi. Similarly, most of the herbivore energy is expended in metabolism or goes into the decomposer system: only 11.4 percent are consumed by carnivores. Once again, the carnivores burn up most of the energy they take in and only 5.5 percent are passed on to top carnivores. The decomposers cycle and recycle all the biomass received from all other trophic levels. Eventually all of the 5,060 kilocalories will appear as heat produced during metabolism. (Decomposers, too, are eventually decomposed.)

This diagram has been deliberately oversimplified. No community is completely isolated from all others. Organisms and materials are constantly dropping into the springs. There is a slow but steady loss of other organisms and materials that flow outward in the stream that leaves the community.

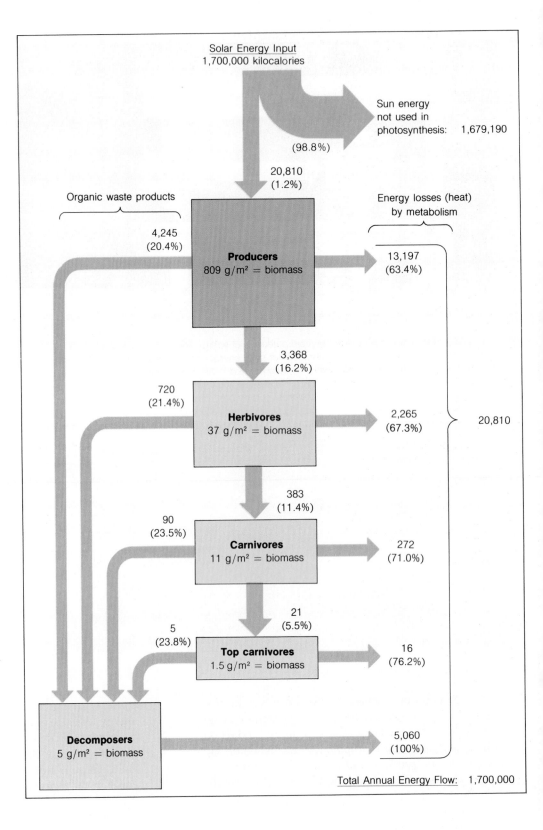

remains in their waste products, and how much they store in body tissues. The energy inputs and outputs are calculated so that energy flow can be expressed per unit of land (or water) *per unit time*. Such calculation is difficult to do, but it has been done in a few studies.

Figure 45.4 shows an energy pyramid constructed as a result of an extended study of a grazing food web of an aquatic ecosystem; Figure 45.5 shows the calculations on which it is based. When we look carefully at the calculations, an important point emerges. Given the met-

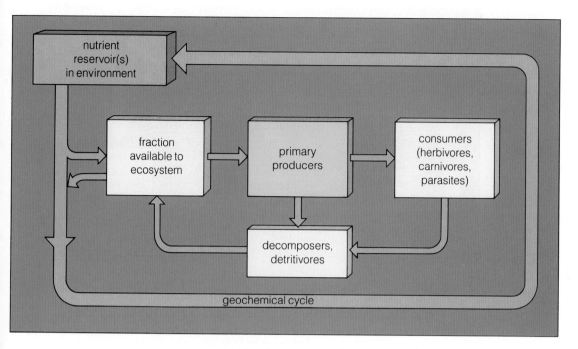

Figure 45.6 Generalized model of nutrient flow through an ecosystem. The overall movement of nutrients from the physical environment, through organisms, and back to the environment constitutes a biogeochemical cycle.

Inside figure:
nutrient reservoir(s) in environment

fraction available to ecosystem

primary producers

consumers (herbivores, carnivores, parasites)

decomposers, detritivores

geochemical cycle

abolic demands of organisms and the considerable amount of energy shunted into organic wastes, *only about six to sixteen percent of the energy entering one trophic level becomes available to organisms at the next level.*

BIOGEOCHEMICAL CYCLES

As we have seen, the availability of energy profoundly affects the trophic structure of ecosystems. So does the availability of nutrients. Besides the carbon, oxygen, and hydrogen they get from water and air, plants generally require about thirteen minerals, including nitrogen and phosphorus, that serve as nutrients. Typically the nutrients become available to plants in ionized forms, such as ammonium (NH_4^+) and phosphorus ($H_2PO_4^-$ or HPO_4^{--}). Some nutrients represent only a few parts per million of the plant's dry weight. Their absence still adversely affects growth, however, because they are essential biochemical constituents in cells (Table 20.2). It follows that nutrient availability can profoundly affect primary productivity and the ecosystem at large.

How do different elements enter and leave ecosystems? Here we must step back and view our planet as a whole. The elements essential for life tend to move in **biogeochemical cycles**, this meaning that they are transferred from the environment, to organisms, then back to the environment. In general, the physical environment serves as a large reservoir through which elements move rather slowly, compared to how rapidly they are exchanged between organisms and the environment.

Broadly speaking, oxygen and hydrogen move in the form of water molecules through the **hydrologic cycle**, and carbon and the mineral elements move through atmospheric and sedimentary cycles. In **atmospheric cycles**, the element occurs in a gaseous phase and a large portion of it exists in the atmosphere. In **sedimentary cycles**, the element does not have a gaseous phase; it moves from land to the seafloor and only "returns" to land when the seafloor is thrust upward and drained by geological forces that act slowly, over immense spans of time (Figure 38.12).

Figure 45.6 shows a generalized model of the relationship between geochemical cycles and most ecosystems. The model is based on four factors:

1. In general, the mineral elements that serve as nutrients for primary producers are in ionized forms.

2. Nutrient reserves of an ecosystem are maintained largely by environmental inputs and by the recycling made possible by the activities of decomposers and detritivores.

3. In most cases, the amount of a nutrient being cycled within the ecosystem is greater than the amount entering or leaving in any given year.

4. Environmental inputs into an ecosystem occur by way of precipitation, metabolic fixation reactions (such as nitrogen fixation), and the weathering and breakdown of rocks. For ecosystems on land, outputs include losses by way of runoff and evaporation.

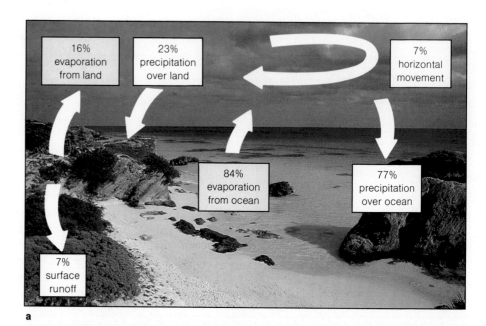

a

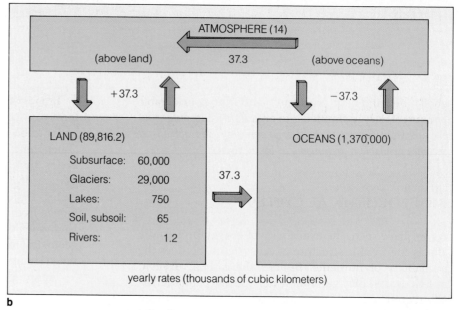

b

Figure 45.7 (**a**) Global water budget. The percentages indicate the annual movements of water into and out of the atmosphere. Of the water entering, 84% is by way of evaporation from oceans. Of that, 7% is carried horizontally to land, which returns it to oceans by way of rivers and streams. The remaining 77% (that is, 84 − 7) leaves the atmosphere as precipitation over oceans.

(**b**) Simplified picture of the hydrologic cycle. Annual net rates of transfer (in thousands of cubic kilometers per year) from one province to another are shown. For example, there is a net transfer of 37.3 × 10^3 km³ of water from the atmospheric to the oceanic province. This is balanced by a comparable net loss from the oceans to the atmosphere, and a net gain by oceans of that amount (through runoff).

Figure 45.8 Effects of different human activities on a forest ecosystem and its biogeochemistry. (**a**) These are experimental watersheds within the Hubbard Brook Valley of New Hampshire. Here, researchers have studied the effects of clear-cutting (foreground), progressive strip cutting (upper left), and deforestation (upper center). One of the gauging weirs used to collect all the water draining from an area under study is shown in (**b**). This watershed had been experimentally deforested, as shown in (**c**), then herbicides were applied to prevent regrowth during three years of studies. Two years after herbicide applications stopped, the vegetation recovered to the extent shown in (**b**).

a

The Hydrologic Cycle

Cold and warm ocean currents, clouds, winds, and rainfall are all part of the global hydrologic cycle. Driven by solar energy (page 766), the waters of the earth move slowly and on a vast scale through the atmosphere, on or through the uppermost layers of land masses, to the oceans, and back again. The following terms are used to define the ways in which water is moved or stored within the provinces of the land, oceans, or atmosphere:

evaporation	*release of water as vapor into the atmosphere*
precipitation	*release of water from the atmosphere as rain or snow*
detention	*temporary storage of water on land or in oceans*
transportation	*movement of water with winds or as surface runoff*

Only a fraction of the water is present in the atmosphere as vapor, clouds, and ice crystals. As Figure 45.7 indicates, the amount of precipitation is greater over oceans than over land. Why? The vast surface area of the oceans places less restriction on evaporation (oceans cover more than seventy percent of the earth's surface).

On the average, a water molecule does not stay aloft for more than about ten days, so the turnover rate is rapid for the airborne part of the hydrologic cycle. Water released as precipitation is detained on land for no more than an average of 10 to 120 days, depending on the season and on where it falls. Some evaporates from land. Some is carried by rivers and streams to the seas where, with large-scale evaporation, the cycle begins again.

In itself, water is essential for life in any ecosystem. *But water is also an important medium by which nutrients move into and out of ecosystems.* In the 1950s, water movements were tracked through the Hubbard Brook Experimental Forest in the White Mountains of New Hampshire. Here, a stream runs down the middle of a valley that has steep slopes and shallow soils (Figure 45.8). Running down the slopes are several smaller streams, each with its own small watershed.

We call a region a **watershed** when all of the rain falling on it is funneled into a single stream or river. A watershed can be any size. The watershed of the Mississippi River extends across roughly one-third of the United States. The watersheds at Hubbard Brook average about fifteen hectares (thirty-six acres).

Water enters a watershed mainly as rain or snow. About eighty to ninety percent of the precipitation passes through the canopy to the forest floor, where it can filter into the soil or run off along the surface (Figure 45.9). In most natural ecosystems, soil infiltration rates are high enough that nearly all water enters the soil. Plant roots absorb water from the shallow soil layer and the plant vascular system carries it to the canopy, where transpiration occurs. "Transpiration," recall, refers to evaporative water loss from plants (page 271). Water also can filter down through the soil (seepage and percolation) to the watertable (groundwater), or it can move laterally to a stream.

At Hubbard Brook, ecologists calculated the water budgets for different watershed "ecosystems." The budgets were based on measurements of precipitation inputs and streamwater and groundwater losses. They also were based on estimates of evaporation and transpiration outputs. The measured amount of precipitation

b

c

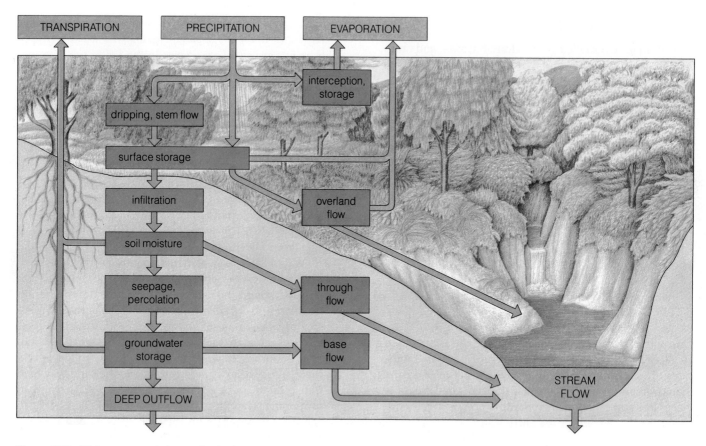

Figure 45.9 Water movements in a watershed.

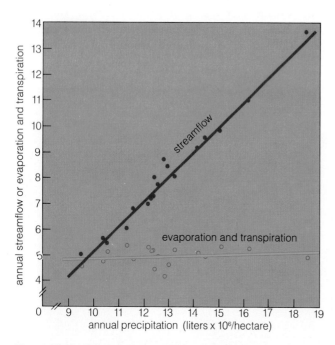

Figure 45.10 Relationship among precipitation, streamflow, evaporation, and transpiration for the Hubbard Brook Experimental Forest during 1956–1974.

changed from year to year, as did the rate of stream flow—but evaporation and transpiration varied only slightly (Figure 45.10). This discovery had practical applications; it indicated how much water can be expected for use in cities that are downstream of watersheds of the Hubbard Brook type.

The Hubbard Brook studies also revealed that plants greatly influence the rate at which nutrients move through the ecosystem phase of biogeochemical cycles. For example, you might think that water draining the watershed would rapidly leach away minerals, including calcium ions. Yet in watersheds with young, undisturbed forests, only about eight kilograms of calcium were lost annually from each hectare. Tree roots were "mining" the soil efficiently, moving calcium upward into growing plant parts and making it available for the food webs. Rain and the weathering of rocks brought calcium replacements into the watershed.

Nutrient outputs were different in experimentally cleared watersheds. In one study, all the vegetation was removed, then herbicides were applied for the next three years to prevent regrowth (Figure 45.8c). The soil itself was not disturbed; no organic material was removed from the site. Yet the loss of calcium in the stream outflow was *six times* greater than in comparable but undisturbed watersheds.

What is the lesson in this? Given how slowly calcium and other nutrients move through geochemical cycles, *stripping the land of vegetation has long-term disruptive effects on nutrient retention for the entire ecosystem.*

Global Movement of Carbon— An Atmospheric Cycle

In the **carbon cycle**, carbon moves from reservoirs in the atmosphere and oceans, through organisms, then back to the reservoirs. As Figure 45.11 indicates, carbon is introduced into the atmosphere by aerobic respiration, fossil fuel burning, and volcanic eruptions (which release carbon from rocks deep in the earth's crust). Atmospheric carbon dioxide exists in gaseous form—mostly carbon dioxide (CO_2). About *half* of all the carbon entering the atmosphere each year will move into two large "holding stations"—that is, into accumulated plant biomass and the oceans.

Each year, photosynthesizers harness airborne or dissolved carbon dioxide and incorporate billions of metric tons of its carbon atoms into organic compounds. However, the average amount of time that a "captured" carbon atom remains in an ecosystem varies greatly. In tropical forests, decomposition and carbon uptake are rapid, so not much carbon is tied up in litter on the soil surface. In bogs, marshes, and other anaerobic settings, organic compounds are not broken down completely and carbon accumulates in forms such as peat (refer to Figure 44.15i).

In aquatic food webs, carbon can become bound as carbonate in shells and other hard parts. When the shelled organisms die, they sink and become buried in bottom sediments of different depths. In deep oceans, carbon can remain buried for millions of years until geologic movements bring it to the surface. Still more carbon is slowly converted to long-standing reserves of gas, petroleum, and coal deep in the earth—reserves that are tapped for use as fossil fuels.

The worldwide burning of fossil fuels is putting more carbon into the atmosphere than can be returned to the global holding stations (oceans and plant biomass). This human activity and others are intensifying the greenhouse effect and may be triggering a global warming. The *Commentary* on the next page describes this effect and its consequences.

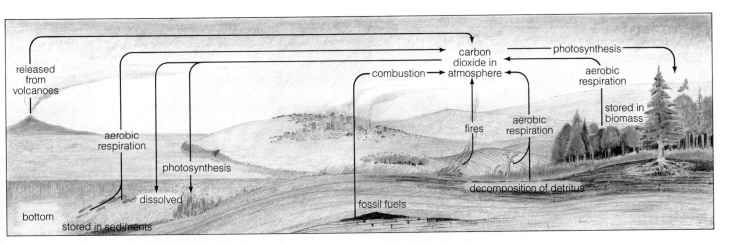

Figure 45.11 Simplified picture of the carbon cycle. Carbon dioxide in the atmosphere is harnessed during the second stage of photosynthesis. Compounds containing carbon and oxygen are assembled into tissues (living biomass). Oxygen is released to the atmosphere during photosynthesis; carbon dioxide is released during aerobic respiration. Conversions to carbon-containing inorganic compounds occur during decomposition. Carbon can also become locked up in peat, coal, oil, and gas, then subsequently released during the combustion of fossil fuels. Some carbon becomes locked up in the bottom sediments of oceans. More carbon enters the atmosphere during volcanic eruptions.

Greenhouse Gases and a Global Warming Trend

The atmospheric concentrations of carbon dioxide, water, ozone, methane, nitrous oxide, and chlorofluorocarbons profoundly influence the average temperature near the earth's surface, and that temperature influences global climates. Collectively, molecules of these gases act somewhat like a pane of glass in a greenhouse (hence their name, "greenhouse gases"). They allow wavelengths of visible light to reach the earth's surface, but they impede the escape of longer, infrared wavelengths—that is, heat—from the earth into space. They absorb infrared wavelengths, much of which get reradiated back toward the earth (Figure a). In short, the greenhouse gases cause heat to build up in the lower atmosphere, a warming action called the **greenhouse effect**.

If there were no greenhouse gases, the earth would be a cold and lifeless planet. But there can be too much of a good thing. Largely as a result of human activities, the levels of greenhouse gases have been increasing (Figure b), and they are contributing to an alarming increase in the global warming.

What is so alarming about a warmer planet? Suppose the temperature of the lower atmosphere were to rise by only 4°C (7°F). Sea levels would rise by about two feet, or 0.6 meter. Why? Ocean surface temperatures would increase—and water expands when heated. Also, global warming could cause partial melting of glaciers and Antarctic ice sheets. Low coastal regions everywhere would flood.

Imagine what a long-term rise in sea level, combined with high tides and storm waves, would do to the waterfronts of Vancouver, Boston, San Diego, Galveston, and other coastal cities. Huge tracts of Florida and Louisiana would face saltwater intrusions. Agricultural lowlands and deltas in India, China, and Bangladesh—where much of the world's rice is grown—would be submerged.

Global warming could affect world agriculture in other ways. Regional patterns of precipitation and temperature would change. Crop yields would decline in currently productive regions (including parts of Canada and the United States) and increase in others. There is speculation that warmer temperatures would promote insect breeding, with the increased population sizes of insect pests leading to more extensive crop losses.

In the late 1950s, a laboratory was set up on a mountaintop in the Hawaiian Islands to measure the concentrations of different greenhouse gases. The remote site was selected because it was free of local contamination and would represent average conditions for the Northern Hemisphere, and the monitoring activities are still going on. Consider what these studies told us about carbon dioxide levels alone.

It turns out that the levels of atmospheric carbon dioxide follow the annual cycle of plant growth in the Northern Hemisphere. The levels are lower during summer, when plants are photosynthesizing most

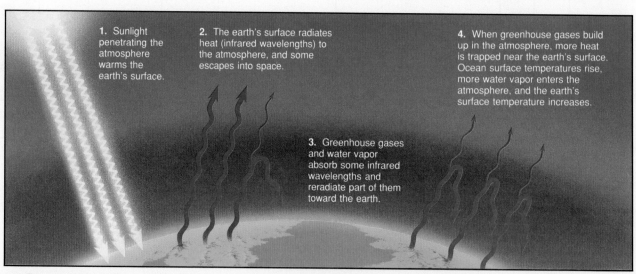

1. Sunlight penetrating the atmosphere warms the earth's surface.

2. The earth's surface radiates heat (infrared wavelengths) to the atmosphere, and some escapes into space.

3. Greenhouse gases and water vapor absorb some infrared wavelengths and reradiate part of them toward the earth.

4. When greenhouse gases build up in the atmosphere, more heat is trapped near the earth's surface. Ocean surface temperatures rise, more water vapor enters the atmosphere, and the earth's surface temperature increases.

a The greenhouse effect.

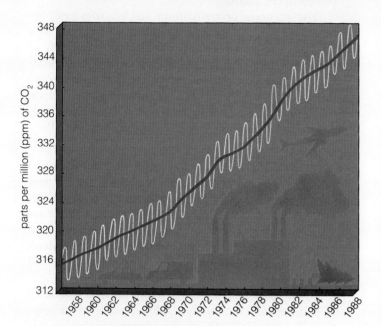

1. **Carbon Dioxide** (CO_2). By the year 2020, the relative contribution of the greenhouse gas CO_2 to the global warming trend is expected to be about fifty percent. Fossil-fuel burning, factory emissions, car exhaust, and deforestation are all contributing to the increased concentration.

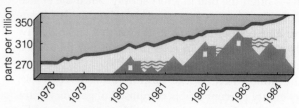

2. **Chlorofluorocarbons** (CFCs). By 2020, this gas will probably be responsible for about twenty-five percent of the greenhouse effect. CFCs are used in plastic foams, air conditioners, refrigerators, and industrial solvents.

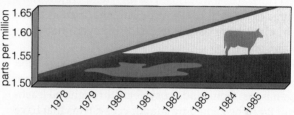

3. **Methane** (CH_4). By 2020, methane may be responsible for fifteen percent of the greenhouse effect. This gas is a natural by-product of bacteria (methanogens), as in feedlots.

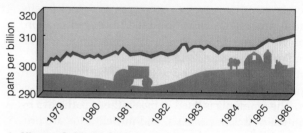

4. **Nitrous Oxide** (N_2O). By 2020, this gas may be responsible for about ten percent of the greenhouse effect. It is a natural by-product of denitrifying bacteria; it is released from fertilizers and animal wastes, as in livestock feedlots.

b Relative contributions of different greenhouse gases to the global warming trend, projected to the year 2020.

rapidly. They are higher in winter, when aerobic respiration continues even while photosynthetic activity declines. The lows and highs are represented by the peaks and troughs around the graph line in Figure b (part 1). *For the first time, scientists could see the integrated effects of the carbon balances of land and water ecosystems of a whole hemisphere.*

More disturbing, the peaks and troughs in the cycle showed a continuous increase. Here was evidence that a buildup of carbon dioxide in the atmosphere may intensify the greenhouse effect over the next century.

Increasing carbon dioxide levels are attributed mostly to the burning of fossil fuels, coal especially, throughout the world. Deforestation is another contributing factor. Today, vast tracts of tropical forests are being cleared and burned at a rapid rate (see, for example, Figure 47.7). Carbon is being released during the wood-burning; and more importantly, the number of plants that absorb carbon dioxide during photosynthesis is plummeting.

Atmospheric concentrations of greenhouse gases are expected to continue increasing into the middle of the twenty-first century, and global warming by several degrees will probably follow. It is doubtful that we can sharply reduce fossil fuel burning and deforestation soon enough to slow down the rate of global warming. There is widespread agreement among scientists that we should begin preparing for the consequences. For example, research in genetic engineering could be intensified to develop drought-resistant and salt-resistant plants (page 244).

Global Movement of Nitrogen—
An Atmospheric Cycle

Of all nutrients that affect the growth of land plants, nitrogen—a component of all proteins and nucleic acids—is often the one in short supply. Since the beginning of life, nitrogen has been abundant in the atmosphere and oceans but not in the earth's crust. Today, nearly all the nitrogen in soils has been put there by a variety of nitrogen-fixing organisms.

The atmosphere is the largest nitrogen reservoir; nearly eighty percent of it is composed of gaseous nitrogen (N_2). The N_2 molecules are held together by stable, triple covalent bonds ($N\equiv N$), and few organisms have the metabolic equipment for breaking them. A limited number of bacteria, volcanic action, and lightning can convert N_2 into usable forms. The high temperatures and pressures of internal combustion engines (such as those in cars) also create usable forms of nitrogen that enter the atmosphere.

Nitrogen is lost from ecosystems through bacteria that use the fixed nitrogen during metabolism. More nitrogen is lost from land ecosystems through soil leaching, but it can be gained by aquatic ecosystems that have nutrient inputs from rivers and streams.

The Cycling Processes. Figure 45.12 shows the nitrogen cycle, which involves six processes. Let's look briefly at those processes, which are called nitrogen fixation, assimilation and biosynthesis, decomposition, ammonification, nitrification, and denitrification.

In **nitrogen fixation**, a few kinds of bacteria convert N_2 to ammonia (NH_3) through a series of reduction reactions. The ammonia is used in biosynthetic pathways by which amino acids, proteins, and nucleic acids are produced. In aquatic ecosystems, *Anabaena*, *Nostoc*, and other cyanobacteria are dominant nitrogen fixers. In terrestrial ecosystems, bacteria such as *Rhizobium* (which live symbiotically with plants) and *Azotobacter* (which live in soil) do the fixing. The bacteria are small in size but mighty in numbers—collectively they fix about 200 million metric tons of nitrogen each year.

How does the fixed nitrogen become available to other organisms in the ecosystem? It is transferred directly into the tissues of legumes (peas, beans, and the like) and other plants that have entered mutually beneficial interactions with the free-living or symbiotic nitrogen fixers (page 269). Also, when the nitrogen-fixers die and undergo decay processes—that is, decomposition—ammonia and other nitrogen-containing substances are released. Such substances dissolve in soil water, from which they can be taken up by the roots of plants. Plants are the only nitrogen source for animals, which feed directly or indirectly on them.

Later, in a process called **ammonification**, the nitrogenous wastes and remains of plants and animals are decomposed by some species of bacteria and fungi. The decomposers use the proteins and amino acids being released for their own growth. They also release the excess as ammonia or ammonium, and some of this is picked up by plants.

The ammonia or ammonium that becomes available in soil also is metabolized by nitrifying bacteria. In a process called **nitrification**, bacteria strip the NH_3 or NH_4^+ of electrons, and nitrite (NO_2^-) is released as a product of the reaction. Still other nitrifying bacteria then use nitrite for energy metabolism, which yields nitrate (NO_3^-) as a product. Nitrification is an example of chemosynthesis, described earlier on pages 109 and 119.

Nitrogen Scarcity. In land ecosystems, the continual production of ammonia by the nitrogen-fixing bacteria would seem to assure plants of plenty of nitrogen. Yet soil nitrogen is scarce. Ammonia, nitrite, and nitrate are all soluble and vulnerable to leaching. Even more fixed nitrogen is lost through a bacterial process called **denitrification**: the reduction of nitrate or nitrite to N_2 and a small amount of nitrous oxide (N_2O).

Most species of the denitrifying bacteria ordinarily rely on aerobic respiration. Under some conditions, though—especially when soil is waterlogged and poorly aerated—they switch to anaerobic electron transport (page 125). Then they use nitrate, nitrite, or nitrous oxide instead of oxygen as the terminal electron acceptor, and in doing so they convert the fixed nitrogen to N_2.

Besides leaching and denitrification, the high metabolic cost of nitrogen fixation also keeps nitrogen from accumulating in soil. Plants engaged in symbiotic interactions with nitrogen-fixing bacteria don't get something for nothing; to gain nitrogen, they give up sugars and other photosynthetic products that can be assembled only with heavy investments of ATP and $NADPH_2$. Although such plants have the competitive edge when soil nitrogen is scarce, they are often displaced from nitrogen-rich soils by species that do not have to pay the high exchange rate, so to speak.

Nitrogen losses are great in agricultural regions. During crop harvests, of course, some nitrogen leaves the fields with the plants. Nitrogen losses also can occur through soil erosion and leaching. Farms in Europe and North America have depended on crop rotation to restore the soil. For example, nitrogen-fixing legumes are planted between plantings of wheat or sugar beet crops. This practice has helped maintain soils in stable and productive condition, in some cases for thousands of years.

Modern agriculture, however, depends on nitrogen-rich fertilizers. With plant breeding, fertilization, and pest control, the crop yields per hectare have doubled and even quadrupled over the past forty years. With intelligent management, it appears that soil can main-

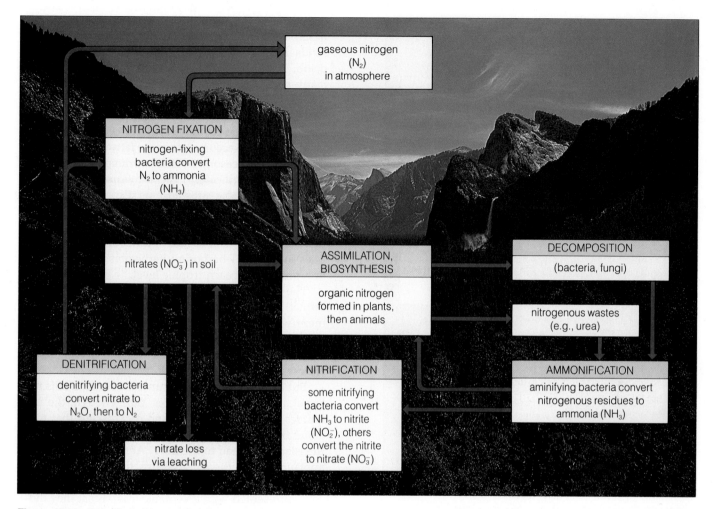

Figure 45.12 Simplified picture of the nitrogen cycle.

tain such high yields indefinitely—as long as water and nitrogen-containing fertilizers are available.

The catch, again, is that we can't get something for nothing. Enormous amounts of energy are needed to produce fertilizer—not energy from the unending stream of sunlight, but energy from petroleum. As long as the supply of oil was viewed as unending, there was little concern about the energetic cost of fertilizer production. In many cases, we have been pouring more energy into the soil (in the form of fertilizer) than we have been getting out of it (in the form of food). Unlike natural ecosystems, in which nutrients such as nitrogen are cycled, our agricultural systems exist only because of constant, massive infusions of fertilizers.

Nevertheless, as any hungry person will tell you, food calories are more basic to survival than are gasoline calories or perhaps, even, than a car. As long as the human population continues to grow exponentially (Chapter Forty-Four), farmers will be engaged in a constant race to supply food to as many individuals as possible. Soil enrichment with nitrogen-containing fertilizers is essential in the race, as it is now being run.

Global Movements of Phosphorus— A Sedimentary Cycle

In the **phosphorus cycle**, phosphorus moves from land, to sediments in the seas, and then back to the land. The earth's crust is the main storehouse for this and other minerals (Figure 45.13).

We can begin our look at the phosphorus cycle with the phosphate in rock formations on land. Through weathering and erosion, phosphorus is washed into rivers and streams, then it moves to the oceans. There, largely on continental shelves, phosphorus accumulates with other minerals as insoluble deposits. Millions of years pass, and geologic forces thrust up and expose the seafloor to form new land surfaces. Then weathering releases phosphorus from the rocks, and the geochemical phase of this sedimentary cycle begins again.

Compared to the slow, long-term geochemical events, the ecosystem phase of the phosphorus cycle is much more rapid. All living organisms require phosphorus, which becomes incorporated in ATP, $NADPH_2$, phospholipids, nucleic acids, and other organic compounds.

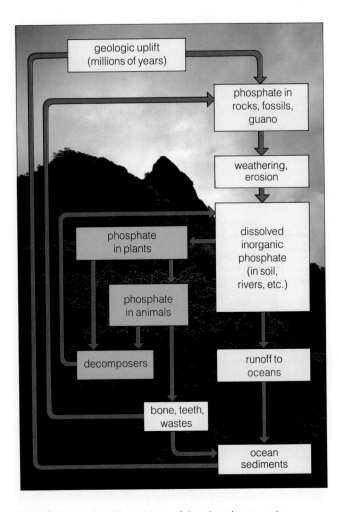

Figure 45.13 Simplified picture of the phosphorus cycle.

Plants have the metabolic means to take up dissolved, ionized forms of phosphorus. Actually, they do this so rapidly and efficiently that they often reduce soil concentrations of phosphorus to extremely low levels. Herbivores obtain phosphorus only by eating the plants; carnivores obtain it by eating herbivores. Herbivores and carnivores excrete phosphorus as a waste product in urine and feces. Phosphorus is also released to the soil by the decomposition of organic matter. The plants then take up phosphorus and so recycle it rapidly within the ecosystem.

Transfer of Harmful Compounds Through Ecosystems

Through our diverse and varied activities, we humans can have major and minor effects on the functioning of ecosystems. We will consider this idea in detail in Chapter Forty-Seven, but for now a simple example will underscore the point.

In 1955, the World Health Organization waged a major campaign to eliminate malaria-transmitting mos-

quitoes from the island of Borneo (now a state of Indonesia). DDT is a chlorinated hydrocarbon compound that sends insects into convulsions, paralysis, and on to death. The compound has been instrumental in bringing many pests, including mosquitoes, more or less under control.

DDT is a relatively stable compound; it is insoluble in water and degrades very slowly. However, it is soluble in fat. It tends to accumulate in the fatty tissues of organisms. For this reason, DDT is a primary candidate for **biological magnification**: the increasing concentration of a nondegradable substance as it moves up through trophic levels. The DDT that becomes concentrated in tissues of herbivores (such as insects) becomes even more concentrated in tissues of carnivores that eat quantities of the DDT-harboring herbivores. The concentration proceeds at each trophic level.

The decision to initiate a DDT-spraying program in Borneo was not made lightly. Nine out of ten people there were afflicted with malaria, an epidemic by any standard. The program worked, insofar as the mosquitoes transmitting this terrible disease were brought almost entirely under control.

But DDT is a broad-spectrum insecticide; it kills nontarget as well as target species. Sure enough, the mosquitoes had company: flies and cockroaches that made a nuisance of themselves in the thatch-roofed houses on the island fell dead to the floor. At first there was much applause. Then the small lizards that also lived in the houses and preyed on flies and cockroaches found themselves presented with a veritable feast. Feast they did—and they died, too. So did the house cats that preyed on the lizards. With the house cats dead, the rat population of Borneo was rid of its main predator, and rats were soon overrunning the island.

The fleas on rats were carriers of still another disease, called sylvatic plague, which can be transmitted to humans. Fortunately, the threat of this new epidemic was averted in time; someone got the inspired idea to parachute DDT-free cats into the remote parts of the island.

But on top of everything else, some of the people of Borneo found themselves sitting under caved-in roofs. The thatch in their roofs was made of certain leaves that happen to be the food resource of a certain caterpillar. The DDT did not affect the caterpillar, but it killed the wasps that were its natural predator. When the predator population collapsed, so did the roofs.

This brief example illustrates that disturbances to one part of an ecosystem can have unexpected effects on other, seemingly unrelated parts. A recent approach to predicting such unforeseen effects is through **ecosystem analysis**, which borrows from the analysis of complex mechanical systems. Ecosystem analysis is a method of extracting crucial bits of information about different components of a system and combining the information through computer programs and models to predict the outcome of disturbances to one or another component.

For example, an analysis of which species feed on which others in the food web shown in Figure 45.2 can be turned into a series of equations describing how much of each species is consumed. The equations can then be used to predict what the effect would be, say, of overharvesting whales. The *Commentary* in the next chapter (page 790) gives a more detailed example of an analysis of long-term interactions in the "global" ecosystem—interactions involving the oceans, the lower atmosphere, and the land.

As we attempt to deal with larger and more complex systems, it becomes more difficult and expensive to run desired experiments in the real ecosystems. The temptation is to analyze them instead on the computer. This is a valid exercise, in that the computer should hold all we know about a system. The danger is that the most important fact may be one that we do not yet know.

SUMMARY

1. An ecosystem is a community of organisms functioning together and interacting with their physical environment through (1) a flow of energy and (2) a cycling of materials.

2. In nearly all ecosystems, photosynthetic autotrophs are the organisms that fix the energy and take up the nutrients used by other members of the community.

3. Ecosystems are *open* systems, with energy and nutrient inputs and outputs. The following variables affect the degree of input and output: the ecosystem size and age, the overall rate of metabolism of its organisms, and the ratio of producer to consumer organisms.

4. Ecosystems are generally more open for inputs and outputs of water, carbon, and energy, whereas nutrients such as nitrogen and phosphorus are mostly recycled within the ecosystem.

5. Energy fixed by photosynthesis passes through grazing food webs and detrital food webs, where it is lost (as heat) through respiration.

6. Biogeochemical cycles include the movement of water, nutrients, and other elements and compounds from the physical environment, to organisms, then back to the environment.

7. Undisturbed ecosystems on land have predictable rates of nutrient losses that generally increase when the ecosystem is disturbed, as by fires or clear-cutting.

8. Fossil fuel burning and conversion of natural ecosystems to cropland or grazing land are contributing to increased atmospheric concentrations of carbon dioxide, and this may be contributing to a global warming trend.

9. Nitrogen availability is often a limiting factor for the total net primary productivity of ecosystems on land. Although gaseous nitrogen (N_2) is abundant in the atmosphere, it must first be converted to forms such as ammonia and nitrates before it can be used by the primary producers (mostly plants). A few species of bacteria, volcanic action, and lightning can cause the conversion.

Review Questions

1. Define ecosystem. Why do autotrophs play such a central role in an ecosystem? *746–747*

2. Define trophic level. Can you name and give examples of some trophic levels in ecosystems? What is the energy source for each level? *747–748*

3. Distinguish between a food chain and a food web. Can you imagine an extreme situation whereby you would be a participant in a food chain? *747–748*

4. Explain the difference between gross primary productivity and net primary productivity. If you were growing a vegetable garden, what variables might affect its net primary production—that is, the amount of energy stored in the organic compounds of plant tissues? *748–749*

5. There are two major pathways of energy flow through ecosystems: grazing food webs and detrital food webs. Can you characterize each? How does energy leave each one? How are the two types of food webs interconnected? *750–751*

6. Describe the greenhouse effect. Make a list of the agricultural products and manufactured goods that you depend on and yet are implicated in the global warming trend that is occurring through amplification of the greenhouse effect. *758–759*

7. What is a biogeochemical cycle? Can you give an example of one and describe the reservoirs and organisms involved? *753–762*

8. Define these terms: nitrogen fixation, nitrification, ammonification, and denitrification. *760*

Readings

Bormann, F., and G. Likens. 1984. *Pattern and Process in a Forested Ecosystem.* New York: Springer-Verlag.

Gibbons, B. September 1984. "Do We Treat Our Soil Like Dirt?" *National Geographic* 166(3):350–388.

Payne, W. 1983. "Bacterial Denitrification: Asset or Defect." *Bioscience* 33(5):319–325. Good summary article.

Pimm, S. 1982. *Food Webs.* London: Chapman-Hall. A look at food webs in community structure. Somewhat advanced.

Price, P. 1984. *Insect Ecology.* Second edition. New York: Wiley. Advanced reading but excellent examples of ecological interactions.

Smith, R. 1986. *Elements of Ecology.* Second edition. New York: Harper & Row.

Waring, R., and W. Schlesinger. 1985. *Forest Ecosystems.* Orlando, Florida: Academic Press.

46

THE BIOSPHERE

When Charles Darwin, Alfred Wallace, and other naturalists made their grand explorations more than a century ago, they became aware of some broad patterns in the world distribution of plants and animals. If you had traveled with them, you would have discovered that many plants and animals are unique to some regions and yet show striking similarities to unrelated species in distant regions.

Suppose you had visited the deserts of Africa and the American Southwest. Flowering plants with spines, extremely reduced leaves, and columnlike, fleshy stems grow in both places (see, for example, Figure 46.2). Yet despite the similarities, those plants evolved from very different ancestors. Suppose you also had explored the hills near the California coast. The woody, many-branched chaparral plants growing there are very much like the woody, many-branched, yet unrelated plants growing near the Mediterranean Sea, in southern Africa, and in Chile.

If you had mapped the geographic locations of those distant yet structurally similar species, you would have found that the American and African desert plants grow about the same distance from the equator. The chaparral plants and their distant look-alikes all tend to grow along western or southern coasts of continents between 30° and 40° latitudes.

Figure 46.1 The biosphere—the thin wrapping of sunlit air, water, and surface land in which organisms live.

Why is it that certain types of plants and animals occur in specific regions of the world and not others? In part, "accidents of history" put many species in particular places. Think about what happened when the supercontinent Pangea broke up more than a million years ago (page 578). The subsequent drifting and jostlings of the huge fragments led to the geographic isolation of species that evolved into (for example) eucalyptus trees, koala bears, and kangaroos. But the distribution of species also is a consequence of climate, topography, and species interactions. We considered the effects of predation, competition, and other species interactions in earlier chapters. Here, we turn to the forces that shape the character of the biosphere itself.

COMPONENTS OF THE BIOSPHERE

The **biosphere** is the entire realm in which organisms live—the waters of the earth, the surface rocks, soils, and sediments of its crust, and the lower region of the surrounding air. The crust (the outer part of the earth's lithosphere) is divided into rigid plates that have been drifting and colliding over evolutionary time (Figure 38.12). The **hydrosphere** is the sum total of all liquid and frozen water on or near the earth's surface. It includes the oceans and smaller bodies of water, groundwater, the polar ice caps, and a small amount of airborne water. The **atmosphere** is a region of gases, airborne particles, and water vapor enveloping the earth. About eighty percent of its total mass is distributed within 17 kilometers of the earth's surface.

GLOBAL PATTERNS OF CLIMATE

The biosphere includes ecosystems that range in size from vast continental forests to tiny ponds. Except for a few remote ecosystems in the deep oceans, they all are influenced profoundly by climate. The word **climate** refers to prevailing weather conditions, including temperature, humidity, wind velocity, degree of cloud cover, and rainfall.

Of the many factors that shape climate, four are especially important:

Climate is shaped by (1) variations in the amount of incoming solar radiation, (2) the earth's daily rotation and its annual path around the sun, (3) the world distribution of continents and oceans, and (4) the elevation of land masses.

Interactions among these factors produce prevailing winds and ocean currents that influence global weather patterns. The weather patterns affect soils and sediments (through erosion, for example). The composition and distribution of soils and sediments influence the growth and distribution of primary producers—and through them, the ecosystem itself.

Mediating Effects of the Earth's Atmosphere

Figure 46.3 shows the layered structure of the atmosphere. When rays from the sun reach the upper atmosphere, ozone (O_3) and oxygen molecules start absorbing

Echinocereus

Euphorbia

Figure 46.2 Convergent evolution of plants that are native to different geographic realms. *Echinocereus*, a member of the cactus family (Cactaceae), grows in deserts of the American Southwest. *Euphorbia*, a member of the spurge family (Euphorbiaceae), grows in deserts of southwestern Africa. Although the plants appear to be similar, they evolved from leafy plants that are not related.

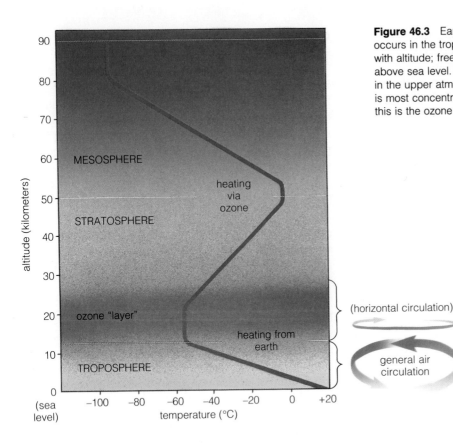

Figure 46.3 Earth's atmosphere. Most of the global air circulation occurs in the troposphere, where temperatures decrease rapidly with altitude; freezing occurs at only 2.3 kilometers (7,000 feet) above sea level. Ultraviolet wavelengths from the sun are absorbed in the upper atmosphere primarily by ozone (O_3) molecules. Ozone is most concentrated between 17 and 25 kilometers above sea level; this is the ozone "layer."

its ultraviolet wavelengths, which are potentially lethal to most forms of life. Of the wavelengths that do pass into the atmosphere, about thirty-two percent are reflected back into space by clouds and airborne particles. Clouds, dust, and water vapor absorb another eighteen percent. In total, about half the solar radiation that might otherwise reach the earth is lost through reflection and absorption in the atmosphere.

The solar radiation that does get through warms the earth's surface, then it indirectly warms the lower atmosphere. How? The earth's surface gives up heat energy by way of radiation, convection, conduction, and evaporation (page 472). Water vapor and carbon dioxide molecules in the atmosphere absorb some of the heat—and they reradiate part of it back toward the earth. The effect is something like heat retention in a greenhouse, which allows the sun's rays to penetrate but retains heat being lost from the plants and soil inside (see *Commentary*, page 758).

Heat energy derived from the sun warms the atmosphere—and that energy drives the earth's weather systems.

Air Currents

Incoming rays from the sun have different heating effects at different latitudes. As Figure 46.4 indicates, the rays are less spread out at the equator than they are at the poles, so air is heated more at the equator. The global pattern of air circulation begins when warm equatorial air rises and spreads northward and southward. Then the pattern becomes modified by two factors that combine to produce worldwide belts of prevailing east and west winds.

First, land absorbs and gives up heat to the air faster than oceans do, and the world's nonuniform distribution of land masses means that some air parcels rise (or sink) faster than others. Atmospheric pressure decreases wherever warm air rises, and the reverse is true when cool air sinks. The pressure differences that are created all around the world give rise to winds, which disrupt the overall air movement from the equator toward the poles. (Those "cold fronts" and "warm fronts" reported by weathermen are simply the leading edges of cold and warm air masses.)

Second, the earth rotates on its axis from east to west, below the air masses. But the earth is ball shaped (it has a declining circumference from the equator to the poles),

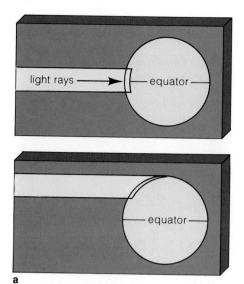

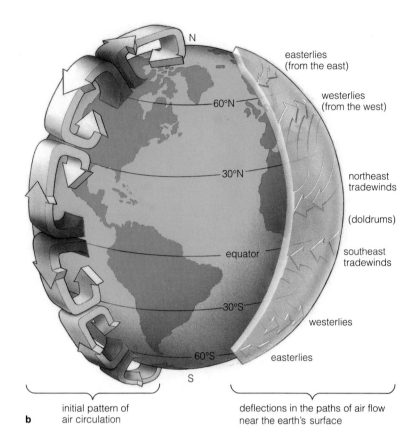

Figure 46.4 Global air circulation patterns. (**a**) Rays of incoming sunlight are more concentrated in equatorial than polar regions. Their unequal heating effects cause air to circulate from the equator to the poles, then back toward the equator. (**b**) The earth's rotation and its declining circumference cause the northward and southward paths of air motion to break up into prevailing east and west winds, in ways described in the text.

initial pattern of air circulation

deflections in the paths of air flow near the earth's surface

so each time the earth makes a full rotation, its surface turns fastest at the equator and slowest near the poles. What this means is that a rising air mass can't really move "straight north" or "straight south."

For example, an air mass moving north from 30° north latitude is being deflected to the east. Air moving south from that same latitude is being deflected to the west. Hold a pencil over Figure 46.4b so that it lines up straight with the earth's poles. Now move it slowly to the right, keeping your eye on 30° north latitude. See how the part of the pencil above that latitude reaches the horizon faster than the part below it? The same things happen to air masses as the earth turns below them.

In sum, three interrelated factors establish the global pattern of air circulation:

1. The different heating effects of the sun's rays in equatorial and polar regions.

2. Pressure variations associated with the nonuniform distribution of land masses and oceans.

3. Easterly and westerly deflections in the paths of air motion as a result of the earth's rotation and its declining circumference.

Global air circulation patterns produce the regional variations in rainfall that are characteristic of the world's

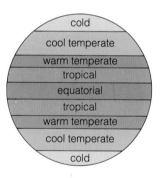

Figure 46.5 The world's climate zones.

climate zones (Figure 46.5). And the amount of rain influences where different ecosystems will occur. At the equator, heated air gives up moisture (as rain) when it rises to cooler altitudes, and where the rain falls it supports luxuriant forests. The drier air continues to move away from the equator and it descends at about 30° latitudes; there, deserts occur. The air again becomes warm, picks up moisture, and ascends at 60° latitudes, then it travels poleward. Almost no precipitation accompanies its descent in polar regions—where you never find luxuriant vegetation.

Figure 46.6 Surface drifts and ocean currents in January. The solid arrows indicate warm-water movements; dashed arrows indicate cold-water movements.

Ocean Currents

The earth's rotation, prevailing surface winds, and variations in water temperature give rise to currents and surface drifts that tend to move parallel with the equator. However, land masses intervene and modify that tendency in the Atlantic, Pacific, and Indian oceans. Two circular water movements, called **gyres**, dominate each of the great oceans. Currents in the two gyres move clockwise in the Northern Hemisphere and counterclockwise in the Southern. An equatorial countercurrent separates the two gyres and carries water away from the western boundaries of the ocean basins (Figure 46.6).

The gyres move warm equatorial waters to the north and to the south. For example, the Gulf Stream flows north from the Caribbean and along the southeastern coast of the United States. From there it flows northeast across the Atlantic. Then it divides into the North Atlantic Drift and the Canary Current (Figure 46.6). Because of the Gulf Stream, the climate of Great Britain and Norway is wetter and milder than you might expect if air currents were the only consideration. The warm water spawns moisture-laden fog and clouds that moderate the temperature and lead to abundant rainfall.

Certain gyres also move water cooled at the poles toward the equator, and others move water along the western coasts of Africa, North America, and South America. The cold currents influence regional climates, particularly near the coasts, which are often shrouded in fog. The mild, wet climate of the Pacific Northwest is largely a result of the California Current. As cool air masses from above those waters move inland, they moderate the temperature even over the hottest desert regions.

Atmospheric and oceanic circulation patterns influence the distribution of different types of ecosystems.

Effects of Topography

Regional climates are influenced by more than patterns of sunlight, atmospheric and oceanic circulation, and rainfall. They also are influenced by the distribution of mountains, valleys, and other topographical features.

Consider how topography can modify the pattern of rainfall. The Sierra Nevada are high mountains that parallel the west coast of California. When prevailing winds reach the mountains, the air rises, cools, and loses moisture (cool air holds less moisture than does warm air). Belts of vegetation at different elevations correspond to the changes in air temperature and moisture.

At the western base of the range, vegetation consists of shrubs adapted to semiarid conditions. Above this is a belt of deciduous and evergreen trees adapted to the moisture levels and cool temperatures of higher elevations. Next is the subalpine belt, with evergreen trees adapted to the rigors of a still cooler climate. The cold alpine belt above this cannot support trees; there, grasses and sedges dominate.

After flowing over the mountain crests, the air descends (Figure 46.7). As it does, it becomes warmer and its moisture-holding capacity increases. The air now draws moisture out of plants and the soil rather than giving it up as rain. In effect, there is a **rain shadow**— a reduction in rainfall on the leeward side of the high

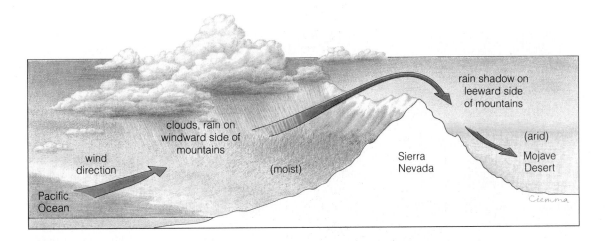

Figure 46.7 Rain shadow effect. As air containing water vapor rises, it cools. Cool air holds less moisture than warm air; hence clouds form on the western side of the Sierra Nevada, and rainfall tends to be heavy on the western flanks of this mountain range. As the air descends on the eastern side, it becomes more compact and warm. Warm air holds more moisture than cooler air; hence it picks up moisture from its surroundings rather than releasing it. This rain shadow effect produces desert conditions on the eastern side of the Sierra Nevada.

mountains—and only plants adapted to arid or semiarid conditions grow there. ("Leeward" means the direction not facing the wind; "windward" means the direction from which the wind is blowing.) Rain shadows also occur in the high mountain ranges of Europe, the Himalayas of Asia, and the Andes of South America. There, too, we find plants adapted to arid and semiarid conditions.

Landforms that interrupt or channel air and water movements influence regional climates.

Seasonal Variations in Climate

The amount of incoming solar radiation changes in the Northern and Southern hemispheres during the year, and the changes cause seasonal variations in climate. The boundary of illumination that divides earth days and nights is always perpendicular to the sun (Figure 46.8). However, the earth's axis is tilted relative to its annual path around the sun. The North Pole is most tilted away from the sun in December, a winter month in the Northern Hemisphere. Then, temperatures are lower than at other times of year because days are shorter and incoming light rays are more spread out. Notice, in Figure 46.8, that winter in the north polar regions is a time of near-perpetual darkness, and that daylight is nearly continuous in summer.

December is a summer month in the Southern Hemisphere; then, temperatures are higher because days are

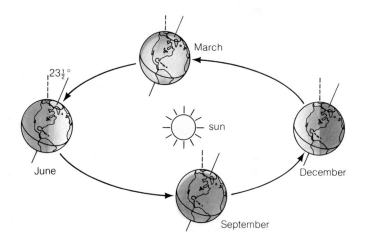

Figure 46.8 Annual variation in the amount of incoming solar radiation. Notice that the northern end of the earth's axis tilts toward the sun in June and away from it in December. Notice also the annual variation in the position of the equator relative to the boundary of illumination between day and night. Such variations in the intensity and duration of daylight lead to seasonal variations in temperature in the two hemispheres. Seasonal change becomes more pronounced with distance from the equator. It is greatest in the central regions of continents, where the moderating effects of oceans are minimal.

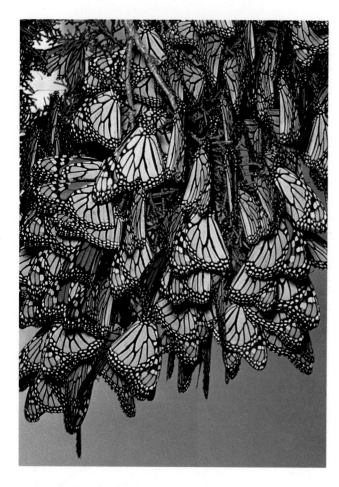

Figure 46.9 Monarch butterflies, migrants that gather in winter in trees of California coastal regions and central Mexico. Monarchs typically travel hundreds of kilometers south to these regions, which are cool and humid in winter; if they stayed in their breeding grounds, they would risk being killed by more severe conditions.

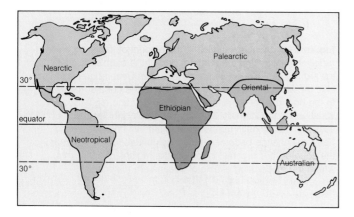

Figure 46.10 Major biogeographic realms, as first proposed by W. Sclater and Alfred Wallace in the 1800s. The scheme is still widely accepted.

longer and light rays are not as spread out. Conditions are reversed for the two hemispheres in June.

Many biological rhythms coincide with annual changes in light, temperature, and rainfall. In essence, biological rhythms permit organisms to anticipate and adjust to seasonal change. In temperate regions (which have moderate climates compared, say, to arctic regions), organisms respond more to seasonal changes in light and temperature than to other factors. In deserts and tropical forests, they respond more to changes in rainfall with corresponding cycles of leafing out, flowering, fruiting, and leaf drop. Many animals, including caribou, bison, and butterflies, follow cycles of breeding and migration (Figure 46.9). In the seas, turtles, seals, whales and other animals also migrate. Such movements correspond with seasonal bursts of primary productivity on land and in the seas.

Cycles of biological activity often coincide with seasonal changes in climate.

THE WORLD'S BIOMES

On the one hand, the climatic factors just described help us understand why vast regions of grasslands, deserts, forests, and tundra occur where they do. They also help explain why some unrelated species that have been geographically isolated from one another over evolutionary time can still show similar adaptations.

On the other hand, it's important to keep in mind that the species in any one of those regions are distinctive in their own right; they have evolved independently of those elsewhere. Barriers such as oceans, mountain ranges, and deserts have kept them more or less isolated from the others by restricting their dispersal. (For example, the kangaroos of Australia have not, on their own, been able to radiate elsewhere.) The naturalist W. Sclater and then Alfred Wallace were the first to propose six **biogeographic realms**, which are broad land regions that contain characteristic types of plants and animals. Figure 46.10 shows the six-realm scheme, which is still widely accepted.

This is not to say that the species *within* each biogeographic realm are monotonously uniform. Each realm is a mosaic of regional climates, landforms, and soils—and each has a distinctive array of organisms. Think about the varied scenery you would encounter if you were to travel across the Nearctic realm, which includes all of Canada and the United States. You would be able to identify regional subdivisions by the predominant plant types—evergreen conifers along the northwest coast, grasses cloaking the interior, and so forth.

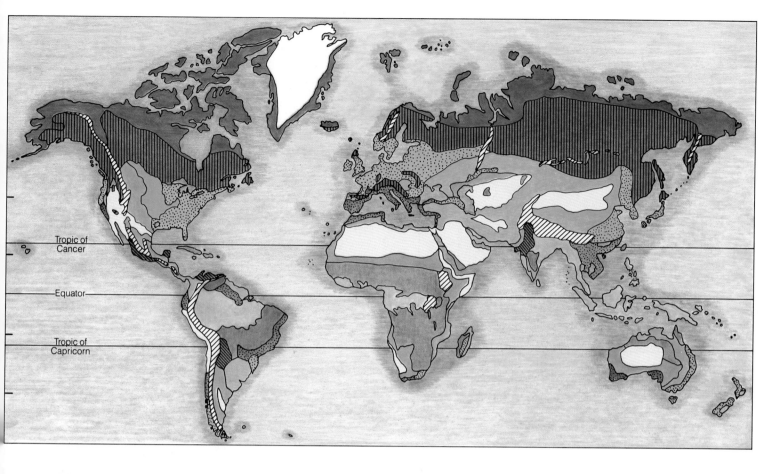

	tundra		sclerophyllous woodlands and shrublands		tropical scrub forest
	evergreen coniferous forest (e.g., boreal forest, montane coniferous forest)		desert		tropical savanna, thorn forest
	temperate deciduous forest		evergreen broadleaf forest (e.g., tropical rain forest)		semidesert, arid grassland
	temperate grassland		tropical deciduous forest		mountains (complex zonation)

Figure 46.11 Simplified picture of the world's major biomes. Arctic tundra and boreal forests are continuous over vast regions and contain ecologically equivalent and taxonomically related species. Other biomes that are isolated but of the same type generally have ecologically equivalent but often taxonomically unrelated species. The overall pattern of biome distribution roughly corresponds with patterns of soil type distributions and climate.

Each broad, vegetational subdivision of a biogeographic realm, together with the animals and other organisms living there, is called a **biome**. Figure 46.11 shows the distribution of the world's major biomes.

The overall pattern of biome distribution generally corresponds not only with climate but also with patterns of soil type. **Soil** is a mixture or rock, mineral ions, and organic matter in some state of physical and chemical breakdown. To varying degrees, water, air, and a variety of organisms are mixed with its components. Figures 46.12 and 46.13 show the characteristics of soils, together with soil profiles for some major types of biomes.

The overall form of the dominant plants in a biome is a useful indicator of whether conditions there are benign or harsh. For example, short plant species tend to be dominant at cold, dry latitudes and at high elevations. Tall, leafy plant species are dominant at tropical latitudes and at low elevations where warm temperatures combine with high rainfall. Such plants are typical of tropical rain forests, which show the highest annual primary productivity (Figure 46.14).

Let's now take a closer look at some features of the major types of biomes—the deserts, shrublands, woodlands, forests, and tundra.

O-Horizon	Surface litter: Freshly fallen leaves and organic debris and partially decomposed organic matter
A-Horizon	Topsoil: Partially decomposed organic matter (humus), living organisms, and some inorganic minerals
E-Horizon	Zone of leaching: Area through which dissolved or suspended materials move downward
B-Horizon	Subsoil: Accumulation of iron, aluminum, and humic compounds, and clay leached down from the A- and E-horizon
C-Horizon	Parent material: Partially broken-down inorganic minerals
R	Bedrock: Impenetrable layer

Figure 46.12 Categories of horizontal layers of soils (also called soil horizons).

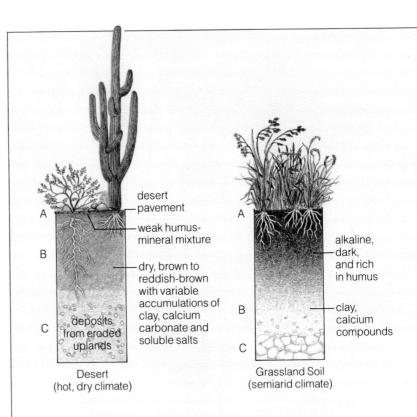

Figure 46.13 Characteristics of soil, together with soil profiles from a few different biomes.

The distribution of different ecosystems on land depends on the composition of regional soils as well as on climate and topography. *Soil* is a mixture of rock, mineral ions, and organic matter in some state of physical and chemical decomposition. Depending on the degree of weathering, the rock component ranges from coarse-grained gravel to sand, silt, and fine-grained clay. Partially decomposed organic matter, called *humus*, helps retain water-soluble ions (such as potassium and ammonium ions) that are released during the activities of soil bacteria and other decomposers.

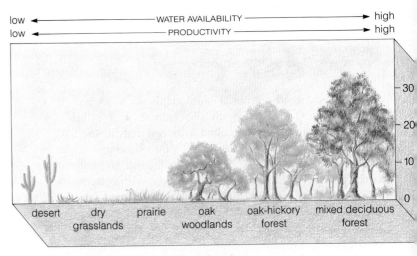

a Moisture gradient for temperate North America

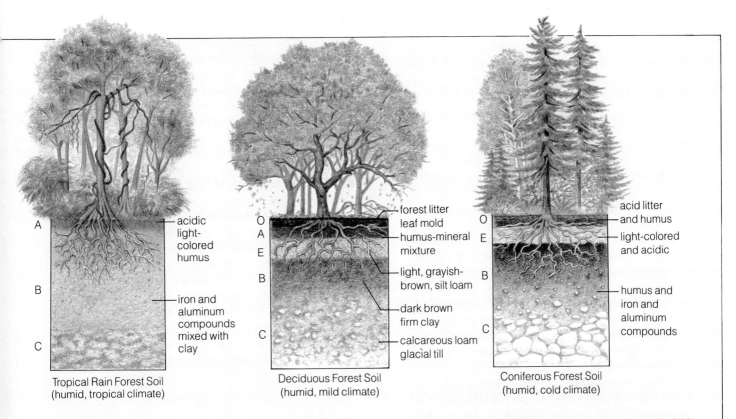

acidic
light-
colored
humus

iron and
aluminum
compounds
mixed with
clay

Tropical Rain Forest Soil
(humid, tropical climate)

forest litter
leaf mold
humus-mineral
mixture
light, grayish-
brown, silt loam
dark brown
firm clay
calcareous loam
glacial till

Deciduous Forest Soil
(humid, mild climate)

acid litter
and humus
light-colored
and acidic

humus and
iron and
aluminum
compounds

Coniferous Forest Soil
(humid, cold climate)

The most productive soil is loam, a mixture of sand, silt, clay, and humus that holds adequate nutrients and mineral ions and that is well aerated. (The roots of most plants, recall, do not do well in poorly aerated, poorly draining soils.) Water percolates too rapidly through soils with a high gravel and sand content; hence these soils are quickly depleted of mineral ions. This process is called *leaching*. Clay soils with small, closely packed particles are poorly aerated and do not drain well; few plants can grow in such soils when they are waterlogged.

Topsoil, the soil layer below any surface litter, ranges in thickness from less than a centimeter on steep slopes to more than a meter in grasslands. Topsoil is the most fertile of all soil layers, the so-

called A-horizon. Biomes differ considerably in the depth of this layer. Most of the world's crops are grown on grassland soils.

When tropical rain forests are cleared (as for agriculture), the torrential, seasonal rains leach most of the nutrients and remaining minerals from the A-horizon of the exposed soil.

About one-third of the area covered by desert soils is not useful for growing crops because of too little rainfall. With proper soil management and extensive irrigation, soils in certain desert areas (such as the Imperial Valley of California) can produce a variety of crops. Such soils, however, often become unproductive from waterlogging and salt buildup when they are irrigated without sufficient drainage.

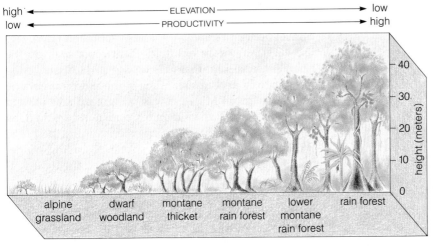

high ← ——————— ELEVATION ——————— → low
low ← ——————— PRODUCTIVITY ——————— → high

alpine
grassland

dwarf
woodland

montane
thicket

montane
rain forest

lower
montane
rain forest

rain forest

b Elevation gradient for tropical South America

Figure 46.14 Changes in plant form along productivity gradients, which are influenced by gradients in water availability, elevation, and other factors, including mean annual temperature and soil drainage characteristics.

Deserts

Deserts occur where the potential for evaporation greatly exceeds rainfall. Such conditions prevail at 30° north and south latitudes, where rainfall is scarce because of the high subtropical air pressure. There we find the deserts of the American Southwest, Chile, Africa, Arabia, and Australia (Figure 46.11). Farther north from the equator, variations in landforms have helped shape other deserts, as through the rain-shadow effects described earlier. Those deserts include the vast Gobi of Asia, the Kyzyl-Kum to the east of the Caspian Sea, and the high deserts of eastern Oregon.

Deserts do not have as much vegetation as other biomes. When rain does fall, it often falls in heavy, brief pulses that cause rapid erosion of the unprotected soil. Because of the low humidity, sunlight penetration through the air is high and the dry ground surface heats rapidly. The surface can cool rapidly at night, when it radiates heat back to the air. Page 473 described some of the behavioral responses that desert animals make to the temperature extremes.

The arid and semiarid conditions generally preclude the growth of tall, leafy plants, but the diversity in form and physiology is still notable. Within the same patch of Arizona desert, you might find evergreen shrubs with extensive roots (creosote bush); fleshy-stemmed, shallow-rooted cacti that are either short (prickly pear) or tall (saguaro); shrubs that drop leaves several times a year and then grow new ones within a week following each new rain (ocotillo); as well as briefly but spectacularly flowering annuals and perennials. Figure 46.15 shows some of these plants. Even very deep-rooted species such as mesquite and cottonwood may be growing along streambeds and other land formations that have a permanent underground water supply.

Today, more than a third of the world's total land area is arid or semiarid. In many parts of the world, prolonged periods of drought, overgrazing, and other factors are leading to **desertification** (conversion of grasslands and other regions to dry wastelands) on a grand scale (Chapter Forty-Seven).

Sclerophyllous Shrublands and Woodlands

Shrubland and woodland biomes prevail in the western or southern coastal regions of continents that lie between 30° and 40° latitudes. Those regions are semiarid; they get more rain than deserts do, but not much more. Most of the rain falls during mild winter months, and summers are long, hot, and dry. The dominant plants are "sclerophyllous," meaning they have hard, tough, evergreen leaves.

Sclerophyllous shrublands occur when the annual rainfall is less than sixty centimeters, and they go by such exotic local names as maquis, fynbos, and chaparral (Figure 46.16). The dominant plant species are woody and multibranched, yet never more than a few meters tall. California has about 2.4 million hectares (about 6 million acres) of chaparral-covered hills.

The plants of sclerophyllous shrublands can grow together into a nearly impenetrable mass. Every so often, however, firestorms (touched off by lightning, for example) sweep through them. Many of the shrubs have highly flammable leaves and their aboveground parts burn rapidly. Yet they are exquisitely adapted to episodes of fire; they resprout from their root crown and the vegetation cover is quickly reestablished. Trees (and suburban housing developments) do not do nearly as well in surviving the firestorms, and the shrubs—which actually feed the fires—have the competitive edge.

Sclerophyllous woodlands occur when the annual rainfall exceeds 100 centimeters. Although the dominant trees do not form a dense, continuous canopy over the whole region, they can be quite tall. The eucalyptus woodlands of southwestern Australia and the oak woodlands of western North America are like this.

Grasslands

The world's great temperate grasslands occur in South Africa, South America, and midcontinental regions of North America and the USSR. This type of biome includes shortgrass prairie, tallgrass prairie, and tropical grasslands. All have several features in common. The land itself is usually flat or rolling and it is dry, with high rates of evaporation. There is enough rainfall to keep the regions from turning into deserts (between four and twelve centimeters, on the average), but not enough to support extensive forests. Grazing and burrowing species are the dominant forms of animal life. At the fringes of many grassland biomes, grazing and periodic fires are other factors that restrict the encroachment of forests.

Shortgrass prairie occurs where winds are strong, rainfall light and infrequent, and evaporation rapid. Figure 46.17 shows an example. Plant roots soak up the brief, seasonal rain that falls on the surface, above the permanently dry subsoil. Much of the shortgrass prairie of the American Midwest was overgrazed and plowed under for wheat, which requires more moisture than the region can provide without irrigation. During the 1930s, the combination of drought, strong prevailing winds, and poor farming practices that removed the tight vegetation cover turned much of the prairie into a Dust Bowl. John Steinbeck's *Grapes of Wrath* and James Michener's *Centennial* are two historical novels that speak eloquently of the disruption of this biome and its consequences.

Tallgrass prairie of the sort shown in Figure 46.18 once extended west from the temperate deciduous for-

Figure 46.15 Warm desert near Tucson, Arizona. The vegetation includes creosote bushes, ocotillo, saguaro cacti, and prickly pear cacti.

ests of North America. In some regions, the species of grasses were actually outnumbered by species of composites (flowering plants that include daisies). Legumes also were abundant, and their nitrogen-fixing activities increased the net primary productivity throughout the biome. Most tallgrass prairie has been converted to farmland.

Tropical grasslands include the broad belts of African *savanna* described in Chapter One. In savanna regions of low rainfall, the main plant species are rapidly growing, tufted grasses. Where there is slightly more rainfall, acacia and other shrubs grow in scattered patches. Where rainfall is high, we find mosaics of tall, coarse grasses, shrubs, and low trees. Other tropical regions have comparable biomes called *monsoon grasslands*. "Monsoon" refers to a season of heavy rainfall that corresponds to a shift in prevailing winds over the Indian Ocean and parts of southern Asia; it alternates with a pronounced dry season.

Figure 46.16 Chaparral-covered hills east of San Diego, California.

Figure 46.17 Rolling shortgrass prairie to the east of the Rocky Mountains.

Figure 46.18 A patch of natural tallgrass prairie.

Forests

All the world's major forest biomes have tall trees that form an extensive and fairly continuous canopy. There are three general types of forest trees, and which type dominates a region depends partly on distance from the equator:

evergreen broadleaf trees	*between 20° north and south latitudes*
deciduous broadleaf trees	*north and south midlatitudes*
evergreen coniferous trees	*Northern Hemisphere, mostly*

As before, keep in mind that the species composition is not entirely uniform through any forest biome; local variations in climate, topography, and other factors can produce patches of distinct communities within its boundaries.

bromeliad

Figure 46.19 Tropical rain forest, showing aerial plants (including bromeliads) and vines in the canopy. Where light breaks through the canopy, luxuriant new growth occurs (see, for example, Figure 44.1).

Evergreen Broadleaf Forests. Forest biomes that are dominated by evergreen broadleaf trees occur in tropical parts of Africa, southwestern Asia, the East Indies, the Malay Archipelago, South America, and Central America. Annual rainfall generally exceeds 200 centimeters and is never less than 130 centimeters.

Where regular, heavy rainfall coincides with high humidity (eighty percent or more) and the annual mean temperature is about 25°C, you will find the highly productive biomes called **tropical rain forests** (Figure 46.19). Many of the trees native to these forests are sold in nurseries; they include the popular ornamental figs (notably *Ficus benjamina* and rubber plants), certain palms, and tree ferns. Unlike other forest biomes, in which perhaps 10 to 20 tree species are present, a tropical rain forest typically has 50 to 70 in a single hectare; sometimes the number exceeds 300!

Some of the evergreen trees produce new leaves and shed old ones throughout the year. Some species near the fringes of the biome are periodically bare, but not for more than a few weeks. Because leaf production and leaf drop are generally continuous, tropical rain forests produce more litter than any other forest biome. How-

ever, decomposition and mineral cycling are rapid in the hot, humid climate, so the soil is not really a significant reservoir of nutrients.

The diversity in tropical rain forests is not confined to tree species. Different kinds of vines twist around tree trunks and grow toward the canopy, where more sunlight is available for photosynthesis. Orchids, mosses, lichens, and bromeliads (plants related to the pineapple) grow on tree branches, absorbing minerals dissolved in tiny pockets of water. (The minerals were released during the decay of bits of leaves, insects, and litter.) Entire communities of exotic insects, spiders, and amphibians live, breed, and die in the small pools of water that collect in the leaves of the aerial plants.

Compared to other biomes, tropical rain forests have the greatest variety of spectacularly plumaged birds, the heaviest insects (rhinoceros beetles), and the plants with the largest flowers (these often being a meter across; see Figure 40.36). There are not as many large herbivores or large predators as in the open grasslands, but those in the forests are splendidly varied; among them are the tapirs, monkeys, and jaguars in some regions and apes, okapi, and leopards in others.

Spring

Autumn

Figure 46.20 The changing character of a temperate deciduous forest in spring, summer, autumn, and winter. The one shown here is south of Nashville, Tennessee.

Deciduous Broadleaf Forests. As we move out from the tropical rain forests, we enter regions where temperatures remain mild but rainfall dwindles during part of the year. These peripheral regions are dominated by semideciduous and deciduous trees that drop leaves during a pronounced dry season. The so-called **monsoon forests** of India and southeastern Asia fall into this category. Farther north, rainfall is lower still and temperatures become cold during the winter, and here we find regions of **temperate deciduous forests** (Figure 46.20).

Conditions are not as favorable for decomposition as they are in humid tropics. Nevertheless, nutrients are not lost; they are conserved in the accumulated litter on the forest floor.

At one time, deciduous broadleaf forests stretched across northeastern North America, Europe, and eastern Asia. Ash, beech, birch, chestnut, elm, and deciduous oak trees dominated Europe and Asia, but they largely disappeared as land was cleared for farming. The same trees were once common in North America, although

species introductions wiped out nearly all chestnuts and many American elms (Table 44.2). Today, maple and beech forests predominate in the northeast. They give way to oak-hickory forests in the west, which grade into scattered woodlands and then into the grasslands of the Midwest.

Evergreen Coniferous Forests. Evergreen conifers are cone-bearing trees, typically with needlelike leaves that can withstand arid conditions. (Each needle has a thick

cuticle and recessed stomata, these being adaptations that help prevent water loss.) This type of tree is the primary producer in a variety of biomes, including the boreal forests, montane coniferous forests, temperate rain forests, and pine barrens.

The sweeping expanses of coniferous trees in northern Europe, Asia, and North America are called **boreal forests**, or **taiga** (meaning "swamp forest"). Most boreal forests occupy glaciated regions that are punctuated by cold lakes and streams (Figure 46.21). Precipitation occurs mainly in the cool summers, and evaporation is low. The winter air is cold and dry. The winters are more severe in the eastern parts of the continents than in the west, where prevailing offshore winds moderate the climate. (That also is why the northern tree line is at lower latitudes in the east.)

In North America, the main primary producers are spruce and balsam fir. Deciduous birches and aspens also are abundant in areas that have been burned (following summer lightning strikes) or heavily logged. To the north, where nutrient-poor soils and bogs grade into the tundra, the boreal forests become considerably less dense.

Montane coniferous forests extend southward through the great mountain ranges. Spruce and fir are major primary producers in mountains that parallel the Pacific coast of Canada and the United States; fir and pine predominate in the southern extensions of the Rockies and in the Sierra Nevada (Figure 46.22). Large coniferous forests once grew around the northern Great Lakes as well, but they were nearly decimated when loggers moved in just before the turn of the century.

Coniferous forests also occur in some temperate lowlands. A **temperate rain forest** extends inward from the coast all the way from Alaska into northern California. That forest has some of the world's tallest trees—Sitka spruce to the north and sequoia redwoods to the south (see, for example, Figure 20.1). Much of New Jersey's coastal plain, which has sandy, nutrient-poor soil, is dominated by **pine barrens**. This is a scrub forest, with grasses and low shrubs growing among open stands of pitch pine and oak trees. The pines recover quickly from the frequent fires in this region. Open stands of pine occur in the coastal plains of North Carolina, South Carolina, Georgia, and Florida, and these forested regions also are maintained by frequent fires.

Tundra

Tundra is a word derived from the Finnish *tuntura*, meaning a treeless plain. **Arctic tundra** lies to the north, between the polar ice cap and the huge belts of boreal forests in North America, Europe, and Asia. It is vast; this biome occupies twenty percent of the earth's land surface. In Alaska, much of the arctic tundra is flat, windswept, and desolate (Figure 46.23). Temperatures average 5°C (41°F) in midsummer, and −32°C (−26°F) in midwinter. The air is too cold to permit formation of much water vapor, so rainfall is sparse. Yet for three summer months, when sunlight is nearly continuous, the short plants grow profusely, flowers bloom, and seeds ripen quickly.

Although the tundra is not completely covered with snow all year long, the summer is too brief to warm much more than surface soil. Just beneath the surface is

Figure 46.21 Example of a boreal forest.

Figure 46.22 (Right, above) The montane coniferous forest of Yosemite Valley in the Sierra Nevada, beneath the first snows of winter.

the **permafrost**, a permanently frozen layer more than 500 meters thick in some regions. Permafrost forms an impenetrable basement beneath stretches of flat terrain, so water drainage is inhibited. The low temperatures and anaerobic conditions of the waterlogged soil have major effects on nutrient cycling. Organic matter cannot completely decompose, and it accumulates in soggy masses (peat). More than ninety-five percent of the carbon in the arctic is locked up in peat.

Alpine tundra occurs at high elevations in mountains throughout the world. The dominant plant species often form cushions and mats that can withstand buffeting from strong winds. Alpine tundra is characterized by low temperatures and shallow, nutrient-poor soils with correspondingly low rates of primary productivity. In contrast to arctic tundra, there is no permafrost layer below the soil surface.

Figure 46.23 A view of Alaska's arctic tundra in summer, looking toward the Brooks Range.

Figure 46.24 Example of a lake basin carved by glacial action in the Canadian Rockies.

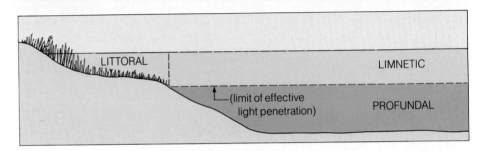

Figure 46.25 Lake zonation. The littoral includes all areas around the edge of the lake, from the shore to the depth where aquatic plants stop growing. The profundal includes areas below the depth of light penetration. Above the profundal are the open, sunlit waters of the limnetic zone.

THE WATER PROVINCES

As vast as the six biogeographic realms are, they cover less than thirty percent of the earth's surface. The water provinces are far more extensive and they, too, encompass diverse ecosystems. Inland are the "lentic" ecosystems (lakes and other bodies of standing fresh water) and the "lotic" ecosystems (rivers and other systems of running fresh water). Also inland are freshwater swamps, marshes, bogs, and similar wetlands, as well as brackish seas. Along the coasts or within the oceans and seas are communities of organisms suspended in the water or tucked into the nooks and crannies of estuaries, rocky and sandy shores, coral reefs, and the ocean floor.

There is no such thing as a "typical" aquatic ecosystem. Lentic ecosystems alone vary in size from small ponds to lakes that cover many thousands of square kilometers. Some are shallow enough to wade across; one is more than 1,700 meters deep. All have gradients in light penetration, temperature, and dissolved gases, but the gradients vary from one ecosystem to another, both daily and seasonally. All we can do here is provide a sampling of the diversity that exists.

LAKE ECOSYSTEMS

The topography, climate, and geologic history of a lake determine the kinds and numbers of organisms found there, how they are distributed, and how nutrients are cycled among them. All lakes form in land basins (Figure 46.24), and most are doomed to become filled with sediments over time.

Lake Zones

Deep lakes have littoral, limnetic, and profundal zones (Figure 46.25). All around the lake's edge is the **littoral**, a shallow zone that extends from the shore to the depth where rooted aquatic plants stop growing. It is the zone with the greatest species diversity. Light penetration usually is good in the shallow waters, and numerous consumers (including snails, snakes, and frogs) as well as decomposers live here, together with plants that are rooted in or attached to the lake bottom.

The **limnetic zone** includes the open, sunlit waters that lie beyond the littoral; it extends to the maximum depth at which significant photosynthesis occurs. Here we find **plankton**: communities of floating or weakly

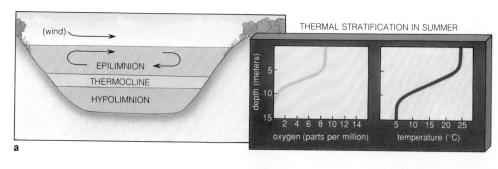

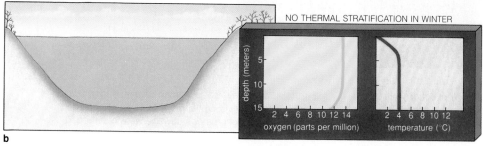

THERMAL STRATIFICATION IN SUMMER

NO THERMAL STRATIFICATION IN WINTER

Figure 46.26 (a) Thermal stratification and restriction on water mixing during summer, as measured in a productive north temperate lake (this one is in Connecticut). (b) Stratification does not occur in winter.

swimming organisms, mostly microscopic. The photosynthetic autotrophs ("phytoplankton") include diatoms, green algae, and cyanobacteria; the heterotrophs ("zooplankton") include rotifers, copepods, and a variety of protistans.

The **profundal zone** is the deep, open water below the depth of effective light penetration. Detritus from the limnetic zone sinks through the profundal to the bottom ooze, which contains communities of bacterial decomposers. Through their activities, the decomposers release mineral ions into the surrounding water and thereby enrich it with nutrients.

Seasonal Changes in Lakes

The surface water temperature rarely rises above 4°C (39.2°F) for lakes in polar regions, and it never falls below 4°C for lakes in the tropics. In north temperate regions, however, where warm summers alternate with cold winters, the deeper lakes undergo seasonal changes in density and temperature from the surface to the bottom.

Water density is greatest at 4°C. Just after the snow and ice melt in spring, the entire lake is almost that cold. As daylength increases and the atmosphere warms up, the lake's surface waters warm, also. By midsummer, the lake has three layers of different densities, as defined by temperature:

epilimnion	*warm upper layer*
thermocline	*middle layer in which temperature declines abruptly with depth*
hypolimnion	*cool layer in the profundal zone*

The vertical layering in north temperate lakes during the summer is a type of **thermal stratification**, and it persists until autumn (Figure 46.26a).

In autumn, the upper layer of water begins to cool, increase in density, and sink—and the thermocline disappears (Figure 46.26b). Now dissolved oxygen is moved downward and nutrients (released by decomposition) are moved up. This thorough mixing of the lake is called the **fall overturn**. In spring, the upper layer of water warms, increases in density, and sinks through the less dense water below. This episode of thorough mixing is called the **spring overturn**. Once more, oxygen is moved downward and nutrients are moved up.

Cycles of primary productivity correspond to the lake's seasonal changes. Following the spring overturn, the increased light and the recycled nutrients support a rapid increase in primary productivity. Through the growing season, phosphorus, nitrogen, and other nutrients necessary for biosynthesis may be utilized faster than replacements can accumulate. All the while, microscopic producers and consumers are dying off, and nutrients tied up in their individually tiny but collectively large remains sink to the deeper waters. As nutrient supplies dwindle, primary productivity in the upper layer of water decreases until, by late summer, only a limited amount of photosynthesis is possible.

After the fall overturn, dissolved nutrients brought up to the surface layer trigger another increase in primary productivity, but this increase is not sustained for long. With autumn's decreasing daylength and declining temperatures, the "burst" of activity often does not occur or is soon over, and primary productivity will not increase again until spring.

Figure 46.27 Field experiment demonstrating the effect of nutrient enrichment on a lake. A plastic curtain was stretched across a narrow channel between two basins of the same lake. Phosphorus, carbon, and nitrogen were added to the basin at the left; only carbon and nitrogen were added to the basin in the right foreground. Within two months, the phosphorus-enriched basin showed accelerated eutrophication, with a dense algal bloom.

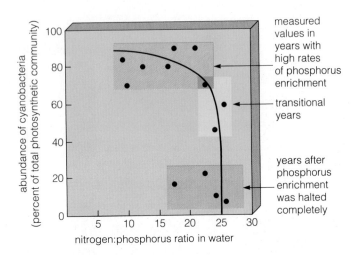

Figure 46.28 Reversal of human-caused eutrophication of Lake Washington in Seattle. From 1941 to 1963, large amounts of phosphorus-rich sewage were discharged into the lake. It promoted large blooms of cyanobacteria, which formed thick mats over the lake each summer and made it useless for recreation. (With phosphorus enrichment, nitrogen became the limiting resource and cyanobacteria—which are superior competitors for nitrogen because of their ability fo fix N_2—became dominant.)

From 1963 to 1968, sewage discharges were cut back and finally stopped. By 1975, the lake had almost fully recovered, and the cyanobacteria were replaced by lower density populations of diatoms and green algae.

Table 46.1	Differences Between Oligotrophic and Eutrophic Lakes	
Oligotrophic	**Eutrophic**
Deep, steeply banked	Shallow with broad littoral zone
Small water volume in epilimnion, compared to hypolimnion	Greater water volume in epilimnion, compared to oligotrophic lakes
Highly transparent	Limited transparency
Water blue or green	Water green to yellow or brownish-green
Low nutrient content	High nutrient content
Oxygen abundant through all levels at all times	Oxygen depleted in hypolimnion in summer
Not much phytoplankton; green algae and diatoms dominant	Abundant, massed phytoplankton; cyanobacteria dominant
Abundant aerobic decomposers in profundal zone	Limited diversity in profundal zone; anaerobic species favored
Low biomass in profundal zone	High biomass in profundal zone
Salmon, trout, other deep-water fishes intolerant of temperature changes	No fishes in hypolimnion that cannot tolerate temperature changes

Data from Cole, 1983; modified from Thienemann, 1925.

Trophic Nature of Lakes

Geologic processes give rise to lakes, as when the grinding action of advancing glaciers carves out basins in mountains. The basins become exposed and then filled with water when the glaciers retreat. Figure 46.24 is an example of a lake formed this way. Other geologic processes, including erosion and sedimentation, change each lake's dimensions over time. The soils of the lake basin and the surrounding regions contribute to the type and amount of dissolved nutrients that will be available for the primary producers of the lake ecosystem.

Interactions among soils, basin shape, and climate produce conditions that range from oligotrophy to marked eutrophy, these being terms that refer to the lake's nutrient content and the resulting primary productivity. **Oligotrophic lakes** are often deep, nutrient-poor, and low in primary productivity. **Eutrophic lakes** are often shallow, nutrient-rich, and high in primary productivity. Table 46.1 summarizes these and other differences between the two extremes in the trophic nature of lakes.

At one time, it was thought that as lake sediments accumulated, conditions always progressed from oligotrophy to eutrophy, then on to a final successional stage corresponding to a completely filled-in basin. However,

Figure 46.29 New England salt marsh, with *Spartina* predominating. This salt marsh grass is the major producer, with its microbial-enriched litter providing the food for consumers in the creeks and sounds.

the trophic nature of lakes may not change as the basin becomes increasingly shallow. Many ancient lakes that are now largely filled in have reverted to oligotrophic conditions. For them, geologic, climatic, and biotic forces are acting in balanced ways.

Many human activities can disrupt the balance, as when sewage is dumped into lakes or when the surrounding land is logged over or cleared for farming. Field experiments dramatically demonstrate the effects of such disruption (Figure 46.27). Even then, however, the disruption may be reversible. Through long-term studies of Lake Washington in Seattle, for example, W. Edmondson showed that human-caused eutrophication was reversed when the sewage discharges that were overloading the lake with nutrients were stopped (Figure 46.28).

MARINE ECOSYSTEMS

Types of Marine Environments

Like freshwater ecosystems, the oceans and seas show variations in physical and chemical properties, including light penetration and water temperature, depth, and salinity.

Some marine ecosystems are partially enclosed by land, yet they have open connections with the sea and are influenced by tides. In these ecosystems, which are called **estuaries**, seawater mixes with fresh water from rivers, streams, and runoff from the land. Along rocky and sandy shores are ecosystems of the **intertidal zone** (also called the littoral), which is subdivided into smaller realms that are influenced in different ways by the tides. Beyond the estuaries and intertidal zones are the **open oceans**.

Estuaries

All estuaries are partially enclosed coastal regions where fresh water and seawater meet. Tides, winds, and the earth's rotation influence their water circulation pattern, in which fresh water generally flows outward over the more dense seawater beneath it.

Estuaries do not all look alike. Some, including Chesapeake Bay and Mobile Bay, are broad and shallow. Others, such as the fjords of Norway, British Columbia, and Alaska, cut deeply into coastal landforms. Still others lie behind long spits of sand and mud; estuaries along the south coast of Texas and the west coast of Florida are like this. Most of the estuaries just described were formed by river or glacial erosion since the last Ice Age. In contrast, San Francisco Bay is an estuary that formed during the geologic upheavals of the Cenozoic.

Estuaries also differ greatly in their physical and chemical properties and in the communities they sustain. Here we will focus on the sheltered estuaries along the New England coast (Figure 46.29). Grassy flatlands called coastal salt marshes are common in the estuaries. *Spartina* and other salt-tolerant marsh plants are equipped to withstand total or partial submergence at high tide. The leaves of *Spartina*, for example, have salt-excreting cells. Its roots, which extend down into oxygen-poor sediments, have hollow tubes that carry oxygen from the leaves to living root cells.

Spartina is a major producer for the estuary, and a large part of its biomass is in the form of cellulose. Few herbivores can digest cellulose and so, as you might expect, grazing food webs do not dominate the estuary. Instead, *Spartina* is the start of detrital food webs, in which bacterial and fungal decomposers are the first to feed. The detrital particles produced by their activities serve as food for nematodes, snails, many different

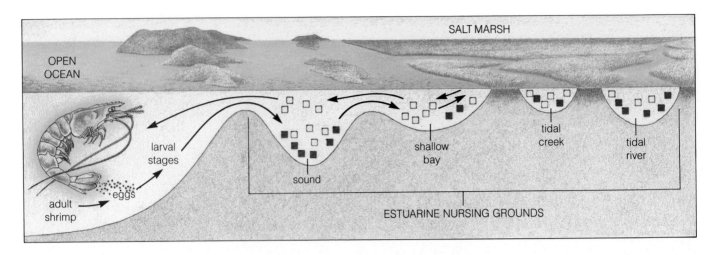

Figure 46.30 Life history of a shrimp that uses estuaries as nursery grounds. Adult shrimp spawn offshore and the larval stages move shoreward. In the partially enclosed estuaries they find the food and protection they need for rapid growth during the juvenile stage (red boxes) and adolescent stages (yellow boxes). As they mature, the shrimp move back into deeper waters of the sounds, then of the open ocean.

Figure 46.31 Vertical zonation in the intertidal zone of a rocky shore.

crabs, and some fish, including mullet. Clams, barnacles, and other filter feeders live on detritus that becomes suspended in the water by rising tides. Many other invertebrates, including the larval stages of shrimp, live in and feed directly on eelgrass; so do some shorebirds and migratory fowl such as ducks and geese.

Like other estuaries, the ones along the New England coast serve not only as feeding grounds but also as nurseries for marine organisms that are essential components of food webs in the seas (see, for example, Figure 46.30). These vital, highly productive ecosystems are rapidly being altered by sewage effluent, agricultural runoff, and diversions of fresh water for farm use—water that normally would drain into estuaries and keep salinity levels within tolerable ranges for the communities.

Life Along the Coasts

The rocky and sandy shores of coastlines are not exactly renowned for their creature comforts. The resident organisms are battered by waves, fiercely so during storms. With the periodic rise and fall of tides, they alternately are submerged and then exposed to air. The higher up they are on the shore, the more they face extreme conditions (such as dehydration, freezing in winter, or baking in summer) and the less food comes their way. The lower they are on the shore, the more they must compete with other species for the limited space available. At low tides, birds, rats, and raccoons move in to feed on them; high tides bring the predatory fishes. Such is life in the intertidal, or littoral, zone.

It is nearly impossible to generalize about life here, because bewildering arrays of habitats are continually

Figure 46.32 The kinds of residents of tide pools along the coast of Vancouver, British Columbia.

being created and then modified as waves and tides erode different substrates. About the only feature that all rocky and sandy shores have in common is a vertical zonation—and even this varies from place to place in its details, with resident species often violating the boundaries we artificially impose on them.

Rocky Shores. The rocky shores of temperate regions often have the following pattern of zonation:

upper littoral	*submerged only during the highest tide of the lunar cycle*
mid-littoral	*submerged during the highest regular tide and exposed during the lowest tide of each day*
lower littoral	*exposed only during the lowest tide of the lunar cycle*

Figure 46.31 shows an example of the type of vertical zonation you might see along the rocky shores of the Pacific Northwest.

Because of the erosive forces at work in all three zones, not much detritus accumulates on rocky shores. Instead, detrital food webs are displaced and we are more apt to find grazing food webs. In the sparsely populated upper littoral, the primary producers are cyanobacteria, green algae (such as *Ulothrix*), and the algal components of lichens, all of which grow in mats or jellylike masses on the infrequently wet rocks. Small snails (*Littorina*) and limpets feed on the producers. Some large, aggressive shore crabs feed on the snails and limpets, and seabirds feed on snails, limpets, and smaller crabs. The upper littoral may also be populated by barnacles that can get by with a few hours of filter-feeding each month, when the highest tides of the lunar cycle carry plankton to them.

Barnacles also are profuse in the regularly submerged mid-littoral, along with mussels and predatory sea stars. Red, brown, and green algae are primary producers here. Tide pools are typical of this zone, and there you may find sea anemones, snails, nudibranchs, hermit crabs, and small fishes (Figure 46.32).

Diversity is greatest in the lower littoral. Seaweeds usually are abundant here, with their leaves forming a canopy for myriad organisms (refer to Figure 40.20a). All major and several minor phyla of animals have members living in the lower littoral.

Sandy and Muddy Shores. Sandy and muddy shores are rather unstable stretches of loose sediments, continually rearranged by waves and currents. Few large plants grow in either place, so you won't find many grazing food webs. However, bits of organic debris from offshore or nearby landforms become trapped among the sediments and form the basis of detrital food webs.

Patterns of zonation are not as distinct on sandy shores, compared to their rocky counterparts, yet zones do occur. Along temperate coasts, for example, blue crabs and sea cucumbers live below the low tide mark. The sands between the high and low tide marks are the domain of burrowing marine worms (polychaetes), crabs, cockles, and different clams. Here also are isopods—carnivores less than a centimeter long that prey on even smaller animals (some of which graze on bacteria and diatoms that form films on individual sand grains). Beach hoppers, ghost crabs, and other crustaceans burrow in at the high tide mark during the day, then bound or lurch about the beach at night after bits of detritus.

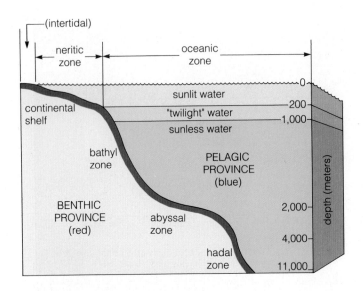

Figure 46.33 Zones of the ocean.

The Open Ocean

Ocean Zonation. Beyond the intertidal are the open oceans, which have two vast provinces (see also Figure 46.33):

1. **Pelagic province.** This includes the entire volume of ocean water. It is subdivided into the **neritic zone** (the relatively shallow waters overlying the continental shelves) and the **oceanic zone** (the water over the ocean basins). The tropical reefs described in Chapter One occur in the neritic zone as well as in the waters around islands in the oceanic zone.

2. **Benthic province.** This includes all the sediments and other substrates of the ocean bottom. It begins with the **continental shelf**, beneath the neritic zone. It extends down through the **bathyl zone** (from 100 to 2,000 meters below the water's surface), then the **abyssal zone** (2,000 down to 4,000 meters), and finally the **hadal zone** of the deep-sea trenches (4,000 down to 11,000 meters).

In the neritic and oceanic zones, photosynthetic activity is restricted to the upper surface waters. There, marine phytoplankton often drift with the currents. Collectively they form vast, suspended pastures for zooplankton that include copepods, arthropods, and shrimplike krill, all of which are food for larger, strong-swimming carnivores such as squids and roving, predatory fishes. Deep ocean waters are largely unexplored, although we have discovered some remarkable examples of diversity here (Figure 46.34).

The remains and wastes of organisms of the upper surface waters sink down through the ocean and serve as nutrients for most communities of the benthic province. A permanent thermal stratification in deep water restricts the circulation of nutrients back to surface waters. Permanent stratification is most pronounced in tropical oceans. There, production is lowest of all, even though light is the most intense and temperatures are warm. Temperate oceans are more productive because they are not permanently stratified; there are spring (and, to some extent) fall overturns that distribute oxygen and nutrients.

Although we have yet to discover huge populations in the abyssal zone, remarkable communities do thrive at **hydrothermal vents**, where water is heated and spewed from fissures between two crustal plates (Figure 46.35). The hydrothermal vent communities are unique in that their primary producers do not use sunlight as the primary energy source. The producers are chemosynthetic bacteria that obtain energy through reactions involving hydrogen sulfide. They are consumed by clams, mussels, marine worms, and other types of organisms.

Upwelling. Wherever currents stir the water and keep nutrients circulating back to the surface waters, primary productivity increases. In this respect, **upwelling** is especially important. The word refers to an upward movement of deep, nutrient-rich water along the margins of continents. Upwelling occurs when winds force surface waters along the coast to move away from the shore. When the surface water moves out, deep water moves in vertically to replace it.

For example, prevailing winds from the south and southeast force surface water away from the west coast of Peru. The cold, deeper water brought to the continental shelf by the Humboldt Current moves toward the surface and brings up tremendous amounts of nitrate and phosphates from below. Phytoplankton growing on the nutrients are the foundation for one of the world's richest fisheries, with schools of sardines and other fishes reaching huge numbers.

Periodically, warm surface waters of the western equatorial Pacific move eastward. The massive displacement of warm water affects the prevailing winds, which accelerate the eastward movement even more. The movement is enough to displace the cooler waters of the Humboldt current and prevent upwelling. This phenomenon, which local fishermen named El Niño, causes productivity to decline. The effects on the populations of marine birds that feed on sardines (as well as on the sardine industry) are catastrophic. The *Commentary* gives a closer look at what goes on during the El Niño phenomenon. With it, we come full circle to a concept presented at the start of this chapter—that interactions among the atmosphere, the oceans, and the land profoundly influence the world of life.

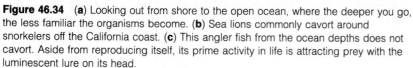

Figure 46.34 (**a**) Looking out from shore to the open ocean, where the deeper you go, the less familiar the organisms become. (**b**) Sea lions commonly cavort around snorkelers off the California coast. (**c**) This angler fish from the ocean depths does not cavort. Aside from reproducing itself, its prime activity in life is attracting prey with the luminescent lure on its head.

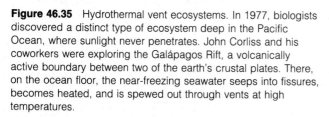

Figure 46.35 Hydrothermal vent ecosystems. In 1977, biologists discovered a distinct type of ecosystem deep in the Pacific Ocean, where sunlight never penetrates. John Corliss and his coworkers were exploring the Galápagos Rift, a volcanically active boundary between two of the earth's crustal plates. There, on the ocean floor, the near-freezing seawater seeps into fissures, becomes heated, and is spewed out through vents at high temperatures.

This hydrothermal outpouring results in deposits of zinc, iron, and copper sulfides as well as calcium and magnesium sulfates (all leached from rocks as pressure forces the heated water upward). In marked contrast to most of the deep ocean floor, these nutrient-rich, warm "oases" support diverse marine communities. Chemosynthetic bacteria and other microbes use the inorganic deposits as energy sources. They are the primary producers in a food web that includes tube worms, clams, sea anemones, crabs, and fishes (**a, b**).

So far, other hydrothermal vent ecosystems have been discovered near Easter Island (off the northwestern United States); in the Gulf of California, about 150 miles south of the tip of Baja California, Mexico; near the Galápagos Islands; and in the Atlantic.

El Niño and Oscillations in the World's Climates

(This Commentary is based on data from Eugene Rasmusson's article, "The Prediction of Drought: A Meteorological Perspective," Endeavour, 1987, volume 11, number 4.)

Oceans cover more than seventy percent of the earth's surface, so it is not difficult to imagine that variations in ocean surface temperatures can have effects on climates around the world. But it is only within the past decade that numerical modeling has begun to show how dramatic the effects may be.

In the winter of 1987–1988, for example, rainfall was below normal all along the California coast, and by spring, strict water conservation and rationing programs were in the works. People were being advised to take quick showers instead of baths, to cut back on flushing the toilet, running the dishwasher and washing machine, and to forget about deep-watering lawns and gardens. By summer, the American Midwest was in the middle of one of its worst droughts.

Drought is a relative term. If you live in Tucson, Arizona or some other desert city, "drought" is what you expect nearly year around. If you live along the western coasts or in the interior of continents, where semiarid conditions prevail, prolonged drought conditions can be disastrous.

Yet it now appears that "abnormally" dry seasons may be part of a recurring feedback relationship between sea surface temperature, drought-related conditions on land, and drought-sustaining atmospheric circulation patterns.

Consider the global climate system called the *El Niño Southern Oscillation* (ENSO). The "El Niño" part of this phenomenon is an irregular but episodic warming of the surface waters in the eastern equatorial Pacific. The "Southern Oscillation" is a global-scale seesaw in atmospheric pressure at the earth's surface—specifically, at Indonesia, northern Australia, and the southeastern Pacific. This area of the Southern Hemisphere is the world's largest reservoir of warm water (Figure **a**), and more warm, moisture-laden air rises here than anywhere else. Rainfall is also heavy here, and it releases much of the heat energy that drives the world's air circulation system.

Every two to seven years, the warm reservoir—and the associated heavy rainfall—moves eastward (Figures **b** and **c**). This causes prevailing surface winds in the western equatorial Pacific to pick up speed. The stronger winds have a more pronounced effect on "dragging" the ocean surface waters eastward. Upper ocean currents are affected to the extent

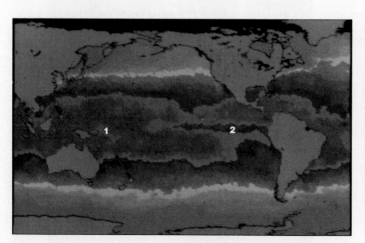

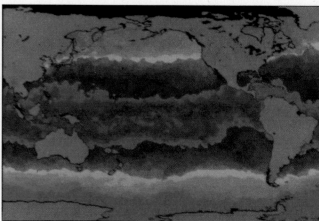

(a) Distribution of ocean surface temperatures in normal years, with the warmest waters found in the western equatorial Pacific (numeral "1"), and a tongue of relatively cold water extending westward along the equator from South America ("2").

(b) Distribution of ocean surface temperatures that were associated with the 1982–1983 ENSO episode.

that the westward transport of water slows down and the eastward transport increases. The outcome? More warm water in the vast reservoir moves east—and so on in a feedback loop between the ocean and the atmosphere.

When ENSO warm episodes do occur, there is a massive dislocation in the pattern of rainfall in the Pacific and Indian oceans. During the 1982–1983 episode, the vital monsoon rains hardly materialized at all over India—and record droughts occurred in Australia and nearby regions as well as in the Hawaiian Islands. Month after month, record rainfall drenched the arid and semiarid coasts of Ecuador and Peru. The ENSO also prolonged a devastating drought that already was under way in Africa, with its consequent and appalling human starvation.

Numerical models are now being used to study these and other episodes of climatic change a few seasons in advance. Will reliable forecasting follow? That will not happen until more and better observations are made of the interrelated systems of the ocean, land, and atmosphere.

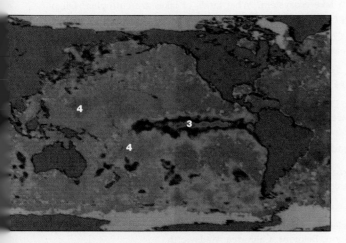

Between 1983 and 1984, temperatures were 2°–6°C warmer than they were in 1984 in the green, yellow, and red areas identified by "3." Also, large regions of the western tropical Pacific were 1°–3°C cooler than normal (dark blues identified by "4"s).

SUMMARY

1. The biosphere is the narrow zone of water, the lower atmosphere, and the fraction of the earth's crust in which organisms live. It is composed of ecosystems, each of which is influenced by the flow of energy and the movement of materials on a global scale.

2. The world distribution of species is partly a result of accidents of history (as when Pangea broke up and some species were carried away from others on the huge land fragments). It also is a result of climate, topography, and species interactions.

3. "Climate" means prevailing weather conditions, including temperature, humidity, wind velocity, degree of cloud cover, and rainfall.

4. Climate is influenced by differences in the amount of solar radiation reaching equatorial and polar regions. It is influenced also by the earth's daily rotation and its annual path around the sun, the distribution of continents and oceans, and the elevation of land masses.

5. Climatic factors interact to produce prevailing winds and ocean currents, which together influence global weather patterns. The weather in turn affects soil composition, sedimentation, and water availability, which influence the growth and distribution of primary producers (mostly photosynthetic plants).

6. Rain shadows are an example of how topography can influence the distribution of different types of organisms. A rain shadow is a region of diminished rainfall on the leeward side of high mountain ranges, and only plants and animals adapted to arid or semiarid conditions will be found on the leeward side.

7. The earth's land masses can be classified as six major biogeographic realms, each with characteristic types of plants and animals and each more or less isolated by oceans, mountain ranges, or desert barriers.

8. Biomes are distinct vegetational subdivisions of the six major realms, created by regional variations in climate, landforms, and soil composition. Each is dominated by plant species adapted to a particular set of conditions. The major types are deserts, sclerophyllous shrublands and woodlands, grasslands, forests, and tundra.

9. *Deserts:* Many deserts occur at latitudes 30° from the equator; others have formed through rain shadow effects. Evaporation greatly exceeds rainfall, vegetational cover is minimal, and plants are drought tolerant.

10. *Sclerophyllous shrublands, woodlands:* These biomes occur in semiarid coastal regions of continents between 30° and 40° latitudes. Winters are cool and moist; summers are long, hot and dry. Plants with hard, tough, evergreen leaves dominate, as in California's chaparral country. Periodic fires help maintain these biomes.

11. *Grasslands*: Grasslands occur in midcontinental regions and other areas where evaporation is high but about four to twelve centimeters of rain fall annually. The land is flat or rolling. Depending on moisture levels, shortgrass or tallgrass species dominate.

12. There are three major categories of *forest* biomes, as defined by the predominant trees:

   a. Evergreen broadleaf forests typically occur between 20° north and south latitudes. Humidity is high and annual rainfall is abundant. Tropical rain forests are an example.

   b. Broadleaf deciduous forests occur in north and south midlatitudes. Rainfall is moderate and temperatures are high during the growing season. An example is the temperate deciduous forest, which grows in regions with pronounced dry or cold winters.

   c. Evergreen coniferous forests occur mostly in the Northern Hemisphere. They include vast boreal forests (taiga), montane coniferous forests, and some forests in temperate coastal lowlands.

13. *Tundra*: Temperatures are low, rainfall is limited, and the growing season is short. Arctic tundra is characterized by permafrost, a permanently frozen layer just beneath the soil surface. Alpine tundra occurs in high mountain ranges.

14. The water provinces, which cover more than seventy percent of the earth's surfaces, include:

   a. Lentic ecosystems (standing fresh water such as ponds and lakes), lotic ecosystems (running fresh water such as streams and rivers), and brackish inland seas.

   b. Marine ecosystems include estuaries, intertidal zones, rocky and sandy shores, coral reefs, and scattered ecosystems of the open oceans.

15. All aquatic ecosystems have gradients in light penetration, temperature, salinity, and dissolved gases, features that vary daily and seasonally in most cases.

16. Many north temperate lakes show thermal stratification in summer (with warm upper waters, a thermocline in which temperature decreases abruptly with depth, and a cool layer in the profundal zone). Thorough mixing occurs only during spring and fall overturns.

17. Estuaries are partially enclosed coastal regions where fresh water mixes with seawater. Many serve as nurseries as well as feeding grounds for organisms that are vital members of marine ecosystems.

18. Rocky and sandy shores show sometimes distinct, sometimes ill-defined vertical zonation based on the tides. Environmental conditions are harsher the higher up organisms live on the shore; competition is greater the lower they are. Like tropical reefs, rocky shores are notable for their species diversity.

19. In the neritic and oceanic zones of the open ocean, photosynthetic activity is greatest in shallow coastal waters and in regions of upwelling along the margins of continents. Upwelling is an upward movement of deep, cooler water that carries nutrients to the surface.

20. A permanent thermal stratification in deep oceans, especially in the tropics, limits primary productivity compared to oceans of temperate regions, which often show spring and fall overturns.

21. The interrelatedness of ocean surface temperatures, the atmosphere, and the land is especially clear through studies of the El Niño Southern Oscillation, a recurring phenomenon that is accompanied by abnormal drought conditions in many parts of the world.

Review Questions

1. Describe the three main physical components of the biosphere. *765*

2. Define climate. What interacting factors influence climate? What does climate in turn influence? *765*

3. How do prevailing air and ocean currents help dictate the distribution of different types of ecosystems? *766–768*

4. Define rain shadow. How does a rain shadow affect ecosystems on both sides of a mountain range? *768–769*

5. Distinguish between biogeographic realm and biome. In what type of biome region would you say you live? *770–771*

6. How does the composition of regional soils affect ecosystem distribution? *771–773*

7. How do climatic conditions affect the character of the following biomes: desert, shrublands, tropical rain forests, temperate deciduous forests, evergreen coniferous forests, and tundra? *774–780*

8. What is the difference between the littoral and profundal zones of a lake ecosystem? What kinds of lakes are eutrophic? Oligotrophic? *782–784*

9. Compare rocky shores and sandy shores in terms of their zonation patterns and types of organisms. *787*

10. Is thermal stratification most pronounced in shallow or deep lakes? In temperate or tropical oceans? *788–790*

11. Define upwelling, and give an example of a region where upwelling has a profound effect on primary productivity. *788*

Readings

Barber, M., J. Burk, and W. Pitts. 1980. *Terrestrial Plant Ecology.* Menlo Park, California: Benjamin/Cummings.

Brewer, R. 1988. *The Science of Ecology.* Philadelphia: Saunders.

Eagleman, J. 1985. *Meteorology: The Atmosphere in Action.* Second edition. Belmont, California: Wadsworth.

Ehrlich, P. and J. Roughgarden. 1987. *The Science of Ecology.* New York: Macmillan.

Sumich, J. 1988. *Biology of Marine Life.* Fourth edition. Dubuque, Iowa: W. C. Brown.

HUMAN IMPACT ON THE BIOSPHERE

ENVIRONMENTAL EFFECTS OF HUMAN POPULATION GROWTH

Of all the concepts introduced in the preceding chapter, the one that should be foremost in your mind is this: Complex interactions between the atmosphere, the oceans, and the land are the engines of the biosphere. Driven by energy from the sun, they produce the world-wide temperatures and circulation patterns on which life ultimately depends. With this chapter, we turn to a related concept of equal importance. Simply put, the human population has been straining the global engines without fully comprehending that engines can crack.

To gain perspective on what is happening, think about something we all take for granted—the air around us. The composition of the present atmosphere is the outcome of geologic and metabolic events, including photosynthesis, that began billions of years ago. At some time between 6 and 4 million years ago, the human species emerged. Like us, the first humans breathed oxygen from an atmosphere of ancient origins. Like us, they were protected from harmful ultraviolet radiation by an ozone shield in the stratosphere. Their population sizes were not much to speak of, and their interactions with the biosphere were not significant. About 10,000 years ago, however, agriculture began in earnest, and it laid the foundation for rapid population growth. With agriculture, and with the medical and industrial revolutions that followed, human population growth became exponential in a mere blip of evolutionary time (page 718).

Today, our burgeoning population is placing unrealistic demands on the biosphere. As we take energy and resources from it, we give back wastes in monumental amounts. In the process, we are destroying the stability of ecosystems on land, contaminating the hydrosphere,

Figure 47.1 How do you view our world and the effects we have on it? Write your own caption for this photograph before reading this chapter; and write another one after the reading is done.

and changing the composition of the atmosphere. Our carbon dioxide wastes alone are contributing to an amplified "greenhouse effect," described on page 758, and this may already be causing an unexpected warming of the entire planet.

In the developed countries of North America, Europe, and elsewhere, the rate of population growth has more or less stabilized, and resource use per individual is no longer increasing by much. But the resource utilization levels are already high. At the same time, population growth and demands for resources are increasing exponentially in the developing countries of Central America, South America, Asia, Africa, and elsewhere—even though millions there are already starving to death and millions more suffer from malnutrition.

Many of the problems we will consider in this chapter are not going to go away soon. It will take decades, even centuries, to reverse some of the trends already in motion, and not everyone is ready to make the effort. An enlightened individual in Michigan or Alberta or New South Wales can make good attempts at conservation—but scattered attempts will not be enough. Individuals of every nation will make a concerted effort to reverse global trends only when they perceive that the dangers of *not* doing so outweigh the personal benefits of ignoring them.

Does this seem pessimistic? Think about the exhaust fumes being released into the atmosphere every time you drive a car. Think about the oil refineries, paper mills, and food-processing plants that supply you with goods but also release chemical wastes into our nation's waterways. Think about Mexico and other developing countries that produce cheap food by using an unskilled labor force and dangerous pesticides—which poison the people who work the land, and the land itself. Who changes behavior first? We have no answer to the question. We suggest, however, that a strained biosphere can rapidly impose the answer upon us.

Table 47.1 Major Classes of Air Pollutants

Carbon oxides:	Carbon monoxide (CO), carbon dioxide (CO_2)
Sulfur oxides:	Sulfur dioxide (SO_2), sulfur trioxide (SO_3)
Nitrogen oxides:	Nitric oxide (NO), nitrogen dioxide (NO_2), nitrous oxide (N_2O)
Volatile organic compounds:	Methane (CH_4), benzene (C_6H_6), chlorofluorocarbons (CFCs)
Photochemical oxidants:	Ozone (O_3), peroxyacyl nitrates (PANs), hydrogen peroxide (H_2O_2)
Suspended particles:	Solid particles (dust, soot, asbestos, lead, etc.), liquid droplets (sulfuric acid, oils, dioxins, pesticides)

CHANGES IN THE ATMOSPHERE

If you were to compare the earth with an apple from the supermarket, the atmosphere would be no thicker than the layer of shiny wax applied to it. Yet this thin, finite wrapping of air around the planet receives more than 700,000 metric tons of pollutants each day in the United States alone. **Pollutants** are substances with which ecosystems have had no prior evolutionary experience, in terms of kinds or amounts, and so have no mechanisms for dealing with them. From the human perspective, pollutants are substances that adversely affect our health, activities, or survival.

Table 47.1 lists the major air pollutants. They are carbon dioxides, sulfur oxides, nitrogen oxides, chlorofluorocarbons (CFCs), and photochemical oxidants.

Local Air Pollution

Whether air pollutants are dispersed throughout the atmosphere or concentrated at their source in a given time period depends on local climate and topography. Consider what happens during a **thermal inversion**, when a layer of dense, cool air becomes trapped beneath a layer of warm air (Figure 47.2). The pollutants cannot be dispersed by winds or rise higher in the atmosphere, so they accumulate to dangerous levels close to the ground. By intensifying a phenomenon known as smog, thermal inversions have contributed to some of the worst local air pollution disasters.

There are two types of smog (gray air and brown air), both of which occur in major cities. **Industrial smog** is gray air that predominates in industrialized cities with cold, wet winters. London, New York, Pittsburgh, and Chicago are examples. These cities use fossil fuel for heating, manufacturing, and producing electric power. The burning fuel releases airborne pollutants, including dust, smoke, ashes, soot, asbestos, oil, bits of heavy metals, and sulfur oxides. The pollutants may reach lethal concentrations when winds and rain do not disperse them. Industrial smog was the cause of London's 1952 air pollution disaster, in which 4,000 people died.

Photochemical smog is a brown and smelly trademark of large cities found in warm climates. When the surrounding land forms a natural basin, as it does around Los Angeles and Mexico City, photochemical smog can reach harmful concentrations. The main culprit is nitric oxide, which is produced chiefly by cars and other vehicles with internal combustion engines. Nitric oxide reacts with oxygen in the air to form nitrogen dioxide. When exposed to sunlight, nitrogen dioxide can react with hydrocarbons (spilled or partly burned gasoline, most often) to form photochemical oxidants. Other components of smog are ozone and PANs (short for *per*oxyacyl *n*itrates). PANs are similar to tear gas; even traces can sting the eyes and irritate the lungs.

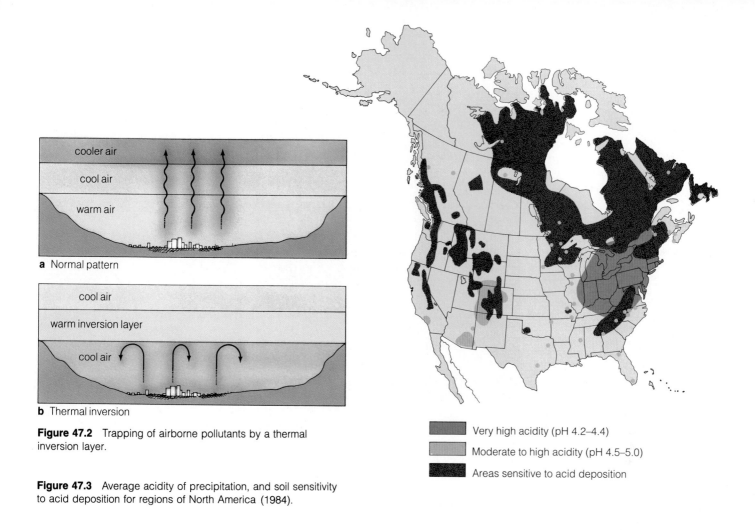

a Normal pattern

b Thermal inversion

Figure 47.2 Trapping of airborne pollutants by a thermal inversion layer.

Figure 47.3 Average acidity of precipitation, and soil sensitivity to acid deposition for regions of North America (1984).

Very high acidity (pH 4.2–4.4)

Moderate to high acidity (pH 4.5–5.0)

Areas sensitive to acid deposition

Acid Deposition

Oxides of sulfur and nitrogen are among the most dangerous air pollutants. Coal-burning power plants, factories, and metal smelters are the main sources of sulfur dioxides. Vehicles, power plants that burn fossil fuels, and nitrogen fertilizers are sources of nitrogen oxides.

Depending on climatic conditions, tiny particles of these substances may be airborne for a while and then fall to earth as **dry acid deposition**. Most sulfur and nitrogen dioxides dissolve in atmospheric water to form weak solutions of sulfuric acid and nitric acid. Winds can distribute them over great distances before they fall to earth in rain and snow; this is called **wet acid deposition**. Acid rain can be four to forty times more acidic than normal rainwater, sometimes as much as lemon juice. The acids attack marble, metals, mortar, rubber, plastic, even nylon stockings. And they are disrupting ecosystems.

Because soils and vegetation are not identical in all watersheds, some regions are more sensitive to acid deposition than others (Figure 47.3). Highly alkaline soils neutralize some of the acids before runoff carries them into lakes, streams, and rivers. Water with high concentrations of carbonates also will help neutralize the acids.

However, in watersheds throughout much of northern Europe, southeastern Canada, and scattered regions of the United States, thin soils overlie solid granite—and such soils provide little buffer against the acids.

The precipitation in much of eastern North America is thirty to forty times more acidic than it was several decades ago, and croplands and forests are suffering (Figure 47.4). All fish populations have been wiped out in 300 lakes of the Adirondack Mountains of New York. Some Canadian biologists predict that within the next two decades, fish will disappear from 48,000 lakes in Ontario. Acidic pollutants originating in industrial regions of England and West Germany are acidifying lakes and streams, and damaging large tracts of forests in northern Europe. They also are emerging as a serious problem in heavily industrialized parts of Asia, Latin America, and Africa.

Researchers confirmed years ago that power plants, factories, and vehicles are the main sources of acid depositions, and that the depositions are indeed damaging the environment. Not much has been done about it. Also, some of the responses to local air pollution standards have contributed to the problem, as when very tall smokestacks are added to power plants and smelting plants. The idea is to dump acid-laden smoke high in

Figure 47.4 Dead spruce trees in the Whiteface Mountains of New York (**a**) and in Germany (**b**). Acid deposition, perhaps in combination with prolonged exposure to other pollutants, seems to be contributing to the rapid destruction of trees in these and other forests. In some cases, prolonged exposure to multiple air pollutants directly damages the trees, especially conifers. In many cases, the pollutants weaken the trees and make them more susceptible to drought and disease.

a

b

the atmosphere so winds can distribute it elsewhere—which winds readily do. The world's tallest smokestack, in Sudbury, Ontario, accounts for about one percent of the annual worldwide emissions of sulfur dioxide.

But Canada cannot be singled out in this issue. Canada presently receives more acid depositions from industrialized regions of the northeastern United States than it sends across its southern border. Most of the acidic pollutants in Finland, Norway, Sweden, the Netherlands, Austria, and Switzerland are blown there from industrialized regions of western and eastern Europe. Prevailing winds do not stop at national boundaries; the problem is of global concern.

Damage to the Ozone Layer

The ozone layer in the lower stratosphere absorbs most of the ultraviolet wavelengths from the sun—a form of radiation that is harmful to organisms (page 112). Yet this layer has been thinning since 1976. Each spring, an ozone "hole" appears over the Antarctic; it extends over an area about the size of the continental United States. Less pronounced thinning also occurs all the way into the midlatitudes.

Satellites and high-altitude planes have been monitoring the ozone hole since 1978. Some of the data are shown in Figure 47.5. By 1987, ozone levels above Antarctica had declined by fifty percent—this compared to the previous worst case of forty percent in 1985.

The reduction in the ozone layer is allowing more ultraviolet radiation to reach the earth's surface, with potentially serious and wide-ranging consequences. Already there has been a dramatic increase in skin cancers, which almost certainly are related to increases in ultraviolet radiation. Cataracts may become more common, and it appears that ultraviolet radiation also can weaken the immune system, making individuals more vulnerable to some viral and parasitic infections. Reduction in the ozone layer also may adversely affect the world's populations of phytoplankton—the basis of food webs in freshwater and marine ecosystems and a factor in maintaining the composition of the atmosphere. (Collectively, these microbial photosynthesizers serve as a sink for carbon dioxide and a source for oxygen.)

The causes of ozone reduction are hotly debated in the scientific community. To be sure, large volcanic eruptions and cyclic changes in solar activity have some effects. But the prime suspects are chlorofluorocarbons (CFCs), which are compounds of chlorine, fluorine, and carbon. These odorless, invisible, and otherwise harmless compounds are widely used as propellants in aerosol spray cans, coolants in refrigerators and air conditioners, and industrial solvents; and they also are used in making plastic foams, including the Styrofoam cups and cartons used for packaging foods, drinks, and other consumer goods. CFCs enter the atmosphere slowly and resist breakdown. By some estimates, about ninety-five percent of the CFCs released between 1955 and 1987 are still making their way up to the stratosphere.

When a CFC molecule absorbs ultraviolet light, it gives up a chlorine atom. The chlorine can react with ozone to form an oxygen molecule and a chlorine monoxide molecule. When the chlorine monoxide reacts with

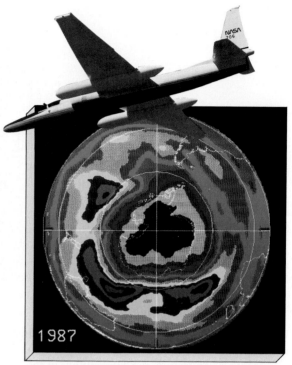

Figure 47.5 Expansion of the ozone hole over Antarctica from 1979 to October 1987, as recorded by special high-altitude planes. The lowest ozone values (the "hole") are indicated by pink colors in the plots for 1979 through 1984. In the plot for 1987, the lowest value ever recorded by that year is indicated by the black area at the center of the plot. The photograph shows the ice clouds over Antarctica that play a role in the formation of an ozone hole each spring.

a free oxygen atom, another chlorine atom is released that can attack another ozone molecule. Each chlorine atom released in the reactions can convert as many as 10,000 molecules of ozone to oxygen!

Recent studies show that chlorine monoxide levels above Antarctica are 100 to 500 times higher than at midlatitudes. Why? High-altitude clouds of ice form there during the frigid winters, and they are isolated from other latitudes by winds that rotate around the South Pole for most of the winter months (Figure 47.5). The ice provides a surface that facilitates the breakdown of chlorine compounds, so that chlorine is free to destroy ozone when the Antarctic air warms somewhat in the spring. (Hence the ozone hole.)

Since 1978, the United States, Canada, and most Scandinavian countries have banned the use of CFCs in aerosol spray cans. Aerosol uses have risen sharply in western Europe, however, as have nonaerosol uses of CFCs throughout the world. In late 1987, an international group assembled by the United Nations Environment Program agreed to a draft treaty to halve CFC emissions by the year 1999. Most nations seem certain to ratify its provisions. The treaty is a step in the right direction, although some feel that it is too little and too late. CFCs already in the air will be there for over a century, before natural processes neutralize them. You, your children, and your grandchildren will be living with their destructive effects. Think about that, the next time you carry a Styrofoam container out of a fast-food restaurant.

CHANGES IN THE HYDROSPHERE

There is a tremendous amount of water in the world, yet two of every ten humans do not have enough water or, if they do, it is contaminated. Most water is saline (too salty) and cannot be used for human consumption or agriculture. For every 1 million liters of water, only about 6 liters are in a readily usable form.

Consequences of Large-Scale Irrigation

Expansion of agriculture is the basis of the exponential growth of the human population—and about half the food being produced today grows on irrigated land.

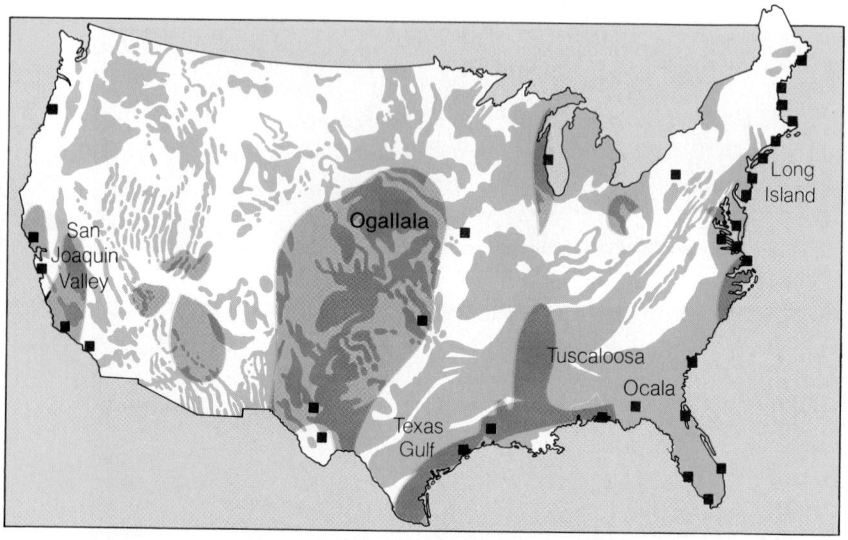

Figure 47.6 Major underground aquifers containing ninety-five percent of all fresh water in the United States. These aquifers are being depleted in many areas and contaminated elsewhere through pollution and saltwater intrusion. Blue areas indicate major aquifers; gold, the areas of groundwater depletion; and the black boxes indicate areas of saltwater intrusion.

Water is piped into agricultural fields from groundwater or from lakes and other sources of surface waters.

Irrigation can change the productivity of the land. If the irrigation water contains quantities of salts and the soil does not drain well, evaporation may cause salt buildup in the soil (salination), which can stunt growth, decrease yields, and eventually kill crop plants. Improperly drained irrigated lands can also become waterlogged. Water accumulating underground gradually raises the water table close to the soil surface, saturating the soil around plant roots with toxic saline water. Salinity and waterlogging can be corrected with proper management of the water-soil system, but the economic cost is high.

Groundwater is used for many purposes, but irrigation is often paramount. Consider what is happening to the Ogallala aquifer in the United States (Figure 47.6). Farmers withdraw so much water from the aquifer that the annual overdraft (the amount of water not replenished) is nearly equal to the annual flow of the Colorado River! As a result, the already low water tables in much of the region are falling rapidly, and stream and underground spring flows are dwindling. Where will the water come from when the aquifer is depleted?

Maintaining Water Quality

Not having enough water is serious enough, but the problem is being compounded by increased pollution of the water that is available. Water becomes unfit to drink (even to swim in) once it contains human sewage and animal wastes, which harbor pathogenic microbes. Agricultural runoff pollutes water with sediments, insecticides, herbicides, and plant nutrients. Power plants and factories pollute water with chemicals, radioactive materials, and excess heat (thermal pollution).

Pollutants accumulate in lakes, rivers, and bays before reaching their ultimate destination, the oceans. The polluted muck is dredged from the bottom of rivers and harbors in order to maintain shipping channels and then is often barged out to sea. Sewage sludge may be barged out, also. For many decades, dredge spoils, industrial wastes, and sewage sludge were dumped into the relatively shallow waters about nineteen kilometers (nearly twelve miles) off the coasts of New York and New Jersey, near the mouth of the Hudson River. There, the ocean floor became coated with a black sludge that is teeming with bacteria and viruses and that is highly contaminated with toxic metals and hospital wastes. The public is now made aware of its presence when storms wash the sludge ashore on Long Island and New Jersey beaches.

The United States has about 15,500 facilities for treating the liquid wastes from about ninety-eight percent of the urban population and from 87,000 industries. Wastes from the remaining population (mostly suburban and rural) are treated in lagoons or septic tanks or discharged—untreated—directly into waterways.

There are three levels of wastewater treatment. In *primary treatment*, screens and settling tanks remove the sludge (coarse suspended solids), which is then burned, dumped in landfills, or treated further. Chlorine often is used to kill pathogens in the water, but it does not kill them all. Also, chlorine may react with certain industrial chemicals to produce chlorinated organic compounds, some of which may cause cancer.

In *secondary treatment*, microbial populations are employed to degrade the organic matter. After primary treatment (but before chlorination), the wastewater is either (1) sprayed and trickled through large beds of exposed gravel in which the microbes live or (2) aerated in tanks and seeded with microbes. The microbial "employees" are sometimes poisoned by toxic substances dissolved in the water. Then the treatment facilities are shut down until populations are reestablished.

Secondary treatment does not remove all of the oxygen-demanding wastes, suspended solids, nitrates, phosphates, viruses, and toxic substances, including heavy metals, pesticides, and industrial chemicals. Sometimes the water gets chlorinated before being released into the waterways, and sometimes not.

Tertiary treatment involves expensive and largely experimental methods of precipitating suspended solids and phosphate compounds, adsorption of dissolved organic compounds, reverse osmosis, stripping nitrogen from ammonia, and disinfecting the water through chlorination or ultrasonic energy vibrations. It may adequately reduce pollution levels, but it is used on only about five percent of the nation's wastewater.

What all this means is that most wastewater is not being properly treated. A typical pattern is repeated thousands of times along our waterways. Water for drinking is removed *upstream* from a city, and wastes from industry and sewage treatment are discharged *downstream*. It takes no great leap of the imagination to see that pollution intensifies as rivers flow toward the oceans. In Louisiana, where waters drained from the central states flow toward the Gulf of Mexico, pollution levels are high enough to be a threat to public health. Water destined for drinking does get treated to remove pathogens—but the treatment does not remove all the toxic wastes from numerous factories upstream. *You may find it illuminating to investigate where your own city's supply of water comes from and where it has been.*

CHANGES IN THE LAND

Solid Wastes

Resources are scarce in the developing countries, and very few materials are discarded. In the more affluent countries, the United States especially, a "throwaway" mentality prevails. Consumers use something once, discard it, and buy another.

Billions of metric tons of solid wastes are dumped, burned, and buried annually in the United States alone. About 60 billion beverage containers are part of it—50 billion of which are nonreturnable cans and bottles. Paper products represent half the total volume of solid wastes. If only half of those paper wastes were recycled, energy as well as trees could be conserved. The energy it takes to produce an equivalent amount of new paper could be diverted to provide electricity to about 10 million homes each year. Also, it takes 150 acres of forest to produce the paper in each Sunday Edition of the *New York Times* alone.

Associated with the throwaway mentality is a problem that is rather unique in the world of life—what to

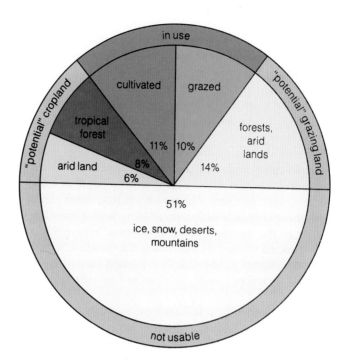

Figure 47.7 Classification of the earth's land. Theoretically, the world's cropland could be doubled in size by clearing tropical forests and irrigating arid lands. But converting this marginal land to cropland would destroy valuable forest resources, cause serious environmental problems, and possibly cost more than it is worth.

do with the solid wastes. Instead of recycling materials, as is done in natural ecosystems, we bulldoze them down the sides of canyons or dump them in wetlands or landfills. There are only so many canyons; what happens when they are filled?

A transition from a throwaway mentality to one based on conservation and reuse is economically and technologically feasible. Consumers can put pressure on manufacturers by refusing to buy goods that are lavishly wrapped, excessively boxed, and designed for one-time use. Individuals can ask the local post office to turn off their daily flow of junk mail, which wastes an astounding amount of paper, time, and energy. Individuals can urge local governments to develop well-designed, large-scale resource recovery centers. Such a center has been operating in Saugus, Massachusetts. With such systems, existing dumps and landfills would be urban "mines" from which some materials might be recovered.

Conversion of Marginal Lands for Agriculture

The dumping of solid wastes is peculiar to affluent societies. There is another type of assault on the land that is occurring throughout the world, and that is the expansion of cultivation and grazing into areas that are only marginally suitable for agriculture (Figure 47.7). Almost twenty-one percent of available land is now being used for agriculture. Another twenty-eight percent is said to be potentially suitable for cropland or grazing land, but its potential productivity is so low that conversion may not be worth the cost.

Asia and some other heavily populated regions suffer severe food shortages, yet more than eighty percent of the productive lands are already being intensively cultivated there. Those countries rely on subsistence agriculture (with energy inputs from sunlight and from human labor) and animal-assisted agriculture (with energy inputs from human labor and the work of draft animals, such as oxen). Valiant efforts have been made to improve crop production on existing land. Under the banner of the so-called **green revolution**, research has been directed toward (1) improving the varieties of crop plants for higher yields and (2) exporting modern agricultural practices and equipment to developing countries.

Modern agriculture is based on massive inputs of fertilizers and pesticides, and ample irrigation to sustain high-yield crops. It is based also on fossil fuel energy to drive the farming machines. Although crop yields are four times as high, the modern practices use up a hundred times more energy and mineral resources.

Besides, the plain truth is that developing countries depend on subsistence farmers who cannot afford to take widespread advantage of the new crop strains. The ones who can afford to make the investment come to depend on industrialized producers of fertilizers and machinery. Of necessity, the costs of fertilizers and machinery are reflected in market food prices—which are too high for much of the country's own population.

Pressures for occupying new farmland are most intense in areas of Central and South America, Asia, the Middle East, and Africa. There, the human population is rapidly expanding into marginal lands, and the repercussions will extend beyond national boundaries, as the following discussion will make clear.

Deforestation

The world's great forests play major roles in the biosphere. As we saw in Chapter Forty-Five, forested watersheds are like giant sponges that absorb, hold, and release water gradually. By influencing the downstream flow of water, forests help control soil erosion, flooding, and sediment buildup in rivers, lakes, and reservoirs. Deforestation, especially on steep slopes, leads to loss of the fragile soil layer and disrupts the watershed. In the tropics, soil loss means long-term fertility loss as nutrients are quickly washed out of the system, leaving nutrient-poor soil behind (see *Commentary*).

More than this, deforestation also can change regional patterns of rainfall as a result of altered rates of evaporation, transpiration, and runoff. For example, between fifty and eighty percent of the water vapor above tropical forests is released from the trees them-

COMMENTARY

Tropical Forests—Disappearing Biomes

Tropical forests contain an incredible variety of organisms. Yet despite the diversity, they are one of the worst places to grow crops. Because of rapid decomposition in the hot, humid climates where such forests occur, there is practically no litter on the forest floor and very little nutrient storage in the subsoil. Minerals released during decomposition are rapidly picked up by roots and mycorrhizae concentrated in the top soil layers, and most become tied up in the standing biomass.

With *slash-and-burn agriculture*, forest biomass can be reduced to nutrient-rich ashes, then the ashes can be tilled into the soil. Even then, heavy rains wash away most of the nutrients from the exposed clay soils. After a few years, cleared plots become infertile and usually are abandoned. Because nutrients are so depleted, successional replacement is extremely slow.

Developing countries in Central America, Africa, and Southeast Asia have been clearing their tropical forests on a massive scale. Human populations are growing rapidly in those countries; by some estimates, ninety percent of the total human population growth is occurring there. The land is being cleared for fuel and farmland (see Figures **a** and **b**). At the present rate of clearing, most tropical rain forests will disappear by the year 2035.

Clearing of the tropical forests means extinction for thousands of species. It will be our loss as well as theirs. Consider that a very small number of crop and livestock species represent most of the world's food supply. However, that base of food production can be broadened and made less vulnerable if we can develop new or hybrid crop plants. Through genetic engineering, the diverse organisms of tropical rain forests can be tapped as genetic resources for developing not only improved varieties of crop plants, but also new antibiotics and vaccines.

Many tropical plants already provide alkaloids that are used in drug treatments of cardiovascular disorders, cancer, and other illnesses. Aspirin, probably the most widely used drug in the world, was formulated by using a chemical "blueprint" of a compound extracted from the leaves of tropical willow trees. Coffee, bananas, cocoa, cinnamon and other spices, sweeteners, Brazil nuts, and many other foods we take for granted originated in the tropics. So did latex, gums, resins, dyes, waxes, and many oils; these

(a) Tropical rain forest in Brazil being burned to clear the land for cattle grazing.

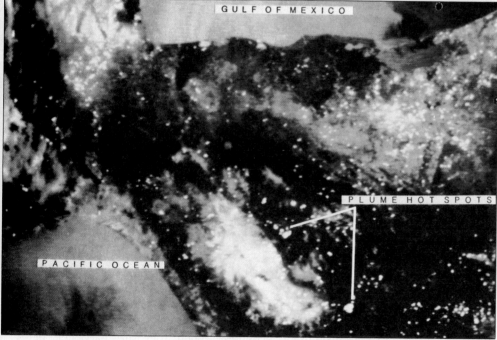

GULF OF MEXICO

PACIFIC OCEAN

PLUME HOT SPOTS

(b) Satellite photograph of southern Mexico and northern Guatemala taken on April 18, 1984. The well-defined white spots are fires, most of which are associated with land clearing for agriculture. The white area near the center of the photograph (the Grijalva Basin) is a major agricultural region; most of the burning here is for clearing previously cultivated land. The areas in the upper left (around Veracruz) and in the lower right (around the Guatemala border) are primarily virgin tropical forest, being cleared at great ecological cost for agriculture. (Satellite sensory devices penetrated the smoke to reveal the underlying fire activity.)

and other substances are used in such diverse products as ice cream, toothpaste, shampoo, condoms, cosmetics, perfumes, records, tires, and shoes.

Several international groups have been working together on a comprehensive plan to preserve tropical forests, and the plan is already being implemented.

A few developing countries have been reevaluating their agricultural policies. Brazil has designated about 100,000 square kilometers of tropical rain forest as unsuitable for agriculture. The region has been set aside for ecological research and for recreation. How many other countries will follow suit?

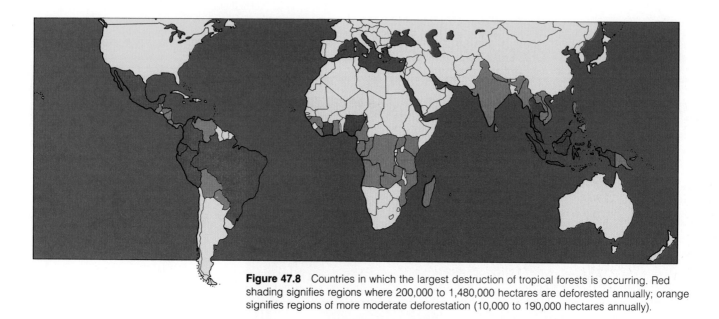

Figure 47.8 Countries in which the largest destruction of tropical forests is occurring. Red shading signifies regions where 200,000 to 1,480,000 hectares are deforested annually; orange signifies regions of more moderate deforestation (10,000 to 190,000 hectares annually).

Figure 47.9 Dust storm approaching Prowers County, Colorado, in 1934. The Great Plains of the American Midwest is normally dry, windy, and subject to severe recurring droughts. Beginning in the 1870s, the land was converted to agriculture. Overgrazing destroyed large regions of natural grassland, leaving the ground bare. In May 1934, the entire eastern portion of the United States was blanketed with a massive dust cloud of topsoil blown off the land, giving the Great Plains a new name—the Dust Bowl. About 9 million acres of cropland were destroyed and 80 million more severely damaged. Today, without massive irrigation and intensive conservation farming, desertlike conditions could prevail in this region.

selves. Without trees, annual precipitation declines and the region gets hotter and drier. Rain that does fall rapidly runs off the bare soil. As the local climate gets hotter and drier, soil fertility and moisture levels decline even more. Eventually, sparse grassland or even desertlike conditions might prevail where there had once been a rich tropical forest.

Clearing large tracts of tropical forests also may have global repercussions. Consider that these forests absorb much of the solar radiation reaching the equatorial regions of the earth's surface. When they are cleared, the land becomes shinier, so to speak, and reflects more incoming energy back into space. Consider also that the

trees of these vast forested regions help maintain the global cycling of carbon and oxygen through their photosynthetic activities. When they are harvested or burned, the carbon stored in their biomass is released to the atmosphere in the form of carbon dioxide—and this may play a role in the amplified greenhouse effect.

Almost half of the world's expanses of tropical forests has already been cleared for cropland, grazing land, timber, and fuelwood. Deforestation is greatest in Brazil, Indonesia, Colombia, and Mexico (Figure 47.8). At present rates of clearing and degradation, only Brazil and Zaire will have large tracts of tropical forest in the year 2010; by 2035, those forests also will be gone.

Desertification

The term **desertification** refers to the conversion of grasslands, rain-fed cropland, or irrigated cropland to desertlike conditions, with a drop in agricultural productivity of ten percent or more. About 900 million hectares (2 billion acres) have turned into deserts over the past fifty years; at least 20 million hectares are still being transformed each year. Prolonged drought may accelerate desertification, as it did in the American Midwest many decades ago (Figure 47.9). Today, however, large-scale desertification is occurring mainly as a result of overgrazing on marginal lands.

In Africa, for example, there are too many cattle in the wrong places. Cattle require more water than the wild herbivores that are native to the region, and so they move back and forth between grazing areas and watering holes. When doing this, they trample grasses and compact the soil surface (Figure 47.10). In contrast, gazelles, elands, and other native herbivores obtain most (if not all) of the water they require from the vegetation they eat. They also are better at conserving water; little is lost in feces, compared to cattle.

In 1978, the biologist David Hopcraft started a ranch composed of antelopes, zebras, giraffes, ostriches, and other native herbivores. He is raising cattle as "control groups" in order to compare costs and meat yields on the same land. So far, results are exceeding expectations. Native herds are increasing steadily and yielding meat. And range conditions are improving, not deteriorating. There are still problems to be overcome. African tribes have their own idea of what constitutes "good" meat, and some tribes view cattle as the symbols of wealth in their society.

Figure 47.10 Desertification in the Sahel, a region of West Africa that forms a belt between the dry Sahara Desert and tropical forests. This is savanna country that is rapidly undergoing desertification as a result of overgrazing and overfarming.

A QUESTION OF ENERGY INPUTS

Paralleling the J-shaped curve of human population growth is a steep rise in energy consumption. The rise is due not only to increased numbers of energy users, but also to extravagant consumption and waste.

For example, in one of the most temperate of all climates, a major university constructed seven- and eight-story buildings with narrow, sealed windows. The windows cannot be opened to catch the prevailing ocean breezes; the windows and the buildings themselves were not designed or aligned to take advantage of the abundant sunlight for passive solar heating and breezes for passive solar cooling. Massive energy-demanding cooling and heating systems are used instead.

Inefficiencies of this sort may be curtailed sooner than might be expected, for current energy supplies are limited. When you hear talk of abundant energy supplies, keep in mind that there is an enormous difference between the total supply and the net amount available. **Net energy** is the energy left over after subtracting the energy used to locate, extract, transport, store, and deliver energy to consumers. In addition, some sources of energy, such as direct solar energy, are renewable; others, such as coal and petroleum, are not. Currently, seventy-nine percent of the energy stores being tapped fall in the second category (Figure 47.11).

Fossil Fuels

Fossil fuels are the carbon-containing remains of plants that lived hundreds of millions of years ago. The plants were buried and compressed in sediments and gradually transformed into coal, petroleum (oil), and natural gas. Often you will see references to our annual "production" rates for fossil fuels. How much oil, coal or natural gas do we really produce each year? None. We simply *extract* it from the earth.

Even with stringent conservation efforts, known petroleum and natural gas reserves may be depleted during the next century. As petroleum and natural gas deposits become depleted in easily accessible areas, we begin to seek new sources, often in wilderness areas such as Alaska and in other fragile environments such as the continental shelves. The *net* energy decreases as costs of extraction and transportation increase; the environmental costs of extraction and transportation escalate.

Colorado, Utah, and Wyoming probably have more potential oil than the entire Middle East. These states have vast deposits of **oil shale**, a type of buried rock containing the hydrocarbon kerogen. However, concentrating and collecting kerogen, then converting it into usable oil, may cost so much that the net energy yield would be negligible. The extraction process may disfigure the land, increase water and air pollution, and tax existing water supplies in regions already facing water shortages.

What about coal? In theory, world reserves can meet the energy needs of the entire human population for at least several centuries. But coal burning has been the largest single source of air pollution; most coal reserves contain low-quality, high-sulfur material. Unless the sulfur is removed before or after burning, sulfur dioxides are released into the air and add to the global problem of acid deposition. Fossil fuel burning also adds carbon dioxide to the atmosphere and so helps amplify the greenhouse effect.

Pressure is on to permit widespread strip-mining of coal reserves close to the earth's surface. Strip mining limits the usefulness of the land for agriculture, grazing, and wildlife. Restoration is difficult and expensive in arid and semiarid lands, where most strip-mining occurs.

Nuclear Energy

As Hiroshima burned in 1945, the world recoiled in horror from the destructive potential of nuclear energy. Optimism replaced horror as nuclear energy became publicized during the 1950s as an instrument of progress. Today, nuclear power plants dot the landscape. Industrialized nations that are poor in energy resources, including France, depend heavily on nuclear power. Yet in most countries, plans to extend reliance on nuclear energy have been delayed or canceled. The cost, effi-

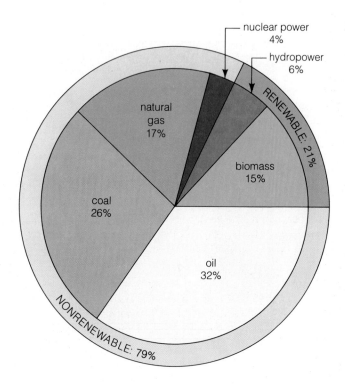

Figure 47.11 World consumption of nonrenewable and renewable energy sources in 1986.

ciency, environmental impact, and safety of nuclear energy are being seriously questioned.

The overall net energy produced by nuclear reactors is relatively low, and the cost of constructing nuclear power plants is high—much higher than initially expected. By 1990, using nuclear energy to generate electricity may cost slightly more than using coal—even if the coal-burning plants are equipped with expensive pollution control devices.

What about safety? Radioactivity escaping from a nuclear plant during normal operation is actually less than the amount released from a coal-burning plant of the same capacity. Also, nuclear plants do not add carbon dioxide to the atmosphere, as coal-burning plants do. However, there is the potential danger of a **meltdown**. As nuclear fuel breaks down, it releases considerable heat, which typically is absorbed by water that is circulating over the fuel. The heated water produces steam, which drives electricity-generating turbines. Should a leak develop in the circulating water system, water levels around the fuel might plummet. The nuclear fuel would heat rapidly, past its melting point. The melting fuel would pour onto the floor of the generator, where it would come into contact with the remaining water and instantly convert it to steam. Formation of enough steam, along with other chemical reactions, could blow the system apart, releasing radioactive material into the surroundings. Also, the overheated reactor core could melt through its thick concrete containment slab and into the earth, causing groundwater contamination.

All reactors have secondary cooling systems designed to flood the reactor if the initial cooling water is lost. In 1979, the Three Mile Island nuclear plant in Pennsylvania underwent partial meltdown and came close to a full meltdown as a result of equipment failure and human error. Some critics believe a meltdown was avoided only by luck. Others believe the incident proved the effectiveness of multiple backup safety systems in nuclear power plants.

In 1986, the potential dangers of nuclear power were brought into sharp focus when a meltdown occurred at the Chernobyl power station in the Soviet Union. Radiation was released into the atmosphere as the plant's containment structures were breached. A number of people died immediately and others died of radiation sickness during the weeks after the accident. Throughout Europe today, people are still concerned about the long-term consequences. How long will the environment be contaminated? What are the risks of cancer from such exposure? We are in the midst of our first major, real-world experiment to find out.

The Chernobyl incident underscores the consequences of nuclear accidents. What about routine nuclear wastes? Nuclear fuel cannot be burned to harmless ashes, like coal. After about three years, the fuel elements of a reactor are spent. They still contain uranium fuel, but they also contain hundreds of new radioactive isotopes produced during the reactor operation. Altogether, the wastes are an enormously radioactive, extremely dangerous collection of materials. As they undergo radioactive decay, they produce tremendous heat. They are immediately plunged into water-filled pools and stored for several months at the power plant. The water cools the wastes and keeps radioactive material from escaping. At the end of the holding period, the remaining isotopes are still lethal, and the decay rates of some of them mean that they must be isolated for 10,000 years. If a certain isotope of plutonium (^{239}Pu) is not removed, the wastes must be kept isolated for a quarter of a million years!

There are plans to seal radioactive wastes in ceramic material, place them in steel cylinders, then bury them deep underground in supposedly stable rock formations that are free from exposure to salt water and earthquakes for at least 10,000 years. No radioactive wastes have yet been put into permanent underground storage in the United States. Such facilities are not expected to be available until after the turn of the century.

The development of another type of reactor is being considered. In theory, breeder reactors would consume a rare isotope of uranium, and, in the process, convert a much greater amount of a common form of uranium (^{238}U) to an isotope of plutonium that can be used as nuclear fuel. However, even though a conventional reactor cannot explode like an atomic bomb, a breeder reactor could undergo a very small nuclear explosion.

A third type of nuclear power source on the drawing boards is fusion power, in which hydrogen atoms are fused to form helium atoms, with considerable release of energy. The process is analogous to the reactions that produce heat energy in the sun. The scientific, technological, and economic problems associated with developing fusion power are so great that, without a major breakthrough, fusion power is not expected to be available to produce electricity on a commercial basis until the last half of the next century, if ever.

One more point should be made here. Earlier, we considered the catastrophic extinctions that marked the boundaries of the great eras of geologic time. We can only speculate on their causes—collisions between asteroids and the earth being at least one of them. Quite probably, nuclear energy has placed in human hands the means of equally catastrophic extinctions. We will not know this until it happens, but several predictions have been made on the basis of detailed computer simulations and analyses.

According to one scenario, a nuclear exchange involving about one-third of the existing American and Soviet arsenals would probably kill outright between forty and sixty-five percent of the human population (along with a good portion of most other forms of life). Those escaping immediate death would have to remain in shelters for a week to three months or more to avoid exposure to dangerous radiation levels. The nuclear detonations would inject a huge, dark cloud of soot and smoke over most of the earth, especially the Northern Hemisphere, and would block out the sun. Much of the planet might be thrown into darkness and temperatures would fall below freezing for months—an effect called **nuclear winter**. If the freezing lasted for a shorter time, we might have a nuclear "autumn"—but even then, the cold temperatures and darkness might be well beyond the tolerance limits of many plant and animal species. Will we cause the "catastrophe" that brings the Cenozoic Era to a close? And what, we can only wonder, will we usher in as a result?

PERSPECTIVE

Molecules, cells, tissues, organs, organ systems, multicelled organisms, populations, communities, ecosystems, the biosphere. These are the architectural systems of life, assembled in increasingly complex ways over the past $3\frac{1}{2}$ billion years. We are latecomers to this immense biological building program. Yet, during the relatively short span of 10,000 years, we have been restructuring the stuff of life at all levels—from recombining DNA of different species to changing the nature of the land, the oceans, and the atmosphere.

It would be presumptuous to think we are the only organisms that have ever changed the nature of living systems. Even during the Proterozoic Era, photosynthetic organisms were irrevocably changing the course of biological evolution by gradually enriching the atmosphere with oxygen. In the present as well as the past, competitive adaptations have assured the rise of some groups, whose dominance has assured the decline of others. Thus, change is nothing new to this biological building program. What *is* new is the accelerated, potentially cataclysmic change being brought on by the human population. We now have the population size, the technology, and the cultural inclination to use energy and modify the environment at frightening rates.

Where will rampant, accelerated change lead us? Will feedback controls begin to operate as they do, for example, when population growth exceeds the carrying capacity of the environment? In other words, will negative feedback controls come into play and keep things from getting too far out of hand?

Feedback control will not be enough, for it operates only when deviation already exists. Our explosive population growth and patterns of resource consumption are founded on an illusion of unlimited resources and a forgiving environment. A prolonged, global shortage of food or the passing of a critical threshold for the global engines can come too fast to be corrected. At some point, such deviations may have too great an impact to be reversed.

What about feedforward mechanisms? Many organisms have early warning systems. For example, skin receptors sense a drop in outside air temperature. Each sends messages to the nervous system, which responds by triggering mechanisms that raise the core temperature before the body itself becomes dangerously chilled. With feedforward control, corrective measures can begin before change in the external environment significantly alters the system.

Even feedforward controls are not enough for us, for they go into operation only when change is under way. Consider, by analogy, the DEW line—the Distant Early Warning system. This system is like a sensory receptor, one that detects intercontinental ballistic missiles that may be launched against North America. By the time the system detects what it is designed to detect, it may be far too late, not only for North America but for the entire biosphere.

It would be naive to assume we can ever reverse who we are at this point in evolutionary time, to de-evolve ourselves culturally and biologically into becoming less complex in the hope of averting disaster. However, there is reason to believe that we can avert disaster by using a third kind of control mechanism, one that is uniquely our own. We have the capacity to anticipate events *before* they happen. We are not locked into responding only after irreversible change has begun. We have the capacity to anticipate the future—it is the essence of our visions of utopia or of nightmarish hell. Thus we all have the

capacity for adapting to a future that we can partly shape. We can, for example, learn to live with less. Far from being a return to primitive simplicity, it would be one of the most complex and intelligent behaviors of which we are capable.

Having that capacity and using it are not the same thing. We have already put the world of life on dangerous grounds because we have not yet mobilized ourselves as a species to work toward self-control. Our survival depends on predicting possible futures. It depends on designing and constructing ecosystems that are in harmony not only with what we define as basic human values but also with the biological models available to us. Human values can change; our expectations can and must be adapted to biological reality. *For the principles of energy flow and resource utilization, which govern the survival of all systems of life, do not change.* It is our biological and cultural imperative that we come to terms at last with these principles, and ask ourselves what our long-term contribution will be to the world of life.

Review Questions

1. Make a list of the advantages you personally enjoy as a member of an affluent, industrialized society. Then list some of the drawbacks. Do you believe that the benefits outweigh the costs? (This is not a trick question.)

2. Describe some of the potential global consequences of acid deposition, reduction of the ozone layer, and amplification of the greenhouse effect. *795–797; see also 758*

3. What are some of the consequences of deforestation? Of desertification? *800–803*

4. It is not likely that humans will turn their back on their technological advances and economic growth as a way of putting the biosphere on a stable footing. Is it possible that massive amounts of money and technology will be put to use to find answers as well as solutions? What prospects and problems do you foresee in making a global effort in this regard?

Readings

Anderson, S. 1985. *Managing Our Wildlife Resources.* Columbus, Ohio: Merrill.

Gribbin, J. 1988. *The Hole in the Sky.* New York: Bantam. An informative little paperback.

Miller, G. T. 1988. *Living in the Environment.* Fifth edition. Belmont, California: Wadsworth.

Mohnen, V. August 1988. "The Challenge of Acid Rain." *Scientific American.* 259(2):30–38.

Mooney, H., P. Vitousek, and P. Matson. 1987. "Exchanges of Materials Between Terrestrial Ecosystems and the Atmosphere." *Science* 238:926–932.

Schneider, S. May 1987. "Climate Modeling." *Scientific American.* 256(2):72–80.

Wilson, E. 1988. *Biodiversity.* Washington, D.C.: The National Academy of Sciences.

With the arrival of spring, a male white-throated sparrow living in a patch of swampy Canadian forest whistles a song that sounds something like "Sam Peabody, Peabody, Peabody." He repeats the Sam-Peabody song thousands of times and with such clarity and consistency, we might well wonder how he does it, and what good it does him. We might also wonder why the male swamp sparrows and white-crowned sparrows living in the same patch of forest eschew the Sam-Peabody song and belt out distinctive songs of their own. Moreover, do all those male birds automatically "know" what they are supposed to sing the first time they do it, or does each one learn something from the environment that influences the way they sing?

Questions of this sort lead us into the world of animal behavior studies. Earlier, we considered various aspects of the external environment and the internal state that stimulate an animal's sensory receptors (page 368). We saw how the nervous system and endocrine system process information about such stimuli and then mobilize the body's effectors (muscles and glands) to bring about appropriate, coordinated responses to the stimulation. Here we will consider specific examples of the internal mechanisms underlying predictable patterns of behavior, as well as examples of learning experiences by which some predictable responses can be modified in novel ways. The following premises provide a framework for the chapter:

1. *Animal behavior* refers to the coordinated neuromotor responses an animal makes to external and internal stimuli.

2. Behavioral responses are outcomes of the integration of sensory, neural, endocrine, and effector components, all of which have a genetic basis and therefore are subject to evolution by natural selection.

3. Heritable, genetically based neural programs provide each new individual with a means of responding to situations that members of its species are likely to encounter in their environment (page 367).

4. Interactions between the animal and its environment can lead to modifications in the neural mechanisms underlying behavior; this can happen while the animal embryo is developing as well as later, when juvenile and adult animals undergo learning experiences.

5. Behavior is adaptive, to the extent that it contributes to the reproductive success of the individual.

48

ANIMAL BEHAVIOR

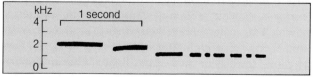

Figure 48.1 A male white-throated sparrow and a sound spectrogram of his territorial song.

MECHANISMS OF BEHAVIOR

Genetic Foundations of Behavior

The motor responses an animal can make to its environment are dictated by the gridwork of neurons in its nervous system and the patterns of activity among them (page 367). Thus, for example, our capacity to control our lips and tongue when speaking resides in the specific design features of our brain (Figure 25.11), just as a white-throated sparrow's ability to sing depends on the coordinated action of millions of neurons in his brain. Genetic instructions are necessary for the growth and "design" of the brain; hence genes contribute in an important way to the development of all behavior.

For example, suppose a researcher takes eggs from the nests of white-throated sparrows and white-crowned sparrows. She places the eggs in an incubator and hand-rears the young after they hatch (which, by the way, is a monumentally difficult task that gives one great respect for the parental abilities of birds). The baby white-throats and white-crowns are both reared under the exact same conditions, fed exactly the same foods, and in all ways treated similarly. As the birds grow up, they are permitted to hear the tape-recorded songs of adult white-throats and white-crowns. Before a year passes, the males will be singing. The white-throats will sing "Sam Peabody, Peabody, Peabody." And the white-crowns will sing a different song, one with all the whistles and buzzy trills characteristic of their species.

This experiment does not demonstrate that bird song is "genetically determined" in the sense that the environment plays no role whatsoever in its development. However, since the white-throated and white-crowned birds were exposed to identical environmental conditions, the *difference* in their singing behavior must be the result of genetic differences between them. The number of genes involved to produce the difference is not important to this argument. Even a one-gene difference could produce this result. One gene-specified enzyme acting at an early stage of embryonic development can have large effects on the sequence of events by which part of the nervous system develops. By analogy, suppose you substitute garlic powder for sugar when following a recipe for chocolate cake. Even if all the other ingredients and all subsequent cake-making steps are identical, that one substitution will have a dramatic effect on the actual cake.

The effects of single genes on behavioral development have been studied in great detail. Studies of fruit flies (*Drosophila*) have been especially revealing. For example, specific mutations can change the frequency at which the males vibrate their wings to produce a courtship "song." Some mutations alter the tendency of flies to move toward or away from light. Other mutations affect the capacity of flies to remember to avoid entering tubes with distinctive odors and wired to give the entrant an electric shock.

Keep in mind that it is incorrect to say there is a "gene for courtship song" in fruit flies. Rather, gene mutation can affect the development of this and other forms of behavior, as can happen when the gene codes for a key enzyme in a developmental pathway that shapes the part of the nervous system concerned with courtship song.

Hormonal Effects on Behavior

Among the proteins specified by genes are *hormones*, the signaling molecules produced by cells in one body region and carried by the bloodstream to target cells elsewhere in the body. The cellular responses to a hormonal message may ultimately have profound behavioral consequences.

Melatonin, recall, is a hormone produced by cells in the pineal gland of many vertebrates (page 358). This hormone serves as a *behavioral primer*; changes in its concentration in the blood prepare the nervous system for changes in reproductive activity by way of its regulatory effect over the gonads (the testes and ovaries). The pineal gland secretes melatonin mostly at night; photoreceptors activated during the day cause a decrease in its secretion. Thus the glandular output varies with the changing daylength of different times of year; it shows **seasonal photoperiodicity**.

In birds, light penetrating directly through the head feathers and the skull influences melatonin production. Imagine what goes on in the brain of a white-throated sparrow during the lengthening days of spring. The pineal puts out less and less melatonin, the gonads are freed from the hormone's inhibitory effects, and the gonads increase in size compared to what they were in the fall. As the gonads grow, they release hormones of their own, thereby setting in motion a sequence of behavioral responses that will include migration, singing and other forms of territorial behavior, and mating. All of these forms of behavior will occur in a predictable sequence, at a time of year that will most favor reproductive success.

But now a new question arises: Why do the males and *not* the females of so many songbirds sing? Anatomical studies of some songbirds revealed differences between the sexes in the structure and size of brain regions that connect with pathways leading to the syrinx, or vocal organ. (The syrinx is a modified region of the bird windpipe, bronchial tubes, or both.) We know now that hormones influence the development of this entire network, the so-called **song system**.

In very young *male* but not female songbirds, estrogen secretions act on embryonic brain cells, initiating the masculinization of the brain and development of the song system. (Interestingly, when researchers implanted tiny pellets of estrogen in nestling *female* zebra finches, the females developed a brain with a malelike song system.)

A masculinized brain is necessary for singing behavior, but in itself it is not enough to cause a bird to sing. At the start of the breeding season in temperate zones, testosterone is manufactured in the enlarged testes of the male bird. Cells of the song system have receptors that can bind this hormone. Binding triggers changes in metabolic activity that enable the bird to sing after it has staked out a territory and must repel other males and attract a female. As you might predict, if female zebra finches receive testosterone implants when they are adults, they will not sing—*unless* they also have been exposed to estrogen secretions early in life (Figure 48.2).

Thus, estrogen *organizes* the development of the song system; then testosterone *activates* the song system and prepares the bird to sing when properly stimulated. Coordinated interactions between nerve cells and hormones are necessary if a white-throated sparrow is ever to sing so much as a single "Sam Peabody, Peabody, Peabody."

INSTINCTIVE BEHAVIOR

When presented with essence of slug, a newly born garter snake flicks its tongue, bringing the chemical scent into its mouth. Even when a cotton swab doused with slug extract is the snake's first encounter with potential food, it will complete this functional response; sometimes the snake goes so far as to strike at the swab. In nature, this is how the garter snake detects and captures small slugs for food without ever having seen another snake do the same thing.

The tongue-flicking response is an example of a behavior that could be called instinctive. In the past, "instinctive" or "innate" behavior was said to be genetically determined, the mistaken idea being that some forms of behavior can arise without environmental influences. However, *no aspect of phenotype—including behavior—can develop without both genetic and environmental contributions*. The tongue-flicking response requires a tongue, chemical receptors, and a nervous system, all of which developed in the snake embryo as a result of complex interactions between genes and the embryonic environment.

Nevertheless, the term **instinct** can be salvaged if we limit its meaning to the capacity of an animal to complete a fairly complex, stereotyped response to a first-time encounter with a key stimulus—in other words, without having had prior experience with the stimulus.

a

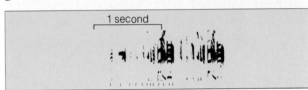

b

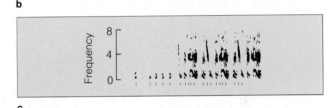

c

Figure 48.2 (a) Zebra finch, with sound spectrograms of the full song of (b) the male and of the song of (c) a female that had been exposed to estrogen as a nestling, then given a testosterone implant as an adult.

What is striking about instinctive behavior is that it seems to be automatically triggered by rather limited environmental cues. When you shake their nests, the baby birds of many species respond with gaping behavior, which is exactly what they do when the (food-carrying) parent lands on the nest rim (see, for example, Figure 34.2d). Male stickleback fish aggressively chase and strike almost any red object placed near their nest even if they have never before seen a red object of any sort. Male sticklebacks happen to have red bellies and they happen to guard nests containing the eggs that they themselves have fertilized, although they will eat the eggs of other sticklebacks in other nests. Under natural

a

b

Figure 48.3 A complex, innate behavioral response by a newly hatched cuckoo to a sign stimulus—spherical objects in its "foster parent's" nest. The European cuckoo lays its eggs in the nests of other species. Even before a newly hatched cuckoo can open its eyes, it responds to the shape of the host's eggs and shoves them out of the nest (**a**). The foster parents continue to feed the usurper, even when it has grown larger than they are (**b**).

conditions, these fish chase away rival red-bellied males. Very young human infants can be induced to smile simply by showing them a face-sized flat mask with two dark spots corresponding to where eyes would be on a face (one "eye" will not do the trick).

Many years ago, Konrad Lorenz, Niko Tinbergen, and others noted that part of a complex object triggers a complete instinctive response in a host of animals. They named the triggering component of the stimulus the **sign stimulus**. They also proposed that the nervous system of the responding animal was organized into neural units, called *innate releasing mechanisms*, each of which controlled a specific response to a specific sign stimulus. Although these ideas have been modified over the years, it is still worth recognizing that the neural organization of many animals promotes effective first-time responses to sign stimuli in their environment (Figure 48.3).

LEARNING

Many animals respond innately to certain objects yet also store information about the *connections* between various experiences and the consequences of their actions. This information storage leads to **learning**, the adaptive modification of behavior in response to specific experiences during the individual's life. Thus, although the initial feeding response of a newly hatched bird is innate, later on it learns from personal experience where it can find food most profitably—in other words, which food items occur in which areas, and even the best time of day to search for a particular kind of food.

Often, learning has been said to be "environmentally determined" as if it depends solely on experience. Yet the ability to learn resides within the nervous system, where information is stored and the neural wiring is modified in ways that lead to altered behavior. Nervous systems depend on genetic instructions for their development, and the neural wiring that results greatly influences what, exactly, its owner can learn. Thus it is just as inaccurate to ignore the genetic component of learning as it is to consider instincts purely genetic.

Categories of Learning

The conditions under which learning takes place are varied. Here we will briefly describe a few of the different categories of learning as reference points for later discussions.

Associative Learning: In this form of behavior, an animal has a capacity to make a connection between a new stimulus and a familiar one. Ivan Pavlov, a physiologist who was interested in digestive juice secretion, provided

a   b

Figure 48.4 (**a**) Human imprinting objects. No one can tell these goslings that Konrad Lorenz is not Mother Goose. (**b**) An imprinted rooster wading out to meet the objects of his affections. During a critical period of the rooster's life, he was exposed to a mallard duck. Although sexual behavior patterns were not yet developing during that period, the imprinting object became fixed in the rooster's mind for life. Then, with the maturation of sexual behavior, the rooster sought out ducks, forsaking birds of his own kind, and lending further support to the finding that imprinting may be one of the reasons why birds of a feather do flock together.

a classic, controlled study of associative learning. Pavlov observed that his laboratory dogs salivated just after he placed a meat extract on their tongues. He interpreted this to be a simple reflex response. Then he found that if he rang a bell just before giving the dogs the extract, the dogs began salivating at the sound of the bell alone. Pavlov called this new response a **conditioned reflex**, for the dogs had come to associate the sound of the bell (a conditioned stimulus) with food (a reinforcing stimulus).

Instrumental conditioning is another form of associative learning. Here, a reinforcing stimulus (either reward or punishment) appears after a particular behavior is performed by chance. The animal learns by trial and error. For example, earthworms can do this in simple T-mazes (which have a base and two arms shaped like a "T"). They enter the maze at the base, and if they turn down one of its arms, they encounter an irritating stimulus such as an electric shock. If they turn down the other arm, they encounter a moist, darkened chamber that approximates the earthworm habitat. After many trials and enough shocks, the "right" response becomes more frequent.

Extinction: In the forms of learning just defined, the behavior persists for as long as the reinforcement persists. However, if the reinforcing stimulus is withdrawn so the animal does not encounter it again, the learned behavior may soon become extinguished. This, too, is a learning process; it is called extinction.

Latent Learning: This term refers to an ability to store information about features of the environment, information that is later used in guiding the animal through its habitat. For example, a rat will learn to find its way through a complicated maze with fewer food-reinforced learning trials *if* it is first given a chance to explore the maze, without any reward when it happens to find the way out.

Insight Learning: It is only among some primates that insight learning has been adequately demonstrated. With this behavior, novel problems can be solved without trial-and-error practice. In a sense, insight learning is a trial-and-error process that goes on in the brain. It is a synthesis of accumulated experiences that can suggest what responses might be appropriate in new situations. For example, some captive chimpanzees will study a banana dangling from the high ceiling of their enclosure, then they will pile up boxes scattered around the enclosure, having perceived that they can climb the boxes to reach the out-of-reach banana.

Imprinting

Among at least some animal species, the capacity to learn specific kinds of information may be especially pronounced during certain early stages of development. This form of learning, which seems to be time dependent, is called **imprinting**.

For example, some birds are able to walk a few hours after hatching and will traipse after their mother as she moves away from the nest. In one of his studies, Lorenz separated newly hatched goslings from their mother and discovered they followed *any* moving object, even a human being (Figure 48.4). The young birds must learn something when they walk after an object, because they form an attachment to their guide and hurry after it whenever it moves off. However, if young chicks or goslings are not offered any object to follow within a couple of days after hatching, they lose their readiness to imprint. *Thus it appears that the young animals are primed during a short sensitive period early in life to form a learned attachment to a moving object, normally their mother.*

Imprinting has significant long-term consequences. When male goslings or ducklings have matured, they direct their sexual behavior toward members of whatever species they followed as hatchlings. Again, in nature this normally would be a female member of their own species. However, sexually mature male goslings that had imprinted on Lorenz early in life preferred to court humans!

Imprinting and Migration

One of the most intriguing examples of imprinting comes from studies of bird migration. Many animals, including a large number of birds, have the capacity to travel from a summer breeding area to a winter refuge of some sort. The length of the trip varies from species to species. The arctic tern makes round-trip journeys of 22,000 miles per year, a truly remarkable achievement. To migrate across unfamiliar terrain, an animal must possess a **compass sense** (an ability to travel in a constant direction) and a **navigational sense** (a sense of its destination).

Stephen Emlen's work with a small songbird, the indigo bunting, revealed that imprinting plays a role in the development of its compass sense. This bird breeds in the northeastern United States but spends its winters in Mexico south to Panama. The young depart on their own for the wintering grounds when they are just a few months old, and they fly at night. The buntings, Emlen suspected, must use the position of the stars as an orientation guide for their nocturnal travels. Emlen took young, hand-reared birds into a planetarium where he could control the position of the apparent stars. Each bird had its own cage with an ink pad on the floor and sloping paper-covered walls (Figure 48.5). In the fall, the buntings became restless at night and attempted to fly. As they jumped from the floor of their cages they landed on the paper walls and left a mark of their chosen direction. When the planetarium sky mimicked the natural night sky of the fall, the marks were clumped at the south end of the cage; clearly the birds were using the position of the stars as a compass. When the planetarium star pattern was shifted 90° from the actual fall pattern, the birds also shifted the orientation of their nightly leaps by 90°.

Interestingly, when the North Star and the stars of the Big Dipper were eliminated from the artificial night sky in the planetarium, the buntings became disoriented. The North Star, which seems to remain stationary all night in the natural sky, was acting as a fixed compass cue.

In nature, young indigo buntings imprint on the visual image of the night sky, and what they learn is essential if they are to migrate successfully at night later on. When they are hand-reared and kept in a completely dark room for several months, or when they grow up in a planetarium under an artificial sky from which the North Star and the Big Dipper are missing, then they cannot orient themselves in the proper migratory direction. *As with sexual imprinting, there is a sensitive period for the selective acquisition of information that has long-term effects on critical aspects of an animal's behavior.*

Song Learning

What does all of this mean for our singing sparrows? Apparently song behavior is learned—but the song that *is* learned requires species-specific wiring in the bird brain. For example, some male white-crowned sparrows were hand-reared in chambers where they were isolated from sounds produced by birds of any species, including their own. They sang at maturity, but the song was only vaguely reminiscent of typical white-crowned sparrow song. However, when a similarly isolated sparrow heard tape recordings of its species song from ten to fifty days after hatching—and long before it would start to sing itself—the bird remembered what it heard. When it started to sing, it gradually came closer and closer to matching its output with its memory of the adult song. Eventually the sparrow produced a nearly perfect copy of the taped song that it had heard months before.

THE ADAPTIVE VALUE OF BEHAVIOR

Foundations of Behavioral Evolution

Now that we have considered some of the mechanisms underlying the behavior of an individual animal, let's turn to evolutionary mechanisms by which diverse forms of behavior come about. We can begin by defining a few terms currently used in animal behavior studies:

1. *Reproductive success:* survival and production of offspring.

2. *Adaptive behavior:* forms of behavior that promote reproductive success and that thereby tend to occur at increased frequency in successive generations.

3. *Selfish behavior:* forms of behavior by which an individual protects or increases its own chance of producing offspring, regardless of the consequences for the group to which it belongs.

4. *Altruistic behavior:* self-sacrificing behavior; the individual behaves in a way that helps others but, in so doing, decreases its own chance to produce offspring.

5. *Natural selection:* differential reproduction among individual members of a group that vary in heritable traits—including many behavioral traits—that promote survival and reproduction.

When a behavioral biologist talks about a "selfish" animal or "altruistic" animal, there is no implication that the animal is consciously aware of what it is doing or that it knows its behavior is related ultimately to reproductive success. Talking about animals as if they knew they were trying to reproduce as much as possible is evolutionary shorthand for what would be a cumbersome mouthful—namely, that the evolution of animal nervous systems (the physical foundation for behavior) is correlated with the production of surviving descendants. A lion does not have to know that eating zebras is good for its survival and reproductive success in order to want to eat zebras.

Before 1966, the idea of "group selection" dominated the thinking of many behavioral researchers. Implicit in this idea is the assumption that individuals of a group (or species) stand ready to sacrifice their own chance of producing offspring if their self-sacrificing behavior will promote the welfare of the group as a whole.

Then the evolutionary biologist George Williams asked a key question. Assuming a population was composed of self-sacrificing types, what would happen if mutation gave rise to a "selfish" individual, able to increase its own reproductive success without regard to the welfare of other members of the group? If the selfish individual out-reproduced the "good-of-the-species" types, then its distinctive genetic makeup would become somewhat more common in the next generation. If the process continued, the good-of-the-species types would inevitably be completely replaced because they were not as successful at leaving surviving descendants as the selfish types over the generations.

A simple example will illustrate Williams's point. Norwegian lemmings disperse when population densities become extremely high, and many perish by acci-

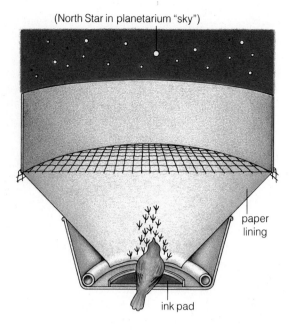

(North Star in planetarium "sky")

paper lining

ink pad

Figure 48.5 Cross-section through the cage used to test the ability of indigo buntings to use the position of the stars as an orientation guide. The bird stands on an ink pad while looking at the night sky through the wire-mesh cage cover. When it jumps up in its attempt to fly, it lands on the paper funnel lining the cage, leaving an ink mark that shows which direction it selected.

dental drowning, starvation, and predatory attack during their travels. The dispersing animals were once thought to be sacrificing themselves in order to prevent overpopulation. After all, an over-large population would destroy its environment, and this could lead to extinction of the species as a whole. The idea was given popular impetus by a nature film that showed masses of lemmings marching over a cliff and drowning in the water below. Lemmings do disperse from areas with dense populations, and some die as a result. But the question is, do they die to help their species?

There is a wonderfully instructive cartoon by Gary Larson in which a legion of lemmings is shown plunging over a cliff into the water, presumably in the act of suicide. However, one member of the group has come equipped with an inflated inner tube about its waist. The point is, if lemming populations really were composed of animals programmed to commit suicide so a few unrelated survivors could carry on the species, then individual selection would favor mutants able to avoid the ultimate sacrifice. Those mutants would be among the few left to exploit remaining resources, produce offspring, and leave copies of their "selfish" genes in surviving "selfish" descendants. Over time, nonsuicidal types obviously would completely replace those with suicidal tendencies.

Figure 48.6 Beneath the hot African sun, a black heron holds its wings over its head, like an umbrella. Minnows drawn to the shade of the umbrella are within range of the heron's bill.

In sum, the current consensus among behavioral researchers is this: *In developing a working hypothesis to explain some behavioral trait, it will almost always be more profitable to use individual selection (not good-of-the-group selection).* A few examples will illustrate how hypotheses consistent with this concept of individual selection can be developed and tested. We will draw these examples from three categories of behavior: feeding, avoidance of predators, and reproductive behavior.

Adaptation and Feeding Behavior

Figure 48.6 shows a black heron holding its wings like an umbrella, creating a patch of shade that may attract minnows nearby in the water. Minnows require shaded cover from bird predators—and the black heron exploits their requirement with often fatal results for the minnows. In terms of individual selection, the bird is behaving in a way that will enhance its own survival and reproduction by providing it with something to eat.

An elegant example of natural selection for feeding behavior is Stevan Arnold's study of garter snakes from two populations in California. Garter snakes living in moist habitats along the coast search for and eat banana slugs, exceptionally slimy animals almost the size of a mouse (Figure 48.7). Inland, the populations of garter snakes will have nothing to do with slugs and besides, the slugs don't live in those relatively dry regions. There, the snakes feed primarily on tadpoles and small fish.

Even newborn garter snakes from coastal and inland habitats react differently to slugs. For their first meal, Arnold offered recently born and isolated garter snakes a (previously frozen) chunk of thawed banana slug.

Coastal newborns almost always approached and ate the slug, but inland baby snakes rarely did. Thus the difference between adults from the two regions could not stem from the effects of feeding experience; different feeding preferences had to be present right from birth.

Could the different reactions to slug cubes be attributable to differences in their responsiveness to slug odor? Arnold found that newborn snakes from inland populations tongue-flicked at cotton swabs drenched in essence of tadpole but were essentially unresponsive to swabs drenched in essence of slug. In contrast, coastal snakes actively tongue-flicked at the slug-treated swabs (Figure 48.7c). Clearly the two populations differed in their sensitivity to the chemicals inherent in banana slugs.

When males and females from the two populations were mated in the laboratory, their offspring were intermediate in responsiveness to slug extracts. Overall, the "hybrid" offspring were much more variable in slug acceptance than either parental population. By maintaining similar environments for the young animals, Arnold was able to show that the behavioral differences among the offspring were largely due to genetic differences. Members of the two parental populations had to differ with respect to genes regulating chemical responsiveness to slug odors, and when they interbred, the outcome was a variety of genotypes (and phenotypes). The chemoreceptor system of the young snakes had been affected during development, hence the wide range of degrees of slug acceptance or slug rejection (Figure 48.8).

In Arnold's view, the genetic basis of slug acceptance by inland populations has been selected against over evolutionary time. Although there are no slugs inland, there are leeches. Slug acceptors also attack and try to

a

b

c

Figure 48.7 (a) Banana slug of the Pacific coastal regions of California, food for garter snakes (b). (c) Newborn garter snake from a coastal population tongue-flicking at a cube of banana slug.

eat leeches, because leeches and slugs share some chemical similarity. But a leech is hard to digest. Some leeches even remain alive in the digestive tracts of their consumers, perhaps feeding upon or damaging the internal organs. If the dangers associated with ingesting leeches are as great as they seem to be, then individual inland snakes that happened to have a genetic predisposition for slug acceptance would leave fewer progeny, on the average, than those endowed with different genes. Similarly, on the California coast, where edible slugs abound and potentially dangerous leeches are absent, those individuals with a preference for slug odors would gain access to a rich food supply, leave more descendants, and help spread the genetic basis for slug acceptance.

Anti-Predator Behavior

It does an animal no good to gather food with great efficiency if it is likely to be attacked and killed in the process. By compromising their foraging efficiency to some extent, most animals improve their chances of escaping detection by predators or of avoiding them even if detected.

The hoary marmot of Alaska is a good example of this point (Figure 48.9). This chunky "woodchuck" feeds on grasses and low-lying vegetation. If maximizing energy intake were its only concern, it should forage in those grassy patches of meadow with the most dense plant growth. But a foraging marmot is at constant risk from golden eagles and wolves. Given a choice between food patches of equal quality, a marmot always selects the one growing closest to safe burrow retreats (often located under rocky slopes on mountainsides). More-

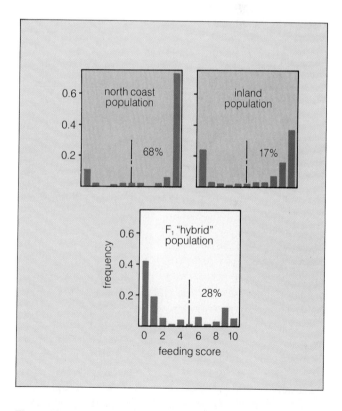

Figure 48.8 Variation in the rate of acceptance of slug cubes of garter snakes that were produced by crossing adult snakes from coastal and inland populations. (Adults from inland populations avoid slugs, for the most part; adults from the coastal regions of California almost always eat them.)

Figure 48.9 Hoary marmot, an animal that compromises foraging efficiency but, in so doing, increases its chances of surviving a predator attack.

Figure 48.10 Predator avoidance behavior by a snake caterpillar. This insect altered the shape of its body in response to being poked. The anterior segments of the caterpillar let go of the vine and puff up like a snake head, which "strikes" at whatever touches it.

over, when it eats, it constantly interrupts the meal and looks around. Although vigilance reduces the rate of food consumption, it prolongs survival; if a marmot can spot an eagle in time, it can race back to its burrow.

Caterpillars of very large tropical moths employ a special behavioral response to predation. When fully grown, these caterpillars are nearly fifteen centimeters long. During the day, they remain motionless on vines. When poked sharply (as by a bird beak), the caterpillar drops part way from the vine and puffs up its anterior segments, so that it looks like a snake (Figure 48.10). The pseudosnake even strikes at whatever touches it. A small bird typically hesitates before continuing an attack. A reluctance to deal with snakes generally prolongs the bird's life, a response that the caterpillar exploits to its own advantage.

Reproductive Behavior

After the rediscovery of the importance of individual selection in the 1960s, it became widely recognized that, *among sexually reproducing animals, the other members of one's species create obstacles to reproductive success.* The intraspecific pressures created as a result of competition for mates and discrimination among potential mates constitutes **sexual selection**, a special reproductive-related category of individual selection. Consider a male white-throated sparrow singing his "Sam-Peabody" song in a Canadian forest. He faces constant challenges from other males that may force him away from the breeding site he has selected. Moreoever, even if he succeeds in holding a small patch of the forest as his own territory, females of his species may refuse to settle there, leaving him with no descendants to show for all his effort. This is the "discriminating mate" aspect of sexual selection. The sparrow is not unique, for male-male competition and female selectivity are extremely widespread. Why?

Among sexually reproducing animals, the male produces smaller gametes (sperm) and vastly more of them compared to the oocytes, or "eggs," produced by the female. It takes a much greater energy investment to produce the larger and far more complex eggs. The disparity of investment in each potential offspring is even more pronounced when the females (but not the males) of the species care for their young. Said another way, females (and their eggs) are a limited resource for the males of most species.

If we measure a male's reproductive success in terms of the number of his descendants, then it will usually be the case that the more females he mates with, the more eggs can be fertilized, and the more successful he will be. In contrast, a female's reproductive success generally will be dictated largely by how many eggs she can produce or how many offspring she can care for. Thus for females, the quality of a mate, not the quantity of

Figure 48.11 Contest between male bighorn sheep for the possession of a cluster of females.

matings, should usually be the prime factor influencing her sexual behavior.

With this simple theoretical background, we can interpret a great deal of male and female reproductive behavior, including the adaptive value of singing by male songbirds. To leave offspring, a male bird must "convince" females that he rather than some other individual should fertilize her (very large and energetically expensive) eggs. The female songbird's reproductive success will depend on how well her offspring are fed.

Nestling white-throated sparrows eat astonishing quantities of insects. Thus the males compete with one another to monopolize an insect-rich patch of forest, which will attract females. In part, they compete by vocally advertising their control over a resource-based territory. (A "territory" is an area in which an animal or group of animals establishes residency and forages and which may be defended against others of the species.) Through many studies, we know that simply hearing an established resident is enough to discourage intruders from settling in an area, for it means they will have to fight for its possession when there may be unoccupied areas elsewhere.

Females also represent a selective pressure by favoring vocal males, which "announce" potentially suitable habitats for them. However, although male song functions as an attractant, the females go on to assess the quality of the territory for themselves.

The point is, *environmental factors influencing the distribution of females are central to male competition for access*

Figure 48.12 A male hangingfly (*Harpobittacus*) holding a prey insect that he will offer to a female if one comes to visit him. He will mate with the female if she accepts his food present.

to mates. If the resources sought by females are clumped in space, then **resource-defense** behavior should evolve, with the winners getting mates and the losers kept from reproducing (at least temporarily).

Resource-defense behavior has evolved independently in many groups besides birds. For example, the females of some dragonflies lay their eggs only in particular habitats on ponds or streams. Males of their species attempt to defend this limited resource, and the winners mate with females that come to the spot suitable for laying eggs.

As another example, sometimes females are not concentrated at discrete patches of a useful resource but they live together, often for defense against predators. In such cases, males usually show **female-defense behavior**.

Male bison, lions, elk, and bighorn sheep compete fiercely as a consequence of the female clustering behavior. The competition favors large males with formidable combat abilities, obvious in the strength of

male lions, the antlers of elk, and the readiness of bison and bighorn sheep to use their horns on each other (Figure 48.11). The return can be great; by winning possession of females, the males of these species have a ready-made harem.

What happens when resources are widely distributed and females do *not* cluster? In such cases, identifying what the males are actually competing for is difficult. For example, the territories defended by male sage grouse of the central United States do not have concentrated resources, nor are the territories centered in places where females live in groups. During the breeding season, males gather in groups on small patches of prairie. Each individual defends his own tiny territory, where he prances about in a truly astonishing display. With tail feathers erect and splayed, head held forward, and neck pouches inflated (Figure 36.9b), a male projects booming calls over the prairie, all the while stamping about like a wind-up toy on its display ground. Females come to the lek, observe the males, and eventually select a partner from among the many. After mating, a female returns to her nest site and rears the young by herself. Many females choose the same male from those at the lek, conferring exceptional reproductive success on the preferred male. What is the basis of this peculiar reproductive strategy? No one knows, although it may be that the males form leks only as a last resort to advertise their qualities, there being no other "workable" and productive mode of mate selection.

So far, we have focused on the advantages that certain forms of behavior bestow on males. Now we conclude with a look at how females may enhance their reproductive success through the judicious selection of a mate. *Two main types of benefits may accrue as a result of her choice: superior genes or superior material benefits.* Consider the male sage grouse, which provide females with the genes in their gametes, nothing more. Do the females "assess" a male's genetic quality by observing his display? Perhaps the nature of his calls, the vigor of his visual display, or his endurance provide cues about his survival ability or foraging skills. To the extent that these behavioral attributes are heritable, a male that puts on a superior show could endow his offspring with useful abilities; and females that could identify such a male and be inseminated by him would enjoy heightened reproductive success. This possibility, like the related issue of the evolution of leks, remains an open question for further research.

One example in which female choice clearly benefits the female is provided by the hangingfly (*Harpobittacus apicalis*). In this species, males capture and kill a fly or a moth and hold it for a female, which they attract by releasing a sex pheromone (Figure 48.12). The size of a "nuptial gift" can vary. If females exercise mate choice in terms of material benefits, they should prefer males that can feed them more, and they do in fact choose males on the criterion of prey size. When a female approaches a male, he offers her his prey; and if she begins to feed on it, she allows mating to begin. However, she will not allow him to transfer *any sperm* until she has eaten for about five minutes. Thereafter, she accepts a steady flow of sperm into her reproductive tract, *but only as long as the food holds out.* At any point up to twenty minutes, she can break off the mating and leave without a full complement of sperm. If so, she will mate again and accept another male's gametes, diluting or replacing her first partner's sperm.

Because the females manufacture a limited resource (eggs), they dictate the rules of male competition, with males attempting to fertilize as many eggs as possible. The improved understanding that has been gained by testing hypotheses on the value of reproductive traits to *individuals,* not to a species as a whole, represents a major triumph of the individual selection approach.

SUMMARY

1. Genetic differences among individuals can lead to differences in the development of the endocrine and nervous systems that control an animal's behavior.

2. Both instinctive and learned behavior are based on genetic information and environmental inputs. Instincts are not purely genetic; learned responses are not purely environmental, as clearly seen in the various forms of imprinting and imprinting-like behavior.

3. The view that behavior is adaptive in advancing individual reproductive success, not group survival, stems from the logic of natural selection and is the basis for most current research on the ultimate causes of behavior.

4. Hypotheses about the adaptive (reproductive) value of a trait can be tested in several ways. For example, there are ways to determine what the "best" response of animals should be to a certain set of obstacles to reproductive success. The animal can then be observed to see if its behavior matches the prediction. Alternatively, one can determine whether different species confronted with the same ecological problem have developed similar ways of dealing with that problem.

5. Both approaches have been used in many studies of feeding, anti-predator, and reproductive behavior. Major advances in the study of sexual behavior have come from the recognition that members of a species create problems for each other as they compete to reproduce.

6. In particular, males are predicted to compete (often aggressively) for access to receptive females because

winners of this competition will tend to have many offspring. Females, on the other hand, produce a limited resource (their large and valuable eggs) that the males need, and therefore they should not often compete for males but instead exercise mate choice to accept the best possible partner from among the many suitors available to them.

Review Questions

1. Rephrase the statement "There is a gene for sexual imprinting by greylag geese." The reworded statement should avoid the implication that genes "make" behavioral traits in a one-to-one relationship. *808*

2. What role does the environment play in the development of an instinct? *809–810*

3. What contributions have behavioral geneticists made to an understanding of the evolutionary basis of behavior? *812–814*

4. Why is song learning by white-throated sparrows a good illustration of the principle that genetic mechanisms contribute to the development of a learned response? *808–809*

5. When an adult female sparrow receives a testosterone implant, why won't she sing the territorial song of males of her species? *809*

6. Give a group selectionist and individual selectionist explanation for territorial behavior in a species, and then criticize the logic of the group selectionist hypothesis. *813–814*

7. You find an insect that looks and behaves like a piece of bark. How could you compare a number of insect species to test the hypothesis that the behavior of the animal is an adaptation to avoid being eaten by a predator? *815–816*

8. Develop a hypothesis for the observation that male lions kill the offspring of females they acquire after they chase away the males that had been the previous pride holders and mates of these females. How would you test your hypothesis? *816–819 (Refer also to the Commentary on page 21.)*

Readings

Alcock, J. 1984. *Animal Behavior: An Evolutionary Approach.* Third edition. Sunderland, Massachusetts: Sinauer Associates. A broad-based survey of the many topics that make up the study of behavior, with a strong emphasis on evolutionary theory.

Dawkins, R. 1976. *The Selfish Gene.* New York: Oxford University Press. An entertainingly written book that explains the basis for the argument that natural selection favors traits that are associated with individual success in propagating genes.

Krebs, J., and N. Davies. 1984. *Behavioral Ecology: An Evolutionary Approach.* Second edition. Sunderland, Massachusetts: Sinauer Associates. A more advanced text consisting of chapters written by experts on many of the issues covered in this chapter.

Williams, G. C. 1966. *Adaptation and Natural Selection.* Princeton, New Jersey: Princeton University Press. The classic book that brought about the revolution in thinking about animal behavior in terms of individual selection.

49

SOCIAL BEHAVIOR

COMMUNICATION AND SOCIAL BEHAVIOR

Consider the Termite

In a forest in southern Queensland, Australia, a long tube about a centimeter across runs up the white trunk of a dead eucalyptus tree. Here, millions of glued-together pellets form a continuous tunnel above the wood's surface. When a fragment of the firm but brittle tube chips away and exposes the interior, several small, nearly white insects scurry away from the light, then a larger number of brown insects quickly line the breach. Those companions of the pale insects look like something out of science fiction; their swollen, eyeless heads taper into long, pointed "noses" (Figure 49.1). When you blow a puff of air at them, each shoots a thin, silvery filament out of its nose!

These insects, one of many species of nasute termites, live in various tropical and subtropical areas around the world. The tunnel is really a covered highway running many meters from an underground nest to areas where the pale workers gather dead wood, which they carry in their mouth to underground gardens. After processing the wood (they cannot digest its tough cellulose fibers), they add it to their gardens, where they are growing a special fungus that *can* digest the cellulose. The pale insects harvest "the mushrooms" (basidiocarps) of the fungus and use them as their food source. The entire colony is defended by the snout-shooters—soldiers that fire off silky strands of glue that entangle and immobilize ants attempting to invade the tunnel.

Perhaps a million or more workers and soldiers in one nasute colony cooperate to build tunnels, create and tend gardens, and defend the colony against predators—and this remarkable cooperation is based on their ability to *communicate*. Despite their pinhead-sized brain, termites use a complex array of communication signals to exchange information in ways that keep the colony running smoothly.

Not one of those million workers and soldiers will ever reproduce. Within the underground bunker beneath the dead eucalyptus are one king and one queen termite, the original fertile founders of the colony, parents to all one million sterile progeny.

With this example of life among the termites, we turn our attention to **social behavior**, the tendency of individual animals to enter into cooperative and interdependent relationships with others of their kind. Because communication signals are the means by which animals manage to be social, we will first explore what "communication" means and how animals use different channels of communication. Then we will turn to the diversity of animal societies, which range from simple aggregations to complex societies.

Communication Defined

When a termite tunnel is breached, the pale workers bang their head against the ceiling and floor of the tunnel, creating vibrations that alert nearby soldiers and cause them to rush to the point of disturbance. The soldiers point their nose in the direction of potential danger, as announced by the scent of ants or a sudden jostling of the tunnel wall. When they shoot out their sticky strands, chemical odors released from the glue attract more soldiers to the site; the odors serve as an alarm scent (a type of pheromone).

The alarm scent of a soldier is clearly a communication signal. Is the scent of the ant one also? In both cases, airborne chemicals associated with one animal change the behavior of another. However, to most biologists, the term **communication signal** means a stimulus produced by one animal that changes the behavior of another individual, in ways that benefit both the signaler and the receiver.

Natural selection has favored the evolution of information exchanges between some signalers and some receivers. When one soldier termite attracts another with an alarm scent, they both benefit through improved defense of their colony. When the original signaler appeared in a population of termites, its scent proved beneficial for itself and its nestmates. Had the scent *harmed* the signaler, then natural selection would have favored nonsignalers, eventually eliminating the "signal" from the population. Similarly, if the respondents had been harmed by reacting to the signal, then selection would have favored types that ignored the signal.

Thus, a communication signal evolves as a benefit for *both* the signaler and the receiver. Such benefits do not accrue in ant-termite interactions. The ant odors did not evolve so that ants could announce their impending attacks on termites; they evolved because they serve useful functions within *ant* colonies.

Communication signals evolved in one context are sometimes exploited by animals of other species. Thus,

Figure 49.1 Nasute termites—highly social insects with a soldier caste that defends the colony by shooting strands of sticky silk from the noselike extension of their individual heads.

from the perspective of the raiding ants, the soldier termites are "illegitimate receivers" of one of their communication signals. Nature also has its share of "illegitimate signalers," which fool the normal receivers. An assassin bug covers its body with some nest material of its termite prey. Thus the bug adopts an accepted odor and is tolerated by living members of the colony, even as it goes about consuming them, one by one.

CHANNELS OF COMMUNICATION

Communication signals are based on diverse sensory modes. For example, nasute termites make use of tactile senses (when they transmit alarm vibrations through wood) and olfactory senses (in the release of alarm pheromones). Humans are more familiar with acoustical signals (as in speech) and visual displays (as in facial expressions like smiles and frowns). Let's consider a few examples of signals in these major sensory modes.

Figure 49.2 Example of the "yawns" by which male baboons threaten each other. Such facial displays often resolve a conflict without actual fighting.

Visual Signals

Visual signals are nearly universal among species of animals that are active by day. Often male baboons "yawn" at each other when they are both interested in a receptive female or some other resource. This facial movement is a visual display, designed to threaten a rival (Figure 49.2), and often precedes an attack on the rival.

How can it be advantageous for a baboon to signal an intention to attack or for a receiver to "accept" this information? Here, the signaler benefits if the receiver backs down after perceiving the threat signal, rather than physically attacking and possibly injuring the signaler. The receiver also benefits if, after mulling over the threat signal, he can accurately judge that his opponent can inflict a real beating unless he gives ground.

Visual communication is essential in the courtship of many animal species. For example, male and female albatross have a battery of visual displays, each with its own message (Figure 49.3). As is true for many birds, the albatross adopt exaggerated and sometimes contorted postures that convey readiness to mate, to take over incubation duties from the partner sitting on the eggs, and so on. Possibly, these highly distinctive poses represent unambiguous messages for the intended receiver and minimize the chance of confusion.

Visual communication is limited at night, but a number of organisms use *bioluminescent* signals (based on special metabolic reactions that are accompanied by a release of light energy). For example, fireflies (beetles, really) have light-generating organs at the tip of the abdomen. Each male produces a signal characteristic of his species and one that a receptive female may answer. After seeing his signal, she waits a standard time and replies with a single flash. A male approaches an answering female, engaging her in a visual "dialogue" until he precisely locates and mates with her.

Females of a predatory species of firefly have "broken the code" of their prey. When they detect a signaling male, they wait the appropriate interval and then give the "come-hither" answering flash. The male approaches and is embraced by the predator, which consumes her victim.

Chemical Signals

Chemical signals abound among the insects, which use them as sex attractants, trail markers, alarm calls, and so on. Male hangingflies lure females to themselves with sex pheromones (page 354). More often, a receptive female insect releases sex pheromones and waits for the male. The male insects have large antennae, packed with specialized receptors. After detecting the pheromone, males fly upwind, tracking the odor in a race to be first to reach the chemically beckoning female (page 371).

Figure 49.3 Courtship behavior of various species of albatross. (**a**) The male spreads his wings as part of a courtship ritual that also includes pointing his head harmlessly at the sky. These birds are at a future nest site in the male's territory (**b–d**). After pair formation is well advanced, the birds begin to touch bills. This contact display precedes copulation.

Chemical signaling is also notable among some of the primates. A female baboon uses chemical and visual signals· to announce her readiness to mate. Her external genitalia become swollen and she produces distinctive sex pheromones that elicit mating behavior from the males. In humans, male pheromones may influence the physiological health of the female's reproductive system.

Tactile Signals

Many animals communicate over short ranges, where tactile signals (distinct patterns of touch) become important.

When a foraging honeybee returns from several successful trips to a rich nectar or pollen source, she may begin to "dance" on the vertical surface of the comb in the darkness of the beehive. The forager moves in circles, jostling her way through the crowded mass of workers on the comb. Other individuals may follow her, keeping in physical contact with the dancer and, in doing so, acquiring information about the distance and direction of the food source.

Karl von Frisch was the first to discern the message encoded in this tactile signaling. In one set of experiments, he trained marked foragers to come to stations baited with concentrated sugar water. He then set out a

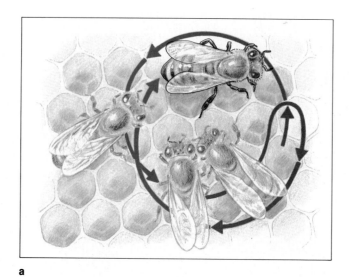

a

b

Figure 49.4 (Above) (a) The round dance of the honeybee. When foragers perform these movements upon their return to the hive, other bees are stimulated to fly out fairly close to the hive in search of the food that the round-dancing bee had found. (b) The waggle dance of the honeybee. Bees that have found a rich source more than eighty meters from the hive perform a waggle dance to recruit other bees in the hive.

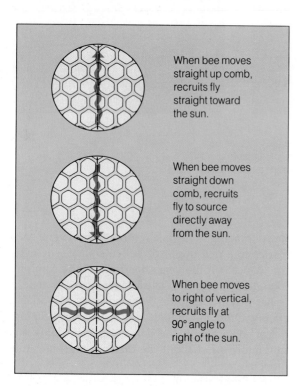

When bee moves straight up comb, recruits fly straight toward the sun.

When bee moves straight down comb, recruits fly to source directly away from the sun.

When bee moves to right of vertical, recruits fly at 90° angle to right of the sun.

Figure 49.5 The manner in which the angle of the run of the waggle dance conveys information about the angle between the food source, the hive, and the sun.

series of stations with identical baits at different distances or directions from the hive. Generally, the station closest to the original training site attracted by far the greatest number of new (unmarked) bees in a short time. The trained bees apparently were acting as recruiters, directing new bees to the general area where they had found so much food.

By observing marked bees and knowing where they had been trained, von Frisch was able to break the code of the dancing bees. He saw that bees trained at stations close to a hive performed a round dance (Figure 49.4a). Worker bees trailing the round-dancing bees were stimulated to fly out of the hive in search of nearby food but they did not learn which direction to fly. If, however, the station was more than eighty meters distant, the foragers trained at this station performed waggle dances (Figure 49.6b). The frequency with which they waggled their abdomen during the straight runs of the dance and the speed with which they completed each loop of the dance conveyed information about the distance to the food. *The more waggles and the faster the dance, the closer the food source* (down to about eighty meters, at which point bees switch to round dancing).

The angle of the straight run with respect to vertical provided information about the direction of the food source relative to the sun and the hive. If a dancer went straight up the comb surface on her straight runs, followers learned that food could be found by flying from the hive in the direction of the sun. If, however, a dancer made her straight run 90° to the right of vertical, she was trained at a food station located at 90° to the right of a line drawn between the hive and the sun (Figure 49.5). Thus dancing bees transpose information about the position of the sun, hive, and food source into a code, and they convey the symbolic message via tactile stimulation to their hivemates.

Figure 49.6 Frog-eating bat (*Trachops cirrhosus*) catching a singing tropical frog (*Physalaemus*). The bat can distinguish between poisonous and edible frogs, as well as locate them, by the frog's calls. The frogs call to attract mates but face the dilemma of how not to attract the bat instead. Bat predation obviously has major influence on frog courtship behavior.

Acoustical Signals

Both chemical and visual signals travel long distances in air (and water), and the same is true for acoustical messages (distinctive sounds). Bird song, described in the preceding chapter, is a familiar example of acoustical signaling. Male frogs also use acoustical signals to communicate with rival males and receptive females. As is true for bird song, calling is primarily a male activity, and each frog or toad species has its own distinctive call.

Frog calling is complicated for many species because (1) some predators use the calls of their amphibian prey to locate victims, and (2) many rival males may be concentrated in the same relatively small area, creating intense competition for space on the "air waves." For example, a tropical frog (*Physalaemus pustulosus*) is a favorite prey of a bat that tracks down calling individuals and snatches them from the water with its formidable jaws (Figure 49.6). The frog call has two parts, a whine followed by a "chuck." A frog calling all by itself often drops the "chuck" which, because of its acoustical properties, gives better information about his location than the initial whine. Although the male thereby makes himself less attractive to females, he is more likely to live to

call another day. He gives the complete call when part of a male chorus; being one of many that could be grabbed that evening by a frog-eating bat, he may live *and* get to mate.

COSTS AND BENEFITS OF SOCIAL LIFE

The forms of communication just described are the foundation for diverse social units. Although there are certain advantages to being highly social (in the sense of living together), some animals live in small family groups, others as loose aggregations of unrelated individuals, and still others in solitude. The degrees of sociality (or its absence) represent different adaptations to different environmental challenges.

The reason for diversity in social life becomes clearer if we consider the risks as well as benefits of social life *in terms of individual reproductive success*. Consider the vast rookeries of egrets, gulls, or terns, in which hundreds of individuals form pairs and nest in apparent harmony

Figure 49.7 Porpoises engaging in social pursuit of schooling fish. Sociality sometimes exists because individuals can forage more effectively for prey when they do so as a group.

a short distance from one another. The appearance of harmony is superficial. A colonial nesting gull does not hesitate to eat its neighbor's eggs or chicks if they are left unguarded for a moment. Moreover, many birds deplete the local food supplies more rapidly than would happen if they were dispersed. Given the crowded conditions, they are all probably more vulnerable than solitary birds would be to contagious diseases.

What does a social animal gain that more than compensates for the price of associating with others? No one overriding benefit applies to all social species. However, lions and porpoises living in groups are able to capture more prey than if they operated alone. Through cooperative hunting, a pride of lions can bring down a giraffe or a Cape buffalo that no one lion would dare attack. When porpoises collectively go after a school of fish, the members of the school break apart in a wild dash to safety. But many escape from one porpoise only to run into the mouth of another (Figure 49.7).

PREDATION AND SOCIALITY

Predation is by far the dominant selection pressure favoring the evolution of social behavior. Here we will consider some of the ways in which group-living animals minimize (at least in theory) the risk of predation.

Dilution Effect

Simply by being part of a large group, a prey animal "dilutes" the risk of being the one animal that predators will select out of the many. For example, the diminutive fairy penguins of Australia become social in the evening when they return to nest burrows on land. The birds tend to come ashore in groups. The surging waves deposit them upon the beach, where they stand about uncertainly before beginning to waddle together inland. Penguins on land are not noted for agility or speed; they are vulnerable to large hawks and eagles patrolling the beaches for prey. But by making their move across the beach only when large numbers of their fellow penguins have assembled, each animal improves its odds of surviving. While an eagle is dispatching one victim, the others can waddle along to safety.

Improved Vigilance

In effect, animals living in groups have many eyes available to spot predators. When the first animal to detect danger flees for cover, others can follow its lead even if they themselves have detected nothing.

In a study of captive starlings, one researcher showed that isolated birds reacted less quickly to an artificial hawk model released over their cages than did a flock of ten starlings. Moreover, the flock flew up sooner after the release of the "hawk," even though any one bird in the group had been looking up *less often* than the isolated starling. By virtue of being a flock member, a starling is not only safer, it has more time to feed because less time is needed for vigilance.

Group Defense

Often, social animals can better repel a predator through coordinated group defense. Nasute termites work as a team to alert soldiers, which then combine forces to spray intruders with their sticky glue. Members of a herd of musk oxen also employ group defense when threatened by attacking wolves (Figure 49.8). Musk oxen have powerful hooked horns, but a solitary animal cannot protect its flanks from a wolf pack. The array of Cape buffalo shown in Figure 1.9 would be formidable even to a pride of lions.

Figure 49.8 Defensive formation of musk oxen. Predation pressure can favor the evolution of social life when members of the group are safer than are solitary animals.

The Selfish Herd

Some animals may live in groups simply to "use" other individuals as living shields against predators. Such groups have been labeled "selfish herds." When hyenas or hunting dogs attack a herd of wildebeest on an African savanna, the prey animals cluster into a tight mass that dashes across the plain as if it were a single animal. Although their escape behavior seems to be a cooperative undertaking, it almost certainly results from the efforts of individuals not to be isolated from the group. Instead they maneuver themselves as closely as they can to the center of the herd, where it is safest. Hyenas and hunting dogs separate an easy victim (from the periphery of the herd) rather than plunge into a mass of flying hooves.

Given the survival advantages, we can predict that members of a selfish herd will compete for the safe central positions. Such competition is evident in nesting colonies of freshwater bluegill fish. The many predators of these small fish include largemouth bass (which can inhale an entire adult) to snails that slip into a nest and eat the fish eggs. Male bluegill fish build their nests together, with each male constructing a depression in the lake bottom to receive the eggs that his mate(s) will deposit there. The largest, most powerful males claim the center of the nesting area, forcing other, smaller males to assemble around them. Thus the smaller males form a living shield against bass and snails, which attack from the edge of the colony inward.

SOCIAL LIFE AND SELF-SACRIFICE

From the preceding discussion, it should be clear that the individuals of many animal societies do not make any personal sacrifice for other members of their group. However, the degree to which individuals tolerate and interact with others is influenced by natural selection. If the environment is such that social members of a species leave fewer descendants on the average than the more solitary types, what will the species become—social or solitary?

Yet if it is true that natural selection influences the evolution of social behavior, then there would never be any cases of self-sacrificing behavior—*unless such behavior enhances the continuity of the genetic line of the altruistic individual.* Let's look at a few examples that will clarify this point.

Parental Behavior

Parenting is the most familiar form of self-sacrificing behavior. A breeding pair of Caspian terns (Figure 49.9) incubate their eggs, shelter the nestlings, bring them food, defend them from predators, and accompany them for some time even after they have begun to fly and feed partly on their own. Such parental activities use up time and energy the adults might spend in ways that would improve their own chance of living to reproduce another time.

Figure 49.9 Male and female Caspian terns, which cooperate in the care of their young. Male parental behavior occurs in most bird species, but overall is rare in the animal kingdom.

However, if a parent can significantly help some of its offspring live to reproductive age, then its sacrifice could increase the frequency of the parental genes in the population. *Because of the genetic similarity between a parent and its offspring, which share fifty percent of their genes, it is possible for an individual to give up some of its own chances for survival and future reproduction—and still have its genes spread through the population.*

The critical point is this: the parent must sacrifice for its *own* offspring, not for those of another. Typically, parents are adept at recognizing their own offspring, which they help, while ignoring or even attacking the offspring of others. Again, many terns and gulls nest at the same time in extraordinarily dense colonies. After eggs have hatched and the young have grown a bit, there may be hundreds or thousands of dependent chicks, all about the same age, in a relatively small area. Yet the parents never feed the chicks of other parents, and they may even eat the "foreign" chicks if they catch them unprotected.

Cooperative Societies

Despite the existence of selfish herds, discriminating parents, and cannibalistic neighbors, there are also numerous examples of apparently self-sacrificing behavior, in which friendly, helpful behavior is not directed exclusively to offspring.

For example, membership in a troop of baboons is stable, all the animals recognize each other as individuals, and they actively seek out the company of their fellows, spending hours grooming a favored companion to remove parasites and burrs. At some risk to themselves, the adult males may join forces and confront an attacking leopard while the smaller females and juveniles flee to safety.

Similarly, the members of a wolf pack do more than merely tolerate each other. They have active greeting ceremonies after they have been separated for a time and they spend long periods in very close association (Figure 49.10). They coordinate their hunting activities and share food in a kill; returning hunters often regurgitate food to pack members that have remained at the den to protect the pups.

This last example raises an interesting question. When a wolf brings back food to pack members that are *not* its offspring, it seems to be sacrificing something (time, food, or both) that could be used to further its own reproductive success. How can we account for such behavior in evolutionary terms? Let's take a look.

Dominance Hierarchies

To gain insight into self-sacrificing behavior, we must first realize that the apparent harmony within even the most cooperative societies is, to some extent, superficial.

Figure 49.10 A dominant male and female member of a wolf pack (**a**). Typically, the dominant male is the only pack member likely to breed successfully. (**b**) Subordinate members of the pack greet a dominant male.

Some "sacrifices" are made simply because some members are less dominant than others, and their aggression and other expressions of individuality are held in check. Thus we see a **dominance hierarchy** in some societies, with some members subordinate to less subordinate ones, which are dominated by other members, and so on up the ladder of who socks it to whom.

The top male in a baboon troop is the so-called alpha member of the hierarchy. Merely by walking toward another baboon, an alpha male can cause it to move quickly to the side, leaving him in charge of a safe sleeping place, choice bits of food, or a receptive female. The second-ranking male can preempt all but the alpha individual. The third-ranking male dominates all but the alpha and beta baboons, and so on down the line.

Because the group members accept their status, aggression is minimized. Subordinates almost never actively challenge more dominant individuals over limited resources. They are often forced to show appeasement gestures if they are to remain in the company of a more dominant group member (Figure 49.11).

What is notable about this behavior is that subordinates almost never reproduce as much as the dominant members do. How, then, can subordinate behavior evolve by way of natural selection—in other words, through pressures that favor individual reproductive success? According to one view, *even dominant members die or become injured, old, or feeble—and a patient subordinate may then move up to a higher position in the social hierarchy, and ultimately reproduce.*

A young, small, or weak animal has three options. First, it could pull out all the stops and compete for dominance—but a challenge to a much stronger individual might lead to injury or death, so any future chance at reproduction would be lost. Second, the subordinate animal could strike out on its own, find a mate, and form a new breeding group. This, too, usually is a prescription for injury or early death. (A wandering, solitary baboon no doubt quickens the pulse of the first leopard to see it.) Third, the subordinate animal could *accept* its low status and thereby secure better protection from predators and more safety when foraging. It may even end up producing offspring if chance events favor its upward movement in the dominance hierarchy.

EVOLUTION OF ALTRUISM

So far, we have seen that often it is not necessary to invoke anything other than individual selection to account for the evolution of helpful behavior. However, a special evolutionary process could, in theory, promote the evolution of altruistic behavior toward individuals other than offspring. To explain this process, we will be using some terms used by behavioral biologists:

Figure 49.11 Appeasement behavior among baboons. Notice the assured position of the dominant animal—and the abject stare and groveling posture of the subordinate one, who is intent on making little conciliatory smacking noises with its lips.

1. *Altruistic behavior:* helpful actions that reduce the individual's production of surviving offspring while increasing the reproductive success of the helped individual.

2. *Cooperation:* helpful behavior that raises the helper's output of its own surviving offspring as well as the number produced by the helped individual.

3. *Self-sacrificing behavior:* helpful actions that reduce the probability of survival of an individual but can, under some circumstances, increase the reproductive success of the individual.

4. *Kin selection:* selection for aid to relatives, which increases an individual's transmission of genes to subsequent generations. The *direct* form of kin selection favors aid given by adults to their offspring when such aid increases the number of their surviving offspring. The *indirect* form favors aid given to relatives (and others with similar genotypes)—not the helper's offspring but surviving relatives that preserve the gene lineage they share with the helpful individual.

Kin Selection

Even if a subordinate never actually succeeds in reproducing, it may still help pass on copies of its genes. *True altruism, even if it involves reproductive self-sacrifice, can be genetically advantageous if the beneficiaries of the sacrifice are relatives of the altruist.* Individuals share a certain proportion of genes not only with their offspring but also with their relatives. Two siblings will have about half the same genotype as the other, because they have inherited some of the same genes from the same parents. Due to common descent, about one-fourth of the genotype of an uncle and nephew is the same. Similarly, about one-eighth of the genotype of cousins is the same. Thus an individual that bears genes leading to the development of helpful behavior can *indirectly* propagate those genes by helping preserve and produce more relatives.

In many animal societies, altruistic behavior is directed strictly to close relatives. For example, a communal flock of scrubjays is composed of a breeding pair and as many as six helpers-at-the-nest. The helpers almost always are older offspring of the pair. They help, feed, and protect younger brothers and sisters. They may never reproduce, yet the sacrificing individuals help perpetuate some "shared" genes when their siblings reproduce.

Suicide, Sterility, and Social Insects

What about animals that are permanently incapable of reproduction? What genetic advantage could be attributed to their altruistic behavior in a group? Consider sterile helpers or workers, which are found among social insects such as honeybees. A honeybee colony is characterized by great sacrifices of the sterile workers (Figure 49.12). A single queen bee is the sole reproductively active female among tens of thousands of bees. Workers force-feed her with food that they collect, groom her, and distribute her socially binding pheromones throughout the colony. They also feed one another, build honeycomb, attend to the queen's eggs, feed larvae, remove dead or diseased pupae, and guard the entrance to the hive.

When a guard bee detects a nest parasite or predator (such as a raccoon), it attacks and stings the intruder. It also releases a chemical attractant that recruits still more defenders, which also attack and sting. Because the loss of a stinger and associated poison glands is fatal, these workers are animals that commit suicide! Suicidal behavior is regularly exhibited by worker ants, wasps, and soldier termites, as well as by worker bees.

How can suicidal behavior and sterility evolve, given that individuals having these traits fail to reproduce? Here, too, close relatives play a role. For example, at some stage a queen bee departs with about half of the work force and starts a new colony. Before she does, drone sons and future queen daughters are produced. All are fertile. One of these future queens inherits the remaining half of the work force, and she stays at the hive. The drones leave the colony and may mate with new queens produced at other colonies. Thus worker bees help the queen mother rear one reproducing sister, and many workers stay with this new queen and help her after their mother leaves. They also help produce a number of potentially reproducing brothers. *When workers sacrifice their own reproductive chances, they increase the number of potentially reproducing siblings—which have a substantial proportion of particular genes in common with them.* This is true of social insects generally.

Sterility, then, can be explained in evolutionary terms as a mechanism that increases the representation of certain genes in certain populations. Modern analysts of worker castes now suggest that worker members do not exist to benefit the species as a whole. Rather, the focus is on how even extreme cooperation in social animals may have evolved through competition at the genetic level.

Human Social Behavior

So far, we have considered examples of how the concept of individual selection can be used to explain animal behavior. Can the concept also be applied to analysis of human behavior? The question is controversial. Since Darwin's time, strong evidence has accumulated in support of the principle of biological evolution, and today there is widespread belief that we share an evolutionary heritage with all other forms of life. Yet there is still a prevailing belief that humans are so evolutionarily advanced that they are totally unique, even compared with our closest relatives in the animal kingdom.

To be sure, our cultural evolution has been so extraordinary that we are unique in many ways. But surely, beneath the elaborate and diverse layerings of culture, there still resides a biological core for human behavior.

Think about some forms of behavior that occur in all human societies. For example, smiling has an innate basis. It starts among newborns, even those born prematurely. Apparently, smiling helps establish emotional ties between infants and the adults who help assure their survival. As another example, beginning at four or five weeks after birth, all infants make strong visual contact with their mothers—even infants who were born blind. Here, too, the behavior apparently helps strengthen social bonds between the protector and the protected. Universally, humans express pleasure, anger, distress, surprise, and rage with the same kinds of facial movements. The meaning of all such behavioral expressions is universally recognized; they are all evolved mechanisms that help promote survival and reproductive success in human social groups.

Some contend that if you were to ask people why they behaved in a particular way, no one would ever say "My genes made me do it." They contend that any behavior in any environmental context is an entirely learned response. Yet conscious awareness of the genetic contribution to some behavior is not a requirement for its expression. If it could talk, a baby egret battering its younger sibling to death (as baby egrets are prone to do) wouldn't say "My genes made me do it" either; it simply has an inherited capacity to behave in ways that help assure its own survival and, ultimately, its reproduction.

Notice the word *capacity*. Having a genetic basis for the development of some behavior does not mean that the trait is biologically determined, in the sense of being impossible to alter. There are no "genes for behavioral traits." There are only genes that code for enzymes and other products—and the synthesis and activity of those products are profoundly influenced by the environment. As pointed out repeatedly in this book, *change the environment, and the course of development can change.* Thus, whatever the extent of its genetic foundation, expression of different forms of behavior that are considered to be unacceptable in a social environment can also be modified.

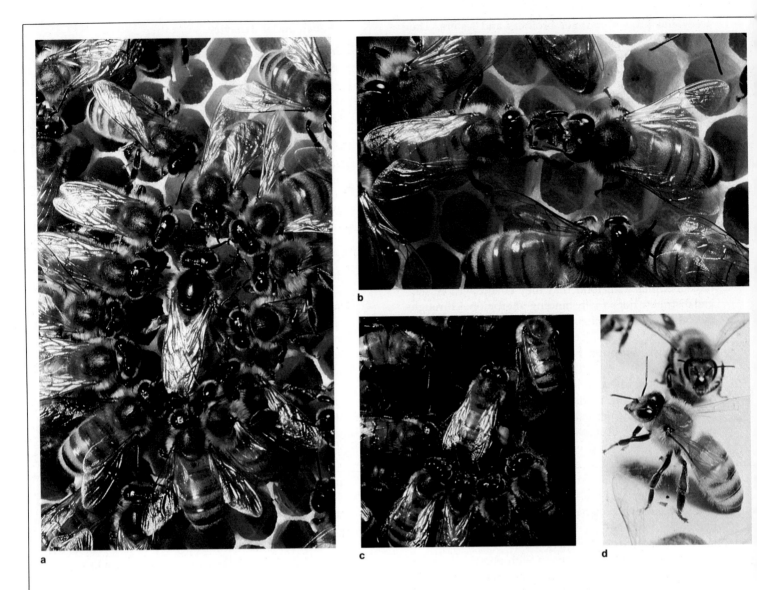

Figure 49.12 Life in a honeybee colony. (**a**) The only fertile female in the hive is the queen bee, shown here surrounded by her court of sterile worker daughters. These individuals feed her and relay her pheromone throughout the hive, regulating the activity of all of its members.

(**b**) Transfer of food from bee to bee, one of the helpful actions within honeybee society.

(**c**) Bee dance. The central bee is in the midst of a complex dance maneuver, which contains information about the direction and distance to a food source. Floral odors on the dancer's body are also useful to the recruits, which follow her movements before flying out in search of the food.

(**d**) Guard bees. Worker females assume a typical stance at the colony entrance. They are quick to repel intruders from the hive.

(**e**) The queen is much larger than the workers, in part because her ovaries are fully developed (unlike those of her sterile daughters).

(**f**) Stingless drones are produced at certain times of the year. They do not work for the colony but instead attempt to mate with queens of other hives.

(**g**) Worker bees forage, feed larvae, guard the colony, construct honeycomb, and clean and maintain the nest. Between 30,000 and 50,000 are present in a colony. They live

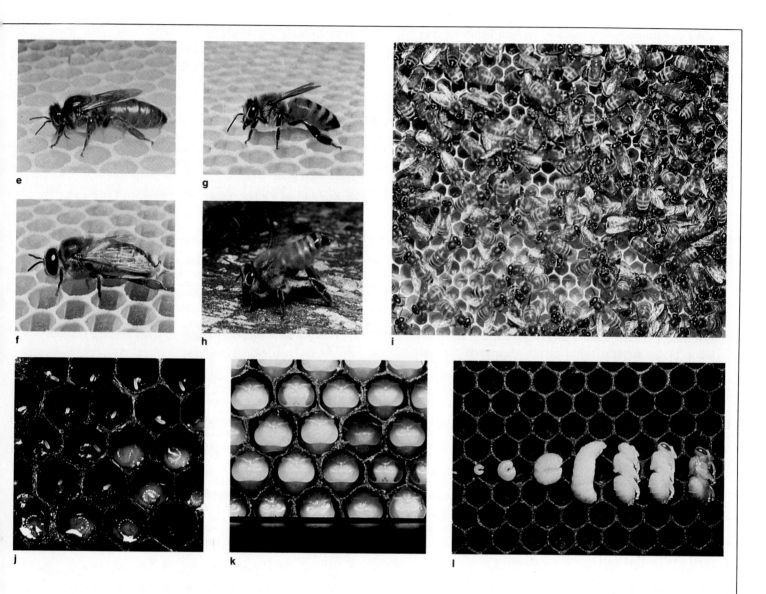

about six weeks in the spring and summer, and can survive about four months in an overwintering colony.

(h) Scent-fanning, another cooperative action by the worker. As air is fanned, it passes over the exposed scent gland of the bee. The pheromones released from the gland help other bees orient to the colony entrance.

(i) Worker bees constructing new honeycomb from wax secretions. Here, honey or pollen may be stored or new generations may be cared for from the egg stage to the emergence of the adult.

(j) The initial stages in the life cycle of a bee. The brood cell caps have been removed, revealing eggs and larvae of various ages. Larvae are fed by young worker bees.

(k) Worker pupae. Once again the cell caps have been removed, exposing pupae that will metamorphose into future workers.

(l) Complete sequence of developmental stages of a worker bee, from the egg (far left) to a six-day-old larva (fourth from left), to a twenty-one-day-old pupa about to become an adult (far right).

An evolutionary approach to human behavior may be under most attack by individuals who fear that what it reveals will be misused. They may think that to suggest a trait such as aggression is adaptive (genetically advantageous) is to imply that it is desirable in moral or social terms. Yet there is a clear difference between trying to explain something (by reference to its possible evolutionary history) and trying to justify it. "Adaptive" does not mean the same thing as "moral." It means *valuable in the transmission of an individual's genes*.

Research into the relationship between evolution, genes, and our behavioral capacity is in its infancy. Yet improved knowledge of ourselves could be used to help resolve many critical issues related to our biology as a species, including overpopulation, exploitation of the environment, and aggression among modern nations. It remains to be seen how quickly this knowledge will be gained, and whether it will actually be employed to alleviate negative aspects of the human condition. In the meantime, perhaps some readers will find that an appreciation of the natural world can be deepened through application of an evolutionary approach to one of its most diverse expressions—human behavior.

SUMMARY

1. Social behavior depends on communication among animals living together. Communication signals are a form of cooperation in which both a signaler and a receiver benefit from the transfer of information between them. Signaling animals employ many different sensory modes as channels of communication.

2. Social species, those that live in groups, are not more highly evolved than solitary species whose members live largely in isolation from each other. Sociality carries with it higher risks of communicable diseases, exploitation by others, and increased competition for scarce resources. But particularly if predation pressure is severe, the various benefits that arise from improved defense against predation can outweigh the costs of social life.

3. Most animal societies are composed of individuals that do not sacrifice their reproductive success to benefit others in their group. In some species, parents gain by living with and helping their offspring. In other species, weaker subordinate animals gain by deferring reproduction temporarily in order to take advantage of the benefits, especially improved safety, that come from belonging to a group.

4. It is theoretically possible for individuals to reduce their net lifetime production of offspring in order to help other kinds of relatives reproduce. Kin selection can lead to the extreme altruism of workers of some social insects;

even though the workers are sterile, they help their reproducing siblings survive, passing on "by proxy" the genes underlying the development of altruism.

Review Questions

1. A hyena places scent marks on vegetation in its territory by releasing specific chemicals from certain glands. What evidence would you need to demonstrate that this action is an evolved communication signal? *821*

2. Explain how a threat display can be considered an example of cooperation. *822*

3. Why don't the members of a "selfish herd" live apart if each member is "trying to take advantage" of the others? *826–827*

4. How can an animal be cooperative and still pass on more of its genes than a noncooperative individual? *827–829*

5. Is parental behavior always adaptive? *827–828*

6. What is true altruism? *830*

7. How can an individual propagate its genes even if it is sterile? *830–831*

Readings

In addition to the readings outlined at the end of the preceding chapter, all of which deal with animal behavior, there are a number of excellent texts that focus primarily on material covered in this chapter, including:

Daly, M., and M. Wilson. 1983. *Sex, Evolution, and Behavior*. Second edition. Boston: Willard Grant Press. This book has two particularly good chapters on the evolutionary analysis of human behavior.

Frisch, K. von. 1961. *The Dancing Bees*. New York: Harcourt Brace Jovanovich. A classic on the natural history and behavior of honeybees.

Trivers, R. 1985. *Social Evolution*. Menlo Park, California: Benjamin/ Cummings. A complete and readable account of all the topics that make up an evolutionary approach to social behavior.

Wilson, E. O. 1975. *Sociobiology: The New Synthesis*. Cambridge, Massachusetts: Harvard University Press. The pivotal book in presenting a new approach to the study of social behavior.

Wittenberger, J. F. 1981. *Animal Social Behavior*. Boston: Willard Grant Press. A book that covers much the same ground as *Social Evolution* by Robert Trivers but at a somewhat more advanced level. The book does an excellent job of laying out competing hypotheses on aspects of social behavior and evaluating the evidence for them.

APPENDIX I
A BRIEF SYSTEM OF CLASSIFICATION

This classification scheme is a composite of several used in microbiology, botany, and zoology. Although major groupings are more or less agreed upon, what to call them and (sometimes) where to place them in the overall hierarchy are not. As Chapter 37 indicated, there are several reasons for this. First, the fossil record varies in its quality and completeness, so certain evolutionary relationships are open to interpretation. Comparative studies at the molecular level are firming up the picture, but this work is still under way.

Second, since the time of Linneaus, classification schemes have been based on perceived morphological similarities and differences among organisms. Although some original interpretations are now open to question, we are so used to thinking about organisms in certain ways that reclassification proceeds slowly. For example, birds and reptiles traditionally are considered separate classes (Reptilia and Aves)—even though there now are compelling arguments for grouping lizards and snakes as one class, and crocodilians, dinosaurs, and birds as another (page 556).

Finally, botanists as well as zoologists have inherited a wealth of literature based on schemes that are peculiar to their fields; and most see no good reason to give up established terminology and so disrupt access to the past. Thus botanists continue to use Division as a major taxon in the hierarchical schemes and zoologists use Phylum in theirs. Opinions are notably divergent with respect to an entire Kingdom (the Protista), certain members of which could just as easily be called single-celled forms of plants, fungi, or animals. Indeed, the term protozoan is a holdover from earlier schemes that ranked the amoebas and some other forms as simple animals.

Given the problems, why do we bother imposing hierarchical schemes on the natural history of life on earth? We do this for the same reason that a writer might decide to break up the history of civilization into several volumes, many chapters, and a multitude of paragraphs. Both efforts are attempts to impart structure to what might otherwise be an overwhelming body of information.

One more point to keep in mind: The classification scheme in this Appendix is primarily for reference purposes, and it is by no means complete (numerous phyla of existing and extinct organisms are not represented). Our strategy is to focus mainly on the organisms mentioned in the text, with numerals referring to some of the pages on which representatives are illustrated or described. A few examples of organisms are also listed under the entries.

SUPERKINGDOM PROKARYOTA. Prokaryotes (single-celled organisms with no nucleus or other membrane-bound organelles in the cytoplasm).

KINGDOM MONERA. Bacteria, either single cells or simple associations of cells; autotrophic and heterotrophic forms. *Bergey's Manual of Systematic Bacteriology*, the authoritative reference in the field, calls this "a time of taxonomic transition" and groups bacteria mainly on the basis of form, physiology, and behavior, not on phylogeny (Table 39.3 gives examples). The scheme presented here does reflect the growing evidence of evolutionary relationships for at least some bacterial groups.

SUBKINGDOM ARCHAEBACTERIA. Methanogens, halophiles, thermoacidophiles. Strict anaerobes, distinct from other bacteria in their cell wall, membrane lipids, ribosomes, and RNA sequences. 576, 594–595

SUBKINGDOM EUBACTERIA. Gram-negative and Gram-positive forms. Peptidoglycan in cell walls. Photosynthetic autotrophs, chemosynthetic autotrophs, and heterotophs. 576, 596

DIVISION GRACILICUTES. Typical Gram-negative, thin wall. Autotrophs (photosynthetic and chemosynthetic) and heterotrophs. *Anabaena, Chlorobium, Escherichia, Shigella, Desulfovibrio, Agrobacterium, Pseudomonas, Neisseria.* 224, 534, 575, 596

DIVISION FIRMICUTES. Typical Gram-positive, thick wall. Heterotrophs. *Staphylococcus, Streptococcus, Clostridium, Bacillus, Actinomyces.* 199, 238, 332, 590, 598

DIVISION TENERICUTES. Gram-negative, wall absent. Heterotrophs (saprobes, parasites, pathogens). *Mycoplasma.* 596

KINGDOM PROTISTA. Mostly single-celled eukaryotes. Some colonial.

PHYLUM GYMNOMYCOTA. Heterotrophs.

Class Acrasiomycota. Cellular slime molds. *Dictyostelium.* 225, 601

Class Myxomycota. Plasmodial slime molds. *Physarum.* 601, 602

PHYLUM EUGLENOPHYTA. Euglenids. Mostly heterotrophic, some photosynthetic. Flagellated. *Euglena.* 602, 603

PHYLUM CHRYSOPHYTA. Golden algae, yellow-green algae, diatoms. Photosynthetic. Some flagellated, others not. 603, 747

PHYLUM PYRROPHYTA. Dinoflagellates. Mostly photosynthetic, some heterotrophs. *Gonyaulax.* 603, 604

PHYLUM SARCOMASTIGOPHORA. Heterotrophs. Free-living, symbiotic, and parasitic forms.

Subphylum Mastigophora. Flagellated protozoans. *Trypanosoma.* 605

Subphylum Sarcodina. Amoeboid protozoans. Amoebas, foraminiferans, heliozoans, radiolarians. 62, 605, 606

PHYLUM APICOMPLEXA. Sporozoans and some other parasitic protozoans, many intracellular. *Plasmodium.* 606

PHYLUM CILIOPHORA. Ciliated protozoans. *Paramecium, Didinium.* 79, 91, 674, 607

KINGDOM FUNGI. Mostly multicelled eukaryotes. Heterotrophs (mostly saprobes, some parasites). All rely on extracellular digestion and absorption of nutrients.

DIVISION MASTIGOMYCOTA. All produce flagellated spores.

Class Chytridiomycetes. Chytrids. 613–614

Class Oomycetes. Water molds and related forms. *Plasmopora, Phytophthora.* 614, 620

DIVISION AMASTIGOMYCOTA. All produce nonmotile spores.

Class Zygomycetes. Bread molds and related forms. *Rhizopus, Pilobilus.* 611, 614, 615

Class Ascomycetes. Sac fungi. Most yeasts and molds; morels, truffles. *Saccharomyces, Morchella.* 615, 616

Class Basidiomycetes. Club fungi. Mushrooms, shelf fungi, bird's nest fungi, stinkhorns. *Agaricus, Amanita.* 616–618

FORM-DIVISION DEUTEROMYCOTA. Imperfect fungi. All with undetermined affiliations because sexual stage unknown; if better known they would be grouped with sac fungi or club fungi. *Verticillium, Candida.* 619, 620

KINGDOM PLANTAE. Nearly all multicelled eukaryotes. Photosynthetic autotrophs, except for a few saprobes and parasites.

DIVISION RHODOPHYTA. Red algae. *Porphyra.* 66, 624, 625

DIVISION PHAEOPHYTA. Brown algae. *Fucus, Laminaria.* 622, 624–627

DIVISION CHLOROPHYTA. Green algae. *Ulva, Spirogyra.* 11, 112, 624, 627

DIVISION CHAROPHYTA. Stoneworts.

DIVISION BRYOPHYTA. Liverworts, hornworts, mosses. *Marchantia, Sphagnum.* 628, 629

DIVISION RHYNIOPHYTA. Earliest known vascular plants; extinct. *Cooksonia, Rhynia.* 581

DIVISION PSILOPHYTA. Whisk ferns. 581

DIVISION LYCOPHYTA. Lycopods, club mosses. *Lycopodium, Selaginella.* 629, 630, 632

DIVISION SPHENOPHYTA. Horsetails. *Equisetum.* 629, 630, 632

DIVISION PTEROPHYTA. Ferns. 629, 631, 632

DIVISION PROGYMNOSPERMOPHYTA. Progymnosperms. Ancestral to early seed-bearing plants; extinct. *Archaeopteris.* 581

DIVISION PTERIDOSPERMOPHYTA. Seed ferns (extinct fernlike gymnosperms).

DIVISION CYCADOPHYTA. Cycads. *Zamia.* 632, 634

DIVISION GINKGOPHYTA. Ginkgo. *Ginkgo.* 632, 634

DIVISION GNETOPHYTA. Gnetophytes. *Ephedra, Welwitschia, Gnetum.* 632, 634–635

DIVISION CONIFEROPHYTA. Conifers. 248, 632, 635*ff.*

Family Pinaceae. Pines, firs, spruces, hemlock, larches, Douglas firs, true cedars. *Pinus.* 635, 636, 727, 740, 796

Family Cupressaceae. Junipers, cypresses, false cedars. 635

Family Taxodiaceae. Bald cypress, redwood, Sierra bigtree, dawn redwood. *Sequoia.* 3, 267, 635

Family Taxaceae. Yews.

DIVISION ANTHOPHYTA. Flowering plants. 248*ff.,* 632, 636*ff.*

Class Dicotyledonae. Dicotyledons (dicots). Some families of several different orders are listed:

Family Magnoliaceae. Magnolias, tulip trees.

Family Ranunculaceae. Buttercups, delphinium. 260

Family Nymphaeaceae. Water lilies. 638

Family Papaveraceae. Poppies, including opium poppy.

Family Brassicaceae. Mustards, cabbage, radishes, turnips. 300

Family Malvaceae. Mallows, cotton, okra, hibiscus.

Family Solanaceae. Potatoes, eggplant, petunias. 77, 118, 620

APPENDIX II
SOLUTIONS TO GENETICS PROBLEMS

Chapter Twelve

1. a. *AB*
   b. *AB* and *aB*
   c. *Ab* and *ab*
   d. *Ab, aB, Ab,* and *ab*

2. a. *AaBB* will occur in all the offspring.
   b. 25% *AABB*; 25% *AaBB*; 25% *AABb*; 25% *AaBb*.
   c. 25% *AaBb*; 25% *Aabb*; 25% *aaBb*; 25% *aabb*.

   d. $\frac{1}{16}$ *AABB* (6.25%)

   $\frac{1}{8}$ *AaBB* (12.5%)

   $\frac{1}{16}$ *aaBB* (6.25%)

   $\frac{1}{8}$ *AABb* (12.5%)

   $\frac{1}{4}$ *AaBb* (25 %)

   $\frac{1}{8}$ *aaBb* (12.5%)

   $\frac{1}{16}$ *AAbb* (6.25%)

   $\frac{1}{8}$ *Aabb* (12.5%)

   $\frac{1}{16}$ *aabb* (6.25%)

3. Yellow is recessive. Because the first-generation plants must be heterozygous and had a green phenotype, green must be dominant over the recessive yellow.

4. a. Mother must be heterozygous for both genes; father is homozygous recessive for both genes. The first child is also homozygous recessive for both genes.
   b. The probability that the second child will not be able to roll the tongue and will have free earlobes is $\frac{1}{4}$ (25%).

5. a. *ABC*
   b. *ABc* and *aBc*
   c. *ABC, aBC, ABc,* and *aBc*
   d. *ABC, aBC, AbC, abC, ABc, aBc, Abc,* and *abc*

6. Because the man can only produce one type of allele for each of his ten genes, he can only produce one type of sperm. The woman, on the other hand, can produce two types of alleles for each of her two heterozygous genes; she can produce 2×2 or 4 different kinds of eggs. As can be observed, as the number of heterozygous genes increases, more and more different types of gametes can be produced. Such variability might prove useful, in that the variant individuals might have better chances of surviving under a variety of environmental conditions.

7. The first-generation plants must all be double heterozygotes. When these plants are self-pollinated, $\frac{1}{4}$ (25%) of the second-generation plants will be doubly heterozygous.

8. The most direct way to accomplish this would be to allow a true-breeding mouse having yellow fur to mate with a true-breeding mouse having brown fur. Such true-breeding strains could be obtained by repeated inbreeding (mating of related individuals; for example, a male and a female of the same litter) of yellow and brown strains. In this way, it should be possible to obtain homozygous yellow and homozygous brown mice.

When true-breeding yellow and true-breeding brown mice are crossed, the progeny should all be heterozygous. If the progeny phenotype is either yellow or brown, then the dominance is simple or complete, and the phenotype reflects the dominant allele. If the phenotype is intermediate between yellow and brown, there is incomplete dominance. If the phenotype shows both yellow and brown, there is codominance.

9. a. The mother must be heterozygous (I^Ai). The man having type B blood could have fathered the child if he were also heterozygous (I^Bi).
   b. If the man is heterozygous, then he *could be* the father. However, because any other type B heterozygous male also could be the father, one cannot say that this particular man absolutely must be. Actually, any male who could contribute an O allele (i) could have fathered the child. This would include males with type O blood (ii) or type A blood who are heterozygous (I^Ai).

10. a. F_1 genotypes and phenotypes: 100% *Bb Cc*, brown progeny.
    F_2 phenotypes: $\frac{9}{16}$ brown $+ \frac{3}{16}$ tan $+ \frac{4}{16}$ albino.

    F_2 genotypes: $\begin{cases} \frac{1}{16} \text{ } BB\text{ } CC + \frac{2}{16} \text{ } BB\text{ } Cc + \frac{2}{16}\text{ } Bb\text{ } CC + \frac{4}{16}\text{ } Bb\text{ } Cc; \\ (\frac{9}{16}\text{ brown}) \\ \frac{1}{16}\text{ } bb\text{ } CC + \frac{2}{16}\text{ } bb\text{ } Cc; (\frac{3}{16}\text{ tan}) \\ \frac{1}{16}\text{ } BB\text{ } cc + \frac{2}{16}\text{ } Bb\text{ } cc + \frac{1}{16}\text{ } bb\text{ } cc; (\frac{4}{16}\text{ albino}) \end{cases}$

    b. Backcross phenotypes: $\frac{1}{4}$ brown $+ \frac{1}{4}$ tan $+ \frac{2}{4}$ albino.

    Backcross genotypes: $\begin{cases} \frac{1}{4}\text{ } Bb\text{ } Cc; (\frac{1}{4}\text{ brown}) \\ \frac{1}{4}\text{ } bb\text{ } Cc; (\frac{1}{4}\text{ tan}) \\ \frac{1}{4}\text{ } Bb\text{ } cc + \frac{1}{4}\text{ } bb\text{ } cc; (\frac{1}{2}\text{ albino}) \end{cases}$

11. The mating is $Ll \times Ll$.
    Progeny genotypes: $\frac{1}{4}\text{ } LL + \frac{1}{2}\text{ } Ll + \frac{1}{4}\text{ } ll$

    Phenotypes: $\begin{cases} \frac{1}{4}\text{ homozygous survivors } (LL) + \\ \frac{1}{2}\text{ heterozygous survivors } (Ll) + \\ \frac{1}{4}\text{ lethal } (ll)\text{ nonsurvivors.} \end{cases}$

    Thus, among the survivors, $\frac{2}{3}$ will be heterozygous.

12. To work this problem, consider the effect of each pair of genes separately.
    a. $Aa \times aa$: The resulting progeny genotypes are $\frac{1}{2}$ *Aa* and $\frac{1}{2}$ *aa*. Thus, half of the progeny will receive an *A* allele, which permits kernel pigmentation.
    b. $cc \times Cc$: Here, too, half of the progeny will receive a dominant *C* allele, which permits kernel pigmentation.
    c. $Rr \times Rr$: In this case, $\frac{3}{4}$ of the progeny will receive at least one *R* allele, which permits kernel pigmentation.

    Remember that in order to have pigmented kernels, at least one dominant allele of each and every one of these three gene loci must be simultaneously present. Since the *A*, *C*, and *R* loci assort independently, to find the overall fraction of kernels that are pigmented, you must multiply together the fraction of pigmented kernels produced separately by the *A*, *C*, and *R* loci, and that is: $\frac{1}{2} \times \frac{1}{2} \times \frac{3}{4} = \frac{3}{16}$.

Chapter Thirteen

1. a. Males inherit their X chromosome from their mothers.
   b. A male can produce two types of gametes with respect to an X-linked gene. One type will lack this gene and possess a Y chromosome. The other will have an X chromosome and the linked gene.
   c. A female homozygous for an X-linked gene will produce just one type of gamete containing an X chromosome with the gene.
   d. A female heterozygous for an X-linked gene will produce two types of gametes. One will contain an X chromosome with the dominant allele, and the other type will contain an X chromosome with the recessive allele.

2. a. Because this gene is only carried on Y chromosomes, females would not be expected to have hairy pinnae because they normally do not have Y chromosomes.
   b. Because sons always inherit a Y chromosome from their fathers and because daughters never do, a man having hairy pinnae will always transmit this trait to his sons and never to his daughters.

3. A 0% crossover frequency means that 50% of the gametes will be *AB* and 50% will be *ab*.

4. The first-generation females must be heterozygous for both genes. The 42 red-eyed, vestigial-winged and the 30 purple-eyed, long-winged progeny represent recombinant gametes from these females. Because the first-generation females must have produced 600 gametes to give these 600 progeny, and because 42 + 30 of these were recombinant, the percentage of recombinant gametes is 72/600, or 12%, which implies that 12 map units separate the two genes.

5. The rare vestigial-winged flies could be explained by a deletion of the dominant allele from one of the chromosomes, due to the action of the x-rays.

6. If the longer-than-normal chromosome 14 represented the translocation of most of chromosome 21 to the end of a normal chromosome 14, then this individual would be afflicted with Down syndrome due to the presence of this attached chromosome 21 as well as two normal chromosomes 21. The total chromosome number, however, would be 46.

7. The initial cells of the hybrid from a cross between *T. turgidum* and *T. tauschii* did have twenty-one chromosomes. These chromosomes were induced to double in the absence of cell division by application of colchicine; the result was the hexaploid, with forty-two chromosomes. Gametes from *T. aestivum* should have twenty-one chromosomes (ABD); when these combine with gametes from *T. turgidum* (fourteen chromosomes, AB), offspring with 21 + 14 = 35 chromosomes (AABBD) would arise. Because there is only one set of D chromosomes, normal meiotic pairing should not occur and these hybrids would be sterile.

8. Using *c* as the symbol for color blindness and *C* for normal vision, then the cross can be diagrammed as follows:

   C(Y) female × Cc male

   In mugwumps, a son receives one sex-linked allele from each of his parents, but a daughter inherits her unpaired sex-linked allele solely from her father. In this cross, half of the sons will be CC and half Cc, but none will be color blind. Of the daughters, half will be C(Y) and half c(Y). Thus, there is a 50% chance that a daughter will be color blind. (Note: this answer is backward from the way it would be in humans; but it is correct not only for mugwumps but also for all birds and Lepidoptera (moths and butterflies), as well as a few other forms. It may be easier for you to work sex-linked problems involving such organisms if you first work them exactly as you would a human or fruit fly problem; but at the end, simply change all progeny called male to female and all female to male. Remember it this way: "Birds is backwards!")

9. In order to produce a female suffering from childhood muscular dystrophy, not only must her mother be a carrier of the disease, but her father must have it. Few, if any, such males who survive to adulthood are capable of having children.

Chapter Fourteen

1. The gene for hemophilia occurs on the X but not the Y chromosome. A male has only one X chromosome. Therefore, it would be impossible for a male simply to be a carrier; the allele associated with hemophilia would always be expressed.

2. Assuming the mother is heterozygous (most individuals with Huntington's disorder are), the woman has a $\frac{1}{2}$ (50%) chance of being heterozygous and therefore of later developing the disorder. Also, if this woman married a normal male, they would have a 50% chance of having a child with the disorder. Thus the *total* probability of their having a child with Huntington's disorder is $(0.5)(0.5) = 0.25$, or $\frac{1}{4}$ (25%).

3. The first child can only be color blind if it is a boy. Why? The probability of this happening is 25%. Similarly, their second child also has a 25% chance of being color blind. The probability that both will be color blind is $(0.25)(0.25) = 0.0625$, or 6.25%.

4. This indicates that genetic information other than that necessary for sex determination must reside on the X chromosome. Such information is necessary for survival regardless of whether one is male or female. Obviously, this is not true for the Y chromosome, in that individuals (females) survive quite nicely in its absence. A major function of the Y chromosome is to change what would have been a female individual into a male.

5. The only parent from whom this child could have received an X chromosome that bears a nonhemophilia allele is the mother. Therefore, nondisjunction must have occurred in the father.

6. A child with Klinefelter syndrome could be produced if a Y-bearing sperm fertilized an egg having two X chromosomes (as a result of nondisjunction during egg development). Such a child could also be produced if a normal egg (with one X chromosome) were fertilized by a sperm having an X and Y chromosome (as a result of nondisjunction during sperm development).

7. The Punnett square for this situation would be as follows:

	X-bearing sperm	Y-bearing sperm
XX-bearing egg	trisomic XXX	Klinefelter XXY
no X in egg	Turner XO	dies before birth; has only Y

8. a. An unaffected female selected at random has 1 chance in 50 (2%) of being heterozygous.
   b. If you selected an unaffected male and unaffected female at random, the probability that both will be heterozygous is $(0.02)(0.02) = 0.0004$, or 0.04%. The probability that a pair of unaffected individuals selected at random could have a child affected by PKU is given by $(0.02)(0.02)(0.25) = 0.0001$, or 0.01%. This is the same thing as 1/10,000, which suggests that about one birth in every 10,000 will be a child with PKU, assuming that only heterozygous individuals have such children (an assumption which is not completely true).

9. There would be 1 chance in 50 (2%) that a randomly selected mate would also be a PKU carrier. There would be a much greater chance that your first cousin would be a PKU carrier, because both of you would share a common ancestor and thus, possibly, a common source for the PKU allele.

APPENDIX III
UNITS OF MEASURE

Length

English		Metric
inch	=	2.54 centimeters
foot	=	0.30 meter
yard	=	0.91 meter
mile (5,280 feet)	=	1.61 kilometer

To convert	multiply by	to obtain
inches	2.54	centimeters
feet	30.00	centimeters
centimeters	0.39	inches
millimeters	0.039	inches

Weight

English		Metric
grain	=	64.80 milligrams
ounce	=	28.35 grams
pound	=	453.60 grams
ton (short) (2,000 pounds)	=	0.91 metric ton

To convert	multiply by	to obtain
ounces	28.3	grams
pounds	453.6	grams
pounds	0.45	kilograms
grams	0.035	ounces
kilograms	2.2	pounds

Volume

English		Metric
cubic inch	=	16.39 cubic centimeters
cubic foot	=	0.03 cubic meter
cubic yard	=	0.765 cubic meters
ounce	=	0.03 liter
pint	=	0.47 liter
quart	=	0.95 liter
gallon	=	3.79 liters

To convert	multiply by	to obtain
fluid ounces	30.00	milliliters
quart	0.95	liters
milliliters	0.03	fluid ounces
liters	1.06	quarts

Area

2.4711 acres	=	1 hectare		
0.386 square mile	=	100 hectares	=	1 square kilometer

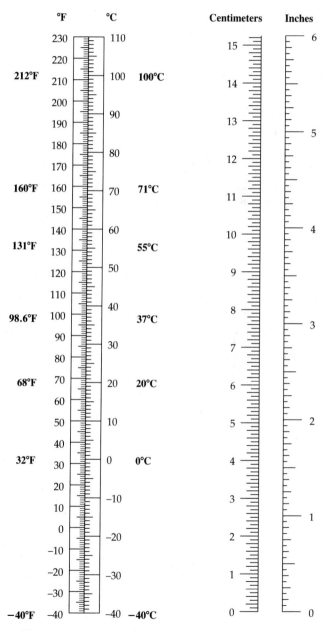

To convert temperature scales:

Fahrenheit to Celsius: $°C = 5/9 (°F - 32)$

Celsius to Fahrenheit: $°F = 9/5 (°C) + 32$

CREDITS

Chapter 1

1.1 James A. Kern / **1.2** (a) Robert Stern; (b) W. Hargreaves and D. Deamer; (c) Robley C. Williams; (d) G. Cohen-Bazire; (e) Michael A. Walsh; (f) M. P. L. Fogden/Bruce Coleman Ltd. / **1.4** Jack de Coningh / **1.5** Jim Doran / **1.6** Douglas Faulkner/ Sally Faulkner Collection / **1.7** Douglas Faulkner/ Sally Faulkner Collection / **1.8** Roger K. Burnard / **1.9** Roger K. Burnard / **1.10** (a) Norman Meyers/ Bruce Coleman Inc.; (b) Timothy Ransom / **1.11** Roger K. Burnard

Chapter 2

2.1 The Bettmann Archive / **2.4** Art by D. & V. Hennings / **2.5** Courtesy George P. Darwin / **2.6** (above) Bianca Lavies, © 1982 National Geographic Society; (below) Charles R. Knight, © Field Museum of Natural History / **2.7** (a) The Bettmann Archive / **2.8** (b) Alan Root/Bruce Coleman Ltd. / **2.9** Encyclopedia of Pigeon Breeds / **2.10** Down House and The Royal College of Surgeons of England / **2.11** (a) Neg. #325288 Courtesy Department Library Services, American Museum of Natural History; (b) John D. Cunningham/Visuals Unlimited

Chapter 3

3.9 Robert D. Schuster / **3.10** William H. Amos / **3.11** Roger K. Burnard

Chapter 4

4.1 Computer Graphics Laboratory, University of California, San Francisco, © Regents, University of California / **4.6** Micrograph E. Frei and R. D. Preston / **4.10** Roland R. Dute / **4.15–4.16** Arthur M. Lesk, *Trends in Biochemical Sciences*, June 1984 / **4.17** Micrograph Jerome Gross; sketch after D. Eyre, *Science*, 207, 21 March 1980, copyright 1980 AAAS

Chapter 5

5.1 Biophoto Associates / **5.4** Jeremy Pickett-Heaps, School of Botany, University of Melbourne / **5.6** (a) Susumu Ito; (b) Kim Taylor/Bruce Coleman Ltd.; (c) Hervé Chaumeton/Agence Nature; (d) McCutcheon/ZEFA / **5.7** (a) Gary Gaard and Arthur Kelman; (b) Micrograph G. Cohen-Bazire / **5.10** Micrograph M. C. Ledbetter, Brookhaven National Laboratory; art by D. & V. Hennings / **5.11** Micrograph G. L. Decker; art by D. & V. Hennings / **5.12** D. Fawcett, *The Cell*, Philadelphia: W. B. Saunders Co., 1966; art by D. & V. Hennings / **5.13** (a) A. C. Faberge, *Cell and Tissue Research*, 151: 403–415, 1974; (b) E. G. Pollock / **5.14** Art by K. Kasnot / **5.15** (b) W. Bloom and D. Fawcett, *A Textbook of Histology*, Philadelphia: W. B. Saunders Co., 1975; (c) Daniel S. Friend, M.D. / **5.16** (a) Gary W. Grimes; (b) Hilton H. Mollenhauer / **5.17** Gary W. Grimes / **5.18** Micrograph Keith R. Porter / **5.19** Micrograph L. K. Shumway / **5.20** (a) J. Victor Small and Gottfried Rinnerthaler; (b) Andrew S. Bajer, University of Oregon; (c) J. E. Heuser and M. Kirschner, *Journal Cell Biology*, 86: 212, 1980, by copyright permission of The Rockefeller University Press / **5.21** (a) Sidney L. Tamm; (b) art by D. & V. Hennings / **5.22** (b) U. W. Goodenough and J. W. Heuser / **5.23** (a) After B. Alberts et al., *Molecular Biology of the Cell*, Garland Publishing Co., 1983; (b)

Diane T. Woodrum and Richard W. Linck / **5.24** Sketch by D. & V. Hennings after P. Raven et al., *Biology of Plants*, Third edition, Worth Publishers, 1981; micrograph P. A. Roelofsen

Chapter 6

6.2 Micrograph K. R. Porter; art by K. Kasnot / **6.3** Micrograph P. Pinto da Silva and D. Branton, *Journal Cell Biology*, 45:598, 1970, by copyright permission of The Rockefeller University Press / **6.5** Micrograph M. Sheetz, R. Painter, and S. Singer, *Journal Cell Biology*, 70:193, 1976, by copyright permission of The Rockefeller University Press / **6.6** (a),(b) Bruce Russell/BioMedia Associates; (c),(d) Thomas Eisner, Cornell University / **6.7** Frank B. Salisbury / **6.11** Micrographs K. W. Jeon / **6.12** M. M. Perry and A. B. Gilbert

Chapter 7

7.1 (left) NASA; (right) Manfred Kage/Peter Arnold / **7.7** W. S. Bennett and T. A. Steitz / **7.10** Art by Victor Royer / **7.11** Michael Connolly, *Science*, 19 August 1983, 221:709–713, copyright 1983 AAAS / **7.14** Art by L. Calver after B. Alberts et al., *Molecular Biology of the Cell*, Garland Publishing Co., 1983.

Chapter 8

8.1 Photograph Sam Zarember/Image Bank / **8.2** (a) Photograph Kjell B. Sandved; art by K. Kasnot; (b),(c) Harry T. Horner; art by Joel Ito / **8.3** Art by Victor Royer / **8.4** Photograph E. R. Degginger / **8.5** Art by L. Calver / **8.6** Art by L. Calver / **8.9** Art by L. Calver / **8.10** Micrograph M. D. Hatch

Chapter 9

9.1 (a) Michael Freeman/Bruce Coleman Ltd.; (b) Adrian Warren/Ardea, London; (c) Tom McHugh / **9.3** Photograph Janeart Ltd./Image Bank / **9.6** (a) Keith R. Porter; art by L. Calver / **9.11** Roger K. Burnard

Chapter 10

10.1 Micrograph J. J. Cardamone, Jr., University of Pittsburgh/BPS / **10.2** E. J. Dupraw, *DNA and Chromosomes*, Holt, Rinehart and Winston, 1970 / **10.3** (a) S. Brecher / **10.4** (bottom) After John D. Jackson and Dallas D. Johnson, *Biology Encounters*, Second edition, Burgess International Group, Inc. 1980 / **10.5** Andrew S. Bajer, University of Oregon / **10.6** Photograph Andrew S. Bajer, University of Oregon; art by K. Kasnot / **10.8** (b) B. R. Brinkley / **10.9** H. Beams and R. G. Kessel, *American Scientist*, 64:279–290, 1976 / **10.10** Art by D. & V. Hennings; (d) B. A. Palevitz and E. H. Newcomb, University of Wisconsin/BPS/Tom Stack & Associates

Chapter 11

11.1 Courtesy of Kirk Douglas/The Bryna Company / **11.2** Art by L. Calver / **11.4** Art by K. Kasnot / **11.5** Art by K. Kasnot / **11.6** Micrograph B. John / **11.8** Art by K. Kasnot / **11.9** (a) Art by K. Kasnot; (b) © 1986 David M. Phillips/Visuals Unlimited / **11.11** Art by K. Kasnot

Chapter 12

12.1 The Granger Collection, New York / **12.2** Jean M. Labat/Ardea, London / **12.11** Jack Carey / **12.13** Tedd Somes / **12.14** David Hosking / **12.15** (b) David M. Phillips/Visuals Unlimited; (c) Bill Longcore/PR / **12.16** Douglas Faulkner/Sally Faulkner Collection / **12.17** After John G. Torrey, *Development in Flowering Plants*, by permission of Macmillan Publishing Company, copyright © 1967 by John G. Torrey / **12.18** (b) F. Blakeslee, *Journal of Heredity*, 1914

Chapter 13

13.1 Photograph Chuck Brown; art by K. Kasnot / **13.2** Photograph Carolina Biological Supply Company / **13.4** Photograph Carolina Biological Supply Company / **13.9** A. D. Stock, City of Hope National Medical Center / **13.10** Photograph courtesy of B. R. Brinkley from D. E. Merry et al., *American Journal of Human Genetics*, 37:425–430, 1985. © 1985 by The American Society of Human Genetics. All rights reserved. / **13.11** Art by K. Kasnot / **13.12** (a) H. P. Olmo, University of California, Davis; (b) Allison L. Hansen / **13.13** After W. Jensen and F. B. Salisbury, *Botany: An Ecological Approach*, Wadsworth, 1972

Chapter 14

14.4 After V. McKusick, *Human Genetics*, Second edition, copyright 1969. Reprinted by permission of Prentice-Hall, Inc., Englewood Cliffs, New Jersey / **14.5** (a) Cytogenetics Laboratory, University of California, San Francisco; (b) after Collmann and Stoller, *American Journal of Public Health*, 52, 1962 / **14.6** (a) Used by permission of Carole Iafrate; (b–d) Courtesy of Peninsula Association for Retarded Children and Adults, San Mateo Special Olympics, Burlingame, CA / **14.10** After A. Emery, *An Introduction to Recombinant DNA*, Wiley, 1984.

Chapter 15

15.2 (a) Lee D. Simon, Waksman Institute of Microbiology, Rutgers University; (b) A. K. Kleinschmidt / **15.3** Art by D. & V. Hennings / **15.6** Photograph Computer Graphics Laboratory, University of California, San Francisco, © Regents, University of California; art by D. & V. Hennings / **15.7** A. C. Barrington Brown, © 1968 J. D. Watson / **15.8** Art by L. Calver / **15.9** Art by L. Calver

Chapter 16

16.6 Micrograph courtesy O. L. Miller, Jr. / **16.9** (b) Sung Hou Kim / **16.12** Art by L. Calver / **16.13** Art by L. Calver

Chapter 17

17.1 Jack Carey / **17.2** Art by Victor Royer / **17.3** Data for sketch contributed by John Sternfeld; (a) John T. Bonner; (b) London Scientific Films; (c–e) Carolina Biological Supply Company / **17.5** (a) E. J. Dupraw; (b) J. G. Gall; (c) B. Hamkalo; (d) V. Foe / **17.6** Micrograph O. L. Miller, Jr. and Steve McKnight / **17.7** I. Paulsen and U. Laemmli / **17.8** Micrograph J. G. Gall; art based on the work of J. G. Gall, H. Callan, H. McGregor, and others / **17.9** W. Beerman / **17.10** Murray L. Barr

Chapter 18

18.1 R. L. Brinster / **18.2** Peter Starlinger / **18.3** Dr. Huntington Potter and Dr. David Dressler, Harvard Medical School / **18.4** C. C. Brinton, Jr. and J. Carnahan / **18.13** Damon Biotech, Inc. / **18.14** Mike Maloney/San Francisco Chronicle / **18.15** W. Merrill

Chapter 19

19.3 (a–c) Photographs Biophoto Associates; (d) Thomas Eisner, Cornell University / **19.4** Micrographs H. A. Core, W. A. Coté, and A. C. Day, *Wood Structure and Identification*, Second edition, Syracuse University Press, 1979 / **19.7** (a) Chuck Brown; (b) G. Shih and R. G. Kessel, *Living Images*, Jones and Bartlett, Publishers, Inc., Boston, © 1982 / **19.8** Art by D. & V. Hennings / **19.9** (left) Art by D. & V. Hennings; (center) Carolina Biological Supply Company; (right) James W. Perry / **19.10** (left) Art by D. & V. Hennings; (center) Ray F. Evert; (right) James W. Perry / **19.11** (a) Robert and Linda Mitchell; (b),(c) Roland R. Dute / **19.12** E. R. Degginger / **19.13** Art by D. & V. Hennings / **19.14** (a) John N. A. Lott, *Scanning Electron Microscope Study of Green Plants*, St. Louis: C. V. Mosby Company, 1976; (b–d) Robert and Linda Mitchell / **19.15** J. Troughton and L. A. Donaldson; art by D. & V. Hennings / **19.16** G. Shih and R. G. Kessel, *Living Images*, Jones and Bartlett, Publishers, Inc., Boston, © 1982 / **19.17** W. Thomson, *American Journal of Botany*, 57(3):316, 1970 / **19.18** John Hodgin / **19.19** Sketch Marian Reeve; photograph E. R. Degginger / **19.20** (a),(b) Chuck Brown; (c),(d) Weier et al., *Botany: An Introduction to Plant Biology*, Sixth edition, Wiley, 1982 / **19.21** Carolina Biological Supply Company / **19.22** Ripon Microslides, Inc. / **19.23** (b) Jerry D. Davis / **19.24** After Marian Reeve / **19.27** (a) Biophoto Associates; (b) H. A. Core, W. A. Coté, and A. C. Day, *Wood Structure and Identification*, Second edition, Syracuse University Press, 1979

Chapter 20

20.1 Dennis Brokaw / **20.2** Adrian P. Davies/Bruce Coleman Ltd. / **20.3** J. Mexal, C. Reid, and E. Burke, *Botanical Gazette*, 140: No. 3, University of Chicago Press, 1979 / **20.4** F. B. Reeves / **20.5** Jean Paul Revel / **20.7** (a),(b) John Troughton and L. A. Donaldson / **20.8** T. A. Mansfield / **20.10** Martin Zimmerman, *Science*, 133:73–79, © AAAS 1961

Chapter 21

21.1 Wardene Wiesser/Ardea, London / **21.5** F. D. Hess / **21.6** Art by D. & V. Hennings / **21.7** (a) Ted Schwartz; (b) R. Taggart; (c) Harlo H. Hadow / **21.8** Thomas Eisner, Cornell University / **21.9** Edward S. Ross / **21.10** (a) R. Taggart; (b) Edward S. Ross; (c) Ted Schwartz / **21.11** (a),(b) Patricia Schulz; (c),(d) Ray F. Evert; (e),(f) Ripon Microslides, Inc. / **21.12** (a) Kjell B. Sandved; (b) Edward S. Ross / **21.13** B. Bracegirdle and P. Miles, *An Atlas of Plant Structure*, Heinemann Educational Books, 1977 / **21.14** Janet Jones / **21.15** (a) B. J. Miller, Fairfax, VA/ BPS; (b) R. Carr/Bruce Coleman Ltd.; (c) Richard H. Gross, Motlow State Community College

Chapter 22

22.1 (a) Carolina Biological Supply Company; (b) Hervé Chaumeton/Agence Nature / **22.6** (a) H. E. Thompson / **22.7** Frank B. Salisbury / **22.8** John Digby and Richard Firn / **22.9** B. E. Juniper / **22.10** Cary Mitchell / **22.11** Frank B. Salisbury / **22.13** Frank B. Salisbury / **22.15** Jan Zeevaart / **22.16** N. R. Lersten, Iowa State University / **22.17** A. C. Leopold et al., *Plant Physiology*, 34:570, 1958 / **22.18** A. C. Leopold and M. Kawase, *American Journal of Botany*, 51:294–298, 1964 / **22.19** R. J. Downs in T. T. Kozlowski, ed., *Tree Growth*, The Ronald Press, 1962 / **22.21** (left) Edward S. Ross; (right) Dennis Brokaw

Chapter 23

23.1 Lennart Nilsson from *Behold Man*, © 1974 by Albert Bonniers Forlag and Little, Brown and Company, Boston / **23.4** Art by L. Calver / **23.5** Art by D. & V. Hennings / **23.6** (a),(b) Manfred Kage/ Bruce Coleman Ltd.; (c) Ed Reschke/Peter Arnold, Inc. / **23.8** Photographs Ed Reschke / **23.9** Art by Joel Ito; micrograph by Bruce Russell/BioMedia Associates / **23.10** Photographs by Ed Reschke / **23.11** Art by Joel Ito / **23.12** Photograph Chaumeton-Lanceau/Agence Nature; sketch after T. Storer et al., *General Zoology*, Sixth edition, McGraw-Hill, 1979

Chapter 24

24.3 Sketch after J. E. Crouch, *Functional Human Anatomy*, Fourth edition, Lea & Febiger, Philadelphia, 1985; Photograph Lennart Nilsson from *Behold Man*, © 1974 by Albert Bonniers Forlag and Little, Brown and Company, Boston / **24.3** Art by D. & V. Hennings / **24.4** (b) A. L. Hodgkin, *Journal of Physiology*, vol. 131, 1956 / **24.6** Art by D. & V. Hennings / **24.8** Art by K. Kasnot / **24.10** (a) Photograph Carolina Biological Supply Company / **24.11** Art by D. & V. Hennings; (d) J. E. Heuser and T. S. Reese / **24.13** Art by K. Kasnot

Chapter 25

25.2 Art by D. & V. Hennings / **25.4** Art by Joel Ito / **25.7** Art by Joel Ito / **25.9** Art by D. & V. Hennings after Romer and others / **25.10** C. Yokochi and J. Rohen, *Photographic Anatomy of the Human Body*, Second edition, Igaku-Shoin Ltd., 1979 / **25.11** Art by Joel Ito / **25.12** After Penfield and Rasmussen, *The Cerebral Cortex of Man*, copyright © 1950 Macmillan Publishing Company, Inc. Renewed 1978 by Theodore Rasmussen / **25.13** (a) Based on A. Vander et al., *Human Physiology*, Third edition, © 1980 by McGraw-Hill / **25.14** After H. Jasper, 1941

Chapter 26

26.1 Art by D. & V. Hennings / **26.2** Art by Joel Ito / **26.3** Art by Joel Ito / **26.4** Art by Joel Ito / **26.7** (a) Syndication International (1986) Ltd.; (b) Mitchell Layton / **26.8** Photographs courtesy of Dr. William H. Daughaday, Washington University School of Medicine. From A. I. Mendelhoff and D. E. Smith, eds., *American Journal of Medicine*, 20:133 (1956) / **26.10** Art by Joel Ito / **26.11** The Bettmann Archive

Chapter 27

27.1 Eric A. Newman / **27.2** From Hensel and Bowman, *Journal of Physiology*, 23:564–568, 1960 / **27.3** After Penfield and Rasmussen, *The Cerebral Cortex of Man*, copyright © 1950 Macmillan Publishing Company, Inc. Renewed 1978 by Theodore Rasmussen / **27.4** (a) After R. Murray and A. Murray, *Taste and Smell in Vertebrates*, Churchill, 1970; (b) Ed Reschke / **27.5** Art by Ron Ervin; photograph Ed Reschke / **27.6** Art by D. & V. Hennings / **27.8** Photograph Douglas Faulkner/ Sally Faulkner Collection / **27.10** (a) Dr. Goran Bredburg, Uppsala, Sweden; (b),(c) Robert E. Preston, courtesy Joseph E. Hawkins, Kresge Hearing Research Institute, University of Michigan Medical School / **27.11** Merlin D. Tuttle, Bat Conservation International / **27.12** After M. Gardiner, *The Biology of Vertebrates*, McGraw-Hill, 1972 / **27.13** (a) Keith Gillett/Tom Stack and Associates; (b) J. Grossauer/ZEFA / **27.14** (a) E. R. Degginger; (b) L. M. Beidler / **27.15** G. A. Mazohkin-Porshnykov (1958). Reprinted with permission from *Insect Vision*, © 1969 Plenum Press / **27.18** Micrograph E. R. Lewis, F. S. Werblin, and Y. Y. Zeevi / **27.19** Sketches after S. Kuffler and J. Nicholls, *From Neuron to Brain*, Sinauer, 1977

Chapter 28

28.1 Douglas P. Wilson / **28.2** D. A. Parry, *Journal of Experimental Biology*, 36:654, 1959 / **28.3** Art by D. & V. Hennings / **28.5** Art by Ron Ervin / **Page 389** Photograph C. Yokochi and J. Rohen, *Photographic Anatomy of the Human Body*, Second edition, 1979 / **28.6** Art by K. Kasnot / **28.7** Photographs courtesy of J. Joseph Prendergast, M.D., Redwood City, CA / **28.8** (b–d) Art by L. Calver / **28.9** (a) Ed Reschke; (b) Micrograph D. Fawcett, *The Cell*, Philadelphia: W. B. Saunders Co., 1966 / **28.11** Inset after L. D. Peachey, *Journal of Cell Biology*, 25:222, 1965 / **28.12** Adapted from R. Eckert and D. Randall, *Animal Physiology: Mechanisms and Adaptations*, Second edition, W. H. Freeman and Co., 1983

Chapter 29

29.1 After M. Labarbera and S. Vogel, *American Scientist*, 70:54–60, 1982 / **29.3** (a) Lennart Nilsson from *Behold Man*, © 1974 by Albert Bonniers Forlag and Little, Brown and Company, Boston; (b) CNRI/ Science Photo Library/Photo Researchers / **29.4** Art by Kathleen Talaro / **29.5** Art by L. Calver / **29.6** (c) After Gerard J. Tortora and Nicholas P. Anagnostakos, *Principles of Anatomy and Physiology*, Fifth edition, Copyright © 1987 by Biological Sciences Textbooks, Inc., A & P Textbooks, Inc., and Elia-Sparta, Inc. Reprinted by permission of Harper & Row, Publishers, Inc. / **29.7** (a) C. Yokochi and J. Rohen, *Photographic Anatomy of the Human Body*, Second edition, Igaku-Shoin Ltd., 1979; (c) art by Joel Ito / **29.12** Art by L. Calver based on A. Spence, *Basic Human Anatomy*, Benjamin-Cummings, 1982 / **29.14** Data from A. Vander, J. Sherman, and D. Luciano, *Human Physiology*, Fourth edition, McGraw-Hill, 1985 / **Page 409** (b) Lester V. Bergman & Associates, Inc.; (c) Lewis L. Lainey / **29.15** After J. A. Gosling et al., *Atlas of Human Anatomy with Integrated Text*, copyright © 1985 by Gower Medical Publishing Ltd. / **29.16** (a) After F. Ayala and J. Kiger, *Modern Genetics*, © 1980 Benjamin-Cummings; (b),(c) Lester V. Bergman & Associates, Inc. / **29.17** After Gerard J. Tortora and Nicholas P. Anagnostakos, *Principles of Anatomy and Physiology*, Fifth edition, Copyright © 1987 by Biological Sciences Textbooks, Inc., A & P Textbooks, Inc. and Elia-Sparta, Inc. Reprinted by permission of Harper & Row, Publishers, Inc. / **29.18** Micrograph Emil Bernstein, *Science*, cover 27 August 1971, copyright 1971 AAAS / **29.19** Art by D. & V. Hennings

Chapter 30

30.1 (a) Arthur J. Olson © 1986 Research Institute of Scripps Clinic; (b) The Granger Collection / **30.5** Art by L. Calver after S. Tonegawa, *Scientific American*, October 1985 / **Page 425** Photographs Dr. Gilla Kaplan / **30.6** (a),(c) Art by L. Calver (b) Arthur J. Olson, © 1986 Research Institute of Scripps Clinic / **30.7** Art by L. Calver after B. Alberts et al., *Molecular Biology of the Cell*, Garland Publishing Company, 1983 / **Pages 430–431** (a–c) Photographs Z. Salahuddin, National Institutes of Health; (d) Art by L. Calver / **Page 432** Lennart Nilsson © Boehringer Ingelheim International GmbH

Chapter 31

31.1 Art by D. & V. Hennings; (a–c) after C. P. Hickman et al., *Integrated Principles of Zoology*, Sixth edition, St. Louis: C. V. Mosby Co., 1979 / **31.2** Ed Reschke / **31.3** Ed Reschke / **31.4** After C. P. Hickman et al., *Integrated Principles of Zoology*, Sixth edition, St. Louis: C. V. Mosby Co., 1979 / **31.5** Art by Victor Royer; micrograph H. R. Duncker, Justus-Liebig University, Giessen, West Germany; diagram after W. L. Bretz and K. Schmidt-Nielsen, *Journal of Experimental Biology*, 56:57–65, 1972 / **31.6** Art by L. Calver / **31.7** After J. Sobotta, *Atlas of Human Anatomy*, Urban & Schwarzenberg, 1978 / **31.8** After A. Vander, J. Sherman, and D. Luciano, *Human*

Physiology, Third edition, McGraw-Hill, 1980 / **31.9** Art by K. Kasnot / **31.11** After J. M. Kinney, *Anesthesiology*, 21:615, 1980 / **31.12** David Steinberg / **31.13** R. G. Kessel and R. H. Kardon, *A Text-Atlas of Scanning Electron Microscopy*, W. H. Freeman and Co., copyright © 1979 / **31.14** Gerard D. McLane

Chapter 32

32.1 Martin W. Grosnick/Ardea, London / **32.3** (a) Kim Taylor/Bruce Coleman Ltd.; (b) P. Morris/ Ardea, London; (c) Wardene Weisser/Ardea, London / **32.6** Art by L. Calver; photograph C. Yokochi and J. Rohen, *Photographic Anatomy of the Human Body*, Igaku-Shoin Ltd., 1979 / **32.7** After A. Vander, J. Sherman, and D. Luciano, *Human Physiology*, Fourth edition, McGraw-Hill, 1985 / **32.9** (a–c) R. G. Kessel and R. H. Kardon, *Tissues and Organs: A Text-Atlas of Scanning Electron Microscopy*, W. H. Freeman and Co., copyright © 1979; (d) J. D. Hoskins, W. G. Henk, and Y. Z. Abdelbaki, *American Journal of Veterinary Research*, 43:10, 1982 / **32.10** Photograph P. Morris/Ardea, London / **32.12** Modified from A. Vander, J. Sherman, and D. Luciano, *Human Physiology*, Fourth edition, McGraw-Hill, 1985

Chapter 33

33.1 David Mcdonald / **33.2** After A. Vander, J. Sherman, and D. Luciano, *Human Physiology*, Fourth edition, McGraw-Hill, 1985 / **33.3** Terry Vaughan / **Table 33.3** Data on human from B. Rose, *Clinical Physiology of Acid-Base and Electrolyte Disorders*, 1984; Data on kangaroo rat from K. Schmidt Nielsen, *Animal Physiology*, 1970 / **33.4** (a) Art by Ron Ervin; (b) art by Joel Ito / **33.5** Art by L. Calver / **33.10** Art by Joel Ito / **33.11** Tom McHugh/ Photo Researchers

Chapter 34

34.1 (a) Hans Pfletschinger; (b–e) John H. Gerard / **34.2** (a) Leonard Lee Rue III; (b) Frieder Sauer/ Bruce Coleman Ltd.; (c) Warren Garst/Tom Stack & Associates; (d) Wisniewski/ZEFA; (e) Jack Dermid; (f) Jean-Paul Ferrero/Ardea, London; (g) Alan Root/ Bruce Coleman Ltd. / **34.4** Adapted from R. G. Ham and M. J. Veomett, *Mechanisms of Development*, St. Louis: C. V. Mosby Co., 1980 / **34.6** R. G. Kessel and C. Y. Shih, *Scanning Electron Microscopy in Biology*, Springer-Verlag, 1976 / **34.7** Photograph Jeff Hardin / **34.8** Photographs Carolina Biological Supply Company; sketches after M. B. Patten, *Early Embryology of the Chick*, Fifth edition, McGraw-Hill, 1971 / **34.9** From L. B. Arey, *Developmental Anatomy*, Philadelphia: W. B. Saunders Co., 1965 / **34.10** Sketches after Willier, Weiss, and Hamburger, *Analysis of Development*, Philadelphia: W. B. Saunders Co., 1955; Photograph Roger K. Burnard / **34.11** Sketches after B. Burnside, *Developmental Biology*, 26:416–441, 1971; micrograph K. W. Tosney / **34.12** Art by Victor Royer / **34.13** From Lennart Nilsson, *A Child Is Born*, © 1966, 1977 Dell Publishing Company, Inc. / **34.14** (a) K. W. Tosney / **34.15** (a–c) After John W. Saunders, Jr., *Patterns and Principles of Animal Development*, copyright © by John W. Saunders, Jr., and Macmillan Publishing Company; (d) S. R. Hilfer and J. W. Yang, *The Anatomical Record*, 197:423–433, 1980 / **34.16** After J. W. Fristrom et al., in E. W. Hanly, ed., *Problems in Biology: RNA Development*, University of Utah Press / **34.17** Sketches after Willier, Weiss, and Hamburger, *Analysis of Development*, Philadelphia: W. B. Saunders Co., 1955; photograph Roger K. Burnard

Chapter 35

35.1 Art by Ron Ervin / **35.2** Art by L. Calver; micrograph R. G. Kessel and R. H. Kardon, *Tissues and Organs: A Text-Atlas of Scanning Electron Microscopy*, W. H. Freeman and Co., copyright ©

1979 / **35.3** Art by Ron Ervin / **35.4** Art by Ron Ervin / **35.6** Art by Ron Ervin / **35.8** Art by Joan Carol / **35.10** (a) From Lennart Nilsson, *A Child Is Born*, © 1966, 1977, Dell Publishing Company, Inc.; (b) From Lennart Nilsson, *Behold Man*, © 1974 by Albert Bonniers Forlag and Little, Brown and Co. (Canada) Ltd. / **35.11** The Carnegie Institution of Washington and Ronan O'Rahilly, M.D., Director / **35.12** Art by L. Calver / **35.13** Art by L. Calver; (c) after A. S. Romer and T. S. Parsons, *The Vertebrate Body*, Sixth edition, Saunders College Publishing, © 1986 CBS College Publishing / **35.14** Art by L. Calver after Bruce Carlson, *Patten's Foundations of Embryology*, Fourth edition, McGraw-Hill, 1981 / **35.15–35.18** From Lennart Nilsson, *A Child Is Born*, © 1966, 1977 Dell Publishing Company, Inc. / **Pages 526–527** Art by Victor Royer; (b) Mills-Peninsula Hospitals / **35.21** Modified from Keith L. Moore, *The Developing Human: Clinically Oriented Embryology*, Fourth edition, Philadelphia: W. B. Saunders Co., 1988 / **Page 534** (a) Cheun-mo To and C. C. Brinton, Jr.; (b) Joel B. Baseman

Chapter 36

36.1 (a) Hans Reinhard/Bruce Coleman Ltd.; (b) Gary Zahm; (c) G. Alan Solem / **36.3** After D. Futuyma, *Evolutionary Biology*, Sinauer, 1979 / **36.5** After M. Karns and L. Penrose, *Annals of Eugenics*, 15:206–233, 1951 / **36.6** (a) Peabody Museum of Natural History, Yale University; (b) Leonard Lee Rue III / **36.7** J. A. Bishop and L. M. Cook / **36.8** Edward S. Ross / **36.9** (a) Charles W. Fowler/ National Marine Fisheries; (b) C. G. Summers/Tom Stack & Associates; (c) G. Sirena/ZEFA / **36.10** After F. Ayala and J. Valentine, *Evolving*, Benjamin-Cummings, 1979 / **36.11** After V. Grant, *Organismic Evolution*, W. H. Freeman and Co., 1977

Chapter 37

37.3 After S. M. Stanley, *Macroevolution: Pattern and Process*, W. H. Freeman and Co., 1979 / **37.4** (a) A. Feduccia, *The Age of Birds*, Harvard University Press, 1980; (b) Donald Baird, Princeton Museum of Natural History / **37.5** Art by Joel Ito / **37.6** (top) D. P. Wilson/Eric & David Hosking; (center) J. Kirschner, Courtesy Department Library Services, American Museum of Natural History; (bottom) E. R. Degginger / **37.7** L. R. Maxson and A. C. Wilson, *Systematic Zoology*, 24(1):1–15, 1975 / **37.8** After C. G. Sibley and J. E. Ahlquist, *Scientific American*, February 1986, copyright © by Scientific American, Inc., all rights reserved / **37.11** After E. O. Dodson and P. Dodson, *Evolution: Process and Product*, Third edition, Prindle, Weber & Schmidt, 1985 / **37.12** Art by Victor Royer; (a) after D. M. Raup and J. J. Sepkoski, Jr., *Science*, 215:1501–1503, 19 March 1982; (b) after J. J. Sepkoski, Jr., *Paleobiology*, 7(1):36–53, 1981

Chapter 38

38.1 Chesley Bonestell / **38.4** (a) Sidney W. Fox; (b) W. Hargreaves and D. Deamer / **38.5** (a) Stanley M. Awramik; (b) J. W. Schopf; (c) M. R. Walter / **38.6** J. W. Schopf / **38.7** Art by Victor Royer / **38.8** David Chase / **38.9** (a) H. Stolp; (b) L. Margulis / **38.11** Jack Carey / **38.12** (a),(b) G. R. Adlington; (c–f) National Museum of Natural History / **38.13** (a) H. P. Banks; (b),(c) Patricia G. Gensel / **38.15** Negative #322871, Courtesy Department Library Services, American Museum of Natural History / **38.16** Art by D. & V. Hennings / **38.17** (a) R. T. Bakker; (b) Negative #109403, Courtesy Department Library Services, American Museum of Natural History; (c) Charles R. Knight, Field Museum of Natural History; (d) Peabody Museum of Natural History, Yale University / **38.18** Data from J. J. Sepkoski, Jr., *Paleobiology* 7(1):36–53 and J. J. Sepkoski, Jr. and M. L. Hulver in Valentine, ed., *Phanerozoic Diversity Patterns: Profiles in Macroevolution*, Princeton University Press, 1985 / **38.19** Smithsonian Museum

Chapter 39

39.1 Tony Brain/Science Photo Library/Photo Researchers / **39.2** (b) Art by L. Calver / **39.3** Micrograph Robley C. Williams; art by L. Calver / **39.4** (a) C. McLaren and F. Siegel/Burroughs Wellcome Co.; (b) Robley C. Williams / **39.5** Breck's / **39.6** Art by L. Calver / **39.7** L. J. LeBeau, University of Illinois Hospital/BPS / **39.8** Paul A. Zahl, © 1967 National Geographic Society / **39.9** (a) John D. Cunningham/Visuals Unlimited; (b) Tony Brain/ Science Source/Photo Researchers; (c) P. W. Johnson and J. McN. Sieburth, University of Rhode Island/ BPS / **39.10** Stanley W. Watson, *International Journal of Systematic Bacteriology*, 21:254–270, 1971 / **39.11** J. R. Norris / **39.12** Richard Blakemore / **39.13** Martin Dworkin, *Journal of Bacteriology*, 154:452–459, 1983 / **39.14** Hans Reichenbach, Gesellschaft für Bio-technologische Forschung, Braunschweig, FRG / **39.16** (a) Art by Joan Carol after H. C. Bold and J. W. LeClaire II, *The Plant Kingdom*, Fifth edition, © 1987. Adapted by permission of Prentice-Hall Inc., Englewood Cliffs, New Jersey; (b) Edward S. Ross; (c) John Shaw/Bruce Coleman Ltd. / **39.17** (a) P. L. Walne and J. H. Arnott, *Planta*, 77:325–354, 1967; (b) G. Shih and R. G. Kessel, *Living Images*, Jones and Bartlett Publishers, Inc., Boston, © 1982 / **39.18** (a) Hans Paerl; (b) G. Shih and R. G. Kessel, *Living Images*, Jones and Bartlett Publishers, Inc., Boston © 1982 / **39.19** Edward Gabrielle / **39.20** (a) Florida Department of Natural Resources, Bureau of Marine Research; (b) C. C. Lockwood / **39.21** (a),(b) From Stanier, Adelberg, and Ingraham, *The Microbial World*, Fourth edition, © 1976 Prentice-Hall Inc. Englewood Cliffs, New Jersey; (c) M. Abbey/Visuals Unlimited / **39.22** John Clegg/Ardea, London / **39.23** Micrograph Steven L'Hernault; illustration based on correspondence with Raul J. Cano / **39.24** (a),(b) Gary W. Grimes and Steven L'Hernault; (c) Thomas Eisner, Cornell University / **39.25** Richard W. Greene / **39.26** Manfred Kage/Peter Arnold, Inc. / **39.27** Laszlo Meszoly in L. Margulis, *Early Life*, Jones and Bartlett, Publishers, Inc., Boston, © 1982

Chapter 40

40.1 John E. Hodgin / **40.2** Art by Joan Carol / **40.4** M. S. Fuller, *Zoosporic Fungi in Teaching and Research*, M. S. Fuller and A. Jaworski (eds.), 1987, Southeastern Publishing Corp., Athens, GA / **40.5** W. Merrill / **40.6** (a) John D. Cunningham/Visuals Unlimited; (b) David M. Phillips/Visuals Unlimited; (c) Art by D. & V. Hennings / **40.7** David M. Phillips/ Visuals Unlimited / **40.8** Sketch after T. Rost et al., *Botany*, Wiley, 1979; photograph Victor Duran / **40.9** (a) Victor Duran; (b) New York State Agricultural Experiment Station, Geneva, NY / **40.10** (a) Jane Burton/Bruce Coleman Ltd.; (b),(c) Thomas J. Duffy / **40.11** (a) Roger K. Burnard; (b),(c) Victor Duran; (d) Edward S. Ross / **40.12** Art by D. & V. Hennings / **40.13** G. Shih and R. G. Kessel, *Living Images*, Jones and Bartlett, Publishers, Inc., Boston, © 1982 / **40.14** (a) N. Allin and G. L. Barron; (b) Gary T. Cole, University of Texas, Austin/BPS; (c) G. L. Barron, University of Guelph / **40.15** After Raven, Evert, and Eichhorn, *Biology of Plants*, Fourth edition, Worth Publishers, New York, 1986 / **40.16** (a) Edward S. Ross; (b) Ken Davis/Tom Stack & Associates / **40.17** Steven C. Wilson/Entheos / **40.18** From Mary Barkworth, Utah State University / **40.19** (a) D. P. Wilson/Eric & David Hosking; (b) Douglas Faulkner/Sally Faulkner Collection / **40.20** (a) J. R. Waaland, University of Washington/BPS; (b) Dennis Brokaw / **40.21** Hervé Chaumeton/Agence Nature; (b) A. Kerstitch/Tom Stack & Associates / **40.22** Photograph D. J. Patterson/Seaphot Ltd: Planet Earth Pictures; art by D. & V. Hennings / **40.23** Carolina Biological Supply Company / **40.24** Photograph Hervé Chaumeton/Agence Nature / **40.25** Photograph Jane Burton/Bruce Coleman Ltd.; art by D. & V. Hennings / **40.26** (a),(b) Kingsley R. Stern; (c) John D. Cunningham/Visuals Unlimited / **40.27** Field Museum of Natural History / **40.28** (a) Edward S. Ross; (b) W. H. Hodge; (c) Kratz/ZEFA

/ **40.29** Jean Paul Ferrero/Ardea, London / **40.30** Photograph Anthony & Elizabeth Bomford/Ardea, London; art by D. & V. Hennings / **40.31** John N. A. Lott, McMaster University/BPS / **40.32** (a), Edward S. Ross; (b),(d) John H. Gerard / **40.33** (a) Edward S. Ross; (b) F. J. Odendaal, Duke University/BPS / **40.34** Photograph Edward S. Ross; art by D. & V. Hennings / **40.35** Art by D. & V. Hennings / **40.36** (a),(b) Edward S. Ross; (c) John D. Cunningham/Visuals Unlimited; (d) E. R. Degginger; (e) Dick Davis/Photo Researchers; (f) Hans Reinhard/Bruce Coleman Ltd.; (g) Edward R. Ross; (h) Heather Angel; (i) Martin W. Grosnick/Ardea, London

Chapter 41

41.1 Kjell B. Sandved / **41.2** Art by L. Calver / **41.3** Art by K. Kasnot / **41.5** David C. Haas/Tom Stack & Associates / **41.6** H. M. Reiswig / **41.8** Frieder Sauer/Bruce Coleman Ltd. / **41.9** Kim Taylor/Bruce Coleman Ltd. / **41.10** (a) Walter Deas/Seaphot Limited: Planet Earth Pictures; (b) Bill Wood/Seaphot Limited: Planet Earth Pictures / **41.11** F. Stuart Westmorland/Tom Stack & Associates / **41.13** Art by Joan Carol after T. Storer et al., *General Zoology*, Sixth edition, © 1979 McGraw-Hill / **41.14** Andrew Mounter/Seaphot Limited: Planet Earth Pictures; art by Raychel Ciemma / **41.15** E. R. Degginger / **41.16** (a) Larry Madin/Seaphot Limited: Planet Earth Pictures; (b) Kathie Atkinson/Oxford Scientific Films / **41.17** Photograph Kim Taylor/Bruce Coleman Ltd.; art by K. Kasnot / **41.18** Photograph Robert L. Calentine / **41.19** C. P. Hickman / **41.20** Cath Ellis, University of Hull/Science Photo Library/Photo Researchers / **41.21** Photograph Carolina Biological Supply Company; art by K. Kasnot / **41.22** Photograph Eugene Kozloff/**41.23** (a) Lorus J. and Margery Milne; (b) Ed Reschke / **41.24** Photograph J. Solliday/BPS; art by Raychel Ciemma / **41.25** Art by D. & V. Hennings / **41.26** Art by Joel Ito / **41.27** (a) Anthony & Elizabeth Bomford/Ardea, London; (b) Kjell B. Sandved / **41.28** Jeff Foott/Tom Stack & Associates / **41.29** (top) Eugene Kozloff; (bottom) Rick M. Harbo / **41.30** Hervé Chaumeton/Agence Nature / **41.31** Art by Laszlo Meszoly and D. & V. Hennings / **41.32** Hervé Chaumeton/Agence Nature / **41.33** © by Chris Newbert / **41.34** Douglas Faulkner/Sally Faulkner Collection / **41.35** Art by Laszlo Meszoly and D. & V. Hennings / **41.36** J. A. L. Cooke/Oxford Scientific Films / **41.37** (a) Hervé Chaumeton/Agence Nature; Jon Kenfield/Bruce Coleman Ltd. / **41.38** Art by Ron Ervin / **41.39** After C. P. Hickman et al., *Integrated Principles of Zoology*, Sixth edition, St. Louis: C. V. Mosby Co., 1979 / **41.40** C. B. & D. W. Frith/Bruce Coleman Ltd. / **41.41** (a) John H. Gerard; (b) Ken Lucas/Seaphot Limited: Planet Earth Pictures; (c) P. J. Bryant, University of California, Irvine/BPS / **41.43** Photograph Hervé Chaumeton/Agence Nature; art by Laszlo Meszoly / **41.44** Agence Nature / **41.45** Kjell B. Sandved / **41.46** Tom McHugh / **41.47** (a) Z. Leszczynski/Animals Animals; (b) Steve Martin/Tom Stack & Associates / **41.48** Art by D. & V. Hennings / **41.49** (a) David Maitland/Seaphot Limited: Planet Earth Pictures; (b–i),(k),(l) Edward S. Ross; (j) C. P. Hickman / **41.50** (a),(b) Douglas Faulkner/Sally Faulkner Collection; (c) Kjell B. Sandved; (d) Ian Took/Biofotos / **41.51** (a),(c) Kjell B. Sandved; (b) John Mason/Ardea, London / **41.52** (a),(d) Hervé Chaumeton/Agence Nature; (b) Jane Burton/Bruce Coleman Ltd.; art by L. Calver / **41.54** (a),(b) After W. D. Russell-Hunter, Macmillan Publishing Company; (c) Rick M. Harbo / **41.55** C. R. Wyttenbach, University of Kansas/BPS / **41.56** Photograph Hervé Chaumeton/Agence Nature; art by Laszlo Meszoly and D. & V. Hennings / **41.57** After A. S. Romer and T. S. Parsons, *The Vertebrate Body*, Sixth edition, Saunders College Publishing, © 1986 CBS College Publishing; art by Laszlo Meszoly and D. & V. Hennings / **41.58** After C. P. Hickman, Jr. and L. S. Roberts, *Integrated Principles*

of Zoology, Seventh edition, St. Louis: Times Mirror/Mosby College Publishing, 1984 / **41.60** Art by Laszlo Meszoly and D. & V. Hennings / **41.61** Peter Scoones/Seaphot Limited: Planet Earth Pictures / **41.62** Heather Angel / **41.63** (a) Allan Power/Bruce Coleman Ltd.; (b) Erwin Christian/ZEFA; (c) Bill Wood/Bruce Coleman Ltd.; (d) Robert and Linda Mitchell / **41.64** (a) E. J. Maruska/Visuals Unlimited; (b) Jerry W. Nagel; (c) John Serraro/Visuals Unlimited / **41.65** (a) Peter Scoones/Seaphot Limited: Planet Earth Pictures; (b) D. Kaleth/Image Bank; (c) W. A. Banaszewski/Visuals Unlimited; (d) C. B. & D. W. Frith/Bruce Coleman Ltd.; (e) W. J. Weber/Visuals Unlimited; (f) Heather Angel; (g) Jane Burton/Bruce Coleman Ltd. / **41.66** Sketch after O. S. Pettinghill, *Ornithology in Laboratory and Field*, Burgess Publishing Company, 1970; photograph Reinhard/ZEFA / **41.67** Christopher Crowley

Chapter 42

42.2 (a) Bruce Coleman Ltd.; (b) Tom McHugh/Photo Researchers; (c) Larry Burrows/Aspect Picture Library, London / **42.3** Art by D. & V. Hennings / **42.6** © Time Inc. 1965/Larry Burrows Collection / **42.8** Art by D. & V. Hennings / **42.9** Dr. Donald Johanson, Institute of Human Origins / **42.10** Louise M. Robbins / **42.12** Art by D. & V. Hennings / **42.13** Photographs by Joan Reader copyright 1981 / **42.14** Art by D. & V. Hennings

Chapter 43

43.1 (left) Robert K. Colwell; (right) William J. Weber/Visuals Unlimited / **43.3** E. R. Degginger / **43.6** Data from V. Scheffer, *Science Monthly*, 73:356–362, 1951; photograph E. Vetter/ZEFA / **43.7** Data from A. J. Nicholson, *Cold Spring Harbor Symposium on Quantitative Biology*, 22:153–173 / **43.8** (d) Eric Crichton/Bruce Coleman Ltd. / **43.10** John Endler / **43.14** After G. T. Miller, *Living in the Environment*, Wadsworth, 1982 / **43.15** Data from Population Reference Bureau

Chapter 44

44.1 (left) Dona Hutchins; (right) Edward S. Ross / **44.2** (a),(c) Harlo H. Hadow; (b) Bob and Miriam Francis/Tom Stack & Associates / **44.3** After G. Gause, 1934 / **44.4** Photograph Clara Calhoun/Bruce Coleman Ltd. / **44.5** John Dominis, Life Magazine,© 1965, Time Inc. / **44.7** Photograph Ed Cesar/NAS/Photo Researchers / **44.8** W. E. Ruth / **44.9** W. M. Laetsch / **44.10** (a–c) Edward S. Ross; (d) James H. Carmichael, Jr. / **44.11** (a) Roger T. Peterson/NAS/Photo Researchers; (b),(c) Thomas Eisner, Cornell University / **44.12** Edward S. Ross / **44.13** Data from P. Price and H. Tripp, *Canadian Entomology*, 104:1003–1016, 1972 / **44.14** After N. Weland and F. Bazzaz, *Ecology*, 56:681–688, © 1975 Ecological Society of America / **44.15** (a–f),(i) Roger K. Burnard; (g),(h) E. R. Degginger / **44.16** (top) Dr. Harold Simon/Tom Stack & Associates; (bottom) After S. Fridriksson, *Evolution of Life on a Volcanic Island*, Butterworth: London, 1975 / **44.17** After J. M. Diamond, *Proceedings of the National Academy of Sciences*, 69:3199–3201, 1972 / **44.18** After M. H. Williamson, *Island Populations*, Oxford University Press, 1981 / **44.19** (a) After F. G. Stehli et al., *Geological Society of America Bulletin*, 78:455–466, 1967; (b) After M. Kusenov, *Evolution*, 11:298–299, 1957; (c) After T. Dobzhansky, *American Scientist*, 38:209–221, 1950

Chapter 45

45.1 Roger K. Burnard / **45.2** Photograph Sharon R. Chester / **45.3** Photograph Alan D. Briere/Tom Stack & Associates / **45.5** After H. T. Odum, "Trophic Structure and Productivity of Silver Springs," *Ecological Monographs*, 27:55–112, copyright © 1957 Ecological Society of America / **45.7** (a) Photograph Steven D. Bach / **45.8** (a)

Photograph by Gene E. Likens from G. E. Likens and F. H. Bormann, *Proceedings First International Congress of Ecology*, pp. 330–335, September 1974, Centre Agric. Publ. Doc. Wageningen, The Hague, The Netherlands; (b),(c) Photographs by Gene E. Likens from G. E. Likens et al., *Ecology Monograph*, 40(1):23–47, 1970 / **45.9** Art by Raychel Ciemma / **41.10** G. Likens et al., *Biogeochemistry of a Forest System*, Springer-Verlag, 1977 / **45.12–45.13** Photographs Jack Carey

Chapter 46

46.1 NASA / **46.2** Edward S. Ross / **46.4** (b) Art by L. Calver / **46.7** Art by Raychel Ciemma / **46.9** Edward S. Ross / **46.11** Art by D. & V. Hennings after G. T. Miller, Jr., *Environmental Science: An Introduction*, Wadsworth, 1986 / **46.13** After Whittaker; Bland; and Tilman / **46.14** Art by Joan Carol / **46.15** Harlo H. Hadow / **46.16** Dennis Brokaw / **46.17** Kenneth W. Fink/Ardea, London / **46.18** Ray Wagner/Save the Tall Grass Prairie, Inc. / **46.19** Thase Daniel / **46.20** Thomas E. Hemmerly / **46.21** Dennis Brokaw / **46.22** Ansel Adams / **46.23** Lynn Eckmann, University of Washington/BPS / **46.24** D. W. MacManiman / **46.26** Art by L. Calver after Edward S. Deevy, Jr., *Scientific American*, October 1951, copyright © 1951 Scientific American Inc., all rights reserved / **46.27** D. W. Schindler, *Science*, 184:897–899 / **46.29** E. R. Degginger / **46.30** Art by D. & V. Hennings / **46.31** Courtesy of J. L. Sumich, *Biology of Marine Life*, Fourth edition, William C. Brown, 1988 / **46.32** (left) H. Clason/Tom Stack & Associates; (right) Phil Degginger / **46.34** (a) Dennis Brokaw; (b) Chuck Niklin; (c) William H. Amos / **46.35** (a) Robert Hessler; (b) Fred Grassle, Woods Hole Institution of Oceanography / **Pages 790–791** Photographs R. Legeckis/NOAA

Chapter 47

47.1 © 1983 Billy Grimes / **47.3** After G. T. Miller, Jr., *Environmental Science: An Introduction*, Wadsworth, 1986 and the Environmental Protection Agency / **47.4** (a) USDA Forest Service; (b) Heather Angel / **47.5** (Bottom: left) National Science Foundation; (Top: left; right) NASA / **47.6** From Water Resources Council / **47.7** Data from G. T. Miller, Jr. / **Page 801** (a) R. Bierregaard/Photo Researchers; (b) National Oceanic and Atmospheric Administration/NESDIS / **47.8** After G. T. Miller, Jr., *Living in the Environment*, Fourth edition, Wadsworth, 1985 / **47.9** USDA Soil Conservation Service/Thomas G. Meier / **47.10** Agency for International Development / **47.11** Data from G. T. Miller, Jr.

Chapter 48

48.1 Photograph John S. Dunning/Ardea, London; sonogram J. Bruce Falls and Tom Dickinson, University of Toronto / **48.2** (a) Hans Reinhard/Bruce Coleman Ltd.; (b),(c) G. Pohl-Apel and R. Sussinka, *Journal für Ornithologie*, 123:211–214 / **48.3** (a) Eric & David Hosking; (b) Stephen Dalton/Photo Researchers / **48.4** (a) Nina Leen in *Animal Behavior*, Life Nature Library (b) F. Schulz / **48.5** After J. Alcock, *Animal Behavior*, Second edition, Sinauer, 1979 / **48.6** Eric Hosking / **48.7** (a) Eugene Kozloff; (b),(c) Stevan Arnold / **48.9** John Alcock / **48.10** Lincoln P. Brower / **48.11** Pat and Tom Leeson/Photo Researchers / **48.12** John Alcock

Chapter 49

49.1 John Alcock / **49.2** Edward S. Ross / **49.3** (a) E. Mickleburgh/Ardea, London; (b–d) G. Ziesler/ZEFA / **49.4** Art by D. & V. Hennings / **49.6** Merlin D. Tuttle, Bat Conservation International / **49.7** Scott Preiss / **49.8** Fred Bruemmer / **49.9** Frank Lane Agency/Bruce Coleman Inc. / **49.10** Patricia Caulfield / **49.11** Timothy Ransom / **49.12** Kenneth Lorenzen

GLOSSARY

abortion. Spontaneous or induced expulsion of the embryo or fetus from the uterus. Spontaneous abortions are also called miscarriages.

abscission (ab-SIH-zhun) [L. *abscissus*, to cut off]. Leaf (or fruit or flower) drop after hormonal action causes a corky cell layer to form where a leaf stalk joins a stem; nutrient and water flow is thereby shut off.

acid [L. *acidus*, sour]. A substance that releases a hydrogen ion (H⁺) in solution.

acoelomate (ay-SEE-la-mate). Type of animal that has no fluid-filled cavity between the gut and body wall.

actin (AK-tin). A protein that functions in contraction; together with myosin, a component of the myofibrils of muscle cells.

action potential. Nerve impulse; a sudden, dramatic reversal of the polarity of charge across the plasma membrane of neurons and some other cells.

activation energy. The minimum amount of collision energy needed to boost reactant molecules to the point at which a reaction will proceed spontaneously.

active site. A crevice in the surface region of an enzyme in which a particular reaction is catalyzed.

active transport. Movement of ions and molecules across a cell membrane, against a concentration gradient, by ATP expenditure. The ion or molecule is moved in a direction other than the one in which simple diffusion would take it.

adaptation [L. *adaptare*, to fit]. An existing structural, physiological, or behavioral trait of an individual that promotes survival and reproduction under prevailing conditions.

adaptive radiation. A burst of evolutionary activity in geologic time, with lineages branching away from one another as they partition the existing environment or invade new ones.

adaptive zone. A way of life, such as "catching insects in the air at night." A lineage must have physical, ecological, and evolutionary access to an adaptive zone to become a successful occupant of it.

adenine (AH-de-neen). A purine; a nitrogen-containing base found in nucleotides.

adenosine diphosphate (ah-DEN-uh-seen die-FOSS-fate). ADP, a molecule involved in cellular energy transfers; typically formed by hydrolysis of ATP.

adenosine triphosphate. ATP, a molecule that is a major carrier of energy (by way of its phosphate groups) from one reaction site to another in all living cells.

aerobic cell (air-OH-bik) [Gk. *aer*, air, + *bios*, life]. A cell that is able to use free oxygen as a final electron acceptor in carbohydrate metabolism.

aerobic respiration. Pathway of carbohydrate metabolism, including glycolysis, the Krebs cycle, and electron transport phosphorylation. The "spent" electrons are transferred finally to oxygen. Much greater net energy yield than from anaerobic pathways.

afferent (AFF-uh-rent) [L. *affere*, to bring to]. Conducting inward to a body part or region, as from sensory receptor endings toward the spinal cord or brain.

allantois (ah-LAN-twahz) [Gk. *allas*, sausage]. Vascularized extraembryonic membrane of reptiles, birds, and mammals that develops as a bladderlike pouch from the primitive gut. Functions in excretion and respiration in reptiles and birds; functions in oxygen transport by way of the umbilical cord in placental mammals.

allele (uh-LEEL). One of two or more alternative forms of a gene at a given gene locus.

allele frequency. The relative abundance of different alleles carried by the individuals of a population. Also called gene frequency, which is something of a non sequitur.

allopatric speciation [Gk. *allos*, different, + *patria*, native land]. Speciation that occurs when geographic separation prevents gene flow and has assured reproductive isolation of different parts of a population or of different populations of the same species.

alternation of generations. In many plant life cycles, the alternation of diploid multicelled bodies with haploid multicelled bodies.

altruistic behavior. Helpful actions that reduce the individual's production of surviving offspring while increasing the reproductive success of the helped individual.

alveolus, plural **alveoli** (ahl-VEE-uh-luss) [L. *alveus*, small cavity]. One of many small, thin-walled pouches in the lungs; sites of gas exchange between air in the lungs and the bloodstream.

amino acid (uh-MEE-no). A molecule having an amino group (NH₂) and an acid group (—COOH); a subunit for protein synthesis.

ammonification (uh-moan-ih-fih-KAY-shun). Decomposition of nitrogenous wastes and remains of organisms by certain bacteria and fungi.

amnion (AM-nee-on). In reptiles, birds, and mammals, an extraembryonic membrane that arises from the inner cell mass of a blastocyst; becomes a fluid-filled sac in which the embryo develops freely.

amyloplast (AM-uh-low-plast) [L. *amylum*, starch, + Gk. *plastos*, formed]. A plastid having no pigments; functions in starch storage.

anaerobic cell (an-uh-ROW-bik) [Gk. *an*, without, + *aer*, air). A cell that either cannot use free oxygen as a final electron acceptor in carbohydrate metabolism or dies upon exposure to it.

anaerobic electron transport. Degradative pathway that does not use oxygen as the final electron acceptor; the "spent" electrons are transferred to inorganic compounds in the environment.

anaphase (AN-uh-faze). In mitosis, the stage when the two sister chromatids of each chromosome are separated and moved to opposite poles of the microtubular spindle.

angiosperm (AN-gee-oh-sperm) [Gk. *angeion*, vessel, + *sperma*, seed]. Flowering plant.

animal. Multicelled heterotroph; except for a few parasites, most ingest food, which is then digested and absorbed.

annual plant. Vascular plant that completes its life cycle in one growing season.

anther [Gk. *anthos*, flower]. In flowering plants, the pollen-bearing part of the male reproductive structure (stamen).

antheridium (an-thuh-RID-ee-um). Protective layer of sterile cells surrounding the haploid sperm in a male gametophyte.

antibody [Gk. *anti*, against]. Any of various Y-shaped proteins, of the immunoglobulin class, produced by B cells. Antibodies bind specific foreign agents invading the body, thus tagging them for destruction by phagocytes or by activating the complement system. Some are bound to the plasma membrane of B cells; others circulate freely in blood and lymph vessels.

antigen [Gk. *anti*, against, + *genos*, race, kind; against "self"]. Any of various specific molecular patterns that are on the surface of foreign agents invading the body and that trigger defense responses.

anus. In some invertebrates and all vertebrates, the terminal opening of the gut through which solid residues of digestion are eliminated.

aorta (ay-OR-tah) [Gk. *airein*, to lift, heave]. Main artery of systemic circulation; carries oxygenated blood away from the heart to all regions except the lungs.

apical meristem (AY-pih-kul MARE-ih-stem) [L. *apex*, top, + Gk. *meristos*, divisible]. In most plants, a mass of self-perpetuating cells at a root or shoot tip that is responsible for primary growth, or elongation, of plant parts.

archegonium (ar-kih-GO-nee-um) [Gk. *archegonos*, first of a kind]. Protective layer surrounding the haploid egg in a female gametophyte.

asexual reproduction. Production of new individuals by any process that does not involve gametes.

atom [Gk. *atomos*, indivisible]. Smallest unit of an element that still retains the properties of that element.

atomic number. A relative number assigned to each kind of element based on the number of protons in one of its atoms.

atomic weight. The weight of an atom of any element relative to the weight of the most abundant isotope of carbon (which is set at 12).

australopith (OHSS-trah-low-pith) [L. *australis*, southern, + Gk. *pithekos*, ape]. Any of the earliest known species of hominids; that is, the first species on the human evolutionary branch.

autonomic nervous system (auto-NOM-ik). Those efferent nerves leading from the central nervous system to cardiac cells, muscle cells, smooth muscle cells (such as those of the stomach), and glands—that is, the visceral portion of the body; generally not under conscious control.

autosome. Any of those chromosomes that are of the same number and kind in both males and females of the species.

autotroph (AH-toe-trofe) [Gk. *autos*, self, + *trophos*, feeder]. An organism able to build all the organic molecules it requires using carbon dioxide (present in air and in water) and energy from the physical environment. Photosynthetic autotrophs use sunlight energy; chemosynthetic autotrophs extract energy from chemical reactions involving inorganic substances. Compare *heterotroph*.

axon. Nerve cell process serving as a through-conducting pathway for action potentials, which are messages that travel rapidly, without alteration, from one region to another.

bacillus, plural **bacilli** (bah-SILL-us, bah-SILL-eye) [L. *baculus*, small staff, rod]. Rodlike form of bacterium.

bacteriophage (bak-TEER-ee-oh-fahj) [Gk. *baktērion*, small staff, rod, + *phagein*, to eat]. Category of viruses that infect and destroy certain bacterial cells.

basal body. A centriole that has given rise to the microtubular system of a cilium or flagellum and that remains attached at the base of the motile structure, just beneath the plasma membrane.

base. Any substance that combines with a hydrogen ion (H^+) in solution.

behavior. Any coordinated neuromotor response that an animal makes to external and internal stimuli. The responses are outcomes of the integration of sensory, neural, endocrine, and effector components, all of which have a genetic basis (hence are subject to natural selection); and they may be modified by learning processes.

biennial (by-EN-ee-ull). Flowering plant that lives two growing seasons.

bilateral symmetry. Body plan by which an animal has an anterior and a posterior end, a left and right side, and a dorsal and ventral surface. The only plane in which the body can be divided into two equivalent halves extends from dorsal to ventral along the midline.

binary fission. Asexual reproduction by division of a body (or cell) into two equivalent parts.

biogeographic realm [Gk. *bios*, life, + *geographein*, to describe the surface of the earth]. In one scheme, one of six major regions having a characteristic array of species that are generally isolated from the other realms by physical barriers that restrict dispersal.

biological clock. Internal timing mechanism that allows organisms to anticipate and adjust to environmental change. In plants, phytochromes may figure in the timing mechanism; in some vertebrates, the pineal gland seems to.

biological magnification. Increasing concentration of a relatively nondegradable substance in body tissues, beginning at low trophic levels and moving up through those organisms that are diners, then are dined upon in food webs.

biomass. The total dry weight of all organisms at a given trophic level of an ecosystem.

biome. A broad, vegetational subdivision of some biogeographic realm, shaped by climate, topography, and the composition of regional soils.

biosphere [Gk. *bios*, life, + *sphaira*, globe]. Narrow zone that harbors life, limited to the waters of the earth, a fraction of its crust, and the lower region of the surrounding air.

biosynthesis [Gk. *bios*, life, + *synthesis*, a putting together]. Assembly of the lipids, carbohydrates, proteins, and nucleic acids that make up a cell.

blastocyst (BLASS-tuh-sist) [Gk. *blastos*, sprout, + *kystis*, pouch]. In mammalian development, a modified blastula stage consisting of a hollow ball of surface cells (trophoblast) having inner cells massed at one end.

blastula (BLASS-chew-lah). In many animal species, an embryonic stage consisting of a hollow, fluid-filled ball of cells one layer thick.

blood pressure. Fluid pressure, generated by heart contractions, that keeps blood circulating. Generally measured at large arteries of systemic circulation.

bronchus, plural **bronchi** (BRONG-cuss, BRONG-kee) [Gk. *bronchos*, windpipe]. Tube-like branchings of the trachea (windpipe) that lead to the lungs.

budding. Asexual reproduction in which some cells differentiate and grow outward from the parent body, then the bud breaks away to form a new individual.

buffer. In living cells, a substance that combines with and/or releases hydrogen ions as a function of pH.

bulk flow. In response to a pressure gradient, a movement of more than one kind of molecule in the same direction in the same medium (gas or liquid).

calorie (KAL-uh-ree) [L. *calor*, heat]. The amount of heat needed to raise the temperature of one gram of water by 1°C. Nutritionists sometimes use "calorie" to mean kilocalorie (1,000 calories), which is a source of much confusion.

Calvin-Benson cycle. Stage of light-independent reactions of photosynthesis in which carbon-containing compounds are used to form carbohydrates (such as glucose) and to regenerate a sugar phosphate (RuBP) required in carbon dioxide fixation. The first product is a three-carbon compound (PGA).

cambium, plural **cambia** (KAM-bee-um). In vascular plants, one of two types of embryonic tissue masses that are responsible for secondary growth (increase in stem or root diameter). Vascular cambium gives rise to secondary xylem and phloem; cork cambium gives rise to periderm.

camouflage. Adaptation in form, coloration, and/or behavior that enables an organism to blend with its background, the advantage being to escape detection. Also known as crypsis.

cancer. Malignancy arising from cells that are characterized by profound abnormalities in the plasma membrane and in the cytoplasm, abnormal growth and division, and diminished capacity for adhesion to substrates.

capillary [L. *capillus*, hair]. Small blood vessel whose thin walls are permeable to many materials; exchange point between blood and interstitial fluid.

carbohydrate [L. *carbo*, charcoal, + *hydro*, water]. Monomer or polymer of a sugar, which is a compound of carbon, hydrogen, and oxygen present in about a 1:2:1 ratio.

carbohydrate metabolism [Gk. *metabolē*, change]. The release of chemical bond energy from carbohydrates by means of phosphorylation and oxidation-reduction reactions.

carbon dioxide fixation. First stage of light-independent reactions of photosynthesis; carbon dioxide from the air is combined with a sugar phosphate to form an intermediate necessary in the synthesis of glucose and other carbon compounds.

carcinogen (CAR-sin-oh-jen). Any agent capable of promoting cancer.

cardiac cycle [Gk. *kardia*, heart, + *kyklos*, circle]. The sequence of muscle contractions and relaxation constituting one heartbeat.

cardiovascular system. Of animals, an organ system composed of blood, one or more hearts, and blood vessels.

carnivore [L. *caro, carnis*, flesh, + *vovare*, to devour]. An animal that eats other animals; a type of heterotroph.

carpel. The central whorl of modified leaves that represents the female reproductive structure of a flower. Typically consists of a stigma, style, and ovary. There may be more than one carpel per flower.

carrying capacity. The equilibrium size at which a particular population in a particular environment will stabilize when its supply of resources (including nutrients, energy, and living space) remains constant.

cell [L. *cella*, small room]. The basic *living* unit. There are large, complex organic molecules below this level of organization, but such molecules by themselves are nonliving.

cell plate. In plant cell division, a partition that forms from vesicles at the equator of the mitotic spindle, between the two newly forming cells.

cell wall. One or more layers of surface deposits outside the plasma membrane that generally provide support and resist mechanical pressure. Cell walls occur among bacteria, protistans, fungi, and plants.

central nervous system. Brain and spinal cord of vertebrates.

centriole. One of a pair of short, barrel-shaped structures that gives rise to the microtubule system of a cilium or a flagellum.

centromere (SEN-troh-meer) [Gk. *kentron*, center, + *meros*, a part]. A special region of the chromosome serving as the attachment site for spindle microtubules during nuclear division.

cephalization (sef-ah-lah-ZAY-shun) [Gk. *kephalikos*, head]. Differentiation of one end of the animal body into a head in which nervous tissue and sensory organs are especially concentrated.

cerebellum (ser-ah-BELL-um) [L. diminutive of *cerebrum*, brain]. Hindbrain region that coordinates motor activity for refined limb movements, maintaining posture, and spatial orientation.

cerebrum (suh-REE-bruhm). In vertebrate forebrain, paired masses of gray matter; cerebral hemispheres overlying thalamus, hypothalamus, and pituitary. Includes primary receiving centers for receptors at body periphery, association centers for coordinating and processing sensory input, and motor centers for coordinating motor responses.

chemical bond. A union between the electron structures of two or more atoms or ions.

chemiosmotic theory (kem-ee-oz-MOT-ik). Concept that an electrochemical gradient across a cell membrane drives ATP synthesis. Operation of electron transport systems builds up the hydrogen ion concentration on one side of the membrane. Then the electrical and chemical force of the H+ flow down the gradient is linked to enzyme machinery that combines ADP with inorganic phosphate to form ATP.

chemoreceptor (KEE-moe-ree-SEP-tur). Sensory cell or cell part that directly or indirectly transforms chemical stimuli into action potentials, the means of communication in nervous systems.

chemosynthetic autotroph (KEE-moe-sin-THET-ik). One of a few kinds of bacteria able to build all the organic molecules it requires using carbon dioxide as the carbon source and certain inorganic substances (such as sulfur) as the energy source.

chiasma, plural **chiasmata** (kai-AZ-mah, kai-az-MAH-tah) [Gk. *chiasma*, cross]. A crossing between two nonsister chromatids during prophase I of meiosis; evidence that breakage and exchange (a crossover) occurred earlier between them.

chlorophyll (KLOR-uh-fill) [Gk. *chloros*, green, + *phyllon*, leaf]. Light-trapping pigment molecule that acts as an electron donor in photosynthesis.

chloroplast. Eukaryotic organelle that houses membranes, pigments, and enzymes of photosynthesis.

chorion (CORE-ee-on). In reptiles, birds, and mammals, outermost membrane around embryo. Its vascularized villi form a nutritional link with the mother; its hormonal secretion (chorionic gonadotropin) helps maintain the corpus luteum (hence the uterine lining) following implantation.

chromatid (CROW-mah-tid) [Gk. *chroma*, color]. One of the two threadlike forms of a replicated chromosome, for as long as they remain attached at the centromere prior to and during nuclear division.

chromosome (CROW-muh-sohme) [Gk. *chroma*, color, + *soma*, body]. A DNA molecule and the proteins intimately associated with it; the vehicle by which hereditary information is transmitted from one generation to the next. For eukaryotes, the word can refer to a single or a duplicated chromosome (with its two sister chromatids).

cilium, plural **cilia** (SILL-ee-um) [L. *cilium*, eyelid]. Short, hairlike process extending from the plasma membrane and containing a regular array of microtubules. Some function as motile structures, others in creating currents of fluids; modified cilia are components of diverse sensory structures.

circulatory system. An organ system consisting of a muscular pump, blood vessels, and blood itself; the means by which materials are transported to and from cells; in many animals, also helps stabilize body temperature and pH.

cleavage (KLEE-vidj). Rapid, successive divisions in an animal zygote that lead to increase in cell number but not in cell size to produce the early embryo (blastocyst).

coccus, plural **cocci** (COCK-us, COCK-eye). Spherical form of bacterium.

codominance. The discernible expression of both alleles of a pair in heterozygotes.

coelom (SEE-lum) [Gk. *koilos*, hollow]. A type of body cavity defined by the peritoneum (a continuous lining of epithelial cells).

coenzyme. A large, nonprotein organic molecule that serves as a carrier of electrons or atoms in metabolic reactions and that is necessary for proper functioning of many enzymes. NAD+ and coenzyme A are examples.

cofactor. A metal ion or coenzyme that helps make substrates bind to an active site of an enzyme or that makes them more reactive. Some are bound tightly to the enzyme, others diffuse freely to and away from it.

cohesion. Condition in which molecular bonds resist rupturing when under tension.

commensalism [L. *com*, together, + *mensa*, table]. Two-species interaction in which one species benefits significantly while the other is neither helped nor harmed to any great degree.

communication signal. A stimulus produced by one animal that changes the behavior of another animal of the same species, in ways that benefit both the signaler and the receiver.

community. The populations of all species that occupy a habitat. Also used to refer to certain groups of organisms (such as the bird community) in a habitat.

competition, interspecific. Two-species interaction in which both species can be harmed as a result of their overlapping niches (that is, they have some requirements or activity in common).

competition, intraspecific. Interaction among individuals of the same species that are competing for the same resources.

competitive exclusion. The concept that if the resources required by two competing species are similar enough, the population growth rate of the one better able to exploit those resources (because of structural, physiological, or behavioral differences) will depress the growth rate of the other and lead to its exclusion from the habitat.

complement system. About twenty proteins, circulating in inactive form in the bloodstream until contact is made with certain bacterial or fungal invaders, whereupon they become activated and contribute to the inflammatory response. They are activated during general defense responses *and* specific immune responses.

concentration gradient. For a given substance, a greater concentration of its molecules in one region of a system than in another. A concentration difference between extracellular fluid and cytoplasm is an example.

condensation. Covalent linkage of small molecules in an enzyme-mediated reaction that can also involve formation of water.

conditioning. Form of learning in which a behavioral response becomes associated (by means of a reinforcing stimulus) with a new stimulus that was not previously associated with the response.

conjugation [L. *conjugatio*, a joining]. In some bacteria, transfer of DNA between two different mating strains that have made cell-to-cell contact.

consumers [L. *consumere*, to take completely]. Any organism that is not self-feeding (it is heterotrophic) and that ingests other (usually) living organisms in whole or in part to obtain organic nutrients. Herbivores, carnivores, omnivores, and parasites are consumers.

continuous variation. Small degrees of phenotypic variation that occur over a more or less continuous range of values in a population.

contractile vacuole (kun-TRAK-till VAK-you-ohl) [L. *contractus*, to draw together]. In some single-celled organisms, a membranous chamber that takes up excess water in the cell body, then contracts, expelling the water through a pore to the outside.

convergence. An outcome of natural selection whereby morphologically dissimilar and only distantly related lineages adopt a similar mode of living and come to resemble one another rather closely.

corpus luteum (CORE-pus LOO-tee-um). A glandular structure that develops from cells of a ruptured ovarian follicle and that secretes progesterone and some estrogen, both of which maintain the endometrium (the lining of the uterus).

cortex [L. *cortex*, bark]. In general, a rindlike layer; the kidney cortex is an example. In vascular plants, ground tissue that makes up most of the primary plant body, supports plant parts, and stores food.

cotyledon (cot-ill-EE-don). "Seed leaf"; often contains stored nutrients that are used in early growth when a seed germinates.

covalent bond (koe-VAY-lunt) [L. *con*, together, + *valere*, to be strong]. A sharing of one or more electrons between atoms or groups of atoms. When electrons are shared equally, it is a nonpolar covalent bond. When they are shared unequally, it is a polar covalent bond.

creatine phosphate (KREE-uh-teen FOSS-fate). Compound that readily gives up phosphate to ADP; important storage form of phosphate that is used in regenerating ATP for muscle contraction.

crossing over. During prophase I of meiosis, the breakage and exchange of corresponding segments of nonsister chromatids (of homologous chromosomes) at one or more sites along their length, resulting in genetic recombination.

cyclic adenosine monophosphate; cyclic AMP (SIK-lik ah-DEN-uh-seen mon-oh-FOSS-fate). A nucleotide that serves as an intracellular mediator of the cellular response to hormonal signals; a type of second messenger.

cyclic photophosphorylation (SIK-lik foe-toe-FOSS-for-ih-LAY-shun). Photosynthetic pathway in which electrons excited by sunlight energy move from a photosystem to a transport chain, then back to the photosystem. Energy released in the transport chain is coupled to ATP formation.

cytochrome (SIGH-toe-krome) [Gk. *kytos*, hollow vessel, + *chrōma*, color]. Iron-containing protein molecule that occurs in electron transport systems used in photosynthesis and aerobic respiration.

cytokinesis (SIGH-toe-kih-NEE-sis) [Gk. *kinesis*, motion]. Cytoplasmic division.

cytomembrane system. Those organelles concerned with modification and distribution of newly formed proteins and lipids, which are used by the cell itself or secreted from it. Includes endoplasmic reticulum, Golgi bodies, lysosomes, microbodies, and transport vesicles.

cytoplasm (SIGH-toe-plaz-um) [Gk. *plassein*, to mold]. In a cell, everything but the plasma membrane and the nucleus (or, in bacteria, the nucleoid). Includes internal membranes and other structures that function in metabolism, biosynthesis, and cell movements. The structures are bathed in cytosol, a semifluid substance.

cytosine (SIGH-toe-seen). A pyrimidine; one of the nitrogen-containing bases in nucleotides.

cytoskeleton. In the cytoplasm of eukaryotic cells, an internal framework of microtubules, microfilaments, and other fine strands by which organelles and other structures are anchored, organized, and moved about.

cytosol. The continuous aqueous portion of cytoplasm, with its dissolved solutes.

decomposers [L. *de*, down, away, + *companere*, to put together]. Mostly heterotrophic bacteria and fungi that obtain organic nutrients by breaking down the remains or products of other organisms; their activities help cycle simple compounds back to autotrophs.

denaturation (deh-NAY-chur-AY-shun). Disruption of bonds holding a protein in its three-dimensional form, such that its polypeptide chain(s) unfolds partially or completely.

dendrite (DEN-drite) [Gk. *dendron*, tree]. Nerve cell process, typically short and slender, which together with the cell body receives and integrates most incoming signals.

denitrification. Reduction of nitrate or nitrite to gaseous nitrogen (N_2) and a small amount of nitrous oxide (N_2O) by soil bacteria.

deoxyribonucleic acid (dee-ox-ee-rye-bow-new-CLAY-ik). DNA; double-stranded helically coiled nucleic acid, in which the hydrogen bonds between strands can be unzipped and DNA's chemical messages exposed to agents of protein synthesis or DNA replication. Overall, a stable molecule in which genetic information is stored, yet which is subject to change in some of its structural details.

detritivore (dih-TRY-tih-vore) [L. *detritus*; after *deterere*, to wear down]. An earthworm, crab, nematode, or other heterotroph that feeds on particles of organic matter, such as would be produced by the partial decomposition of plant and animal tissues.

deuterostome (DYEW-ter-oh-stome) [Gk. *deuteros*, sond, + *stoma*, mouth]. Any of those bilateral animals, including echinoderms and chordates, in which the first opening that appears in the gastrula stage of embryonic development bomes the anus and the sond opening, the mouth. Also characterized by radial cleavage and by formation of the coelom from pouches in the embryonic gut wall.

diaphragm (DIE-uh-fram) [Gk. *diaphragma*, to partition]. Muscular partition between the thoracic and abdominal cavities, the contraction and relaxation of which contribute to breathing. Also, a contraceptive device used temporarily to close off and thus prevent sperm from entering the uterus during sexual intercourse.

dicot (DIE-kot) [Gk. *di*, two, + *kotylēdōn*, cup-shaped vessel]. Short for dicotyledon; class of flowering plants characterized primarily by seeds having embryos with two cotyledons (seed leaves), generally net-veined leaves, and floral parts generally arranged in fours, fives, or multiples of these.

differential reproduction. The tendency of bearers of adaptive traits to reproduce successfully more than bearers of less adaptive traits. Because their offspring tend to make up an increasingly greater proportion of the reproductive base for each new generation, the adaptive traits increase in frequency also.

differentiation. Processes by which eukaryotic cells of identical genetic makeup become structurally and functionally different from one another, according to the genetically controlled developmental program of the species.

diffusion. Tendency of like molecules to move from their region of greater concentration to a region where they are less concentrated; occurs through random energetic movements of individual molecules, which tend to become dispersed uniformly in a given system.

digestive system. Some form of body cavity or tube by which food is ingested and prepared for absorption into the internal environment and from which the residues are eliminated.

diploid (DIP-loyd). Having two chromosomes of each type (that is, homologous chromosomes) in the somatic cells of sexually reproducing species. Except for sex chromosomes, the homologues resemble their partner in length, shape, and which genes they carry. Compare *haploid*.

directional selection. Mode of natural selection that moves the frequency distribution of alleles in a steady, consistent direction, such that the phenotypic character of a population shifts as a whole.

disaccharide (die-SAK-uh-ride) [Gk. *di*, two, + *sakcharon*, sugar]. A carbohydrate; two monosaccharides covalently bonded.

disruptive selection. Mode of natural selection that increases the frequency of two or more alleles that give rise to extreme forms of a trait, such that intermediate forms are selected against.

divergence [L. *dis*, apart, + *vergere*, to incline]. A buildup of differences in allele frequencies between reproductively isolated populations of a species or local breeding units of the same population. In *morphological* divergence, selection leads to departures from the ancestral species form.

diversity, organismic. Sum total of variations in form, functioning, and behavior that have accumulated in different lineages. Those variations generally are adaptive to prevailing conditions or were once adaptive to conditions that existed in the past.

dominance hierarchy. Social ranking of members of a group.

dominant allele. In a diploid cell, an allele whose expression masks the expression of its partner at the same gene locus on the homologous chromosome.

dormancy [L. *dormire*, to sleep]. Cessation of growth under physical conditions that could be quite suitable for growth.

ecology [Gk. *oikos*, home, + *logos*, reason]. Study of the interactions of organisms with one another and with the physical and chemical environment.

ecosystem. A community and its physical and chemical environment. Its biotic (living) component includes producers, consumers, decomposers, and detritivores. Its abiotic (nonliving) component includes soils, temperature, and rainfall.

ectoderm [Gk. *ecto*, outside, + *derma*, skin]. In an animal embryo, an outermost cell layer that gives rise to the outer layer of skin and to tissues of the nervous system.

effector. A muscle (or gland) that responds to nerve signals by producing movement (or chemical change) that helps adjust the body to changes in internal and/or external conditions.

efferent (EFF-uh-rent) [L. *effere*, to carry outward]. Conducting away from a body part or region, as from the brain or spinal cord to muscles or glands by way of motor neurons.

electric charge. A property of matter that enables ions, atoms, and molecules to attract or repel one another.

electron. Negatively charged particle that orbits the nucleus of an atom.

electron transport system. In a cell membrane, electron carriers and enzymes positioned in an organized array that enhances oxidation-reduction reactions. Such systems function in the release of energy that is used in ATP formation and other reactions.

element. Any substance that cannot be decomposed into substances with different properties.

embryo (EM-bree-oh) [Gk. *en*, in, + probably *bryein*, to swell]. In animals generally, the early stages of development (including cleavage, gastrulation, organogenesis, and morphogenesis) after fertilization. In most plants, the young sporophyte, from the first cell divisions after fertilization until germination.

endergonic reaction (en-dur-GONE-ik). Chemical reaction showing a net gain in energy.

endocrine element (EN-doe-krin) [Gk. *endon*, within, + *krinein*, to separate]. Cell or gland that produces and/or secretes hormones.

endocrine system. System of cells, tissues, and organs functionally linked to the nervous system and whose chemical secretions (hormones) help control body functioning.

endocytosis (EN-doe-sigh-TOE-sis). The process by which a region of the plasma membrane encloses substances (or cells, in the case of phagocytes) at or near the cell surface,

then pinches off to form a vesicle that transports the substances into the cytoplasm.

endoderm [Gk. *endon*, within, + *derma*, skin]. In an animal embryo, the innermost cell layer, which differentiates into the inner lining of the gut and organs derived from it.

endometrium (EN-doh-MEET-ree-um) [Gk. *metrios*, of the womb]. Inner lining of the uterus, consisting of connective tissues, glands, and blood vessels.

endoplasmic reticulum, or **ER** (EN-doe-PLAZ-mik reh-TIK-you-lum). A system of membranous tubes, channels, and flattened sacs that form compartments within the cytoplasm of eukaryotic cells and that function in the processing of proteins destined for secretion from the cell, and in the manufacture of the protein and lipid components of most organelles. *Rough* ER has ribosomes attached to the side of the membrane facing the cytoplasm; *smooth* ER does not.

endoskeleton [Gk. *endon*, within, + *skleros*, hard, stiff]. In chordates, the internal framework of bone, cartilage, or both. Together with skeletal muscle, supports and protects other body parts, helps maintain posture, and moves the body.

endosperm [Gk. *endon*, within, + *sperma*, seed]. Mass of tissue that surrounds embryo in a seed; in monocots, storage site for nutrients needed after seed germination.

energy. Capacity to make things happen, to cause change, to do work.

entropy (EN-trohp-ee). A measure of the degree of disorganization of a system—that is, how much energy in a system has become so dispersed (usually as evenly distributed, low-quality heat) that it is no longer available to do work.

enzyme. Any of a class of proteins that catalyze reactions (they greatly enhance the rate at which a reaction approaches equilibrium) by lowering the required activation energy.

epidermis. Outermost tissue layer of a multicelled animal or plant.

epithelium (EP-ih-THEE-lee-um). Sheet of cells, one or more layers thick, lining internal or external surfaces of the multicelled animal body.

equilibrium, dynamic [Gk. *aequus*, equal, + *libra*, balance]. The point at which a chemical reaction runs forward as fast as it runs in reverse, so that there is no net change in the concentrations of products or reactants.

erythrocyte (eh-RITH-row-site) [Gk. *erythros*, red, + *kytos*, vessel]. Red blood cell.

estrus (ESS-truss) [Gk. *oistrus*, frenzy]. For mammals generally, the cyclic period of a female's sexual receptivity to the male.

estuary (ESS-chew-airy). A region where fresh water from a river or stream mixes with salt water from the sea.

eukaryote (yoo-CARRY-oht) [Gk. *eu*, good, + *karyon*, nut]. Having a true nucleus; a cell that has membranous organelles, most notably the nucleus.

evaporation [L. *e-*, out, + *vapor*, steam]. The changes by which a substance is converted

from a liquid state into (and carried off in) vapor.

evolution [L. *evolutio*, act of unrolling]. In biology, successive changes in allele frequencies in a population, as brought about by such occurrences as mutation, genetic drift, gene flow, and selection pressure.

excretion. Elimination of excess water and excess or harmful solutes from the internal environment, as by kidneys.

exergonic reaction (EX-ur-GONE-ik). A chemical reaction that shows a net loss in energy.

exocrine gland (EK-suh-krin) [Gk. *ex*, out of, + *krinein*, to separate]. Secretory structure whose products travel through ducts that empty at a free epithelial surface.

exocytosis (EK-so-sigh-TOE-sis). The process by which substances are moved out of a cell. The substances are transported in cytoplasmic vesicles, the surrounding membrane of which merges with the plasma membrane in such a way that the substances are dumped outside.

exoskeleton [Gk. *exō*, out, + *skleros*, hard, stiff]. An external skeleton, as in arthropods.

exponential growth (EX-poe-NEN-shul). Pattern of population growth in which the number of individuals increases in doubling increments (2, 4, 8, 16, 32 . . .). Occurs when the birth rate remains even slightly above the death rate.

extinction, background. The steady rate of species turnover that characterizes lineages through most of their histories.

extinction, mass. An abrupt increase in the rate at which higher taxa disappear, with several higher taxa being affected simultaneously.

extracellular fluid. In animals generally, the medium through which substances are continuously exchanged between cells and between body regions. In vertebrates most is interstitial fluid; the rest is blood plasma.

facilitated diffusion. The movement of specific solutes across a plasma membrane in the direction that diffusion would take them, but with the passive assistance of carrier proteins that span the lipid bilayer of the membrane.

fat. A lipid with one, two, or three fatty acid tails attached to a glycerol backbone.

fatty acid. A long, unbranched hydrocarbon with a —COOH group at the end.

feedback inhibition. Control mechanism whereby an increase in some substance or activity inhibits the very process leading to (or allowing) the increase.

fermentation [L. *fermentum*, yeast]. Degradative pathway that begins with glycolysis and ends with the "spent" electrons being transferred back to one of the breakdown products or intermediates.

fertilization [L. *fertilis*, to carry, to bear]. Fusion of sperm nucleus with egg nucleus. In flowering plants, an additional sperm nucleus fuses with two other nuclei present in the ovule, forming a single triploid nucleus that will divide and give rise to endosperm; this is *double fertilization*.

first law of thermodynamics [Gk. *thermē*, heat, + *dynamikos*, powerful]. The total amount of energy in the universe remains constant; more energy cannot be created and existing energy cannot be destroyed. What already exists can only undergo conversion from one form to another.

flagellum, plural **flagella** (fluh-JELL-um) [L. *flagellum*, whip]. A motile structure that some cells use to move rapidly through the environment. Longer and less numerous than cilia; contains system of microtubules.

fluid mosaic membrane structure. Current model of membrane structure, in which diverse proteins are embedded in a lipid bilayer or attached to one of its two surfaces. The lipids give the membrane its basic structure and its relative impermeability to water-soluble molecules; packing variations and movements of lipids impart fluidity to the membrane. The proteins carry out most membrane functions, such as transport, enzyme action, and reception of chemical signals or substances.

follicle (FOLL-ih-kul). In a mammalian ovary, one of the spherical chambers containing an oocyte on the way to becoming a mature ovum (egg).

food chain. Linear sequence of who eats whom in an ecosystem.

food web. Network of many interlinked food chains, encompassing primary producers, consumers, decomposers, and detritivores.

fruit [L. after *frui*, to enjoy]. In flowering plants, the ripened ovary of one or more carpels, sometimes with accessory structures incorporated.

functional group. Atom or groups of atoms bonded covalently to the carbon backbone of an organic molecule and contributing to its characteristic structure and properties.

fungus [probably modification of Gk. *spongos*, sponge]. A heterotrophic organism the cells of which secrete enzymes that promote digestion of large organic molecules *outside* the cell, which then absorbs the breakdown products.

gamete (GAM-eet) [Gk. *gametēs*, husband, and *gametē*, wife]. Mature haploid cell (sperm or egg) that functions in sexual reproduction.

gametogenesis (gam-EET-oh-JEN-ih-sis). Formation of gametes by way of meiosis.

gametophyte (gam-EET-oh-fight) [Gk. *phyton*, plant]. Haploid, multicelled, gamete-producing phase in the life cycle of most plants.

ganglion (GANG-lee-un) [Gk. *ganglion*, a swelling]. A clustering of cell bodies of neurons into a distinct structure in body regions other than the brain or spinal cord. (Such clusterings in the brain or spinal cord are called nuclei.)

gastrula (GAS-truh-lah). Stage of animal development in which elaborate patterns of cell migrations in the embryo bring about the formation of two or three embryonic tissue layers (which, in most animals, are the endoderm, mesoderm, and ectoderm).

gene (jeen) [short for German *pangen*, after Gk. *pan*, all, + *-genēs*, to be born]. The basic unit of inheritance; a specific region of DNA coding for a tRNA, rRNA, or mRNA molecule, with the translation product of the mRNA being a polypeptide chain (the basic structural unit of proteins).

gene flow. Change in allele frequencies due to immigration (new individuals join the population), emigration (some individuals leave), or both.

gene frequency. More precisely, allele frequency: the relative abundance of different alleles carried by the individuals of a population.

gene locus. Particular location on a chromosome for a given gene.

gene pair. In diploid cells, the two alleles at a given gene locus on homologous chromosomes.

gene pool. Sum total of all genotypes in a given population. More accurately, allele pool.

genetic code [after L. *genesis*, to be born]. Basic language of protein synthesis, by which nucleotide triplets in DNA (and then mRNA) call for specific amino acids used in protein synthesis.

genetic drift. Random fluctuation in allele frequencies over time, due to chance occurrence alone.

genetic engineering. Altering the information content of DNA through use of recombinant DNA technology.

genetic equilibrium. Stability of allelic and genotypic ratios in a population over succeeding generations. Reference point, signifying zero evolution, for measuring rates of evolutionary change.

genetic recombination. Presence of a new combination of alleles in a DNA molecule compared to the parental genotype. The outcome of events such as crossing over at meiosis, chromosomal aberrations, gene mutation, and recombinant DNA technology.

genotype (JEEN-oh-type). Genetic constitution of an individual. Can mean a single gene pair or the sum total of the individual's genes. Compare *phenotype*.

genus, plural **genera** (JEEN-us, JEN-er-ah) [L. *genus*, race, origin]. A taxon (that is, a category of relationship based on phenotypic similarities, descent, or both) in which all species sharing (or exhibiting) certain characteristics are grouped.

germ cell. Animal cell that may develop into gametes. Compare *somatic cell*.

germ layer. Embryonic tissue layer that gives rise to the tissues and organs of the adult. Ectoderm, mesoderm, and endoderm are examples.

gill. A respiratory organ, typically with a moist, thin, vascularized layer of epidermis that functions in gas exchange.

glomerulus (glow-MARE-yoo-luss) [L. *glomus*, ball]. Cluster of capillaries in Bowman's capsule of the nephron, the functional unit of the kidney.

glucagon (GLUE-kuh-gone). Animal hormone secreted by pancreatic cells and essential in breakdown of glycogen (a polysaccharide) to glucose subunits during the post-absorptive state of organic metabolism.

glycerol (GLISS-er-ohl) [Gk. *glykys*, sweet, + L. *oleum*, oil]. Three-carbon molecule with three hydroxyl groups attached; combines with fatty acids to form fat or oil.

glycogen (GLY-kuh-jen). In animals, a starch that is a main food reserve; can be readily broken down into glucose subunits during the post-absorptive state of organic metabolism.

glycolysis (gly-CALL-ih-sis) [Gk. *glykys*, sweet, + *lysis*, loosening or breaking apart]. In carbohydrate metabolism, the initial breaking apart of sugar molecules such as glucose, with the release of energy.

gonad (GO-nad). Primary reproductive organ in which gametes are produced.

graded potential. At chemical synapses and receptors, a change in membrane permeability characteristics that can vary in magnitude, depending on the stimulus. Many graded potentials acting at the same time can so change the voltage difference across the membrane that they initiate an action potential. Compare *action potential*.

granum, plural **grana**. In chloroplasts, stacked membrane system where sunlight energy is actually trapped and where ATP is formed.

gravitropism (GRAV-ih-TROPE-izm) [L. *gravis*, heavy, + Gk. *trepein*, to turn]. Directional growth of a coleoptile, root, or shoot in response to gravity. Also called *geotropism*.

ground meristem (MARE-ih-stem) [Gk. *meristos*, divisible]. A primary meristem that produces ground tissue (hence the bulk of the plant body).

gymnosperm [Gk. *gymnos*, naked, + *sperma*, seed]. One of two divisions of seed plants, characterized by having their seeds borne on surfaces of reproductive structures, without protective tissue layers. Conifers such as pines are examples.

habitat [L. *habitare*, to live in]. Place where an individual or population of a given species lives; its "mailing address."

haploid (HAP-loyd). State in which a nucleus contains half the parental number of chromosomes characteristic of the somatic cells of a species; brought about by meiosis, which is necessary for gamete formation. Each gamete ends up with one of each pair of homologous chromosomes present in the parental nucleus.

heart. Muscular pump that keeps blood circulating through the animal body.

hemoglobin (HEEM-oh-glow-bin) [Gk. *haima*, blood, + L. *globus*, ball]. Iron-containing protein that gives red blood cells their color; functions in oxygen transport.

hemorrhage. Bulk flow of blood from damaged vessels.

herbivore [L. *herba*, grass, + *vovare*, to devour]. Plant-eating animal.

heterotroph (HET-er-oh-trofe) [Gk. *heteros*, other, + *trophos*, feeder]. Organism that obtains carbon and all metabolic energy from organic molecules that have already been assembled by autotrophs. Animals, fungi, many protistans, and most bacteria are heterotrophs.

heterozygote (HET-er-oh-ZYE-gote) [Gk. *zygoun*, join together]. Individual having non-identical alleles at a given gene locus.

histone. Any of a class of structural proteins complexed with DNA in the eukaryotic chromosome.

homeostasis (HOE-me-oh-STAY-sis) [Gk. *homo*, same, + *stasis*, standing]. For multicelled organisms, maintaining the internal environment (that is, the extracellular fluid environment) within some tolerable range throughout the life cycle even when conditions change.

homeostasis, dynamic. Maintaining the living state by adjusting the organism's form and behavior over time, as a function of the genetically prescribed developmental program of the species.

hominid [L. *homo*, man]. A member of the genus *Homo*; modern humans and their most recent ancestors. *H. sapiens* is the only living representative.

hominoid [after Gk. *eidos*; resembling the form of]. Ape or human species.

homologous chromosome (huh-MOLL-uh-gus) [Gk. *homologia*, agreement; correspondence]. In the nucleus of a somatic cell, one of a pair of chromosomes that resemble each other in length, shape, and which genes they carry and that synapse at meiosis. Sex chromosomes, which differ morphologically in males and females, also function as homologues. (Typically the two chromosomes of a homologous pair are derived from two different parents, but exceptions do occur, as in the case of self-fertilizing plants.)

homologous recombination. Crossing over and the exchange of segments between homologous chromosomes during meiosis I. An exchange can occur almost anywhere along the chromosomes, the exchange is reciprocal, and a fairly long, staggered joint forms between the interacting DNA strands.

homozygote [Gk. *homos*, same, + *zygoun*, join together]. Individual having two identical alleles at a given gene locus.

hormone [Gk. *hormōn*, to stir up, set in motion]. A secretion from an endocrine cell or gland, transported by the bloodstream to nonadjacent target cells, these being any cells having receptors to which the hormone can bind.

hydrogen bond. Type of chemical bond in which an electronegative atom interacts weakly with a hydrogen atom that is already participating in a polar covalent bond.

hydrolysis (high-DRAWL-ih-sis) [L. *hydro*, water, + Gk. *lysis*, loosening or breaking apart]. Reaction in which covalent bonds between parts of molecules are broken and an H^+ ion and an OH group derived from water become attached to the fragments.

hydrophilic [Gk. *philos*, loving]. Having an attraction for water molecules; refers to a polar substance that readily dissolves in water.

hydrophobic [Gk. *phobos*, dreading]. Repelled by water molecules; refers to a nonpolar substance that does not readily dissolve in water.

hypha, plural **hyphae** [Gk. *hyphē*, web]. A filament of a mycelium; composed of elongated cells that have chitin-reinforced walls.

hypothalamus [Gk. *hypo*, under, + *thalamos*, inner chamber or possibly *tholos*, rotunda]. Region of vertebrate forebrain concerned with neural-endocrine control of visceral activities (e.g., salt-water balance, temperature control, reproduction).

immune system, vertebrate. Three types of white blood cells (phagocytes, T lymphocytes, and B lymphocytes) and their products, all of which make a *specific* response to a particular invader of the body (as opposed to a general attack response) and which are characterized by *memory* (an ability to mount a rapid attack when the same type of invader returns). The complement system and other components of the general, inflammatory response are also activated during immune responses.

independent assortment. Mendelian principle that the alleles of two (or more) gene pairs located on nonhomologous chromosomes tend to be assorted independently of one another into gametes.

inflammation. Nonspecific defense response involving mobilization of phagocytic cells and the complement system; a series of homeostatic events that restore damaged tissues and intercellular conditions.

instinctive behavior. The capacity of an animal to complete a fairly complex, stereotyped response to a first-time encounter with a key stimulus (without having had prior experience with that stimulus).

integration, neural [L. *integrare*, to coordinate]. Moment-by-moment summation of all excitatory and inhibitory synapses acting on a neuron; occurs at each level of synapsing in a nervous system.

interneuron. Main component of integrating centers such as the brain and spinal cord; integrates information arriving from sensory neurons, then influences other neurons in turn.

interphase. Time interval (differs among species) in which a cell increases its mass, approximately doubles the number of its structures and organelles, and finally replicates its DNA, prior to nuclear division.

interstitial fluid (IN-ter-STISH-ul) [L. *interstitus*, to stand in the middle of something]. In vertebrates, that portion of the extracellular fluid occupying spaces between cells and tissues. (The remaining portion is blood plasma.)

ion, negatively charged. An atom or a compound that has gained one or more electrons, hence has acquired an overall negative charge.

ion, positively charged. An atom or a compound that has lost one or more electrons, hence has acquired an overall positive charge.

ionic bond. An association between ions of opposite charge.

isogamy (EYE-soh-gam-ee) [Gk. *isos*, equal, + *gametēs*, husband, and *gametē*, wife]. In some sexual reproductive modes, gametes that are all identical (no differentiation into sperm and eggs).

isolating mechanism. Some aspect of structure, functioning, or behavior that prevents interbreeding between populations that are undergoing or have undergone speciation. Also applies to local breeding units within a population.

isotope (EYE-so-tope). Individual atom that contains the same number of protons as other atoms of a given element, but that has a different number of neutrons.

kidney. Organ of salt and water regulation; its nephron/capillary units are concerned with filtration of water and other noncellular components of blood, selective reabsorption of solutes and most of the water, and tubular secretion (through active transport) of certain substances from the capillaries.

Krebs cycle. Stage of aerobic respiration in which pyruvate fragments are completely broken down into carbon dioxide; molecules reduced in the process can be used in forming many electron carriers for use in the last stage of the aerobic pathway.

larva, plural **larvae**. A sexually immature, free-living and free-feeding animal that grows and develops into the sexually mature adult form.

larynx (LARE-inks). Tube that leads to the lungs. In humans, contains vocal cords, the production site of sound waves used in speech.

lateral meristem. Either vascular cambium or cork cambium, the meristems responsible for secondary growth (increases in diameter) in most plants.

learning. Modification of behavior, arising from specific experiences during the lifetime of an animal.

leucoplast (LEW-kuh-plast). In some plant cells, colorless plastid in which starch grains and other substances may be stored.

life cycle. For any species, the genetically programmed sequence of events by which individuals are produced, grow, develop, and themselves reproduce.

light-dependent reactions. First stage of photosynthesis, concerned with harnessing sunlight and using it as an energy source for synthesizing ATP alone (the cyclic pathway) or ATP and NADPH (the noncyclic pathway).

light-independent reactions. Second stage of photosynthesis, in which sugars and other organic compounds are assembled using the ATP and NADPH produced during the first stage.

linkage. The tendency of genes physically located on the same chromosome to be inherited together instead of undergoing independent assortment.

lipid [Gk. *lipos*, fat]. A hydrocarbon (mostly) that generally does not dissolve in water but will dissolve in nonpolar substances. Some lipids have fatty acid components (e.g., oils, waxes), others do not (e.g., steroids).

lymph vascular system [L. *lympha*, water, + *vasculum*, a small vessel]. Network of vessels that supplements the blood circulation system; reclaims water that has entered interstitial regions from the bloodstream; also transports fats from small intestine to bloodstream. Fluid in its vessels is called *lymph*.

lymphocyte. Any of various white blood cells that take part in vertebrate immune responses.

lymphoid organs. Those organs (and some tissue regions) that function as blood cell production centers and as sites for some defense responses; bone marrow, thymus, lymph nodes, spleen, appendix, tonsils, adenoids, and patches of small intestine.

lymphokine. Any of a class of proteins by which the cells of the vertebrate immune system communicate with one another.

lysosome. Membrane-bound organelle containing hydrolytic enzymes that can break down all polysaccharides, nucleic acids, and proteins as well as some lipids. Central in the cell's materials recycling and biosynthesis programs.

macroevolution. Large-scale rates, trends, and patterns of change among groups of species since the beginning of life.

mantle [L. *mantellum*, a loose external garment]. In mollusks, a body wall surrounding internal parts; secretes substances that form the molluscan shell.

mass number. Total number of protons and neutrons in the nucleus of atoms of a given element.

mechanoreceptor. Sensory cell or cell part that detects mechanical energy associated with changes in pressure, position, or acceleration.

medusa (meh-DOO-sah) [Gk. *Medousa*, one of three sisters in Greek mythology having snake-entwined hair; this image probably evoked by the tentacles and oral arms extending from the medusa]. Free-swimming, bell-shaped stage in cnidarian life cycles.

megaspore. In seed plants, a meiospore that develops into a female gametophyte.

meiosis (my-OH-sis) [Gk. *meioun*, to diminish]. Two-stage nuclear division process in which the parental number of chromosomes in each daughter nucleus becomes haploid (half of what it was in the parent nucleus). Basis of gamete and meiospore formation. Compare *mitosis*.

meiospore. Haploid cell that divides by mitosis and differentiates into multicelled haploid bodies (gametophytes).

menopause (MEN-uh-pozz) [L. *mensis*, month, + *pausa*, stop]. End of the period of a human female's reproductive potential.

menstrual cycle. The cyclic reproductive capacity of female humans and other primates.

menstruation. Periodic sloughing of the blood-enriched lining of the uterus when pregnancy does not occur.

meristem (MARE-ih-stem) [Gk. *meristos*, divisible]. In most plants, a mass of self-perpetuating cells not yet committed to developing into a specialized cell type. Compare *apical*, *lateral*, and *primary meristems*.

mesoderm (MEH-so-derm) [Gk. *mesos*, middle, + *derm*, skin]. In most animal embryos, a tissue layer between ectoderm and endoderm; gives rise to muscle, the organs of circulation, reproduction, and excretion, most of the internal skeleton (when present), and connective tissue layers of the gut and body covering.

metabolic pathway. In a cell, breakdown or synthesis reactions that occur in sequential, stepwise fashion.

metabolic reaction. Some form of internal energy change in a cell.

metabolism (meh-TAB-oh-lizm) [Gk. *metabolē*, change]. All activities by which organisms extract and transform energy from their environment, and use it in manipulating materials in ways that assure maintenance, growth, and reproduction.

metamorphosis (met-ah-more-FOE-sis) [Gk. *meta-*, change, + *morphē*, form]. For animals that undergo indirect development, the reactivation of development from the larval stage to the adult form.

metaphase. In mitosis, stage when microtubules increase in number and become organized into mitotic spindle, which is responsible for separating the sister chromatids of chromosomes from each other prior to cell division.

metazoan (MET-ah-ZOE-un). Multicelled animal.

microfilament [Gk. *mikros*, small, + L. *filum*, thread]. Component of the cytoskeleton; involved in cell shape, motion, and growth.

microspore. In seed plants, a meiospore that develops into a male gametophyte.

microtubular spindle. An array of microtubules that helps establish the polarity necessary for chromosome movements during nuclear division.

microtubule. Hollow cylinder of (mostly) tubulin subunits; involved in cell shape, motion, and growth; functional unit of cilia and flagella.

microvillus (MY-crow-VILL-us) [L. *villus*, shaggy hair]. A slender, cylindrical extension of the animal cell surface that functions in absorption or secretion.

migration. A cyclic movement between two distant regions at times of year corresponding to seasonal change.

mimicry (MIM-ik-ree). Situation in which one species (the mimic) bears deceptive resemblance in color, form, and/or behavior to another species (the model) that enjoys some survival advantage.

mitochondrion, plural **mitochondria** (MY-toe-KON-dree-on). Eukaryotic organelle that specializes in aerobic respiration.

mitosis (my-TOE-sis) [Gk. *mitos*, thread]. Nuclear division in which the parental number of chromosomes is maintained from one cell generation to the next. Basis of reproduction of single-celled eukaryotes; basis of physical growth (through cell divisions) in multicelled eukaryotes.

molecule [diminutive of L. *moles*, mass]. A unit of two or more atoms of the same or different elements bonded together.

monocot. Short for monocotyledon; a flowering plant in which seeds have only one cotyledon, whose floral parts generally occur in threes (or multiples of threes), and whose leaves typically are parallel-veined. Compare *dicot*.

monomer (MON-oh-mur). Any of numerous individual subunits that become incorporated into polymers.

monosaccharide (MON-oh-SAK-ah-ride) [Gk. *mono*, alone, single, + *sakcharon*, sugar]. A sugar monomer, typically with a backbone of three to seven carbon atoms, an aldehyde or ketone group, and two or more hydroxyl groups.

morphogenesis (MORE-foe-GEN-ih-sis) [Gk. *morphē*, form, + *genesis*, origin]. The growth, shaping, and arrangement of body parts according to genetically predefined patterns. The extent, direction, and rate of morphogenesis depend on genetic controls and environmental factors.

motor neuron. Type of neuron that relays information away from integrating centers (such as the brain and spinal cord) to the body's effectors (muscles and glands).

multiple allele system. More than two forms of alleles that can occur at a given gene locus.

muscle fiber. Contractile cell of skeletal, smooth, or cardiac muscle tissue.

mutation [L. *mutatus*, a change, + *-ion*, result of a process or an act]. A heritable change in the kind, structure, sequence, or number of the component parts of a DNA molecule.

mutualism [L. *mutuus*, reciprocal]. A major type of community interaction from which both members of a pair of species benefit clearly and directly. When such a mutually beneficial relationship involves continuous, intimate contact, it is called symbiosis.

mycelium, plural **mycelia** (my-SEE-lee-um) [Gk. *mykēs*, fungus, mushroom, + *hēlos*, callus]. A multicelled structure, in the form of a mesh of branching, microscopic filaments, that forms during the life cycle of most fungi.

NAD⁺. Nicotinamide adenine dinucleotide, oxidized form.

NADH. Nicotinamide adenine dinucleotide, reduced form.

NADP⁺. Nicotinamide adenine dinucleotide phosphate, oxidized form.

NADPH. Nicotinamide adenine dinucleotide phosphate, reduced form.

natural selection. The result of differential reproduction of genotypes within a population, this being one of the most important mechanisms bringing about evolutionary change. See also *differential reproduction.*

negative feedback mechanism. Homeostatic control whereby an increase in some substance or activity sooner or later inhibits the very processes leading to (or allowing) the increase.

nematocyst (NEM-add-uh-sist) [Gk. *nēma,* thread, + *kystis,* bladder, pouch]. A stinging capsule that assists in capturing prey and that may serve in protection; a distinguishing feature of cnidarians such as jellyfishes.

nephridium, plural **nephridia** (neh-FRID-ee-um). In invertebrates such as earthworms, a system for regulating water and solute levels.

nephron (NEH-frohn) [Gk. *nephros,* kidney]. One of more than a million long, slender tubules in the kidney in which urine is formed by processes of filtration, reabsorption, and secretion.

nerve. Cordlike communication line of nervous systems, composed of axons of sensory or motor neurons (or both) packed tightly in bundles within connective tissue. In the brain and spinal cord, such bundles are called nerve pathways or tracts.

nerve impulse. Action potential.

nerve net. Cnidarian nervous system consisting of nerve cells associated with the gastrodermis and epidermis and concerned primarily with feeding behavior.

nervous system. Constellations of neurons oriented relative to one another in precise message-conducting and information-processing pathways.

neuroglia (NUR-oh-GLEE-uh). Cells intimately associated with neurons and functioning in their structural and metabolic support, maintenance, and in some cases as axonal sheaths. In vertebrates they represent at least half of the volume of the nervous system.

neuromuscular junction. The synapses between the splayed-out axon terminals of a motor neuron and a muscle cell. The terminals are positioned in troughs (in the muscle cell membrane) called the motor end plate.

neuron. In most animals, a cell that responds to specific chemical, electrical, or mechanical stimuli in three ways: it can *integrate* different incoming signals, *propagate* excitation as a pulse of information along its plasma membrane, and *transmit* information about change to other neurons, muscles, or glands.

neutron. Subatomic particle of about the same size and mass as a proton but having no electric charge.

niche (nitch) [L. *nidus,* nest]. The full range of abiotic and biotic conditions under which a particular species can live and reproduce.

nicotinamide adenine dinucleotide, or **NAD⁺.** A local electron carrier that transfers hydrogen atoms and electrons *within* metabolic pathways. NAD⁺ is a free-moving carrier, not membrane-bound in a transport system.

nicotinamide adenine dinucleotide phosphate. NADPH. Together with ATP, a major coupling agent *between* degradative and biosynthetic pathways.

nitrification. Process by which certain soil bacteria strip ammonia or ammonium of electrons, and nitrite (NO_2^-) is released as a reaction product, then other soil bacteria use nitrite for energy metabolism, yielding nitrate (NO_3^-).

nitrogen fixation. Among some bacteria, assimilation of gaseous nitrogen (N_2) from the air; through reduction reactions, electrons (and associated H^+) become attached to the nitrogen, thereby forming ammonia (NH_3) or ammonium (NH_4^+).

noncyclic photophosphorylation (non-SIK-lik foe-toe-FOSS-for-ih-LAY-shun) [L. *non,* not, + Gk. *kylos,* circle]. Photosynthetic pathway in which new electrons derived from water molecules flow through two photosystems and two transport chains, the result being formation of ATP and NADPH.

notochord (NO-toe-kord) [Gk. *nōtos,* back, + L. *chorda,* cord]. A stiffened but flexible supporting structure, just beneath the nerve cord, that is present during at least some stages of the life cycle of all chordates. During embryonic development of complex vertebrates, it is replaced by a vertebral column.

nuclear envelope. Double membrane forming the surface boundary of a eukaryotic nucleus.

nucleic acid (new-CLAY-ik). Long-chain, single- or double-stranded nucleotide; DNA and RNA are examples.

nucleoid. In bacterial cells, the irregularly shaped region in which DNA is concentrated but not bound by a membrane.

nucleolus (new-KLEE-oh-lus) [L. *nucleolus,* a little kernel]. Within the nucleus of a nondividing cell, a mass of proteins, RNA, and other material used in ribosome synthesis.

nucleotide (NEW-klee-oh-tide). Molecule containing a five-carbon sugar (ribose), a nitrogen-containing base (either purine or pyrimidine), and a phosphate group. Structural unit of adenosine phosphates, nucleotide coenzymes, and nucleic acids.

nucleus (NEW-klee-us) [L. *nucleus,* a kernel]. In atoms, the central core of one or more positively charged protons and (in all but hydrogen) electrically neutral neutrons. In eukaryotic cells, the membranous organelle that houses the DNA.

omnivore [L. *omnis,* all, + *vovare,* to devour]. An organism able to obtain energy from more than one source rather than being limited to one trophic level.

oogenesis (oo-oh-JEN-uh-sis). Formation of a female gamete, from a germ cell (oogonium) to a mature haploid ovum.

organ. A structure of definite form and function that is composed of more than one tissue and the character of which is influenced by the type, combination, and arrangement of those tissues.

organelle. Any of various membranous sacs, envelopes, and other compartmented portions of cytoplasm that separate different, often incompatible metabolic reactions in the space of the cytoplasm and in time (through specific reaction sequences).

osmosis (oss-MOE-sis) [Gk. *ōsmos,* act of pushing]. Movement of water molecules across a differentially permeable membrane in response to a concentration and/or pressure gradient.

ovary. The primary female reproductive organ in which oogenesis occurs.

oviduct (OH-vih-dukt). Passageway through which ova travel from the ovary to the uterus.

ovule (OH-vewl) [L. *ovum,* egg]. In seed-bearing plants, the structure destined to become the seed; includes the nucellus, integuments, and the stalk attaching them to the ovarian wall. Cell divisions in the nucellus give rise to the female gametophyte, with its egg cell and endosperm mother cell.

oxidation. The loss of one or more electrons from an atom or molecule.

oxidation-reduction reaction. An electron transfer from one atom or molecule to another. Often hydrogen is also transferred along with the electron or electrons.

oxidative phosphorylation. Use of electron energy being released during oxidation reactions to phosphorylate (tack a phosphate group onto) a molecule such as ADP (which yields energy-rich ATP).

parapatric speciation [Gk. *para,* alongside, + *patria,* native land]. Speciation that occurs when adjoining populations undergo divergence despite some gene flow.

parasite [Gk. *para,* alongside, + *sitos,* food]. A type of heterotroph that lives on or in a living host organism during some part of its life cycle, obtains nutrients from the host's tissues, and may or may not end up killing the host as a consequence of the association (but usually produces an intermediate level of negative effects).

parasitoid. A type of larva that kills a host insect by completely consuming its soft tissues before the host metamorphoses into an adult.

parasympathetic nerves. A division of the autonomic nervous system, consisting of those efferent nerves concerned more with slowing down overall body activity and diverting energy to basic housekeeping tasks. Compare *sympathetic nerves.*

passive transport. Movement of a substance across a cell membrane without any direct energy outlay.

pathogen (PATH-oh-jen) [Gk. *pathos,* suffering, + *-genēs,* origin]. Disease-causing organism.

penis. Component of the male reproductive system of many species; the copulatory organ by which sperm are deposited into a specialized duct of the female reproductive system.

perennial [L. *per-*, throughout, + *annus*, year]. A plant that lives year after year.

pericycle (PARE-ih-sigh-kul) [Gk. *peri-*, around, + *kyklos*, circle]. One or more layers of parenchyma cells, just inside the endodermis of the root vascular column, that maintain the potential for meristematic activity. Gives rise to lateral roots and, in species showing secondary growth, contributes to the formation of vascular cambium and cork cambium.

periderm. In stems and roots of gymnosperms and flowering plants, a protective covering that replaces epidermis during secondary growth.

peripheral nervous system (per-IF-ur-uhl) [Gk. *peripherein*, to carry around]. In vertebrates, the nerves leading into and out from the spinal cord and brain and the ganglia along those communication lines.

pH. Whole number referring to the number of hydrogen ions present in a liter of a given fluid.

phagocytosis (FAG-uh-sigh-TOE-sis) [Gk. *phagein*, to eat, + *kytos*, hollow vessel]. Engulfment of foreign cells or substances by amoebas and some white blood cells, by means of endocytosis.

pharynx (FAR-inks). A muscular tube by which food is taken into the gut. In humans, the gateway to the digestive tract and to the windpipe (trachea).

phenotype (FEE-no-type) [Gk. *phainein*, to show, + *typos*, image]. Observable trait or traits of an individual; arises from interactions between genes, and between genes and the environment.

pheromone (FARE-oh-moan) [Gk. *phero*, to carry, + *-mone*, as in hormone]. A chemical secreted by an exocrine gland that serves as a communication signal between individuals of the same species.

phloem (FLOW-um). The food-conducting tissue of vascular plants, the main components of which are sieve tube members and companion cells (in flowering plants) or sieve cells (in gymnosperms).

phospholipid. A key component of cell membranes in plants and animals; a molecule with a glycerol backbone, two fatty acid tails, and a phosphate group to which an alcohol is attached.

phosphorylation (FOSS-for-ih-LAY-shun). Addition of one or more phosphate groups to a molecule.

photolysis (foe-TALL-ih-sis) [Gk. *photos*, light, + *-lysis*, breaking apart]. First step in noncyclic photophosphorylation, when water is split into oxygen, hydrogen, and associated electrons; photon energy indirectly drives the reaction.

photoreceptor. Light-sensitive sensory cell.

photosynthesis. The trapping of solar energy and its conversion to chemical energy (ATP, NADPH, or both), which is used in manufacturing food molecules from carbon dioxide and water.

photosynthetic autotroph. An organism able to build all of the organic molecules it requires using carbon dioxide as the carbon source and sunlight as the energy source. All plants, some protistans, and a few bacteria are photosynthetic autotrophs.

photosystem. Functional light-trapping unit in photosynthetic membranes; contains pigment molecules and enzymes.

phototropism [Gk. *photos*, light, + *trope*, turning, direction]. Growth toward or away from light shining mainly on one side of a plant.

phytochrome. Light-sensitive pigment molecule whose activation and inactivation trigger hormone activities governing leaf expansion, stem branching, stem length, and, in many plants, seed germination and flowering.

phytoplankton (FIE-toe-PLANK-tun) [Gk. *phyton*, plant, + *planktos*, wandering]. Community of photosynthetic microorganisms in freshwater or saltwater environments.

pinocytosis (PIN-oh-sigh-TOE-sis) [Gk. *pinein*, to drink, + *kytos*, vessel]. "Cell-drinking"; engulfment of liquid droplets.

placenta (play-SEN-tuh). In the uterus, an organ made of extensions of extraembryonic membranes (the chorion especially) and the endometrium. Through this composite of embryonic and maternal tissues and vessels, nutrients reach the embryo and wastes are carried away.

plant. Generally, multicelled autotroph able to build its own food molecules through photosynthesis.

plasma. Liquid component of blood in which numerous plasma proteins, ions, simple sugars, amino acids, vitamins, hormones, and gases are dissolved.

plasma membrane. Outermost membrane of a cell. Its surface has molecular regions that detect changes in external conditions. Spanning the membrane are passageways through which substances move inward and outward in controlled ways. Transport vesicles form from or fuse with its lipid bilayer.

plasmid. In some bacteria, a small circle of DNA in addition to the bacterial chromosome.

plasmodesma (PLAZ-moe-DEZ-muh). In a multicelled plant, a junction between the linked walls of adjacent cells through which nutrients and other substances are transported.

plastid. In some plant cells, a storage organelle; some plastids also function in photosynthesis.

platelet. Component of blood that functions in clotting.

pleiotropism (PLEE-oh-TROE-pizm) [Gk. *pleōn*, more, + *trope*, direction]. Multiple phenotypic effect of a single gene; the action of the gene affects many developmental or maintenance activities.

pollen grain [L. *pollen*, fine dust]. In gymnosperms and flowering plants, the immature male gametophyte (gamete-producing body).

pollination. The transfer of pollen grains to the female gametophyte.

pollutant. Any substance with which an ecosystem has had no prior evolutionary experience, in terms of kinds or amounts, and that can accumulate to disruptive or harmful levels. Can be naturally occurring or synthetic.

polymer (POH-lih-mur) [Gk. *polus*, many, + *meris*, part]. A molecule composed of from three to millions of subunits of relatively low molecular weight that may or may not be identical.

polymorphism (poly-MORE-fizz-um) [Gk. *polus*, many, + *morphe*, form]. In a population, the persistence of two or more forms of a trait, at a frequency that is greater than can be maintained by newly arising mutations alone; and that frequency, if changed, will return to its former value over several generations.

polyp (POH-lip). Vase-shaped, sedentary stage of cnidarian life cycles.

polypeptide. Chain of amino acids linked by peptide bonds, which form through condensation reactions.

polyploid. Having three or more sets of chromosomes in the somatic cells of a species.

polyribosome. During protein synthesis, a clustering of ribosomes engaged in translation of a messenger RNA molecule.

polysaccharide [Gk. *polus*, many, + *sakcharon*, sugar]. Three or more monosaccharides bonded together covalently.

population. Group of individuals of the same species occupying a given area.

predator [L. *prehendere*, to grasp, seize]. Any of various organisms that get food from other living organisms (their prey), but they do not live on or in the prey and they may or may not kill it. Compare *parasite*.

primary growth. Following seed germination, the cell divisions, elongation, and differentiation that produce the primary plant body.

primary productivity, gross. The total rate at which the autotrophs of an ecosystem fix energy in organic compounds (as plants do in photosynthesis).

primary productivity, net. The rate of energy storage in the tissues of autotrophs in excess of the rate of respiration during a measured time interval.

prion (PRY-on). Infectious agent apparently consisting of protein only and typically causing slow but fatal diseases of the central nervous system.

procambium (pro-KAM-bee-um). A primary meristem (formed from apical meristem) that gives rise to the vascular tissues of the primary plant body.

producer. An autotrophic organism; able to build its own complex organic molecules from simple inorganic substances in the environment.

prokaryote (pro-CARRY-oht) [L. *pro*, before, + Gk. *karyon*, kernel]. Single-celled organism that has no membrane-bound nucleus or other internal organelles; all bacteria are prokaryotes.

prokaryotic fission. Form of bacterial cell reproduction; membrane growth divides the replicated DNA and the cytoplasm into daughter cells.

prophase. In mitosis, the stage when chromatin coils into compact chromosome bodies. In meiosis I, the stage when crossing over occurs.

protein. Molecule composed of one or more chains of amino acids (polypeptide chains).

protistan (pro-TISS-tun) [Gk. *prōtistos*, primal, very first]. Single-celled eukaryote.

proton. Positively charged unit of energy that is found in the atomic nucleus.

protostome (PRO-toe-stome) [Gk. *proto*, first, + *stoma*, mouth]. Any of those bilateral animals, including annelids, arthropods, and mollusks, in which the first opening that appears in the gastrula stage of development becomes the mouth and the second opening, the anus. Also characterized by spiral cleavage and formation of the coelom by the splitting of two masses of mesoderm on either side of the gut.

pseudocoelomate (SOO-doe-SEE-la-mate) [Gk. *pseudos*, false, + *koilos*, a hollow]. Type of invertebrate in which the body lacks a continuous peritoneal lining (which occurs in animals having a coelom).

pseudopod (SOO-doe-pod). "False foot"; a nonpermanent cytoplasmic extension of the cell body.

pulmonary circulation. Pathways of blood flow leading to and from the lungs.

purine. Nucleotide base having a double ring structure. Examples are adenine and guanine.

pyrimidine (pih-RIM-ih-deen). Nucleotide base having a single ring structure. Cytosine and thymine are examples.

pyruvate (PIE-roo-vate). Three-carbon compound produced by the initial breakdown of a glucose molecule during glycolysis.

radial symmetry. Body plan in which the body can be divided into four or more pieces that are equal with respect to the structures they contain.

radicle. In plant seeds, the embryonic root; typically, longitudinal growth of its cells gives rise to a slender primary root.

receptor. Sensory cell or cell part that may be activated by a specific stimulus in the internal or external environment.

recessive allele [L. *recedere*, to recede]. In the heterozygous state, an allele whose expression is fully or partially masked by expression of its partner. Recessive alleles can be fully expressed in the homozygous state and in haploid organisms.

recombinant DNA. Whole molecules or fragments that incorporate parts of different parent DNA molecules, as formed by natural recombination mechanisms or by recombinant DNA technology.

reduction. The gaining of one or more electrons by an atom or molecule.

reflex [L. *reflectere*, to bend back]. A simple, stereotyped, and repeatable motor action that is elicited by a sensory stimulus.

reproduction, asexual. Production of new individuals by any process that does not involve gametes.

reproduction, sexual. Process of reproduction that begins with meiosis, proceeds through gamete formation, and ends at fertilization.

reproductive isolating mechanism. Any aspect of structure, functioning, or behavior that prevents successful interbreeding (hence gene flow) between populations or between local breeding units within a population.

respiration [L. *respirare*, to breathe]. In most animals, the overall exchange of oxygen from the environment and carbon dioxide wastes from cells by way of circulating blood. Compare *aerobic respiration*.

resting membrane potential. Steady voltage difference that exists across the plasma membrane of a neuron (or some other excitable cell) that is not being stimulated.

rhizoid (RYE-zoid) [Gk. *rhiza*, root]. Long, single cell or filament that anchors some gametophytes to the ground or some other substrate.

ribonucleic acid (RYE-bow-new-CLAY-ik). RNA; a category of nucleotides used in translating the genetic message of DNA into actual protein structure.

ribosome. In both prokaryotic and eukaryotic cells, a structure made of RNA and proteins and the site of protein synthesis.

salt. Ionic compound (such as NaCl) that forms by the reaction between an acid and a base; usually dissociates into positive and negative ions in water.

saprobe. Heterotroph that obtains nutrients from nonliving organic matter. Most fungi are saprobes.

sarcolemma (SAR-koe-LEM-uh) [Gk. *sarx*, the flesh, + *lemma*, husk]. The plasma membrane surrounding a muscle fiber.

sarcomere (SAR-koe-meer). Fundamental unit of contraction in skeletal muscle; repeating bands of actin and myosin that appear between two Z-lines.

sarcoplasmic reticulum (SAR-koe-PLAZ-mik reh-TIK-you-lum). A continuous system of membrane-bound chambers that surrounds myofibrils within a muscle fiber and that stores calcium ions necessary for the mechanism of muscle contraction.

second law of thermodynamics. When left to itself, any system along with its surroundings undergoes energy conversions, spontaneously, to less organized forms. When that happens, some energy gets randomly dispersed in a form (often evenly distributed, low-grade heat) that is not as readily available to do work.

secondary growth. In vascular plants, an increase in stem and root diameter, made possible by the activity of two types of lateral meristems (vascular cambium and cork cambium).

seed. In gymnosperms and flowering plants, a fully mature ovule (contains the plant embryo), with its integuments forming the seed coat.

segmentation. In many animal species, a series of body units that may be externally similar to or quite different from one another.

segregation, allelic [L. *se-*, apart, + *grex*, herd]. Mendelian principle that two units of heredity (alleles) exist for a trait, and that during gamete formation, the two units of each pair are separated from each other and end up in different gametes.

semen (SEE-mun) [L. *serere*, to sow]. Sperm-bearing fluid expelled from the penis during ejaculation.

semiconservative replication [Gk. *hēmi*, half, + L. *conservare*, to keep]. Manner in which a DNA molecule is reproduced; formation of a complementary strand on each of the unzipping strands of a DNA double helix, the outcome being two "half-old, half-new" molecules.

senescence (seh-NESS-sens) [L. *senescere*, to grow old]. Sum total of processes leading to death of a plant or any of its organs; cause appears to be built into the life cycle of the species.

sensory neuron. Type of neuron that carries signals about changing conditions into integrating centers (such as the central nervous system).

sex chromosomes. In most animals and some plants, chromosomes that differ in number or kind between males and females but that still function as homologues during meiosis. All other chromosomes are called autosomes.

sex-linked gene. A gene located only on a female X chromosome; has no allelic partner on the male Y chromosome.

sexual selection. Mode of natural selection based on any trait that gives the individual a preferential advantage in mating and in producing offspring.

sodium-potassium pump. A transport protein spanning the lipid bilayer of the plasma membrane. When the protein receives an energy boost from ATP, its shape changes in such a way that it selectively transports sodium ions *out* of the cell and potassium ions *in*. This active transport process maintains the ion distributions (hence the voltage difference) across the membrane.

solute (SOL-yoot) [L. *solvere*, to loosen]. Any substance dissolved in some solution. In water, this means its individual molecules are surrounded by spheres of hydration that keep their charged parts from interacting, so the molecules remain dispersed in the water.

solvent. Fluid in which one or more substances is dissolved.

somatic cell (so-MAT-ik) [Gk. *sōma*, body]. Any cell of the animal body that is not a germ cell (which develops by meiosis into sperm or eggs).

somatic nervous system. Those efferent nerves leading from the central nervous system to skeletal muscles.

species (SPEE-sheez) [L. *species*, a kind]. For sexually reproducing organisms, one or more populations whose members interbreed under natural conditions and produce fertile offspring, and who are reproductively isolated from other such groups.

sperm [Gk. *sperma*, seed]. Mature male gamete.

spermatogenesis (sperm-AT-oh-JEN-ih-sis). Formation of a mature sperm from a germ cell (spermatogonium).

sphere of hydration. Through positive or negative interactions, a clustering of water molecules around the individual molecules of a substance placed in water. Compare *solute*.

sphincter (SFINK-tur). Ring of muscle that serves as a gate between regions of a tubelike system (as between the stomach and small intestine).

sporangium, plural **sporangia** (spore-AN-gee-um) [Gk. *spora*, seed]. Protective layer of sterile cells surrounding haploid spores in a sporophyte.

spore. In plants, a meiospore that gives rise to one or more haploid gametophytes. Among fungi, an asexual reproductive cell that gives rise to new hyphae of the fungal mat (mycelium).

sporophyte [Gk. *phyton*, plant]. Diploid, spore-producing stage of plant life cycles.

sporozoite. Infective, sporelike stage of sporozoan life cycles.

stabilizing selection. Mode of natural selection that decreases the frequency of alleles giving rise to extreme forms of a trait, such that intermediate forms already well adapted to prevailing conditions are favored.

stamen (STAY-mun). In flowering plants, the male reproductive structure; commonly consists of pollen-bearing structures (anthers) positioned on single stalks (filaments).

steroid (STAIR-oid). A lipid with a backbone of four carbon rings. Steroids differ in the number and location of double bonds in the backbone and in their number, position, and type of functional groups.

stimulus [L. *stimulus*, goad]. Any form of energy that the body is able to detect by means of its receptors.

stoma, plural **stomata** (STOW-muh) [Gk. *stoma*, mouth]. An opening, defined by two guard cells, across the epidermis of a leaf or stem, through which water vapor moves out of the plant and carbon dioxide moves in, in amounts governed by controls over stomatal widening and closure.

stroma [Gk. *strōma*, bed]. In chloroplasts, the semifluid matrix, surrounding the grana, where complex organic molecules are assembled.

substrate. Molecule or molecules of a reactant on which an enzyme acts.

succession, primary (suk-SESH-un) [L. *succedere*, to follow after]. The orderly changes in species composition from the pioneer species that inhabit an area previously devoid of life to the more or less constant array of species that constitutes the climax community.

succession, secondary. Reestablishment of a climax community that has been disrupted in whole or in part.

surface-to-volume ratio. In cells, a physical constraint on increased size: as the cell's linear dimensions grow, its surface area does not increase at the same rate as its volume (hence each unit of plasma membrane is called upon to serve increasing amounts of cytoplasm).

symbiosis (sim-by-OH-sis) [Gk. *syn*, together, + *bios*, life, mode of life]. A mutually beneficial relationship involving continuous, intimate contact between interacting species. Compare *mutualism*.

sympathetic nerves. A division of the autonomic nervous system, consisting of those efferent nerves concerned with slowing down the body's housekeeping tasks and with increasing overall body activity during times of stress, danger, excitement, and heightened awareness. Compare *parasympathetic nerves*.

sympatric speciation (sim-PAT-rik) [Gk. *syn*, together, + *patria*, native land]. The origin of a species as a result of ecological, behavioral, or genetic barriers that arise *within* the boundaries of a single population. The emergence of a new polyploid species of plant is an example.

synapse, chemical (SIN-aps) [Gk. *synapsis*, union]. A junction between two neurons, or between a neuron and a muscle or gland cell, that are separated by a small gap. At an *excitatory* synapse, a transmitter substance released from the first neuron produces changes in the receiving cell that bring its membrane closer to threshold. At an *inhibitory* synapse, a transmitter substance released from the first neuron produces changes in the receiving cell that drive membrane potential away from threshold.

synapsis (sin-AP-sis). At prophase I of meiosis, the point-by-point alignment of the two sister chromatids of each chromosome with the two sister chromatids of its homologue.

systemic circulation. Pathways of blood flow leading to and from all body parts except the lungs.

telophase (TEE-low-faze). Final stage of mitosis, during which the separated chromosomes decondense and become enclosed within newly forming daughter nuclei.

testis, plural **testes**. Male gonad; primary reproductive organ in which male gametes and sex hormones are produced.

threshold value. The minimum voltage change across the plasma membrane necessary to produce an action potential in neurons and some other excitable cells.

thymine. Nitrogen-containing base found in some nucleotides.

tissue. A group of cells and intercellular substances functioning together in a specialized activity.

tonicity. The relative concentrations of solutes in the fluid inside and outside the cell. When solute concentrations are *isotonic* (equal in both fluids), water shows no net osmotic movement in either direction. When one of the fluids is *hypotonic* (has less solutes than the other), the other is *hypertonic* (has more solutes) and is the direction in which water tends to move.

trachea, plural **tracheae** (TRAY-kee-uh). A tube for breathing; in land vertebrates, the windpipe that carries air between the larynx and bronchi.

tracheid (TRAY-kid). Typically elongated cell, dead at maturity, that passively conducts water and solutes in xylem.

transcription [L. *trans*, across, + *scribere*, to write]. The assembly of an RNA strand on one of the two strands of a DNA double helix; the resulting transcript has a nucleotide sequence that is complementary to the DNA region on which it is assembled.

translation. The interaction of rRNA, tRNA, and mRNA in converting the DNA instructions encoded in the mRNA molecule into a particular sequence of amino acids to form a polypeptide chain.

translocation. In vascular plants, the transport of soluble food molecules (mostly sucrose) from one plant organ to another by way of the phloem tissue.

transmitter substance. Chemical secretion, released in tiny amounts from a neuron, that triggers change in the membrane potential of an adjacent cell.

transpiration. Evaporative water loss from stems and leaves.

transposition. Genetic recombination involving nonhomologous base sequences in which transposable elements move from one site in the DNA to another. Transposable elements are also called jumping genes.

trophic level (TROE-fik) [Gk. *trophos*, feeder]. All organisms that are the same number of energy transfers away from the original source of energy (e.g., sunlight) that enters an ecosystem.

turgor pressure (TUR-gore) [L. *turgere*, to swell]. Internal pressure on a cell wall caused by osmotic movement of water into the cell body.

uracil. Nitrogen-containing base found in RNA molecules; can base-pair with adenine.

urinary system. An organ system concerned with regulating water and solute levels in the body.

uterus (YOU-tur-us) [L. *uterus*, womb]. A chamber in which the developing embryo is contained and nurtured during pregnancy.

vacuole, central (VAK-you-ohle). In plant cells, a membrane-bound, fluid-filled sac that may take up most of the cell interior; main function is to increase cell size and surface area, thereby enhancing absorption of relatively dilute concentrations of nutrients from the external environment.

vagina. Part of the female reproductive system that receives sperm from the male penis; forms part of the birth canal, and acts as a channel to the exterior for menstrual flow.

vascular cambium. In vascular plants, one of the lateral meristems that increase stem or root diameter.

vernalization [L. *ver*, spring]. Cold-temperature stimulation of the flowering process.

vertebra, plural **vertebrae.** One of a series of hard bones that form the backbone in most chordates.

vertebrate. Animal having a backbone made of bony segments called vertebrae.

vesicle (VESS-ih-kul) [L. *vesicula*, little bladder]. In cytoplasm, a small, membrane-bound sac in which various substances may be transported or stored.

vessel element. Typically elongated cell, dead at maturity, that passively conducts water and solutes in xylem.

viroid. An infectious nucleic acid that has no protein coat; a tiny rod or circle of single-stranded RNA.

virus. Infectious agent consisting of nucleic acid encased in protein; incapable of metabolism or reproduction without a host cell, hence is often not considered alive.

vision. Precise light focusing onto a layer of photoreceptive cells that is dense enough to sample details concerning a given light stimulus, followed by image formation in the brain.

water potential. The sum of two opposing forces (osmosis and turgor pressure) that can cause the directional movement of water into or out of a walled cell.

white matter. Mainly axons of interneurons, so named because of the glistening myelin sheaths around them.

wild-type allele. The normal or most common allele at a given gene locus.

xylem (ZYE-lum) [Gk. *xylon*, wood]. In vascular plants, a tissue that transports water and solutes through the plant body.

zygote (ZYE-gote). In most multicelled eukaryotes, the first diploid cell formed after fertilization (fusion of nuclei from a male and a female gamete).

INDEX

and population growth, 717,
728–731, *729, 730*
Predator
defined, 728
versus parasite, 728
Predator-prey interaction
Amoeba-ciliate, *95*
ant-termite, 820–821
bacteria as, 600, *600*
bat-frog, 825, *825*
camouflage in, *732, 733*
Didinium-Paramecium, *607*, 608
frog-snake, 683
fungus-nematode, 619, *619*
leopard-baboon, *729*
lion-zebra, *16*
lynx-hare, *730*
models of, 728, *729*
mouse-beetle, *733*
sea star–scallop, *660*
sea star–sea anemone, 647, *647*
spider-fly, *732*
Venus flytrap-fly, 255–257, *256*
Pregnancy
and AIDS, 432
and cigarette smoking, 446, 528
and Down syndrome, 193, *193*
and drugs, 528
and fetal alcohol syndrome (FAS),
528
first trimester of, 522
first week of, 520, *520*
and infection, 525
and nutrition, 525
among primates, 695
and Rh blood typing, 412-413, *413*
second trimester of, 522
and syphilis, 534
third trimester of, 524
weight gain during, 525
Premature birth, human, 524
Premolar, 692
Premotor cortex, 345, *345*
Prenatal diagnosis, 196, *196*
Prenatal period, defined, *518*, 519
Pressure flow theory, 275–276, *276*
Pressure gradient, defined, 89
Presynaptic cell, 329–330, *329*
Prey defense, 731–734
Price, P., 736
Prickly pear cactus, 774, *775*
Primary follicle, 513, *514*
Primary growth, defined, 249
Primary immune response, *421*,
422–425, 428
Primary meristematic tissue, 253, *253*
Primary oocyte, 156, *157*
Primary productivity, 748, 750, 770,
771, 773, 783
Primary receiving center, cortex, 345,
345
Primary root, *248*, 259, *292, 293, 294,
296*
Primary spermatocyte, 156, *157*, 510,
511
Primary structure, protein, 56–57
Primary succession, 739, *740*
Primary wall, 81, *81, 148, 250, 251,
251*
Primate (Primates)
characteristics, 510, 515, 692–696
classification, *691*, 692
evolution, 696*ff.*
groups, *691*, 692
habitats, 692
Primer (DNA synthesis), *206*, 241,
241
Primitive streak, *480*
Principle, scientific, 18, *19–22*
Prion, 593
PRL (prolactin), 225, *356, 357, 357*
Probability, defined, 166
Proboscis, 6, 653, *654*

Procambium, 253, *253, 265*
Process, neuron, 323
Prochlorobacteria, *596*
Prochloron, *596*, 608
Producer
defined, 707, 746
roles in biosphere, 17, *18*
and trophic levels, 747–748, *748*
Profundal zone, 782–783, *782, 784*
Progesterone
and the Pill, 530
and pregnancy, 524
source of, *358, 513, 515*
targets of, *358*, 517
and uterine function, *516, 517, 517*
Proglottid, 652, *653*
Progymnosperm, *581*
Prokaryote (*see also* Bacterium;
Moneran)
body plan, 67, *67*
defined, 67
versus eukaryote, 67
origin of, 575–576
reproduction, *138, 139*
Prokaryotic fission, *138, 139, 150*
Prolactin (PRL), 225, *356, 357, 357*
Proline, *58, 59, 213, 213*
Promoter, 211–212, 222–223, *223,
224, 244*
Propagation, vegetative, 290
Prophase
meiosis, 152, 153, *153–154, 155,
160*
mitosis, 143, *143, 144, 146, 160*
Prosimian, *691, 692, 693*
Prostaglandin, 55, 363, 428, 512, 517
Prostate gland, *509, 510, 511, 512*
Prosthetic group, 102
Protein
abiotic formation of, 586
activator, 223
behavior in cellular fluid, 47
bonding patterns, 57–59, *56, 57,
58, 59*
and cell structure, 139
channel, 86, *87, 92, 310, 324, 325*
chromosomal, 71–72, 140
complement, 419, *419*, 420
and cytomembrane system, 73–75,
73, 212
denaturation, 58
digestion, 75, 454–458, *455, 461*
electron-transfer, 86–87, *87*, 105
fibrous, *58, 60*
function, 55–56, 139
gated, 86, *324, 325, 326, 326*
globular, *58, 60*
hormones, *363, 364, 365*
and human nutrition, 461–462
and liver, 465
membrane, 86–87, *87, 92–94, 93*
metabolism, 121, *132–133, 133,
360, 362*
oncogene-encoded, 231–232
plasma (*see* Plasma protein)
primary structure, 56–57
pump, 86, *87, 93–94, 324, 325,
480, 481*
quaternary structure of, 58, *58*
receptor, 86–87, *87*
recognition, *87, 231, 312, 501*
R groups, 56, *56, 57, 58*
repressor, 223
roles of, 55–56
secondary structure, 57–58
summary of, *60*
tertiary structure, 58
utilization, human, 461–462, *462*
Protein synthesis
and cancer, *232*
and cell function, 71, 73–74,
132–133
control of, 132–133, 222*ff.*

and DNA transcription, 210,
211–212, *211, 212, 218,
222–223, 226ff., 226*
and DNA translation, 213–217,
216, 218, 226, 226
genetic code for, 213, *213, 214*
hormone stimulation of, 224–225,
357–359, 364, 364
location of, 67, 209
overview of, 210–211
role of cytomembrane system,
73–75, *73, 217*
role of DNA in, 71, 73, 210
role of RNA in, 210–211, 212
summary of, *218*
Proterozoic Era, 562, *562, 578,
587*
Prothoracic gland, *484*
Prothrombin, 399, *399*
Protistan (Protista)
characteristics, *83*, 142
classification, *33*, 600, *601*
evolution, 576, 577, 608–609
size, 590
summary of groups, *601*
Protoavis, 584
Protoderm, 253, *253, 259, 265*
Proton (*see also* Hydrogen ion), 4,
37, *104–105, 123, 124*
Protonephridium, 650, *650, 655,
655*
Proto-oncogene, *232*
Protostome, *643*, 656, *656*
Prototheria, 639
Protozoan
amoeboid, *601*, 605
ciliated, *601*, 607–608, *607*
classification, *601, 604*
evolution, 586, 604
flagellated, *601, 605, 605*
Proximal tubule, *478, 479, 481, 481*
Prunus, 278, *281*
Pseudocoel, *642*
Pseudocoelomate, *642, 643, 644,
653, 654, 655*
Pseudomonas, 67, *138*, 243
Pseudopod, defined, 501
Psilophyton, *581*
Psychedelic, 349
Psychoactive drug, 348–350
Psychosis, 349
Ptarmigan, *746*
Pterophyta (ferns), 623, 629, *629,
630, 631, 632*
Pterosaur, 559, *559*, 583, *585*
Ptychodiscus brevis, 604
Puberty, 7, 362, 510, 514, 518
Puffball, 616
Pulmonary artery, *402, 403*
Pulmonary circulation, 402, *402*, 405
Pulmonary pleura, 441
Pulmonary vein, *402, 403*
Pulmonary ventilation, 440
Pulse pressure, 406
Punctuation, evolutionary model, 557,
557
Punnett, R., 171
Punnett-square method, 166, *166,
167*
Pupa, 6, *7, 504*
Pupil, *340, 377, 379*
Purine, 202
Purple bacteria, *596*, 598
Pyloric sphincter, *453, 457*
Pyridoxine, *463*
Pyrimidine, 202
Pyrrophyta, *601, 624, 624*
Pyrsonympha, 578
Pyruvate
destinations, *123,* 124*ff.*
formation of, 124, 126, *127*
structural formula, *125*
Python, 25, 367–368, *367*

T